Holt Algebra 1

Examination Guide

A complete course for students of all abilities provides everything you need—

- **The Step-by-Step problem-solving process** helps you teach algebraic concepts and methods.

  Chapters 1–5 include lessons that emphasize representing quantities with variables (p. 125), writing equations (p. 129), and inequalities (p. 168). Chapters 6–8 explore polynomial equations and their applications (p. 229). Chapters 10–12 focus on graphing in a coordinate plane (p. 373), exploring the slope of a line (p. 413), and solving systems of equations (p. 455).

- **Exercises** progress in complexity to give you a choice of assignments.

  Exercises are plentiful with Classroom Exercises (p. 92) for oral discussion and Written Exercises (p. 93) that progress from easy to more challenging. Examples correspond to each type of written exercise for levels A and B (pp. 90–91).

- **Reviews, quizzes, and tests** make it easy to integrate assessment with instruction.

  Each lesson includes a Prerequisite Quiz (p. 89) and a post-lesson Checkpoint (p. 92) in the side column of the *Annotated Teacher's Edition* and as blackline masters in the Quick Quizzes of the testing package. Assessments in the textbook include Mixed Reviews (p. 93) and a Midchapter review (p. 100) as well as Chapter Reviews and Tests (pp. 120–122) and College Prep Tests (p. 123) at the end of each chapter and Cumulative Reviews in alternate chapters (pp. 162–163).

- **Extensive features and worksheets** help you meet the NCTM *Standards* for the 90s!

  * **Technology—calculators and computers**
    pp. 54, 68, 71, 628
    Investigating Algebra with the Computer Software–
    Teacher's ResourceBank™

  * **Math connections**
    In Teacher's side column for each lesson
    Applications features (pp. 61, 66)
    Application Worksheets–
    Teacher's ResourceBank™

  * **Communication about math**
    Focus on Reading (p. 104)
    Written Exercises–essays (p. 100, #56)

  * **Models, manipulatives, projects**
    Investigations that begin each chapter
    Project Worksheets and Manipulatives
    Worksheets—*Teacher's ResourceBank*™

  * **Problem solving strategies**
    pp. 35, 101, 157, 185
    Problem Solving Worksheets—
    Teacher's ResourceBank™

  * **Other valuable timesavers for you**
    Reteaching and Practice Worksheets and
    Tests booklet with Tests, Form A and
    Form B—*Teacher's ResourceBank*™
    Instructional Transparencies and
    Solution Key

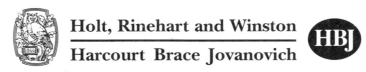

Holt, Rinehart and Winston
Harcourt Brace Jovanovich

HOLT
ALGEBRA 1
Annotated Teacher's Edition

Eugene D. Nichols
Mervine L. Edwards
E. Henry Garland
Sylvia A. Hoffman
Albert Mamary
William F. Palmer

Holt, Rinehart and Winston, Inc.

Harcourt Brace Jovanovich, Inc.

HBJ

Austin • Orlando • San Diego • Chicago • Dallas • Toronto

About the Authors

Eugene D. Nichols
Distinguished Professor of Mathematics Education
Florida State University
Tallahassee, Florida

Mervine L. Edwards
Chairman, Department of Mathematics
Shore Regional High School
West Long Branch, New Jersey

E. Henry Garland
Head of Mathematics Department
Developmental Research School
DRS Professor
Florida State University
Tallahassee, Florida

Sylvia A. Hoffman
Resource Consultant in Mathematics
Illinois State Board of Education
State of Illinois

Albert Mamary
Superintendent of Schools for Instruction
Johnson City Central School District
Johnson City, New York

William F. Palmer
Professor of Education and Director
Center for Mathematics and Science Education
Catawba College
Salisbury, North Carolina

123456   036   98765432   ISBN 0-03-005422-2

Acknowledgments

Reviewers

Jeanette Gann
Mathematics Supervisor
High Point Schools
High Point, North Carolina

Patrice Gossard, Ph.D.
Mathematics Teacher
Cobb County School District
Marietta, Georgia

Linda Harvey
Mathematics Teacher
Reagan High School
Austin, Texas

Gerald Lee
Chairman, Mathematics Department
McArthur High School
Lawton, Oklahoma

Janet Page
Mathematics Teacher
Hoffman Estates High School
Hoffman Estates, Illinois

Photo Credits

Illustration

Chapter Contents

1 | Introduction to Algebra | xiv

Holt Algebra 1

MEETING THE STANDARDS OF TOMORROW—STARTING TODAY

The role that today's teachers play in the education of our students will affect our world for generations to come. At Holt, Rinehart and Winston, we recognize the importance of education and are dedicated to providing today's teachers and students with the finest textbooks available.

Holt Algebra 1 helps you move toward the NCTM *Standards,* while also providing you with a practical text that meets your standards. A variety of supplementary materials has been developed that incorporates the newest technology and current curriculum standards in mathematics education into a traditional mathematics program.

"TODAY, WHAT HAPPENS in America's classrooms is being given lots of attention and scrutiny. The reactions and responses to the recent reports on education offer the mathematics education community a rare opportunity to shape school mathematics during the next decade. Public interest and concern, when combined with changing technology and a growing body of research-based knowledge, are the ingredients necessary for real reform. The NCTM *Standards* is a vehicle that can serve as a basis for improving the teaching and learning of mathematics in America's schools." (NCTM *Standards,* p. 254)

A COMPLETE SELECTION OF ANCILLARY MATERIALS ENRICHES THE PROGRAM.

The materials described below provide a wealth of activities that incorporate important topics in mathematics education. Computer activities help students feel comfortable using today's technology to perform real-life tasks.

Teacher's ResourceBank™
Chapter Worksheets and Tests

The *Teacher's ResourceBank* contains a wealth of blackline-master booklets, organized by chapter, that provide materials for practice, reteaching, review, enrichment, and an extensive testing program.

> "**N**OR DO WE BELIEVE that textbooks should drive instruction. Rather, other materials, . . . such as manipulatives and courseware, must be developed, in addition to new textbooks." (NCTM *Standards*, p. 252)

- *Reteaching Worksheets* (one per lesson)
- *Practice Worksheets* (one per lesson)
- *Application Worksheets* (one per chapter)
- *Manipulative Worksheets* (one per chapter)
- *Problem Solving Worksheets* (one per chapter)
- *Project Worksheets* (one per chapter)
- *Prerequisite Quizzes* (one per lesson)

- *Checkpoint Quizzes* (one per lesson)
- *Chapter Tests,* Forms A and B (one test per chapter)
- *Cumulative Tests,* Forms A and B (one test for every fourth chapter)
- *Semester Tests,* Forms A and B (one test for every eighth chapter)

Also Included in the Teacher's ResourceBank™

Investigating Algebra with the Computer software (Apple® and IBM®)*
SAT®/ACT Practice Tests

*Also available for sale separately.

Other Ancillary Materials

The *Instructional Transparencies* provide over forty colorful transparencies (at least one transparency per chapter) for teachers who use an overhead projector in their classroom.

The *SAT®/ACT Practice Tests* contain 4 sample SAT tests and 4 sample ACT tests which prepare students for the mathematics sections of these college entrance tests.

The *Investigating Algebra with the Computer* software is designed to be used with the lessons that are found at the back of the *Pupil's Edition.* The software is particularly useful because it allows students to explore algebraic concepts visually.

The *Test Generator* uses a random number generator with a bank of algorithms to allow the teacher to create a virtually unlimited number of unique tests.

The *Solution Key* contains worked-out answers for all exercises and end-of-chapter activities in the *Pupil's Edition.*

FLEXIBLE TEACHING STRATEGIES ALLOW YOU TO CUSTOMIZE INSTRUCTION.

The teaching strategies, which are interleaved in the *Annotated Teacher's Edition* before each chapter, help you focus lessons on problem-solving strategies, computer and calculator technology, and special features that enrich and extend students' learning.

In both the **Overview** and the **Objectives,** learning goals are defined in terms of what students will *do*—emphasizing the value of active, not passive, learning.

3 SOLVING EQUATIONS

OVERVIEW

In this chapter, students solve equations by applying the Addition, Subtraction, Multiplication, and Division Properties of Equality. They solve equations with the variable on one side, with the variable on both sides, and with parentheses. Students apply equation-solving skills to solving word problems and to using formulas.

The last lesson in the chapter provides a summary of all the properties discussed to this point. Students then use these properties as reasons that justify statements in algebraic proofs.

OBJECTIVES

- To solve equations with one variable
- To solve equations containing parentheses
- To solve word problems by first translating them into equations
- To write missing reasons in proofs

PROBLEM SOLVING

This chapter introduces a fundamental problem-solving strategy for algebra— *Writing an Equation.* This same strategy, as well as the strategies of *Using a Formula* and *Using Logical Reasoning,* is applied in the *Applications* on pages 93, 109, and 110–113. The problem-solving strategy lesson on page 101 focuses on the second step of the four-step Polya model, *Developing a Plan.* Provision for maintaining problem-solving skills is included on page 119 of this chapter and at the end of every third chapter throughout the textbook.

READING AND WRITING MATH

Throughout this chapter, students read a problem stated in words and then rewrite it as an algebraic statement. Exercise **37** on page 97, Exercise 56 on page 100, as well as Exercise 13 in the Chapter Review, ask students to write explanations of how they reach a conclusion. The Classroom Exercises on page 99 ask students to verbalize a plan before they solve the equation.

TECHNOLOGY

Calculator: Students may find a calculator helpful in solving Written Exercises 35–37 on page 93, Exercises 45–47 on page 100, Exercises 1–20 on page 112, and in the Applications on pages 109 and 114.

SPECIAL FEATURES

Mixed Review pp. 93, 97, 105, 108, 113, 118
Application: Mathematics in Typing p. 93
Midchapter Review p. 100
Problem Solving Strategies: Developing a Plan p. 101
Focus on Reading p. 104
Application: Thunder and Lightning p. 109
Application: The Shock Wave p. 114
Summary of Properties p. 115
Mixed Problem Solving p. 119
Key Terms p. 120
Key Ideas and Review Exercises pp. 120–121
Chapter 3 Test p. 122
College Prep Test p. 123

Reading and Writing Math identifies places in the chapter where students are asked to explain mathematical concepts in their own words.

Lessons in which the calculator can be used to explore mathematical concepts and facilitate routine computations are included in **Technology.**

A **Special Features** list allows you to decide in advance which applications and additional activities to incorporate into lessons.

> **"T**HESE ALTERNATIVE methods of instruction will require the teacher's role to shift from dispensing information to facilitating learning, from that of director to that of catalyst and coach." (NCTM *Standards,* p. 128)

Problem Solving identifies the features and exercises in the chapter that allow students to develop strategies for solving problems.

TIME-SAVING PLANNING GUIDES HELP YOU ORGANIZE INSTRUCTION.

A quality curriculum requires versatile and complete lesson plans and the lesson plans in *Holt Algebra 1* are organized in a format that is both functional and convenient.

The **Planning Guide** recommends appropriate classroom and written exercises for students of all ability levels, suggests when to incorporate special features into the lessons, and reminds you of the availability of other components.

> **"A** VARIETY OF INSTRUC-tional methods should be used in classrooms in order to cultivate students' abilities to investigate, to make sense of, and to construct meanings from new situations; to make and provide arguments for conjectures; and to use a flexible set of strategies to solve problems . . ." (NCTM *Standards,* p. 125)

PLANNING GUIDE

Lesson	Basic	Average	Above Average	Resources
3.1 pp. 92–93	CE 1–21 WE 1–29 odd, Application	CE 1–21 WE 9–37 odd, Application	CE 1–21 WE 17–47 odd, Application	Reteaching 19 Practice 19
3.2 pp. 96–97	CE 1–16 WE 1–25 odd	CE 1–16 WE 9–33 odd	CE 1–16 WE 13–29, 37–43 odd	Reteaching 20 Practice 20
3.3 pp. 99–101	CE 1–12 WE 1–21, 31–35, 39–43 odd, Problem Solving Strategies	CE 1–12 WE 13–29, 33–47 odd, Problem Solving Strategies	CE 1–12 WE 23–29, 33–35 odd, 56 Problem Solving Strategies	Reteaching 21 Practice 21
3.4 pp. 104–105	FR all CE 1–12, WE 1–17, 25–31, 35–39 odd	FR all CE 1–12, WE 9–19 25–41 odd	FR all CE 1–12, WE 17–23 29–49	Reteaching 22 Practice 22
3.5 pp. 107–109	CE 1–15, WE 1–13, 19–23 odd, Application	CE 1–15, WE 7–25 odd, Application	CE 1–15 WE 13–31 odd, Application	Reteaching 23 Practice 23
3.6 pp. 111–114	CE 1–8 WE 1–9 odd, 12, 13, Application	CE 1–8 WE 5–11 odd, 12, 13, 15, Application	CE 1–8 WE 9, 11, 12, 13–19 odd, Application	Reteaching 24 Practice 24
3.7 pp. 117–119	CE 1–9 WE 1–4 all, Mixed Problem Solving	CE 1–9 WE 2–4 all, 5–11 odd. Mixed Problem Solving	CE 1–9 WE 2–4 all, 5–7 odd, 13 Mixed Problem Solving	Reteaching 25 Practice 25
Chapter 3 Review pp. 120–121	all	all	all	
Chapter 3 Test p.122	all	all	all	
College Prep Test p. 123	all	all	all	

CE = Classroom Exercises   WE = Written Exercises   FR = Focus on Reading
NOTE: For each level, all students should be assigned all Try This and all Mixed Review Exercises.

ATTRACTIVE CHAPTER OPENERS CONNECT MATH TO THE WORLD OUTSIDE THE CLASSROOM.

Each **Chapter Opener** contains a colorful photograph of a person who uses applications of mathematics in their profession.

The **Investigation** is a hands-on activity that helps students give concrete meaning to abstract mathematical concepts. Students work cooperatively, in small groups, acquiring interactive problem-solving skills they will use for a lifetime. Follow-up questions help students form conjectures and draw conclusions about the activity.

■ INVESTIGATION

Project: In this investigation, students explore a physical model of a balanced equation. They experiment with the difference between the *weight* of a coin and its value, and between the *number* of coins and their total value.

Materials: Each student will need a copy of Project Worksheet 3. Each group of students needs a balance scale, some coins (at least 10 pennies, 2 nickels, and 2 dimes), and 6 jumbo paper clips. Students can easily make a simple balance scale using a ruler, stick or dowel about 12 inches long, a rubber band, two smaller disposable drinking cups, a textbook, transparent tape, and four jumbo paper clips as shown below.

Make one scale (see the figure above) as a model. Then ask students to construct their own scales.

Students follow the steps given in Project Worksheet 3.

When students have completed the worksheet, encourage them to explore further. Have the class discuss their conclusions about what keeps the two sides in balance, and why one nickel does not balance five pennies. Ask a student volunteer to write a sentence or two on the chalkboard that summarizes the findings of the class. Ask other students to amend the statement until everyone agrees that it is accurate and complete.

3 SOLVING EQUATIONS

For almost 20 years, Dian Fossey and her students collected data on the daily lives of mountain gorillas inhabiting forest regions in Africa. The book she wrote reporting her findings, *Gorillas in the Mist,* was made into a movie.

More About the Photo

The data collected by Dian Fossey and her colleagues includes recordings of gorilla vocalizations which were then displayed graphically as measurements of the tonality, rhythm, amplitude, quality, and frequencies of the sound. The observers also plotted the birth and death rates in various gorilla groups. Data from autopsies on gorillas who died included measures of the lengths and diameters of bones, the size and weight of various organs, and the number and size of parasites. All of this data continues to help those working to preserve this endangered species.

More About the Photo provides in-depth background information about the person featured in the **Chapter Opener**. The accomplishments and innovations of this person provide an opportunity to discuss the value and usefulness of mathematics in our world.

EASY-TO-FOLLOW LESSONS COMMUNICATE LEARNING GOALS CLEARLY AND INTELLIGIBLY.

Easy-to-follow lessons—featuring succinct objectives, brief introductions, and clear-cut objectives—make mathematics accessible to your students.

3.1 Solving Equations by Adding or Subtracting

Objective

To solve equations by using the Addition or Subtraction Property of Equality

The scales at the right are balanced. If two objects of equal weight are placed on each side of the scale, the scale will remain balanced. This shows the following.

If $x = y$, then $x + 2 = y + 2$.

Suppose Bill and Jean are the same age. Then in 4 years they will still be the same age. Let b = Bill's age now in years and j = Jean's age now in years. This situation can be expressed as follows.

If $b = j$, then $b + 4 = j + 4$.

This reasoning suggests one of the equality properties. These properties can be used to find solutions of equations.

Consider the equation $x = 6$. Its only solution is 6; its solution set is {6}. Suppose that you add the same number to each side of $x = 6$.

$$x = 6$$
Add 7 to each side. $$x + 7 = 6 + 7$$
$$x + 7 = 13$$

The resulting equation, $x + 7 = 13$, also has {6} as its solution set. Equations with the same solution set, such as $x = 6$ and $x + 7 = 13$, are called **equivalent equations**.

Addition Property of Equality

For all real numbers a, b, and c, if $a = b$, then $a + c = b + c$. This means that adding the same number to each side of an equation produces an equivalent equation.

Recall that the *replacement set* for a variable is the set of numbers that can replace the variable. For the rest of the equations in this book, *assume that the replacement set is the set of all real numbers*, unless it is otherwise stated.

3.1 Solving Equations by Adding or Subtracting   **89**

" "N MATHEMATICS, AS IN ANY field, knowledge consists of information plus know-how. Know-how in mathematics that leads to mathematical power requires the ability to use information to reason and think creatively and to formulate, solve, and reflect critically on problems." (NCTM *Standards*, p. 205)

▰▰ GETTING STARTED

Prerequisite Quiz

Name the opposite of each number.

1. -9 9
2. $1\frac{3}{5}$ $-1\frac{3}{5}$

Add or subtract as indicated.

3. $-6 + 6$ 0
4. $-1\frac{1}{3} + 1\frac{1}{3}$ 0
5. $-7.9 - 7.9$ -15.8
6. $6.3 - (-6.3)$ 12.6

Motivator

Remind students that they have already found the solution set of an equation by substituting the members of a finite replacement set for the variable (see Lesson 1.8). As a review, have them find the solution set of $-12 + x = 8$ for the replacement set $\{-1, 2, 4\}$.

Highlighting the Standards

Standard 5a: In this lesson, students represent situations involving variables. They learn how to relate words and algebraic representations and they practice this in the Exercises.

89

Teaching Resources provides a list of all supplementary components available for use with the lesson. These materials contain practice, review, reteaching and enrichment activities.

The **Prerequisite Quiz** helps you assess what students have learned in previous lessons and determine the overall readiness of the class for the new lesson.

Through a series of questions, the **Motivator** helps students form a hypothesis. As students complete the lesson, they will discover if their conjectures were correct and will be able to draw mathematical conclusions about the concepts presented in the lesson.

Highlighting the Standards identifies how each lesson addresses the NCTM *Standards*.

TEACHING SUGGESTIONS INTEGRATE A VARIETY OF TOPICS INTO THE LESSONS.

The **Teaching Suggestions** relate each lesson to current issues in mathematics education. These features, located in the side-column of the text, provide a convenient way to enrich the lessons and make them meaningful to students.

Examples lead students through the problem-solving process by showing them how to formulate a Plan and then move step-by-step toward a solution.

The **Lesson Note** helps prepare students for the exercises by identifying concepts and skills from earlier lessons that may need to be reviewed.

Math Connections relates algebra to material taught previously, other areas of mathematics, real-life situations, and the history of mathematics.

"WE NOW CHAL-lenge educators to integrate mathematics topics across courses so that students can view major mathematical ideas from more than one perspective . . . " (NCTM *Standards*, p. 252)

■ TEACHING SUGGESTIONS

Lesson Note

A balance scale is a helpful model for demonstrating the Addition Property of Equality. When the same amount is added to both sides of a scale, the balance is maintained.

Point out to students that the solution for Example 1 is 21. The solution set is {21}. After solving the equation in Example 4, ask students whether the answer is reasonable. For example, $5,300 would not be a reasonable selling price for a tape recorder.

Math Connections

Life Skills: Translating verbal descriptions into algebraic equations is a problem-solving strategy that is used in everyday life. For example, the amount of profit that you can make by selling a number of items for more than they cost you, and the number of miles that you have left to travel before reaching your destination, can be found by solving algebraic equations.

To **solve** an equation means to find its solution.

EXAMPLE 1   Solve $x - 12 = 9$. Check the solution.

Plan   Notice that x is not alone on the left side of $x - 12 = 9$, since 12 is subtracted from x. To get x alone on the left, add 12 to each side.

Solution
$$x - 12 = 9$$
Add 12 to each side. $x - 12 + 12 = 9 + 12$
Simplify. $x + 0 = 21$ ⟵ $-12 + 12 = 0$
$$x = 21$$

All of these equations are equivalent. They all have the same solution.

Check   Check 21 in the original equation.
$$x - 12 = 9$$
Substitute 21 for x. $21 - 12 \overset{?}{=} 9$
$$9 = 9 \text{ True}$$

Thus, the solution is 21.

TRY THIS   1. Solve $x - 6 = -3.5$. Check the solution.   2.5

Since subtracting a number is the same as adding its opposite, the following property follows from the Addition Property of Equality.

Subtraction Property of Equality

For all real numbers, a, b, and c, if $a = b$, then $a - c = b - c$.

In other words, subtracting the same number from each side of an equation produces an equivalent equation.

This property is used in solving the equation in the next example.

EXAMPLE 2   Solve $-17 = y + 5$. Check.

Plan   The opposite of adding 5 is subtracting 5.
To get y alone on the right, subtract 5 from each side.

Solution
$$-17 = y + 5$$
Subtract 5 from each side. $-17 - 5 = y + 5 - 5$
Simplify. $-22 = y$

90   Chapter 3 Solving Equations

Additional Example 1
Solve $19 = a - 7$. Check.   26

Additional Example 2
Solve $-6\frac{2}{3} = x + 1\frac{1}{3}$. Check. $-7\frac{4}{5}$

AN ABUNDANCE OF SAMPLE PROBLEMS HELPS STUDENTS ASSESS THEIR UNDERSTANDING OF THE LESSON.

Sample problems encourage students to become self-directed learners while providing the opportunity to correct student mistakes early in the learning process.

Try This exercises reinforce the lesson by providing the opportunity for immediate practice of newly learned skills.

Critical Thinking Questions develop higher-order thinking skills such as analysis, synthesis, and evaluation.

Check

Substitute -22 for y.

$$-17 = y + 5$$
$$-17 \stackrel{?}{=} -22 + 5$$
$$-17 = -17 \text{ True}$$

Thus, the solution is -22.

TRY THIS   2. Solve $y + 2\frac{1}{2} = -6$. Check the solution. $-8\frac{1}{2}$

Many word problems can be solved by using equations such as the ones illustrated in this lesson. Recall how you translate English phrases into algebraic expressions.

4 more than a number	16 less than a number	A number decreased by 16
4 more than n	16 less than n	n decreased by 16
$n + 4$	$n - 16$	$n - 16$

EXAMPLE 3   The $37-selling price of a tape recorder is $16 less than the original price. Find the original price.

Solution   You are asked to find the original price.
Let $x =$ the original price in dollars.

Thirty-seven is 16 less than the original price.
37 is 16 less than x.

$$37 = x - 16$$
$$37 + 16 = x - 16 + 16$$
$$53 = x$$

Check the answer, $53, in the original problem.

Check   Is it true that $37 is $16 less than $53? Yes, because $53 - 16 = 37$.

Thus, the original price is $53.

TRY THIS   3. A number has been decreased by 12. The result is -5. Find the original number. Check the solution.   7

Notice that in Example 3, the answer was checked in the *original problem*, not in the equation. This is important since you could have made an error in writing an equation. For example, you may have translated

37 is 16 less than x incorrectly as $37 - 16 = x$.

The solution to the equation $37 - 16 = x$ is 21, a solution that checks in the equation but *not* in the word problem itself.

Thirty-seven dollars is *not* $16 less than $21.

3.1 Solving Equations by Adding or Subtracting   **91**

Critical Thinking Questions

Analysis: Ask students to compare the solutions for $x + 3 = 7$ and $3 - x = 7$. (4 and -4 respectively). Then have students write an algebraic equation involving addition whose solution set is -2, and an equation with the same numbers but involving subtraction and having a solution set of 2. Answers may vary.   Sample answers are $x + 6 = 4$ and $6 - x = 4$.

Common Error Analysis

Error: After students have checked the solution of an equation, they often state that the solution is the number named on each side of the equation. For example, after completing the check in Example 2, they may conclude that the solution is -17.

Remind students to look back in Example 2 to the solution they found originally, -22.

Common Error Analysis diagnoses potential problem areas and prescribes corrective strategies that focus students' attention on correct procedures.

Additional Example 3

On a canoe trip, the campers paddled 8.6 miles on the first day. That was 2.7 mi farther than they paddled the second day. How many miles did they paddle the second day?   5.9 mi

---

"THE ASSESSMENT OF students' mathematics learning should enable educators to draw conclusions about their instructional needs . . . and the effectiveness of a mathematics program."
(NCTM *Standards*, p. 193)

Additional Examples may be used for reteaching or as a quiz to evaluate students' understanding of the lesson.

T8

CLASSROOM AND WRITTEN EXERCISES CLARIFY AND REINFORCE NEWLY-LEARNED CONCEPTS AND SKILLS.

An extensive variety of exercises covers all topics from the lesson and challenges students of all ability levels.

Checkpoint provides a quick way to determine what students have learned from the lesson and what concepts and skills need to be reinforced.

Closure questions help students summarize, communicate, and form conclusions about the major concepts of the lesson.

Checkpoint

Solve and check.

1. $x - 9 = -31$ -22
2. $7 + y = -23$ -30
3. $-6 = 14 + z$ -20
4. $0.9 = -7.4 + y$ 8.3

Translate into an equation and solve.

5. Eighteen less than a number is -46. Find the number. $x - 18 = -46;$ -28

Closure

Ask students to explain which property should be used to solve an equation such as $-3 + x = 14$ and why. Do the same for $19 = y + 8$. Add Prop of Eq, Add 3 to each side to get x alone; Subt. Prop of Eq, Add (-8) to each side to get y alone.

Classroom Exercises

What number would you add to or subtract from each side of the equation to solve it?
Add 8.
1. $x - 12 = -32$ Add 12.   2. $46 = y + 7$ Subtract 7.   3. $-3 = x - 8$
4. $p + 1 = 46$ Subtract 1.   5. $36 + y = 42$ Subtract 36.   6. $27 = y - 9$
7. $4.2 + x = 5.7$ Subtract 4.2.   8. $y - \frac{2}{3} = -4$ Add $\frac{2}{3}$.   9. $-2 = z + 1\frac{1}{4}$

Subtract $1\frac{1}{4}$.
10–18. Solve and check the equations of Classroom Exercises 1–9.
10. -20 11. 39 12. 5 13. 45 14. 6 15. 36 16. 1.5 17. $-3\frac{1}{3}$ 18. $-3\frac{1}{4}$ 6. Add 9.
For Exercises 19–21, let x represent the number you are asked to find. Write an equation to find the number. Then find the number.
19. Twenty is 5 more than a number. Find the number. $20 = 5 + x;$ 15
20. Seven less than a number is 23. What is the number? $x - 7 = 23;$ 30
21. Twenty-seven increased by a number is 33. Find the number.
    $27 + x = 33;$ 6

Written Exercises

Solve and check.

1. $x - 23 = 30$ 53   2. $y + 7 = 63$ 56   3. $-18 = x - 35$ 17
4. $a + 8 = -4$ -12 5. $14 = x + 9$ 5   6. $c - 9 = 3$ 12
7. $6 = 15 + x$ -9 8. $b - 10 = 8$ 18   9. $7 + x = 5$ -2
10. $t + 8 = -3$ -11 11. $9 = 10 + x$ -1 12. $4 + n = 4$ 0
13. $x - 9 = -9$ 0   14. $-12 + y = 18$ 30   15. $5 = 8 + a$ -3
16. $n - 6 = 13$ 19   17. $4 = x + 8$ -4 18. $x - 5 = 9$ 14
19. $5 + y = 11$ 6   20. $9 = c + 15$ -6 21. $-4 + x = -4$ 0

Translate each problem into an equation and solve. Other equations are possible.

22. The $44 selling price of a sweater is the cost increased by $18. Find the cost. $44 = c + 18;$ $26
23. Nine more than a number is 42. Find the number. $n + 9 = 42;$ 33
24. Eight less than a number is 56. Find the number. $n - 8 = 56;$ 64
25. The price of a radio decreased by $35 is the discount price of $85. Find the original price. $p - 35 = 85;$ $120

Solve and check.

26. $0.6 + x = 1.4$ 0.8   27. $y - 9.2 = 4.7$ 13.9   28. $-5.2 = z - 0.9$ -4.3

92   Chapter 3 Solving Equations

Enrichment

Have students solve this puzzle. Four unequal weights have a total weight of 22 oz. A scale balances when two weights are placed on the left pan and two are placed on the right pan. The smaller weight on the right pan is 2 oz heavier than the smaller weight on the left pan. If the smaller weight on the left pan is moved to the right pan, 6 oz will have to be added to the left pan to balance the scale.

Find each of the four weights, and how they were placed originally on the scale. 8 oz and 3 oz on the left pan; 6 oz and 5 oz on the right pan

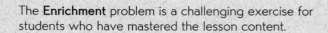

The **Enrichment** problem is a challenging exercise for students who have mastered the lesson content.

FREQUENT REVIEW OPPORTUNITIES HELP YOU EVALUATE STUDENTS' UNDERSTANDING.

The **Mixed Review,** which occurs at the end of most lessons, reinforces newly learned concepts, and maintains skills from previous lessons. For the convenience of the student, exercises are referenced to the chapter and lesson where the related skill or concept was taught.

Follow-up suggests *Guided Practice* exercises that students can do in class to ensure a successful homework experience.

C-level exercises, numbered in red, require students to use critical-thinking skills to solve more challenging problems.

The **Mixed Review** reinforces concepts, skills, and problem-solving skills from the material in earlier lessons.

Application problems provide students with the opportunity to use problem-solving skills in real-world situations.

---

29. $x + 3.4 = -2.6$ -6.0 **30.** $-4.3 + y = -6.7$ -2.4 **31.** $z - 0.5 = 4.8$ 5.3

32. $x + \frac{1}{2} = 5$ $4\frac{1}{2}$ **33.** $y - \frac{2}{3} = 18$ $18\frac{2}{3}$ **34.** $z + 5\frac{1}{3} = -10$ $-15\frac{1}{3}$

Solve.

35. $x - 246.8 = 656.32$ 903.12 **36.** $254.78 + x = -234.5$ -489.28 **37.** $115.3 + x = 889.36$ 774.06

38. Explain in writing how the Subtraction Property of Equality follows from the Addition Property of Equality. To subtract a number is to add its opposite.

39. To solve the equation $5 = x + 7$, explain why you would subtract 7 from each side rather than subtract 5 from each side. The new equation, $0 = x + 2$, does not have x alone on one side.

Find the solution set.

40. $x + |-8| = 17$ {9} **41.** $|x| - 3 = -1$ {2, -2} **42.** $16 = |x| + 7$ {9, -9}

43. $x + 8 = 8 + x$ **44.** $x + 6 = x - 9$ ∅ **45.** $x + 3 = -7 + x$ ∅

46. $|7 - [18 ÷ (-6)]| = x + 3$ {7} **47.** $x + 4 = |-16|$ {12}

Mixed Review

Simplify. 2.4–2.7, 2.9, 2.10

1. $-\frac{2}{3} \cdot \frac{3}{5}$ $-\frac{2}{5}$ **2.** $-1.6 + 5.4$ 3.8 **3.** $6t - 11t$ $-5t$

4. $-3 - (7 + 5)$ -15 **5.** $-(x + 8) - 9x$ **6.** $35 ÷ (-7)$ -5
   $-10x - 8$

 Application: *Mathematics in Typing*

A secretary applying for a job may have to take a typing test to determine typing speed in words per minute. The formula below is often used to determine typing speed. Note that a deduction is made for errors, since accuracy in typing is so important.

$$s = \frac{w - 10e}{m}$$

where s = speed in words per minute
w = number of words typed
e = number of errors
m = number of minutes of typing

1. On an 8-minute typing test, Jane typed 380 words with 4 errors. Find her speed in words per minute. 42.5 words/min

2. The formula above can be rewritten as $e = \frac{w - sm}{10}$. Use the formula to find the number of errors Barry made if he typed 320 words in 5 minutes and got a speed of 42 words per minute. 11 errors

3.1 Solving Equations by Adding or Subtracting **93**

■ **FOLLOW UP**

Guided Practice

Classroom Exercises 1–21
Try This all

Independent Practice

A Ex. 1–25, **B** Ex. 26–39, **C** Ex. 40–47
Basic: WE 1–29 odd, Application
Average: WE 9–37 odd, Application
Above Average: WE 17–47 odd, Application

Additional Exercises

Written Exercises

43. The set of real numbers

Independent Practice groups exercises according to A, B, and C difficulty levels for basic, average, and above-average students.

"THE MAIN PURPOSE OF evaluation, as described in these standards, is to help teachers better understand what students know and make meaningful instructional decisions." (NCTM *Standards,* p.189)

IMPLEMENTING THE NCTM *STANDARDS* WITH HOLT ALGEBRA 1

EVERYONE WHO TEACHES mathematics in the 1990s is concerned about how to implement the *Curriculum and Evaluation Standards for School Mathematics*, published by the National Council of Teachers of Mathematics. The professional articles in this section are designed to help you use *Holt Algebra 1* to implement the NCTM *Standards* in your classroom.

NCTM's vision of the future "sees students studying much of the same mathematics currently taught but with quite a different emphasis" and builds on the premise that "*what* a student learns depends to a great degree on *how* he or she has learned it." (p. 5, NCTM *Standards*)

How will your students learn algebra? The six articles that follow are designed to help you apply these six new approaches in your classroom.

Applications and Connections
Cooperative Learning Groups
Problem-Solving Strategies
Critical Thinking Questions
Reading, Writing, and Discussing Math
Technology

Each of these articles can help you to achieve these five general goals for all students as listed in the NCTM *Standards*. (p. 5-6, NCTM *Standards*)
"
1. *Learning to value mathematics*
2. *Becoming confident in one's own ability*
3. *Becoming a mathematical problem solver*
4. *Learning to communicate mathematically*
5. *Learning to reason mathematically*"

These goals suggest new approaches to much of the traditional content of an algebra course. **Standard 5: Algebra** (p. 150, NCTM *Standards*) states (in part) that the curriculum should include topics such that the students can
"
- *represent situations that involve variable quantities with expressions, equations, inequalities, and matrices;*
- *use tables and graphs as tools to interpret expressions, equations, and inequalities;*
- *operate on expressions and matrices, and solve equations and inequalities;*
- *appreciate the power of mathematical abstraction and symbolism.* "

These topics, of course, are addressed not only in *Holt Algebra 1*, but also in *Holt Pre-Algebra* and in *Holt Algebra with Trigonometry* as well. You will find ample material that relates directly to these topics as you survey *Holt Algebra 1*, as well as content that relates to the other Standards. Pages 659 through 663 of the *Annotated Teacher's Edition* repeat the text of all 14 Standards and correlate each lesson in *Holt Algebra 1* to the individual Standards. ■

APPLICATIONS AND CONNECTIONS

What are applications and connections?

A PERSON GATHERS, discovers, or creates knowledge in the course of some activity having a purpose. . . . instructions should persistently emphasize 'doing' rather than 'knowing that.'" (p. 7, NCTM *Standards*) Applications and connections are the ways in which something is done with mathematics. Applications are the purposeful activities that students do in order to apply, or make use of, the algebra they have learned. Connections are the activities that relate the algebra learned today to that which was previously learned, or to other branches of mathematics, or to other disciplines such as chemistry, physics, or biology.

It is true that a person can keep a checkbook and fill out a tax form without using algebra. Students therefore sometimes have difficulty seeing the relevance of what they are learning—"When am I ever going to use this?" They need to see how algebra and the skills they learn in class relate to getting a job, making decisions, and solving problems in the world outside the classroom.

Students in the United States drop out of mathematics classes at an alarming rate, and minorities drop out at a disproportionally greater rate. "Because mathematics is a key to leadership in our technological society, uneven preparation in mathematics contributes to unequal opportunity for economic power." (p. 18, *Educational Leadership,* Sept. 1989). Applications and connections give students a reason to continue their study of mathematics as they recognize its utility and importance.

How do I use applications and connections with my students?

Take every opportunity to relate mathematics and algebra to what is happening in the world today. Use newspapers, television programs, and magazines to relate mathematics to what is happening outside the classroom. Encourage students to bring in materials that show the use of mathematics in science and other disciplines. Invite them to share with the class the math they may be doing in some other course in school. Assign an interview so that students ask their parents and other adults what preparation in mathematics they had for their careers, and what mathematics they now wish they had taken in school. Use guest speakers to bring the applications of mathematics into the classroom.

For more formal experiences, use the worksheets in the *Teacher's ResourceBank.*™ The *Application Worksheets* and the *Project Worksheets* guide the students in ways to use the skills they have learned. The *Manipulative Worksheets* help students see concrete models for the abstract algebra in the lessons. All of these help to convince the students that algebra is both real and useful.

Take a few minutes to discuss the photographs that open each chapter. Share with the students the additional information that is in the *Annotated Teacher's Edition*. The *Annotated Teacher's Edition* also gives Connections that you can share with the class. Use the Applications sections that occur several times in each chapter to stimulate discussion of how math is used.

How will applications and connections help me implement the *Standards*?

The *Standards* (p. 125, NCTM *Standards*)

“ *call for a shift in emphasis from a curriculum dominated by memorization of isolated facts and procedures and by proficiency with paper-and-pencil skills to one that emphasizes conceptual understandings, multiple representations and connections, mathematical modeling, and mathematical problem solving.* ”

Applications and connections help you and your students take the step from solving equations with paper and pencil to applying algebra and other mathematics to the world outside the classroom. Manipulatives and applications devices can help stimulate that all-important student comment, so rewarding to the teacher, “Oh, now I see!” ∎

COOPERATIVE LEARNING GROUPS

What are cooperative learning groups?

THE *STANDARDS* POINT out that “instructional settings that encourage investigation, cooperation, and communication foster problem posing as well as problem solving.” (p. 138, NCTM *Standards*) Cooperative learning groups are such an instructional setting.

Cooperative learning means more than simple working in teams or groups. Some of the distinguishing characteristics of true cooperative learning are these.

- Groups are heterogeneous to increase tolerance and understanding.
- Students are responsible for their own participation and for making sure that others participate.
- Help is sought from within the group and the teacher is asked for help only when the whole group agrees to ask a question.
- Consensus on the answer is required from the whole group.
- The group evaluates their own strategies and ideas rather than relying on the teacher for this evaluation.
- The teacher provides help by means of questions rather than by giving hints.
- The students who learn at a faster pace do NOT do the task alone and then help other students.
- One student does NOT do all the work and then have the others sign off on it.
- Participants encourage each other to explain answers and how they arrived at them.
- Ideas, not people, are criticized.
- The logic of an idea, not peer pressure or majority rule, determines its value.

Rather than working in competition with other students, or as an individual working for a personal best, students in cooperative learning groups are working together towards shared goals; the success of each student in the group depends on the success of the group as a whole, which in turn depends on the success of each member. Each individual is accountable for mastering the material, and also for helping everyone in the group to master the material. Thus cooperative learning groups foster interdependence.

The benefits of this learning structure incorporate those of peer tutoring (in which the tutor also learns by helping), increased understanding of diverse viewpoints, and simulation of the way work is done on the job.

How do I use cooperative learning groups with my students?

Students may not be familiar with interdependent groups in a classroom setting, though they will usually have experienced the team approach in games. The first time you organize cooperative learning groups, discuss a set of procedural rules for the groups based on the characteristics in the foregoing list. The first time, let students form their own groups of four or five. Later, after they have begun to learn how to work together, you can use random methods to form groups that are heterogeneous.

For the first try, assign a familiar task such as producing study notes for a test. Tell the students that the goal of the activity includes learning the process (cooperation) as well as producing the product (the notes). Encourage each student to take responsibility for

a) participating and contributing,
b) ensuring that all other members of the group participate,
c) staying on task, and
d) producing a high quality final product.

Once the procedures are understood (though the groups will probably need much practice before they can adhere effectively to them), use the appropriate worksheets (Projects,

Applications, Manipulatives, and Problem Solving), and the Investigations that begin each chapter, to provide algebra-related tasks for the groups.

You can also use cooperative learning groups with the quizzes before the lesson to verify needed skills (Prerequisite Quiz, see also in the *Annotated Teacher's Edition*) and after the lesson to verify understanding (Checkpoint, see also in the *Annotated Teacher's Edition*). The Classroom Exercises that follow each lesson can also be done in cooperative groups to prepare students for the Written Exercises.

Conclude each session with a discussion about how the procedure worked, and how they can improve the process.

How will groups help me implement the *Standards*?

Research indicates that "cooperative learning experiences tend to promote higher achievement than do competitive and individualistic learning experiences." (p. 15, *Circles of Learning: Cooperation in the Classroom,* by Johnson, Johnson, Holubec, and Roy.) This results in part because the discussion process promotes the development of critical thinking skills and the group setting increases student motivation. The peer support and the oral repetition of information that occurs in the group also contribute to efficient learning.

The use of cooperative learning groups clearly belongs in an effective variety of instructional methods as called for by the *Standards*. (p. 125, 128, NCTM *Standards*)

" *A variety of instructional methods should be used in classrooms in order to cultivate students' abilities to investigate, to make sense of, and to construct meanings from new situations; to make and provide arguments for conjectures; and to use a flexible set of strategies to solve problems from both within and outside mathematics.* " ∎

PROBLEM-SOLVING STRATEGIES

What are problem-solving strategies?

THE NCTM *STANDARDS* recommend increased attention to "problem solving as a means as well as a goal of instruction." (p. 129 of the *Standards*) As a means of instruction, problem-solving strategies are taught overtly so that students can choose from many possible ways to begin solving a given problem.

Some students, especially those who have been unsuccessful in mathematics, have developed, on their own, only one problem-solving strategy—guess and give up. Teaching problem-solving strategies overtly shows students that there are many procedures used by successful problem-solvers to solve a problem. When you use problem solving as a means of instruction, students begin to realize that getting the right answer is not merely a matter of luck, innate talent, or magic. They begin to see that they too can experience success at mathematics.

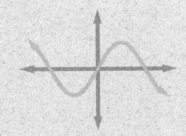

In order to create an atmosphere of successful learning, change the emphasis in your classroom from correct answers to effective process and procedures. Pay attention to the reasons and steps that go into solving a problem. Give credit for the process. Ask for, and value, the explanations students give to tell how they reached a solution. Emphasize that there may be more than one way to arrive at a correct answer. Encourage discussions about the advantages (such as efficiency and generalizability) of various methods. For some students, it may be appropriate to point out that the easiest and most efficient way for a human to solve a problem may not be the most efficient way for a computer program to solve that same problem.

How do I use these strategies with my students?

To use problem-solving strategies with your students, follow these guidelines:

1. Teach problem-solving strategies directly by using the *Problem Solving Worksheets* and the Problem-Solving Strategies features in the text.
2. When you discuss the Examples in the Lesson, point out the step where the student formulates a Plan for the problem.
3. Ask students to tell the class how they arrived at an answer.
4. Discuss with the class whether another strategy would have worked; compare different strategies that students may have used for a given problem.
5. Give recognition and credit for correct procedures as well as for correct answers; teach students that there may be only one correct answer but several correct ways to obtain that answer.
6. In reteaching students, ask them to explain their thoughts as they work the problem. This gives you a chance to help students to think logically and to suggest more effective problem-solving strategies.

How will problem solving help me implement the *Standards*?

" *The importance of problem solving to all education cannot be overestimated. To serve this goal effectively, the mathematics curriculum must provide many opportunities for all students to meet problems that interest and challenge them and that, with appropriate effort, they can solve.* " (p. 139, NCTM *Standards*) ∎

CRITICAL THINKING QUESTIONS

What are critical thinking questions?

CRITICAL THINKING questions ask the students to engage in mathematical thinking and to construct, symbolize, apply, and generalize mathematical ideas. This thinking can include investigations, constructing meanings from new situations, making and providing arguments for conjectures, working cooperatively, and creative and self-directed learning.

"Critical thinking" refers to types of thinking that are of a higher order in Bloom's taxonomy; these include application, analysis, synthesis, and evaluation.

Application involves applying concepts and ideas to new situations and is often signalled by these verbs: apply, build, choose, solve, plan, develop, construct, demonstrate, show.

Analysis means finding the underlying structure and breaking it down into stages or processes. This frequently involves looking for patterns and classifying examples and is often signalled by these verbs: relate, classify, compare, contrast, diagram, analyze, recognize.

Synthesis means bringing together data from various sources to come up with a new conclusion and is often signalled by these verbs and phrases: design, create, develop, make up, what happens if, invent, write a formula for.

Evaluation means judging the quality or worth of something, for example, which method works, or is more efficient. It is often signalled by these verbs and phrases: prove or disprove, evaluate, conclude, defend, choose, select, which is better, do you agree, judge, what do you think.

For example, in your algebra class, you might ask the students to show that one

expression is equivalent to another, and then discuss alternative approaches. You could ask questions such as these: "If substitution is used, how might you use another approach, such as simplifying? Can you think of several different ways to solve this problem? Which way is best, and why?" Critical thinking also includes investigating and exploring to derive properties, generalizations, and procedures.

How do I use these questions with my students?

Critical thinking questions appear throughout this *Holt Algebra 1* text. The Investigation that introduces each chapter and the Motivator that introduces each lesson (in the *Annotated Teacher's Edition*) often include such questions.

The Teaching Suggestions for each lesson include Critical Thinking questions pertinent to that lesson. Many of the worksheets end with critical thinking questions.

How will critical thinking help me implement the *Standards*?

The *Standards* call for "an environment that encourages students to explore, formulate and test conjectures, prove generalizations, and discuss and apply the results of their investigations." (p. 128, NCTM *Standards*)

Critical thinking questions are a vital part of such an environment. ■

READING, WRITING, AND DISCUSSING MATH

What is the role of reading, writing, and discussing math?

THE FOURTH GENERAL goal set forth by the *Standards* is "learning to communicate mathematically." In the past, we have sometimes allowed students to spell mathematical terms incorrectly, or to read aloud "x two" instead of "x-squared" or "x to the power of two" because, as students hastened to explain, "This isn't English class." We now recognize that it is just as important that correct English be spoken in math class as it is that correct math be used in science class. These disciplines are inter-related and teach skills that will be integrated on the job. When we require students to use their newly-acquired skills only in a particular class, we participate in an artificial separation of knowledge into school courses.

We also now realize that precise and accurate communication about mathematics is closely connected to doing precise and accurate mathematics. For the classroom teacher, communication about mathematics brings an extra benefit. When a student writes or speaks about a problem, the teacher can often identify a mistaken or incomplete understanding that leads to an error in problem solving.

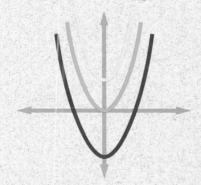

How do I encourage my students to communicate mathematically?

Ask a student to read an exercise (from the Classroom Exercises, for example) aloud and then answer the question. This helps you assess whether the student knows the vocabulary and the meaning of the various symbols.

Ask a student to put an exercise from the previous night's homework (in the Written Exercises) on the chalkboard and then to explain the steps and process to the class. You could then ask if anyone else followed a different procedure to reach the same result. After the second student has explained the alternate method, ask the class to discuss the differences and the advantages and drawbacks of each method. This works particularly well with word problems and solving equations, as there are often two equally valid ways to reach the same conclusion.

Ask students frequently to justify their conclusions, *before* you say whether or not the answer is correct. This encourages them to look beyond a correct answer to think about and defend their procedures and reasoning.

Encourage students to verbalize conjectures *before* they read in the text about certain facts and relationships. Discuss the closure questions that are given in *Holt Algebra 1* at the end of Investigations that begin each chapter in the *Annotated Teacher's Edition* and the various worksheets in the *Teacher's ResourceBank*™. Ask students to keep a math journal in which they record their questions, ideas, and reactions to the activities and discussions in class and a summary of what they read in the text.

Students learn to value those activities that they see you value. If you take time to listen to them as they learn to communicate mathematically, and if you allow their reading, writing, and discussing mathematics to contribute to your on-going evaluation of their achievements, they will know that these things really do count.

How will communicating mathematically help me meet the *Standards*?

The *Standards* (p. 6) put it this way:
" *The development of a student's power to use mathematics involves learning the signs, symbols, and terms of mathematics. This is best accomplished in problem situations in which students have an opportunity to read, write, and discuss ideas in which the use of the language of mathematics becomes natural. As students communicate their ideas, they learn to clarify, refine, and consolidate their thinking.* " ■

TECHNOLOGY
What is the role of technology?

THE NCTM *STANDARDS* speak of "removing the 'computational gate' to the study of high school mathematics" (p. 130, NCTM *Standards*) and goes on to say this:

" *By assigning computational algorithms to calculator or computer processing, this curriculum seeks not only to move students forward but to capture their interest.* "
(p. 130, NCTM *Standards*)

It is particularly important to note that the non-college-intending student must have better preparation for the jobs of tomorrow.

" *The ever-increasing role of technology in our society further argues for a curriculum that moves all students beyond computation.* "
(p. 130, NCTM *Standards*)

Technology—the use of the calculator, computer, and graphing calculator—enables students to "study mathematics that is more interesting and useful and not characterized as remedial" and this in turn "will enhance students' self-concepts as well as their attitudes

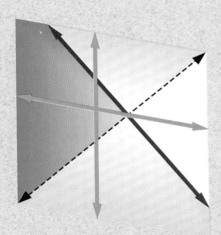

toward, and interest in, mathematics."
(p. 131, NCTM *Standards*)

How do I use technology with my students?

Almost everyone today uses a computer at work. The use of computers on the job and at home continues to grow. Make sure that *all* your students have access to whatever technology is available. The emphasis in today's curriculum is not on writing programs, but on investigation and foreshadowing of mathematical ideas and applications by means of the computer.

The Computer Investigation pages in *Holt Algebra 1* pose problems that can be solved by using the *Investigating Algebra with the Computer* software included in the *Teacher's ResourceBank*.

In the *Holt Algebra 1*, when problems are especially suited to the calculator, the text shows what calculator keys to press. However, you may want to encourage students to use hand-held calculators whenever they wish. This simulates mathematics as it is really used, both on the job and in scientific applications. Frequent use of calculators also helps students learn the importance of *estimating* before they calculate, and *checking* afterwards to be sure that answers are reasonable. "Appropriate use of calculators enhances children's understanding and mastery of arithmetic," according to *Everybody Counts, A Report to the Nation on the*

Future of Mathematics Education, published by the National Academy of Sciences. The *Standards* also make this point.

"*Contrary to the fears of many, the availability of calculators and computers has expanded students' capability of performing calculations. There is no evidence to suggest that the availability of calculators makes students dependent on them for simple calculations.*"
(p. 8, NCTM *Standards*)

The Computer Investigations pages help students to graph equations and systems easily with the help of technology. With these programs, they can make predictions and draw inferences about patterns.

Specific technology resources are listed for each chapter in *Holt Algebra 1* within a Technology paragraph on the introductory interleaf pages. Read this introduction to see how technology can be used in the following chapter.

How will technology help me implement the *Standards*?

The *Standards* propose increased attention to—

"*The use of calculators and computers as tools for learning and doing mathematics.*"
(page 129, NCTM *Standards*)

The technology resources in *Holt Algebra 1* help you use the computer in this way. ■

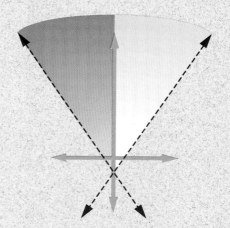

SUMMARY

AT THE BEGINNING OF this section, we listed the five general goals for all students from the NCTM *Standards*.

1. Learning to value mathematics
2. Becoming confident in one's own ability
3. Becoming a mathematical problem solver
4. Learning to communicate mathematically
5. Learning to reason mathematically

These articles have explained how various instructional techniques used with the textbook, special features, and worksheets from *Holt Algebra 1* can help you achieve those goals.

1. Cooperative learning groups help students value mathematics, as do the Applications worksheets and features in *Holt Algebra 1,* and the Math Connections discussed in the *Annotated Teacher's Edition.*
2. Manipulatives, cooperative learning groups, problem-solving strategies, critical thinking questions, and technology used with *Holt Algebra 1* and the accompanying worksheets all help students to become confident in their own abilities.
3. Problem-solving strategies, and the opportunities to practice and increase them provided by the Problem Solving features in the textbooks, and by the Problem Solving Worksheets, help students become mathematical problem solvers.

4. Cooperative learning groups, closure questions for each lesson in the *Annotated Teacher's Edition,* the questions that end the worksheets, and the features called Focus on Reading all help students learn to communicate mathematically.
5. The entire *Holt Algebra 1,* with its accompanying materials, helps the students learn to reason mathematically. The extensive practice with problem-solving strategies strengthens their reasoning abilities and their skills in communicating that reasoning to others. The chapter on analytic geometry (Chapter 11: *Analytic Geometry*) and the lessons on graphing throughout the text help students learn to reason from an analytic perspective.

The *Standards* make this comment in the context of program evaluation:

"*The mathematics classroom envisioned in the* Standards *is one in which calculators, computers, courseware, and manipulative materials are readily available and regularly used in instruction.*"
(p. 243, NCTM *Standards*)

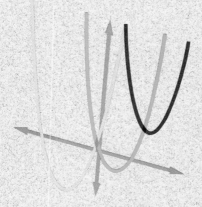

Holt Algebra 1 provides you with materials, in the textbook, in the *Annotated Teacher's Edition,* and in the *Teacher's ResourceBank*™ to help you meet these goals and standards as you teach your students. ■

1 INTRODUCTION TO ALGEBRA

OVERVIEW

The emphasis in this chapter is on simplifying and evaluating numerical and algebraic expressions. Students translate word descriptions into algebraic expressions and use the rules for order of operations, including parentheses and exponents, to simplify and evaluate expressions. These skills are applied to using formulas for perimeter, area, and volume. Then students identify and use the Commutative, Associative, and Distributive Properties to simplify expressions. In the last lesson, students find a solution set for an open sentence from a replacement set.

OBJECTIVES

- To write and evaluate algebraic expressions
- To simplify numerical and algebraic expressions
- To use formulas
- To identify and use the Commutative, Associative, and Distributive Properties
- To find solution sets of open sentences

PROBLEM SOLVING

Using a Formula, an important problem-solving strategy, is introduced on page 10, Example 5, and then used in all of Lesson 1.4 and in the Applications on pages 21 and 29. The strategy of Interpreting Key Words is illustrated in the chart on page 2. Finding More Than One Way to solve a problem by using the Commutative, Associative, and Distributive Properties is illustrated in Example 2 on page 18 and in Example 1 on page 22. The Polya four-step plan for problem-solving is formally introduced on page 35. The main focus of this lesson is on the first step of the plan, Understand the Problem.

READING AND WRITING MATH

The Focus on Reading on pages 11 and 33 reviews the math vocabulary used in this chapter. The Application on page 21 applies a mathematical formula in a real-world situation. Several exercises throughout the chapter ask the student to explain why or how a particular mathematical process works. These are on page 8 (Exercises 44 and 49), page 12 (Exercise 50), page 20 (Exercise 46), and page 37 (Exercise 28).

TECHNOLOGY

Calculator: Students can become familiar with how to enter problems correctly on the calculator starting with Exercises 30–37 on page 7. Calculators will also be helpful for Exercises 59–64 on page 12, Exercises 34–39 on page 16, and for the Application on page 21.

SPECIAL FEATURES

Mixed Review pp. 8, 12, 21, 25, 29, 34
Application: Temperature Humidity Index p. 8
Focus on Reading pp. 11, 33
Midchapter Review p. 16
Application: Heat Transfer p. 21
Application: Costs and Profit p. 29
Problem Solving Strategies: Understanding the Problem p. 35
Key Terms p. 36
Key Ideas and Review Exercises pp. 36–37
Chapter 1 Test p. 38
College Prep Test p. 39

PLANNING GUIDE

Lesson	Basic	Average	Above Average	Resources
1.1 pp. 3–4	CE 1–18, WE 1–39 odd	CE 1–18, WE 9–47 odd	CE 1–18, WE 13–53 odd	Reteaching 1 Practice 1
1.2 pp. 7–8	CE 1–12, WE 1–37 odd. Application	CE 1–12, WE 7–43 odd, Application	CE 1–12, WE 17–43 odd, 44–49 Application	Reteaching 2 Practice 2
1.3 pp. 11–12	FR all, CE 1–14, WE 1–45 odd, 50	FR all, CE 1–14, WE 13–49 odd, 50, 51–57 odd	FR all, CE 1–14, WE 19–49 odd, 50, 51–63 odd	Reteaching 3 Practice 3
1.4 pp. 14–16	CE 1–8, WE 1–23, 29–39 odd Midchapter Review	CE 1–8, WE 7–39 odd, Midchapter Review	CE 1–8, WE 13–39 odd, 41–44 Midchapter Review	Reteaching 4 Practice 4
1.5 pp. 19–21	CE 1–19, WE 1–41, Application	CE 1–19, WE 5–45 odd, Application	CE 1–19, WE 11–45 odd, 46–48 Application	Reteaching 5 Practice 5
1.6 pp. 24–25	CE 1–8, WE 1–33 odd, 40–43	CE 1–8, WE 9–49 odd	CE 1–8, WE 11–23, 29–49 odd, 50–53	Reteaching 6 Practice 6
1.7 pp. 28–29	CE 1–15, WE 1–25, 31–43 odd, Application	CE 1–15, WE 9–47 odd, Application	CE 1–15, WE 13–49 odd, 50, Application	Reteaching 7 Practice 7
1.8 pp. 33–35	FR all, CE 1–17, WE 1–29 odd Problem Solving Strategies	FR all, CE 1–17, WE 7–37 odd Problem Solving Strategies	FR all, CE 1–17, WE 11–43 odd Problem Solving Strategies	Reteaching 8 Practice 8
Chapter 1 Review pp. 36–37	all	all	all	
Chapter 1 Test p. 38	all	all	all	
College Prep Test p. 39	all	all	all	

CE = Classroom Exercises   WE = Written Exercises   FR = Focus on Reading

NOTE: For each level, all students should be assigned all Try This and all Mixed Review Exercises.

◼◼ INVESTIGATION

Project: This investigation is designed to help students develop number sense so that they can better evaluate the reasonableness of answers.

Materials: Each student will need a copy of Project Worksheet 1.

Before you distribute the worksheets, ask each student to estimate the answer to the following question and to write the estimate on a piece of paper.

> How many miles per hour does human hair grow?

Collect the papers (with no names) and write each estimate on the chalkboard. Point out the range of answers to the students and tell them that they will compare their estimates with their calculated answers later.

Distribute the worksheets and have the students work in small groups to complete the answers to all the questions. Students will need time to complete the calculations.

After completing the worksheets, have two groups meet together and compare both their answers and their problem-solving methods. Some groups may wish to rework the problem because of what they have learned from the discussion.

Ask the larger groups to report their final answers to the class. Compare these answers to the original estimates. The answer is about 0.0000001, or 10^{-8} mph.

Here are some further questions for student research.

1. How many seconds has rock music been around? About one billion seconds
2. How many years were dinosaurs on the earth? About 120 million years
3. How many years has modern man been on earth? About 250,000 years How many seconds? Less than 10 trillion seconds

In very dry forest areas, fire spotters of the U.S. Forest Service watch for smoke from towers high above the trees. Using this firefinder table, they can relay the exact direction to the nearest firefighting unit.

More About the Photo

Fire spotters point the long arm on the circular table towards the smoke or fire and use the crosshairs on the arm to pinpoint the direction. They then read an exact bearing (between zero and 360 degrees) from the table. The map on the wall gives the precise location of the tower within the assigned area. If a second bearing is relayed from another tower, the fire spotters plot two readings and calculate both the direction and the distance to the fire.

1.1 Algebraic Expressions

Objectives

To evaluate algebraic expressions for given values of the variables
To write algebraic expressions for word descriptions

The study of algebra involves numbers and operations. A **numerical expression** contains one or more numbers and may also contain one or more operations. The following are examples of numerical expressions.

$$12 \qquad 7.6 \qquad 5 + 9 \qquad 14 - 7 \times 2$$

In algebra, letters are often used to represent numbers. For example, when you buy movie tickets at $5 per ticket, the cost depends on the number of tickets you buy.

Number of tickets bought	Cost in dollars
1	5×1, or 5
2	5×2, or 10
3	5×3, or 15
4	5×4, or 20
5	5×5, or 25
⋮	⋮
n	$5 \times n$

In the chart above, the letter n is a *variable* that represents the number of tickets. A **variable** is a symbol that represents two or more numbers.

An **algebraic expression** contains one or more variables and usually one or more operations. It may also contain numbers. The algebraic expression $5 \times n$ can be written in several other ways:

$$5 \cdot n \qquad 5(n) \qquad (5)n \qquad 5n$$

To **evaluate an expression,** replace each variable with a number and find the numerical value. To evaluate $5n$ for $n = 6$, substitute 6 for n. The result is

$$5n = 5(6)$$
$$= 30$$

Thus, the value of $5n$ is 30 for $n = 6$.

Teaching Resources

Project Worksheet 1
Transparency 1
Quick Quizzes 1
Reteaching and Practice Worksheets 1

GETTING STARTED

Prerequisite Quiz

Add.

1. $4.6 + 5.9$ 10.5
2. $2.36 + 5.47$ 7.83
3. $23.5 + 4.62$ 28.12

Subtract.

4. $7.6 - 3.7$ 3.9
5. $4.23 - 0.51$ 3.72
6. $16.8 - 5.74$ 11.06

Motivator

Write these four expressions on the chalkboard. Have students complete the first three expressions.

1. $25 \times ? = 100$ 4
2. $2,450 \div \square = 35$ 70
3. $610 + \bigcirc = 720$ 110
4. $y - 15 = 60$ 75

Ask students what the first three expressions have in common.   A symbol or placeholder Ask them what the symbol or placeholder stands for.   An unknown number Ask students how the y in the last expression is like the symbols in the first three expressions.   It stands for a number. Ask them to name the number that y stands for.   75

Highlighting the Standards

Standard 1c: In Classroom Exercises 1–6, students formulate their own problems to fit the mathematical expressions.

Lesson Note

In Chapter 1, many Prerequisite Quizzes will provide a review of computations with fractions and decimals.

In this lesson, students evaluate algebraic expressions and also translate them into, and from, English expressions. Point out that 3 + 6 and 6 + 3 have the same value, but that their English translations are different.

> 3 + 6: the sum of 3 and 6
> 6 + 3: the sum of 6 and 3

Emphasize that a numerical expression names a unique (just one) number. Therefore, when a number is substituted for each variable in an algebraic expression, the expression names a unique number.

Review the meaning of the words sum, difference, product, and quotient.

Math Connections

History: The ÷ symbol for division first appeared in print in 1659 in an algebra book by Johann Heinrich Rahn (1622–1676). Discuss how this symbol seems to represent a fraction with dots replacing the numerator and denominator.

EXAMPLE 1 Evaluate $2xy$ for $x = 4$ and $y = 3$.

Solution

$2 \cdot x \cdot y$ ⟵ The expression $2xy$ means $2 \cdot x \cdot y$.
$\underbrace{2 \cdot 4 \cdot 3}$ ⟵ Substitute 4 for x and 3 for y.
$\underbrace{8 \cdot 3}$
$\quad 24$

TRY THIS

1. Evaluate $3mn$ for $m = \frac{2}{3}$ and $n = \frac{1}{2}$. 1

The chart below shows algebraic expressions with the variable x.

Algebraic expression	English expressions	Value for $x = 9$
$x + 28$	x plus 28 28 added to x x increased by 28 28 more than x the *sum* of x and 28	$x + 28$ $9 + 28$ 37
$42 - x$	42 minus x x subtracted from 42 x less than 42 the *difference*, $\quad$ 42 decreased by x	$42 - x$ $42 - 9$ 33
$16 \cdot x$, or $16x$	16 times x 16 multiplied by x the *product* of 16 and x	$16x$ $16 \cdot 9$ 144
$18 \div x$, or $\frac{18}{x}$	18 divided by x the *quotient*, $\quad 18 \div x$	$18 \div x$ $18 \div 9$ 2

Compare: x *decreased by* 16 and x *less than* 16

$\qquad x - 16 \qquad\qquad\qquad 16 - x$

EXAMPLE 2 Evaluate $\frac{x}{y}$ for $x = 12.8$ and $y = 4$.

Solution

$\frac{x}{y} = \frac{12.8}{4} = 3.2$ ⟵ Substitute 12.8 for x and 4 for y.

Thus, $\frac{x}{y} = 3.2$ for $x = 12.8$ and $y = 4$.

TRY THIS

2. Evaluate $\frac{2a}{b}$ for $a = 3.9$ and $b = 6$. 1.3

Additional Example 1

Evaluate $4xy$ for $x = 5$ and $y = 2.8$. 56

Additional Example 2

Evaluate $\frac{x}{y}$ for $x = 2.72$ and $y = 8$. 0.34

Algebraic expressions can represent word descriptions.

EXAMPLE 3 Write an algebraic expression for each word description.

   a. the cost in cents of p pounds of tomatoes at 79¢ per pound
   b. y years less than x years
   c. $\frac{3}{4}$ the number of students s
   d. twice the number of people p
   e. the cost of one pencil, in dollars, if n pencils cost \$1.46

Solutions   a. $79p$ b. $x - y$ c. $\frac{3}{4}s$ d. $2p$ e. $\frac{1.46}{n}$

TRY THIS   3. Write an algebraic expression for x dollars more than ten dollars. $(10 + x)$ dollars

Classroom Exercises

Give the meaning for each expression. Answers may vary (see chart on page 2).

   1. $x + 4$ 2. $5 - y$ 3. $6b$ 4. $\frac{a}{7}$ 5. cd 6. $3 \cdot y$

Evaluate.

   7. $3x$ for $x = 12$ 36   8. $5y$ for $y = 9$ 45   9. $a + 6$ for $a = 13$ 19
   10. $7 + z$ for $z = 17$ 24   11. $q - 5$ for $q = 23$ 18   12. $21 - r$ for $r = 6$ 15
   13. $\frac{x}{3}$ for $x = 18$ 6   14. $\frac{21}{y}$ for $y = 7$ 3   15. $p + q$ for $p = 9$ and $q = 3$ 12
   16. $-10a$ for $a = \frac{2}{5}$ -4 17. $30 - t$ for $t = 2.3$ 27.7 18. pq for $p = 10$ and $q = -3.4$
-34

Written Exercises

Evaluate.

   1. $7p$ for $p = 9$ 63   2. $x + y$ for $x = 12$ and $y = 9$ 21
   3. $\frac{16}{y}$ for $y = 2$ 8   4. $\frac{x}{9}$ for $x = 63$ 7
   5. $a + b$ for $a = 7$ and $b = 13$ 20   6. $8z$ for $z = 15$ 120
   7. $c - d$ for $c = 18$ and $d = 3$ 15   8. mn for $m = 3$ and $n = 15$ 45
   9. $3rs$ for $r = 2$ and $s = 7$ 42   10. $p - q$ for $p = 28$ and $q = 14$ 14
   11. $\frac{a}{b}$ for $a = 52$ and $b = 13$ 4   12. $\frac{c}{d}$ for $c = 48$ and $d = 3$ 16

Write an algebraic expression for each word description.

   13. the cost, in dollars, of y pounds of   14. \$45 decreased by the amount spent, s,
   peaches at 89¢ per pound $0.89y$ in dollars $45 - s$

Additional Example 3

Write an algebraic expression for each word description.

   a. the cost, in cents, of x kilograms of potatoes at 60¢ per kilogram $60x$
   b. y inches less than 14 inches $14 - y$
   c. $\frac{3}{5}$ of the original price P $\frac{3}{5}P$

   d. 8 times the number of miles, m $8m$
   e. The cost, in cents, of one marble if 24 marbles cost n¢ $\frac{n}{24}$

Ask students how an algebraic expression differs from a numerical expression. See p. 1. What does it mean to evaluate an expression, e.g., to evaluate $3n$ for $n = 12$? 36

◼◼◼FOLLOW UP

Guided Practice

Classroom Exercises 1–18
Try This all

Independent Practice

A Ex. 1–28, **B** Ex. 29–44, **C** Ex. 45–54

Basic: WE 1–39 odd

Average: WE 9–47 odd

Above Average: WE 13–53 odd

15. the total number of weeks y in x days $\frac{x}{7}$

16. the cost, in dollars, of each sticker if n stickers cost $\$1.35$ $\frac{1.35}{n}$

17. the number of sunny days s subtracted from 365 $365 - s$

18. the cost, in cents, of 5 pounds of beans at c cents per pound $5c$

19. $\frac{2}{3}$ of the number of girls g $\frac{2}{3}g$

20. 5 less than the number of boys b in the class $b - 5$

21. 3 more than the number of cars c in the parking lot $c + 3$

22. twice the number of record albums r $2r$

23. $\frac{3}{4}$ of the number of seats s $\frac{3}{4}s$

24. $1\frac{1}{2}$ times the regular pay rate r $1\frac{1}{2}r$

25. $\frac{1}{3}$ the length b of a bat $\frac{1}{3}b$

26. x inches less than 8 inches $8 - x$

27. two-tenths of the population p $0.2p$

28. 5 times the number of books b $5b$

Evaluate.

29. $x + 8$ for $x = 22.7$ 30.7

30. $24.3 - n$ for $n = 9.7$ 14.6

31. $n - 2.4$ for $n = 33$ 30.6

32. $n \div 36$ for $n = 7.2$ 0.2

33. $x + \frac{1}{2}$ for $x = 2\frac{1}{2}$ 3

34. $y - \frac{1}{4}$ for $y = 5\frac{3}{4}$ $5\frac{1}{2}$

35. $\frac{3.6}{p}$ for $p = 3$ 1.2

36. $1.2w$ for $w = 8$ 9.6

37. ab for $a = 0.4$ and $b = 13$ 5.2

38. $b - a$ for $a = 8$ and $b = 13.6$ 5.6

39. $\frac{m}{n}$ for $m = 54$ and $n = 0.6$ 90

40. $\frac{3}{5}y$ for $y = \frac{2}{3}$ $\frac{2}{5}$

41. $\frac{m}{n}$ for $m = 3.51$ and $n = 100$ 0.0351

42. $x - y$ for $x = 0.54$ and $y = 0.37$ 0.17

43. $\frac{x}{3.2}$ for $x = 13.12$ 4.1

44. $\frac{23.22}{m}$ for $m = 4.3$ 5.4

45. xy for $x = \frac{1}{3}$ and $y = 2\frac{1}{4}$ $\frac{3}{4}$

46. $a + b$ for $a = 1.25$ and $b = 1.5$ 2.75

47. $c - d$ for $c = \frac{3}{5}$ and $d = \frac{1}{4}$ $\frac{7}{20}$

48. rs for $r = \frac{3}{8}$ and $s = 1\frac{1}{3}$ $\frac{1}{2}$

49. $\frac{m}{n}$ for $m = 6.2$ and $n = 0.4$ 15.5

50. $\frac{a}{b}$ for $a = \frac{2}{3}$ and $b = \frac{3}{9}$ 2

Write an algebraic expression for each word description.

51. A number x is 4 less than another number. Write an algebraic expression for the larger number. $x + 4$

52. Jim's height h is twice Bill's height. Write an algebraic expression for Bill's height. $\frac{1}{2}h$

53. The number of dollars d in Carey's bank is one-third the number of dollars in Jane's bank. Write an algebraic expression for the number of dollars in Jane's bank. $3d$

54. The number of fans f is 18 more than the number of players. Write an algebraic expression for the number of players. $f - 18$

4 Chapter 1 Introduction to Algebra

Enrichment

Have students look up the word "variable" in the dictionary and record the different meanings given. Then have students try to write a one-word synonym for "variable." Possible answers: changeable, alterable, mutable, and so on

1.2 Grouping Symbols

Teaching Resources

Quick Quizzes 2
Reteaching and Practice
  Worksheets 2
Problem Solving Worksheet 1
Transparency 2

Objectives   To simplify numerical expressions by using the rules for order of operations
To evaluate algebraic expressions

To **simplify an expression** such as $4 + 5 + 8$, complete the addition and write the sum 17. Examine how $36 - 5 \cdot 4$ might be simplified.

Multiply first; then subtract.   Subtract first; then multiply.

$$36 - 5 \cdot 4$$
$$36 - \quad 20$$
$$16 \longleftarrow \text{Correct}$$

$$36 - 5 \cdot 4$$
$$31 \quad \cdot 4$$
$$124 \quad \longleftarrow \text{Incorrect}$$

It would be confusing for a numerical expression to name more than one number. So the following order of operations is agreed upon.

Order of Operations
1. Do all multiplications and divisions in order from left to right.
2. Do all additions and subtractions in order from left to right.

EXAMPLE 1   Simplify: **a.** $16 + 8 \cdot 9$ **b.** $18 - 8 \div 4$

Solutions   **a.** Multiply first; then add.   **b.** Divide first; then subtract.

$$16 + 8 \cdot 9$$
$$16 + \quad 72$$
$$88$$

$$18 - 8 \div 4$$
$$18 - \quad 2$$
$$16$$

TRY THIS   Simplify: **1.** $24 - 6 \div 3$ 22   **2.** $12 + 4 \cdot 3$ 24

Parentheses ( ) and brackets [ ] are called *grouping symbols*. Do operations within grouping symbols first. A multiplication symbol may be omitted when it occurs next to a grouping symbol. For example,

$$3 \cdot (5 + 2) = 3(5 + 2) = 3(7).$$

If there is more than one set of grouping symbols, operate within the innermost symbols first.

$$5[8 + (7 - 3)] = 5[8 + 4] = 60$$

GETTING STARTED

Prerequisite Quiz

Multiply.

1. 16×2.5 40
2. 23×7.21 165.83
3. 7.9×6.1 48.19
4. 2.05×4.6 9.43

Divide.

5. $25.4 \div 2$ 12.7
6. $56.4 \div 8$ 7.05
7. $42.93 \div 5.3$ 8.1
8. $6.386 \div 2.06$ 3.1

Motivator

Ask students to evaluate $36 + 12 \div 4 - 2 \cdot 7$. Many will get an answer of 70; others, who know the order of operations, may get the correct answer of 25. Without using grouping symbols, talk through the procedure of obtaining other answers, such as $(36 + 12) \div (4 - 2) \cdot 7 = 168$ or $36 + [12 \div (4 - 2)] \cdot 7 = 78$. Ask students to suggest a solution to this problem. Lead them to a realization of the need for rules relating to order of operations.

Highlighting the Standards

Standard 2c: Students learn to read and write mathematical notation.

Lesson Note

Emphasize that the rules for order of operations represent an agreement that is made to ensure that a given numerical expression will name a unique number. Point out that the fraction in Example 4 is another way of writing $2(13 - 7) \div (15 - 6 \cdot 2)$ and that is why the numerator and denominator are simplified first.

Students need to be aware that some calculators do not follow the standard rules for order of operations and that using parentheses or storing intermediate results in memory is necessary for arriving at the correct answer.

Suggest that the phrase "My Dear Aunt Sally" gives the operations in their correct order.

Explain that to "simplify" usually means to make as short as possible or to express as concisely as possible.

Math Connections

Computers: When writing mathematical expressions on the computer, nested parentheses are used. For example, $18 - 6[4 \div (9 - 7)]$ must be entered as $18 - 6 \cdot (4 / (9 - 7))$. Discuss the difference between notation that is easier for humans to read and consistency of symbols that is easier for a non-seeing computer to read.

EXAMPLE 2 Simplify.

    a. $11 + 21 \div (16 - 9)$ **b.** $18 - 6[4 \div (9 - 7)]$

Solutions

a. $11 + 21 \div \underline{(16 - 9)}$

   $11 + \underline{21 \div 7}$

   $\underline{11 + 3}$

       14

b. $18 - 6[4 \div \underline{(9 - 7)}]$

   $18 - 6[\underline{4 \div 2}]$

   $18 - \underline{6[2]}$

   $\underline{18 - 12}$

       6

EXAMPLE 3 Evaluate $11 - 8x$ for $x = \frac{3}{4}$.

Solution

$11 - 8x = 11 - \underline{8 \cdot \frac{3}{4}} \longleftarrow 8 \cdot \frac{3}{4} = \frac{\overset{2}{\cancel{8}}}{1} \cdot \frac{3}{\underset{1}{\cancel{4}}} = \frac{6}{1} = 6$

       $= \underline{11 - 6}$

       $= 5$

TRY THIS
3. Simplify $15 - 2[3(7 - 5)]$. 3
4. Evaluate $20 - 2y$ for $y = 4.5$. 11

A fraction bar is also a grouping symbol. Perform operations within the numerator and the denominator before simplifying further.

EXAMPLE 4 Simplify $\dfrac{2(13 - 7)}{15 - 6 \cdot 2}$.

Solution

$\dfrac{2(13 - 7)}{15 - 6 \cdot 2} = \dfrac{2 \cdot 6}{15 - 12} = \dfrac{12}{3}$, or 4

EXAMPLE 5 Evaluate $x + 5(8 - y)$ for $x = 9$ and $y = 3$.

Solution

$x + 5(8 - y) = 9 + 5(8 - 3)$

          $= 9 + 5 \cdot 5$

          $= 9 + 25 = 34$

TRY THIS
5. Simplify $\dfrac{3(15 - 7)}{12 - 3 \cdot 2}$. 4   6. Evaluate $3(x + 5) - y$ for $x = 2$ and $y = 4$. 17

6   Chapter 1 Introduction to Algebra

Additional Example 2

Simplify.

a. $36 + 18 \div (4 + 5)$ 38
b. $70 - 5[18 - (9 - 3)]$ 10

Additional Example 3

Evaluate $7 + 24x$ for $x = \frac{5}{6}$.   27

Additional Example 4

Simplify $\frac{4(1 + 8)}{18 - 3 \cdot 2}$.   3

Classroom Exercises

What is the first operation to perform in simplifying each expression?

1. $6 \cdot 8 - 5$
Mult

2. $4 + 3 \cdot 9$
Mult

3. $7 - 8 \div 2$
Div

4. $\dfrac{16}{5 + 3}$ Add

Simplify.

5. $14 - 5 \cdot 2$ 4

6. $(14 - 5) \cdot 2$ 18

7. $12 + 6 \div 3$ 14

8. $(12 + 6) \div 3$ 6

Evaluate.

9. $5 + 2x$ for $x = 3$ 11

10. $14 \div c - 3$ for $c = 2$ 4

11. $7 - (8 - a)$ for $a = 6$ 5

12. $x \div (9 - 7)$ for $x = 4$ 2

Written Exercises

Simplify.

1. $15 \cdot 7 + 9$ 114

2. $15 + 7 \cdot 9$ 78

3. $20 - 4 \cdot 3$ 8

4. $24 \div 3 + 5$ 13

5. $18 - 2 \cdot 9$ 0

6. $24 - 48 \div 2$ 0

7. $43 + 140 \div 70$ 45

8. $24 \div 8 \cdot 2$ 6

9. $24 \div 12 + 8 - 6$ 4

10. $17 - 4 + 2 \cdot 8$ 29

11. $12 + 13 - (12 + 4)$ 9

12. $12 + (13 - 12) + 4$ 17

13. $(12 + 13) - (12 + 4)$ 9

14. $12 - (13 - 12 + 4)$ 7

15. $12 - (13 - 12 \div 4)$ 2

16. $12 + 3(4 \cdot 5)$ 72

17. $12 - (15 \div 3)$ 7

18. $12 - 2(15 \div 3)$ 2

19. $\dfrac{18 - 12}{13 - 7}$ 1

20. $\dfrac{18 - 7 \cdot 2}{3 - 1}$ 2

21. $\dfrac{3(10 - 4)}{15 - 3 \cdot 3}$ 3

Evaluate.

22. $16 - 3x$ for $x = 4$ 4

23. $12x + 9$ for $x = 3$ 45

24. $2a - 6 \div 2$ for $a = 8$ 13

25. $14 + y \div 5$ for $y = 30$ 20

26. $17 - 18 \div x$ for $x = 6$ 14

27. $5a + 3 \cdot 4$ for $a = 2$ 22

28. $15 - 9d$ for $d = \frac{2}{3}$ 9

29. $4y + 2 \cdot 7$ for $y = 3.5$ 28

Simplify.

30. $4.2 + 3.1 \times 4$ 16.6

31. $5.8 - 2.1 \div 7$ 5.5

32. $8.4 \div (6.3 \div 3)$ 4

33. $8.1 - 2(4.3 - 1)$ 1.5

34. $\dfrac{4.02 + 3.21}{3}$ 2.41

35. $10(5.2 - 1.4)$ 38

36. $9 - 2(4.9 - 3.2)$ 5.6

37. $5[8 - (6.2 + 1.8)]$ 0

1.2 Grouping Symbols **7**

Critical Thinking Questions

Application: Insert parentheses in each expression so that it has the value indicated.

1. $24 + 8 \cdot 12 \div 4 - 2 = 94$

2. $24 + 8 \cdot 12 \div 4 - 2 = 72$

3. $24 + 8 \cdot 12 \div 4 - 2 = 28$
$(24 + 8) \cdot (12 + 4) - 2$; $24 + 8 \cdot$
$[12 \div (4 - 2)]$; $(24 + 8 \cdot 12) \div 4 - 2$

Common Error Analysis

Error: Some students will want to change the order of operations into four steps instead of two, thinking that all multiplication must be done before division.

Reinforce the *two* steps by showing that a computer will read through the following line twice from left to right: $18 \div 3 + 6 \cdot 4 - 8$. On the first pass it will find the $\div$ and $\cdot$ signs; it will then go back and pick up the $+$ and $-$ signs in order from left to right.

Checkpoint

Simplify.

1. $16 \div 8 + 9 \cdot 7$ 65

2. $8 - [14 \div (2 + 5)]$ 6

3. $7.3 - 8.4 \div 1.4$ 1.3

4. Evaluate $x + y(3 - x)$ for $x = 1$ and $y = 7$. 15

Additional Example 5

Evaluate $\dfrac{y - 2(x + 3)}{y - 3x}$ for $x = 3$ and $y = 12$. 0

Closure

Ask students why it is important to have a standard set of rules for order of operations.   To avoid confusion
Then ask: How many operations are contained in the expression $23 - 3 \cdot 5$? 2
Is there more than one correct solution?   No   What does it mean to simplify a mathematical expression?   To express it more concisely

■■■FOLLOW UP

Guided Practice

Classroom Exercises 1–12
Try This all

Independent Practice

A Ex. 1–29 **B** Ex. 30–44 **C** Ex. 45–49
Basic: WE 1–37 odd, Application
Average: WE 7–43 odd, Application
Above Average: WE 17–43 odd, 44–49, Application

Additional Answers

Written Exercises

44. Agreement on the order of operations is needed so that a numerical expression will name only one number.
49. Some calculators obey the rules for order of operations; others do not.

Evaluate.

38. $a + 2(b - 4)$ for $a = 5$ and $b = 9$ 15

39. $x - 4(5 - y)$ for $x = 20$ and $y = 3$ 12

40. $m(n + 1) - (m + 1)$ for $m = 5$ and $n = 2$ 9

41. $c + 2[d + d(c - 1)]$ for $c = 4$ and $d = 2$ 20

42. $x - (y - 1)$ for $x = 20$ and $y = 3$ 18

43. $(x + 2) - (y + 3)$ for $x = 7$ and $y = 2$ 4

44. Write an explanation of the need for an agreement on the order of operations.

Write an algebraic expression for each word description.

45. the amount, in dollars, that will remain if you have \$360 now and you spend \$8 a week for w weeks $360 - 8w$

46. the repair bill, in dollars, for x hours if a man charges a base fee of \$25 plus \$15 per hour $25 + 15x$

47. the total cost, in cents, of p pencils at 10¢ each and t tablets at 65¢ each $10p + 65t$

48. 10 lb less than twice the weight y of a package $2y - 10$

49. After you enter 10 $\boxed{+}$ 2 $\boxed{\times}$ 3 $\boxed{=}$, some calculators display 36 and others display 16. Explain why.

Mixed Review

Evaluate.   1.1

1. $15 - y$ for $y = 8$ 7

2. $3x$ for $x = 2.1$ 6.3

3. $4d$ for $d = \frac{3}{4}$ 3

4. $\frac{21}{c}$ for $c = 3$ 7

5. $\frac{a}{b}$ for $a = 28, b = 7$ 4

6. $\frac{m}{n}$ for $m = 28, n = 0.8$ 35

7. Write an algebraic expression for the cost, in dollars, of p pounds of grapes at \$1.09 per pound. $1.09p$

Application: Temperature Humidity Index (THI)

Weather forecasters sometimes report a Temperature Humidity Index, (THI). The THI is a number that measures the degree of discomfort you may feel because of the amount of water vapor in the air. The higher the index, the greater the discomfort. You can find the THI by using the formula, THI $= 0.4(t + s) + 15$, where t is the temperature of the air and s is the temperature of a thermometer with a moistened cloth on its bulb.

For example, if $t = 84$ and $s = 71$,

$$\text{THI} = 0.4(84 + 71) + 15 = 0.4(155) + 15 = 77.$$

Find the THI for each pair of Fahrenheit temperatures.

1. $t = 80, s = 65$ 73

2. $t = 86, s = 74$ 79

3. $t = 100, s = 81$ 87.4

8   Chapter 1 Introduction to Algebra

Enrichment

Have the students write this expression.

$$2 + 5 \cdot 4 - 4 \div 2$$

Then have them rewrite the expression in as many ways as they can by inserting *one* set of parentheses. Have them evaluate each of the resulting expressions, as well as the original. HINT: There are 10 expressions, including the original.

$2 + 5 \cdot 4 - 4 \div 2$ 20
$(2 + 5) \cdot 4 - 4 \div 2$ 26
$2 + (5 \cdot 4) - 4 \div 2$ 20
$2 + 5 \cdot (4 - 4) \div 2$ 2
$2 + 5 \cdot 4 - (4 \div 2)$ 20
$(2 + 5 \cdot 4) - 4 \div 2$ 20
$2 + (5 \cdot 4 - 4) \div 2$ 10
$2 + 5 \cdot (4 - 4 \div 2)$ 12
$(2 + 5 \cdot 4 - 4) \div 2$ 9
$2 + (5 \cdot 4 - 4 \div 2)$ 20

1.3 Exponents

Objectives   To simplify expressions containing exponents
To evaluate expressions containing exponents

Teaching Resources

Quick Quizzes 3
Reteaching and Practice
Worksheets 3

If a car is going 40 mi/h on dry concrete, it takes about 60 ft to stop after the brakes are applied. If it is going 55 mi/h, the braking distance is about twice as long. The braking distance can be estimated by using the *formula* $d = 0.04 \times s \times s$. A simpler way to write this formula uses *exponents*: $d = 0.04s^2$.

In the product $7 \cdot 3$, 7 and 3 are *factors*. The product $5 \cdot 5 \cdot 5$ consists of the factor 5 used three times. Another way to write $5 \cdot 5 \cdot 5$ is to use an exponent. The *exponent* indicates the number of times the *base* is used as a factor.

base $\longrightarrow$ 5^3 $\longleftarrow$ exponent

5^3 is read *5 to the third power* or *5 cubed*. Other expressions containing exponents are shown in the table.

Exponential expression	English expression	Base	Exponent	Meaning
7^2	7 to the second power, or 7 squared	7	2	$7 \cdot 7$, or 49
4^3	4 to the third power, or 4 cubed	4	3	$4 \cdot 4 \cdot 4$, or 64
6^1	6 to the first power	6	1	6
x^5	x to the fifth power	x	5	$x \cdot x \cdot x \cdot x \cdot x$

The rules for simplifying an expression can be extended to include expressions containing exponents.

Order of Operations
1. Operate within grouping symbols first. Work from the inside to the outside.
2. Simplify powers.
3. Multiply and divide from left to right.
4. Add and subtract from left to right.

1.3 Exponents   **9**

■■■ GETTING STARTED

Prerequisite Quiz

Multiply.

1. $3 \times 3 \times 3$ 27
2. $4 \cdot 4 \cdot 4 \cdot 4$ 256
3. $2 \cdot 2 \cdot 2 \cdot 2 \cdot 2$ 32
4. $3 \cdot 3 \cdot 6 \cdot 6$ 324
5. $5 \cdot 6 \cdot 2 \cdot 2$ 120
6. $9 \cdot 9 \cdot 9$ 729

Motivator

Remind students that *square* and *cube* are related to their geometric counterparts, *area* and *volume*. *Ask*: Why is area always given in square units?   Because you multiply a length and a width, such as 9 ft and 3 ft, to get area, and ft × ft = ft². *Ask*: Why is volume given in cubic units?   Because volume is the product of three factors, such as 3 ft × 2 ft × 1 ft, and ft × ft × ft = ft³.

Highlighting the Standards

Standards 1d, 2a: In the opening problem and in Classroom Exercises 51–58, students are shown how to apply the process of mathematical modeling to braking a car. In Exercise 50, they clarify their thinking about exponents through writing.

Lesson Note

Emphasize the distinction between an exponent and a factor. Have students compare expressions such as $5 \cdot 2$ and 5^2. Point out also that $5 \cdot 2 = 2 \cdot 5$, but $2^5 \neq 5^2$. The phrase Please Excuse My Dear Aunt Sally can be used as a mnemonic for the order of operations: (P)arentheses, (E)xponents, (M)ultiplication and (D)ivision, (A)ddition and (S)ubtraction. Emphasize the difference between $4x^3$ and $(4x)^3$. Tell students that an exponent operates on only the one number or variable directly before it unless parentheses are used. That is, $4x^3$ means $4 \cdot x \cdot x \cdot x$ while $(4x)^3$ means $(4x) \cdot (4x) \cdot (4x)$.

Math Connections

Science: Show students one of the most famous formulas of this century: $E = mc^2$. Point out that E stands for energy, m for mass, and c for the speed of light, which is approximately 186,000 mi/sec or 300,000 km/sec. Squaring such a large number demonstrates the "power" of Einstein's formula.

EXAMPLE 1 Simplify: **a.** $6 \cdot 5^2$ **b.** $4^2 \cdot 1^3 + 8$

Solutions
a. $6 \cdot 5^2$
$\quad 6 \cdot 5 \cdot 5$
$\quad 6 \cdot 25$
$\quad 150$

b. $4^2 \cdot 1^3 + 8$
$\quad 4 \cdot 4 \cdot 1 \cdot 1 \cdot 1 + 8$
$\quad 16 + 8$
$\quad 24$

EXAMPLE 2 Simplify: $\dfrac{2^3 + 4^2}{(5 - 3)^2}$

Solution $\dfrac{2^3 + 4^2}{(5 - 3)^2} = \dfrac{2 \cdot 2 \cdot 2 + 4 \cdot 4}{(2)^2} = \dfrac{8 + 16}{2 \cdot 2} = \dfrac{24}{4} = 6$

TRY THIS Simplify: **1.** $2^4 \cdot 4$ 64 **2.** $\dfrac{3^3 + 1}{(3 - 1)^2}$ 7

Example 3 shows that $4x^3$ and $(4x)^3$ do not mean the same thing.

EXAMPLE 3 Evaluate: **a.** $4x^3$ for $x = 2$ **b.** $(4x)^3$ for $x = 2$

Solutions
a. $4x^3 = 4 \cdot 2^3$
$\qquad = 4 \cdot 2 \cdot 2 \cdot 2$
$\qquad = 32$

b. $(4x)^3 = (4 \cdot 2)^3$
$\qquad = 8^3$
$\qquad = 8 \cdot 8 \cdot 8 = 512$

EXAMPLE 4 Evaluate $3x^2 - 2x + 1$ for $x = 4$.

Solution $3x^2 - 2x + 1 = 3 \cdot 4^2 - 2 \cdot 4 + 1$
$\qquad\qquad\qquad\quad = 3 \cdot 16 - 8 + 1$
$\qquad\qquad\qquad\quad = 48 - 8 + 1$
$\qquad\qquad\qquad\quad = 40 + 1 = 41$

EXAMPLE 5 The distance an object falls if dropped from any height is found by the formula $d = 5t^2$, where d is the distance in meters (m) and t is the time in seconds. Find d if $t = 5.2$ seconds. Round the answer to the nearest whole number.

Solution Substitute 5.2 for t and evaluate. **Calculator Steps:**
$d = 5t^2 = 5 \times (5.2)^2 = 5 \times 5.2 \times 5.2$ $5 \;\boxed{\times}\; 5.2 \;\boxed{x^2}\; \boxed{=}\; 135.2$
$\qquad\qquad\qquad\qquad\qquad = 135.2$

Thus, the distance is about 135 m.

TRY THIS **3.** Evaluate $2x^2 - x + 3$ **4.** Evaluate $d = 5t^2$
$\qquad\qquad$ for $x = 3$. 18 $\qquad$ for $t = 0.3$. 0.45

10 Chapter 1 Introduction to Algebra

Additional Example 1

Simplify.

a. $4 \cdot 3^2$ 36
b. $5^2 \cdot 2^3 - 10$ 190
c. $(3 \cdot 4)^2 - 5^2$ 119

Additional Example 2

Simplify $\dfrac{3^3 - 4^2}{(7 + 1)^2}$ $\dfrac{11}{64}$

Additional Example 3

Evaluate.

a. $2x^3$ for $x = 6$ 432
b. $(2x)^3$ for $x = 6$ 1,728

Focus on Reading

Use one of the words at the right to complete each sentence.

1. In x^3, x is the __e__ .
2. In y^4, 4 is the __c__ .
3. 5^4 means 5 to the fourth __f__ .
4. In 6^5, 6 is used as a(n) __d__ 5 times.
5. $7y^3$ means the __a__ of 7 and y cubed.
6. $4z^2$ means 4 times z __b__ .

   a. product
   b. squared
   c. exponent
   d. factor
   e. base
   f. power

Classroom Exercises

Use exponents to rewrite the expressions in Exercises 1–6.

1. $5 \times 5 \times 5$ 5^3
2. $3 \cdot 3 \cdot 2 \cdot 2 \cdot 2$ $3^2 2^3$
3. $x \cdot x \cdot x \cdot y$ $x^3 y$
4. $x \cdot x + 1$ $x^2 + 1$
5. $a \cdot a - b \cdot b + c \cdot c$ $a^2 - b^2 + c^2$
6. $\dfrac{3 \cdot c \cdot c + 2 \cdot d \cdot d}{c \cdot c - d \cdot d \cdot d}$ $\dfrac{3c^2 + 2d^2}{c^2 - d^3}$

Simplify.

7. 6^2 36
8. 10^3 1,000
9. 2^4 16
10. $3^3 - 2$ 25
11. $2 \cdot 5^2$ 50
12. $7^2 + 1$ 50
13. $6 \cdot 5^2$ 150
14. $100^2 + 3$ 10,003

Written Exercises

Simplify.

1. 3^4 81
2. 2^6 64
3. 10^5 100,000
4. $5 \cdot 2^3$ 40
5. $3^2 \cdot 6$ 54
6. $(5 \cdot 2)^3$ 1,000
7. $5^3 \cdot 2^3$ 1,000
8. $4 \cdot 10^3$ 4,000
9. $3 + 7^2$ 52
10. $(3 + 7)^2$ 100
11. $(45 - 5)^2$ 1,600
12. $45 - 5^2$ 20
13. $(5 \cdot 4)^2$ 400
14. $5 \cdot 10^3$ 5,000
15. $5^2 - 4^2$ 9
16. $(5 - 4)^2$ 1
17. $2^6 + 2^4 + 2^2$ 84
18. $3^5 - 3^4 + 3^3 - 3^2$ 180
19. $9^2 - 3 \cdot 5$ 66
20. $6^2 + 2 \cdot 3^2$ 54
21. 4.6×10^3 4,600
22. 10.3×2^3 82.4
23. $(1^3 + 2^2 + 3) \div 4$ 2
24. $(4^3 \div 2^5) + 5$ 7
25. $8^2 \div (3^2 - 1^2)$ 8
26. $5^3 - 3 \cdot 5^2 + 2 \cdot 5$ 60
27. $(2^5 - 3^3)^2 - 5^2$ 0
28. $4^2 + 8(16 - 3^2)$ 72
29. $\dfrac{8^2 - 6^2}{10 - 6}$ 7
30. $\dfrac{4^3 - 2}{4^2 + 15}$ 2
31. $\dfrac{10^2 - 5^2}{(9 - 4)^2}$ 3

Use the formula $d = 5t^2$ to find d for each value of t. Round your answer to the nearest meter.

32. $t = 0$ 0 m
33. $t = 1$ 5 m
34. $t = 10$ 500 m
35. $t = 6$ 180 m
36. $t = 0.5$ 1 m
37. $t = 1.5$ 11 m

1.3 Exponents **11**

Critical Thinking Questions

Application: Remind students that $3^2 \neq 2^3$ and, in general, $x^y \neq y^x$. However, there are instances when $x^y = y^x$, where x and y are positive integers. Ask them to find as many such instances as they can.   Sample answer: $x = 2$, $y = 4$

Common Error Analysis

Error: Students may evaluate $4 \cdot 3^2$ as $12^2 = 144$.

Remind them that the base in this problem is 3 and that, according to the order of operations, powers are simplified before multiplication is completed. Compare this problem with $(4 \cdot 3)^2$

Checkpoint

Simplify.

1. $(6 \cdot 3)^2$ 324
2. $6^2 - (4 + 3^2)$ 23
3. $\dfrac{2^3 + 1^5}{(5 + 2)^2}$ $\dfrac{9}{49}$
4. Evaluate $6x^2 - x$ for $x = 4$. 92
5. Use the formula $S = 12.56r^2$ to find S for $r = 5$. 314

Additional Example 4

Evaluate $5x^2 + 6x - 8$ for $x = 3$. 55

Additional Example 5

Use the formula $d = 5t^2$ to find d for $t = 1.2$ seconds. 7.2 m, or about 7 m

Ask students how exponentiation differs
from multiplication.   In exponentiation, the
factors are all the same; this is not true for
multiplication.   Remind them that a number
having an exponent of 2 is said to be
squared, and a number having an exponent
of 3 is said to be cubed. Ask them to make
a conjecture about how these operations
(squaring and cubing) got their names.
area of square and volume of cube

■■■FOLLOW UP

Guided Practice

Classroom Exercises 1–14
Try This all

Independent Practice

A Ex. 1–45, **B** Ex. 46–58, **C** Ex. 59–64

Basic: FR all, WE 1–45 odd, 50

Average: FR all, WE 13–49 odd, 50,
51–57 odd

Above Average: FR all, WE 19–49 odd,
50, 51–63 odd

Additional Answers

Written Exercises, page 11

18. 180
23. 2

Evaluate.

38. $3x^2 - 2x$ for $x = 4$ 40

39. $x^2 - 2x + 1$ for $x = 4$ 9

40. $x^2 + 2x + 1$ for $x = 3$ 16

41. $m^2 - m - 6$ for $m = 4$ 6

42. $3y^2 + y$ for $y = 4$ 52

43. $(3 \cdot y)^2 + y$ for $y = 4$ 148

44. $36 + 3n - n^2$ for $n = 2$ 38

45. $x^3 + x^2 - 6x$ for $x = 2$ 0

46. $(x^2 - y^2)^3$ for $x = 3$ and $y = 2$ 125

47. $a(ab)^2 - a^3$ for $a = 3$ and $b = 4$ 405

48. $\dfrac{m^2 - n^2}{(m - n)^2}$ for $m = 5$ and $n = 4$ 9

49. $\dfrac{x^2 + 2xy + y^2}{(x + y)^2}$ for $x = 3$ and $y = 4$ 1

50. Explain in writing why $2x^3$ is not equivalent to $(2x)^3$. Answers will vary.

The braking distance of a car can be estimated by the formula $d = 0.04s^2$, where d is the braking distance in feet and s is the speed in miles per hour. Find d for each value of s.

51. $s = 50$ 100 ft

52. $s = 10$ 4 ft

53. $s = 20$ 16 ft

54. $s = 30$ 36 ft

55. $s = 40$ 64 ft

56. $s = 90$ 324 ft

57. $s = 70$ 196 ft

58. $s = 80$ 256 ft

Solve each problem. Round your answer to the nearest whole number. (Ex. 59–62)

59. Use the formula $A = 3.14r^2$ to find A for $r = 2.1$. 14

60. Use the formula $V = 4.19r^3$ to find V for $r = 2.1$. 39

Use the formula $I = ar^{n-1}$ to find I for the given values of a, r, and n.

61. $a = 2$, $r = 3$, and $n = 5$ 162

62. $a = 1.6$, $r = 2.1$, and $n = 3$ 7

Use the formula $y = \left(\dfrac{1}{a}\right)^{kt}$ to find y for the given values of a, k, and t.

63. $a = 2$, $k = 3$, and $t = 2$ $\frac{1}{64}$

64. $a = 2.5$, $k = 3$, and $t = 1$ 0.064

Mixed Review

Simplify *1.2*

1. $15 + 5 \cdot 7 + 14 \cdot 2$ 78

2. $7(8 - 3) - 15 \div 5$ 32

3. $\dfrac{8 + 4 \cdot 3}{7 - 3}$ 5

4. $\dfrac{9 \cdot 3 + 4 \cdot 7}{4 \cdot 17 - 3 \cdot 19}$ 5

Evaluate. *1.2, 1.3*

5. $3r^2 - 5$ for $r = 3$ 22

6. $(x - 5)(9 + y)$ for $x = 12$ and $y = 2$ 77

12   Chapter 1 Introduction to Algebra

Enrichment

Explain to the students that banks pay compound interest on deposited money. If the compounding is done only once a year, for example, interest is paid each year on the beginning balance, which is the initial deposit, in addition to interest earned in previous years. If P dollars is deposited for n years in an account paying a rate of interest i compounded yearly, the account will contain a total of $A = P(1 + i)^n$ dollars at the end of n years. For quarterly compounding, the amount is $A = P(1 + \frac{i}{4})^{4n}$. Have the students use a calculator to compute how much they would have at the end of 10 years if they invest $1,000 in an account paying 10% per year with yearly compounding and with quarterly compounding. $2,593.74 with annual compounding; $2,685.06 with quarterly compounding

1.4 Formulas from Geometry

Objective   To use perimeter, area, and volume formulas

The distance around a rectangle is called its **perimeter**. The formula for the perimeter of a rectangle is

$$p = 2l + 2w$$

where p is the perimeter, l is the length, and w is the width.

EXAMPLE 1   Find the perimeter of a rectangle with length 13.5 ft and width 2.7 ft.

Solution   Evaluate the formula $p = 2l + 2w$ for $l = 13.5$ and $w = 2.7$.

$$
\begin{aligned}
p &= 2l + 2w \\
&= 2(13.5) + 2(2.7) \quad \longleftarrow \; l = 13.5 \text{ and } w = 2.7 \\
&= 27 + 5.4 \\
&= 32.4
\end{aligned}
$$

Thus, the perimeter is 32.4 feet.

Calculator Steps:

2 $\boxed{\times}$ 13.5 $\boxed{+}$ 2 $\boxed{\times}$ 2.7 $\boxed{=}$ 32.4

TRY THIS   1. Find the perimeter of a rectangle with length 6.8 yd and width 2.3 yd. 18.2

The **area** of a geometric figure is the number of square units it contains. Area is measured in *square units*, such as square centimeters (cm^2), square meters (m^2), or square inches ($in.^2$). Two area formulas are shown below.

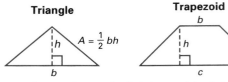

Triangle $A = \frac{1}{2}bh$

Trapezoid $A = \frac{1}{2}(b+c)h$

For the two figures above, the *height* or *altitude* is h. For the triangle, the *base* is b. For the trapezoid, the *bases* are b and c.

Teaching Resources

Application Worksheet 1
Quick Quizzes 4
Reteaching and Practice Worksheets 4

▮▮▮ GETTING STARTED

Prerequisite Quiz

Evaluate each expression for $r = 7$ and $t = 5.2$.

1. $8t$ 41.6
2. $3r + t$ 26.2
3. r^2 49
4. $3(r + t)$ 36.6
5. $\frac{1}{2}rt$ 18.2
6. $\frac{1}{2}t(r + 1)$ 20.8

Motivator

Lead students to a discussion of *length, area,* and *volume* by asking them to associate one of these words with each of the following descriptions.

1. The amount of wallpaper needed to cover the walls of a room   area
2. The amount of solution in a fire extinguisher   volume
3. The distance from home to school   length
4. The amount of ribbon it would take to tie a ribbon around the earth at the equator   length
5. The amount of air taken in when a person inhales   volume

Then ask them to suggest appropriate units of measure for each of the descriptions.
Sample answers: 1. ft^2, m^2; 2. gal, qt, cm^3; 3. mi, m, km; 4. mi, ft, yd, m, km; 5. ml, in^3

Additional Example 1

Use the formula $p = 2l + 2w$ to find the perimeter of a rectangular park that is 5,700 ft long and 4,800 ft wide. Then estimate how long it would take to bicycle around the park at 8 mi/h.   21,000 ft; $\frac{1}{2}$ h

Highlighting the Standards

Standards 7a, 5a, 6b: In Example 3, students interpret a three-dimensional object. In Exercise 40, they represent a situation with a formula, and in Exercises 41–44, they analyze relationships.

Lesson Note

Memorizing formulas is not an objective for this lesson. However, a little time spent on showing the logic behind them is good preparation for geometry. Demonstrate that the area of a triangle can be thought of as half that of a rectangle as in the following diagram.

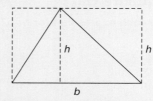

The formulas in this lesson show applications of algebra to some geometric problems.

Point out that the area formula for a trapezoid can be written as $A = \frac{1}{2}(b + c) \cdot h$. Since $\frac{1}{2}(b + c)$ represents the average of the lengths of the bases, the formula can be interpreted as finding the area of a trapezoid by multiplying the average of the lengths of the bases by the height.

Math Connections

Life Skills: Calculating perimeter, area, and volume are skills that can be used around the house. Fencing a yard requires calculation of perimeter. Calculating area is necessary for carpeting a floor or painting a room. Buying soil or fertilizer for a garden involves estimating in cubic feet or cubic yards.

EXAMPLE 2 Find the area of a trapezoid with $b = 5$ in., $c = 7$ in., and $h = 9$ in.

Solution $A = \frac{1}{2}(b + c)h = \frac{1}{2}(5 + 7)(9) = \frac{1}{2} \cdot 12 \cdot 9 = 6 \cdot 9 = 54$

Thus, the area is 54 in^2.

TRY THIS 2. Find the area of a triangle with $b = 2$ ft and $h = 7.8$ ft. 7.8 ft^2

The **volume** of a solid figure is the number of cubic units it contains. The formula for the volume of a *rectangular solid* is $V = lwh$, where l is the length, w is the width, and h is the height of the solid. Volume is measured in *cubic units*, such as cubic centimeters (cm^3), cubic meters (m^3), or cubic inches (in.^{3}).

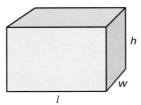

EXAMPLE 3 Use the formula $V = lwh$ to find the volume of a rectangular solid if $l = 4.2$ cm, $w = 3.5$ cm, and $h = 6.7$ cm. Round your answer to the nearest whole number.

Solution
$$V = lwh$$
$$= (4.2)(3.5)(6.7)$$
$$= (14.7)(6.7)$$
$$= 98.49$$

Calculator Steps: 4.2 $\boxed{\times}$ 3.5 $\boxed{\times}$ 6.7 $\boxed{=}$

Thus, the volume is about 98 cm^3.

TRY THIS 3. Find the volume of a rectangular solid where $l = 9$ m, $w = 2.4$ m, and $h = 4.8$ m. About 104 m^3

Classroom Exercises

Give a formula for each of the following.

1. area of a square with side s $A = s^2$
2. perimeter of a square with side s $p = 4s$
3. area of a rectangle of length l and width w $A = lw$
4. perimeter of a rectangle of length l and width w $p = 2l + 2w$
5. area of a triangle with base b and height h $A = \frac{1}{2}bh$
6. area of a trapezoid of height h with bases of length b and c $A = \frac{1}{2}(b + c)h$
7. Use $p = 4s$ to find the perimeter p of a square where $s = 7$ in. 28 in
8. Use $V = e^3$ to find the volume V of a cube where $e = 3$ cm. 27 cm^3

Additional Example 2

Use the formula $A = \frac{1}{2}bh$ to find the area of a triangle where $b = 9$ cm and $h = 6$ cm. 27 cm^2

Additional Example 3

Use the formula $V = e^3$ to find the volume of a cube where $e = 1.5$ in. Round your answer to the nearest whole number. 3 in^3

Written Exercises

Use $p = 4s$ to find the perimeter p of each square where s is given.

1. $s = 9$ ft 36 ft   **2.** $s = 15$ in. 60 in   **3.** $s = 5.2$ cm 20.8 cm

Use the formula $p = 2l + 2w$ to find the perimeter p of each rectangle where l and w are given.

4. $l = 8$ yd, $w = 3$ yd   **5.** $l = 12$ ft, $w = 9$ ft   **6.** $l = 6.2$ m, $w = 5.8$ m
22 yd   42 ft   24 m

Use $A = s^2$ to find the area A of each square where s is given.

7. $s = 8$ yd 64 yd²   **8.** $s = 3.20$ cm 10.24 cm²   **9.** $s = \frac{1}{2}$ in. $\frac{1}{4}$ in²

Use the formula $A = lw$ to find the area A of each rectangle where l and w are given.

10. $l = 6$ in., $w = 5$ in.   **11.** $l = 16$ m, $w = 11$ m   **12.** $l = 6.80$ cm,
30 in²   176 m² $w = 2.40$ cm 16.32 cm²

Use the formula $A = \frac{1}{2}bh$ to find the area A of each triangle where b and h are given. Round your answer to the nearest whole number.

13. $b = 5$ ft, $h = 4$ ft   **14.** $b = 9$ yd, $h = 5$ yd   **15.** $b = 9.8$ cm, $h = 5$ cm
10 ft²   23 yd²   25 cm²

Use the formula $V = lwh$ to find the volume V of each rectangular solid where l, w, and h are given.

16. $l = 5$ in., $w = 3$ in., and $h = 6$ in.   **17.** $l = 7$ ft, $w = 4$ ft, and $h = 2$ ft 56 ft³
90 in³

Use the formula $A = \frac{1}{2}(b + c)h$ to find the area A of each trapezoid where b, c, and h are given.

18. $b = 6$ ft, $c = 8$ ft, and $h = 4$ ft   **19.** $b = 6$ m, $c = 14$m, and $h = 6$ m
28 ft²   60 m²

Use the formula $V = e^3$ to find the volume V of a cube where e, the length of an edge, is given.

20. $e = 5$ cm 125 cm³   **21.** $e = 7$ ft 343 ft³
22. $e = 9$ m 729 m³   **23.** $e = 0.8$ cm 0.512 cm³

In Exercises 24–44, round your answer to the nearest whole number.

24. Find the perimeter and area of a square with each side 6.5 cm. $p = 26$ cm,
                                                         $A = 42$ cm²
25. Find the perimeter and area of a rectangle with length 16.3 m and width 8.4 m. $p = 49$ m, $A = 137$ m²
26. Find the area of a triangle with base 7.5 m and altitude 6.2 cm.   23 cm²
27. Find the area of a trapezoid with bases 3.7 m and 5.9 m and altitude 4.2 m. $A = 20$ m²

Application: Have students imagine that they need to draw a line 6 in. long but they do not have a ruler. However, they do have notebook paper which is $8\frac{1}{2}$ in. wide and 11 in. long. Have them describe how they could use a sheet of notebook paper to measure 6 in. Add the measure of 2 paper widths (17 in.). Subtract one paper length (11 in.). The difference is exactly 6 in.

Common Error Analysis

Error: Some students may try to apply a false Distributive Property to $\frac{1}{2}bh$, taking half of b and half of h before multiplying.

Since the Distributive Property has not been discussed yet, point out that the derivation of the area formula for a triangle, $\frac{1}{2}bh$, is actually half the area of a rectangle whose area is bh.

Checkpoint

Use the formula $p = 2l + 2w$ to find the perimeter p of each rectangle where l and w are given.

1. $l = 7$ cm, $w = 5$ cm 24 cm
2. $l = 4.2$ cm, $w = 3.1$ cm 14.6 cm

Use the formula $A = \frac{1}{2}(b + c)h$ to find the area A of each trapezoid where h, b, and c are given.

3. $h = 7$ ft, $b = 5$ ft, $c = 11$ ft 56 ft²
4. $h = 12$ in., $b = 14$ in., $c = 15$ in. 174 in.²

Ask students these questions. Do two-dimensional objects have volume?   No   Can three-dimensional objects have area?   Yes   Why are square units used to measure area, and cubic units to measure volume?   Square units cover a surface with no gaps; cubic units fill volume with no gaps. Then ask students how they should determine the area of a triangle.   Multiply the product of the base and height by $\frac{1}{2}$.   The perimeter of a rectangle?   Find the sum of twice the length and the width.   The volume of a rectangular solid?   Find the product of the length, width, and height.

◼◼◼FOLLOW UP

Guided Practice

Classroom Exercises 1–10
Try This all

Independent Practice

A Ex. 1–23, **B** Ex. 24–39, **C** Ex. 40–44

Basic: WE 1–23, 29–39 odd, Midchapter Review

Average: WE 7–39 odd, Midchapter Review

Above Average: WE 13–39 odd, 41–44 Midchapter Review

The circumference of a circle is the distance around it. Use the formula $C = 2\pi r$ to find the circumference C of each circle where r, the length of a radius, is given. Use 3.14 for π. Round answers to the nearest whole number.

28. $r = 4$ m 25 m   **29.** $r = 9$ in. 57 in   **30.** $r = 1.6$ cm 10 cm

31. $r = 3$ yd 19 yd   **32.** $r = 6$ ft 38 ft   **33.** $r = 10$ in. 63 in

Use the formula $A = \pi r^2$ to find the area A of each circle where r is given. Use 3.14 for π. Round answers to the nearest square unit.

34. $r = 7.0$ m 154 m²   **35.** $r = 9.0$ in. 254 in²   **36.** $r = 5.2$ cm 85 cm²

37. $r = 2.0$ yd 13 yd²   **38.** $r = 6.0$ in. 113 in²   **39.** $r = 12$ ft 452 ft²

40. In a regular pentagon, all five sides are the same length. Write a formula for the perimeter p of a regular pentagon with s the length of each side. Use it to find the perimeter of a regular pentagon with each side 6.2 cm. $p = 5s, p = 31$ cm

41. If each side of a square is doubled in length, what will happen to its perimeter? It will double.

42. If each side of a square is doubled in length, what will happen to its area? It will be multiplied by 4.

43. If each edge of a cube is doubled in length, what will happen to its volume? It will be multiplied by 8.

44. If the radius of a circle is doubled in length, what will happen to its area? It will be multiplied by 4.

Midchapter Review

Simplify. *1.2, 1.3*

1. $7 \cdot 5 - 18$ 17   **2.** $8 + 2 \cdot 7$ 22   **3.** $3(4 + 5)$ 27   **4.** $\dfrac{5 + 4}{3 - 2}$ 9

Evaluate. *1.1–1.3*

5. $4x$ for $x = 7$ 28   **6.** $5y^3$ for $y = 2$ 40

7. $a + 2b$ for $a = 2$ and $b = 4$ 10   **8.** $(a + 2)b$ for $a = 2$ and $b = 4$ 16

Write an algebraic expression for each word description. *1.1*

9. 7 more than x $x + 7$ **10.** 6 less than the number n $n - 6$

11. the cost, in cents, of y note pads at 79¢ per pad 79y   **12.** $\frac{4}{5}$ of the population p $\frac{4}{5}p$

Find the perimeter and the area of each rectangle with the given dimensions. *1.4*

13. $l = 4$ in., $w = 7$ in.
  $p = 22$ in, $A = 28$ in²

14. $l = 2.3$ cm, $w = 6.0$ cm
  $p = 16.6$ cm, $A = 13.8$ cm²

16 Chapter 1 Introduction to Algebra

Enrichment

Have the students solve these problems.

1. The area of a rectangle is 100 cm². If the length of the rectangle is doubled, and its width is halved, what will be the area of the resulting rectangle? 100 cm²

2. A large circle has a radius that is 5 times as great as the radius of a smaller circle. How many times as great is the area of the larger circle? 25 times

1.5 The Commutative and Associative Properties

Objectives

To determine which property is illustrated by a mathematical sentence

To simplify expressions by using the Commutative and Associative Properties

Suppose you want to calculate $4 \cdot 39 \cdot 25$ mentally. Multiplying 4 by 25 first makes the mental calculation easier. You can multiply any two numbers first because of some properties of multiplication. These properties and others have been given special names. The reason for giving them names is to make it easier to refer to them.

Property	Meaning of property with examples
Commutative Property of Addition For all numbers a and b, $a + b = b + a$.	The order in which two numbers are added does not affect the sum. $6 + 8 = 8 + 6$ $42.7 + 3.1 = 3.1 + 42.7$ $7\frac{1}{2} + 5\frac{1}{4} = 5\frac{1}{4} + 7\frac{1}{2}$
Commutative Property of Multiplication For all numbers a and b, $a \cdot b = b \cdot a$.	The order in which two factors are multiplied does not affect the product. $9 \cdot 7 = 7 \cdot 9$ $\frac{1}{2} \cdot \frac{3}{4} = \frac{3}{4} \cdot \frac{1}{2}$ $12(0.25) = (0.25)12$
Associative Property of Addition For all numbers a, b, and c, $(a + b) + c = a + (b + c)$.	The sum of three numbers is the same no matter how you group the numbers. $(6 + 3) + 8 = 6 + (3 + 8)$ $(2 + 9.3) + 0.7 = 2 + (9.3 + 0.7)$ $\left(3 + \frac{1}{2}\right) + 0.5 = 3 + \left(\frac{1}{2} + 0.5\right)$
Associative Property of Multiplication For all numbers a, b, and c, $(a \cdot b) \cdot c = a \cdot (b \cdot c)$.	The product of three factors is the same no matter how you group the factors. $(3 \cdot 7) \cdot 10 = 3 \cdot (7 \cdot 10)$ $\left(3 \cdot 22\frac{1}{2}\right) \cdot 2 = 3 \cdot \left(22\frac{1}{2} \cdot 2\right)$ $[2(0.125)]8 = 2[(0.125)8]$

From the associative properties, it follows that $a + b + c$ names just one number and abc names just one number.

Teaching Resources

Quick Quizzes 5
Reteaching and Practice
    Worksheets 5

■■■ GETTING STARTED

Prerequisite Quiz

Simplify.

1. $120 + (26 + 4)$ 150
2. $(120 + 26) + 4$ 150
3. $3 \cdot (6 \cdot 5)$ 90
4. $(3 \cdot 6) \cdot 5$ 90
5. $2 \cdot (18 \cdot 5)$ 180
6. $(2 \cdot 5) \cdot 18$ 180
7. $37 + 33 + 16$ 86
8. $37 + 16 + 33$ 86

Motivator

Ask students to give a definition for the word "commute."   Interchange or exchange

Ask them to give a definition for the word "associate."   Connect, combine, join together   Ask them whether parentheses are used in algebra to commute or to associate.   To associate

Highlighting the Standards

Standards 4b, 14b: Examples 4 and 5 show students how to relate procedures in algebra to the use of properties. In Exercises 9–14 and 29–37, they demonstrate their understanding of the logic of algebraic procedures.

Lesson Note

Point out that the Commutative and Associative Properties of Addition and Multiplication are applied when numbers are added in any order, or multiplied in any order.

The dictionary definitions of *associate* and *commute,* as well as the everyday use of these terms, should suggest to the students the reasons why the Associative and Commutative Properties are so named. Point out that addition and multiplication are *binary* operations because our minds can only combine two numbers at one time. The Associative Property states that it does not matter which two we start with. Remind students that when multiplying fractions by whole numbers, such as $\frac{2}{3} \cdot 12$, the 12 can be written as the fraction $\frac{12}{1}$.

Before assigning Written Exercise 47, tell students that a *counterexample* is an example that shows that a statement is false.

Math Connections

Life Skills: Often students are unimpressed by the Commutative Property since it seems so obvious. Ask them to give examples of everyday processes that are not commutative; e.g., putting on shoes and socks.

EXAMPLE 1   Which property of multiplication is illustrated?

a. $(4 \cdot 39) \cdot 25 = 4 \cdot (39 \cdot 25)$ Associative Property of Multiplication

b. $4 \cdot (39 \cdot 25) = 4 \cdot (25 \cdot 39)$ Commutative Property of Multiplication

c. $4 \cdot (25 \cdot 39) = (4 \cdot 25) \cdot 39$ Associative Property of Multiplication

---

TRY THIS   1. Name the property: $(4 + 25) + 39 = (25 + 4) + 39$
Commutative Property of Addition
You can use the commutative and associative properties together.

EXAMPLE 2   Simplify $4 \cdot 39 \cdot 25$ mentally. Explain your thinking.

Solution   You can multiply 4 by 25 first because of the Commutative and Associative Properties of Multiplication, as shown in Example 1 above.

$$4 \cdot 39 \cdot 25 = 4 \cdot 25 \cdot 39$$
$$= 100 \cdot 39$$
$$= 3,900$$

EXAMPLE 3   Which property is illustrated?

a. $(x + 5) + 2 = x + (5 + 2)$ Associative Property of Addition

b. $(5p) \cdot 3 = 3 \cdot (5p)$ Commutative Property of Multiplication

c. $(7 \cdot x) \cdot y = 7 \cdot (x \cdot y)$ Associative Property of Multiplication

d. $7t + 9y = 9y + 7t$ Commutative Property of Addition

TRY THIS   2. Name the property: $(5 + a) + 3 = 3 + (5 + a)$
Commutative Property of Addition
The commutative and associative properties can be used to simplify algebraic expressions, as shown in the next examples. When you actually do such simplifications, you need not write the property.

EXAMPLE 4   Simplify $\frac{2}{3}(12c)$.

Solution $\frac{2}{3}(12c) = \left(\frac{2}{3} \cdot 12\right)c$ ⟵ Associative Property of Multiplication

$= 8c$ ⟵ $\frac{2}{\underset{1}{3}} \cdot \frac{\overset{4}{12}}{1} =$

TRY THIS   Simplify   3. $5\left(\frac{2}{5}a\right)$ $2a$ 4. $\frac{1}{2}(2b)$ b

---

Additional Example 1

Which property is illustrated?

a. $20 \cdot (5 \cdot 9) = (20 \cdot 5) \cdot 9$
   Assoc Prop Mult

b. $20 \cdot (5 \cdot 9) = 20 \cdot (9 \cdot 5)$
   Comm Prop Mult

c. $(20 \cdot 9) \cdot 5 = (9 \cdot 20) \cdot 5$
   Comm Prop Mult

Additional Example 2

Simplify $20 \cdot 9 \cdot 5$ mentally. Explain your work.   Use the Commutative and Associative Properties of Multiplication:
$20 \cdot 9 \cdot 5 = 20 \cdot 5 \cdot 9 = 100 \cdot 9 = 900$

Additional Example 3

Which property is illustrated?

a. $4a + (5b + c) = (4a + 5b) + c$
   Assoc Prop Add

b. $8 + 5y = 5y + 8$ Comm Prop Add

c. $t \cdot (3 \cdot n) = (3 \cdot n) \cdot t$ Comm Prop Mult

d. $2 \cdot (9y) = (2 \cdot 9)y$ Assoc Prop Mult

Example 4 shows that when you simplify $\frac{2}{3}(12c)$, you get $8c$. This means that $\frac{2}{3}(12c) = 8c$ is true for every number c. Two algebraic expressions are said to be **equivalent** if they name the same number for all values of their variables for which the expressions have meaning.

The two expressions $\frac{2}{3}(12c)$ and $8c$ are *equivalent*.

EXAMPLE 5   Simplify $(7x)5$.

Solution $(7x)5 = 5(7x)$ ⟵ Commutative Property of Multiplication
$= (5 \cdot 7)x$ ⟵ Associative Property of Multiplication
$= 35x$

TRY THIS   Simplify: **5.** $(3a)\frac{2}{3}$ *2a*   **6.** $(0.5b)6.2$ *3.1b*

Classroom Exercises

Which property is described?

1. When multiplying two numbers, you may interchange the numbers without changing the result.   Comm Prop Mult
2. When adding three numbers, you get the same result whether you add the third to the sum of the first two or add the first to the sum of the last two.   Assoc Prop Add

Which property is illustrated?

3. $8 \cdot 4 = 4 \cdot 8$ **4.** $a + b = b + a$ **5.** $(7 + 6) + 9 = 7 + (6 + 9)$
6. $19 + 45 = 45 + 19$ **7.** $(3b) \cdot 2 = 2 \cdot (3b)$ **8.** $(x \cdot 3) \cdot 2 = x \cdot (3 \cdot 2)$

Simplify mentally. Explain your thinking.

9. $199 + 47 + 1$ **10.** $25 \cdot 59 \cdot 4$ **11.** $25 + 20.9 + 75$
12. $\frac{1}{3} \cdot 37 \cdot 3$ **13.** $\frac{1}{2} + 3\frac{7}{8} + \frac{1}{2}$ **14.** $99 + 197 + 1 + 3$

Simplify.

15. $3(5a)$ *15a*   **16.** $(4x)2$ *8x*   **17.** $7(6m)$ *42m*

18. Show that $6(3a)$ is equivalent to $2(9a)$. $6(3a) = 18a$ and $2(9a) = 18a$

19. Which expressions are equivalent?   b, c, d

   a. $\left(\frac{1}{3} \cdot 3\right)c$ and $c + \left(\frac{1}{3} \cdot 3\right)$ **b.** $(10t)\frac{2}{5}$ and $\frac{1}{8}(32t)$

   c. $3 + (a + 1)$ and $(a + 1) + 3$ **d.** $3 \cdot (24 \cdot r)$ and $(3 \cdot 24) \cdot r$

Critical Thinking Questions

Synthesis: Subtraction and division are not commutative operations; $5 - 3 \neq 3 - 5$; $8 \div 2 \neq 2 \div 8$. Ask students how problems of this type could be rewritten such that the commutative property would be valid.
Rewrite $5 - 3$ as $5 + (-3)$; then $5 + (-3) = -3 + 5$; rewrite $8 \div 2$ as $8 \cdot \frac{1}{2}$; then $8 \cdot \frac{1}{2} = \frac{1}{2} \cdot 8$

Checkpoint

Which property is illustrated?

1. $(5 \cdot 9)2 = 5(9 \cdot 2)$ Assoc Prop Mult
2. $6 + 2a = 2a + 6$ Comm Prop Add
3. $t + (r + s) = (t + r) + s$ Assoc Prop Add
4. $8(5 + 9) = 8(9 + 5)$ Comm Prop Add

Simplify.

5. $1 + 16 + 199$ 216
6. $0.6 \cdot 5 \cdot 10$ 30
7. $\frac{2}{3} + 9 + 1\frac{1}{3}$ 11
8. $\frac{1}{5} \cdot 2 \cdot 5x$ 2x

Closure

Ask the students to explain which property is used in each of the following.

$(2 \cdot 8) \cdot 5 = (8 \cdot 2) \cdot 5$ Comm Prop Mult
$(8 \cdot 2) \cdot 5 = 8 \cdot (2 \cdot 5)$ Assoc Prop Mult

Then ask:
What is the difference between these two properties?   See p. 17 (order vs. grouping.)
What is meant by equivalent algebraic expressions?   See p. 19.

Additional Example 4

Simplify $\frac{3}{5}(30y)$.   18y

Additional Example 5

Simplify $(4t)7$.   28t

Guided Practice

Classroom Exercises 1–19
Try This all

Independent Practice

A Ex. 1–24, **B** Ex. 25–44, **C** Ex. 45–48

Basic: WE 1–41 odd, Application
Average: WE 5–45 odd, Application
Above Average: WE 11–45 odd, 46–48, Application

Additional Answers

Classroom Exercises, page 19

3. Comm Prop Mult
4. Comm Prop Add
5. Assoc Prop Add
6. Comm Prop Add
7. Comm Prop Mult
8. Assoc Prop Mult
9. $(199 + 1) + 47 = 200 + 47 = 247$
10. $(25 \cdot 4) \cdot 59 = 100 \cdot 59 = 5900$
11. $(25 + 75) + 20.9 = 100 + 20.9 = 120.9$
12. $(\frac{1}{3} \cdot 3) \cdot 37 = 1 \cdot 37 = 37$
13. $(\frac{1}{2} + \frac{1}{2}) + 3\frac{7}{8} = 1 + 3\frac{7}{8} = 4\frac{7}{8}$
14. $(99 + 1) + (197 + 3) = 100 + 200 = 300$

Written Exercises

5. $(299 + 1) + 53 = 300 + 53 = 353$
6. $(25 \cdot 4) \cdot 73 = 100 \cdot 73 = 7,300$
7. $(15 \cdot 2) \cdot 7 = 30 \cdot 7 = 210$
8. $(57 + 3) + 29 = 60 + 29 = 89$
9. $(50 \cdot 2) \cdot 14 = 100 \cdot 14 = 1,400$

Written Exercises

Which property is illustrated?

1. $49 + 73 = 73 + 49$ Comm Prop Add
2. $(2 + 9) + 3 = 2 + (9 + 3)$ Assoc Prop Add
3. $(3 \cdot 7)8 = 3(7 \cdot 8)$ Assoc Prop Mult
4. $(7k)5 = 5(7k)$ Comm Prop Mult

Simplify mentally. Explain your thinking.

5. $299 + 73 + 1$
6. $25 \cdot 73 \cdot 4$
7. $15 \cdot 7 \cdot 2$
8. $57 + 29 + 3$
9. $50 \cdot 14 \cdot 2$
10. $18 + 42 + 2$
11. $20 \cdot 43 \cdot 5$
12. $4 \cdot 23 \cdot 5$
13. $\frac{1}{4} + 18 + \frac{3}{4}$
14. $\frac{1}{2} \cdot 317 \cdot 2$
15. $\frac{2}{3} \cdot 17 \cdot 3$
16. $\frac{3}{5} + 23\frac{1}{2} + \frac{2}{5}$

Simplify.

17. $8(4m)$ $32m$
18. $3(7b)$ $21b$
19. $(5k)3$ $15k$
20. $(8m)9$ $72m$
21. $6(7y)$ $42y$
22. $14(3p)$ $42p$
23. $16(4d)$ $64d$
24. $(9x)7$ $63x$

Which property is illustrated?

25. $(3 + 5a) + 4a = 3 + (5a + 4a)$ Assoc Prop Add
26. $7(3a) = (7 \cdot 3)a$ Assoc Prop Mult
27. $5m + 3m^2 = 3m^2 + 5m$ Comm Prop Add
28. $(3k)(7b) = (7b)(3k)$ Comm Prop Mult

Simplify mentally. Explain your thinking.

29. $299 + 57 + 3 + 1$
30. $2 + 175 + 138 + 5$
31. $688 + 289 + 12 + 11$
32. $0.07 + 0.18 + 0.93$
33. $(0.5) \cdot 27 \cdot 2$
34. $5 \cdot 23 \cdot 4 \cdot 5$
35. $2\frac{3}{7} + 399 + 3\frac{4}{7} + 1$
36. $\frac{1}{3}(22.5) \cdot 6$
37. $\frac{1}{2} \cdot 25 \cdot 12 \cdot 4$

Simplify.

38. $3(5k)4$ $60k$
39. $8y(4)3$ $96y$
40. $7(3a)6$ $126a$
41. $2 \cdot 13 \cdot 50g$ $1,300g$
42. $5t(9.3)20$ $930t$
43. $25 \cdot \frac{3}{5}b \cdot 4$ $60b$

44. Show that $4(5a)$ is equivalent to $2(10a)$. $4(5a) = 20a$ and $2(10a) = 20a$

45. Is there a number x for which $6(3x) = 18x$ is not true? Why?

46. Write an explanation of the difference between the commutative and associative properties of addition. Answers will vary.

47. Is division of nonzero numbers commutative? That is, is "$a \div b = b \div a$" true for all nonzero numbers a and b? If not, give an example to show that it is not. No; $6 \div 3 \neq 3 \div 6$

48. Is division of nonzero numbers associative? That is, is "$(a \div b) \div c = a \div (b \div c)$" true for all nonzero numbers a, b, and c? If not, give an example to show that it is not. No; $(27 \div 9) \div 3 \neq 27 \div (9 \div 3)$

Mixed Review

Evaluate. *1.2, 1.3*

1. $14 - 6x$ for $x = \frac{2}{3}$ 10
2. $x^2 - 4x$ for $x = 6$ 12
3. $(x - 4)x$ for $x = 6$ 12

4. Write an algebraic expression for the cost, in cents, of one pencil if k pencils cost \$1.20.   *1.1* $\frac{120}{k}$

5. Use the formula $A = \frac{1}{2}(b + c)h$ to find the area of a trapezoid for $b = 6$ cm, $c = 10$ cm, and $h = 4$ cm.   *1.4* 32 cm²

Application: Heat Transfer

When the temperature outside your house is colder than it is inside, heat will escape through the windows. The rate of heat transfer is measured in British Thermal Units, or BTUs, per hour. One BTU is the amount of heat needed to raise the temperature of one pound of water one degree Fahrenheit. The formula for the rate of heat transfer through glass is

$$H = \frac{A \times T}{R}.$$

In this formula, H is the rate of heat transfer in BTUs, A is the window area in square feet, T is the difference in inside and outside temperatures in degrees Fahrenheit, and R is the thermal resistance. Thermal resistance is a measure of a material's resistance to the flow of heat through it. The R value for a single-pane window is 0.88. If a storm window is added, the R value of the double-pane window is 1.79.

To calculate H for a single-pane window 2 ft by 3 ft when the inside temperature is 72° and the outside temperature is 30°, begin by writing the formula. Then substitute the values you know.

$$H = \frac{A \times T}{R}$$

$$H = \frac{(2 \cdot 3) \cdot (72 - 30)}{0.88} \quad \longleftarrow \text{ Simplify. Use a calculator.}$$

$$H = 290 \text{ BTU/h (nearest 10 BTUs)}$$

Solve each problem. Round your answers to the nearest 10 BTUs.

1. Calculate H for the same conditions if the single-pane window is replaced by a double-pane window.   140 BTU/h

2. Suppose that in Exercise 1 the inside temperature is lowered to 70°. How many BTUs would be saved in 1 hour? How many in a day (24 h)?
10 BTU/h; 240 BTU/d

Application: Heat Transfer   **21**

---

10. $(18 + 2) + 42 = 20 + 42 = 62$
11. $(20 \cdot 5) \cdot 43 = 100 \cdot 43 = 4{,}300$
12. $(4 \cdot 5) \cdot 23 = 20 \cdot 23 = 460$
13. $(\frac{1}{4} + \frac{3}{4}) + 18 = 1 + 18 = 19$
14. $(\frac{1}{2} \cdot 2) \cdot 317 = 1 \cdot 317 = 317$
15. $(\frac{2}{3} \cdot 3) \cdot 17 = 2 \cdot 17$
16. $(\frac{3}{5} + \frac{2}{5}) + 23\frac{1}{2} = 1 + 23\frac{1}{2} = 24\frac{1}{2}$
29. $(299 + 1) + (57 + 3) = 300 + 60 = 360$
30. $(2 + 138) + (175 + 5) = 140 + 180$
$= 320$
31. $(688 + 12) + (289 + 11) = 700 + 300$
$= 1{,}000$
32. $(0.07 + 0.93) + 0.18 = 1.0 + 0.18$
$= 1.18$
33. $(0.5 \cdot 2) \cdot 27 = 1 \cdot 27 = 27$
34. $(5 \cdot 4) \cdot (23 \cdot 5) = 20 \cdot 115 = 2{,}300$
35. $(2\frac{3}{7} + 3\frac{4}{7}) + (399 + 1) = 6 + 400 = 406$
36. $(\frac{1}{3} \cdot 6) \cdot (22.5) = 2(22.5) = 45$
37. $(\frac{1}{2} \cdot 12) \cdot (25 \cdot 4) = 6 \cdot 100 = 600$
45. No. By the Assoc Prop of Mult, $6(3x) = (6 \cdot 3)x$. Thus, the two expressions are equivalent and are equal for all values of x.

Enrichment

Challenge the students to write the whole numbers from 1 through 10 using four 3s and the operations of addition, subtraction, multiplication, and division. For example, $\frac{3}{3} \cdot \frac{3}{3} = 1$.   Answers may vary. One set of answers is shown.

$1 = \frac{3}{3} \cdot \frac{3}{3}$ $3 = (3 + 3 + 3) \div 3$

$5 = \frac{3 + 3}{3} + 3$ $7 = 3 + 3 + \frac{3}{3}$

$9 = \frac{3 \cdot 3 \cdot 3}{3}$ $2 = \frac{3}{3} + \frac{3}{3}$

$4 = \frac{3 + 3 \cdot 3}{3}$ $6 = \frac{3(3 + 3)}{3}$

$8 = 3 \cdot 3 - \frac{3}{3}$ $10 = 3 \cdot 3 + \frac{3}{3}$

■GETTING STARTED

Prerequisite Quiz

Simplify.

1. $3 \cdot 8 + 3 \cdot 2$ 30
2. $3(8 + 2)$ 30
3. $8(8 + 100)$ 864
4. $63(7) + 63(3)$ 630
5. $\frac{1}{8}(80 - 8)$ 9

Motivator

Have students simplify each pair of
expressions.

1. $45(8)$ and $40 \cdot 8 + 5 \cdot 8$ 360
2. $7\frac{4}{5}(10)$ and $7 \cdot 10 + \frac{4}{5} \cdot 10$ 78
3. $9(58)$ and $9 \cdot 50 + 9 \cdot 8$ 522
4. $2\frac{3}{4}(80)$ and $2 \cdot 80 + \frac{3}{4} \cdot 80$ 220

Ask: Do both expressions in each pair have
the same value?   Yes   Lead students in a
discussion of which expression in each pair
is easier to simplify.

1.6 The Distributive Property

Objective   To rewrite expressions by using the Distributive Property

How do you find the area of the
swimming pool at the right?
Compare the following two
methods.

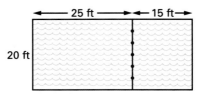

Method 1
Multiply the total length by the
width.
$20(25 + 15)$

$20 \cdot 40$
800

Method 2
Add the areas of the two small
rectangles.
$20 \cdot 25 + 20 \cdot 15$
$500 + 300$
800

So the area of the pool is 800 square feet (800 ft²).

Notice that $20(25 + 15) = 20 \cdot 25 + 20 \cdot 15$. This illustrates that
multiplication is *distributive* over addition.

Property	Meaning of Property with Examples
Distributive Property of Multiplication over Addition For all numbers a, b, and c, $a(b + c) = ab + ac$ and $(b + c)a = ba + ca$.	To multiply a sum by a factor, you can multiply each addend by the factor and then add the products. $8(300 + 2) = 8 \cdot 300 + 8 \cdot 2$ $(3 + 0.2)4 = 3 \cdot 4 + (0.2)4$ $5\left(4 + \frac{1}{2}\right) = 5 \cdot 4 + 5 \cdot \frac{1}{2}$

This property can help you do mental calculations.

EXAMPLE 1   Compute $8 \cdot 52$ mentally. Explain your thinking.

Solution   Think: $52 = 50 + 2$

Thus, $8 \cdot 52 = 8(50 + 2)$
$= 8 \cdot 50 + 8 \cdot 2$ ← Distributive Property of Multiplication over Addition
$= 400 + 16 = 416$

TRY THIS   1. Compute mentally: $5 \cdot 308$ 1,540

Highlighting the Standards

Standards 3a, 3b: In Exercises 50–53,
students test conjectures and formulate
counterexamples.

Additional Example 1

Compute $9 \cdot 83$ mentally. Explain your work.
$9 \cdot 83 = 9(80 + 3) = 9 \cdot 80 + 9 \cdot 3$
$= 720 + 27 = 747$

The Distributive Property of Multiplication over Addition can also be used in reverse to rewrite $7 \cdot 9 + 7 \cdot 5$.

$$7 \cdot 9 + 7 \cdot 5 = 7(9 + 5)$$

EXAMPLE 2 Rewrite by using the Distributive Property of Multiplication over Addition. Then simplify.

a. $7(8 + 5)$ b. $6 \cdot 4 + 6 \cdot 3$

Solutions

a. $7(8 + 5) = 7 \cdot 8 + 7 \cdot 5$ b. $6 \cdot 4 + 6 \cdot 3 = 6(4 + 3)$
$= 56 + 35$ $= 6 \cdot 7$
$= 91$ $= 42$

TRY THIS Use the Distributive Property of Multiplication over Addition to rewrite each expression. Then simplify.

2. $4(7 + 3)$ 40   3. $5 \cdot 9 + 5 \cdot 1$ 50

The Distributive Property of Multiplication over Addition works with any number of addends.

EXAMPLE 3 Rewrite by using the Distributive Property of Multiplication over Addition. Then simplify.

a. $\frac{1}{3}(15 + 9 + 3)$ b. $6 \cdot 7 + 1 \cdot 7 + 4 \cdot 7$

Solutions

a. $\frac{1}{3}(15 + 9 + 3)$ b. $6 \cdot 7 + 1 \cdot 7 + 4 \cdot 7$

$\frac{1}{3} \cdot 15 + \frac{1}{3} \cdot 9 + \frac{1}{3} \cdot 3$ $(6 + 1 + 4)7$

$5 + 3 + 1$ $11 \cdot 7$

9 77

TRY THIS Use the Distributive Property of Multiplication over Addition to rewrite each expression. Then simplify.

4. $1.2(10 + 5 + 1)$ 19.2   5. $7 \cdot \frac{1}{2} + 4 \cdot \frac{1}{2} + 9 \cdot \frac{1}{2}$ 10

Multiplication is also distributive over subtraction. For example, $2(9 - 4) = 2 \cdot 9 - 2 \cdot 4$, as shown below.

$$2(9 - 4) \stackrel{?}{=} 2 \cdot 9 - 2 \cdot 4$$
$$2 \cdot 5 \stackrel{?}{=} 18 - 8$$
$$10 = 10 \quad \text{True}$$

The Distributive Property of Multiplication over Addition and the Distributive Property of Multiplication over Subtraction are often referred to simply as the **Distributive Property**.

Lesson Note

Explain to the students that they use the Distributive Property when they multiply two numbers using the usual paper and pencil method. For example:

54		50 + 4	
× 3		× 3	
162		12	3 × 4
		+ 150	3 × 50
		162	

Point out that the Distributive Property involves both the operations of addition and multiplication in the same expression.

Math Connections

Arithmetic: The Distributive Property is used to add fractions. When adding $\frac{1}{2} + \frac{3}{4} + \frac{5}{8}$, we rewrite them with a common denominator: $\frac{4}{8} + \frac{6}{8} + \frac{5}{8}$. This is the same as $4 \cdot \frac{1}{8} + 6 \cdot \frac{1}{8} + 5 \cdot \frac{1}{8} = (4 + 6 + 5) \cdot \frac{1}{8} = \frac{15}{8}$.

Critical Thinking Questions

Application: Ask students to use the Distributive Property to multiply $2\frac{3}{4}$ and 8 without changing the mixed number to an improper fraction. Have the students demonstrate how this pattern could be used to multiply $8\frac{1}{2}$ and $2\frac{1}{4}$.

$2\frac{3}{4}(8) = 2 \cdot 8 + \frac{3}{4} \cdot 8 = 16 + 6 = 22$; $8\frac{1}{2}(2\frac{1}{4}) = 8\frac{1}{2}(2) + 8\frac{1}{2}(\frac{1}{4}) = [8 \cdot 2 + \frac{1}{2} \cdot 2] + [8 \cdot \frac{1}{4} + \frac{1}{2} \cdot \frac{1}{4}] = 16 + 1 + 2 + \frac{1}{8} = 19\frac{1}{8}$

Additional Example 2

Rewrite by using the Distributive Property of Multiplication over Addition. Then simplify.

a. $4(9 + 6)$ $4(9 + 6) = 4 \cdot 9 + 4 \cdot 6 = 36 + 24 = 60$
b. $5 \cdot 3 + 5 \cdot 6$ $5 \cdot 3 + 5 \cdot 6 = 5(3 + 6) = 5 \cdot 9 = 45$

Additional Example 3

Rewrite by using the Distributive Property of Multiplication over Addition. Then simplify.

a. $\frac{1}{6}(6 + 18 + 24)$ $\frac{1}{6} \cdot 6 + \frac{1}{6} \cdot 18 + \frac{1}{6} \cdot 24 = 1 + 3 + 4 = 8$
b. $8 \cdot 2 + 4 \cdot 2 + 1 \cdot 2$ $(8 + 4 + 1)2 = 13 \cdot 2 = 26$

Common Error Analysis

Error: Some students may try to apply the Distributive Property to an expression such as 2(5 · 7) and get 2 · 5 · 2 · 7.

Remind them that this property always involves *two* operations, such as multiplication and addition.

Checkpoint

Rewrite by using the Distributive Property. Then simplify.

1. $(40 + 3)5$ $40 \cdot 5 + 3 \cdot 5 = 200 + 15 = 215$
2. $\frac{1}{3}(45 - 6)$ $\frac{1}{3} \cdot 45 - \frac{1}{3} \cdot 6 = 15 - 2 = 13$
3. $(25 + 10 + 20)\frac{2}{5}$ $25 \cdot \frac{2}{5} + 10 \cdot \frac{2}{5} + 20 \cdot \frac{2}{5} = 10 + 4 + 8 = 22$
4. $5 \cdot 8 + 3 \cdot 8 + 10 \cdot 8$ $(5 + 3 + 10)8 = 18 \cdot 8 = 144$
5. $2(15x - 3)$ $2 \cdot 15x - 2 \cdot 3 = 30x - 6$
6. $(24y - 16)\frac{5}{8}$ $24y \cdot \frac{5}{8} - 16 \cdot \frac{5}{8} = 15y - 10$

Closure

Have students state the property or fact that leads from one line to the next in the following.

1. $37 \cdot 6 + 4 \cdot 37$
   a. $37 \cdot 6 + 37 \cdot 4$ comm., mult.
   b. $37(6 + 4)$ distr.
2. $37(10) + 3 \cdot (2 \cdot 5)$
   a. $37 \cdot 10 + 3 \cdot 10$ mult. fact
   b. $(37 + 3)10$ distr.
   c. $40 \cdot 10$ add. fact
   d. 400 mult. fact

EXAMPLE 4   Rewrite by using the Distributive Property. Then simplify.

    a. $5(6a - 8)$ **b.** $(12a + 9)\frac{2}{3}$

Solutions   **a.** $5(6a - 8) = 5 \cdot 6a - 5 \cdot 8$ **b.** $(12a + 9)\frac{2}{3} = 12a \cdot \frac{2}{3} + 9 \cdot \frac{2}{3}$
                   $= 30a - 40$ $= 8a + 6$

TRY THIS   Rewrite by using the Distributive Property. Then simplify.

    6. $2(3c - 5)$ $6c - 10$ **7.** $(4b - 2)3$ $12b - 6$

You need to be able to distinguish the Commutative, Associative, and Distributive Properties in order to use them correctly.

EXAMPLE 5   Which property is illustrated?

    a. $7(4 \cdot 6) = (7 \cdot 4)6$ Associative Property of Multiplication
    b. $2(1 + 5) = 2(5 + 1)$ ·   Commutative Property of Addition
    c. $(a + 5) + 2 = a + (5 + 2)$ Associative Property of Addition
    d. $8 \cdot 15 + 8 \cdot 5 = 8(15 + 5)$ Distributive Property
    e. $(x + 2)4 = 4(x + 2)$ Commutative Property of Multiplication

TRY THIS   **8.** Which property is illustrated? $(3 + 4) + 6 = 6 + (3 + 4)$
               Commutative Property of Addition

Classroom Exercises

Rewrite by using the Distributive Property.

1. $8(4 + 5)$ 2. $(6 - 2)9$ 3. $5 \cdot 6 - 5 \cdot 3$ 4. $4(2a + 5)$
   $8 \cdot 4 + 8 \cdot 5$ $6 \cdot 9 - 2 \cdot 9$ $5(6 - 3)$ $4 \cdot 2a + 4 \cdot 5$

Rewrite by using the Distributive Property. Then simplify.

5. $10 \cdot 3 - 6 \cdot 3$ 6. $7 \cdot 4 + 13 \cdot 4$ 7. $6(5x - 2)$ 8. $(4y + 5)2$
   $(10 - 6)3; 12$ $(7 + 13)4; 80$ $6 \cdot 5x - 6 \cdot 2; 30x - 12$ $4y \cdot 2 + 5 \cdot 2; 8y + 10$

Written Exercises

Rewrite by using the Distributive Property. Then simplify.

1. $8(7 + 6)$ 2. $6(5 - 3)$ 3. $(4 + 8)3$
4. $(12 - 4)2$ 5. $\frac{1}{2}(8 + 4 + 2)$ 6. $\frac{2}{3}(9 - 6)$
7. $\frac{1}{4}(28 - 16)$ 8. $(10 + 5 + 15)\frac{1}{5}$ 9. $8 \cdot 7 + 8 \cdot 3$
10. $6 \cdot 4 - 6 \cdot 2$ 11. $8 \cdot 4 + 12 \cdot 4$ 12. $9 \cdot 3 - 7 \cdot 3$
13. $4 \cdot 7 + 5 \cdot 7 + 8 \cdot 7$ 14. $6 \cdot 9 + 7 \cdot 9 + 5 \cdot 9$

Additional Example 4

Rewrite by using the Distributive Property. Then simplify.

a. $8(3b - 1)$ $8(3b - 1) = 8 \cdot 3b - 8 \cdot 1 = 24b - 8$

b. $(12d + 60)\frac{3}{4}$ $(12d + 60)\frac{3}{4} = 12d \cdot \frac{3}{4} + 60 \cdot \frac{3}{4} = 9d + 45$

Additional Example 5

Which property is illustrated?

a. $3(6 + 2) = 3(2 + 6)$ Comm Prop Add
b. $\frac{1}{2}(8 \cdot 5) = (\frac{1}{2} \cdot 8)5$ Assoc Prop Mult
c. $7 \cdot 10 + 7 \cdot 9 = 7(10 + 9)$ Distr Prop
d. $(y + 3)5 = 5(y + 3)$ Comm Prop Mult
e. $x + (11 + 7) = (x + 11) + 7$
    Assoc Prop Add

15. $4(3a - 9)$ **16.** $4(5b + 3)$ **17.** $(x - 5)4$ **18.** $(3k + 7)6$

19. $\frac{3}{4}(8x + 12)$ **20.** $(10y - 15)\frac{1}{5}$ **21.** $\frac{2}{3}(6x + 12)$ **22.** $(14y - 21)\frac{3}{7}$

23. $1.5(2x + 3)$ **24.** $(2x - 1.6)4$ **25.** $3(4.1x - 6)$

26. $3(2a + 4b + 6)$ **27.** $(4c - 2b + 1)1.5$ **28.** $2(3a + 6x - 1.3d)$

Compute mentally. Explain your thinking.

29. $6 \cdot 43$ **30.** $5 \cdot 307$ **31.** $7 \cdot 49$

32. $4 \cdot 598$ **33.** $49 \cdot 3 + 49 \cdot 7$ **34.** $96 \cdot 294 + 4 \cdot 294$

35. $\frac{1}{2} \cdot 47 + \frac{1}{2} \cdot 3$ **36.** $40 \cdot 52 - 40 \cdot 2$ **37.** $15 \cdot \frac{3}{4} + 5 \cdot \frac{3}{4}$

38. Show that $3(4y + 5) = 12y + 15$ for at least 4 values of y.   Answers will vary.

39. Show that $4(2b - 6) = 8b - 24$ for at least 4 values of b.   Answers will vary.

Which property is illustrated?

40. $17\frac{1}{2} \cdot 32 = 32 \cdot 17\frac{1}{2}$ **41.** $13 \cdot 21 + 13 \cdot 9 = 13(21 + 9)$

42. $14 + (9 + 6) = (14 + 9) + 6$ **43.** $14 + (9 + 6) = 14 + (6 + 9)$

44. $6(5.7 + 9) = 6(5.7) + 6(9)$ **45.** $7(5 \cdot 2\frac{1}{2}) = (7 \cdot 5)2\frac{1}{2}$

46. $6(x + y) = 6x + 6y$ **47.** $4(5x) = (4 \cdot 5)x$

48. $(4a)(b + c) = (b + c)(4a)$ **49.** $(4a)(b + c) = 4ab + 4ac$

50. Is it true that $(a + b) \div c = (a \div c) + (b \div c)$ for all nonzero numbers a, b, and c? If not, give an example to show that it is not. Yes

51. Is it true that $c \div (a + b) = (c \div a) + (c \div b)$ for all nonzero numbers a, b, and c? If not, give an example to show that it is not. No

52. Is it true that $(a + b)^2 = a^2 + b^2$ for all numbers a and b? No

53. Is it true that $(a \cdot b)^2 = a^2 \cdot b^2$ for all numbers a and b? Yes

Mixed Review

Evaluate.   *1.1–1.3*

1. $y - x$ for $x = 4.1$ and $y = 7$ 2.9

2. $c - (d + 1)$ for $c = 9$ and $d = 6$ 2

3. $x^2 - 5x$ for $x = 9$ 36

4. $\dfrac{x^2 + y^2}{(x + y)^2}$ for $x = 2$ and $y = 4$ $\frac{5}{9}$

5. Find the perimeter and the area of a square with sides of length 2.1 cm. *1.4*
8.4 cm; 4.41 cm²

6. Use the formula $A = \frac{1}{2}bh$ to find the area of a triangle with base 3 cm and height 5.4 cm. *1.4* 8.1 cm²

1.6 The Distributive Property   **25**

◼◼◼◼FOLLOW UP

Guided Practice

Classroom Exercises 1–8
Try This all

Independent Practice

A Ex. 1–28, **B** Ex. 29–49, **C** Ex. 50–53

Basic: WE 1–33 odd, 40–43

Average: WE 9–49 odd

Above Average: WE 11–23, 29–49 odd, 50–53

Additional Answers

Written Exercises

1. $8 \cdot 7 + 8 \cdot 6$; 104
2. $6 \cdot 5 - 6 \cdot 3$; 12
3. $4 \cdot 3 + 8 \cdot 3$; 36
4. $12 \cdot 2 - 4 \cdot 2$; 16
5. $\frac{1}{2} \cdot 8 + \frac{1}{2} \cdot 4 + \frac{1}{2} \cdot 2$; 7
6. $\frac{2}{3} \cdot 9 - \frac{2}{3} \cdot 6$; 2
7. $\frac{1}{4} \cdot 28 - \frac{1}{4} \cdot 16$; 3
8. $10 \cdot \frac{1}{5} + 5 \cdot \frac{1}{5} + 15 \cdot \frac{1}{5}$; 6
9. $8(7 + 3)$; 80
10. $6(4 - 2)$; 12
11. $(8 + 12)4$; 80
12. $(9 - 7)3$; 6
13. $(4 + 5 + 8)7$; 119
14. $(6 + 7 + 5)9$; 162
15. $4 \cdot 3a - 4 \cdot 9$; $12a - 36$
16. $4 \cdot 5b + 4 \cdot 3$; $20b + 12$
17. $x \cdot 4 - 5 \cdot 4$; $4x - 20$
18. $3k \cdot 6 + 7 \cdot 6$; $18k + 42$
19. $\frac{3}{4} \cdot 8x + \frac{3}{4} \cdot 12$; $6x + 9$
20. $10y \cdot \frac{1}{5} - 15 \cdot \frac{1}{5}$; $2y - 3$

See page 36 for the answers to Ex. 21–49.

Enrichment

Explain the ancient "Egyptian" method of multiplying two whole numbers. The method below for multiplying two even whole numbers involves only two skills, doubling and dividing by 2.

Multiply: 8 × 36

Rewrite the product. Divide the left factor by 2 and multiply the right factor by 2. Continue until the left factor is 1. The final product equals 8·36. Have the students choose other even numbers and multiply them by the Egyptian method.

Teaching Resources

Manipulative Worksheet 1
Quick Quizzes 7
Reteaching and Practice
  Worksheets 7
Transparency 3
Teaching Aid 1

◼◼◼GETTING STARTED

Prerequisite Quiz

Complete.

1. $5(8 + 3 + 1) = 40 + 15 + \underline{\ ?\ }$ 5
2. $6(5 + \underline{\ ?\ }) = 30 + 6$ 1
3. $(2 + 9) \cdot \underline{\ ?\ } = 10 + 45$ 5

Simplify.

4. $5 + 2(6 - 1) + 9$ 24
5. $\frac{1}{2}(8 + 22) + \frac{2}{3}(15 + 3)$ 27

Motivator

Model: Draw the following coins on the chalkboard or place the coins on the overhead and ask students to calculate the total amount of money.

5 quarters, 4 dimes, 3 nickels, and 4 pennies

Ask what might be the fastest way to get the total. Tell students that combining "like" coins is a parallel to combining like terms to simplify an expression.

1.7 Combining Like Terms

Objective | To simplify expressions by combining like terms

The expression $7x + 13xy^2 + 18$ has three *terms*, $7x$, $13xy^2$, and 18. The parts of an expression connected by addition or subtraction are called **terms** of the expression. The term 18, with no variable, is called a *constant term*, or **constant**.

Like terms have the same variables and the corresponding variables have the same exponents. Also, all constant terms are like terms.

Like terms	Unlike terms
$6x$ and $4x$	$5x$ and $5y$
$7a^2b$ and $\frac{1}{2}a^2b$	$8t$ and 8
3 and 2.7	$3m^2n$ and $2mn$

In the term $7x$, 7 is the *numerical coefficient* of x. The numerical coefficient of a term is usually called simply the **coefficient** of the term. For example, the numerical coefficient of the term $13xy^2$ is 13.

You can use the Distributive Property to simplify expressions with like terms. For example, $3t + 7t + 5t = (3 + 7 + 5)t$, or $15t$. You can combine like terms by adding or subtracting coefficients in one step.

EXAMPLE 1 | Simplify.

　　　a. $8y - 3y$　　　　　　　　　b. $6x + 4x + 5x$

Plan | Combine like terms.

Solutions | a. $8y - 3y = 5y$　　　　　　b. $6x + 4x + 5x = 15x$

TRY THIS | 1. Simplify $5b - 2b + 3b$. 6b

Remember, only like terms can be combined. Thus, to simplify the expression $9a + 4b + 3a + 5$:

1. Rearrange the terms so that　　　　$9a + 4b + 3a + 5$
   like terms are grouped together.　　$(9a + 3a) + 4b + 5$
2. Combine like terms.　　　　　　　　$12a + 4b + 5$

The expression $12a + 4b + 5$ cannot be simplified since the terms, $12a$, $4b$, and 5, are unlike.

26　　Chapter 1 Introduction to Algebra

Highlighting the Standards

Standards 2d, 1d: In the opening paragraphs, students practice reading written presentations of mathematics. In the Application, they apply the mathematics to making predictions in business situations.

Additional Example 1

Simplify.

a. $11y - 4y$ 7y
b. $24x + 9x + 3x$ 36x

The product of 1 and any number is that number. Therefore, 1 is called the *identity element for multiplication*, or the **multiplicative identity**.

Identity Property for Multiplication
For any number a, $1 \cdot a = a$ and $a \cdot 1 = a$.

Since $y = 1 \cdot y$, the coefficient of y is understood to be 1.

EXAMPLE 2 Simplify $8y + y - 3y$.

Plan Rewrite y as $1 \cdot y$, or $1y$.

Solution $8y + y - 3y = 8y + 1y - 3y$
$$= 6y$$

TRY THIS **2.** Simplify $7y - y + 2y$. $8y$

To simplify $9 + 3(4a + 2) + a$, multiply 3 and $4a + 2$ first.

EXAMPLE 3 Simplify $9 + 3(4a + 2) + a$.

Solution
Rewrite a as $1a$.	$9 + 3(4a + 2) + 1a$
Use the Distributive Property.	$9 + 3 \cdot 4a + 3 \cdot 2 + 1a$
Multiply.	$9 + 12a + 6 + 1a$
Group the like terms.	$(12a + 1a) + (9 + 6)$
Combine the like terms.	$13a + 15$

EXAMPLE 4 Simplify $2(3x + 5) + \frac{2}{3}(6x - 3)$. Then evaluate for $x = 5$.

Solution
$2(3x + 5) + \frac{2}{3}(6x - 3) = 2 \cdot 3x + 2 \cdot 5 + \frac{2}{3} \cdot 6x - \frac{2}{3} \cdot 3$
$$= 6x + 10 + 4x - 2$$
$$= (6x + 4x) + (10 - 2)$$
$$= 10x + 8$$

Now substitute 5 for x and simplify.
$10x + 8 = 10 \cdot 5 + 8$
$$= 50 + 8$$
$$= 58$$

TRY THIS **3.** Simplify $(4.5x - 8)2 + x$. Then evaluate for $x = 2$. $10x - 16$; 4

1.7 Combining Like Terms **27**

Lesson Note

Emphasize the use of the Distributive Property in combining like terms. This will help students to avoid errors such as writing $2a + 3b$ as $5ab$ or $6ab$. Examples 2 and 3 illustrate two important points: (a) the need to recognize that the coefficient of an expression such as y is 1, and (b) the use of the order of operations in simplifying an expression with sums and products.

Ask students what is meant by "like" fractions. The sum or difference of fractions can only be simplified if they have the same (like) denominators.

Math Connections

Life Skills: When balancing a checkbook, it is generally simpler to collect like items. You find the totals for deposits, checks, teller machine withdrawals, outstanding checks, and service charges, and then add or subtract these from the previous balance to find the new balance.

Additional Example 2

Simplify $15x - 9x + x$. $7x$

Additional Example 3

Simplify $y + 6(2y + 4) + 8$. $13y + 32$

Additional Example 4

Simplify $\frac{3}{5}(10x + 25) + 4(2x + 3)$. Then evaluate for $x = 4$. $14x + 27$; 83

Application: Look back at Example 4 on page 27. Ask students to evaluate the expression given for $x = 5$ *before* it is simplified. Why is the answer the same? It is the same expression.

Evaluation: Ask students whether it is easier to evaluate the above expression before or after it is simplified.   After

Ask why it might be useful to evaluate both expressions.   As a check

Common Error Analysis

Error: Expressions such as $9 + 3(4a + 2) + a$ in Example 3 may confuse some students. They may first combine the 9 and 3 and write $12(4a + 2) + a$.

Emphasize the need to follow the rules for order of operations and multiply $3(4a + 2)$ before combining any terms.

Checkpoint

Simplify, if possible.

1. $9x + 55x$ $64x$
2. $14n - n$ $13n$
3. $8b + b$ $9b$
4. $11t + 4 + t + 7$ $12t + 11$
5. $4 + 2(5x + 1)$ $10x + 6$
6. $7a + 4(a + 4) + 4$ $11a + 20$
7. $\frac{3}{4}(12x + 8) + 5x + 2$ $14x + 8$
8. $18x + 5 + x + 2$ $19x + 7$

Closure

Ask students these questions: How does grouping the terms of an expression allow you to simplify it? How do you know when an expression can be simplified no further? Combining like terms; terms are unlike

Classroom Exercises

Identify the like terms. Give the coefficient of each variable term.

1. $3x + 5x + 7$
2. $y + 4 + 7y$
3. $8m + 4 + m$

Simplify, if possible.

4. $13y - 6y$ $7y$
5. $8x + 4x + x$ $13x$
6. $7 + 4y + y$ $7 + 5y$
7. $3k + 5 + k + 2$ $4k + 7$
8. $x + 5 + 2x - 4$ $3x + 1$
9. $3a + 2 + a$ $4a + 2$
10. $3x + 7$ Not possible
11. $b + 7 + 4b + 1$ $5b + 8$
12. $2(3a + 1) + 4$ $6a + 6$
13. $3(4a + 5) + 1$ $12a + 16$
14. $5(a + 3) + 4$ $5a + 19$
15. $7x + 3(4 - 2x)$ $x + 12$

Written Exercises

Simplify, if possible.

1. $7m - 2m$ $5m$
2. $4k + 5k$ $9k$
3. $2 + 5u$ Not possible
4. $b + 3b$ $4b$
5. $7x + 5x - 2x$ $10x$
6. $3a + 5a + 2a$ $10a$
7. $3a + 4a + 7$ $7a + 7$
8. $6m + 2 + 5m$ $11m + 2$
9. $7y + y + 4y$ $12y$
10. $3 + 5 + m$ $8 + m$
11. $a + 5a + 9a$ $15a$
12. $4 + 5x + 3y$ Not possible
13. $7 + a + 13 + 6a$ $20 + 7a$
14. $5m + 13 + m + 4$
15. $2x + 7y + 9x + 3y$
16. $t + 7g + t + 8g$ $2t + 15g$
17. $3m + 4(2m + 3) - 6$
18. $3(4x + 5) + x$ $13x + 15$
19. $3x + 4(6 + 2x) + 5x$
20. $9 + 2(3m + 4) - m$
21. $y + 4(2y + 3) + 2$
22. $\frac{2}{3}(6y + 9) + 5y - 2$
23. $\frac{1}{2}(2b + 8) + b + 1$
24. $\frac{2}{3}(9 + 12x) + x$
25. $\frac{1}{4}(8x - 4) - 5x$
26. $(9x - 12)\frac{2}{3} - 8$
27. $15 + \frac{3}{5}(5x - 10)$
28. $2\frac{1}{2}m + 3m + 1\frac{1}{2}m$
29. $8\frac{3}{5}c + 1\frac{1}{5}c - 2\frac{3}{5}c$
30. $7\frac{1}{6}p - 2\frac{5}{6}p + 3\frac{1}{6}p$
31. $8\frac{3}{4}m + 6 + 5\frac{1}{2}m + 4\frac{1}{4}$ $14\frac{1}{4}m + 10\frac{1}{4}$
32. $7\frac{5}{6}a + 9\frac{1}{4} + 2\frac{2}{3}a + 3\frac{1}{2}$ $10\frac{1}{2}a + 12\frac{3}{4}$
33. $4(3x + 8) + 4(2x - 3)$ $20x + 20$
34. $5(3x + 4) + 2(7x + 2)$ $29x + 24$
35. $6(x + 4) + 3(2x + 6)$ $12x + 42$
36. $7(3a + 5) - 4a + 2(3a - 5)$ $23a + 25$
37. $\frac{1}{2}(4x + 10) + \frac{2}{3}(9x + 6)$ $8x + 9$
38. $\frac{2}{5}(5v + 10) + \frac{3}{4}(12 + 8v)$ $8v + 13$
39. $\frac{1}{2}(2a + 12) + \frac{3}{7}(7a + 14)$ $4a + 12$
40. $\frac{3}{4}(16m + 20) + \frac{1}{6}(12 + 24m)$ $16m + 17$

Simplify. Then evaluate for the given value of the variable.

41. $5(2a + 3) + 4(3a + 1)$ for $a = 5$
42. $4(6 + 3x) + \frac{3}{4}(8x + 12)$ for $x = 3$
43. $5(3b + 5) + \frac{3}{5}(10b + 25)$ for $b = 4$
44. $\frac{2}{3}(6 + 12p) + \frac{1}{2}(2p - 4)$ for $p = 6$
45. $13y + \frac{1}{4}(8y + 12) + 3 + 7y + \frac{3}{7}(28 + 7y)$ for $y = 9$ $25y + 18; 243$

Simplify.

46. $3a + 2[41 + 5(4a + 6)]$ $43a + 142$ **47.** $3(2x + 3) + 7[12 + 4(7x + 2)]$

48. $2(6y + 4) + 3[8 + 2(5y + 1)]$ **49.** $k + 6[3k + 5 + 5(3 + 7k)]$

50. Find the perimeter of the figure in terms of x. Simplify the result. $10x + 80$

(figure: rectangle with top labels $2(x+5)$, 5, $2(x+5)$; interior 10; left side x)

Mixed Review

Evaluate for the given value of the variable. *1.2, 1.3*

1. $7 - 12x$ for $x = \frac{1}{4}$ 4   **2.** $3x^2$ for $x = 4$ 48   **3.** $2x^2 + 5x$ for $x = 2$ 18

Which property is illustrated? *1.5, 1.6*

4. $a + b = b + a$ **5.** $(x + y) + z = x + (y + z)$ **6.** $a(b + c) = ab + ac$

7. Find the perimeter and area of a rectangle with length 14 cm and width 12 cm. *1.4* 52 cm, 168 cm²

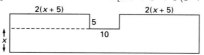

Application: Costs and Profit

In business, accounting, and many other fields, formulas are frequently used to make predictions about costs or profit.

Solve each problem.

1. The manager of a restaurant figures the cost of making coffee is

$$m = 0.049x,$$

where m is the cost of making x cups. Find the cost of making 600 cups of coffee. $29.40

2. Mr. Irvington owns and operates Appealing Apple Pies. A consultant gave him the following formula to analyze his profits

$$p = 2x - 0.01x^2$$

where p is the profit and x is the number of pies baked. Find the profit on 60 pies. $84

3. The following formula is used by some taxi companies to figure how much to charge for a ride,

$$C = f + rd,$$

where C is the charge for the ride, f is a fixed charge, r is the rate per tenth of a mile, and d is the distance of the ride in tenths of a mile. Find the charge if f is $1.10, r is $0.10, and the ride is 25 tenths of a mile. $3.60

Enrichment

Most students should be familiar with *magic squares*. If necessary, remind them that the numbers in each row, in each column, and along each diagonal have the same sum. Then have the students complete the following array so that it will be a magic square.

$x + 3a$	$x + 8a$	$x + a$
$\underline{?}$	$\underline{?}$	$\underline{?}$
$x + 7a$	$?$	$x + 5a$

Challenge the students to use variables and numbers to construct magic squares of their own. Then pairs of students can exchange papers and check each other's work.
$x + 2a$; $x + 4a$; $x + 6a$; x

1.8 Solution Sets of Sentences

Objectives To determine whether a sentence with no variables is true or false
To find the solution set of an open sentence

Prerequisite Quiz

Evaluate for the given value of the variable.

1. $2x + 5; x = 3$ 11
2. $7y - 2; y = 4$ 26
3. $9 - 4x; x = 1$ 5
4. $7(3x - 1); x = 2$ 35
5. $3(p + 2); p = 4$ 18
6. $2(3x - 1) - 4; x = 3$ 12
7. $18 + 5y - 3; y = 6$ 45
8. $\frac{x + 2}{5}; x = 13$ 3

Motivator

Ask students to write an algebraic statement for these sentences: The sum of 6 and 2 equals 8 and 5 subtracted from x equals 12. $6 + 2 = 8; x - 5 = 12$ Ask them whether the first statement is true or false.   T   Ask them whether they can tell whether the second statement is true or false and why. Can't tell; the value for x is unknown.   Ask them whether the second statement is true for $x = 9$.   No   Ask whether it is true for $x = 17$?   Yes

A balance scale can be used to compare the weights of two objects. The result can be written using an inequality symbol.

$4 < 8$	$8 > 4$	$4 \neq 8$
4 is less than 8.	8 is greater than 4.	4 is not equal to 8.

On a horizontal number line, the smaller of two numbers is at the left of the larger number. Thus, 4 is at the left of 8 because $4 < 8$.

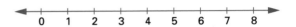

Equations and inequalities are two types of *mathematical sentences*. An **equation** consists of two expressions with an equals sign between them. An **inequality** consists of two expressions with an inequality symbol such as $<$, $>$, or $\neq$ between them. A mathematical sentence can be true or false.

EXAMPLE 1 True or false?

   a. $8(7 - 4) = 23$ **b.** $26 - 7 \cdot 3 > 2 \cdot 6 - 8$

Solutions

a. $8(7 - 4) = 23$ **b.** $26 - 7 \cdot 3 > 2 \cdot 6 - 8$
     $8 \cdot 3 \overset{?}{=} 23$ $26 - 21 \overset{?}{=} 12 - 8$
        $24 \neq 23$ False $5 > 4$ True

TRY THIS True or false?   **1.** $(7 + 5)6 = 37$ F
                                2. $8 \cdot 4 < 2 \cdot 16$ F

The mathematical sentence $6x + 5 = 17$ contains a variable and is called an **open sentence**. An *open sentence* is neither true nor false until the variable or variables are replaced with numbers. The set of numbers that can replace a variable is called the **replacement set** for the variable.

30   Chapter 1 Introduction to Algebra

Highlighting the Standards

Standards 1a, 1b: In Problem Solving Strategies, students practice their problem-solving approaches and apply strategies to problems outside mathematics.

Additional Example 1

True or false?

a. $9(5 + 20) = 225$ T
b. $36 - 5.2 < 41 - 3 \cdot 6$ F

Definition	The **solution set** of an open sentence is the set of numbers from the given replacement set that makes the sentence true.

You can use braces, { }, to indicate a set. In Example 2, {0, 1, 2} is the set with members 0, 1, and 2.

EXAMPLE 2 Find the solution set of $6x + 5 = 17$. The replacement set for x is {0, 1, 2, 3}.

Plan Replace x with each number in the replacement set.

Solution

Replace x with 0.

$$6x + 5 = 17$$
$$6 \cdot 0 + 5 \stackrel{?}{=} 17$$
$$5 \stackrel{?}{=} 17$$
$$5 \neq 17 \quad \text{False}$$

Replace x with 1.

$$6x + 5 = 17$$
$$6 \cdot 1 + 5 \stackrel{?}{=} 17$$
$$11 \stackrel{?}{=} 17$$
$$11 \neq 17 \quad \text{False}$$

Replace x with 2.

$$6x + 5 = 17$$
$$6 \cdot 2 + 5 \stackrel{?}{=} 17$$
$$12 + 5 \stackrel{?}{=} 17$$
$$17 \stackrel{?}{=} 17$$
$$17 = 17 \quad \text{True}$$

Replace x with 3.

$$6x + 5 = 17$$
$$6 \cdot 3 + 5 \stackrel{?}{=} 17$$
$$18 + 5 \stackrel{?}{=} 17$$
$$23 \stackrel{?}{=} 17$$
$$23 \neq 17 \quad \text{False}$$

Thus, the solution set is {2}.

EXAMPLE 3 Find the solution set of $2x > x + 5$. The replacement set is {2, 4, 6}.

Solution

Replace x with 2.

$$2x > x + 5$$
$$2 \cdot 2 \stackrel{?}{>} 2 + 5$$
$$4 \stackrel{?}{>} 7$$
$$4 \not> 7 \quad \text{False}$$

Replace x with 4.

$$2x > x + 5$$
$$2 \cdot 4 \stackrel{?}{>} 4 + 5$$
$$8 \stackrel{?}{>} 9$$
$$8 \not> 9 \quad \text{False}$$

Replace x with 6.

$$2x > x + 5$$
$$2 \cdot 6 \stackrel{?}{>} 6 + 5$$
$$12 \stackrel{?}{>} 11$$
$$12 > 11 \quad \text{True}$$

Thus, the solution set is {6}.

TRY THIS 3. Find the solution set of $2x - 3 < x + 5$. The replacement set for x is {7, 8, 9}. {7}

The set containing no members is called the **empty set**, or the **null set**, and is written $\varnothing$. If no number in the replacement set makes an open sentence true, then the solution set of the sentence is the empty set, $\varnothing$.

Additional Example 2

Find the solution set of $4x - 3 = 5$. The replacement set is {1, 2, 3}. {2}

Additional Example 3

Find the solution set of $16 - 2y > 3y$. The replacement set is {1, 3, 4}. {1, 3}

Application: Given a replacement set of all positive integers, ask students to give the solution set for $x + 3 = x + 2$. $\varnothing$ Then ask them to write an open sentence which is an inequality, has a replacement set of all positive integers, and has the empty set for the solution set.   Sample answer: $x + 1 > x + 5$

Ask students to give examples of open sentences with a replacement set of all positive integers, and with an infinite number of members in the solution set.   Sample answers: $2x + 3 = 5x$ and $7x + 5 > 7x$

Common Error Analysis

Error: It is common for students to find one solution and then stop.

Remind them that the solution set of an open sentence can be empty, have one, two, or more members, or even be all members of the replacement set.

EXAMPLE 4   Find the solution set of $1.2x + 3x = 4.1x$. The replacement set is $\{2, 5\}$.

Solution

Replace x with 2.

$$1.2x + 3x = 4.1x$$
$$1.2(2) + 3(2) \stackrel{?}{=} 4.1(2)$$
$$2.4 + 6 \stackrel{?}{=} 8.2$$
$$8.4 \stackrel{?}{=} 8.2$$
$$8.4 \neq 8.2 \quad \text{False}$$

Replace x with 5.

$$1.2x + 3x = 4.1x$$
$$1.2(5) + 3(5) \stackrel{?}{=} 4.1(5)$$
$$6 + 15 \stackrel{?}{=} 20.5$$
$$21 \stackrel{?}{=} 20.5$$
$$21 \neq 20.5 \quad \text{False}$$

Thus, the solution set is $\varnothing$.

EXAMPLE 5   Find the solution set of $2x + 1 > 3x + 1$. The replacement set is $\{0, 1, 2, 3, 4\}$.

Solution   Replace x with 0. The result is $1 > 1$ (false). Replacing x with numbers greater than 1 will also result in a false sentence. Thus, the solution set is $\varnothing$.

TRY THIS   **4.** Find the solution set of $3x + x \neq 4x$. The replacement set is $\{3, 4, 5, 6, 7\}$. $\varnothing$

Sometimes open sentences involve exponents.

EXAMPLE 6   Find the solution set of $3x^2 + 2x > 7$. The replacement set is $\{0, 1, 3, 5\}$.

Solution

Replace x with 0.

$$3x^2 + 2x > 7$$
$$3 \cdot 0^2 + 2 \cdot 0 \stackrel{?}{>} 7$$
$$3 \cdot 0 + 0 \stackrel{?}{>} 7$$
$$0 + 0 \stackrel{?}{>} 7$$
$$0 \not> 7 \quad \text{False}$$

Replace x with 1.

$$3x^2 + 2x > 7$$
$$3 \cdot 1^2 + 2 \cdot 1 \stackrel{?}{>} 7$$
$$3 \cdot 1 + 2 \stackrel{?}{>} 7$$
$$3 + 2 \stackrel{?}{>} 7$$
$$5 \not> 7 \quad \text{False}$$

Replace x with 3.

$$3x^2 + 2x > 7$$
$$3 \cdot 3^2 + 2 \cdot 3 \stackrel{?}{>} 7$$
$$3 \cdot 9 + 6 \stackrel{?}{>} 7$$
$$33 > 7 \quad \text{True}$$

Replace x with 5.

$$3x^2 + 2x > 7$$
$$3 \cdot 5^2 + 2 \cdot 5 \stackrel{?}{>} 7$$
$$3 \cdot 25 + 10 \stackrel{?}{>} 7$$
$$85 > 7 \quad \text{True}$$

Thus, the solution set is $\{3, 5\}$.

TRY THIS   **5.** Find the solution of $x^2 + 1 > (x + 1)^2 - 1$. The replacement set is $\{0, 1, 2, 3\}$. $\{0\}$

Additional Example 4

Find the solution set of $1.5x = 1.1x + 4x$. The replacement set is $\{1, 3, 10\}$. $\varnothing$

Additional Example 5

Find the solution set of $4x + 2 < 3x + 1$. The replacement set is $\{1, 3, 5\}$. $\varnothing$

Additional Example 6

Find the solution set of $2y^2 + y > 25$. The replacement set is $\{2, 3, 4\}$. $\{4\}$

Focus on Reading

Match each term at the left with the appropriate definition at the right.

1. solution set d
2. replacement set a
3. open sentence e
4. empty set c
5. inequality f
6. equation b

a. the set of numbers that can replace the variable in an open sentence

b. a sentence that states that two expressions name the same number

c. the set containing no members

d. the set of numbers from the given replacement set that makes an open sentence true

e. a sentence containing one or more variables

f. a sentence containing $\neq$, $>$, or $<$

Classroom Exercises

True or false?

1. $5 > 4$ T
2. $8 + 2 > 13$ F
3. $3 + 2 \neq 5$ F
4. $7 < 4 + 6$ T
5. $4 \cdot 4 < 5$ F
6. $3^2 \neq 6$ T
7. $7 > 5 + 6$ F
8. $2(3 + 5) > 13$ T

Tell whether the given number is a solution of the open sentence.

9. $15 = x + 7$; 8 Yes
10. $14 + y > 16$; 12 Yes
11. $a \neq 5.9$; 5.9 No
12. $x - 5 < 2$; 10 No
13. $16 - x > 8$; 8 No
14. $2c - 6 > 1$; 4 Yes
15. $x^2 > x$; 2 Yes
16. $x^2 > x$; $\frac{1}{2}$ No
17. $x > x$; 2.7 No

Written Exercises

True or False?

1. $83 + 0 = 83$ T
2. $23 \cdot 1 < 23$ F
3. $18 + 5 < 5 + 18$ F
4. $6(4 + 3) < 90$ T
5. $20 - 6 < 4(3 + 2)$ T
6. $10(3 + 8) = 30 + 80$ T
7. $3(10 - 3) \neq 4(7 - 1)$ T
8. $32 - 4 \cdot 6 < 12 + 5$ T
9. $24 - 5 \cdot 3 > 10 - 3$ T

Find the solution set. The replacement set is {0, 1, 2, 3}.

10. $3x + 1 > 5$ {2, 3}
11. $4x + 1 < 22$ {0, 1, 2, 3}
12. $3x + 4 = 4$ {0}
13. $3x + 5 = x + 8$ $\varnothing$
14. $\frac{x + 3}{2} = 2$ {1}
15. $\frac{3x + 2}{4} \neq 2$ {0, 1, 3}

Find the solution set. The replacement set is {1, 3, 5}.

16. $3x + 2 = x + 8$ {3}
17. $2x - 2 = x + 3$ {5}
18. $3(x + 1) = 3x + 3$ {1, 3, 5}
19. $2x > x + 1$ {3, 5}
20. $5x - 5 < 4x$ {1, 3}
21. $4(x + 2) > 3(x + 5)$ $\varnothing$

Checkpoint

True or false?

1. $36 \cdot 0 < 1$ T
2. $5(2 + 3) = 3 \cdot 5 + 9$ F
3. $3(9 + 5) \neq 32$ T

Find the solution set. The replacement set is {0, 2, 3}.

4. $2x + 4 = 10$ {3}
5. $3(y + 5) > 15$ {2, 3}
6. $6x + 2 > 20 + x$ $\varnothing$
7. $\frac{1}{2}x + 3 = 2x$ {2}
8. $x^2 + 5x < 20$ {0, 2}

Closure

Ask for the term used to describe a mathematical expression containing a variable or variables, but for which no solution set has been defined.   Open sentence   Ask how a replacement set and a solution set differ. Ask how they are similar.   See p. 31.

Guided Practice

Classroom Exercises 1–17
Try This all

Independent Practice

A Ex. 1–21, **B** Ex. 22–33, **C** Ex. 34–43

Basic: FR all, WE 1–29 odd
Average: FR all, WE 7–37 odd
Above Average: FR all, WE 11–43 odd

Additional Answers

Written Exercises

35. F; if $x = 1$ and $y = 2$, $x - y \neq y - x$.
37. F; if $x = \frac{1}{2}$, $x^2 < x$.

Find the solution set. The replacement set is $\{0, 5, 10\}$.

22. $1.1x = x + 0.5$ {5}   **23.** $0.4x = 0.2x + 0.5 + 0.1x$ {5}
24. $x^2 - 3x < 10$ {0}   **25.** $x^2 + 4x = 45$ {5}
26. $x^2 - 5x < 100$ {0, 5, 10}   **27.** $x^2 + 7x = 3x + 45$ {5}
28. $\frac{3x}{4} + x = \frac{x + 5}{2} + x$ {10}   **29.** $\frac{1}{2}(x + 2) < 2x + 1 + \frac{1}{2}x$ {5, 10}
30. $0.15x > 0.2 + 0.1 + 0.1x$ {10}   **31.** $1.6x < x + 1 + 0.5x$ {0, 5}
32. $x < x^2$ {5, 10}   **33.** $x^2 + 4x = 8x + 5$ {5}

An open sentence can contain more than one variable. For each sentence, tell whether it is true for all numbers. If not, give an example showing it is not.

34. $x + y = y + x$ T   **35.** $x - y = y - x$ **36.** $x + y - y = x$ T   **37.** $x^2 > x$

38. $x(y - z) = xy - xz$ T   **39.** $x \cdot x < x^2 + 1$ T

Find the solution set. The replacement set is $\{1, 2, 3, 4\}$.

40. $x + 3[4 + 2(5x + 1)] = 30x + 19$ {1}   **41.** $\frac{x^2 + 7x}{2x + 5} = \frac{1}{2}x + 1$ {2}
42. $2x + 3x + 4x = x + 8x$ {1, 2, 3, 4}   **43.** $x + 3x + 4x = x + 8x$ $\varnothing$

Mixed Review

Write an algebraic expression for each word description.   *1.1, 1.2*

1. 19 less than x $x - 19$ **2.** $\frac{1}{4}$ of the number n of students enrolled $\frac{1}{4}n$

3. the total bill for x hours if a person charges \$12 per hour. $12x$

Simplify.   *1.3, 1.5–1.7*

4. $3(2y)4$ $24y$ **5.** $2 \cdot 4^2$ 32 **6.** $2(3x + 7)$ $6x + 14$ **7.** $3a + 4(2a + 5) + a$ $12a + 20$

For Exercises 8–9, refer to the rectangle below.   *1.4*

8. What is the perimeter? 20 cm
9. What is the area? 24 cm²

4 cm

6 cm

$p = 12.8$ cm, $A = 10.24$ cm²

10. Find the perimeter and the area of a square with a side of 3.2 cm.

11. Find the area of a trapezoid with bases 3.5 cm and 2.7 m and an altitude of 4.6 m. 14.26 m²

34   Chapter 1 Introduction to Algebra

Enrichment

Have the students use the replacement set $\{0, 1, 2, 3, 4, 5\}$ to find the solution of each open sentence.

1. $x^3 - 1 = 0$ {1}
2. $x^3 = x^2$ {0, 1}
3. $\frac{x^2 + 1}{5} = 2$ {3}
4. $x^2 - 1 > 0$ {2, 3, 4, 5}

5. $x^2 = 5x$ {0, 5}
6. $x^2 - x < 2$ {0, 1}
7. $x^2 = (x + 1)^2$ $\varnothing$
8. $(\frac{1}{2}x)^2 < x$ {1, 2, 3}

Problem Solving Strategies

Understanding the Problem

Everyone solves problems every day,—which shirt to wear, scheduling time for homework assignments and for TV, budgeting money for a class trip, and so on. Whether in personal life, at school, or on the job, you can often solve problems that appear difficult at first glance by using a series of steps. Here is a four-step process that many people have found useful for solving problems.

To Solve a Problem
1. Understand the Problem
2. Develop a Plan
3. Carry Out the Plan
4. Look Back

To understand a problem (Step 1), do these things.
a. Read the problem carefully. Make a list of all new words.
b. Look up the meaning of the words that are new to you.
c. Decide what given information is needed to solve the problem, and what is not needed.
d. Identify what you are asked to find. Identify the units (if any) that you will need to write the answer.

Exercises

Follow the steps (a–d) to understand the problem.
(a) List the information you need to solve the problem; (b) write the words from the problem that tell you what you are looking for, and (c) without solving the problem, write the unit for the answer.

1. Five people are planning a 123-mile trip to the beach. They estimate that the total cost will be $285. About how much is each person's share?

2. Raysa has saved $17 for a new bicycle seat. She needs about $50 more to cover the cost and the sales tax. If Raysa earns money by mowing lawns at $10 each, how many more lawns does she need to mow to earn the amount she needs?

3. For a science club experiment, Ed, Carla, and Ramon are counting the minutes that a hamster spends on its exercise wheel during three 12-hour days. The wheel has a diameter of 5 in. If each student shares the observation time for the experiment equally, how many hours must Ed observe the experiment?

21. $\frac{2}{3} \cdot 6x - \frac{2}{3} \cdot 12; 4x + 8$

22. $14y \cdot \frac{3}{7} - 21 \cdot \frac{3}{7}; 6y - 9$

23. $1.5 \cdot 2x + 1.5 \cdot 3; 3x + 4.5$

24. $2x \cdot 4 - 1.6 \cdot 4; 8x - 6.4$

25. $3 \cdot 4.1x - 3 \cdot 6; 12.3x - 18$

26. $3 \cdot 2a + 3 \cdot 4b + 3 \cdot 6; 6a + 12b + 18$

27. $4c \cdot 1.5 - 2b \cdot 1.5 + 1 \cdot 1.5;$
$6c \cdot 3b + 1.5$

28. $2 \cdot 3a + 2 \cdot 6x - 2 \cdot 1.3d; 6a + 12x - 2.6d$

29. $6 \cdot 40 + 6 \cdot 3; 258$

30. $5 \cdot 300 + 5 \cdot 7 = 1{,}500 + 35 = 1{,}535$

31. $7 \cdot 50 - 7 \cdot 1 = 350 - 7 = 343$

32. $4 \cdot 600 - 4 \cdot 2 = 2{,}400 - 8 = 2{,}392$

33. $49(3 + 7) = 49 \cdot 10 = 490$

34. $(96 + 4)294 = 100 \cdot 294 = 29{,}400$

35. $\frac{1}{2}(47 + 3) = \frac{1}{2} \cdot 50 = 25$

36. $40(52 - 2) = 40 \cdot 50 = 2{,}000$

37. $(15 + 5)\frac{3}{4} = 20 \cdot \frac{3}{4} = 15$

38. Answers will vary.

39. Answers will vary.

40. Comm Prop Mult

41. Distr Prop

42. Assoc Prop Add

43. Comm Prop Add

44. Distr Prop

45. Assoc Prop Mult

46. Distr Prop

47. Assoc Prop Mult

48. Comm Prop Mult

49. Distr Prop

Chapter 1 Review

Key Terms

algebraic expression (p. 1)
area (p. 13)
Associative Properties (p. 17)
coefficient (p. 26)
Commutative Properties (p. 17)
Distributive Property (p. 22)
empty set (p. 31)
equation (p. 30)
equivalent expressions (p. 19)
evaluate (p. 1)
exponent (p. 9)
factors (p. 9)
formula (p. 9)
grouping symbols (p. 5)

Identity Property for Multiplication (p. 27)
inequality (p. 30)
like terms (p. 26)
mathematical sentence (p. 30)
multiplicative identity (p. 27)
numerical expression (p. 1)
open sentence (p. 30)
perimeter (p. 13)
power (p. 9)
replacement set (p. 30)
solution set (p. 31)
term (p. 26)
variable (p. 1)
volume (p. 14)

Key Ideas and Review Exercises

1.1   To evaluate an algebraic expression, replace the variables with the numbers that are given and then simplify.

Evaluate.

1. $7x$ for $x = 9$ 63

2. $a - b$ for $a = 11$ and $b = 7$ 4

Write an algebraic expression for each word description.

3. 31 decreased by the number of cloudy days c $31 - c$

4. the cost, in dollars, of one bracelet if n bracelets cost \$6 $\frac{6}{n}$

1.2–1.4   To simplify a numerical expression:
  1. Operate within grouping symbols. Begin with the innermost symbols.
  2. Simplify powers.
  3. Multiply or divide from left to right.
  4. Add or subtract from left to right.

Simplify.

5. $30 + 5 \cdot 4$ 50

6. $4 + 6^2$ 40

7. $(3 \cdot 5)^2$ 225

8. $3 \cdot 5^2$ 75

9. $35 \div 7 + 8$ 13

10. $6 + [13 - (18 - 9)]$ 10

Evaluate.

11. $x^3 - x^2 + 3x$ for $x = 2$ 10

12. $\dfrac{x^2 + y^2}{x - y}$ for $x = 4$ and $y = 3$ 25

Use the formula $p = 2l + 2w$ to find the perimeter of each rectangle with the given dimensions.

13. $l = 9$ in. and $w = 7$ in. 32 in   **14.** $l = 5.8$ cm and $w = 3.9$ cm 19.4 cm

1.5 For all numbers a, b, and c,
$a + b = b + a$ and $a \cdot b = b \cdot a$ (Commutative Properties)
$(a + b) + c = a + (b + c)$ and $(a \cdot b)c = a(b \cdot c)$ (Associative Properties)

Which property is illustrated?

15. $65 + 43 = 43 + 65$ Comm Prop Add   **16.** $(3a)2 = 2(3a)$ Comm Prop Mult

Simplify.

1.7 To simplify the expression $a + 3(2a - 4)$, first use the Distributive Property.

1.6 For all numbers a, b, and c, the Distributive Property states:
$a(b + c) = ab + ac$ $(b + c)a = ba + ca$
$a(b - c) = ab - ac$ $(b - c)a = ba - ca$

Rewrite by using the Distributive Property. Then simplify.

17. $5(7 + 2)$ $5 \cdot 7 + 5 \cdot 2$; 45   **18.** $8 \cdot 5 + 8 \cdot 7$ 8(5 + 7); 96
19. $4(7a - 3)$ $4 \cdot 7a - 4 \cdot 3$; 28a − 12   **20.** $6x + 5x + 7x$ (6 + 5 + 7)x; 18x

1.7 To combine like terms, use the Distributive Property. To simplify the expression $a + 3(2a - 4)$, first multiply 3 by $2a - 4$.

Simplify. (Exercises 25–27)

21. $7y - 4y$ 3y   **22.** $17 + 4(6x - 2) + x$ 25x + 9
23. Simplify $4(2p + 5) + (6p + 12)\frac{2}{3}$. Then evaluate for $p = 2$. 12p + 28; 52

1.8 The solution set of an open sentence is the set of numbers from the replacement set that makes the sentence true.

Find the solution set. The replacement set is {1, 3, 5}.

24. $4x - 3 = 17$ {5}   **25.** $7 + 4y < 6$ ∅
26. $2x + 8 > 3x$ {1, 3, 5}   **27.** $x^2 + 4x > 12$ {3, 5}
28. Explain in your own words the meaning of the symbols $>$ and $<$. Show how $a > b$ can be read in two different ways. Can you use the shape of the inequality symbol to describe a way to remember which number is larger? Answers will vary.

Chapter 1 Test

A-Level Exercises: 1, 2, 5–10, 12–14, 16–22, 24–27, 29–32
B-Level Exercises: 3–4, 11, 15, 23, 28
C-Level Exercises: 33–34

Evaluate.

1. $6s$ for $s = 12$ 72

2. $x^2 + 2x + 5$ for $x = 4$ 29

3. $\dfrac{4.5}{c}$ for $c = 3$ 1.5

4. $x^3(x + y)$ for $x = 3$ and $y = 7$ 270

Write an algebraic expression for each word description.

5. the cost, in cents, of p pencils if each pencil costs 39¢ 39p

6. 6 less than Julio's age j $j - 6$

Simplify.

7. $20 - 5 \cdot 3$ 5

8. $3 \cdot 2^3$ 24

9. $(3 \cdot 2)^3$ 216

10. $10 + 2(7 - 4)$ 16

11. Use the formula $A = lw$ to find the area of a rectangle of length 11.5 cm and width 6.2 cm. 71.3 cm²

Which property is illustrated?

12. $5(7 + 2) = 5 \cdot 7 + 5 \cdot 2$ Distr Prop

13. $3 \cdot 10 = 10 \cdot 3$ Comm Prop Mult

14. $(4 + 5) + 1 = 4 + (5 + 1)$ Assoc Prop Add

15. $(3 + 9) + 7 = (9 + 3) + 7$ Comm Prop Add

Rewrite by using the Distributive Property. Then simplify.

16. $7(10 - 6)$ $7 \cdot 10 - 7 \cdot 6$; 28

17. $5 \cdot 7 + 5 \cdot 9 + 5 \cdot 3$ $5(7 + 9 + 3)$; 95

18. $4(9x - 5)$ $4 \cdot 9x - 4 \cdot 5$; $36x - 20$

19. $(20a + 15)\dfrac{3}{5}$ $20a \cdot \dfrac{3}{5} + 15 \cdot \dfrac{3}{5}$; $12a + 9$

Simplify.

20. $(7m)4$ 28m

21. $9(21y)$ 189y

22. $7(2x)3$ 42x

23. $16 \cdot \dfrac{3}{4}k \cdot 2$ 24k

24. $5x + x + 7x$ 13x

25. $8m - 2m$ 6m

26. $6x + 3 + x + 2$ $7x + 5$

27. $8 + 2(3x + 2) + x$ $7x + 12$

28. Simplify $5(2y + 4) + (4y + 8)\dfrac{3}{4}$. Then evaluate for $y = 5$. $13y + 26$; 91

Find the solution set. The replacement set is {1, 2, 3}.

29. $3x + 8 = 17$ {3}

30. $4y - 1 > 3y$ {2, 3}

31. $5x + 6 \neq 7x + 2$ {1, 3}

32. $x^2 + 4x < 10x + 1$ {1, 2, 3}

33. Simplify $(5 + 3x)2 + 3[9 + 5(2x + 1)]$ $36x + 52$

34. Use the formula $V = \dfrac{\pi r^2 h}{3}$ to find V for $r = 10$ cm and $h = 5$ cm.

Use 3.14 for π. Round your answer to the nearest whole number. 523 cm³

College Prep Test

Choose the best answer to each question or problem.

1. If $r = 3.8$ and $s = 1.9$, then $r \div s$ is __?__. C
(A) $\frac{1}{2}$ (B) $\frac{1}{20}$ (C) 2
(D) 20   (E) None of these

2. If each apple costs c cents, what is
E   the cost in dollars for a dozen apples?
(A) $\frac{c}{100}$ (B) $\frac{c}{12}$ (C) $12c$
(D) $\frac{12}{c}$ (E) $\frac{12c}{100}$

3. If change for a dollar consists of
C   some quarters and some nickels, what is the least number of coins possible?
(A) 4   (B) 6   (C) 8
(D) 12   (E) 20

4. $(9x)^3$ and $9x^3$ have the same value if x is equal to __?__. A
(A) 0   (B) 1   (C) 0 or 1
(D) 9   (E) None of these

5. How many numbers between 300
B   and 600 begin or end with 5?
(A) 100   (B) 120   (C) 130
(D) 140   (E) 200

6. Find the next number in the sequence 1, 2, 3, 5, 8, 13, __?__. D
(A) 15   (B) 18   (C) 20
(D) 21   (E) 25

7. Thirty-cent cans of fruit juice can be
C   bought for $3.12 per dozen. What would be the total savings on 48 cans of juice?
(A) $1.20   (B) $1.44
(C) $1.92   (D) $2.40
(E) $2.88

8. The value of $\frac{5}{x}$ is greatest if x equals __?__. A
(A) $\frac{1}{5}$ (B) $\frac{1}{2}$ (C) 1
(D) 2   (E) 5

9.

If each side of the square is multiplied by 3, then the area will be __?__. D
(A) unchanged
(B) tripled
(C) multiplied by 6
(D) multiplied by 9
(E) multiplied by 12

10. A taxi meter registers $1.25 for the
E   first $\frac{1}{6}$ mile and 10¢ for each additional $\frac{1}{6}$ mile. If the meter registers $3.25, how many miles was the trip?
(A) $2\frac{5}{6}$ (B) $3\frac{1}{6}$ (C) $3\frac{1}{4}$
(D) $3\frac{1}{3}$ (E) $3\frac{1}{2}$

11. An algebraic expression for y in. less than 5 ft is __?__. A
(A) $60 - y$ (B) $y - 5$
(C) $5 - y$ (D) $y + 5$
(E) $12y - 5$

12. If a number between 1 and 2 is squared, the result is a number between __?__. D
(A) 0 and 1   (B) 2 and 3
(C) 2 and 4   (D) 1 and 4
(E) None of these

13. If the postage rate is 25¢ for the first ounce and 20¢ for each additional ounce, then the weight of a package costing $2.45 is __?__. E
(A) 8 oz   (B) 9 oz   (C) 10 oz
(D) 11 oz   (E) 12 oz

2 OPERATIONS WITH REAL NUMBERS

OVERVIEW

This chapter starts with identifying coordinates on a number line and applying real numbers to real-life situations. Students add numbers on a number line and then progress to the rules for all four operations with real numbers. Throughout the chapter, the students are introduced to the Field Properties not covered in Chapter 1. They use these rules and properties to simplify and evaluate algebraic expressions.

OBJECTIVES

- To compare real numbers
- To add, subtract, multiply, and divide real numbers
- To use the properties of real numbers to simplify arithmetic and algebraic expressions
- To evaluate algebraic expressions

PROBLEM SOLVING

In Lesson 2.1, students apply the strategy of Making a Graph to illustrating positive and negative real numbers. Looking for Patterns is the problem-solving technique used in Lesson 2.6 on page 62 to introduce the rules for multiplying real numbers. Problems throughout the chapter and in the Application on page 61 encourage students to apply the rules for operations with real numbers in everyday problem-solving situations by making the Problem Concrete. The strategy of Using a Formula is applied to automobile rental rates in the Application on page 66.

READING AND WRITING MATH

The Focus on Reading exercises on pages 64 and 73 require students to review vocabulary associated with real number properties. The exercises in each lesson contain problems which ask students to write expressions relating positive and negative real numbers to real-life situations. Exercise 79 on page 48 asks the student to express in writing the connection between absolute value and the opposite of a number.

TECHNOLOGY

Calculator: Entering negative numbers into the calculator is demonstrated in Example 4 on page 54. Finding a reciprocal with or without the reciprocal key found on some calculators is shown on page 71.

SPECIAL FEATURES

Mixed Review pp. 44, 48, 52, 56, 66, 71, 74, 78, 81
Midchapter Review p. 60
Application: Windchill p. 61
Focus on Reading pp. 64, 73
Application: Automobile Rental Rates p. 66
Using the Calculator p. 71
Brainteaser p. 78
Key Terms p. 82
Key Ideas and Review Exercises pp. 82–83
Chapter 2 Test p. 84
College Prep Test p. 85
Cumulative Review (Chapters 1–2) pp. 86–87

PLANNING GUIDE

Lesson	Basic	Average	Above Average	Resources
2.1 pp. 43–44	CE 1–12 WE 1–55 odd	CE 1–12 WE 9–63 odd	CE 1–12 WE 17–63 odd, 65–68	Reteaching 9 Practice 9
2.2 pp. 47–48	CE 1–16 WE 1–65 odd	CE 1–16 WE 13–77 odd	CE 1–16 WE 29–79 odd, 80–87	Reteaching 10 Practice 10
2.3 pp. 51–52	CE 1–22 WE 1–28	CE 1–22 WE 9–35	CE 1–22 WE 13–39	Reteaching 11 Practice 11
2.4 pp. 55–56	CE 1–16 WE 1–41 odd	CE 1–16 WE 9–45 odd, 47–48	CE 1–16 WE 17–45 odd, 49–54	Reteaching 12 Practice 12
2.5 pp. 58–61	CE 1–12 WE 1–61 odd Midchapter Review Application	CE 1–12 WE 17–77 odd Midchapter Review Application	CE 1–12 WE 29–83 odd, 84–89 Midchapter Review Application	Reteaching 13 Practice 13
2.6 pp. 64–66	FR all, CE 1–20, WE 1–57 odd Application	FR all, CE 1–20, WE 11–67 odd Application	FR all, CE 1–20, WE 21–79 odd Application	Reteaching 14 Practice 14
2.7 pp. 69–71	CE 1–15 WE 1–37 odd UC	CE 1–15 WE 7–43 odd UC	CE 1–15 WE 13–43 odd, 44–47 UC	Reteaching 15 Practice 15
2.8 pp. 73–74	FR all, CE 1–8, WE 1–31 odd	FR all, CE 1–8, WE 5–35 odd	FR all, CE 1–8, WE 11–41 odd	Reteaching 16 Practice 16
2.9 pp. 76–78	CE 1–21 WE 1–41 odd, 50–52, 57 Brainteaser	CE 1–21 WE 11–61 odd Brainteaser	CE 1–21 WE 21–71 odd Brainteaser	Reteaching 17 Practice 17
2.10 pp. 80–81	CE 1–10 WE 1–29 odd	CE 1–10 WE 11–39 odd	CE 1–10 WE 21–53 odd	Reteaching 18 Practice 18
Chapter 2 Review pp. 82–83	all	all	all	
Chapter 2 Test p. 84	all	all	all	
College Prep Test p. 85	all	all	all	
Cumulative Review pp. 86–87	1–67 odd	1–67 odd	1–67 odd	

CE = Classroom Exercises   WE = Written Exercises   FR = Focus on Reading   UC = Using the Calculator
NOTE: For each level, all students should be assigned all Try This and all Mixed Review Exercises

INVESTIGATION

Project: This investigation gives students a chance to explore three different methods of subtraction and to explain how each works.

Materials: Each student will need a copy of Project Worksheet 2.

Before distributing the worksheets, tell students that they are going to compare three methods of subtraction.

1. **Decomposition Method** This is the method they have learned.

2. **Equal addends**

```
      13
73    73 ← Add 10 to the 3.
   →   6
− 58  − 58 ← Compensate by adding
     15    10 to 58.
```

3. **Addition of nines complements**

```
  73         73   Add complements:
−  58  →   + 41 ← 9 − 5 = 4; 9 − 8 = 1
          114 ← Add the carried
            1    number to the ones.
           15
```

Distribute the worksheets and divide students into groups of three. Have all students use the first set of problems on the worksheet to practice all three methods. Then have each student in a group choose a different method and have a contest to see who can complete the problems in the first row of Set 2 the fastest. Students in each group then choose a different method to complete the next row of problems in Set 2, and another method to complete the third row in Set 2. Have each group decide which method seems to be most efficient.

Finally, have each group work together to show how each of the three methods works. Choose groups to present their conclusions to the class.

2 OPERATIONS WITH REAL NUMBERS

To turn mud into beautiful ceramics, potters make use of many formulas. They mix clay in different proportions, create glazes that melt into various colors, and fire the dried pieces to precise temperatures.

More About the Photo

Potters vary the mixture of clay and the temperature at which it is fired to produce various types of pottery, such as earthware, stoneware, and porcelain. After the clay is mixed, potters first push the damp clay down until the mass is evenly distributed around the center of the turning wheel. Then they push and pull gently on the clay to form the solid mass into a hollow shape. The air-dried shapes are painted with glazes made from glass and metal oxides. The glazed pottery is then fired in a kiln to a temperature that will fuse the molten glaze to the hardened clay.

2.1 The Set of Real Numbers

Objectives

To graph a point on a number line, given the coordinate of the point
To use real numbers to describe real-life situations
To compare real numbers

A number line may be used to show the relationships between numbers. A starting point, labeled 0, is called the *origin*. Numbers to the right of 0 are *positive*, and numbers to the left of 0 are *negative*. Zero is neither positive nor negative.

Different sets of numbers can be represented by listing their members. The three dots at the end of each list means *go on forever*.

Natural or **counting numbers:** $N = 1, 2, 3, 4, \cdots$
Whole Numbers: $W = 0, 1, 2, 3, 4, \cdots$
Integers: $I = \cdots -3, -2, -1, 0, 1, 2, 3, 4, \cdots$

The positive integers, 1, 2, 3, . . . are read as positive 1, positive 2, positive 3, and so on. The negative integers $-1, -2, -3, \ldots$ are read as negative 1, negative 2, negative 3, and so on.

These sets can be graphed on a number line. The shaded arrows mean that the graph goes on forever.

Graph of N: Natural numbers

Graph of W: Whole numbers

Graph of I: Integers

On a number line, the number that corresponds to a point is called the *coordinate* of the point. The point is called the *graph* of the number. In the graph of the integers above, point A is the graph of the number 3. The coordinate of point A is 3.

Teaching Resources

Application Worksheet 2
Project Worksheet 2
Quick Quizzes 9
Reteaching and Practice Worksheets 9

▰▰▰ GETTING STARTED

Prerequisite Quiz

Insert < or > to make each sentence true.

1. 9 _?_ 4   >
2. 0 _?_ 5   <
3. $\frac{1}{2}$ _?_ $\frac{2}{3}$ <
4. 2.5 _?_ 1.7   >

Find the number that is halfway between each pair of numbers.

5. 0 and $\frac{1}{3}$ $\frac{1}{6}$
6. $3\frac{1}{2}$ and 4 $3\frac{3}{4}$

Motivator

Ask students to describe situations, other than math class, where negative numbers are used.   Answers will vary.   Show students that zero and negative numbers are both necessary and useful for labeling certain quantities.

Highlighting the Standards

Standards 14a, 3d: The opening discussion compares the structure of three subsystems of the real number system. In Exercises 57–68, students judge the validity of statements about real numbers.

TEACHING SUGGESTIONS

Lesson Note

Some of the examples and exercises in this lesson use rational numbers such as -2.7 and $\frac{3}{4}$. Point out that these include the common fractions and decimals that the students are familiar with from arithmetic. Irrational numbers include square roots that are not exact decimals. However, a thorough discussion of rational numbers and irrational numbers is deferred until Chapter 13.
If students have difficulty understanding that $2 > -5$, use an example such as $2°$ being a higher temperature than $-5°$.

Math Connections

Medicine: The words "positive" and "negative" are often used to classify the results of medical tests. A negative test result can be good news for someone who has just been tested for a medical problem.

Critical Thinking Questions

Application: Have students draw a number line, leaving a large space between 0 and 1. Have them divide this distance as closely as they can into 12 equal parts. Then have them order the following fractions from smallest to largest by graphing them on the number line. $\frac{2}{3}, \frac{1}{2}, \frac{3}{4}, \frac{1}{6}, \frac{5}{12}, \frac{5}{6}, \frac{7}{12}, \frac{1}{4}.$
$\frac{1}{6}, \frac{1}{4}, \frac{5}{12}, \frac{1}{2}, \frac{7}{12}, \frac{2}{3}, \frac{3}{4}, \frac{5}{6}$

There are numbers on the number line that are not integers.
Numbers like $\frac{2}{5}, \frac{15}{4}, -\frac{1}{2}$, and 3 $\left(\text{or } \frac{3}{1}\right)$ are called *rational numbers*.

On the graph below, the coordinate of point D is 0.423. The graph of $-1\frac{2}{3}$ is B.

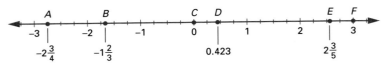

EXAMPLE 1 Give the coordinates of A, C, and E on the graph above.

Solution $-2\frac{3}{4}$ is the coordinate of A; 0 is the coordinate of C; $2\frac{3}{5}$ is the coordinate of E.

EXAMPLE 2 Graph the points with coordinates -4, -2.6, and $3\frac{3}{4}$ on a number line. Label the points A, B, and C.

Solution

Point A is the graph of -4. Point B is the graph of -2.6.
Point C is the graph of $3\frac{3}{4}$.

TRY THIS 1. Give the coordinate of F on the first number line above. 3

Any number that is positive, negative, or zero is called a **real number.** Real numbers are often used to represent real-life situations.

EXAMPLE 3 Write a real number to represent each situation.
  a. 25° above zero   25
  b. a loss of 5 yd in football -5
  c. withdrawal of $400.25 from a checking account -400.25
  d. a gain of $1\frac{3}{8}$ on the stock market $+1\frac{3}{8}$

TRY THIS 2. Write a real number to represent a temperature drop of four degrees. -4

Additional Example 1

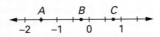

Give the coordinates of the points A, B, and C in the number line above. $A, -1\frac{1}{2}$; $B, -\frac{1}{4}$; $C, \frac{3}{4}$

Additional Example 2

Graph the points with coordinates $-1\frac{3}{4}$, -0.5, and 1.3 on a number line. Label the points D, E, and F, respectively.

Additional Example 3

Give a real number to represent each situation.
  a. A profit of $26 $+26$
  b. A descent of 1,000 ft $-1,000$
  c. A temperature drop of 6° -6
  d. An increase of $45.60 $+45.60$

Numbers on a horizontal number line increase from left to right.

Therefore, $-4 < 3$ *or* $3 > -4$.

EXAMPLE 4   Use $<$ or $>$ to make a true statement.

    a. -5 ___ -2 **b.** $4\frac{1}{2}$ ___ -6 **c.** -5.3 ___ -5.4

Solutions **a.** -5 is to the *left* of -2. Thus, $-5 < -2$.
    b. $4\frac{1}{2}$ is to the *right* of -6. Thus, $4\frac{1}{2} > -6$.
    c. -5.3 is to the *right* of -5.4. Thus, $-5.3 > -5.4$.

TRY THIS   **3.** Use $<$ or $>$ to make a true statement: 2.25 _?_ -2.35. $>$

Classroom Exercises

Give the approximate coordinate of each point graphed below.

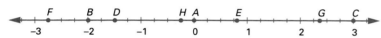

1. A 0 **2.** B -2 **3.** C 3 **4.** D $-1\frac{1}{2}$ **5.** E $\frac{4}{5}$ **6.** F $-2\frac{3}{4}$ **7.** G $2\frac{1}{3}$ **8.** H $-\frac{1}{4}$

Use $<$ or $>$ to make each statement true.

9. 1 _?_ 4 $<$ **10.** 2 _?_ -3 $>$ **11.** -3.7 _?_ -3.6 $<$ **12.** $2\frac{1}{3}$ _?_ $-3\frac{2}{3}$ $>$

Written Exercises

Give the approximate coordinate of each point graphed below.

1. A -3 **2.** B $-\frac{1}{2}$ **3.** C 2 **4.** D $-2\frac{1}{2}$ **5.** E $\frac{1}{4}$ **6.** F $2\frac{2}{3}$ **7.** G $-1\frac{3}{4}$ **8.** H $1\frac{3}{5}$

On a number line, graph each point whose coordinate is given.

9. A: -1 **10.** B: 4 **11.** C: 1.5 **12.** D: -2.1
13. E: $1\frac{2}{5}$ **14.** F: -4.7 **15.** G: $2\frac{3}{4}$ **16.** H: $-3\frac{2}{3}$

Common Error Analysis

Error: Students make the mistake of writing $-5 > -2$.

Remind them that numbers increase from left to right on the number line. Using an example such as $-2°$ is a slightly warmer temperature than $-5°$ may help.

Checkpoint

Give a real number to represent each situation.

1. 2,500 ft above sea level $+2,500$
2. A pay cut of $2 per hour -2
3. A $72,500 increase in export sales $+72,500$

Use $<$ or $>$ to make each sentence true.

4. -8.2 _?_ 3.7 $<$
5. $5\frac{1}{2}$ _?_ $5\frac{3}{8}$ $>$
6. 0 _?_ -0.6 $>$
7. -3.4 _?_ -2.5 $<$

Closure

Ask students what it means to give the coordinate of a point on the number line. Ask: "What is the difference between the coordinates of points to the left of zero as opposed to those that are to the right of zero? How can a number line be used to determine which of two numbers is greater in value?" Negative values as opposed to positive values; numbers increase in values from left to right.

Additional Example 4

Use $<$ or $>$ to make a true sentence. Refer to the number line in the textbook.

a. -1 _?_ -4 $>$
b. $-2\frac{1}{2}$ _?_ 0 $<$
c. 1.4 _?_ -3.2 $>$

◼◼ FOLLOW UP

Guided Practice

Classroom Exercises 1–12
Try This   all

Independent Practice

A Ex. 1–44, **B** Ex. 45–64, **C** Ex. 65–68
Basic: WE 1–55 odd
Average: WE 9–63 odd
Above Average: WE 17–63 odd, 65–68

Additional Answers

Written Exercises, page 43

9.–16.

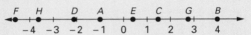

Write a real number to represent each situation.

17. a gain of 20 yd 20
18. a temperature drop of 6° −6
19. 8 wins 8
20. no gain, no loss 0
21. 3 below par −3
22. a deposit of $136.52 136.52
23. a weight loss of $3\frac{1}{2}$ pounds $-3\frac{1}{2}$
24. a loss of $2\frac{3}{8}$ points $-2\frac{3}{8}$
25. a profit of $236.25 236.25
26. a withdrawal of $9.42 −9.42
27. 4.2 km below sea level −4.2
28. a weight gain of 2.1 kg 2.1

Use < or > to make each sentence true.

29. $7 \underline{>} 3$
30. $0 \underline{<} 5$
31. $-2 \underline{<} 1$
32. $4 \underline{>} -4$
33. $-8 \underline{<} -6$
34. $2 \underline{>} -6$
35. $-7 \underline{>} -9$
36. $-3 \underline{<} 0$
37. $2\frac{1}{2} \underline{<} 3$
38. $-4\frac{1}{4} \underline{<} -4$
39. $3\frac{1}{3} \underline{>} 3$
40. $-8\frac{3}{4} \underline{<} 8\frac{3}{4}$
41. $0.2 \underline{>} 0.1$
42. $-0.6 \underline{<} 0.6$
43. $-2.4 \underline{<} -2.3$
44. $-6.1 \underline{<} 0$
45. $\frac{1}{2} \underline{>} \frac{1}{3}$
46. $-\frac{1}{5} \underline{>} -\frac{1}{4}$
47. $\frac{2}{3} \underline{>} -1\frac{1}{3}$
48. $\frac{3}{4} \underline{>} \frac{1}{2}$
49. $4.62 \underline{>} 4.6$
50. $-3.81 \underline{<} 3.81$
51. $7.94 \underline{<} 7.95$
52. $-8.3 \underline{>} -8.31$
53. $3.6 \underline{>} 3\frac{1}{2}$
54. $5.4 \underline{>} -5\frac{1}{2}$
55. $6.7 \underline{<} 6\frac{3}{4}$
56. $-8.2 \underline{>} -8\frac{1}{4}$

True or false?

57. 0 is an integer. T
58. Every integer is a real number. T
59. Every real number is an integer. F
60. $-6\frac{1}{2}$ is not an integer. T
61. 0 is not positive or negative. T
62. -6 is less than $-6\frac{1}{2}$. F
63. Every negative integer is less than every positive integer. T
64. On a number line, -2 lies to the right of $-3\frac{1}{2}$. T
65. On a number line, x and $-x$ are each the same distance from 0. T
66. For all real numbers x and y, if $x < y$, then $y > x$. T
67. For all real numbers x and y, if $x > 0$ and $y < 0$, then $x < y$. F
68. For all real numbers x, y, and z, if $x < y$ and $y < z$, then $x < z$. T

Mixed Review

Simplify.   *1.2, 1.3*

1. $19 - 3 \cdot 6$ 1
2. $(19 - 3)6$ 96
3. $2 \cdot 4^2$ 32
4. $(2 \cdot 4)^2$ 64

Find the solution set. The replacement set is $\{0, 1, 2, 3\}$.   *1.8*

5. $x + 2 < 4$ {0, 1}
6. $2x + 3 = 4$ ∅
7. $4(y + 2) = 4y + 8$ {0, 1, 2, 3}
8. Use the formula $A = \frac{1}{2}(b + c)h$ to find the area of a trapezoid with $b = 10$ cm, $c = 6$ cm, and $h = 4$ cm.   *1.4* 32 cm²

44   Chapter 2 Operations with Real Numbers

Enrichment

Have students use one or more of the following to describe each number.

a. whole number   **b.** integer
c. rational number   **d.** real number

1. -15 **2.** $\frac{1}{5}$ **3.** 0   **4.** $-\frac{9}{7}$
1. b, c, d   2. c, d   3. a, b, c, d
4. c, d

2.2 Opposites and Absolute Value

Objective To simplify expressions involving opposites and absolute values

In football, a 2-yd gain is represented by the number 2. The opposite of a 2-yd gain is a 2-yd loss, which is represented by the number -2. The numbers 2 and -2 are called **opposites**. On a number line they are the same distance from 0, and they lie on opposite sides of 0. Every real number has an opposite.

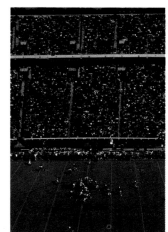

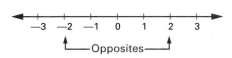

The opposite of 2 is -2.
The opposite of -2 is 2.
The opposite of 0 is 0.

So far, a dash has been used to indicate a negative number. Since -5 is the opposite of 5, a dash can also mean *the opposite of*. For example, -5 means *negative 5* or *the opposite of 5*.

You can indicate the opposite of a negative number also. For example, $-(-4)$ means *the opposite of* -4. Note that $-(-4) = 4$. This is read: *The opposite of negative 4 is 4.*

For any real number a, $-a$ means the opposite of a, and $-(-a) = a$.

EXAMPLE 1 Simplify each of the following expressions.

Expression	Meaning	Value
a. $-\left(-\frac{1}{2}\right)$	the opposite of negative $\frac{1}{2}$	$\frac{1}{2}$
b. -0	the opposite of 0	0
c. $-(7.8 - 3.2)$	the opposite of the difference: $7.8 - 3.2 = 4.6$	-4.6

TRY THIS 1. Simplify $-(6.9 - 4.7)$. -2.2

Additional Example 1

Simplify each expression.

a. $-(0.5)$ -0.5
b. $-(-1.8)$ 1.8
c. $-(10.5 - 8.1)$ -2.4

Teaching Resources

Quick Quizzes 10
Reteaching and Practice Worksheets 10
Transparency 4

■■■ GETTING STARTED

Prerequisite Quiz

A B C D E F G H
-2 0 2 4

Give the coordinate of each point.

1. A $-3\frac{1}{2}$ **2.** C $-1\frac{1}{2}$
3. E 1 **4.** G $2\frac{1}{2}$

What point corresponds to the given coordinate?

5. 3.9 H **6.** 0 D
7. $-\frac{5}{2}$ B **8.** $1\frac{1}{2}$ F

Motivator

Ask students for the *opposite* of earning $25, of gaining 5 yards on the football field, and of a temperature drop of 10 degrees. Owing $25, losing 5 yards, and a temperature gain of 10 degrees Then ask what gaining 5 yards and losing 5 yards have in common. A distance of 5 yd This last question can lead to an explanation of absolute value, i.e., magnitude without regard to direction.

Highlighting the Standards

Standards 2f, 2c: In the explanation of the definition, students can see the power of the absolute value notation. In Exercise 79, they express mathematical ideas in writing.

Lesson Note

Both *opposite* and *absolute value* are presented in terms of distances from 0 on a number line. An alternate definition of absolute value is also given.

$|x| = x$, if $x \geq 0$
$|x| = -x$, if $x < 0$

The second part of the definition is confusing to those who think that $-x$ is always a negative number. Point out that $-x$ means "the opposite of x," and $-x$ is positive if x is negative ($x < 0$). For example, if $x = -4$, then $-x$ is the opposite of -4, or 4.

In expressions, such as $-|16-7|$, absolute value bars are treated somewhat like parentheses. Operations inside the absolute value symbols are completed before taking the absolute value.

Math Connections

Music: A person who has absolute pitch (sometimes called perfect pitch) is one who can identify the exact pitch of a note without reference to any other pitch. This can be compared to the distance between two points without reference to direction.

Critical Thinking Questions

Analysis: Have students study the following to find a pattern relating the number of negative signs to the value of such an expression.

$-4 = -4$
$-(-4) = 4$
$-(-(-4)) = -4$
$-(-(-(-4))) = 4$ Odd number of negative signs gives negative value; even number gives positive value.

The distance between 4 and 0 on the number line is 4 units. The distance between -4 and 0 is also 4 units.

The distance of each number from zero is its *absolute value*.

The absolute value of 4 is equal to 4.

$|4| = 4$

The absolute value of -4 is equal to 4.

$|-4| = 4$

Definition

> The **absolute value** of a real number x is the distance between x and 0 on a number line. The symbol $|x|$ means the absolute value of x.

$|x| = x$, if $x \geq 0$ ($\geq$ means *is greater than or equal to*.)
$|x| = -x$, if $x < 0$

In general, if x is a positive number or zero, $|x| = x$. If x is a negative number, $|x| = -x$. In this case, $-x$ is positive.

EXAMPLE 2   Simplify.

a. $|7|$ b. $|-2.3|$ c. $|0|$ d. $|8 + 3|$

Solutions   a. $|7| = 7$ b. $|-2.3| = 2.3$ c. $|0| = 0$ d. $|8 + 3| = |11| = 11$

EXAMPLE 3   Simplify. First rewrite the expressions without absolute value symbols.

a. $-|-4|$ b. $|-8| + |5|$ c. $-|13 - 7|$

Solutions   a. $-|-4| = -(4)$ b. $|-8| + |5| = 8 + 5$ c. $-|13 - 7| = -(6)$
$= -4$ $= 13$ $= -6$

EXAMPLE 4   Compare. Use $>$, $<$, or $=$.

a. $|2.5| \underline{\ ?\ } |-2.5|$ b. $|-6| \underline{\ ?\ } -|-9|$ c. $-(-5) \underline{\ ?\ } -|-5|$

Solutions   a. $|2.5| \underline{\ ?\ } |-2.5|$ b. $|-6| \underline{\ ?\ } -|-9|$ c. $-(-5) \underline{\ ?\ } -|-5|$
2.5 $\underline{\ ?\ }$ 2.5   6 $\underline{\ ?\ }$ $-(9)$ 5 $\underline{\ ?\ }$ $-(5)$
2.5 $=$ 2.5   6 $>$ -9 5 $>$ -5

TRY THIS   Simplify: **2.** $|3| + |-2|$ 5   **3.** $-|3 - 2|$ -1
4. Compare. Use $<$ or $>$. $-(-4) \underline{\quad} -|6 - 2|$ $>$

Additional Example 2

Simplify each expression.

a. $|3\frac{1}{2}|$ $3\frac{1}{2}$
b. $|-6|$ 6
c. $|3 - 3|$ 0
d. $|6 + 2|$ 8

Additional Example 3

Simplify each expression.

a. $-|-8|$ -8
b. $|-14| - |10|$ 4
c. $-|12 - 10|$ -2

Additional Example 4

Compare. Use $>$, $<$, or $=$.

a. $|-7.2| \underline{\ ?\ } |7.2|$ $=$
b. $-|-6| \underline{\ ?\ } |-6|$ $<$
c. $|-7| \underline{\ ?\ } -|-11|$ $>$

Classroom Exercises

Give the opposite of each number.

1. 9 -9 **2.** -6.4 6.4   **3.** 1.876 -1.876 **4.** -0.72 0.72

Give the absolute value of each number.

5. 9 9   **6.** -5.2 5.2   **7.** $-\frac{3}{5}$ $\frac{3}{5}$ **8.** 5.384 5.384

Simplify.

9. $-\left(-6\frac{1}{2}\right)$ $6\frac{1}{2}$ **10.** $-(7 + 9)$ -16 **11.** $-\left(\frac{18}{6}\right)$ -3 **12.** $-(-8.2)$ 8.2

13. $|15|$ 15   **14.** $|6 - 7|$ 1   **15.** $|2 + 6|$ 8   **16.** $|7| + |-8|$ 15

Written Exercises

Give the opposite of each number.

1. 2 -2 **2.** -5.6 5.6   **3.** 0 0   **4.** 40 -40

5. -3 3   **6.** $-\frac{4}{5}$ $\frac{4}{5}$ **7.** 0.5 -0.5 **8.** $\frac{1}{3}$ $-\frac{1}{3}$

Simplify.

9. $-(-10)$ 10   **10.** $-(-4.7)$ 4.7   **11.** $-(5 - 3)$ -2 **12.** $-(3 \cdot 8)$ -24

13. $-(14 \div 2)$ -7 **14.** $-(8 \cdot 0)$ 0   **15.** $-(17 + 0)$ -17 **16.** $-(6.8 - 4.9)$ -1.9

17. $|3|$ 3   **18.** $|-20|$ 20   **19.** $|-1|$ 1   **20.** $|6|$ 6

21. $|1|$ 1   **22.** $|-18|$ 18   **23.** $|0.8|$ 0.8   **24.** $|-1.9|$ 1.9

25. $-|3|$ -3 **26.** $-|-3|$ -3 **27.** $|-(-3)|$ 3   **28.** $-|-(-3)|$ -3

29. $|8 - 5|$ 3   **30.** $-|4 \cdot 2|$ -8 **31.** $\left|\frac{20}{4}\right|$ 5   **32.** $-|8 - 7|$ -1

33. $|6 \cdot 9|$ 54   **34.** $-|6 + 10|$ -16 **35.** $-|30 \div 5|$ -6 **36.** $|1| + |5|$ 6

37. $|7| + |-4|$ 11   **38.** $|-9| + |-1|$ 10   **39.** $|6| + |-6|$ 12   **40.** $|3| - |-3|$ 0

41. $|-8| + |4|$ 12   **42.** $|-6| + |-10|$ 16   **43.** $|-6| \cdot |8|$ 48   **44.** $|-4| \cdot |0|$ 0

45. $|-9| \cdot |-2|$ 18   **46.** $|-4| \cdot |-3|$ 12   **47.** $|-1| \cdot |8|$ 8   **48.** $|-38| \cdot |1|$ 38

Write a positive number or a negative number for each phrase. Then write the opposite phrase and the corresponding opposite number.

    Example: depositing $40
    Answer: 40; withdrawing $40; -40

49. losing 5 lb   **50.** a profit of $80   **51.** temperature falling 8°

52. 5 mi/h gain in speed   **53.** 12 ft above sea level   **54.** a $15 price markdown

Common Error Analysis

Error: Some students interpret the distance between 0 and -4 as -4 since they think of a move in a negative direction.

Remind students that distance is the number of units between two points, expressed as a positive number or 0. The distance between two points is never less than zero. Thus, the distance between 0 and -4 is $|-4|$, or 4 units.

Checkpoint

Simplify.

1. $-(-9)$ 9
2. $-(16 - 4)$ -12
3. $-|-7|$ -7
4. $-|14|$ -14

Compare. Use $>$, $<$, or $=$.

5. $-|7| \underline{\ ?\ } |-7|$ $<$
6. $|-9| \underline{\ ?\ } |-1|$ $>$
7. $|4| \underline{\ ?\ } |-5|$ $<$

Closure

Ask students to evaluate $-x$ and $|x|$ for $x = 5$. $-5, 5$ Then have them replace the $\underline{\ ?\ }$ with $=$, $<$, or $>$ in the following.
$-x \underline{\ ?\ } x$ and $|x| \underline{\ ?\ } x$ $<$, $=$

Repeat this procedure for $x = -8$. $>$, $=$

Guided Practice

Classroom Exercises 1–16
Try This   all

Independent Practice

A Ex. 1–54, **B** Ex. 55–79, **C** Ex. 80–87

Basic: WE 1–65 odd
Average: WE 13–77 odd
Above Average: WE 29–79 odd, 80–87

Additional Answers

Written Exercises, pages 47–48

49. -5; gaining 5 lb; 5
50. 80; $80 loss; -80
51. -8; 8° increase; 8
52. 5; slows 5 mi/h; -5
53. 12; 12 ft below sea level; -12
54. -15; $15 increase; 15
55. F, $-6 < -2$
56. T, $4 = 4$
57. T, $-2 < 2$
58. F, $-3 < 3$
60. T, $16.2 > 6.2$
61. F, $\frac{3}{4} > \frac{1}{2}$
63. T, $16 = 16$
79. If $x < 0$, then $|x| = -x$. The absolute value of a number may equal the opposite of the number.

True or false? Justify your answer.

55. $-6 > -(2)$ **56.** $-(-4) = 4$ **57.** $-2 < |2|$ **58.** $-(3) = |-3|$
59. $|-8| = 8$ T, 8 = 8   **60.** $|16.2| > 6.2$ **61.** $\left|-\frac{3}{4}\right| < \frac{1}{2}$ **62.** $|0| < -1$ F, 0 > -1
63. $|-16| = |16|$ **64.** $\left|\frac{1}{2}\right| < \left|\frac{1}{4}\right|$ F, $\frac{1}{2} > \frac{1}{4}$ **65.** $|0| < |-1|$ T, 0 < 1 **66.** $|9| > |-9|$ F, 9 = 9

Compare. Use $>$ **,** $<$ **, or** $=$ **.**

67. $-6 \underline{\;=\;} -(6)$ **68.** $0 \underline{\;>\;} -(5)$ **69.** $-8 \underline{\;\leq\;} -(-8)$ **70.** $-(3) \underline{\;\leq\;} |-3|$
71. $|5| \underline{\;=\;} |-5|$ **72.** $|-6| \underline{\;=\;} |6|$ **73.** $|-1| \underline{\;\geq\;} -(1)$ **74.** $|2| \underline{\;\leq\;} |-8|$
75. $|0| \underline{\;=\;} -|0|$ **76.** $\left|-\frac{1}{3}\right| \underline{\;\leq\;} \left|-\frac{1}{2}\right|$ **77.** $|6| \underline{\;\geq\;} -|-6|$ **78.** $-|-4| \underline{\;\leq\;} |4|$

79. Explain in writing why the absolute value of a number may be the same as the opposite of that number.

True or false? If false, justify your answer.

80. If x is a positive number, then $-x$ is a negative number. T
81. If $-x$ is a positive number, then x is a negative number. T
82. If x is a negative number, then $-(-x)$ is a positive number. F; $-(-x) = x$ and $x < 0$ (given)
83. If x is any real number, then $|x| = x$. F; for x < 0, |x| = -x.
84. If x is any real number, then $|x| = |-x|$. T
85. There are two real numbers whose absolute value is 8. T
86. If $x > 0$, then $|x| = x$. T
87. If $x < 0$, then $x > -x$. F; for x < 0, x < -x.

Mixed Review

Evaluate for $x = 2$, $y = 3$, **and** $z = 4$.   *1.1, 1.3*

1. $yz - xy$ 6
2. $\dfrac{x + y + z}{x + y - z}$ 9
3. xy^2 18
4. $x(7z - xy - 6)$ 32

Which property is illustrated?   *1.5, 1.6*

5. $y(z - x) = yz - yx$ Distr Prop
6. $(9y + 2z) + 6x = 6x + (9y + 2z)$ Comm Prop Add
7. $(10y)z = 10(yz)$ Assoc Prop Mult
8. $4x + 5x = (4 + 5)x$ Distr Prop

Simplify.   *1.7*

9. $7x + 5x + 4x$ 16x
10. $7y + 4 + 3y$ 10y + 4
11. $3k + 5b + 7k$ 10k + 5b
12. $5a + 4b + 7a + 9b$ 12a + 13b

48   Chapter 2 Operations with Real Numbers

Enrichment

Suppose x represents any integer from -6 to 6. Ask the students which numbers, if any, will make each of the following sentences true.

5. $x < |x|$ $-6, -5, -4, -3, -2, -1$
6. $|x| < x$ None

1. $|x| > 4$ $-6, -5, 5, 6$
2. $4|x| = x$ 0
3. $|x| < 5$ $-4, -3, -2, -1, 0, 1, 2, 3, 4$
4. $|x| - 2 = 0$ $-2, 2$

2.3 Addition on a Number Line

Objective   To add real numbers using a number line

The result of two successive weight gains can be expressed as the sum of two positive numbers. Similarly, the result of two successive weight losses can be expressed as the sum of two negative numbers.

The number line can be used as a model when numbers are added.

EXAMPLE 1   Add 3 + 2.

Solution   Start at 3. Move 2 units to the right.
Stop at 5.

Thus, 3 + 2 = 5.

TRY THIS   1. How can you use the number line to add 2 + 4?
Start at 2. Move 4 units to the right. Stop at 6.

The sum of two negative numbers, such as −2 and −4 is written

$$-2 + (-4).$$

The parentheses are used to separate the addition symbol (+) from the negative symbol (−) that is part of the numeral, −4.

EXAMPLE 2   Add −2 + (−4).

Solution   Start at −2. Move 4 units to the left.
Stop at −6.

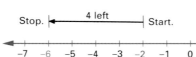

Thus, −2 + (−4) = −6.

TRY THIS   2. How can you use a number line to add −5 + (−2)?
Start at −5; move 2 units to the left; stop at −7.

An application of adding two negative numbers is finding a total weight loss when dieting. The result of a loss of 2 lb last week, followed by a loss of 4 lb this week, can be represented as follows.

$$-2 + (-4) = -6 \qquad \text{The total loss is 6 lb.}$$

Teaching Resources

Quick Quizzes 11
Reteaching and Practice Worksheets 11
Transparency 5

▰▰▰ GETTING STARTED

Prerequisite Quiz

Add.

1. 6.9 + 4.8   11.7
2. 5.6 + 9.4 + 2.7   17.7
3. 24 + 0 + 26 + 17   67
4. $\frac{3}{7} + \frac{2}{7}$ $\frac{5}{7}$
5. $1\frac{1}{4} + 3\frac{2}{3}$ $4\frac{11}{12}$

Motivator

Show students how a thermometer can be thought of as a vertical number line. Ask them to relate the addition of −3 and −5 to temperature on the thermometer.   A temperature of −3° followed by a drop of 5°.

Additional Example 1

Add 2 + 5.   7

Additional Example 2

Add −1 + (−5).   −6

Highlighting the Standards

Standards 1d, 2b: In the opening problem, the number line is used as a model for weight gains and losses. In Exercises 37 and 38, students express generalizations in writing.

49

Lesson Note

Another way to illustrate addition of a positive and a negative number is to introduce positively- and negatively-charged particles as a model for addition. For example, to add 4 and -3, imagine 4 positive particles and 3 negative particles.

+ + − − + + −

Each positive charge counterbalances a negative charge, so there is 1 positive charge left. Thus, $4 + (-3) = 1$.

Point out that the Commutative Property of Addition still applies. Therefore, $-3 + 8 = 8 + (-3)$. Show this on the number line.

Math Connections

Life Skills: As an aid for remembering the term "inverse," compare it to "reverse." Putting the car in *reverse* allows you to back out of a particular spot; it's the *opposite* of moving forward. Compare this to the additive *inverse* of a number; this is the *opposite* of the given number.

Critical Thinking Questions

Synthesis: Ask students to compare $5 - 2$ with $5 + (-2)$. Give other examples but limit the answers to positive numbers. Ask if someone can state a rule that relates subtraction to adding the opposite of a number. *Subtracting a number is the same as adding its opposite.* Then ask them to rewrite a problem such as $3 - 7$ so that it can be solved on the number line. $3 + (-7)$

A number line can be used to add numbers that have unlike signs.

EXAMPLE 3 Add $-4 + 3$.

Solution Start at -4. Move 3 units to the right. Stop at -1.

Thus, $-4 + 3 = -1$.

EXAMPLE 4 Add $-2 + 5$.

Solution Start at -2. Move 5 units to the right. Stop at 3.

Thus, $-2 + 5 = 3$.

TRY THIS Add: **3.** $2 + (-5)$ -3 **4.** $-3 + 8$ 5

A number line can be used to add more than two numbers. To find the sum of $(-7) + 6 + (-3)$, add $(-7) + 6$ first. Then add the result to (-3).

EXAMPLE 5 Add $-7 + 6 + (-3)$.

Solution Start at -7. Move 6 units to the right. Then move 3 units to the left. Stop at -4.

Thus, $-7 + 6 + (-3) = -4$.

TRY THIS **5.** Add $4 + (-7) + 3$. 0 **6.** Add $(-1) + 5 + (-6)$. -2

Adding 0 to a number on the number line is the same as moving 0 units. The sum $-3 + 0$ means to start at -3 and move 0 units. The sum $3 + 0$ means to start at 3 and move 0 units. Thus,

$$-3 + 0 = -3 \quad \text{and} \quad 3 + 0 = 3$$

The sum of any given number and 0 is equal to the given number. This illustrates the Property of Zero for Addition of real numbers. Thus, 0 is called the *identity element for addition.*

> **Identity Property for Addition**
> For all real numbers a, $a + 0 = a$ and $0 + a = a$.

Additional Example 3

Add $-6 + 2$. -4

Additional Example 4

Add $-5 + 11$. 6

Additional Example 5

Add $5 + (-8) + (-1)$. -4

Think of adding a pair of opposites, such as 5 and -5 on a number line.

Start at 5.

Then move 5 units to the left to 0.

Therefore, $5 + (-5) = 0$.

Similarly, it can be shown that
$$-5 + 5 = 0.$$

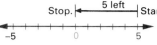

Another name for opposite is *additive inverse*. Thus, the additive inverse, or opposite, of -7 is 7. The following property relates to the sum of two additive inverses.

Additive Inverse Property

For each real number a, there is exactly one real number $-a$ such that
$$a + (-a) = 0 \quad \text{and} \quad -a + a = 0.$$

Classroom Exercises

Tell how to add these numbers using a number line. Then add.

1. $5 + 6$ 11   **2.** $-7 + (-2)$ -9 **3.** $8 + (-10)$ -2 **4.** $-5 + 0$ -5
5. $4 + (-4)$ 0   **6.** $8 + 1$ 9   **7.** $0 + (-7)$ -7 **8.** $10 + (-2)$ 8
9. $7 + (-9)$ -2 **10.** $-5 + (-4)$ -9 **11.** $-4 + 5$ 1   **12.** $3 + (-3)$ 0
13. $-6 + (-1)$ -7 **14.** $2 + (-7)$ -5 **15.** $-8 + (-3)$ -11 **16.** $-11 + 2$
-9

Give the additive inverse of each number.

17. -6 6   **18.** -32 32   **19.** 24 -24 **20.** -13 13   **21.** 17 -17 **22.** 64 -64

Written Exercises

Add.

1. $3 + 7$ 10   **2.** $8 + 5$ 13   **3.** $-6 + (-9)$ -15 **4.** $-7 + (-5)$ -12
5. $0 + (-3)$ -3 **6.** $-4 + 0$ -4 **7.** $7 + (-7)$ 0   **8.** $11 + (-3)$ 8

Error: Some students may have heard the statement, "Two negatives make a positive." Point out that this does *not* pertain to addition of two negative real numbers.

Direct students to the notion of beginning at zero on a number line and making two moves in a negative direction. The result is a negative number.

Checkpoint

Add.

1. $-8 + (-3)$ -11
2. $-5 + (-5)$ -10
3. $11 + (-11)$ 0
4. $-30 + 4$ -26
5. $8 + (-24)$ -16
6. $15 + (-9)$ 6
7. $16 + (-4) + (-6) + 3$ 9
8. Find the value of $-[(-6) + 2]$. 4

Closure

Ask students to explain and demonstrate on a number line the following statement: The sum of any real number and its additive inverse is the identity element for addition.

Guided Practice

Classroom Exercises 1–22
Try This   all

Independent Practice

🅐 Ex. 1–25, 🅑 Ex. 26–35, 🅒 Ex. 36–39

Basic: WE 1–28

Average: WE 9–35

Above Average: WE 13–39

Additional Answers

Written Exercises

37. Let a and b be any positive real number; $-a + (-b) = -(a + b)$, -161

38. Let a and b be any real numbers with different signs; if $a > b$ then $a + b = a - |b|$. If $a < b$ then $a + b = b - |a|$. If $a = 0$ and b is the opposite, $a + b = 0$; 39

39. No, let $a = 2$ and $b = -1$; then $|a + b| = |2 + (-1)| = |1| = 1$. Yet $|a| + |b| = |2| + |-1| = 2 + 1 = 3$. Not true for all numbers.

9. $-9 + 8$ -1 **10.** $-1 + 9$ 8   **11.** $-5 + (-5)$ -10 **12.** $8 + (-12)$ -4

13. $8 + (-8)$ 0   **14.** $-14 + 1$ -13 **15.** $-9 + 9$ 0 -17 **16.** $-14 + (-3)$

17. $-5 + 4 + (-2)$ -3 **18.** $5 + 4 + (-2)$ 7   0 **19.** $6 + (-2) + (-4)$

20. $-8 + 8 + (-1)$ -1 **21.** $9 + 5 + (-14)$ 0 -5 **22.** $3 + (-3) + (-5)$

23. $-5 + (-8) + 4$ -9 **24.** $9 + (-5) + (-2)$ 2   **25.** $-7 + (-7) + 14$

26. $-8 + (-3) + (-2) + 4$ -9 **27.** $12 + (-2) + (-5) + (-3)$ 2   0

28. $-16 + 5 + (-7) + 16$ -2 **29.** $-25 + 50 + 50 - 75$ 0

30. $12 + (-14) + (-9) + 5$ -6 **31.** $100 + (-60) + 10 + (-40)$ 10

32. Find the additive inverse of the sum $-5 + 4$. 1

33. Find the value of $-7 + (-9)$. -16

Evaluate for the given values of the variables.

34. $x + y$, for $x = -5$ and $y = 4$ -1

35. $-x + y$, for $x = -3$ and $y = -6$ -3

36. For what values of a will $-a < 0$ be true? $a > 0$

37. Write a rule for adding two negative numbers without using a number line. Then use this rule to find the sum $-43 + (-118)$.

38. Write a rule for adding two numbers with different signs. Then use this rule to find the sum $-47 + 86$.

39. Is $|a + b| = |a| + |b|$ true for all numbers a and b? If not, give a *counterexample*. A **counterexample** is an example that shows that a statement is false.

Mixed Review

Simplify.   *1.2, 1.3, 1.7*

1. $4 \cdot 3 + 18$ 30   **2.** $8 \cdot 4^3$ 512   **3.** $4y + 8y$ $12y$ **4.** $3x + 5 + x$ $4x + 5$

5. Write an algebraic expression for 6 less than the number x of workdays in a year. *1.1* $x - 6$

6. Write an algebraic expression for \$62 increased by d dollars saved. $62 + d$

7. Write an algebraic expression for the number of yards in x inches. $\frac{x}{36}$

8. Find the perimeter of a rectangle with length 8 m and width $6\frac{1}{4}$ m. Use the formula $p = 2l + 2w$. *1.4* $28\frac{1}{2}$ m

9. Find the solution set of $3x + 2 = 11$. The replacement set is $\{1, 3, 7\}$. *1.8* {3}

10. Find the solution set of $2x + 5 = 15$. The replacement set is $\{0, 2, 5\}$. {5}

Enrichment

Challenge the students to arrange the integers from -4 through 4 in a triangular array so that the four numbers on each side have a sum of 0.   Answers may vary. One possible answer:

```
              0
         - 2      2
       3            - 3
   - 1     4    - 4      1
```

2.4 Adding Real Numbers

Objective

To add real numbers using the rules for addition

In the last lesson a number line was used to add integers. This is not very practical for finding sums of numbers such as $4\frac{1}{2} + \left(-6\frac{1}{3}\right)$.

Examine the sums below.

$$2 + 4 = 6 \qquad -3 + (-2) = -5$$
$$3 + 2 = 5 \qquad -2 + (-4) = -6$$

The sums suggest the following rule.

To add two real numbers with the same sign:

1. Add their absolute values.
2. Determine the sign of the sum:
   a. If both numbers are positive, then the sum is positive.
   b. If both numbers are negative, then the sum is negative.

EXAMPLE 1 Add $-12 + (-19)$.

Solution

Add the absolute values. $|-12| + |-19| = 12 + 19 = 31$
The sum is negative. Thus, $-12 + (-19) = -31$.

TRY THIS Add: **1.** $-7 + (-3)$ -10 **2.** $4 + 11$ 15

Examine the sums below.

$$-4 + 3 = -1 \qquad\qquad -2 + 5 = 3$$

The sums suggest the following rule.

To add two real numbers with unlike signs, if the numbers are not opposites:

3. Subtract their absolute values, the smaller from the larger.
4. The sign of the sum will be the same as the sign of the number with the greater absolute value.

2.4 Adding Real Numbers **53**

■ **GETTING STARTED**

Prerequisite Quiz

Add.

1. $-5 + (-9)$ -14
2. $5 + (-7)$ -2
3. $8 + (-3)$ 5
4. $7.6 + 1.4$ 9.0

Subtract.

5. $1\frac{3}{4} - \frac{1}{4}$ $1\frac{1}{2}$
6. $8.5 - 4.9$ 3.6

Motivator

Remind students that the last lesson introduced addition of numbers using a number line. Ask for some instances when it is impractical to use a number line for such problems.   When the addends are large numbers.

Additional Example 1

Add $-5.4 + (-6.7)$. -12.1

Highlighting the Standards

Standards 14d, 1d: The discussion in this lesson continues to develop the number system. Throughout this lesson and the next, students apply mathematics to problem situations of daily life.

53

Lesson Note

Remind students that the answer to a subtraction problem is called the difference. When adding two numbers with "different" signs, you take the "difference" of the absolute values.

In Example 2, it might help to suggest that a positive number and a negative number "pull" in opposite directions and that since there is more "negative pull" than positive, the result is negative.

Demonstrate the use of the ⊞/⊟ key on the calculator for changing the sign of a number. Remind students that this key is used *after* the number is entered.

Math Connections

Life Skills: Maintaining a checking account requires keeping the total of the positive amounts (deposits) larger than the negative amounts (checks, service charges, withdrawals). Here the different transactions must be taken in chronological order so that at no time is the balance allowed to drop below zero.

EXAMPLE 2 Add $-16 + 12$.

Solution The signs are unlike, and -16 has the greater absolute value.
Subtract the absolute values. $|-16| - |12| = 16 - 12 = 4$
$|-16| > |12|$; sum is negative.   Thus, $-16 + 12 = -4$.

EXAMPLE 3 Evaluate $x + y$ for $x = -4.5$ and $y = 6.3$.

Solution The signs are unlike, and 6.3 has the greater absolute value.

Subtract the absolute values. $|6.3| - |-4.5| = 6.3 - 4.5 = 1.8$
$|6.3| > |-4.5|$; sum is positive.   Thus, $-4.5 + 6.3 = 1.8$.

TRY THIS   **3.** Evaluate $x + y$ for $x = 5$ and $y = -8$. -3

Notice that when you add two real numbers the sum is also a real number. We say that the set of real numbers is *closed* under addition.

The commutative and associative properties allow you to group numbers in any convenient way and add in any order. Adding more than two real numbers is sometimes easier when numbers with like signs are grouped together. This is shown in Example 4.

EXAMPLE 4   Add $3 + (-2) + 4 + (-6)$.

Solution

Method 1
Add the numbers in order from left to right.

$$3 + (-2) + 4 + (-6)$$
$$\underbrace{1} + 4 + (-6)$$
$$\underbrace{5} + (-6)$$
$$-1$$

Method 2
Group the positive numbers and the negative numbers. Then add.

$$3 + (-2) + 4 + (-6)$$
$$\underbrace{3 + 4} + \underbrace{(-2) + (-6)}$$
$$7 + (-8)$$
$$-1$$

Here are calculator steps for Method 1.

3 ⊞ 2 ⊞/⊟ ⊞ 4 ⊞ 6 ⊞/⊟ ⊟ -1

TRY THIS   **4.** Add $-3 + 5 + (-8) + 9$. 3

Additional Example 2

Add $-14 + 9$. -5

Additional Example 3

Add $-5.2 + 7.1$.   1.9

Additional Example 4

Add $7 + (-4) + (-8) + 15$.   10

EXAMPLE 5 On Tuesday afternoon the temperature was $-3°F$. That night it dropped 8° and the next morning it rose 15°. What was the temperature on Wednesday morning?

Plan Translate the problem into a sum of real numbers and add.

Solution

Tuesday afternoon		Tuesday night		Wednesday morning
-3	$+$	(-8)	$+$	15
	-11		$+$	15
		4		

Thus, the temperature was $4°F$ on Wednesday morning.

TRY THIS 5. Translate: Nancy is three years younger than Harry. Harry is 17. How old is Nancy? $17 + (-3)$

Classroom Exercises

Add.

1. $5 + 8$ 13
2. $-4 + (-6)$ -10
3. $-8 + (-2)$ -10
4. $0 + 11$ 11
5. $-7 + 4$ -3
6. $9 + (-10)$ -1
7. $-8 + 3$ -5
8. $-4 + 9$ 5
9. $-6 + 13$ 7
10. $8 + (-13)$ -5
11. $-9 + (-7)$ -16
12. $15 + (-15)$ 0
13. $-7 + (-3)$ -10
14. $7 + (-8)$ -1
15. $-4 + (-5)$ -9
16. $-13 + (-8)$ -21

Written Exercises

Add.

1. $7 + 6$ 13
2. $9 + 5$ 14
3. $-4 + (-9)$ -13
4. $-10 + (-5)$ -15
5. $5 + (-7)$ -2
6. $-11 + (-8)$ -19
7. $-12 + 4$ -8
8. $-13 + (-6)$ -19
9. $-16 + 8$ -8
10. $32 + (-15)$ 17
11. $-17 + 0$ -17
12. $23 + (-16)$ 7
13. $4.8 + (-4.8)$ 0
14. $-3.7 + (-2.8)$ -6.5
15. $-7.4 + 3.8$ -3.6
16. $2.9 + 4.7$ 7.6
17. $16.2 + (-8.5)$ 7.7
18. $-8.7 + 7.9$ -0.8
19. $-5.4 + (-5.4)$ -10.8
20. $7.3 + (-2.6)$ 4.7
21. $-2.8 + 9.6 + (-1.5)$ 5.3
22. $-6 + 2 + (-4)$ -8
23. $19 + (-4) + 3$ 18
24. $-2 + 12 + (-14)$ -4

25. A football team had a 4-yd gain followed by a 7-yd loss. Find the resulting gain or loss. 3-yd loss
26. The temperature was $-5°F$ at midnight. It dropped 7° by 3:00 A.M. Find the temperature at that time. $-12°F$
27. In one month Mike gained 3 lb. The next month he lost 5 lb. He lost 2 more pounds the third month. Find the net gain or loss. 4-lb loss
28. A team scored 5 points. Then there was a 2-point penalty. This was followed by a 3-point score. Find the net gain or loss. 6-point gain

Critical Thinking Questions

One student had the following grades on major quizzes: 96, 90, 89, 92, 97, 100. The student found the average of these mentally by adding $4 + 10 + 11 + 8 + 3 + 0$ and dividing this sum by 6. Ask students to relate this second list of numbers to the original grades and have them explain how the student used them to find the average quiz grade. Subtract each grade from 100; subtract average of second group of numbers (6) from 100 to get the quiz average (94)

Common Error Analysis

Error: Some students get into the habit of saying "subtract the smaller number from the larger number" when referring to a problem such as $-7 + 2$.

Remind them that $2 > -7$ and that the rule says to subtract the number with the smaller *absolute value* from the number with the larger absolute value.

Checkpoint

Add.

1. $-8 + (-6)$ -14
2. $-4 + 7$ 3
3. $16 + (-16)$ 0
4. $-9 + 5$ -4
5. $\frac{1}{3} + \left(-\frac{2}{3}\right)$ $-\frac{1}{3}$
6. $-7 + 1\frac{3}{4}$ $-5\frac{1}{4}$
7. $9.2 + (-8.5)$ 0.7
8. $18 + (-3) + (-10) + 7$ 12

Additional Example 5

Helen bought a stock on Monday. On Tuesday, it dropped $\frac{1}{2}$ point. On Wednesday it gained $\frac{1}{4}$ point. What was the net gain or loss between Monday and Wednesday?

$\frac{1}{4}$ point loss

Closure

Ask students for the most efficient way to find the sum of the following numbers.

23	−17	−43	28
16	−32	−8	−50

Answers will vary.

■■■ FOLLOW UP

Guided Practice

Classroom Exercises 1–16
Try This all

Independent Practice

A Ex. 1–42, **B** Ex. 43–50, **C** Ex. 51–54

Basic: WE 1–41 odd

Average: WE 9–45 odd, 47, 48

Above Average: WE 17–45 odd, 49–54

Add.

29. $-3\frac{1}{3} + \left(-3\frac{2}{3}\right)$ −7   **30.** $7\frac{5}{6} + \left(-2\frac{1}{6}\right)$ $5\frac{2}{3}$ **31.** $7 + \left(-2\frac{4}{5}\right)$ $4\frac{1}{5}$

32. $-6\frac{7}{10} + \left(-8\frac{3}{5}\right)$ $-15\frac{3}{10}$ **33.** $8\frac{1}{6} + \left(-3\frac{1}{3}\right)$ $4\frac{5}{6}$ **34.** $-5\frac{2}{3} + \left(-4\frac{1}{2}\right)$ $-10\frac{1}{6}$

Evaluate for $x = -8$, $y = 18$, and $z = -13$.

35. $x + 5$ −3   **36.** $-6 + y$ 12   **37.** $z + (-4)$ −17   **38.** $x + 8$ 0

39. $x + y$ 10   **40.** $x + z$ −21   **41.** $y + z$ 5   **42.** $x + y + z$ −3

43. $x + (-y)$ −26   **44.** $-x + y$ 26   **45.** $|x| + |y|$ 26   **46.** $|x| + |z|$ 21

47. Evaluate $a + b$ for $a = 7.5$ and $b = -0.9$. 6.6

48. Evaluate $a + b$ for $a = -\frac{3}{4}$ and $b = -1\frac{1}{8}$. $-1\frac{7}{8}$

49. Evaluate $a + b + c$ for $a = 2.6$, $b = -1.4$, and $c = -0.7$. 0.5

50. Evaluate $a + b + c$ for $a = -\frac{2}{3}$, $b = 2\frac{1}{3}$, and $c = -\frac{5}{6}$. $\frac{5}{6}$

51. The lowest point in Death Valley, California, is 276 ft below sea level. A hot-air balloon was floating 83 ft above this point. The balloon climbed 68 ft higher and then dropped 72 ft. Find the altitude of the balloon relative to sea level. 197 ft below sea level

52. Mary took out a loan to open a small business. In the first year the business lost $1,700. In the second year the business made a profit of $2,845. In the third year the business made a profit of $3,650. When Mary paid off the loan and a total of $1,125 in interest, she had a net profit of $1,140. How much was the loan? $2,530

53. José left on a flight from London at 2:42 P.M. The flight was due to land in New York 7 h 35 min later. In New York, the time is 5 h earlier than it is in London. Give the time in New York when the plane was due to land. 5:17 P.M.

54. If the balance in Bill's checking account falls below $600.00 during a month, he is charged a $7.00 service charge for that month. At the beginning of January his balance was $1,426.51. At the end of January his balance was $582.70. During January he wrote checks totaling $1,872.85. How much did he deposit during January? $1,036.04

Mixed Review

Compare. Use $<$, $>$, or $=$. *2.1*

1. $2 \underline{\ >\ } -3$

2. $-7 \underline{\ <\ } -4$

Simplify. *2.2*

3. $-(8 - 3)$ −5

4. $9 - 2$ 7

Enrichment

Have the students make each sentence true by inserting one or more absolute value signs.

1. $(-3) + 5 = 8$ $|(-3)| + 5 = 8$

2. $2 + (-4) + 1 = 7$ $2 + |(-4)| + 1 = 7$

3. $2 + (-4) + 1 = 5$ $2 + |(-4) + 1| = 5$

4. $10 + (-12) + (-3) + (-2) = 1$
$|10 + (-12)| + (-3) + |(-2)| = 1$

5. $(-3) + (-7) + 5 + (-12) = 11$
$(-3) + |(-7) + 5| + |(-12)| = 11$

6. $(-8) + 2 + (-1) + (-6) + 1 = 12$
$|(-8) + 2 + (-1)| + |(-6) + 1| = 12$

2.5 Subtraction of Real Numbers

Objectives
To subtract real numbers
To simplify and evaluate expressions involving addition and subtraction of real numbers

Consider the following examples.

$$9 - 3 = 6 \qquad 7 - 3 = 4 \qquad 3 - 3 = 0$$
$$9 + (-3) = 6 \qquad 7 + (-3) = 4 \qquad 3 + (-3) = 0$$

In all three cases, subtracting 3 and adding -3 gave the same result. Subtracting a number gives the same result as adding the opposite of that number. Thus, $1 - 3 = 1 + (-3) = -2$.

Definition

Subtraction
For all real numbers a and b, $a - b = a + (-b)$.
That is, to subtract a number, add its opposite.

EXAMPLE 1   Subtract.
  a. $2 - 5$ Subtract 5 by adding -5. $2 - 5 = 2 + (-5) = -3$
  b. $6 - (-9)$ Subtract -9 by adding 9. $6 - (-9) = 6 + 9 = 15$
  c. $-2.3 - (-1.8)$ Subtract -1.8 by adding 1.8. $-2.3 - (-1.8) =$
                                                        $-2.3 + 1.8 = -0.5$

TRY THIS   Subtract: **1.** $3 - 8$ -5 **2.** $-7 - (-10)$ 3

Since subtraction is defined in terms of addition, the expression $6 - 9 + 4 - 1$ can be thought of as the sum $6 + (-9) + 4 + (-1)$.

You can then use the properties of addition to group the numbers in a convenient way, as shown in Example 2.

EXAMPLE 2   Simplify $6 - 9 + 4 - 1$.

Plan   Group the positive numbers and the negative numbers.

Solution
$$6 - 9 + 4 - 1 = 6 + (-9) + 4 + (-1)$$
$$= 6 + 4 + (-9) + (-1)$$
$$= \underbrace{10} + \underbrace{-10} \qquad 0$$

TRY THIS   **3.** Simplify $-4 + 7 - 9 + 2$. -4

Additional Example 1

Subtract.

a. $8 - 12$ -4
b. $7 - (-3)$ 10
c. $-7.5 - (-4.6)$ -2.9

Additional Example 2

Simplify $8 - 3 + 2 - 7$. 0

Teaching Resources

Manipulative Worksheet 2
Quick Quizzes 13
**Reteaching and Practice
    Worksheets** 13
Teaching Aid 1
Transparency 5

▬▬ GETTING STARTED

Prerequisite Quiz

Give the opposite of each number.

1. -8 8
2. 1.2 -1.2
3. $-1\frac{1}{3}$ $1\frac{1}{3}$
4. 0 0

Add.

5. $-6 + 8$ 2
6. $-9 + (-7)$ -16
7. $2.8 + (-5.7)$ -2.9
8. $-\frac{1}{8} + \frac{3}{4}$ $\frac{5}{8}$

Motivator

Ask students to compare $7 + 3$ and $7 - 3$ on the number line. They should notice that to find the second "answer," you move 3 units in the opposite direction to that used in the first expression. In other words, to subtract 3 you move in a direction opposite to that used to add 3. Then have them compare this to $7 + (-3)$ on the number line.

Highlighting the Standards

Standards 14a, 5b: In Exercises 84–89, students continue to compare and contrast the structure of real number subsystems. In the Application, they begin the process of using tables as part of their problem-solving process.

57

Lesson Note

Since subtracting is defined as adding the opposite, emphasize that the first step in subtracting is to rewrite the subtraction as addition of the opposite number. Then complete the solution by applying for addition.

Math Connections

Life Skills: Subtracting a negative number is similar to a double negative in English. The statement, "I don't have no money", literally means that the person does have money.

Critical Thinking Questions

Analysis: Ask these questions: Since $8 - 5$ is not equal to $5 - 8$, which property of addition does not apply to subtraction? Can $8 - 5$ be rewritten so that the commutative property would hold?   Commutative; rewrite as $8 + (-5)$

Common Error Analysis

Error: In problems such as Example 3, some students may evaluate $-z$ as -9 instead of $-(-9)$.

Remind them that subtracting the value of z is the same as adding the opposite of the value assigned to z and that $-(-9)$ equals 9.

EXAMPLE 3 Evaluate $x + y - z$ for $x = -7$, $y = 8$, and $z = -9$.

Solution
$$x + y - z = -7 + 8 - (-9)$$
$$= -7 + 8 + 9 \quad \longleftarrow \text{Subtracting } -9 \text{ is the same as adding 9.}$$
$$= -7 + \underbrace{17}$$
$$= 10$$

Thus, $x + y - z = 10$ for $x = -7$, $y = 8$, and $z = -9$.

TRY THIS   4. Evaluate $x - y + z$ for $x = -2$, $y = -6$, and $z = -3$. 1

Subtraction of real numbers can be used to solve problems.

EXAMPLE 4 The temperature rose from $-3°F$ to $16°F$. What was the temperature change?

Solution
$$16 - (-3) = 16 + 3 \quad \longleftarrow \text{Subtract } -3 \text{ from 16.}$$
$$= 19$$

Thus, the temperature rose $19°$.

EXAMPLE 5 An elevator on the 8th floor goes up 3 floors, then down 5 floors, then up 2 floors, and then down 7 floors. At what floor is the elevator after these moves are completed?

Solution
THINK: $8 + 3 + (-5) + 2 + (-7)$
WRITE: $8 + 3 - 5 + 2 - 7$ Then group positive and negative terms together.

$$\underbrace{-5 - 7} + \underbrace{8 + 3 + 2}$$
$$-12 \quad + \quad 13$$
$$1$$

The elevator is at the first floor

TRY THIS   5. An elevator started three floors below the ground and stopped four floors above the ground. How many floors did the elevator move?
7 floors

Classroom Exercises

State each subtraction problem as an equivalent addition problem. Then simplify.

1. $6 - 8$ $6 + (-8), -2$ 2. $-3 - 9$ 3. $4 - (-7)$ $4 + 7, 11$ 4. $10 - 2$ $10 + (-2), 8$
5. $16 - 20$ 6. $-9 - (-5)$ 7. $-8 - 6$ 8. $7 - (-12)$
9. $-8 - (-6)$ 10. $4 - 11$ 11. $21 - 3$ 12. $-6 - (-7)$

Additional Example 3

Evaluate $x - y - z$ for $x = 9$, $y = -1$, and $z = 4$.   6

Additional Example 4

The temperature fell from 3°C to $-4°C$. How many degrees did it fall?   7°

Additional Example 5

During a five-month period, Joe had the following gains and losses in weight: -2, -3, 1, 1, and -2. What was his net gain or loss?   A net loss of 5 pounds

Written Exercises

Subtract.

1. $10 - 4$ 6   **2.** $4 - 10$ −6   **3.** $-5 - 3$ −8   **4.** $3 - (-5)$ 8

5. $7 - (-2)$ 9   **6.** $-8 - (-3)$ −5   **7.** $-7 - (-1)$ −6   **8.** $4 - (-6)$ 10

9. $-7 - 8$ −15   **10.** $18 - 19$ −1   **11.** $16 - 16$ 0   **12.** $4 - (-4)$ 8

13. $-8 - 8$ −16   **14.** $-9 - (-9)$ 0   **15.** $6 - (-12)$ 18   **16.** $-14 - (-3)$ −11

17. $12 - 12$ 0   **18.** $15 - (-5)$ 20   **19.** $-18 - 16$ −34   **20.** $32 - 0$ 32

21. $-14 - 0$ −14   **22.** $-6 - 5$ −11   **23.** $15 - (-9)$ 24   **24.** $0 - (-6)$ 6

25. $17 - (-23)$ 40   **26.** $0 - 8$ −8   **27.** $-42 - 26$ −68   **28.** $87 - 54$ 33

29. $-3.8 - 2.7$ −6.5   **30.** $16 - 4.2$ 11.8   **31.** $-2.9 - (-5.1)$ 2.2   **32.** $4.8 - (-3.6)$ 8.4

33. $6.8 - 4.2$ 2.6   **34.** $4.2 - 6.8$ −2.6   **35.** $-3.4 - (-7.2)$ 3.8   **36.** $-7.2 - (-3.4)$

37. $\frac{1}{2} - \frac{1}{4}$ $\frac{1}{4}$ **38.** $\frac{1}{4} - \frac{1}{2}$ $-\frac{1}{4}$ **39.** $-\frac{1}{2} - \left(-\frac{1}{8}\right)$ $-\frac{3}{8}$ **40.** $\frac{1}{6} - \left(-\frac{2}{3}\right)$ −3.8 $\frac{5}{6}$

Simplify.

41. $-7 + 8 - 3$ −2   **42.** $-5 + 12 - 3$ 4   **43.** $17 + 8 - 15$ 10

44. $-6 - 5 - (-3)$ −8   **45.** $-31 + 17 - 2$ −16   **46.** $-12 + 13 - 1$ 0

47. $-7 + 5 - 2 - 9$ −13   **48.** $2 - 7 + 6 - (-8)$ 9   **49.** $-13 + 5 - 2 + 7$ −3

50. $-31 + 17 - 2 + 5$ −11   **51.** $-6 - 3 + 20 - (-4)$ 15   **52.** $5 + 8 - 4 - (-9)$ 18

53. $62 - 26 + 51 + 14$ 101   **54.** $-52 - 48 + 53 + 17$ −30   **55.** $-71 + 82 - 46 - 5$ −40

Evaluate for $x = -3$, $y = 9$, and $z = -7$.

56. $x - y$ −12   **57.** $y - x$ 12   **58.** $x + z$ −10

59. $z + y$ 2   **60.** $x - x$ 0   **61.** $x + y$ 6

62. $x - y + z$ −19   **63.** $-x - y - z$ 1   **64.** $-(x + y + z)$ 1

65. $-(-x - y - z)$ −1   **66.** $|x + y + z|$ 1   **67.** $|x| + |y| + |z|$ 19

Solve each problem.

68. A sea gull started 24 ft above the sea, and dove into the sea to catch a fish 8 ft beneath the surface. How far did the gull plunge from its starting point? 32 ft

69. In April, Jane weighed 1.5 kg less than she weighed in May. In June she weighed 0.75 kg more than she weighed in May. What is the difference between her weight in April and her weight in June? 2.25 kg

70. A rock cliff is 265 ft high. Jim climbed up 38 ft from the bottom. Heidi climbed down 56 ft from the top. What is the distance between Heidi and Jim? 171 ft

71. The formula $C = K - 273$ can be used to convert temperatures from degrees Kelvin to degrees Celsius. Oxygen solidifies at 54°K. What is this temperature in degrees Celsius? −219°C

2.5 Subtraction of Real Numbers   **59**

Subtract.

1. $5 - 9$ −4
2. $-7 - 6$ −13
3. $-5 - (-6)$ 1
4. $0 - (-7)$ 7
5. $6.3 - 9.2$ −2.9
6. $4 - 7 - (-9) + 6$ 12

Evaluate for $x = -2$, $y = -7$, and $z = 3$.

7. $z - y$ 10
8. $-(x + y - z)$ 12

Closure

Have a student explain how to simplify $-9 - (-5)$ by stating and using the rule for subtraction followed by the appropriate rule for addition.

■■■ FOLLOW UP

Guided Practice

Classroom Exercises 1–12
Try This   all

Independent Practice

A Ex. 1–61, **B** Ex. 62–83, **C** Ex. 84–89

Basic: WE 1–61 odd, Midchapter Review, Application

Average: WE 17–77 odd, Midchapter Review, Application

Above Average: WE 29–83 odd, 84–89, Midchapter Review, Application

Additional Answers

Classroom Exercises, page 58

2. $-3 + (-9), -12$
5. $16 + (-20), -4$
6. $-9 + 5, -4$
7. $-8 + (-6), -14$
8. $7 + 12, 19$
9. $-8 + 6, -2$
10. $4 + (-11), -7$
11. $21 + (-3), 18$
12. $-6 + 7, 1$

Written Exercises

84. No. If $a = 3$ and $b = 7$, then $a - b = 3 - 7$, or -4, and $b - a = 7 - 3$, or 4. Since $-4 \neq 4$, $3 - 7 \neq 7 - 3$. Thus, since "$a - b = b - a$" is not true for $a = 3, b = 7$, it is not true for all real numbers.

86. No. If $a = 3$, $b = 7$, and $c = 12$, then $(a - b) - 12 = (3 - 7) - 12 = -4 - 12$, or -16 and $a - (b - c) = 3 - (7 - 12) = 3 - (-5)$, or 8. Since $-16 \neq 8$, $(3 - 7) - 12 \neq 3 - (7 - 12)$. Since "$(a - b) - c = a - (b - c)$" is not true for $a = 3, b = 7, c = 12$, it is not true for all real numbers.

87. No. If $a = 2$, then $2 - e = e - 2$, only if $e = 2$. Since e is unique, its value must be 2 for all values of a. If $a = 5, 5 - 2 = 3$ and $2 - 5 = -3$. Since $3 \neq -3$, $5 - 2 \neq 2 - 5$, that is, $5 - e \neq e - 5$. Since "$a - e = e - a$" is not true for $a = 5$, it is not true for all real numbers.

See page 61 for answers to Written Ex. 88 and 89.

Simplify.

72. $|16 - 9|$ 7
73. $|9 - 16|$ 7
74. $|-16| - |9|$ 7
75. $|9| - |-16|$ -7
76. $|-5 + 12|$ 7
77. $|12 - 5|$ 7
78. $|-5| - |12|$ -7
79. $|12| - |-5|$ 7
80. $|-11 - 8|$ 19
81. $|-8 - 11|$ 19
82. $|-11| - |-8|$ 3
83. $|-8| - |-11|$ -3

84. Is subtraction of real numbers commutative? That is, is $a - b = b - a$ true for all real numbers a and b? Explain.

85. If a and b are real numbers, what is the relationship between $a - b$ and $b - a$? They are opposites (additive inverses).

86. Is subtraction associative? That is, is $(a - b) - c = a - (b - c)$ true for all real numbers a, b, and c? Explain.

87. Is there an identity element for subtraction? That is, is there a real number e such that $a - e = a$ and $e - a = a$ for all real numbers a? Explain.

88. Is the set of real numbers closed under subtraction? That is, is $a - b$ a real number for all numbers a and b? Explain.

89. Is the set of whole numbers closed under subtraction? That is, is $a - b$ a whole number for all whole numbers a and b? Explain.

Midchapter Review

Simplify.   *2.2–2.5*

1. $|-5|$ 5
2. $4 - 8$ -4
3. $-8 + 12$ 4
4. $-15 + (-12)$ -27
5. $-(-15)$ 15
6. $-[6 + (-10)]$ 4
7. $-6.8 - 7.5$ -14.3
8. $\frac{1}{5} - \left(-\frac{1}{10}\right)$ $\frac{3}{10}$

Evaluate.   *2.2–2.5*

9. $a + b$, for $a = -7$ and $b = 4$ -3
10. $-x$, for $x = -8$ 8
11. $s + (-t)$, for $s = 25$ and $t = -17$ 42
12. $|r + p|$, for $r = -9$ and $p = -1$ 10
13. $h - j$, for $h = -4$ and $j = 11$ -15
14. $l - m$, for $l = 9$ and $m = -5$ 14

Give the integer that describes each situation.   *2.1*

15. five wins   5
16. eight degrees below zero -8
17. a bank withdrawal of $15 -15

18. On the first sale of the day, Groggin Small Appliances lost $3.52. On the next sale there was a profit of $2.75. Then there were profits of $2.99 and $1.08 followed by losses of $3.01 and $2.76. Is the total result a profit or loss?   *2.4* Loss

19. The summit of Mt. Everest is 29,002 ft above sea level. The lowest point of the Marianas Trench is 36,198 ft below sea level. Find the difference between their altitudes.   *2.5* 65,200 ft

Enrichment

Have students imagine adding hours on a standard clock marked with the numerals 1 through 12. It is easy to see that 3 o'clock plus 5 hours would be $3 + 5 = 8$, or 8 o'clock. But 9 o'clock plus 8 hours $= 9 + 8 = 17$. The answer should be 5 for 5 o'clock. Ask students to devise a rule for adding hours on the clock that will always give the correct answer. Have them demonstrate their rule for 8 o'clock plus 36 hours. Add and then divide by 12; the remainder will give the new time.

Application: *Windchill*

How cold you feel depends on more than just the actual temperature. It also depends on the speed of the wind. Scientists in Antarctica developed the idea of a second temperature, called the *equivalent temperature*, that includes this windchill factor. For example, at 20°F with a wind speed of 15 mi/h, the equivalent temperature is −5°F. This means that under these conditions your body would lose heat as quickly as on a calm day when the temperature is −5°F.

Windchill

Wind speed in mi/h	Actual temperature (°F)								
	50	40	30	20	10	0	−10	−20	−30
	Equivalent temperature (°F)								
0	50	40	30	20	10	0	−10	−20	−30
5	48	37	27	16	6	−5	−15	−26	−36
10	40	28	16	4	−9	−21	−33	−46	−58
15	36	22	9	−5	−18	−31	−45	−58	−72
20	32	18	4	−10	−25	−39	−53	−67	−82
25	30	16	0	−15	−29	−44	−59	−74	−88
30	28	13	−2	−18	−33	−48	−63	−79	−94

Use the windchill table to solve each of the following exercises.

1. If the actual temperature is 30°F and the wind speed is 5 mi/h, what is the equivalent temperature? 27°F

2. If the actual temperature is 20°F but it feels more like −10°F, what is the wind speed? 20 mi/h

3. If the equivalent temperature is −21°F and the wind speed is 10 mi/h, what is the actual temperature? 0°F

4. Give two cases where the equivalent temperature is 16°F.

5. Give two cases where the equivalent temperature is −18°F.

6. The actual temperature dropped from 20°F to 10°F in a day.
   a. If the wind speed was constant at 20 mi/h, how did the equivalent temperature change?
   b. Compare the drop in the actual temperature to the drop in the equivalent temperature.

■■■ GETTING STARTED

Prerequisite Quiz

Multiply.

1. $16 \cdot 0$ 0
2. $8.2(0.5)$ 4.1
3. $7.3(9)$ 65.7
4. $\frac{1}{3} \cdot \frac{3}{4}$ $\frac{1}{4}$
5. $\frac{5}{6} \cdot \frac{3}{5}$ $\frac{1}{2}$
6. $8 \cdot 2 \cdot 0$ 0
7. $2^2 \cdot 3^3$ 108
8. $1^8 \cdot 2^5$ 32

Motivator

Ask students to give the connection between multiplication and addition both in words and by examples.   Answers will vary.   Then point out that multiplication by a whole number can be thought of as repeated addition and demonstrate this by several examples.

2.6 Multiplication of Real Numbers

To multiply real numbers
To evaluate expressions involving multiplication and exponents

The formula $F = \frac{9}{5}C + 32$ can be used to convert temperatures given in degrees Celsius to degrees Fahrenheit. If the temperature is $-5°C$, you need to multiply $\frac{9}{5}$ and -5, and then add 32, to find the temperature in degrees Fahrenheit.

From arithmetic you know that the product of two positive numbers is positive. Also, the product of any real number and zero is zero. For example, $-8 \cdot 0 = 0$ and $0 \cdot (-8) = 0$.

Property of Zero for Multiplication
For all real numbers a, $a \cdot 0 = 0$ and $0 \cdot a = 0$

Now consider the product of a positive number and a negative number. Examine this pattern.

$$
\left.
\begin{array}{l}
3 \cdot 3 = 9 \\
3 \cdot 2 = 6 \\
3 \cdot 1 = 3 \\
3 \cdot 0 = 0
\end{array}
\right\}
\begin{array}{l}
\text{The product} \\
\text{decreases by 3} \\
\text{each time.}
\end{array}
$$

Continue the pattern.
$$3 \cdot (-1) = -3$$
$$3 \cdot (-2) = -6$$
$$3 \cdot (-3) = -9$$

The pattern suggests that the product of a positive number and a negative number is negative. Because of the Commutative Property for Multiplication, "the product of a positive number and a negative number" refers, for example, to both $3 \cdot (-1)$ and $-1 \cdot 3$.

To multiply two real numbers with opposite signs:
1. Multiply their absolute values.
2. The sign of the product is negative.

negative $\cdot$ positive = negative   positive $\cdot$ negative = negative
$(-)$ $\cdot$ $(+)$ $=$ $(-)$ $(+)$ $\cdot$ $(-)$ $=$ $(-)$

Highlighting the Standards

Standards 2b, 2a: The lesson asks students to perceive and to continue patterns leading to mathematical generalizations. In the Focus on Reading, students clarify their thinking through examples followed by generalizations.

EXAMPLE 1   Multiply: **a.** $9(-8)$ **b.** $(-6.2)5$

Solutions   **a.** $9(-8) = -72$ ← pos · neg = neg

b. $(-6.2)5 = -31$ ← neg · pos = neg

TRY THIS   Multiply: **1.** $(-51)4$ -204 **2.** $7(-3)$ -21

We already know the product of two positive numbers is positive. To find the product of two negative numbers, examine the following pattern.

$$-3 \cdot 3 = -9$$
$$-3 \cdot 2 = -6$$
$$-3 \cdot 1 = -3$$
$$-3 \cdot 0 = 0$$

The product increases by 3 each time.

Continue the pattern.
$$-3 \cdot (-1) = 3$$
$$-3 \cdot (-2) = 6$$
$$-3 \cdot (-3) = 9$$

The pattern suggests that the product of two negative numbers is positive.

To multiply two real numbers with the same sign:
3. Multiply their absolute values.
4. The sign of the product is positive.

negative · negative = positive   positive · positive = positive
$(-)$ · $(-)$ = $(+)$ $(+)$ · $(+)$ = $(+)$

EXAMPLE 2   Multiply: **a.** $\frac{3}{4} \cdot 20$ **b.** $-6.8(-5)$

Solutions   **a.** $\frac{3}{4} \cdot 20 = 15$ ← pos · pos = pos

b. $-6.8(-5) = 34$ ← neg · neg = pos

TRY THIS   Multiply: **3.** $-8(-5)$ 40 **4.** $\frac{2}{3} \cdot 12$ 8

If any two real numbers are multiplied, their product is a real number. That is, the set of real numbers is *closed* under multiplication.

The commutative and associative properties allow you to multiply numbers in any order, as shown in Example 3 on the next page.

2.6 Multiplication of Real Numbers **63**

▬ TEACHING SUGGESTIONS

Lesson Note

An alternate method for showing the product of a positive number and a negative number is as follows.

$$7(-2 + 2) = 7(0)$$
$$= 0$$

But, by the Distributive Property,

$$7(-2 + 2) = 7(-2) + 7(2)$$
$$= ? \quad + 14$$

Ask what must be added to 14 to give the correct result of 0. (-14) Therefore, $7(-2) = -14$.

When demonstrating Example 4, remind students that exponents are calculated before multiplication. Emphasize that $(-5)^2 = 25$ is not the same as $-5^2 = -25$.

Math Connections

Science: Use the formula $C = \frac{5}{9}(F - 32)$ to convert $-4°$ Fahrenheit to degrees Celsius. Then convert $-40°F$ to degrees Celsius.
$-20°C$, $-40°C$

Additional Example 1

Multiply.

a. $7(-5)$ -35
b. $(-7.3)6$ -43.8

Additional Example 2

Multiply.

a. $\frac{2}{7}(56)$ 16
b. $-9.2(-7)$ 64.4

Critical Thinking Questions

Synthesis: After showing examples such as $-1(8) = -8$, $-1(\frac{2}{3}) = -\frac{2}{3}$, and $-1(-5) = 5$, ask students what happens when you multiply a real number by -1. What is $-1(a)$? We call this the Property of -1 for Multiplication. Now ask them to give a rule that justifies each step in the following and to express the result in their own words.

$$(-a)(-b) = (-1 \cdot a)(-1 \cdot b)$$
$$= (-1)(-1)ab$$
$$= 1ab$$
$$= ab$$

Prop. of -1 for Mult.; Comm. & Assoc. Prop. of Mult.; Prop. of -1 for Mult.; Identity Prop. for Mult.; The product of two opposites is the same as the product of the original numbers.

Common Error Analysis

Error: Students often confuse the rules for adding real numbers with the rules for multiplying real numbers.

Once students have learned the rules for multiplication, it is helpful to compare the two sets of rules.

EXAMPLE 3 Multiply $-2(29)(-5)(-3)$.

Solution Regroup to simplify computations. $\underbrace{-2(-5)}\underbrace{(-3)(29)}$
$$10(-87)$$
$$-870$$

EXAMPLE 4 Evaluate $-5a^3b$, for $a = -2$ and $b = -4$.

Solution $-5a^3b$
$-5(-2)^3(-4)$ $\longleftarrow$ Only the -2 is raised to the third power, not -5.
$-5\underbrace{(-2)(-2)(-2)}(-4)$
$\underbrace{-5(-8)}(-4)$
$40(-4) = -160$

TRY THIS 5. Evaluate $3x^2y$ for $x = -5$ and $y = -2$. -150

Focus on Reading

To complete each statement, choose a, b, c, or d at the right.

1. The product $0 \cdot (-1)$ is __c__.   **a.** positive
2. The product $(-4.2)(-7.5)$ is __a__.   **b.** negative
3. The product $7(-81)$ is __b__.   **c.** zero
4. The product $(3.1)(5.4)$ is __a__.   **d.** either positive, negative, or zero
5. The product of a negative number and zero is __c__.
6. The product of two numbers with the same sign is __a__.

Classroom Exercises

Multiply.

1. $5 \cdot 6$ 30   **2.** $-7 \cdot 9$ -63 **3.** $6(-3)$ -18 **4.** $(-8)(-8)$ 64
5. $16 \cdot 0$ 0   **6.** $7 \cdot 7$ 49   **7.** $-8 \cdot 0$ 0   **8.** $4(-9)$ -36
9. $(-6)(-9)$ 54   **10.** $-4 \cdot 6$ -24 **11.** $-9 \cdot 7$ -63 **12.** $-5 \cdot 1$ -5
13. $-3(9)(-2)$ 54   **14.** $1(15)(-3)$ -45 **15.** $0(15)(-5)$ 0   **16.** $-8(-1)4$ 32

Evaluate for $x = -4$ and $y = -5$.

17. xy 20   **18.** x^2y -80 **19.** $x^2 + y^2$ 41   **20.** $x^2 - y^2$ -9

64 Chapter 2 Operations with Real Numbers

Additional Example 3

Multiply $3(-4)(-6)(-10)$. -720

Additional Example 4

Evaluate $3x^2y$ for $x = -2$ and $y = -4$. -48

Written Exercises

Multiply.

1. $-1 \cdot 0$ 0
2. $10(-1)$ -10
3. $-6(-5)$ 30
4. $7 \cdot 8$ 56

5. $0(-12)$ 0
6. $6(-2)$ -12
7. $-9 \cdot 9$ -81
8. $-5(-5)$ 25

9. $6 \cdot 4$ 24
10. $5^2(-6)$ -150
11. $-4(-12)$ 48
12. $-9 \cdot 4$ -36

13. $-\frac{3}{5}(5)$ -3
14. $\frac{3}{4}\left(-\frac{4}{3}\right)$ -1
15. $-\frac{1}{8}(-16)$ 2
16. $-\frac{3}{8}\left(-\frac{8}{3}\right)$ 1

17. $1(3.2)$ 3.2
18. $-1(8.6)$ -8.6
19. $(9.3)(0)$ 0
20. $-1(-9.7)$ 9.7

21. $-1(4)(-2)$ 8
22. $3(6)(-1)$ -18
23. $5(-9)(-1)^2$ -45
24. $(-2)^2(1)(4)$ 16

25. $-6(3)(-5)$ 90
26. $8(-2)^2(-3)$ -96
27. $-6(8)(-3)$ 144
28. $(-5)^2(-2)$ -50

Evaluate for $x = -2$, $y = -5$, and $z = 0$.

29. $3x$ -6
30. $-3x$ 6
31. $4y$ -20
32. $-7z$ 0

33. y^2 25
34. $3x^2$ 12
35. $(3x)^2$ 36
36. xyz 0

37. $6x^2y$ -120
38. $(-x)^3$ 8
39. $-2xy^2$ 100
40. $-2(xy)^2$ -200

41. $6x - 7$ -19
42. $6y + x$ -32
43. $x - y$ 3
44. $y - x$ -3

45. $(-2xy)^2$ 400
46. $5x^2y$ -100
47. $(5x)^2y$ -500
48. $(y - x)^2$ 9

49. $(x + y)^2$ 49
50. $x^2 + y^2$ 29
51. $(x - y)^2$ 9
52. $x^2 - y^2$ -21

Multiply.

53. $9(-7)(-3)(5)$ 945
54. $6(0)(-8)(9)$ 0
55. $-5(-2)(-3)(4)$ -120

56. $(-4)^3 \cdot 2^2$ -256
57. $(-2)^3(-5)$ 40
58. $-2(-4)^3$ 128

59. $(-3)^3(-1)^3$ 27
60. $6^2(-1)^5$ -36
61. $(-2)^5 \cdot 3^3$ -864

62. $(-1)^8(-3)^3(-2)$ 54
63. $(-6)^2(-1)(-2)^3$ 288
64. $-5^2(-3)^2(-4)$ 900

65. The formula $F = \frac{9}{5}C + 32$ is used to convert temperatures from degrees Celsius (°C) to degrees Fahrenheit (°F). When the temperature is -10°C, what is the corresponding temperature in degrees Fahrenheit? 14°F

66. The formula $C = \frac{5}{9}(F - 32)$ is used to convert temperatures from degrees Fahrenheit (°F) to degrees Celsius (°C). When the temperature is -4°F, what is the corresponding Celsius temperature? -20°C

Simplify.

67. $(-1)^{12}8^2(-2)^3$ -512
68. $-1^{15}(-2 \cdot 3)^2(-4)$ 144

69. $(-5 + 3 - 8)(-7 + 3 - 4)$ 80
70. $(-7 + 3)(7 - 8)(-5 + 2)(-7 + 7)$ 0

71. $(-3 \cdot 4)[-2 + (-5)(-4)]$ -216
72. $(-5 + 8)[-7 + (-4 \cdot 3) + (-6)]$ -75

73. $(-8 + 6)^2[-9 + (-4)2]$ -68
74. $(-2 - 1)^3(7 - 8 - 1)^5$ 864

Multiply.

1. $-1(-8)$ 8
2. $-7(0)$ 0
3. $-8 \cdot 6$ -48
4. $12(-4)$ -48
5. $-6.2(-0.5)$ 3.1
6. $-\frac{1}{3}\left(\frac{4}{5}\right)$ $-\frac{4}{15}$
7. $6(-4)(-1)$ 24
8. $7(-3)(-2)^3$ 168

Evaluate for $x = -3$, $y = 7$ and $z = -1$.

9. $4x - y$ -19
10. $3x^2z$ -27

Closure

Ask students to state the rules for multiplication of real numbers.   See page 63.   Then ask them to compare these to the rules for addition. Have them give examples to explain their answers.   Answers will vary.

◼◼◼ FOLLOW UP

Guided Practice

Classroom Exercises 1–20
Try This   all, FR   all

Independent Practice

A Ex. 1–36, **B** Ex. 37–66, **C** Ex. 67–79
Basic: WE 1–57 odd, Application
Average: WE 11–67 odd, Application
Above Average: WE 21–79 odd, Application

Enrichment

Ask students to determine which of the following are true for all real numbers, and to give numerical examples for expressions that are not true.

1. $|a| + |b| = |a + b|$
2. $|a + b| = |b + a|$ T.

3. $|a| - |b| = |a - b|$
4. $|a - b| = |b - a|$ T
5. $|a||b| = |ab|$ T
6. $|ab| = |ba|$ T

1. F;$|-1| + |1| \neq |-1 + 1|$
3. F;$|1| - |-1| \neq |1 - (-1)|$

True or false?

75. When a negative number is squared, the result is negative.

76. When a negative number is cubed, the result is negative.

77. When a negative number is raised to an even power, the result is positive.

78. When a negative number is raised to an odd power, the result is negative.

79. When a number of factors are multiplied and one factor is zero, the result is negative. F; result is always 0.

Mixed Review

Simplify. *1.2, 1.7, 2.4*

1. $15 + 7 \cdot 2$ 29

2. $8 \cdot 3 - 14 \div 7$ 22

3. $a + 5 + 7a + 2$ $8a + 7$

4. $4(7x + 12) + 2$ $28x + 50$

5. $x + 4(x + 5)$ $5x + 20$

6. $-17 + 18 + (-3)$ -2

Find the solution set. The replacement set is $\{-4, -2, 0, 2, 4\}$. *1.8*

7. $x + 2 = 0$ $\{-2\}$

8. $x + x < x + (-1)$ $\{-4, -2\}$

9. $\dfrac{x + 6}{2} = 3$ $\{0\}$

10. The temperature was $-8°F$ in the morning and then rose $10°$ by afternoon. Find the afternoon temperature. *2.4* $2°F$

Application: *Automobile Rental Rates*

The Jacksons plan a 3-week trip with a budget of $600 for renting a car. Car rental rates are $85/wk plus $0.17/km. The following formula shows the number of kilometers the Jacksons can travel.

$$\text{number of km} = \frac{(\text{total cost}) - (\text{flat rate})}{\text{cost per km}}$$

Substituting the values for the Jacksons' trip gives the following.

$$\text{number of km} = \frac{600 - 3(85)}{0.17} = 2{,}029 \text{ km (nearest kilometer)}$$

Solve. Round answers to the nearest kilometer.

1. Earl budgeted $315 for car rental. How far can he travel if the rate is $75 plus $0.12/km? 2,000 km

2. Julie budgeted $200 for car rental. How far can she travel if the rate is $65 plus $0.15/km? 900 km

3. Moira rents a car for $65 plus $0.12/km. How far can she travel, to the nearest kilometer, on a budget of $260? 1,625 km

4. The Brodskys rent a camper for a week for $275 plus $15 insurance plus $0.13/km. How far can they go on a budget of $500? 1,615 km

2.7 Division of Real Numbers

Objectives

To divide real numbers
To find the arithmetic mean of a set of real numbers

Multiplication and division are related operations. Thus, the rules for division depend upon the rules for multiplication.

Example	Conclusion
$12 \div 4 = 3$ since $3 \cdot 4 = 12$	pos $\div$ pos = pos
$-30 \div (-6) = 5$ since $5 \cdot (-6) = -30$	neg $\div$ neg = pos
$-56 \div 8 = -7$ since $-7 \cdot 8 = -56$	neg $\div$ pos = neg
$36 \div (-9) = -4$ since $-4 \cdot (-9) = 36$	pos $\div$ neg = neg

Dividing Nonzero Real Numbers
1. The quotient of two real numbers with the same sign is positive.
2. The quotient of two real numbers with unlike signs is negative.

EXAMPLE 1 Divide.

a. $-15 \div (-5)$ b. $\frac{-27}{3}$ c. $4.8 \div (-0.6)$

Solutions

a. $-15 \div (-5) = 3$ neg $\div$ neg = pos

b. $\frac{-27}{3} = -9$ $\frac{\text{neg}}{\text{pos}}$ = neg

c. $4.8 \div (-0.6) = -8$ pos $\div$ neg = neg

EXAMPLE 2 Simplify $-54 \div [15 \div (-5)]$.

Solution Divide within the brackets first. $-54 \div [15 \div (-5)]$
                                                        $-54 \div (-3)$
                                                        18

TRY THIS Divide: **1.** $-18 \div 6$ -3 **2.** $-12 \div (-3)$ 4

3. Simplify $24 \div (-12 \div 2)$. -4

Teaching Resources

Problem Solving Worksheet 2
Quick Quizzes 15
**Reteaching and Practice
   Worksheets** 15

▪▪▪ GETTING STARTED

Prerequisite Quiz

Divide.

1. $56 \div 8$ 7
2. $0 \div 15$ 0
3. $72 \div 9$ 8
4. $63 \div 63$ 1
5. $14.4 \div 1.2$ 12
6. $16.5 \div 1$ 16.5
7. $\frac{3}{8} \div \frac{3}{2}$ $\frac{1}{4}$
8. $\frac{5}{6} \div \frac{1}{4}$ $3\frac{1}{3}$

Motivator

Show the following examples to the students.

$8 \div 2 = 4$ and $8 \cdot \frac{1}{2} = 4$
$27 \div 3 = 9$ and $27 \cdot \frac{1}{3} = 9$

Ask them in what way multiplication and division are related operations. Ask what operations are involved in finding the average of a set of numbers.   Inverses; addition and division

---

Additional Example 1

Divide.

a. $-28 \div (-4)$ 7
b. $\frac{42}{-7}$ -6
c. $-7.2 \div 0.9$ -8

Additional Example 2

Compute $56 \div [24 \div (-3)]$. -7

Highlighting the Standards

Standard 4a: As they examine division by zero, students see various representations of the same idea, and use the calculator to expand their understanding.

Lesson Note

The fact that multiplication and division are inverse operations can be demonstrated by reminding students that multiplication can be used to check the answer to a division problem. Emphasize that the rules for determining the sign of the answer are the same for both multiplication and division. Many students have difficulty understanding the relationship between a fraction and its reciprocal since the reciprocal of a number a is described as $\frac{1}{a}$. Have them consider an example such as the following.

Let $a = \frac{2}{3}$. Then $\frac{1}{a} = \frac{1}{\frac{2}{3}} = 1 \div \frac{2}{3} = 1 \cdot \frac{3}{2} = \frac{3}{2}$.

Thus, the reciprocal of $\frac{2}{3}$ is $\frac{3}{2}$.

Math Connections

Previous Algebra: In Lesson 2.3, the students learned that the sum of any real number and its *additive inverse* (or *opposite*) is the *identity* element for addition, i.e., $a + (-a) = 0$. Ask the students to write a similar sentence for multiplication.

$a \cdot \frac{1}{a} = 1$

The rules for 0 in division are related to the Property of Zero in Multiplication. Consider the following three cases.

$\frac{0}{8} = 0$ If 0 is divided by any nonzero real number such as 8, the quotient is 0, since $0 \cdot 8 = 0$.

$\frac{5}{0}$ is undefined.   There is *no* number n such that $n \cdot 0 = 5$, or any other nonzero number.

$\frac{0}{0}$ is undefined.   Since $n \cdot 0 = 0$ is true for *every* number n, mathematicians agree to leave $0 \div 0$ undefined.

Thus, if 0 is divided by a nonzero number, the quotient is 0, but division by 0 is undefined. Verify this on your calculator.

$$0 \boxed{\div} 6 \boxed{=} \qquad\qquad 6 \boxed{\div} 0 \boxed{=}$$

The *reciprocal* of a number is often used in division. Two numbers are called reciprocals, or *multiplicative inverses*, of each other if their product is 1. Notice that each product below is 1.

$4 \cdot \frac{1}{4} = \frac{4}{1} \cdot \frac{1}{4}$ Thus, $\frac{1}{4}$ is the reciprocal of 4, and

$\qquad = \frac{4}{4}$, or 1   4 is the reciprocal of $\frac{1}{4}$.

$-\frac{5}{6} \cdot \left(-\frac{6}{5}\right) = \frac{30}{30}$ Thus, $-\frac{6}{5}$ is the reciprocal of $-\frac{5}{6}$, and

$\qquad\qquad = 1$ $-\frac{5}{6}$ is the reciprocal of $-\frac{6}{5}$.

Note that 0 has no reciprocal since $\frac{1}{0}$ is undefined.

Multiplicative Inverse Property
For each nonzero number a, there is exactly one number $\frac{1}{a}$ such that $a \cdot \frac{1}{a} = 1$ and $\frac{1}{a} \cdot a = 1$.

The number $\frac{1}{a}$ is called the **reciprocal** or **multiplicative inverse** of a.

Division by a nonzero real number can be defined as multiplication by the reciprocal of that number, as illustrated below.

$$\text{Division: } 12 \div 4 = 3 \qquad \text{Multiplication: } 12 \cdot \frac{1}{4} = 3$$

If a and b are real numbers and $b \neq 0$, $a \div b = a \cdot \frac{1}{b}$.
That is, to divide by a nonzero number, multiply by its reciprocal.

EXAMPLE 3 Divide.

a. $\frac{8}{9} \div \left(-\frac{2}{3}\right)$ b. $-2\frac{4}{5} \div (-7)$

         reciprocals   reciprocals

Solutions   a. $\frac{8}{9} \div \left(-\frac{2}{3}\right) = \frac{8}{9} \cdot \left(-\frac{3}{2}\right)$ b. $-2\frac{4}{5} \div (-7) = -\frac{14}{5} \cdot \left(-\frac{1}{7}\right)$

                   $= -\frac{24}{18}$, or $-\frac{4}{3}$ $= \frac{14}{35}$, or $\frac{2}{5}$

TRY THIS   Divide: 4. $-8 \div \frac{4}{5}$ -10 5. $-\frac{3}{7} \div \left(-2\frac{1}{3}\right)$ $\frac{9}{49}$

The **arithmetic mean**, or *average*, of a set of numbers is the sum of the members of the set divided by the number of members.

EXAMPLE 4 The table below shows the gains and losses made by a football team in eight successive plays. Find the average gain or loss per play.

Play	1	2	3	4	5	6	7	8
Gain or loss	5	-8	-2	22	3	-15	0	-13

Plan   Add the eight numbers and divide by 8.

Solution

$$5 - 8 - 2 + 22 + 3 - 15 + 0 - 13$$
$$\underbrace{-8 - 2 - 15 - 13}_{-38} + \underbrace{5 + 22 + 3}_{30} \quad \leftarrow \text{Group like signs together before adding.}$$
$$-8$$

Divide the sum by 8. $\frac{-8}{8} = -1$

Therefore, the average *loss* was 1 yd per play.

TRY THIS   6. Find the average gain or loss for the first 5 plays.   4

Classroom Exercises

Divide, if possible.

1. $\frac{-9}{-3}$ 3   2. $\frac{8}{4}$ 2   3. $\frac{-16}{2}$ -8 4. $\frac{18}{-9}$ -2 5. $-\frac{-6}{0}$ Undef **6.** $\frac{0}{10}$ 0

7. $-42 \div 7$ -6 8. $6.2 \div (-0.2)$ -31 9. $-1\frac{2}{3} \div \left(-\frac{5}{6}\right)$ 2

Give the reciprocal of the number, if it exists.

10. $\frac{3}{4}$ $\frac{4}{3}$ 11. 7 $\frac{1}{7}$ 12. 0   None   13. $-\frac{1}{3}$ -3 14. 1   1   15. $-2\frac{3}{4}$ $-\frac{4}{11}$

2.7 Division of Real Numbers   **69**

Ask students questions such as the following.

1. Why is the rule for determining the sign of the answer the same for division as for multiplication?   Multiplication and division are inverse operations.
2. What is another term for *multiplicative inverse*?   Reciprocal
3. What is the reciprocal of the reciprocal of $\frac{1}{3}$? $\frac{1}{3}$
4. Why does 0 not have a reciprocal? Cannot divide by 0.

◼◼◼ FOLLOW UP

Guided Practice

Classroom Exercises 1–15
Try This   all

Independent Practice

A Ex. 1–37, **B** Ex. 38–43, **C** Ex. 44–47

Basic: WE 1–37 odd, Using the Calculator

Average: WE 7–43 odd, Using the Calculator

Above Average: WE 13–43 odd, 44–47, Using the Calculator

Written Exercises

Divide, if possible.

1. $\frac{8}{2}$ 4
2. $\frac{-12}{3}$ -4
3. $\frac{32}{-8}$ -4
4. $\frac{-72}{-8}$ 9
5. $\frac{-40}{10}$ -4
6. $\frac{28}{7}$ 4
7. $\frac{-45}{9}$ -5
8. $\frac{9}{-9}$ -1
9. $\frac{-100}{4}$ -25
10. $\frac{80}{-8}$ -10
11. $\frac{0}{5}$ 0
12. $\frac{21}{0}$ Undef
13. $-60 \div 10$ -6
14. $32 \div (-8)$ -4
15. $1.4 \div 0.07$ 20
16. $12 \div 0$ Undef
17. $3.4 \div (-1.7)$ -2
18. $-22 \div 0.002$ $-11,000$
19. $-32 \div 0$ Undef
20. $-0.4 \div (-0.2)$ 2
21. $\frac{4}{5} \div \left(-\frac{8}{15}\right)$ $-\frac{3}{2}$
22. $\frac{7}{8} \div (-14)$ $-\frac{1}{16}$
23. $-\frac{5}{9} \div 3\frac{1}{3}$ $-\frac{1}{6}$
24. $-1\frac{1}{2} \div \left(-1\frac{1}{8}\right)$ $1\frac{1}{3}$
25. $-15 \div \frac{5}{7}$ -21
26. $3\frac{1}{3} \div \frac{5}{9}$ 6
27. $-1\frac{1}{3} \div \left(-2\frac{2}{3}\right)$ $\frac{1}{2}$

Simplify.

28. $-48 \div [60 \div (-30)]$ 24
29. $(-36 \div 18) \div (-2)$ 1
30. $72 \div [-24 \div (-8)]$ 24
31. $-45 \div (-15 \div 3)$ 9
32. $14 \div [28 \div (-2)]$ -1
33. $-27 \div [-18 \div (-6)]$ -9
34. A town's daily low temperatures for the week are shown below. Find the average low temperature for the week. $-1°$

Sun.	Mon.	Tues.	Wed.	Thurs.	Fri.	Sat.
$-9°$	$-2°$	$6°$	$-3°$	$-3°$	$-7°$	$11°$

35. The table below shows the profits and losses of a small business for the four quarters of a year. What was the average profit or loss per quarter? $675 profit

1st quarter	2nd quarter	3rd quarter	4th quarter
$5,200	$-$1,400	$-$2,700	$1,600

36. The table below shows Maria's weight gain or weight loss in pounds at the end of the month for six months. What was the average gain or loss per month? $\frac{2}{3}$-lb gain

Jan.	Feb.	Mar.	Apr.	May	June
5	-3	-3	4	-1	2

37. The table below shows the gain or loss in the price of a stock for a five-day period. What was the average change per day? $-\frac{3}{40}$ per day

Mon.	Tues.	Wed.	Thurs.	Fri.
$-\frac{1}{4}$	$+\frac{3}{8}$	No change	$+\frac{1}{8}$	$-\frac{5}{8}$

Enrichment

Have students solve this problem.

Thirty families in a survey are numbered 1–30. The average number of children in families 1–10 is 4. The average number of children in families 16–30 is 5. If the average number of children in the entire group of 30 families is 4, what is the average number of children in families 11–15?   1

Simplify.

38. $(-8 + 3 - 7 + 4) \div (-7 - 8 + 13)$ 4

39. $(-9 - 5 + 4 - 4) \div (6 - 8 - 4 + 3)$ $4\frac{2}{3}$

40. $(-7 - 4 + 2)(-5 + 3)(-8 - 4 + 12)$ 0

41. $(-8 + 6 - 4 - 2 + 20) \div (2 \cdot 4 - 3 \cdot 6)$ $-1\frac{1}{5}$

42. $(8 - 6 + 14 + 6) \div [(-7 + 15) - (2 - 6)]$ $1\frac{5}{6}$

43. $[(24 - 32) \div (8 - 12)] \div (1 - 5 + 6 - 8 + 4)$ -1

Give an example to show that each statement is false.

44. Division of real numbers is commutative. Ex. 44.–47. Answers will vary.

45. Division of real numbers is associative.

46. The identity element for division is 0.

47. The set of real numbers is closed under division.

Mixed Review

Simplify.   *1.2, 1.3, 1.7, 2.4–2.6*

1. $7 + 5 \cdot 2$ 17

2. $39 - 3 \cdot 6 - 6 \cdot 2$ 9

3. $5 + 2 \cdot 4^2$ 37

4. $7b + 5 + b$ $8b + 5$

5. $2c + 3(4c + 1)$ $14c + 3$

6. $-7 + 5$ -2

7. $2 - (-3)$ 5

8. $-4 \cdot 5$ -20

9. $-6(-2)$ 12

10. At Spearfish, South Dakota, one day in January 1943, the temperature rose from $-4°F$ to $45°F$ in two minutes! What was the change in temperature? *2.5* 49°

/Using the Calculator

A calculator can be used to find a decimal value for the reciprocal of a positive number (or at least an approximation of the reciprocal). Some calculators have a reciprocal key marked $\boxed{\frac{1}{x}}$. If yours does, you can enter a number x and press the $\boxed{\frac{1}{x}}$ key to find the reciprocal of x. If a calculator has no $\boxed{\frac{1}{x}}$ key, you can find the reciprocal as follows.

$$\boxed{1} \;\; \boxed{\div} \;\; \boxed{x} \;\; \boxed{=}$$

Find a decimal for the reciprocal of each number.

1. 5 0.2

2. 0.2 5

3. 8 0.125

4. 0.125 8

5. 0.4 2.5

6. 2.5 0.4

7. 0.16 6.25

8. 6.25 0.16

Find the reciprocal of the reciprocal of each number.

0.00128, 781.25

9. 0.8 1.25, 0.8

10. 31.25 0.032, 31.25

11. 64 0.015625, 64

12. 781.25

▉ GETTING STARTED

Prerequisite Quiz

Simplify.

1. $6 + (-11)$ -5
2. $-5(2) + 3$ -7
3. $2(-3)^2$ 18
4. $-1(12)^2$ -144
5. $-3(-4)(-5)$ -60
6. $-18\left(-\frac{5}{9}\right)$ 10

Motivator

Tell students that this lesson is a review of the rules they have learned to this point. Take time to have students restate the rules for the four operations with real numbers and the order of operations. Then ask them to evaluate the following expression.

$[-8 + (-2)] + [-8 - (-2)] +$
$[-8 \div (-2)] + [-8 \div (2)]$ -16

▉ TEACHING SUGGESTIONS

Lesson Note

Encourage students to write out each step as illustrated in the Examples. Accuracy is important; it only takes one incorrect sign to obtain an incorrect answer. Remind students that a fraction bar, as in Example 4, is also a grouping symbol.

2.8 Mixed Operations

Objectives   To simplify numerical expressions
To evaluate algebraic expressions

To simplify the expressions in this lesson, you will be using all of the rules for operating with real numbers. Be careful not to confuse the rules. For example,

The *sum* of two negative numbers is *negative*.
The *product* of two negative numbers is *positive*.

You also need to recall the rules for the order of operations (see Lesson 1.3). For example, an expression such as $-6 - 2 \cdot 5$ involves two operations and the multiplication $2 \cdot 5$ is performed first.

EXAMPLE 1   Simplify.

a. $-6 - 2 \cdot 5$ b. $8 - (-7)^2$

Solutions   a. $-6 - 2 \cdot 5 = -6 - 10$ b. $8 - (-7)^2 = 8 - 49$
$\qquad\qquad\qquad = -6 + (-10)$ $\qquad\qquad\qquad = 8 + (-49)$
$\qquad\qquad\qquad = -16$ $\qquad\qquad\qquad = -41$

TRY THIS   Simplify: 1. $-8 + 4(-5)$ -28 2. $6 + (-2)^3$ -2

The order of operations must be followed when you evaluate expressions involving two or more operations.

EXAMPLE 2   Evaluate $-8 - 4x$ for $x = -3$.

Solution $-8 - 4x = -8 - 4(-3)$
Multiply first. $= -8 - (-12)$
$\qquad\qquad\qquad\qquad = -8 + 12 = 4$

EXAMPLE 3   Evaluate $-8e^2 + 7f - 5g$, for $e = -3, f = 4$, and $g = -2.1$.

Solution $-8e^2 + 7f - 5g = -8(-3)^2 + 7 \cdot 4 - 5(-2.1)$
$\qquad\qquad\qquad\qquad = -8 \cdot 9 + 7 \cdot 4 - (-10.5)$
$\qquad\qquad\qquad\qquad = -72 + 28 + 10.5$
$\qquad\qquad\qquad\qquad = \underbrace{-72 + 28} + 10.5$
$\qquad\qquad\qquad\qquad = -44 + 10.5 = -33.5$

TRY THIS   3. Evaluate $5c^2 - 3d - 2$ for $c = -2$ and $d = -4$.   30

Highlighting the Standards

Standard 2a: The Focus on Reading asks students to think about the mathematical ideas and then evaluate generalizations about various relationships.

Additional Example 1

Simplify.

a. $-9 - 5 \cdot 8$ -49
b. $2 - (-4)^2$ -14

Additional Example 2

Evaluate $7 + 3x$ for $x = -6$. -11

EXAMPLE 4   Evaluate $\dfrac{a - b}{-4}$ for $a = 3.6$ and $b = -1.2$.

Solution
$$\frac{a - b}{-4} = \frac{3.6 - (-1.2)}{-4}$$
$$= \frac{3.6 + 1.2}{-4}$$
$$= \frac{4.8}{-4} = -1.2$$

TRY THIS   4. Evaluate $3a - 5b$ for $a = 0.5$ and $b = -0.7$. 5

Focus on Reading

True or false?

1. If two nonzero numbers are opposites, then their quotient is -1. T
2. The quotient of two nonzero numbers with the same sign always has that same sign. F
3. If $a \cdot b$ is negative, and a is negative, then b is negative. F
4. Every real number has an additive inverse. T
5. Every real number has a multiplicative inverse. F
6. If a is negative, then a^3 is negative. T
7. For every real number a, $a \div 0$ is undefined. T
8. The sum of two negative numbers is positive. F
9. The multiplicative inverse of a negative number is positive. F
10. If $a + b = 0$, then a is the additive inverse of b. T
11. If $a + b$ is positive, and a is positive, then b must be positive. F
12. The product of a number and its multiplicative inverse is 1. T

Classroom Exercises

Simplify.

1. $-7 \cdot 4 + 9$ -19
2. $-6 - 3 \cdot 4$ -18
3. $8 - 2 \cdot 3$ 2
4. $6 + (-8)^2$ 70

Evaluate for the given values of the variables.

5. $-7 - 5x$, for $x = -3$ 8
6. $6 - 3y$, for $y = -6$ 24
7. $\dfrac{5x - y}{5}$, for $x = -2$ and $y = 5$ -3
8. $2x - 4y^2$, for $x = 3$ and $y = -2$ -10

2.8 Mixed Operations   **73**

History: The Hindus in India were calculating with negative numbers by 700 A.D. Their names for positive and negative were derived from the words for credit and debit.

Critical Thinking Questions

Analysis: Evaluate $(x + 3)(x - 1)$ for $x = -3$. Then evaluate the same expression for $x = 1$. 0, 0   For what values of x will the expression $(x - 5)(x + 4)$ have a value of 0? 5, -4 For what values of x will the expression $x(x + 1)(x - 8)(x + 2)$ have a value of 0? 0, -1, 8, -2

Common Error Analysis

Error: Some students will fail to use parentheses when evaluating an expression such as $-8 - 4x$ for $x = -3$. They will write $-8 - 4 - 3$ and get an incorrect result.

Emphasize that a number directly preceding a variable indicates multiplication, and that when the variable is replaced with a number, they must indicate the multiplication in some other way. This is usually best accomplished with parentheses.

Checkpoint

Evaluate for the given values of the variables.

1. $4 - 8x$ for $x = 6$ -44
2. $10 - 3y$ for $y = -4$ 22
3. $3a - 15$ for $a = -2$ -21
4. $10x - 7y$ for $x = -\dfrac{3}{5}$ and $y = -2$ 8
5. $3d^2 - 5e + 3f$ for $d = -2$, $e = -4$, and $f = -\dfrac{2}{3}$ 30

Additional Example 3

Evaluate $3r - 2s^2 + 5t$ for $r = -2$, $s = -3$, and $t = 1.5$ -16.5

Additional Example 4

Evaluate $\dfrac{a + b}{8}$ for $a = -9.6$ and $b = 2.4$. -0.9

Closure

Ask a student to summarize as briefly but accurately as possible, the rules for adding, subtracting, multiplying, and dividing real numbers. See if other students can improve on this response. When the class has reached some kind of consensus, write the concise rules on the chalkboard so that everyone can use them.

■■■ FOLLOW UP

Guided Practice

Classroom Exercises 1–8
Try This   all, FR   all

Independent Practice

A Ex. 1–28, **B** Ex. 29–38, **C** Ex. 39–42
Basic: WE 1–31 odd
Average: WE 5–35 odd
Above Average: WE 11–41 odd

Written Exercises

Simplify.

1. $-2 + 3 \cdot 9$ 25
2. $-5 \cdot 8 - 7$ −47
3. $8 - 6 \cdot 5$ −22
4. $4 + 3(-9)$ −23
5. $6 - 7^2$ −43
6. $-3 + (-8)^2$ 61
7. $5^2 - (-1)^2$ 24
8. $13 - (-2^3)$ 21
9. $-3 \cdot 6 + 4(-7) - 8(-5)$ −6
10. $6(-2) + (-4)^2 - 9 \cdot 3$ −23

Evaluate for the given values of the variables.

11. $a - 5$, for $a = -4$ −9
12. $7 - 5y$, for $y = -2$ 17
13. $-2x - 11$, for $x = -2$ −7
14. $-5x - 8$, for $x = -2$ 2
15. $-2x + 6$, for $x = -3$ 12
16. $-8 - 6x$, for $x = -6$ 28
17. $6 - 3x$, for $x = 2$ 0
18. $-3m - 10$, for $m = -3$ −1
19. $6 - 9k$, for $k = 0$ 6
20. $5b - 3$, for $b = -7$ −38
21. $-6 - 5t$, for $t = -6$ 24
22. $6 - 7b$, for $b = -5$ 41
23. $3x + 2y$, for $x = 5$ and $y = -3$ 9
24. $7a - 5b$, for $a = -3$ and $b = 2$ −31
25. $-6c + 3d$, for $c = -1$ and $d = 4$ 18
26. $4p - 9q$, for $p = -8$ and $q = -7$ 31
27. $\frac{a + 2b}{6}$, for $a = 4$ and $b = -5$ −1
28. $\frac{4x - y}{x - 3}$, for $x = 7$ and $y = -8$ 9
29. $2x - 4y$, for $x = -3$ and $y = -1.5$ 0
30. $4r - 7s$, for $r = 3$ and $s = 0.2$ 10.6
31. $\frac{5x - 8y}{-x}$, for $x = -3$ and $y = -\frac{3}{4}$ −3
32. $\frac{15x + 2y}{3x}$, for $x = -\frac{2}{3}$ and $y = -3$ 8
33. $|3a - b^2|$, for $a = -2$ and $b = 4$ 22
34. $|-4c + d^2|$, for $c = 2$ and $d = -5$ 17
35. $|2x^2 - (5y)^2|$, for $x = -3$ and $y = 2$ 82
36. $|4p^2 - 3q^3|$, for $p = -3$ and $q = -1$ 39
37. $-4p^2 - 9q + 6r$, for $p = -3$, $q = -9$, and $r = -5$ 15
38. $|-6b - 5c^2 + 15d|$, for $b = -2$, $c = -6$, and $d = -\frac{3}{5}$ 177

Simplify. Then evaluate for $x = -3$, $y = -3$, and $z = -4$.

39. $4[3x + 5(7y - 8)]$ −616
40. $-4x + 6[2z + 3(5 + 2y)]$ −54
41. $8x \div [2(5 + 3y)] - 2$ 1
42. $(10z - 4y^3) \div [2(4 + 7x)]$ −2

Mixed Review

Simplify.   *1.7, 2.2, 2.5, 2.6*

1. $5a + 3 + a + 2a$ $8a + 3$
2. $2x + 4(3x + 7)$ $14x + 28$
3. $|-5|$ 5
4. $8 - (-3)$ 11
5. $-7 - 2 + 4$ −5
6. $3(-3)^3$ −81

True or false?   *2.1*

7. $-7 > -5$ F
8. $-3 < 2$ T
9. $2.3 < 2.19$ F
10. $-\frac{1}{4} < -\frac{1}{5}$ T

Enrichment

For what value of the variable will the expression in the left column have the value given in the right column?

$3y - 8$	1	3
$2a + 5$	23	9
$x^2 + 7$	43	6
$-3g + 2$	−13	5
$12 - 4b$	48	−9
$3k - 7k$	4	−1

2.9 Like Terms: Real Number Coefficients

Objectives

To simplify expressions by combining like terms
To evaluate expressions after combining like terms

In Lesson 1.7 you learned to simplify an expression such as $5x - 2x$ using the Distributive Property.

$$5x - 2x = (5 - 2)x = 3x$$

In this lesson you will learn to simplify expressions such as $-5x + 2x$. Now that you have learned to add positive and negative numbers, you can combine like terms as shown below.

$$-5x + 2x = (-5 + 2)x \qquad -6xy - 4xy = -6xy + (-4xy)$$
$$= -3x \qquad\qquad\qquad = [-6 + (-4)]xy$$
$$= -10xy$$

EXAMPLE 1 Simplify.
   a. $-9y + 2y$ b. $6cd - 9cd$

Solutions a. THINK: $-9 + 2 = -7$ b. THINK: $6 - 9 = -3$
   $-9y + 2y = -7y$ $6cd - 9cd = -3cd$

TRY THIS Simplify: **1.** $-4t + 7t$ _3t_   **2.** $-5ab - 6ab$ _-11ab_

EXAMPLE 2 Write a formula for the perimeter of the rectangle at the right. Simplify the formula.

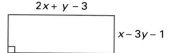

$2x + y - 3$

$x - 3y - 1$

Plan Use the formula $p = 2l + 2w$, for $l = 2x + y - 3$ and $w = x - 3y - 1$.

Solution
$$p = 2(2x + y - 3) + 2(x - 3y - 1)$$
$$p = 4x + 2y - 6 + 2x - 6y - 2$$

Group like terms. $p = 4x + 2x + 2y - 6y - 6 - 2$
Combine like terms. $p = 6x - 4y - 8$

TRY THIS **3.** Write the formula for the perimeter of a rectangle with $l = a - b$ and $w = 2a + b$. Simplify. _p = 6a_

You know that $1a = a$, or $a = 1a$, from the Identity Property for Multiplication. A similar property states that the product of -1 and any real number is the opposite of that number. For example,

$$-1 \cdot 3 = -3 \qquad \text{and} \qquad -1(-4) = 4.$$

▉▉▉ GETTING STARTED

Prerequisite Quiz

Add.

1. $-8 + (-6)$ -14
2. $12 + (-4)$ 8
3. $-7 + 16$ 9
4. $-5 + (-12)$ -17
5. $4 + (-16)$ -12
6. $-9 + 15$ 6
7. $-0.7 + 1.2$ 0.5
8. $-6.4 + (-8.2)$ -14.6

Motivator

Ask students to name the Multiplicative Identity element. What is $1 \cdot a$? $-1 \cdot a$? Have them complete this sentence: Multiplying a real number by -1 gives _?_ .
$1; -1;$ the opposite of the number.

▉▉▉ TEACHING SUGGESTIONS

Lesson Note

This lesson is an extension of Lesson 1.7. Here, combining like terms involves addition and subtraction of real numbers.

The multiplication properties of 1 and -1 enable students to combine like terms when no numerical coefficient appears with a term. Many students avoid errors by inserting the "hidden" coefficient of 1 or -1 before combining like terms.

Additional Example 1

Simplify.

a. $-8x + 5x$ $-3x$
b. $9rs - 15rs$ $-6rs$

Additional Example 2

$x + 2y - 4$

$3x - y + 2$

Write a formula for the perimeter of this rectangle. Simplify the formula.
$p = 8x + 2y - 4$

Highlighting the Standards

Standards 7a, 6b: In Exercises 60 and 61, students interpret three-dimensional objects and then write formulas for their total surface area.

Math Connections

Life Skills: Finding the perimeter of a section of several blocks in your neighborhood can give you an estimate for the distance you want to jog. You can plan your route for neighborhoods which have blocks that are all the same size by using a map with a scale. You can use the dimensions of one block and a little arithmetic to plan the route.

Critical Thinking Questions

Application: Ask students to give an expression for the perimeter of a rectangle having a width represented by $2x + y$. The rectangle is 3 times as long as it is wide.
$16x + 8y$

Synthesis: Use the multiplication properties of 1 and -1 to show that a real number and its additive inverse have a sum of 0.
$1a + (-1)a = [1 + (-1)]a = 0a = 0$

Common Error Analysis

Error: Some students may simplify $2x + 8y - y$ to $2x + 8$.

Tell them to write the original expression as $2x + 8y - 1y$ to avoid this error.

Checkpoint

Simplify.

1. $-7x - 5x$ $-12x$
2. $8a - a$ $7a$
3. $-6x + 3 + x$ $-5x + 3$
4. $8z - 9z - z$ $-2z$
5. $5c + 3 - 4c - 12$ $c - 9$
6. $-8r - 9s + r - 5s$ $-7r - 14s$
7. $6a - 4b - a + 2 - 5b$ $5a - 9b + 2$
8. $-1.2x + 0.3y - 5x + 5.9y + 3$
   $-6.2x + 6.2y + 3$

Property of -1 for Multiplication
For any real number a, $-1a = -a$ and $a(-1) = -a$

EXAMPLE 3 Simplify.
  a. $-x - 7 + 6x + 4$ b. $4a - 3b - 5a + 5b$

Solutions a. $-x - 7 + 6x + 4 = -1x - 7 + 6x + 4$ ⟵ $-x = -1x$
  $ = -1x + 6x - 7 + 4$
  $ = 5x - 3$

  b. $4a - 3b - 5a + 5b = 4a - 5a - 3b + 5b$
  $ = -1a + 2b$
  $ = -a + 2b$ ⟵ $-1a = -a$

EXAMPLE 4 Simplify $6y - 5z - 7y - 4 + 6z$. Evaluate for $y = -8$ and $z = 5$.

Solution $6y - 5z - 7y - 4 + 6z = \underbrace{6y - 7y}_{} + \underbrace{-5z + 6z}_{} - 4$

  $ = \underbrace{-1y}_{} + \underbrace{1z}_{} - 4$
  $ = -y + z - 4$ ⟵ $-1y = -y$ and $1z = z$

  Now substitute -8 for y and 5 for z.
  $-y + z - 4 = -(-8) + 5 - 4$
  $ = 8 + 5 - 4$
  $ = 9$

  Thus, $6y - 5z - 7y - 4 + 6z = 9$, for $y = -8$ and $z = 5$.

TRY THIS 4. Simplify $3x - 2y - x$ for $x = -1$ and $y = -2$. 2

Classroom Exercises

Simplify.

1. $5a + a$ 6a
2. $-7m + 8m$ m
3. $-3m - m$ $-4m$
4. $9t - 8t$ t
5. $-a + 3a$ 2a
6. $-3x + 4x$ x
7. $4t - t$ 3t
8. $-k - 2k$ $-3k$
9. $-7j - j$ $-8j$
10. $-a + a$ 0
11. $k + 7k$ 8k
12. $g + 5g$ 6g
13. $5y - 6y$ $-y$
14. $-x + 6x$ 5x
15. $10b - 11b$ $-b$
16. $3xy - 8 - 9xy$ $-6xy - 8$
17. $-4xy + 3xy - 7$ $-xy - 7$
18. $2ac - 3ac - 5ac$ $-6ac$

Simplify. Then evaluate for $x = 3$, $y = -2$, and $z = -1$.

19. $2x + y + 3x - 2y$ 17
20. $3xy + 2xz - 3xy$ -6
21. $8xyz - 2xyz$ 36

Additional Example 3

Simplify.

a. $-3z + 8 + z - 6$ $-2z + 2$
b. $7c + 2d - 8c - d$ $-c + d$

Additional Example 4

Simplify $2y + 2x + 6 - 3y - x$.
Then evaluate for $x = -3$ and $y = 5$.
$x - y + 6; -2$

Written Exercises

Simplify.

1. $8y - 3y$ 5y
2. $-7x + 9x$ 2x
3. $2a - 8a$ −6a

4. $-6b + 8b$ 2b
5. $4a - 7a$ −3a
6. $-5y + 9y$ 4y

7. $-9x + 2x$ −7x
8. $2c - 2c$ 0
9. $-8r + 3r$ −5r

10. $-5y - 5y$ −10y
11. $-4c - 7c$ −11c
12. $-7x - 3x$ −10x

13. $-7b + 9b$ 2b
14. $-4k - k$ −5k
15. $15c - 4c$ 11c

16. $7x + x$ 8x
17. $-x - 5x$ −6x
18. $4y - y$ 3y

19. $5c - 3.4c$ 1.6c
20. $6x - 7 - 3x$ 3x − 7
21. $4z - 10 - 9z$

22. $-5r - 2 + 2r - 8$
23. $9y - 3 - 7y + 4$ 2y + 1
24. $9z - 8z + 3z$ 4z

25. $-7 + 4b + 3 - b$ 3b − 4
26. $-2a - 7 + a$ −a − 7
27. $9 - b - 7 - 3b$

28. $-3x - 7 - x - 8$
29. $-2b - 3 + b - 4$ −b − 7
30. $8y - 4 - y + 2$ 7y − 2

31. $5x - x + 4x$ 8x
32. $7b - b - b$ 5b
33. $3a - 7a - a$ −5a

34. $5b - 6b - b$ −2b
35. $3b - 4 - b + 8$ 2b + 4
36. $-a - 13 - a + 4$

37. $b - 4 - 2b + 8$ −b + 4
38. $6 - z - 4z + 5$ −5z + 11
39. $8c + 9 - 9c - 4$

40. $-4y - b + 4y + b$ 0
41. $-3k - 2m + 2k + m$
42. $10c - 7 - 11c - 3$

43. $8x - 10 - 7x + 4$
44. $-4a + 5b - a - 4b$
45. $-7r + 8s + 2s - 9s$

46. $-6r - 10s + 8r + 4s - 6$ 2r − 6s − 6
47. $-y + 6 - 5y - 4 + 3y - 2$ −3y

48. $-a + 5.2b + 3a + 3.8b - 10b$ 2a − b
49. $7g - 4t - 6.9g + 3t - g$ −0.9g − t

21. −5z − 10
27. −4b + 2
36. −2a − 9
39. −c + 5

Simplify. Then evaluate for the given values of the variable.

50. $9x + 7y - 2y$ for $x = 2$ and $y = -3$ 3
51. $-2a + 5c - 3c$ for $a = -1$ and $c = 2$ 6
52. $-4x - 8 + 5y - 4x - 2$ for $x = 2$ and $y = -1$ −31
53. $6c - 4b - 8 + 5b - 3c$ for $b = -3$ and $c = 2$ −5
54. $8a - 2.1y + 11a + 1.4y$ for $a = -2$ and $y = 5$ −41.5
55. $-2 - 5x + 2y + 4x - y - 8 + 7x$ for $x = 3$ and $y = -2$ 6
56. $-t + 3 - t - 4n + 8 + 3t - 5 + 6n$ for $n = -\frac{1}{2}$ and $t = 3$ 8

Write a formula for the perimeter of each rectangle below. Simplify.

57.
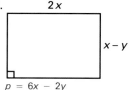
$p = 6x - 2y$

58.

$p = 10y + 2x$

59.
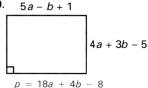
$p = 18a + 4b - 8$

2.9 Like Terms: Real Number Coefficients   **77**

64. $-a + b$, $1\frac{5}{8}$

66. $-2x + y$, $\frac{4}{5}$

67. $-a - b$, $-3\frac{1}{2}$

68. $1.2x - 2.292y - 6$, -11.73

69. $-0.06x - 4.1z + 4.3$, 4.2836

The total surface area of a geometric solid is the sum of the areas of the faces. Write a formula for the total surface area of each figure below. Combine like terms.

60. rectangular solid

$A = 2wl + 2wh + 2lh$

61. square pyramid

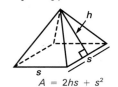

$A = 2hs + s^2$

Simplify. Then evaluate for the given values of the variables.

62. $a - 4b - 2a + 5b$; $a = 7.2$, $b = -8$
$-a + b$, -15.2

63. $-x + 8y + 2x - 6y$; $x = -1.4$, $y = 5$ $x + 2y$; 8.6

64. $5a - 9b - 6a + 10b$; $a = -\frac{3}{4}$; $b = \frac{7}{8}$

65. $7x + 9 - 8x$; $x = 2\frac{3}{5}$ $-x + 9$; $6\frac{2}{5}$

66. $-x - y - x + 2y$; $x = 1\frac{3}{5}$, $y = 4$

67. $6a - b - 7a$; $a = -\frac{1}{2}$, $b = 4$

Simplify. Then evaluate for $x = 0$, $y = 2.5$, and $z = 0.004$.

68. $1.4x - 6 + 0.008y - 2.3y - 0.2x$

69. $4.2 - 0.06x - 8.1z + 0.1 + 4z$

70. $3.8z - 2.4x + 0.4y - 0.1z + 7y$
$-2.4x + 3.7z + 7.4y$; 18.5148

71. $2.4 - 0.08x + 0.3y - 0.1 + 3z$
$-0.08x + 0.3y + 3z + 2.3$; 3.062

Mixed Review

Simplify.   *1.7, 2.4–2.7*

1. $3x + 2(4x + 3)$ $11x + 6$

2. $3a + 2(5a + b) + 11a$ $24a + 2b$

3. $-8 + 6$ -2

4. $-3(-4)^2$ -48

5. $-4 \div 2$ -2

6. $6 \div (7 - 9)$ -3

Evaluate for $x = -5$.   *2.8*

7. $4 + 2x$ -6

8. $17(x + 6)$ 17

9. $17 - 4x$ 37

10. $12(4x - 7)$ -324

11. $(4x)^3$ $-8,000$

12. $4x^3$ -500

13. $\dfrac{2x + 6}{x + 3}$ 2

14. $\dfrac{x + 3}{x + 5}$ Undef

/Brainteaser

A 120-min cassette tape plays for 60 min on each side. As it plays, the tape moves at $1\frac{7}{8}$ in/s. Is the tape longer or shorter than a 100-yd football field?   187.5 yards longer

Enrichment

Nine square blocks of the same size are formed by 4 streets running north and south and 4 intersecting streets running east and west. If each block is $\frac{1}{8}$ of a mile long, what is the shortest distance from the southwest corner of this arrangement to the northeast corner? You must follow the streets. $\frac{3}{4}$ mile

2.10 Removing Parentheses: Negative Factors

Objective   To simplify and evaluate expressions containing negative factors

In this lesson you will simplify expressions such as $-2(3a + 4)$, where multiplication by a negative number is indicated.

EXAMPLE 1   Simplify $-4(5x + 7)$.

Solution
$$-4(5x + 7) = -4 \cdot 5x + (-4)7 \quad \longleftarrow \text{Distributive Property}$$
$$= -20x - 28 \quad \longleftarrow \text{THINK: } -20x + (-28) = -20x - 28$$

EXAMPLE 2   Simplify $-(7x - 2y + 8)$.

Solution
$$-(7x - 2y + 8) = -1(7x - 2y + 8) \quad \longleftarrow -a = -1 \cdot a$$
$$= -1 \cdot 7x + (-1)(-2y) + (-1)8 \quad \longleftarrow \text{Distributive Property}$$
$$= -7x + 2y - 8$$

TRY THIS   Simplify.   **1.** $-3(4x - 5)$ $-12x + 15$ **2.** $-(6x + 3y - 4)$
$-6x - 3y + 4$

To simplify an expression such as $4x - (7 + 5x)$, you need to remove the parentheses first and then combine like terms.

EXAMPLE 3   Simplify $4x - (7 + 5x)$.

Solution
$$4x - (7 + 5x) = 4x - 1(7 + 5x)$$
$$= 4x + (-1)(7 + 5x) \quad \longleftarrow \text{Definition of subtraction}$$
$$= 4x + (-7 - 5x) \quad \longleftarrow \text{Distributive Property}$$
$$= 4x - 7 - 5x$$
$$= 4x - 5x - 7$$
$$= -1x - 7$$
$$= -x - 7$$

TRY THIS   **3.** Simplify $-6x - (8 - 9x)$. $3x - 8$

The solution shown in Example 3 can be shortened by writing
$$4x - 1(7 + 5x) \text{ as } 4x - 7 - 5x.$$
Such a shortcut is used in Example 4 on page 80.

■■■ GETTING STARTED

Prerequisite Quiz

Write without parentheses. Use the Distributive Property.

1. $3(x + 2)$ $3x + 6$
2. $7(y + 5)$ $7y + 35$
3. $4(a + 3)$ $4a + 12$
4. $6(1 + z)$ $6 + 6z$
5. $5(4 + c)$ $20 + 5c$
6. $(x + 2)7$ $7x + 14$
7. $(y + 8)3$ $3y + 24$

Motivator

Remind students that the opposite of a is $-a$. Ask them how they would write the opposite of $(x + y)$. Then ask them to use the Distributive Property to remove the parentheses from $-1(a + b - c)$.
$-(x + y)$; Answers will vary.

■■■ TEACHING SUGGESTIONS

Lesson Note

As you discuss Examples 2–4, be sure to emphasize the use of the Multiplication Property of -1. When the parentheses are removed in an expression such as $-(7x - 2y + 8)$, the result is the opposite of each term inside the parentheses.

Additional Example 1

Simplify $-7(3x + 2)$. $-21x - 14$

Additional Example 2

Simplify $-(5y + 6z - 1)$. $-5y - 6z + 1$

Highlighting the Standards

Standard 3c: The Mixed Review prepares students for the next chapter in which they follow the reasoning in logical arguments.

Math Connections

Life Skills: When recording checks in a checkbook register, each check amount is subtracted to give a new balance. If several checks have been written, you can also add all the check amounts first and then subtract this total from the beginning balance.

Critical Thinking Questions

Application: If the opposite of $a + b + c$ is $-(a + b + c)$, ask students for the reciprocal of $a + b + c$. $\frac{1}{(a + b + c)}$

Application: Ask students to give the product of a real number and the opposite of its reciprocal. -1

Common Error Analysis

Error: Students often make the mistake of rewriting $-(3x - 7)$ as $-3x - 7$.

Remind them that the leading negative sign changes all terms to their opposites when the parentheses are removed.

Checkpoint

Simplify.

1. $4x - 3(7 - 2x)$ $10x - 21$
2. $-(2x - 4) + 8x$ $6x + 4$
3. $-(3 - y) + 4(6 - 2y)$ $-7y + 21$
4. $2(3x + 7) - (4x - 2)$ $2x + 16$
5. $-(7c - 9) + 2(5 - 4c)$ $-15c + 19$

Simplify. Then evaluate for $x = -1$.

6. $-3(2x - 7) + 5(8 - 4x)$ $-26x + 61; 87$

EXAMPLE 4   Simplify $-7(2 + 3x) - (9 - 3x)$. Then evaluate for $x = -1$.

Solution
$$-7(2 + 3x) - (9 - 3x) = -7(2 + 3x) - 1(9 - 3x)$$
$$= -14 - 21x - 9 + 3x$$
$$= -21x + 3x - 14 - 9$$
$$= -18x - 23$$

Now substitute -1 for x in $-18x - 23$ and simplify.
$$-18x - 23 = -18(-1) - 23$$
$$= 18 - 23 = -5$$

Thus, $-7(2 + 3x) - (9 - 3x) = -5$ for $x = -1$.

TRY THIS   **4.** Simplify $-(5 - 3x) - 2x$. Then evaluate for $x = -2$. -7

Classroom Exercises

Simplify.

1. $6(5x - 1)$ $30x - 6$
2. $-2(4y + 6)$ $-8y - 12$
3. $-(7a + 3)$ $-7a - 3$
4. $-(-2 + 8b)$ $2 - 8b$
5. $-(-6 + 8p - q)$ $6 - 8p + q$
6. $-(3z + 4y - 1)$ $-3z - 4y + 1$
7. $x - (5 - x)$ $2x - 5$
8. $-3(a + b) + \frac{1}{2}(4a - 8b)$ $-a - 7b$
9. $6a - 5(8 - 3a)$ $21a - 40$
10. $4x - 6(2x + 8)$ $-8x - 48$

Written Exercises

Simplify.

1. $-7(3 + 4x)$ $-21 - 28x$
2. $-3(-9 + 2y)$ $27 - 6y$
3. $-(-a + 2b + 8)$ $a - 2b - 8$
4. $-(15x + y - 7)$ $-15x - y + 7$
5. $5(2x - 4) - 3x$ $7x - 20$
6. $6(4 - 2y) + 9$ $-12y + 33$
7. $3p - 2(4p + 7) - 5$ $-5p - 19$
8. $8r - 2(2r - 5) + 6$ $4r + 16$
9. $10 + 8(4 - 3d) + 5d$ $-19d + 42$
10. $-4(7 - 4y) + 7y$ $23y - 28$
11. $6x - (x + 5)$ $5x - 5$
12. $8 - (4a - 10)$ $-4a + 18$
13. $-7 - (a - 5)$ $-a - 2$
14. $(7y - 4) + 2y$ $9y - 4$
15. $z - (8 - 6z) - 4$ $7z - 12$
16. $-(b + 8) - 6b + 5$ $-7b - 3$
17. $5 - (7 - a) - 4a$ $-3a - 2$
18. $-6x - (9 - x) - 9$ $-5x - 18$
19. $-4(3x - 5) - 3(2 + 7x)$ $-33x + 14$
20. $-5(3y - 7) - 2(6 + 4y)$ $-23y + 23$
21. $-5(6 + 3y) - 7(2y - 9)$ $-29y + 33$
22. $-3(5x - 4) + 8(-3x + 2)$ $-39x + 28$
23. $-3(5x + 2) - 6(7 + 2x)$ $-27x - 48$
24. $-3(8 - 7z) - 9(4 - 3z)$ $48z - 60$

Additional Example 3

Simplify $-2y - (9 + 8y)$. $-10y - 9$

Additional Example 4

Simplify $-2(3 + 5x) - (3x + 2)$.
Then evaluate for $x = -2$. $-13x - 8; 18$

Simplify. Then evaluate for the given value of the variable. (Exercises 25–38)

25. $-(-7x + 2) - 8x + 1$, for $x = 5$

26. $-8y - (5y + 6) + 12y$, for $y = -1$

27. $5z - 9 - (7z - 6)$, for $z = 0.2$

28. $7a - (8 - a) + 12$, for $a = -0.8$

29. $-(7 + 3y) - 5(4y - 8)$, for $y = -1$

30. $-6(7x - 5) - (4 - x)$, for $x = 4$

31. $-(5z - 7) - (9 - z)$, for $z = -0.2$

32. $-(2a - 4) - (8 - a)$, for $a = 3.4$

33. $6(3z - 7) - \frac{1}{2}(-8z - 6)$, for $z = -2$ $22z - 39; -83$

34. $-\frac{1}{4}(12 - 8d) - 2(-6d + 5)$, for $d = 3$ $14d - 13; 29$

35. $-(-6r + 8) - (-7 - r)$, for $r = -2.1$ $7r - 1; -15.7$

36. $6b - 3(b + 7) - (5 - 4b)$, for $b = 0.04$ $7b - 26; -25.72$

37. $-2(3 - n) - (5n + 4) - 7n$, for $n = -\frac{4}{5}$ $-10n - 10; -2$

38. $-(-c + 5) - (-6 - c)$, for $c = -3\frac{1}{2}$ $2c + 1; -6$

39. On Monday, Jerry jogged for m minutes. On four other days that week he jogged 9 more minutes than he jogged on Monday. On two other days he jogged 7 fewer minutes than he jogged on Monday. Write an algebraic expression in simplest form for the total number of minutes he jogged that week. $7m + 22$

40. Amy bought x headbands and decorated them. It cost her $3 to make each headband. She sold 8 of them at a profit of $4 and the rest at a loss of $1. Write an algebraic expression in simplest form for her total profit or loss. $40 - x$

Simplify.

41. $x - [3(x - 2) - (4 - 5x)]$ $-7x + 10$

42. $-2y - [-2(1 - 7y) - (5 - 3y)]$ $-19y + 7$

43. $-5 - [5 - (3 - x)] + 3x$ $2x - 7$

44. $-x - [-(4 - x) - 2(3 - x)]$ $-4x + 10$

45. $-[6 - 3(2x + 4) - x] + 8$ $7x + 14$

46. $7y - [-(2 + 3y) - 4(6 - y)]$ -26 $6y$

47. $-3[-(2 - y) + 4] - 5(-y + 8)$ $2y - 46$

48. $-[-x + 6(3 - x)] - [-2(x - 3)]$ $9x - 24$

49. $a(x - y) + 2ax$ $3ax - ay$

50. $ax - (ax - ay)$ ay

51. $ay - a(x + y)$ $-ax$

52. $-ax - ay - (ax + ay)$ $-2ax - 2ay$

53. $ce + (c - d)e$ $2ce - de$

54. $2ce - 3de - (de + 2ce)$ $-4de$

Mixed Review

Which property is illustrated? *1.5, 1.6, 2.3, 2.9*

1. $-7x + 7x = 0$ Add Inverse Prop

2. $-t + 8t = -1t + 8t$ Prop of -1 for Mult

3. $4a + 0 = 4a$ Identity Prop for Add

4. $-3 \cdot 4 = 4(-3)$ Comm Prop Mult

5. $-4(m + n) = -4m + (-4n)$ Distr Prop

6. $[3 + (-6)] + (-4) = 3 + [(-6) + (-4)]$ Assoc Prop Add

7. $5m^2 + 3m^4 = 3m^4 + 5m^2$ Comm Prop Add

8. $-4(6c) = (-4 \cdot 6)c$ Assoc Prop Mult

2.10 Removing Parentheses: Negative Factors **81**

Closure

Ask students how $x + y - z$ compares to $-(x + y - z)$. opposite Then ask them to evaluate $2a + b - 3c - (2a + b - 3c)$ for $a = -5$, $b = 3$, and $c = -1$. 0

▰ FOLLOW UP

Guided Practice

Classroom Exercises 1–10
Try This all

Independent Practice

🅐 Ex. 1–24, 🅑 Ex. 25–40, 🅒 Ex. 41–54

Basic: WE 1–29 odd

Average: WE 11–39 odd

Above Average: WE 21–53 odd

Additional Answers

Written Exercises

25. $-x - 1; -6$
26. $-y - 6; -5$
27. $-2z - 3; -3.4$
28. $8a + 4; -2.4$
29. $33 - 23y; 56$
30. $-41x + 26; -138$
31. $-4z - 2; -1.2$
32. $-a - 4; -7.4$

Enrichment

Students have been removing parentheses by using the Distributive Property. Challenge them to reverse the process and write each of the following expressions in the form $a(x + b)$.

1. $3x + 21$ $3(x + 7)$

2. $5x - 45$ $5(x - 9)$

3. $3x - 2$ $3(x - \frac{2}{3})$

4. $4 - 5x$ $-5(x - \frac{4}{5})$

5. $\frac{2}{3}x - \frac{5}{4}$ $\frac{2}{3}(x - \frac{15}{8})$

6. $0.25x + 1$ $0.25(x + 4)$

81

Chapter 2 Review

Key Terms

absolute value   (p. 46)
additive inverse   (p. 51)
Additive Inverse Property   (p. 51)
arithmetic mean   (p. 69)
average   (p. 69)
coordinate   (p. 41)
counting number   (p. 41)
graph   (p. 41)
identity element for addition   (p. 50)
Identity Property for Addition   (p. 51)
integer   (p. 41)

multiplicative inverse   (p. 68)
Multiplicative Inverse Property   (p. 68)
opposite   (p. 45)
natural number   (p. 41)
Property of -1 for Multiplication   (p. 76)
Property of Zero for Multiplication
    (p. 62)
real number   (p. 42)
reciprocal   (p. 68)
whole number   (p. 41)

Key Ideas and Review Exercises

2.1   On a horizontal number line, the numbers increase from left to right.

Use $<$ or $>$ to make each sentence true.

1. $-5 \underline{\ <\ } 6$ **2.** $0 \underline{\ >\ } -2$ **3.** $8 \underline{\ >\ } -8$ **4.** $-6 \underline{\ <\ } -4$

5. $0.5 \underline{\ >\ } 0.4$ **6.** $-3\frac{1}{2} \underline{\ <\ } -3$ **7.** $-2.1 \underline{\ <\ } 2$ **8.** $-\frac{1}{2} \underline{\ <\ } -\frac{1}{4}$

2.2   Two numbers are opposites if they are the same distance from 0 on a number line and they are on opposite sides of 0. The opposite of 0 is 0. The symbol $-a$ means *the opposite of a.*

The absolute value of a real number x is the distance between x and 0 on a number line. The symbol $|x|$ means *the absolute value of x.*

Simplify.

9. $-(12 - 7)$ -5 **10.** $-(-4)$ 4 **11.** $|5|$ 5 **12.** $|-4|$ 4

Compare. Use $>$, $<$, or $=$.

13. $0 \underline{\ <\ } |6|$ **14.** $|3| \underline{\ =\ } |-3|$ **15.** $-(-4) \underline{\ >\ } -4$ **16.** $-\left|\frac{1}{2}\right| \underline{\ <\ } \left|\frac{1}{3}\right|$

2.3,   To add two real numbers, follow the rules in Lesson 2.4.
2.4

Add.

17. $-8 + 0$ -8 **18.** $7 + (-7)$ 0 **19.** $15 + (-8)$ 7 **20.** $-9 + 7$ -2

21. $-4 + (-11)$ -15 **22.** $8 + 9$ 17 **23.** $6.2 + (-9.1)$ -2.9 **24.** $-2\frac{3}{4} + \left(-1\frac{1}{2}\right)$ $-4\frac{1}{4}$

2.5   To subtract a number, add its opposite.

Subtract.

25. $6 - 9$ -3 **26.** $-5 - 8$ -13 **27.** $4 - (-7)$ 11   **28.** $-6.3 - (-9.8)$
3.5

Simplify.

29. $-28 + 18 - (-4)$ -6 **30.** $-6 + 7 - 9 - (-5)$ -3

31. Give two numerical examples to show that subtraction is not a commutative operation. Use at least one negative number in each example. Answers will vary.

2.6,   The product (quotient) of two numbers with like signs is positive.
2.7   The product (quotient) of two numbers with unlike signs is negative.

Multiply.

32. $7 \cdot 9$ 63   **33.** $-8 \cdot 5$ -40 **34.** $-6 \cdot 0$ 0   **35.** $(-4)(-7)$ 28
36. $4(-8)$ -32 **37.** $-1(-5.2)$ 5.2   **38.** $16 \cdot 5^3$ 2,000   **39.** $4(-8)(-1)$ 32

Divide.

40. $\frac{63}{-7}$ -9 **41.** $\frac{0}{-4}$ 0   **42.** $\frac{9}{0}$ Undef   **43.** $\frac{-49}{-7}$ 7
44. $-16 \div 2$ -8 **45.** $5.2 \div (-1.3)$ -4 **46.** $-\frac{3}{8} \div (-6)$ $\frac{1}{16}$ **47.** $-\frac{5}{6} \div \frac{2}{3}$ $-\frac{5}{4}$

2.8   To simplify expressions involving more than one operation, follow the rules for the order of operations.

Simplify.

48. $6 - 4 \cdot 8$ -26 **49.** $12 - (-6)^2$ -24

Evaluate.

50. $7 - 5x$, for $x = 2$ -3 **51.** $-8 - 9y$, for $y = -3$ 19
52. $2x - 3y$, for $x = -6$ and $y = -3$ -3 **53.** $-5c + d^2$, for $c = 3$ and $d = -5$ 10

2.9,   To simplify the algebraic expression $4x - 3(x - 4)$, apply the Distributive
2.10   Property. Then combine like terms.

Simplify.

54. $-2a + 9a$ 7a   **55.** $-7y - y$ $-8y$ **56.** $x - (5 + x)$ -5 **57.** $-(x - 2)$
$$$-x + 2$
58. $7a - 2(3 - a) + 6$ 9a   **59.** $-4(7 + 2y) - (y + 9)$ $-37 - 9y$

Chapter 2 Test

A-Level Exercises: 1–6, 9–14, 16, 19–23, 25–28, 31–37
B-Level Exercises: 7–8, 15, 17–18, 24, 29–30
C-Level Exercises: 39–40

Simplify.

1. $-(-6)$ 6

2. $-(7 - 5)$ -2

3. $|-8|$ 8

4. $-|-9|$ -9

Compare. Use > , < , or =.

5. $-7 \underline{\le} -6$

6. $0.1 \underline{\ge} -0.2$

7. $|-8| \underline{=} |8|$

8. $|\frac{1}{4}| \underline{\ge} -|-\frac{1}{2}|$

Add.

9. $-9 + 5$ -4

10. $16 + (-16)$ 0

11. $-8 + (-9)$ -17

12. $8.2 + (-6.5)$ 1.7

Subtract.

13. $8 - (-5)$ 13

14. $-4 - 7$ -11

15. $|-15| - |-1|$ 14

16. $5.6 - 9.2$ -3.6

Evaluate for $x = 7$, $y = -7$, and $z = -12$.

17. $x - y + z$ 2

18. $|x| + |z|$ 19

19. $2z$ -24

20. $-x - 2y$ 7

Multiply.

21. $-8(-9)$ 72

22. $\frac{1}{6}(-30)$ -5

23. $-8(12)$ -96

24. $4^2(-3)^2$ 144

Divide, if possible.

25. $\frac{-48}{-8}$ 6

26. $\frac{0}{-2}$ 0

27. $18 \div 0$ Undef

28. $-\frac{3}{5} \div \frac{4}{5}$ $-\frac{3}{4}$

29. Evaluate $2a^2 + b^3$, for $a = -4$ and $b = -3$. 5

30. Evaluate $\dfrac{x - 25y}{-5y}$, for $x = -6$ and $y = \frac{3}{5}$. 7

Simplify.

31. $-8 + 5 - 9 - (-2)$ -10

32. $-8 + 6(-5)$ -38

33. $-8y + 8y$ 0

34. $-6c + 5c + d$ $-c + d$

35. $-4a + 3b - 6a - 2b$ $-10a + b$

36. $-3c - 2(4 - 5c)$ $7c - 8$

37. $-9y - (5 - y) + 8$ $-8y + 3$

38. $4.2x - 3.7y + 8.2 - 9.5x + 2.9y$

39. $(-9 + 4)(8 - 10 \cdot 2 - 3)$ 75

40. $5y - [-(3 + 4y) - 2(7 - y)] - 4$

84 Chapter 2 Test

Test-Taking Strategy

When a test is given, there is often a time limit for completing it. It is important not to spend too much time on any one problem. If you find this happening, leave the problem and continue with the rest of the test. If there is time remaining, go back to the problem later on.

Choose the best answer to each question or problem.

1. $|6 - 10| = $ __C__.
  (A) 16   (B) -16 (C) 4
  (D) -4 (E) None of these

2. Find the missing number in the sequence: $-10, -7, -4, -1, \underline{\quad}$. D
  (A) -3 (B) 1 (C) 0 (D) 2
  (E) 3

3. A relationship exists between Figure I and Figure II.

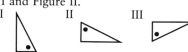

A similar relationship exists between Figure III and which of the following figures? A

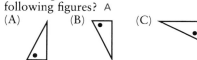

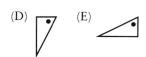

4. At 9:00 P.M. the temperature reached a low of $-15°F$. Then it rose $4°$ each hour. It reached $13°F$ at __C__.
  (A) 2:00 A.M.   (B) 3:00 A.M.
  (C) 4:00 A.M.   (D) 5:00 A.M.
  (E) 6:00 A.M.

5. Which of the following is greater than $\frac{1}{2}$? E
  (A) $-\frac{1}{2}$ (B) $\left(\frac{1}{2}\right)^2$ (C) $\left(\frac{1}{2}\right)^3$
  (D) 0.5 (E) None of these

6. If a water tank is leaking at a rate of 6 liters per 24 h, how much will it leak in 2 h? D
  (A) 1 liter   (B) 2 liters
  (C) 0.25 liter (D) 0.5 liter
  (E) 0.75 liter

7. For which value of x does y have the greatest value for $y = \dfrac{-16}{x}$? D
  (A) $x = 2$ (B) $x = -2$
  (C) $x = \frac{1}{2}$ (D) $x = -\frac{1}{2}$
  (E) $x = \frac{1}{20}$

8. How much more than $-5\frac{1}{2}$ is $3\frac{3}{4}$? A
  (A) $9\frac{1}{4}$ (B) $-9\frac{1}{4}$ (C) $1\frac{1}{4}$
  (D) $-1\frac{1}{4}$ (E) $3\frac{3}{4}$

9. What is the maximum total weight of a dozen apples if four of them weigh 90 to 110 g each and the rest weigh 115 to 120 g each? B
  (A) 1,500 g   (B) 1,400 g
  (C) 230 g   (D) 120 g
  (E) 3

10. Find the value of $0^5 \cdot 1^4 + 1^3$. C
  (A) -1 (B) 0 (C) 1 (D) 2
  (E) None of these

Cumulative Review (Chapters 1–2)

Evaluate.

1. $6x$, for $x = 9$ 54   *1.1*

2. $7 - y$, for $y = 3$ 4

3. $\frac{a}{b}$, for $a = 72$ and $b = 8$ 9

4. mn, for $m = 2.1$ and $n = 3$ 6.3

5. $r + s$, for $r = 5.6$ and $s = 9.8$ 15.4

6. $x(y + 1) - (x - 2)$, for $x =$ *1.2*
4 and $y = 7$ 30

7. $4x^2 - 3x$, for $x = 2$ 10   *1.3*

8. $(2y)^2 + y$, for $y = 12$ 588

9. $(a^2 - 2b)^2$, for $a = 4$ and $b = 3$ 100

Use < or > to make each statement true.

10. -5 ___ -4 <   *2.1*

11. 3 ___ -2 >

Simplify, if possible.

12. $|6|$ 6   *2.2*

13. $|-9|$ 9

14. $-|-12|$ -12

15. $|-(-7)|$ 7

16. $-9 + (-8)$ -17 *2.4*

17. $-16 + 7$ -9

18. $19 + (-19)$ 0

19. $-4.2 + 0$ -4.2

20. $1.3 + (-0.6)$ 0.7

21. $16 - 9$ 7

22. $-5 - 9$ -14

23. $8 - (-4)$ 12

24. $-9 - (-7)$ -2

25. $\frac{1}{6} - \frac{2}{3}$ $-\frac{1}{2}$

26. $-14 + 2 + 7 - (-8)$ 3

27. $-1(-15)$ 15   *2.6*

28. $2 \cdot (-9)$ -18

29. $-9 \cdot 7$ -63

30. $0(-10)$ 0

31. $(2)(-9)(-3)$ 54

32. $(3^2)(-2)^3(-1)$ 72

33. $\frac{-16}{8}$ -2 *2.7*

34. $\frac{0}{-19}$ 0

35. $42 \div 0$ Undef

36. $-18 \div (-3)$ 6

37. $-54 \div [24 \div (-4)]$ 9

Evaluate for $x = -2$, $y = 4$, and $z = -1$.

38. $x + (-y)$ -6 *2.4*

39. $|x - 3z|$ 1   *2.8*

40. $|4z - 7y|$ 32

Simplify.

41. $6x - 11x$ $-5x$ *2.9*

42. $3x - 4y - 7x$ $-4x - 4y$

43. $-a + 3 - 7a - 5 + 4a - 2$ $-4a - 4$

44. $-(4x - 7y) + 2x$ $-2x + 7y$ *2.10*

45. $-3a - 2(7b + a) - 4b$ $-5a - 18b$

46. $3(2x - y) - 4(y - x)$ $10x - 7y$

Solve each problem.

47. Use the formula $d = 5t^2$. Find   *1.3*
the distance d in meters that an
object falls in a time t of 5 sec.

48. Use the formula $d = 0.04s^2$ to
find d for $s = 60$. 144

49. Use the formula $p = 2l + 2w$ *1.4*
to find p for $l = 7$ ft, $w = 5$ ft.

50. Use the formula $A = \frac{1}{2}bh$ to
find A for $b = 7$ yd, $h = 8$ yd.

51. Use the formula $p = 4s$ to find
p for $s = 3.2$ cm. 12.8 cm

52. Use the formula $A = s^2$ to find
A for $s = 4.1$ m. 16.81 m^2

53. Use the formula $A = lw$ to find A for $l = 9$ in. and $w = 6$ in. 54 in²

54. Use the formula $V = lwh$ to find V for $l = 10$ cm, $w = 6$ cm, and $h = 4$ cm. 240 cm³

55. Use $A = \frac{1}{2}(b + c)h$ to find A for $b = 5$ in., $c = 8$ in., and $h = 12$ in. 78 in²

For Exercises 56 and 57, round your answers to the nearest whole number.

56. Use the formula $C = 2\pi r$ to find C for $r = 3.6$ in. Use 3.14 for π. 23 in

57. Use the formula $A = \pi r^2$ to find A for $r = 6$ cm. Use 3.14 for π. 113 cm²

Choose the one best answer to each question or problem.

58. $7x^2y$ and $6xy^2$ are __?__. C   1.7
(A) constants
(B) like terms
(C) unlike terms
(D) numerical coefficients
(E) none of these

59. The sum of a positive number and a negative number can be __?__. D   2.4
(A) positive   (B) negative
(C) zero   (D) all of these
(E) none of these

60. Subtracting a number is the same as adding its __?__. A   2.5
(A) opposite
(B) absolute values
(C) difference
(D) multiplicative inverse
(E) none of these

61. The sum of two opposites is __?__. C   2.4
(A) positive   (B) negative
(C) zero   (D) the additive inverse
(E) none of these

62. Dividing by a nonzero number is the same as multiplying by its __?__. A   2.7
(A) reciprocal
(B) opposite
(C) absolute value
(D) additive inverse
(E) none of these

63. During one year, the highest temperature recorded in Utah was 102°F. The lowest temperature was -15°F. What was the difference between the two temperatures? 117°F   2.5

64. One day Len withdrew $512.32 from his checking account. The next day he deposited $326.45. What real number describes the net change in his account? -185.87 2.4

65. A submarine was at a depth of 245 m below sea level. It rose 75 m. Use a real number to give its new position with respect to sea level. -170 m   2.6

66. If an elephant lost $1\frac{1}{2}$ lb per week for 8 weeks and then gained $\frac{3}{4}$ lb per week for 5 weeks, what is its net gain or loss at the end of the 13 weeks? $-8\frac{1}{4}$ lb   2.6

67. Alvin set up a monthly budget to help him manage his finances. For six months he determined whether he spent more or less than his budget. 2.4
   June: $18.20 over
   July: $1.50 under
   Aug.: $2.45 over
   Sept.: $15.30 over
   Oct.: $20.65 under
   Nov.: $0.50 over
By how much was he over or under his budget at the end of the six months? $14.30 over

3 SOLVING EQUATIONS

OVERVIEW

In this chapter, students solve equations by applying the Addition, Subtraction, Multiplication, and Division Properties of Equality. They solve equations with the variable on one side, with the variable on both sides, and with parentheses. Students apply equation-solving skills to solving word problems and to using formulas.

The last lesson in the chapter provides a summary of all the properties discussed to this point. Students then use these properties as reasons that justify statements in algebraic proofs.

OBJECTIVES

- To solve equations with one variable
- To solve equations containing parentheses
- To solve word problems by first translating them into equations
- To write missing reasons in proofs

PROBLEM SOLVING

This chapter introduces a fundamental problem-solving strategy for algebra— *Writing an Equation.* This same strategy, as well as the strategies of *Using a Formula* and *Using Logical Reasoning,* is applied in the *Applications* on pages 93, 109, and 110–113. The problem-solving strategy lesson on page 101 focuses on the second step of the four-step Polya model, *Developing a Plan.* Provision for maintaining problem-solving skills is included on page 119 of this chapter and at the end of every third chapter throughout the textbook.

READING AND WRITING MATH

Throughout this chapter, students read a problem stated in words and then rewrite it as an algebraic statement. Exercise 36 on page 97, Exercise 56 on page 100, as well as Exercise 13 in the Chapter Review, ask students to write explanations of how they reach a conclusion. The Classroom Exercises on page 99 ask students to verbalize a plan before they solve the equation.

TECHNOLOGY

Calculator: Students may find a calculator helpful in solving Written Exercises 35–37 on page 93, Exercises 45–47 on page 100, Exercises 1–20 on page 112, and in the Applications on pages 109 and 114.

SPECIAL FEATURES

Mixed Review pp. 93, 97, 105, 108, 113, 118
Application: Mathematics in Typing p. 93
Midchapter Review p. 100
Problem Solving Strategies: Developing a Plan p. 101
Focus on Reading p. 104
Application: Thunder and Lightning p. 109
Application: The Shock Wave p. 114
Summary of Properties p. 115
Mixed Problem Solving p. 119
Key Terms p. 120
Key Ideas and Review Exercises pp. 120–121
Chapter 3 Test p. 122
College Prep Test p. 123

PLANNING GUIDE

Lesson	Basic	Average	Above Average	Resources
3.1 pp. 92–93	CE 1–21 WE 1–29 odd, Application	CE 1–21 WE 9–37 odd, Application	CE 1–21 WE 17–47 odd, Application	Reteaching 19 Practice 19
3.2 pp. 96–97	CE 1–16 WE 1–25 odd	CE 1–16 WE 9–33 odd	CE 1–16 WE 13–29, 37–45 odd	Reteaching 20 Practice 20
3.3 pp. 99–101	CE 1–12 WE 1–21, 31–35, 39–43 odd, Midchapter Review Problem Solving Strategies	CE 1–12 WE 13–29, 33–47 odd, Midchapter Review Problem Solving Strategies	CE 1–12 WE 23–29, 33–35 odd, 56 Midchapter Review Problem Solving Strategies	Reteaching 21 Practice 21
3.4 pp. 104–105	FR all CE 1–12, WE 1–17, 25–31, 35–39 odd	FR all CE 1–12, WE 9–19 25–41 odd	FR all CE 1–12, WE 17–23 29–49	Reteaching 22 Practice 22
3.5 pp. 107–109	CE 1–15, WE 1–13, 19–23 odd, Application	CE 1–15, WE 7–25 odd, Application	CE 1–15 WE 13–31 odd, Application	Reteaching 23 Practice 23
3.6 pp. 111–114	CE 1–8 WE 1–9 odd, 12, 13, Application	CE 1–8 WE 5–11 odd, 12, 13, 15, Application	CE 1–8 WE 9, 11, 12, 13–19 odd, Application	Reteaching 24 Practice 24
3.7 pp. 117–119	CE 1–9 WE 1–4 all, Mixed Problem Solving	CE 1–9 WE 2–4 all, 5–11 odd. Mixed Problem Solving	CE 1–9 WE 2–4 all, 5–7 odd, 13 Mixed Problem Solving	Reteaching 25 Practice 25
Chapter 3 Review pp. 120–121	all	all	all	
Chapter 3 Test p.122	all	all	all	
College Prep Test p. 123	all	all	all	

CE = Classroom Exercises   WE = Written Exercises   FR = Focus on Reading

NOTE: For each level, all students should be assigned all Try This and all Mixed Review Exercises.

▰▰ INVESTIGATION

Project: In this investigation, students explore a physical model of a balanced equation. They experiment with the difference between the *weight* of a coin and its value, and between the *number* of coins and their total value.

Materials: Each student will need a copy of Project Worksheet 3. Each group of students needs a balance scale, some coins (at least 10 pennies, 2 nickels, and 2 dimes), and 6 jumbo paper clips. Students can easily make a simple balance scale using a ruler, stick or dowel about 12 inches long, a rubber band, two smaller disposable drinking cups, a textbook, transparent tape, and four jumbo paper clips as shown below.

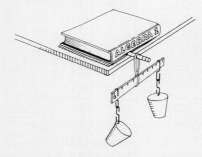

Make one scale (see the figure above) as a model. Then ask students to construct their own scales.

Students follow the steps given in Project Worksheet 3.

When students have completed the worksheet, encourage them to explore further. Have the class discuss their conclusions about what keeps the two sides in balance, and why one nickel does not balance five pennies. Ask a student volunteer to write a sentence or two on the chalkboard that summarizes the findings of the class. Ask other students to amend the statement until everyone agrees that it is accurate and complete.

For almost 20 years, Dian Fossey and her students collected data on the daily lives of mountain gorillas inhabiting forest regions in Africa. The book she wrote reporting her findings, *Gorillas in the Mist,* was made into a movie.

More About the Photo

The data collected by Dian Fossey and her colleagues includes recordings of gorilla vocalizations which were then displayed graphically as measurements of the tonality, rhythm, amplitude, quality, and frequencies of the sound. The observers also plotted the birth and death rates in various gorilla groups. Data from autopsies on gorillas who died included measures of the lengths and diameters of bones, the size and weight of various organs, and the number and size of parasites. All of this data continues to help those working to preserve this endangered species.

3.1 Solving Equations by Adding or Subtracting

Objective

To solve equations by using the Addition or Subtraction Property of Equality

The scales at the right are balanced. If two objects of equal weight are placed on each side of the scale, the scale will remain balanced. This shows the following.

If $x = y$, then $x + 2 = y + 2$.

Suppose Bill and Jean are the same age. Then in 4 years they will still be the same age. Let b = Bill's age now in years and j = Jean's age now in years. This situation can be expressed as follows.

If $b = j$, then $b + 4 = j + 4$.

This reasoning suggests one of the equality properties. These properties can be used to find solutions of equations.

Consider the equation $x = 6$. Its only solution is 6; its solution set is $\{6\}$. Suppose that you add the same number to each side of $x = 6$.

Add 7 to each side.
$$x = 6$$
$$x + 7 = 6 + 7$$
$$x + 7 = 13$$

The resulting equation, $x + 7 = 13$, also has $\{6\}$ as its solution set. Equations with the same solution set, such as $x = 6$ and $x + 7 = 13$, are called **equivalent equations**.

Addition Property of Equality
For all real numbers a, b, and c, if $a = b$, then $a + c = b + c$. This means that adding the same number to each side of an equation produces an equivalent equation.

Recall that the *replacement set* for a variable is the set of numbers that can replace the variable. For the rest of the equations in this book, *assume that the replacement set is the set of all real numbers*, unless it is otherwise stated.

Teaching Resources

Manipulative Worksheet 3
Project Worksheet 3
Quick Quizzes 19
Reteaching and Practice Worksheets 19
Transparency 6
Teaching Aid 1

▰▰▰ GETTING STARTED

Prerequisite Quiz

Name the opposite of each number.

1. -9 9
2. $1\frac{3}{5}$ $-1\frac{3}{5}$

Add or subtract as indicated.

3. $-6 + 6$ 0
4. $-1\frac{1}{3} + 1\frac{1}{3}$ 0
5. $-7.9 - 7.9$ -15.8
6. $6.3 - (-6.3)$ 12.6

Motivator

Remind students that they have already found the solution set of an equation by substituting the members of a finite replacement set for the variable (see Lesson 1.8). As a review, have them find the solution set of $-12 + x = 8$ for the replacement set $\{-1, 2, 4\}$.

Highlighting the Standards

Standard 5a: In this lesson, students represent situations involving variables. They learn how to relate words and algebraic representations and they practice this in the Exercises.

■ TEACHING SUGGESTIONS

Lesson Note

A balance scale is a helpful model for demonstrating the Addition Property of Equality. When the same amount is added to both sides of a scale, the balance is maintained.

Point out to students that the solution for Example 1 is 21. The solution set is {21}. After solving the equation in Example 4, ask students whether the answer is reasonable. For example, $5,300 would not be a reasonable selling price for a tape recorder.

Math Connections

Life Skills: Translating verbal descriptions into algebraic equations is a problem-solving strategy that is used in everyday life. For example, the amount of profit that you can make by selling a number of items for more than they cost you, and the number of miles that you have left to travel before reaching your destination, can be found by solving algebraic equations.

To **solve** an equation means to find its solution.

EXAMPLE 1 Solve $x - 12 = 9$. Check the solution.

Plan Notice that x is not alone on the left side of $x - 12 = 9$, since 12 is subtracted from x. To get x alone on the left, add 12 to each side.

Solution
$$x - 12 = 9$$
Add 12 to each side. $\quad x - 12 + 12 = 9 + 12$
Simplify. $\qquad\qquad x + 0 = 21 \quad \longleftarrow -12 + 12 = 0$
$$x = 21$$

All of these equations are equivalent. They all have the same solution.

Check Check 21 in the original equation.
$$x - 12 = 9$$
Substitute 21 for x. $\quad 21 - 12 \overset{?}{=} 9$
$$9 = 9 \text{ True}$$

Thus, the solution is 21.

TRY THIS 1. Solve $x - 6 = -3.5$. Check the solution. 2.5

Since subtracting a number is the same as adding its opposite, the following property follows from the Addition Property of Equality.

> **Subtraction Property of Equality**
>
> For all real numbers, a, b, and c, if $a = b$, then $a - c = b - c$.
>
> In other words, subtracting the same number from each side of an equation produces an equivalent equation.

This property is used in solving the equation in the next example.

EXAMPLE 2 Solve $-17 = y + 5$. Check.

Plan The opposite of adding 5 is subtracting 5.
To get y alone on the right, subtract 5 from each side.

Solution
$$-17 = y + 5$$
Subtract 5 from each side. $\quad -17 - 5 = y + 5 - 5$
Simplify. $\qquad\qquad\qquad -22 = y$

90 Chapter 3 Solving Equations

Additional Example 1

Solve $19 = a - 7$. Check. 26

Additional Example 2

Solve $-6\frac{3}{5} = x + 1\frac{1}{5}$. Check. $-7\frac{4}{5}$

Check

Substitute -22 for y.

$$-17 = y + 5$$
$$-17 \stackrel{?}{=} -22 + 5$$
$$-17 = -17 \text{ True}$$

Thus, the solution is -22.

TRY THIS   **2.** Solve $y + 2\frac{1}{2} = -6$. Check the solution. $-8\frac{1}{2}$

Many word problems can be solved by using equations such as the ones illustrated in this lesson. Recall how you translate English phrases into algebraic expressions.

4 more than a number	16 less than a number	A number decreased by 16
4 more than n	16 less than n	n decreased by 16
$n + 4$	$n - 16$	$n - 16$

EXAMPLE 3   The $37-selling price of a tape recorder is $16 less than the original price. Find the original price.

Solution   You are asked to find the original price.
Let $x =$ the original price in dollars.

Thirty-seven is 16 less than the original price.
37 is 16 less than x.

$$37 = x - 16$$
$$37 + 16 = x - 16 + 16$$
$$53 = x$$

Check the answer, $53, in the original problem.

Check   Is it true that $37 is $16 less than $53? Yes, because $53 - 16 = 37$.

Thus, the original price is $53.

TRY THIS   **3.** A number has been decreased by 12. The result is -5. Find the original number. Check the solution. 7

Notice that in Example 3, the answer was checked in the *original problem*, not in the equation. This is important since you could have made an error in writing an equation. For example, you may have translated

37 is 16 less than x incorrectly as $37 - 16 = x$.

The solution to the equation $37 - 16 = x$ is 21, a solution that checks in the equation but *not* in the word problem itself.

Thirty-seven dollars is *not* $16 less than $21.

Critical Thinking Questions

Analysis: Ask students to compare the solutions for $x + 3 = 7$ and $3 - x = 7$. (4 and -4 respectively). Then have students write an algebraic equation involving addition whose solution set is -2, and an equation with the same numbers but involving subtraction and having a solution set of 2. Answers may vary.   Sample answers are $x + 6 = 4$ and $6 - x = 4$.

Common Error Analysis

Error: After students have checked the solution of an equation, they often state that the solution is the number named on each side of the equation. For example, after completing the check in Example 2, they may conclude that the solution is -17.

Remind students to look back in Example 2 to the solution they found originally, -22.

Additional Example 3

On a canoe trip, the campers paddled 8.6 miles on the first day. That was 2.7 mi farther than they paddled the second day. How many miles did they paddle the second day?   5.9 mi

Solve and check.

1. $x - 9 = -31$ -22
2. $7 + y = -23$ -30
3. $-6 = 14 + z$ -20
4. $0.9 = -7.4 + y$ 8.3

Translate into an equation and solve.

5. Eighteen less than a number is -46. Find the number. $x - 18 = -46$; -28

Closure

Ask students to explain which property should be used to solve an equation such as $-3 + x = 14$ and why. Do the same for $19 = y + 8$. Add Prop of Eq, Add 3 to each side to get x alone; Subt. Prop of Eq, Add (-8) to each side to get y alone.

Classroom Exercises

What number would you add to or subtract from each side of the equation to solve it?

Add 8.

1. $x - 12 = -32$ Add 12.
2. $46 = y + 7$ Subtract 7.
3. $-3 = x - 8$
4. $p + 1 = 46$ Subtract 1.
5. $36 + y = 42$ Subtract 36.
6. $27 = y - 9$
7. $4.2 + x = 5.7$ Subtract 4.2.
8. $y - \frac{2}{3} = -4$ Add $\frac{2}{3}$.
9. $-2 = z + 1\frac{1}{4}$

10–18. Solve and check the equations of Classroom Exercises 1–9. Subtract $1\frac{1}{4}$.
10. -20 **11.** 39 **12.** 5 **13.** 45 **14.** 6 **15.** 36 **16.** 1.5 **17.** $-3\frac{1}{3}$ **18.** $-3\frac{1}{4}$ **6.** Add 9.

For Exercises 19–21, let x represent the number you are asked to find. Write an equation to find the number. Then find the number.

19. Twenty is 5 more than a number. Find the number. $20 = 5 + x$; 15
20. Seven less than a number is 23. What is the number? $x - 7 = 23$; 30
21. Twenty-seven increased by a number is 33. Find the number. $27 + x = 33$; 6

Written Exercises

Solve and check.

1. $x - 23 = 30$ 53
2. $y + 7 = 63$ 56
3. $-18 = x - 35$ 17
4. $a + 8 = -4$ -12
5. $14 = x + 9$ 5
6. $c - 9 = 3$ 12
7. $6 = 15 + x$ -9
8. $b - 10 = 8$ 18
9. $7 + x = 5$ -2
10. $t + 8 = -3$ -11
11. $9 = 10 + x$ -1
12. $4 + n = 4$ 0
13. $x - 9 = -9$ 0
14. $-12 + y = 18$ 30
15. $5 = 8 + a$ -3
16. $n - 6 = 13$ 19
17. $4 = x + 8$ -4
18. $x - 5 = 9$ 14
19. $5 + y = 11$ 6
20. $9 = c + 15$ -6
21. $-4 + x = -4$ 0

Translate each problem into an equation and solve. Other equations are possible.

22. The $44 selling price of a sweater is the cost increased by $18. Find the cost. $44 = c + 18$; $26
23. Nine more than a number is 42. Find the number. $n + 9 = 42$; 33
24. Eight less than a number is 56. Find the number. $n - 8 = 56$; 64
25. The price of a radio decreased by $35 is the discount price of $85. Find the original price. $p - 35 = 85$; $120

Solve and check.

26. $0.6 + x = 1.4$ 0.8
27. $y - 9.2 = 4.7$ 13.9
28. $-5.2 = z - 0.9$ -4.3

92 Chapter 3 Solving Equations

Enrichment

Have students solve this puzzle.
Four unequal weights have a total weight of 22 oz. A scale balances when two weights are placed on the left pan and two are placed on the right pan. The smaller weight on the right pan is 2 oz heavier than the smaller weight on the left pan. If the smaller weight on the left pan is moved to the right pan, 6 oz will have to be added to the left pan to balance the scale.

Find each of the four weights, and how they were placed originally on the scale. 8 oz and 3 oz on the left pan; 6 oz and 5 oz on the right pan

29. $x + 3.4 = -2.6$ -6.0 **30.** $-4.3 + y = -6.7$ -2.4 **31.** $z - 0.5 = 4.8$ 5.3

32. $x + \frac{1}{2} = 5$ $4\frac{1}{2}$ **33.** $y - \frac{2}{3} = 18$ $18\frac{2}{3}$ **34.** $z + 5\frac{1}{3} = -10$ $-15\frac{1}{3}$

Solve.

35. $x - 246.8 = 656.32$ 903.12 **36.** $254.78 + x = -234.5$ -489.28 **37.** $115.3 + x = 889.36$ 774.06

38. Explain in writing how the Subtraction Property of Equality follows from the Addition Property of Equality. To subtract a number is to add its opposite.

39. To solve the equation $5 = x + 7$, explain why you would subtract 7 from each side rather than subtract 5 from each side.
The new equation, $0 = x + 2$, does not have x alone on one side.

Find the solution set.

40. $x + |-8| = 17$ {9}   **41.** $|x| - 3 = -1$ {2, -2}   **42.** $16 = |x| + 7$ {9, -9}

43. $x + 8 = 8 + x$ **44.** $x + 6 = x - 9$ ∅   **45.** $x + 3 = -7 + x$ ∅

46. $|7 - [18 \div (-6)]| = x + 3$ {7}   **47.** $x + 4 = |-16|$ {12}

Mixed Review

Simplify. *2.4–2.7, 2.9, 2.10*

1. $-\frac{2}{3} \cdot \frac{3}{5}$ $-\frac{2}{5}$ **2.** $-1.6 + 5.4$ 3.8 **3.** $6t - 11t$ $-5t$

4. $-3 - (7 + 5)$ -15 **5.** $-(x + 8) - 9x$ **6.** $35 \div (-7)$ -5
$-10x - 8$

Application: Mathematics in Typing

A secretary applying for a job may have to take a typing test to determine typing speed in words per minute. The formula below is often used to determine typing speed. Note that a deduction is made for errors, since accuracy in typing is so important.

$$s = \frac{w - 10e}{m}$$

where s = speed in words per minute
w = number of words typed
e = number of errors
m = number of minutes of typing

1. On an 8-minute typing test, Jane typed 380 words with 4 errors. Find her speed in words per minute. 42.5 words/min

2. The formula above can be rewritten as $e = \dfrac{w - sm}{10}$. Use the formula to find the number of errors Barry made if he typed 320 words in 5 minutes and got a speed of 42 words per minute. 11 errors

Guided Practice

Classroom Exercises 1–21
Try This   all

Independent Practice

A Ex. 1–25, **B** Ex. 26–39, **C** Ex. 40–47
Basic: WE 1–29 odd, Application
Average: WE 9–37 odd, Application
Above Average: WE 17–47 odd, Application

Additional Exercises

Written Exercises

43. The set of real numbers

Teaching Resources

Problem Solving Worksheet 3
Quick Quizzes 20
Reteaching and Practice
   Worksheets 20
Transparency 7

■ GETTING STARTED

Prerequisite Quiz

Give the reciprocal of each number.

1. 6 $\frac{1}{6}$

2. $-\frac{1}{3}$ -3

3. -1 -1

Multiply.

4. $-\frac{1}{6}(-6)$ 1

5. $(-\frac{3}{4})(-\frac{3}{4})$ $\frac{9}{16}$

6. $(0.6)(-0.6)$ -0.36

Motivator

Ask students if it is possible to solve the equation $8x = -96$ by the Addition or Subtraction Property of Equality.   No
Then ask them if they can guess the solution (-12). Have one or more students tell how they found the solution.   Answers will vary.

3.2 Solving Equations by Multiplying or Dividing

Objective To solve equations by using the Multiplication or Division Property of Equality

When doing a number puzzle, Ralph and Sarah were each asked to pick a number and double it. If they both picked the same number, then they would both have the same result when they doubled it. Let r = Ralph's number and s = Sarah's number. Then the situation described above can be shown as follows.

$$\text{If } r = s, \text{ then } 2r = 2s.$$

This reasoning suggests the following property.

Multiplication Property of Equality
For all real numbers a, b, and c, if $a = b$, then $ac = bc$.
Multiplying each side of an equation by the same nonzero number produces an equivalent equation.

EXAMPLE 1   Solve $\frac{x}{4} = -7$. Check.

Plan   The variable x is not alone on the left side, since x is divided by 4. To get x alone on the left, multiply each side by 4.

Solution

$$\frac{x}{4} = -7$$

Multiply each side by 4. $4\left(\frac{x}{4}\right) = 4(-7)$

Simplify. $1x = -28$ ⟵ $4 \cdot \frac{x}{4} = 4 \cdot \frac{1}{4}x = 1x$

$x = -28$

Check

$$\frac{x}{4} = -7$$

Substitute -28 for x. $\frac{-28}{4} \overset{?}{=} -7$

$-7 = -7$ True

Thus, the solution is -28.

TRY THIS   **1.** Solve $\frac{x}{-2} = 15$. Check. -30

94   Chapter 3 Solving Equations

Highlighting the Standards

Standards 1b, 2c: In Exercises 17–21, students solve application problems and in Exercise 37 they express mathematical ideas in writing.

Additional Example 1

Solve $8 = \frac{y}{-3}$. Check. -24

If the numerical coefficient of the variable is a fraction, multiply each side of the equation by the reciprocal of the fraction.

EXAMPLE 2 Solve $6 = -\frac{2}{3}y$. Check.

Plan The reciprocal of $-\frac{2}{3}$ is $-\frac{3}{2}$, since $\left(-\frac{2}{3}\right) \cdot \left(-\frac{3}{2}\right) = 1$. Therefore, multiply each side of the equation by $-\frac{3}{2}$.

Solution
$$6 = -\frac{2}{3}y$$

Multiply each side by $-\frac{3}{2}$. $\quad -\frac{3}{2} \cdot 6 = -\frac{3}{2} \cdot \left(-\frac{2}{3}\right)y$

Simplify.
$$-9 = 1y$$
$$-9 = y$$

Check
$$6 = -\frac{2}{3}y$$

Substitute -9 for y. $\qquad 6 \overset{?}{=} -\frac{2}{3}(-9)$

$$6 = 6 \quad \text{True}$$

Thus, the solution is -9.

TRY THIS 2. Solve $-\frac{3}{5}x = -30$. Check. 50

Since dividing by a nonzero number is the same as multiplying by its reciprocal, each side of an equation can be divided by such a number.

Division Property of Equality

For all real numbers a, b, and c, if $a = b$, and $c \neq 0$, then $\frac{a}{c} = \frac{b}{c}$.

Dividing each side of an equation by the same nonzero number produces an equivalent equation.

EXAMPLE 3 Solve $-6x = -42$.

Solution
$$-6x = -42$$

Divide each side by -6. $\qquad \frac{-6x}{-6} = \frac{-42}{-6}$

Simplify. $\qquad 1x = 7 \quad \longleftarrow \frac{-6x}{-6} = \frac{1}{-6}(-6)x = 1x$

$$x = 7$$

Thus, the solution is 7. The check is left for you.

TRY THIS 3. Solve $72 = -8x$. Check. -9

95

Common Error Analysis

Error: For equations such as $16 = -2x$, where the variable occurs on the right side, some students divide both sides of the equation by the 16 on the left.

Remind students that, to get x alone, they should divide each side of the equation by the numerical coefficient of the variable, which is -2.

Checkpoint

Solve and check.

1. $-7x = 56$ -8
2. $-42 = 6y$ -7
3. $\frac{1}{3}z = -17$ -51
4. $\frac{y}{-5} = -13$ 65
5. $0.6x = 0.54$ 0.9

Translate into an equation and solve.

6. Molly gave 18 stickers to Carl. This was $\frac{3}{8}$ of her sticker collection. How many stickers did Molly have before she gave some to Carl?   48 stickers

Closure

Have students summarize the four equation properties that have been covered in Lessons 3.1 and 3.2.   See pages 89, 90, 94, and 95.   Have them state an equation that can be solved by using each of the four properties.   Answers will vary.

The word *of* is frequently associated with multiplication. Thus,

$$\tfrac{2}{3} \text{ of } 12 \quad \text{means} \quad \tfrac{2}{3} \cdot 12.$$

EXAMPLE 4   Joe jogged 3 mi. He jogged $\frac{2}{5}$ of the distance Mike jogged. How far did Mike jog?

Solution   Let x = number of miles Mike jogged.

THINK: Three miles that Joe jogged is $\frac{2}{5}$ of the distance that Mike jogged.

$$3 = \tfrac{2}{5} \cdot x$$
$$\tfrac{5}{2} \cdot 3 = \tfrac{5}{2} \cdot \tfrac{2}{5}x$$
$$\tfrac{15}{2} = x, \text{ or } x = 7\tfrac{1}{2}$$

Check   Check the answer, $7\frac{1}{2}$ mi, in the original problem.

$$\tfrac{2}{5} \cdot 7\tfrac{1}{2} = \tfrac{\overset{1}{2}}{\underset{1}{5}} \cdot \tfrac{\overset{3}{15}}{\underset{1}{2}} = \tfrac{3}{1} = 3$$

Thus, Mike jogged $7\frac{1}{2}$ mi .

TRY THIS

4. Robin used $\frac{3}{4}$ of her savings to buy a new bicycle. If Robin spent $144, how much had Robin saved? $192

Classroom Exercises

By what number would you multiply or divide each side of the equation to solve it?

1. $6x = 12$ 6; div
2. $\frac{1}{3}x = -2$ 3; mult
3. $18 = -2x$ -2; div
4. $-\frac{1}{9}x = 5$ -9; mult
5. $-6 = \frac{3}{4}x$ $\frac{4}{3}$; mult
6. $-6x = -18$ -6; div
7. $-\frac{2}{3}x = 14$ $-\frac{3}{2}$; mult
8. $\frac{2}{9}x = -8$ $\frac{9}{2}$; mult

9–16. Solve and check the equations of Classroom Exercises 1–8.

9. 2   10. -6 11. -9 12. -45 13. -8 14. 3   15. -21 16. -36

Written Exercises

Solve and check.

1. $4x = 20$ 5
2. $-3y = 18$ -6
3. $-24 = 2z$ -12
4. $-5a = -30$ 6
5. $\frac{1}{3}c = 2$ 6
6. $7 = -\frac{1}{5}x$ -35
7. $\frac{1}{4}y = -6$ -24
8. $-\frac{1}{7}z = -3$ 21

96   Chapter 3 Solving Equations

Additional Example 4

Carmen spent $\frac{3}{4}$ of her allowance on a notebook. The notebook cost $2.25. How much was her allowance? $3.00

9. $\frac{2}{3}y = 14$ 21   **10.** $-\frac{3}{4}x = 15$ −20   **11.** $\frac{5}{8}z = -25$ −40   **12.** $16 = -\frac{2}{5}c$ −40

13. $3 = \frac{x}{8}$ 24   **14.** $\frac{x}{7} = 4$ 28   **15.** $\frac{y}{9} = -2$ −18   **16.** $\frac{z}{5} = -6$ −30

Translate each problem into an equation and solve. Other equations are possible.

17. Seven times a number is -84. Find the number. $7n = -84$; −12

18. Two-thirds of a number is -26. Find the number. $\frac{2}{3}n = -26$; −39

19. Bill worked 15 h building scenery for a play. He worked 5 times as long as Maria did. How long did Maria work? Explain your reasoning.

20. Jesse spent $\frac{2}{5}$ of her savings on a new coat. The coat cost $60. How much were her savings before she bought the coat? Explain your reasoning.

21. A class is collecting empty bottles to return for 5¢ each. The average rate of collection is 43 bottles a day. At this rate, how many days would it take to reach $100? Explain your reasoning. $5(43)n = 10{,}000$; 47 days

Solve.

22. $-5x = \frac{3}{4}$ $-\frac{3}{20}$ **23.** $\frac{2}{3} = 7y$ $\frac{2}{21}$ **24.** $-6z = -\frac{5}{8}$ $\frac{5}{48}$

25. $-\frac{3}{4} = -9c$ $\frac{1}{12}$ **26.** $1\frac{1}{3}x = 8$ 6   **27.** $-2\frac{1}{4}y = 18$ −8

28. $15 = 1\frac{2}{3}z$ 9   **29.** $-4\frac{1}{2}c = -27$ 6   **30.** $0.2a = 3$ 15

31. $-0.7x = 14$ −20   **32.** $2.4z = -24$ −10   **33.** $-5.2 = -1.3y$ 4

34. $-1 = -\frac{2}{3}n$ $\frac{3}{2}$, or $1\frac{1}{2}$ **35.** $-\frac{4}{5}x = 0$ 0   **36.** $-\frac{1}{8} = 8t$ $-\frac{1}{64}$

37. Explain in writing how the Division Property of Equality follows from the Multiplication Property of Equality.
To divide by a number is to multiply by its reciprocal.

Find the solution set.

38. $3|x| = 15$ {5, −5}   **39.** $\frac{2}{3}|x| = 18$ {27, −27}

40. $x \cdot 1 = x$ The set of real numbers   **41.** $0 \cdot x = 7$ ∅

42. $\frac{|x|}{7} = 2$ {−14, 14}   **43.** $3 = \frac{1}{5}|x|$ {−15, 15}

44. $|x| = x$ The set of real numbers ≥ 0   **45.** $|x| = -x$ The set of real numbers < 0

Mixed Review

Evaluate. Simplify first for Exercises 2, 4, and 5. *1.2 1.7, 2.6, 2.8*

1. $3(2a + 4)$ for $a = 6$ 48   **2.** $6x - 8 - x - 4$ for $x = -3$ −27   **3.** $2a^3$ for $a = -4$ −128

4. $5x - 4 - x$ for $x = -3$ −16   **5.** $8x + 8 + 4x + 3$ for $x = 2\frac{1}{2}$ 41   **6.** $-x^5$ for $x = -3$ 243

7. At 11:00 P.M. the temperature was 6°F. That night it dropped 9° and then rose 4° by 6:00 A.M. What was the temperature at 6:00 A.M.? *2.4* 1°F

Guided Practice

Classroom Exercises 1–16
Try This   all

Independent Practice

A Ex. 1–16, **B** Ex. 17–37, **C** Ex. 38–45

Basic: WE 1–25 odd
Average: WE 9–33 odd
Above Average: WE 13–29, 37–43 odd

Additional Answers

Written Exercises

19. $5x = 15$; 3 h
20. $\frac{2}{5}x = 60$; $150

Enrichment

Explain that the half-life of a radioactive isotope is the time during which half of it will decay. Then have the students solve this problem. The half-life of a certain isotope is 4 days. Suppose that 2 grams of this isotope are present today.

1. How many grams will be present in 20 days?   0.0625 or $\frac{1}{16}$ g

2. How many grams were present 20 days ago?   64 g

▰ GETTING STARTED

Prerequisite Quiz

Simplify.

1. $-8 + 9$ 1
2. $13 - (-5)$ 18
3. $2.6 - 8.4$ -5.8
4. $-\frac{3}{4} - \frac{1}{2}$ $-1\frac{1}{4}$
5. $-3.5 + 4.9$ 1.4
6. $-8 \cdot 7$ -56
7. $-\frac{1}{3}(-\frac{2}{5})$ $\frac{2}{15}$
8. $-5.4 \div 0.25$ -21.6

Motivator

In Lesson 1.8, students solved equations such as $4x + 9 = -7$ where the replacement set was finite. Have them solve this equation for the replacement set $\{-5, -4, -3\}$. Point out that after studying this lesson, they will be able to solve equations where the replacement set is the set of all real numbers.

3.3 Using Two Properties of Equality

Objective To solve equations by using two properties of equality

A plumber charges $42 for each hour he works plus $35 to come to the house. The Batistas were sent a bill for $119. The equation

$$42n + 35 = 119$$

can be solved for n to find the number of hours the plumber worked.

EXAMPLE 1 Solve the equation above. How long did the plumber work?

Plan To get n alone on the left side, first subtract 35 from each side. Then divide each side by 42.

Solution

	$42n + 35 = 119$
Subtract 35 from each side.	$42n + 35 - 35 = 119 - 35$
Simplify.	$42n = 84$
Divide each side by 42.	$\frac{42n}{42} = \frac{84}{42}$
Simplify.	$n = 2$ ⟵ 2 hours

Check Did the plumber work 2 h?
Yes, because $42 \times 2 + 35 = 84 + 35 = 119$.
Thus, the plumber worked 2 h.

EXAMPLE 2 Solve $5 = -16 - \frac{3}{4}y$. Check.

Solution

$$5 = -16 - \frac{3}{4}y$$

Add 16 to each side.

$$5 + 16 = -16 - \frac{3}{4}y + 16$$

$$21 = -\frac{3}{4}y$$

Multiply each side by $-\frac{4}{3}$.

$$-\frac{4}{3} \cdot \overset{7}{\underset{1}{\cancel{21}}} = -\frac{4}{3}\left(-\frac{3}{4}y\right)$$

$$-28 = y$$

Thus, the solution is -28. The check is left for you.

TRY THIS 1. Solve $-\frac{2}{3}x + 7 = -5$. Check. 18

Highlighting the Standards

Standards 5a, 1c: In Exercises 31–38, students represent situations with equations and in Problem Solving Strategies they see how to choose strategies for problems outside the classroom.

Additional Example 1

Solve $2a - 8 = 4$. Check.   6

Additional Example 2

Solve $-3 = \frac{3}{5}x + 9$. Check. -20

Some equations need to be simplified before a property of equality is used. In the next example, like terms should be combined first.

EXAMPLE 3 Solve $8 - (-3) = 7 + 5z - 6z$. Check.

Solution

$$8 - (-3) = 7 + 5z - 6z$$

Rewrite $8 - (-3)$ as $8 + 3$. $8 + 3 = 7 + 5z - 6z$
Simplify. $11 = 7 - 1z$
Subtract 7 from each side. $11 - 7 = 7 - 1z - 7$
Simplify. $4 = -1z$
Divide each side by -1. $\dfrac{4}{-1} = \dfrac{-1 \cdot z}{-1}$
Simplify. $-4 = z$

Thus, the solution is -4. The check is left for you.

TRY THIS   2. Solve $3x + \frac{1}{8} - 7x = \frac{5}{8}$. Check. $-\frac{1}{8}$

Classroom Exercises

To solve the equation, what number would you add to or subtract from each side? Then, by what number would you multiply or divide each side?

1. $4x - 3 = -15$ 3, add; 4, div   **2.** $-2y + 7 = 3$ 7, subt; -2, div

3. $9 - \frac{2}{3}z = 1$ 9, subt; $-\frac{2}{3}$, div   **4.** $5 = 16 - x$ 16, subt; -1, div

5. $-x - 9 = -4$ 9, add; -1, div   **6.** $-1 = 8 - 3c$ 8, subt; -3, div

7–12. Solve and check the equations of Classroom Exercises 1–6.
      7. -3 **8.** 2 **9.** 12 **10.** 11 **11.** -5 **12.** 3

Written Exercises

Solve and check.

1. $5x - 1 = -26$ -5 **2.** $4y - 2 = 14$ 4   **3.** $2z + 4 = 8$ 2

4. $6 + 2a = 10$ 2   **5.** $9z - 5 = 4$ 1   **6.** $4y - 7 = 21$ 7

7. $6 = -4x - 2$ -2 **8.** $4 - 3y = 13$ -3 **9.** $-4 = -8 - 2z$ -2

10. $-7 = 3 + 5a$ -2 **11.** $-6x + 25 = -11$ 6   **12.** $18 = 2c + 10$ 4

13. $17 - 3y = -10$ 9   **14.** $8n - 14 = -22$ -1 **15.** $-x + 3 = -4$ 7

16. $5 - m = 12$ -7 **17.** $-6 - z = -2$ -4 **18.** $-4 = 6 - 2x$ 5

19. $\frac{x}{5} + 9 = 13$ 20   **20.** $-7 = 9 + \frac{x}{3}$ -48 **21.** $5 + \frac{y}{3} = -4$ -27

22. $\frac{1}{3}x - 7 = 2$ 27   **23.** $6 - \frac{3}{5}y = 9$ -5 **24.** $17 = -10 + \frac{3}{4}z$ 36

25. $32 = 7x + 8 - 5x$ 12   **26.** $6y + 8 - 5y = -11$ -19 **27.** $-3 = 8z + 8 - 9z$ 11

28. $5 = 5x - 7x + 25$ 10   **29.** $3y + 7 - 5y = -9$ 8   **30.** $-7 = 3p - 9 - 7p$ $-\frac{1}{2}$

Lesson Note

Before introducing two-step equations, have students perform these steps. Pick a number. Multiply by 4. Add 5. Now ask what steps can be taken to get back to the original number.   Subtract 5. Divide by 4.   Then show a generalized form using x for the number: $4x + 5$. Show that subtracting 5 and then dividing by 4 gives the original number. Next, introduce a similar example using division and subtraction.

Math Connections

Science: Environmentalists spend much of their time studying balances in nature. When man introduces some new element into the environment, problems can result unless measures are taken to compensate for the change that has occurred on one side of the "equation of nature." Equations are used as models to study these effects.

Critical Thinking Questions

Analysis: In solving an equation such as $5x - 3 = 17$, it is usually easier to start by adding 3 than to start by dividing by 5. Ask students to compare this to the rules for the Order of Operations and to suggest a method for deciding which Property of Equality to use first in solving such an equation.   If the side containing the variable indicates a certain order of operations, use the inverse operations in the opposite order.

Additional Example 3

Solve $4x - (-8) - 9x = 13$. Check. -1

Common Error Analysis

Error: In an equation such as the one in Example 3, some students will use Properties of Equality before simplifying each side of the equation.

Remind students that the methods learned for simplifying expressions in the previous two chapters should be applied before the Properties of Equality.

Checkpoint

Solve and check.

1. $-2x + 7 = -19$ 13
2. $13 = 5x - 2$ 3
3. $\frac{x}{7} + 9 = 1$ -56
4. $4 - \frac{3}{5}x = -2$ 10
5. $-5 = 6y + 4 - 7y$ 9
6. $-0.5 - 0.3x = 7.6$ -27

Closure

Have students write problems involving real-world situations that can be solved by writing and solving equations that require the use of two properties of equality.

◼◼◼ FOLLOW UP

Guided Practice

Classroom Exercises 1–12
Try This   all

Independent Practice

A Ex. 1–38, **B** Ex. 39–53, **C** Ex. 54–56

Basic: WE 1–21, 31–35, 39–43 odd, Problem Solving Strategies

Average: WE 13–29, 33–47 odd, Problem Solving Strategies

Above Average: WE 23–29, 33–55 odd, 56, Problem Solving Strategies

---

Translate each problem into an equation and solve.

31. Thirty-four is 6 less than 5 times a number. Find the number. 8
32. 4 increased by 3 times a number is 22. Find the number. 6
33. The $347 selling price of a stereo is $35 more than 3 times the cost. Find the cost. $104
34. Two less than 4 times the temperature is $-20°$. Find the temperature. $-4.5°$
35. 5 times Mary's age decreased by 17 is 28. How old is she? 9
36. Seven pounds less than twice Jason's weight is 219 lb. What is Jason's weight? 113 lb
37. For babysitting Mary charges $3 per hour plus an additional $2 for transportation. One evening she was paid $17. How many hours did she babysit? 5 h
38. A car rental company charges $19 a day plus $0.15 for each mile driven. Mr. Aboud's bill for a one-day rental was $38.50. How many miles did he drive? 130 mi

Solve.

39. $0.6c - 1.5 = 3.9$ 9
40. $2.3 - 0.7a = 7.2$ -7
41. $0.3 + \frac{x}{0.6} = 2.8$ 1.5
42. $\frac{3}{5} = \frac{4}{5} - 3x$ $\frac{1}{15}$
43. $\frac{1}{3} + 4y = \frac{1}{6}$ $-\frac{1}{24}$
44. $\frac{1}{4} = \frac{1}{2} - \frac{3}{4}b$ $\frac{1}{3}$
45. $-5x + 1.7 = 1.8$ -0.02
46. $1.4 - 0.4z = 0.9$ 1.25
47. $-0.8 = 6x + 0.4$ -0.2
48. $5 = 2x + x + 7$ $-\frac{2}{3}$
49. $13 = -3y + 7y + 10$ $\frac{3}{4}$
50. $3z + z - 8 = -2$ $1\frac{1}{2}$
51. $19 = 20 - 7a - a$ $\frac{1}{8}$
52. $-7z - 5 + 4z = -4$ $-\frac{1}{3}$
53. $-12y - 5 - 2y = 8$ $-\frac{13}{14}$
54. $x - 2x = |-3 + 8|$ -5
55. $|-9 + 8| = -5x + 4x + 1$ 0

56. Solve the equation $4x - 3 = 9$ by first using the Addition Property, and then the Multiplication Property. Then solve it by first using the Multiplication Property and then the Addition Property. Explain why the first method is simpler than the second.

Midchapter Review

Solve and check.   *3.1–3.3*

1. $x - 4 = 8$ 12
2. $4 = 6 - x$ 2
3. $x + 5 = 12$ 7
4. $-4 = -8 + \frac{4}{x}$
5. $-2x = 12$ -6
6. $7 = \frac{1}{5}x$ 35
7. $-8 = -2x$ 4
8. $-6 = \frac{2}{3}x$ -9
9. $3x - 2 = 8$ $3\frac{1}{3}$
10. $4 = -2 - \frac{1}{2}x$ -12
11. $2x - 4 - 3x = -8$ 4
12. $6 = 2\frac{1}{2}x - 9$ 6

Solve each problem. Then check in the problem.   *3.1, 3.2*

13. Eighteen is $\frac{1}{3}$ of a number. Find the number. 54
14. Five less than Naomi's age is 16. How old is Naomi? 21 yr old

100   Chapter 3 Solving Equations

Enrichment

Ask the students to start with any number and follow these steps, in order: Multiply by 6, subtract 5, multiply by 2, add 10, divide by 12. They should get the number they started with. Then ask if anyone can show why this will always work. $[2(6x - 5) + 10] \div 12 = x$

Problem Solving Strategies

Developing a Plan

In a baseball game, the home team has runners on second and third base when its best hitter comes up to bat. What plan, or strategy, will the manager of the opposing team use?

Part of solving a problem is choosing the strategy to use. Choosing a strategy that is effective and efficient for you is an important focus of the second step, *Develop a Plan*, in the problem-solving process.

To Solve a Problem

1. Understand the Problem
2. Develop a Plan
3. Carry Out the Plan
4. Look Back

Exercises

Column 1 contains six questions that suggest problem situations. Column 2 lists helpful problem-solving strategies. Beside the number of each question in Column 1, write the letter of one or more strategies that you think you could use to solve the related problem. Explain your choice(s). Answers will vary.

Column 1	Column 2
1. What time will we arrive at the contest site?	**a.** Drawing a Diagram
	b. Making a Model
2. Which made more profit, the popcorn stand or the soft drink stand?	**c.** Using a Formula
	d. Making a Graph
3. Is shoe size related to the height of a person?	**e.** Using Logical Reasoning
	f. Working Backwards
4. How many computer desks will fit in the lab?	
5. What is the shortest route to deliver all the news-papers?	
6. Why are winter days shorter than summer days?	

Additional Answers

Written Exercises, page 100

56. In Method 2, mult each side by $\frac{1}{4}$; then add 3 to each side. Fractions are involved, Distr Prop is needed.

3.4 Equations with the Variable on Both Sides

Objective   To solve equations with the variable on both sides

■ GETTING STARTED

Prerequisite Quiz

Simplify, if possible.

1. $3x - 5x$ $-2x$
2. $-2a - 7b$ Cannot simplify
3. $-x + 19x$ $18x$
4. $-40y - y$ $-41y$
5. $-3y + 8 - 9y - 14$ $-12y - 6$
6. $-6 - c + 9 + 10c$ $9c + 3$

Motivator

Have students solve the equation $5x + 2x = 28$. Then have them use this solution (4) as a replacement for x in the equation $5x = 28 - 2x$. Point out that 4 must be used to replace the variable both times it occurs in the equation. Then have the students show how the second equation can be changed into the first. Add 2x to both sides.

Until now, you have solved only equations with the variable on one side. The equation $6x - 5 = 2x - 21$ has a variable on both sides. Solving such an equation requires the additional step of getting the variable on one side only. This can be done in two different ways.

EXAMPLE 1   Solve $6x - 5 = 2x - 21$. Check.

Plan   Subtract either $2x$ or $6x$ from each side. Try both ways.

Solution

Method 1
To get x alone on the left, first subtract $2x$ from each side. Then add 5 to each side.

$$6x - 5 = 2x - 21$$
$$6x - 5 - 2x = 2x - 21 - 2x$$
$$4x - 5 = -21$$
$$4x - 5 + 5 = -21 + 5$$
$$4x = -16$$
$$\frac{4x}{4} = \frac{-16}{4}$$
$$x = -4 \quad \longleftarrow \text{Same result} \longrightarrow$$

Method 2
To get x alone on the right, first subtract $6x$ from each side. Then add 21 to each side.

$$6x - 5 = 2x - 21$$
$$6x - 5 - 6x = 2x - 21 - 6x$$
$$-5 = -4x - 21$$
$$-5 + 21 = -4x - 21 + 21$$
$$16 = -4x$$
$$\frac{16}{-4} = \frac{-4x}{-4}$$
$$-4 = x$$

Check   Substitute -4 for x.

$$6x - 5 = 2x - 21$$
$$6(-4) - 5 \stackrel{?}{=} 2(-4) - 21$$
$$-24 - 5 \stackrel{?}{=} -8 - 21$$
$$-29 = -29 \text{ True}$$

Thus, the solution is -4.

TRY THIS   1. Solve $8 - x = 5x - 4$. Check.   2

There are equations that have no solution at all. There are also equations for which *any* real number is a solution.

$x = x + 1$ $\frac{1}{2}(2x) = x$

(True for *no* value of x)   (True for *every* value of x)

Highlighting the Standards

Standard 4b: The discussion in this lesson continues to develop the number system. Throughout this lesson and the next, students apply mathematics to problem situations related to daily life.

Additional Example 1

Solve $-7y + 2 = -3y - 30$. Check.   8

EXAMPLE 2   Find the solution set of $-17 - x = 8x + 6 - 9x$.

Solution

Combine $8x$ and $-9x$. $-17 - x = 6 - x$
Add x to each side. $-17 - x + x = 6 - x + x$
Simplify. $-17 = 6$ $\longleftarrow$ False statement

The equation $-17 - x = 8x + 6 - 9x$ is equivalent to the false statement $-17 = 6$. No value of x makes the equation true. Thus, the solution set is the empty set, $\varnothing$.

EXAMPLE 3   Find the solution set of $7x - 4 = -2x + 1 + 9x - 5$.

Solution

Combine $-2x$ and $9x$. $7x - 4 = 7x - 4$
Subtract $7x$ from each side. $7x - 7x - 4 = 7x - 7x - 4$
Simplify. $-4 = -4$ $\longleftarrow$ true statement

The equation $7x - 4 = -2x + 1 + 9x - 5$ is equivalent to the true statement $-4 = -4$. All values of x makes the equation true.

The solution set is the set of all real numbers.

TRY THIS   2. Find the solution set of $5 - 2x = 4x + 8 - 6x - 3$.
The set of all real numbers

The equation $\frac{1}{2}(2x) = x$ is an example of an *identity*.

An **identity** is an equation that is true for each value from the replacement set of the variable.

EXAMPLE 4   Jane has $30 and is saving at the rate of $7 per week. Susan has $50 and is saving at the rate of $3 per week. When will they have the same amount of money?

Solution

Let $n =$ the number of weeks until they have the same amount.

After n weeks, Jane will have $30 + 7n$ dollars and Susan will have $50 + 3n$ dollars. These two expressions are equal.

$$30 + 7n = 50 + 3n$$
Subtract $3n$ from each side. $30 + 4n = 50$
Subtract 30 from each side. $4n = 20$
                                        $n = 5$

Check

After 5 weeks, Jane will have $30 + 7 \cdot 5$ dollars, which is $65.
After 5 weeks, Susan will have $50 + 3 \cdot 5$ dollars, which is $65.

Thus, they will have the same amount after 5 weeks.

TRY THIS   3. In 18 years, John's age will be 3 times his present age. How old is John now?   9 years old

Lesson Note

When solving equations such as $-5x + 9 = 3x - 7$, many students will get the variable on the left side.

$$-8x = -16 \text{ and } x = 2$$

Point out that in cases such as this, getting the variable on the right will result in a positive coefficient of x. That is, $16 = 8x$ and $2 = x$, or $x = 2$. Emphasize that when no real number makes an equation true, we say that the solution set is the empty set and use the symbol, $\varnothing$, to indicate that set. When all values of the variable make an equation true, we say that the solution set is all real numbers. This can also be written as {real numbers}.

Math Connections

History: As late as the 18th century, the mathematician Leonhard Euler (1707–1783) declared that the solution to the equation $x^2 = -1$ was "impossible" or, as we would say, the empty set. Today the solution of this equation is known as the imaginary unit, i, and is a very important number in higher mathematics.

Additional Example 2

Find the solution set of $-3 - 5y = 4 + 3y + 7 - 8y$. $\varnothing$

Additional Example 3

Find the solution set of $x + 4 - 6x = 6 - 5x - 2$.   The set of real numbers

Additional Example 4

Six more than 4 times a number is the same as twice the number. Find the number.
-3

Critical Thinking Questions

Evaluation: The following statement in the computer language BASIC is used to increase the value of the variable x by 1.

LET X = X + 1

Ask students why this type of statement might be criticized as misleading.

Common Error Analysis

Error: In solving equations such as the one in Example 1, some students who use Method 2 will inadvertently omit the negative sign at some point and get an incorrect answer.

Suggest that it might be safer to use the method that will maintain a positive coefficient for the variable.

Checkpoint

Solve and check.

1. $15m - 19 = -4m$ 1
2. $-3c = 21 + 4c$ -3
3. $5d - 9 = 2d - 6 + 3d$ $\varnothing$
4. $16y - 5 - 9y = 13 + 7y - 18$
   The set of real numbers

Translate into an equation and solve.

5. Five times the temperature is the same as the temperature decreased by 8. Find the temperature. -2

Focus on Reading

1. Write, in order, the letters of a sequence of steps for solving the equation $-7x + 8 = -3x - 6$ to get the variable alone on the *left*.   d, e, c or e, d, c

2. Repeat Exercise 1 above, but get the variable alone on the *right*.   b, a, f or a, b, f

  a. Add 6 to each side.
  b. Add $7x$ to each side.
  c. Divide each side by -4.
  d. Add $3x$ to each side.
  e. Subtract 8 from each side.
  f. Divide each side by 4.

Classroom Exercises

What operation would you perform first to get the variable on the right side of the equation?

1. $5x - 6 = 2x$ Subt $5x$.
2. $3y + 7 = -4y$ Subt $3y$.
3. $-5c = -14 + c$ Add $5c$.
4. $6 - x = x + 2$ Add x.
5. $3a + 8 = 5a - 14$ Subt $3a$.
6. $y + 5 = 9 - 4y$ Subt y.

7–12. Solve and check the equations of Classroom Exercises 1–6.
  7. 2   **8.** -1 **9.** $2\frac{1}{3}$ **10.** 2   **11.** 11   **12.** $\frac{4}{5}$

Written Exercises

Solve and check.

1. $9y - 18 = 3y$ 3
2. $7c - 9 = 8c$ -9
3. $8n - 12 = 5n$ 4
4. $-11m = 14 - 9m$ -7
5. $-6x = 10 - 4x$ -5
6. $-4z = 35 - 9z$ 7
7. $8p = -5p + 65$ 5
8. $-84 + 15r = 3r$ 7
9. $11c + 36 = 8c$ -12
10. $7z - 9 = 3z + 19$ 7
11. $6 + 10t = 8t + 12$ 3
12. $3x + 7 = 16 + 6x$ -3
13. $18 + 3y = 5y - 4$ 11
14. $11a + 8 = -2 + 9a$ -5
15. $9x - 5 = 6x + 13$ 6
16. $5 - x = x + 9$ -2
17. $14 + 3n = n - 14$ -14
18. $7 - x = 5 + 3x$ $\frac{1}{2}$
19. $4y + 2 = 2y + 4 + 3y$ -2
20. $8c - 12 = 15c - 4c$ -4
21. $5x - 3 = 7x + 7 + 3x$ -2
22. $y + 11 = -2y + 6$ $-1\frac{2}{3}$
23. $-2y + 3 - y = 11 + y$ -2
24. $16 - x = 4x + 8 + 3x$ 1

Translate each problem into an equation and solve. Other equations are possible.

25. Courtney's allowance increased by $30 is the same as twice her allowance. Find her allowance. $x + 30 = 2x$; \$30

26. Five times Mary's age is the same as 28 less than 7 times her age. Find her age. $5x = 7x - 28$; 14 yr old

27. Ten more than twice a number is the same as 4 times the number. Find the number. $2x + 10 = 4x$; 5

28. The temperature, increased by 80 degrees, is the same as 6 times the temperature. Find the temperature. $x + 80 = 6x$; $16°$

29. Three less than 5 times Jerry's age is the same as 3 times his age increased by 37. Find his age.
$5x - 3 = 3x + 37$; 20

30. The perimeter of a triangle increased by 3 cm is the same as 35 cm decreased by 7 times the perimeter. Find the perimeter of the triangle.

31. An automobile salesman is offered a choice of two compensation plans. The first consists of a weekly salary of $120 and a $150 commission on each car he sells. The second consists of no salary and a $180 commission on each car sold. How many cars would the salesman have to sell in one week to make the same money under both plans? $120 + 150x = 180x$; 4 cars

32. Two men are planting tulip bulbs. One man has already planted 57 bulbs and is planting 44 bulbs per hour. The other man has already planted 96 bulbs and is planting 32 bulbs per hour. In how many hours will each man have planted the same number of bulbs? What is the number of bulbs? $57 + 44x = 96 + 32x$; $3\frac{1}{4}$h; 200

33. A number decreased by 30 is the same as 14 decreased by 3 times the number. Find the number. $x - 30 = 14 - 3x$; 11

Find the solution set.

The set of real numbers

34. $-6x = 5 - 6x$ ∅

35. $4y + 7 = 7 + 4y$

36. $5y - 4 + 6 - 7y = 3 + 6y$ $\{-\frac{1}{8}\}$

37. $8p - 5 - 9p = 4 + p - 6$ $\{-1\frac{1}{2}\}$

38. $2a - 5a + 4 = 4 - 3a$

39. $6 - 3c = -3c + 2$ ∅

40. $-7 + 0.6a + 2 = 15 - 1.4a$ $\{10\}$

41. $0.2x + 3.2 = 0.4x - 0.2x + 3.2$

42. $3(x - 2) = 7 - (4 - x)$ $\{4\frac{1}{2}\}$

43. $3x + 4(3 - x) = 7 - x$ ∅

44. $3|x| - 7 = 2|x| + 5$ $\{12, -12\}$

45. $5|x| - 9 = 5 - 2|x|$ $\{2, -2\}$

46. $8 - 4|x| = 2|x| - 9$ $\{2\frac{5}{6}, -2\frac{5}{6}\}$

47. $4|2x| - 5 = 7|2x| + 3$ ∅

48. $6y = 8 - 9 + 6y$ ∅

49. $5 = 3a - 3a + 2$ ∅

Mixed Review

Simplify, if possible. Then evaluate for the given value of the variable. *2.6, 2.8, 2.10*

1. $-3x^3$ for $x = -4$ 192

2. $3x - (4 - x)$ for $x = -5$ -24

3. $\dfrac{2a - 3}{3a + 4}$ for $a = -2$ $3\frac{1}{2}$

Insert < or > to make a true statement. *2.1*

4. $-7 \underline{\ ?\ } -3$ <

5. $5 \underline{\ ?\ } -15$ >

6. $0 \underline{\ ?\ } -12$ >

7. $-2.5 \underline{\ ?\ } -1.75$ <

8. Harry bought 14 records from Melissa. That was $\frac{2}{3}$ of her records. How many records did Melissa have before she sold some to Harry? *3.2* 21 records

Closure

Have students list the types of solution sets they have encountered thus far. One or more real numbers, empty set, the set of all real numbers. Then have them give examples of equations for each type of solution set. Answers will vary.

◼◼◼ FOLLOW UP

Guided Practice

Classroom Exercises 1–12
Try This all, FR all

Independent Practice

A Ex. 1–28, **B** Ex. 29–41, **C** Ex. 42–49

Basic: WE 1–17, 25–31, 35–39 odd

Average: WE 9–19, 25–41 odd

Above Average: WE 17–23, 29–49

Additional Answers

Written Exercises

30. $x + 3 = 35 - 7x$; 4 cm
38. The set of real numbers
41. The set of real numbers

Enrichment

Have each student write a set of directions, similar to those given in the Enrichment for Lesson 3.3, such that the result will always be 5. One possible answer: Start with a nonzero number, multiply by 10, subtract 8, divide by 2, add 4, divide by the original number. $[(10x - 8) \div 2 + 4] \div x = 5$

Teaching Resources

Quick Quizzes 23
Reteaching and Practice
Worksheets 23

■ GETTING STARTED

Prerequisite Quiz

Simplify.

1. $3y - (16 - 2y)$ $5y - 16$
2. $-4 + 2(x + 7)$ $2x + 10$
3. $5d - 3(6 - 7d)$ $26d - 18$
4. $-(4x - 3) + 2(6 - 5x)$ $-14x + 15$
5. $4a - (3 + 2a) + 6a$ $8a - 3$
6. $2(5x - 3) - 4(7 - 4x)$ $26x - 34$

Motivator

Tell students that the Distributive Property provides the basis for removing parentheses before solving an equation. Have them simplify each of these expressions and compare their answers.

1. $5(-4 + 3x)$ $-20 + 15x$
2. $4(4 - 3x)$ $16 - 12x$
3. $-(3x - 4)$ $-3x + 4$
4. $-2(-3x - 4)$ $6x + 8$

■ TEACHING SUGGESTIONS

Lesson Note

Review the following procedure for solving equations containing parentheses.

1. Remove parentheses.
2. Simplify each side of the equation by collecting like terms.
3. Use the properties of equality to solve the resulting equation.

3.5 Equations with Parentheses

Objective   To solve equations that contain parentheses

To solve an equation that contains parentheses, first use the Distributive Property. Then solve the resulting equation and check.

EXAMPLE 1   Solve $8x - 3(2 - 5x) = 40$. Check.

Solution

Use the Distributive Property.
Combine like terms.
Add 6 to each side.

Divide each side by 23.

$$8x - 3(2 - 5x) = 40$$
$$8x - 6 + 15x = 40$$
$$23x - 6 = 40$$
$$23x - 6 + 6 = 40 + 6$$
$$23x = 46$$
$$x = 2$$

Check

Substitute 2 for x.

$$8x - 3(2 - 5x) = 40$$
$$8 \cdot 2 - 3(2 - 5 \cdot 2) \stackrel{?}{=} 40$$
$$16 - 3(2 - 10) \stackrel{?}{=} 40$$
$$16 - 3(-8) \stackrel{?}{=} 40$$
$$16 + 24 \stackrel{?}{=} 40$$
$$40 = 40 \text{ True}$$

Thus, the solution is 2.

TRY THIS   1. Solve $12 = 5x + 3(4 - x)$. Check.   0

In Example 2, two sets of parentheses must be removed before like terms are combined.

EXAMPLE 2   Solve $6y - (4 - 2y) = 3(7 + 2y)$.

Solution

$$6y - (4 - 2y) = 3(7 + 2y)$$

Use the property $-a = -1a$.
Remove parentheses.
Combine like terms.
Subtract $6y$ from each side.
Add 4 to each side.

Divide each side by 2.

$$6y - 1(4 - 2y) = 3(7 + 2y)$$
$$6y - 4 + 2y = 21 + 6y$$
$$8y - 4 = 21 + 6y$$
$$8y - 4 - 6y = 21 + 6y - 6y$$
$$2y - 4 = 21$$
$$2y = 25$$
$$y = \frac{25}{2}, \text{ or } 12\frac{1}{2}$$

Thus, the solution is $12\frac{1}{2}$. The check is left for you.

TRY THIS   2. Solve. $2(5 - 2y) = -1 + 3(y - 1)$. Check.   2

106   Chapter 3 Solving Equations

Highlighting the Standards

Standards 1c, 1d: Students continue to work on problems from everyday situations in Exercises 19–26, and, in the Application, they apply mathematical modeling to problems about thunder and lightning.

Additional Example 1

Solve $-5x - (3 + 2x) = 25$. Check. -4

Additional Example 2

Solve $2(-4y - 5) = 6y - 3(y - 4)$. -2

| EXAMPLE 3 | The long-distance telephone rate to Iowa is 21¢ for the first minute and 15¢ for each additional minute or part thereof. Judy's call to Iowa cost $1.11. How long was Judy's call? |

Plan
Let x = the maximum number of minutes.
The first minute cost 21¢.
The remaining $x - 1$ minutes cost 15¢ each.
The call cost $1.11, which is 111¢.

Solution
$$21 + (x - 1)15 = 111$$
$$21 + 15x - 15 = 111$$
$$15x + 6 = 111$$
$$15x = 105$$
$$x = 7$$

Check
Does 1 min at 21¢ and 6 min at 15¢ cost $1.11?
Yes, because $21 + 6 \cdot 15$ equals $21 + 90$, or 111.

Thus, the call lasted at least 6 minutes but not more than 7 minutes.

TRY THIS

3. Twice a number subtracted from 6 is the same as that number added to 3. What is the number? 1

Classroom Exercises

Use the Distributive Property to remove parentheses.

1. $2(3x + 4)$ 6x + 8   **2.** $3(7y - 2)$ 21y − 6   **3.** $4(n - 1)$ 4n − 4   **4.** $-7(x + 2)$ −7x − 14

5. $-5(2x + 3)$ **6.** $(6 + a)3$ 18 + 3a   **7.** $(4 - 3x)2$ 8 − 6x   **8.** $(6t + 2)4$

9. $-1(3x - 5)$ −3x + 5   **10.** $-(-4 + 2y)$ 4 − 2y   **11.** $-(1 + 4z)$ −1 − 4z   **12.** $(4a - 2)(-3)$ −12a + 6

5. −10x − 15   **8.** 24t + 8

Solve and check.

13. $3(x + 4) = 15$ 1   **14.** $-3(a - 4) = 39$ −9   **15.** $4(y + 1) = 14 - y$ 2

Written Exercises

Solve and check.

1. $2(y + 7) = 16$ 1   **2.** $3(x - 2) = 18$ 8   **3.** $-5(a + 2) = 30$ −8

4. $x + 9 = 2(x - 3)$ 15   **5.** $2(y + 3) = 12 - y$ 2   **6.** $25 - 5a = 3(2a + 1)$ 2

7. $-2(3 - 2c) = 10 - 4c$ 2   **8.** $23 = 12 - (6 + c)$ −17

9. $5(x - 1) = 2x + 4(x - 1)$ −1   **10.** $13 - (2x - 5) = 2(x + 2) + 3x$ 2

11. $-(3 - 2n) + 7n = (n + 3)3$ 2   **12.** $-(y + 8) - 5 = 4(y + 2) - 6y$ 21

Math Connections

History: The vinculum, from the Latin word which means "to bind," was originally used instead of parentheses. For example,
$2 \cdot \overline{x + 4}$ meant the same as $2(x + 4)$.

Critical Thinking Questions

Evaluation: In reading the expression $3(x + 4)$, someone might say "three times x plus four." Ask students whether this is an unambiguous way to translate the algebraic expression. Ask them how the expression could be translated to avoid any ambiguity. One possible answer: "three times the quantity x plus four"

Common Error Analysis

Error: When translating word problems, students often misinterpret the required order of operations. See Example 3, where parentheses are needed to clarify the order of operations.

Encourage students to think about the required order of operations as they read each problem.

Checkpoint

Solve and check.

1. $4(x - 3) = 4$ 4
2. $8 - y = 5(y - 2)$ 3
3. $3(7 - y) + 7y = -(y + 4)$ −5
4. $5(3c - 2) + 1 = -(c + 4)$ $\frac{5}{16}$
5. $-(-6 + 2a) = 4a - 3(6 - 2a)$ 2

Translate into an equation and solve.

6. Eight times the difference between a number and 3 is 56. Find the number. 10

Additional Example 3

Three times the sum of Edward's age and 4 is 51 years. Find Edward's age.   13 years old

Closure

Have students write an equation with parentheses where the solution set is the set of all real numbers. Have them repeat the process where the solution set is the empty set.   Answers will vary.

FOLLOW UP

Guided Practice

Classroom Exercises   1–16
Try This   all

Independent Practice

A Ex. 1–20, **B** Ex. 21–26, **C** Ex. 27–32
Basic: WE 1–13, 19–23 odd, Application
Average: WE 7–25 odd, Application
Above Average: WE 13–31 odd, Application

Additional Answers

Written Exercises

20. $8(x - 9) = 48$; 15 yr
21. $x + 5 = 4(x + 8)$; -9
22. $3(x + 4) = 18 + x$; 3
24. $25 - 20(x - 1) = 165$; 8 oz

---

13. $(c + 4)3 - 6c = 2(4 - 2c)$ -4
14. $8y - 3(4 - 2y) = 6(y + 1) - 2$ 2
15. $-2(3 - 4z) + 7z = 12z - (z + 2)$ 1
16. $-3(6 - 2x) + 4x = -(2x - 6)$ 2
17. $7x - (9 - 4x) = 3(x - 11)$ -3
18. $7r + 3(7 - r) = -(r + 4)$ -5

Translate each problem into an equation and solve.   Other equations are possible.

19. Five times the sum of a number and 3 is 35. Find the number.
$5(x + 3) = 35$; 4

20. If Jim's age in years is decreased by 9, and that difference is multiplied by 8, the result is 48 years. Find Jim's age.

21. Five more than a number is the same as 4 times the sum of the number and 8. Find the number.

22. Three times the sum of a number and 4 is the same as 18 increased by the number. Find the number.

23. If Amy's age in years is increased by 5 years and then multiplied by 2, the result is 38 more than Amy's age. Find her age.
$2(x + 5) = x + 38$; 28 yr old

24. First-class postage costs 25¢ for the first ounce and 20¢ for each additional ounce or part thereof. If you are charged $1.65 to mail a package, what is the maximum weight?

25. Rectangular tables that seat 6 people are being placed end-to-end for a banquet. Five people will be placed at each of the 2 end tables and 4 at each of the other tables. How many tables are needed to seat 50 people in one line of tables? Explain your reasoning. $10 + 4(x - 2) = 50$; 12 tables

26. Mr. Parsons is planning to take his 12-year-old son, Jamie, and some of his friends to a show. Adult tickets are $7; children's are $4. Jamie's two younger sisters will join them at the theatre. How many friends can Jamie invite if Mr. Parsons can spend $35 and if he is the only adult?
$7 + (x + 3)4 = 35$; 4 friends

Solve and check. (HINT: Remove the parentheses first. Then remove the brackets.)

27. $5 - n = n - [4 + 7(2n - 1)]$ $-\frac{1}{6}$
28. $-3[5 - (2 + c)] = 6c + 1$ $-3\frac{1}{3}$
29. $3[5 - 3(y - 4)] = 2y + 7$ 4
30. $-9x - [2(1 + 3x) + 6] = 5x$ $-\frac{2}{5}$
31. $6z - 2[7(z + 1) + 4] = 10z$ $-1\frac{2}{9}$
32. $-[-8 + 2(1 - 4r)] = 1 - 7r$ $-\frac{1}{3}$

Mixed Review

Solve and check.   3.2, 3.3, 3.4, 3.5

1. $-2x + 4 = 7$ $-1\frac{1}{2}$
2. $-\frac{3}{5}y = 18$ -30
3. $13 = 2x - 7$ 10
4. $4y + 8 - y = -13$ -7
5. $3(4a - 1) = 21$ 2
6. $5x - 8 - 2x = 7 + 3x$ $\varnothing$

Simplify.   2.8

7. $-26 - (-2 + 6)$ -30
8. $-(-12 - 4) - 18$ -2
9. $16 + (3 - 6)$ 13
10. $(10 - 3) - (2 + 4)$ 1
11. $4 - 8 - (2 - 3)$ -3
12. $-(3 - 2) - 4$ -5

Enrichment

Have the students solve this puzzle.

Sally and Sue are twins who keep pennies in a tin box on their dresser. On Monday morning, Sally puts 10 pennies into the box, but on Monday afternoon Sue takes out half the number of pennies in the box. On Tuesday morning, Sally puts another 10 pennies into the box; but on Tuesday afternoon, Sue again takes out half the pennies in the box. On Wednesday morning, Sally puts 10 more pennies into the box, and in the afternoon Sue takes out a dollar's worth of pennies. If this left 65 pennies in the box, how many pennies were there originally?   590 pennies

Application: *Thunder and Lightning*

During a thunderstorm there is a difference in time between when you see the lightning flash and when you hear the thunder. Both the flash and the sound are created at essentially the same time and place in the storm, but sound travels slower than light and, therefore, arrives after you have seen the flash. If you count the seconds between the flash of lightning and the sound of thunder, you can compute how far away the lightning is.

EXAMPLE   Twelve seconds elapse between the time you see lightning and then hear the thunder. How far away was the lightning? The speed of sound is 1,127 feet per second.

Use the formula, distance = rate · time.

$d = rt$
$d = 1,127 \cdot 12$
$d = 13,524$ feet

The lightning was about 13,500 feet away.

(NOTE: Since light travels at approximately 186,000 miles per second, it takes about 0.000014 second for light to travel 13,524 feet. This time is so much less than 12 seconds that it can be ignored.)

Solve.

1. How far away in feet is the lightning if the elapsed time after the time you hear the thunder is 1 second? 3 seconds? 6 seconds?
 1,127 ft; 3,381 ft; 6,762 ft

2. There are 5,280 feet in a mile. Use your answers to Exercise 1 to find out how many miles away the lightning is for each of the three cases. 0.21 mi; 0.64 mi; 1.3 mi

3. Make a chart that organizes the data in the Example and in Exercises 1 and 2.

4. What rule could you use to approximate mentally the number of miles between you and a storm if you know the elapsed time in seconds? To approximate the miles, divide the number of seconds by five.

Application

3. seconds	ft	mi
1	1,127	0.21
3	3,381	0.64
6	6,762	1.2
12	13,524	2.56

3.6 Problem Solving: Using Formulas

Objective   To solve problems involving formulas

Suppose a school dance will cost $800 for a band and other expenses. If tickets cost $5 each, and only 100 are sold, will there be a profit? You can use the following relationship:

profit = number of tickets × price per ticket − expenses
profit =   100   ×   5   − 800
profit =   500   − 800
profit = −300

A negative profit is a *loss*, so the sale of 100 tickets is a loss of $300.

Profit = number of tickets × price per ticket − expenses, can be expressed as a *formula*: $p = nt - e$.

Notice that in a formula, different quantities are represented by different letters. Thus, t (not p) is used to represent "price per ticket" since "p" has already been selected to represent "profit."

EXAMPLE 1   Suppose that your school dance hires a band for $800.

a. If you expect to sell 200 tickets, what should be the price of each ticket in order to make a profit of $1,000?

b. If you plan to charge $5 per ticket and wish to break even (have a zero profit), how many tickets should you sell?

Solutions   **a.**

$$p = nt - e$$
$$1{,}000 = 200t - 800$$
$$1{,}000 + 800 = 200t - 800 + 800$$
$$1{,}800 = 200t$$
$$9 = t$$

b.

$$p = nt - e$$
$$0 = n \cdot 5 - 800$$
$$0 + 800 = 5n - 800 + 800$$
$$800 = 5n$$
$$160 = n$$

Check   The total receipts from the sale of 200 tickets, at $9 each, are $1,800. After $800 in expenses is deducted, the profit is $1,000.

The price should be $9 each.

The total receipts from the sale of 160 tickets, at $5 apiece, are $800, which equals the expenses.

Thus, 160 tickets must be sold to break even.

TRY THIS   1. Use the formula $p = nt - e$ to find the expenses, e, if 420 tickets sell for $6 each and the profit is $1,950. $570

Distance, rate, and time are related by the formula $d = rt$, where d is the distance, r is the rate (speed), and t is the time.

GETTING STARTED

Prerequisite Quiz

Find the value of *x* for *a* = 30, *b* = 5, and *c* = 15.

1. $x = \dfrac{ab}{c}$ 10
2. $x = \dfrac{a + b}{c}$ $2\frac{1}{3}$
3. $x = abc$ 2,250
4. $ax = bc$ $2\frac{1}{2}$
5. $a = bx - c$ 9
6. $a = bc - x$ 45

Motivator

Remind students of the formula $p = 2l + 2w$ for finding the perimeter of a rectangle. Ask them to find the perimeter when $l = 8.5$ and $w = 5$.   27   Then ask them to draw a rectangle with a width of 7 cm and a perimeter of 34 cm. This will require them to use the formula to solve for l. $l = 10$

Highlighting the Standards

Standards 2b, 1d: In the opening discussion, students read about how to use a formula to represent a relationship. In Exercise 20, students write their own formula for a problem.

Additional Example 1

Suppose that expenses for putting on your school play total $160.

a. If you plan to charge $3 per ticket, how many tickets should you sell in order to make a profit of $500?   220

b. If you expect to sell 200 tickets, what should the price of each ticket be if you wish to break even (have a zero profit)? $.80

110

EXAMPLE 2 Gail plans to ride her bike 6 mi in 45 min. At what rate in miles per hour (mi/h) must she travel?

Plan Use the formula $d = rt$. You are asked to give your answer in miles per hour. Thus, d must be given in miles and t must be given in hours.

Solution Since $d = 6$ mi and $t = 45$ min, change minutes to hours.

$$45 \text{ min} = \frac{45}{60} \text{ h} = \frac{3}{4} \text{ h. So } t = \frac{3}{4} \text{ h.}$$

$$d = rt$$

Substitute 6 for d and $\frac{3}{4}$ for t. $6 = r \cdot \frac{3}{4}$

Multiply each side by $\frac{4}{3}$. $6 \cdot \frac{4}{3} = r \cdot \frac{3}{4} \cdot \frac{4}{3}$

$$8 = r$$

Thus, Gail must travel at a rate of 8 mi/h The check is left for you.

TRY THIS 2. It took Mr. Greenberg $4\frac{1}{2}$ hours to drive 270 miles. What is his average speed? 60 mi/h

The surface area of a rectangular solid is the total area of its six sides. The formula for the surface area, A, of a rectangular solid with length l, width w, and height h, is $A = 2(lw + hw + lh)$.

EXAMPLE 3 The surface area of a box is 94 in². Find the height of a box that is 3 in. wide and 5 in. long.

Solution $A = 94, w = 3,$ and $l = 5.$ Find h.
$A = 2(lw + hw + lh)$
$94 = 2(5 \cdot 3 + h \cdot 3 + 5 \cdot h)$
$94 = 2(15 + 8h)$
$94 = 30 + 16h$
$64 = 16h$
$4 = h$ Thus, the height of the box is 4 in

TRY THIS 3. Find the width of a box with a surface area of 46 in.², a length of 4 in., and a height of 2 in. $2\frac{1}{2}$ in.

Classroom Exercises

1. Using $d = rt$, find d for $r = 45$ mi/h and $t = 11$ h. 495 mi
2. Using $d = rt$, find t for $d = 330$ mi and $r = 55$ mi/h. 6 h

Lesson Note

A formula is a compact or shorthand way of stating a relationship. Emphasize that, in a formula, different quantities are represented by different letters. Since the formula $d = rt$ will be used again later in the text, be sure that students understand what the formula means. Use simple examples such as: At a rate of 50 mi/h, what distance can you travel in 4 hours? 200 mi Use a box as a model to demonstrate how the formula $A = 2(lw + hw + lh)$ was derived.

Math Connections

Life Skills: When planning a trip, you can estimate the time it will take if you know the distance and the approximate average speed at which you will be able to travel.

Critical Thinking Questions

Application: The directions for Written Exercises 15 and 16 indicate a limit on the size of a package that you can mail. If the volume of a package is found by the formula $V = lwh$, ask students to use a calculator to find the volume of the largest cube (all 3 dimensions the same) that could be mailed. $21.6^3 = 10,077.696$ in³

Additional Example 2

If Ken rows his boat at an average rate of $12\frac{1}{2}$ mi/h, how many minutes will it take him to row 10 mi? 48 min

Additional Example 3

Find the length of a box, having a width of 10 cm, a height of 8 cm and a surface area of 700 cm². 15 cm

Common Error Analysis

Error: Given the total surface area of a rectangular solid and asked to find l, w, or h, students forget to include the unit of measure in the answer.

Remind students that a correct answer always includes the correct unit of measure.

Checkpoint

1. A bus travels 60 mi in 1 h 20 min. What is its average speed in miles per hour? Use the formula $d = rt$.   45 mi/h
2. Use the formula $A = 2(lw + hw + lh)$ to find h for $l = 6$, $w = 6$ and $A = 216$.   6
3. The school choral group will give a concert to raise $1,600 for a tour. If expenses for the concert total $150, and they expect to sell 500 tickets, what should they charge for each ticket? Use the formula $p = nt - e$. $3.50

Closure

Ask students how a formula is unlike the equations they have solved in previous lessons.   Has more than one variable Ask how many of the variables in a formula must have values assigned to them in order to use the formula to solve a problem.   All but one

3. Using $F = \frac{9}{5}C + 32$, find F for $C = 25$. 77
4. Barry jogs at a rate of 6 mi/h. How many hours must he jog in order to cover a distance of 15 mi? $2\frac{1}{2}$ h

A school orchestra is giving a concert. Its expenses will be $250. Use this information to answer Exercises 5–8.

5. How many tickets must be sold at $2 each to make a $1,000 profit? 625 tickets
6. If the orchestra expects to sell approximately 500 tickets, what should it charge for each ticket to make a $1,000 profit? $2.50
7. If the orchestra sells 550 tickets at $2 each, what will be its profit? $850.00
8. If the orchestra expects to sell only 500 tickets at $2.25 each, by how much would it have to lower expenses to make a $1,000 profit? $125

Written Exercises

Use the formula $p = nt - e$ for Exercises 1–4.

1. Find n for $p = \$1,578$, $t = \$8$, and $e = \$150$. 216 tickets
2. Find t for $p = \$2,210$, $n = 410$, and $e = \$250$. $6
3. A day-care center needs a profit of $600 for its variety show. Expenses are $75. How many tickets must be sold if the price of each ticket is $1.50? 450
4. If the day-care center expects to sell about 350 tickets, expenses are $75, and a $600 profit is the goal, what should be the price of each ticket, to the nearest quarter? $2.00

Use the formula $d = rt$ for Exercises 5–8.

5. Find r for $d = 175$ miles and $t = 3.5$ h. 50 mi/h
6. Find t for $d = 1,716$ mi and $r = 264$ mi/h. 6.5 h
7. Kate lives 2 mi from school. She runs at a rate of 10 mi/h. How long must she run to get from her home to school? Use $d = rt$. $\frac{1}{5}$ h, or 12 min
8. How fast must a train travel in order to cover a distance of 130 mi in $2\frac{1}{2}$ h? Use $d = rt$. 52 mi/h

Use the formula $A = 2(lw + hw + lh)$ for Exercises 9–12.

9. Find l for $A = 118$ in.^{2}, $w = 5$ in., and $h = 7$ in. 2 in
10. Find w for $A = 78$ in.^{2}, $l = 4$ in., and $h = 6$ in.   1.5 in
11. Find h for $A = 81.4$ in.^{2}, $l = 4$ in., and $w = 5$ in. 2.3 in
12. Nancy is painting the surface of a wooden crate that has a surface area of 108 in.^{2} If the length and width both equal 6 in., what is the height of the crate? $1\frac{1}{2}$ in.

The formula $\frac{W}{4A} = p$ represents the tire pressure p in pounds per square inch of a tire on a car. W is the weight of the car in pounds and A is the area in square inches of each tire's contact with the ground.

13. Find p for $W = 4,000$ lb and $A = 50$ in.^{2} 20 lb/in.^{2}

14. Find W for $A = 31.25$ in.^{2} and $p = 8.1$ lb/in^2. 1012.5 lb

The maximum size for a package accepted by the United States Postal Service is one that satisfies the condition $l + 2w + 2h = 108$ in., where the length l, the width w, and the height h are all expressed in inches.

15. If you wish to mail a package with $l = 48$ in. and $w = 22$ in., what is its maximum permitted height? 8 in

16. If you wish to mail a package with $w = h = 1.5$ ft, what is its maximum permitted length? 36 in

On a 160-mi trip, Ray averages 50 mi/h for the first $2\frac{1}{2}$ h. After a 10-min rest, he continues driving and arrives at his destination in another 45 min.

17. What is his average speed during the last 45 min? Approximately 46.7 mi/h

18. What is the average speed for the entire trip? Approximately 46.8 mi/h

In a certain city, the streets are parallel, 200 ft apart, and approximately 35 ft wide. There are 5,280 ft in 1 mi.   Approximately 18.3 blocks

19. How many blocks can you walk in 14 min at the rate of 3.5 mi/h?

20. Write a formula that relates the number of blocks walked to the rate in mi/h and the time in minutes. $n = \frac{5,280rt}{60(235)}$

Mixed Review

Evaluate for the given values of the variables.   1.3, 2.6, 2.8

1. $8x^2$ for $x = \frac{1}{4}$ $\frac{1}{2}$

2. $(8x)^2$ for $x = \frac{1}{4}$ 4

3. $3a^4b$ for $a = 2$, $b = -5$ -240

4. $-3x - 4$ for $x = -6$ 14

Simplify.   2.2, 2.6, 2.9, 2.10

5. $|-16|$ 16

6. $-x + 5 - x - 9$ $-2x - 4$

7. $2x - (5 - x)$ $3x - 5$

8. $4(-4)^3$ -256

Solve.

9. Five feet less than twice the length of a board is the same as its length increased by 12 ft. Find the length of the board. *3.4* 17 ft

Guided Practice

Classroom Exercises 1–8
Try This   all

Independent Practice

A Ex. 1–14, **B** Ex. 15–16, **C** Ex. 17–20

Basic: WE 1–9 odd, 12, 13, Application

Average: WE 5–11 odd, 12, 13, 15, Application

Above Average: WE 9, 11, 12, 13–19 odd, Application

Enrichment

Ask students to use any strategy they wish to solve this puzzle problem. Evita drew a rectangle using just the grid lines on graph paper. The length of the rectangle was 1 unit more than twice the width. If the area of the rectangle was more than 50 square units but less than 150 square units, what are the possible dimensions?   11 units, 5 units; 13 units, 6 units; 15 units, 7 units; 17 units, 8 units

The Shock Wave roller coaster is located at Six Flags Great America in Gurnee, Illinois, outside of Chicago. It is the tallest roller coaster in the world and has three vertical loops, two corkscrew loops, and two boomerang loops.

Solve. Recall that 5,280 feet equal 1 mile.

1. The Shock Wave is 170 ft high at its highest point. What part of a mile is this? Give the answer to the nearest hundredth of a mile. 0.03

2. The track length of the Shock Wave is 3,900 ft. Is this closer to $\frac{2}{3}$ mile or $\frac{3}{4}$ mile? Justify your answer. Closer to $\frac{3}{4}$ mi, since 3,900 ÷ 5,280 = 0.74.

3. A ride on the Shock Wave takes about 2 min 20 s. What is the average rate of speed in feet per second? Give the answer to the nearest tenth. 27.9 ft/s

4. Jerry stood in line for 42 min waiting to ride on the Shock Wave. If he could have used that time to take one ride after another without stopping, how many rides could he have taken? 18

5. The Shock Wave has 7 cars per train with 4 passengers per car. How many train loads would it take for 1,176 passengers to ride the Shock Wave? 42

6. If one trainload of passengers departs every 5 min, how many hours would it take for the 1,176 passengers to ride? $3\frac{1}{2}$ h

3.7 Proving Statements (Optional)

Objectives To give a reason for each step in the solution of an equation
To prove theorems about real numbers

Here is a summary of the properties of operations that have been introduced so far in this book. Unless otherwise stated, a property applies to all real numbers a, b, and c.

Name of Property	Statement of Property
Closure for Addition	$a + b$ is a real number.
Closure for Multiplication	ab is a real number.
Commutative of Addition	$a + b = b + a$
Commutative of Multiplication	$ab = ba$
Associative of Addition	$(a + b) + c = a + (b + c)$
Associative of Multiplication	$(ab)c = a(bc)$
Distributive for Multiplication over Addition	$a(b + c) = ab + ac$ $(b + c)a = ba + ca$

Additive Identity	There is exactly one real number, 0, such that for each real number a, $a + 0 = a$ and $0 + a = a$.
Multiplicative Identity	There is exactly one real number, 1, such that for each real number a, $a \cdot 1 = a$ and $1 \cdot a = a$.
Additive Inverse	For each real number a, there is exactly one real number $-a$ such that $a + (-a) = 0$ and $-a + a = 0$.
Multiplicative Inverse	For each nonzero real number a, there is exactly one real number $\frac{1}{a}$ such that $a \cdot \frac{1}{a} = 1$ and $\frac{1}{a} \cdot a = 1$.

The following *properties of equality* are also true.

Properties of Equality
For all real numbers a, b, and c,

Reflexive Property	$a = a$ (A number is equal to itself.)
Symmetric Property	If $a = b$, then $b = a$.
Transitive Property	If $a = b$ and $b = c$, then $a = c$.
Substitution Property	If $a = b$, then a can be replaced by b, and b by a.

Teaching Resources

Quick Quizzes 25
Reteaching and Practice Worksheets 25

■ GETTING STARTED

Prerequisite Quiz

What property is illustrated?

1. $(a \cdot b) \cdot c = a \cdot (b \cdot c)$ Assoc Prop Mult
2. $a + [b + (-b)] = a + 0$ Add Inv Prop
3. $a \cdot (b \cdot \frac{1}{b}) = a \cdot 1$ Mult Inv Prop
4. $b \cdot 1 = 1 \cdot b$ Comm Prop Mult
5. $1 \cdot b = b$ Identity Prop Mult

Motivator

Have students list the properties of operations they have studied so far in the textbook. Have them give a numerical example to illustrate each property they list. Answers will vary.

Highlighting the Standards

Standards 3c, 3d, 3e: Throughout this lesson, students follow logical arguments (Examples 1–3), judge the validity of arguments (Exercises), and construct their own arguments (Written Exercises 9–14).

115

Lesson Note

This lesson is optional and may be assigned as enrichment for more capable students. You may wish to point out to the students that the format for proofs is similar to the format used for two-column geometric proofs. The Mixed Problem Solving on page 119 is appropriate for all students.

Math Connections

Law: To prove a case in court, a lawyer must use logical arguments, containing statements in a logical sequence that are based on facts that can be proved and on the laws of the state and country in which the case is being tried.

Critical Thinking Questions

Analysis: Ask students which, if any, of the four Properties of Equality will be true if the = sign were replaced with the symbol, <. Only the Transitive Property

---

When you solve an equation, it is possible to give a reason for each step by stating one of these properties or an equation property.

EXAMPLE 1   Solve $7x + 9 = -5$. Give a reason for each step.

Proof

	Statement		Reason
1.	$7x + 9 = -5$	1.	Given
2.	$7x + 9 - 9 = -5 - 9$	2.	Subt Prop of Eq
3.	$7x + 0 = -14$	3.	Add Inverse Prop
4.	$7x = -14$	4.	Add Identity Prop
5.	$\frac{7x}{7} = \frac{-14}{7}$	5.	Div Prop of Eq
6.	$1 \cdot x = -2$	6.	Mult Inverse Prop $\left[\frac{7}{7} = 7 \cdot \frac{1}{7}\right] = 1$
7.	$x = -2$	7.	Mult Identity Prop

TRY THIS   1. Solve $x - 3 = 2$. Give a reason for each step.   Check students' proofs.

A **theorem** is a statement that has been proved. In statements of theorems, variables represent real numbers.

EXAMPLE 2   Prove $(a + b) + (-a) = b$.

Proof

	Statement	Reason
1.	$(a + b) + (-a) = (b + a) + (-a)$	1. Comm Prop of Add
2.	$= b + [a + (-a)]$	2. Assoc Prop of Add
3.	$= b + 0$	3. Add Inverse Prop
4.	$= b$	4. Add Identity Prop
5.	$(a + b) + (-a) = b$	5. Trans Prop of Eq

TRY THIS   2. Prove: $a + (b + c) = (b + a) + c$ Check students' proofs.

Recall that the Multiplication Property of Equality states that for all real numbers a, b, and c, if $a = b$, then $ac = bc$. Note that the property contains an *if-clause* (if $a = b$) and a *then-clause* (then $ac = bc$).

A statement that is written in if-then form is called a **conditional**. The if-clause is called the **hypothesis**, and the then-clause is called the **conclusion**.

Some conditionals, such as the Multiplication Property of Equality, can be proved to be true. To prove a conditional, you first assume that the hypothesis is true. Then you show by a series of logical steps that the conclusion must follow. For each step, or *statement*, in the proof, you must give a *reason* that has been previously accepted as true. A proof of the Multiplication Property of Equality is shown in Example 3.

116   Chapter 3 Solving Equations

Additional Example 1

Solve $\frac{1}{2}x - 3 = 11$. Give a reason.

Statements	Reasons
1. $\frac{1}{2}x - 3 = 11$	1. Given
2. $\frac{1}{2}x - 3 + 3 = 11 + 3$	2. Add Prop Eq
3. $\frac{1}{2}x + 0 = 14$	3. Add Inv Prop
4. $\frac{1}{2}x = 14$	4. Iden Prop Add
5. $2 \cdot \frac{1}{2}x = 2 \cdot 14$	5. Mult Prop Eq
6. $1 \cdot x = 28$	6. Mult Inv Prop
7. $x = 28$	7. Iden Prop Mult

Additional Example 2

Prove: $a \cdot 0 = 0$

Statements	Reasons		
1. $a \cdot 0 = a(0 + 0)$	1. Iden Prop Add	5. $0 = 0 + a \cdot 0$	5. Add Inv Prop
2. $a \cdot 0 = a \cdot 0 + a \cdot 0$	2. Distr Prop	6. $0 = a \cdot 0$	6. Iden Prop Add
3. $-(a \cdot 0) + a \cdot 0 = -(a \cdot 0) + [a \cdot 0 + a \cdot 0]$	3. Add Prop Eq	7. $a \cdot 0 = 0$	7. Sym Prop Eq
4. $-(a \cdot 0) + a \cdot 0 = [-(a \cdot 0) + a \cdot 0] + a \cdot 0$	4. Assoc Prop Add		

116

EXAMPLE 3   Prove: If $a = b$, then $ac = bc$.

Proof

Statement	Reason
1. $a = b$	1. Given
2. $ac = ac$	2. Reflex Prop
3. $ac = bc$	3. Sub(b is substituted for a.)

Thus, if $a = b$, then $ac = bc$.

TRY THIS   3. Prove: If $a = b$ then $ca = cb$. Check students' proofs.

Classroom Exercises

Give a reason for each step.

1.

Statement	Reason
1. $3y - 4 = -19$	1. Given
2. $3y - 4 + 4 = -19 + 4$	2. ____?____ Add Prop Eq
3. $3y + 0 = -15$	3. ____?____ Add Inverse Prop
4. $3y = -15$	4. ____?____ Add Identity Prop
5. $\frac{3y}{3} = \frac{-15}{3}$	5. ____?____ Div Prop Eq
6. $1 \cdot y = -5$	6. ____?____ Mult Inverse Prop
7. $y = -5$	7. ____?____ Mult Identity Prop

Give a reason for each statement.

2. If $x = y$ and $y = z$, then $x = z$.

3. If $x = y$, then $y = x$. Sym Prop

4. $(x + y) + (-y) = x + [y + (-y)]$

5. $x \cdot y = x \cdot y$ Reflex Prop

6. $x + [y + (-y)] = x + 0$

7. $x + y = x + y$ Reflex Prop

8. $y \cdot \left(x \cdot \frac{1}{x}\right) = y \cdot 1$ Mult Inverse Prop

9. $(x \cdot y) \cdot \frac{1}{y} = x \cdot \left(y \cdot \frac{1}{y}\right)$ Assoc Prop Mult

Written Exercises

Give a reason for each step.

1.

Statement	Reason
1. $\frac{2}{3}x + 5 = -9$	1. Given
2. $\frac{2}{3}x + 5 - 5 = -9 - 5$	2. ____?____ Subt Prop Eq
3. $\frac{2}{3}x + 0 = -14$	3. ____?____ Add Inverse Prop
4. $\frac{2}{3}x = -14$	4. ____?____ Add Identity Prop
5. $\frac{3}{2} \cdot \frac{2}{3}x = \frac{3}{2}(-14)$	5. ____?____ Mult Prop Eq
6. $1 \cdot x = -21$	6. ____?____ Mult Inverse Prop
7. $x = -21$	7. ____?____ Mult Identity Prop

Common Error Analysis

Error: Some students confuse the Additive Identity with the Additive Inverse.

Remind them that "identical" means "the same," and that another word for "inverse" is "opposite."

Checkpoint

1. Solve $2(3x - 2) = 38$. Give a reason.

Statements	Reasons
1. $2(3x - 2) = 38$	1. Given
2. $6x - 4 = 38$	2. Distr Prop
3. $6x - 4 + 4 = 38 + 4$	3. Add Prop Eq
4. $6x + 0 = 42$	4. Add Inv Prop
5. $6x = 42$	5. Add Iden Prop
6. $\frac{6x}{6} = \frac{42}{6}$	6. Div Prop Eq
7. $1 \cdot x = 7$	7. Mult Inv Prop
8. $x = 7$	8. Iden Prop Mult

2. Prove: If $a = b$, then $a - c = b - c$. (Subtraction Property of Equality)

Statements	Reasons
1. $a = b$	1. Given
2. $a - c = a - c$	2. Reflex Prop Eq
3. $a - c = b - c$	3. Sub

Closure

List the names of the 11 Field Properties on the chalkboard or overhead projector. Have students close their books and then call on volunteers to fill in the statements for each of the properties.

Additional Example 3

Prove: If $c \neq 0$ and $a = b$, then $\frac{a}{c} = \frac{b}{c}$.

Statements	Reasons
1. $a = b, c \neq 0$	1. Given
2. $\frac{a}{c} = \frac{a}{c}$	2. Reflex Prop Eq
3. $\frac{a}{c} = \frac{b}{c}$	3. Sub

Guided Practice

Classroom Exercises 1–9
Try This all

Independent Practice

A Ex. 1–4, **B** Ex. 5–12, **C** Ex. 13–14

Basic: WE 1–4 all, Mixed Problem Solving

Average: WE 2–4 all, 5–11 odd, Mixed Problem Solving

Above Average: WE 2–4 all, 5–7 odd, 13, Mixed Problem Solving

Additional Answers

Classroom Exercises

2. Trans Prop
4. Assoc Prop Add
6. Add Inverse Prop

Written Exercises

5.
Statements	Reasons
1. $5x - 3 = 18$	1. Given
2. $5x - 3 + 3 = 18 + 3$	2. Add Prop Eq
3. $5x + 0 = 21$	3. Add Inv Prop
4. $5x = 21$	4. Add Iden Prop
5. $\frac{5x}{5} = \frac{21}{5}$	5. Div Prop Eq
6. $1 \cdot x = 4\frac{1}{5}$	6. Mult Inv Prop
7. $x = 4\frac{1}{5}$	7. Mult Iden Prop

6.
Statements	Reasons
1. $\frac{3}{4}y + 8 = -1$	1. Given
2. $\frac{3}{4}y + 8 - 8 = -1 - 8$	2. Subt Prop Eq
3. $\frac{3}{4}y + 0 = -9$	3. Add Inv Prop
4. $\frac{3}{4}y = -9$	4. Add Iden Prop
5. $\frac{4}{3} \cdot \frac{3}{4}y = \frac{4}{3} \cdot (-9)$	5. Mult Prop Eq
6. $1 \cdot y = -12$	6. Mult Inv Prop
7. $y = -12$	7. Mult Iden Prop

See page 119 for the answers to Ex. 7–10.

Write the missing reasons in each proof. All variables represent real numbers.

2. Prove: If $a = b$, then $a + c = b + c$. (Add Prop of Eq)

Statement	Reason
1. $a = b$	1. Given
2. $a + c = a + c$	2. _____?_____ Reflex Prop
3. $a + c = b + c$	3. _____?_____ Sub (b for a)

3. Prove: If $a + c = b + c$, then $a = b$.

Statement	Reason
1. $a + c = b + c$	1. Given
2. $(a + c) + (-c) = (b + c) + (-c)$	2. _____?_____ Add Prop of Eq
3. $a + [c + (-c)] = b + [c + (-c)]$	3. _____?_____ Associative Prop Add
4. $a + 0 = b + 0$	4. _____?_____ Add Inverse Prop
5. $a = b$	5. _____?_____ Add Identity Prop

4. Prove: If $x \neq 0$, then $(xy)\frac{1}{x} = y$.

Statement	Reason
1. $x \neq 0$	1. _____?_____ Given
2. $(xy)\frac{1}{x} = (yx)\frac{1}{x}$	2. _____?_____ Comm Prop Mult
3. $\quad = y\left(x \cdot \frac{1}{x}\right)$	3. _____?_____ Assoc Prop Mult
4. $\quad = y \cdot 1$	4. _____?_____ Mult Inverse Prop
5. $\quad = y$	5. _____?_____ Mult Identity Prop
6. $(xy)\frac{1}{x} = y$	6. _____?_____ Trans Prop of Eq

Solve each equation. Give a reason for each step.

5. $5x - 3 = 18$ **6.** $\frac{3}{4}y + 8 = -1$ **7.** $7 - 5a = 4$ **8.** $3(x + 2) = -15$

Write a proof for each statement. All variables represent real numbers.

9. If $a = b$, then $a - c = b - c$.
10. If $a - c = b - c$, then $a = b$.
11. $(ax + b) + ay = a(x + y) + b$
12. $x + (3 + y) = 3 + (y + x)$
13. $mx + (a + x) = a + (m + 1)x$
14. $a(b - c) = ab - ac$

Mixed Review

Evaluate. *2.8, 2.10*

1. $5x + 8$ for $x = 3.2$ 24
2. $-9 - 2z$ for $z = 5.4$ -19.8
3. $-6y - 4.8$ for $y = -8.1$ 43.8
4. $-6.8 + 4a$ for $a = 6.8$ 20.4
5. $-5 - 4(2x + 5)$ for $x = 3.5$ -53
6. $3(8 - 4y) + 2.6$ for $y = 3.1$ -10.6

Enrichment

Have the students try this problem. In the Kingdom of Omo, the unit of currency is the moo. One moo equals 3 blobs and 1 blob equals 5 drubs. If an Omoan buys a hat that costs 2 moos, in how many different ways might he pay for it? 12

	Moo	Blob	Drub
1.	2	0	0
2.	1	3	0
3.	1	2	5
4.	1	1	10
5.	1	0	15
6.	0	6	0
7.	0	5	5
8.	0	4	10
9.	0	3	15
10.	0	2	20
11.	0	1	25
12.	0	0	30

Mixed Problem Solving

Recall that the first step in problem solving is to understand the problem. Read each problem carefully to identify the information you need. Think about what you are asked to find and the units you will use in writing the final answer. After solving the problem, ask: "Is there another way to solve this problem?"

1. Jean received time-and-a-half pay for working on her day off. This means that she received $1\frac{1}{2}$ times her normal pay. If she received $9.60/h working on her day off, what is her normal hourly pay? $6.40/h

2. The long-distance telephone rate for calling California is 32¢ for the first minute and 20¢ for each additional minute. Judy's call to California cost $2.32. How long did she talk? 11 min

3. Jane and Kai are both selling their bicycles. Jane is charging $72. Her price is $18 less than Kai's price. How much is Kai charging for his bicycle? $90

4. For babysitting, Jerry charges a $2.50 transportation fee plus $3.75 an hour. One evening he earned $28.75. How many hours did he babysit? 7 h

5. On Sunday, 12,450 people came to a concert. Sunday's crowd was $\frac{3}{4}$ as large as Saturday's crowd. How many people came on Saturday? 16,600

6. Morris bought 8 pairs of socks and a T-shirt. His bill was $18.50 without tax. The T-shirt was $6.50. What was the price of each pair of socks? $1.50

7. Martha bought 4 doz cookies. She put 3 cookies aside and divided the rest equally among 5 people. How many cookies did each of the 5 people receive? 9

8. On Tuesday morning, Justin's stock opened at $24\frac{3}{4}$. It had dropped $\frac{3}{8}$ of a point since the closing price on Monday. What was the stock's closing price on Monday? $25\frac{1}{8}$

9. To buy beach tokens at Paradise Beach, one must pay $25.00 for a family membership plus $3 additional for each child's token. Mrs. Breckel spent $37 for a membership and tokens. How many child's tokens did she buy? 4

10. Bob was paid $15 for proofreading an article plus 75¢ for each error he could find. He earned a total of $24.75 for his work. How many errors did he find? 13

11. After Jim withdrew $24.50 from his savings account, his balance was $276.83. How much was in his account before he made the withdrawal? $301.33

12. First-class postage costs 25¢ for the first ounce and 20¢ for each additional ounce. Joe spent $2.05 to mail a package first class. How much did it weigh? 10 oz

7.

Statements	Reasons
1. $7 - 5a = 4$	1. Given
2. $7 + (-5a) = 4$	2. Def Subt
3. $-7 + 7 + (-5a) = -7 + 4$	3. Add Prop Eq
4. $0 + (-5a) = -3$	4. Add Inv Prop
5. $-5a = -3$	5. Add Iden Prop
6. $\frac{-5a}{-5} = \frac{-3}{-5}$	6. Div Prop Eq
7. $1 \cdot a = \frac{3}{5}$	7. Mult Inv Prop
8. $a = \frac{3}{5}$	8. Mult Iden Prop

8.

Statements	Reasons
1. $3(x + 2) = -15$	1. Given
2. $3x + 6 = -15$	2. Distr Prop
3. $3x + 6 - 6 = -15 - 6$	3. Subt Prop Eq
4. $3x + 0 = -21$	4. Add Inv Prop
5. $3x = -21$	5. Add Iden Prop
6. $\frac{3x}{3} = \frac{-21}{3}$	6. Div Prop Eq
7. $1 \cdot x = -7$	7. Mult Inv Prop
8. $x = -7$	8. Mult Iden Prop

9.

Statements	Reasons
1. $a = b$	1. Given
2. $a - c = a - c$	2. Reflex Prop
3. $a - c = b - c$	3. Sub Prop

10.

Statements	Reasons
1. $a - c = b - c$	1. Given
2. $a - c + c = b - c + c$	2. Add Prop Eq
3. $a + 0 = b + 0$	3. Add Inv Prop
4. $a = b$	4. Add Iden Prop

See page 650 for the answers to Ex. 11–14.

Chapter 3 Review

Key Terms

Addition Property of Equality (p. 89)
conclusion (p. 116)
conditional (p. 116)
Division Property of Equality (p. 95)
equivalent equations (p. 89)

hypothesis (p. 116)
identity (p. 103)
if-clause (p. 116)
Multiplication Property of Equality (p. 94)
Subtraction Property of Equality (p. 90)
then-clause (p. 116)
theorem (p. 116)

Key Ideas and Review Exercises

3.1, 3.2 Adding or subtracting the same number to each side of an equation produces an equivalent equation. Also, multiplying by the same number or dividing by the same nonzero number on each side of an equation produces an equivalent equation.

Solve and check.

1. $x - 9 = -7$ 2
2. $8 = 14 + y$ -6
3. $-8 + z = 12$ 20

4. $x - \frac{2}{3} = -\frac{1}{6}$ $\frac{1}{2}$
5. $3x = -24$ -8
6. $19 = 5x$ $3\frac{4}{5}$

7. $15 = -\frac{3}{5}d$ -25
8. $2 = \frac{-x}{4}$ -8
9. $-5 + x = -5$ 0

10. $-7 = -\frac{2}{3}c$ $10\frac{1}{2}$
11. $1\frac{2}{3}d = -20$ -12
12. $3.2 + y = -7.4$
  -10.6

13. Explain in writing why you would multiply each side of $-\frac{1}{3}x = 7$ by -3 in order to solve it. Answers will vary.

3.3, 3.4 To solve the equation $-6x + 7 = -5$, first subtract 7 from each side. Then divide each side by -6.

To solve the equation $8x - 7 = 2x + 11$, you can
(1) subtract $2x$ from each side to get $6x - 7 = 11$,
(2) add 7 to each side to get $6x = 18$, and
(3) divide each side by 6 to get $x = 3$.

Solve and check.

14. $4x - 7 = 5$ 3
15. $\frac{x}{8} + 6 = 7$ 8
16. $\frac{2}{3}c + 5 = 9$ 6
17. $7x - 30 = 2x$ 6

18. $-4y = 6y + 20$
19. $16 = -3y - 2$
20. $7 + y = 9 - y$
21. $2.5x + 0.4 = -4.6$

22. $-3x + 2 - 5x = -14$ 2
23. $3x - 1 = 4x + 7 - x$ ∅

Translate each problem into an equation and solve.

24. Three times Tara's age decreased by 2 years is the same as twice her age increased by 13 years. How old is Tara? 15

25. The length of a rope increased by 5 ft is the same as 3 ft less than twice the length of the rope. Find the length of the rope. 8 ft

3.5   To solve an equation containing parentheses, use the Distributive Property to remove the parentheses. Combine like terms on each side, if necessary. Then solve the equation.

Solve and check.

26. $3(x - 4) = 18$ 10

27. $x + 10 = 3(2 - x)$ −1

28. $-(-2a + 3) + 18 = 3(a - 5)$ 30

29. $-(-6 + 2y) - 4y = -3(6 - 2y)$ 2

30. $3r - 5(7 - r) = 5r - (2r - 10)$ 9

31. $-7x - (8 - 12x) = 2x$ $2\frac{2}{3}$

Translate into an equation and solve.

32. Six times the difference between a two-digit number and 9 is 24. Find the number. 13

3.6   Some formulas can be used to solve problems using the equation properties. Substitute known values into the formula and solve for the variable that does not have a given value.

33. Using $d = rt$, find d for $r = 150$ km/h and $t = 11$ h. 1,650 km

34. Using $p = nt - e$, find the price of a carnival ticket t if the expenses e are $800, the profit p is $2,500, and if the number of sold tickets n is 2,000. $1.65

3.7   To prove a theorem, show that it is true by a series of logical steps. Give a reason for each step.

35. Write the missing reasons in the proof. All variables represent real numbers.

Statement	Reason
1. $a \cdot 0 = a \cdot 0 + 0$	1. ____?____ Add Identity Prop
2. $\quad = a \cdot 0 + a + (-a)$	2. ____?____ Add Inverse Prop
3. $\quad = a \cdot 0 + a \cdot 1 + (-a)$	3. ____?____ Mult Identity Prop
4. $\quad = a(0 + 1) + (-a)$	4. ____?____ Distr Prop
5. $\quad = a \cdot 1 + (-a)$	5. ____?____ Add Identity Prop
6. $\quad = a + (-a)$	6. ____?____ Mult Identity Prop
7. $\quad = 0$	7. ____?____ Add Inverse Prop
8. $a \cdot 0 = 0$	8. ____?____ Trans Prop of Eq

Chapter 3 Review   **121**

Chapter 3 Test

A-Level Exercises: 1–5, 7, 10–12, 15–19, 21
B-Level Exercises: 6, 8–9, 13–14, 20

C-Level Exercises: 23

Solve and check.

1. $x - 14 = -5$ 9

2. $12 = -4 + x$ 16

3. $-\frac{2}{3}x + 5 = 3$ 3

4. $-2x + 5 - x = 17 + x$ −3

5. $-6 + x = -4$ 2

6. $0.4 + x = -3.7$ −4.1

7. $6x = -54$ −9

8. $-3x = 3\frac{1}{3}$ $-1\frac{1}{9}$

9. $-x - 4 = 4x - 5 - 5x + 1$

10. $12 - 5y = -3$ 3

11. $4(x + 2) = 20$ 3

12. $2(z - 3) = z + 5$ 11

13. $z - \frac{2}{3} = -\frac{5}{6}$ $-\frac{1}{6}$

14. $6 + \frac{-x}{3} = 15$ −27

Translate each problem into an equation and solve.

15. Eight more than a number is 42. Find the number. 34

16. Jim worked 12 h on a science project. He worked $\frac{3}{4}$ as long as Jessie worked. How long did Jessie work? 16 h

17. Three less than twice the temperature is −17. Find the temperature. −7 degrees

18. Bob's salary increased by $98 is the same as three times his salary. Find his salary. $49

19. Six times the difference between Margot's age and 2 is 54. Find Margot's age. 11 yr old

20. Use the formula $T = 3(ab - c)$ to find b for $T = 18$, $a = 3$, and $c = 15$. 7

21. If a number is decreased by 6 and then multiplied by −5, the result is 12 less than twice the number. Find the number. 6

22. Write the missing reasons in this proof. All variables represent real numbers.

Prove: If $a = b$ and $c \neq 0$, then $\frac{a}{c} = \frac{b}{c}$.

Statement	Reason
1. $a = b$ and $c \neq 0$	1. _____?_____ Given
2. $\frac{a}{c} = \frac{a}{c}$	2. _____?_____ Reflex Prop Eq
3. $\frac{a}{c} = \frac{b}{c}$	3. _____?_____ Substitution

Thus, if $a = b$ and $c \neq 0$, then $\frac{a}{c} = \frac{b}{c}$.

23. Solve $-5x - [2(1 - 3x) + 4] = 7x$. −1

Test-taking Strategy

Since time is an important factor in test-taking, it is important to look for properties that offer shortcuts in finding answers. Consider the following example.

If $5x + 10 = 30$, find the value of $x + 2$.

By the Distributive Property, $5x + 10 = 5(x + 2)$. Instead of solving $5x + 10 = 30$ for x, solve it for $x + 2$.

Use the Distributive Property. $5(x + 2) = 30$

Divide each side by 5. $\dfrac{5(x + 2)}{5} = \dfrac{30}{5}$

$$x + 2 = 6$$

Thus, the value of $x + 2$ is 6.

Choose the one best answer to each question or problem.

1. If $3x - 12 = 18$, find the value of $x - 4$.
   (A) 10   (B) 6   (C) 9
   (D) 8   (E) None of these   B

2. If $2x - 8 = 7$, find the value of $10x - 40$.
   (A) $\frac{7}{2}$ (B) 14   (C) 35
   (D) 56   (E) None of these   C

3. If $x + y = 3.9 + x$, find the value of y.
   (A) 0   (B) -3.9 (C) 3.9
   (D) x (E) None of these   C

4. If $5x - 9 = 23$, find the value of $5x - 7$.
   (A) 25   (B) -2 (C) 21
   (D) -23 (E) None of these   A

5. If $9y + 8 = -16$, find the value of $9y - 2$.
   (A) 0   (B) 10   (C) -6
   (D) -24 (E) None of these   E

6. If $\frac{c}{9} = 2$, find the value of $\frac{c}{2}$.
   (A) $\frac{1}{9}$ (B) $\frac{1}{2}$ (C) 2
   (D) 9   (E) None of these   D

7. If $3(2 - 4x) = 17$, find the value of $6 - 12x$.
   (A) 6   (B) 8   (C) 17
   (D) 51   (E) None of these   C

8. If $\frac{3}{4}x = -9$, find the value of $\frac{1}{4}x$.
   (A) -3 (B) 3   (C) $\frac{1}{2}$
   (D) $-\frac{1}{2}$ (E) None of these   A

9. If a number is decreased by 5 and then multiplied by 3, the result is 26. Find the result if the same number is decreased by 5 and then multiplied by 6.
   (A) 12   (B) 13   (C) 21
   (D) 52   (E) None of these   D

10. The formula for the area of a circle with radius r is $A = \pi r^2$. Find the area of the ring shaded below.

   (A) 4π (B) 25π (C) 8π
   (D) 24π (E) None of these   E

OVERVIEW

This chapter provides students with opportunities to apply the equation solving skills they have learned to solving various types of word problems. Problems include those involving two or more numbers, consecutive integers, perimeter, and angle measure. After solving equations involving fractions and decimals, students apply these skills to percent problems dealing with sales tax, profit and loss, discount, and commission.

OBJECTIVES

- To solve problems involving two or more numbers
- To solve problems involving perimeter and angle measure
- To solve equations that contain fractions and decimals
- To solve problems involving percent and its applications

PROBLEM SOLVING

In this chapter, the students continue to write equations to solve problems. Using a Formula is the basis for the lessons on perimeter, angle measure, and the various applications of percent. Interpreting Key Words is essential to all types of problems presented. The focus of the problem-solving strategy lesson on page 157 is on the third step of the Polya model, Carry Out the Plan.

READING AND WRITING MATH

Lesson 4.1 gives students practice in writing expressions to represent relationships between numbers. This is also the emphasis of the Focus on Reading on page 137. Every lesson in the remainder of the chapter provides practice with reading a stated problem and rewriting it as an algebraic statement.

TECHNOLOGY

Calculator: Several of the lessons, particularly those involving decimals and percents, offer an opportunity for students to check their solutions with a calculator.

SPECIAL FEATURES

Mixed Review pp. 128, 134, 138, 146, 149, 153, 156
Brainteaser p. 134
Focus on Reading p. 137
Midchapter Review p. 143
Application: Gear Depths p. 143
Problem Solving Strategies: Carrying Out the Plan p. 157
Key Terms p. 158
Key Ideas and Review Exercises p. 158
Chapter 4 Test p. 160
College Prep Test p. 161
Cummulative Review (Chapters 1–4) p. 162

PLANNING GUIDE

Lesson	Basic	Average	Above Average	Resources
4.1 pp. 126–128	CE 1–9 WE 1–17 all	CE 1–9 WE 5–22 all	CE 1–9 WE 9–27 all	Reteaching 26 Practice 26
4.2 pp. 132–134	CE 1–14 WE 1–21 odd Brainteaser	CE 1–14 WE 9–29 odd Brainteaser	CE 1–14 WE 17–37 odd Brainteaser	Reteaching 27 Practice 27
4.3 pp. 137–138	FR all CE 1–23 WE 1–10, 15–17 all	FR all CE 1–23 WE 1–13, odd 15–21 all	FR all CE 1–23 WE 9–13 odd, 15–25 all	Reteaching 28 Practice 28
4.4 pp. 141–143	CE 1–7 WE 1–10 all Midchapter Review Application	CE 1–7 WE 5–14 all Midchapter Review Application	CE 1–7 WE 1–20 all Midchapter Review Application	Reteaching 29 Practice 29
4.5 pp. 145–146	CE 1–8 WE 1–12 all, 15	CE 1–8 WE 1–20 odd	CE 1–8 WE 5–21 odd, 22, 23	Reteaching 30 Practice 30
4.6 pp. 148–149	CE 1–12 WE 1–19 odd, 25	CE 1–12 WE 7–29 odd	CE 1–12 WE 11–33 odd	Reteaching 31 Practice 31
4.7 pp. 152–153	CE 1–19 WE 1–13, 19–27 odd	CE 1–19 WE 9–33 odd	CE 1–19 WE 11–17, 25–37 odd, 38	Reteaching 32 Practice 32
4.8 pp. 155–157	CE 1–6 WE 1–10 all Problem Solving Strategies	CE 1–6 WE 4–13 all Problem Solving Strategies	CE 1–6 WE 7–16 all Problem Solving Strategies	Reteaching 33 Practice 33
Chapter 4 Review pp. 158–159	all	all	all	
Chapter 4 Test p. 160	all	all	all	
College Prep Test p. 161	all	all	all	
Cumulative Review (Chaps. 1–4) pp. 162–163	all	all	all	

CE = Classroom Exercises   WE = Written Exercises   FR = Focus on Reading

Note: For each level, all students should be assigned all Try This and all Mixed Review Exercises.

■ INVESTIGATION

Project: This investigation shows students how to use a calculator to find decimal values for common fractions to any number of decimal places.

Materials: Each pair of students will need a copy of Project Worksheet 4 and a calculator.

Have students use a calculator to calculate the decimal value of $\frac{5}{12}$. 0.41$\overline{6}$ Point out that this is a repeating decimal and show them the correct way to write the answer. Then ask students to find $\frac{9}{17}$ on the calculator. Have them work along with you as you demonstrate the following method for finding the exact decimal value of $\frac{9}{17}$.

$9 \div 17 = 0.5294\ 1176$ (rounded to 8 places)

Step 1: Multiply the last four digits rewritten as 0.1176 by the divisor, 17.

$$0.1176 \times 17 = 1.9992$$

Step 2: Round this result to a whole number and divide by 17.

$$2 \div 17 = 0.1176\ 4706$$

Now there are 12 decimal places.

$$0.5294\ 1176\ 4706$$

Repeat steps 1 and 2 until the decimal terminates or starts to repeat. For $\frac{9}{17}$, this will happen after two more repetitions. So $\frac{9}{17} = 0.\overline{5294\ 1176\ 4705\ 8823}$.

Work one or two more examples from the worksheet. Then have students work in pairs so that one of them can record the digits four at a time as the other works with the calculator.

Jim Abbott is a pitcher with the California Angels baseball team. Born with only one hand, Jim taught himself to handle a baseball and glove. After pitching for the USA Olympic team in 1988, he went directly to the major leagues.

More About the Photo

In baseball, pitching performance is measured by a ratio called earned-run average. Earned-run average (abbreviated ERA) is the average number of runs a pitcher allows in nine innings. A pitcher's ERA is calculated by first finding the ratio of earned runs to innings pitched and then multiplying the result by nine. In his rookie season, Jim Abbott's ERA was 3.93. He pitched 181 innings and gave up 79 earned runs. The sports page of the newspaper lists other baseball statistics such as percent of games won and batting averages.

4.1 Translating English to Algebra

Objective

To represent two or more numbers in terms of one variable, given the relationship between the numbers

Frequently, a word problem will contain an English sentence that describes a relationship between two or more numbers. To write an equation used to solve the problem, you will have to determine:

(1) what numbers you are asked to find,
(2) which *one* of the numbers will be represented by a variable, such as x, and
(3) how to represent other numbers in terms of that variable.

EXAMPLE 1

The larger of two numbers is twice the smaller. Write representations of the two numbers.

Plan

There is a larger number and a smaller number. Use the *smaller* as the basis of comparison (larger . . . is twice smaller).

Solution

Let s = the *smaller* number.
Then $2s$ = the larger number. $\longleftarrow$ twice the smaller number

EXAMPLE 2

The price of a book is $3 less than twice the price of a tape. Answer each of the following questions.

Solutions

a. What are the two unknown numbers being compared? The unknown numbers are the book price and the tape price.

b. Which unknown number is the basis of comparison? (THINK: The book price is 3 *less than* twice the *tape* price.) The tape price is the basis of comparison.

c. Represent the two unknowns in terms of one variable.

Choose a variable to represent the number that is the basis of comparison. Represent the other number in terms of that variable.

Let x = the tape price in dollars. $\longleftarrow$ basis of comparison

Then $2x - 3$ = the book price in dollars. $\longleftarrow$ 3 less than twice the *tape* price.

TRY THIS

1. Juan's age is 5 years more than three times Manuel's age. Let a represent Manuel's age. Represent Juan's age in terms of a. $3a + 5$

2. The width of a rectangle is 8 cm less than twice its length. Let l represent the length. Represent the width of the rectangle in terms of l. $2l - 8$

Teaching Resources

Project Worksheet 4
Quick Quizzes 26
Reteaching and Practice
   Worksheets 26
Transparency 8

■■■ GETTING STARTED

Prerequisite Quiz

Write an algebraic expression for each word description.

1. five more than a number y $y + 5$
2. eight less than n $n - 8$
3. x divided by 2 $\frac{x}{2}$
4. three more than twice y $2y + 3$
5. six less than 5 times x $5x - 6$
6. four less than half of z $\frac{1}{2}z - 4$
7. five more than 3 times $(c + 6)$
   $3(c + 6) + 5$

Motivator

Have students consider these questions: "Susan has saved twice as much as Chris. If Chris has saved $150, how much has Susan saved? $300   If Chris has saved x dollars, what expression represents how much Susan has saved?" $2x$ Point out that Chris' savings are used as the basis of comparison.

Additional Example 1

Jack's age is 1 year less than twice Paul's age. Use this information to answer each question.

a. What are the two unknown ages being compared?   Jack's age, Paul's age
b. Which age is the basis of comparison?
   Paul's age
c. Represent the two ages in terms of one variable. $x, 2x - 1$

Additional Example 2

The smaller of two numbers is 7 less than the larger. Write algebraic representations for the two numbers. $x, x - 7$

Highlighting the Standards

Standard 4c: The applications problems throughout this lesson connect algebra to geometry and to various areas such as sports and business.

Lesson Note

Emphasize finding the quantity that can be used as a basis of comparison. Ask students to give suggestions for deciding which quantity this is. Point out the difference in these expressions.

$3 less x $3 less than x
$3 - x$ $x - 3$

Suggest to the students that if more than two quantities are involved, they should first establish a relationship between two quantities before considering other relationships.

Math Connections

Life Skills: Most grocery stores label items with unit prices as a basis of comparison. If the price of two boxes of rice is compared by the price per ounce, the buyer can determine the better buy regardless of the size of each box.

Critical Thinking Questions

Application: Ask students to consider the following: If a 15 oz. can of peas has a price of 60¢ and another brand of peas comes in a 32 oz. can at a price of $1.12, which is the better buy?   32 oz. can   What other factors might determine which can you will buy?   brand name, amount you need, amount of money you have

---

Recall that the word "of" often indicates *multiplication*.

EXAMPLE 3   This year, the price of a movie ticket increased by one-fifth of the price last year. Let l represent the price last year. Represent the price this year in terms of l.

Solution   Let l = the price of a ticket last year, in dollars.

THINK: price last year increased by $\frac{1}{5}$ of price last year

$$l \qquad + \qquad \frac{1}{5} \times \qquad l$$

Thus, $l + \frac{1}{5}l$ = the price of a ticket this year, in dollars.

EXAMPLE 4   The first side of a triangle is 3 times as long as a second side. The third side is 5 cm shorter than the first side. Represent the lengths of the three sides in terms of one variable.

Plan   In the first sentence, a first side is compared with a *second* side. The *second* side is the basis of comparison.

Solution   Let s = the length of the *second* side in cm.
Then $3s$ = the length of the first side in cm. ←— 3 times length of *second*
So $3s - 5$ = the length of the third side in cm. ←— 5 less than the *first*

TRY THIS   3. Ling has twice as many nickels as dimes. She has 5 more quarters than nickels. If d represents the number of dimes, represent the number of nickels and quarters in terms of d.
$2d$ = number of nickels; $2d + 5$ = number of quarters

Classroom Exercises

Complete each statement.

1. Marcie's age is 2 years more than 3 times Lou's age. Let l = Lou's age. Then __?__ is Marcie's age. $3l + 2$
2. The base of a triangle is 5 cm less than twice the height. Let h = the height in centimeters. Then __?__ is the base in centimeters. $2h - 5$
3. Bill has twice as many basketball points as George. Let g = the number of George's basketball points. Then __?__ is the number of Bill's points. $2g$

In Classroom Exercises 4 and 5, James worked 5 h longer than Henry.

4. How much longer did James work? 5 hours longer
5. Let h = the number of hours worked by Henry. Represent, in terms of h, the number of hours that James worked. $h + 5$

Additional Example 3

This year, the price of a calculator increased by one-tenth of the price last year. Let x represent the price last year. Represent the price this year in terms of x. $x + \frac{x}{10}$

Additional Example 4

The second of three numbers is twice the first number. The third number is 6 more than the second number. Represent the three numbers in terms of x. $x, 2x, 2x + 6$

In Classroom Exercises 6–7, the selling price of a skirt is $4 less than 3 times the cost.

6. What are the two unknown numbers in the sentence? Price and cost
   Which is less, the selling price or the cost? Cost

7. Let c = the cost. Represent, in terms of c, the selling price of the skirt. $3c - 4$

8. Bob is 4 years older than Mary. Write representations for each age in terms of one variable, m. m = Mary's age, $m + 4$ = Bob's age

9. The larger of two numbers is 3 less than 5 times the smaller. Write representations for each number in terms of one variable, s.
   s = smaller number, $5s - 3$ = larger number

Written Exercises

In Exercises 1–4, first choose a variable to represent the smaller of the two unknowns. Then represent the other unknown in terms of the same variable.

1. Jeanie and Bill were working together on an art project. Bill worked 2 h less than Jeanie worked.

2. This week Jack worked $6\frac{1}{2}$ more hours at the supermarket than he worked last week.

3. Kate has half as many fish as Bob has.

4. The larger of two numbers is 3 times the smaller.

Complete each statement.

5. Robert's age is 2 years less than 3 times Matt's age. Let m = Matt's age in years. Then ___?___ is Robert's age in years. $3m - 2$

6. This year Kyle's salary increased by one-fifth of his salary last year. Let l = his salary last year. Then ___?___ is his salary this year. $l + \frac{1}{5}l$

7. A piece of wire is to be cut into two pieces so that one piece is 1 cm longer than 3 times the second piece. Let s = the shorter length in centimeters. Then ___?___ is the greater length in centimeters. $3s + 1$

8. This year a company's earnings decreased by $\frac{1}{3}$ of its earnings last year. Let l = the company's earnings last year. Then ___?___ is the company's earnings this year. $l - \frac{1}{3}l$

Choose a variable to represent one unknown. Then represent the other unknown in terms of that variable.

9. Jason scored one-third as many points in a game as Ron scored.

10. Bobbie has 2 less than 8 times as many stamps as Selise has.

11. Kyle earned $5 more than twice what Josie earned.

12. The length of a rectangle is 2 cm less than 5 times its width.

13. This year the price of a ticket to a baseball game has increased by one-fifth of the price last year.

14. The length of one piece of wire is 1 cm greater than half the length of a second piece of wire.

Common Error Analysis

Error: After having worked with formulas in Chapter 3, some students may try to use more than one variable for a particular problem.

Remind students that they have learned to solve equations with one variable only. Solving equations involving two variables will be studied in a later chapter.

Checkpoint

Kevin has a collection of coins, consisting of nickels, dimes, and quarters. He has 6 fewer dimes than nickels. He has twice as many quarters as nickels.

1. Which coin does he have the most of, nickels, dimes, or quarters? quarters
2. Which coin does he have the least of, nickels, dimes, or quarters? dimes
3. Let n represent the number of nickels Kevin has. Represent the number of dimes in terms of n. $n - 6$
4. Represent the number of quarters in terms of n. $2n$

Closure

Have students summarize their methods for choosing which quantity will serve as the basis of comparison in a word problem.
Answers will vary.

Enrichment

Have students work in pairs. Have each student write five word descriptions, each of which gives the relationship of two or more numbers. Have the students exchange papers and write algebraic expressions in one variable that represent their partner's word descriptions.

Guided Practice

Classroom Exercises 1–9
Try This   all

Independent Practice

A Ex. 1–14, **B** Ex. 15–23, **C** Ex. 24–27

Basic: WE 1–17 all
Average: WE 5–22 all
Above Average: WE 9–27 all

Additional Answers

Written Exercises

1. Let x = the amount of time Jeanie worked. Then $x - 2$ = time Bill worked.
2. Let x = the amount of time Jack worked last week. Then $x + 6.5$ = time worked this week.
3. Let x = the number of Bob's fish. Then $\frac{1}{2}x$ = number of Kate's fish.
4. Let x = the smaller number. Then $3x$ = the larger number.
9. Let x = Ron's points. Then $\frac{1}{3}x$ = Jason's points.
10. Let x = Selise's stamps. Then $8x - 2$ = Bobbie's stamps.
11. Let x = Josie's earnings. Then $2x + 5$ = Kyle's earnings.
12. Let w = width. Then $5w - 2$ = length.
13. Let l = last year's price. Then $l + \frac{1}{5}l$ = this year's price.
14. Let x = length of the second piece. Then $\frac{1}{2}x + 1$ = length of the first piece.

Sarah has a collection of coins, consisting of pennies, nickels, and dimes. She has 6 more pennies than nickels. She has the same number of nickels as dimes.

15. What coin does she have the most of? Pennies
16. Let n = the number of nickels she has. Represent the number of pennies in terms of n. $n + 6$
17. Represent the number of dimes in terms of n. n

A sausage pizza costs \$3.25 more than a cheese pizza and \$4.50 less than a combination pizza.   Combination

18. Which pizza costs the most?   **19.** Which pizza costs the least? Cheese
20. Let s = the price of the sausage pizza. Represent the price of the cheese pizza in terms of s. $s - 3.25$
21. Represent the price of the combination pizza in terms of s. $s + 4.50$

Choose a variable to represent one unknown. Then represent the other unknown(s) in terms of that variable. (Exercises 22–23)

22. On a test, a true-false problem is worth 2 points less than a short-answer problem. An essay problem is worth 1 point more than three times as much as a short-answer problem.
23. One side of a triangle is 3.2 cm longer than the second side. The third side is 1.2 cm shorter than the first side.
24. Acme Car Rental Agency charges \$21 a day plus \$0.19 a mile for a mid-sized car. Represent the number of miles driven in a day and the cost for that day in terms of one variable.
25. To buy a car, Bill plans to save \$80 a month for 15 months and after that to save \$60 a month for the next n months. Represent the total amount saved at the end of n months. $1,200 + 60n$
26. Shira's grandmother is 7 years older than Shira's grandfather, and 6 years older than 10 times Shira's age. Represent Shira's grandfather's age in terms of Shira's age.
27. A town is planning emergency action for a hurricane that is now 330 mi away, but moving towards the town at 15 mi/h. Represent the distance of the hurricane from the town t hours from now.
$330 - 15t$

Mixed Review

Solve and check. *3.1–3.3* $y = 10$
1. $x + 4 = 1$ $x = -3$ **2.** $2t = -8$ $t = -4$ **3.** $\frac{a}{4} = 7$ $a = 28$ **4.** $y - 7 = 3$
5. $2n + 3 = 11$ $n = 4$ **6.** $3n - 4 = 6$ $n = 3\frac{1}{3}$ **7.** $4x + 5 = 1$ $x = -1$ **8.** $2x + 3 = -5$
9. Bill's 24 hits this baseball season were 3 times his number of hits $x = -4$
last season. How many hits did he make last season? *3.2* 8

---

22. Let x = value of short answer problem. Then value of a true/false problem = $x - 2$ and the value of an essay problem is $3x + 1$.
23. Let x = length of the second side. Then $x + 3.2$ = length of the first side and $x + 2$ = length of the third side.

24. Let x = miles driven in a day. Then $21 + 0.19x$ = cost for the day.
26. Let s = Shira's age. Then $10s - 1$ = grandfather's age.

4.2 Problem Solving: Two or More Numbers

Objective   To solve problems involving two or more numbers

In Example 1 of the previous lesson, you used the sentence "The price of a book is $3 less than twice the price of a tape" to represent the price of a book as x and the price of a tape as $2x - 3$. If you have more information, you may be able to write an equation that can be solved to find the actual prices, as shown below.

New information: Al spent a total of $63 for a book and a tape.

Equation: $\overbrace{63}$ $=$ x $+$ $(2x - 3)$

$63 = x + (2x - 3)$
$63 = 3x - 3$
$66 = 3x$
$22 = x$
$x = 22$ and $2x - 3 = 2 \cdot 22 - 3 = 41$

The book cost $22 and the tape cost $41.

The step-by-step method outlined below can guide you through the entire problem-solving process.

The larger of two numbers is 17 more than the smaller. The sum of the numbers is 59. Find the two numbers.

What are you to find?	Two numbers, a larger and a smaller
What is given?	The larger number is 17 more than the smaller. The sum of the numbers is 59.
Choose a variable.	Let n = one number. Then $n + 17$ = the larger number.
Write an equation.	The sum of the two numbers is 59. $n + (n + 17) = 59$
Solve the equation.	$2n + 17 = 59$ $2n = 42$ $n = 21$ ⟵ smaller $n + 17 = 21 + 17 = 38$ ⟵ larger

Teaching Resources

Quick Quizzes 27
Reteaching and Practice Worksheets 27

▬▬ GETTING STARTED

Prerequisite Quiz

In each exercise, choose a variable to represent one number. Then represent the other number in terms of that variable.

1. The larger of two numbers is 9 more than the smaller. Smaller, x; larger, $x + 9$
2. The smaller of two numbers is $\frac{4}{5}$ of the larger. Larger, x; smaller, $\frac{4}{5}x$
3. The first of two weights is 3 times as heavy as the second weight. Second, x; first, $3x$
4. The length of a rectangle is 4 cm less than 3 times the width. Width, x; length, $3x - 4$

Motivator

Give the students this problem. "The price of a book is $3 less than twice the price of a tape. Find the price of each." Students should realize that not enough information has been given. Then proceed with the lesson as given in the text.

Highlighting the Standards

Standards 1a, 1b, 1c, 1d: The focus of this entire lesson is problem-solving strategies and their applications.

Lesson Note

The step-by-step method for problem-solving given on this page will be helpful to students as they solve word problems. Write these steps on the chalkboard or overhead and follow them as you explain each Example. You may wish to relate Polya's four-step problem-solving model (see page 35) to the method outlined on page 129.

Understand the Problem	What are you to find? What is given? Choose a variable. What does it represent?
Develop a Plan	Write an equation.
Carry out the Plan	Solve the equation.
Look Back	Check in the original problem. State the answer.

Usually one or more sentences in a word problem give the comparison(s) of the numbers involved. Another sentence provides the basis for the equation that models the problem. You may wish to have students identify this "equation sentence" in the Examples and several of the Written Exercises.

Math Connections

Statistics: When taking a poll of voters, only a small sample is used. If the sample indicates that twice as many will vote for candidate A as for candidate B, then the number of votes for a particular candidate out of a large number of votes cast can be predicted by using an equation.

Check in the original problem.	Is the larger number 17 more than the smaller? Yes, because $21 + 17$ equals 38.
	Is the sum of the two numbers 59? Yes, because $21 + 38$ equals 59.
State the answer.	The numbers are 21 and 38.

Think about the step-by-step method as you work the examples below.

EXAMPLE 1   A board that is 81 cm long is cut into two pieces. The first piece is 7 cm less than 3 times the second. Find the length of each piece.

Solution

Let s = length of second piece, in cm. ⟵ smaller length
Then $3s - 7$ = length of first piece, in cm.
THINK: Can the answers be negative? Can they be more than 81 cm?

The combined lengths must total 81 cm.

$$s + (3s - 7) = 81$$
$$4s - 7 = 81$$
$$4s = 88$$
$$s = 22 \quad ⟵ \text{second length: 22 cm}$$
$$3s - 7 = 3 \cdot 22 - 7 = 59 \quad ⟵ \text{first length: 59 cm}$$

Check

Is the length of the first piece 7 cm less than 3 times the length of the second? Yes, because $3 \cdot 22 - 7$ equals $66 - 7$, or 59.

Is the sum of the lengths 81 cm? Yes, because $22 + 59$ equals 81. Thus, the lengths are 59 cm and 22 cm

EXAMPLE 2   Bob has 3 times as many records as Jim. Kacey has 5 fewer records than Bob. Together the boys have 16 records. How many records does each boy have?

Plan

Bob's number is expressed in terms of Jim's. Kacey's is expressed in terms of Bob's. You can express Bob's and Kacey's in terms of Jim's.

Solution

Let j = the number that Jim has.
Then $3j$ = the number that Bob has,
and $3j - 5$ = the number that Kacey has.

$$j + 3j + (3j - 5) = 16 \quad ⟵ \text{Together they have 16 records.}$$
$$7j - 5 = 16$$
$$7j = 21$$
$$j = 3 \quad ⟵ \text{Jim: 3 records}$$
$$3j = 3 \cdot 3 = 9 \quad ⟵ \text{Bob: 9 records}$$
$$3j - 5 = 3 \cdot 3 - 5, \text{ or } 4 \quad ⟵ \text{Kacey: 4 records}$$

Additional Example 1

A board that is 89 cm long is separated into two pieces so that the length of the first piece is 5 cm more than twice the length of the second piece. Find the length of each piece. First piece, 61 cm; second piece, 28 cm

Additional Example 2

The second of three numbers is 4 more than the first. The third is 6 less than twice the second. The sum of the numbers is 74. Find the three numbers. 17, 21, 36

Check   Bob has 3 times as many records as Jim: $9 = 3 \cdot 3$

Kacey has 5 fewer records than Bob: $9 - 5 = 4$

Together they have 16 records: $3 + 9 + 4 = 16$

Thus, Jim has 3 records, Bob has 9, and Kacey has 4.

TRY THIS

1. Solve.

Tim's age is 4 times Gina's age. John is 6 years younger than Tim. The sum of their ages is 57. Find their ages.   Gina: 7; Tim: 28; John: 22

Example 3 is based on Example 3 on page 126.

EXAMPLE 3   This year, the price of a movie ticket increased by one-fifth of the price last year. The price this year is $3.00. Find the price of a movie ticket for last year.

Solution   Let l = the price of a movie ticket last year, in dollars.

Then $\frac{1}{5}l$ = the price increase in dollars $\longleftarrow \frac{1}{5}$ of the price last year

$l + \frac{1}{5}l = 3$ $\longleftarrow$ last year's price + increase = this year's price

$1l + \frac{1}{5}l = 3$

$1\frac{1}{5}l = 3$

$\frac{6}{5}l = 3$

$\frac{5}{6} \cdot \frac{6}{5}l = \frac{5}{6} \cdot 3$

$l = \frac{5}{2}$, or 2.5 $\longleftarrow$ last year's price: $2.50

Check   Is the amount of increase equal to $\frac{1}{5}$ of last year's price?

Yes, because $\frac{1}{5}(2.50)$ is 0.50 and $2.50 + 0.50 = 3.00$.

Thus, the price of a movie ticket last year was $2.50.

TRY THIS

2. Solve.

This year, the price of a stamp increased by one-fifth of the price last year. The price this year is 30¢. Find the price of the stamp last year.   25¢

Critical Thinking Questions

Evaluation: Have students consider these two ways of representing Tom's age and Sally's age if Sally is twice as old as Tom.

Method 1: Tom's age: x; Sally's age: $2x$

Method 2: Sally's age: x; Tom's age: $\frac{x}{2}$

Ask students to give reasons why one method might be better than the other. Answers will vary.

Common Error Analysis

Error: Students sometimes forget to write all the answers to a problem. Remind students to reread the question in the original problem before stating the answers.

Remind them to label their answers and to check whether the answers are reasonable.

Checkpoint

Solve each problem.

1. One number is 3 times another. The sum of the numbers is 172. Find the numbers.   129, 43
2. Casey's age is 3 years less than twice Paul's age. The sum of their ages is 30. Find the age of each.   Paul, 11; Casey 19
3. Bob has 6 more coins than Dirk. Jane has twice as many coins as Bob. Together they have 46 coins. How many coins does each person have?   Bob, 13; Dirk, 7, Jane, 26

Closure

Have students relate the steps for problem solving outlined on page 129 in the Polya model.   See the Lesson Note.

Additional Example 3

This year, the price of a special calendar has increased by one-sixth of the price last year. The price this year is $5.60. Find the price of the calendar last year. $4.80

Guided Practice

Classroom Exercises 1–14
Try This   all

Independent Practice

A Ex. 1–21, **B** Ex. 22–30, **C** Ex. 31–38
Basic: WE 1–21 odd, Brainteaser
Average: WE 9–29 odd, Brainteaser
Above Average: WE 17–37 odd,
Brainteaser

Additional Answers

Classroom Exercises

1. m = Mary's age; $2m$ = Jim's age
2. x = larger number, $x - 6$ = smaller number
3. x = shorter ribbon, $x + 9$ = longer ribbon
4. x = smaller number, $3x - 5$ = larger number
7. $m + 2m = 39$; $m = 13$; Mary: 13; Jim: 26
8. $x + (x - 6) = 47$; $x = 26.5$; $x - 6 = 20.5$
9. $x + (x + 9) = 215$; $x = 103$, $x + 9$ $= 112$
10. $x + (3x - 5) = 19$; $x = 6$; 6, 13
11. $l + (2l - 3) = 4.5$; $l = 2.5$, $w = 2$
12. $x + (2x + 5) = 365$; $x = \$120$; Caroline: $245

Classroom Exercises

Using one variable, represent the numbers described in each exercise.

1. Jim's age is twice Mary's age.

2. One number is 6 less than another.

3. One ribbon is 9 inches longer than another ribbon.

4. The larger of two numbers is 5 less than 3 times the smaller.

5. The width of a rectangle is 3 cm less than twice the length.
l = length; $2l - 3$ = width

6. Caroline's weekly salary is $5 more than twice Sarah's salary.
x = Sarah's salary; $2x + 5$ = Caroline's salary

For Classroom Exercises 1–6, the following additional relationship between the two quantities is known. Write an equation, solve it, and find the quantities.

7. For Exercise 1, the sum of the ages is 39.

8. For Exercise 2, the sum of the numbers is 47.

9. For Exercise 3, there are 215 in. of ribbon in all.

10. For Exercise 4, the sum of the numbers is 19.

11. For Exercise 5, the *semiperimeter* (half the perimeter) is $4\frac{1}{2}$ cm.

12. For Exercise 6, Caroline and Sarah together earn $365 each week.

Solve.

13. The larger of two numbers is 3 times the smaller. The sum of the numbers is 28. Find the numbers. 7, 21

14. Joe's age is 3 less than twice Bill's age. The sum of their ages is 21. Find the age of each. Bill: 8; Joe: 13

Written Exercises

Solve.   First, 60; second, 12

1. One number is 4 times another. The sum of the numbers is 65. Find the numbers. 13, 52

2. Seventy-two students are separated into two groups. The first group is 5 times as large as the second. How many students are in each group?

Brian, 16; Gus, 23

3. Gus has 7 more tapes than Brian. Together they have 39 tapes. How many tapes does each boy have?

4. One number is 7 less than another. The sum of the numbers is 35. Find the numbers. 14, 21   18.5 m, 26.5 m

5. Clyde worked 8 hours longer than Barry worked. Together they worked 40 hours. How long did each boy work? Barry, 16 h; Clyde, 24 h

6. A board that is 45 m long is separated into two pieces. The longer piece is 8 m longer than the shorter. Find the length of each piece.

7. Aaron's bowling score is 12 less than twice Dan's score. The sum of their scores is 258. Find each score.
Dan, 90; Aaron, 168

8. Juan's age is 5 years more than twice Margo's age. The sum of their ages is 29 years. Find their ages.
Margo: 8; Juan: 21

Enrichment

Review the concept of *prime numbers*, which should be familiar to students from their work in earlier grades. Then ask students to find the solutions, if any exist, for these problems.

1. The second of three prime numbers is 5 more than the first. The third is 3 times the second. The sum of the three numbers is 30.   No solution

2. The second of three prime numbers is 6 more than the first. The third is 1 more than twice the second. The sum of the three numbers is 39.   5, 11, 23

9. A fish tank contained 23 fish which were either guppies or swordtails. The number of guppies was 1 less than 3 times the number of swordtails. How many of each kind of fish were there?

10. The larger of two numbers is 2 more than 4 times the other. Their sum is 33. Find the two numbers. 6.2, 26.8

Written Exercises

11. James has saved $560 in two years. He saved 3 times as much during the first year as during the second year. How much did he save each year?

12. In a game, Bob's score was 3 times Otto's score. Together they scored 72 points. Find their scores. Otto: 18; Bob: 54

13. A salesperson's profit on a portable TV set is $10 more than the cost. The set sells for $130. Find the cost and the profit. (HINT: selling price = cost + profit) Cost: $60; profit: $70

14. Last month, a salesman earned $560 more than he did this month. Total earnings for the two months were $8,720. Find each month's earnings. Last month: $4,640; this month: $4,080

15. Bessie's age is $\frac{1}{2}$ of Jean's age. The sum of their ages is 27 years. How old is each girl? Bessie: 9; Jean: 18

16. A house costs $3\frac{1}{2}$ times as much as the lot. Together they sold for $135,000. Find the cost of each.

17. If the cost of a shirt is increased by one-fourth of the cost, the result is the selling price of $35. Find the cost. $28

18. Separate 53 people into two groups so that the first group has 7 fewer than 4 times the number of people in the second group. 41, 12

19. An air pump and a gang valve for an aquarium cost $31. The air pump costs 1 dollar less than 3 times the cost of the gang valve. Find the cost of each.

20. This year, Gregg bought 3 more than twice the number of tapes he bought last year. He bought 11 tapes last year. How many did he buy this year?

21. This year, Christy has 15 fish in her tank. She has one fish fewer than twice the number she had last year. How many fish did she have last year? 8

22. Hisako has 2 records fewer than 3 times the number Gretchen has. Together they have 42 records. How many records does Gretchen have? 11

23. Last month's phone bill was $2.51 more than this month's bill. The bill for the two months totaled $44.45. Find the bill for each month.

24. It costs $88.35 to buy a pair of slacks and a jacket. The cost of the jacket is $1.80 more than twice the cost of the slacks. Find the cost of each.

25. The price of an antique book is $91, which is two-fifths more than the cost to the book dealer. Find the cost. $65

26. The larger of two numbers is 5 times the smaller. The difference between them is 36. Find the numbers. 9, 45

27. Amy's part-time salary is 4 times Katie's part-time salary. The difference between their salaries is $126. Find each girl's salary. Amy: $168; Katie: $42

28. A number is decreased by $\frac{1}{3}$ of the number. The result is 62. Find the number. 93

29. After Sten gave away $\frac{1}{8}$ of his record collection, he had 14 records left. How many records did he have before he gave some away? 16

30. This year, a theater ticket costs $5 less than twice what it cost last year. This year 4 tickets cost $36. What did 4 tickets cost last year? $28

Written Exercises

9. 6 swordtails, 17 guppies
11. First, $420; second, $140
16. House: $105,000; lot: $30,000
19. Pump: $23; valve: $8
20. 25
23. Last month: $23.48; this month: $20.97
24. Slacks: $28.85; jacket: $59.50

36. Mon: 3; Tues: 7; Wed: 8

31. Horace's mother is three times as old as her only child. Horace's grandmother is 35 years older than Horace's mother. If Horace's grandmother were 7 years younger, she would be 5 times as old as Horace. How old is Horace's grandmother? 77 yr

32. The second of three numbers is 4 times the first. The third is 2 more than the second. If the second number is decreased by twice the third, the result is 28. Find the three numbers. $-8, -32, -30$

33. The first of three numbers is 4 times the second. The third is 15 more than the first. The average of the 3 numbers is -1. Find the three numbers. $-8, -2, 7$

34. The second of three numbers is 8 less than 3 times the first. The third number is 18 less than 6 times the first. If twice the first number is decreased by the third number, the result is -2. Find the three numbers. 5, 7, 12

35. Jack bought a shirt, a tie, and a sport jacket. The cost of the shirt is 3 dollars more than twice the cost of the tie. The cost of the sport jacket is 8 dollars less than 4 times the cost of the shirt. The combined cost of the three items is $84. Find the cost of each. Shirt: $17, tie: $7, jacket: $60

36. Jessie went jogging on Monday. On Tuesday she jogged 1 mi farther than twice the number of miles she jogged on Monday. On Wednesday she jogged 1 mi less than 3 times the number of miles she jogged on Monday. She jogged a total of 18 mi in the three days. How many miles did she jog on each of the three days?

37. A double bed costs $80 more than a twin bed. Three twin beds cost as much as two double beds. Find the cost of each bed. Twin: $160; double: $240

38. On a social studies test, Charlayne scored 85 points. On a science test, she scored 6 points more than on a mathematics test. Her average score for the three tests was 89 points. What did she score on the mathematics test? 88

Mixed Review

Simplify. *2.2, 2.6, 2.9, 2.10*

1. $-2(-3)^3$ 54 **2.** $3x - (4 - x)$ $4x - 4$ **3.** $7 - a - 5 - 2a$ $2 - 3a$ **4.** $|-7|$ 7

Solve. *3.2, 3.5*

5. $9 = \frac{3}{4}x$
$x = 12$

6. $3x - (4 - x) = 2x + 8$
$x = 6$

7. $3(x + 2) = 3x + 5$
No solution

8. $-2y = 3$
$y = -\frac{3}{2}$

▰▰/*Brainteaser*

A cyclist completed an 84-km course in three hours. In the first hour she covered $\frac{4}{7}$ of the distance; in the second hour she covered $\frac{1}{2}$ the distance covered in the first hour; in the third hour she covered the remaining distance. How far did she travel in the third hour? 12 km

4.3 Problem Solving: Consecutive Integer Problems

Objective   To solve problems involving consecutive integers

The years 1991, 1992, and 1993 are examples of three consecutive integers. Here are four other examples of consecutive integers.

$$1, 2, 3, 4, \qquad -5, -4, -3 \qquad 29, 30 \qquad -1, 0, 1, 2, 3$$

If n is the first of two or more consecutive integers, then those integers can be represented by $n, n + 1, n + 2$, and so on.

EXAMPLE 1   Find two consecutive integers with a sum of 63.

Solution   Let n = the first integer. Then $n + 1$ = the next consecutive integer. The sum is 63. Use this fact to write an equation.

$$n + (n + 1) = 63$$
$$2n + 1 = 63$$
$$2n = 62$$
$$n = 31 \quad \longleftarrow \text{first integer: 31}$$
$$n + 1 = 31 + 1, \text{ or } 32 \quad \longleftarrow \text{second integer: 32}$$

Check   Are 31 and 32 consecutive integers? Yes.
Is their sum 63? Yes, because $31 + 32 = 63$.

Thus, the integers are 31 and 32.

EXAMPLE 2   Find two consecutive integers with a sum of 32.

Solution   Let n = the first integer. Then $n + 1$ = the second integer.

$$n + (n + 1) = 32 \quad \longleftarrow \text{The sum is 32.}$$
$$2n + 1 = 32$$
$$2n = 31$$
$$n = 15\tfrac{1}{2}$$

But $15\tfrac{1}{2}$ is not an integer.

Thus, no two consecutive integers have a sum of 32.

TRY THIS   **1.** Find two consecutive integers with a sum of 55.   27, 28

An **even integer** is an integer that is divisible by 2. Examples of even integers are $-10, -4, 0, 20$, and 100.

Teaching Resources

Quick Quizzes 28
Reteaching and Practice Worksheets 28

▰▰ GETTING STARTED

Prerequisite Quiz

1. What integer is 1 more than -6? -5
2. What integer is 1 more than -1?   0
3. What integer is 4 less than 1? -3

Write the next two integers in the pattern.

4. 4, 6, 8,   10, 12
5. $-9, -7, -5$ $-3, -1$
6. $-6, -4, -2$ 0, 2

Motivator

The word "consecutive" means following one after the other in order without gaps. Have students give several everyday uses of the word. For example, Monday, Tuesday, and Wednesday are three consecutive days and June, July, and August are three consecutive months.   Answers will vary.

Additional Example 1

Find two consecutive integers with a sum of -57. $-29, -28$

Additional Example 2

Find two consecutive integers with a sum of -18.   There are no two consecutive integers with a sum of -18.

Highlighting the Standards

Standards 14a, 14c, 3e: The Examples and Exercises lead students to consider the structure of odd, even, and consecutive numbers and, in Exercises 22–25, to construct arguments that prove generalizations on this topic.

Lesson Note

Stress the algebraic representation of these sets of integers.

a. Three consecutive integers
b. Three consecutive even integers
c. Three consecutive odd integers

Some students are confused by consecutive integers that are negative. A number line can be used to demonstrate that consecutive integers increase by 1 from left to right.

Math Connections

Previous Math: The sign of the power of a negative number can be determined by whether the exponent is even or odd. An odd exponent, as in $(-1)^9 = -1$, will give a negative value; an even exponent will always give a positive value.

An **odd integer** is an integer that is not divisible by 2. Examples of odd integers are -7, -1, 13, and 121.

Three consecutive odd integers can be represented algebraically by n, $n + 2$, and $n + 4$ where n is an odd integer. Similarly, three consecutive even integers can also be represented by n, $n + 2$, and $n + 4$, where n is an even integer.

EXAMPLE 3 Find three consecutive even integers such that the third is 8 less than twice the second.

Solution

Let n = the first even integer.
Then $n + 2$ = the second even integer,
and $n + 4$ = the third even integer.
The third is 8 less than twice the second.

$$n + 4 = 2(n + 2) - 8$$
$$n + 4 = 2n + 4 - 8$$
$$n + 4 = 2n - 4$$
$$8 = n, \text{ or } n = 8 \quad \longleftarrow \text{first even integer: 8}$$
$$n + 2 = 10 \quad \longleftarrow \text{second even integer: 10}$$
$$n + 4 = 12 \quad \longleftarrow \text{third even integer: 12}$$

Thus, the three consecutive even integers are 8, 10, and 12

EXAMPLE 4 Is the sum of two consecutive odd integers *never* an even integer, *sometimes* even, or *always* even? Use the following two facts to justify your answer.

Fact 1: An integer is divisible by 2 if and only if it is even.
Fact 2: The sum of two integers is an integer.

Solution Look at specific numbers. Then use algebra to find a general answer.

$$9 + 11 = 20 \qquad 15 + 17 = 32 \qquad 41 + 45 = 86$$

These sums are even integers. So the answer is either *sometimes* even or *always* even.

Next, let n = *any* odd integer.
Then, $n + 2$ = the next consecutive odd integer.
The sum of these two integers is an integer, namely
$$n + (n + 2) = 2n + 2 = 2(n + 1).$$

Since $2(n + 1)$ is divisible by 2, it is even. Thus, the sum of two consecutive odd integers is *always* even

TRY THIS 2. Find three consecutive odd integers such that the second is 20 less than 3 times the third integer. 5, 7, 9

136 Chapter 4 Applying Equations

Additional Example 3

Find three consecutive odd integers such that three times the second is 9 more than twice the third. 11, 13, 15

Additional Example 4

Is the sum of two consecutive even integers *never* even, *sometimes* even, or *always* even? Use algebraic representations to justify your answer. Always even: Let n = any even integer. Then $n + 2$ is the next consecutive even integer, and $n + (n + 2) = 2n + 2 = 2(n + 1)$. Since $2(n + 1)$ is divisible by 2, it is even.

Focus on Reading

Match each item at the left with one expression or phrase at the right.

1. $-4, -3, -2$ c
2. $5, 7, 9$ e
3. $24, 26, 28$ a
4. consecutive integers if x is an integer d
5. consecutive odd integers if x is odd b
6. consecutive even integers if x is even b

a. three consecutive even integers
b. $x, x + 2, x + 4$
c. three consecutive integers
d. $x, x + 1, x + 2$
e. three consecutive odd integers
f. $x, x + 1, x + 3$

Classroom Exercises

Give the next three consecutive integers.

1. 7 8, 9, 10
2. -21
3. 78 79, 80, 81
4. -2 -1, 0, 1
5. n
6. $n + 5$

Give the next three consecutive odd integers.

7. 1 3, 5, 7
8. -3 -1, 1, 3
9. -33 -31, -29, -27
10. 5 7, 9, 11
11. k
12. $k + 6$

Give the next three consecutive even integers.

13. 10 12, 14, 16
14. -4 -2, 0, 2
15. 56 58, 60, 62
16. -42
17. t
18. $t + 8$

Suppose n is the first of three consecutive odd integers. Represent each of the following with an algebraic expression. (Exercises 19 and 20)

19. the sum of the three integers
    $n + (n + 2) + (n + 4)$
20. the sum of the second and the third integers $(n + 2) + (n + 4)$

21. Find three consecutive integers with a sum of 42. 13, 14, 15
22. Find two consecutive even integers with a sum of 82. 40, 42
23. Find three consecutive odd integers with a sum of 171. 55, 57, 59

Written Exercises

Solve each problem. If there is no solution, so indicate. (Exercises 1–14)

1. The sum of two consecutive integers is 75. Find the integers. 37, 38
2. The sum of two consecutive integers is -63. Find the integers. $-32, -31$
3. Find two consecutive even integers with a sum of 78. 38, 40
4. Find two consecutive odd integers with a sum of -56. $-29, -27$
5. Find two consecutive integers with a sum of 52. No solution
6. Find two consecutive even integers with a sum of 41. No solution
7. Find two consecutive odd integers with a sum of 65. No solution
8. The sum of three consecutive integers is -51. Find the integers. $-18, -17, -16$

4.3 Problem Solving: Consecutive Integer Problems **137**

Closure

To review the concepts in this lesson, have students suggest a word problem that would lead to writing and solving the equation

$$x + (x + 1) + (x + 2) = 90.$$

Have them suggest a second problem for this equation.

$$x + (x + 2) + (x + 4) = 135.$$

Sample answers: The sum of three consecutive integers is 90, or the sum of three consecutive odd integers is 135.

◼◼FOLLOW UP

Guided Practice

Classroom Exercises 1–23
Try This   all, Focus on Reading   all

Independent Practice

A Ex. 1–14, **B** Ex. 15–21, **C** Ex. 22–25

Basic: WE 1–10, 15–17 all

Average: WE 1–13 odd, 15–21 all

Above Average: WE 9–13 odd, 15–25 all

Additional Answers

Classroom Exercises

2. $-20, -19, -18$
5. $n + 1, n + 2, n + 3$
6. $n + 6, n + 7, n + 8$
11. $k + 2, k + 4, k + 6$
12. $k + 8, k + 10, k + 12$
16. $-40, -38, -36$
17. $t + 2, t + 4, t + 6$
18. $t + 10, t + 12, t + 14$

---

9. Find three consecutive odd integers with a sum of 273. 89, 91, 93
10. Find three consecutive even integers with a sum of -126. $-44, -42, -40$
11. Find four consecutive integers with a sum of 113. No solution
12. Find five consecutive integers with a sum of -45. $-11, -10, -9, -8, -7$
13. The sum of four consecutive even integers is 4. Find the integers. $-2, 0, 2, 4$
14. The sum of four consecutive odd integers is -8. Find the integers. $-5, -3, -1, 1$

In Exercises 15–20, complete the statement by writing *never*, *sometimes*, or *always*.

15. The sum of any two consecutive integers is __?__ even. never
16. The sum of any two consecutive even integers is __?__ even. always
17. The sum of any two consecutive odd integers is __?__ odd. never
18. The sum of any three consecutive even integers is __?__ even. always
19. The sum of any three consecutive odd integers is __?__ odd. always
20. The sum of any four consecutive integers is __?__ odd. never
21. Find three consecutive odd integers such that their sum decreased by 18 is equal to the first integer. No solution
22. Show that for all integers n, if $n + (n + 1) = S$, then $n = \dfrac{S - 1}{2}$.

    Is it true that if you know the sum of any two consecutive integers, you can find the first integer by subtracting 1 from the sum and dividing the result by 2? Yes
23. Show that for all integers n, if $n + (n + 2) + (n + 4) = S$, then

    $n = \dfrac{S - 6}{3}$. Is it true that if you know the sum of any three consecutive odd integers, you can find the first integer by subtracting 6 from the sum and dividing the result by 3? Yes
24. Give a rule for finding the first of four consecutive odd integers if you know the sum of the four integers. n = integer; s = sum; $n = \frac{s - 12}{4}$
25. Give a rule for finding the first of four consecutive integers if you know the sum of the four integers. n = integer; s = sum; $n = \frac{s - 6}{4}$

Mixed Review

Evaluate for the given values of the variable.   1.3

1. $(8x)^2$, for $x = \frac{1}{4}$ 4
2. x^3y^2, for $x = 1$ and $y = 7$ 49
3. $\dfrac{x^2 - y^2}{x - y}$, for $x = 5$ and $y = 2$ 7
4. $3a^4b$, for $a = 2$ and $b = 5$ 240

Simplify.   2.2

5. $|-16|$ 16
6. $|-2| + |8|$ 10
7. $|5 - 7|$ 2
8. $-|6| \cdot |-6|$ -36

Enrichment

Challenge students to demonstrate algebraically that the following statement is true.

The sum of any three consecutive integers is divisible by 3.   Let n = the first integer. Then $n + (n + 1) + (n + 2) = 3n + 3 = 3(n + 1)$, and $3(n + 1)$ is divisible by 3.

Then have them determine whether the sum of any four consecutive integers is divisible by 4.   No.   Whether the sum of any five consecutive integers is divisible by 5.   Yes.

4.4 Problem Solving: Perimeter and Angle Measure

Objectives

To solve problems about perimeter
To solve problems about angle measure

In this lesson, you will find it helpful to draw figures and label them.

In designing a quilt, Jackie needs to cut rectangular pieces of cloth. The length of each rectangular piece must be 2 cm less than 3 times the width. The perimeter must be 44 cm. How can you find the length and width of each rectangular piece?

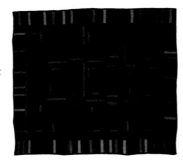

What are you to find?	The length and width of a rectangle
Draw and label a figure.	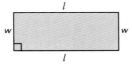
What is given?	The length is 2 less than 3 times the width. The perimeter is 44 cm.
Choose a variable.	Let w = the width, in cm. Then $3w - 2$ = the length, in cm. ← 2 less than 3 times the width
Write an equation.	$p = 2l + 2w$ ← the perimeter of a rectangle $44 = 2(3w - 2) + 2w$ ← Substitute 44 for p and $3w - 2$ for l.
Solve the equation.	$44 = 6w - 4 + 2w$ $44 = 8w - 4$ $48 = 8w$ $6 = w$ ← width: 6 cm $l = 3w - 2 = 3 \cdot 6 - 2 = 16$ ← length: 16 cm
Check in the original problem.	The check is left for you.
State the answer.	Thus, the length is 16 cm; the width is 6 cm.

4.4 Problem Solving: Perimeter and Angle Measure   **139**

Teaching Resources

Manipulative Worksheet 4
Problem Solving Worksheet 4
Quick Quizzes 29
Reteaching and Practice Worksheets 29
Transparencies 9A, 9B, 9C, 9D

■■■ GETTING STARTED

Prerequisite Quiz

Find the perimeter of each figure.

1. A square with sides 8 cm long   32 cm
2. A rectangle with length 12 in and width 7 in   38 in
3. A triangle with sides that have these lengths: 4 ft, 7 ft, 9 ft   20 ft
4. A triangle with sides that have these lengths: 4.8 cm, 4.8 cm, 5.5 cm   15.1 cm
5. A triangle with sides that are each 6.4 cm long   19.2 m

Highlighting the Standards

Standards 7b, 7d, 4c: In Exercises 1–20, students use the properties of geometric figures to solve problems that connect algebra and geometry.

Motivator

Remind students that a circle contains 360 degrees and that the sides of a straight angle form a straight line. Thus, the measure of a straight angle is 180°. Then have each student carefully cut a piece of paper to form a triangular shape. Direct them to tear off the three corners and place the points together so as to form a straight line. The fact that different students with different triangles get the same result will demonstrate that the sum of the measures of the angles of a triangle must be 180 degrees.

▰TEACHING SUGGESTIONS

Lesson Note

Review the concept of perimeter. In the review, use drawings of a square, a rectangle that is not a square, a triangle with no two sides having the same length (scalene triangle), an isosceles triangle, and an equilateral triangle. Elicit the perimeter formulas from the class.

Emphasize the importance of making a drawing when solving problems such as those in Examples 1 and 2.

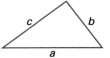

The *perimeter* of a triangle is the sum of the lengths of its sides. Therefore, the formula for the perimeter of a triangle with sides a, b, and c is

$$p = a + b + c.$$

EXAMPLE 1 The first side of a triangle is 3 cm longer than the second side. The third side is 4 cm shorter than twice the length of the second side. The perimeter is 31 cm. How long is each side?

Plan Draw and label a figure (see below). Represent the first and third sides in terms of the *second* side.

Solution Let s = the length of the second side, in cm.
Then $s + 3$ = the length of the first side, in cm, ← 3 cm longer than s
and $2s - 4$ = the length of the third side, in cm. ← 4 cm shorter than twice s

Use $p = a + b + c$.

$$31 = (s + 3) + s + (2s - 4)$$
$$31 = 4s - 1$$
$$32 = 4s$$
$$8 = s, \text{ or}$$
$$s = 8 \quad \longleftarrow \text{second side: 8 cm}$$
$$s + 3 = 11 \quad \longleftarrow \text{first side: 11 cm}$$
$$2s - 4 = 2 \cdot 8 - 4 = 16 - 4 = 12 \quad \longleftarrow \text{third side: 12 cm}$$

Check The check is left for you.

Thus, the lengths of the sides are 8 cm, 11 cm, and 12 cm.

TRY THIS 1. The length of a rectangle is 7 cm longer than 3 times the width. The perimeter is 78 cm. Find the length and the width.
Width: 8 cm; length: 31 cm

For any triangle, the sum of the degree measures of the angles is 180.

$$x + y + z = 180$$

In an **isosceles triangle**, two sides have the same length. These two sides are called the **legs** and the third side is called the **base**. An isosceles triangle also has two angles with the same measure. These two angles are called the **base angles** and the third angle is called the **vertex angle**.

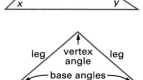

Isosceles triangle

140 Chapter 4 Applying Equations

Additional Example 1

One side of a triangle is twice as long as a second side. The third side is 3 m longer than the second side. The perimeter is 35 m. Find the length of each side.
16 m, 8 m, 11 m

EXAMPLE 2 The degree measure of the vertex angle of an isosceles triangle is 30 more than that of a base angle. Find the degree measure of each angle of the triangle.

Solution Let b = the measure of each base angle, in degrees.

Then $b + 30$ = the measure of the vertex angle, in degrees.

The sum of the degree measures of the angles is 180.

$$b + b + (b + 30) = 180$$
$$3b + 30 = 180$$
$$3b = 150$$
$$b = 50 \longleftarrow \text{degree measure of each base angle: 50}$$
$$b + 30 = 50 + 30 = 80 \longleftarrow \text{degree measure of vertex angle: 80}$$

Check Does the vertex angle have a degree measure that is 30 more than that of a base angle? Yes, because $50 + 30$ equals 80. Is the sum of the measures of the three angles 180? Yes, because $50 + 50 + 80$ equals 180. Thus, the degree measure of each base angle is 50 and the degree measure of the vertex angle is 80.

TRY THIS 2. The degree measure of the vertex angle of an isosceles triangle is 4 times that of a base angle. Find the degree measure of each angle of the triangle. 30, 30, 120

Classroom Exercises

Find the indicated measure for each geometric figure. (Exercises 1–3)

1. Rectangle: length, $2x$; width, x; perimeter, 24 in.
   Find the length and the width. l = 8 in, w = 4 in

2. Rectangle: length, $2x - 4$; width, x; perimeter, 28 m
   Find the length and the width. l = 8m; w = 6m

3. Square: each side of length x; perimeter, 40 ft
   Find the length of each side. 10 ft

Solve each problem.

4. The length of a rectangle is 3 times the width. The perimeter is 42 cm. Find the length and width. l = $15\frac{3}{4}$ cm; w = $5\frac{1}{4}$ cm

5. The perimeter of an isosceles triangle is 40 ft. The length of the base is 10 ft. Find the length of each leg. 15 ft

6. The base of an isosceles triangle is 5 ft longer than either of the other two sides. The perimeter is 41 ft. Find the lengths of all three sides. Base: 17 ft; legs: 12 ft

7. The degree measure of the vertex angle of an isosceles triangle is 42. Find the measure of each base angle. 69°

Math Connections

Geometry: In a triangle, the longest side is opposite (across from) the largest angle and the shortest side is opposite the smallest angle. This can be proved in geometry and it leads to the conclusion that if a triangle has two sides of equal length, it must also have two angles of the same measure.

Critical Thinking Questions

Analysis: Ask students if a triangle can be constructed with sides that measure 3 cm, 4 cm, and 8 cm. Have them use a ruler and try to draw such a triangle. No. Then ask if someone can make a statement concerning restrictions on the lengths of the three sides of a triangle. The measure of each side must be less than the sum of the measures of the other two sides.

Common Error Analysis

Error: Some students will fail to account for all the sides or all the angles in their equation.

Remind them to draw a sketch of the figure for each problem.

Additional Example 2

The degree measure of each base angle of an isosceles triangle is 22 less than 3 times the degree measure of the vertex angle. Find the degree measure of each angle of the triangle. Degree measures: each base angle 74, vertex angle 32

Checkpoint

1. The length of a rectangle is 7 cm less than 3 times the width. The perimeter is 58 cm. Find the length and the width.
   $l = 20$ cm, $w = 9$ cm

2. One side of a triangle is 4 ft longer than a second side. The third side is 5 ft shorter than twice the length of the second side. The perimeter is 47 ft. Find the length of each side. 16 ft, 12 ft, 19 ft

3. The degree measure of the vertex angle of an isosceles triangle is 8 more than twice the degree measure of each base angle. Find the degree measure of each angle of the triangle. Degree measures: each base angle 43, vertex angle 94

Closure

Ask students to define and give an example of each of these terms.
   perimeter
   isosceles triangle
   vertex angle of an isosceles triangle
   base angle of an isosceles triangle
See p. 140. Then ask them to give a formula for the perimeter of a rectangle and for the perimeter of a triangle. See pp. 139–140.

Written Exercises

Solve each problem.

1. The length of a rectangle is 3 times the width. The perimeter is 52 cm. Find the length and the width.

2. The width of a rectangle is one-fourth of the length. The perimeter is 130 cm. Find the length and the width.

3. The width of a rectangle is 7 ft less than the length. The perimeter is 34 ft. Find the length and the width.

4. The length of a rectangle is 3 m more than 5 times the width. The perimeter is 126 m. Find the length and width.

5. The width of a rectangle is 2 ft more than twice the length. The perimeter is 28 ft. Find the length and width.

6. The degree measure of the vertex angle of an isosceles triangle is 46. Find the measure of each base angle.

7. For an *equiangular* triangle, all three angles have the same degree measure. Find the degree measure of each angle of an equiangular triangle. 60

8. The degree measure of the vertex angle of an isosceles triangle is 20 more than twice the measure of a base angle. Find the degree measure of each angle. Vertex: 100; bases: 40

9. The length of each leg of an isosceles triangle is 4 km longer than the base. The perimeter is 35 km. Find the length of each side.

10. The length of each leg of an isosceles triangle is 3 times the length of the base. The perimeter is 77 m. Find the length of each side.

11. The perimeter of a triangle is 45 in. The first side is 5 in. shorter than twice the second side. The third side is 2 in. longer than the second side. Find the length of each side. 19 in., 12 in., 14 in.

12. The first side of a triangle is 2 cm longer than the second side. The third side is 5 cm shorter than twice the second side. The perimeter is 49 cm. Find the length of each side. 15 cm; 13 cm; 21 cm

13. One side of a triangle is 5 yd shorter than a second side. The remaining side is 3 yd longer than the second side. The perimeter is 52 yd. Find the length of each side. 13 yd, 18 yd, 21 yd

14. One side of a triangle is twice as long as a second side. The remaining side is 5 m longer than the second side. The perimeter is 75 m. Find the length of each side. 35 m, 17.5 m, 22.5 m

15. In triangle ABC, the measure of angle A is 1 less than twice the measure of angle B. The measure of angle C is 22 more than the measure of angle A. Find the measure of each angle.
    $A: 63°$; $B: 32°$; $C: 85°$

16. For triangle XYZ, the measure of angle Z is 3 more than half the measure of angle Y. The degree measure of angle X is 17 less than 4 times the measure of angle Z. Find the degree measure of each angle.

17. A rectangle's length is 1 cm less than twice its width. If the length is decreased by 3 cm and the width is decreased by 2 cm, the perimeter will be 36 cm. Find the length and width of the original rectangle.
    $l = 15$ cm; $w = 8$ cm

18. A rectangular field is 3 times as long as it is wide. If the length is decreased by 3 m and the width is increased by 2 m, the perimeter will be 54 m. Find the length and width of the original field.
    $l = 21$ m; $w = 7$ m

142   Chapter 4   Applying Equations

In Exercises 19 and 20, a rectangular house lies on a rectangular lot. The owner plans to enclose the lot with a fence from a corner of the house (Point A) around the lot and finishing at another corner of the house (Point B). Find the length of the fence. All dimensions are in feet.

19.

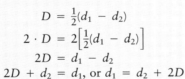

121   3x   x+20   x   house   B   232 ft   A

perimeter of house: 200

20.

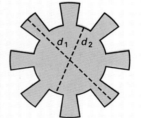

2y+20   25 house   30   y   B   A   220 ft

perimeter of property: 280
perimeter of house: 170

Midchapter Review

1. Find three consecutive odd integers whose sum is 117.   **4.4** 37, 39, 41

2. The length of a rectangle is twice the width. The perimeter is 72 cm. Find the length and the width.   **4.2** $l = 24$ cm; $w = 12$ cm

3. The degree measure of the vertex angle of an isosceles triangle is 10 less than twice the measure of each base angle. Find the degree measure of each angle of the triangle.   **4.2** Vertex: 85°; bases: 47.5°

4. Lola's age is 14 more than 6 times Juan's age. The sum of their ages is 35. How old is each?   **4.3** Juan: 3; Lola: 32

Application: Gear Depths

The depth of a gear tooth is found by multiplying the difference of the larger and smaller diameters of the gear by $\frac{1}{2}$.

Use the formula $D = \frac{1}{2}(d_1 - d_2)$, where D is the depth of the tooth, d_1 is the larger diameter, and d_2 is the smaller diameter. The formula can be solved for d_1 as follows.

$$D = \tfrac{1}{2}(d_1 - d_2)$$

Multiply each side by 2. $2 \cdot D = 2\left[\tfrac{1}{2}(d_1 - d_2)\right]$

Simplify. $2D = d_1 - d_2$

Add d_2 to each side. $2D + d_2 = d_1$, or $d_1 = d_2 + 2D$

Exercises

1. Find D if $d_1 = 8.2$ cm and $d_2 = 7.6$ cm.   0.3 cm

2. Find d_1 if $D = 5.1$ mm and $d_2 = 23.5$ mm.   33.7 mm

3. Solve the formula for d_2. $d_2 = d_1 - 2D$

4. Find d_2 if $D = 12.8$ mm and $d_1 = 61.3$ mm.   35.7 mm

Enrichment

Show students how *diagonals* can be drawn from one vertex of a polygon to partition the figure into triangles.

Have them find the sum of the degree measure of the angles of a

a. 4-sided polygon   360
b. 5-sided polygon   540
c. 6-sided polygon   720
d. 7-sided polygon   900
e. 8-sided polygon   1080
f. 9-sided polygon   1,260

Then ask them to generalize and write a formula for S, the sum of the degree measures of the angles of a polygon with n sides. $S = (n - 2)180$

Finally, have them use the formula to compute the sum of the degree measures of the angles of a polygon with 100 sides. 17,640

4.5 Equations with Fractions

Objective | To solve equations that contain fractions

You know that an equation containing fractions, such as $\frac{4}{5}x + \frac{3}{10} = \frac{1}{2}$, can be solved by applying the properties of equality. However, there is a much more efficient method for solving this equation. First, multiply each side by the *least* number that is divisible by all three denominators, 5, 10, and 2. That number is 10, the **least common multiple** (LCM) of 5, 10, and 2. The number 10 is also referred to as the **least common denominator** (LCD) of the three fractions $\frac{4}{5}$, $\frac{3}{10}$, and $\frac{1}{2}$. If you multiply each side of $\frac{4}{5}x + \frac{3}{10} = \frac{1}{2}$ by 10, the resulting equation will not contain fractions.

EXAMPLE 1 | Solve $\frac{4}{5}x + \frac{3}{10} = \frac{1}{2}$. Check the solution.

Solution

Multiply each side by 10. $10\left(\frac{4}{5}x + \frac{3}{10}\right) = 10 \cdot \frac{1}{2}$

Use the Distributive $10 \cdot \frac{4}{5}x + 10 \cdot \frac{3}{10} = 10 \cdot \frac{1}{2}$
Property.

$$8x + 3 = 5 \quad \longleftarrow \overset{2}{\cancel{10}} \cdot \frac{4}{\cancel{5}}x = 8x$$

$$8x = 2$$

$$x = \frac{2}{8}, \text{ or } \frac{1}{4}$$

Check

Substitute $\frac{1}{4}$ for x.

$$\frac{4}{5}x + \frac{3}{10} = \frac{1}{2}$$

$$\frac{4}{5} \cdot \frac{1}{4} + \frac{3}{10} \overset{?}{=} \frac{1}{2}$$

$$\frac{1}{5} + \frac{3}{10} \overset{?}{=} \frac{1}{2}$$

$$\frac{2}{10} + \frac{3}{10} \overset{?}{=} \frac{1}{2}$$

$$\frac{1}{2} = \frac{1}{2} \text{ True}$$

Thus, the solution is $\frac{1}{4}$.

EXAMPLE 2 | Solve $\frac{a + 3}{3} - \frac{a - 2}{5} = \frac{a}{4}$.

Plan | Multiply each side by 60, the LCD of the three fractions.

144   Chapter 4 Applying Equations

Solution

$$60\left(\frac{a+3}{3} - \frac{a-2}{5}\right) = 60 \cdot \frac{a}{4}$$

$$60 \cdot \frac{a+3}{3} - 60 \cdot \frac{a-2}{5} = 60 \cdot \frac{a}{4}$$

$$20(a+3) - 12(a-2) = 15a$$

$$20a + 60 - 12a + 24 = 15a$$

$$8a + 84 = 15a$$

$$84 = 7a$$

$$12 = a$$

Thus, the solution is 12.

TRY THIS 1. Solve $\frac{x+8}{4} - \frac{x-5}{8} = \frac{x}{5}$ 35

Recall that the *average* or *arithmetic mean* of a set of numbers is the *sum* of the numbers divided by the number of numbers.

EXAMPLE 3 Mary received the following test scores: 70, 80, and 90. What grade must she earn on a fourth test to have a passing average of 70?

Solution Let s = the score on the fourth test.
The sum of four scores divided by 4 is 70.

$$\frac{70 + 80 + 90 + s}{4} = 70$$

$$4 \cdot \frac{70 + 80 + 90 + s}{4} = 4 \cdot 70$$

$$240 + s = 280$$

$$s = 40$$

Thus, Mary must score at least 40 on the fourth test to pass the course.

TRY THIS 2. Joe received the following test scores: 86, 92, 78, and 95. What grade must he earn on a fifth test to have an average of 90?   99

Classroom Exercises

For each equation, by what number would you multiply each side to eliminate the fractions?

1. $\frac{1}{5}x - \frac{1}{3} = -\frac{8}{15}$ 15   2. $\frac{5}{6}y + \frac{1}{2} = \frac{7}{3}$ 6   3. $\frac{5}{12} = \frac{2}{3}c + \frac{1}{4}$ 12   4. $\frac{5}{6}y - 8 = \frac{1}{3}y + \frac{1}{8}$ 24

5–8. Solve the equations of Exercises 1–4 above. 5. $x = -1$ 6. $y = 2\frac{1}{5}$ 7. $c = \frac{1}{4}$ 8. $y = 16\frac{1}{4}$

Lesson Note

Point out that, in Example 1, multiplying both sides by 10 requires that each term be multiplied by 10. When the correct number to multiply by is chosen, and all multiplications are performed correctly, the resulting equation will contain no fractions. In Example 3, remind students that a fraction bar is a grouping symbol, similar to parentheses.

Math Connections

Music: In printed music, a $\frac{3}{4}$ time signature means that each measure consists of 3 quarter notes or their equivalent. The fractional equivalents of the notes in each measure must have a total value of $\frac{3}{4}$.

Critical Thinking Questions

Analysis: After reviewing the definition of GCF with students, have them complete a table such as the following. The table should include at least 5 pairs of numbers.

Numbers	GCF	LCM	
6, 9	?	?	3, 18
8, 12	?	?	4, 24

Ask students to find a way to determine the LCM, given two numbers and their GCF.   LCM = product of numbers ÷ GCF.

Common Error Analysis

Error: Some students will fail to multiply all the terms of an equation such as those in Examples 1 and 2 by the LCD, especially if any term is already a whole number.

Remind them to use parentheses and the Distributive Property.

Additional Example 3

Dinah received the following test scores on the first four math tests of the year: 80, 92, 86, and 78. After the fifth math test, her average was exactly 85. What score did she receive on the fifth test?   89

Checkpoint

Solve.

1. $\frac{1}{6}x + 9 = \frac{2}{3}x$ $x = 18$
2. $\frac{x}{4} = \frac{2}{5}x - 3$ $x = 20$
3. $\frac{x-5}{2} + \frac{x+1}{3} = 9$ $x = 13\frac{2}{5}$
4. In his first two games, Jim bowled 136 and 145. What score must he bowl on the third game to have an average of 150?   169

Closure

Ask students to summarize how to clear an equation of all fractions.   See p. 144.
Have them summarize the method for finding the average of a given number of numbers.   See p. 145.

◼◼◼FOLLOW UP

Guided Practice

Classroom Exercises 1–8
Try This   all

Independent Practice

A Ex. 1–14, **B** Ex. 15–20, **C** Ex. 21–23

Basic: WE 1–12 all, 15
Average: WE 1–20 odd
Above Average: WE 5–21 odd, 22, 23

Written Exercises

Solve each equation.

1. $\frac{3}{4}c = 15$ 20
2. $\frac{3}{5}y = \frac{2}{15}$ $\frac{2}{9}$
3. $\frac{5}{9} = \frac{2}{3}c$ $\frac{5}{6}$
4. $\frac{1}{4}c + \frac{2}{3} = \frac{1}{3}$ $-1\frac{1}{3}$
5. $\frac{3}{4}y - \frac{1}{2} = \frac{5}{3}$ $2\frac{8}{9}$
6. $\frac{5}{8} = \frac{3}{2}a + \frac{1}{4}$ $\frac{1}{4}$
7. $\frac{2}{3}x + 7 = 9$ 3
8. $9 = \frac{1}{2}x + 6$ 6
9. $\frac{1}{2}x + 6 = \frac{3}{4}x - 3$ 36
10. $\frac{1}{2}y - 4 = \frac{1}{5}y + 7$ $36\frac{2}{3}$
11. $\frac{3}{2}x - 3 = \frac{2}{3}x + 2$ 6
12. $\frac{1}{3}r - 1 = \frac{2}{5}r + 2$ -45

Solve each problem.

13. Seven years more than $\frac{3}{5}$ of Juan's age is 25 years. How old is Juan?   30 yr
14. Six less than $\frac{2}{3}$ of a number equals the number. Find the number. $x = -18$
15. The income for 7 weeks of football-game refreshments were: $200, $350, $500, $375, $400, $625, and $250. What income from the 8th and last game will give an average income of $400? $500
16. Megan's batting averages for her first 3 years of high school are .225, .325, and .300. What batting average for the senior year will she need in order to have a yearly batting average of .300 for the four years?   .350

Solve.

17. $\frac{3a+4}{12} - \frac{5}{3} = \frac{2a-1}{2}$ $a = -1\frac{1}{9}$
18. $\frac{2x-3}{7} - \frac{x}{2} = \frac{x+3}{14}$ $x = -2\frac{1}{4}$
19. $\frac{3y}{4} - \frac{2y-9}{3} = \frac{y+1}{5}$ $y = 24$
20. $\frac{3b}{4} - \frac{2b-1}{2} = \frac{b-7}{6}$ $b = 4$

21. When the second of two consecutive even integers is divided by 2 and added to the first even integer, the sum is 25. What are the two integers?   16 and 18
22. The price of a football is decreased by one-third to give the discount price of $38. Find the original price. $57
23. Solve for x: $\frac{2}{3}(x - 5) - \frac{3x-2}{4} = \frac{2}{3}[1 - (4 - x)] - 5\frac{1}{3}$. $x = 6$

Mixed Review

Simplify.   *1.3, 2.2, 2.6, 2.9, 2.10*

1. $|-8 + 3|$ 5
2. $\left(-\frac{3}{4}\right)^3$ $-\frac{27}{64}$
3. $-y - 4 - y + \frac{3}{4}$ $-2y - \frac{13}{4}$
4. $7 - (6 - b) - 3b$ $-2b + 1$
5. $(-1)^3$ -1

6. Find the number of meters a bowling ball drops in 6.3 seconds. Use $d = 5t^2$.   *1.3* 198.45 m

Enrichment

Challenge students to solve this problem.

In a bowling game, Bob's score was $1\frac{1}{5}$ times as great as Ann's score. Carol's score was 198, which was $1\frac{1}{10}$ times as great as Bob's score. What was Ann's score?   150

4.6 Equations with Decimals

Objective

To solve equations that contain decimals

The equation $1.2x - 0.04 = 0.8$ contains *decimals*. It can be solved by adding 0.04 to each side and then dividing each side by 1.2. However, it is often easier to solve an equation such as this by first finding an equivalent equation that does not contain decimals.

Recall that multiplying a decimal by a power of ten, such as 10, 100, or 1,000, may result in an integer.

$10 \times 1.2 = 12$	(Move the decimal point one place to the right.)
$100 \times 0.04 = 4$	(Move the decimal point two places to the right.)
$1,000 \times 0.800 = 800$	(Move the decimal point three places to the right.)

EXAMPLE 1

Solve $1.2x - 0.04 = 0.8$.

Plan

First, find the number of digits at the right of each decimal point.

$$1.2x - 0.04 = 0.8$$
$$\text{one} \quad \text{two} \quad \text{one}$$

The greatest number of digits at the right of any decimal point is two. Thus, you should multiply each side of the equation by 10^2, or 100.

Solution

$$1.2x - 0.04 = 0.8$$

Multiply each side by 100.　　　$100(1.2x - 0.04) = 100(0.8)$

Use the Distributive Property.　$100(1.2x) - 100(0.04) = 100(0.8)$

Simplify.

$$120x - 4 = 80$$
$$120x = 84$$
$$x = \frac{84}{120}, \text{ or } 0.7$$

Check

Substitute 0.7 for x.

$$1.2x - 0.04 = 0.8$$
$$1.2(0.7) - 0.04 \overset{?}{=} 0.8$$
$$0.84 - 0.04 \overset{?}{=} 0.8$$
$$0.80 = 0.8 \text{ True}$$

Thus, the solution is 0.7.

TRY THIS

1. Solve $0.05x + 4.27 = 4.3$. 0.6

Additional Example 1

Solve $0.003x - 7.4 = 0.16$. 2,520

Teaching Resources

Application Worksheet 4
Quick Quizzes 31
Reteaching and Practice Worksheets 31

■■■ GETTING STARTED

Prerequisite Quiz

Perform the indicated operation.

1. $52.6 + 3.05$　55.65
2. $6.025 - 3.4$　2.625
3. 10×0.7　7
4. $1,000 \times 1.9$　1,900
5. $6.8 \div 0.4$　17
6. $16.5 \div 0.15$　110

Motivator

Have students multiply both sides of each equation by the given numbers.

a. $0.5x = 1.8$ by 10　$5x = 18$
b. $0.25y = 13.75$ by 100　$25y = 1375$
c. $0.03t = 0.018$ by 1000　$30t = 18$

Then ask them to describe the result of their multiplications.　The equations have been cleared of decimals.

Highlighting the Standards

Standard 14a: In this lesson and the next, students compare the structure and ways of representing decimals and percents, and solve application problems involving decimals.

Lesson Note

In discussing Examples 1 and 2, emphasize that to solve this type of equation involving decimals, students should determine the greatest number of decimal places in the terms of the equation. Then they should multiply each side of the equation by a power of 10 whose exponent is the same as the number of decimal places.

- For 1 decimal place, multiply by 10^1, or 10.
- For 2 decimal places, multiply by 10^2, or 100.
- For 3 decimal places, multiply by 10^3, or 1,000.

Math Connections

Life Skills: In European countries, the use of the period and comma in a decimal numeral differs from the way we use them. In Europe, 23,500.125 is written as 23.500,125.

Critical Thinking Questions

Evaluation: Ask students to evaluate this description: "To multiply by 100, just add two zeros." Ask them whether this is an accurate description of multiplying by 100 and have them rewrite the sentence to make it more accurate.
Adding two zeros: $23 + 0 + 0 = 23$
Attaching two zeros: $23.8 = 23.800$

Common Error Analysis

Error: Some students forget that a whole number is a decimal number, but that the decimal point is usually omitted.

Remind them that the decimal point in a whole number follows immediately after the last digit.

EXAMPLE 2 This year, Jim's wages increased by 0.125 of last year's wages. If his hourly wage is now $9.45, what was his wage last year?

Solution Let w = Jim's wage last year, in dollars.

wage last year   increased by 0.125 of wage last year   = wage this year

$$w \qquad + \qquad 0.125w \qquad = 9.45$$

$$w + 0.125w = 9.45$$
$$1{,}000(w + 0.125w) = 1{,}000(9.45) \quad \longleftarrow \text{Greatest number of digits to the}$$
$$\text{right of a decimal point: 3;}$$
$$10^3 = 1{,}000$$

$$1{,}000w + 125w = 9{,}450$$
$$1{,}125w = 9{,}450$$
$$w = 8.40 \quad \longleftarrow \text{Jim's wage last year: \$8.40}$$

Check Is last year's wage increased by 0.125 of last year's wage equal to this year's wage? Yes, because $8.40 + 0.125(8.40)$ is $8.40 + 1.05$, or 9.45.

Thus, Jim's wage last year was $8.40.

TRY THIS 2. A number decreased by 0.79 of the number is 6.3. What is the number? 30

Classroom Exercises

For each equation, by what number would you multiply each side to eliminate the decimals?

1. $4.1x + 2 = 14.3$ 10
2. $1.2y + 0.05 = 8.45$ 100
3. $0.005z + 0.02 = 0.1z - 1.5$ 1,000
4. $0.3y - 7.5 = 0.6$ 10
5. $9.8 = 0.02 - 0.1x$ 100
6. $0.004 + 5x = 7.1$ 1,000

7–12. Solve the equations of Classroom Exercises 1–6 above.

Written Exercises

Solve each equation.

1. $0.2x = 1.8$ $x = 9$
2. $2.6x = -7.8$ $x = -3$
3. $1.3y - 1.7 = 7.4$ $y = 7$
4. $0.05m = 7.45$ $n = 149$
5. $8.2 - 3.2c = -17.4$ $c = 8$
6. $0.25q = 8.75$ $q = 35$
7. $0.10d = 3.40$ $d = 34$
8. $0.036x = -1.08$ $x = -30$
9. $0.03r = 0.018$ $r = 0.6$
10. $1.3 = 0.15x - 3.2$
11. $0.02d - 2.6 = 0.84$
12. $0.05n + 1.45 = 10.20$
13. $0.25q - 1.50 = 13.75$ $q = 61$
14. $1.75 + 0.25q = 15.50$ $q = 55$
15. $0.006x - 7.3 = 0.14$ $x = 1{,}240$
16. $0.08 - 0.2y = -4.4$ $y = 22.4$
17. $0.7z - 0.1071 = 0.07z$ $z = 0.17$
18. $-0.009 = 5.2 - 0.1x$ $x = 52.09$
19. $0.112y + 2 = 0.012y - 4$ $y = -60$
20. $0.7x - 0.11 = 5 - 0.03x$ $x = 7$

Additional Example 2

This year, a company's earnings increased by 0.375 of last year's earnings. If the company's earnings are $682,000 this year, what were the earnings last year? $496,000

21. $0.5n + 0.02 = -0.2 - 0.6n$ $n = -0.2$ **22.** $0.23z + 119.7 = 0.8z$ $z = 210$

23. $1.2x + 0.004 = 1.4x - 2$ $x = 10.02$ **24.** $0.09 - 5.1 = 1.5 - 0.24c$

$c = 27.125$

Solve each problem.

25. A number increased by 0.8 of the number is 16.2. Find the number. 9

26. A number decreased by 0.37 times the number is 12.6. Find the number. 20

27. This year a company's earnings increased by 0.28 of last year's earnings. If the company's earnings are $135,680 this year, what were the earnings last year? $106,000

28. The greater of two numbers is 3.2 more than twice the smaller. If 0.3 times the smaller is added to the larger, the result is 11.48. Find the numbers. 3.6, 10.4

29. One number is 0.9 of another. Find the numbers if their sum is 0.038. 0.02, 0.018

30. A taxicab driver charges $1.75 for the first mile and $12\frac{1}{2}$¢ ($0.125) for each additional tenth of a mile. How far can you go for $5.00? 3.6 mi

31. An adult's ticket to a movie theater costs 2.5 times as much as a child's ticket. If John, an adult, pays $18.75 for tickets for himself and 5 children, how much did he pay in all for the children's tickets? $12.50

32. Jan has 3.5 times as much money as Lucia who has $75.60 less than Ken. Jan has $113.10 more than Lucia and Ken together. How much money does Jan have? $440.30

On the Kelvin temperature scale, a temperature of 0 ("absolute zero") is the lowest temperature that can be reached anywhere in the universe. To convert from the Kelvin scale to the Fahrenheit scale (or vice versa), use the formula $1.8K = F + 459.67$.

33. Convert the temperature of outer space, $-454.27°F$, to degrees Kelvin. 3°K

34. Is a temperature of 300°K comfortable? Explain your answer. Yes; 300°K = 80.33°F

Mixed Review

Solve each equation. *3.3, 3.5, 4.5*

1. $6 - 2x = 8$ −1   **2.** $4 - 3x = 7 - x$ −1.5   **3.** $\frac{2}{3}x - \frac{1}{2} = \frac{5}{6}$ 2   **4.** $7 - (3 - x) = 1$ −3

Simplify. *2.2, 2.6, 2.9, 2.10*

5. $|8 - 9|$ 1   **6.** $-5(-2)^3$ 40   **7.** $7 - (4 - x) - 3x + 1$ 4 − 2x

8. Bill is one-third as old as his dad. Bill is 12 years old. How old is his dad? *3.2* 36 yr old

4.6 Equations with Decimals   **149**

Enrichment

Ask students to solve this problem.
A mouse decides to journey to a cheese factory that is *n* miles away. On the first day, it travels 0.5 of the distance. On the second day, it travels of 0.5 of the remaining distance. On the third day, it travels 0.5 of the remaining distance. If the mouse continues to travel in this pattern, how many days will the journey to the cheese factory take? The journey will never be completed.

Checkpoint

Solve each equation. (Exercises 1–4)

1. $0.4y = 0.3$ 0.75
2. $0.6a - 0.32 = 0.004$ 0.54
3. $1.18 = 0.3y - 2.6$ 12.6
4. $0.04z + 0.024 = 0.6z - 0.2$ 0.4
5. One number is 0.6 of another number. Their sum is 0.512. Find each number. 0.32, 0.192

Closure

Have students summarize their method of determining by what number to multiply each side of an equation in order to produce an equivalent equation that contains only integers. See p. 147.

▰▰▰ FOLLOW UP

Guided Practice

Classroom Exercises 1–12
Try This   all

Independent Practice

A Ex. 1–24, **B** Ex. 25–30, **C** Ex. 31–34
Basic: WE 1–19 odd, 25
Average: WE 7–29 odd
Above Average: WE 11–33 odd

Additional Exercises

Classroom Exercises

7. $x = 3$
8. 7
9. $z = 16$
10. $y = 27$
11. $x = -97.8$
12. $x = 1.4192$

Written Exercises

10. $x = 30$
11. $d = 172$
12. $n = 175$

■■■ GETTING STARTED

Prerequisite Quiz

Write as a percent.

1. 0.05   5%
2. 0.125   12.5%
3. 1.4   140%
4. $\frac{2}{5}$ 40%
5. $\frac{2}{3}$ $66\frac{2}{3}$%

Write as a decimal.

6. 46%   0.46
7. 0.3%   0.003
8. $\frac{4}{5}$ 0.8

Motivator

Tell students that the Latin word "centum" means "hundred," from which we get English words such as century, centimeter, centigrade, cent, and percent. Remind the students that percent means per hundred. Therefore, 23% $= \frac{23}{100} = 0.23$.

4.7 Percent Problems

Objective   To solve problems involving percent

Algebra can be used to deal with many situations involving percent. **Percent**, written %, means *per hundred*, or *hundredths*. A percent can be expressed as a fraction or as a decimal, as illustrated below.

$$38\% = \frac{38}{100} = 0.38 \qquad 2\% = \frac{2}{100} = 0.02$$

$$150\% = \frac{150}{100} = 1.5 \qquad 6\frac{1}{2}\% = 6.5\% = \frac{6.5}{100} = 0.065$$

There are three basic types of percent problems. Each involves an equation with two known numbers and one unknown number.

EXAMPLE 1   *Finding a percent of a number*
What number is 65% of 132?

Plan   Translate the statement of the problem into an equation. Use the fact that "of" means "multiplied by."

Solution   Let n = the number.
What number is 65% of 132?

$$n = 0.65 \cdot 132$$
$$n = 85.8 \longleftarrow \text{Calculator steps: } 0.65 \boxed{\times} 132 \boxed{=} 85.8$$

Thus, the number is 85.8.

EXAMPLE 2   *Finding what percent one number is of another*
3 is what percent of 8?

Solution   Let p = the percent.
3 is what percent of 8?

$$3 = p \cdot 8$$
$$\frac{3}{8} = p, \text{ or } p = \frac{3}{8} = 0.375$$

Write as a percent.   0.375 = 37.5%.

Thus, 3 is 37.5% of 8.

TRY THIS   1. What number is 76% of 112?   85.12   2. 4 is what percent of 5?
                                                                              80%

Highlighting the Standards

Standards 4b, 4d, 5a: These examples use decimals to work percent problems. In the Classroom Exercises, students write mathematical formulas related to retail selling.

Additional Example 1

What number is 7% of 150?   10.5

Additional Example 2

Fourteen is what percent of 42? $33\frac{1}{3}$%

EXAMPLE 3 *Finding the number when a percent of it is known*
72 is 150% of what number?

Solution

Let n = the number.
72 is 150% of what number?

Write 150% of n as 1.50n.
$$72 = 1.5 \cdot n$$

Multiply each side by 10.
$$10(72) = 10(1.5n)$$
$$720 = 15n$$
$$48 = n$$

Thus, 72 is 150% of 48.

TRY THIS   **3.** 44 is 110% of what number? 40

When the price of a T-shirt increases from $5 to $6, you subtract to find the amount of increase: $6 - 5 = 1$. To find the percent increase, you compare the amount of increase, $1, to the original price, $5.

$$\text{percent increase} = \frac{\text{amount of increase}}{\text{original amount}} = \frac{1}{5} = 0.2 = 20\%$$

So, the percent increase is 20%.

If the price of a T-shirt decreases from $6 to $5, you subtract to find the amount of decrease: $6 - 5 = 1$.

$$\text{percent decrease} = \frac{\text{amount of decrease}}{\text{original amount}} = \frac{1}{6} = 0.16\frac{2}{3} = 16\frac{2}{3}\%$$

So, the percent decrease is $16\frac{2}{3}\%$.

EXAMPLE 4 Susan's salary was raised from $7 to $9 per hour. Find the percent increase, to the nearest whole percent.

Solution Amount of increase: $9 - $7 = $2

$2 is what percent of $7? Let p = the percent.

$$2 = \frac{p}{100} \cdot 7$$

$$\frac{2}{7} = \frac{p}{100} \quad \longleftarrow \text{Calculator steps: } 2 \boxed{\div} 7 \boxed{=} \ 0.2857142$$

$$p = \frac{200}{7} \approx 29 \quad \longleftarrow \approx \text{ means } \textit{is approximately equal to.}$$

Thus, the percent increase is 29%, to the nearest whole percent.

TRY THIS   **4.** David's salary was decreased from $8 to $6 per hour. Find the percent decrease, to the nearest whole percent. 25%

TEACHING SUGGESTIONS

Lesson Note

Some students have solved percent problems by writing proportions. Show them that writing equations in the form demonstrated here is an alternate method and lends itself well to what they have learned about solving equations. As you discuss the three cases of percent, encourage students to translate each problem into an equation, as shown in Examples 1–3. Point out that *is* indicates equality, and *of* indicates multiplication. In Example 2, remind students that they are asked to find a percent, so they must write the answer in the form of a percent, *not* as a fraction or a decimal.

Emphasize that in computing both percent of increase and percent of decrease, the amount of change is compared to the original amount.

Math Connections

History: In the 15th century, the abbreviation for percent was per $\frac{o}{c}$. In the 17th century this became per $\frac{o}{o}$. Later the "per" was dropped and the horizontal bar became a slash, %.

Additional Example 3

Sixteen is 5% of what number? 320

Additional Example 4

In one year, the enrollment at Kelly School dropped from 800 to 777. Find the percent of decrease to the nearest whole percent. 3%

Common Error Analysis

Error: In problems involving *percent of increase* or *percent of decrease*, students often divide the wrong quantities.

Before assigning Exercises 24, 27, 28, and 29 for homework, have students write the equation for each exercise in class. Have volunteers write each equation on the chalkboard and have the class determine whether the equations are correct.

Checkpoint

1. Twenty-four is what percent of 30? 80%
2. Thirty-five is 5% of what number? 700
3. What number is 28% of 420? 117.6
4. A baseball team won 9 games and lost 6. What percent of their games did the team win? 60%
5. The price of a barrette rose from $1.60 to $1.99. Find the percent increase to the nearest whole percent. 24%

Sometimes interest can be computed using the formula for *simple interest*,

$$i = prt$$

where i is the interest earned, p is the principal (the amount invested), r is the annual rate of interest, and t is the time in years. This formula is used when interest is not paid on interest previously earned.

EXAMPLE 5   Mrs. King invested $1,000 in municipal bonds at a simple-interest rate of 5% per year for 3 years. Find the interest earned.

Solution $p = 1,000$, $r = 5\% = 0.05$, and $t = 3$
interest = principal · rate · time, or $i = prt$
     $= 1,000(0.05)(3) = 150$

Thus, the interest earned was $150.

TRY THIS   5. If $850 is invested at a simple-interest rate of 6% per year for 5 years, how much interest is earned? $255

Classroom Exercises

Give a decimal for each percent.

1. 20%   2. 45%   3. 3%   4. 12.5%   5. 0.5%   6. $7\frac{1}{2}\%$ 7. $8\frac{3}{4}\%$
   0.20   0.45   0.03   0.125   0.005   0.075   0.0875

Give a percent for each decimal or fraction.   10. 12.5%

8. 0.26 26%   9. 0.05 5%   10. 0.125   11. $\frac{2}{5}$ 40%   12. 1.4 140%   13. $\frac{3}{8}$ 37.5%   14. 0.003
     0.3%

15. 24 percent of 6 is what number? 1.44   16. What is $33\frac{1}{3}\%$ of 81? 27

17. 12% of what number is 24? 200   18. 3 is what percent of 9? $33\frac{1}{3}\%$

19. A price is lowered from $5 to $3. What is the percent decrease? 40%

Written Exercises

1. What number is 15% of 60? 9   2. What number is 8% of 50? 4
3. 16 is what percent of 80? 20%   4. 13 is what percent of 52? 25%
5. 20 is 40% of what number? 50   6. 8 is 5% of what number? 160
7. What number is 32% of 80? 25.6   8. What number is 5% of 142? 7.1
9. 24 is what percent of 42? $57\frac{1}{7}\%$ 10. 32 is what percent of 48? $66\frac{2}{3}\%$
11. 84 is 12% of what number? 700   12. 48 is $37\frac{1}{2}\%$ of what number? 128
13. What number is 43% of 78? 33.54   14. 38 is 25% of what number? 152
15. 15 is what percent of 40? 37.5%   16. 18 is 150% of what number? 12
17. What number is $3\frac{1}{2}\%$ of 150? 5.25   18. 30 is what percent of 25? 120%

Additional Example 5

If $5,000 is invested at a simple interest rate of 6% per year for 2 years, how much interest is earned? $600

19. A football team won 15 out of 20 games. What percent of their games did they win? 75%

20. If $600 is invested at a simple-interest rate of 5% per year for 3 years, how much interest is earned? $90

21. If there are 30 points possible on a quiz and a student scored 24 points, what percent did she get right? 80%

22. Roberto got 6 hits in 24 times at bat. The number of hits is what percent of the number of times at bat? 25%

23. A band gets 6% of the total sales of a record that it made. If it got $67,500, what were the total sales? $1,125,000

24. The price of a stock dropped from $72 a share to $63 a share. What was the percent decrease in price? 12.5%

25. If $8,000 is invested at a simple-interest rate of 6.2% per year for 4 years, how much interest is earned? $1,984

26. If $4,000 is invested at a simple-interest rate of 6% for 5 years, how much interest is earned? $1,200

27. A price is decreased from $10 to $8. What is the percent decrease? 20%

28. A price is increased from $8 to $10. What is the percent increase? 25%

29. The price of an airline ticket was $320. Then the price was increased 15%. What was the new price? $368

30. A man flipped a coin 6,000 times. He got heads 3,069 times. What percent of the time did he get heads? $51\frac{3}{20}$%

31. How much money must be invested at a simple-interest rate of 7.2% to earn $1,440 in interest in 5 years? $4,000

32. At what simple-interest rate should $25,000 be invested to earn interest of $13,000 in 8 years? 6.5%

33. A price is raised from $6 to $7. What is the increase, to the nearest whole percent? 17%

34. A price is lowered from $7 to $6. What is the decrease, to the nearest whole percent? 14%

35. There are 2,374 students in a school. If 6% are absent, how many students are present, to the nearest whole number? 2,232

36. Seven percent of a country is ambidextrous. If that represents 1,200,000 people, what is the total population of the country, to the nearest 100,000? 17,100,000

37. In 1987, Bob earned $280. In 1988, his earnings increased 15%. In 1989, his earnings decreased 15% from his earnings in 1988. Was $280 his earnings for 1989? Show why or why not.

38. In 1987, Jane earned d dollars. In 1988, her earnings decreased p percent. In 1989, her earnings increased p percent from her 1988 earnings. Express her 1989 earnings in terms of d and p. $d - dp^2$

Mixed Review

Solve. **3.4–3.6**

1. $5x - 8 = 3x + 12$
   $x = 10$

2. $2(3x - 2) - 4 = 6x + 12$
   No solution

3. $0.6a - 1 = 0.5a + 0.03$
   $a = 10.3$

Evaluate for the given values of the variables. **2.6, 2.8**

4. $-8x^3$, for $x = -5$
   1,000

5. $3x - 4y$, for $x = -3$ and $y = -6$ 15

6. $\dfrac{a - b}{-6}$, for $a = -2$, $b = 4$ 1

Teaching Resources

Quick Quizzes 33
Reteaching and Practice
    Worksheets 33
Transparency 10
Transparency 11

■■■**GETTING STARTED**

Prerequisite Quiz

Solve.

1. $x + 0.5x = 45$ 30
2. $y - 0.8y = 48$ 240
3. $z + 0.04z = 56.16$ 54
4. $60 + 0.05s = 640$ 11,600
5. $r - 0.25r = 24$ 32
6. $560 + 0.02x = 820$ 13,000

Motivator

Discuss with students how percent is used in daily life. For example, a sales tax of 8% means that for every dollar spent, 8¢ or $.08 must be added for tax. Have students offer other examples of uses of percent.
Answers will vary.

■■■**TEACHING
SUGGESTIONS**

Lesson Note

Point out to students that the solution to the problem in Example 1 can also be found by writing one equation.

$$\$32.99 + .08(\$32.99) = \text{total price}$$

4.8 Problem Solving: Using Percent

Objective   To solve problems about sales tax, profit, discount, and commission

When you buy a sweater with a marked price and a known sales tax, you can find the total price by using the following relationship.

marked price + sales tax = total price

EXAMPLE 1   Ahmed bought a sweater with a marked price of $32.99. If the sales tax was 8%, what was the total price of the sweater?

Solution   sales tax = 8% of marked price
            $= 0.08 \times \$32.99 = \$2.6392$
    sales tax = $2.64   ←—— to the nearest cent

    marked price + sales tax = total price
            $32.99 + $2.64 = total price
                    $35.63 = total price

Thus, the total price was $35.63.

TRY THIS   1. Jack bought a jacket with a marked price of $89.99. If the sales tax was 6%, what was the total price of the sweater? $95.39

If a book dealer buys a book at a *cost* of $10 and wants to make a *profit* of $6, then he must add the profit to the cost to determine his *selling price*. So, his selling price must be 10 + 6, or $16.

cost + profit = selling price

The profit is often determined as a percent of the cost.

EXAMPLE 2   A store manager lists the selling price of a bicycle at $78. This includes a 20% profit on her cost. Find the cost of the bicycle.

Solution   Let c = the cost, in dollars.
    Then the profit = 20% of cost = $0.20c$, or $0.2c$, in dollars.

                    cost + profit = selling price
                        $c + 0.2c = 78$
    Multiply each side by 10. $10(c + 0.2c) = 10 \cdot 78$
                        $10c + 2c = 780$
                            $12c = 780$
                            $c = 65$ ←—— Cost: $65

Thus, the cost of the bicycle is $65.

TRY THIS   2. The selling price of a stereo is $124. This includes a 25% profit on the cost. Find the cost of the stereo. $99.20

Highlighting the Standards

Standards 1a, 1b: In the Problem Solving Strategies, the students examine and apply another problem-solving approach.

Additional Example 1

Jessie bought a jacket with a marked price of $67.50. If the sales tax was 9%, what was the total price of the jacket? $73.58

Additional Example 2

The selling price of a keyboard is $280. This includes a profit of 40% on the cost. Find the cost. $200

Stores often sell items at a *discount*, stated as a percent of the *regular price*. The *sale price* is the regular price minus the *amount of discount*.

regular price − amount of discount = sale price

EXAMPLE 3 A radio is on sale for $57.85. The discount is 35%. Find the regular price of the radio and the amount of discount.

Solution Let r = the regular price, in dollars.
Then the amount of discount = 35% of regular price = $0.35r$, in dollars.
regular price − amount of discount = sale price
$$r - 0.35r = 57.85$$
Multiply each side by 100. $100r - 35r = 5785$
$$65r = 5785$$
$$r = 89 \longleftarrow \text{regular price } \$89.00$$
Amount of discount = $0.35r = 0.35(89) = 31.15$

Thus, the regular price is $89.00 and the amount of discount is $31.15.

TRY THIS 3. A shirt is on sale for $24.48. The discount is 20%. What is the regular price of the shirt and the amount of discount?
$30.60; $6.12

Salespeople often receive a percent of the amount they sell. The payment is called a *commission*. A regular salary may also be paid.

salary + commission = total earnings

EXAMPLE 4 Hector earned $350 last week plus a 2% commission on each car he sold. If his total earnings were $710, what were his sales?

Solution Let s = the amount of sales needed, in dollars.
Then the commission = 2% of sales = $.02s$, in dollars.
salary + commission = total earned
$$350 + 0.02s = 710$$
$$0.02s = 360$$
$$2s = 36{,}000$$
$$s = 18{,}000 \quad \text{Thus, Hector's sales were } \$18{,}000.$$

TRY THIS 4. Tamara earned $300 last week plus a 4% commission on sales. If her total earnings were $500, what were her sales? $5,000

Classroom Exercises

State the formula that relates the three items.

1. cost, selling price, profit
2. sales tax, total price, marked price
3. commission, total earned, salary
   salary + commission = total earned
4. amount of discount, regular price, sale price   regular price − discount = sale price

4.8 Problem Solving: Using Percent   **155**

Closure

Have each student write a problem involving sales tax, profit (or loss), discount, and commission (4 problems in all). Have students exchange problems with a classmate, solve the problems, and check each other's work.

◼◼◼FOLLOW UP

Guided Practice

Classroom Exercises 1–6
Try This   all

Independent Practice

A Ex. 1–10, **B** Ex. 11–14, **C** Ex. 15–16

Basic: WE 1–10 all, Problem Solving Strategies

Average: WE 4–13 all, Problem Solving Strategies

Above Average: WE 7–16 all, Problem Solving Strategies

Additional Answers

Classroom Exercises

1. cost + profit = selling price
2. marked price + sales tax = total price

Written Exercises

8. $6,054.75

Find the missing item.

5. salary: $175 plus 2% of sales
   total earnings: $355
      total sales: __?__ $9,000

6. marked price: $12.95
   sales tax: 7%
      total price: __?__ $13.86

Written Exercises

1. A jacket is marked $159. If the tax is $6\frac{1}{2}\%$, what is the total price? $169.34

2. If a record is marked $8.99 and the tax is 7%, what is the total price? $9.62

3. The marked price of a T-shirt is $9.00. If it is on sale for 15% off, what is the sale price? $7.65

4. The regular price of a tape is $8. If the discount is 20%, what is the sale price? $6.40

5. The selling price of a TV set is $360. The profit is 20% of the cost. Find the cost of the set. $300

6. Matt was paid $125 last week plus a 5% commission on his sales of $6,720. What were his total earnings? $461

7. The selling price of a turntable is $156. The profit is 30% of the cost. Find the cost. $120

8. Mr. King bought 150 shares of stock at $39 per share plus $3\frac{1}{2}\%$ commission. How much did he pay in all?

9. A book is on sale for $10.20. The rate of discount is 25%. Find the regular price and the amount of discount. $13.60; $3.40

10. A camera's selling price is $260. Find the dealer's cost if the profit is 30% of the cost. $200

11. Marcus is paid an 8% commission on all sales plus $3.50 an hour for 40 hours a week. What must be his total sales to earn a total of $220? $1,000

12. Sofia is paid $125 plus a commission of 15% on all sales over $300. What must her total sales be if she is to earn a total of $185? $700

13. Gary is paid a 15% commission on his sales plus $6.50 an hour. One week his sales were $1,240. How many hours did he work to earn $459? 42 h

14. The regular price of a video-cassette recorder is $480. It is on sale for $390. Find the discount. $18\frac{3}{4}\%$

15. The original price of a TV set allowed for a profit of 30% of the cost. The new price is $195, an increase of 25%. Find the dealer's cost. $120

16. The sale price s of an item with a regular price r, sold at a 15% discount, is given by the formula $s = kr$. Find the value of k. Justify your answer. 0.85

Mixed Review

Simplify.   2.10

1. $9^2 \div (5^2 - 4^2)$ 9

2. $2^2(125 - 8^2)$ 244

3. $(12 - 7)^2$ 25

4. $12^2 - 7^2$ 95

Simplify.   1.3

5. $-4n(n - 10)$
   $-4n^2 + 40n$

6. $3 - (a + b)$
   $3 - a - b$

7. $-(c - 8) + 9$
   $-c + 17$

8. $a - (-a + 1)$
   $2a - 1$

156   Chapter 4 Applying Equations

Enrichment

Have the students solve for n mentally by using the Distributive Property.

a. 11% of 1,500 = n 165
b. 15% of $20.00 = n $3.00
c. 15% of $14.20 = n $2.13
d. 16% of 350 = n 56
e. 33% of 1,200 = n 396
f. 55% of 620 = n 341

156

Problem Solving Strategies

Carrying Out the Plan

Carrying Out the Plan in problem solving involves these things.

a. Estimating the answer and thinking about the appropriate units for the answer.

b. Applying the strategy or strategies you have chosen. A good strategy helps you to find a correct solution in an efficient way.

Example Jean wants to buy some cassettes that are on sale for $2.59 each. How many can she buy with a $20 bill?

Plan Use the guess-and-check strategy.

Solution THINK: The number of cassettes will be a small whole number. So try 10 as a first guess. Use a calculator to check.

Guess	Check	Does it check?
10	$10 \times \$2.59 = \$25.90$	No. Too many
8	$8 \times \$2.59 = \$20.72$	No. Too many
7	$7 \times \$2.59 = \$18.14$	Reasonable answer

So Jean can buy 7 cassettes with the $20 bill.

Exercises

Solve. Use the guess-and-check strategy.

1. A tree trimmer charges $35 plus $18 per hour for trimming trees. Mrs. Danoff pays $143 for his work. How long did the tree trimmer work for Mrs. Danoff? 6 h

2. Travis has $1.00 in 7 coins of two different kinds. What are the coins and how many of each does he have? 2 quarters; 5 dimes

3. Find two consecutive odd integers whose product is 323. 17 and 19

4. A veterinarian has a rectangular pen with an area of 96 ft². It is surrounded by a fence 44 feet long. What are the dimensions of the pen? 6 ft by 16 ft

5. Chris, Mike, and John play on the Wheelchair Whiz basketball team. The numbers on their jerseys are consecutive even integers. The sum of the integers is 42. Find the numbers. 12, 14, and 16

Key Terms

even integer   (p. 135)
isosceles triangle   (p. 140)
  base angles of,   (p. 140)
  base of,   (p. 140)
  legs of,   (p. 140)
  vertex angles of,   (p. 140)

least common denominator (LCD)   (p. 144)
least common multiple (LCM)   (p. 144)
odd integer   (p. 136)
percent   (p. 150)
perimeter of a triangle   (p. 140)

Key Ideas and Review Exercises

**4.1,
4.2**
If a word problem describes a relationship between two or more numbers, choose a variable and let it represent the number that is the *basis of comparison*. Then represent the other numbers in terms of that variable. Write an equation and solve it. Find the numbers, and check them in the original problem.

Bonita picked 6 more red apples than green apples and 8 more green apples than yellow apples. Let y be the number of yellow apples Bonita picked. (Exercises 1–4)

1. Represent the number of green apples in terms of y. $8 + y$

2. Represent the number of red apples in terms of y. $6 + (8 + y)$, or $14 + y$

3. The total number of apples picked was 115. Write an equation and solve it. $3y + 22 = 115$; $y = 31$

4. How many apples of each color were picked?   31 yellow, 39 green, 45 red

Solve each problem.

5. One number is 6 times another. Their sum is 48. Find the numbers. $6\frac{6}{7}$, $41\frac{1}{7}$

6. Separate $71 into 2 parts such that the second part is $5 less than 3 times the first part. $19; $52

4.3
To solve problems involving consecutive integers, remember that consecutive integers can be represented by n, $n + 1$, $n + 2$, and so on, and that consecutive odd integers and consecutive even integers can be represented by n, $n + 2$, $n + 4$, and so on.

Solve each problem.

7. Find two consecutive integers with a sum of -25. -13, -12

8. Find three consecutive even integers such that their sum decreased by 34 is equal to the first integer.   14, 16, 18

4.4 To solve problems involving perimeter and angle measure, first draw and label a figure. Then follow the usual problem-solving process.

9. The length of a rectangle is 10 ft less than 4 times the width. The perimeter is 50 ft. Find the length and the width. $l = 18$ ft; $w = 7$ ft

10. Each base angle of an isosceles triangle has a degree measure that is 5 more than twice that of the vertex angle. Find the measure of each angle of the triangle. Bases: 73; vertex: 34

4.5 To solve an equation that contains fractions, multiply each side by the LCD of all the fractions. Then solve in the usual way.

Solve each equation.

11. $\frac{2}{3}y - \frac{1}{2} = \frac{5}{6}$ 2

12. $\frac{3}{8} + \frac{1}{4}c = \frac{1}{2}c$ $\frac{3}{2}$

13. $\frac{4x - 1}{3} - \frac{x}{4} = \frac{11x + 4}{12}$ 4

4.6 To solve an equation with decimals, multiply each side by a power of ten that will give you an equivalent equation with integers in place of the decimals. Then solve in the usual way.

Solve each equation.

14. $0.03y = 0.6$ $y = 20$

15. $10.4 - 0.2c = -0.018$ $c = 52.09$

16. $0.6y + 0.002 = 0.7y - 1$ $y = 10.02$

4.7 To solve any of the three basic types of percent problems, write an equation and solve as in Examples 1, 2, and 3 of Lesson 4.7.

To find the percent increase or decrease from one number to another:
a. Subtract to find the amount of increase or decrease.
b. Find what percent this is of the original number.

17. What number is 62% of 150? 93

18. Forty-one is 25% of what number? 164

19. Pierre got 12 hits in 40 times at bat. The number of hits is what percent of the number of times at bat? 30%

20. Last year Jake was paid $7.50 an hour. This year he is paid $8.00 an hour. What is the percent increase in his wages? $6\frac{2}{3}\%$

4.8 For business problems involving percents, use these relationships.

cost + profit = selling price
regular price − amount of discount = sale price
salary + commission = total earned

21. The selling price of a radio is $70. The profit is 40% of the cost. Find the cost. $50

22. Jane is paid $140 a week plus 5% commission on all sales. Find the total sales needed to make her total earnings $350. $4,200

Solve each problem.

1. A 49-cm piece of ribbon is cut into two parts. The first part is 5 cm shorter than twice the second part. Find the length of each part. 31 cm; 18 cm

2. The larger of 2 numbers is 3 times the smaller. The difference between the numbers is 14. Find the numbers. 7, 21

3. The length of a rectangle is 3 ft greater than 4 times the width. The perimeter is 86 ft. Find the length and the width. l = 35 ft; w = 8 ft

4. Each base angle of an isosceles triangle measures 12 degrees less than the vertex angle. Find the measure of each angle of the triangle. Bases: 56°; vertex: 68°

5. The sum of two consecutive odd integers is −72. Find the integers. −37, −35

6. Find 3 consecutive integers such that 4 times the middle integer plus the largest integer is 9 less than 6 times the smallest. 15, 16, 17

Solve each equation.

7. $\frac{1}{2}y = \frac{1}{3}y + 8$ 48

8. $\frac{6a}{7} - \frac{a}{2} = 5$ 14

9. $3.78 = 0.06x - 4.2$ 133

10. $1.4x - 0.2142 = 0.14x$ 0.17

Solve each problem.

11. 15 is 20% of what number? 75

12. 54 is what percent of 60? 90%

13. The price of a jacket rose from $60 to $70. What was the percent increase in the price? $16\frac{2}{3}$%

14. Olga is paid $160 a week plus a commission of 4% of her sales. How much must she sell to earn $250 in a week? $2,250

15. The regular price of a radio is $64. It is on sale for $56. Find the rate of discount. $12\frac{1}{2}$%

16. Mr. Golden bought a shipment of paperweights for a total of $576. He will make a profit of 60% of that amount when he sells them. He priced each paperweight at $4.80. How many paperweights were there? 192

17. The selling price s of an item when there is a cost c and a profit of 40% of the cost is given by a formula of the form $s = kc$. Find the value of k. k = 1.40

College Prep Test

Choose the one best answer to each question or problem.

1. At Harper School, which grade had the largest percent increase in the number of students? A

Grade	8	9	10	11	12
1989	60	55	65	62	60
1990	80	62	72	72	70

   (A) 8   (B) 9   (C) 10
   (D) 11   (E) 12

2. If n is an odd integer, which of the following represents the sum of n and the next two consecutive odd integers? D
   (A) $3n$ (B) $3n + 3$
   (C) $3n + 4$ (D) $3n + 6$
   (E) None of these

3. If m and n are integers and m is divisible by 5, which of the following is always true? E
   (A) $m + n$ is odd.
   (B) $m + n$ is even.
   (C) $m + n$ is divisible by 5.
   (D) mn is odd.
   (E) mn is divisible by five.

4. If 12% of a class of 25 students failed an exam, how many students passed the exam? A
   (A) 22   (B) 20   (C) 13
   (D) 12   (E) 3

5. Evaluate $\dfrac{x^2 + x^3}{x - 1}$ if $x^3 = 8$. C
   (A) $\dfrac{16}{7}$ (B) 4   (C) 12
   (D) 16   (E) 24

6. A truck was originally priced at $7,000. The price was reduced 20% and then raised 5%. What was the net reduction in price? D
   (A) $5,950   (B) $5,880
   (C) $1,400   (D) $1,120
   (E) $1,050

7. A rectangle is 3 units longer than it is wide. Find its length if the perimeter is 38 units. D
   (A) 4   (B) 6   (C) 8   (D) 11
   (E) 19

8. How many 25-cent stamps may be purchased for d dollars? E
   (A) $\dfrac{d}{25}$ (B) $\dfrac{25}{d}$ (C) $25d$
   (D) $\dfrac{25}{100d}$ (E) $\dfrac{100d}{25}$

9. Mary receives D dollars for a 5-day work week. What is her daily salary after receiving a $10.00 per week raise? B
   (A) $D + 10$ (B) $\dfrac{D}{5} + 2$
   (C) $\dfrac{D}{5} + 10$ (D) $5D + 2$
   (E) $5D + 10$

10. The sum of 4 consecutive odd integers exceeds twice the largest by 22. Find the sum of the 4 numbers. C
   (A) 11   (B) 48   (C) 56
   (D) 60   (E) 88

11. Find the perimeter of the figure. D

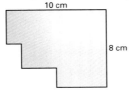

   10 cm
   8 cm

   (A) 18 cm   (B) 28 cm   (C) 30 cm
   (D) 36 cm   (E) 80 cm

12. Bob is 5 times as old as Jean, who is 2 years younger than Mike. If Mike is 5 years old, how old is Bob? C
   (A) 27   (B) 25   (C) 15
   (D) 10   (E) 3

For additional standardized test practice, see the SAT/ACT test booklet for cumulative tests Chapters 1–4.

Cumulative Review (Chapters 1–4)

Each of Exercises 1–8 has five choices for answers. Choose the best answer.

1. Evaluate $6.1x$ for $x = 4$.   C   *1.1*
   (A) 244   (B) 2.44
   (C) 24.4   (D) 6.5
   (E) None of these

2. Simplify $\dfrac{3 + 8 \cdot 6}{17}$.   B   *1.2*
   (A) 1   (B) 3   (C) $\frac{1}{17}$
   (D) $3\frac{15}{17}$ (E) None of these

3. In the formula $p = 2l + 2w$, find p for $l = 3.4$, $w = 4.1$.   A   *1.4*
   (A) 15   (B) 7.5   (C) 0.7
   (D) 30   (E) None of these

4. Solve $3x - 9 = 2(4 + x)$.   E   *3.5*
   (A) $\frac{17}{4}$ (B) -1 (C) 1
   (D) -17 (E) None of these

5. Simplify $5(2x - 6) - (3 - 6x)$.   *2.10*
   (A) $4x - 9$ (B) $4x - 33$ D
   (C) $16x - 9$ (D) $16x - 33$ (E) None of these

6. Simplify $-6 - (-5)$.   B   *2.5*
   (A) 11   (B) -1
   (C) -11 (D) -30
   (E) None of these

7. Simplify $-15 + 2 - (-8) - 3$.   A
   (A) -8 (B) -24
   (C) -28 (D) 8
   (E) None of these

8. Rewrite $(5x - 9)4$ without parentheses.   D   *2.10*
   (A) $5x - 36$ (B) $5x + 36$
   (C) $20x + 36$ (D) $20x - 36$
   (E) None of these

Solve each equation.

9. $x - (-4) = 7$ 3   *3.1*

10. $46 = -2y$ -23 *3.2*

11. $\frac{2}{3}x = -22$ -33

12. $-5y + 2 = -18$ 4   *3.3*

13. $2(5y + 7) = -y - 8$ -2 *3.5*

14. $3x - 7 = 4 - (x + 1)$ $\frac{5}{2}$

15. $5.6 - 2.4y = 7.2$ $-\frac{2}{3}$ *4.6*

16. $3(2 - a) = 4(a + 5)$ -2 *3.5*

Simplify.

17. $6a - 9(2a + 3) - 7$ $-12a - 34$ **2.10**

18. $-5x + 3(2 + x) - (x - 1)$ $-3x + 7$

19. $3c + 2d - 5(c - 4d)$ $-2c + 22d$

Evaluate. (Exercises 20–24)

20. x^3, for $x = -3$ -27 *2.6*

21. $4b^4$, for $b = -1$ 4

22. $-7 - 3x$, for $x = 2$ -13 *2.8*

23. $6a^2 - b$, for $a = -2$, $b = -9$ 33

24. $6(x - 9)$, for $x = 4$ -30 *1.2*

25. If Jim's age is increased by 3 years more than twice his age, the result is 51 years. Find Jim's age.   24 yr   *4.2*

26. This year, a ring selling for $53 costs 7 dollars less than twice what it cost last year. What did the ring cost last year? $30

27. Phyllis has 28 records in her collection. She has $\frac{2}{3}$ as many as Josie has. How many records does Josie have?   42   *3.2*

28. The temperature at 6:00 P.M. was 5°F. By 1:00 A.M. it had dropped 9°. By 4:00 A.M. it had risen 6°. What was the temperature at 4:00 A.M.? 2°F   *2.4*

29. The lowest point in a valley is 160 ft below sea level. A helicopter was flying 2,200 ft above this point. The helicopter dropped 500 ft. Find the altitude of the helicopter relative to sea level. 1,540 ft   *2.4*

30. Use $C = \frac{5}{9}(F - 32)$ to find the temperature in degrees Celsius (C) when a thermometer reads 50°F. 10°C   *1.3*

31. The daily high temperatures for the first 5 days of January are shown below. What was the average high temperature for the 5 days? 0   *2.7*

Jan. 1	Jan. 2	Jan. 3	Jan. 4	Jan. 5
−3	10	6	−4	−9

32. On Monday, Katie ran the 100-yard dash in 12.3 seconds. On Tuesday she improved her time by 0.9 s. What was her time for the 100-yard dash on Tuesday? 11.4 s

33. After jogging 4 mi from home, Jordan was $\frac{2}{3}$ of the way to the shopping center. How far was the shopping center from his home? 6 mi   *3.2*

34. The $56 price of a coat is $1 less than 3 times the cost. Find the cost of the coat. $19   *4.2*

35. After Jean gave away $\frac{2}{5}$ of her fish, she had 12 fish left. How many fish did she have before she gave some away? 20

36. This year a party favor costs $0.60 less than twice what one cost last year. This year a favor costs $1.10. What did it cost last year? $0.85   *4.2*

37. If a number is increased by 8 and then multiplied by −3, the result is 21. Find the number. −15   *4.2*

38. For the square below, write a formula for the perimeter in terms of k. Express the formula in simplest form. Perimeter : 12k   *2.9*

3 k

39. Separate 31 into two parts so that the larger is 5 less than 3 times the smaller. 9, 22   *4.2*

40. The sum of two consecutive integers is 75. Find the two integers. 37, 38   *4.3*

41. Find three consecutive odd integers whose sum is −39. −15, −13, −11

42. In a rectangle, the length is 3 m shorter than twice the width. The perimeter is 36 m. Find the length and the width.   *4.4*

w : 7 m; l : 11 m

w
2w − 3

43. What number is 8% of 40? 3.2   *4.7*

44. The price of a tape deck is $156. The profit is 30% of the cost. Find the cost. $120   *4.8*

45. What percent of 90 is 15? $16\frac{2}{3}$%   *4.7*

46. Six is 5% of what number? 120

47. Use the formula $i = prt$ to find the amount of simple interest received when $2,500 is invested for 3 years at an interest rate of $6\frac{1}{2}$%. $487.50

5 INEQUALITIES AND ABSOLUTE VALUE

OVERVIEW

The first two lessons of Chapter 5 deal with the properties needed for solving and graphing solution sets of inequalities. Then a lesson on conjunctions and disjunctions prepares students for solving combined inequalities, and for applying these skills in problem situations related to everyday life. Finally, equations and inequalities involving absolute value are introduced and applied in problem-solving contexts.

OBJECTIVES

- To solve inequalities and combined inequalities and to graph their solution sets
- To determine whether a conjunction or a disjunction is true or false
- To solve and graph equations and inequalities containing absolute value
- To solve problems using inequalities

PROBLEM SOLVING

The strategy of Making a Graph is emphasized in this chapter as a method for solving inequalities and open sentences involving absolute value. Interpreting Key Words is covered in relationship to writing inequalities. Making a Table allows the student to decide the truth value of conjunctions and disjunctions. The problem-solving strategy on page 185 emphasizes the importance of the Looking Back step in problem solving. The Mixed Problem Solving exercises on page 195 ask students to apply the strategy of Drawing a Diagram as they review various types of problems presented thus far in the textbook.

READING AND WRITING MATH

The Focus on Reading on page 182 asks students to match conjunctions and disjunctions with their graphs. Throughout the chapter, students translate phrases into conjunctions and disjunctions and sentences into equations and inequalities. The students are asked to compare a mathematical phrase and a sentence in Exercise 21 on page 189.

TECHNOLOGY

Calculator: The Mixed Problem Solving exercises on page 195 offer opportunities for students to use a calculator as part of this problem-solving process.

SPECIAL FEATURES

Mixed Review pp. 169, 174, 178, 189, 191, 194
Focus on Reading p. 182
Midchapter Review p. 184
Brainteaser p. 184
Problem Solving Strategies: Looking Back p. 185
Application: Triangle Inequality p. 189
Mixed Problem Solving p. 195
Key Terms p. 196
Key Ideas and Review Exercises p. 196
Chapter 5 Test p. 198
College Prep Test p. 199

PLANNING GUIDE

Lesson	Basic	Average	Above Average	Resources
5.1 pp. 168–169	CE 1–14 WE 1–19 odd	CE 1–14 WE 5–23 odd	CE 1–14 WE 9–27 odd	Reteaching 34 Practice 34
5.2 pp. 173–174	CE 1–9 WE 1–29 odd	CE 1–9 WE 9–37 odd	CE 1–9 WE 15–43 odd	Reteaching 35 Practice 35
5.3 pp. 177–178	CE 1–8 WE 1–14 all	CE 1–8 WE 1–25 odd	CE 1–8 WE 7–31 odd	Reteaching 36 Practice 36
5.4 pp. 182–185	FR all CE 1–20 WE 1–25 odd 35–37 all Midchapter Review Brainteaser Problem Solving Strategies	FR all CE 1–20 WE 7–37 odd Midchapter Review Brainteaser Problem Solving Strategies	FR all CE 1–20 WE 19–51 odd Midchapter Review Brainteaser Problem Solving Strategies	Reteaching 37 Practice 37
5.5 pp. 187–189	CE 1–8 WE 1–10 all, 21 Application	CE 1–8 WE 5–14 all, 21 Application	CE 1–8 WE 5–15 odd, 17–21 all Application	Reteaching 38 Practice 38
5.6 p. 191	CE 1–8 WE 1–12 all	CE 1–8 WE 1–11 odd 13–18 all	CE 1–8 WE 5–11 odd 13–20 all	Reteaching 39 Practice 39
5.7 pp. 194–195	CE 1–10 WE 1–10 all 21, 23 Mixed Problem Solving	CE 1–10 WE 1–23 odd Mixed Problem Solving	CE 1–10 WE 11–33 odd Mixed Problem Solving	Reteaching 40 Practice 40
Chapter 5 Review pp. 196–197	all	all	all	
Chapter 5 Test p. 198	all	all	all	
College Prep Test p. 199	all	all	all	

CE = Classroom Exercises   WE = Written Exercises   FR = Focus on Reading
NOTE: For each level, all students should be assigned all Try This and all Mixed Review Exercises.

■ INVESTIGATION

Project: Students use data collected from their classmates to write compound statements using *or* and *and*.

Materials: Each group of students will need a data sheet and a copy of Project Worksheet 5.

The day before doing this Investigation, have students complete a data sheet with the following headings: Birthday (Month/Day), Height in Inches, House Number, Number of Letters in Last Name, and Shoe Size. Have each student initial his or her line of data.

On the day of the activity, discuss with students which birth dates would or would not fit the description: "All students born after March 31 and before August 1." Discuss which heights would fit this description: "Students less than 64 inches or greater than 70 inches tall."

Distribute a worksheet and a data sheet to each group. Have students, working in small groups, use the data sheet to write five compound statements, some with *or* and some with *and,* on a separate sheet of paper. Have the groups fill in the initials of all students that fit each sentence they have written on the worksheeet. Then direct the groups to exchange worksheets only and write a sentence with *and* or *or* that will fit the initials they have been given for each category. Finally, have the groups check each other's answers for correctness. Remind students that an answer may be correct even though it does not match the original sentence exactly.

These surveyors study the movement of glaciers in the Juneau icefield in Alaska. Most glaciers move, or "flow" at less than one foot per day. Scientists, however, have measured a flow of more than 50 feet in one day.

More About the Photo

Ancient Egyptians developed the techniques of surveying, or measuring the Earth's surface in three dimensions. Surveying is based on the laws of simple geometry. Surveyors measure distances and angles, establish reference points, and apply triangulation. They measure distances by reflecting from a target a laser or radar beam, or a beam of infra-red light. Modern versions of the theodolite, which measures angles in both the vertical and horizontal plane, greatly increase accuracy but the basic principles, used in ancient times to build the pyramids, remain the same.

5.1 The Addition and Subtraction Properties of Inequality

Objective

To solve and graph the solution set of an inequality by using the Addition or Subtraction Property of Inequality

Examine the chart below to review the two inequality symbols, $<$ and $>$.

Inequality	Read	Meaning
$a < b$	a is less than b.	a is to the *left* of b on a number line.
		a \qquad\qquad b
$c > d$	c is greater than d.	c is to the *right* of d on a number line.
		d \qquad\qquad c

The open sentence $x < -2$ is an example of an *inequality*. An **inequality** contains at least one variable and consists of two expressions with an inequality symbol such as $<$, $>$, or $\neq$ between them. To **solve an inequality** means to find its solution set. The replacement set for the variable is assumed to be the set of all real numbers. To solve the inequality $x < -2$, you must find all the real numbers that make the open sentence true. Some real numbers make $x < -2$ a true statement and some real numbers make $x < -2$ a false statement.

$x < -2$		$x < -2$	
$-3 < -2$	true	$2 < -2$	false
$-5.4 < -2$	true	$0 < -2$	false
$-100 < -2$	true	$-2 < -2$	false

It is not possible to list all the solutions of $x < -2$. However, a number line can be used to graph the solution set.

The graph of the solution set of $x < -2$ includes every point with a coordinate that is less than -2. To indicate that -2 itself is not a solution, an open circle appears at -2. A heavy arrow to the *left* of the open circle indicates that the coordinate of every point to the left of -2 is a solution.

$$-5 \quad -4 \quad -3 \quad -2 \quad -1 \quad 0 \quad 1 \quad 2$$

graph of the solution set of $x < -2$

Teaching Resources

Project Worksheet 5
Quick Quizzes 34
Reteaching and Practice Worksheets 34
Transparency 12
Teaching Aid 2

◼◼◼ GETTING STARTED

Prerequisite Quiz

Write $<$, $=$, or $>$.

1. $7 \ \underline{\ ?\ } \ -8 \quad >$
2. $0 \ \underline{\ ?\ } \ -3 \quad >$
3. $0.5 \ \underline{\ ?\ } \ 0.6 \quad <$
4. $-1.2 \ \underline{\ ?\ } \ -1.5 \quad >$
5. $2(5 - 8) \ \underline{\ ?\ } \ -4 \quad <$
6. $20 - 4 \ \underline{\ ?\ } \ 2(10 - 2) \quad =$

Motivator

Point out to students that inequalities involve comparisons. Have them give examples of comparisons from real life that involve the words, "equal, less than, and greater than."

Highlighting the Standards

Standards 2f, 4a, 3a, 3b, 3e: Students learn the inequality notation and to solve and graph inequalities. In Exercises 24–27 they test conjectures, formulate counterexamples, and construct proofs.

TEACHING SUGGESTIONS

Lesson Note

Extensive use of the number line will help students understand the operational properties of inequalities. It also provides the best way of visualizing the solution sets of open sentences containing inequality symbols. Students should perceive that these are infinite sets, in contrast to the finite solution sets of equations.

Suggest that the students use a different color of pencil or pen to show their graphs on the number lines that they have drawn in pencil.

Math Connections

Science: The time it takes the earth to revolve around the sun is greater than 365 days. In other words, 1 year > 365 days. A leap year every fourth year compensates for this discrepancy.

Critical Thinking Questions

Analysis: Ask students this question. Given that $a < b$, which of the following are always true?

1. $a^2 < b^2$
2. $|a| < |b|$
3. $\frac{1}{a} < \frac{1}{b}$ none

Additional Answers

Try This

1.
  −2 −1 0 1 2 3 4

EXAMPLE 1   Graph the solution set of $x > 0$.

Solution   Make an open circle at 0. Then draw a heavy arrow to the right of the circle to show all the points with coordinates that are greater than 0.

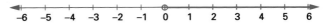
  −6 −5 −4 −3 −2 −1 0 1 2 3 4 5 6

TRY THIS   1. Graph the solution set of $x < 3$.

The Addition and Subtraction Properties of Equality allow you to add or subtract the same number from each side of an equation to obtain an equivalent equation. Examine the chart below to determine whether there are similar properties for inequalities.

True Inequality	Operate on each side	New Inequality	True or False?
$2 < 6$	Add 5.	$2 + 5 < 6 + 5$ $7 < 11$	True
$3 > -7$	Subtract 2.	$3 - 2 > -7 - 2$ $1 > -9$	True
$-8 < -2$	Subtract 4.	$-8 - 4 < -2 - 4$ $-12 < -6$	True

In each case above, the new inequality is true. These cases suggest the properties below.

Definition   Open inequalities with the same solution set are called **equivalent inequalities**.

Addition Property of Inequality
For all real numbers a, b, and c,
  if $a < b$, then $a + c < b + c$, and
  if $a > b$, then $a + c > b + c$
That is, adding the same number to each side of an inequality produces an *equivalent inequality*.

Subtraction Property of Inequality
For all real numbers a, b, and c,
  if $a < b$, then $a - c < b - c$, and
  if $a > b$, then $a - c > b - c$
That is, subtracting the same number from each side of an inequality produces an equivalent inequality.

166   Chapter 5 Inequalities and Absolute Value

Additional Example 1

Graph the solution set of $y < 1$.

  1

EXAMPLE 2   Solve $x - 8 < -11$. Graph the solution set.

Plan   Use the Addition Property of Inequality.

Solution
$$x - 8 < -11$$
Add 8 to each side. $x - 8 + 8 < -11 + 8$
$$x < -3$$

Check   You cannot check every solution. Try numbers less than -3 and greater than -3. Also try -3 itself. Check in the *original* inequality.

Let $x = -4$.	Let $x = -3$.	Let $x = -2$.
$x - 8 < -11$	$x - 8 < -11$	$x - 8 < -11$
$-4 - 8 \overset{?}{<} -11$	$-3 - 8 \overset{?}{<} -11$	$-2 - 8 \overset{?}{<} -11$
$-12 < -11$	$-11 \not< -11$	$-10 \not< -11$
(True)	(False)	(False)

Thus, the solution set is the set of all real numbers less than -3.

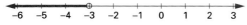

EXAMPLE 3   Solve $7 < 5 - \left(\frac{1}{2} - x\right)$. Graph the solution set.

Plan   First simplify the expression on the right side of the inequality. Then find an equivalent inequality that has x alone on the right.

Solution
$$7 < 5 - \left(\frac{1}{2} - x\right)$$
$$7 < 5 - \frac{1}{2} + x$$
$$7 < 4\frac{1}{2} + x$$
Subtract $4\frac{1}{2}$ from each side. $7 - 4\frac{1}{2} < 4\frac{1}{2} + x - 4\frac{1}{2}$
$$2\frac{1}{2} < x, \quad \text{or} \quad x > 2\frac{1}{2}$$

Thus the solution set is the set of all real numbers greater than $2\frac{1}{2}$.

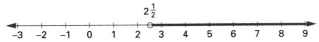

TRY THIS   2. Solve $-8 > -2 - (4 - x)$. $x < -2$

Additional Example 2

Solve $y - 6 > 2$. Graph the solution set.

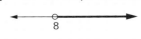

Additional Example 3

Solve $1 < 2 - (\frac{1}{4} - x)$. Graph the solution set.

(graph: open circle at $-\frac{3}{4}$)

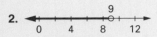
EXAMPLE 4   After Mary paid $8.36 for a tape, she had less than $2.50 left. How much money did she have originally?

Solution   Let m = the original amount of money.
Then $m - 8.36$ = the amount of money after the purchase of a tape.
The amount of money after purchase is less than $2.50.

Inequality: $m = 8.36$ $<$ 2.50

Solve the inequality. $m - 8.36 + 8.36 < 2.50 + 8.36$

$m < 10.86$

Thus, the original amount of money was less than $10.86.

TRY THIS   3. After Bill paid $7.21 at the movies, he had less than $1.75 left. How much money did he have originally?   less than $8.96

Classroom Exercises

Complete each statement. Use $<$, $>$, or $=$.

1. $7 > 6$. Therefore, $7 - 2$ _?_ $6 - 2$ $>$
2. $-3 < 0$. Therefore, $-3 + 3$ _?_ $0 + 3$ $<$
3. $-9 + 4 = -5$. Therefore, $-9 + 4 - 4$ _?_ $-5 - 4$ $=$
4. $-8 < -3 - 4$. Therefore, $-8 + 8$ _?_ $-3 - 4 + 8$ $<$
5. If $x + 6 < 2$, then $x + 6 - 6$ _?_ $2 - 6$ $<$

Match each open sentence with the correct graph.

a.
   -2 -1 0 1 2 3 4 5

b.
   -2 -1 0 1 2 3 4 5

6. $x < 3$ b
7. $x > 3$ a
8. $3 < x$ a
9. $3 > x$ b
10. $x - 5 < -2$ b
11. $7 < x + 4$ a

Solve each inequality. Graph the solution set.

12. $y - 4 < 2$ $y < 6$
13. $x + 7 > 0$ $x > -7$
14. $c - 5 < -1$ $c < 4$

Written Exercises

Solve each inequality. Graph the solution set.

1. $x + 8 > 5$ $x > -3$
2. $y - 7 < 2$ $y < 9$
3. $z + 3 < -7$ $z < -10$
4. $c - 5 > -2$ $c > 3$
5. $r + 7 < 0$ $r < -7$
6. $x - 8 > 1$ $x > 9$

Additional Example 4

Ricky bought tickets to a game for little less than $14. His change from the purchase was $2.80. Could he have paid for the tickets with a $20 bill?   No

7. $y - \frac{1}{4} < 2$ $y < 2\frac{1}{4}$ **8.** $z + 0.5 > -4$ $z > -4.5$ **9.** $d - 2\frac{1}{3} > -6$

10. $5 < x - 3$ $x > 8$ **11.** $-2 > y + \frac{1}{2}$ $y < -2\frac{1}{2}$ **12.** $-6 < -2.8 + c$

13. $6.2 + y < -8.1$ $y < -14.3$ **14.** $-5.7 + a > 3.6$ $a > 9.3$ **15.** $-7\frac{1}{2} < 3\frac{1}{4} + z$

12. $c > -3.2$

15. $z > -10\frac{3}{4}$

Write the inequality for each problem. Then solve the problem.

16. If Sarah gains 6 lb, she will weigh more than 113 lb. How much does Sarah weigh now?
$w + 6 > 113; w > 107$

17. After Jim deposited $6.25 in his savings account, he had less than $50 in the account. How much was in the account before he made the deposit?

18. After Jason gave 8 tapes away, he had more than 35 tapes. How many tapes did he have before he gave 8 away?
$t - 8 > 35; t > 43$

19. After selling 13 copies of the Daily Gazette, a newsdealer had fewer than 90 copies left. How many copies did the newsdealer have originally?
$s - 13 < 90; s < 103$

Solve each inequality. Graph the solution set.

20. $5x - (1 + 4x) > -4$ $x > -3$ **21.** $-2x - 3(4 - x) < 14$ $x < 26$

22. $1.5x - 0.5(x + 4) < 12$ $x < 14$ **23.** $6.5 - 4.2c + 5.2(c - 1) > 3.7$
$c > 2.4$

24. The Symmetric Property of Equality states that for the equality symbol, $=$, if $x = y$, then $y = x$. Is there a similar property for the inequality symbol, $<$? If so, state it. If not, give a numerical example that disproves it. No. $2 < 3$ yet $3 \not< 2$

25. The Reflexive Property of Equality states that for the equality symbol, $=$, $x = x$. Is there a reflexive property for the inequality symbol, $>$? If so, state it. If not, give an example to disprove it. No, $2 \not> 2$

26. Begin with a true inequality, such as $5 > -1$. Use a number line to show that the Addition Property of Inequality holds for $>$ when you add either a positive or a negative number to each side of the inequality.

27. Use the Addition Property of Inequality to prove the Subtraction Property of Inequality. (HINT: In the Addition Property, let $c = -d$.)

Mixed Review

Solve.   *3.3, 3.4, 3.5, 4.6, 4.7*

1. $2x - 5 = 9$ 7   **2.** $5y - y = 11 + 3y$ 11

3. $5c - (4c - c) = 3 + 5c + 2$ $-1\frac{2}{3}$ **4.** $1.6x = 0.72 - 0.8x$ 0.3

5. What number is 6% of 50? 3   **6.** Forty-two is 150% of __?__ . 28

7. Fifteen is what percent of 75? 20%   **8.** Twelve is __?__ % of 3? 400%

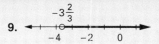

9. (number line from -4 to 0, open circle at $-3\frac{2}{3}$, $d > -3\frac{2}{3}$)

10. (number line from 0 to 12, open circle at 8)

11. (number line, open circle at $-2\frac{1}{2}$)

12. (number line, open circle at -3.2)

13. (number line -18 to 0, open circle at -14.3)

14. (number line 0 to 12, open circle at 9.3)

15. (number line -12 to 0, open circle at $-10\frac{3}{4}$)

17. $s + 6.25 < 50; s < \$43.75$ where $s =$ amount in savings before

20. (number line -6 to 6, open circle at -3)

21. (number line 0 to 52, open circle at 26)

22. (number line -14 to 28, open circle at 14)

23. (number line 0 to 4, open circle at 2.4)

See page 650 for the answers to Ex. 26, 27.

Enrichment

Present this situation to students.
Mary is older than Cheryl. Cheryl is younger than Jane. Then ask which of the following statements, if any, can be stated with certainty based on the given facts?

1. Mary is younger than Jane.   No
2. Jane is younger than Mary.   No

3. Mary's age minus Jane's age is less than Cheryl's age.   No
4. Mary's age plus Jane's age is greater than Cheryl's age.   Yes
5. Cheryl's age plus Jane's age is greater than Mary's age.   No

▰▰▰GETTING STARTED

Prerequisite Quiz

Solve.

1. $-8y = 72$ -9
2. $-\frac{3}{8}a = 45$ -120
3. $\frac{1}{7}r = -7$ -49
4. $7x + 4 = -2x - 14$ -2
5. $-4a + 6 = 2(3 - 2a)$ All real numbers
6. $3 - 2x = x - 9$ 4

Motivator

Write these inequalities on the chalkboard.

    $1 < 6$ $-1 > 2$ $4 > -6$

Have students multiply both sides of each inequality by 3, then by -4, and insert the correct inequality symbol. Have students summarize the results of this experiment. Multiplying both sides of an inequality by a negative number changes the direction of the inequality symbol.

5.2 The Multiplication and Division Properties of Inequality

Objective   To solve and graph the solution set of an inequality by using the Multiplication or Division Property of Inequality

The Multiplication and Division Properties of Equality allow you to multiply or divide each side of an equation by the same number. There are similar properties for inequalities.

True Inequality	Operate on each side.	New Inequality	True or False?
$3 < 4$	Multiply by 5. (5 is *positive*.)	$3 \cdot 5 < 4 \cdot 5$ $15 < 20$	True
$-4 > -20$	Divide by 2. (2 is *positive*.)	$\frac{-4}{2} > \frac{-20}{2}$ $-2 > -10$	True
$-5 < -3$	Multiply by -1. (-1 is *negative*.)	$-5(-1) < -3(-1)$ $5 < 3$	False. Reverse the order of the inequality. $5 > 3$ True
$18 > -6$	Divide by -3. (-3 is *negative*.)	$\frac{18}{-3} > \frac{-6}{-3}$ $-6 > 2$	False. Reverse the order of the inequality. $-6 < 2$ True

Notice that multiplying (or dividing) each side of a true inequality by a *negative* number produces a *false* inequality. The order of the inequality must be *reversed* to make the new inequality true. These results suggest the following property.

Multiplication Property of Inequality
For all real numbers a, b, and c,
        if $a < b$ and $c > 0$, then $ac < bc$, and
        if $a > b$ and $c > 0$, then $ac > bc$
That is, multiplying each side of an inequality by the same *positive* number produces an equivalent inequality.
For all real numbers a, b, and c,
        if $a < b$ and $c < 0$, then $ac > bc$, and
        if $a > b$ and $c < 0$, then $ac < bc$
That is, multiplying each side of an inequality by the same *negative* number and *reversing the order of the inequality* produces an equivalent inequality.

Highlighting the Standards

Standards 5a, 5c: In this lesson, students see alternate methods for solving inequalities and, in the Exercises, represent situations by means of inequalities.

EXAMPLE 1 Solve each open inequality. Graph the solution set.

  a. $7x < -56$ b. $-\frac{2}{3}x > 16$

Plan Multiply each side of the inequality by the *reciprocal* of the coefficient of x.

Solutions a. $7x < -56$

   $\frac{1}{7}(7x) < \frac{1}{7}(-56)$

   $x < -8$

$\xleftarrow{\quad\oplus\;|\;|\;|\;|\quad}$
$\;-12\;\;-8\;\;-4\;\;\;0\;\;\;\;4$

The solution set is the set of all real numbers less than -8.

  b. $-\frac{2}{3}x > 16$

   $\left(-\frac{3}{2}\right)\left(-\frac{2}{3}x\right) < \left(-\frac{3}{2}\right)16$ $\longleftarrow$ $-\frac{3}{2}$ is negative; change $>$ to $<$. 24

   $x < -\frac{48}{2}$, or $x < -24$

$\xleftarrow{\quad\circ\;|\;|\;|\;|\quad}$
$\;-30\;-20\;-10\;\;\;0\;\;\;10$

The solution set is the set of all real numbers less than -24.

TRY THIS   1. Solve $-4 < -2x$. Graph the solution set. $x < 2$

The second step of Example 1a could be

   $\frac{7x}{7} < \frac{-56}{7}$ rather than $\frac{1}{7}(7x) < \frac{1}{7}(-56)$.

This illustrates the *Division Property of Inequality.*

Division Property of Inequality
For all real numbers a, b, and c,

   if $a < b$ and $c > 0$, then $\frac{a}{c} < \frac{b}{c}$, and

   if $a > b$ and $c > 0$, then $\frac{a}{c} > \frac{b}{c}$.

That is, dividing each side of an inequality by the same *positive* number produces an equivalent inequality.

For all real numbers a, b, and c,

   if $a < b$ and $c < 0$, then $\frac{a}{c} > \frac{b}{c}$, and

   if $a > b$ and $c < 0$, then $\frac{a}{c} < \frac{b}{c}$.

That is, dividing each side of an inequality by the same *negative* number and *reversing the order of inequality* produces an equivalent inequality.

Lesson Note

The effect of multiplying or dividing both sides of an inequality by a negative factor will require special attention. Drawings such as this will help make it clear that whereas $2 > 1$, $2\cdot(-1)$ or -2 is less then $1\cdot(-1)$ or -1.

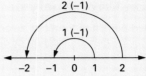

$\;-2\;\;-1\;\;\;0\;\;\;1\;\;\;2$

Remind students that $c > 0$ indicates that c is a positive number and that $c < 0$ indicates that c is a negative number.

Math Connections

History: Thomas Harriot (1560–1621) was the first to use the signs $>$ and $<$ for "is greater than" and "is less than." His work in the field of algebra did much toward setting the standards for algebra textbooks.

Additional Answers

Try This

1.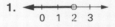
   $\;0\;\;1\;\;2\;\;3$

Additional Example 1

Solve each inequality. Graph the solution set.

a. $-4y > 28$ $\xleftarrow{\quad\circ\quad}$ $y < -7$
$\qquad\qquad\qquad\quad\;-7$

b. $-\frac{1}{4}c < -4$ $\xleftarrow{\;\circ\quad\rightarrow}$ $c > 16$
$\qquad\qquad\qquad\;16$

171

Synthesis: Point out to students that the Multiplication Property of Inequality says nothing about what would happen if $c = 0$. Ask students what effect this would have. The result would be $0 = 0$, which is not an equivalent inequality. Ask what would happen in the Division Property of Inequality if $c = 0$. Division by 0 is not possible.

Common Error Analysis

Error: Students often forget to reverse the inequality sign when multiplying or dividing by a negative number.

Remind them that this is the major difference between the properties they have learned for equations and inequalities.

Checkpoint

Solve each inequality. Graph the solution set.

1. $2x > -14$

-7

2. $-3y < 15$

-5

3. $-4a > -24$

6

4. $\frac{1}{-4}c > 5$

-20

5. $2x + 5 + x > -9 + 3x$
All real numbers

6. $3(-2y + 1) < -7 - 6y$ $\varnothing$

A calculator may be helpful in solving problems involving inequalities.

EXAMPLE 2 If Jill sells more than $100 worth of peanut brittle, she will win a radio. Each box of peanut brittle sells for $2.75. How many boxes must she sell to win the radio?

Solution   Let p = the number of boxes Jill must sell.

$$2.75p > 100$$
$$p > \frac{100}{2.75}$$
$$p > 36.36\overline{36} \quad \text{(THINK: She must sell a whole number of boxes.)}$$

Jill must sell more than 36 boxes.
Thus, she must sell 37 or more boxes.

EXAMPLE 3   Solve $-3x + 6 < -5$.

Plan   Use the Subtraction Property of Inequality first. Then use the Division Property of Inequality.

Solution   Subtract 6 from each side. $-3x + 6 < -5$
Divide each side by -3 $-3x < -11$
(or multiply by $-\frac{1}{3}$). $\frac{-3x}{-3} > \frac{-11}{-3}$ ⟵ Change < to >.
                                                  $x > \frac{11}{3}$, or $x > 3\frac{2}{3}$

Thus, the solution set is the set of real numbers greater than $3\frac{2}{3}$.

EXAMPLE 4   Solve $5 - 4x < 2x - 7$.

Solution

Method 1	Method 2
$5 - 4x < 2x - 7$	$5 - 4x < 2x - 7$
$5 < 6x - 7$	$5 - 6x < -7$
$12 < 6x$	$-6x < -12$
$2 < x$, or $x > 2$	$x > 2$

Thus, the solution set is the set of all real numbers greater than 2.

TRY THIS   **2.** Solve $6x - 5 < 8x + 3$. $x > -4$

Additional Example 2

Tim earns $7.50 for each magazine subscription he sells. If he earns more than $50, he will receive a bonus. How many subscriptions must he sell to earn the bonus?   7 or more

Additional Example 3

Solve $-2x + 1 > 8$. $x < -3\frac{1}{2}$

Additional Example 4

Solve $7 - 2x > x - 5$. $x < 4$

It is possible for the solution set of an inequality to be the empty set, $\varnothing$, or to be the set of all real numbers.

EXAMPLE 5   Solve $-\frac{3}{2}x + 4 > 7 - \frac{3}{2}x$. Graph the solution set.

Solution
$$-\frac{3}{2}x + 4 > 7 - \frac{3}{2}x$$
Add $\frac{3}{2}x$ to each side. $-\frac{3}{2}x + 4 + \frac{3}{2}x > 7 - \frac{3}{2}x + \frac{3}{2}x$
$$4 > 7 \quad \longleftarrow \text{False}$$

When solving an inequality results in a false inequality (involving no variable), then *no* value of x will make the original inequality true. Thus, the solution set is the empty set $\varnothing$.

(No points are graphed.)

EXAMPLE 6   Solve $-2(2x + 1) + 5x < x + 5$. Graph the solution set.

Solution
$$-2(2x + 1) + 5x < x + 5$$
$$-4x - 2 + 5x < x + 5 \quad \longleftarrow \text{Distributive Property}$$
$$x - 2 < x + 5$$
Add $-x$ to each side. $-x + x - 2 < -x + x + 5$
$$-2 < 5 \quad \longleftarrow \text{True}$$

When solving an inequality results in a true inequality (involving no variable), then *every* real-number value of x will make the original inequality true. Thus, the solution set is the set of all real numbers

graph of the set of all real numbers (All points are graphed.)

TRY THIS   **3.** Solve $-5\left(2 - \frac{4}{5}x\right) + 3x > -4\left(\frac{1}{4} - x\right)$. $x > 3$

Classroom Exercises

Complete each statement. Use $<$, $>$, or $=$.

1. If $8 > 5$, then $8 \cdot 3 \underline{\ ?\ } 5 \cdot 3$ $>$
2. $-7 < 0$. So, $-7(-4) \underline{\ ?\ } 0(-4)$ $>$
3. If $4x < 12$, then $x \underline{\ ?\ } 3$ $<$
4. If $-3y > 15$, then $y \underline{\ ?\ } -5$ $<$
5. If $-20 < -5x$, then $4 \underline{\ ?\ } x$, or $x \underline{\ ?\ } 4$ $>, <$
6. If $0 > -7x$, then $0 \underline{\ ?\ } x$, or $x \underline{\ ?\ } 0$ $<, >$

Solve each open inequality. Graph the solution set.

7. $4a > 12$ $a > 3$
8. $-2c < 18$ $c > -9$
9. $3x - 5 > 22$
$x > 9$

Closure

Have students summarize the four properties of inequality introduced in Lessons 5.1 and 5.2. Emphasize that the only time the inequality symbol is reversed is when both sides are either multiplied by, or divided by, a negative number.

■■■■**FOLLOW UP**

Guided Practice

Classroom Exercises 1–9
Try This   all

Independent Practice

A Ex. 1–30, **B** Ex. 31–40, **C** Ex. 41–43
Basic: WE 1–29 odd
Average: WE 9–37 odd
Above Average: WE 15–43 odd

Additional Answers

Classroom Exercises

7.
8.
9.

1. $y < 6$;
   -6 0 6

2. $x < -4$;
   -6 -4 -2 0

3. $z > \frac{2}{7}$;
   0 2 4

4. $y < -8$;
   -12 -8 -4 0

5. $a < -15$;
   -20 -10 0

6. $k < -18$;
   -36 -18 0

7. $x < -1$;
   -2 0 2

8. $y < 1$;
   -2 0 2

9. $z > 2$;
   0 2 4

10. $a < -1\frac{4}{7}$;
    -4 -2 0

11. $y > -4\frac{1}{2}$;
    -6 -4 -2 0

12. $k < 6$;
    -6 0 6

See page 650 for the answers to Ex. 13–26, 42, 43.

Written Exercises

Solve each inequality. Graph the solution set.

1. $3y < 18$ 2. $-4x > 16$ 3. $-7z < -2$ 4. $-16 > 2y$

5. $-\frac{1}{3}a > 5$ 6. $\frac{2}{3}k < -12$ 7. $2x + 3 < 1$ 8. $-3y - 4 > -7$

9. $-2 + 5z > 8$ 10. $3 < -7a - 8$ 11. $-6 < 2y + 3$ 12. $9 - 4k > -15$

13. $\frac{1}{3}x + 7 < -2$ 14. $-\frac{3}{5}y - 1 > -4$ 15. $5 < \frac{1}{2} + \frac{3}{2}x$ 16. $x + 8 > x - 3$

17. $4y - 4 > 7 + 4y$ ∅   18. $-6x - 6 < 3x - 27$ $x > 2\frac{1}{3}$

19. $7 - 5x > 9 - 4x$ $x < -2$ 20. $17 - 2z < 3z - 8$ $z > 5$

21. $10c - 16 < 12c - 18$ $c > 1$ 22. $2x - 7 > -7x + 20$ $x > 3$

23. $5x - 9 > 8 + 5x$ ∅   24. $35 - 4y > -9 + 7y$ $y < 4$

25. $9c - 7 < 4c + 18$ $c < 5$ 26. $-6z + 8 > -3 - 6z$
                                             All real numbers

Write an inequality for each problem. Then solve the problem.

27. Bert and Juan made more than $63. If they divided the money equally, how much did each boy receive?
$2x > 63$; $x > \$31.50$

28. Morey paid less than $5.10 for a package of three golf balls. What was the cost of each golf ball? $3x < 5.10$; $x < \$1.70$

29. Dana has saved $62 toward the cost of a ten-speed bike. This is less than $\frac{1}{3}$ of the cost of the bike. What is the cost of the bike? $\frac{1}{3}x > 62$; $x > \$186$

30. Helena and Grace must take in more than $150 to make a profit on some T-shirts they are selling. If each T-shirt costs $6.50, how many must be sold to make a profit? $6.50x > 150$; $x > 23$

Solve each inequality. $x > 16$

31. $2x + 4 < 3(x - 4)$ 32. $5(x - 2) > 7 + 5x$ ∅   33. $-(3 - 2c) > 10c + 5$ $c < -1$

34. $\frac{1}{2}y - 8 > \frac{2}{5}y - 7$ $y > 10$ 35. $\frac{1}{2}c + 1 < \frac{3}{4}c - 1$ $c > 8$ 36. $6 + \frac{1}{2}z > 5 + \frac{2}{3}z$ $z < 6$

37. $8x - 2(5x - 2) < 7 + 3x$ $x > -\frac{3}{5}$ 38. $3(4y + 1) + 7 > 3y - 2$ $y > -1\frac{1}{3}$

39. $2 + 5(y + 1) > y - 3(2y + 1)$ $y > -1$ 40. $x + 2 < 5x - 7 - (3 + x)$ $x > 4$

Solve for x. (HINT: Where appropriate, consider more than one case.)

41. $x + a < b$ 42. $ax < b, a \neq 0$ 43. $\frac{ax + bx}{c} < d, c \neq 0, a \neq -b$
     $x < b - a$

Mixed Review

Simplify.   *1.2, 1.3, 2.2, 2.8*

1. $18 - 5 \cdot 3$ 3   2. $6^2 - 5^2$ 11   3. $|-4.6|$ 4.6   4. $-(-1)^8$ -1   5. $14 - 3(-5)$
                                                                            29
6. The length of a rectangle is 10 m less than twice the width. The perimeter is 112 m. Find the length and the width of the rectangle.   *4.7* $l = 34$ m, $w = 22$ m

Enrichment

Write these two statements on the chalkboard.

If $a < b$, then $\frac{1}{a} > \frac{1}{b}$.

If $a > b$, then $\frac{1}{a} < \frac{1}{b}$.

Have students test the truth of each statement by substituting different values for a and b. Ask if there are any conditions under which the statements would not be true. The restrictions are that neither a nor b could equal zero, and both a and b must have the same sign.

5.3 Conjunctions and Disjunctions

Objective

To determine whether a conjunction or a disjunction is true or false

The words "and" and "or" occur frequently in everyday life. These two familiar words also occur in mathematical settings, for example, when forming combined, or *compound*, mathematical sentences.

A **conjunction** is composed of two sentences connected by the word *and*. Whether a conjunction is true (T) or false (F) depends upon the two sentences. Examine the chart below.

Conjunctions

First sentence	T or F?	Second sentence	T or F?	Conjunction	T or F?
$5 < 8$	T	$2 + 9 = 11$	T	$5 < 8$ and $2 + 9 = 11$	T
$6 - 4 = 2$	T	$-3 > 1$	F	$6 - 4 = 2$ and $-3 > 1$	F
$2 < 0$	F	$8 = 5 + 4$	F	$2 < 0$ and $8 = 5 + 4$	F

A conjunction is true only if *both* of the sentences are true.

EXAMPLE 1

Determine whether each conjunction is true or false.

a. $5 - 2 = 7$ and $6 - 4 > -1$
b. $-7 + 3 < -2$ and $6 - 3 > -1$

Plan

For each sentence of the conjunction, determine whether it is true or false.

Solutions

a. $5 - 2 = 7 \quad$ and $\quad 6 - 4 > -1$
$\qquad 3 \neq 7 \qquad\qquad\quad 2 > -1$
$\qquad\quad$ F $\qquad\qquad\qquad$ T

One sentence is false. Thus, the conjunction is false.

b. $-7 + 3 < -2 \quad$ and $\quad 6 - 3 > -1$
$\qquad -4 < -2 \qquad\qquad\quad 3 > -1$
$\qquad\quad$ T $\qquad\qquad\qquad$ T

Both sentences are true. Thus, the conjunction is true.

TRY THIS

Determine whether the conjunction is true or false.
1. $6 + 5 < 2 \quad$ and $\quad 2 - 3 > 1$ False
2. $5 - 8 < 0 \quad$ and $\quad 8 - 5 > 0$ True

■ GETTING STARTED

Prerequisite Quiz

Match each inequality with the graph of its solution set.

1. $x - 6 > -1$ a. ←——○—→ -2

2. $6 - x > -1$ b. ←——○—→ -1

3. $x - 3 < -5$ c. ←——○—→ -8

4. $3 - x < -5$ d. ←○——→ 5

5. $x - 3 > 2x + 5$ e. ←——○—→ 8

6. $2 - x > 2x + 5$ f. ←——○—→ 7
   1d, 2f, 3a, 4e, 5c, 6b

Motivator

Have students review the everyday meaning of the words "and" and "or." Have them give examples of compound English sentences using these connectors. Answers will vary.

Additional Example 1

Determine whether the given conjunction is true or false.

a. $-2 < -3 + 2$ and $8 + (-5) = -3$ F
b. $6 + (-6) > 0$ and $-9 > 3 - 13$ F

Highlighting the Standards

Standards 2f, 5c: Students learn to write compound sentences with mathematical notation and apply this to solving problems.

Lesson Note

Venn diagrams which are used in the algebra of sets can be used to illustrate conjunction and disjunction. The shaded regions indicate the set.

p <u>and</u> q   p <u>or</u> q

The diagram on the right shows that the disjunctive *or* includes *and*. This is to say that in mathematical logic, *or* is used inclusively, not exclusively. A disjunction is true if either or both of the components can be said to be true. You may wish to assign the study of truth tables to the above average students only.

Math Connections

Geometry: Tell students that an obtuse angle is one whose measure is greater than 90 *and* less than 180. Ask students if these angle measures satisfy the conjunction: 60°, 90°, 180°, 210°.   No

Critical Thinking Questions

Application: Ask students to describe in words all possible values for *x* in each of the following.

1) $x > -2$ *and* $x < 4$ All numbers between -2 and 4
2) $x < 6$ *and* $x > 6$ Empty set
3) $x > -2$ *and* $x = -2$ Empty set
4) $x > 3$ *or* $x < 0$ All numbers greater than 3 or less than 0
5) $x > 2$ *or* $x > 5$ All numbers greater than 2

A **disjunction** is composed of two sentences connected by the word *or*. Whether a disjunction is true or false depends upon the two sentences.

Disjunctions

First sentence	T or F?	Second sentence	T or F?	Disjunction	T or F?
$5 > -2$	T	$8 + 2 = 10$	T	$5 > -2$ or $8 + 2 = 10$	T
$5 \cdot 8 = 40$	T	$9 - 10 = 1$	F	$5 \cdot 8 = 40$ or $9 - 10 = 1$	T
$6 < -4$	F	$3 \cdot 5 = 10$	F	$6 < -4$ or $3 \cdot 5 < 10$	F

A disjunction is false only if *both* sentences are false.

EXAMPLE 2   Determine whether each disjunction is true or false.

  a. $6 > 3 + 5$ *or* $9 = 12 - 3$
  b. $5^3 < 100$ *or* $6 \cdot 8 > 50$

Plan   Determine whether *either* sentence of the disjunction is true.

Solutions   a. $6 > 3 + 5$ *or* $9 = 12 - 3$
                $6 \not> 8$ $9 = 9$ One sentence is true.
                   F   T   Thus, the disjunction is true.

  b. $5^3 < 100$ *or* $6 \cdot 8 > 50$
       $125 \not< 100$ $48 \not> 50$ Both sentences are false.
          F   F   Thus, the disjunction is false.

TRY THIS   3. Determine whether the disjunction is true or false.

  $9 + 6 > 5$ *or* $4^2 < 24$ True

The truth of a conjunction and a disjunction can be summarized in **truth tables**, as shown below. The letters *p* and *q* represent sentences.

Conjunction

p	q	p and q
T	T	T
T	F	F
F	T	F
F	F	F

A conjunction is true only if both *p* and *q* are true.

Disjunction

p	q	p or q
T	T	T
T	F	T
F	T	T
F	F	F

A disjunction is false only if both *p* and *q* are false.

Additional Example 2

Determine whether the given disjunction is true or false.

a. $8 > 9 + (-2)$ *or* $6 - 4 = 2$ T
b. $-8 - (-5) < -14$ *or* $8 - 9 > 0$ F

EXAMPLE 3   In a laboratory stress test, a tube of material is found to fail if it is stretched to a length of more than 21.3 cm or compressed to a length of less than 19.7 cm. Use a conjunction or disjunction to describe the complete failure zone for the material.

Plan   Write two inequalities, one for each failure zone. Then combine into one open sentence using "and" or "or."

Solution   Let x = the length of the tube in a failure zone.

Then $x < 19.7$ describes the failure zone under compression, and $x > 21.3$ describes the failure zone under stretching.

The material fails if its length is less than 19.7 cm *or* greater than 21.3 cm.

This is described by the *disjunction* $x < 19.7 \text{ or } x > 21.3$.

TRY THIS   4. Write a conjunction or a disjunction to describe the problem. Mark weighs more than 160 lb but less than 168 lb.

$x > 160 \quad and \quad x < 168$

Classroom Exercises

Determine whether the given sentence is *true* or *false*.

1. $7 \neq 5 + 2$ F
2. $8 = -1 + 9$ T
3. $7 \neq 5 + 2 \text{ and } 8 = -1 + 9$ F
4. $7 \neq 5 + 2 \text{ or } 8 = -1 + 9$ T
5. $5 < -6^2$ F
6. $0 > 7 - 2^3$ T
7. $5 < -6^2 \text{ and } 0 > 7 - 2^3$ F
8. $5 < -6^2 \text{ or } 0 > 7 - 2^3$ T

Written Exercises

Determine whether the given conjunction or disjunction is *true* or *false*.

1. $-2 < 5 \text{ and } 6 = 2 + 4$ T
2. $5 > 1 \text{ or } 3 = -2$ T
3. $7 \neq 2 \text{ and } 5 = -1 + 4$ F
4. $8 > -2 \text{ and } -5 < -2$ T
5. $7 < 3 + 3 \text{ or } 6 \neq 4 + 2$ F
6. $8 < 10 - 1 \text{ or } 4^2 = 16$ T
7. $10 - 5 \cdot 2 < 0 \text{ or } 5^3 > 100$ T
8. $-3 > 2 \text{ and } (-6)^2 = 36$ F
9. $5 \cdot 6 > 30 \text{ or } 7 \cdot 2 < 15$ T
10. $8 \neq 5 + 2 \text{ and } 9 > -9$ T

For each problem, use x for the variable and write a conjunction or a disjunction to describe the problem.

11. Mary would like to weigh more than 110 lb but less than 115 lb.

$x > 110 \text{ and } x < 115$

12. The new addition to Fairview Stadium must have 6 or more gates. $x = 6 \text{ or } x > 6$

13. Jack hopes to save at least $150.

$x = 150 \text{ or } x > 150$

14. When Mark makes a cake, he bakes it no more than 50 min.

$x = 50 \text{ or } x < 50$

5.3 Conjunctions and Disjunctions   **177**

177

Guided Practice

Classroom Exercises 1–8
Try This   all

Independent Practice

A Ex. 1–10, **B** Ex. 11–26, **C** Ex. 27–32

Basic: WE 1–14 all
Average: WE 1–25 odd
Above Average: WE 7–31 odd

Additional Answers

Written Exercises

28.

p	$\sim p$	$\sim(\sim p)$
T	F	T
F	T	F

29.

p	q	$\sim p$	$\sim q$	$\sim p$ or $\sim q$
T	T	F	F	F
T	F	F	T	T
F	T	T	F	T
F	F	T	T	T

30.

p	q	$\sim q$	p and $\sim q$
T	T	F	F
T	F	T	T
F	T	F	F
F	F	T	F

31.

p	q	$\sim p$	$\sim p$ or q
T	T	F	T
T	F	F	F
F	T	T	T
F	F	T	T

32.

p	q	$\sim q$	p and $\sim q$	$\sim(p$ and $\sim q)$
T	T	F	F	T
T	F	T	T	F
F	T	F	F	T
F	F	T	F	T

Give a value of x that will make each conjunction or disjunction true. If there is no such value, state this. (For each exercise, the same value of x must be used in both sentences.) Possible answers are given.

15. $x = 3$ and $x > 2$ 3
16. $x < 2$ or $x < -7$ 1
17. $x > -3$ or $x = -3$ -2
18. $x < 5$ and $x = 5$ No sol.
19. $x > -1$ or $x < 2$ 0
20. $x > -2$ and $x < 3$ -1

21–26. Give a value of x that will make each conjunction or disjunction of Exercises $15-20$ *false*. If there is no such value, state this.

21. 4 **22.** 3 **23.** -4 **24.** 4 **25.** None **26.** 3

27. The *negation* of sentence p is *not p*, written $\sim p$. If p is true, then $\sim p$ is false. If p is false, then $\sim p$ is true. See the truth table at the right.
Complete the truth table below.

Negation

p	$\sim p$
T	F
F	T

p	q	$\sim p$	$\sim q$	$\sim p$ and $\sim q$
T	T	F	F	F
T	F	F	T	F
F	T	T	F	F
F	F	T	T	T

Write a truth table for each conjunction, disjunction, or negation.

28. $\sim(\sim p)$ **29.** $\sim p$ or $\sim q$ **30.** p and $\sim q$ **31.** $\sim p$ or q **32.** $\sim(p$ and $\sim q)$

Mixed Review

Simplify.   *1.2, 2.8*

1. $3y - 7 - 4y - 9$ $-y - 16$
2. $11x - (7 - x) - 13x$ $-x - 7$

Solve.   *3.3, 3.5*

3. $3(x + 1) = 5 - (4 - x)$ -1
4. $13 = -\frac{1}{2}x + 2$ -22

Evaluate.   *2.8, 2.9*

5. $5 - 3z$, for $z = -7$ 26
6. $14c + 1.7$, for $c = 0.2$ 4.5

7. Mr. Rosenfeld sells dishwashers for $360. The profit is 20% of the cost. How many dishwashers must he sell to make a total profit of $900?   *4.8* 15

178   Chapter 5 Inequalities and Absolute Value

Enrichment

Have students construct a truth table with these 8 columns: p, q, $\sim p$, $\sim q$, $\sim(p$ and $q)$, $\sim p$ or $\sim q$, $\sim(p$ or $q)$, $\sim p$ and $\sim q$. Then ask them to identify the columns that have the same truth values.   Columns 7 and 8

5.4 Combining Inequalities

Objective   To graph the solution sets of conjunctions and disjunctions of inequalities

The disjunction $x < 2$ *or* $x = 2$ can be written as $x \leq 2$ (*x is less than or equal to 2*).

Recall that a disjunction is true if either or both of its sentences are true. Thus, the solution set of a disjunction contains all the solutions of the two sentences.

Sentence   Graph

$x < 2$

$x = 2$

$x < 2$ *or* $x = 2$ $(x \leq 2)$

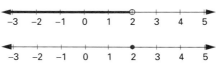

In the first graph above, an open circle at 2 indicates that 2 is not a member of the solution set of $x < 2$. In the third graph, the heavy dot at 2 indicates that 2 *is* a member of the solution set of $x \leq 2$.

When finding the solution set of a disjunction containing $\leq$ or $\geq$, you can use the inequality properties of Lessons 5.1 and 5.2.

EXAMPLE 1   Graph the solution set of $11 - 3x \leq 26$.

Solution
$$11 - 3x \leq 26$$

Subtract 11 from each side. $-3x \leq 15$

Divide each side by -3 and $x \geq -5$
reverse the order of the inequality.

Thus, the solution set is the set of all real numbers greater than or equal to -5.

   ⟵ Note the heavy dot at -5.

TRY THIS   Graph the solution set of $2y + 1 \leq 7$.

▮▮▮ GETTING STARTED

Prerequisite Quiz

Classify the following compound statements as true or false.

1. $2 + 2 = 4$ *and* $1 > -1$ T
2. $2 < 1$ *or* $3 > 2$ T
3. $4 = 3 + 1$ *and* $-1 < -2$ F
4. $10 > 3 - 2$ *or* $6 > 1 + 2$ T
5. $5 + 1 < 6$ *or* $2 + 1 > 3$ F
6. $2^2 = 4$ *and* $3^3 = 27$ T

▮▮▮ TEACHING SUGGESTIONS

Lesson Note

After explaining that $\geq$ and $\leq$ are disjunctions ("or" statements), you may wish to refer to these simply as "closed dot" (as opposed to "open dot") problems so that the disjunctions in problems like Examples 2, 3, and 4 do not confuse the students by using two or more "or's." When graphing combined inequalities, many students find it helpful to use three lines, one for each component, and then to mentally slide the lines together to graph the disjunction or intersection on the third line as shown in the examples.

Additional Answers

Try This

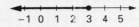

Additional Example 1

Graph the solution set of $-9 - 2y \geq 17$.

-13

Highlighting the Standards

Standards 14b, 4a: This lesson takes students further into the logic of algebraic procedures and into representing algebraic statements as number line graphs.

Math Connections

Language Arts: Stating that someone must be between the ages of 15 and 18 allows the possibility that the person could be 15 or 18. In mathematics, $15 < x < 18$ does not include either 15 or 18 as a value for x.

Critical Thinking Questions

Synthesis: In geometry, if point P is between points A and C, then A, P, and C must all lie on the same line. Have the students write an equation for this situation using AP, PC, and AC as the distances between pairs of points. $AP + PC = AC$ Ask students if the equation will be true if P does not lie between A and B? No

Common Error Analysis

Error: After drawing two separate graphs for the component inequalities, many students have difficulty in deciding how to combine them.

Have them think of the graph of an *and* statement as the overlap of two component graphs. Have them think of an *or* statement as the union, or uniting, of two component graphs.

Recall that a conjunction is true only if both sentences are true. Thus, the solution set contains only the solutions common to both sentences.

EXAMPLE 2 Graph the solution set of $x \le -2$ or $x > 3$. Then graph the solution set of $x \le -2$ and $x > 3$.

Plan First, graph the solution set of $x \le -2$ and the solution set of $x > 3$. Then use the definition of *disjunction* and of *conjunction* to draw the other two graphs.

Solutions $x \le -2$

$x > 3$

$x \le -2$ or $x > 3$

(graph of all numbers that are all less than –2, equal to –2, or greater than 3)

$x \le -2$ and $x > 3$

(no points)

In Example 2, there are no solutions common to both $x \le -2$ and $x > 3$. Therefore, the solution set of "$x \le -2$ *and* $x > 3$" is the empty set.

EXAMPLE 3 Graph the solution set of $x > -1$ or $x < 4$. Then graph the solution set of $x > -1$ and $x < 4$.

Solutions $x > -1$

$x < 4$

$x > -1$ or $x < 4$

(entire number line since *every* number is less than –1 or greater than 4)

$x > -1$ and $x < 4$

(only the numbers that are both greater than –1 and less than 4)

180 Chapter 5 Inequalities and Absolute Value

Additional Example 2

Graph the solution set of $x \le 0$ or $x > 1$.

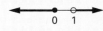

Then graph the solution set of $x < 0$ and $x > 1$. Empty set

Additional Example 3

Find the solution set of $x > 0$ or $x < 1$.
All real numbers
Then graph the solution set of $x \ge 0$ and $x < 1$.

EXAMPLE 4   Graph the solution set of $x - 1 \geq -3$ *or* $4 < 2x - 2$. Then graph the solution set of $x - 1 \geq -3$ *and* $4 < 2x - 2$.

Plan   Solve $x - 1 \geq -3$. Then solve $4 < 2x - 2$. Graph their solution sets on separate number lines. Finally, graph the solution set of the disjunction and the solution set of the conjunction.

Solutions
$$x - 1 \geq -3 \qquad\qquad 4 < 2x - 2$$
$$x \geq -2 \qquad\qquad\quad 6 < 2x$$
$$\qquad\qquad\qquad\qquad 3 < x, \text{ or } x > 3$$

$x \geq -2$

$x > 3$

$x \geq -2$ or $x > 3$

$x \geq -2$ and $x > 3$

The conjunction $x > -1$ *and* $x < 4$ can be written as $-1 < x < 4$. This can be read in three ways.

-1 is less than x and x is less than 4.

x is greater than -1 and x is less than 4.

x is between -1 and 4.

EXAMPLE 5   Graph the solution set of $-3 < 2x + 5 \leq 9$.

Plan   Write the conjunction as $-3 < 2x + 5$ *and* $2x + 5 \leq 9$. Solve each sentence and graph the solution set of the resulting conjunction.

Solution
$$-3 < 2x + 5 \quad and \quad 2x + 5 \leq 9$$
$$-8 < 2x \qquad\qquad\qquad 2x \leq 4$$
$$-4 < x \qquad\qquad\qquad\quad x \leq 2$$
$$\qquad -4 < x \leq 2$$

$\longleftarrow$ x is between -4 and 2, including 2.

Additional Example 4

Graph the solution set of $3 > y + 8$ *or* $3y - 7 \leq 5$.

4

Then graph the solution set of $3 > y + 8$ *and* $3y - 7 \leq 5$.

-5

Additional Example 5

Graph the solution set of $1 \leq -3x + 2 < 8$.

$-2 \qquad \frac{1}{3}$

181

◼️◼️ FOLLOW UP

Guided Practice

Classroom Exercises 1–20
Try This   all

Independent Practice

🅰 Ex. 1–26, 🅱 Ex. 27–37, 🅲 Ex. 38–51

Basic: WE 1–25 odd, 35–37 all,
Midchapter Review, Brainteaser, Problem
Solving Strategies

Average: WE 7–37 odd, Midchapter
Review, Brainteaser, Problem Solving
Strategies

Above Average: WE 19–51 odd,
Midchapter Review, Brainteaser, Problem
Solving Strategies

Additional Answers

Classroom Exercises

1. $x < 5 \, or \, x = 5$
2. $x < 7 \, and \, x > -2$
3. $y > -6 \, or \, y = -6$
4. $y < 12 \, and \, y > 0$
5. $x > -9 \, or \, x = -9$
6. $x < 9 \, and \, x > -8$
7. $x > 18 \, or \, x = 18$
8. $x < 9 \, and \, x > -9$

15.
16.
17.
18.

EXAMPLE 6 This year, Jim's weight fluctuated within a 7-lb range. At times he weighed as little as 125 lb. At other times he weighed almost 132 lb. Write a conjunction that describes the range of values for Jim's weight.

Solution Let x = Jim's weight.

If Jim weighed "as little as" 125 lb, then sometimes he weighed 125 lb and sometimes he weighed more than 125 lb. This is expressed by the compound inequality

$$125 = x \text{ or } 125 < x; \text{ that is, } 125 \leq x$$

If Jim weighed "almost 132 lb" then sometimes he weighed less than 132 lb, but never actually 132 lb. This is expressed by the inequality $x < 132$.

Jim's weight in the 7 lb range is expressed by this conjunction.

$$125 \leq x \text{ and } x < 132, \text{ or } 125 \leq x < 132.$$

◼️◼️ Focus on Reading

Match each conjunction or disjunction
with its graph.

1. $x \geq -5 \, and \, x < 0$ d
2. $x \leq -5 \, or \, x > 0$ e
3. $x \geq -5 \, or \, x > 0$ a
4. $x \leq -5 \, and \, x > 0$ c
5. $x \geq -5 \, or \, x < 0$ b

a.
b.
c.
d.
e.

Classroom Exercises

Rewrite using *and* or *or*.

1. $x \leq 5$
2. $-2 < x < 7$
3. $y \geq -6$
4. $0 < y < 12$
5. $-2 \leq 7 + x$
6. $-8 < x < 9$
7. $x - 3 \geq 15$
8. $-7 < x + 2 < 11$

182 Chapter 5 Inequalities and Absolute Value

Additional Example 6

Rosine wants to spend $5 or more for a pair of earrings, but she has only $12. Write a conjunction that describes her spending range. $5 \leq x \leq 12$

Rewrite without *and* or *or*.

9. $x > -5$ and $x < 7$ $-5 < x < 7$

10. $x = -8$ or $x > -8$ $x \geq -8$

11. $x < 0$ or $x = 0$ $x \leq 0$

12. $x \geq -6$ and $x < 9$ $-6 \leq x < 9$

13. $x > 0$ and $x < 10$ $0 < x < 10$

14. $x > 5$ or $x = 5$ $x \geq 5$

Graph the solution set.

15. $2x - 3 = 5$

16. $-7 \leq 5 + 4y$

17. $y \geq 5$ or $y < -1$

18. $x > -2$ and $x \leq -1$

19. $-3 < x + 1 \leq 0$

20. $-2x > 6$ or $x - 7 < -2$

Written Exercises

Graph the solution set.

1. $3x - 8 \geq 7$

2. $2x + 15 \leq 21$

3. $-17 \geq 9c - 8$

4. $-15 \leq -3a + 12$

5. $-3d + 4 \leq -11$

6. $8r - 8 \leq -24$

7. $9x - 7 \geq 18 + 4x$

8. $2y - 5 \geq 2y + 7$

9. $3z + 2 \leq 4 + 3z$

10. $7a - 9 \geq 35 - 4a$

11. $x < -2$ or $x > 1$

12. $y > -3$ and $y < -1$

13. $z > -5$ or $z < 7$

14. $-1 < x < 3$

15. $c \leq -3$ and $c > 0$

16. $x \geq -2$ or $x > 5$

17. $0 \leq x < 6$

18. $y < 3$ or $y \geq -1$

19. $c > -2$ or $c > 2$

20. $a \geq 5$ and $a < 5$

21. $x \leq 3$ or $x > 4$

22. $-5 \leq x \leq 0$

23. $t > 7$ or $t < 7$

24. $7 < t < 7$

25. $c \leq 2$ or $c < -8$

26. $x \leq 4$ and $x < -2$

27. $-2y + 5 < 13$ or $-2 < 2y + 4$

28. $5 + x > 6$ and $3x + 5 > 20$

29. $3z + 4 \leq 19$ and $z + 5 > 8$

30. $-3x > -27$ or $2x - 8 > 6$

31. $-11 \leq 3y - 2 < 18$

32. $3 \leq 2x - 5 < 1$

33. $5 < -z - 8 \leq 7$

34. $4 > -2 - x > -5$

Solve.

35. The high temperature today was 56° Fahrenheit, while the low was 42°. If t represents the temperature, then $\underline{\ ?\ } \leq t \leq \underline{\ ?\ }$. 42, 56

36. In one year the least amount Jane had in her savings account was $102.55. The greatest amount was $221.42. Let s represent the amount in her savings account, and write a conjunction. $102.55 \leq s \leq 221.42$

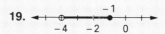

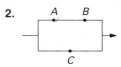

5.4 Combining Inequalities   **183**

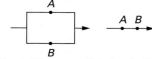

Enrichment

Conjunctions and disjunctions can be visualized in terms of electrical circuits.

On the left is a *parallel* circuit, through which current will flow if either switch A or switch B is closed. On the right is a *series* circuit, through which current will flow only if both switches are closed. Ask the students which illustrates conjunction and which illustrates disjunction.

parallel — disjunction
series — conjunction

Now have the students draw circuits which illustrate these compound statements.

1. *A and (B or C)*
2. *(A and B) or C*

11.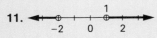
$-2 \quad 0 \quad 2$ (with 1 marked above, open circles)

12.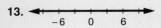
$-4 \quad -2 \quad 0$ (with −3 and −1 marked above)

13.
$-6 \quad 0 \quad 6$

14.
$-2 \quad 0 \quad 2 \quad 4$ (with −1 and 3 marked above)

15.
$-4 \quad 0 \quad 4$

16.
$-2 \quad 0 \quad 2$

17.
$0 \quad 4 \quad 8$ (with 6 marked above)

18.
$-2 \quad 0 \quad 2$

19.
$-2 \quad 0 \quad 2$

20.
$-2 \quad 0 \quad 2$

21.
$0 \quad 2 \quad 4$ (with 3 marked above)

22.
$-8 \quad -4 \quad 0 \quad 4$ (with −5 marked above)

See page 650 for the answers to Ex. 23–26, page 651 for the answers to Ex. 27–51, and for the answers to the Midchapter Review, Ex. 1–6, 11, 12, 13.

37. Riley took his car to United Car Dealers for repairs. The mechanic estimated that repairs would cost $200. The final bill came within $30 of the estimate. Let d = the cost of the repairs and write a conjunction to describe the range of values for d. Explain your reasoning. $170 < d < 230;\ 200 \pm 30$

Graph the solution set.

38. $7 \le 2x - 1 \le 17\ and\ 8 > 2x - 4 > 6$ **39.** $4y > 20\ and\ 2y - 4 > 6$

40. $2z + 6 < 2z + 4 < 2z - 1$ **41.** $3x - 2 < 3x + 5 < 3x + 7$

42. $2y + 5 < 10\ or\ 4y - 1 \ge y + 11$ **43.** $a - 5 < 2a + 4\ or\ -3a > 27$

44. $9 < 2c + 1 < 15\ or\ 2c - (4 - 6) \ge 11$ **45.** $2 < x \le 4\ and\ 6x - (4 - x) < 17$

46. Graph the solution set of $-a < x\ and\ x < a$, where a is positive.

47. Graph the solution set of $-a < x\ and\ x < a$, where a is negative.

48. Graph the solution set of $x > a\ or\ x < b$, where $a > b$.

49. Graph the solution set of $x > a\ and\ x < b$, where $a > b$.

50. Graph the solution set of $x > a\ or\ x < b$, where $a < b$.

51. Graph the solution set of $x > a\ and\ x < b$, where $a < b$.

Midchapter Review

Solve each open inequality. Graph the solution set.   **5.1, 5.2**

1. $4x < 24$ $x < 6$ **2.** $-2y > 14$ $y < -7$ **3.** $-3 > -x - 5$ $x > -2$

4. $0.2x - 0.9 > 1.5$ **5.** $2y - 9 \le 4 + 3y$ **6.** $3c - 9 < 2(-3 + c) - 6$
    $x > 12$ $y \ge -13$ $c < -3$

For each conjunction or disjunction, determine whether it is *true* or *false*.   **5.3**

7. $3 - 2 < 1\ or\ 6^2 = 36$ T   **8.** $7 + (-4) = 3\ and\ 6 < -8$ F

9. $14 < 9 + 4\ or\ -9 = 9 + 0$ F   **10.** $3 > -4 + 1\ and\ -7 < 0$ T

Graph the solution set.   **5.4**

11. $y - 9 > -2$ **12.** $5 < x\ or\ x < -3$ **13.** $-4 < x + 7 < 2$

▰▰/*Brainteaser*

A bug was trying to crawl to the top of an 18-ft drain pipe. Each day it climbed 4 ft, but each night it slipped back 3 ft. How many days did it take it to crawl out of the pipe?   15 days

Looking Back

Looking Back is the fourth step in problem solving. As you look back after solving a problem, check these points.

a. Did you use all the necessary information?
b. Are your calculations correct?
c. Does the answer fit your estimate?
d. Does your answer make sense?
e. Does your answer respond to the question asked?
f. Is your answer written in correct form and with the correct units?

Exercises

In these exercises, the person who worked the problem forgot to *Look Back* after arriving at a solution. Find what is wrong with each solution and explain what was forgotten. Then write the correct answer to the problem.

1. A circle has a radius of 2 feet. Find the area.

   Solution: Since $A(\text{circle}) = \pi r^2$ and $r = 2$, $A = \pi(2)^2$.
   So the area of the circle is 4π. Needs correct units; $4\pi\text{ft}^2$

2. The sum of Jaime's and Cheryl's ages is 26. Cheryl is two years older than Jaime. How old is Cheryl?

   Solution: Let $x =$ Jaime's age. Then $x + 2 =$ Cheryl's age.
   So $x + (x + 2) = 26$ Did not answer
   $\qquad\qquad 2x + 2 = 26$ the question;
   $\qquad\qquad\quad 2x = 24$ Cheryl is 14.
   $\qquad\qquad\quad\ x = 12$ Thus, Jaime is 12 years old.

3. Chad uses a calculator for this problem.
   $256.2 + 312.6 + 613 = \underline{\ ?\ }$

   Solution: 630.1 Answer doesn't make sense; 1181.8

4. Jennifer buys 5 yards of ribbon at 60¢ per yard and 2 yards of ribbon at 70¢ per yard. How much change will she get from a $5-bill?

   Solution: $5(0.60) + 2(0.70) = 3.00 + 1.50 = \$4.40$
   Did not answer the question, and the calculation is incorrect; $0.60

■■■ GETTING STARTED

Prerequisite Quiz

1. José is paid $130 a week plus 7% commission on all sales. What must be his weekly sales if he wishes to earn a total of $200 per week? $1,000
2. Mr. Rosenfeld sells dishwashers for $360 each. The price includes a profit of 20% of the cost. How many dishwashers must he sell to make a total profit of $900?   15

Motivator

In this lesson students are introduced to additional English words that can be translated into inequality symbols. Ask them to give the meanings of *at least, not less than, at most, not greater than,* and *between.* Have them suggest sentences that illustrate the meaning of each.   Answers will vary.

5.5 Problem Solving: Using Inequalities

Objective   To solve problems by using inequalities

The chart below lists sentences that can be translated into inequalities.

English sentence	Inequality
x is greater than a.	$x > a$
x is less than a.	$x < a$
x is greater than or equal to a. x is at least a. x is not less than a. ($x \not< a$)	$x \geq a$
x is less than or equal to a. x is at most a. x is not greater than a. ($x \not> a$)	$x \leq a$
x is between a and b.	$a < x < b$ (if $a < b$)

> **Trichotomy Property**
> For all real numbers a and b, exactly one of the following three statements is true.
> $$a < b \qquad a = b \qquad a > b$$

Suppose that the number of records in Jane's collection *is at least* 30. Then the number of records is not less than 30. By the Trichotomy Property, there are two remaining possibilities. The number is equal to 30 or is greater than 30. Thus, *is at least* is translated as *is greater than or equal to,* that is, as "≥." Similarly, if the number of records *is at most* 30, then Jane has 30 records or less than 30 records. Thus, *is at most* is translated as *is less than or equal to,* that is, as "≤."

EXAMPLE 1   Translate each sentence into an inequality.

    a. Roberto weighs no less than 125 lb.   **b.** Jesse has between 20 and 30 books.

Solutions   **a.** Let r = Roberto's weight.   **b.** Let b = the number of books.
           Then $r \geq 125$ Then $20 < b < 30$.

TRY THIS   1. Translate this sentence into an inequality.
           John has at most 28 records. $x \leq 28$

Highlighting the Standards

Standards 5c, 1b, 7d: In the Exercises, students apply the solving of inequalities to problem situations. In the Application, they explore the triangle inequality from geometry.

Additional Example 1

Translate each sentence into an inequality.

a. Jeremy has at least 200 stamps in his collection. $j \geq 200$
b. Barbara has at most $400 in her savings account. $b \leq 400$

EXAMPLE 2   Mr. Johnson rented a car for $39 a day plus $0.20 a mile. How far can he drive in one day if he wants to spend at most $100.

What are you to find?	The distance Mr. Johnson can drive in a day
What is given?	The rental charge is $39 a day plus $0.20 a mile. At most, $100/day can be spent.
Choose a variable. What does it represent?	Let m = the number of miles Mr. Johnson can drive in a day.
Write an inequality.	The one-day cost *is at most* $100.
Solve the inequality.	$39 + 0.20m \leq 100$ $0.20m \leq 61$ $m \leq 305$ ⟵ 305 mi
Check.	The check is left for you.
State your answer.	Mr. Johnson can drive at most 305 mi.

EXAMPLE 3   The sum of two consecutive positive odd integers is at most 24. Find the integers.

Solution   Let x and $x + 2$ represent the two odd integers.
Their sum *is at most* 24, they are positive, and they are odd integers.

$$x + (x + 2) \leq 24 \quad and \quad\quad x > 0 \quad and \quad x \text{ is an}$$
$$2x \leq 22 \quad\quad\quad\quad\quad\quad\quad\quad\quad\quad\quad\quad \text{odd integer.}$$

First integer: $x \leq 11$ *and* $x > 0$ *and* x is odd.
Second integer: $x + 2 \leq 13$ *and* $x + 2 > 2$ *and* $x + 2$ is odd.

All possible pairs of integers are shown in the chart below.

First integer (x)	1	3	5	7	9	11
Second integer $(x + 2)$	3	5	7	9	11	13

TRY THIS   2. The result of 5 more than 4 times a number is between 17 and 41. What real numbers are possible solutions? $3 < x < 9$

Classroom Exercises

Translate each sentence into an inequality.

1. Mary has at least 8 goldfish. $x \geq 8$

2. Bob weighs between 110 and 120 lb. $110 < x < 120$

3. A number is greater than 50. $x > 50$

4. Bill has at most 6 cats. $x \leq 6$

5. Twice a number is not greater than 15. $2x \leq 15$

6. Josh has at most $5 to spend. $x \leq 5$

Lesson Note

Translating word problems into inequalities is more challenging than translating word problems into equations because of the variety of relationships and symbols used to represent them. Have students give numerical examples to illustrate each inequality in the chart at the top of page 186. Proceed with word problems only when it is certain that the students have mastered translations such as those found in Example 1.

Math Connections

Life Skills: Income tax tables have two columns with headings of "at least" and "but less than." For earnings of $23,435, the amount of tax is on the line that reads "at least 23,400, but less than 23,450."

Critical Thinking Questions

Analysis: Have students write the corresponding algebraic statement for each phrase. Use x as the variable.

1. between $10 and $20 $10 < x < $20
2. at least $10, but less than $20 $10 \leq x < $20
3. more than $10, but at most $20 $10 < x \leq $20

Additional Example 2

Jim is paid $35 a week plus a commission of $2 for each sweatshirt he sells. How many sweatshirts must he sell each week in order to earn not less than $100 a week?   At least 33 sweatshirts

Additional Example 3

Nicole has 7 more bracelets than Molly has. Together they have fewer than 12 bracelets. How many bracelets can each girl have? Give all possible answers.

Molly	Nicole
2	9
1	8
0	7

Error: Students often confuse the symbols for "at least" and "at most."

Suggest that students think of short sentences, such as "Jim has at least $10," which means that Jim has $10 or more, or $j \geq 10$. A similar short sentence can be used for "at most."

Checkpoint

Solve each problem.

1. Rebecca wants to make a total profit of at least $36 by selling painted hair ribbons. How many ribbons must she sell if she makes a profit of $1.50 on each ribbon?   At least 24 ribbons

2. Mr. Rivera rented a car for $29 a day plus $0.15 a mile. How far can he drive in a day if he can spend not more than $50?   Not more than 140 miles

3. The sum of two consecutive positive odd integers is at most 12. Find the numbers. Give all possible solutions. 5, 7; 3, 5; 1, 3

Closure

Ask students to explain why "is at least" is translated as "$\geq$" and "is at most" is translated as "$\leq$".   See page 186.

Use an inequality to solve each problem.

7. The Sports Club raises money by selling boxes of nuts at a profit of $2 on each box. How many boxes must they sell to make a profit of at least $350? $x \geq 175$

8. Margaret rented a car for $25 a day plus 15¢ a mile. How far can she travel in one day if she can spend at most $55? $m \leq 200$

Written Exercises

Translate each sentence into an inequality.

1. Five more than 3 times a number is at least 95. $3x + 5 \geq 95$

2. Two less than 5 times a number is not less than 60. $5x - 2 \geq 60$

3. The sum of a number and 4 times the number is at most 40. $x + 4x \leq 40$

4. Half of Juanita's age is not equal to 16 years. $\frac{1}{2}x \neq 16$

Solve each problem by using an inequality.

5. A store makes a profit of $6 on each book sold. How many books must be sold to make a profit of at least $450? $x \geq 75$

6. A freight elevator can carry 2,000 lb safely. Shipping crates weigh 70 lb each. At most, how many crates can be safely carried on the elevator? $x \leq 28$

7. Michael is paid $175 a week plus a commission of $25 on each TV he sells. How many sets must he sell to earn not less than $400 a week? $x \geq 9$

8. Melinda rented a car for $60 a week plus 12¢ a kilometer. How far can she drive in a week if she can spend not more than $120? $x \leq 500$

Give all possible solutions.

9. The sum of two consecutive positive integers is less than 15. Find the numbers. 1, 2; 2, 3; 3, 4; 4, 5; 5, 6; 6, 7

10. Barry has 6 more tropical fish than Marcel. Together they have fewer than 28. How many fish can each have?

11. The sum of two consecutive positive even integers is at most 14. Find the numbers. 6, 8; 4, 6; 2, 4

12. The sum of two consecutive positive odd integers is not greater than 16. Find the numbers. 7, 9; 5, 7; 3, 5; 1, 3

13. Lightweight boxers must weigh more than 58 kg but no more than 60 kg. Rocky weighs 53 kg. How many kilograms could he possibly gain to be in the lightweight class? Explain. $5 < x \leq 7$; Solve $58 < 53 + x \leq 60$.

14. If a number is doubled, the result is between 2 and 17. What real numbers are possible solutions? $1 < x < 8.5$

15. If a number is tripled, the result is between -6 and 15. What real numbers are possible solutions? $-2 < x < 5$

16. If a number is multiplied by -4, the result is between -1 and 8. What real numbers are possible solutions? $-2 < x < \frac{1}{4}$

17. Helena earns $14,000 a year in salary plus an 8% commission on her sales. How much must her sales be if her annual income is to be between $20,000 and $25,000? $75,000 < x < 137,500$

18. Five less than 3 times a number is between -7 and 16. What real numbers are possible solutions? $-\frac{2}{3} < x < 7$

19. Seven more than half of a number is between -10 and 10. What real numbers are possible solutions? $-34 < x < 6$

20. The reciprocal of a number is between $\frac{1}{3}$ and 10. What real numbers are possible solutions? $\frac{1}{10} < x < 3$

21. Compare the English words "3 less than x" and "3 is less than x" and translate each into algebraic symbols. Are the algebraic expressions the same? Explain your reasoning.

Mixed Review

Simplify. *1.2, 1.3, 2.2, 2.9, 2.10*

1. $(2 \cdot 5)^2$ 100
2. $2 \cdot 5^2$ 50
3. $25 + 6(3 - 4)$ 19
4. $7x - 11x$ $-4x$
5. $3(2y - 1) + (4y - 8)\frac{3}{4}$ $9y - 9$
6. $|-3| + |3|$ 6
7. Seven more than 5 times a number is -18. Find the number. *3.3* -5
8. At 8:00 P.M. the temperature was 5°C. That night it dropped 12°. By 6:00 A.M. it had risen 8°. What was the temperature at 6:00 A.M.? *2.4* 1°C

Application: *Triangle Inequality*

Suppose a triangle is to be constructed from three pieces of wood, with one piece being 10 in. long. In order to form a triangle, the sum of the lengths of the other two pieces must be greater than 10 in. In other words, $x + y > 10$. Similarly, $10 + x > y$ and $10 + y > x$. This relationship, called the **Triangle Inequality,** states that the sum of the lengths of two sides of a triangle is greater than the length of the third side.

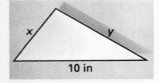

1. Can a triangle have sides with the three given lengths?
   a. 6 cm, 2 cm, 4 cm No
   b. 5 ft, 5 ft, 5 ft Yes
   c. 3 m, 8 m, 4 m No
2. State three inequalities that apply to a triangle with sides having lengths a, b, and c. $a + b > c, b + c > a, c + a > b$
3. Suppose two sides of a triangular lot measure 75 ft and 62 ft. Then the length of the third side must be between what lengths?
   13 ft and 137 ft

5.5 Problem Solving: Using Inequalities **189**

Guided Practice

Classroom Exercises 1–8
Try This   all

Independent Practice

A Ex. 1–12, **B** Ex. 13–16, **C** Ex. 17–21

Basic: WE 1–10 all, 21, Application

Average: WE 5–14 all, 21, Application

Above Average: WE 5–15 odd, 17–21 all, Application

Additional Answers

Written Exercises

10. (10,16); (9,15); (8,14); (7,13); (6,12); (5,11); (4,10); (3,9); (2,8); (1,7)
21. "3 less than x" written as $x - 3$. "3 is less than x" written as $3 < x$. Symbols are not the same.

Enrichment

Challenge students to solve this problem. There are no more than 40 birds perched on three branches of a tree. On the middle branch there are 3 times as many birds as there are on the top branch. On the bottom branch there are 9 fewer birds than there are on the middle branch. If no branch is empty, name all the possible number combinations of the birds on the branches.

Top	Middle	Bottom
7	21	12
6	18	9
5	15	6
4	12	3

Prerequisite Quiz

Evaluate if $x = -1$, $y = 3$, and $z = -4$.

1. $|3z + 5|$ 7
2. $|8 - 2x|$ 10
3. $|8 - 3y|$ 1
4. $|2x - 2|$ 4

Motivator

Have students explain how distance is related to absolute value. Ask them if distance can be represented by a negative number. See page 190. No, distance is always positive.

Lesson Note

Emphasize that equations involving absolute value must be written as disjunctions.

Math Connections

Science: On the Celsius thermometer scale, the freezing point of water, 0°C, is the point of reference for all other temperatures, which are either above (positive) or below (negative) this point.

Critical Thinking Questions

Application: Ask students to substitute different values for a and b to complete $|a + b|$? $|a| + |b|$ with $=$, $\leq$, or $\geq$. $\leq$

5.6 Equations with Absolute Value

Objective | To solve equations containing absolute value

Recall (Lesson 2.2) that the absolute value of a real number is the distance between the number and 0 on a number line. Thus, 5 and -5 both have an absolute value of 5, since both are 5 units from 0.

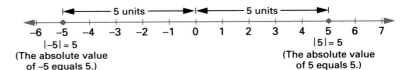

|−5| = 5
(The absolute value of −5 equals 5.)

|5| = 5
(The absolute value of 5 equals 5.)

The equation $|x| = 5$ can be written as the disjunction $x = 5$ or $x = -5$. It has two solutions, 5 and -5. Since an absolute value cannot be negative, an equation such as $|x| = -3$ has no solution.

EXAMPLE 1 | Solve each equation: **a.** $|x - 1| = 3$ **b.** $|9 - 2x| = 15$

Plan | Write the equation as a disjunction and solve.

Solutions | **a.** $x - 1 = 3$ or $x - 1 = -3$
$x = 4$ $x = -2$

b. $9 - 2x = 15$ or $9 - 2x = -15$
$-2x = 6$ $-2x = -24$
$x = -3$ $x = 12$

Checks | **a.** $|x - 1| = 3$ **b.** $|9 - 2x| = 15$

$|x - 1| = 3$ $|x - 1| = 3$ $|9 - 2(-3)| \overset{?}{=} 15$ $|9 - 2x| = 15$
$|4 - 1| \overset{?}{=} 3$ $|-2 - 1| \overset{?}{=} 3$ $|9 + 6| \overset{?}{=} 15$ $|9 - 2(12)| \overset{?}{=} 15$
$|3| \overset{?}{=} 3$ $|-3| \overset{?}{=} 3$ $|15| \overset{?}{=} 15$ $|9 - 24| \overset{?}{=} 15$
$3 = 3$ $3 = 3$ $15 = 15$ $|-15| \overset{?}{=} 15$
$15 = 15$

Thus, the solutions are -2 and 4.   Thus, the solutions are -3 and 12.

TRY THIS | Solve each equation.   **1.** $|x + 5| = 8$ $x = 3$ or $x = -13$
2. $|4x - 6| = 46$ $x = 13$ or $x = -10$

Highlighting the Standards

Standards 2f, 4a, 5c: In the Exercises, students relate absolute value notation to graphing and to solving equalities as compound sentences.

Additional Example 1

Solve each equation.

a. $|3y - 7| = 5$ $4, \frac{2}{3}$
b. $|3y - 7| = -5$ $\varnothing$

EXAMPLE 2   Solve $-2|3y + 5| + 4 = 2$.

Solution

$$-2|3y + 5| + 4 = 2$$

Subtract 4 from each side. $-2|3y + 5| = -2$
Divide each side by -2. $|3y + 5| = 1$

Write a disjunction. $3y + 5 = 1 \quad or \quad 3y + 5 = -1$

$$3y = -4 \qquad\qquad 3y = -6$$
$$y = -\frac{4}{3} \qquad\qquad y = -2$$

Thus, the solutions are $-\frac{4}{3}$ and -2. The check is left for you.

TRY THIS   3. Solve $8 + 5|6 - 2y| = 68$. *$y = -3$ or $y = 9$*

Classroom Exercises

Solve each equation.

1. $|x| = 6$ $6, -6$ 2. $|y| = -1$ $\varnothing$ 3. $|z| = 0$ 0 4. $|d| + 1 = 4$ $3, -3$
5. $2|b| = -8$ $\varnothing$ 6. $|y - 6| = 10$ $16, -4$ 7. $|2y| = 10$ $5, -5$ 8. $|4x - 1| = 5$ $-1, 1\frac{1}{2}$

Written Exercises

Solve each equation.

1. $|x| = 4$ $4, -4$ 2. $|x| = -4$ $\varnothing$ 3. $2|x| = 6$ $3, -3$ 4. $-3|x| = -12$ $4, -4$
5. $|x - 4| = 5$ $9, -1$ 6. $|y - 3| = 7$ $10, -4$ 7. $|y + 8| = -3$ $\varnothing$ 8. $|5 - z| = 3$ $2, 8$
9. $|-5c| = 15$ $3, -3$ 10. $5|-c| = 15$ $3, -3$ 11. $|3y + 1| = -2$ $\varnothing$ 12. $2|d + 2| = 14$
13. $5|x - 2| - 13 = 7$ $6, -2$ 14. $4|y - 3| - 2 = 10$ $0, 6$ 15. $3|z + 5| + 5 = 8$
16. $-6 - 3|4x - 2| = 8$ $\varnothing$ 17. $10 = 6|a - 7| - 14$ $3, 11$ 18. $1 = 8 - |2y - 5|$
12. $5, -9$
15. $-6, -4$
18. $-1, 6$

Complete each sentence. Use $>$, $<$, or $=$.

19 The solution set of $|x| = a$ is $\{0\}$ if $a \underline{\ ?\ } 0$. $=$
20. The solution set of $|x| = a$ is $\varnothing$ if $a \underline{\ ?\ } 0$. $<$

Mixed Review

Solve. *3.2, 3.5, 4.6* 1. $\frac{2}{3}$

1. $\frac{1}{2}x = \frac{1}{3}$ 2. $4(x + 2) = 3x - 9$ -17 3. $3x + 1 = -(4 - x)$ $-2\frac{1}{2}$ 4. $-0.2 = y + 4.9$ -5.1
5. Last year a particular car cost \$10,000. This year the same model costs \$10,600. What is the percent increase in the cost? *4.7* 6%

Common Error Analysis

Error: Students decide that equations such as $|3x + 5| - 9 = -4$ have no solution because the right side of the equation is negative.

Remind them that this is true only for equations of the form, $|3x + 5| = -4$.

Checkpoint

Solve each equation.

1. $|x - 7| = 14$ 21, -7
2. $|7 - c| = 7$ 0, 14
3. $|-6y| = 36$ 6, -6
4. $|4 + 2e| = 16$ -10, 6

Closure

Have students explain why an equation such as $|x + 7| = 4$ has two solutions. Since $|x + 7|$ is 4 units from 0 on the number line, $x + 7 = 4$ or $x + 7 = -4$. So the solutions are -3 and -11.

▰▰FOLLOW UP

Guided Practice

Classroom Exercises 1–8
Try This   all

Independent Practice

A Ex. 1–12, **B** Ex. 13–18, **C** Ex. 19–20

Basic: WE 1–12 all
Average: WE 1–11 odd, 13–18 all
Above Average: WE 5–11 odd, 13–20 all

Additional Example 2

Solve $7 = -4|3 - 3x| + 11$.   1, 2

Enrichment

Have the students find the solutions for the equation $|x + 8| = -x$. -4
Then have them find the solutions for the equation $|x - 8| = -x$. $\varnothing$

▰▰ GETTING STARTED

Prerequisite Quiz

Graph the solution set of each inequality.

1. $2x - 5 < 9$

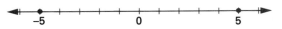

2. $6 - 3a \geq 15$

3. $3x - 2 < -4 \ or \ 3x - 2 > 4$

4. $7 - 2y > -5 \ and \ 7 - 2y < 5$

Motivator

Have students graph all points that are
less than 6 units from 2 on the number
line. Relate their graphs to the inequality
Vx − 2V < 6. Do the same for points
greater than 6 units from 2. (Vx − 2V > 6)

5.7 Inequalities with Absolute Value

Objective   To graph solution sets of inequalities containing absolute value

One winter night, the temperature
remained within 5° of 0 degrees
Fahrenheit (0°F). If x represents the
temperature in °F at any time that
night, then these temperature conditions can be described by the
inequality $|x| < 5$.

The situation just described can be represented by a graph on a number
line. First, recall from the previous lesson that the equality $|x| = 5$ has
two solutions, 5 and -5.

graph of solution
set of $|x| = 5$

It follows that the solutions of $|x| < 5$ will be *less than* 5 units from 0.

all points less than 5 units from 0

graph of solution
set of $|x| < 5$

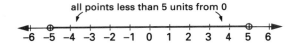

From the graph above, you can see that $|x| < 5$ is equivalent to the
conjunction $x > -5 \ and \ x < 5$. (This can be written as $-5 < x < 5$.)
Similarly, the solutions of $|x| > 5$ will be *greater than* 5 units from 0 on
a number line.

all points greater than 5 units from 0

graph of solution
set of $|x| > 5$

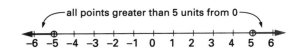

$|x| > 5$ is equivalent to the disjunction $x < -5 \ or \ x > 5$.

Highlighting the Standards

Standards 5a, 5c, 1a, 1b: Absolute value
notation is extended to the graphing and
solving of inequalities. In Mixed Problem
Solving, students practice strategies and
applications.

EXAMPLE 1   Graph the solution set of $|5 - 3x| \geq 9$.

Plan   All values of x that make $5 - 3x$ greater than or equal to 9 units from 0 on a number line are solutions of $|5 - 3x| \geq 9$. Solve the equivalent disjunction $5 - 3x \leq -9$ or $5 - 3x \geq 9$.

Solution

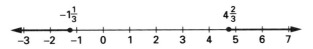

$$|5 - 3x| \geq 9$$

$$5 - 3x \leq -9 \quad or \quad 5 - 3x \geq 9$$
$$-3x \leq -14 \qquad\qquad -3x \geq 4$$
$$x \geq \frac{-14}{-3} \qquad\qquad x \leq -\frac{4}{3} \quad \longleftarrow \text{When dividing by a}$$
$$x \geq 4\frac{2}{3} \qquad\qquad x \leq -1\frac{1}{3} \quad \begin{array}{l}\text{negative number, reverse}\\ \text{the order of inequality.}\end{array}$$
$$x \leq -1\frac{1}{3}$$

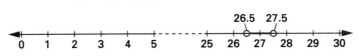

EXAMPLE 2   In an automobile assembly plant, a certain kind of connecting rod is accepted for installation in an automobile only if the length of the rod is within 0.5 cm of 27 cm. For the acceptable lengths of the rod, write an inequality with absolute value and graph the solution set of the inequality.

Solution   Let x = the length of the connecting rod.

$$|x - 27| < 0.5$$

$$x - 27 > -0.5 \quad and \quad x - 27 < 0.5$$
$$x > 26.5 \qquad\qquad x < 27.5$$
$$26.5 < x < 27.5$$

TRY THIS

1. Graph the solution set of $|4y - 1| \geq 7$. $y \leq -1\frac{1}{2}$ or $y \geq 2$

2. Yesterday the temperature stayed within 6° of 70°C. Graph the solution set. $64 < x < 76$

■■■■ **TEACHING SUGGESTIONS**

Lesson Note

Show students a comparison of the following three open sentences, their solutions, and their *graphs*.

$|x| = 4$ $\qquad\qquad$ $x = 4$ *or* $x = -4$
$|x| < 4$ $\qquad\qquad$ $x < 4$ *and* $x > -4$
$\qquad\qquad\qquad$ $(-4 < x < 4)$
$|x| > 4$ $\qquad\qquad$ $x > 4$ *or* $x < -4$

Tell them that all exercises in the lesson can be related to one of these three types.

Math Connections

Industry: Manufactured parts are allowed to vary from a standard only by a specified amount, called **tolerance**. For example, the standard diameter for a valve might be 3 cm and a customer may specify that it may not vary from this standard by more than 0.001 cm. In other words, the tolerance is $\pm$ 0.001 cm. The customer's specifications may be expressed as $|x - 3| \leq 0.001$, when x is the diameter of the valve.

Critical Thinking Questions

Analysis: Ask students this question.

Can $|x - y| = |-x|$? Explain.   Yes.   Suggest that the students substitute values for x and y, such as $x = -9$ and $y = 5$, $x = 9$ and $y = -5$, $x = -9$ and $y = -5$, and so on.

Additional Answers

Try This

1.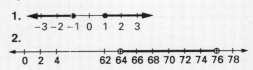

2.

Additional Example 1

Graph the solution set of $|2 + 5y| \geq 12$.

$-2\frac{4}{5}$ $\qquad$ -2

Additional Example 2

A shop owner wants the amount in her cash register to stay within $15 of $50. Write this as an inequality with absolute value, and graph the solution set. $35 < c < 65$

Checkpoint

Graph the solution set of each inequality.

1. $|x - 2| > 1$

2. $|2y + 5| < 3$

3. $|3x + 7| \leq 2$

4. $|5 - 4y| \geq 7$

5. $|6a + 3 - 4a| < 9$

Closure

Have students summarize the three basic patterns for equations and inequalities that involve absolute values. See page 192.

▰▰FOLLOW UP

Guided Practice

Classroom Exercises 1–10
Try This   all

Independent Practice

A Ex. 1–24, **B** Ex. 25–30, **C** Ex. 31–33

Basic: WE 1–10 all, 21, 23, Mixed Problem Solving

Average: WE 1–23 odd, Mixed Problem Solving

Above Average: WE 11–33 odd, Mixed Problem Solving

Additional Answers

See pages 651–652 for the answers to Classroom Ex. 1–10 and Written Ex. 1–33.

Classroom Exercises

For each inequality, give an equivalent conjunction or disjunction.

1. $|x| > 6$ **2.** $|x| \leq 3$ **3.** $|y| \geq 1$ **4.** $|4 - 7x| > 2$ **5.** $|3c + 1| \leq 0$

6–10. Graph the solution set of each inequality of Classroom Exercises 1–5.

Written Exercises

Graph the solution set of each inequality.

1. $|x| < 4$ **2.** $|x| \geq 4$ **3.** $|x| > 0$ **4.** $|x - 5| < 3$

5. $|y + 4| < 7$ **6.** $|z - 2| \geq -4$ **7.** $|2y - 8| > 6$ **8.** $|3x + 3| \leq 9$

9. $|4x - 8| \geq 20$ **10.** $|7z - 14| < 7$ **11.** $|c + 2| < 5$ **12.** $|6k - 18| \leq 12$

13. $|5 + 2x| > 7$ **14.** $|-3 + 2y| \geq 9$ **15.** $|-8 + 4y| \leq 28$ **16.** $|4 - 7x| \geq -2$

17. $|2y - 5| < 6$ **18.** $|3x + 4| < 2$ **19.** $|5 - 4x| \leq 3$ **20.** $|3 - 2z| \leq 0$

For each problem, use x for the variable and write an inequality with absolute value. Then graph the solution set of the inequality. $|x - 75| < 3$

21. Last night, the temperature stayed within 8° of 0°C. $|x| < 8$

22. Marian's golf score stayed within 3 strokes of par (75 strokes).

23. The monthly profits or losses for the Spedini Corporation were within $2,000 of the goal of $10,000.
$|x - 10{,}000| < 2{,}000$

24. Last week the price of Enton Company's common stock changed less than $1\frac{1}{8}$ points from its opening price of $10. $|x - 10| < 1\frac{1}{8}$

Graph the solution set of each inequality.

25. $|x - (5 - x)| > 3$ **26.** $|6 - (4 - y)| \leq 5$ **27.** $|5z - (3z - 6)| < 4$

28. $|2y - (y - 6)| \geq 3$ **29.** $|x - 3| - 2 < 4$ **30.** $4 + |y - 2| > 3$

31. $7 - 2|x - 4| \leq 3$ **32.** $4|c - 2| + 3 < 5$ **33.** $3 - 2|2x + 7| \leq -5$

Mixed Review

1. Use the formula $A = \frac{1}{2}bh$ to find A if $b = 9$ cm and $h = 12$ cm.   **1.4** 54 cm²

Evaluate.   **2.6, 2.8**

2. x^5, for $x = -2$ –32

3. $-9 - 5x$, for $x = -3$ 6

4. The length of a rectangle is 3 cm more than twice the width. The perimeter is 36 cm. Find the length and the width.   **4.2** Length: 13 cm; width: 5 cm

5. José is paid $130 a week plus a commission of 7% on all sales. What must be his weekly sales if José is to earn a total of $200/week?   **4.8** $1,000

Enrichment

Draw these graphs on the chalkboard.

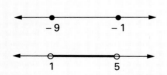

Challenge students to write equations or inequalities involving absolute value for these graphs.

1. $|x + 5| = 4$
2. $|x - 3| < 2$
3. $|x - 4| > 3$
4. $|x + 6| < 5$

Mixed Problem Solving

The algebra skills you have acquired up to this point provide you with tools for problem solving. As part of a problem-solving plan, you choose a variable and use it to represent the unknown quantity or quantities, translate the words of the problem into an equation, and solve the equation.

For some of the problems on this page, try the strategy of *Drawing a Diagram* to represent the problem situation. For example, a diagram or sketch may help you in solving Exercises 3, 5, and 9.

Solve each problem.

1. The table below shows the profits and losses of a small business for four successive months. What is the average profit or loss per month? $1,730 profit

Jan	Feb	March	April
+ $6,700	+ $2,300	− $1,600	− $480

2. Bonnie needs $\frac{3}{4}$ yd of ribbon to make each hair bow. To be safe, she bought an extra $\frac{1}{2}$ yd of ribbon. How many hair bows can she make if she bought $7\frac{1}{4}$ yd of ribbon? 9 bows

3. Between midnight and 4:00 A.M. the temperature dropped 8°. If the temperature was −3°F at 4:00 A.M., what was the temperature at midnight? 5°F

4. On Tuesday Mary swam $\frac{2}{3}$ as many laps as she swam on Monday. If she swam 48 laps on Tuesday, how many laps did she swim on Monday? 72 laps

5. Yesterday Mark jogged 1.2 km farther than he jogged today. If he jogged 5.9 km yesterday, how far did he jog today? 4.7 km

6. The selling price of a book is $\frac{1}{3}$ more than the cost. If a book sells for $18, find the cost. $13.50

7. Martha has $100 and is saving at a rate of $5/week. Kate has $50 and is saving at a rate of $10/week. After how many weeks will they have the same amount of money? 10 weeks

8. Jackie showed a profit of $560 in her business in January. In February she showed a profit of $490. What was the percent decrease in her profit from January to February? 12.5%

9. The vertex angle of an isosceles triangle measures 62. Find the measure of each base angle. 59°

10. The sum of two consecutive integers is −77. Find the integers. −39, −38

11. Of three consecutive odd integers, x represents the second integer. Give an algebraic expression for the first integer and for the third integer. $x - 2, x + 2$

12. A collection of nickels and dimes is worth $2.75. There are 7 more nickels than dimes. How many of each type of coin are there? 16 dimes; 23 nickels

13. An electrician earns $6.50 more per hour than his apprentice. For a 6-h job, their total earnings were $267. How much does each make per hour? Apprentice: $19; electrician: $25.50

1.
$$\xleftarrow{\qquad \overset{\circ}{|} \;\;|\;\;|\;\;|\;\;|\;\;|\;\;|\;\;|\;\;|\;\;|\;\; \rightarrow}$$
$$\quad\;\; -4 \quad\;\; -2 \quad\;\;\; 0$$

2.
$$\overset{\textstyle -5}{\xleftarrow{\;|\;\;\overset{\circ}{|}\;\;|\;\;|\;\;|\;\;|\;\;|\;\;|\;\;|\;\;|\;\;}\rightarrow}$$
$$\;\; -6 \quad -4 \quad -2 \quad\;\; 0$$

3.
$$\overset{\textstyle -5}{\xleftarrow{\;|\;\;\overset{\circ}{|}\;\;|\;\;|\;\;|\;\;|\;\;|\;\;|\;\;|\;\;|\;\;}\rightarrow}$$
$$\;\; -6 \quad -4 \quad -2 \quad\;\; 0$$

4.
$$\overset{\textstyle 1.3}{\xleftarrow{\;|\;\overset{\circ}{|}\;\;|\;\;|\;\;|\;\;|\;\;|\;\;|\;\;|\;\;}\rightarrow}$$
$$\quad\;\; 0 \quad\;\; 2 \quad\;\; 4$$

5.
$$\overset{\textstyle -\frac{3}{4}}{\xleftarrow{\;|\;\;|\;\;|\;\;\overset{\circ}{|}\;\;|\;\;|\;\;|\;\;}\rightarrow}$$
$$\;\; -4 \quad\; -2 \quad\;\; 0$$

6.
$$\xleftarrow{\;|\;\;|\;\;|\;\;|\;\;|\;\;\overset{\circ}{|}\;\;|\;\;}\rightarrow$$
$$\quad\;\; 0 \quad\;\; 2 \quad\;\; 4$$

7.
$$\overset{\textstyle 5}{\xleftarrow{\;|\;\;|\;\;|\;\;\overset{\circ}{|}\;\;|\;\;|\;\;|\;\;}\rightarrow}$$
$$\;\; -10 \quad\; 0 \quad\;\; 10$$

8.
$$\xleftarrow{\;|\;\;|\;\;\overset{\circ}{|}\;\;|\;\;|\;\;|\;\;|\;\;}\rightarrow$$
$$\;\; -12 \quad -6 \quad\;\; 0$$

9.
$$\xleftarrow{\;|\;\;\overset{\circ}{|}\;\;|\;\;|\;\;|\;\;|\;\;|\;\;|\;\;}\rightarrow$$
$$\;\; -10 \quad\; 0 \quad\;\; 10$$

10.
$$\xleftarrow{\;|\;\;|\;\;|\;\;|\;\;|\;\;|\;\;|\;\;|\;\;|\;\;}\rightarrow$$
$$\;\; -4 \quad\;\; 0 \quad\;\; 4$$

11.
$$\xleftarrow{\;|\;\;|\;\;|\;\;|\;\;|\;\;|\;\;\overset{\circ}{|}\;\;|\;\;}\rightarrow$$
$$\;\; -4 \quad\;\; 0 \quad\;\; 4$$

12.
$$\overset{\textstyle -3}{\xleftarrow{\;|\;\;\overset{\circ}{|}\;\;|\;\;|\;\;|\;\;|\;\;|\;\;}\rightarrow}$$
$$\;\; -6 \quad\;\; 0 \quad\;\; 6$$

Chapter 5 Review

Key Terms

Addition Property of Inequality (p. 166)
conjunction (p. 175)
disjunction (p. 176)
Division Property of Inequality (p. 171)
equivalent inequalities (p. 166)

Multiplication Property of Inequality (p. 170)
Subtraction Property of Inequality (p. 166)
Trichotomy Property (p. 186)
truth table (p. 176)

Key Ideas and Review Exercises

5.1 Adding (or subtracting) the same number to (or from) an inequality produces an equivalent inequality.

Solve each inequality. Graph the solution set.

1. $x + 6 < 2$ $\;x < -4$

2. $y - 4 > -9$ $\;y > -5$

3. $-2 < x + 3$ $\;x > -5$

4. $0.4 + x > 1.7$ $\;x > 1.3$

5. $-2\frac{1}{2} > y - 1\frac{3}{4}$ $\;y < -\frac{3}{4}$

6. $3x - 2(x + 7) < -10$ $\;x < 5$

5.2 Multiplying (or dividing) each side of an inequality by the same positive number produces an equivalent inequality. Multiplying (or dividing) each side of an inequality by the same negative number and reversing the order of the inequality produces an equivalent inequality.

Solve each inequality. Graph the solution set.

7. $4x < 20$ $\;x < 5$

8. $-3z > 18$ $\;z < -6$

9. $\frac{2}{5}k > -4$ $\;k > -10$

10. $x + 7 < x - 1$ $\;\varnothing$

11. $7 - 3x > 4x - 21$ $\;x < 4$

12. $4a + 5 < -(2 - 3a) - 4a - 8$ $\;a < -3$

5.3 A *conjunction* (two sentences connected by *and*) is true only if both of the sentences are true. Otherwise it is false. A *disjunction* (two sentences connected by *or*) is false only if both of the sentences are false. Otherwise it is true.

Determine whether the given conjunction or disjunction is *true* or *false*.

13. $-2 < -1$ *or* $5 > 7$ T

14. $3 > -1$ *and* $6 < -6$ F

15. $5^2 = 25$ *and* $8 \neq 5 + 1$ T

16. $4 = 9 - 5$ *or* $7 > 5 + 3$ T

5.4 To graph the solution set of a *disjunction* of two sentences, graph the set that contains all the solutions of the two sentences. To graph the solution set of a *conjunction* of two sentences, graph the set that contains only the solutions common to both sentences.

196 Chapter 5 Review

Graph the solution set.

17. $4a - 5 \geq -2 + 3a$

18. $-3c + 6 \leq -4c - 2$

19. $x > 2$ or $x \leq 0$

20. $-5 < x < -1$

21. Write a short explanation of why the solution set of $x \geq a$ and $x \leq a$ is $\{a\}$.

5.5   The inequality $x \geq a$ can mean *x is greater than or equal to a, x is at least a*, or *x is not less than a*. The inequality $x \leq a$ can mean *x is less than or equal to a, x is at most a*, or *x is not greater than a*. The inequality $a < x < b$ means *x is between a and b* with $a < b$.

Solve each problem.

22. Jake rented a car for \$21/day plus 18¢/mi. How far can he travel in one day if he can spend at most \$62.40?   At most 230 mi

23. The sum of two positive, consecutive odd integers is at most 20. Find all possible pairs of integers.   9, 11; 7, 9; 5, 7; 3, 5; 1, 3

5.6   To solve an equation involving absolute value, solve a disjunction.

     For example, to solve $|2x - 5| = 11$, solve the disjunction $2x - 5 = 11$ *or* $2x - 5 = -11$.

Solve each equation.

24. $|x - 2| = 3$ 5, −1

25. $|4 - y| = 6$ −2, 10

26. $|c + 3| = -1$ ∅

27. $|2y - 4| = 14$ 9, −5

28. $4 = |5a - 2|$ $1\frac{1}{5}, -\frac{2}{5}$

29. $-2|4x - 2| - 8 = 16$ ∅

5.7   To solve an inequality with absolute value that contains $<$ or $\leq$, solve a conjunction. For example, to solve $|3x + 6| \leq 9$, solve the conjunction $3x + 6 \geq -9$ *and* $3x + 6 \leq 9$.

     To solve an inequality with absolute value that contains $>$ or $\geq$, solve a disjunction. For example, to solve $|-2y - 1| \geq 7$, solve the disjunction $-2y - 1 \leq -7$ *or* $-2y - 1 \geq 7$.

Graph the solution set of each inequality.

30. $|x - 4| < 2$

31. $|y + 3| \geq 5$

32. $|5 + 2x| > 9$

33. $|-4 + 3y| \leq 2$

34. $|-2y + 1| < -1$

35. $9 - 3|x - 2| < 6$

Chapter 5 Review   **197**

17. (number line: solid dot at 3; labeled 0, 6, 12)

18. (number line: solid dot at −8; labeled −16, −8, 0)

19. (number line: solid dot at 0, open dot at 2; labeled −2, 0, 2)

20. (number line: open dot at −5, open dot at −1; labeled −6, −4, −2, 0)

21. *a* is the only number where $x \geq a$ and $x \leq a$ are true.

30. (number line: open dots at 2 and 6; labeled 0, 2, 4, 6)

31. (number line: solid dots at −8 and 2; labeled −8, 0, 8)

32. (number line: open dots at −7 and 2; labeled −8, −4, 0, 4)

33. (number line: solid dots at $\frac{2}{3}$ and 2; labeled −2, 0, 2)

34. (number line: labeled −4, 0, 4)

35. (number line: open dots at 1 and 3; labeled 0, 2, 4)

1. (number line, open circle at 1) 0 2 4

2. (number line, open circle at -3) -4 -2 0

3. -7 (number line) -8 -4 0 4

4. (number line, open circle at 3) 0 2 4

5. (empty number line) -4 0 4

6. -1 (number line, open circle) -2 0 2

9. (number line, closed dot) -12 0 12

10. -5, -1 (number line) -6 -4 -2 0

11. (empty number line) -4 0 4

12. -5, 1 (number line) -8 -4 0 4

23. -5, 2 (number line) -8 -4 0 4

24. 1, 3 (number line) 0 2 4

25. (number line) -4 0 4

26. (empty number line) -4 0 4

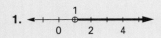
Solve each inequality. Graph the solution set.

1. $x - 4 > -3$ $x > 1$

2. $7 > y + 9$ $y < -2$

3. $-5x < 35$ $x > -7$

4. $-3c + 4 > -8$ $c < 4$

5. $2p - 9 > 16 + 2p$ $\varnothing$

6. $-2(4x + 5) < 1 + 3x$ $x > -1$

Determine whether the given conjunction or disjunction is *true* or *false*.

7. $7 = 3 + 5 \ or \ 6 \neq 5 + 2$ T

8. $-3 > -2 + 1 \ and \ (-6)^2 > 30$ F

Graph the solution set.

9. $-2x - 11 \leq 13$

10. $-5 \leq x < -1$

11. $-3y > -24 \ or \ 2y - 6 > 8$

12. $3x - 1 < 2 \ and \ x + 4 > -1$

Solve each problem.

13. Mr. Heisler makes a profit of $7 on each shirt he sells. How many shirts must he sell to make a profit of at least $224? $x \geq 32$

14. An elevator can carry up to 1,800 lb safely. At most, how many 80-lb crates can be carried in the elevator without overloading it? $x \leq 22$, where x is a natural number.

15. Eight less than 4 times a number is between -5 and 12. What real numbers are possible solutions? $\frac{3}{4} < x < 5$

16. The selling price of a stereo is $270. The profit is 35% of the cost. Find the cost. $200

17. The first side of a triangle is 4 cm longer than twice the second. The third side is 2 cm shorter than the second side. The perimeter is 30 cm. Find the length of each side. 18 cm, 7 cm, 5 cm

18. The second of three numbers is one more than twice the first. The third is 10 less than the second. The sum of the three numbers is 27. Find each. 7, 15, 5

Solve each equation.

19. $|x - 7| = 2$ $x = 9 \ or \ x = 5$

20. $-|x + 1| = -3$ $x = 2 \ or \ x = -4$

21. $|4x + 1| = -5$ $\varnothing$

22. $3 - 4|3y - 1| = -5$ $y = 1 \ or \ y = -\frac{1}{3}$

Graph the solution set of each inequality.

23. $|2x + 3| < 7$

24. $|6 - (4 + c)| < 1$

25. $|r| < 4 \ or \ |r| > 4$

26. $3c + 4 < 6 \ and \ 5c - 1 > c + 7$

27. Solve for x: $ax + b < c$, where $a < 0$. $x > \frac{c - b}{a}$

In each item you are to compare a quantity in Column 1 with a quantity in Column 2. Write the letter of the correct answer from these choices:

A—The quantity in Column 1 is greater than the quantity in Column 2.
B—The quantity in Column 2 is greater than the quantity in Column 1.
C—The quantity in Column 1 is equal to the quantity in Column 2.
D—The relationship cannot be determined from the information given.

NOTES: Information centered over both columns refers to one or both of the quantities to be compared. A symbol that appears in both columns has the same meaning in each column, and all variables represent numbers.

Sample Question $2x + 1 > 7$	Answer
Column 1 **Column 2**	The solution to $2x + 1 > 7$ is $x > 3$. Therefore, the quantity in Column 2 is greater. The answer is B.
3 x	

	Column 1	**Column 2**		**Column 1**	**Column 2**
1.	\multicolumn{2}{c}{$x + 2 < 7$}	**8.**	\multicolumn{2}{c}{$0 < x < 1$}
	x	5 B		x^2	1 B
2.	\multicolumn{2}{c}{$5y < 0$}	**9.**	\multicolumn{2}{c}{$-1 < y < 0$}
	y	1 B		y^2	1 B
3.	\multicolumn{2}{c}{$x < y$}	**10.**	\multicolumn{2}{c}{$x > 0, y < 0$}
	$x - y$	$y - x$ B		0	xy A
4.	\multicolumn{2}{c}{$0 < x < 8$ $0 < y < 10$}	**11.**	\multicolumn{2}{c}{$x = 0, y < 0$}
	x	y D		0	xy C
5.	$\lvert x - y \rvert$	$\lvert y - x \rvert$ C	**12.**	\multicolumn{2}{c}{$x > 0, y < 0$}
				$\dfrac{1}{x}$	$\dfrac{1}{y}$ A
6.	\multicolumn{2}{c}{a is a real number.}	**13.**	\multicolumn{2}{c}{$-1 < x < 1, x \neq 0$}
	a	$-a$ D		$\dfrac{1}{x}$	0 D
7.	\multicolumn{2}{c}{$x > 0, y < 0$}	**14.**	\multicolumn{2}{c}{$x > 1$}
	$\dfrac{x}{y}$	$\dfrac{y}{x}$ D		$\dfrac{1}{x}$	1 B

6 POWERS AND POLYNOMIALS

OVERVIEW

The first two lessons of this chapter extend the concept of exponents introduced in Chapter 1 to multiplying and dividing monomials. Negative and zero exponents are then introduced and immediately applied to working with very large and very small numbers expressed in scientific notation. Sections 6–6 through 6–8 cover operations with polynomials. The special products treated in the last two lessons show students how to multiply certain binomials more efficiently.

OBJECTIVES

- To multiply and divide monomials using the properties of powers
- To simplify monomials containing zero and negative exponents
- To work with numbers in scientific notation
- To add, subtract, and simplify polynomial expressions
- To multiply polynomials and learn special product forms

PROBLEM SOLVING

Looking for Patterns is the strategy used to introduce the properties of powers to the students. In the Application on page 217, students use formulas and scientific notation to solve problems involving very large numbers. Using Formulas is also the strategy applied in the Application dealing with compound interest on page 225.

READING AND WRITING MATH

Students are asked to use the properties they have learned to explain why particular open sentences are true or false (see p. 213, Exercise 34, p. 221, Exercise 37, and p. 228, Exercise 36). The Focus on Reading exercises on pages 220, 230, and 234 help students to review essential vocabulary and check their understanding of basic concepts and procedures by identifying equivalent expressions.

TECHNOLOGY

Calculator: Students are introduced to the $\boxed{\text{EE}}$ and $\boxed{\text{EXP}}$ keys on page 215. Examples of how to use a calculator in computations are also presented. Opportunities for using a calculator may be found in Section 6.5, in the Application on page 217, and in the Application on page 225.

SPECIAL FEATURES

Mixed Review pp. 203, 206, 209, 213, 216, 225, 228, 231, 235
Application: The Earth's Hydrosphere p. 217
Focus on Reading p. 220, 230, 234
Midchapter Review p. 221
Application: Compound Interest p. 225
Brainteaser p. 235
Key Terms p. 236
Key Ideas and Review Exercises pp. 236-237
Chapter 6 Test p. 238
College Prep Test p. 239
Cumulative Review (Chapters 1–6) pp. 240-241

PLANNING GUIDE

Lesson	Basic	Average	Above Average	Resources
6.1 pp. 202–203	CE 1–12 WE 1–21 odd, 28–30	CE 1–12 WE 7–27 odd, 28–30	CE 1–12 WE 11–37 odd	Reteaching 41 Practice 41
6.2 pp. 205–206	CE 1–16 WE 1–39 odd	CE 1–16 WE 13–39 odd, 41–46 all	CE 1–16 WE 17–39 odd, 41–46 all, 47–51 odd	Reteaching 42 Practice 42
6.3 p. 209	CE 1–8 WE 1–19 odd	CE 1–8 WE 9–27 odd	CE 1–8 WE 13–31 odd	Reteaching 43 Practice 43
6.4 pp. 212–213	CE 1–14 WE 1–21 odd, 34	CE 1–14 WE 5–25, odd, 34	CE 1–14 WE 13–33 odd, 34	Reteaching 44 Practice 44
6.5 pp. 215–217	CE 1–8 WE 1–20 all, Application	CE 1–8 WE 5–24 all, Application	CE 1–8 WE 1–19 odd, 21–30 all, Application	Reteaching 45 Practice 45
6.6 pp. 220–221	FR all, CE 1–10 WE 1–25 odd, 37 Midchapter Review	FR all, CE 1–10 WE 9–33 odd, 37 Midchapter Review	FR all, CE 1–10 WE 13–35 odd, 36–37 Midchapter Review	Reteaching 46 Practice 46
6.7 pp. 224–225	CE 1–10 WE 1–16 all, Application	CE 1–10 WE 1–6 all, 7–25 odd, Application	CE 1–10 WE 1–23 odd, 24–27, Application	Reteaching 47 Practice 47
6.8 pp. 227–228	CE 1–9 WE 1–27 odd, 36	CE 1–9 WE 9–35 odd, 36	CE 1–9 WE 13–39 odd	Reteaching 48 Practice 48
6.9 pp. 230–231	FR all, CE 1–17 WE 1–24 all	FR all, CE 1–17 WE 1–33 odd, 34–39	FR all, CE 1–17 WE 1–47 odd	Reteaching 49 Practice 49
6.10 pp. 234–235	FR all, CE 1–24 WE 1–20 all Brainteaser	FR all, CE 1–24 WE 1–39 odd Brainteaser	FR all, CE 1–24 WE 9–47 odd Brainteaser	Reteaching 50 Practice 50
Chapter Review pp. 236–237	all	all	all	
Chapter 6 Test p. 238	all	all	all	
College Prep Test p. 239	all	all	all	
Cumulative Review (Chapters 1–6) pp. 240–241	all	all	all	

CE = Classroom Exercises   WE = Written Exercises   FR = Focus on Reading

Note: For each level, all students should be assigned all Try This and all Mixed Review Exercises.

■■■INVESTIGATION

Manipulative: In this *Investigation,* students use a grid of squares and rectangles to illustrate multiplication of binomials.

Materials: Each student will need a copy of Manipulative Worksheet 6.

Before handing out the worksheet, use the following illustration to show $(a + b)(a + b)$ as the sum of four areas. Point out that $(a + b)^2$.

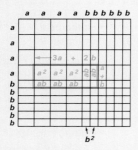

Then help students draw and complete the first exercise on the worksheet. Students may have to experiment to find which part of the grid to mark.

After students have had a chance to complete the worksheet, ask them if they can see any patterns in the answers that might allow them to do these problems without using a grid. Have them try to complete the following multiplications without a grid.

1. $(5a + 2b)(3a + 7b)$ $15a^2 + 41ab + 14b^2$
2. $(7a + b)(9a + b)$ $63a^2 + 16ab + b^2$
3. $(6a + 5b)^2$ $36a^2 + 60ab + 25b^2$

Marine biologists study animal and plant life in the Antarctic to help monitor the air and water quality of the planet. Computer models, built from sets of equations, make possible early warnings of possible pollution problems.

More About the Photo

The world's coldest temperature was recorded in Antarctica as minus 88 degrees Celsius. Antarctica's largest land creature is an insect, related to the housefly, that is less than $\frac{1}{2}$ inch long. The waters of the Antarctic are home to the largest mammal that ever lived, the blue whale, which has a length of about 100 feet. Antarctica is uninhabited by native people, but scientists from several nations have set up laboratories there. A 1959 treaty says that no individual nation can claim territory on Antarctica for 30 years.

6.1 Multiplying Monomials

Objective To multiply monomials

The power 3^6 can be broken into factors in many ways. For example,

$$3^6 = (3 \cdot 3) \cdot (3 \cdot 3 \cdot 3 \cdot 3) = 3^2 \cdot 3^4$$
$$3^6 = (3 \cdot 3 \cdot 3) \cdot (3 \cdot 3 \cdot 3) = 3^3 \cdot 3^3$$
$$3^6 = (3 \cdot 3 \cdot 3 \cdot 3 \cdot 3) \cdot (3) = 3^5 \cdot 3^1$$

In each case the sum of the exponents of the factors is equal to 6. This idea is used in multiplying *monomials*.

Each of the following is an example of a monomial.

-7	x	$4x^3$	$-5xy$
a constant	a variable	a product of a constant and one or more variables	

A **monomial** is an algebraic expression that is either a constant, a variable, or a product of a constant and one or more variables.

The monomials a^3 and a^4 have the same base a but different exponents, 3 and 4. To multiply a^3 and a^4, write a as a factor 3 times and a as a factor 4 more times, as shown below.

$$a^3 \cdot a^4 = (a \cdot a \cdot a) \cdot (a \cdot a \cdot a \cdot a) = a^{3+4} = a^7$$

The result is a as a factor 7 times. Similarly,

$$y^3 \cdot y^3 = (y \cdot y \cdot y) \cdot (y \cdot y \cdot y) = y^6$$
$$x \cdot x^4 = x^1 \cdot x^4 = x \cdot (x \cdot x \cdot x \cdot x) = x^5$$

This suggests that to multiply powers with the same base, you add the exponents and keep the same base. In general,

$$b^m \cdot b^n = \overbrace{(b \cdot b \ldots b)}^{m \text{ factors}} \cdot \overbrace{(b \cdot b \ldots b)}^{n \text{ factors}} = b^{m+n}$$

Product of Powers
For all real numbers b and all positive integers m and n,

$$b^m \cdot b^n = b^{m+n}$$

GETTING STARTED

Prerequisite Quiz

Write with exponents.

1. $5 \cdot 5 \cdot 5$ 5^3
2. $(7 \cdot 7)(7 \cdot 7 \cdot 7)$ 7^5
3. $(-4)(-4)$ $(-4)^2$
4. $c(c \cdot c \cdot c \cdot c)$ c^5

Evaluate.

5. $-2y^5$ for $y = -1$ 2
6. $5x^2y$ for $x = 3$ and $y = -7$ -315
7. $-4ab^3$ for $a = 4$ and $b = -2$ 128

Motivator

Tell students that the height of a stack of newspapers for recycling doubles every day after the first. Ask them to use powers of 2 to indicate the height of the stack for the first 5 days if the stack was 2 inches high on the first day. $2^1, 2^2, 2^3, 2^4, 2^5$. Ask them when the stack will be over 6 ft high. On the 7th day.

Highlighting the Standards

Standards 3b, 2d, 2f: Students read a written presentation about exponential notation and see examples of how economically this notation denotes products. In the Written Exercises, they provide counterexamples for false statements.

Lesson Note

When introducing the Product of Powers Property, emphasize that exponents cannot be added unless the bases are the same. In Example 2, emphasize that the coefficients are *multiplied,* but the exponents are *added.*

Math Connections

Biology: Amoebas reproduce by dividing in half. Given a suitable environment, this process of splitting in half will continue indefinitely. Thus, if you start with one amoeba, the number of amoebas, *n,* present after *t* days is given by $n = 2^t$. The growth indicated in the formula is an example of the exponential growth of a population.

Critical Thinking Questions

Synthesis: Ask students to write each expression as the product of two factors.   Answers will vary.

1. 5^6 $5^a \cdot 5^b$, where $a + b = 6$.
2. $(xy)^3$ $(xy)^a \cdot (xy)^b$, where $a + b = 3$.
3. 9^{3m} $9^m \cdot 9^{2m}$

Common Error Analysis

Error: When multiplying monomials, some students multiply the exponents.

Suggest that they write powers in factored form until they understand that the exponents must be added.

EXAMPLE 1   Simplify each product, if possible.

a. $r^7 \cdot r$ b. $c^2 \cdot d^2$

Solutions   a. Rewrite r as r^1. Since the bases are the same, add the exponents.

b. The bases, c and d, are different. Thus, $c^2 \cdot d^2$ cannot be simplified.

$r^7 \cdot r = r^7 \cdot r^1 = r^{7+1} = r^8$

TRY THIS   Simplify each product, if possible.

1. $x^4 \cdot y^6$ x^4y^6 2. $a^2 \cdot a^3$ a^5

To simplify a product of monomials such as $(-3c^2d)(7cd^3)$, use the Commutative and Associative Properties of Multiplication to group the constant coefficients together and the like bases together. Then multiply the coefficients and apply the Product-of-Powers Property to the factors with the same base.

EXAMPLE 2   Simplify: a. $-3c^2d(7cd^3)$ b. $(-5x^2y)(-xz^3)$

Solutions   a. $-3c^2d(7cd^3)$
$= (-3 \cdot 7)(c^2 \cdot c^1)(d^1 \cdot d^3)$
$= -21 \cdot c^{2+1} \cdot d^{1+3}$
$= -21c^3d^4$

b. $(-5x^2y)(-xz^3)$
$= (-5 \cdot x^2y^1)(-1 \cdot x^1z^3)$
$= [-5(-1)] \cdot (x^2 \cdot x^1) \cdot y^1 \cdot z^3$
$= 5 \cdot x^{2+1} \cdot y \cdot z^3 = 5x^3yz^3$

EXAMPLE 3   Simplify $x^2y(x^2y^2)$. Then evaluate for $x = -3$ and $y = -1$.

Solution $x^2y(x^2y^2) = x^2 \cdot x^2 \cdot y^1 \cdot y^2 = x^{2+2} \cdot y^{1+2} = x^4y^3$

If $x = -3$ and $y = -1$, then $x^4y^3 = (-3)^4(-1)^3$
$= (81)(-1)$, or -81

TRY THIS   3. Simplify $(3a^5b^4c)(-2b^3c^8)$. $-6a^5b^7c^9$
4. Simplify $(xy^3)(x^2y^3)$. Then evaluate for $x = -5$ and $y = 2$.
x^3y^6; -8000

Classroom Exercises

Simplify, if possible.

1. $(-1)^4$ 1 2. $(-1)^{21}$ -1 3. $x^2 \cdot x^3$ x^5 4. $y^5 \cdot z^2$ y^5z^2
5. $(-2)^3 \cdot (-2)^2$ -32 6. $-2^3 \cdot [-(2^2)]$ 32 7. $-a^4 \cdot a$ $-a^5$ 8. $5y(7y)$ $35y^2$
9. $6c(-5d)$ $-30cd$ 10. $-ab(-ab)$ a^2b^2 11. $3a^2b(5ab^2)$ $15a^3b^3$ 12. $-7x^4y(4xy)$
$-28x^5y^2$

Additional Example 1

Simplify each product.

a. $x \cdot x^5$ x^6
b. $r \cdot s^3$ Cannot be simplified

Additional Example 2

Simplify each product.

a. $-3a^2b(-6bc^3)$ $18a^2b^2c^3$
b. $-cd^2(3c^3d)$ $-3c^4d^3$

Additional Example 3

Simplify $(-x)(3x^2)(5x)$. $-15x^4$
Then evaluate for $x = -2$. -240

Written Exercises

Simplify, if possible.

1. $2x \cdot x^3$ $2x^4$
2. $-3y^4 \cdot y^2$ $-3y^6$
3. $4a^4 \cdot b^2$ $4a^4b^2$
4. $-y \cdot 2y^2$ $-2y^3$
5. $(-z^2) \cdot (-z^2)$ z^4
6. $-z^2 \cdot (-z)^2$ $-z^4$
7. $4a^2(3a^5)$ $12a^7$
8. $-6x^3y^2$ $-6x^3y^2$
9. $5y(-10z^3)$ $-50yz^3$
10. $(-3c^4)(4d)$ $-12c^4d$
11. $(2b)(-6b^7)$ $-12b^8$
12. $(-5x^2)(-4x^3)$ $20x^5$
13. $(5x^2y)(-2xy^2)$ $-10x^3y^3$
14. $(3c^2d)(7c^2d^4)$ $21c^4d^5$
15. $-xy(-xy)$ x^2y^2
16. $(2a^2b)(3b^2c)$ $6a^2b^3c$
17. $(2x^4z)(-3y^3z)$ $-6x^4y^3z^2$
18. $(-5x^2y^2z)(2xz^3)$
19. $3x(2x^2)(5x^3)$ $30x^6$
20. $(5y)(-4y^2)(-y)$ $20y^4$
21. $-3z^2(-7z)(-2z^4)$ $-42z^7$
22. $\frac{1}{3}a^2(-6ab^2)$ $-2a^3b^2$
23. $-\frac{3}{4}xy^3(4xyz)$ $-3x^2y^4z$
24. $(-0.5c^2d)(2d^3c)$ $-c^3d^4$
25. $(-12x)(\frac{2}{3}x^2y)(-4y)$ $32x^3y^2$
26. $(-a^2b)(-ab^2)(-ab^4)$ $-a^4b^7$
27. $(16x^3yz)(-\frac{3}{4}y^2z)$

18. $-10x^3y^2z^4$

27. $-12x^3y^3z^2$

Simplify. Then evaluate for $x = -1$ and $y = -2$.

28. $x^5 \cdot x^3$ x^8; 1
29. $x^2y \cdot xy$ x^3y^2; -4
30. $-x^4y^3(x^3y)$ $-x^7y^4$; 16

Simplify. Assume that all exponents are positive integers.

31. $x^a \cdot x^b$ x^{a+b}
32. $y^n \cdot y^{4n}$ y^{5n}
33. $z^{x+2} \cdot z^{x-3}$ z^{2x-1}

Find the value of n that makes each sentence true.

34. $x^{2n+3} \cdot x^{2n-3} = x^{12}$ 3
35. $y^{5n-2} \cdot y^{-3n+4} = y^8$ 3
36. $c^{4n-1} \cdot c^{n+2} = c^{6n-5}$ 6

For each statement, tell whether it is true or false. If not true, give an example showing that it is not true. Example answers may vary.

37. For all real numbers x, $x^8 - x^5 = x^{8-5}$ F; $x = 2$
38. An odd power of a negative number is always negative. T

Mixed Review

Evaluate. *2.6, 2.8*

1. $16 - y$ for $y = -3$ 19
2. $5x - 2y$ for $x = 2$ and $y = -6$ 22

Solve. *3.4, 5.2*

3. $3x + 7 = 5x - 9$ 8
4. $-3x + 12 < -15$ $x > 9$
5. $3x - 6 > 4x - 2$ $x < -4$
6. Mary and Bob bought new jackets. Mary paid \$49. Her jacket cost \$19 less than twice Bob's jacket. How much did Bob pay? *3.3* \$34

Checkpoint

Simplify.

1. $-x \cdot 3x^4$ $-3x^5$
2. $2a \cdot 3b$ $6ab$
3. $-4c(-2c^3)$ $8c^4$
4. $-ab(-ab)$ a^2b^2
5. $7x^2yz(-5xz^5)$ $-35x^3yz^6$
6. $-2a(7b)(-3ac)$ $42a^2bc$

Closure

Ask students to give three examples of a monomial. Answers will vary.
Ask them if $-5 + x$ is a monomial. No.

■■■ FOLLOW UP

Guided Practice

Classroom Exercises 1–12
Try This all

Independent Practice

A Ex. 1–21, **B** Ex. 22–30, **C** Ex. 31–38
Basic: WE 1–21 odd, 28–30 all
Average: WE 7–27 odd, 28–30 all
Above Average: WE 11–37 odd

Enrichment

Cross-Monomial Puzzle

Complete the puzzle.

1 -6	2 b^4		
3 12	p^4	a	4 t^3
5 r^8	o^5	s^4	y
	6 w^2	e^2	

Across
1. $6b \cdot -b^3$
3. $2ap \cdot 2pt \cdot 3p^2t^2$
5. $r^3o^5y \cdot r^5s^4$
6. $-we \cdot -we$

Down
1. $-2pw^2 \cdot 3p^3o^5$
2. $ab^2e \cdot b^2es^4$
3. $-3r^6 \cdot -4r^2$
4. $t^2 \cdot ty$

203

Teaching Resources

Project Worksheet 6
Quick Quizzes 42
Reteaching and Practice
    Worksheets 42
Transparency 15

▰▰▰ GETTING STARTED

Prerequisite Quiz

Simplify.

1. $(2^2)^3$ 64
2. $(-3^3)^2$ 729
3. $(-1)^{12}$ 1
4. $-(-3^2)^3$ 729
5. $(-2 \cdot 3)^3$ -216

Motivator

Challenge students to evaluate $(x^4)^3$, $(x^3)^5$, and $(x^2y)^3$ x^{12}, x^{15}, x^6y^3
If students are not successful at first, cover the expression inside the parentheses with a finger, and ask them the meaning of the exponent outside the parentheses. Then help them simplify $(x^4)^3$ as $x^4 \cdot x^4 \cdot x^4 = x^{4+4+4} = x^{12}$, and so on.

6.2 Powers of Monomials

Objective | To simplify a power of a power and a power of a product

Suppose that a monomial such as a^3 is raised to the fourth power. The result, $(a^3)^4$, is called a *power of a power*. You can simplify $(a^3)^4$ as shown below.

$$(a^3)^4 = a^3 \cdot a^3 \cdot a^3 \cdot a^3 \quad \text{(Definition of exponent)}$$
$$= a^{3+3+3+3} \quad \quad \text{(Product of Powers)}$$
$$= a^{3 \cdot 4}$$
$$= a^{12}$$

The result is a as a factor 12 times. Similarly,

$$(x^5)^2 = x^5 \cdot x^5 = x^{5+5} = x^{5 \cdot 2} = x^{10}$$
$$(r^6)^3 = r^6 \cdot r^6 \cdot r^6 = r^{6+6+6} = r^{6 \cdot 3} = r^{18}$$

This suggests that to raise a power to a power, you multiply the exponents and keep the same base. In general,

$$\overbrace{\qquad}^{n \text{ factors}} \qquad \overbrace{\qquad}^{n \text{ terms}}$$
$$(b^m)^n = b^m \cdot b^m \cdot \ldots \cdot b^m = b^{m+m+\ldots+m} = b^{mn}.$$

> **Power of a Power**
> For all real numbers b and all positive integers m and n,
> $$(b^m)^n = b^{mn}.$$

EXAMPLE 1 | Simplify.

     a. $(c^2)^5$ **b.** $(y^3)^{10}$

Solutions | **a.** $(c^2)^5 = c^{2 \cdot 5} = c^{10}$ **b.** $(y^3)^{10} = y^{3 \cdot 10} = y^{30}$

TRY THIS | Simplify.   **1.** $(x^3)^3$ x^9 **2.** $(a^{21})^2$ a^{42}

Now suppose that a product such as $2x$ is raised to the third power. The result, $(2x)^3$, is called a *power of a product*. You can simplify $(2x)^3$, as follows.

$$(2x)^3 = 2x \cdot 2x \cdot 2x = (2 \cdot 2 \cdot 2)(x \cdot x \cdot x) = 2^3 \cdot x^3 = 8x^3$$

Highlighting the Standards

Standards 2f, 5c: This lesson further expands exponential notation by finding the power of a power. In the Written Exercises, students explore possible combinations of bases and exponents.

Additional Example 1

Simplify.

a. $(x^3)^6$ x^{18}
b. $(a^5)^3$ a^{15}

Notice that both 2 and x are raised to the third power. This suggests that to find a power of a product, raise each factor to that power.

$$\overbrace{(ab)^m = (ab)(ab) \ldots (ab)}^{m \text{ factors}} = \overbrace{(a \cdot a \cdot \ldots \cdot a)}^{m \text{ factors}} \overbrace{(b \cdot b \cdot \ldots \cdot b)}^{m \text{ factors}} = a^m b^m.$$

Power of a Product
For all real numbers a and b and all positive integers m,
$$(ab)^m = a^m b^m.$$

Do not confuse $(2x)^3$ and $2x^3$. $(2x)^3$ means "$2x \cdot 2x \cdot 2x$," while $2x^3$ means "$2 \cdot x \cdot x \cdot x$."

EXAMPLE 2 Simplify.

a. $(-4c)^3$ b. $(-xy^2)^4$ c. $-4x\,(5x^3)^2$

Solutions

a. $(-4c)^3$
$= (-4)^3 c^3$
$= -64c^3$

b. $(-xy^2)^4$
$= (-1x)^4(y^2)^4$
$= (-1)^4 x^4 y^{2 \cdot 4}$
$= 1 \cdot x^4 y^8$, or $x^4 y^8$

c. $-4x(5x^3)^2$
$= -4x^1 \cdot 5^2 \cdot (x^3)^2$
$= -4 \cdot 25 \cdot x^1 \cdot x^6$
$= -100x^{1+6}$, or $-100x^7$

TRY THIS Simplify. 3. $(-2a^4b^6)^5$ $-32a^{20}b^{30}$ 4. $-y^2(4y^7)^2$ $-16y^{16}$

Classroom Exercises

Simplify.

1. $(a^2)^4$ a^8
2. $(y^3)^2$ y^6
3. $(c^5)^2$ c^{10}
4. c^5c^2 c^7
5. $[-2(-1)]^2$ 4
6. $-2(-1)^2$ -2
7. $-[2(-1)]^2$ -4
8. $(-2)^2(-1)^2$ 4
9. $(cd)^2$ c^2d^2
10. $(4x)^2$ $16x^2$
11. $4(x)^2$ $4x^2$
12. $-(2a)^3$ $-8a^3$
13. $(x^2y)^3$ x^6y^3
14. $-(2xy)^4$ $-16x^4y^4$
15. $(3b^2)^3$ $27b^6$
16. $-(2a^2bc^3)^2$ $-4a^4b^2c^6$

Written Exercises

Simplify.

1. $(y^4)^3$ y^{12}
2. $(x^3)^5$ x^{15}
3. $y^4 \cdot y^3$ y^7
4. $x^3 \cdot x^5$ x^8
5. $(-3c)^2$ $9c^2$
6. $-(3c)^2$ $-9c^2$
7. $(-3^2)^3$ -729
8. $[(-3)^2]^3$ 729
9. $(-r)^3$ $-r^3$
10. $(-y)^6$ y^6
11. $-(c)^5$ $-c^5$
12. $(xy)^4$ x^4y^4

Lesson Note

Contrast $x^2 \cdot x^3$ with $(x^2)^3$ by showing each in factored form. Point out the different number of factors in each expression: $x^2 \cdot x^3$ has 5 factors; $(x^2)^3$ has 6 factors. When presenting Example 2b, point out to the students that the rewriting of $-x$ as $-1x$ will help them avoid making a mistake with the sign of the answer.

Math Connections

Measurement: Exponents are used to indicate units of area and volume. A cm^2 indicates an area of 1 cm · 1 cm = (1 cm)^2, or 1 cm^2. Likewise, for volume, 1 cm · 1 cm · 1 cm = 1 cm^3.

Critical Thinking Questions

Application: Ask students to evaluate $(-1)^n$ for various positive integral values of n. Which values of n give a result of -1? Which values of n give a result of 1? odd; even

Common Error Analysis

Some students simplify an expression such as $(-3a^2b)^4$ by writing $-12a^8b^4$.

If so, have them write the preliminary step as follows:
$(-3a^2b)^4 = (-3)^4 \cdot (a^2)^4 \cdot (b)^4$
$= 81a^8b^4$
Emphasize that they are finding the fourth power of 3, or 3^4.

Additional Example 2

Simplify.

a. $(-3c^2d)^5$ $-243c^{10}d^5$
b. $(-a^2b)^3$ $-a^6b^3$
c. $-(2y^4)^5$ $-32y^{20}$

Checkpoint

1. $(-y^2)^3$ $-y^6$
2. $(-2a^2)^5$ $-32a^{10}$
3. $(xy^3)^2$ x^2y^6
4. $-(a^2b^3)^4$ $-a^8b^{12}$
5. $(-3a^2bc^3)^4$ $81a^8b^4c^{12}$
6. $2x(\frac{1}{2}x)^2$ $\frac{1}{2}x^3$
7. Simplify $(ab)^3(a^2b)^2$.
   Then evaluate if $a = -1$ and $b = 2$.
   a^7b^5; -32

Closure

Ask students to simplify each pair of expressions.

1. $x^3 \cdot x^4$ and $(x^3)^4$ x^7, x^{12}
2. $x^3 \cdot y^4$ and $(x^3y)^4$ x^3y^4, $x^{12}y^4$
3. $(-1)^5$ and $(-1)^{20}$ -1, 1

◼◼◼ FOLLOW UP

Guided Practice

Classroom Exercises 1–16
Try This   all

Independent Practice

Ⓐ Ex. 1–24, Ⓑ Ex. 25–44, Ⓒ Ex. 45–52

Basic: WE 1–39 odd

Average: WE 13–39 odd, 41–46 all

Above Average: WE 17–39 odd, 41–46 all, 47–51 odd

Additional Answers

Written Exercises

21. $32a^{20}b^5c^{10}$
22. $-27r^3s^6c^9$
23. $-125x^6y^9z^3$
24. $-36a^4b^4c^6$
31. $-24c^4d^3$

---

13. $(5y^3)^2$ $25y^6$
14. $(-3c^4)^3$ $-27c^{12}$
15. $(a^2b)^4$ a^8b^4
16. $(r^3s^2)^6$ $r^{18}s^{12}$
17. $(-xy)^4$ x^4y^4
18. $-(a^2b)^6$ $-a^{12}b^6$
19. $(x^2yz^3)^2$ $x^4y^2z^6$
20. $(-a^4b^2c)^4$ $a^{16}b^8c^4$
21. $(2a^4bc^2)^5$
22. $(-3rs^2c^3)^3$
23. $(-5x^2y^3z)^3$
24. $-(-6a^2b^2c^3)^2$
25. $x(2x^3)^2$ $4x^7$
26. $4a(3a)^3$ $108a^4$
27. $a^4(a^2b)^2$ a^8b^2
28. $4x(-2xy)^3$
29. $3a(\frac{1}{3}a)^2$ $\frac{1}{3}a^3$
30. $(-2x)^3(\frac{1}{2}x)^2$ $-2x^5$
31. $3c(-2cd)^3$
32. $(\frac{1}{2}x)^3(2x)^4$ $2x^7$
33. $(2x^3)^3(3x^4)^2$ $72x^{17}$
34. $(-x^2)^3(4x^2)^3$ $-64x^{12}$
35. $-3ab^3(a^2b)^4$ $-3a^9b^7$
36. $0.05xy(2x^2y)^3$
   28. $-32x^4y^3$
   36. $0.4x^7y^4$

Simplify, if possible. Then evaluate for $x = -1$ and $y = 3$.

37. $(x^2)^5 \cdot y^2$ $x^{10}y^2$; 9
38. $(xy)^2(x^3)^4$ $x^{14}y^2$; 9
39. $(2x^4y)^3$ $8x^{12}y^3$; 216
40. $3xy(\frac{1}{3}xy)^2$

For the given conditions, tell whether x^n is positive, negative, or zero. $\frac{x^3y^3}{3}$; -9

41. x is negative and n is even. Positive
42. x is positive and n is odd. Positive
43. x is zero and n is odd. Zero
44. x is negative and n is odd. Negative

For the given conditions, tell whether $(x^m)^n$ is positive, negative, or zero.

45. x is negative, m is odd, and n is even. Positive
46. x is negative, m is odd, and n is odd. Negative

For each sentence, find the value of a that makes it true.

47. $(x^4)^{2a-3} = x^4$ 2
48. $x^{2a+6} = (x^{2a+1})^2$ 2
49. $(y^{a+1})^5 = y^{3a+9}$ 2
50. $(x^{3a+5})^2(x^a)^4 = x^{8a+12}$ 1

Solve.

51. The exponent of a power of 2 is 3 more than the exponent of a power of 4. Find the exponents. 3, 6
52. The exponent of a power of 3 is one more than twice the exponent of a power of 27. What are the exponents? 1, 3

Mixed Review

1. Use the formula $A = \frac{1}{2}(b + c)h$ to find A for $b = 9$ cm, $c = 12$ cm, and $h = 14$ cm.   *1.4* 147 cm²

Simplify.   *2.9, 2.10*

2. $3x + 2y - x + y$ $2x + 3y$
3. $4a + 3b + 2a - 2b$ $6a + b$
4. $6x - 7(y - 4x) + 1$ $34x - 7y + 1$
5. Solve for a: $2a + 6 = 11$ *3.3* $\frac{5}{2}$
6. Jackie bought $\frac{3}{4}$ yd of fabric for each stuffed animal she was making. To be safe, she bought an extra $\frac{1}{2}$ yd of fabric, for a total of $6\frac{1}{2}$ yd. How many stuffed animals did she plan to make?   *4.5* 8

Enrichment

12^4 can be expressed in many different ways. Have the students use the set {1, 2, 3, 4, 5, 6, 7, 8, 9} to replace the variable in the following expressions so that each expression equals 12^4.

1. $12^3 \cdot 12^a$ 1
2. $3^4 \cdot c^4 \cdot c^4$ 2
3. $(3^y \cdot 2^4)^2$ 2
4. $6^3 \cdot 3 \cdot 2^a$ 5
5. $b^4 \cdot (2^4)^2$ 3
6. $3^4 \cdot b^4$ 4

Have pairs of students create an exercise set similar to the one at the left. Then pairs can exchange papers and solve each other's exercises.

6.3 Dividing Monomials

Objective   To divide monomials

You can divide the numerator and denominator of a fraction by the same nonzero number. For example,

$$\frac{12}{20} = \frac{\overset{1}{\cancel{4}} \cdot 3}{\underset{1}{\cancel{4}} \cdot 5} = \frac{3}{5} \qquad\qquad \frac{7}{14} = \frac{\overset{1}{\cancel{7}} \cdot 1}{\underset{1}{\cancel{7}} \cdot 2} = \frac{1}{2}$$

To divide monomials you can use the same cancellation property.

Cancellation Property of Fractions
For all real numbers a, b, and c, such that $b \neq 0$ and $c \neq 0$,

$$\frac{a \cdot c}{b \cdot c} = \frac{a}{b}.$$

Now consider the quotient $\dfrac{a^5}{a^2}$, where $a \neq 0$.

$$\frac{a^5}{a^2} = \frac{\overbrace{\cancel{a} \cdot \cancel{a} \cdot a \cdot a \cdot a}^{5 \text{ factors}}}{\underbrace{\cancel{a} \cdot \cancel{a}}_{2 \text{ factors}}} = a \cdot a \cdot a = a^3$$

Similarly, for $x \neq 0$ and $y \neq 0$,

$$\frac{x^4}{x^6} = \frac{\overset{1}{\cancel{x}} \cdot \overset{1}{\cancel{x}} \cdot \overset{1}{\cancel{x}} \cdot \overset{1}{\cancel{x}}}{\underset{1}{x} \cdot \underset{1}{\cancel{x}} \cdot \underset{1}{\cancel{x}} \cdot \underset{1}{\cancel{x}} \cdot \underset{1}{\cancel{x}} \cdot x} = \frac{1}{x^2} \text{ and } \frac{y^3}{y^3} = \frac{\overset{1}{\cancel{y}} \cdot \overset{1}{\cancel{y}} \cdot \overset{1}{\cancel{y}}}{\underset{1}{\cancel{y}} \cdot \underset{1}{\cancel{y}} \cdot \underset{1}{\cancel{y}}} = 1$$

These three cases suggest the following property for the *quotient of powers*, which you can use without having to write the factored forms.

Quotient of Powers
For all real numbers b, $(b \neq 0)$, and all positive integers m and n,

if $m > n$, then	if $m < n$, then	if $m = n$, then
$\dfrac{b^m}{b^n} = b^{m-n}.$	$\dfrac{b^m}{b^n} = \dfrac{1}{b^{n-m}}.$	$\dfrac{b^m}{b^n} = \dfrac{b^m}{b^m} = 1.$
(Case 1)	(Case 2)	(Case 3)

Teaching Resources

Quick Quizzes 43
Reteaching and Practice
  Worksheets 43

■■■ GETTING STARTED

Prerequisite Quiz

1. $\dfrac{6^3}{6}$ 36

2. $\dfrac{2^6}{2^2}$ 16

3. $\dfrac{3^4}{3}$ 27

4. $\dfrac{5}{5^3}$ $\dfrac{1}{25}$

5. $\dfrac{(-4)^2}{(-4)^5}$ $\dfrac{1}{-64}$

Motivator

Have students evaluate each pair of expressions.

1. $\dfrac{3^4}{3}$ and 3^3 27, 27

2. $\dfrac{5^3}{5}$ and 5^2 25, 25

3. $\dfrac{(-2)^6}{(-2)^4}$ and $(-2)^2$ 4, 4

Ask students what they notice when they evaluate each pair.   They are equal. Challenge them to formulate a rule for dividing a^m by a^n. $a^m \div a^n = a^{m-n}$

Lesson Note

When presenting Example 2, point out that, since the highest power of a is in the numerator, a will be in the numerator of the answer; likewise, since the highest power of b is in the denominator, b will be in the denominator of the answer. Emphasize that, when dividing mononials, although exponents are subtracted, coefficients are divided.

Additional Example 1

Simplify.

a. $\dfrac{a^7}{a}$, $(a \neq 0)$ a^6

b. $\dfrac{c^4}{c^4}$, $(c \neq 0)$ 1

c. $\dfrac{x^3}{x^8}$, $(x \neq 0)$ $\dfrac{1}{x^5}$

Highlighting the Standards

Standard 5c: Students practice operating on algebraic expressions. The "simplest form" for a quotient is defined.

Math Connections

Science: The force of attraction between two bodies can be found by dividing the monomial kMm by r^2, or $\frac{kMm}{r^2}$, where M and m are the masses of the two bodies, r is the distance between them, and k is constant for all bodies. This force of attraction is called *gravity*.

Critical Thinking Questions

Synthesis: First, ask students to simplify $\frac{x^3}{x^5} \cdot \frac{1}{x^2}$. Then tell them to assume that the restrictions on the relative values of m and n are removed, such that $b^m \div b^n = b^{m-n}$ regardless of whether m is greater than, less than, or equal to n. Then ask them to simplify $x^3 \div x^5$. x^{-2} Finally, ask them to compare the two answers and make a conjecture about the meaning of x^{-2}. $\frac{1}{x^2}$

Common Error Analysis

Error: Some students may incorrectly simplify $6^3 \div 6$ to 1^2, because $6 \div 6 = 1$. Have them rewrite the problem as $\frac{6 \cdot 6 \cdot 6}{6}$ and then simplify.

EXAMPLE 1   Simplify.

   a. $\dfrac{x^8}{x^5}$ $(x \neq 0)$ **b.** $\dfrac{y^9}{y^9}$ $(y \neq 0)$ **c.** $\dfrac{z}{z^5}$ $(z \neq 0)$

Plan   Use the three different cases of the Quotient-of-Powers Property.

Solutions   **a.** $\dfrac{x^8}{x^5} = x^{8-5} = x^3$ **b.** $\dfrac{y^9}{y^9} = 1$ **c.** $\dfrac{z}{z^5} = \dfrac{z^1}{z^5} = \dfrac{1}{z^{5-1}} = \dfrac{1}{z^4}$

           by Case 1.   by Case 3.   by Case 2.

EXAMPLE 2   Simplify $\dfrac{-15a^2b^3}{5ab^8}$ $(a \neq 0, b \neq 0)$.

Solution $\dfrac{-15a^2b^3}{5ab^8} = \dfrac{-3 \cdot 5 \cdot a^2 \cdot b^3}{5 \cdot a^1 \cdot b^8} = \dfrac{-3a^{2-1}}{b^{8-3}} = \dfrac{-3a^1}{b^5} = \dfrac{-3a}{b^5}$

EXAMPLE 3   Simplify $\dfrac{18x^3y^4z}{27x^3y^2z^5}$ $(x \neq 0, y \neq 0, z \neq 0)$.

Solution $\dfrac{18x^3y^4z}{27x^3y^2z^5} = \dfrac{2 \cdot 9 \cdot x^3 \cdot y^4 \cdot z^1}{3 \cdot 9 \cdot x^3 \cdot y^2 \cdot z^5} = \dfrac{2y^{4-2}}{3z^{5-1}} = \dfrac{2y^2}{3z^4}$

TRY THIS   Simplify.

  1. $\dfrac{a^3}{a^9}$ $\frac{1}{a^6}$ **2.** $\dfrac{-13x^8y}{26x^4y^5}$ $-\frac{x^4}{2y^4}$ **3.** $\dfrac{-24a^7bc^3}{36abc^6}$ $\frac{-2a^6}{3c^3}$

It may be necessary to simplify the numerator or the denominator, or both, before using the Quotient-of-Powers Property.

EXAMPLE 4   Simplify $\dfrac{(3x^5y)^2}{6x^4y^3}$ $(x \neq 0, y \neq 0)$.

Solution $\dfrac{(3x^5y)^2}{6x^4y^3} = \dfrac{3^2 \cdot x^{5 \cdot 2} \cdot y^2}{6 \cdot x^4 \cdot y^3} = \dfrac{3^2 \cdot x^{10} \cdot y^2}{6 \cdot x^4 \cdot y^3} = \dfrac{3 \cdot 3 \cdot x^{10-4}}{3 \cdot 2 \cdot y^{3-2}} = \dfrac{3x^6}{2y}$

Notice that a quotient *in simplest form* contains no parentheses, each base appears just once, and numerical coefficients do not have a common factor (other than 1 or -1).

Additional Example 2

Simplify.

$\dfrac{-18x^2y}{-3xy^4}$, $(x \neq 0, y \neq 0)$. $\frac{6x}{y^3}$

Additional Example 3

Simplify.

$\dfrac{4a^2b^3c}{-6ab^5c^4}$ $(a \neq 0, b \neq 0, c \neq 0)$. $\frac{2a}{-3b^2c^3}$

Additional Example 4

Simplify.

$\dfrac{(5x^2y)^3}{15xy^4}$, $(x \neq 0, y \neq 0)$. $\frac{25x^5}{3y}$

Classroom Exercises

Simplify. Assume that no variable is equal to zero.

1. $\dfrac{6^4}{6}$ 216

2. $\dfrac{8^2}{8^3}$ $\dfrac{1}{8}$

3. $\dfrac{5^4}{5^4}$ 1

4. $\dfrac{y^4}{y^8}$ $\dfrac{1}{y^4}$

5. $\dfrac{a^5}{a^5}$ 1

6. $\dfrac{a^2 b}{ab^2}$ $\dfrac{a}{b}$

7. $\dfrac{r^3 s}{s^2 t}$ $\dfrac{r^3}{st}$

8. $\dfrac{4xy^3}{8x^2 y}$ $\dfrac{y^2}{2x}$

Written Exercises

Simplify. Assume that no variable is equal to zero.

1. $\dfrac{5^3}{5}$ 25

2. $\dfrac{7^2}{7^6}$ $\dfrac{1}{2{,}401}$

3. $\dfrac{x^3}{x^3}$ 1

4. $\dfrac{9^{12}}{9^{10}}$ 81

5. $\dfrac{z^2}{z^{10}}$ $\dfrac{1}{z^8}$

6. $\dfrac{x^2 y}{xy}$ x

7. $\dfrac{c^2 d^3}{c^3 d}$ $\dfrac{d^2}{c}$

8. $\dfrac{3x^5}{5x^2}$ $\dfrac{3x^3}{5}$

9. $\dfrac{5a^2 b}{10a^2 b}$ $\dfrac{1}{2}$

10. $\dfrac{6mn}{6m^2 n}$ $\dfrac{1}{m}$

11. $\dfrac{-2r^2 s}{18rs}$ $\dfrac{-r}{9}$

12. $\dfrac{14x^2 y}{14x^2 y}$ 1

13. $\dfrac{(7x)^2}{(7x)^5}$ $\dfrac{1}{343x^3}$

14. $\dfrac{(9y)^3}{9y}$ $81y^2$

15. $\dfrac{13xy}{26x^2 y}$ $\dfrac{1}{2x}$

16. $\dfrac{-5z^2}{15yz}$ $\dfrac{-z}{3y}$

17. $\dfrac{-18ab^2}{6a^2 b^3}$ $\dfrac{-3}{ab}$

18. $\dfrac{-12x^2 yz^3}{9xy^4 z^3}$ $\dfrac{-4x}{3y^3}$

19. $\dfrac{-4a^2 b^3 c}{-28abc}$ $\dfrac{ab^2}{7}$

20. $\dfrac{16xy^2 z^3}{18x^2 y^2 z^2}$ $\dfrac{8z}{9x}$

21. $\dfrac{(2a^2 b)^3}{6ab^4}$ $\dfrac{4a^5}{3b}$

22. $\dfrac{4xy^2(-3x^2 y)}{10x^3 y}$ $\dfrac{-6y^2}{5}$

23. $\dfrac{(3r^2)^3(2rs)}{6rs}$ $9r^6$

24. $\dfrac{5a^2 b^2}{10b(2ab)^3}$ $\dfrac{1}{16ab^2}$

25. $\dfrac{(4x^5 y^2)^2}{16x^{10} y^4}$ 1

26. $\dfrac{-3ab(6a^2 b^4)}{9a^3 b^2}$ $-2b^3$

27. $\dfrac{(8 \cdot 10^2)^2}{16 \cdot 10^5}$ $\dfrac{2}{5}$

28. $\dfrac{(0.5)^8}{[(0.5)^2]^3}$ 0.25

Simplify. Assume that no variable equals zero and that all exponents are positive integers.

29. $\dfrac{y^{3n}}{y^n}$ y^{2n}

30. $\dfrac{x^n y^{2m}}{xy^m}$ $x^{n-1} y^m$

31. $\dfrac{(3x^n y^{m-1})^2}{6x^n y^{m+2}}$ $\dfrac{3x^n y^{m-4}}{2}$

32. $\dfrac{7c^{2n}(2c)}{4c^n}$ $\dfrac{7c^{n+1}}{2}$

Mixed Review

Which property of real numbers is illustrated?   *1.5*

   Commutative for mult

1. $4(2 \cdot 6) = (4 \cdot 2)6$ Associative for mult

2. $(7 + x)3 = 3(7 + x)$

3. $(4 + y) + z = 4 + (y + z)$

4. $(3 + a) + b = b + (3 + a)$

5. Barry's salary is 4 times Jack's salary. The difference of their salaries is $126. Find each boy's salary.   *4.2* Barry: $168; Jack: $42

Checkpoint

Simplify. Assume that no variable equals zero and that all exponents are positive integers.

1. $\dfrac{8^3}{8^5}$ $\dfrac{1}{64}$

2. $\dfrac{x^4}{x^4}$ 1

3. $\dfrac{y}{y^6}$ $\dfrac{1}{y^5}$

4. $\dfrac{a^2 b}{ab^4}$ $\dfrac{a}{b^3}$

5. $\dfrac{-21x^2 y}{-7x^5 y}$ $\dfrac{3}{x^3}$

6. $\dfrac{(3ab^2)^2}{12a^3 b}$ $\dfrac{3b^3}{4a}$

Closure

Ask students to describe how to find the quotient of two monomials having the same base. Write the base, subtract the exponents. Ask them also to describe when a quotient is in simplest form. Quotient has no parentheses, each base appears just once, numerical coefficients have no common factors other than 1 or −1.

■■■FOLLOW UP

Guided Practice

Classroom Exercises 1–8
Try This   all

Independent Practice

A Ex. 1–20, **B** Ex. 21–28, **C** Ex. 29–32
Basic: WE 1–19 odd
Average: WE 9–27 odd
Above Average: WE 13–31 odd

Additional Answers

Mixed Review

3. Assoc Prop Add
4. Comm Prop Add

Enrichment

Refer to the formula given in the Math Connections and show the students that if the distance between the bodies is doubled, the force of attraction is represented by $\dfrac{kMm}{(2r)^2}$, or $\dfrac{kMm}{4r^2}$. Point out that as the distance is doubled, the force of attraction is decreased to one-fourth its original force.

Have the students represent the force of attraction, in simplest form, between the above bodies in each of the following cases.

1. Triple their distance apart. $\dfrac{kMm}{9r^2}$
2. Double their distance apart and double the mass of one body. $\dfrac{kMm}{2r^2}$

3. Triple their distance apart and triple the mass of each body. $\dfrac{kMm}{r^2}$
4. Half the mass of one body and half their distance apart. $\dfrac{2kMm}{r^2}$
5. Half the mass of each body and double their distance apart. $\dfrac{kMm}{16r^2}$

▰▰▰ GETTING STARTED

Prerequisite Quiz

Simplify.

1. $2^4 \cdot 2^1$ 32
2. $3^2 \cdot 3^2$ 81
3. $\dfrac{2^5}{2^2}$ 8
4. $\dfrac{3^4}{3^2}$ 9
5. $\dfrac{4 \cdot 4^3}{4^5}$ $\dfrac{1}{4}$

Motivator

Ask students to tell under what circumstances the Quotient-of-Powers Property will result in a negative exponent. When $\dfrac{b^m}{b^n}$ is written as b^{m-n} and $n > m$. Ask them under what circumstances, the Quotient-of-Powers Property will result in an exponent of 0. For $\dfrac{b^m}{b^n}$ where $m = n$.

6.4 Negative Exponents

Objectives   To simplify expressions with integral exponents
To evaluate monomials with integral exponents

Thus far, you have been working with positive exponents only. **Zero** and **negative exponents** can be defined by extending the Quotient of Powers Property. For example,

$$\frac{5^3}{5^3} = 5^{3-3} = 5^0 = 1$$

For this reason, mathematicians have made the following agreement.

Definition

> **Zero Exponent**
> For each nonzero real number b,
> $$b^0 = 1. \qquad (0^0 \text{ is undefined.})$$

EXAMPLE 1   Simplify.

   a. $4x^0, x \neq 0$ **b.** $(5m)^0, m \neq 0$

Solutions   **a.** $4x^0 = 4(1) = 4$ **b.** $(5m)^0 = 1$
              Thus, $4x^0 = 4$ and $(5m)^0 = 1$.

TRY THIS   Simplify.   **1.** $(xy)^0$ 1   **2.** $(7c^6)^0$ 1

Next consider the quotient $\dfrac{6^2}{6^5}$.

You know that $\dfrac{6^2}{6^5} = \dfrac{6 \cdot 6}{6 \cdot 6 \cdot 6 \cdot 6 \cdot 6} = \dfrac{1}{6^3}$.

If $\dfrac{6^2}{6^5} = 6^{2-5} = 6^{-3}$, then by substitution $6^{-3} = \dfrac{1}{6^3}$.

Now suppose that the negative exponent appears in the denominator.

$$\frac{1}{4^{-3}} = \frac{1}{\frac{1}{4^3}} = 1 \div \frac{1}{4^3} = 1 \cdot \frac{4^3}{1} = 4^3$$

These cases suggest a definition for negative exponent.

Highlighting the Standards

Standards 2b, 4a: This lesson develops the definition of special exponents from the rules for quotients. In the Written Exercises, students write quotients in equivalent forms.

Additional Example 1

Simplify.

a. $-7y^0, y \neq 0$ -7
b. $-(6q)^0, q \neq 0$ -1

Definition

Negative Exponent

For each nonzero real number b and for each positive integer n,

$$b^{-n} = \frac{1}{b^n} \text{ and } \frac{1}{b^{-n}} = b^n.$$

EXAMPLE 2 Simplify. Use positive exponents only. No variable equals zero.

a. $3^5 \cdot 3^{-8}$ **b.** $(ab^{-7})(-b^{-3})$

Solutions   **a.** $3^5 \cdot 3^{-8} = 3^{5+(-8)} = 3^{-3} = \frac{1}{3^3} = \frac{1}{27}$

b. $(ab^{-7})(-b^{-3}) = -1(ab^{-7})(b^{-3}) = -ab^{-7+(-3)}$
$$= -ab^{-10} = \frac{-a}{b^{10}}$$

EXAMPLE 3 Evaluate $3x^{-4}$ for $x = -2$.

Solution $3x^{-4} = 3 \cdot \frac{1}{x^4} = \frac{3}{x^4} = \frac{3}{(-2)^4} = \frac{3}{16}$

TRY THIS   **3.** Simplify $(x^{-8}y^{-5})(x^{10}y^{-2})$. Use positive exponents only. $\frac{x^2}{y^7}$

An expression such as $\frac{3x^{-5}}{7y^{-2}}$ can be simplified so that there are no negative exponents.

Use $\frac{a}{b} = \frac{a \cdot c}{b \cdot c}$. $\frac{3x^{-5}}{7y^{-2}} = \frac{3x^{-5} \cdot x^5y^2}{7y^{-2} \cdot x^5y^2}$.

Use $b^m \cdot b^n = b^{m+n}$. $= \frac{3x^{-5+5}y^2}{7x^5y^{-2+2}}$

$$= \frac{3x^0y^2}{7x^5y^0}$$

Use $b^0 = 1$. $= \frac{3 \cdot 1 \cdot y^2}{7 \cdot x^5 \cdot 1}$, or $\frac{3y^2}{7x^5}$

This shows that $\frac{3x^{-5}}{7y^{-2}}$ is equivalent to $\frac{3y^2}{7x^5}$.

As a shortcut, many of the steps such as those shown above are often omitted, as in Example 4 on page 212.

TEACHING SUGGESTIONS

Lesson Note

It helps some students to show them that a negative exponent indicates that a *factor* can be moved to the other member (numerator or denominator) of a fraction by changing the sign of its exponent. After presenting the Examples in this lesson, point out that the Quotient-of-Powers Property is now stated as just one rule rather than as 3 different cases.

Math Connections

Previous Math: The reciprocal of a number can be indicated by using an exponent of -1. Since $a \cdot a^{-1} = a^0 = 1$, a and a^{-1} are reciprocals of one another.

Critical Thinking Questions

Analysis: The definition of a Zero Exponent states that 0^0 is undefined. Ask students to look at the development prior to this definition and give a reason for this exclusion.   Since division by 0 is undefined, $0^n \div 0^n$ is undefined.

Additional Example 2

Simplify. Use positive exponents only. No variable equals zero.

a. $9^{-4} \cdot 9^6$ 9^2
b. $(-x^{-5}y^2)(x^{-4}y^{-2})$ $\frac{-1}{x^9}$

Additional Example 3

Evaluate $7a^{-3}$ for $a = -3$. $-\frac{7}{27}$

Common Error Analysis

Error: Some students may evaluate x^0 as 0.

Have students simplify several expressions such as these: $\frac{9}{9}$, $\frac{3^3}{3^3}$, $\frac{6^2}{6^2}$, and so on. Then have them apply the Quotient-of-Powers Property to simplify the expressions and compare the two sets of answers.

Checkpoint

Simplify. Use positive exponents only. Assume that no variable equals zero.

1. $-5x^0$ -5
2. $\frac{3a^{-2}}{6a^{-5}}$ $\frac{a^3}{2}$
3. $\frac{4x^2y^{-3}z}{-14x^0yz^{-5}}$ $\frac{2x^2z^6}{-7y^4}$
4. $\left(\frac{5x^2}{2a}\right)^{-3}$ $\frac{8a^3}{125x^6}$
5. Evaluate $-6m^{-2}$ for $m = -8$. $-\frac{3}{32}$

Closure

Ask students to express each of the following so that the exponent is positive or zero. In each case, n is a positive integer and $a \neq 0$.

1. $\frac{1}{a^{-n}}$ a^n
2. a^{-n} $\frac{1}{a^n}$
3. $\frac{a^{-n}}{a^{-n}}$ a^0, or 1
4. $a^n \cdot \frac{1}{a^n}$ a^0, or 1

EXAMPLE 4   Simplify $\frac{6x^{-4}y^2}{-3xy^{-5}}$, $x \neq 0$, $y \neq 0$. Use positive exponents only.

Plan   To write with positive exponents, use $a^{-n} = \frac{1}{a^n}$ and $\frac{1}{a^{-n}} = a^n$.

Solution   Write 6 as $-2(-3)$ and x as x^1.

$$\frac{6x^{-4}y^2}{-3xy^{-5}} = \frac{-2(-3)x^{-4}y^2}{-3x^1y^{-5}}$$

$$= \frac{-2y^2y^5}{x^1x^4} = \frac{-2y^7}{x^5}$$

EXAMPLE 5   Simplify $\left(\frac{2}{3}\right)^{-4}$.

Solution

$$\left(\frac{2}{3}\right)^{-4} = \frac{1}{\left(\frac{2}{3}\right)^4} = \frac{1}{\left(\frac{16}{81}\right)} = \frac{81}{16}$$

TRY THIS   4. Simplify $\frac{-5x^{-5}y^{-1}}{10x^2y^{-6}}$, $x \neq 0$, $y \neq 0$. Use positive exponents only. $\frac{-y^5}{2x^7}$

In general, for all nonzero real numbers a and b, and all integers n,

$$\left(\frac{a}{b}\right)^{-n} = \left(\frac{b}{a}\right)^n.$$

The following properties hold for all integral exponents m and n.

Product of Powers	Quotient of Powers ($b \neq 0$)	Power of a Power	Power of a Product
$b^m \cdot b^n = b^{m+n}$	$\frac{b^m}{b^n} = b^{m-n}$	$(b^m)^n = b^{mn}$	$(ab)^m = a^m b^m$

Classroom Exercises

Evaluate each expression for the given value of the variable.

1. $2a^{-1}$; $a = -5$ $\frac{2}{-5}$ 2. $-3n^4$; $n = -2$ -48

Simplify. Use positive exponents only. No variable equals zero.

3. $-3x^0$ -3
4. $(-3x)^0$ 1
5. 3^{-2} $\frac{1}{9}$
6. $6^7 \cdot 6^{-8}$ $\frac{1}{6}$
7. $d^{-2} \cdot d^4$ d^2
8. $4a^{-5} \cdot 2a^3$ $\frac{8}{a^2}$
9. $3x^{-2}y \cdot 5xy^3$ $\frac{15y^4}{x}$
10. $\left(\frac{1}{3}\right)^{-3}$ 27
11. $\frac{(2y)^0}{(3y)^0}$ 1
12. $\frac{4x^{-2}y^3}{-2xy^{-1}}$ $\frac{-2y^4}{x^3}$
13. $\frac{1}{2^{-3}}$ 8
14. $\left(\frac{-12c^2}{3a^{-1}}\right)^{-2}$ $\frac{1}{16a^2c^4}$

Additional Example 4

Simplify. $\frac{-4a^3b^{-2}}{10a^{-3}b^2}$, ($a \neq 0$, $b \neq 0$). Use positive exponents only.

$$\frac{-2a^6}{5b^4}$$

Additional Example 5

Simplify. $\left(-\frac{1}{5}\right)^{-3}$ -125

Written Exercises

Evaluate each expression for the given value of the variable.

1. $2a^{-5}$; $a = 3$ $\frac{2}{243}$

2. y^{-2}; $y = 8$ $\frac{1}{64}$

3. $-5m^{-1}$; $m = -105$ $\frac{1}{21}$

4. $7n^{-9}$; $n = -1$ -7

5. x^0; $x = -93$ 1

6. $-5b^{-4}$; $b = -2$ $\frac{-5}{16}$

Simplify. Use positive exponents only. Assume that no variable is equal to zero.

7. 4^{-3} $\frac{1}{64}$

8. $5m^{-4}$ $\frac{5}{m^4}$

9. $-3a^0 b$ $-3b$

10. $2m^{-5}n^3$ $\frac{2n^3}{m^5}$

11. $m^{-3}m^{-8}$ $\frac{1}{m^{11}}$

12. $(-2x^{-3})(-3x^4)$ $6x$

13. $(6y^{-1})(-xy)$ $-6x$

14. $(-4b^0)(-2ab^{-3})$ $\frac{8a}{b^3}$

15. $\dfrac{8x}{4^0 x^{-4}}$ $8x^5$

16. $\dfrac{(7x)^0}{4x^0}$ $\frac{1}{4}$

17. $\dfrac{5a^{-2}}{15a^3}$ $\frac{1}{3a^5}$

18. $\dfrac{8x^{-3}y^4}{-2xy^{-5}}$ $\frac{-4y^9}{x^4}$

19. $\left(\dfrac{1}{3}\right)^5$ $\frac{1}{243}$

20. $\left(-\dfrac{1}{4}\right)^2$ $\frac{1}{16}$

21. $\left(-\dfrac{1}{2}\right)^{-3}$ -8

22. $\left(\dfrac{3}{4}\right)^{-2}$ $\frac{16}{9}$

23. $\dfrac{-16a^{14}b^{-6}}{-8a^{-10}b^2}$ $\frac{2a^{24}}{b^8}$

24. $\dfrac{-6m^{-4}n^3}{2m^{-4}n^{-3}\cdot 5a^0}$ $\frac{-3n^6}{5}$

25. $\dfrac{8c^{-7}d^{10}}{12c^{-4}d^{-9}}$ $\frac{2d^{19}}{3c^3}$

26. $\left(\dfrac{6a^2b^3}{-2ab^{-4}}\right)^{-2}$ $\frac{1}{9a^2b^{14}}$

Solve each of the following for n.

27. $x^{-3n-6} = (x^3)^{n-6}$ 2

28. $8^{-n+6}\cdot(8^2)^{-n} = (8^{4n})^{-1}$ -6

29. $\left(\dfrac{a}{4}\right)^{-n}\cdot\left(\dfrac{a}{2}\right)^{2n} = a^0$ $(a\neq 0)$ 0

30. $x^{5n}\cdot x^{-2n} = \dfrac{1}{x^{-2}}$ $(x\neq 0)$ $\frac{2}{3}$

31. $5^{6n-1}\cdot 5^{n+4} = (-5)^{2n}$ $-\frac{3}{5}$

32. $\dfrac{1}{x^{3n}} = x^{2n}\cdot x^{-10}$ $(x\neq 0)$ 2

33. Show that $\left(\dfrac{a}{b}\right)^m = \dfrac{a^m}{b^m}$, $a\neq 0$, $b\neq 0$. (HINT: Rewrite $\left(\dfrac{a}{b}\right)^m$ as $(ab^{-1})^m$.)

34. Explain why $x^{-1} > 1$ if $0 < x < 1$.

Mixed Review

1. Simplify $8a - (-6 - 5a) - 2a$ **2.10** $11a + 6$

Evaluate. **1.3, 2.6, 2.8**

2. $(3m)^3$ for $m = \frac{1}{2}$ $3\frac{3}{8}$

3. $\frac{1}{3}x + 6x^2$ for $x = -6$ 214

4. $(-a)^3$ for $a = -6$ 216

5. $-m + 3n$ for $m = -5$ and $n = -2$ -1

Solve. **3.3, 3.5**

6. $\frac{x}{4} - 1 = 8$ 36

7. $2x - (-2 + x) = 6$ 4

◼◼FOLLOW UP

Guided Practice

Classroom Exercises 1–14
Try This   all

Independent Practice

A Ex. 1–22, **B** Ex. 23–26, **C** Ex. 27–34
Basic: WE 1–21 odd, 34
Average: WE 5–25 odd, 34
Above Average: WE 13–33 odd, 34

Additional Answers

Written Exercises

33. $\left(\dfrac{a^m}{b}\right) = (ab^{-1})^m$ (Def Neg Exp)

$= a^m(b^{-1})^m$ (Pow of a Prod Prop)

$= a^m b^{-m}$ (Pow of a Prod Prop)

$= \dfrac{a^m}{b^m}$ (Def Neg Exp)

34. $x^{-1} = \dfrac{1}{x}$. If x is between 0 and 1, x^{-1} gets smaller and closer to 1 as x gets closer to 1. As x gets closer to 0, x^{-1} gets larger and larger. Thus, x^{-1} is always greater than 1.

Enrichment

Tell students that an equation of the form $y = 2^x$ is an exponential equation, and that the equation can be graphed as a smooth curve in the coordinate plane. Then have them complete these exercises.

1. Find the values for y when $x = 0, 1, 2, 3,$ and 4.   1, 2, 4, 8, 16

2. Find the values for y when $x = -1, -2, -3,$ and -4. $\frac{1}{2}, \frac{1}{4}, \frac{1}{8}, \frac{1}{16}$

3. For what values of x is $y > 0$? $x > 0$

4. For what values of x is $y < 1$? $x < 0$

5. For what values of x is $y < 0$? No values

6. Choose the best way to complete this sentence.
For $y = 2^x$, as x increases in value:
a. y increases at an even rate.
b. y decreases at an even rate.
c. y increases more and more rapidly.   c

Teaching Resources

Application Worksheet 6
Quick Quizzes 45
Reteaching and Practice
   Worksheets 45
Transparency 16

◼◼◼ GETTING STARTED

Prerequisite Quiz

Multiply.

1. 0.594 × 10   5.94
2. 0.047 × 100   4.7
3. 0.0065 × 1,000   6.5
4. 0.0007 × 10,000   7

Divide.

5. 186.94 ÷ 100   1.8694
6. 78,297 ÷ 10,000   7.8297
7. 600,000 ÷ 100,000   6

Motivator

Ask students to give examples of very large
distances (distance to the sun and other
stars) and very small measurements
(diameter of an atom). Then ask them to
determine whether these pairs of numbers
are equivalent.

1. 95,672 and 9.5672×10^4
2. 956.72 and 9.5672×10^2
3. 0.008631 and 8.631×10^{-3}
4. 0.00008631 and 8.631×10^{-5}
   All are equivalent.

6.5 Scientific Notation

Objective | To convert ordinary notation to scientific notation and vice versa

Scientific Notation is a compact way of writing very large and very
small numbers. It enables you to see the magnitude of a number at a
glance. The distance that light travels in one year is about
9,500,000,000,000 kilometers. In scientific notation this number is
written as 9.5×10^{12}, which the eye can grasp quickly. The exponent
12 is a positive integer. However, the exponents used in scientific notation may be positive, negative, or zero.

Definition

> A number is written in **scientific notation** when it is in the form
> $$a \times 10^n,$$
> where $1 \leq a < 10$ and n is an integer.

Study the table below. Notice that for numbers greater than 10, the
exponent is positive; for numbers between 0 and 1, the exponent is
negative.

Number	Decimal-Point Shift	Scientific Notation
430,000,000	left 8 places	4.3×10^8
517,65	left 2 places	5.1765×10^2
0,0000956	right 5 places	$\frac{9.56}{100,000} = 9.56 \times 10^{-5}$
0,0471	right 2 places	$\frac{4.71}{100} = 4.71 \times 10^{-2}$

A number such as 1.432, between 1 and 10, is equal to 1.432×10^0.
Also, $1 = 1 \times 10^0$ and $10 = 1 \times 10^1$.

EXAMPLE 1 Express each number in scientific notation.

   a. 8654.32   b. 0.00346

Solutions a. $8654.32 = 8_{\,}6\,5\,4_{,}3\,2 = 8.65432 \times 10^3$

   b. $0.00346 = 0_{,}0\,0\,3_{,}4\,6 = 3.46 \times 10^{-3}$

TRY THIS Express each number in scientific notation.

   1. 0.00002013 2.013×10^{-5} 2. 54179.82 5.417982×10^4

214   Chapter 6 Powers and Polynomials

Highlighting the Standards

Standards 4c, 4d: This lesson connects the
use of exponents to scientific notation. In the
Application, students apply this to the earth's
hydrosphere.

Additional Example 1

Express in scientific notation.

a. 60,400 6.04×10^4
b. 0.02 2.0×10^{-2}

To change a number from scientific notation to ordinary notation, perform the indicated multiplication by the power of 10, as in Example 2.

EXAMPLE 2   Express each number in ordinary notation.

a. 4.602×10^5 **b.** 3.76×10^{-6}

Solutions   **a.** 4.60200×10^5 **b.** 000003.76×10^{-6}

$= 460{,}200$ $= 0.00000376$

TRY THIS   Express each number in ordinary notation.

3. 8.0032×10^4 80,032   **4.** 6.415×10^{-3} 0.006415

A scientific calculator with an exponential shift key, $\boxed{\text{EE}}$ or $\boxed{\text{EXP}}$, can be used to perform multiplication and division of numbers written in scientific notation.

EXAMPLE 3   Simplify. Express the answer in ordinary notation.

a. $(5.2 \times 10^2)(8.6 \times 10^4)$ **b.** $(6.1 \times 10^{-2}) \div (4.5 \times 10^{-1})$

Solution   **a.** $5.2 \boxed{\text{EE}} 2 \boxed{\times} 8.6 \boxed{\text{EE}} 4 = 4.472 \quad 07 \longleftarrow 4.472 \times 10^7$
In ordinary notation, $4.472 \times 10^7 = 44{,}720{,}000.$

b. $6.1 \boxed{\text{EE}} \boxed{+/_{-}} 2 \boxed{\div} 4.5 \boxed{\text{EE}} \boxed{+/_{-}} 1 = 1.3556 \quad -01$
In ordinary notation, $1.3556 \times 10^{-1} = 0.13556.$

To perform computations involving very large or very small numbers, first express the numbers in scientific notation. Then use a calculator.

$$\frac{(48{,}200{,}000)(0.0042)}{0.0000012} = \frac{(4.82 \times 10^7)(4.2 \times 10^{-3})}{1.2 \times 10^{-6}}, \text{ or}$$

$4.82 \boxed{\text{EE}} 7 \boxed{\times} 4.2 \boxed{\text{EE}} \boxed{+/_{-}} 3 \boxed{\div} 1.2 \boxed{\text{EE}} \boxed{+/_{-}} 6 \boxed{=} \qquad 1.687 \quad 11$

So, the result is 1.687×10^{11}, or 168,700,000,000.

Classroom Exercises

Express each number in scientific notation.

1. 463 4.63×10^2 **2.** 0.0034 3.4×10^{-3} **3.** 18×10^5 1.8×10^6 **4.** 0.513×10^{-5}
5.13×10^{-6}

Express each number in ordinary notation.

5. 5.3×10^6 **6.** 7.24×10^{-4} **7.** 546×10^{-4} **8.** 0.025×10^6
5,300,000   0.000724   0.0546   25,000

Lesson Note

Be sure that students understand that, in scientific notation, the first factor is a number between 1 and 10, and that the exponent of 10, the second factor, indicates how many places the decimal point moves when changing from one notation to the other. You may wish to decide which problems and how many to assign depending on whether scientific calculators are available for student use.

Math Connections

Science: A *light year* is the distance that light travels in one year. It is known that light travels at approximately 1.86×10^5 miles per hour or 3×10^5 kilometers per hour.

Critical Thinking Questions

Synthesis: Large numbers are constantly used in the news when referring to the state or national budget, but few people really comprehend the size of these numbers. Ask students to give examples of ways to explain the concept of a million. How much is it?   Possible answer: it takes a million millimeters (about the thickness of a dime) to equal 1 kilometer (about $\frac{5}{8}$ of a mile)

Common Error Analysis

Error: Some students make the mistake of thinking that a negative exponent indicates a negative quantity.

Remind them of the definition of b^{-n}.

Additional Example 2

Express each number in ordinary notation.

a. 6.95×10^6 6,950,000
b. 5.102×10^{-8} 0.00000005102

Additional Example 3

Simplify. Express answers in scientific notation.

a. $(1.86 \times 10^5)(3.6 \times 10^3)$ 6.696×10^8
b. $(1.75 \times 10^3) \div (2.5 \times 10^{-2})$ 7×10^4

Checkpoint

Express in scientific notation.

1. 4679.5 4.6795×10^3
2. 0.00824 8.24×10^{-3}
3. 1,350,000 1.35×10^6
4. The number of seconds in a week
      6.048×10^5

Express in ordinary notation.

5. 0.078×10^8 7,800,000
6. 0.67×10^{-5} 0.0000067

Closure

Ask students to explain in their own words how to write a very large and a very small number in scientific notation. Ask them to explain the advantages of using scientific notation and to give examples of situations where this type of notation is used.
Answers will vary.

■■■ FOLLOW UP

Guided Practice

Classroom Exercises 1–8
Try This   all

Independent Practice

A Ex. 1–20, **B** Ex. 21–24, **C** Ex. 25–30

Basic: WE 1–20 all, Application

Average: WE 5–24 all, Application

Above Average: WE 1–19 odd, 21–30 all, Application

Written Exercises

Express each number or measure in scientific notation.

1. 5,620,000
2. 0.0301 3.01×10^{-2}
3. 5.002 5.002×10^0
4. 6,300 6.3×10^3
5. 0.0000067
6. 8 8.0×10^0
7. 91,420,000
8. 0.0000008
9. the speed of light (approximately 186,000 mi/s) 1.86×10^5 mi/s
10. the wavelength of red light (approximately 0.00000068 m) 6.8×10^{-7} m
11. the wavelength of a TV signal (approximately 77,250,000 m)
    7.725×10^7 1. 5.62×10^6
    5. 6.7×10^{-6}
12. the mass of one atom of oxygen (0.0000000000000000000000265 g)
    7. 9.142×10^7
    8. 8.0×10^{-7}
    12. 2.65×10^{-23}

Express each number in ordinary notation.

13. 3.1×10^{-4}
14. 5.62×10^7
15. 1×10^{-2} 0.01
16. 8.04×10^5 804,000
17. 1×10^3 1,000
18. 7.03×10^{-6}
19. 2×10^1 20
20. 4.6×10^{-3} 0.0046
13. 0.00031   14. 56,200,000   0.00000703

Simplify. Express the answer in ordinary notation.

21. $(4.6 \times 10^3)(7.9 \times 10^2)$ 3,634,000
22. $(6.1 \times 10^{-3})(8.6 \times 10^{-2})$ 0.0005246
23. $(8.8 \times 10^{-1}) \div (2.5 \times 10)$ 0.0352
24. $(2.25 \times 10^2) \div (9 \times 10^{-4})$ 250,000

Simplify. Express the answer in scientific notation.

25. $\dfrac{(216,000,000)(0.00000168)}{0.00000084}$ 4.32×10^8
26. $\dfrac{69,300,000}{(16,500)(0.0000015)}$ 2.8×10^9
27. $\dfrac{(0.525)(7,820)}{22.5}$ 1.8247×10^2
28. $\dfrac{(7,500,000)(0.05)}{150,000,000}$ 2.5×10^{-3}
29. $\dfrac{(9,060,000)(0.00447)}{0.00042}$ 9.6424×10^7
30. $\dfrac{(0.000212)(0.588)}{57,700,000}$ 2.1604×10^{-12}

Mixed Review

Evaluate for $x = -3$, $y = -1$, and $z = 4$.   2.5, 2.6

1. $-(x - y - z)$ 6
2. x^2yz -36
3. Write an algebraic expression for 5 less than the product of 3 and x.   4.1 $3x - 5$
4. Solve: $\frac{1}{2}x + \frac{1}{4} = -\frac{3}{5}$ 4.5 $x = -1\frac{7}{10}$
5. The degree measure of the vertex angle of an isosceles triangle is 20 less than twice the degree measure of each base angle of the triangle. What is the measure of each angle?   4.4   50°, 50°, 80°

Enrichment

Johann Kepler (1571–1630) was a scientist who believed in a harmonious universe where the planets followed patterns that could be described mathematically. Have students complete the table at the right by finding the quotient of the square of each planet's time, T, to revolve once around the sun, and the cube of its distance, D, from the sun.

Students should find the pattern that Kepler discovered.

$T^2 \div D^3$ has the same value for each planet; $\frac{T^2}{D^3} =$ approx 3×10^{-19}

Planet	Time (T) in seconds around sun	Distance (D) in meters from sun	$\dfrac{T^2}{D^3}$
Mercury	7.60×10^6	5.79×10^{10}	?
Venus	1.94×10^7	1.08×10^{11}	?
Earth	3.15×10^7	1.49×10^{11}	?
Mars	5.94×10^7	2.28×10^{11}	?
Jupiter	3.73×10^8	7.78×10^{11}	?
Saturn	9.35×10^8	1.43×10^{12}	?
Uranus	2.64×10^9	2.87×10^{12}	?
Neptune	5.22×10^9	4.50×10^{12}	?
Pluto	7.82×10^9	5.91×10^{12}	?

Application: The Earth's Hydrosphere

How many drops of water are there in the ocean?

You cannot know the exact answer, but you can get an idea of the size of the answer, or its *order of magnitude*. To estimate the number of drops of water in the hydrosphere, made up of the seas and the oceans, follow this plan.

1. Find the surface area of the earth in square meters (m^2).
2. Find the part of the surface area that is water in m^2.
3. Find the total volume of the water in the hydro-sphere in m^3.
4. Divide the total volume by the volume of a drop of water to find the number of drops of water.

The first three exercises below help to estimate the total volume of the water. Use a scientific calculator for all the exercises (refer to page 215). Round answers to the nearest whole number.

1. The surface area of a sphere is given by the formula, $A = 4\pi r^2$, where r, the average radius of the earth, is about 6,370,000 m. Find the surface area of the earth. Use 3.1416 or the calculator key for π. 5×10^{14} m^2

2. Since water covers approximately 70% of the earth, multiply your answer to Exercise 1 by 70% to find the surface area of the hydro-sphere. 4×10^{14} m^2

3. To find the total volume of the water in cubic meters, multiply your answer to Exercise 2 by 3,790 meters, the average depth of the earth's seas and oceans. 1×10^{18} m^3

4. The volume of a large drop of water is about 5.0×10^{-8} m^3 (or 50 billionths of a cubic meter). Divide your answer to Exercise 3 by this volume to approximate how many drops of water there are in the oceans. Give your answer in scientific notation using as many digits as your calculator will show.   About 2×10^{25}

5. The *order of magnitude* of an estimate is indicated by the power of 10 when the estimate is written in scientific notation.
   Complete: The order of magnitude of the number of drops of water in the hydrosphere is __?__. 10^{25}

◼◼ GETTING STARTED

Prerequisite Quiz

Simplify.

1. $3x + 2y - x - y$ $2x + y$
2. $4a + 3b + 2a - 2b$ $6a + b$
3. $6x - 7y - 4x + 1$ $2x - 7y + 1$
4. $-7a - 3b - a + b$ $-8a - 2b$
5. $-x - y + 5y - 2x$ $-3x + 4y$
6. $3c - d - c + 2d - 2c$ d

Motivator

Ask several students to give an example of an expression with like terms. Then ask them to explain how the Distributive Property can be used to simplify their expression. Answers will vary.

6.6 Simplifying Polynomials

Objectives   To classify polynomials according to the number of terms
To simplify a polynomial and determine its degree
To write a polynomial in descending order and in ascending order

The sculpture shown here has the shape of two joined cubes. Its twelve faces are squares. If x and y represent the lengths of the sides of the large and small cube, respectively, then the total surface area can be represented by the *polynomial* $6x^2 + 6y^2$.

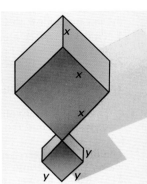

$6x^2$ and $6y^2$ are *monomials*.

A **polynomial** is a monomial or the sum of two or more monomials. The monomials in a polynomial are called *terms* of the polynomial. In the polynomial $5x^2 - 4x + 2$, the terms are $5x^2$, $-4x$, and 2.

Polynomials are classified according to their number of terms. A **binomial** has two terms and a **trinomial** has three terms.

Number of Terms	Name	Example
1	*mono*mial	$-5x^3$
2	*bino*mial	$a^2 - 2ab$
3	*trino*mial	$x^2 + 2x - 1$

Like terms are terms with the same variable(s) raised to the same power(s). A polynomial is *in simplest form* if no two of its terms are like terms.

EXAMPLE 1   Simplify.

a. $3y^2 - 5y - y^2$ b. $5x^2 - 7x + 2$

Solutions   a. $3y^2 - 5y - y^2$
= $3y^2 - y^2 - 5y$
= $2y^2 - 5y$

b. $5x^2 - 7x + 2$ has no like terms.
Thus, $5x^2 - 7x + 2$ is in simplest form.

TRY THIS   Simplify: 1. $10y + 4y^2 - 5y^2$ $10y - y^2$ 2. $3x^2 + 2x^2 - 1$ $5x^2 - 1$

Highlighting the Standards

Standards 6a, 7a: A three-dimensional sculpture illustrates a polynomial function (total surface area) to introduce students to the vocabulary of polynomials.

Additional Example 1

Simplify.

a. $6a - 8a^2 + a^3$ Cannot be simplified
b. $-4x^3 + 6x + 3x^3 - x$ $-x^3 + 5x$

The **degree of a monomial** is the sum of the exponents of all of its variables, as illustrated in the table below.

Monomial	-7	y	$6x^3y^2$	$-5a^4b^2c$
Degree	0	1	3 + 2, or 5	4 + 2 + 1, or 7

The **degree of a polynomial** is the same as the degree of its term with the greatest degree when the polynomial is in simplest form.

Polynomial	$8x^2 + 2x^3 + 1$			$9a^3b^2 + 6b^4$		$6 + 5c^2 - 3c^4 + c$			
Degree of terms	2	3	0	5	4	0	2	4	1
Degree of polynomial		3			5		4		

EXAMPLE 2 Simplify $3x^3 - 5x^2 + 6x^2 - 3x^3 - 7$. Then give the degree of the polynomial.

Solution
$$3x^3 - 5x^2 + 6x^2 - 3x^3 - 7 = (3x^3 - 3x^3) + (6x^2 - 5x^2) - 7$$
$$= 0 + x^2 - 7$$
$$= x^2 - 7$$

The term with the greatest degree is x^2. Its degree is 2. Thus, the degree of the given polynomial is 2.

TRY THIS
3. Simplify $4y^2 - 6y + 7y^4 - y^2 + y^4$. Give the degree of the polynomial. $8y^4 + 3y^2 - 6y; 4$

A polynomial in one variable is in **descending order** when the terms are in order from greatest to least degree. Thus, $7x^4 - 6x^3 + 4x + 2$ is in descending order.

EXAMPLE 3 Simplify $5x^4 + 7x - 1 - 3x^3 - 5x^4 - 3x + 2x^3$. Write the result in descending order.

Plan First group terms in descending order. Then combine like terms.

Solution
$$5x^4 + 7x - 1 - 3x^3 - 5x^4 - 3x + 2x^3$$
$$= (5x^4 - 5x^4) + (-3x^3 + 2x^3) + (7x - 3x) - 1$$
$$= -x^3 + 4x - 1 \text{ (descending order)}$$

TRY THIS
4. Simplify $8x^5 - 6x^2 + 4x^4 + 3x^2 - 2x^5 + x^4 + 2x - 7$. Write the result in descending order. $6x^5 + 5x^4 - 3x^2 + 2x - 7$

Additional Example 2

Simplify $-c^3 + 3c^2 + 2 + 4c^3 - c^2 - 9$. Then give the degree of the polynomial.
$3c^3 + 2c^2 - 7; 3$

Additional Example 3

Simplify $-5 + 3d^4 - 2d + 8 - 4d^4 + 6d^2$. Then write the result in descending order. What is the degree of the polynomial?
$-d^4 + 6d^2 - 2d + 3; 4$

Common Error Analysis

Students tend to rename $5x^3 + 4x^2$ as $9x^5$, confusing it with multiplication of monomials.

Point out that addition of monomials, like addition of fractions, requires <u>like terms</u>. $5x^3$ and $4x^2$ are <u>not</u> like terms since the variables do not have the same exponents. Thus, $5x^3$ and $4x^2$ <u>cannot</u> be combined unless a value for x is known.

Checkpoint

Simplify. Write the results in descending order.

1. $3y^2 - 2y + 1 - 5y^2 - y$ $-2y^2 - 3y + 1$
2. $-r + 6r^3 + 8r^2 - 5r^3 + r^2$ $r^3 + 9r^2 - r$
3. $4x^5 - 5x + 2x^4 - 3x^5 + x - 1$
   $x^5 + 2x^4 - 4x - 1$
4. $-a^4 + 6a^3 + a - a^3 + a - 3$
   $-a^4 + 5a^3 + 2a - 3$
5. $3a^2b - 9ab^2 + 2a^2b + 7ab - 3a^2b$
   $2a^2b - 9ab^2 + 7ab$

Closure

Ask students these questions:

1. How many terms can a polynomial have?   Any number
2. When is a polynomial in simplest form? When no two of its terms are like terms
3. Are $3y^2$ and $3y^3$ like terms? Explain. No; the exponents are not the same.
4. How do you simplify a polynomial? Write terms in descending order and combine like terms.

Focus on Reading

To complete each sentence, write a letter from the column at the right.

1. $7 - 3x^2$ is a(n) $\underline{\ ?\ }$. b
2. The degree of the polynomial $6x^3 - 2x + 1$ is $\underline{\ ?\ }$. g
3. In $7x^3 - 2x - 1$, -1 is a(n) $\underline{\ ?\ }$. h
4. In $4x^2 - 3x + 1$, 2 is a(n) $\underline{\ ?\ }$. i
5. The degree of the monomial -1 is $\underline{\ ?\ }$. d
6. $9y^2 - 2y + 3$ is a(n) $\underline{\ ?\ }$. c
7. In the polynomial $3c^3 - 2c + 1$, c is the only $\underline{\ ?\ }$. j
8. $-7x^3y^4$ is a(n) $\underline{\ ?\ }$. a
9. The degree of the polynomial $7 - 4x$ is $\underline{\ ?\ }$. e
10. The $\underline{\ ?\ }$ of a monomial is the sum of the exponents of all of its variables. k
11. The degree of the polynomial $7x^3 + 2x^2 - 7x$ is $\underline{\ ?\ }$. g

a. monomial
b. binomial
c. trinomial
d. 0
e. 1
f. 2
g. 3
h. constant
i. exponent
j. variable
k. degree

Classroom Exercises

Classify each polynomial as a monomial, a binomial, or a trinomial.

1. 4^2 Monomial   **2.** $6a^4 - 2a^3$ Binomial   **3.** $3y^2x^3$ Monomial   **4.** $7x^2 - 3x^3 + 1$
                                                                     Trinomial

Simplify. Then give the degree of the polynomial.

5. $4y^3 - 3y + 2y^2 - 5y^3 + y$
   $-y^3 + 2y^2 - 2y; 3$
6. $-3a + 2a^2 - 7a + 6a^3 - a^2$
   $6a^3 + a^2 - 10a; 3$

Give each polynomial in descending order and ascending order.

7. $2y^4 - 3y + 7y^3$ **8.** $3z^2 - z^4 + 7z$ **9.** $c^3 + 4 - c^2$ **10.** $-x + 2 - x^2$

Written Exercises

Tell whether the given expression is a polynomial.

1. $x - 2x^2 + 5$ Yes   **2.** $(0.5)^{-1}$ Yes   **3.** $n^2 + \frac{n}{2}$ Yes   **4.** $3y^2 - 3y^{-1}$ No

Classify each polynomial as a monomial, a binomial, or a trinomial.

5. $a^2 - b^2$ Binomial   **6.** $2a + b - 3c$ **7.** $-6xy^4$ Monomial   **8.** $-3x + 5y^2$
                                             Trinomial   Binomial

Simplify. Then give the degree of the polynomial. $2c^3 + 6c^2 - c + 5; 3$

9. $5y^3 - 3y^2 + 2y - 5y^3 + 1$ **10.** $3c + 2c^3 - 4c + 6c^2 + 5$
11. $8d^4 - 5d^3 + 4d^3 - 3d^4 - 5d^4$ **12.** $7 - 3y^4 + 2y^3 - y^4 + 5y^3 - 2y^4$
9. $-3y^2 + 2y + 1; 2$ $-d^3; 3$ $-6y^4 + 7y^3 + 7; 4$

13. $7x^2 - 2x^3 - 4x^2 - x^3 + 3x + 1$ $-3x^3 + 3x^2 + 3x + 1; 3$

14. $-5z^2 - 2z + 4z^2 + 3z + z^2$ $z; 1$

15. $3y^6 - 2y^4 + y^6 + 2y^4 + 3y + 7$ $4y^6 + 3y + 7; 6$

16. $x^4 - 2x + 5x^2 + 7x - x^4 + 2x^2$ $7x^2 + 5x; 2$

Simplify. Write the result in descending order and in ascending order.

17. $7a + 2a^2 + 3a - 6$

18. $-5c + 1 - 4c^2 + 3c^2$

19. $3 + 7x^2 - 2x + 4x^2 - 1$

20. $-r^2 + 3r - 5 + 2r - 7r^2$

21. $7x^4 - 3x^2 + 2x^3 - x^4 - 2x^3$

22. $6y^5 - 4y^3 + y^4 - 7y + 1$

23. $-3 + 2r^4 + 2r - 6r^4 + 2r + 1$

24. $x^3 - 8x - 4x^2 + 8x + 3x^2 + 9x^3$

25. $-5a + 4a^3 - 3a^2 + 2 + 8a^3 - 9a + 5a - 6$ $12a^3 - 3a^2 - 9a - 4; -4 - 9a - 3a^2 + 12a^3$

26. $-2 + 5x^2 - 4x - 3x^2 + x^2 + 7x + 2x^3 + 4$ $2x^3 + 3x^2 + 3x + 2; 2 + 3x + 3x^2 + 2x^3$

27. $y^4 - 3y^3 - 8y + y^2 - 7y^3 + 9y + y^2 - 5 + 4y^4$ $5y^4 - 10y^3 + 2y^2 + y - 5; -5 + y + 2y^2 - 10y^3 + 5y^4$

Simplify. (Exercises 28–32)

28. $7xy - 2yz + 3xz - 4yz + 4x - 7y$ $7xy - 6yz + 3xz + 4x - 7y$

29. $5b^2 - 6a^2 - 2b^2 + 4a^2 + 7a + 6b + 3a$ $-2a^2 + 3b^2 + 10a + 6b$

30. $2a^3b - 3ab^3 + 4a^3 - a^3b - 7ab^3$ $a^3b - 10ab^3 + 4a^3$

31. $(-3a)(4b^2)(2a) + (2a^2)(4b^2) - (5ab^2)(2ab)$ $-16a^2b^2 - 10a^2b^3$

32. $(2xy)(3x^2)(-x^2y) + (3y^2)(2x)(5x^4) + (x^2y)(-x^3y)$ $23x^5y^2$

33. Construct a polynomial of degree 3 with one variable and four terms.   Answers will vary.

34. Construct a trinomial of degree 5 with one variable.   Answers will vary.

35. Construct a monomial of degree 5 with two variables.   Answers will vary.

36. Solve $x(x^2 - 2x + 3) + 2x^2 + 3 = x(1 + x^2) + 6.$ $\frac{3}{2}$

37. Explain why "$x^3 + x^2 = x^5$ for all real values of x" is false.

Midchapter Review

Simplify. Use positive exponents only. No denominator equals zero.   *6.1, 6.2, 6.3, 6.4*

1. $2y^4 \cdot y$ $2y^5$

2. $(x^6)^2$ x^{12}

3. $-2x^0y^{-1}$ $\frac{-2}{y}$

4. $4cd^2(6cd)$ $24c^2d^3$

5. $(ab^2)^3$ a^3b^6

6. $(-2a^2bc)^4$

7. $6c^2d^3(-8bc)$

8. $0.5xy(-2x^2y^3)$

9. $\dfrac{y^3z}{yz^2}$ $\dfrac{y^2}{z}$

10. $\dfrac{-3ab^2}{6b^3c}$ $\dfrac{-a}{2bc}$

11. $\dfrac{7x^{-1}}{14x^2}$ $\dfrac{1}{2x^3}$

12. $5x\left(\dfrac{1}{5}x\right)^2$ $\dfrac{x^3}{5}$

13. $\dfrac{(5x^2)^3}{7x^4}$ $\dfrac{125x^2}{7}$

14. $\left(\dfrac{4}{5}\right)^{-2}$ $\dfrac{25}{16}$

15. $\left(\dfrac{3xy}{2y}\right)^{-1}$ $\dfrac{2}{3x}$

16. $\dfrac{15r^{-2}s^2}{5r^{-3}s^{-1}}$ $3rs^3$

Express each number in scientific notation.   *6.5*

17. 32,000,000 3.2×10^7

18. 0.000017 1.7×10^{-5}

19. 0.0065 6.5×10^{-3}

20. 405,000,000,000 4.05×10^{11}

6.6 Simplifying Polynomials   **221**

Enrichment

Operations $\boxed{}$, $\triangle$, and $\bigcirc$ are defined as follows:

$a \boxed{} b = 2a^2 + b$; $a \triangle b = ab$;
$a \bigcirc b = \dfrac{2b}{a^2}$

Have students use the definition to simplify each expression below.

1. $4 \boxed{} 6$ 38
2. $4 \triangle 6$ 24
3. $4 \bigcirc 6$ $\frac{3}{4}$
4. $(2 \triangle 3) \boxed{} 6$ 78
5. $(3 \bigcirc 18) \triangle 2$ 8

6. $d \boxed{} c$ $2d^2 + c$
7. $d \boxed{} d^2$ $3d^2$
8. $(e \bigcirc e^4) \triangle 2e^2$ $4e^4$

■■ GETTING STARTED

Prerequisite Quiz

Rewrite without parentheses.

1. $-(2a + 3b)$ $-2a - 3b$
2. $-(-4x + 5y)$ $4x - 5y$
3. $-(7r - 5s)$ $-7r + 5s$
4. $-(-7a - 4b - 9c)$ $7a + 4b + 9c$
5. $-(-3x + 2y - 4z)$ $3x - 2y + 4z$

Motivator

Ask students to name the operations they have applied to positive and negative numbers. $+$, $-$, $\times$, $\div$, evaluating powers Tell them that they will now apply these same operations to polynomials.

■■ TEACHING SUGGESTIONS

Lesson Note

In discussing the vertical form for adding and subtracting polynomials, remind students to align like terms and to leave a "hole" in the polynomial that does not have a term like one in the other polynomials being added or subtracted. Focus student's attention on the way in which Example 6 is worded. Illustrate with "subtract 5 from 12," which means $12 - 5$.

6.7 Addition and Subtraction of Polynomials

Objectives
To add and subtract polynomials
To write the indicated sum of polynomials in descending order.

The *perimeter* of a polygon is the sum of the lengths of its sides. To find the perimeter of this triangle, use vertical notation and add like terms.

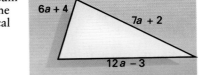

$$6a + 4$$
$$7a + 2$$
$$\underline{12a - 3}$$
$$25a + 3$$ The perimeter is $(25a + 3)$ units.

EXAMPLE 1   Add $7x^2 - 2x + 1$ and $4x^2 + 3x - 6$.

Plan   Write the sum. Then group like terms in descending order and simplify.

Solution $(7x^2 - 2x + 1) + (4x^2 + 3x - 6) = 7x^2 + 4x^2 - 2x + 3x + 1 - 6$
$$= 11x^2 + x - 5$$

EXAMPLE 2   Add $9y^3 - 7y + 8 + y^2$ and $7y - y^3 - 3y^4 - 3$.

Plan   When simplifying the sum, begin with the term with the greatest degree, $-3y^4$. Continue grouping like terms in descending order.

Solution $(9y^3 - 7y + 8 + y^2) + (7y - y^3 - 3y^4 - 3)$
$$-3y^4 + 9y^3 - y^3 + y^2 - 7y + 7y + 8 - 3$$
$$-3y^4 + 8y^3 + y^2 + 5$$

TRY THIS   1. Add $5x^2 + 7x^5 - 8$ and $-9x^2 + x - 2 - 6x^5$. $x^5 - 4x^2 + x - 10$

You can also use a vertical form to add polynomials.

EXAMPLE 3   Add. $(3x^5 - 2x^3 + 4x + 7) + (6x^3 - x^2 - 4x + 1)$

Solution
$$3x^5 - 2x^3 \qquad + 4x + 7$$
$$\underline{+ \; 6x^3 - x^2 - 4x + 1}$$
$$3x^5 + 4x^3 - x^2 + 0 \quad + 8, \text{ or } 3x^5 + 4x^3 - x^2 + 8$$

TRY THIS   2. Add. $(5x^4 + 6x^3 - 3x^2 + x) + (2x^4 - x^3 - x^2 - x)$
$$7x^4 + 5x^3 - 4x^2$$

Highlighting the Standards

Standards 4c, 4d: Several problems in this lesson connect geometric figures to the addition of polynomials. The Application shows a connection to compound interest.

Additional Example 1

Add $3y^2 + 5y - 2$ and $-4y^2 - y - 6$.
$-y^2 + 4y - 8$

Additional Example 2

Add $5a^5 - 6a^3 + 7a^2 - 2a$ and $3a^4 + 2a^3 - 6a + 1$.
$5a^5 + 3a^4 - 4a^3 + 7a^2 - 8a + 1$

Additional Example 3

Add $(6x^4 - 5x + 3x^3 - 1) + (5x^2 - 7x^3 + 4x - 9)$.
$6x^4 - 4x^3 + 5x^2 - x - 10$

Recall that $-a = -1 \cdot a$. This property is used in finding the opposite (additive inverse) of a polynomial.

EXAMPLE 4   Simplify $-(-5x^2 + 2x - 1)$.

Solution
$$-(-5x^2 + 2x - 1) = -1 \cdot (-5x^2 + 2x - 1) \longleftarrow -a = -1 \cdot a$$
$$= 5x^2 - 2x + 1 \longleftarrow \text{Distributive Property}$$

TRY THIS   **3.** Simplify $-(6x^3 - 4x^2 - 8x + 5)$. $\quad -6x^3 + 4x^2 + 8x - 5$

Example 5 illustrates that the opposite of a polynomial is found by *changing the sign of each term* of the polynomial. That is,

$$-(x - 2) = -x + 2,$$
$$-(y^2 - y + 3) = -y^2 + y - 3, \quad \text{and}$$
$$-(a - b + c) = -a + b - c.$$

EXAMPLE 5   Subtract. $(5a^2 - 3a + 6) - (2a^2 - 3a - 2)$

Plan   Use $a - b = a + (-b)$. (Definition of subtraction)
Add the opposite of $(2a^2 - 3a - 2)$ to $(5a^2 - 3a + 6)$.

Solution
$$(5a^2 - 3a + 6) - (2a^2 - 3a - 2) = (5a^2 - 3a + 6) + (-2a^2 + 3a + 2)$$
$$= \underbrace{5a^2 - 2a^2}_{} \underbrace{- 3a + 3a}_{} \underbrace{+ 6 + 2}_{}$$
$$= \quad 3a^2 \qquad + 0 \qquad + 8$$
$$= \quad 3a^2 + 8$$

TRY THIS   **4.** Subtract. $(9y^3 + 4y - 3) - (8y^3 - 2y + 5)$ $\quad y^3 + 6y - 8$

You can also use a vertical form in subtraction.

EXAMPLE 6   Subtract $0.2x^3 - 0.5x + 0.1$ from $0.4x^2 + 0.3x - 0.7$.

Plan   "Subtract a from b" means add $(-a)$ to b.
Therefore, find the opposite of $0.2x^3 - 0.5x + 0.1$.
Add it to $0.4x^2 + 0.3x - 0.7$.

Solution
$$\begin{array}{ll} -(0.2x^3 - 0.5x + 0.1) & \quad 0.4x^2 + 0.3x - 0.7 \\ = -0.2x^3 + 0.5x - 0.1 & \underline{-0.2x^3 \qquad + 0.5x - 0.1} \\ & -0.2x^3 + 0.4x^2 + 0.8x - 0.8 \end{array}$$

TRY THIS   **5.** Subtract $0.7x^4 - 0.3x^2 - 0.2$ from $0.2x^4 + 0.5x^2 - 0.1$.
$-0.5x^4 + 0.8x^2 + 0.1$

6.7 Addition and Subtraction of Polynomials   **223**

Math Connection

History: The French mathematician François Viète (1540–1603) was the first to use the same letter of the alphabet for different powers of a quantity. He would write $a^3 - 7a^2b + 2ab$ as "A cub − B 7 in A quad + B plano 2 in A."

Critical Thinking Questions

Application: Ask students to give the result when a polynomial is added to its additive inverse.   0   Have them give examples to illustrate the result.

Common Error Analysis

Error: When subtracting polynomials, students tend to change the sign of only the first term in the subtrahend, as follows:

$6x^2 - 5 - (4x^2 + 2) =$
$6x^2 - 5 - 4x^2 + 2$

Remind students to use the Distributive Property with -1 as follows:

$6x^2 - 5 - (4x^2 + 2) =$
$6x^2 - 5 - 1(4x^2 + 2) =$
$6x^2 - 5 - 4x^2 - 2 = 2x^2 - 7$

Additional Example 4
Simplify $-(-7a^2 + a - 5)$. $7a^2 - a + 5$

Additional Example 5
Subtract $(2c^2 - 6c + 5) - (5c^2 - 6c - 3)$.
$-3c^2 + 8$

Additional Example 6
Subtract $0.3y^3 - 0.7y^2 + 0.2y$ from $0.4y^3 + 0.6y - 0.1$.
$0.1y^3 + 0.7y^2 + 0.4y - 0.1$

Checkpoint

Add.

1. $-x^2 + 5x - 2$ and $9x^2 - 4x + 7$
   $8x^2 + x + 5$
2. $2y^3 - 3y + 8$ and $-7y^2 - 4y - 9$
   $2y^3 - 7y^2 - 7y - 1$

Subtract.

3. $(3x^2 - 7x + 2) - (7x^2 + 4x - 1)$
   $-4x^2 - 11x + 3$
4. $(5a^3 - 6a - 3) - (5a^3 - 4a^2 + 3)$
   $4a^2 - 6a - 6$
5. Subtract $-x^2 - 7x + 1$ from
   $2x^2 + 5$. $\quad 3x^2 + 7x + 4$

Closure

Ask students to sketch a rectangle and a triangle and label the lengths of the sides with polynomials in a manner similar to the triangle at the top of page 222. Have them find the perimeter of each figure and then subtract the perimeter of the triangle from the perimeter of the rectangle. Have them exchange papers with a classmate and check each other's answers.

Classroom Exercises

Add.

1. $\begin{array}{r} 5c^2 - 3c + 2 \\ -4c^2 - c - 7 \\ \hline c^2 - 4c - 5 \end{array}$

2. $\begin{array}{r} -x^3 + 2x^2 + 1 \\ -9x^2 - 2x + 3 \\ \hline -x^3 - 7x^2 - 2x + 4 \end{array}$

3. $\begin{array}{r} 2b^4 - 3b^2 + 5b \\ 6b^3 - 4b \\ \hline 2b^4 + 6b^3 - 3b^2 + b \end{array}$

Give the additive inverse of each polynomial.

4. $2x + 8 \quad -2x - 8$
5. $-y^3 - 6y + 1 \quad y^3 + 6y - 1$
6. $3d^4 - 7d^2 + d - 1$
   $-3d^4 + 7d^2 - d + 1$

Simplify.

7. $-(3x - 2) \quad -3x + 2$
8. $-(5x^4 - 7x^3 + 3x) \quad -5x^4 + 7x^3 - 3x$
9. $-(-6c^5 + 3c^3 + 1)$
   $6c^5 - 3c^3 - 1$
10. Subtract: $(3c^2 - 7c + 2) - (c^2 + 8c + 5) \quad 2c^2 - 15c - 3$

Written Exercises

Add the given polynomials.

1. $5x^2 - 7x$ and $3x^2 + 2x \quad 8x^2 - 5x$
2. $4y^2 + 6y + 1$ and $2y^2 + 3y - 2 \quad 6y^2 + 9y - 1$
3. $-4a^2 - 9a + 6$ and $3a^2 - 2a + 5$
   $-a^2 - 11a + 11$
4. $x^2 - 7x + 2$ and $4x^2 + 2x - 9$
   $5x^2 - 5x - 7$

Add.

5. $(-3b^2 - 6b + 2) + (-4b^2 - b - 7) \quad -7b^2 - 7b - 5$
6. $(6a^4 - 7a^3 + a) + (5a^3 - 3a^2 - 2a) \quad 6a^4 - 2a^3 - 3a^2 - a$
7. $(x^3 - 5x^4 + 2x^2 + x) + (-x^4 + 6x^2 - 5x + 2) \quad -6x^4 + x^3 + 8x^2 - 4x + 2$

Simplify.

8. $-(5c^2 - 3c) \quad -5c^2 + 3c$
9. $-(y^3 - 3y^2 + 1) \quad -y^3 + 3y^2 - 1$
10. $-(-8t^2 + 3t) \quad 8t^2 - 3t$
11. $-(2x^3 + 6x^2 + 3) \quad -2x^3 - 6x^2 - 3$

Perform the indicated operations.

12. $(6x - 4) - (2x + 7) \quad 4x - 11$
13. $(3y + 8) - (3y - 9) \quad 17$
14. $(7x^2 + 2x - 4) - (2x^2 - 2x + 6)$
15. $(5r^2 + 7r + 2) - (-5r^2 + 7r - 3) \quad 10r^2 + 5$
16. Subtract $x^2 - 3x + 2$ from $5x^2 + 4$.
17. Subtract $-x^2 - 4x$ from $-x^3 - x^2 - 9x$.
18. $\left(\frac{1}{2}x^2 + \frac{1}{4}x - \frac{1}{5}\right) + \left(-\frac{1}{4}x^2 - x + \frac{4}{5}\right)$
19. $\left(\frac{2}{3}y^3 + \frac{1}{6}y - \frac{1}{4}\right) + \left(\frac{1}{3}y^3 - \frac{5}{8}y^2 - \frac{5}{6}y + \frac{1}{2}\right)$
20. $0.5x^2 - 0.8x + 0.2) + (-0.9x^2 + 1.4x - 3.8) \quad -0.4x^2 + 0.6x - 3.6$
21. $(5.04x^3 - 6.2x^2 + 0.7) + (8.37x^3 - 4.9x^2 + 6.3x) \quad 13.41x^3 - 11.1x^2 + 6.3x + 0.7$
22. $(0.6y^3 - 5.3y - 6.1) - (1.9y^3 + 4.9y^2 - 7.8y) \quad -1.3y^3 - 4.9y^2 + 2.5y - 6.1$
23. $(8.5x^5 - 3.8x^3 + 2.7x) - (4.2x^4 + 3.8x^3 - 7.4x + 0.2)$
    $8.5x^5 - 4.2x^4 - 7.6x^3 + 10.1x - 0.2$

Express the perimeter of each figure as a polynomial in simplest form.

24.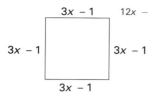

3x − 1   12x − 4

3x − 1   3x − 1

3x − 1

25.

5y − 3   19y − 2

4y + 2   4y

6y − 1

26. Subtract the sum of $5y^2 - 3y + 2$ and $7y^3 + 2y - 1$ from $8y^4 - 7y^2 + 6y - 1$. $8y^4 - 7y^3 - 12y^2 + 7y - 2$

27. Subtract $6x^2y + xy^2$ from $3x^2y - 8y^3$ and add the difference to $x^2y - 5xy^2$. $-8y^3 - 6xy^2 - 2x^2y$

Mixed Review

Simplify. *1.6, 2.2, 2.10, 6.1, 6.2, 6.3*

1. $-|7 - 9|$ -2
2. $3(4x + 8)$ $12x + 24$
3. $-7(3a + 2b - 1)$ $-21a - 14b + 7$
4. $-1(-5x - 2y + 3z)$
5. $-y - (6 + 3y)$
6. $-(2x + 4) - 5(x - 2)$
7. $7c - (2 + 6c - 5)$ $c + 3$
8. $x^2 \cdot x^4$ x^6
9. $2y(-3y^2)$ $-6y^3$ $-4y - 6$ $-7x + 6$
10. $-x(-4xy^3)$ $4x^2y^3$
11. $(3a^2b)^3$ $27a^6b^3$
12. $\dfrac{4x^2y}{12xy^5}$ $\dfrac{x}{3y^4}$

4. $5x + 2y - 3z$

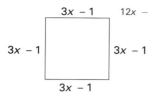

Application: *Compound Interest*

In January, Jackie started a savings account into which she deposited $100. Her bank pays 8% yearly interest compounded monthly according to this formula: $A = p(1 + r)^n$. Use your scientific calculator with these exercises to find how much money Jackie will have in her account at the end of the year.

1. To find r, divide the yearly rate of interest by the number of interest periods (12). Write your answer with as many places as your calculator will show. 0.0066666

2. Find A, the amount of the new balance, when p, the principal or amount deposited, is $100 and n, the number of interest periods per year, is 12. Here are the calculator steps:

    100 ⎡×⎤ ⎡(⎤ 1.0066666 ⎡y^x⎤ 12 ⎡)⎤ ⎡=⎤ 108.29986

    Round the answer to the nearest cent. How much money will Jackie have in her account at the end of the year? $108.30

3. How much money would she have if the bank only paid simple interest of 8% per year? $108.00

6.7 Addition and Subtraction of Polynomials   **225**

FOLLOW UP

Guided Practice

Classroom Exercises 1–10
Try This   all

Independent Practice

A Ex. 1–17, **B** Ex. 18–25, **C** Ex. 26–27

Basic: WE 1–16 all, Application

Average: WE 1–6 all, 7–25 odd, Application

Above Average: WE 1–23 odd, 24–27 all, Application

Additional Answers

Written Exercises

14. $5x^2 + 4x - 10$
16. $4x^2 + 3x + 2$
17. $-x^3 - 5x$
18. $\frac{1}{4}x^2 - \frac{3}{4}x + \frac{3}{5}$
19. $y^3 - \frac{5}{8}y^2 - \frac{2}{3}y + \frac{1}{4}$

Enrichment

The method of writing expressions and equations containing polynomials has changed over the last few centuries. Assign interested students to do some research in the library and to write a short essay or design a poster illustrating an older method of writing algebraic expressions.

▄▄▄ GETTING STARTED

Prerequisite Quiz

Simplify.

1. $-3a(2a)$ $-6a^2$
2. $-5y^2(-7y^2)$ $35y^4$
3. $(-4a^2)2a$ $-8a^3$
4. $(-d^3) - d$ d^4
5. $4b(2b^5)$ $8b^6$
6. $(-6r^4)2r$ $-12r^5$

Motivator

Ask students whether $2x$ is a monomial or a binomial.   Monomial   Ask them whether $3x^2 + 5$ is a monomial or a binomial.   Binomial   Ask them what property could be used to find the product, $2x(3x^2 + 5)$.   Distributive Property   Ask them to complete this product.

$2x(3x^2 + 5) = (2x)(3x^2) + (2x)(?)$ 5
$= 6x^3 + ?$ 10x

▄▄▄ TEACHING SUGGESTIONS

Lesson Note

Emphasize that it is the Distributive Property that is used when students multiply a monomial and a polynomial. *Each* term in the polynomial must be multiplied by the monomial.

6.8 Multiplying a Polynomial by a Monomial

Objective   To multiply a polynomial by a monomial

The area of a rectangle is the product of its width and length. To find the product of $(a + 3)$ and $2a$, begin by multiplying a by $2a$.

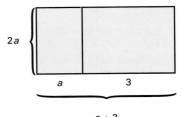

$$\begin{array}{ll} a + 3 & \text{Length} \\ \underline{2a} & \text{Width} \\ 2a^2 + 6a & \text{Area} \end{array}$$

Thus, the area of the rectangle is $(2a^2 + 6a)$ square units.

EXAMPLE 1   Multiply $3x(4x - 1)$.

Solution $3x(4x - 1) = (3x)(4x) + (3x)(-1) = 12x^2 - 3x$

EXAMPLE 2   Multiply $(3y^2 - y + 6)(-2y^2)$.

Solution $(3y^2 - y + 6)(-2y^2) = (3y^2)(-2y^2) + (-1y)(-2y^2) + (6)(-2y^2)$
$= -6y^4 + 2y^3 - 12y^2$

TRY THIS   Multiply: **1.** $9y^2(2y - 7)$ $18y^3 - 63y^2$ **2.** $y^4(-y - 2)$ $-y^5 - 2y^4$

Sometimes you may have to apply the Distributive Property within an expression.

EXAMPLE 3   Simplify $a^4 - 6a^2 - 4a^2(3a^2 - 7a + 2)$.

Solution
$a^4 - 6a^2 - 4a^2(3a^2 - 7a + 2)$
$= a^4 - 6a^2 + (-4a^2)(3a^2) + (-4a^2)(-7a) + (-4a^2)(2)$
$= a^4 - 6a^2 - 12a^4 + 28a^3 - 8a^2$
$= a^4 - 12a^4 + 28a^3 - 6a^2 - 8a^2 = -11a^4 + 28a^3 - 14a^2$

TRY THIS   **3.** Simplify $a^3 + 2a^5 - 5a^4(4a - 7a^2 - 6)$. $35a^6 - 18a^5 + 30a^4 + a^3$

Highlighting the Standards

Standards 4c, 7a, 2c: Again in this lesson, geometric figures (in two and three dimensions) are used to illustrate polynomials. Exercise 36 asks students to explain a mathematical idea in writing.

Additional Example 1

Multiply $-5y(2y - 8)$. $-10y^2 + 40y$

Additional Example 2

Multiply $(6x^2 - 3x + 2)(-x^2)$.
$-6x^4 + 3x^3 - 2x^2$

Classroom Exercises

Multiply.

1. $3a(2a + 1)$ $6a^2 + 3a$

2. $(4x - 2)6x$ $24x^2 - 12x$

3. $5y(-3y - 9)$
$-15y^2 - 45y$

4. $-x(4x^2 - 5x - 2)$
$-4x^3 + 5x^2 + 2x$

5. $(5t^2 - 3t + 4)2t$
$10t^3 - 6t^2 + 8t$

6. $-6z^2(-3z^2 + 2z - 1)$
$18z^4 - 12z^3 + 6z^2$

Simplify.

7. $5y^2 + 2y(y - 6)$
$7y^2 - 12y$

8. $3x^3 - (4x^2 + 3)5x$
$-17x^3 - 15x$

9. $2(-5x^2 + 7) - 8x^2$
$-18x^2 + 14$

Written Exercises

Multiply.

1. $5y(3y - 2)$ $15y^2 - 10y$

2. $c(-4c + 2)$ $-4c^2 + 2c$

3. $-r(-3r - 7)$ $3r^2 + 7r$

4. $3a(a^2 - 2a + 1)$

5. $-2x(x^2 + 4x - 1)$

6. $4c(2c^2 - c + 3)$

7. $x^2(7x^2 - 4x + 1)$

8. $(-3t^2 + 4t - 7)(-2t)$

9. $-a^2(3a^2 - a + 1)$

10. $4x(5x^2 + 3x - 2)$

11. $-2y^2(4y^3 - 3y^2 - y)$

12. $(-2x^2 + 3x - 1)(-3x^3)$

13. $2b^3(4b^2 - 3b + 1)$
$8b^5 - 6b^4 + 2b^3$

14. $-3b^4(2b^2 - 6b + 2)$
$-6b^6 + 18b^5 - 6b^4$

15. $(-y^2 - 2y - 1)4y^3$
$-4y^5 - 8y^4 - 4y^3$

Express the area of each figure as a polynomial in simplest form.

16.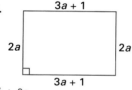

$6a^2 + 2a$

17.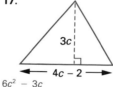

$6c^2 - 3c$

18.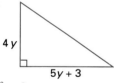

$10y^2 + 6y$

19.

$16c^2$

Simplify.

20. $-3y^2 - 4y + 6 + 2(y^2 + 7y - 1)$
$-y^2 + 10y + 4$

21. $-a^2 + 5a - 4 - 3(2a^2 - a - 5)$
$-7a^2 + 8a + 11$

22. $3x^2 + 2x(x - 5)$ $5x^2 - 10x$

23. $5y^2 + 2(y - 3) - 6y$ $5y^2 - 4y - 6$

24. $4a^2 + 3a + (a + 7)2a$ $6a^2 + 17a$

25. $x^2 + 5x + 4 - 3x(-x + 1)$ $4x^2 + 2x + 4$

26. $x^3 + (5x^2 - 6x - 1)x + 7x^2$
$6x^3 + x^2 - x$

27. $3y^3 - 2y(4y^2 - 3y + 7) - 6y^2$
$-5y^3 - 14y$

Additional Example 3

Simplify $7c^4 - 6c^3 + (c^2 - 2c + 1)3c$.
$7c^4 - 3c^3 - 6c^2 + 3c$

228

Closure

Ask students to find the errors in these products.

1. $r(r^2 - 9) = r^3 - 9$ $r^3 - 9r$
2. $7y(-6y - 9) = -42y^2 + 9y$
     $-42y^2 - 63y$
3. $5 - 2(8x + 3) = 5 - 16x - 6$
     $-16x - 1$
4. $x^3 + (2x^2 - 1)2x = x^3 + 4x^2 - 2$
     $5x^3 - 2x$.

■ FOLLOW UP

Guided Practice

Classroom Exercises 1–9
Try This   all

Independent Practice

A Ex. 1–27, **B** Ex. 28–35, **C** Ex. 36–39

Basic: WE 1–27 odd, 36
Average: WE 9–35 odd, 36
Above Average: WE 13–39 odd, 36

Additional Answers

Written Exercises

4. $3a^3 - 6a^2 + 3a$
5. $-2x^3 - 8x^2 + 2x$
6. $8c^3 - 4c^2 + 12c$
7. $7x^4 - 4x^3 + x^2$
8. $6t^3 - 8t^2 + 14t$
9. $-3a^4 + a^3 - a^2$
10. $20x^3 + 12x^2 - 8x$
11. $-8y^5 + 6y^4 + 2y^3$
12. $6x^5 - 9x^4 + 3x^3$
30. $-6x^4y^2 + 2x^3y^3 - 8x^2y^4$
32. $-6c^3d^3 + 4c^3d^4 - 2c^2d^4$

See page 235 for the answers to Ex. 33, 36.

Multiply. $5x^3y + 10x^2y^2 + 5xy^3$

28. $5xy(x^2 + 2xy + y^2)$
29. $-3ab(a^2 - 2ab + b^2)$ $-3a^3b + 6a^2b^2 - 3ab^3$
30. $-2x^2y^2(3x^2 - xy + 4y^2)$
31. $-r^2s(r^2 + 2rs - 5s^2)$ $-r^4s - 2r^3s^2 + 5r^2s^3$
32. $-2cd^2(3c^2d - 2c^2d^2 + cd^2)$
33. $(-2a^3 - 5ab + b^3)(3a^2b^2)$
34. $(a^3 + 2a^2b - ab^2 + 3b^3)(4ab)$
     $4a^4b + 8a^3b^2 - 4a^2b^3 + 12ab^4$
35. $-2xy(x^3 + 3x^2y - 5xy^2 + 4y^3)$
     $-2x^4y - 6x^3y^2 + 10x^2y^3 - 8xy^4$

36. Explain in writing why $(x - 2)(x + 4) = (x - 2)x + (x - 2)4$, for all real numbers x is true.

37. Express in simplest polynomial form the total surface area of the rectangular solid below. $38x^2 + 28x$

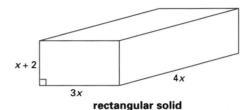

$x + 2$
$3x$
$4x$
rectangular solid

38. Add the product of m^3n^2 and $(5m^2 + 3mn + n^2)$ to the product of m^3n^2 and $(m^2 - 3mn + 2n^2)$.
     $6m^5n^2 + 3m^3n^4$

39. Add the product of $3ab$ and $(a^2 - 5ab + 4b^2)$ to the product of $-2ab$ and $(6a^2 - 7ab - 8b^2)$.
     $-9a^3b - a^2b^2 + 28ab^3$

Mixed Review

1. Evaluate $18a^3$ for $a = -\frac{2}{3}$.   **2.6** $-5\frac{1}{3}$

Solve.   **4.5, 5.7**

2. $\frac{2}{3}y - \frac{1}{6} = \frac{5}{6}$ $1\frac{1}{2}$ **3.** $|2a - 4| < 8$ $-2 < a < 6$

4. Use the formula $p = 2l + 2w$ to find p if $l = 7.5$ in. and $w = 4.5$ in.   **1.4** 24 in

5. 6 is what percent of 24?   **4.7** 25%

6. 12 is 75% of what number?   **4.7** 16

7. For the school play, the number of student tickets sold was 300 more than the number of adult tickets. Find the number of tickets of each type sold if the total number of tickets was 1,500.   **3.6**
Adult: 600; student: 900

Enrichment

Have the students find the total surface area and volume of a rectangular solid whose dimensions are x, $2x$, and $3x + 1$. Then have them double each of these dimensions and find the volume and surface area of the new solid. Ask what happens to the total surface area when each dimension is doubled.   4 times as great   What happens to the volume?   8 times as great

6.9 Multiplying Binomials

Objectives
To multiply two binomials
To multiply a trinomial by a binomial

A patio that measures 3 m by 5 m has an area of 3×5, or 15 m². Suppose the patio is to be expanded in two directions by a certain amount, say by x meters. Then the new dimensions of the patio are $(x + 3)$ m and $(x + 5)$ m. The area of the expanded patio, in square meters, can be represented by the product of two binomials, $(x + 3)(x + 5)$.

The Distributive Property is used twice in multiplying $(x + 3)$ by $(x + 5)$. In the horizontal multiplication below, $(x + 3)$ is first treated as a single number.

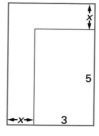

Horizontal Multiplication

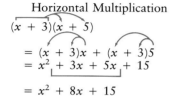

$(x + 3)(x + 5)$

$= (x + 3)x + (x + 3)5$
$= x^2 + 3x + 5x + 15$

$= x^2 + 8x + 15$

or

Vertical Multiplication

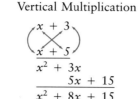

$x + 3$

$x + 5$

$x^2 + 3x$

$\underline{\qquad 5x + 15}$

$x^2 + 8x + 15$

Thus, the area, in square meters, is represented by $x^2 + 8x + 15$.

Two binomials can be multiplied mentally by using the FOIL method.

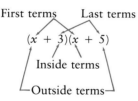

First terms   Last terms

$(x + 3)(x + 5)$

Inside terms

Outside terms

First terms: x and x

Outside terms: x and 5

Inside terms: 3 and x

Last terms: 3 and 5

Remember these four steps when you multiply two binomials.

First terms	Outside terms	Inside terms	Last terms
$x \cdot x$	$x \cdot 5$	$3 \cdot x$	$3 \cdot 5$
x^2	$5x$	$3x$	15
F	O	I	L

Then combine like terms: $x^2 + 8x + 15$.

Teaching Resources

Manipulative Worksheet 6
Problem Solving Worksheet 6
Quick Quizzes 49
Reteaching and Practice Worksheets 49

■ GETTING STARTED

Prerequisite Quiz

Simplify.

1. $x \cdot x + x \cdot 5 + 2 \cdot x + 2 \cdot 5$
   $x^2 + 7x + 10$
2. $y \cdot y + y \cdot (-3) + 7 \cdot y + 7(-3)$
   $y^2 + 4y - 21$
3. $r \cdot r + r(-4) - 6 \cdot r - 4(-6)$
   $r^2 - 10r + 24$
4. $3a(2a) + 3a(-3) - 8(2a) - 8(-3)$
   $6a^2 - 25a + 24$
5. $x(2x) + x \cdot 5 - 5(2x) - 5 \cdot 5$
   $2x^2 - 5x - 25$

Motivator

Show an example of multiplying two binomials by using the Distributive Property. For example,

$(x + 3)(x + 5) = (x + 3)x + (x + 3)5 = (x^2 + 3x) + (5x + 15) = x^2 + 8x + 15.$

Then tell them that they will learn a shortcut method that will allow them to find such products more efficiently and, with practice, by using mental computation.

Highlighting the Standards

Standards 4b, 4c: Students see three ways to multiply binomials, and geometric figures provide examples for this multiplication.

229

Lesson Note

The figure at the top of page 229 can be used to illustrate the result of multiplying the two binomials, $x + 3$ and $x + 5$, as shown below.

The areas of the four regions are x^2, $3 \cdot x$, $5 \cdot x$, and 15. The area of the large rectangle $(x + 3)(x + 5)$ is the same as the sum of the areas of the four regions.

$$(x + 3)(x + 5) = x^2 + 3x + 5x + 15$$
$$= x^2 + 8x + 15$$

Math Connections

Previous Math: The lesson continues the study of multiplying polynomials from the previous lesson and applies the FOIL method as well as the Distributive Property to finding binomial products.

Critical Thinking Questions

Synthesis: Have students explain how to apply the FOIL to finding products such as these.

1. 21^2 **2.** 35^2 **3.** 49^2

$(20 + 1)(20 + 1) = 400 + 40 + 1 = 441$
$(30 + 5)(30 + 5) = 900 + 300 + 25 = 1,225$
$(50 - 1)(50 - 1) = 2500 - 100 + 1 = 2401$

EXAMPLE 1   Multiply: $(2x + 1)(7x - 4)$

Solution
$$\qquad\qquad\qquad\quad F \qquad O \qquad I \qquad L$$
$$(2x + 1)(7x - 4) = 2x(7x) + 2x(-4) + 1(7x) + 1(-4)$$
$$= 14x^2 - 8x + 7x - 4$$
$$= 14x^2 - x - 4$$

TRY THIS   Multiply: **1.** $(5y - 2)(2y + 4)$ **2.** $(a - 3)(a + 4)$
$\qquad\qquad\qquad\qquad\qquad 10y^2 + 16y - 8 \qquad\qquad\qquad\qquad a^2 + a - 12$
Some binomials contain more than one variable.

EXAMPLE 2   Multiply: $(5a - b)(-2a + b)$

Solution
$$\qquad\qquad\qquad\qquad F \qquad\quad O \qquad\quad I \qquad\quad L$$
$$(5a - b)(-2a + b) = 5a(-2a) + 5a(b) - b(-2a) - b(b)$$
$$= -10a^2 + 5ab + 2ab - b^2$$
$$= -10a^2 + 7ab - b^2$$

TRY THIS   Multiply: **3.** $(6x - y)(-x - y)$ $-6x^2 - 5xy + y^2$

Focus on Reading

Use one of the expressions at the right below to complete each product.

1. $(x + 1)(x + 9) = x^2 + \underline{\ ?\ } + 9$ c
2. $(x - 3)(x + 4) = x^2 + x + \underline{\ ?\ }$ e
3. $(x - 2)(x - 8) = x^2 + \underline{\ ?\ } + 16$ d
4. $(3x - 7)(-2x + 5) = \underline{\ ?\ } + 29x - 35$ g

a. $-12x$ **e.** -12
b. $6x^2$ **f.** $9x$
c. $10x$ **g.** $-6x^2$
d. $-10x$

Classroom Exercises

Give the missing term in each product.

1. $(y + 3)(y + 2) = y^2 + 5y + \underline{\ ?\ }$ 6   **2.** $(x - 4)(x + 7) = x^2 + 3x - \underline{\ ?\ }$ 28
3. $(c - 9)(c - 1) = c^2 - 10c + \underline{\ ?\ }$ 9   **4.** $(a - 8)(a - 3) = a^2 - \underline{\ ?\ } + 24$ 11a
5. $(x + 8)(x - 9) = x^2 - \underline{\ ?\ } - 72$ x   **6.** $(r - 3)(r + 7) = r^2 + \underline{\ ?\ } - 21$ 4r
7. $(3y + 2)(y - 4) = \underline{\ ?\ } - 10y - 8$ **8.** $(4x + 7)(5x - 1) = \underline{\ ?\ } + 31x - 7$
$\qquad\qquad\qquad\qquad 3y^2$ $\qquad\qquad\qquad\qquad 20x^2$

Multiply.
$\qquad\qquad\qquad\qquad\qquad\qquad\qquad\qquad\qquad\qquad\qquad\qquad\qquad x^2 + x - 2$
9. $(x + 2)(x + 4)$ $x^2 + 6x + 8$ **10.** $(y - 3)(y - 2)$ $y^2 - 5y + 6$ **11.** $(x + 2)(x - 1)$
12. $(a - 4)(a + 5)$ $a^2 + a - 20$ **13.** $(r - 8)(r - 3)$ $r^2 - 11r + 24$ **14.** $(c + 1)(c - 4)$
15. $(2x + 3)(x - 1)$ $\qquad\qquad\qquad$ **16.** $(3y - 8)(2y - 3)$ $\qquad\qquad$ **17.** $(4a + b)(3a - b)$
$\qquad 2x^2 + x - 3$ $\qquad\qquad\qquad\qquad 6y^2 - 25y + 24$
$\qquad\qquad\qquad\qquad\qquad\qquad\qquad\qquad\qquad\qquad\qquad\qquad\qquad$ **14.** $c^2 - 3c - 4$
$\qquad\qquad\qquad\qquad\qquad\qquad\qquad\qquad\qquad\qquad\qquad\qquad\qquad$ **17.** $12a^2 - ab - b^2$

Additional Example 1

Multiply $(4y - 1)(2y + 5)$. $8y^2 + 18y - 5$

Additional Example 2

Multiply $(-3x + y)(2x - 7y)$.
$-6x^2 + 23xy - 7y^2$

Written Exercises

Multiply.

1. $(x + 8)(x - 9)$ $x^2 - x - 72$
2. $(a + 5)(a + 7)$ $a^2 + 12a + 35$
3. $(y - 4)(y - 10)$ $y^2 - 14y + 40$
4. $(r + 5)(r + 5)$ $r^2 + 10r + 25$
5. $(c - 6)(c - 6)$
6. $(x + 8)(x - 8)$ $x^2 - 64$
7. $(2x + 5)(x + 1)$
8. $(3c + 4)(c + 5)$
9. $(y - 3)(2y - 4)$
10. $(3x + 4)(4x + 5)$
11. $(2y + 5)(3y + 1)$
12. $(2c - 3)(3c - 7)$
13. $(4x - 6)(2x + 1)$
14. $(3d - 7)(d + 7)$
15. $(r - 8)(2r + 5)$
16. $(3y + 5)(3y - 5)$ $9y^2 - 25$
17. $(2y - 3)(y - 4)$
18. $(3x - 2)(3x - 2)$
19. $(x + 2y)(4x - y)$
20. $(3a + 2b)(a - b)$
21. $(r + s)(2r - 3s)$
22. $(7a + b)(3a - b)$
23. $(5y - z)(2y + 3z)$
24. $(4c - d)(-2c + d)$
25. $(x + 6)(x^2 + 4x - 1)$
26. $(2x - 1)(3x^2 - x + 4)$
27. $(3x - 2)(4x^2 - 1)$
28. $(5p - 1)(p^2 - 2p + 1)$
29. $(7y + 2)(2y^2 - 3y + 1)$
30. $(6a - 1)(3a^2 - a)$
31. $(x^2 - 2)(x^2 + 5)$
32. $(y^2 - 7)(y^2 - 4)$
33. $(c^2 + 2)(c^2 - 6)$
34. $(3x^2 - 1)(2x^2 + 5)$ $6x^4 + 13x^2 - 5$
35. $(4y^3 + 1)(4y^3 - 1)$ $16y^6 - 1$
36. $(2c^3 - 3)(2c^3 - 3)$ $4c^6 - 12c^3 + 9$

Express the area of each figure as a polynomial in simplest form.

37.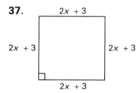
$4x^2 + 12x + 9$

38.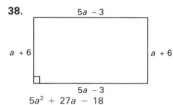
$5a^2 + 27a - 18$

39.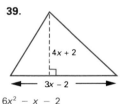
$6x^2 - x - 2$

Simplify.

40. $(2x^4 - 3y)(4x^4 + 2y)$ $8x^8 - 8x^4y - 6y^2$
41. $(3y^3 + 2x^2)(7y^3 - x^2)$ $21y^6 + 11x^2y^3 - 2x^4$
42. $(a + b)(x + y)$ $ax + ay + bx + by$
43. $(a + b)(x - y)$ $ax - ay + bx - by$
44. $(x^2 + y^2)(2x^2 - 3xy - y^2)$
45. $(a^2 - b^2)(7a^2 - 4ab + 6b^2)$
46. $(4x - 1)(4x + 1)(2x - 3)$
47. $(a + b + 3)(a + b - 3)$ $a^2 + 2ab + b^2 - 9$
48. Subtract the product of $(4y - 1)$ and $(3y + 2)$ from the product of $(6y - 5)$ and $(y - 8)$. $-6y^2 - 58y + 42$

Mixed Review

1. Evaluate $x^2 - 5$ for $x = -9$.   **2.8**   76

Solve.   **3.1, 3.3, 5.2, 5.7**

2. $x - 8 = -52$ -44
3. $2y + 7 = 1$ -3
4. $\frac{a}{9} + 4 = -3$ -63
5. $13 < 2x + 5$ $x > 4$
6. $|x + 2| < -4$ $\varnothing$
7. $|x - 1| > 7$ $x > 8 \text{ or } x < -6$

Common Error Analysis

Error: Since only one step of the FOIL method involves addition, students may try to multiply, instead of add, the two like terms.

Review the long form of multiplying two binomials to show where the two like products must be added.

Checkpoint

Multiply.

1. $(x + 6)(x + 1)$ $x^2 + 7x + 6$
2. $(y - 4)(y - 9)$ $y^2 - 13y + 36$
3. $(a + 3)(a - 7)$ $a^2 - 4a - 21$
4. $(r - 2)(r + 5)$ $r^2 + 3r - 10$

Closure

Ask students to summarize how they can multiply binomials mentally.
Answers will vary.

◼◼◼ FOLLOW UP

Guided Practice

Classroom Exercises 1–17
Try This   all, FR   all

Independent Practice

A Ex. 1–24, **B** Ex. 25–39, **C** Ex. 40–48

Basic: WE 1–24 all
Average: WE 1–33 odd, 34–39 all
Above Average: WE 1–47 odd

Additional Answers

See page 236 for the answers to Written Ex. 5, 7–33, 44–46.

Enrichment

Have students find the area of each rectangle below.

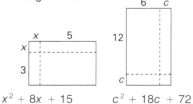

$x^2 + 8x + 15$ $c^2 + 18c + 72$

Then have them draw a rectangle representing each of the following products. Remind them to label the lengths of sides. Students may find it helpful to use graph paper.

1. $(x + 7)(x + 3)$
2. $(a + 6)(a + 1)$
3. $(c + 20)(c + 1)$
4. $(y + 4)(z + 2)$

▰▰▰ GETTING STARTED

Prerequisite Quiz

Multiply.

1. $(x + 2)(x + 2)$ $x^2 + 4x + 4$
2. $(y + 5)(y + 5)$ $y^2 + 10y + 25$
3. $(y - 3)(y - 3)$ $y^2 - 6y + 9$
4. $(x - 6)(x - 6)$ $x^2 - 12x + 36$
5. $(r + 2)(r - 2)$ $r^2 - 4$
6. $(d + 7)(d - 7)$ $d^2 - 49$

Motivator

Have the students multiply several pairs of binomials that fit the pattern $(a + b)(a - b)$. This will allow them to see that this particular type of product is a binomial rather than a trinomial. Ask them to explain why this happens.

▰▰▰ TEACHING SUGGESTIONS

Lesson Note

Tell the students that of the three methods shown for squaring a sum or difference, the last method will serve them best as they continue in mathematics courses. Once they have mastered this process mentally, they will be able to complete complex problems more efficiently.

6.10 Special Products

Objectives   To square a binomial
To find products of the form $(a + b)(a - b)$

To square a binomial means to multiply the binomial by itself.

Square of a Sum

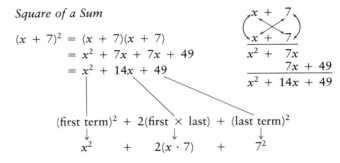

$$(x + 7)^2 = (x + 7)(x + 7)$$
$$= x^2 + 7x + 7x + 49$$
$$= x^2 + 14x + 49$$

$$(\text{first term})^2 + 2(\text{first} \times \text{last}) + (\text{last term})^2$$
$$x^2 \quad + \quad 2(x \cdot 7) \quad + \quad 7^2$$

Square of a Difference

$$(3y - 5)^2 = (3y - 5)(3y - 5)$$
$$= 9y^2 - 15y - 15y + 25$$
$$= 9y^2 - 30y + 25$$

$$(\text{first term})^2 - 2(\text{first} \times \text{last}) + (\text{last term})^2$$
$$(3y)^2 \quad - \quad 2(3y \cdot 5) \quad + \quad 5^2$$

Notice that $(x + 7)^2 \neq x^2 + 7^2$,
and $(3y - 5)^2 \neq (3y)^2 - (5)^2$.

When you square a binomial the product always contains **three** terms.

In general, the **square of a binomial** can be expressed as a trinomial.

Square of a Binomial
$$(a + b)^2 = a^2 + 2ab + b^2$$
$$(a - b)^2 = a^2 - 2ab + b^2$$

The square of a binomial is the square of the first term, plus or minus twice the product of the two terms, plus the square of the last term.

Highlighting the Standards

Standards 4a, 5b: Squaring binomials is represented in several different equivalent ways, and students operate on binomial expressions.

EXAMPLE 1   Multiply.

  a. $(x - 8)(x - 8)$ **b.** $(2y + 1)^2$

Solutions   **a.** $(x - 8)(x - 8)$ **b.** $(2y + 1)^2$
  $x^2 - 2(x \cdot 8) + 8^2$ $(2y)^2 + 2(2y \cdot 1) + 1^2$
  $x^2 - 16x + 64$ $4y^2 + 4y + 1$

TRY THIS   Multiply: **1.** $(3x - 2)^2$ $9x^2 - 12x + 4$
  2. $(5a + 1)^2$ $25a^2 + 10a + 1$

Now observe the result of multiplying *the sum and the difference* of the same two terms.

$$(x + 6)(x - 6) = x^2 - 6x + 6x - 36$$
$$= x^2 \quad - \quad 36$$

$$\underset{\text{(first term)}^2}{\big|} - \underset{\text{(last term)}^2}{\big|}$$
$$\underset{x^2}{\downarrow} - \underset{6^2}{\downarrow}$$

$$\begin{array}{r} x + 6 \\ x - 6 \\ \hline x^2 + 6x \\ -6x - 36 \\ \hline x^2 \quad\quad - 36 \end{array}$$

Notice that the product of the sum and difference of the same two terms results in an expression containing **two** terms.

In general, the **product** $(a + b)(a - b)$ can be expressed as a binomial.

Product of $(a + b)(a - b)$
$$(a + b)(a - b) = a^2 - b^2$$

The product of the sum and the difference of the same two terms is the square of the first term minus the square of the last term.

EXAMPLE 2   Multiply.

  a. $(c - 1)(c + 1)$ **b.** $(4x - 3)(4x + 3)$ **c.** $(3x + 2y)(3x - 2y)$

Solutions   **a.** $(c - 1)(c + 1)$ **b.** $(4x - 3)(4x + 3)$ **c.** $(3x + 2y)(3x - 2y)$
  $= c^2 - 1^2$ $= (4x)^2 - 3^2$ $= (3x)^2 - (2y)^2$
  $= c^2 - 1$ $= 16x^2 - 9$ $= 9x^2 - 4y^2$

TRY THIS   Multiply: **3.** $(2x + 2)(2x - 2)$ $4x^2 - 4$
  4. $(6x - 4)(6x + 4)$ $36x^2 - 16$

Math Connections

Trigonometry: In trigonometry, special products often yield interesting and useful results. For example,
$(1 + \sin \Theta)(1 - \sin \Theta) = 1 - \sin^2 \Theta = \cos^2 \Theta$

Critical Thinking Questions

Application: Ask students to compute the following mentally by using the patterns covered in this lesson.

  1. $52 \cdot 48$ **2.** 23^2
  $(50 + 2)(50 - 2) = 2500 - 4 = 2496$
  $(20 + 3)^2 = 400 + 2(60) + 9 = 529$

Common Error Analysis

Error: Students will confuse $(a + b)^2$ with $(ab)^2$ and write $a^2 + b^2$.

Remind them that the square of a binomial results in a trinomial.

Checkpoint

Multiply.

  1. $(x + 9)^2$ $x^2 + 18x + 81$
  2. $(y - 4)^2$ $y^2 - 8y + 16$
  3. $(x + 3)(x - 3)$ $x^2 - 9$
  4. $(3y - 1)(3y + 1)$ $9y^2 - 1$
  5. $(5x - 2)^2$ $25x^2 - 20x + 4$
  6. $(6a + 3b)^2$ $36a^2 + 36ab + 9b^2$

Closure

Have students summarize the patterns for squaring binomials and for finding the product of the sum and difference of the same two terms. Have them illustrate each pattern with a specific example.   Answers will vary.

Additional Example 1

Multiply.

a. $(3 + y)(3 + y)$ $9 + 6y + y^2$
b. $(6 + 2c)^2$ $36 + 24c + 4c^2$

Additional Example 2

Multiply.

a. $(4 + r)(4 - r)$ $16 - r^2$
b. $(2a - b)(2a + b)$ $4a^2 - b^2$
c. $(4a + 5b)(4a - 5b)$ $16a^2 - 25b^2$

Guided Practice

Classroom Exercises 1–24
Try This   all, FR   all

Independent Practice

A Ex. 1–24, **B** Ex. 25–39, **C** Ex. 40–47
Basic: WE 1–20 all, Brainteaser
Average: WE 1–39 odd, Brainteaser
Above Average: WE 9–47 odd,
Brainteaser

Additional Answers

Classroom Exercises

13. $x^2 + 10x + 25$
14. $y^2 - 2y + 1$
15. $x^2 + 6x + 9$
16. $a^2 - 18a + 81$
17. $x^2 + 4x + 4$
18. $c^2 - 10c + 25$
19. $4x^2 - 4x + 1$
20. $9y^2 + 24y + 16$

Written Exercises

5. $y^2 + 20y + 100$
6. $d^2 - 16$
9. $-x^2 + 49$
12. $16c^2 - 25$
13. $100d^2 - 4$
15. $-36y^2 + 1$
16. $c^2 + 2cd + d^2$
18. $9y^2 - 16z^2$
19. $4p^2 - 9q^2$
20. $9a^2 + 30ab + 25b^2$
21. $49x^2 - y^2$
22. $25x^2 + 20xy + 4y^2$

Some equations contain products of binomials.

EXAMPLE 3   Solve $(x - 5)^2 = (x - 8)(x + 8)$.

Solution

$$(x - 5)^2 = (x - 8)(x + 8)$$
$$x^2 - 10x + 25 = x^2 - 64$$
$$-10x + 25 = -64 \quad \longleftarrow \text{Subtract } x^2 \text{ from each side.}$$
$$-10x = -89$$
$$x = 8.9$$

The check is left for you.

TRY THIS   **5.** Solve $(x - 2)(x + 2) = (x - 4)^2$. 2.5

■■ Focus on Reading

Match each expression with one expression lettered a–e.

1. $(x + 6)^2$ e
2. $(x - 6)^2$ d
3. $(x + 6)(x - 6)$ b
4. $(6 - x)^2$ d

a. $x^2 + 36$
b. $x^2 - 36$
c. $x^2 + 6x + 36$
d. $x^2 - 12x + 36$
e. $x^2 + 12x + 36$

Classroom Exercises

Give the missing term in each product.

1. $(x + 1)^2 = x^2 + \underline{?} + 1$ 2x
2. $(y - 4)^2 = y^2 - \underline{?} + 16$ 8y
3. $(r + 5)(r - 5) = r^2 - \underline{?}$ 25
4. $(d - 2)(d + 2) = \underline{?} - 4$ d^2
5. $(c + 9)(c + 9) = c^2 + \underline{?} + 81$ 18c
6. $(y - 10)(y - 10) = y^2 - 20y + \underline{?}$ 100
7. $(p + q)(p - q) = \underline{?} - q^2$ p^2
8. $(3 - x)(3 + x) = 9 - \underline{?}$ x^2
9. $(1 + a)(1 + a) = 1 + 2a + \underline{?}$ a^2
10. $(x + y)^2 = x^2 + \underline{\quad?\quad} + y^2$ 2xy
11. $(r - t)^2 = r^2 + \underline{\quad?\quad} + t^2$ $-2rt$
12. $(1 + 2b)^2 = 1 + 4b + \underline{\quad?\quad}$ $4b^2$

Multiply.

13. $(x + 5)^2$
14. $(y - 1)^2$
15. $(x + 3)(x + 3)$
16. $(a - 9)(a - 9)$
17. $(x + 2)^2$
18. $(5 - c)^2$
19. $(2x - 1)^2$
20. $(3y + 4)^2$
21. $(4c + 1)(4c - 1)$ $16c^2 - 1$
22. $(3a - 7b)(3a + 7b)$ $9a^2 - 49b^2$
23. $(3x - 2y)(3x + 2y)$ $9x^2 - 4y^2$
24. $(9 + 4p)(9 - 4p)$ $81 - 16p^2$

234   Chapter 6 Powers and Polynomials

Additional Example 3

Solve $(y - 2)(y + 8) = (y + 6)^2$.
$\frac{-26}{3}$, or $-8\frac{2}{3}$

Written Exercises

Multiply.

1. $(y - 6)^2$ $y^2 - 12y + 36$
2. $(x + 2)^2$ $x^2 + 4x + 4$
3. $(c - 8)(c - 8)$ $c^2 - 16c + 64$
4. $(x - 3)(x + 3)$ $x^2 - 9$
5. $(y + 10)(y + 10)$
6. $(d - 4)(d + 4)$
7. $(r - 12)(r + 12)$ $r^2 - 144$
8. $(6 + y)(6 - y)$ $-y^2 + 36$
9. $(7 + x)(7 - x)$
10. $(2x + 3)(2x - 3)$ $4x^2 - 9$
11. $(7y + 1)(7y - 1)$ $49y^2 - 1$
12. $(4c - 5)(4c + 5)$
13. $(10d - 2)(10d + 2)$
14. $(3r + 8)(3r - 8)$ $9r^2 - 64$
15. $(1 + 6y)(1 - 6y)$
16. $(c + d)(c + d)$
17. $(2r - s)^2$ $4r^2 - 4rs + s^2$
18. $(3y + 4z)(3y - 4z)$
19. $(2p - 3q)(2p + 3q)$
20. $(3a + 5b)^2$
21. $(7x + y)(7x - y)$
22. $(5x + 2y)^2$
23. $(4r + s)(4r + s)$
24. $(6a - b)^2$
25. $(3 - 5c)(5c + 3)$ $-25c^2 + 9$
26. $(3x + 2y)(2y + 3x)$
27. $(7x - y)(-y + 7x)$
28. $(x^2 + 2)^2$ $x^4 + 4x^2 + 4$
29. $(y^2 - 5)^2$ $y^4 - 10y^2 + 25$
30. $(z^2 + 1)(z^2 - 1)$
31. $(3x^2 - 2)^2$ $9x^4 - 12x^2 + 4$
32. $(5a^2 + 1)(5a^2 + 1)$
33. $(7y^3 + 2)(7y^3 - 2)$
34. $\left(y + \frac{1}{2}\right)^2$ $y^2 + y + \frac{1}{4}$
35. $\left(x - \frac{1}{4}\right)^2$ $x^2 - \frac{1}{2}x + \frac{1}{16}$
36. $\left(a - \frac{1}{3}\right)\left(a + \frac{1}{3}\right)$ $a^2 - \frac{1}{9}$
37. $(x - 0.1)^2$ $x^2 - 0.2x + 0.01$
38. $(a + 0.2)(a - 0.2)$ $a^2 - 0.04$
39. $(y + 0.5)^2$ $y^2 + y + 0.25$

Solve each equation.

40. $(x + 2)(x - 7) = (x - 4)^2$ 10
41. $(x + 3)^2 = (x - 1)^2$ -1
42. $(y + 5)(y - 5) = (y + 2)(y - 7)$ $\frac{11}{5}$
43. $a^2 = (a + 8)^2$ -4
44. $(x + 4)^2 = (x + 2)(x - 4)$ $-\frac{12}{5}$
45. $(y - 6)^2 = y^2$ 3
46. $(x + 5)^2 - x^2 = 0$ $-\frac{5}{2}$
47. $(2r - 4)(r + 5) = (2r + 5)(r - 1)$ 5

Mixed Review

1. Evaluate $3x^2 - 6y$ for $x = -2$ and $y = -5$.   **2.8**   42

Solve.   **3.3, 3.5, 5.2, 5.6, 5.7**

2. $-\frac{2}{5}y + 42 = 36$ 15
3. $3y - (4 - y) = 12$ 4
4. $5y - 7 > 3y - 5$ $y > 1$
5. $6x - 4 < -3 + 5x$ $x < 1$
6. $|d| + 8 = 10$ $2, -2$
7. $|4x + 2| < -7$ $\varnothing$
8. Graph the solution set: $8 + x < 12$ *and* $x \geq -2$ **5.4**

/Brainteaser

Maggie Mae buys pencils at 3 for 46¢. She sells them at 5 for 80¢.
How many pencils must she sell to make a profit of $1.00?   150

---

23. $16r^2 + 8rs + s^2$
24. $36a^2 - 12ab + b^2$
26. $9x^2 + 12xy + 4y^2$
27. $49x^2 - 14xy + y^2$
30. $z^4 - 1$
32. $25a^4 + 10a^2 + 1$
33. $49y^6 - 4$

Mixed Review

8.

Additional Answers, page 228

33. $-6a^5b^2 - 15a^3b^3 + 3a^2b^5$
36. $(x - 2)(x + 4) = x^2 - 2x + 4x - 8 = x^2 + 2x - 8$ and $(x - 2)x + (x - 2)4 = x^2 - 2x + 4x - 8 = x^2 + 2x - 8$. So $(x - 2)(x + 4) = (x - 2)x + (x - 2)4$ for all real numbers x.

Enrichment

The measure of each side of a square is $(a + b + c)$. Have students find the area of this square.
$a^2 + b^2 + c^2 + 2ab + 2ac + 2bc$
Then have them use this *pattern* to find the areas of squares having sides of each given measure.

a. $x + y + z$
$x^2 + y^2 + z^2 + 2xy + 2xz + 2yz$
b. $m + n + 3$
$m^2 + n^2 + 2mn + 6m + 6n + 9$
c. $x^2 + x + 5$
$x^4 + 2x^3 + 11x^2 + 10x + 25$

235

9. $(x^2)^3 = x^{2 \cdot 3}$ (Pow of Pow Prop)
   $= x^{3 \cdot 2}$ (Comm Prop Mult)
   $= (x^3)^2$ (Pow of Pow Prop)

Additional Answers, page 231

V tten Exercises

5. $c^2 - 12c + 36$
7. $2x^2 + 7x + 5$
8. $3c^2 + 19c + 20$
9. $2y^2 - 10y + 12$
10. $12x^2 + 31x + 20$
11. $6y^2 + 17y + 5$
12. $6c^2 - 23c + 21$
13. $8x^2 - 8x - 6$
14. $3d^2 + 14d - 49$
15. $2r^2 - 11r - 40$
17. $2y^2 - 11y + 12$
18. $9x^2 - 12x + 4$
19. $4x^2 + 7xy - 2y^2$
20. $3a^2 - ab - 2b^2$
21. $2r^2 - rs - 3s^2$
22. $21a^2 - 4ab - b^2$
23. $10y^2 + 13yz - 3z^2$
24. $-8c^2 + 6cd - d^2$
25. $x^3 + 10x^2 + 23x - 6$
26. $6x^3 - 5x^2 + 9x - 4$
27. $12x^3 - 8x^2 - 3x + 2$
28. $5p^3 - 11p^2 + 7p - 1$
29. $14y^3 - 17y^2 + y + 2$
30. $18a^3 - 9a^2 + a$
31. $x^4 + 3x^2 - 10$
32. $y^4 - 11y^2 + 28$
33. $c^4 - 4c^2 - 12$
44. $2x^4 - 3x^3y + x^2y^2 - 3xy^3 + y^4$
45. $7a^4 - 4a^3b - a^2b^2 + 4ab^3 - 6b^4$
46. $32x^3 - 48x^2 - 2x + 3$

Chapter 6 Review

Key Terms

binomial   (p. 218)
Cancellation Property for Fractions   (p. 207)
degree of a monomial   (p. 219)
degree of a polynomial   (p. 219)
descending order   (p. 219)
FOIL method   (p. 229)
monomial   (p. 201)
negative exponent   (p. 211)
polynomial   (p. 218)

Power-of-a-Power Property   (p. 204)
Power-of-a-Product Property   (p. 205)
Product-of-Powers Property   (p. 201)
Quotient-of-Powers Property   (p. 207)
scientific notation   (p. 214)
square of a binomial   (p. 232)
trinomial   (p. 218)
zero exponent   (p. 210)

Key Ideas and Review Exercises

6.1,
6.2   To multiply powers with the same base, add the exponents and keep the same base.

Simplify.
                                                                    $-21c^3d^5$

1. $5x \cdot x^4$ $5x^5$ 2. $-6x \cdot 2x^2 \cdot 3x$ $-36x^4$ 3. $0.5(6c^3d)$ $3c^3d$ 4. $(3c^2d)(-7cd^4)$
5. $(x^4)^2$ x^8 6. $(3c^2)^3$ $27c^6$ 7. $(-xy^2z^3)^5$ $-x^5y^{10}z^{15}$ 8. $(5c^3)^2(2c)^2$
9. Explain why "$(x^2)^3 = (x^3)^2$ for all real numbers x" is true. $100c^8$

6.3   To simplify $\dfrac{b^m}{b^n}$ for any nonzero real number b and positive integers m and n, use the Quotient-of-Powers Property. See page 207.

Simplify.
                                                                                              $\dfrac{-1}{5y^2}$
10. $\dfrac{x^5}{x^2}$ x^3 11. $\dfrac{c^2d}{cd^3}$ $\dfrac{c}{d^2}$ 12. $\dfrac{(4ab^2)^3}{2a^2b}$ $32ab^5$ 13. $\dfrac{-2xy(x^3y^2)}{10x^4y^5}$

6.4   To simplify expressions with zero and negative exponents, use $b^0 = 1$, $b^{-n} = \dfrac{1}{b^n}$, $\dfrac{1}{b^{-n}} = b^n$, and $\left(\dfrac{a}{b}\right)^{-m} = \left(\dfrac{b}{a}\right)^m$ where a and b are nonzero real numbers and n is an integer.

Simplify. Use positive exponents only. No variable equals zero.

14. $5c^{-2}$ $\dfrac{5}{c^2}$ 15. $\left(\dfrac{1}{3}\right)^{-3}$ 27 16. $\dfrac{12x^{-1}y^{-2}}{-9xy^{-1}}$ $\dfrac{4}{-3x^2y}$ 17. $\left(\dfrac{3x^2y}{6y}\right)^{-2}$ $\dfrac{4}{x^4}$
18. Evaluate $\dfrac{(5y)^0}{y^{-2}}$ for $y = -4$. 16

6.5 To write a number in scientific notation, write it in the form $a \times 10^n$, where $1 \le a < 10$ and n is an integer.

Write each answer in scientific notation.

19. Find the number of seconds in a 24 h day. 8.64×10^4

20. Simplify $\dfrac{20.8(16{,}000{,}000)}{0.00004}$.
8.32×10^{12}

6.6,
6.7 To simplify a polynomial, combine like terms.

To write a polynomial in descending (ascending) order, arrange the terms in order from greatest to least (least to greatest) degree. See the definition of degree of a polynomial (Lesson 6.6).

To add polynomials, combine their like terms.

To subtract polynomials, add the opposite of the polynomial that is being subtracted to the other polynomial.

Simplify. Write the result in descending order and in ascending order. Give the degree of the polynomial.

21. $8x^2 + 7x - 5x^4 - 7x^2 + x$
$-5x^4 + x^2 + 8x;\ 8x + x^2 - 5x^4;\ 4$

22. $16y - 8y^5 + 3y^2 - 4y - y^2 + y^5$
$-7y^5 + 2y^2 + 12y;\ 12y + 2y^2 - 7y^5;\ 5$

Add or subtract.

23. $(2y^3 - 3y + 1) + (-2y^3 + 4y - 7)$ $y - 6$

24. $(3x^2 - 7x + 2) - (-4x^2 + 3x + 5)$ $7x^2 - 10x - 3$

25. Add $-2a^5 + 6a^3 - 3a + 1$ and $4a^4 - 6a^3 + 2a^2 - 8a$. $-2a^5 + 4a^4 + 2a^2 - 11a + 1$

26. Subtract $c^3 - 5c$ from $c^4 - c^3 + 5c - 2$. $c^4 - 2c^3 + 10c - 2$

6.8,
6.9,
6.10 To multiply a polynomial by a monomial, multiply each term in the polynomial by the monomial.

To multiply two polynomials, multiply each term of the first polynomial by each term of the second. To multiply two binomials mentally, use the FOIL method (Lesson 6.9). Two special products are (1) the square of a binomial: $(a + b)^2 = a^2 + 2ab + b^2$, or $(a - b)^2 = a^2 - 2ab + b^2$, and (2) the product of the sum and difference of two terms: $(a + b)(a - b) = a^2 - b^2$.

Multiply.

27. $2y(3y^3 - 6y + 1)$ $6y^4 - 12y^2 + 2y$

28. $(x + 8)(x - 7)$ $x^2 + x - 56$

29. $(2x + y)(3x + 4y)$ $6x^2 + 11xy + 4y^2$

30. $(a - 5)(2a^2 + 3a - 1)$ $2a^3 - 7a^2 - 16a + 5$

31. $(y + 5)(y + 5)$ $y^2 + 10y + 25$

32. $(2a - 7)^2$ $4a^2 - 28a + 49$

33. $(x - 3)(x + 3)$ $x^2 - 9$

34. $(y^2 + 6)(y^2 - 6)$ $y^4 - 36$

35. Simplify $-x^2 - 4x + (2x^2 + 3x - 1)5x$. $10x^3 + 14x^2 - 9x$

9. $-28a^4b^6$
23. 4.39×10^{-4}
24. 5.24×10^4
31. $7a^4 - 3a^3 - 2a + 2$
32. $-c^3 + 2c^2 - 9c + 1$
35. $-12x^4y^2 + 4x^3y^2 - 36x^3y^3$
42. $c^2 - 25c$

Chapter 6 Test

A Exercises: 1–12, 14, 17–26, 28–29, 31–32, 34, 36–39, 42
B Exercises: 13, 15–16, 27, 30, 33, 35, 40–41, 44
C Exercise: 43

Classify each polynomial as a monomial, a binomial, or a trinomial.

1. $a + b - 1$ Trinomial **2.** $4xy^2$ Momomial **3.** $2n + 5$ Binomial **4.** -100
Monomial

Simplify, if possible. Assume that no denominator equals zero.

5. $4x \cdot x$ $4x^2$ **6.** $-8y^2(-y^4)$ $8y^6$ **7.** $(-a)^6$ a^6 **8.** $5c^2(2cd)(7cd^3)$ $70c^4d^4$

9. $(4a^3b)(-7ab^5)$ **10.** $(4x)^3$ $64x^3$ **11.** $-(7xy^3)^2$ $-49x^2y^6$ **12.** $(-2x^3y^2)^3$ $-8x^9y^6$

13. $4c(-3c^2)^3$ $-108c^7$ **14.** $\dfrac{-5x^2y}{15xy^3}$ $\dfrac{-x}{3y^2}$ **15.** $9x^4yz^2\left(-\dfrac{1}{3}yz\right)$ $-3x^4y^2z^3$ **16.** $\dfrac{(2a^2b)^3}{(4ab^2)(3a)}$ $\dfrac{2a^4b}{3}$

Simplify. Use positive exponents only. Assume that no variable equals zero.

17. $6x^0$ 6 **18.** $10^2 \cdot x^{-1}$ $\dfrac{100}{x}$ **19.** $\left(\dfrac{2}{3}\right)^{-4}$ $\dfrac{81}{16}$ **20.** $\dfrac{-12a^4b^{-3}c^0}{-6a^{-7}b^{-2}}$ $\dfrac{2a^{11}}{b}$

21. Evaluate $6n^3$ for $n = -2$. -48 **22.** Evaluate $-4t^{-2}$ for $t = 6$. $\dfrac{-1}{9}$

Express each number in scientific notation.

23. 0.000439 **24.** $52,400$ **25.** $900,000$ 9.0×10^5 **26.** 0.084 8.4×10^{-2}

27. Simplify $\dfrac{(186,000,000)(0.00000051)}{0.0034}$ and express the answer
in scientific notation. 2.79×10^4

Simplify. Then give the degree of the polynomial.

28. $5c^3 - 7c^2 - 3c^3 + 2c - 2c^3$ $-7c^2 + 2c$; 2   **29.** $-4y^4 + 3y^3 + 6y - 3y^3 - 7y + 2$ $-4y^4 - y + 2$; 4

30. Simplify $-3 + 7a^2 + 2a - 3a^2 + a^2 - 7a + 6a^3 - 4$.
Write the result in descending order and in ascending order.
$6a^3 + 5a^2 - 5a - 7$; $-7 - 5a + 5a^2 + 6a^3$

Add or subtract.

31. $(7a^4 - 6a^3 + 1) + (3a^3 - 2a + 1)$ **32.** $(c^3 + 2c^2 - 5c) - (2c^3 + 4c - 1)$

33. Add $0.4x^2 - 0.3x + 0.1$ to $-0.9x^2 + 0.4x - 0.6$. $-0.5x^2 + 0.1x - 0.5$

Multiply.

34. $2a^2(6a^2 - 4a - 2)$ $12a^4 - 8a^3 - 4a^2$ **35.** $4x^2y(-3x^2y + xy - 9xy^2)$

36. $(x + 3)(x - 7)$ $x^2 - 4x - 21$ **37.** $(2x + 3)(5x + 1)$ $10x^2 + 17x + 3$

38. $(a + 8)^2$ $a^2 + 16a + 64$ **39.** $(2x + 1)(2x - 1)$ $4x^2 - 1$

40. $(x + 4)(x^2 - x + 3)$ $x^3 + 3x^2 - x + 12$ **41.** $(r^2 - 2)^2$ $r^4 - 4r^2 + 4$

42. Simplify. $-5c^2 - 7c + (c - 3)6c$ **43.** Simplify. $c^n \cdot c^{3n+2}$ c^{4n+2}

44. Solve. $7 + x(x + 3) = x^2 - (5 - 4x)$ 12

Choose the one best answer to each question or problem.

1. The statement $(x + y)^2 = (x - y)^2$ is true if __?__ . D
   (A) $x < y$ (B) $x > y$ (C) $x = y$
   (D) $x = 0$ and $y = 0$
   (E) $x = 1$ and $y = -1$

2. How many centimeters are there in 12 mm? D
   (A) 1,200   (B) 120   (C) 12
   (D) 1.2   (E) 0.12

3. One ft is what percent of one yd? E
   (A) $\frac{1}{3}$%   (B) $\frac{2}{3}$%   (C) 3%
   (D) 300%   (E) $33\frac{1}{3}$%

4. What is the area of the shaded region in the figure below? B

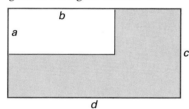

   (A) $ab - cd$ (B) $cd - ab$
   (C) ab (D) $ab + cd$
   (E) $ac - bd$

5. If $0.6x = 4$, find $0.3x$. C
   (A) 0.4   (B) 1   (C) 2   (D) 2.4
   (E) 0.8

6. Find the next term in the sequence:
   0, 1, 3, 7, 15, 31, __?__ . D
   (A) 42   (B) 52   (C) 62
   (D) 63   (E) 73

7. If $1^m = 2^{m-1}$, then $m =$ __?__ . C
   (A) -1 (B) 0   (C) 1   (D) 2
   (E) none of these

8. Which expression is equivalent to $x^2 + y^2 + 18 - 11y$? E
   (A) $(x + y)^2 - 11y + 18$
   (B) $(x + y - 9)(x + y - 2)$
   (C) $(x + 18)(y - 11)$
   (D) $y^2 + (x + 9)(x - 2)$
   (E) $(y - 9)(y - 2) + x^2$

9. What is the perimeter of the figure below? A

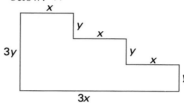

   (A) $6x + 6y$ (B) $3(x + y)$
   (C) $3xy$ (D) xy (E) None of these

10. If $0 < x < 1$, then what is true of x^{-1}? C
    (A) $x^{-1} < 0$ (B) $0 < x^{-1} < 1$
    (C) $x^{-1} > 1$ (D) $x^{-1} = 0$
    (E) It cannot be determined from the information given.

11. What part of a dollar is the value of 1 quarter, 1 dime, and 4 nickels? B
    (A) $.55   (B) $\frac{11}{20}$ (C) 50%
    (D) $\frac{1}{55}$ (E) 5.5

12. If $0.00009 \times 0.02 = 1.8 \times 10^n$, then $n =$ __?__ . C
    (A) -4 (B) -5 (C) -6
    (D) -7 (E) -8

39.

40.

41.

42.

43.

44.

45.

Cumulative Review (Chapters 1–6)

Simplify.

1. $16 - 2 \cdot 5$ 6   *1.2*

2. $\left(\frac{2}{3}\right)\left(\frac{3}{4}\right) - \frac{1}{2}$ 0

3. 2.8×10^4 28,000   *1.3*

4. $-|-3| \cdot |6|$ −18   *2.2*

5. $-7 + (-9)$ −16   *2.4*

6. $-6.8 + 6.8$ 0

7. $6 - (-11)$ 17   *2.5*

8. $-1.8 - 5.6$ −7.4

9. $\frac{1}{2}(-16)$ −8   *2.6*

10. $(-8)(2)^3$ −64

11. $0 \div (-18)$ 0   *2.7*

12. $-\frac{3}{4} \div \frac{6}{7}$ $-\frac{7}{8}$

13. $3x - 5x$ −2x   *2.9*

14. $-4x + 2y - 7x - 9y$ −11x − 7y

15. $6(a + 3) - 4(2a - 1)$ −2a + 22   **2.10**

16. $-2x^2y(-5xy^3)$ 10x³y⁴   *6.1*

17. $(3c^2d)^4$ 81c⁸d⁴   *6.2*

18. $5a(-2ab)^3$ −40a⁴b³

19. $\frac{-2x^2y}{14xy^4}$ $\frac{-x}{7y^3}$ *6.3*

20. $\frac{(-2x^3)^2}{8x^5}$ $\frac{x}{2}$

21. $-3x^2 - 2x + 3 - x^2 + 6x$ −4x² + 4x + 3   *6.6*

22. Simplify $\frac{3a^0b^{-1}c}{15bc^{-3}}$. Use positive exponents only. $\frac{c^4}{5b^2}$ *6.6*

Evaluate.

23. $6.2n$ for $n = 4$ 24.8   *1.1*

24. $\frac{x}{y}$ for $x = -5.5$ and $y = 2.7$ −2.037

25. $16 - 3y$ for $y = 3$ 7   *1.2*

26. $(4y)^2 - y$ for $y = 2$ 62   *1.3*

27. $-x + y$ for $x = -7$ and $y = 4$ 11   *2.5*

28. x^2y for $x = -2$ and $y = -12 \div -3$ *2.6*

29. $\frac{6x^2}{y}$ for $x = 9$ and $y = 2$ 243   *1.3*

30. $3x^{-2}$ for $x = 5$ $\frac{3}{25}$ *6.4*

Solve.

31. $-6 + a = 14$ 20   *3.1*

32. $x + 9 = -12$ −21

33. $-4y = 42$ $-10\frac{1}{2}$ *3.2*

34. $-\frac{3}{5}x = 51$ −85

35. $\frac{x}{5} - 7 = -20$ −65   *3.3*

36. $--4c + 3 = 17$ $-3\frac{1}{2}$

37. $3y - 4 = 28 - y$ 8   *3.4*

38. $2(x + 9) = 5 - 7(2 + x)$ −3   *3.5*

Graph the solution set.

39. $x - 7 < -2$ *5.1*

40. $2x + 3 > -7$ *5.2*

41. $-7y + 8 \le 2 - 6y$ *5.4*

42. $-3 < c < 1$

43. $|x - 7| \ge 3$ *5.7*

44. $|x + 2| < 4$

45. $|y - 3| < -5$

Add or subtract.

46. $(3t^2 - 5) + (-t^2 + 7 - 3)$ 2t² − 1   *6.7*

47. $(y^2 - y - 4) - (-2y^2 + y + 6)$ 3y² − 2y − 10

Multiply.

48. $3x(7x - 1)$ 21x² − 3x   *6.8*

49. $(4y - 5)2y^2$ 8y³ − 10y²

50. $(x + 8)(2x - 1)$ 2x² + 15x − 8   *6.9*

51. $(a - 6)(a - 10)$ a² − 16a + 60

52. $(y - 8)^2$ y² − 16y + 64   *6.10*

53. $(3x + 1)^2$ 9x² + 6x + 1

54. $(2x - 5)(2x + 5)$ 4x² − 25

Solve each problem.

55. Use the formula $A = s^2$ to find A if $s = 1.2$ cm. 1.44 cm² *4.2*

56. Use the formula $p = 2l + 2w$ to find p if $l = 7$ m and $w = 5$ m. 24 m

57. Use the formula $V = lwh$ to find V if $l = 6$ in., $w = 9$ in., and $h = 2$ in. 108 in³

58. After the temperature dropped 17°, the thermometer read $-6°$. What did the thermometer read before the temperature dropped? 11° *3.6*

59. If Mr. Berensen travels at an average rate of 45 mi/h, how many hours must he travel to cover 105 mi? $2\frac{1}{3}$ h *4.4*

60. The perimeter of a rectangular room is 72 ft. The width is $\frac{1}{3}$ of the length. Find the width and the length of the room. *4.2* l: 27 ft w: 9 ft

61. Each leg of an isosceles triangle is 3 cm longer than the base. The perimeter of the triangle is 45 cm. Find the length of each side. 13 cm; 16 cm; 16 cm

62. Use the formula $A = \frac{1}{2}(b + c)h$ to find the area A of a trapezoid if the height h is 12 cm and the bases b and c measure 7 cm and 10 cm, respectively. 102 cm²

63. Bill has 3 times as much money as Jim, and Jim has $8 less than Lynn. Together they have $268. How much money does each of them have? *4.2* Jim: $52 Bill: $156 Lynn: $60

64. The sum of two consecutive odd integers is -92. Find the integers. -45, -47 *4.3*

65. Find three consecutive integers such that their sum, decreased by 60, is equal to the second integer. 29, 30, 31

66. Erika has half as many quarters as dimes. How many coins of each type does she have if the total number of coins is 4 more than the number of quarters? 2 quarters; 4 dimes *4.5*

67. This year Juan's earnings increased $\frac{2}{3}$ over last year's. This year he earned $605. How much did he earn last year? $363 *4.2*

68. After Bob lent $\frac{2}{5}$ of his pocket money to Marilee, he had $4.80 left. How much pocket money did he have at first? $8 *4.2*

69. Mrs. Danoff rented a car for $35 per day plus $0.12 per mile. How far can she travel in one day if she can spend at most $52, not including taxes or insurance? $x \le 141\frac{2}{3}$ mi *5.5*

70. Jason bought 5 pads of writing paper. The total was $3.79, but $0.24 of this was for tax. What was the price of each pad before the tax? $0.71 *4.8*

71. This year a book costs $1.90 less than twice what it cost last year. The book now costs $10. What did it cost last year? $5.95 *3.3*

72. The greater of two numbers is 6.8 more than 3 times the smaller. The difference between the two numbers is 21.6. Find the two numbers. 7.4, 29 *4.2*

73. What number is 62% of 40? 24.8 *4.7*

74. Fourteen is what percent of 84? $16\frac{2}{3}$%

75. Seventy-five is $62\frac{1}{2}$% of what number? 120

7 FACTORING POLYNOMIALS

OVERVIEW

After reviewing the concepts of prime factor and greatest common factor, students identify the greatest common factor of a polynomial. Students then learn to factor trinomials and to identify special factorizations, such as factoring perfect square trinomials and the difference of two squares. In the last three lessons, students solve quadratic equations by factoring and then solve word problems which can be modeled by quadratic equations.

OBJECTIVES

- To use prime factorizations to find the greatest common factor of two numbers and two or more monomials
- To factor polynomials using several basic patterns
- To solve quadratic equations by factoring
- To solve word problems involving quadratic equations

PROBLEM SOLVING

Looking for Patterns is the basic strategy applied in this chapter to factoring polynomials. Making a Table of values and Guessing and Checking are also used as aids in factoring certain trinomials. Using a Formula, Drawing a Diagram, and Writing an Equation are additional strategies useful for solving the word problems in this chapter. Several Mixed Problem Solving Exercises on page 283 emphasize the usefulness of the Estimating Before Solving strategy.

READING AND WRITING MATH

Students are asked to explain in writing how or why a particular procedure works for solving Exercise 68 on p. 247, Exercise 46 on p. 258, Exercise 48 on p. 262, and Exercise 25 on p. 275. The Focus on Reading exercises on pages 264 and 269 ask students to demonstrate an understanding of the vocabulary presented by answering short questions and giving examples.

TECHNOLOGY

Calculator: Using a calculator to find prime factors is demonstrated on p. 244. Example 5 on p. 249 and the Application on p. 255 as well as the related problems in the Written Exercises provide good opportunities for the use of a calculator.

SPECIAL FEATURES

Mixed Review pp. 247, 251, 255, 258, 266, 270, 276, 281
Brainteaser p. 251, 276
Application: Gas Mileage p. 255
Midchapter Review p. 262
Focus on Reading pp. 264, 269
Application: Commission Sales p. 266
Application: Boiling Point of Water p. 271
Problem Solving Strategies p. 282
Mixed Problem Solving p. 283
Key Terms p. 284
Key Ideas and Review Exercises pp. 284–285
Chapter 7 Test p. 286
College Prep Test p. 287

PLANNING GUIDE

Lesson	Basic	Average	Above Average	Resources
7.1 pp. 246–247	CE 1–21 WE 1–57 odd	CE 1–21 WE 11–67 odd	**CE 1–21** **WE 17–71 odd**	Reteaching 51 Practice 51
7.2 pp. 250–251	CE 1–20 WE 1–47 odd Brainteaser	CE 1–20 WE 11–57 odd Brainteaser	CE 1–20 WE 17–63 odd Brainteaser	Reteaching 52 Practice 52
7.3 pp. 254–255	CE 1–12 WE 1–21 all Application	CE 1–12 WE 7–27 all Application	CE 1–12 WE 1–37 odd Application	Reteaching 53 Practice 53
7.4 pp. 257–258	CE 1–18 WE 1–27 odd, 40	CE 1–18 WE 11–39 odd, 40	**CE 1–18** **WE 21–49 odd**	Reteaching 54 Practice 54
7.5 pp. 261–262	CE 1–15 WE 1–24 all, 42 Midchapter Review	CE 1–15 WE 1–41 odd, 42–44, 48 Midchapter Review	CE 1–15 WE 1–47 odd, 48 Midchapter Review	Reteaching 55 Practice 55
7.6 pp. 265–266	FR all, CE 1–24 WE 1–10 all, 11–27 odd Application	FR all, CE 1–24 WE 1–39 odd Application	FR all, CE 1–24 WE 13–51 odd Application	Reteaching 56 Practice 56
7.7 pp. 270–271	FR all, CE 1–12 WE 1–8 all, 9–23 odd Application	FR all, CE 1–12 WE 7–39 odd Application	FR all, CE 1–12 WE 19–51 odd, 52 Application	Reteaching 57 Practice 57
7.8 pp. 275–276	CE 1–18 WE 1–27 odd Brainteaser	CE 1–18 WE 7–33 odd Brainteaser	CE 1–18 WE 11–37 odd Brainteaser	Reteaching 58 Practice 58
7.9 pp. 280–283	CE 1–10 WE 1–11 odd Problem Solving Strategies Mixed Problem Solving	CE 1–10 WE 5–15 odd Problem Solving Strategies Mixed Problem Solving	CE 1–10 WE 9–19 odd Problem Solving Strategies Mixed Problem Solving	Reteaching 59 Practice 59
Chapter 7 Review pp. 284–285	all	all	all	
Chapter 7 Test p. 286	all	all	all	
College Prep Test p. 287	all	all	all	

CE = Classroom Exercises   WE = Written Exercises   FR = Focus on Reading
Note: For each level, all students should be assigned all Try This and all Mixed Review exercises.

◾◾◾ INVESTIGATION

Project: This investigation leads students to discover how many factors a number has.

Materials: Each group of students will need a copy of Project Worksheet 7. A calculator will be helpful with the exercises involving larger numbers.

Before handing out the worksheet, have students review the meaning of factors of a whole number and then give them a few minutes to find as many factors of 720 as they can. When you have worked with them to be sure they have found all 30 factors, tell them that the purpose of the worksheet is to help them discover a much faster way of determining how many factors a number has.

Hand out the Project worksheet. Some students may need a quick review of what prime factors are and how to write them using exponents. Suggest that the group divide up the work needed to complete all the exercises and share their results.

The rule for determining the number of factors can be stated as: Add 1 to the exponent of each prime factor. Then find the product of these numbers. For example, the number of factors of $2^3 \cdot 3^2$ is $(3 + 1)(2 + 1) = 12$. When the students have completed the worksheet, ask:

1. What does your rule tell you about how many factors the prime number 43 has? $43 = 43^1$; $1 + 1 = 2$; 43 has 2 factors.
2. What kind of number has an odd number of factors?   A perfect square, such as 36 or 49
3. Name a number that has exactly 15 factors.   Possible answer: $2^4 \cdot 3^2 = 144$

George R. Carruthers, Ph.D., is an astrophysicist who designs and builds instruments to study outer space which have been used in Apollo, Skylab, and space shuttle missions. He was the first to detect interstellar molecular hydrogen.

More About the Photo

Dr. Carruthers' interest in outer space began before high school, as he read science fiction comic books and astronomy texts. In high school, he entered science fairs and was a member of the astronomy and space clubs. After receiving his doctorate, he went to the Naval Research Laboratory in Washington, D.C. where he is now a senior astrophysicist. There he works on detecting ultraviolet radiation on deep space missions. To do this he has designed instruments using a *charge coupled device,* a specially prepared semiconductor also used in most television cameras.

7.1 Introduction to Factoring

Teaching Resources

Project Worksheet 7
Quick Quizzes 51
Reteaching and Practice
  Worksheets 51

Objectives

To find the greatest common factor (GCF) of two or more integers
To find the greatest common factor (GCF) of two or more monomials
To find the missing factor, given a monomial and one of its factors

How can you arrange 36 square tiles to form a rectangle? The formula for the area of a rectangle is $A = lw$ (area = length × width), so look for two factors of 36.

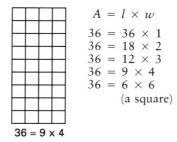

$A = l \times w$
$36 = 36 \times 1$
$36 = 18 \times 2$
$36 = 12 \times 3$
$36 = 9 \times 4$
$36 = 6 \times 6$
     (a square)

36 = 9 x 4

There are five ways to arrange the tiles. In each product, 36 is said to be *factored*. To **factor an integer** means to express it as the product of two or more integers.

There are several ways to factor 36. For example,

$36 = 1 \times 36$ $36 = 2 \times 3 \times 6$
$36 = 4 \times 9$ $36 = 2 \times 2 \times 3 \times 3$

In $36 = 2 \times 2 \times 3 \times 3$, none of the factors can be factored further without using the factor 1. The only positive factors of 2 are 2 and 1; the only positive factors of 3 are 3 and 1. The numbers 2 and 3 are called *prime numbers*.

Definition

A **prime number** is an integer greater than 1 whose only positive factors are 1 and itself.

Other prime numbers are 5, 7, 11, 13, 17, 19, 23, 29, 31, and so on.

Definition

A **composite** number is a positive integer that has two or more positive factors other than 1.

Other composite numbers are 4, 6, 8, 9, 10, 12, 14, 15, and so on. The number 1 is neither prime nor composite.

7.1 Introduction to Factoring   **243**

GETTING STARTED

Prerequisite Quiz

Multiply.

1. $2^3 \cdot 3^2 \cdot 5$ 360
2. $2^2 \cdot 3 \cdot 5$ 60
3. $x^3 \cdot x^6$ x^9
4. $-8a^5 \cdot 2a$ $-16a^6$
5. $\dfrac{a^5}{a^2}$ a^3
6. $\dfrac{-12m^8}{-4m^4}$ $3m^4$

Motivator

Ask students how 500 people at a convention can be divided into equal groups of 2 or more persons.   Groups of 2, 4, 5, 10, 20, 25, 50, 100, 125, and 250   Then ask these questions.

1. How are the group numbers related to 500?   All are factors of 500.
2. Which of the factors are prime numbers?   2, 5
3. How can you write 500 as a product of prime factors? $2 \cdot 2 \cdot 5 \cdot 5 \cdot 5$, or $2^2 \cdot 5^3$

Highlighting the Standards

Standards 2a, 14a: The process of factoring integers is modeled with tiles in a rectangle. Students compare primes and composite numbers and learn several ways to factor.

Lesson Note

Emphasize that prime numbers are *positive* and that 1 is not a prime number. You may wish to have some students use the calculator ideas on page 244 as the basis for writing a computer program for finding prime factorizations. When presenting Example 4, remind students that the answer is a factor, but not necessarily a prime factor.

Math Connections

History: Eratosthenes (c. 274–194 B.C.) was a Greek geographer and mathematician who calculated the circumference of the earth 2000 years ago. His estimate was only about 150 miles off the true value. The Sieve of Eratosthenes can be used to find prime numbers.

Critical Thinking Questions

Analysis: Remind students that the prime factorization of an integer must be unique; that is, there is exactly one prime factorization for a number. Then ask what connection this has to the fact that 1 is not considered to be a prime number. If 1 were prime, the prime factorization of an integer would not be unique; e.g., $10 = 2 \cdot 5$ and $10 = 1 \cdot 2 \cdot 5$.

A factorization such as $36 = 2 \times 2 \times 3 \times 3$ is called a **prime factorization** because all the factors are prime numbers.

EXAMPLE 1 Give the prime factorization of 90.

Plan Choose two positive factors of 90. Continue factoring until all the factors are prime numbers.

Solution

Method 1	Method 2	Method 3
$90 = 45 \cdot 2$	$90 = 6 \cdot 15$	$90 = 30 \cdot 3$
$\quad = 9 \cdot 5 \cdot 2$	$\quad = 2 \cdot 3 \cdot 3 \cdot 5$	$\quad = 6 \cdot 5 \cdot 3$
$\quad = 3 \cdot 3 \cdot 5 \cdot 2$		$\quad = 3 \cdot 2 \cdot 5 \cdot 3$

Each method gives the same prime factors, but in a different order. In each case, 3 is a factor twice, and 5 and 2 are the other factors.

Thus, the prime factorization of 90 is $2 \cdot 3 \cdot 3 \cdot 5$, or $2 \cdot 3^2 \cdot 5$.

TRY THIS 1. Give the prime factorization of 100. $2^2 \cdot 5^2$

Example 1 illustrates that the prime factorization of an integer is *unique*. Only the order of the prime factors can differ.

A systematic way of finding the prime factorization of a positive integer is to divide by 2 as often as 2 is a factor, then by 3, and so on. You can use this method with your calculator. If a decimal appears in the display, return to the previous step. Then divide by the next larger prime number. Find the prime factorization of 660.

$660 \;\boxed{\div}\; 2 \;\boxed{=}\; 330$	$660 = 2 \cdot 330$
$330 \;\boxed{\div}\; 2 \;\boxed{=}\; 165$	$660 = 2 \cdot 2 \cdot 165$
$165 \;\boxed{\div}\; 3 \;\boxed{=}\; 55$	$660 = 2 \cdot 2 \cdot 3 \cdot 55$
$55 \;\boxed{\div}\; 5 \;\boxed{=}\; 11$	$660 = 2 \cdot 2 \cdot 3 \cdot 5 \cdot 11$

Thus, $660 = 2^2 \cdot 3 \cdot 5 \cdot 11$.

The integer 6 is a factor of *both* 30 and 12.

$$30 = 6 \cdot 5 \qquad 12 = 6 \cdot 2$$

Therefore, 6 is a **common factor** of 30 and 12. Notice that 6 is the *greatest common factor* of 30 and 12.

Definition The **greatest common factor** (GCF) of two or more integers is the largest integer that is a factor of all the integers.

Additional Example 1

Give the prime factorization of 48.

$2 \cdot 2 \cdot 2 \cdot 2 \cdot 3$, or $2^4 \cdot 3$

EXAMPLE 2   Find the GCF of 72 and 84.

Plan   To find the GCF of 72 and 84, factor both integers into primes. For each common factor of 72 and 84, compare its powers and choose the smaller power. The product of these powers is the GCF.

Solution $72 = 12 \cdot 6 = 2 \cdot 2 \cdot 3 \cdot 2 \cdot 3 = 2^3 \cdot 3^2$
$84 = 14 \cdot 6 = 2 \cdot 7 \cdot 2 \cdot 3 \quad = 2^2 \cdot 3 \cdot 7$

smaller power of 2 $\longrightarrow$ 2^2 3^1 $\longleftarrow$ smaller power of 3

So, the GCF of 72 and 84 is $2^2 \cdot 3$, or 12.

TRY THIS   **2.** Find the GCF of 36 and 90.   18

It may be necessary to find the GCF of two or more monomials whose factors are variables. The monomials x^3, x^2, and x^5 can be written as follows.

$$x^3 = x \cdot x \cdot x, \qquad x^2 = x \cdot x, \qquad x^5 = x \cdot x \cdot x \cdot x \cdot x$$

The product $x \cdot x$ is a common factor of these monomials. Thus, the GCF of x^3, x^2, and x^5 is x^2, the *smallest* power of x.

EXAMPLE 3   Find the GCF of $12x^4$, $-28x^3$, and $120x^2$.

Solution   Rewrite $-28x^3$ as $-1 \cdot 28 \cdot x^3$.

$12x^4 = 2 \cdot 2 \cdot 3 \cdot x^4 \qquad\qquad = \quad\quad 2^2 \cdot 3 \cdot \quad x^4$
$-28x^3 = -1 \cdot 2 \cdot 2 \cdot 7 \cdot x^3 \quad = -1 \cdot 2^2 \cdot 7 \cdot \quad x^3$
$120x^2 = 2 \cdot 2 \cdot 2 \cdot 3 \cdot 5 \cdot x^2 \quad = \quad\quad 2^3 \cdot 3 \cdot 5 \cdot x^2$

smallest power of 2 times smallest power of x $\longrightarrow$ $2^2 \cdot$ x^2

Thus, $2^2 \cdot x^2$, or $4x^2$, is the GCF of $12x^4$, $-28x^3$, and $120x^2$.

TRY THIS   **3.** Find the GCF of $84y^3$, $98y^6$, and $-196y^5$. $14y^3$

Monomials such as $3a$ and $5b^2$ have no common factors other than 1. Thus, the GCF of $3a$ and $5b^2$ is 1.

Sometimes you need to find a missing factor, as in $x^3 \cdot \underline{\ ?\ } = x^7$. To do this, divide the given product by the given factor.

$$\frac{x^7}{x^3} = x^4 \qquad \text{Thus, the missing factor is } x^4. \qquad x^3 \cdot x^4 = x^7$$

Additional Example 2
Find the GCF of 18 and 54.   18

Additional Example 3
Find the GCF of $9y^5$, $-3y^6$, and $36y^4$. $3y^4$

Guided Practice

Classroom Exercises 1–21
Try This   all

Independent Practice

A Ex. 1–57, **B** Ex. 58–67, **C** Ex. 68–72
Basic: WE 1–57 odd
Average: WE 11–67 odd
Above Average: WE 17–72 odd

EXAMPLE 4   Find the missing factor.

a. $a^5 \cdot \underline{\ ?\ } = a^9$ **b.** $\underline{\ ?\ } \cdot 5a = -30a^{10}$ **c.** $3a^3b^5 \cdot \underline{\ ?\ } = 24a^4b^7$

Solutions   **a.** $\dfrac{a^9}{a^5} = a^4$ **b.** $\dfrac{-30a^{10}}{5a} = -6a^9$ **c.** $\dfrac{24a^4b^7}{3a^3b^5} = 8ab^2$

So, the missing factors are a^4, $-6a^9$, and $8ab^2$, respectively.

TRY THIS   **4.** Find the missing factor for $13a^4b^7 \cdot \underline{\ \ \ } = 52\,a^9b^8$. $_{4a^5b}$

Classroom Exercises

Give the prime factorization of each number.

1. 4 $_{2^2}$ **2.** 8 $_{2^3}$ **3.** 6 $_{2\cdot3}$ **4.** 9 $_{3^2}$ **5.** 14 $_{2\cdot7}$ **6.** 10 $_{2\cdot5}$
7. 15 $_{3\cdot5}$ **8.** 21 $_{3\cdot7}$ **9.** 12 $_{2^2\cdot3}$ **10.** 18 $_{2\cdot3^2}$ **11.** 20 $_{2^2\cdot5}$ **12.** 24 $_{2^3\cdot3}$

Give the GCF of each pair of monomials.

13. $2^5 \cdot 5^3$ and $2^3 \cdot 5^6$ $_{1,000}$ **14.** $3^2 \cdot 7^3$ and $3^4 \cdot 7$ $_{63}$ **15.** x^4 and x^3 $_{x^3}$
16. $3x^4$ and $9x^3$ $_{3x^3}$ **17.** $5a^2$ and $3a^5$ $_{a^2}$ **18.** $4n^7$ and $5t^3$ $_1$

Find the missing factor.

19. $x^2 \cdot \underline{\ ?\ } = x^7$ $_{x^5}$ **20.** $-a^3 \cdot \underline{\ ?\ } = -5a^7$ $_{5a^4}$ **21.** $3b^4 \cdot \underline{\ ?\ } = -12b^5$ $_{-4b}$

Written Exercises

Give the prime factorization of each number.

1. 40 $_{2^3\cdot5}$ **2.** 16 $_{2^4}$ **3.** 45 $_{3^2\cdot5}$ **4.** 42 $_{2\cdot3\cdot7}$ **5.** 49 $_{7^2}$ **6.** 51$_{3\cdot17}$
7. 28   **8.** 32 $_{2^5}$ **9.** 60   **10.** 180   **11.** 200   **12.** 360
$\quad\ _{2^2\cdot7}$ $\qquad\qquad\qquad\quad _{2^2\cdot3\cdot5}$ $_{2^2\cdot3^2\cdot5}$ $\quad _{2^3\cdot5^2}$ $\ _{2^3\cdot3^2\cdot5}$

Find the greatest common factor (GCF).

13. 12, 15 $_3$ **14.** 8, 20 $_4$ **15.** 20, 32 $_4$ **16.** 27, 81 $_{27}$ **17.** 30, 18$_6$ **18.** 36, 54 ^{18}
19. 81, 144 $_9$ **20.** 45, 90 $_{45}$ **21.** 18, 24 $_6$ **22.** 26, 78 $_{26}$ **23.** 34, 51$_{17}$ **24.** 48, 32
25. 4, 10, 6 $_2$ **26.** 14, 21, 35 $_7$ **27.** 12, 10, 15 $_1$ ^{16}
28. 14, 15, 16 $_1$ **29.** 6, 8, 4, 20 $_2$ **30.** 18, 12, 24 $_6$
31. 60, 48, 84, 12 $_{12}$ **32.** $3x^3, 9x^5$ $_{3x^3}$ **33.** $6m^5, 8m^4$ $_{2m^4}$
34. $12a^3, 18a$ $_{6a}$ **35.** $24x^3, 33y^4$ $_3$ **36.** $6x^2, 15x^3$ $_{3x^2}$
37. $14x^3, -21x^4$ $_{7x^3}$ **38.** $-30y^3, 20y$ $_{10y}$ **39.** $3x^3, 6x^2, -9x$ $_{3x}$
40. $4a^4m, 6a^3, 12a^2$ $_{2a^2}$ **41.** $6a^5, 8a^4, 2a^3$ $_{2a^3}$ **42.** $5a, -b^5, 8c^4$ $_1$
43. $10y^3, 5y^2, 20y, -40y^2$ $_{5y}$ **44.** $4a^5, -12a^4, 28a^3$ $_{4a^3}$ **45.** $3x^3, 6x^4, 5x^2, 2x$ $_x$

Additional Example 4

Find the missing factor.

a. $a^5 \cdot \underline{\ ?\ } = a^{12}$ $_{a^7}$
b. $7x^2 \cdot \underline{\ ?\ } = 28x^4$ $_{4x^2}$
c. $-4x^2y \cdot \underline{\ ?\ } = 36x^5y^5$ $_{-9x^3y^4}$

46. $6ab^2$, $6a^2$, $12ab^3$ $6a$ **47.** $2m^3n$, $8m^2n^2$, $16m^2$ $2m^2$ **48.** $3mn$, $6n$, $3m$ 3

Find the missing factor.

49. $a^4 \cdot \underline{a^6} = a^{10}$ **50.** $m^3 \cdot \underline{m^5} = m^8$ **51.** $\frac{b^7}{n} \cdot n = b^7$ $\frac{2y^{10}}{m^6}$

52. $\underline{x^8} \cdot x^6 = x^{14}$ **53.** $7m^3 \cdot \underline{-3m^5} = -21m^8$ **54.** $9m^6 \cdot \underline{\quad} = 18y^{10}$

55. $\underline{-3a^7} \cdot 4a^6 = -12a^{13}$ **56.** $4x^5 \cdot \underline{8x} = 32x^6$ **57.** $7c^3 \cdot \underline{9c^2} = 63c^5$

58. $8a^3b^7 \cdot \underline{9a^6b^3} = 72a^9b^{10}$ **59.** $-6x^4y^6 \cdot \underline{-16y^6} = 96x^4y^{12}$

60. $\underline{15x^7y^2} \cdot 5xy^{10} = 75x^8y^{12}$ **61.** $5a^3b^4 \cdot \underline{-6a^3b} = -30a^6b^5$

62. $6a^2m^7 \cdot \underline{4a^3m^2} = 24a^5m^9$ **63.** $\underline{-13m} \cdot 3a^4m^6 = -39a^4m^7$

64. The monomial $45x^4y^3$ was factored into two factors. One factor was $-9xy^3$. What was the other factor? $-5x^3$

65. The product of $4a^3b^5$ and what monomial is $80a^4b^6$? $20ab$

66. A band director has a formation designed for 118 members, but 2 more members arrive. What are the possible rectangular formations?

67. An artist has 98 square tiles and wishes to make a rectangular design with at least 4 tiles in a column. What arrangements are possible? 7×14

68. Explain in writing why $2x^2$ is the greatest common factor of $(2xy)^2$ and $18x^6$.

In Exercises 69–72, use the formula for the area of a square, $A = s^2$, or the formula for the area of a rectangle, $A = lw$.

69. If the area of a rectangle is $72a^mb^{2n}$ and the length is $4a^2b^2$, what is the width? $18a^{m-2}b^{2n-2}$

70. If the area of a rectangle is $144x^{2a}y^{4b}$ and the width is $6x^{4a}y^{3b}$, what is the length? $24x^{-2a}y^b$

71. If the area of a square is $36x^{4a}y^{6b}$, what is the length of each side? $6x^{2a}y^{3b}$

72. If the area of a square is $81a^{8c}b^{12d}$, what is the length of each side? $9a^{4c}b^{6d}$

Mixed Review

Simplify. *1.6, 2.10, 6.2, 6.9*

1. $5(6a - 4)$ $30a - 20$ **2.** $4y + 6 - (8 - y)$ $5y - 2$ **3.** $3x - 7 - 5(x + 2)$ $-2x - 17$

4. $(-4x^2y)^3$ $-64x^6y^3$ **5.** $(3x - 2)(4x + 5)$ $12x^2 + 7x - 10$ **6.** $(x + 5)(x^2 - 4x + 2)$ $x^3 + x^2 - 18x + 10$

Solve. *5.6, 5.7*

7. $|x| = 8$ $x = 8 \text{ or } x = -8$ **8.** $|x - 3| < 9$ $-6 < x < 12$ **9.** $|x + 5| > 3$ $x > -2 \text{ or } x < -8$

10. The smaller of two numbers is 4 less than the greater. If the greater number is decreased by twice the smaller, the result is -10. Find the two numbers. *4.2* $14, 18$

Additional Answers

Written Exercises

66. {1 × 120; 2 × 60; 3 × 40; 4 × 30; 5 × 24; 6 × 20; 8 × 15; 10 × 12}

68. $(2xy)^2 = 2 \cdot 2 \cdot x \cdot x \cdot y \cdot y$; $18x^6 = 2 \cdot 3 \cdot 3 \cdot x \cdot x \cdot x \cdot x \cdot x \cdot x$; GCF: $2 \cdot x \cdot x$, or $2x^2$

Enrichment

Have students start with a 10 by 10 grid of the integers from 1 to 100. Then have them find all the prime numbers less than 100 by using the Sieve of Eratosthenes. Strike out 1. Circle 2, the first prime number, but strike out all the other multiples of 2: 4, 6, 8, 10, 12. Circle 3, the next number that has not been struck out. Then strike out all the other multiples of 3. Continue in this fashion until reaching a prime that is greater than the square root of 100: 11. All numbers that have not been struck out are prime. Have students count the number of primes less than 100. 25

Teaching Resources

Application Worksheet 7
Quick Quizzes 52
Reteaching and Practice
    Worksheets 52
Transparency 17

7.2 Greatest Common Monomial Factor

▰▰▰GETTING STARTED

Prerequisite Quiz

Multiply.

1. $3(3a^3 + 4a^2 - 2)$ $9a^3 + 12a^2 - 6$
2. $x^3(6x^5 + 3x^4 - 2x^2)$ $6x^8 + 3x^7 - 2x^5$
3. $2y(-3y + 5y^2 - 6y^4)$
   $-6y^2 + 10y^3 - 12y^5$
4. $-5b^4(-3b^4 + b - 8)$
   $15b^8 - 5b^5 + 40b^4$

Motivator

Ask students to give the greatest common factor for each pair.

1. 9 and 27   9
2. 18 and 27   9
3. 18 and $27x$ 9
4. $18x$ and $27x$ $9x$
5. $18x^2$ and $27x^3$ $9x^2$

Objective	To factor out the greatest common monomial factor from a polynomial

The area of the large rectangle is

$$5x + 5y.$$

The Distributive Property can be used to rewrite the area as

$$5(x + y).$$

The Distributive Property can be used to multiply a polynomial having any number of addends by a monomial. For example,

$$2(x^2 - 3x - 5) = 2x^2 - 6x - 10.$$

The Distributive Property can also be used to factor polynomials. For example,

$$3x^2 - 9x + 6 \qquad 3(x^2 - 3x + 2).$$

The **GCF of a polynomial** is the greatest common factor of its terms.

EXAMPLE 1   Factor out the GCF from $5y^2 + 10y + 20$.

Plan   First find the GCF of the terms $5y^2$, $10y$, and 20.

$5y^2 = 5 \cdot y^2$, $10y = 5 \cdot 2y$, $20 = 5 \cdot 4$ The GCF is 5.

Write each term as a product of 5 and a monomial. Then use the Distributive Property.

Solution $5y^2 + 10y + 20 = 5(y^2) + 5(2y) + 5(4)$
$= 5(y^2 + 2y + 4)$

EXAMPLE 2   Factor out the GCF from $t^5 - 8t^4 + t^3$.

Solution   The terms are $1t^5$, $-8t^4$, and $1t^3$.
The GCF of the numerical coefficients is 1.
The GCF of the variables is t^3, the smallest power of t.
Write each term as a product of t^3 and a monomial.

$$t^5 - 8t^4 + t^3 = t^3(t^2) + t^3(-8t) + t^3(1)$$

Thus, $t^5 - 8t^4 + t^3 = t^3(t^2 - 8t + 1)$.

TRY THIS   1. Factor out the GCF from $2t^4 + 4t^3 - 5t^2$. $t^2(2t^2 + 4t - 5)$

248   Chapter 7 Factoring Polynomials

Highlighting the Standards

Standards 4c, 14b: Students use factoring in connection with various geometric figures and find the GCF of polynomials.

Additional Example 1

Factor out the GCF from $14x^2 + 10x - 18$.
$2(7x^2 + 5x - 9)$

Additional Example 2

Factor out the GCF from $t^5 - 6t^3 + t^2$.
$t^2(t^3 - 6t + 1)$

Some terms include constants and variables as factors.

EXAMPLE 3 Factor out the GCF from $2b^4 - 10b^3 + 8b^2$.

Plan The GCF of 2, -10, and 8 is 2.
The GCF of b^4, b^3, and b^2 is b^2. The GCF of the polynomial is $2b^2$.
Write each term as a product of $2b^2$ and a monomial. Then factor.

Solution $2b^4 - 10b^3 + 8b^2 = 2b^2(b^2) + 2b^2(-5b) + 2b^2(4)$
$$= 2b^2(b^2 - 5b + 4)$$

EXAMPLE 4 Factor out the GCF from $24a^2b^5 - 8a^7b^6 + 12a^4b^3$.

Solution The GCF of 24, -8, and 12 is 4.
The GCF of a^2, a^7, and a^4 is a^2.
The GCF of b^5, b^6, and b^3 is b^3. The GCF of the polynomial is $4a^2b^3$.
Write each term as a product of $4a^2b^3$ and a monomial. Then factor.

$$24a^2b^5 - 8a^7b^6 + 12a^4b^3 = 4a^2b^3(6b^2) + 4a^2b^3(-2a^5b^3) + 4a^2b^3(3a^2)$$
$$= 4a^2b^3(6b^2 - 2a^5b^3 + 3a^2)$$

TRY THIS 2. Factor out the GCF from $30x^5y + 12x^6y^2 - 6x^4y^4$. $6x^4y(5x + 2x^2y - y^3)$

Factoring can be used to simplify computational work with many area problems.

EXAMPLE 5 Find the area of the shaded region.

Write the result in factored form.

Solution First, by drawing two radii, you can
see that the side of the square is $2r$.

Area of shaded region = area of square − area of circle
$$A = 2r \cdot 2r - \pi r^2$$
$$= 4r^2 - \pi r^2$$
$$= r^2(4 - \pi)$$

Sometimes the GCF of a polynomial is 1 as in $5t^2 - 12t + 3$.

7.2 Greatest Common Monomial Factor **249**

■ **TEACHING SUGGESTIONS**

Lesson Note

Point out to students that, after factoring out
the GCF, they should check two things.
(1) The GCF of the terms of the polynomial
remaining within the parentheses must be 1.
(2) The product indicated in the answer must
give the original polynomial. (Students
should mentally multiply the factors to check
this.) When presenting Example 5, remind
students that π is a number, not a variable.
However, it is not an integer and cannot be
factored into prime factors. Give the value
of r as 12 ft., and have students use a
calculator to find the shaded area in
Example 5. Use 3.14 for π.   123.84

Math Connections

Geometry: The surface area of a cylinder is
$2\pi r^2 + 2\pi rh$. The factored form, $2\pi r(r + h)$
is much easier to use especially when
entering a problem into a calculator.

Critical Thinking Questions

Application: Tell students that if the GCF of
two numbers is 1, then the numbers are said
to be *relatively prime,* even though one or
both of the numbers may not be prime. For
example, 14 and 15 are relatively prime. Ask
students how this concept can be applied to
deciding whether a fraction is in lowest
terms.   A fraction is in lowest terms if the
numerator and denominator are relatively
prime.

Additional Example 3

Factor out the GCF from $6f^5 - 3f^4 - 12f^3$.
$3f^3(2f^2 - f - 4)$

Additional Example 4

Factor out the GCF from $16a^4b^2 + 12a^2b^3 - 20a^3b^5$. $4a^2b^2(4a^2 + 3b - 5ab^3)$

Additional Example 5

Factor out the GCF from $c + rc$. Then
evaluate for $c = \$30$ and $r = 0.06$.
$c(1 + r)$; $\$30(1 + 0.06) = \$31.80$

249

Common Error Analysis

Error: Sometimes students will factor $x^3 - 3x^2 + x$ incorrectly as $x(x^2 - 3x)$.

This can be avoided if students rewrite $x^3 - 3x^2 + x$ as $x(x^2) - x(3x) + x(1)$ before using the Distributive Property.

Checkpoint

Factor out the GCF from each polynomial.

1. $15a^2 - 25a + 30$ $5(3a^2 - 5a + 6)$
2. $3x^4 + x^3 - 2x^2$ $x^2(3x^2 + x - 2)$
3. $12m^4n^3 - 20m^2n + 24m^3n^2$
   $4m^2n(3m^2n^2 - 5 + 6mn)$

Closure

Remind students that they should check their answers by multiplication. Then ask what is wrong with the following factorization:
$6x^4 - 15x^3 + 9x^2 + 3x =$
$3x(2x^3 - 5x^2 + 3x)$.   The last term, 1, inside the parentheses is missing.

▰▰▰FOLLOW UP

Guided Practice

Classroom Exercises 1–20
Try This   all

Independent Practice

Ⓐ Ex. 1–48, **Ⓑ** Ex. 49–57, **Ⓒ** Ex. 58–63

Basic: WE 1–47 odd

Average: WE 11–57 odd

Above Average: WE 17–63 odd

Classroom Exercises

State the GCF of each polynomial.

1. $3x^2 - 6$ 3
2. $5a - 10$ 5
3. $3y - 15$ 3
4. $x^3 + x^2$ x^2
5. $5x^3 - 2x^2$ x^2
6. $3x^4 - x^2$ x^2
7. $3x^2 + 9x$ $3x$
8. $4x^3 + 12x$ $4x$
9. $7a^2 - 14a + 21$ 7
10. $3x^3 + 2x^2 + 4x$ x
11. $4a^2 - 2a + 16$ 2
12. $3a^4 - 9a^2 + 120$ 3

Factor out the GCF from each polynomial.

13. $3x(x^2) + 3x(5x) + 3x(7)$ $3x(x^2 + 5x + 7)$
14. $2x^2(5x^2) - 2x^2(7x) + 2x^2(9)$
15. $5x(2x^2) + 5x(7x) - 5x(3)$ $5x(2x^2 + 7x - 3)$
16. $4a^2(3a^2) - 4a^2(13a) - 4a^2(11)$
17. $m^3 + 5m^2 - 8m$ $m(m^2 + 5m - 8)$
18. $5x^3 - 15x^2 + 10x$ $5x(x^2 - 3x + 2)$
19. $14y^2 + 28y - 7$ $7(2y^2 + 4y - 1)$
20. $15t^4 - 20t^2 + 25t$ $5t(3t^3 - 4t + 5)$

Written Exercises

Factor out the GCF from each polynomial. If the GCF is 1, just write the polynomial.

1. $2x^2 - 8x - 6$ $2(x^2 - 4x - 3)$
2. $3a^2 + 9a - 15$ $3(a^2 + 3a - 5)$
3. $5t^2 - 15t + 25$ $5(t^2 - 3t + 5)$
4. $5b^2 + 35b - 60$
5. $4x^2 - 12x - 36$
6. $3a^2 + 7a + 2$
7. $28y^2 - 20y - 24$
8. $30y^2 - 35y - 45$
9. $2x^2 + 11x - 4$
10. $2x^2 - 10x + 12$
11. $8a^2 + 16a + 8$
12. $12x^2 - 10x - 12$
13. $6x^2 - 21x - 12$
14. $18a^2 + 21a - 9$
15. $4b^2 + 26b - 14$
16. $2x^2 + 10x - 28$
17. $5b^2 + 12b - 60$
18. $5y^2 - 45y - 110$
19. $a^3 + 3a^2 - a$
20. $2y^3 - y^2 + y$
21. $m^4 - 3m^3 - 7m^2$
22. $7r^4 - 2r^3 + 8$ $7r^4 - 2r^3 + 8$
23. $a^6 - a^4 - 2a^2$
24. $y^7 - 2y^6 + 20$
25. $9a^3 - 3a^2 + 6a$
26. $4x^3 - 12x^2 + 16x$
27. $3a^2 - 21a$ $3a(a - 7)$
28. $4x^4 - 24x^2$ $4x^2(x^2 - 6)$
29. $5r^2 - 36$ $5r^2 - 36$
30. $6a^3 - 18a^2 - 24a$
31. $4y^3 - 20y^2 + 24y$
32. $7y^3 - 21y^2 + 14y$
33. $9x^5 + 81x^3 - 27x$
34. $51p^2q + 3pq$
35. $3x^3 - 33x^2 + 84x$
36. $6z^3 - 3z^2 - 30z$
37. $5a^5 - b^5$ $5a^5 - b^5$
38. $42r - 14r^3$ $14r(3 - r^2)$
39. $2y^3 - 4y^2 - 48y$
40. $x^4 - 9x^2$ $x^2(x^2 - 9)$
41. $6f^2 - 15f^3$ $3f^2(2 - 5f)$
42. $n^2 + 5n + 7$
43. $2t^3 - 128t^5$ $2t^3(1 - 64t^2)$
44. $3x^2 + 21x^4$ $3x^2(1 + 7x^2)$
45. $8a^2 + b^2 + ab$
46. $ab^2 - ab - 72a$
   $a(b^2 - b - 72)$
47. $8a^4 + 4a^3 - 12a^2$
   $4a^2(2a^2 + a - 3)$
48. $st^2 - st - 20s$
   $s(t^2 - t - 20)$

250   Chapter 7 Factoring Polynomials

Write a formula in factored form for the area of each shaded region.

49.

4 quarter-circles in a square $r^2(4 - \pi)$

50.
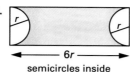
←—— 6r ——→
semicircles inside a rectangle
$r^2(12 - \pi)$

51.

←2r→
square with semicircle ends
$r^2(4 + \pi)$

Factor out the GCF from each polynomial.

52. $8a^3b + 12a^4b^3 - 4ab^2$

53. $6x^2y^3 - 20x^4y^5 + 36x^7y^6$

54. $14m^4n^7 + 21m^3n^8 - 35m^9n^6$

55. $8x^9y^4 - 20x^7y^8 + 12x^5y^9$

56. $100a^{10}b^4 - 50a^8b^8 + 75a^6b^9$

57. $35c^4d^2 + 45c^3d^3 - 50c^2d^4$

58. $x^{n+3} + x^n$ $x^n(x^3 + 1)$

59. $2y^{n+1} + 4y^n$ $2y^n(y + 2)$

60. $4p^{n-1} + 6p^{n+1}$ $2p^{n-1}(2 + 3p^2)$

61. $9w^{2n} + 21w^{2n+1}$ $3w^{2n}(3 + 7w)$

62. $t^{3n+21} + 2t^{2n+14}$ $t^{2n+14}(t^{n+7} + 2)$

63. $3r^{9n-27} - 13r^{6n-18}$
$r^{6n-27}(3r^{3n} - 13r^9)$

Mixed Review

Solve.   *3.4, 5.2, 5.6*

1. $3x - 4 = -x + 20$
6

2. $3x - 4 < 5x + 8$
$x > -6$

3. $|8 - 2b| = 10$
$-1, 9$

Simplify. Use positive exponents only. Assume that no variable equals zero.   *6.2, 6.3, 6.9, 6.10*

4. $(-2x^4y^2)^2$ $4x^8y^4$

5. $\dfrac{16a^4b}{-8a^3b^9}$ $-\dfrac{2a}{b^8}$

6. $(x + 5)(x + 8)$
$x^2 + 13x + 40$

7. $(x + 7)(x - 7)$
$x^2 - 49$

▰▰/Brainteaser

There is a legend that the man who invented the game of chess taught it to the king. The king was so pleased that he offered to give the man anything he wished as payment. "All I wish for payment, sire, is the amount of grain that it would take to do the following. Place one grain of wheat on the first square of this chessboard. Then place two grains on the second, four grains on the third, and so on so that each square has twice as many as the previous square."

1. How many grains of wheat should be on the last square? 2^{63}

2. How many grains of wheat should the man receive altogether? About 1.8×10^{19}

3. Suppose that you use pennies instead of grains of wheat. For which square will the payment reach more than one billion dollars? 37th square

Enrichment

Have students research the method that Eratosthenes used for estimating the circumference of the earth. Have them write a report on this method.

Teaching Resources

Manipulative Worksheet 7
Quick Quizzes 53
**Reteaching and Practice
Worksheets** 53
Teaching Aid 3

7.3 Factoring Trinomials: $x^2 + bx + c$

GETTING STARTED

Prerequisite Quiz

Multiply.

1. $(x - 7)(x + 2)$ $x^2 - 5x - 14$
2. $(x - 3)(x - 4)$ $x^2 - 7x + 12$
3. $(x + 4)(x + 1)$ $x^2 + 5x + 4$
4. $(2x - 1)(x + 2)$ $2x^2 + 3x - 2$
5. $(2x - 5)(3x - 2)$ $6x^2 - 19x + 10$

Motivator

Use the coefficient of the middle term and the last term of 5 or more of the Written Exercises on page 254 to fill in the first two columns of a chart as shown below on the chalkboard or on an overhead projector.

$a + b$	ab	a	b
3	-4	?	?
2	-15	?	?
-9	14	?	?

Ask the students to find two integers, a and b, whose sum is the number given in the $a + b$ column and whose product is the number given in the ab column. $a = 4$, $b = -1$; $a = 5$, $b = -3$; $a = -2$, $b = -7$

Objective   To factor trinomials of the form $x^2 + bx + c$

The product of two binomials is usually a trinomial. For example,

$$(x + 3)(x - 2) = x^2 + x - 6.$$

To factor a trinomial means to express it as a product of two binomials. Thus, factoring can be thought of as undoing the result of multiplication; that is, to factor $x^2 + x - 6$ is to rewrite it as $(x + 3)(x - 2)$.

To factor a trinomial, the FOIL method of multiplying can be used in reverse. For example, to factor $x^2 - 2x - 48$, the first step is to think of x^2 as $x \cdot x$ and write

$$x^2 - 2x - 48 = (x \quad)(x \quad).$$

Now examine the sign of the constant term, -48. Since the sign of the constant term is negative, the sign of the second terms in the binomials must be opposites. Thus you can write

$$x^2 - 2x - 48 = (x + \quad)(x - \quad).$$

Apply the guess and check strategy to find the numbers needed to complete the binomial factors. However, the FOIL method does tell us that the product of those numbers must be -48 and their sum must be -2.

Guess	Check	Result
1. $4, -12$	$4 \cdot -12 = -48$; $4 + -12 = -8$	Does not work.
2. $8, -6$	$8 \cdot -6 = -48$; $8 + -6 = 2$	Does not work.
3. $-8, 6$	$-8 \cdot 6 = -48$; $-8 + 6 = -2$	It works.

The numbers are -8 and 6. The binomial factors can be completed.

$$x^2 - 2x - 48 = (x + 6)(x - 8)$$

The FOIL method indicates which signs to choose in factoring $x^2 + bx + c$. If the constant term c is negative, then the second terms in the binomial factors will have opposite signs.

On the other hand, if the constant term c is positive, then the second terms in the binomial factors must have the same sign, both negative or both positive.

$(x - 4)(x - 6) = x^2 - 10x + 24$ $(x + 7)(x + 2) = x^2 + 9x + 14$

| same sign (negative) | positive constant | same sign (positive) | positive constant |

EXAMPLE 1   Factor $x^2 - 13x + 40$.

Solution

Trial 1
Try -10 and -4 as factors of 40: $(x - 10)(x - 4)$.

Check: $(x - 10)(x - 4) = x^2 - 4x - 10x + 40$
$= x^2 - 14x + 40$

This combination does *not* work.

$$x^2 - 14x + 40 \neq x^2 - 13x + 40$$

Trial 2
Try -5 and -8 as factors of 40: $(x - 5)(x - 8)$.

Check: $(x - 5)(x - 8) = x^2 - 8x - 5x + 40$
$= x^2 - 13x + 40$

This combination *works*, so $x^2 - 13x + 40 = (x - 5)(x - 8)$

TRY THIS

1. Factor $x^2 + 11x + 24$. $(x + 3)(x + 8)$
2. Factor $x^2 - 15x + 56$. $(x - 7)(x - 8)$

By now you may have discovered an easier way to factor a polynomial of the form $x^2 + bx + c$, where the coefficient of x^2 is 1. Examples 1 and 2 suggest a pattern. Look for two numbers whose product is c and whose sum is b.

$$x^2 - 2x - 48 = (x - 8)(x + 6)$$
Product: $-8(6) = -48$
Sum: $-8 + 6 = -2$

$$x^2 - 13x + 40 = (x - 5)(x - 8)$$
Product: $-5(-8) = 40$
Sum: $-5 + (-8) = -13$

EXAMPLE 2   Factor $x^2 - x - 20$.

Solution

$x^2 - x - 20 = x^2 - 1x - 20$

Try factors of -20. Check to see whether their sum is -1.

$-20 = -10(2)$ $-10 + 2 = -8$ Sum is *not* -1.
$-20 = -4(5)$ $-4 + 5 = 1$ Sum is *not* -1.
$-20 = 4(-5)$ $4 + (-5) = -1$ Sum *is* -1.

So, $x^2 - x - 20 = (x + 4)(x - 5)$.

TRY THIS

3. Factor $x^2 - 2x - 24$. $(x - 6)(x + 4)$

7.3 Factoring Trinomials: $x^2 + bx + c$ **253**

Lesson Note

Many students believe that mathematics is an exact science, but as these examples show, there is often much trial and error in the processes they must master. Since there are a limited number of pairs of factors that give a particular product, have the students start with the last term rather than trying to guess pairs of numbers that give the required sum of the middle term.

Math Connections

Vocabulary: A polynomial is classified as reducible if it can be expressed as a product of polynomials of lower degree. If this is not possible, the polynomial is classified as irreducible.

Critical Thinking Questions

Analysis: The factors of $x^2 + 10x + 24$ are $x + 4$ and $x + 6$. Ask students to give the factors of $x^2 + 10x - 24$. $x + 12$ and $x - 2$ Tell students that polynomials of the forms $x^2 + bx + c = 0$ that can be factored when $c > 0$ and when $c < 0$ are called *twin polynomials*. Ask them what conditions for c are needed in twin polynomials.   The sum of one set of factors for c must equal the difference of a second set of factors of c. Then have students find other examples of twin polynomials.   Some examples are $x^2 + 5x \pm 6$, $x^2 + 13x \pm 30$, $x^2 + 15x \pm 54$.

Common Error Analysis

Error: Students often forget to check the sign of the middle term. For example, they may factor $x^2 - 8x + 12$ as $(x + 6)(x + 2)$.

Emphasize the importance of *checking* the sign of the middle term.

Additional Example 1

Factor $x^2 - 11x + 28$. $(x - 4)(x - 7)$

Additional Example 2

Factor $x^2 + x - 42$. $(x + 7)(x - 6)$

Checkpoint

Factor each trinomial.

1. $x^2 + 7x + 10$ $(x + 2)(x + 5)$
2. $x^2 - 10x + 21$ $(x - 3)(x - 7)$
3. $x^2 - 2x - 24$ $(x + 4)(x - 6)$
4. $x^2 + 3x - 40$ $(x - 5)(x + 8)$

Closure

▮▮FOLLOW UP

Guided Practice

Classroom Exercises 1–12
Try This   all

Independent Practice

A Ex. 1–24, **B** Ex. 25–28, **C** Ex. 29–37

Basic: WE 1–21 all, Application

Average: WE 7–27 all, Application

Above Average: WE 1–37 odd, Application

▬▬ Summary

To factor a trinomial of the form $x^2 + bx + c$, look for binomial factors $(x + r)(x + s)$ such that $r \cdot s = c$ and $r + s = b$.
  If c is *negative*, then r and s have *opposite signs*.
  If c is *positive* and b is *negative*, then r and s are *both negative*.
  If c is *positive* and b is *positive*, then r and s are *both positive*.

Classroom Exercises

For each trinomial, predict the signs contained in its binomial factors. State all possible factors of the constant term.

1. $x^2 + 5x + 6$
2. $a^2 + a - 20$
3. $y^2 + 9y + 14$
4. $y^2 - 2y - 35$
5. $t^2 - 19t + 34$
6. $s^2 + s - 16$
7. $b^2 + 5b + 4$
8. $k^2 - k - 30$
9. $j^2 + 5j - 14$

Factor each trinomial.

10. $x^2 - 5x + 4$
$(x - 1)(x - 4)$
11. $x^2 - x - 6$
$(x - 3)(x + 2)$
12. $a^2 + 6a + 5$
$(a + 5)(a + 1)$

Written Exercises

Factor each trinomial.

1. $x^2 + 3x - 4$ $(x + 4)(x - 1)$ 2. $y^2 + 2y - 15$ $(y + 5)(y - 3)$ 3. $a^2 - 9a + 14$ $(a - 7)(a - 2)$
4. $b^2 - 4b - 21$
5. $r^2 + r - 6$ $(r + 3)(r - 2)$
6. $x^2 - 10x + 9$
7. $t^2 - 2y - 24$ $(t - 6)(t + 4)$ 8. $a^2 - a - 42$ $(a - 7)(a + 6)$ 9. $y^2 - 9y + 22$
10. $s^2 - 11s + 28$
11. $b^2 - 6b - 27$
12. $m^2 - 10m + 25$
13. $x^2 + 7x + 12$
14. $t^2 + 10t + 24$
15. $y^2 - 18y + 17$
16. $a^2 - 11a - 60$
17. $x^2 - 16x + 48$
18. $m^2 - 6m - 40$
19. $y^2 + 17y + 72$
20. $y^2 - 16y + 64$
21. $a^2 + 15a + 54$
22. $x^2 - 15x + 44$
23. $c^2 - c - 56$
24. $w^2 - 3w - 54$

25. The formula for the area of a rectangle is $A = lw$. Find possible algebraic expressions for l and w if $A = x^2 + 13x + 36$. $(x + 4), (x + 9)$

26. The formula for the area of a parallelogram is $A = bh$. Find possible algebraic expressions for b and h if $A = x^2 - 3x - 10$. $(x - 5), (x + 2)$

27. The formula for the area of a square is $A = s^2$. Find a possible algebraic expression for s if $A = x^2 - 8x + 16$. $(x - 4)$

28. The formula for the area of a triangle is $A = \frac{1}{2}bh$. Find possible algebraic expressions for b and h if $A = \frac{1}{2}(x^2 - 11x + 30)$. $(x - 5), (x - 6)$

Factor. Assume that a, b, c, d, e, and n are positive integers.

29. $x^2 + dx + ex + de$ **30.** $x^2 - dx - cx + dc$ **31.** $x^2 + nx + 3x + 3n$

32. $a^{10n} + 3a^{5n} - 130$ **33.** $x^{6a} + 21x^{3a} + 108$ **34.** $-98 - 3t^{4d} + t^{8d}$

35. $y^{6b} + 27y^{3b} + 180$ **36.** $a^{4c} + 17a^{2c} - 38$ **37.** $t^{8n} - 27t^{4n} - 160$

$(y^{3b} + 15)(y^{3b} + 12)$ $(a^{2c} - 2)(a^{2c} + 19)$ $(t^{4n} - 32)(t^{4n} + 5)$

Mixed Review

Simplify. *1.2, 1.3, 6.6*

1. $3.8 + 4.1 \times 2.3$ 13.23
2. $1.8(4.4 - 1.6)$ 5.04
3. 4^3 64
4. 2^5 32
5. $(4 \cdot 3)^2$ 144
6. $3^3 + 3^2 + 3^4$ 117
7. $8x^2 + 3x + 4x - 2x^2 - 8$ $6x^2 + 7x - 8$
8. $9y^3 - 3y + 8y^2 - 2y^3 - y + 2y^2$
$7y^3 + 10y^2 - 4y$

Application: Gas Mileage

Anita is planning to tour the Northwest and needs to estimate her gasoline costs. To make this estimate, she writes a proportion.

$$\frac{\text{miles traveled on one gallon}}{\text{one gallon of gasoline}} = \frac{\text{total miles for the trip}}{\text{total gallons needed}}$$

Anita knows that her car usually travels about 23 miles on one gallon of gasoline. Since she plans a trip of about 3,500 miles, Anita substitutes the known values into the proportion and solves for the total number of gallons g.

$$\frac{23}{1} = \frac{3500}{g} \quad \longleftarrow \text{If } \frac{a}{b} = \frac{c}{d}, \text{ then } ad = bc.$$
$$23g = 3500$$
$$g = \frac{3500}{23}$$

1. Use your calculator to find the total numbers of gallons of gasoline, rounded to the nearest gallon. 152 gallons

2. If the average price for gasoline is $1.17/gal, what is Anita's total cost for gasoline? $177.84

3. If the average price falls to $1.13/gal, how much less will Anita's total cost be? $6.08

4. If her car only averages 21 mi/gal, how many more gallons will Anita need? 15 gallons

5. Len budgets $50 for fuel for his vacation trip of 1500 miles. His car averages 32 mi/gal. If gasoline costs $1.20 gal, has he budgeted enough? No

Factoring Trinomials: $x^2 + bx + c$ **255**

Enrichment

Write the trinomial $x^2 + bx + 16$ on the chalkboard. Ask the students to find all the integers that could be substituted for b to make the trinomial factorable. Have them explain their answers. $b = 8, -8, 10, -10, 17,$ or -17; b can be the sum of any two integers whose product is 16.

Repeat this procedure with the trinomial $x^2 + cx - 12$. $c = 1, -1, 4, -4, 11,$ or -11; c can be the sum of any two integers whose product is -12. Have the students create exercises of a similar nature and challenge each other with them.

▮▮GETTING STARTED

Prerequisite Quiz

Multiply.

1. $(2x + 3)(3x - 1)$ $6x^2 + 7x - 3$
2. $(5a + 2)(2a + 3)$ $10a^2 + 19a + 6$
3. $(3y - 4)(4y - 1)$ $12y^2 - 19y + 4$
4. $(2t - 5)(3t + 2)$ $6t^2 - 11t - 10$
5. $(7m + 5)(2m - 3)$ $14m^2 - 11m - 15$
6. $(9s - 7)(5s - 6)$ $45s^2 - 89s + 42$

Motivator

Have students replace the _?_ to make each
sentence true.

1. $6x^2 + 11x + 10 = (2x + 5)(3x + \underline{?})$
   2
2. $2x^2 + 7x + 3 = (x + \underline{?})(2x + 1)$ 3
3. $12x^2 - 11x + 2 = (3x - 2)(\underline{?} - 1)$
   $4x$
4. $3x^2 - 10x + 3 = (x - 3)(3x \underline{?} 1)$ −
5. $3x^2 - 5x - 2 = (\underline{?} + 1)(x - 2)$ $3x$
6. $3x^2 - x - 2 = (3x \underline{?} 2)(x - 1)$ +

7.4 Factoring Trinomials: $ax^2 + bx + c$

Objectives | To factor trinomials of the form $ax^2 + bx + c$, where $a > 1$

For the trinomial $ax^2 + bx + c$, the signs of b and c give clues for
finding the binomial factors, if there are any. To determine the binomial
factors of a trinomial such as $3x^2 - 14x + 8$, trial factors must include pairs of factors of $3x^2$ as well as of 8.

EXAMPLE 1 | Factor $3x^2 - 14x + 8$.

Plan | Note that b is negative (-14) and c is positive (8).
So, the trial factors of 8 must both be negative.

Solution | **Trial 1**
Try $3x$ and x as factors of $3x^2$; -4 and -2 as factors of 8.

$$(3x - 4)(x - 2) = 3x^2 - 6x - 4x + 8 = 3x^2 - 10x + 8$$

This combination does *not* work: $3x^2 - 10x + 8 \neq 3x^2 - 14x + 8$

Trial 2
Interchange the factors -4 and -2. Try -2 and -4.

$$(3x - 2)(x - 4) = 3x^2 - 12x - 2x + 8 = 3x^2 - 14x + 8.$$

This combination *works* so $3x^2 - 14x + 8 = (3x - 2)(x - 4)$.

EXAMPLE 2 | Factor $4x^2 - x - 5$.

Solution | The trial factors of -5 must have opposite signs.

Trial 1
Try $2x$ and $2x$ as factors of $4x^2$; -5 and 1 as factors of -5.

$$(2x - 5)(2x + 1) = 4x^2 + 2x - 10x - 5 = 4x^2 - 8x - 5$$

This combination does *not* work: $4x^2 - 8x - 5 \neq 4x^2 - x - 5$.

Trial 2
Try $4x$ and $1x$ as factors of $4x^2$ and keep -5 and 1 as factors of -5.

$$(4x - 5)(1x + 1) = 4x^2 + 4x - 5x - 5 = 4x^2 - 1x - 5.$$

This combination *works* so $4x^2 - x - 5 = (4x - 5)(x + 1)$.

TRY THIS | 1. Factor $4x^2 - 2x - 6$.   2. Factor $5x^2 + 7x - 6$.
             $(4x - 6)(x + 1)$ $(5x - 3)(x + 2)$

Highlighting the Standards

Standards 1a, 1b: Students apply
the problem-solving strategy of
Guess-and-Check to factoring some
trinomials.

Additional Example 1

Factor $5a^2 - 21a + 4$.
$(5a - 1)(a - 4)$

Additional Example 2

Factor $3x^2 - 11x - 4$.
$(3x + 1)(x - 4)$

It may be necessary to try several combinations before finding the correct pair of binomial factors.

EXAMPLE 3   Factor $6x^2 + 7x - 24$.

Solution   **Trial 1**
Try $6x$ and x as factors of $6x^2$; 6 and -4 as factors of -24.

$$(6x + 6)(x - 4) = 6x^2 - 24x + 6x - 24 = 6x^2 - 18x - 24$$

This combination does *not* work: $6x^2 - 18x - 24 \neq 6x^2 + 7x - 24$.

Trial 2
Interchange the factors of 6 and -4. Try -4 and 6.

$$(6x - 4)(x + 6) = 6x^2 + 36x - 4x - 24 = 6x^2 + 32x - 24$$

This combination does *not* work: $6x^2 + 32x - 24 \neq 6x^2 + 7x - 24$

Trial 3
Try $2x$ and $3x$ as factors of $6x^2$; -3 and 8 as factors of -24.

$$(2x - 3)(3x + 8) = 6x^2 + 16x - 9x - 24 = 6x^2 + 7x - 24$$

This combination *works*. So $6x^2 + 7x - 24 = (2x - 3)(3x + 8)$.

TRY THIS   **3.** Factor $4x^2 - 12x - 7$.   **4.** Factor $10x^2 - 19x + 6$.
$(2x + 1)(2x - 7)$ $(5x - 2)(2x - 3)$

Some trinomials cannot be factored into two binomials. For example, the only possible binomial factors of $2x^2 + 3x + 5$ are shown below.

$$(2x + 1)(x + 5) = 2x^2 + 11x + 5 \neq 2x^2 + 3x + 5$$
$$(2x + 5)(x + 1) = 2x^2 + 7x + 5 \neq 2x^2 + 3x + 5$$

A polynomial that cannot be factored into polynomials of lower degree is said to be an **irreducible polynomial.** Thus, $2x^2 + 3x + 5$ is an irreducible trinomial.

Classroom Exercises

For each trinomial $ax^2 + bx + c$, state all possible factors of a and of c.

1. $8x^2 + 14x - 15$ **2.** $10a^2 - 11a - 6$ **3.** $6t^2 + 31t - 35$
4. $14a^2 - 55a + 21$ **5.** $12t^2 - 3t - 20$ **6.** $18x^2 + 39x + 20$
7. $16m^2 + 34m - 15$ **8.** $27x^2 + 6x - 8$ **9.** $20x^2 + 88x - 9$
10. $4y^2 + 13y - 35$ **11.** $6a^2 - 43a + 72$ **12.** $15y^2 - y - 2$

7.4 Factoring Trinomials: $ax^2 + bx + c$ **257**

■■■ TEACHING SUGGESTIONS

Lesson Note

Point out to the students that if a or c is a prime number, then the factoring will be easier because there will not be as many possible combinations of factors. In Example 1 the only possible factors for $3x^2$ are x and $3x$.

Math Connections

Alternate Method: The following is another way to factor by making $a = 1$.
$3x^2 - 7x - 6$
$\frac{1}{3} \cdot 3(3x^2 - 7x - 6)$
$\frac{1}{3}(3 \cdot 3x^2 - 3 \cdot 7x - 3 \cdot 6)$
$\frac{1}{3}[(3x)^2 - 7(3x) - 18]$

Let $y = 3x$.
$\frac{1}{3}(y^2 - 7y - 18)$
$\frac{1}{3}(y - 9)(y + 2)$

Replace y with $3x$.
$\frac{1}{3}(3x - 9)(3x + 2)$
$(x - 3)(3x + 2)$

Critical Thinking Questions

Evaluation: Ask students to consider how they would factor $12x^2 - 60x + 72$. What is the most efficient way to proceed?   Factor out a 12 first.

Common Error Analysis

Error: Some students may rewrite $10 - t - 3t^2$ as $-3t^2 - t + 10$ and then make mistakes in assigning the correct signs to the factors.

Suggest that they factor the polynomial in the given order.

Additional Example 3

Factor $6t^2 - t - 15$. $(2t + 3)(3t - 5)$

257

Checkpoint

Factor each trinomial. If not possible, write irreducible.

1. $2x^2 + x - 6$ $(2x - 3)(x + 2)$
2. $6x^2 + 23x + 15$ $(x + 3)(6x + 5)$
3. $4x^2 + 4x + 3$ Irreducible
4. $8x^2 - 18x + 9$ $(2x - 3)(4x - 3)$

Closure

Have students list all possible pairs of factors for $3x^2 + bx + 8$ and give the value of b that each will produce.

$(3x + 1)(x + 8)$; $b = 25$
$(3x + 2)(x + 4)$; $b = 14$
$(3x + 4)(x + 2)$; $b = 10$
$(3x + 8)(x + 1)$; $b = 11$
$(3x - 1)(x - 8)$; $b = -25$
$(3x - 2)(x - 4)$; $b = -14$
$(3x - 4)(x - 2)$; $b = -10$
$(3x - 8)(x - 1)$; $b = -11$

Emphasize that for a given value of b there is only *one* correct factorization of this trinomial.

▰▰▰ FOLLOW UP

Guided Practice

Classroom Exercises 1–18
Try This   all

Independent Practice

A Ex. 1–27, **B** Ex. 28–45, **C** Ex. 46–49

Basic: WE 1–27 odd, 40
Average: WE 11–39 odd, 40
Above Average: WE 21–49 odd, 40

Additional Answers

See page 271 for the answers to Classroom Ex. 1–12 and Written Ex. 1–18. See page 276 for the answers to Written Ex. 19–46.

Factor each trinomial.

13. $2x^2 + x - 1$ $(2x - 1)(x + 1)$
14. $3a^2 + 13a - 10$ $(3a - 2)(a + 5)$
15. $6m^2 + m - 1$ $(3m - 1)(2m + 1)$
16. $9y^2 + 3y - 2$ $(3y - 1)(3y + 2)$
17. $6y^2 - 17y + 12$ $(2y - 3)(3y - 4)$
18. $21x^2 + 5x - 6$ $(3x + 2)(7x - 3)$

Written Exercises

Factor each trinomial. If not possible, write *irreducible*.

1. $2x^2 + 5x - 3$
2. $3a^2 + 10a + 3$
3. $5x^2 - 11x + 2$
4. $3y^2 + 2y - 5$
5. $2b^2 - 5b + 3$
6. $3a^2 - 7a - 6$
7. $4f^2 - 7f + 2$
8. $2x^2 - x - 6$
9. $10a^2 + 13a - 3$
10. $2x^2 - x - 15$
11. $3b^2 - b - 2$
12. $6y^2 - 7y + 3$
13. $6y^2 + 5y - 6$
14. $4a^2 - 11a - 3$
15. $6a^2 + a - 3$
16. $3x^2 + 3x - 4$
17. $6x^2 - 7x - 5$
18. $4b^2 + 4b - 3$
19. $6t^2 - t - 15$
20. $27t^2 - 12t - 7$
21. $4x^2 - 12x + 9$
22. $28a^2 - 15a + 2$
23. $8m^2 + 34m + 35$
24. $2t^2 - 17t + 30$
25. $15a^2 - 9a - 6$
26. $3x^2 + 17x + 40$
27. $36c^2 + 12c - 35$
28. $4a^2 + 17a - 15$
29. $2b^2 - b - 45$
30. $14y^2 - 39y + 10$
31. $8t^2 + 42t + 49$
32. $8y^2 + 45y - 18$
33. $10a^2 - 91a + 9$
34. $10 - t - 3t^2$
35. $16 - 34a - 15a^2$
36. $8 - 26m + 21m^2$
37. $18 - 3x - 10x^2$
38. $9 + 6y - 8y^2$
39. $20 + 44b - 15b^2$
40. $18 - 9y - 35y^2$
41. $8x^2 - 35x + 12$
42. $2x^2 - 5x - 3$
43. $8x^2 - 10x + 3$
44. $14t^2 - 57t - 27$
45. $40d^2 + 39d - 40$
46. Write an explanation of how to factor $8x^2 + 31x - 4$.

Factor each polynomial. Simplify each factor of the product.

47. $4(a + 3)^2 - 6(a + 3) - 10$ $(2a + 8)(2a + 1)$
48. $12(t - 1)^2 + 8(t - 1) - 15$ $(6t - 11)(12t + 1)$
49. If $8x^2 + (3b - 1)x + 45 = (4x - 5)(2x - 9)$, what is the value of b? $b = -15$

Mixed Review

Solve and check.   *3.1, 3.3, 3.5, 4.1, 4.5, 4.6*

1. $15 = -6 + x$ $x = 21$
2. $z - 4 = -3$ $z = 1$
3. $x + \frac{3}{4} = 4$ $x = 3\frac{1}{4}$
4. $4y - 8 = 12$ $y = 5$
5. $8 - \frac{1}{3}x = 10$ $x = -6$
6. $0.3b - 2.6 = 5.3$ $b = 26\frac{1}{3}$
7. $-3(2 - 5x) = 3(4 - 3x)$ $x = \frac{3}{4}$
8. $-6(6 - 4y) + 3y = (y - 4)5$ $y = \frac{8}{11}$
9. Five less than twice a number is 13 less than 3 times the number. Find the number.   *4.2* 8

Enrichment

Tell students that sums and differences of cubes can be factored and write these factoring patterns on the chalkboard.
$x^3 + y^3 = (x + y)(x^2 - xy + y^2)$
$x^3 - y^3 = (x - y)(x^2 + y + y^2)$

1. Have students verify the patterns by multiplying $(x + y)(x^2 + xy + y^2)$ and $(x - y)(x^2 + xy + y^2)$.

Then have students follow the patterns to factor these binomials.

2. $x^3 - 8$ $(x - 2)(x^2 + 4x + 4)$
3. $n^3 + 27$ $(n + 3)(n^2 - 3n + 9)$
4. $1 - 64t^3$ $(1 - 4t)(1 + 4t + 16t^2)$
5. $1 + x^{3a}$ $(1 + x^a)(1 - x^a + x^{2a})$

7.5 Two Special Cases of Factoring

Objectives
To factor the difference of two squares
To factor a perfect-square trinomial

Let a^2 and b^2 represent areas of two squares. The shaded region at the right represents the *difference* of the two squares, where b^2 is subtracted from a^2. Now see how the region is separated into two parts (I and II) and reassembled, as shown below.

  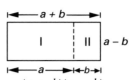

Then, $a^2 - b^2 = a(a - b) + b(a - b) = (a + b)(a - b)$
Region I + Region II

This formula can be used to factor binomials of the form $a^2 - b^2$.

Difference of Two Squares

$$a^2 - b^2 = (a + b)(a - b)$$

EXAMPLE 1   Factor $a^2 - 16$.

Plan   Since 16 is the square of 4, factor $a^2 - 16$ as $a^2 - 4^2$.

Solution $a^2 - 16 = a^2 - 4^2 = (a + 4)(a - 4)$

TRY THIS   1. Factor $y^2 - 36$. $(y - 6)(y + 6)$ 2. Factor $x^2 - 81$. $(x - 9)(x + 9)$

Sometimes both the coefficients and the variables are perfect squares.

EXAMPLE 2   Factor each of the following.
  a. $144b^2 - 49y^2$ **b.** $(x - 3)^2 - 25y^2$

Solution   **a.** $144b^2 - 49y^2$ **b.** $(x - 3)^2 - 25y^2$
          $= (12b)^2 - (7y)^2$ $= (x - 3)^2 - (5y)^2$
          $= (12b + 7y)(12b - 7y)$ $= (x - 3 + 5y)(x - 3 - 5y)$

7.5 Two Special Cases of Factoring   **259**

Lesson Note

Emphasize the importance of writing the middle step where the squares are shown explicitly before completing the factoring.
Example: $y^2 - 64 = y^2 - 8^2$
$$= (y + 8)(y - 8)$$
Show students that the last term of a perfect square trinomial must be positive. Therefore, a trinomial such as $4x^2 - 12x - 9$ does not fit the pattern.

Math Connections

Arithmetic: The perfect square trinomial pattern can be used to square a two-digit number mentally.
$53^2 = (50 + 3)^2 = 50^2 + (2)(50)(3) + 3^2 = 2500 + 300 + 9 = 2809.$

Critical Thinking Questions

Synthesis: Ask students to try to factor $x^2 + 9$. When they begin to suspect that this is not possible, ask them to show why they think so.   Possible answer: Since $(x + 3)^2$, $(x - 3)^2$, and $(x + 3)(x - 3)$ give other products, this leaves no possible pair of factors that will give $x^2 + 9$.

EXAMPLE 3   The figure at the right shows a square inside a square. Write a formula for the shaded region. Then evaluate for $r = 6.2$ and $t = 4.3$.

Solution   The area of the shaded region is $(3r)^2 - (2t)^2$.

$$(3r)^2 - (2t)^2 = (3r + 2t)(3r - 2t)$$
$$= (18.6 + 8.6)(18.6 - 8.6)$$
$$= 272 \text{ square units}$$

TRY THIS   **3.** Factor $25a^2 - 100b^2$.   **4.** Factor $(a + 4)^2 - 64b^2$.
(5a + 10b)(5a − 10b)   (a + 4 + 8b)(a + 4 − 8b)
Write a formula for each shaded region.

5.   **6.**

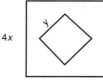

(4x + y)(4x − y)   (5a + 3c)(5a − 3c)

Since $(a + b)^2 = a^2 + 2ab + b^2$ and $(a - b)^2 = a^2 - 2ab + b^2$, the resulting trinomials are called *perfect-square trinomials*. To factor a perfect square trinomial, you must first recognize its form: *square of the first term $\pm$ twice the product of the two terms $+$ square of the second term*.

Perfect Square Trinomials

$$a^2 + 2ab + b^2 = (a + b)^2 \qquad a^2 - 2ab + b^2 = (a - b)^2$$

EXAMPLE 4   Determine whether $4x^2 - 12x + 9$ is a perfect square trinomial. If so, factor the trinomial.

Solution   Express the first and last terms as squares.

$$4x^2 - 12x + 9 = (2x)^2 - 12x + (3)^2$$

Check whether the middle term is $2(2x \cdot 3)$ or $-2(2x \cdot 3)$.

$$-12x = -2(2x \cdot 3) \ ✔$$
So, $4x^2 - 12x + 9 = (2x)^2 - 2(2x \cdot 3) + (3)^2$, a perfect square.
$$4x^2 - 12x + 9 = (2x - 3)^2$$

TRY THIS   **7.** Determine whether $9x^2 + 42x + 49$ is a perfect square trinomial. If so, factor the trinomial. $(3x + 7)^2$

260   Chapter 7 Factoring Polynomials

Additional Example 3

The figure shows a square inside a square. Write a formula in factored form for the area of the shaded region. Then evaluate for $a = 2.5$ and $b = 0.5$. $(a + 3b)(a - 3b)$; 4 square units

Additional Example 4

Determine whether $36x^2 - 6x + 1$ is a perfect square. If so, factor it.   The trinomial is not a perfect square; the middle term is not $12x$ or $-12x$.

Classroom Exercises

Express as the square of a monomial.

1. 100 $(10)^2$

2. $16c^2$ $(4c)^2$

3. $36b^2$ $(6b)^2$

4. $49y^2$ $(7y)^2$

5. $81w^2$ $(9w)^2$

6. $144k^2$ $(12k)^2$

Factor as the difference of two squares.

7. $(3x)^2 - (4y)^2$

8. $(4a)^2 - (11y)^2$

9. $(5t)^2 - (8j)^2$

10. $t^2 - 4$ $(t-2)(t+2)$

11. $a^2 - 1$ $(a-1)(a+1)$

12. $36 - x^2$ $(6-x)(6+x)$

Tell how to fill the blank so that the trinomial will be a perfect square.

13. $(2x)^2 + \underline{\ ?\ } + (5)^2$ $20x$

14. $(5x)^2 - \underline{\ ?\ } + (3)^2$ $30x$

15. $(a)^2 + \underline{\ ?\ } + (3b)^2$ $6ab$

Written Exercises

Factor.

1. $a^2 - 25$ $(a-5)(a+5)$

2. $m^2 - 49$ $(m-7)(m+7)$

3. $t^2 - 36$ $(t-6)(t+6)$

4. $100 - y^2$ $(10-y)(10+y)$

5. $25 - r^2$ $(5-r)(5+r)$

6. $f^2 - 81$ $(f-9)(f+9)$

7. $9 - g^2$ $(3-g)(3+g)$

8. $9k^2 - 64$ $(3k-8)(3k+8)$

9. $49u^2 - 25$

10. $25b^2 - 4$ $(5b-2)(5b+2)$

11. $49 - 16y^2$ $(7-4y)(7+4y)$

12. $1 - 64r^2$ $(1-8r)(1+8r)$

13. $4e^2 - 1$ $(2e-1)(2e+1)$

14. $9w^2 - 64$ $(3w-8)(3w+8)$

15. $144 - h^2$ $(12-h)(12+h)$

Determine whether the given trinomial is a perfect square. If so, factor it.

16. $x^2 + 8x + 16$

17. $m^2 - 8m + 49$ No

18. $y^2 + 12y + 36$

19. $t^2 - 4t + 4$

20. $k^2 - 14k + 49$

21. $a^2 - 12a + 9$ No

22. $4x^2 + 4x + 1$

23. $9a^2 - 6a + 1$

24. $25x^2 - 12x + 1$ No

25. $25a^2 + 30a + 4$ No

26. $16y^2 - 12y + 1$ No

27. $49u^2 - 14u + 1$
Yes; $(7u-1)^2$

Factor.

28. $121a^2 - 169b^2$ $(11a-13b)(11a+13b)$

29. $49t^2 - 225y^2$ $(7t-15y)(7t+15y)$

30. $25x^2 - 196y^2$ $(5x-14y)(5x+14y)$

31. $256a^2 - 225b^2$

32. $(2a+5)^2 - 121r^2$

33. $4y^2 - (2x-5)^2$

34. $36a^2 - 60ab + 25b^2$

35. $81t^2 + 90tv + 25v^2$

36. $4p^2 - 44pq + 121q^2$

37. $49r^2 + 42rt + 9t^2$

38. $9k^2 - 30kh + 25h^2$

39. $4y^2 + 28yx + 49x^2$

40. Find an algebraic expression for the length of a side of a square of area $25p^2 + 110pq + 121q^2$. What is the length for $p = 5$ and $q = 4$?
$l = 5p + 11q;\ 69$

41. Find an algebraic expression for the length of a side of a square of area $9a^2 - 24ab + 16b^2$. What is the length for $a = 5$ and $b = 3$?
$l = 3a - 4b;\ 3$

7.5 Two Special Cases of Factoring **261**

Classroom Exercises

7. $(3x - 4y)(3x + 4y)$

8. $(4a - 11y)(4a + 11y)$

9. $(5t - 8j)(5t + 8j)$

Written Exercises

9. $(7u - 5)(7u + 5)$

16. Yes; $(x + 4)^2$

18. Yes; $(y + 6)^2$

19. Yes; $(t - 2)^2$

20. Yes; $(k - 7)^2$

22. Yes; $(2x + 1)^2$

23. Yes; $(3a - 1)^2$

31. $(16a - 15b)(16a + 15b)$

32. $(2a + 5 - 11r)(2a + 5 + 11r)$

33. $(2y - 2x + 5)(2y + 2x - 5)$

34. $(6a - 5b)^2$

35. $(9t + 5v)^2$

36. $(2p - 11q)^2$

37. $(7r + 3t)^2$

38. $(3k - 5h)^2$

39. $(2y + 7x)^2$

48. Let $10t + 5$ represent any two-digit number ending in 5, where $1 \le t \le 9$.
$(10t + 5)^2 = 100t^2 + 100t + 25$
$= (t^2 + t)100 + 25$
$= t(t + 1)100 + 25$
Hence, a short cut for mentally squaring any two-digit number ending in 5 is using the formula $t(t + 1)100 + 25$ where t equals the first digit.

Midchapter Review

10. $(x - 5)(x + 3)$

11. $(a + 5)(a - 4)$

12. $(t + 1)(t + 7)$

13. $(2a + 5)(a - 1)$

14. $(2x - 1)(3x - 4)$

15. $(4t - 1)(4t + 1)$

Write a formula in factored form for the area of each shaded region. Then evaluate for the given values of the variables. Use 3.14 for π. Give answers to the nearest integer.

42.

$a = 10$ and $b = 6$
$(a + b)(a - b)$; 64

43.

$R = 12$ and $r = 5$
$\pi(R^2 - r^2)$; 374

44.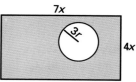

$r = 5$ and $x = 8$
$28x^2 - \pi 9r^2$; 1,086

Factor as the difference of two squares. Then simplify.

45. $(5t + 6)^2 - (t - 3)^2$ $(4t + 9)(6t + 3)$

46. $(x - y)^2 - (4x + 3y)^2$
$(-3x - 4y)(5x + 2y)$

47. Express $(a + b)^4 - 81$ as the product of three factors.
$(a + b - 3)(a + b + 3)(a^2 + 2ab + b^2 + 9)$

48. There is a shortcut for mentally squaring any two-digit number ending in 5. Notice the pattern in the following examples.

$$(65)^2 = 6(6 + 1)100 + 25 = 6(7)(100) + 25 = 4,225$$
$$(85)^2 = 8(8 + 1)100 + 25 = 8(9)(100) + 25 = 7,225$$

Explain this shortcut by letting $10t + 5$ represent any two-digit number ending in 5, where $1 \le t \le 9$. (HINT: $(10t + 5)^2 = ?$)

Midchapter Review

Find the greatest common factor (GCF). *7.1*

1. 12, 22 2

2. 15, 27 3

3. 15, 18, 30 3

Find the missing factor. *7.1*

4. $x^3 \cdot \underline{\ ?\ } = x^6$ x^3

5. $9a^2 \cdot \underline{\ ?\ } = 36a^8$ $4a^6$

6. $5x^2y^3 \cdot \underline{\ ?\ } = 60x^5y^6$
$12x^3y^3$

Factor out the GCF from each polynomial. *7.2*

7. $3a^2 - 9a + 15$
$3(a^2 - 3a + 5)$

8. $15x^4 - 20x^2 + 10$
$5(3x^4 - 4x^2 + 2)$

9. $8a^4b^2 - 48a^2b^3 + 4a^3b$
$4a^2b(2a^2b - 12b^2 + a)$

Factor each polynomial. If not possible, write _irreducible_. *7.3, 7.4, 7.5*

10. $x^2 - 2x - 15$

11. $a^2 + a - 20$

12. $t^2 + 8t + 7$

13. $2a^2 + 3a - 5$

14. $6x^2 - 11x + 4$

15. $16t^2 - 1$

16. $9a^2 - 25$
$(3a - 5)(3a + 5)$

17. $a^2 + 6a + 8$
$(a + 2)(a + 4)$

18. $4t^2 - 20t + 25$
$(2t - 5)^2$

Enrichment

Point out to the students that finding the product of two numbers can sometimes be simplified if the numbers fall into the pattern of $(a + b)(a - b)$. For example, $31 \cdot 29$ is equivalent to $(30 + 1)(30 - 1)$, which equals $30^2 - 1^2$, or 899. Challenge students to find the following products mentally by applying this strategy.

1. $(19)(21)$ $20^2 - 1^2 = 399$

2. $(51)(49)$ $50^2 - 1^2 = 2,499$

3. $(32)(28)$ $30^2 - 2^2 = 896$

4. $(72)(68)$ $70^2 - 2^2 = 4,896$

5. $(98)(102)$ $100^2 - 2^2 = 9,996$

6. $(1,003)(997)$ $1,000^2 - 3^2 = 999,991$

7.6 Combined Types of Factoring

Teaching Resources

Problem Solving Worksheet 7
Quick Quizzes 56
Reteaching and Practice
Worksheets 56

Objectives   To factor polynomials completely
To factor polynomials by grouping terms

The polynomial $3x^2 - 27$ can be factored in two steps.

$3x^2 - 27 = 3(x^2 - 9)$ 3 is the GCF.
$\qquad\qquad = 3(x + 3)(x - 3)$ Each factor is irreducible.

When a polynomial is factored into irreducible polynomials that have
no common factor, then it is *factored completely*.

Steps in Factoring Completely	Examples
1. Look for the GCF.	$8t^4 - 32t^3 + 40t = 8t(t^3 - 4t - 5)$
2. Look for special cases.	
a. difference of two squares	$4x^2 - 9y^2 = (2x)^2 - (3y)^2$
	$\qquad\qquad\quad = (2x + 3y)(2x - 3y)$
b. perfect-square trinomial	$x^2 + 8x + 16 = x^2 + 8x + 4^2$
	$\qquad\qquad\qquad\quad = (x + 4)^2$
3. If a trinomial is not a perfect square, look for two different binomial factors.	$6x^2 + 11x - 10 = (2x + 5)(3x - 2)$

EXAMPLE 1   Factor $8x^3 + 12x^2 + 4x$ completely.

Solution
1. The GCF is $4x$.
$8x^3 + 12x^2 + 4x = 4x(2x^2 + 3x + 1)$

2. $2x^2 + 3x + 1$ can be factored into binomials.
$2x^2 + 3x + 1 = (2x + 1)(x + 1)$

So, $8x^3 + 12x^2 + 4x = 4x(2x + 1)(x + 1)$.

TRY THIS
1. Factor $6x^3 + 14x^2 + 8x$. $2x(3x + 4)(x + 1)$

The factoring in Example 2 uses the GCF, as well as the special case of
the difference of two squares, applied twice.

EXAMPLE 2   Factor $3x^5 - 243x$ completely.

Solution $3x^5 - 243x = 3x(x^4 - 81)$ ⟵ GCF: $3x$
$\qquad\qquad\quad = 3x[(x^2)^2 - 9^2)]$
$\qquad\qquad\quad = 3x(x^2 + 9)(x^2 - 9) = 3x(x^2 + 9)(x + 3)(x - 3)$

GETTING STARTED

Prerequisite Quiz

Factor.

1. $25a^2 - 15a$ $5a(5a - 3)$
2. $y^3 + 7y^2 - 13y$ $y(y^2 + 7y - 13)$
3. $6x^2 + 11x - 10$ $(2x + 5)(3x - 2)$
4. $64t^2 - 25$ $(8t + 5)(8t - 5)$
5. $9y^2 + 12xy + 4x^2$ $(3y + 2x)^2$

Motivator

Ask students what is meant by prime
factorization of numbers.   A number written
as a product of prime-number factors or of
powers of prime-number factors   Ask them
what is meant by prime factorization of
polynomials.   A polynomial expressed as
a product of prime polynomials   Then ask
them when a polynomial is prime.   When it
cannot be factored into polynomials of lower
degree

Additional Example 1

Factor $6x^3 + 21x^2 - 12x$ completely.
$3x(2x - 1)(x + 4)$

Additional Example 2

Factor $5a^5 - 405a$ completely.
$5a(a^2 + 9)(a + 3)(a - 3)$

Highlighting the Standards

Standards 2d, 1d: The Focus on Reading
asks students to read written presentations
of mathematics and answer questions and
give examples. The Application connects
mathematics to commission sales.

Lesson Note

Direct students to look *first* for a GCF and factor it out. Then they should systematically check for the difference of two squares, perfect square trinomials, and trinomials of the form $ax^2 + bx + c$. Tell the students that since factoring by grouping is a new method, they should be sure to check such factorizations by multiplying the factors to determine whether the product is the original polynomial.

Math Connections

Cryptography: Very long numbers which can be factored into large prime numbers are used to develop secret codes. One method of breaking a number into primes involves finding two numbers, x and y, that satisfy the equation $x^2 - y^2 = N$, where N is the large number to be factored.

Critical Thinking Questions

Application: Show students that $4x^4 + 1$ is equivalent to $4x^4 + 4x^2 + 1 - 4x^2$ which is equivalent to $(2x^2 + 1)^2 - (2x)^2$. Then ask students how this last expression could be factored. $(2x^2 + 1 + 2x)(2x^2 + 1 - 2x)$ Finally, have students multiply these two factors together to obtain the original expression.

Common Error Analysis

Error: Sometimes students forget to factor out the GCF before attempting to factor into binomials, or they may forget to include the GCF as part of the final factorization.

Remind them to check their answers by multiplication.

EXAMPLE 3   Factor $2m^2 - 20mn + 50n^2$ completely.

Plan   Factor out 2, the GCF. Then factor the trinomial into a perfect square.

Solution
$$2m^2 - 20mn + 50n^2 = 2(m^2 - 10mn + 25n^2)$$
$$= 2[m^2 - 2(m)(5n) + (5n)^2]$$
$$= 2(m - 5n)^2$$

TRY THIS   2. Factor $5x^4y - 80x^2y^3$ completely. $5x^2y(x + 4y)(x - 4y)$

To be factorable, a polynomial need not be a binomial or a trinomial. In the polynomial $x(r + s) + y(r + s)$, notice that $r + s$ is common to each product. So, $r + s$ can be factored from the polynomial.

$$x(r + s) + y(r + s) \text{ can be factored as } (x + y)(r + s).$$

Sometimes the possibility of factoring this way is less obvious, as in $5x - 5y + ax - ay$. However, 5 is common to the first two terms and a is common to the last two terms. This suggests grouping those terms to find a common binomial factor. This method of factoring is called **factoring by grouping.**

EXAMPLE 4   Factor $5x - 5y + ax - ay$.

Plan   Group the terms in either of the two ways shown to factor out the common monomial factor.

Solution

Method 1	Method 2
$(5x - 5y) + (ax - ay)$	$(5x + ax) + (-5y - ay)$ ← $-y$ is common
$5(x - y) + a(x - y)$	$x(5 + a) - y(5 + a)$ to the last two terms.
$(5 + a)(x - y)$	$(x - y)(5 + a)$

TRY THIS   3. Factor $2a - 4c + ax - 2cx$. $(2 + x)(a - 2c)$

Focus on Reading

When it is factored into irreducible polynomials having no common factor
1. When is a polynomial factored completely?
2. One factor of a perfect square trinomial is $(a - 2b)$. What is the other factor? $(a - 2b)$
3. Give an example of a binomial that has three factors. Answers will vary.
4. Give an example of a polynomial having a common binomial factor.
Answers will vary.

Additional Example 3

Factor $24x^2 + 4xy - 48y^2$ completely.
$4(3x - 4y)(2x + 3y)$

Additional Example 4

Factor $ax - 3x + ay - 3y$. $(a - 3)(x + y)$

Classroom Exercises

Indicate whether each of the following can be factored (1) as the product of two different binomials, (2) as a perfect square trinomial, (3) as the difference of two squares, or (4) by grouping.

1. $x^2 - 25$ (3)   **2.** $x^2 - 8x + 12$ (1)   **3.** $x^2 + 6x + 9$ (2)

4. $ax + ay + 3x + 3y$ (4)   **5.** $x^2 - 20x - yx + 20y$ (4)   **6.** $4a^2 - 121r^2$ (3)

State the GCF, if there is one other than 1.

7. $3x^2 + 3x - 6$ 3   **8.** $2a^2 - 2a - 12$ 2   **9.** $3y^2 - 27$ 3

10. $2a^3 - 8a^2$ $2a^2$ **11.** $12x^2 + 20x - 8$ 4   **12.** $5a^2 + 15a + 10$ 5

13. $3y^3 + 12y^2 - 15y$ $3y$ **14.** $5t^3 - 20t$ $5t$ **15.** $9x^2y - 36y$ $9y$

16–24. Factor the polynomials in Classroom Exercises 7–15 completely.

Written Exercises

Factor completely.

1. $2x^2 + 10x + 8$ **2.** $5x^2 - 10x - 15$ **3.** $2a^2 - 14a + 24$

4. $6a^3 + 3a^2 - 18a$ **5.** $3y^3 + 15y^2 + 12y$ **6.** $4a^3 - 12a^2 - 16a$

7. $3x^2 - 75$ **8.** $5y^3 - 20y$ **9.** $98 - 2a^2$

10. $2a^2 + 48a - 50$ **11.** $3x^2 + 12x + 12$ **12.** $2x^3 + 12x^2 + 18x$

13. $9p^2 + 33pq - 12q^2$ **14.** $9a^2 - 9b^2$ **15.** $6x^2 - 18xy - 60y^2$

16. $2a^2 - 6ab - 20b^2$ **17.** $50y^2 - 98x^2$ **18.** $8a^2 - 24ab + 18b^2$

19. $8m^2 - 18mn - 26n^2$ **20.** $8x^2 + 14xy - 30y^2$ **21.** $8a^2 - 8ab + 2b^2$

22. $8a + 8b + ca + cb$ **23.** $2x + 2y + bx + by$ **24.** $ac + bc + 5a + 5b$

25. $5ax + 3bx + 5ay + 3by$ **26.** $a^2 + ac - 7a - 7c$ **27.** $tr - 3r + st - 3s$

28. $ax^2 - 9a + bx^2 - 9b$ **29.** $p^2y - 25y - 2p^2 + 50$ **30.** $x^2a + x^2b - 16a - 16b$

31. $12r^2 - 45r - 12$ **32.** $48t^3 - 24t^2 + 3t$ **33.** $-3f^2 - 6f + 24$

34. $6x^2y - 39xy - 210y$ **35.** $288a - 50a^3$ **36.** $a^3b - ab^3$

37. $30s^3 + 22s^2 - 28s$ **38.** $x^3 - 6x^2y - 5xy^2$ **39.** $20y^3 - 6y^2 - 8y$

40. $ax^2 - 9a + bx^2 - 9b$ **41.** $x^2m + x^2 - 4m - 4$ **42.** $2x^3 - 8y^2x + 4x^2 - 16y^2$

43. $-4a^4y - 64a^2y^3 + 12a^3y^2 + 192ay^4$ **44.** $45x^2y - 5x^4y - 45xy^2 + 5x^3y^2$

45. $a^4 - (a - 2)^2$ **46.** $(r^2 - 1)^2 - (r - 1)^2$

47. $3(2x - s)^2 + 5(2x - s) - 12$ **48.** $x(x + 1)(4x - 5) - 6(x + 1)$

49. $4x^2 - 25y^2 + 2x - 5y$ **50.** $(x^2 - 5)^2 - 7(x^2 - 5) + 12$

51. Factor $64a^4 + 1$ by writing it as $(64a^4 + 16a^2 + 1) - 16a^2$, the difference of two squares. $(8a^2 - 4a + 1)(8a^2 + 4a + 1)$

Checkpoint

Factor completely.

1. $6x^3 + 26x^2 - 20x$ $2x(3x - 2)(x + 5)$

2. $3t^5 - 48t$ $3t(t^2 + 4)(t + 2)(t - 2)$

3. $24s^3 + 4s^2t - 48st^2$ $4s(2s + 3t)$
   $(3s - 4t)$

4. $2ax - 4x + ay - 2y$ $(2x + y)(a - 2)$

Closure

Ask students to give the order of the steps to use for factoring completely.   See page 263.   Then ask them what method they have learned for factoring certain polynomials which have four terms.
By grouping two terms at a time

■■■■FOLLOW UP

Guided Practice

Classroom Exercises 1–24
Try This   all; FR   all

Independent Practice

A Ex. 1–27, **B** Ex. 28–42, **C** Ex. 43–51

Basic: WE 1–10 all, 11–27 odd, Application

Average: WE 1–39 odd, Application

Above Average: WE 13–51 odd, Application

Classroom Exercises

16. $3(x - 1(x + 2)$
17. $2(a + 2)(a - 3)$
18. $3(y - 3)(y + 3)$
19. $2a^2(a - 4)$
20. $4(3x - 1)(x + 2)$
21. $5(a + 2)(a + 1)$
22. $3y(y + 5)(y - 1)$
23. $5t(t - 2)(t + 2)$
24. $9y(x - 2)(x + 2)$

Written Exercises

1. $2(x + 4)(x + 1)$
2. $5(x - 3)(x + 1)$
3. $2(a - 4)(a - 3)$
4. $3a(2a - 3)(a + 2)$
5. $3y(y + 4)(y + 1)$
6. $4a(a - 4)(a + 1)$
7. $3(x + 5)(x - 5)$
8. $5y(y - 2)(y + 2)$
9. $-2(a - 7)(a + 7)$
10. $2(a + 25)(a - 1)$
11. $3(x + 2)^2$
12. $2x(x + 3)^2$
13. $3(3p - q)(p + 4q)$
14. $9(a + b)(a - b)$
15. $6(x + 2y)(x - 5y)$
16. $2(a + 2b)(a - 5b)$
17. $2(5y - 7x)(5y + 7x)$
18. $2(2a - 3b)^2$
19. $2(4m - 13n)(m + n)$

See page 283 for the answers to Ex. 20–50.

Mixed Review

Solve. *3.2, 4.5, 5.4*

1. $\frac{3}{5}x - 3 = \frac{4}{3}x$ $-4\frac{1}{11}$
2. $3a - 2(3 - a) = 2a + 9$ $a = 5$
3. $3.2y - 1.7y = 3(0.02y + 13)$ $27\frac{1}{12}$
4. $8 - 2x \geq 10 - (4x - 8)$ $x \geq 5$

Simplify. *6.2, 2.10, 6.7*

5. $(p^3r)^7$ $p^{21}r^7$
6. $(-2ab^3)^4$ $16a^4b^{12}$
7. $(5a^2b^3c^4)^3$ $125a^6b^9c^{12}$
8. $(-3p^4q^3r^5)^3$ $-27p^{12}q^9r^{15}$
9. $-(-3a^2 - 4a + 6)$ $3a^2 + 4a - 6$
10. $9a - 3b - (7a + b - 10)$ $2a - 4b + 10$
11. $(8x^3 + 3x^2 - 8) - (-x^3 - 2x^2 + 9)$
    $9x^3 + 5x^2 - 17$
12. $(14a^4 - 12a^2 + 8) - (15a^3 - 12a^2 + 4a)$
    $14a^4 - 15a^3 - 4a + 8$

Application: Commission Sales

Example

Jill works as a salesperson in an appliance store. She earns $140 per week plus a commission of 15% on the selling price of the television sets she sells. Find the total of Jill's television sales, s, if she earns $260 for the week.

Solution

Let s = the amount of sales needed.

Regular salary plus commission equals total earned.

$$140 \quad + \quad 15\% \text{ of } s \quad = \quad 260$$
$$140 + 0.15s = 260 \quad \longleftarrow 15\% \text{ of } s \text{ means } 0.15s.$$
$$100(140 + 0.15s) = 100 \cdot 260 \quad \longleftarrow \text{ Multiply each side}$$
$$100 \cdot 140 + 100 \cdot 0.15s = 26{,}000 \qquad \text{by the LCM, 100.}$$
$$14{,}000 + 15s = 26{,}000$$
$$15s = 12{,}000$$
$$s = 800$$

Thus, Jill must sell $800 worth of television sets.

1. Diego is paid $165 a week plus a 10% commission on all camera sales. How much must his sales be for him to earn a total of $215 for the week? $500
2. Janell earns only a 30% commission of her total sales. What must her total sales be to earn $270? $900
3. Mary earns $175 a week plus a 2% commission on each car she sells. What must her total sales be to have total earnings of $355? $9,000

Enrichment

Have students, working in small groups, research the connection between secret codes (cryptography) and mathematics and report on their research to the class.

7.7 Solving Quadratic Equations by Factoring

Objectives

To solve quadratic equations by factoring
To solve cubic equations by factoring

Recall that the Property of Zero for Multiplication states that the product of any number and 0 is 0.

$$\text{If } a = 0 \text{ or } b = 0, \text{ then } a \cdot b = 0.$$

The converse of the Property of Zero for Multiplication is also true.

$$\text{If } a \cdot b = 0, \text{ then } a = 0 \text{ or } b = 0.$$

You can use the words *if and only if* to combine a conditional and its converse.

Zero-Product Property

For all real numbers a and b, $ab = 0$ if and only if $a = 0$ or $b = 0$.

EXAMPLE 1   Solve $(x - 5)(2x + 6) = 0$.

Plan   By the Zero-Product Property, if $(x - 5)(2x + 6) = 0$, then $x - 5 = 0$ or $2x + 6 = 0$. Solve each of these equations.

Solution
$$(x - 5)(2x + 6) = 0$$
$$x - 5 = 0 \text{ or } 2x + 6 = 0$$
$$x = 5 \qquad 2x = -6$$
$$x = -3$$

Check

For $x = 5$:

$$(x - 5)(2x + 6) = 0$$
$$(5 - 5)\,[2(5) + 6] \overset{?}{=} 0$$
$$0 \cdot 16 \overset{?}{=} 0$$
$$0 = 0 \text{ True}$$

For $x = -3$:

$$(x - 5)(2x + 6) = 0$$
$$(-3 - 5)\,[2(-3) + 6] \overset{?}{=} 0$$
$$-8 \cdot 0 \overset{?}{=} 0$$
$$0 = 0 \text{ True}$$

So, the solutions of $(x - 5)(2x + 6) = 0$ are 5 and −3.

TRY THIS   1. Solve $(x + 4)(3x - 15) = 0$. −4, 5

Additional Example 1

Solve $(x - 1)(3x + 18) = 0$.   {1, −6}

Teaching Resources

Quick Quizzes 57
Reteaching and Practice Worksheets 57
Transparency 19

▬▬GETTING STARTED

Prerequisite Quiz

Factor.

1. $x^2 + 4x - 12$ $(x + 6)(x - 2)$
2. $8a^2 + 14a - 15$ $(2a + 5)(4a - 3)$
3. $t^3 - 25t$ $t(t + 5)(t - 5)$
4. $y^2 + 9y + 14$ $(y + 7)(y + 2)$

Motivator

Write this equation on the chalkboard: $a \cdot b = 0$. Ask: If you know that $a \neq 0$, what is the value of b? 0   Ask: If you know that $b \neq 0$, what is the value of a? 0   Ask: If you do not know either that $a = 0$, or $b = 0$, what are the possible solutions of the equation, $a \cdot b = 0$? $a = 0$, or $b = 0$, or $a = 0$ and $b = 0$

▬▬TEACHING SUGGESTIONS

Lesson Note

Tell students that the word *quadratic* is an adjective that means square, as in a square shape, and that quadratic equations are equations in which the highest degree of the variable is 2.

Highlighting the Standards

Standards 2a, 2d, 1b: In Focus on Reading, students read mathematics and clarify their thinking about equations. The Application connects mathematics to the boiling point of water.

In Example 1, the solution set of $(x - 5)(2x + 6) = 0$ is $\{-3, 5\}$. If the factors are multiplied, the equation becomes $2x^2 - 4x - 30 = 0$. You can check by substitution that this equation has the same solution set, $\{-3, 5\}$.

Equations such as $2x^2 - 4x - 30 = 0$ are called *quadratic equations*.

Definition

> A **quadratic equation** is an equation that can be written in the form $ax^2 + bx + c = 0$, where a, b, and c are real numbers and $a \neq 0$.

Some quadratic equations can be solved by factoring the polynomial and then applying the Zero-Product Property, as shown in Example 2.

EXAMPLE 2 Solve $3m^2 + 13m - 10 = 0$.

Solution

Factor.
Set each factor equal to 0.
Solve for m.

$3m^2 + 13m - 10 = 0$
$(3m - 2)(m + 5) = 0$
$3m - 2 = 0$ or $m + 5 = 0$
$3m = 2 \qquad\qquad m = -5$
$m = \frac{2}{3}$

Check:

For $m = \frac{2}{3}$:

$3m^2 + 13m - 10 = 0$
$3\left(\frac{2}{3}\right)^2 + 13\left(\frac{2}{3}\right) - 10 \stackrel{?}{=} 0$
$3\left(\frac{4}{9}\right) + \frac{26}{3} - 10 \stackrel{?}{=} 0$
$\frac{4}{3} + \frac{26}{3} - 10 \stackrel{?}{=} 0$
$\frac{30}{3} - 10 \stackrel{?}{=} 0$
$10 - 10 \stackrel{?}{=} 0$
$0 = 0$ True

For $m = -5$:

$3m^2 + 13m - 10 = 0$
$3(-5)^2 + 13(-5) - 10 \stackrel{?}{=} 0$
$3(25) - 65 - 10 \stackrel{?}{=} 0$
$75 - 65 - 10 \stackrel{?}{=} 0$
$10 - 10 \stackrel{?}{=} 0$
$0 = 0$ True

So, the solution set of $3m^2 + 13m - 10 = 0$ is $\left\{\frac{2}{3}, -5\right\}$.

TRY THIS

2. Solve $2a^2 - 7a - 4 = 0$. $\left\{-\frac{1}{2}, 4\right\}$

A quadratic equation has at most two solutions, or **roots**. If the polynomial of a quadratic equation is a perfect square, then its factors will be the same. In this case, there will be only one root.

Additional Example 2

Solve $12x^2 - 5x - 3 = 0$. $\left\{\frac{3}{4}, -\frac{1}{3}\right\}$

EXAMPLE 3   Solve $4a^2 - 12a + 9 = 0$.

Solution
$$4a^2 - 12a + 9 = 0$$
$$(2a - 3)(2a - 3) = 0$$
$$2a - 3 = 0 \quad or \quad 2a - 3 = 0$$
$$2a = 3 \qquad\qquad 2a = 3$$
$$a = \frac{3}{2} \qquad\qquad a = \frac{3}{2}$$ The check is left for you.

So, the solution set of $4a^2 - 12a + 9 = 0$ is $\left\{\frac{3}{2}\right\}$.

TRY THIS   3. Solve $16y^2 + 24y + 9 = 0$. $\left\{-\frac{3}{4}\right\}$

In Example 3, the solution $\frac{3}{2}$ is called a **double root** since $2a - 3$ appears twice as a factor.

The equation $y^3 - 9y = 0$ is a *third-degree* or **cubic equation**. To solve this cubic equation, extend the Zero-Product Property to three factors.

$a \cdot b \cdot c = 0$ if and only if $a = 0$ or $b = 0$ or $c = 0$.

EXAMPLE 4   Solve $y^3 - 9y = 0$.

Solution

	$y^3 - 9y = 0$
Factor out the GCF.	$y(y^2 - 9) = 0$
Factor $y^2 - 9$.	$y(y + 3)(y - 3) = 0$
Set each factor equal to 0.	$y = 0 \quad or \quad y + 3 = 0 \quad or \quad y - 3 = 0$
Solve each equation.	$y = 0 \qquad\qquad y = -3 \qquad\qquad y = 3$

The solution set is $\{-3, 0, 3\}$

TRY THIS   4. Solve $x^3 - 36x = 0$. $\{-6, 0, 6\}$

◢ *Focus on Reading*

1. Give another name for the *solutions* of an equation. Roots
2. What is the degree of the equation $x^2 - 12x + 20 = 0$? 2
3. How does a cubic equation differ from a quadratic equation?
4. When will a quadratic equation have a double root? When the polynomial is a perfect square
5. Can the equation $ax^2 + bx + c = 0$ be quadratic if b equals 0? Yes

---

Checkpoint

Solve each equation.

1. $t^2 + 8t + 15 = 0$ $\{-3, -5\}$
2. $y^2 - 9y + 20 = 0$ $\{4, 5\}$
3. $16a^2 - 24a + 9 = 0$ $\left\{\frac{3}{4}\right\}$
4. $y^3 - 100y = 0$ $\{0, 10, -10\}$

Closure

Ask these questions.

1. How many solutions would you expect for the equation $x^3 - 5x^2 + 10x = 0$? 3
2. What is another name for the solutions of an equation? roots
3. How many different solutions does the equation $x^2 + 10x + 25 = 0$ have? 1
4. How many different solutions does the equation $9t^2 - 25 = 0$ have? 2

◢ FOLLOW UP

Guided Practice

Classroom Exercises 1–12
Try This all; FR all

Independent Practice

A Ex. 1–24, **B** Ex. 25–39, **C** Ex. 40–52
Basic: WE 1–8 all, 9–23 odd, Application
Average: WE 7–39 odd, Application
Above Average: WE 19–51 odd, 52, Application

Additional Answers

Focus on Reading

3. Quadratic equation has at most 2 roots; cubic equation has at most 3 roots.

Additional Example 3

Solve $9t^2 + 24t + 16 = 0$. $\left\{-\frac{4}{3}\right\}$

Additional Example 4

Solve $a^3 - 49a = 0$. $\{0, 7, -7\}$.

269

9. $\{-\frac{1}{3}, 2\}$ **12.** $\{-\frac{1}{3}, 3\}$

16. $\{4, -11\}$ **17.** $\{-1, \frac{1}{4}\}$

18. $\{-5, -2\}$ **19.** $\{\frac{2}{3}, -7\}$

20. $\{\frac{1}{5}, 9\}$ **21.** $\{\frac{1}{2}, 7\}$

22. $\{\frac{3}{7}, 7\}$ **23.** $\{\frac{2}{3}\}$

24. $\{\pm\frac{5}{2}\}$ **25.** $\{\frac{1}{4}, -6\}$

26. $\{-\frac{2}{3}, 8\}$ **27.** $\{\frac{5}{2}, -4\}$

28. $\{-4\}$ **29.** $\{-\frac{5}{3}, -8\}$

30. $\{\frac{5}{6}, -7\}$ **31.** $\{-\frac{8}{3}, -\frac{1}{2}\}$

32. $\{-\frac{13}{3}\}$ **33.** $\{0, \pm 6\}$

34. $\{0, \pm 5\}$ **35.** $\{0, \pm 2\}$

36. $\{0, -1, -\frac{1}{3}\}$ **37.** $\{0, \frac{1}{2}, -2\}$

38. $\{0, \frac{3}{2}, 1\}$ **39.** $\{0, \frac{7}{4}\}$

40. $\{0, 5, 4\}$ **41.** $\{0, \pm 11\}$

42. $\{0, \frac{2}{3}, \frac{5}{2}\}$ **43.** $\{0, \pm 2, \pm 4\}$

44. $\{\pm\frac{1}{3}, \pm\frac{3}{2}\}$ **45.** $\{0, \pm\frac{3}{2}\}$

46. $\{0, \pm\frac{3}{2}, \pm 1\}$ **47.** $\{0, \pm 2, \pm 5\}$

48. $\{0, \pm 2, \pm 3\}$ **49.** $\{0, \pm\frac{5}{2}, \pm 1\}$

50. $\{0, \pm 1, \pm 3\}$ **51.** $\{0 \pm 7, \pm 1\}$

Classroom Exercises

Find the solution set of each equation.

1. $(x - 2)(x + 1) = 0$ $\{2, -1\}$ **2.** $(x + 5)(x - 6) = 0$ $\{-5, 6\}$ **3.** $(x - 9)(x - 2) = 0$ $\{9, 2\}$

4. $(x - 8)(x - 8) = 0$ $\{8\}$ **5.** $x(x - 7) = 0$ $\{0, 7\}$ **6.** $x(x + 7) = 0$ $\{0, -7\}$

7. $(2x - 6)(x - 5) = 0$ $\{5, 3\}$ **8.** $x(2x + 1)(x - 4) = 0$ **9.** $x(x + 5)(3x - 1) = 0$

10. $x^2 + 4x - 5 = 0$ $\{1, -5\}$ **11.** $x^2 + 2x - 3 = 0$ $\{1, -3\}$ **12.** $2x^2 - 5x + 3 = 0$

8. $\{4, 0, -\frac{1}{2}\}$

9. $\{\frac{1}{3}, 0, -5\}$

12. $\{\frac{3}{2}, 1\}$

Written Exercises

Find the solution set of each equation.

$\{-1, 3\}$

1. $(x - 2)(x - 3) = 0$ $\{2, 3\}$ **2.** $(x + 4)(x + 1) = 0$ $\{-4, -1\}$ **3.** $p^2 - 2p - 3 = 0$

4. $r^2 + 9r + 20 = 0$ $\{-4, -5\}$ **5.** $a^2 - 16 = 0$ $\{4, -4\}$ **6.** $y^2 - 144 = 0$ $\{12, -12\}$

7. $x^2 - 12x = 0$ $\{0, 12\}$ **8.** $2t^2 - 3t - 2 = 0$ $\{-\frac{1}{2}, 2\}$ **9.** $3x^2 - 5x - 2 = 0$

10. $m^2 + 7m = 0$ $\{0, -7\}$ **11.** $2m^2 - m - 1 = 0$ $\{-\frac{1}{2}, 1\}$ **12.** $3t^2 - 8t - 3 = 0$

13. $4a^2 + a = 0$ $\{0, -\frac{1}{4}\}$ **14.** $3y^2 - 7y + 2 = 0$ $\{\frac{1}{3}, 2\}$ **15.** $x^2 - 49 = 0$ $\{7, -7\}$

16. $x^2 + 7x - 44 = 0$ **17.** $4x^2 + 3x - 1 = 0$ **18.** $y^2 + 7y + 10 = 0$

19. $3x^2 + 19x - 14 = 0$ **20.** $5y^2 - 46y + 9 = 0$ **21.** $2g^2 - 15g + 7 = 0$

22. $7a^2 - 52a + 21 = 0$ **23.** $9x^2 - 12x + 4 = 0$ **24.** $4b^2 - 25 = 0$

25. $4a^2 + 23a - 6 = 0$ **26.** $3t^2 - 22t - 16 = 0$ **27.** $2a^2 + 3a - 20 = 0$

28. $4x^2 + 32x + 64 = 0$ **29.** $3t^2 + 29t + 40 = 0$ **30.** $6x^2 + 37x - 35 = 0$

31. $6y^2 + 19y + 8 = 0$ **32.** $9y^2 + 78y + 169 = 0$ **33.** $x^3 - 36x = 0$

34. $5t^3 - 125t = 0$ **35.** $4a^3 - 16a = 0$ **36.** $3t^3 + 4t^2 + t = 0$

37. $2x^3 + 3x^2 - 2x = 0$ **38.** $2y^3 - 5y^2 + 3y = 0$ **39.** $16t^3 - 56t^2 + 49t = 0$

40. $a^3 - 9a^2 + 20a = 0$ **41.** $5y^3 - 605y = 0$ **42.** $6b^3 - 19b^2 + 10b = 0$

43. $x^5 - 20x^3 + 64x = 0$ **44.** $36a^4 - 85a^2 + 9 = 0$ **45.** $8t^4 - 18t^2 = 0$

46. $4x^5 - 13x^3 + 9x = 0$ **47.** $b^5 - 29b^3 + 100b = 0$ **48.** $3a^5 - 39a^3 + 108a = 0$

49. $8b^5 - 58b^3 + 50b = 0$ **50.** $t^5 - 10t^3 + 9t = 0$ **51.** $2y^6 - 100y^4 + 98y^2 = 0$

52. Solve $x^2 - 14x + 49 - 16y^2 = 0$ for x in terms of y.

$x = 4y + 7$ or $x = -4y + 7$

Mixed Review

Multiply.   *6.9, 6.10, 4.4*

1. $6x^2 - 11x + 3$
2. $2x^3 + x^2 - 22x + 3$ $6y^3 + y^2 - 27y - 20$

1. $(2x - 3)(3x - 1)$ **2.** $(x - 3)(2x^2 + 7x - 1)$ **3.** $(3y + 5)(2y^2 - 3y - 4)$

4. $(a - b)(a + b)$ $a^2 - b^2$ **5.** $(5t + 7)(5t - 7)$ $25t^2 - 49$ **6.** $(2n - 11)(2n - 11)$

7. The length of a rectangle is 2 more than 3 times the width. The $4n^2 - 44n + 121$
perimeter of the rectangle is 36. Find the length and the width.   *4.4*
$l = 14,\ w = 4$

Enrichment

Show students how to use the equation solving steps in reverse to find an equation whose roots are -3 and 5.

$x = -3$ or $x = 5$

$x + 3 = 0$ or $x - 5 = 0$

$(x + 3)(x - 5) = 0$

$x^2 - 2x - 15 = 0$

Then have them write equations which would have the following solution sets.

1. $\{7, -2\}$ $x^2 - 5x - 14 = 0$
2. $\{5, -5\}$ $x^2 - 25 = 0$
3. $\{0, -3, -4\}$ $x^3 + 7x^2 + 12x$

Application: Boiling Point of Water

If you live in Houston and like your breakfast soft-boiled egg cooked for four minutes, you will have to adjust the timing when you go camping near Denver, the "mile-high city."

In Houston, where the altitude is approximately at sea level, water boils when it reaches a temperature of 100°C or 212°F.

In Denver, where the altitude is about 5,280 feet above sea level, there is less atmospheric pressure. So water boils at a lower temperature. This means that you have to cook an egg longer at the higher altitude to obtain the same degree of "doneness."

It is the temperature, and not the boiling, that determines how fast food cooks. Since the temperature of water remains the same after it reaches the boiling point, water that is boiling vigorously doesn't cook eggs any faster than water that is boiling gently.

A convenient rule-of-thumb for campers who often cook foods at different altitudes is that the boiling point of water decreases about 1°C for each 1,000-foot increase in altitude.

Exercises

1. Estimate the boiling point of water in Celsius degrees in Denver. Round your answer to the nearest degree. 95°C

2. If the boiling point of water is about 98°C in Mountain City, what is the approximate altitude there? 2,000 ft

3. Estimate the boiling point of water at a location that has an altitude of 7,000 feet. 93°C

4. Find out the altitude of your city or town. What boiling point will the rule-of-thumb predict for this altitude? Perform an experiment to determine the temperature in Celsius degrees at which water boils in the area where you live. Compare your experimental result with the estimate. Answers will vary.

7.8 Standard Form of a Quadratic Equation

Objectives To write quadratic equations in standard form
To solve word problems involving quadratic equations

The quadratic equation $3x^2 + 14x - 5 = 0$ is in **standard form** because

1. the polynomial, $3x^2 + 14x - 5$, is set equal to zero, and
2. the terms $3x^2$, $14x$, and -5 are in descending order of exponents.

Notice also that the coefficient of the square term is positive.

■ GETTING STARTED

Prerequisite Quiz

Solve.

1. $3x(x - 20) = 0$ $\{0, 20\}$
2. $(2x - 5)(x - 3) = 0$ $\{\frac{5}{2}, 3\}$
3. $x^2 - 12x + 27 = 0$ $\{3, 9\}$
4. $2x^2 + 9x - 5 = 0$ $\{\frac{1}{2}, -5\}$

Motivator

Have students write *Yes* if the given equation satisfies these conditions. If it does not, write *No*.

a. The terms of the polynomial are in descending order of exponents.
b. The equation is set equal to 0.

1. $6 - p^2 = p$ No
2. $25r^2 + 81 = 90x$ No
3. $6a^2 - 11a - 72 = 0$ Yes
4. $9k^2 - 4k = 0$ Yes
5. $0 = 4s^2 - 4s + 1$ Yes

EXAMPLE 1 Write $x^2 - 18 = 7x$ in standard form.

Solution

Subtract $7x$ from each side.
Arrange terms in descending order.

$$x^2 - 18 = 7x$$
$$x^2 - 18 - 7x = 0$$
$$x^2 - 7x - 18 = 0$$

EXAMPLE 2 Solve $11x = -x^2 - 28$. Check the solutions.

Plan Add $1x^2$ to each side so that the coefficient of the x^2 term will be positive. Then write the equation in standard form and solve.

Solution

	$11x = -1x^2 - 28$
Add $1x^2$ to each side.	$1x^2 + 11x = -28$
Add 28 to each side.	$1x^2 + 11x + 28 = 0$ ⟵ standard form
Factor.	$(x + 7)(x + 4) = 0$
Set each factor equal to 0.	$x + 7 = 0$ or $x + 4 = 0$
Solve for x.	$x = -7$ $x = -4$

Check

For $x = -7$:

$$11x = -x^2 - 28$$
$$11(-7) \overset{?}{=} -(-7)^2 - 28$$
$$-77 \overset{?}{=} -49 - 28$$
$$-77 = -77 \text{ True}$$

For $11x = -4$:

$$11x = -x^2 - 28$$
$$11(-4) \overset{?}{=} -(-4)^2 - 28$$
$$-44 \overset{?}{=} -16 - 28$$
$$-44 = -44 \text{ True}$$

The roots are -7 and -4. So the solution set is $\{-7, -4\}$.

TRY THIS 1. Solve $x^2 = 30 - x$. Check the solutions. $\{-6, 5\}$

Highlighting the Standards

Standards 2d, 1a, 1b: This lesson helps students to read words and translate them into mathematical statements. In the Written Exercises, students practice stating and solving problems.

Additional Example 1

Write $x^2 = 16x - 48$ in standard form.
$x^2 - 16x + 48 = 0$

Additional Example 2

Solve $x = -x^2 + 56$. $\{-8, 7\}$

In the next two examples, it is more convenient to have the polynomial on the right side of the equation and 0 on the left side.

EXAMPLE 3   Solve $-x^2 = 7x$.

Solution

$$-x^2 = 7x$$

Add x^2 to each side. $\quad 0 = x^2 + 7x$

Factor out the GCF, x. $\quad 0 = x(x + 7)$

Set each factor equal to 0. $\quad x = 0 \ or \ x + 7 = 0$

Solve. $\quad x = 0 \qquad\qquad x = -7$

The roots are 0 and -7. The check is left for you.

So, the solution set is $\{-7, 0\}$

EXAMPLE 4   Solve $6 - 11x = 2x^2$.

Plan   The x^2 term is on the right and its coefficient is positive. Write the equation in standard form with the polynomial on the right and 0 on the left side of the equation. Solve by factoring.

Solution

$$6 - 11x = 2x^2$$
$$0 = 2x^2 + 11x - 6$$
$$0 = (2x - 1)(x + 6)$$
$$2x - 1 = 0 \ or \ x + 6 = 0$$
$$x = \frac{1}{2} \qquad\qquad x = -6$$

Check

For $x = \frac{1}{2}$:

$$6 - 11x = 2x^2$$
$$6 - 11\left(\frac{1}{2}\right) \stackrel{?}{=} 2\left(\frac{1}{2}\right)^2$$
$$6 - 5\frac{1}{2} \stackrel{?}{=} 2\left(\frac{1}{4}\right)$$
$$\frac{1}{2} = \frac{1}{2} \ \text{True}$$

For $x = -6$:

$$6 - 11x = 2x^2$$
$$6 - 11(-6) \stackrel{?}{=} 2(-6)^2$$
$$6 + 66 \stackrel{?}{=} 2(36)$$
$$72 = 72 \ \text{True}$$

The roots are $\frac{1}{2}$ and -6. So, the solution set is $\left\{\frac{1}{2}, -6\right\}$

TRY THIS   2. Solve $22x - 5x^2 = 8$. $\left\{\frac{2}{5}, 4\right\}$

Sometimes a word problem will lead to a quadratic equation. A quadratic equation may have two different roots.

7.8 Standard Form of a Quadratic Equation   **273**

Additional Example 3

Solve $-3x^2 = 15x$. $\{0, -5\}$

Additional Example 4

Solve $3 - x = 4x^2$. $\left\{\frac{3}{4}, -1\right\}$

Lesson Note

Tell students that if the coefficient of the x^2-term is positive, it is left on the side of the equation where it is. If the coefficient of the x^2-term is negative, then they can use the Addition Property of Equality to move it to the other side of the equation. Remind students that they should find all possible solutions. Since a quadratic equation has two roots, they should expect to find two solutions. For the Brainteaser, suggest that they let n equal some number, e.g., $n = 7$ as an aid in solving Exercise 1. Then suggest that they use the solutions for Exercises 1 and 2 to solve Exercise 3. If they need a further hint, point out that $\left(1 - \frac{1}{3}\right)\left(1 + \frac{1}{3}\right) = 1 - \left(\frac{1}{3}\right)^2$.

Math Connections

Previous Math: Finding the square roots of 25 is equivalent to solving the equation $x^2 = 25$, which is easily solved by factoring if the constant term is a perfect square.

Critical Thinking Questions

Evaluation: Ask students why the standard form of a quadratic equation requires that the x^2-term have a positive coefficient.
Easier to factor   Have them factor
$-x^2 + 7x - 12.$ $(-x + 3)(x - 4)$
Compare this to factoring $x^2 - 7x + 12.$
$(x - 3)(x - 4) = -(-x + 3)(x - 4)$

Common Error Analysis

Error: When solving equations involving a quadratic trinomial, some students will have difficulty when $a < 0$.

Point out that multiplying both sides of the equation by -1 will result in $a > 0$, thus making the trinomial easier to factor.
Example: Solve $-x^2 + 3x + 10 = 0$
Solution: $-1(-x^2 + 3x + 10) = -1(0)$
$$x^2 - 3x - 10 = 0$$
$$(x - 5)(x + 2) = 0$$
$$x - 5 = 0 \qquad x + 2 = 0$$
$$x = 5 \qquad x = -2$$

Checkpoint

Solve each equation.

1. $65 - 18x = -x^2$ $\{5, 13\}$
2. $7x = 3x^2$ $\{0, \frac{7}{3}\}$
3. $4x^2 = -27x + 7$ $\{\frac{1}{4}, -7\}$
4. Twelve less than 7 times a number is the same as the square of the number. Find the number.   3 or 4

EXAMPLE 5   The square of a number is 12 less than 8 times the number. Find the number.

What are you to find?	A number
What is given?	The square of the number is 12 less than 8 times the number.
Choose a variable. What does it represent?	Let x = the number.
Write an equation.	*Square of x* is 12 less than *8 times x*.

$$x^2 = 8x - 12$$

Solve the equation.
$$x^2 - 8x + 12 = 0$$
$$(x - 6)(x - 2) = 0$$
$$x - 6 = 0 \text{ or } x - 2 = 0$$
$$x = 6 \qquad x = 2$$

Check in the original problem.
Is the square of 6, which is 36, 12 less than 8 times 6? Yes, because $36 = 8(6) - 12 = 48 - 12$.

Is the square of 2, which is 4, 12 less than 8 times 2? Yes, because $4 = 8(2) - 12 = 16 - 12$.

State the answer.   There are two answers, 6 and 2.

EXAMPLE 6   The product of a number and 3 more than twice the number is 44. Find the number.

Solution   Let x = the number. The *product* of x and *3 more than twice x* is 44.

$$x(2x + 3) = 44$$
$$2x^2 + 3x = 44$$
$$2x^2 + 3x - 44 = 0$$
$$(2x + 11)(x - 4) = 0$$
$$2x + 11 = 0 \qquad or \quad x - 4 = 0$$
$$2x = -11 \qquad\qquad x = 4$$
$$x = -\frac{11}{2}$$

Check   Show that both $-\frac{11}{2}$ and 4 are correct answers.

TRY THIS   3. The product of a number and 7 less than 3 times the number is 20. $-\frac{5}{3}$ and 4

274   Chapter 7 Factoring Polynomials

Additional Example 5

The square of a number is 14 more than 5 times the number. Find the number.
7 or -2

Additional Example 6

The product of a number and 1 less than 3 times the number is 140. Find the number.
7 or $-\frac{20}{3}$

Classroom Exercises

State each equation in standard form with a positive coefficient for the square term.

$x^2 - 4x + 3 = 0$

1. $x^2 = 4x - 3$
2. $-a^2 = 13a$ $a^2 + 13a = 0$
3. $-14 = a^2 - 9a$ $a^2 - 9a + 14 = 0$
4. $-t^2 - 5t = 4$
5. $p^2 = 11p - 30$
6. $4 = d^2$ $d^2 - 4 = 0$
7. $5 + g^2 = -6g$
8. $8x = -x^2$ $x^2 + 8x = 0$
9. $4x - x^2 = 0$ $x^2 - 4x = 0$

10–18. Solve the equations in Exercises 1–9.

10. $\{1, 3\}$ 11. $\{-13, 0\}$ 12. $\{2, 7\}$ 13. $\{-4, -1\}$ 14. $\{5, 6\}$
15. $\{-2, 2\}$ 16. $\{-5, -1\}$ 17. $\{-8, 0\}$ 18. $\{0, 4\}$

Written Exercises

Solve each equation.

1. $13y = -y^2 - 40$ $\{-5, -8\}$
2. $-x^2 = -3x$ $\{0, 3\}$
3. $5c = -c^2$ $\{-5, 0\}$
4. $8x = -x^2 - 16$ $\{-4\}$
5. $21 + 4m = m^2$ $\{-3, 7\}$
6. $10 - 3u = u^2$ $\{2, -5\}$
7. $x^2 = 8x + 20$ $\{-2, 10\}$
8. $2y = -y^2$ $\{0, -2\}$
9. $-4a + 4 = -a^2$ $\{2\}$
10. $64 = b^2$ $\{8, -8\}$
11. $8 - 10a = 3a^2$ $\{\frac{2}{3}, -4\}$
12. $4k^2 = 9$ $\{\frac{3}{2}, -\frac{3}{2}\}$
13. $n^2 - 3n = 28$ $\{-4, 7\}$
14. $2x^2 + 15 = -11x$
15. $3a^2 = -5a + 2$
16. $5x - 25 = -2x^2$ $\{\frac{5}{2}, -5\}$
17. $49 = 25t^2$ $\{\frac{7}{5}, -\frac{7}{5}\}$
18. $3a^2 - 4 = a$ $\{\frac{4}{3}, -1\}$
19. $31x + 42 = -4x^2$
20. $12x^2 = 13x + 35$ $\{\frac{7}{3}, -\frac{5}{4}\}$
21. $-2b = 3 - 8b^2$
22. $-10 = 7y - 12y^2$ $\{-\frac{2}{3}, \frac{5}{4}\}$
23. $21 = 11x + 6x^2$ $\{-3, \frac{7}{6}\}$
24. $10x^2 - 6 = -11x$

25. Write an explanation of how to solve the equation $4x^3 + 9x = 15x^2$.

21. $\{-\frac{1}{2}, \frac{3}{4}\}$ 24. $\{\frac{2}{5}, -\frac{3}{2}\}$

Solve each problem.

26. The square of a number is 6 less than 5 times the number. Find the numbers.   2, 3

27. The product of a number and 5 more than twice the number is 75. Find the numbers. $5, -\frac{15}{2}$

28. The sum of the square of a number and 6 times the number is 40. Find the numbers. $4, -10$

29. Six less than 5 times the square of a number is the same as the number. Find the numbers. $\frac{6}{5}, -1$

30. Twice the square of a number is 15 less than 11 times the number. Find the numbers. $\frac{5}{2}, 3$

31. Three times the square of a number is 5 less than 16 times the number. Find the numbers. $\frac{1}{3}, 5$

32. The product of 1 more than twice a number and 3 less than 4 times the same number is 25. Find the numbers. $2, -\frac{7}{4}$

Closure

Ask students to give the three requirements for a quadratic equation to be in standard form.   One side equal to 0; terms in descending order of exponents; coefficient of squared term positive

◼◼ FOLLOW UP

Guided Practice
Classroom Exercises 1–18
Try This   all

Independent Practice
A Ex. 1–18, **B** Ex. 19–32, **C** Ex. 33–37
Basic: WE 1–27 odd
Average: WE 7–33 odd
Above Average: WE 11–37 odd

Additional Answers
Classroom Exercises

4. $t^2 + 5t + 4 = 0$
5. $p^2 - 11p + 30 = 0$
7. $g^2 + 6g + 5 = 0$

Written Exercises

14. $\{-\frac{5}{2}, -3\}$
15. $\{-2, \frac{1}{3}\}$
19. $\{-\frac{7}{4}, -6\}$
25. Write in standard form: $4x^3 - 15x^2 - 9x = 9$; Factor: $x(4x - 3)(x - 3) = 0$; Set each factor equal to zero and solve for x: $x = 0$ or $x = \frac{3}{4}$ or $x = 3$. The solution set is $\{0, \frac{3}{4}, 3\}$.

Enrichment

Have the students imagine that they have won a lottery, and that they have the choice of two prizes.

1. Receive $100,000 now, and each year that follows receive half of the amount received the previous year for life.

2. Receive $20,000 now, and $20,000 every year for life.

If the second prize is chosen, how long will a student have to live to collect more than if the student had chosen Prize 1?   10 years

19. $(3t - 5)(2t + 3)$
20. $(3t + 1)(9t - 7)$
21. $(2x - 3)^2$
22. $(4a - 1)(7a - 2)$
23. $(4m + 7)(2m + 5)$
24. $(2t - 5)(t - 6)$
25. $(3a - 3)(5a + 2)$
26. Irreducible
27. $(6c + 7)(6c - 5)$
28. $(4a - 3)(a + 5)$
29. $(2b + 9)(b - 5)$
30. $(2y - 5)(7y - 2)$
31. $(4t + 7)(2t + 7)$
32. $(8y - 3)(y + 6)$
33. $(10a - 1)(a - 9)$
34. $(-3t + 5)(t + 2)$
35. $(-5a + 2)(3a + 8)$
36. $(3m - 2)(7m - 4)$
37. $(2x + 3)(-5x + 6)$
38. $(-2y + 3)(4y + 3)$
39. $(5b + 2)(-3b + 10)$
40. $(-5y + 3)(7y + 6)$
41. $(8x - 3)(x - 4)$
42. $(2x + 1)(x - 3)$
43. $(4x - 3)(2x - 1)$
44. $(7t + 3)(2t - 9)$
45. $(8d - 5)(5d + 8)$
46. For a trinomial of the form $ax^2 + bx + c$, the possible factors of $ax^2 (= 8x^2)$ are 1 and 8, and 2 and 4. Since $c (-4)$ is negative, the signs contained in the binomial factors will be opposites. Use trial and error to test possible combinations of factors. The factors are $(x + 4)(8x - 1)$.

Solve each equation.

33. $(3a - 4)(2a + 3) = (a - 1)(5a + 4)$ $\{-4, 2\}$
34. $6(x + 2)^2 - 5(x + 2) = 6$ $\{-\frac{1}{2}, -\frac{8}{3}\}$
35. $(y + 4)(3y - 2) = -y - 14$ $\{-\frac{2}{3}, -3\}$
36. $-13x^2 + 36 = -x^4$ $\{-2, 2, -3, 3\}$
37. Solve for x: $x^2 - cx = ac - ax$ $\{-a, c\}$

Mixed Review

Simplify. *2.10, 6.2, 6.8, 6.9*

1. $3x - [4 - (2 - x)]$ $2x - 2$
2. $(-4a^2)^3$ $-64a^6$
3. $(3x - 2)(4x - 3)$ $12x^2 - 17x + 6$
4. $(x + 2)(x^2 - 4x + 3)$ $x^3 - 2x^2 - 5x + 6$
5. $(x - 5)(x + 2) - (x^2 + 5x)$ $-8x - 10$
6. $(2a + 3)^2$ $4a^2 + 12a + 9$

Solve. *7.7*

7. $x^2 - 2x = 0$ $\{0, 2\}$
8. $16x^2 - 1 = 0$ $\{\frac{1}{4}, -\frac{1}{4}\}$
9. $x^2 - 6x + 5 = 0$ $\{1, 5\}$
10. The sum of three consecutive even integers is 48. Find the integers. *4.3* 14, 16, 18

/Brainteaser

1. When simplified, the product,

$$\left(1 - \tfrac{1}{3}\right)\left(1 - \tfrac{1}{4}\right)\left(1 - \tfrac{1}{5}\right)\left(1 - \tfrac{1}{6}\right) \cdots \left(1 - \tfrac{1}{n}\right),$$

becomes which of the following? b

   a. $\dfrac{1}{n}$ b. $\dfrac{2}{n}$ c. $\dfrac{2(n - 1)}{n}$ d. $\dfrac{3}{n(n + 1)}$

2. When simplified, the product,

$$\left(1 + \tfrac{1}{3}\right)\left(1 + \tfrac{1}{4}\right)\left(1 + \tfrac{1}{5}\right)\left(1 + \tfrac{1}{6}\right) \cdots \left(1 + \tfrac{1}{m}\right),$$

becomes which of the following? c

   a. $\dfrac{4}{m}$ b. $\dfrac{m + 1}{m}$ c. $\dfrac{m + 1}{3}$ d. $\dfrac{m + 1}{3m}$

3. Simplify.

$$\left[1 - \left(\tfrac{1}{3}\right)^2\right]\left[1 - \left(\tfrac{1}{4}\right)^2\right]\left[1 - \left(\tfrac{1}{5}\right)^2\right]\left[1 - \left(\tfrac{1}{6}\right)^2\right] \cdots \left[1 - \left(\tfrac{1}{n}\right)^2\right]$$ $\dfrac{2(n + 1)}{3n}$

7.9 Problem Solving: Using Quadratic Equations

Teaching Resources

Quick Quizzes 59
Reteaching and Practice
  Worksheets 59

Objectives   To solve consecutive integer problems involving quadratic equations
To solve area problems involving quadratic equations

Some problems lead to quadratic equations. They can be solved by solving their corresponding equations.

EXAMPLE 1   Find two consecutive odd integers such that the square of the second, decreased by the first, is 14.

What are you to find?	Two consecutive odd integers
What is given?	The square of the second integer, decreased by the first integer, is 14.
Choose a variable.	Let x = the first odd integer.
What does it represent?	Then $x + 2$ = the second odd integer.

Write an equation.
$$(x + 2)^2 - x = 14$$
$$x^2 + 4x + 4 - x = 14$$
$$x^2 + 3x - 10 = 0$$
$$(x + 5)(x - 2) = 0$$
$$x + 5 = 0 \quad or \quad x - 2 = 0$$
$$x = -5 \qquad x = 2$$
$$x + 2 = -3 \qquad x + 2 = 4 \longleftarrow \text{Not odd integers}$$

Check in the original problem.   Are -5 and -3 consecutive odd integers? Yes.

Does the square of -3, decreased by -5, equal 14? Yes, because
$$(-3)^2 - (-5) = 9 + 5 = 14$$

State the answer.   The consecutive odd integers are -5 and -3.

EXAMPLE 2   The product of the first and third of three consecutive integers is 1 less than 8 times the second. Find the three integers.

Solution   Let x, $x + 1$, and $x + 2$ represent the three consecutive integers. Use the given information to write an equation.

Product of 1st and 3rd is 1 less than 8 times 2nd.
$$x(x + 2) = 8(x + 1) - 1$$

Additional Example 1

Find two consecutive even integers such that the square of the first, decreased by the second, is 130.   12 and 14

Additional Example 2

The product of the second and third of three consecutive integers is 200 more than 10 times the first. Find the three integers.   18, 19, 20 or $-11, -10, -9$

Highlighting the Standards

Standards 7d, 4c, 1b: Students find various mathematical facts about geometric figures and practice solving problems that involve geometric figures. In Problem Solving Strategies, they practice organizing the possibilities in a problem.

Lesson Note

Point out the importance of the parentheses around $x + 2$ in Example 1. The fact that one of the solutions to the equation for Example 1 does not fit the conditions of the original problem (odd integers), should emphasize the importance of checking each solution in the original problem rather than in the equation. In Example 3, point out that dividing the equation $0 = 2(x - 6)(x + 4)$ by 2 gives an equivalent equation.

Math Connections

Life Skills: When replanting a rectangular flower garden, you may wish to double its area but retain the the same ratio of length to width. The task of finding the new dimensions can be solved with a quadratic equation.

Critical Thinking Questions

Application: Ask students what happens to the area of a rectangular garden if both the length and the width are doubled.   The area is 4 times as great.   Then ask them to show this algebraically. $(2x)(2x) = 4x^2$ Ask them to find the increase in area of a rectangular garden with length x and width y in area if x is increased by 5 and y is increased by 7. $7x + 5y + 35$

$$x^2 + 2x = 8x + 8 - 1$$
$$x^2 - 6x - 7 = 0$$
$$(x - 7)(x + 1) = 0$$
$$x - 7 = 0 \ \text{or} \ x + 1 = 0$$
$$x = 7 \qquad\qquad x = -1$$
$$x + 1 = 8 \qquad x + 1 = 0$$
$$x + 2 = 9 \qquad x + 2 = 1$$

Check   Are 7, 8, and 9 consecutive integers?   Yes.

Is the product of 7 and 9 equal to 1 less than 8 times 8?
Yes, because $7(9) = 8(8) - 1$, or $63 = 64 - 1$

Are -1, 0, and 1 consecutive integers?   Yes.

Is the product of -1 and 1 equal to 1 less than 8 times 0?
Yes, because $-1(1) = 8(0) - 1$, or $-1 = 0 - 1$

Thus, there are two groups of such integers: $7, 8, 9$ and $-1, 0, 1$.

TRY THIS   1. The product of the first and second of three consecutive integers is 22 more than 5 times the third. Find the three integers. $-4, -3, -2$
and 8, 9, 10

Only positive solutions of an equation can represent measurements such as the length and width of a rectangle, as shown in Example 3.

EXAMPLE 3   The area of a rectangle is 48 cm². The length is 4 cm less than twice the width. Find the length and width.

What are you to find?   length and width of a rectangle

Draw and label a figure.

$A = lw$, with sides labeled w and l

What is given?   length = 4 cm less than 2 times width
area = 48 cm²

Choose a variable.   Let $x = $ the width.

What does it represent?   Then $2x - 4 = $ the length.

Write an equation. $A = lw$
$48 = (2x - 4)x$

Solve the equation. $48 = 2x^2 - 4x$
$0 = 2x^2 - 4x - 48$
$0 = 2(x - 6)(x + 4)$
$x - 6 = 0 \ \text{or} \ x + 4 = 0$
$x = 6 \qquad\qquad x = -4$

Additional Example 3

The area of a rectangle is 240 cm². The length is 6 cm more than 3 times the width. Find the length and the width. $w = 8$ cm;
$l = 30$ cm

<table>
<tr><td>Check in the original problem.</td><td>If $x = 6$, the width is 6 cm and the length is $2(6) - 4$, or 8 cm. Does $A = lw = 48$? Yes.</td></tr>
<tr><td></td><td>Reject -4. A width must be a positive number.</td></tr>
<tr><td>State the answer.</td><td>The width is 6 cm and the length is 8 cm.</td></tr>
</table>

EXAMPLE 4 A rectangular patio is 6 m long and 4 m wide. The area of the patio is to be increased by 39 m² by increasing the length and width by the same amount. Find the amount by which each side must be increased.

Solution Let $t =$ the amount of increase of each side.
Then $t + 6 =$ the new length and
$t + 4 =$ the new width.

$A = 6 \cdot 4 + 39$ $t + 4$

$t + 6$

New area = original area + increase in area

$(t + 6)(t + 4) = \quad 6 \cdot 4 \quad + \quad 39$

$t^2 + 10t + 24 = 24 + 39$
$t^2 + 10t - 39 = 0$
$(t + 13)(t - 3) = 0$
$t + 13 = 0 \quad or \quad t - 3 = 0$
$\qquad t = -13 \qquad\qquad t = 3$

Check Since the amount of increase must be a positive number, reject -13.

If each side is increased by 3 m, the new width is 7 m and the new length is 9 m. Is the new area 39 m² more than the original area?

Yes, since $9(7) = 39 + 6(4)$, or $63 = 39 + 24$.

Thus, each side is increased by 3 m.

TRY THIS 2. The area of a rectangle is 36 cm². The length is 3 cm less than 3 times the width. Find the width and length. $w = 4$ cm, $l = 9$ cm

Common Error Analysis

Error: Students sometimes make errors in writing consecutive negative integers. For example, if $x = -8$, they may mistakenly write $x + 1$ and $x + 2$ as -9 and -10, instead of as -7 and -6.

Remind them that values increase as you move from the left to the right on the number line.

Checkpoint

1. Find two consecutive integers whose product is 210. 14, 15, or $-15, -14$
2. Find two consecutive even integers whose product is 2,600. 50, 52, or $-52, -50$
3. Find three consecutive odd integers such that the square of the second, increased by the square of the first, is 7 less than the square of the third. 5, 7, 9, or $-1, 1, 3$
4. A rectangular flag is 6 in. longer than it is wide. Its area is 216 in². Find the length and the width of the flag. $w = 12$ in; $l = 18$ in

Additional Example 4

A rectangle is 14 cm long and 11 cm wide. If each dimension is decreased by the same amount, the area is decreased by 84 cm². Find the amount by which each side is decreased. 4 cm

1. Adding 2 to an integer produces an even integer.   Sometimes
2. If a word problem leads to a quadratic equation, then there will be two solutions for the original *word* problem.
   Sometimes
3. Adding 1 to an integer results in an odd integer.   Sometimes
4. Two consecutive odd integers differ by 2.   Always
5. The sum of two odd integers is an odd integer.   Never

◢◣◤FOLLOW UP

Guided Practice

Classroom Exercises 1–10
Try This   all

Independent Practice

A Ex. 1–10, **B** Ex. 11–15, **C** Ex. 16–19

Basic: WE 1–11 odd, Problem Solving Strategies, Mixed Problem Solving

Average: WE 5–15 odd, Problem Solving Strategies, Mixed Problem Solving

Above Average: WE 9–19 odd, Problem Solving Strategies, Mixed Problem Solving

Classroom Exercises

Let n, $n + 2$, and $n + 4$ represent three consecutive odd integers. Write an algebraic expression for each of the following. Do not simplify.

1. the product of the first two consecutive integers $n(n + 2)$
2. the square of the second integer decreased by twice the third integer $(n + 2)^2 - 2(n + 4)$
3. the sum of the squares of the second and the third integers $(n + 2)^2 + (n + 4)^2$
4. the product of the first and the third integers, decreased by the second integer $n(n + 4) - (n + 2)$

Give, in simplest form, an algebraic expression for the area of each rectangle with the given length and width.

5. length: $w + 5$
   width: w $w^2 + 5w$
6. width: w $2w^2 - 3w$
   length: $2w - 3$
7. width: $w + 1$ $w^2 + 3w + 2$
   length: $w + 2$
8. width: $w - 5$ $w^2 - 25$
   length: $w + 5$
9. Find two consecutive even integers whose product is 120.
   -10, -12 and 10, 12
10. Find two consecutive odd integers whose product is 143.
    -11, -13 and 11, 13

Written Exercises

Solve each problem.

1. Find two consecutive even integers whose product is 48. -6, -8 and 6, 8
2. Find two consecutive odd integers whose product is 63. -7, -9 and 7, 9
3. Find two consecutive even integers such that the square of the second, decreased by the first, is 58.   6, 8
4. Find two consecutive odd integers such that the square of the second, increased by the first, is 88.   7, 9
5. Find two consecutive odd integers such that the sum of their squares is 34.   3, 5 and -3, -5
6. Find two consecutive even integers such that the sum of their squares is 52.   4, 6 and -4, -6
7. The width of a rectangle is 3 in. less than the length. The area is 54 in^2. Find the length and width. $l = 9$ in, $w = 6$ in
8. The length of a rectangle is 4 yd more than the width. The area is 60 yd^2. Find the length and width.
9. The length of a rectangle is 3 cm more than twice its width. The area is 44 cm^2. Find its dimensions.
10. The length of a rectangle is 5 m more than 3 times its width. Find the dimensions if the area is 42 m^2.
11. Find three consecutive odd integers such that the square of the second integer, increased by the first, is 54.   5, 7, 9
12. Find four consecutive integers such that the sum of the squares of the first and the third integers is 130.
    -9, -8, -7, -6 and 7, 8, 9, 10

280   Chapter 7 Factoring Polynomials

13. Find three consecutive even integers such that the product of the first and the third integers is 20 more than 5 times the second. 6, 8, 10

14. If each of the dimensions of a 7 ft by 10 ft rectangle is increased by the same amount, the area is increased by 38 ft². Find the amount of increase in each dimension. 2 ft

15. A movie screen measures 8 m by 6 m. Each dimension is to be increased by the same amount to increase the area by 72 m². Find the amount of increase in each dimension. 4 m

16. Find three consecutive integers such that the product of all three, decreased by the cube of the first, is 33. 3, 4, 5

17. A room has dimensions 11 ft by 12 ft. A rug with area 90 ft² is placed in the center of the room leaving bare edges of the same width all the way around. Find the dimensions of the rug. 9 ft × 10 ft

18. A photograph, 14 in. by 11 in., is to have a white border of uniform width around it. The total area of the picture, including the border, is 270 in². Find the width of the border. 4 in.

19. Squares 2 in. wide are cut out from the corners of a square sheet of metal. The sides of the remaining sheet are then turned up to form an open box. The volume of the box is 32 in³. Find the original length of each side of the square. 8 in.

Mixed Review

Evaluate. *1.1, 1.2, 1.3*

1. $n - 3.5$ for $n = 14.1$ 10.6

2. $\frac{x}{12}$ for $x = 4.8$ 0.4

3. $27 - 5y$ for $y = 6$ −3

4. $m(n - 3) - (m - 2)$ for $m = -2$ and $n = -1$ 12

5. $3a^2 - 4a$ for $a = 8$ 160

6. $\frac{x^2 - y^2}{(x - y)^2}$ for $x = 6$ and $y = 3$ 3

Simplify. *6.1, 6.2*

7. $3y \cdot y^6$ $3y^7$

8. $-5a^4(2a^3)$ $-10a^7$

9. $(6x^3y^2)(-3xy^4)$

10. $(y^6)^2$ y^{12}

9. $-18x^4y^6$ **12.** $-32a^5b^{10}c^{20}$

11. $-(a^5)^3$ $-a^{15}$

12. $(-2ab^2c^4)^5$

13. $(3c^4)^3(2c^2)^4$ $432c^{20}$

14. $-m^2(m^3n)^4$ $-m^{14}n^4$

15. John has 6 more quarters than dimes. He has 18 coins in all. Find the number of coins of each type. *4.2* 6 dimes, 12 quarters

7.9 Problem Solving: Using Quadratic Equations **281**

Enrichment

Have students solve this problem. A rectangle with an area of 180 cm² contains 5 smaller rectangles as shown below.

Each of the smaller rectangles has length of $(2x + 2)$ cm and width of $(x - 2)$ cm.

1. Find the value of x. 5
2. Find the dimensions of each of the 5 smaller rectangles. $l = 12$ cm; $w = 3$ cm
3. Find the dimensions of the largest rectangle. $l = 15$ cm; $w = 12$ cm

3. First game matches: *AB* in one gym, *CD* in other. Only 4 distinct possible pairs for second game.

Problem Solving Strategies

Organizing the Possibilities

In some problems you need to keep track of possibilities. One way to help manage a large number of possibilities is to list or diagram the possibilities in a systematic way. Organizing the possibilities can help you make sure you have them all and that you have not counted the same one twice.

Example A rock band wants to add a new twist to its act by featuring a different duet in each song of a performance. How many songs will the 5-member band need to play so that each member sings one duet with every other member? Start by representing each member with the letters *A–E* to make it easier to list duets. Next, list all possible duets on a chart. Begin with all duets that include band member *A*, then those that include *B*, then *C*, and so on. Then cross out any repeats, such as *AB* and *BA*.

With *A*	With *B*	With *C*	With *D*	With *E*
AB	B̸A̸	C̸A̸	D̸A̸	E̸A̸
AC	BC	C̸B̸	D̸B̸	E̸B̸
AD	BD	CD	D̸C̸	E̸C̸
AE	BE	CE	DE	E̸D̸

Since each of the 5 band members can be in 4 duets, there are 4(5), or 20 possible duets. But half of these 20 are repeats, so there are only 10 different duets. The band must play 10 songs.

Solve each problem.

1. As head of a dance club with a membership of 8 girls and 8 boys, you are planning a "Mix-Up" Dance at which each boy has 1 complete dance with each girl. How many dances must you allow for? 8

2. Four teams play basketball at the same time in 2 adjacent gyms. After each game, each team plays a different team. What is the fewest number of games that will allow each team to play each other team the same number of times and allow each team to play in each gym the same number of times? 12

3. An organizer for the teams in Exercise 2 wants them to play 3 games each and to play a different team each time. He does not want a team to play all 3 of its games in the same gym. Show that such a schedule is impossible.

4. How much money would it take to display all the possible combinations that can be made using 3 different denominations at one time from these 5 coins: half-dollar, quarter, dime, nickel, penny? Each combination must have a different monetary value. $5.46

Mixed Problem Solving

20. $2(4x - 5y)(x + 3y)$
21. $2(2a - b)^2$
22. $(8 + c)(a + b)$
23. $(2 + b)(x + y)$
24. $(c + 5)(a + b)$
25. $(5a + 3b)(x + y)$
26. $(a - 7)(a + c)$
27. $(r + s)(t - 3)$
28. $(a + b)(x - 3)(x + 3)$
29. $(y - 2)(p - 5)(p + 5)$
30. $(x + 4)(x - 4)(a + b)$
31. $3(4r + 1)(r - 4)$
32. $3t(16t^2 - 8t + 1) = 3t(4t - 1)^2$
33. $-3(f + 4)(f - 2)$
34. $3y(2x + 7)(x - 10)$
35. $2a(12 - 5a)(12 + 5a)$
36. $ab(a - b)(a + b)$
37. $2s(5s + 7)(3s - 2)$
38. $x(x^2 - 6xy - 5y^2)$
39. $2y(5y - 4)(2y + 1)$
40. $(a + b)(x - 3)(x + 3)$
41. $(x - 2)(x + 2)(m + 1)$
42. $2(x + 2)(x + 2y)(x - 2y)$
43. $4ay(3y - a)(a^2 + 16y^2)$
44. $5xy(x - y)(3 + x)(3 - x)$
45. $(a + 2)(a - 1)(a^2 - a + 2)$
46. $r(r + 2)(r - 1)^2$
47. $(6x - 3s - 4)(2x - s + 3)$
48. $(x + 1)(4x + 3)(x - 2)$
49. $(2x - 5y)(2x + 5y + 1)$
50. $(x + 3)(x - 3)(x^2 - 8)$

As you work Exercises 1, 2, 3, 5, 6, 7, and 12, apply the strategy of *Estimating Before Solving.* For each problem, write an estimate (including the units) of the final answer before you begin to solve the problem.

If your final answer is not close to your estimate, think about how you might improve your estimating skills.

1. Find the perimeter and the area of a square with each side 8.7 cm.
   $p = 34.8$ cm; $a = 75.69$ cm²

2. Find the area of a triangle with a base of 8.3 cm and altitude 4.8 cm. 19.92 cm²

3. Willa has saved $290 in two years. She saved 4 times as much the first year as she saved the second year. How much did she save each year? First: $232; second: $58

4. Find the perimeter of a triangle if the perimeter, decreased by 5 cm, is the same as 43 cm decreased by 7 times the perimeter. 6 cm

5. For the February opera, 2,452 people attended. This audience was $\frac{4}{5}$ as large as the October audience. How many attended the opera in October? 3,065

6. Sue lives $2\frac{1}{2}$ miles from school. She walks at the rate of 3 mi/h. How long does it take her to walk to school? 50 min

7. Luis drove 234 mi at an average rate of 52 mi/h. If he stopped for a 40-min lunch break, how long did the trip take? 5 h 10 min

8. The length of a rectangle is 12 cm more than 3 times the width. The perimeter is 96 cm. Find the length and width. $l = 39$ cm, $w = 9$ cm

9. A triangle has a perimeter of 76 cm. The first side is 8 cm longer than the second side. The third side is 5 cm longer than the second side. Find the length of each side. 29 cm, 21 cm, 26 cm

10. One side of a triangle is 3 cm shorter than the second side. The remaining side is 8 cm longer than the second side. The perimeter is 95 cm. Find the length of each side. 30 cm, 27 cm, 38 cm

11. Forty-eight students are separated into two groups. The first group is 3 times as large as the second. How many students are in each group? 36, 12

12. On a video game, Wilma and Alex scored 2,895 points. Wilma's score was 2 times Alex's score. Find their scores. Alex: 965, Wilma: 1,930

13. The vertex angle of an isosceles triangle measures 16 more than twice a base angle. Find the measure of each angle of the triangle. vertex: 98°; base angles: 41°

14. One typist worked 8 h longer than a second. A third typist worked 3 h less than the second. How long did each typist work if together they worked 95 h? 38 h, 30 h, 27 h

15. Last month George earned $620 more than this month. He earned a total of $5,420 for the two months. Find his earnings for each month. $3,020; $2,400

16. Find three consecutive integers if 3 times the square of the smallest equals the product of the other two. 2, 3, 4

Key Terms

composite number   (p. 243)
cubic equation   (p. 269)
difference of two squares   (p. 259)
double root   (p. 269)
factoring by grouping   (p. 264)
greatest common factor (GCF)   (p. 244)
irreducible polynomial   (p. 257)
perfect square trinomial   (p. 260)

prime factorization   (p. 244)
prime number   (p. 243)
quadratic equation   (p. 268)
quadratic polynomial   (p. 268)
roots   (p. 268)
standard form of a quadratic
   equation   (p. 272)
Zero-Product Property   (p. 267)

Key Ideas and Review Exercises

7.1   To find the greatest common factor (GCF) of two or more monomials, factor the integers into primes. Then multiply the smallest powers of prime numbers and of variables that are common to the monomials. To find a missing factor, divide the product by the given factor.

Find the greatest common factor (GCF).

1. 18, 45 9

2. $30a^3, 35a^2$ $5a^2$

3. $72b^5, 135b, 54ab^2$ $9b$

Find the missing factor.

4. $a^6 \cdot \underline{\ ?\ } = a^{12}$ a^6

5. $-5n^4 \cdot \underline{\ ?\ } = 20n^{10}$ $-4n^6$

6. $8a^3b^2 \cdot \underline{\ ?\ } = 64a^6b^8$ $8a^3b^6$

7.2   To factor out the greatest common monomial factor from a polynomial, find the GCF of the terms. Then use the Distributive Property to factor out the GCF.

Factor out the GCF from each polynomial.

7. $5x^2 - 15x + 20$
$5(x^2 - 3x + 4)$

8. $a^5 + a^3 - 3a^2$
$a^2(a^3 + a - 3)$

9. $3x^4 - 9x^3 + 27x$
$3x(x^3 - 3x^2 + 9)$

7.3   To factor trinomials of the form $x^2 + bx + c$, find binomial factors $(x + r)(x + s)$ such that $r \cdot s = c$ and $r + s = b$.

Factor each trinomial.

10. $x^2 - 8x + 15$
$(x - 3)(x - 5)$

11. $a^2 + 3a - 28$
$(a - 4)(a + 7)$

12. $y^2 + 14y + 24$
$(y + 2)(y + 12)$

7.4   To factor trinomials of the form $ax^2 + bx + c$, where $a > 1$, form trial binomial factors using factors of ax^2 for the first terms and factors of c for the second terms. Then multiply the binomials to check that the middle term is bx.

Factor each trinomial. **13.** $(3a + 5)(a - 2)$

$(5a + 3)(2a - 1)$

$(3x + 4)(2x + 3)$

13. $3a^2 - 11a + 10$ **14.** $10a^2 + a - 3$ **15.** $6x^2 + 17x + 12$

16. $2a^2 + 13a + 15$ **17.** $2m^2 - 7m + 5$ **18.** $4a^2 - 20a + 25$

$(2a + 3)(a + 5)$ $(2m - 5)(m - 1)$ $(2a - 5)^2$

7.5   To factor the difference of two squares, use $a^2 - b^2 = (a + b)(a - b)$.
To factor a perfect square trinomial, use
$a^2 + 2ab + b^2 = (a + b)^2$ or $a^2 - 2ab + b^2 = (a - b)^2$.

Factor.

$(x - 20)(x + 20)$

$(6y + 5)(6y - 5)$

19. $x^2 - 400$ **20.** $y^2 + 18y + 81$ $(y + 9)^2$ **21.** $36y^2 - 25$

22. $a^2 + 8a + 16$ **23.** $x^2 - 12x + 36$ **24.** $9x^2 + 30xy + 25y^2$

$(a + 4)^2$ $(x - 6)^2$ $(3x + 5)^2$

7.6   To factor a polynomial by grouping, look for the GCF in pairs of terms.
Factor out the GCF. Then factor out the common binomial.

Factor by grouping.

25. $5x + ax - 5y - ay$ **26.** $ax + ay + bx + by$ **27.** $x^2y - 25y - 3x^2 + 75$

28. $xm + xb + tm + tb$ **29.** $ta + 6t + a^2 + 6a$ **30.** $x^2m + x^2 - 4m - 4$

31. $ax + ay + 3x + 3y$ **32.** $ax + bx - ay - by$ **33.** $2axy - 2ay + 3bx - 3b$

7.6   To factor a polynomial completely, first factor out any GCF greater than 1.
Continue factoring until polynomial factors are irreducible.

Factor completely.

34. $2a^2 + 2a - 12$ **35.** $24x^2 + 4x - 8$ **36.** $100a^2 - 4b^2$

$2(a + 3)(a - 2)$ $4(3x + 2)(2x - 1)$ $4(5a + b)(5a - b)$

7.7, 7.8,   To solve quadratic equations that are in standard form, factor the poly-
7.9   nomial and apply the Zero-Product Property.

Solve each equation or problem.   **37.** $\{-5, -3\}$

$\{-\frac{5}{3}, \frac{3}{2}\}$

37. $a^2 + 8a + 15 = 0$ **38.** $t^2 - 196 = 0$ $\{-14, 14\}$ **39.** $y - 15 = -6y^2$

40. If a number is multiplied by 5 more than twice that number, the
result is 12. Find the number. $\frac{3}{2}, -4$

41. Find three consecutive integers such that the square of the third
integer is 9 times the first integer. 1, 2, 3 and 4, 5, 6

42. Find the solution of $(3y - 6)^2 - 12(3y - 6) + 36 = 0$ 4

43. The square of a number is 24 more than twice the number. Find
the number. $-4, 6$

25. $(x - y)(5 + a)$

26. $(a + b)(x + y)$

27. $(y - 3)(x - 5)(x + 5)$

28. $(x + t)(m + b)$

29. $(a + 6)(a + t)$

30. $(m + 1)(x + 2)(x - 2)$

31. $(a + 3)(x + y)$

32. $(x - y)(a + b)$

33. $(2ay + 3b)(x - 1)$

13. $(x + 1)(x - 7)$
14. $(a + 4)(a + 3)$
15. $(y - 5)(y - 1)$
16. $(a - 6)(a + 15)$
17. $(4a - 1)(a + 2)$
18. $3(3t - 1)(2t - 1)$
19. $(a + 10)(a - 10)$
20. $(3x - 5y)^2$
21. $(3x + y)(2a - b)$
22. $2(x - 7)^2$
23. $9(3x - 1)(3x + 1)$
24. $4a(2a + 1)(3a + 2)$
25. $2x(x - 3)(x + 5)$
26. $xy(x - 6y)(x + 6y)$
27. $(x - 1)(x - 3y)(x + 3y)$

Chapter 7 Test

Give the prime factorization of each number.

1. 48 $2^4 \cdot 3$
2. 81 3^4
3. 250 $2 \cdot 5^3$

Find the greatest common factor (GCF).

4. 27, 45 9
5. $18y^3, -24y$ $6y$
6. $x^4, 27x^2, 36x^3$ x^2

Find the missing factor.

7. $a^5 \cdot \underline{\ ?\ } = a^{10}$ a^5
8. $-4x^4 \cdot \underline{\ ?\ } = 24x^9$ $-6x^5$
9. $\underline{\ ?\ } \cdot 5b^3c^7 = -20b^7c^{12}$ $-4b^4c^5$

Factor out the GCF from each trinomial.

10. $5x^2 - 20x - 15$ $5(x^2 - 4x - 3)$
11. $a^4 - 5a^3 + 6a$ $a(a^3 - 5a^2 + 6)$
12. $18a^3b^4 - 30a^4b^3 - 24a^4b^2$ $6a^3b^2(3b^2 - 5ab - 4a)$

Factor each polynomial completely.

13. $x^2 - 6x - 7$
14. $a^2 + 7a + 12$
15. $y^2 - 6y + 5$
16. $a^2 + 9a - 90$
17. $4a^2 + 7a - 2$
18. $18t^2 - 15t + 3$
19. $a^2 - 100$
20. $9x^2 - 30xy + 25y^2$
21. $6ax - 3bx + 2ay - by$
22. $2x^2 - 28x + 98$
23. $81x^2 - 9$
24. $24a^3 + 28a^2 + 8a$
25. $2x^3 + 4x^2 - 30x$
26. $x^3y - 36xy^3$
27. $x^3 - 9xy^2 - x^2 + 9y^2$

Solve each equation.

28. $x^2 - 7x + 12 = 0$ $3, 4$
29. $a^2 - 16a = 0$ $0, 16$
30. $3 - 5t = 2t^2$ $-3, \frac{1}{2}$

Solve each problem.

31. The square of a number is 4 more than 3 times the number. Find the number. 4 and -1

32. If 4 times a number is added to the square of the number, the result is 21. Find the number. -7 and 3

33. The area of a wedding photograph is 300 cm². The length is 5 cm longer than the width. Find the length and width. $l = 20$ cm, $w = 15$ cm

34. Find three consecutive even integers if the product of the first and the second integers is 8 more than 10 times the third integer. $-4, -2, 0$ and 12, 14, 16

Factor each polynomial completely. (Exercises 35–36)

35. $x^{6a} - 11x^{3a} + 30$ $(x^{3a} - 5)(x^{3a} - 6)$
36. $32a^4 - 1{,}250b^4$ $2(2a - 5b)(2a + 5b)(4a^2 + 25b^2)$
37. Solve the equation $x^5 - 17x^3 + 16x = 0$ for all values of x. $\{0, -1, 1, -4, 4\}$

Choose the one best answer to each question or problem.

1. The sum of an even number and an
D odd number is __?__ .
   (A) always composite
   (B) always prime
   (C) always even
   (D) always odd
   (E) always divisible by 3

2. If $5a - 15 = 40$, then $a - 3 =$
C __?__
   (A) 11   (B) 5   (C) 8   (D) 40
   (E) 3

3. The symbol $\begin{vmatrix} a & b \\ c & d \end{vmatrix}$ means $ad - bc$.
B What is the value of x^2 if $\begin{vmatrix} x & 3 \\ 3 & x \end{vmatrix} = 0$?
   (A) -9 (B) 9   (C) 3
   (D) -3 (E) 0

4. If r, s, and t are positive numbers
B such that $3r = 4s$ and $2s = 5t$,
   arrange the numbers in order from
   least to greatest.
   (A) r, t, s (B) t, s, r (C) r, s, t
   (D) s, r, t (E) t, r, s

5. $ACIG$ is a square. $AB = 6$ cm and
E $HI = 5$ cm.
   The area of the
   shaded region is
   75 cm². What is
   the area of $BCFE$?
   (A) 25 cm²
   (B) 121 cm²
   (C) 61 cm²
   (D) 66 cm²
   (E) 46 cm²

6. The perimeter of a rectangle is given
B by the formula $p = 2l + 2w$, where
   l is the length and w is the width. If
   $3a - 2$ represents the width and

$18a + 12$ represents the perimeter,
then the length is __?__ .
(A) $6a + 3$ (B) $6a + 8$
(C) $12a + 16$ (D) $15a + 14$
(E) $15a + 10$

7. If $0.5a + 1.5 = 3.0$, find the value
E of $3a - 10$.
   (A) 3   (B) -3 (C) 17
   (D) 1   (E) -1

8. Six consecutive odd integers are
A given. The sum of the first three in-
   tegers is 3. What is the sum of the
   last three?
   (A) 21   (B) 0   (C) 9
   (D)   4 (E)   3

9. If $x - 5 = 2$ and $x^2 - y^2 = 0$,
C then $y =$ ___
   (A) 7   (B) -7 (C) 7 or -7
   (D) 49   (E) -49

10. If the roots of the equation
D $(x - 5)(2x + c) = 0$ are integers,
    then c cannot equal ___ .
    (A) -6 (B) -4 (C) 0
    (D) 5   (E) 10

11. When is $(x - y)^2 = x^2 + y^2$
C   true?
    (A) Always   (B) Never
    (C) Sometimes   (D) only with
    decimals   (E) It cannot be deter-
    mined from the information given.

12. A rectangle has a width of 8 cm and
B   a length of 10 cm. If the length and
    width are each increased by 2 cm,
    what is the percent increase in the
    area of the rectangle?
    (A) $33\frac{1}{3}\%$ (B) 50%   (C) 80%
    (D) 120%   (E) 150%

8 RATIONAL EXPRESSIONS

OVERVIEW

This chapter enables students to extend the procedures they have learned for working with fractions to operating with rational expressions. The students add, subtract, multiply, and divide rational expressions and express the results in simplest form. A final lesson on simplifying complex fractions and complex rational expressions serves as a review of all the techniques they have learned.

OBJECTIVES

- To simplify rational expressions and tell for what values they are undefined
- To add, subtract, multiply, and divide rational expressions
- To divide polynomials by monomials and by binomials
- To simplify complex fractions and complex rational expressions

PROBLEM SOLVING

Looking for Patterns is the technique used to encourage students to write polynomials in "convenient form" in order to factor them more easily. Making a List of factors aids the student in finding the LCD. Solving a Simpler Problem in arithmetic helps the student understand the processes used for rational expressions. The application on page 324 applies the strategy of Testing Conditions to solving problems related to everyday, real-life situations.

READING AND WRITING MATH

In Exercise 42 on page 292 and in Exercises 33 and 34 on page 314, students are asked to give written explanations of procedures related to operations with rational expressions. The Focus on Reading Exercises on pages 295 and 313 ask students to identify certain patterns and to justify the answers they give.

TECHNOLOGY

Calculator: The calculator can be used to show what happens when you try to divide by zero, as illustrated on page 289. Exercise 26 on page 292 is a good problem to work with a calculator.

SPECIAL FEATURES

PLANNING GUIDE

Lesson	Basic	Average	Above Average	Resources
8.1 pp. 291–293	CE 1–16 WE 1–29 all, 42	CE 1–16 WE 1–29 odd, 34–48 all	CE 1–16 WE 1–55 odd	Reteaching 60 Practice 60
8.2 pp. 295–296	FR all, CE 1–12 WE 1–21 odd	FR all, CE 1–12 WE 5–25 odd	FR all, CE 1–12 WE 11–31 odd	Reteaching 61 Practice 61
8.3 pp. 300–301	CE 1–12 WE 1–30 odd	CE 1–12 WE 11–40 odd	CE 1–12 WE 17–40 odd, 41–43 all	Reteaching 62 Practice 62
8.4 pp. 303–305	CE 1–14 WE 1–33 odd	CE 1–14 WE 9–41 odd	CE 1–14 WE 19–51 odd	Reteaching 63 Practice 63
8.5 pp. 308–310	CE 1–10 WE 1–29 odd Midchapter Review	CE 1–10 WE 9–37 odd Midchapter Review	CE 1–10 WE 13–41 odd Midchapter Review	Reteaching 64 Practice 64
8.6 pp. 313–315	FR all, CE 1–18 WE 1–22 all	FR all, CE 1–18 WE 1–10 all, 11–37 odd	FR all, CE 1–18 WE 1–43 odd	Reteaching 65 Practice 65
8.7 pp. 318–319	CE 1–12 WE 1–27 odd	CE 1–12 WE 11–37 odd	CE 1–12 WE 15–41 odd	Reteaching 66 Practice 66
8.8 pp. 322–324	CE 1–6 WE 1–17 odd Problem Solving Strategies	CE 1–6 WE 1, 3, 9–21 odd Problem Solving Strategies	CE 1–6 WE 1, 3, 17–29 odd Problem Solving Strategies	Reteaching 67 Practice 67
8.9 pp. 328–329	CE 1–16 WE 1–12 all	CE 1–16 WE 1–23 odd	CE 1–16 WE 7–27 odd, 28	Reteaching 68 Practice 68
Chapter 8 Review pp. 330–331	all	all	all	
Chapter 8 Test p. 332	all	all	all	
College Prep Test p. 333	all	all	all	
Cumulative Review (Chapters 1–8) pp. 334–335	all	all	all	

CE = Classroom Exercises WE = Written Exercises FR = Focus on Reading
Note: For each level, all students should be assigned all Try This and all Mixed Review Exercises.

■■■ INVESTIGATION

Project: In this Investigation, students review fraction concepts while finding solutions for $\frac{1}{a} + \frac{1}{b} = \frac{1}{c}$.

Materials: Each student will need a copy of Project Worksheet 8.

Hand out the worksheet and have students work in small groups.

The first exercise asks students to find positive integers for a and b such that $\frac{1}{a} + \frac{1}{b} = \frac{1}{8}$, where $a \leq b$.

Help students by suggesting that they replace a with 9, giving $\frac{1}{9} + \frac{1}{b} = \frac{1}{8}$.

Ask how they could find a value for b. $\frac{1}{b} = \frac{1}{8} - \frac{1}{9}$; $\frac{1}{b} = \frac{1}{72}$. Then b is the reciprocal of $\frac{1}{b}$; that is, $b = 72$.

Then have students replace a with 10 to find another solution for b. 40 Next, have them try to find a solution when $a = 11$. Take time to explain that the solution, $a = \frac{88}{3}$, does not fit the condition that a and b must be positive integers. Ask students what would be the largest number they should try for a. Help them to see that if $a = 16$, then $b = 16$ also. Any further increase in the value of a will require a smaller value for b and that would contradict the requirement that $a \leq b$. Therefore, 16 is the maximum value for a.

The solutions for Exercise 1 are those found by letting $a = 8 + n$, where n is a factor of 8. In this case, $n = 1, 2, 4,$ and 8. There are four different solutions. See if students can discover this pattern on their own.

Belinda Campos is a champion at the game of billiards. To be a winner, she must judge the force and direction of her shots carefully. As she hits the ball, Belinda also uses her knowledge of angles and reflections.

More About the Photo

The game of billiards is popular now in Japan, England, and the United States. Skill in all variations depends on a knowledge of how balls rebound. If a ball is hit at 45 degrees from a corner of a table with sides of integral measure (say 4 × 9, 5 × 10, or 8 × 4), it will eventually rebound into a corner pocket. Number of Rebounds = Width + Length − 2. Thus, on a 4 × 9 table, the ball rebounds 11 times and then lands in a corner pocket. Since the angle of incidence is always equal to the angle of reflection, the ball traces a series of 45-degree right triangles.

8.1 Simplifying Rational Expressions

Teaching Resources

Application Worksheet 8
Project Worksheet 8
Quick Quizzes 60
Reteaching and Practice
   Worksheets 60

Objectives
To find the values, if any, for which rational expressions are undefined
To simplify rational expressions
To evaluate rational expressions

An expression such as $\dfrac{5}{x + 5}$ is called a *rational expression*. A **rational expression** is a polynomial or the quotient of two polynomials.

Each of the following expressions is a rational expression.

$$\frac{3x}{10} \qquad \frac{2x - 5}{x^2 - 3x + 5} \qquad 2a \qquad \frac{0}{3x + 2} \qquad x^2 - 9$$

A rational expression is undefined when the denominator is equal to 0.

EXAMPLE 1 For what value or values of the variable is the given rational expression undefined?

a. $\dfrac{a - 3}{3a + 9}$ b. $\dfrac{5x + 2}{x^2 - 7x + 12}$

Plan Find the values of the variable that will make the denominator 0.

Solutions
a. Let $3a + 9 = 0$.
$$3a = -9$$
$$a = -3$$

Thus, $\dfrac{a - 3}{3a + 9}$ is undefined for $a = -3$

b. Let $x^2 - 7x + 12 = 0$.
$$(x - 4)(x - 3) = 0$$
$$x - 4 = 0 \ or \ x - 3 = 0$$
$$x = 4 \ or \qquad x = 3$$

Thus, $\dfrac{5x + 2}{x^2 - 7x + 12}$ is undefined for $x = 4$ and for $x = 3$

TRY THIS 1. For what value or values of the variable is the rational expression $\dfrac{6x - 4}{x^2 - 10x + 25}$ undefined? 5

In Example 1a, if a is replaced by -3, then the numerator of $\dfrac{a - 3}{3a + 9}$ is $-3 - 3$, or -6. The denominator is 0. Use a calculator to find $\dfrac{-6}{0}$.
What does the display show? E, for error

Assume that if a value of a variable causes a denominator in an expression to be zero, then the expression is undefined for that value.

■■■■ **GETTING STARTED**

Prerequisite Quiz

Solve.

1. $x^2 - 36 = 0$ 6, -6
2. $6x^2 + 13x - 5 = 0$ $\frac{1}{3}$, $-\frac{5}{2}$
3. $5a - a^2 = 0$ 0, 5
4. $81y - y^3 = 0$ 0, 9, -9

Motivator

Ask students what is meant by simplifying or reducing a fraction. Expressing it in lowest terms Ask how they know when a fraction is expressed in lowest terms. When the numerator and denominator have no common factors other than 1. Ask them to describe the first step in simplifying fractions. Look for common factors of the numerator and denominator.

Additional Example 1

For what value(s) of the variable is the given rational expression undefined?

a. $\dfrac{a + 5}{4a + 8}$ -2

b. $\dfrac{x - 1}{x^2 + 2x - 8}$ 2 and -4

Highlighting the Standards

Standards 5a, 5c, 7a: Students work with rational expressions to find when they are undefined and, in the Exercises, they represent various aspects of three-dimensional objects by writing equations.

Lesson Note

Students should be aware that the restrictions on a variable will not always be given. However, students should also realize that these restrictions exist. In Example 3 on page 290, emphasize that the a in $(a + 3)$ and $(a - 3)$ cannot be cancelled or divided out to give $\frac{3}{-3} = -1$. Replace a with some value, such as 5, and show students that $\frac{(5 + 3)}{(5 - 3)}$ is not equal to -1. In Example 5, point out that to evaluate after simplifying is easier than evaluating before simplifying. However, one way to check for careless errors in simplifying is to choose some value for the variable and evaluate before and after simplifying.

Math Connections

Number Theory: The largest prime number found by a super-computer as of September, 1989, is $391{,}581 \times 2^{216{,}193} - 1$, which has 65,087 digits.

EXAMPLE 2 Simplify $\dfrac{18a^2}{12a}$.

Solution

$$\frac{18a^2}{12a} = \frac{2 \cdot 3 \cdot 3 \cdot a \cdot a}{2 \cdot 2 \cdot 3 \cdot a}$$

$$= \frac{\overset{1}{\cancel{2}} \cdot \overset{1}{\cancel{3}} \cdot 3 \cdot \overset{1}{\cancel{a}} \cdot a}{\cancel{2} \cdot 2 \cdot \cancel{3} \cdot \cancel{a}}_{1 \quad 1 \quad 1} = \frac{3a}{2}, \text{ or } \frac{3}{2}a$$

TRY THIS   2. Simplify $\dfrac{30x^4}{36x^2}$. $\frac{5}{6}x^2$

A rational expression is said to be *simplified* when it is written as a polynomial or as a quotient of polynomials with 1 as the greatest common factor (GCF) of the numerator and denominator.

EXAMPLE 3 Simplify $\dfrac{(a + 3)(a + 4)}{(a - 3)(a + 4)}$.

Solution

$$\frac{(a + 3)(a + 4)}{(a - 3)(a + 4)} = \frac{(a + 3)\overset{1}{\cancel{(a + 4)}}}{(a - 3)\underset{1}{\cancel{(a + 4)}}}$$

$$= \frac{(a + 3)1}{(a - 3)1}$$

$$= \frac{a + 3}{a - 3}$$

EXAMPLE 4 Simplify $\dfrac{x^2 + 2x - 15}{2x^2 - 7x + 3}$.

Plan Factor the numerator and the denominator. Then divide out the common factors.

Solution

$$\frac{x^2 + 2x - 15}{2x^2 - 7x + 3} = \frac{(x - 3)(x + 5)}{(2x - 1)(x - 3)} \quad \longleftarrow \text{ The GCF is } x - 3.$$

$$= \frac{\overset{1}{\cancel{(x - 3)}}(x + 5)}{(2x - 1)\underset{1}{\cancel{(x - 3)}}}$$

$$= \frac{1(x + 5)}{(2x - 1)1} \quad \longleftarrow \text{ The GCF is now 1.}$$

$$= \frac{x + 5}{2x - 1}$$

TRY THIS   3. Simplify $\dfrac{2x^2 - 5x - 12}{x^2 - 11x + 28}$. $\frac{2x + 3}{x - 7}$

290   Chapter 8 Rational Expressions

Additional Example 2

Simplify $\dfrac{21c^4}{28c^2}$. $\frac{3c^2}{4}$

Additional Example 3

Simplify $\dfrac{(y - 5)(y + 8)}{(y - 2)(y - 5)}$. $\frac{y + 8}{y - 2}$

Additional Example 4

Simplify $\dfrac{x^2 - 2x - 8}{3x^2 - 13x + 4}$. $\frac{x + 2}{3x - 1}$

EXAMPLE 5 Simplify $\dfrac{3x^3 - 12x}{6x^4 - 12x^3}$.

Solution
$$\dfrac{3x^3 - 12x}{6x^4 - 12x^3} = \dfrac{3x(x^2 - 4)}{6x^3(x - 2)}$$
$$= \dfrac{3x(x + 2)(x - 2)}{6x^3(x - 2)} \quad \longleftarrow \text{The GCF is } 3x(x - 2).$$
$$= \dfrac{\overset{1}{\cancel{3x}}(x + 2)\cancel{(x - 2)}}{\underset{2x^2}{\cancel{6x^3}}\underset{1}{\cancel{(x - 2)}}}$$
$$= \dfrac{x + 2}{2x^2} \quad \longleftarrow \text{The GCF is now 1.}$$

EXAMPLE 6 Evaluate the rational expression $\dfrac{2x^2 - 5}{x^2 - 3x + 9}$ for $x = -2$.

Solution
$$\dfrac{2x^2 - 5}{x^2 - 3x + 9} = \dfrac{2(-2)^2 - 5}{(-2)^2 - 3(-2) + 9}$$
$$= \dfrac{2(4) - 5}{4 + 6 + 9}$$
$$= \dfrac{8 - 5}{19}, \text{ or } \dfrac{3}{19}$$

Thus, the value of the given rational expression for $x = -2$ is $\dfrac{3}{19}$

TRY THIS   4. Simplify $\dfrac{10x^4 + 40x^3}{2x^3 - 32x}$. $\frac{5x^2}{x-4}$

Classroom Exercises

For what values of the variable is the rational expression undefined?

1. $\dfrac{10m}{m - 3}$ 3
2. $\dfrac{x - 6}{y}$ 0
3. $\dfrac{7a}{a - 4}$ 4
4. $\dfrac{10}{5 - y}$ 5

5. $\dfrac{8 - x}{6 - x}$ 6
6. $\dfrac{m - 3}{3 - m}$ 3
7. $\dfrac{x - 4}{2x - 10}$ 5
8. $\dfrac{3 - y}{y^2 - 9}$ -3,3

Simplify, if possible. If not possible, write NP.

9. $\dfrac{110}{225}$ $\frac{22}{45}$
10. $\dfrac{a}{a^6}$ $\frac{1}{a^5}$
11. $\dfrac{-52y^2}{39y^2}$ $\frac{-4}{3}$
12. $\dfrac{10m^{15}}{22m^5}$ $\frac{5m^{10}}{11}$

13. $\dfrac{x + 4}{(x + 4)(x - 3)}$ $\frac{1}{x-3}$
14. $\dfrac{(5x - 3)(x + 9)}{(x + 1)(5x - 3)}$ $\frac{x+9}{x+1}$
15. $\dfrac{9x - 2}{2x + 9}$ NP
16. $\dfrac{x^3 + x^2}{2x^3 - 2x}$ $\frac{x}{2x-2}$

8.1 Simplifying Rational Expressions   **291**

Additional Example 5

Simplify $\dfrac{x^3 - 9x^2}{x^4 - 81x^2} \cdot \dfrac{1}{x + 9}$

Additional Example 6

Evaluate $\dfrac{y - y^2 + 12}{28 - 7y}$ for $y = -3$. 0

Closure

Ask students these questions.

1. What is a rational expression?
See p. 289.
2. For what values of the variables is a rational expression not defined?
For values that make any denominator equal to 0
3. When is a rational expression in simplest form? See p. 290.

◼◼◼◼ FOLLOW UP

Guided Practice

Classroom Exercises 1–16
Try This all

Independent Practice

A Ex. 1–29, **B** Ex. 30–44, **C** Ex. 45–56

Basic: WE 1–29 all, 42
Average: WE 1–29 odd, 34–48 all
Above Average: WE 1–55 odd, 42, 46, 48

Written Exercises

For what value or values of x is the rational expression undefined?

1. $\dfrac{x - 3}{2x + 6}$ -3

2. $\dfrac{-12}{5x - 10}$ 2

3. $\dfrac{x}{2x - 8}$ 4

4. $\dfrac{x - 14}{3x - 15}$ 5

5. $\dfrac{x}{x^2 + 5x + 6}$ $-3, -2$

6. $\dfrac{x - 3}{x^2 + 3x - 4}$ $-4, 1$

7. $\dfrac{4x + 5}{x^2 - 4x + 3}$ $1, 3$

8. $\dfrac{x + 7}{x^2 - 25}$ $-5, 5$

Simplify, if possible. If not possible, write NP.

9. $\dfrac{14k}{21}$ $\frac{2k}{3}$

10. $-\dfrac{9}{33}x$ $-\frac{3}{11}x$

11. $\dfrac{42t}{70t}$ $\frac{3}{5}$

12. $-\dfrac{15x}{25x^3}$ $-\frac{3}{5x^2}$

13. $\dfrac{2x - 10}{x - 5}$ 2

14. $\dfrac{a - 6}{a^2 - 6a}$ $\frac{1}{a}$

15. $\dfrac{y^2 - 49}{y^2 + 4y - 21}$

16. $\dfrac{2a + 6}{3a - 15}$ NP

17. $\dfrac{a^2 + 3a - 10}{2a^2 + 11a + 5}$

18. $\dfrac{2x^2 - x - 1}{2x^2 - 5x + 3}$

19. $\dfrac{2y^2 - y}{2y^2 - 5y - 3}$ NP

20. $\dfrac{m^2 + 4m}{m^2 + m - 12}$

21. $\dfrac{x^2 - 7x + 18}{x^2 - 12x + 27}$ NP

22. $\dfrac{2a^2 + 7a - 4}{a^2 - 16}$

23. $\dfrac{3m^2 - 9m - 30}{6m - 30}$

24. $\dfrac{5a - 20}{a^2 - 4a}$ $\frac{5}{a}$

25. $\dfrac{t - 3}{8.96t - 26.88}$ $\frac{1}{8.96}$

26. $\dfrac{3m^2 + 21m - 54}{0.341m - 3.069}$ NP

Evaluate for the given value of the variable.

27. $\dfrac{3a - 5}{2a + 1}$ for $a = -1$ 8

28. $\dfrac{5x - 3}{x^2 - 3}$ for $x = -2$ -13

29. $\dfrac{m^2 - 9}{3m + 5}$ for $m = 3$ 0

Simplify, if possible.

30. $\dfrac{3a^4 + 7a^3 - 20a^2}{6a^4 - 7a^3 - 5a^2}$ $\frac{a + 4}{2a + 1}$

31. $\dfrac{2p^3 - 14p^2 + 20p}{4p^4 - 8p^3 - 60p^2}$ $\frac{p - 2}{2p^2 + 6p}$

32. $\dfrac{a^3 + 10a^2 + 25a}{a^5 - 3a^4 - 40a^3}$ $\frac{a + 5}{a^3 - 8a^2}$

33. $\dfrac{3y^3 - 15y^2 - 12y}{6y^3 - 42y^2}$

34. $\dfrac{3a^3 - a^2 - 14a}{2a^4 + 3a^3 - 2a^2}$

35. $\dfrac{3x^2 + 18x - 21}{15x^2 - 15}$

36. $\dfrac{2x^3 - 4x^2 - 6x}{4x^4 - 12x^3 - 16x^2}$

37. $\dfrac{3x^3 + 24x^2 + 48x}{9x^4 - 144x^2}$

38. $\dfrac{2a^2 - ab - 3b^2}{2a - 3b}$

39. $\dfrac{a^3b^7(2x^2 + 9x - 5)}{a^2b^9(2x^2 + 7x - 15)}$

40. $\dfrac{a^3b^4(y^2 + 7y + 10)}{a^6b^2(y^2 + y - 20)}$

41. $\dfrac{m^2n(2x^2 - 8x + 6)}{mn^2(12x - 36)}$ $\frac{mx - m}{6n}$

42. Write an explanation of how to determine the values of the variable for which a rational expression is undefined. Give an example of a rational expression that is always defined and an example of a rational expression that is undefined for certain values of the variable.

43. The area of a certain rectangle is represented by $x^2 + 2x - 3$ and its length by $x - 1$. Find its width. $x + 3$

44. The area of a certain rectangle is represented by $x^2 - 9$ and its width by $x - 3$. Find its length. $x + 3$

The volume of the cube at the right is determined by multiplying the area of its square base by its height. The total surface area of the cube is determined by multiplying the area of one face by 6.

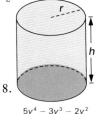

45. Find an expression for the volume of the cube. e^3

46. Find an expression for the total surface area of the cube. $6e^2$

47. Find the quotient of the total surface area divided by the volume. $\frac{6}{e}$

48. Find the value of the quotient in Exercise 47 for $e = 4$. $\frac{3}{2}$

The volume of the right circular cylinder at the right is $\pi r^2 h$ and its surface area is $2\pi r^2 + 2\pi rh$.

49. Find the quotient of the volume of the cylinder divided by its surface area. $\dfrac{rh}{2r + 2h}$

50. Find the value of the quotient of Exercise 49 for $r = 3$ and $h = 8$. $\frac{12}{11}$

Simplify. $\dfrac{x^2 - x - 6}{x}$

51. $\dfrac{x^4 - 13x^2 + 36}{x^3 + x^2 - 6x}$

52. $\dfrac{9p^6 - 145p^4 + 16p^2}{3p^2 + 11p - 4}$ $3p^4 - 11p^3 - 4p^2$

53. $\dfrac{50y^6 - 58y^4 + 8y^2}{30y^2 + 18y - 12}$ $\dfrac{5y^4 - 3y^3 - 2y^2}{3}$

54. $\dfrac{x^4 - 10x^2 + 9}{3x^2 - 27}$ $\dfrac{x^2 - 1}{3}$

55. $\dfrac{a^2 + 4ab - 21b^2}{a^2 + 7ab - 3a - 21b}$ $\dfrac{a - 3b}{a - 3}$

56. $\dfrac{ax^2 - ay^2 + 3x^2 - 3y^2}{ax - ay + 3x - 3y}$ $x + y$

Mixed Review

Multiply. **6.9. 6.10**

1. $(2x + 1)(x - 3)$ $2x^2 - 5x - 3$
2. $(2y - 3)(3y + 5)$ $6y^2 + y - 15$
3. $(m - 1)(3m + 5)$ $3m^2 + 2m - 5$
4. $(x - 9)(x + 9)$
5. $(y - 3)(y + 3)$ $y^2 - 9$
6. $(3a + 4)(3a - 4)$
7. $(2m + 5)(m - 3)$ $2m^2 - m - 15$
8. $(2b + 3)^2$ $4b^2 + 12b + 9$
9. $(3a - 1)^2$ $9a^2 - 6a + 1$

Simplify. **6.1, 6.2, 6.3**

10. $(-3x^2)^4$ $81x^8$
11. $(-2x^2y)(5x^5y^4)$ $-10x^7y^5$
12. $\dfrac{-18a^5b}{12ab^8}$ $\dfrac{-3a^4}{2b^7}$

Find the solution set. **7.9**

13. $x^2 - 9 = 0$ $\{-3, 3\}$
14. $x^2 + 2x - 3 = 0$ $\{-3, 1\}$
15. $2x^3 - 7x^2 - 15x = 0$ $\{-\frac{3}{2}, 0, 5\}$

16. Use the formula $A = \frac{1}{2}h(b + c)$ to find the area A of a trapezoid, where $b = 14$ cm, $c = 18$ cm, and $h = 8$ cm. **1.4** 128 cm²

Additional Answers

Written Exercises

15. $\dfrac{y - 7}{y - 3}$
17. $\dfrac{a - 2}{2a + 1}$
18. $\dfrac{2x + 1}{2x - 3}$
20. $\dfrac{m}{m - 3}$
22. $\dfrac{2a - 1}{a - 4}$
23. $\dfrac{m + 2}{2}$
33. $\dfrac{y^2 - 5y - 4}{2y^2 - 14y}$
34. $\dfrac{3a - 7}{2a^2 - a}$
35. $\dfrac{x + 7}{5x + 5}$
36. $\dfrac{x - 3}{2x^2 - 8x}$
37. $\dfrac{x + 4}{3x^2 - 12x}$
38. $a + b$
39. $\dfrac{2ax - a}{2b^2x - 3b^2}$
40. $\dfrac{b^2y + 2b^2}{a^3y - 4a^3}$
42. To determine the values of the variable for which a rational expression is undefined, find the values that will make the denominator zero.

Examples: $\dfrac{2x}{3}$, always defined;

$\dfrac{2 + x}{x^2 - 3x}$, undefined for $x = 0$ and $x = 3$.

Mixed Review

4. $x^2 - 81$
6. $9a^2 - 16$

Enrichment

Tell students that each expression on the left is simplified incorrectly. Have them find the error.

1. $\dfrac{3(x + y)}{a + 5(x + y)} = \dfrac{3}{x + 5}$ $x + y$ is not a common factor of the numerator and denominator.

2. $\dfrac{a + b}{a^2 + b^2} = \dfrac{1}{a + b}$ $a + b$ is not a factor of $a^2 + b^2$.

3. $\dfrac{x + y + z}{x + y} = z$ $x + y$ is not a factor of $x + y + z$.

4. $\dfrac{(m + n)}{(m + n)(m + n)} = m + n$ When the numerator and the denominator are divided by $m + n$, the correct answer is $\dfrac{1}{m + n}$.

5. $\dfrac{2c + d}{c + d} = 2$ $c + d$ is not a factor of $2c + d$.

8.2 Simplifying Rational Expressions: Convenient Form

Objective	To simplify rational expressions by writing polynomials in convenient form

To simplify $\dfrac{14 + a - 3a^2}{-3a + 2 - 2a^2}$, it is necessary first to factor both the numerator and denominator. To get the polynomial $14 + a - 3a^2$ in a more convenient form, follow these steps.

1. Write in descending order of the exponents.

$$-3a^2 + a^1 + 14, \text{ or } -3a^2 + a + 14$$

2. Factor out -1 in order to obtain a first coefficient that is positive.

$$-1(3a^2 - a - 14)$$

The convenient form of $14 + a - 3a^2$ is $-1(3a^2 - a - 14)$. The denominator $-3a + 2 - 2a^2$ can also be written in convenient form following the same two steps.

EXAMPLE 1	Simplify $\dfrac{a - 3}{9 - a^2}$.
Plan	The numerator, $a - 3$, is already in convenient form. The denominator, $9 - a^2$, is not. Rewrite the denominator in convenient form.

$$9 - a^2 = -1a^2 + 9 = -1(a^2 - 9)$$

Solution

$$\frac{a - 3}{9 - a^2} = \frac{a - 3}{-1(a^2 - 9)}$$

$$= \frac{a - 3}{-1(a + 3)(a - 3)} = \frac{\overset{1}{\cancel{(a - 3)}}}{-1(a + 3)\underset{1}{\cancel{(a - 3)}}} = \frac{1}{-(a + 3)}$$

TRY THIS	1. Simplify $\dfrac{c^2 - 36}{-c - 6}$. $\quad -(c - 6)$

The following expressions are equivalent.

$$-\frac{x}{y} \qquad \frac{-x}{y} \qquad \frac{x}{-y}$$

Thus, $\dfrac{1}{-(a + 3)}$ in Example 1 can also be written as $-\dfrac{1}{a + 3}$,

Prerequisite Quiz

Factor out -1 in each of the following polynomials. State answers in order of descending powers.

1. $-x^2 + 3x - 5 \quad -1(x^2 - 3x + 5)$
2. $25 - b^2 \quad -1(b^2 - 25)$
3. $16 - 4m \quad -1(4m - 16)$
4. $36 - y^4 \quad -1(y^4 - 36)$
5. $6 - x - x^2 \quad -1(x^2 + x - 6)$

Motivator

Ask students these questions.

1. Does $a - 8 = 8 - a$? No
2. What is the opposite of $8 - a$?
 $-(8 - a)$
3. Does $a - 8 = -(8 - a)$? Yes
4. Does $a - 8 = -1 \cdot (8 - a)$? Yes

Then ask them to suggest a way to simplify $\dfrac{4 - t}{t - 4}$. $\quad \dfrac{4 - t}{t - 4} = \dfrac{-1(t - 4)}{(t - 4)} = -1$

Highlighting the Standards

Standards 2d, 4a: In Focus on Reading, students read about and provide their own equivalent forms of rational expressions.

Additional Example 1

Simplify $\dfrac{x - 2}{4 - x^2}$. $\quad -\dfrac{1}{x + 2}$

or as $\dfrac{-1}{a + 3}$. Although any one of the three forms, $-\dfrac{x}{y}, \dfrac{-x}{y}, \dfrac{x}{-y}$ is acceptable, the form $-\dfrac{x}{y}$ is used most frequently.

EXAMPLE 2 Simplify $\dfrac{14 + a - 3a^2}{-3a + 2 - 2a^2}$.

Plan Rewrite the numerator and denominator in convenient form.

Solution
$$\dfrac{14 + a - 3a^2}{-3a + 2 - 2a^2} = \dfrac{-3a^2 + a + 14}{-2a^2 - 3a + 2}$$
$$= \dfrac{-1(3a^2 - a - 14)}{-1(2a^2 + 3a - 2)}$$
$$= \dfrac{\overset{1}{\cancel{-1}}(3a - 7)\overset{1}{\cancel{(a + 2)}}}{\underset{1}{\cancel{-1}}(2a - 1)\underset{1}{\cancel{(a + 2)}}} = \dfrac{3a - 7}{2a - 1}$$

TRY THIS 2. Simplify $\dfrac{25 + 4x^2 - 20x}{15 - x - 2x^2}$. $\quad -\frac{2x - 5}{x + 3}$

Focus on Reading

Determine which polynomials are in convenient form. Explain.
1. $3x^2 - 5x + 2$ 2. $-4x^2 - 3x + 1$ 3. $-1(x^2 - 7x - 12)$
4. List the steps for putting $-4 + 5x - x^2$ in convenient form.
5. List two other forms for expressing $-\dfrac{x + 5}{2x - 7}$. $\quad \frac{-x - 5}{2x - 7}, \frac{x + 5}{-2x + 7}$

Classroom Exercises

Give each expression in convenient form.
1. $-5x - 2$ $\;-1(5x + 2)$ 2. $-x^2 + 9$ $\;-1(x^2 - 9)$ 3. $25 - m^2$ $\;-1(m^2 - 25)$ 4. $-2x^2 + 8x$ $\;-1(2x^2 - 8x)$
5. $-x^2 + 4x + 12$ 6. $42 + y - y^2$ 7. $-c^2 + 8c + 20$ 8. $6 - x - 2x^2$
$\;-1(x^2 - 4x - 12)$ $-1(y^2 - y - 42)$ $-1(c^2 - 8c - 20)$ $-1(2x^2 + x - 6)$

Simplify.
9. $\dfrac{9 - x}{3x - 27}$ $\;-\frac{1}{3}$ 10. $\dfrac{a - 2}{6 - 3a}$ $\;-\frac{1}{3}$ 11. $\dfrac{a - 4}{16 - a^2}$ $\;-\frac{1}{a + 4}$ 12. $\dfrac{x^2 - 5x + 6}{3 - x}$
$\qquad\qquad\qquad\qquad\qquad\qquad\qquad\qquad\qquad\qquad\qquad\qquad\qquad -x + 2$

Additional Example 2

Simplify $\dfrac{12 - 2x^2 + 5x}{-6 + 2x^2 - x}$. $\quad -\frac{x - 4}{x - 2}$

Closure

Have students write an equivalent expression having a denominator of $x - y$.

1. $\dfrac{-2}{y-x}$ $\dfrac{2}{x-y}$
2. $\dfrac{x-y}{y-x}$ $\dfrac{y-x}{x-y}$
3. $\dfrac{a-b}{y-x}$ $\dfrac{b-a}{x-y}$
4. $\dfrac{5}{y-x}$ $-\dfrac{5}{x-y}$, or $\dfrac{-5}{x-y}$

◼◼◼ FOLLOW UP

Guided Practice

Classroom Exercises 1–12
Try This all, FR all

Independent Practice

A Ex. 1–20, **B** Ex. 21–28, **C** Ex. 29–32

Basic: WE 1–21 odd

Average: WE 5–25 odd

Above Average: WE 11–31 odd

Additional Answers

Focus on Reading

1. Convenient; exponents in descending order, first coefficient pos.
2. Not convenient; first coefficient neg.
3. Convenient; exponents in descending order, first coefficient pos.
4. **1.** Write in descending exponent order: $-x^2 + 5x - 4$. **2.** Factor out -1 to obtain pos first coefficient: $-1(x^2 - 5x + 4)$.

See page 301 for the answers to Written Ex. 9–19.

Written Exercises

Simplify.

1. $\dfrac{x-5}{25-x^2}$ $\quad -\dfrac{1}{x+5}$
2. $\dfrac{8-b}{3b-24}$ $\quad -\dfrac{1}{3}$
3. $\dfrac{a+6}{36-a^2}$ $\quad \dfrac{-1}{a-6}$
4. $\dfrac{a-1}{1-a^2}$ $\quad \dfrac{-1}{a+1}$

5. $\dfrac{x^2 - 7x + 10}{5 - x}$ $\quad -x+2$
6. $\dfrac{18 - 3x}{x^2 - 36}$ $\quad \dfrac{-3}{x+6}$
7. $\dfrac{x^2 - x - 2}{2 - x}$ $\quad -x-1$
8. $\dfrac{x^2 - 3x}{9 - x^2}$ $\quad \dfrac{-x}{x+3}$

9. $\dfrac{x-4}{12+x-x^2}$
10. $\dfrac{8-x}{x^2 - 6x - 16}$
11. $\dfrac{2-c}{c^2 - 11c + 18}$
12. $\dfrac{5 - 2b}{2b^2 - b - 10}$

13. $\dfrac{12 - x - x^2}{2x^2 + 5x - 12}$
14. $\dfrac{2a^2 - 5a - 3}{4 + 7a - 2a^2}$
15. $\dfrac{p^2 - 8p - 20}{-p^2 + 12p - 20}$
16. $\dfrac{16 - y^2}{y^2 - 5y + 4}$

17. $\dfrac{-2b^2 - b + 6}{b^2 - 2b - 8}$
18. $\dfrac{4k - 12}{-9 + 6k - k^2}$
19. $\dfrac{x^2 + x - 30}{36 - x^2}$
20. $\dfrac{13x - 15 - 2x^2}{2x^2 - x - 3}$ $\quad \dfrac{-x+5}{x+1}$

21. $\dfrac{2m^2 + 8m - 64}{28 - 3m - m^2}$ $\quad \dfrac{-2m - 16}{m+7}$
22. $\dfrac{42 + a - a^2}{2a^2 + 10a - 12}$ $\quad \dfrac{-a+7}{2a-2}$

23. $\dfrac{3x^2 - 3x - 60}{96 - 6x^2}$ $\quad \dfrac{-x+5}{2x-8}$
24. $\dfrac{3x^3 - 12x^2 + 9x}{18x - 6x^2}$ $\quad \dfrac{-x+1}{2}$

25. $\dfrac{m^8 p^6 (y^2 + 9y + 20)}{m^{11} p^5 (-4 - y)}$ $\quad \dfrac{-py - 5p}{m^3}$
26. $\dfrac{18x - 6x^2}{2x^4 + x^3 - 15x^2}$ $\quad \dfrac{-6x + 18}{2x^3 + x^2 - 15x}$

27. $\dfrac{14 - 17s - 6s^2}{3s^3 - 20s^2 + 12s}$ $\quad \dfrac{-2s - 7}{s^2 - 6s}$
28. $\dfrac{-x^3 - 2x^2 + 15x}{x^4 + 2x^3 - 15x^2}$ $\quad \dfrac{-1}{x}$

29. $\dfrac{6a^3 + 10a^2}{100a - 36a^3}$ $\quad \dfrac{-a}{6a - 10}$
30. $\dfrac{a^3 + 6a^2 - 4a - 24}{-a^3 + 2a^2 + 36a - 72}$ $\quad \dfrac{-a - 2}{a - 6}$

31. $\dfrac{x^2 - 6x + 9 - 4y^2}{2y + 3 - x}$ $\quad -x + 3 - 2y$
32. $\dfrac{2 - (x + y) - (x + y)^2}{x + y + 2}$ $\quad -x - y + 1$

Mixed Review

Solve each equation or inequality. *3.2, 3.4, 3.5, 4.5, 5.6, 5.7*

1. $15 - (7 - x) = 10$ $\quad x = 2$
2. $-7x - 2(3 - 4x) = -15$ $\quad x = -9$
3. $7y - 12 = 3y + 4$ $\quad y = 4$
4. $-6a + 15 = 12 - 9a$ $\quad a = -1$
5. $\dfrac{3}{4}a = \dfrac{5}{2}$ $\quad a = \dfrac{10}{3}$
6. $\dfrac{2}{3}x - \dfrac{1}{3} = \dfrac{1}{4}x + \dfrac{1}{5}$ $\quad x = \dfrac{32}{25}$
7. $|3x - 4| = 12$ $\quad x = \dfrac{16}{3}$ or $x = \dfrac{-8}{3}$
8. $|5x + 4| \le 7$ $\quad \dfrac{-11}{5} \le x \le \dfrac{3}{5}$
9. $|x + 8| > 16$ $\quad x < -24$ or $x > 8$
10. $|x + 8| < 16$ $\quad -24 < x < 8$

Enrichment

Have students simplify this expression.

$$\left[\left[\left[\dfrac{r^4 t^4 + 8r^2 t^2 + 12}{r^4 t^4 - 36} \right]^{-2} \right]^0 \right]^7 \quad 1$$

8.3 Multiplying Rational Expressions

Objective   To multiply rational expressions

Multiplying rational expressions is similar to multiplying fractions.

$$\frac{2}{3} \cdot \frac{4}{5} = \frac{2 \cdot 4}{3 \cdot 5} = \frac{8}{15} \qquad \frac{2}{x} \cdot \frac{y}{5} = \frac{2 \cdot y}{x \cdot 5} = \frac{2y}{5x}$$

Rule for Multiplying Rational Expressions

If $\frac{a}{b}$ and $\frac{c}{d}$ are rational expressions, $b \neq 0$ and $d \neq 0$, then

$$\frac{a}{b} \cdot \frac{c}{d} = \frac{a \cdot c}{b \cdot d}.$$

EXAMPLE 1   Multiply $\frac{3}{10} \cdot \frac{15x}{9}$.

Plan   Write the numerator and denominator of the product in factored form. Then use the Cancellation Property of Fractions.

Solution   Use the Rule for Multiplying Rational Expressions.

$$\frac{3}{10} \cdot \frac{15x}{9} = \frac{3 \cdot 15x}{10 \cdot 9}$$

Factor the numerator and the denominator.

$$= \frac{3 \cdot 3 \cdot 5 \cdot x}{2 \cdot 5 \cdot 3 \cdot 3}$$

Use the Cancellation Property of Fractions (Lesson 6.3).

$$= \frac{\overset{1}{\cancel{3}} \cdot \overset{1}{\cancel{3}} \cdot \overset{1}{\cancel{5}} \cdot x}{2 \cdot \underset{1}{\cancel{5}} \cdot \underset{1}{\cancel{3}} \cdot \underset{1}{\cancel{3}}}$$

$$= \frac{x}{2}$$

TRY THIS   1. Multiply $\frac{6}{10x} \cdot \frac{5}{2}$. $\frac{3}{2x}$

The Cancellation Property can be used *before* multiplying.

$$\frac{3}{10} \cdot \frac{15x}{9} = \frac{3}{2 \cdot 5} \cdot \frac{3 \cdot 5 \cdot x}{3 \cdot 3}$$

$$= \frac{\overset{1}{\cancel{3}}}{2 \cdot \underset{1}{\cancel{5}}} \cdot \frac{\overset{1}{\cancel{3}} \cdot \overset{1}{\cancel{5}} \cdot x}{\underset{1}{\cancel{3}} \cdot \underset{1}{\cancel{3}}}$$

$$= \frac{1}{2} \cdot \frac{x}{1} = \frac{1 \cdot x}{2 \cdot 1} = \frac{x}{2}$$

Teaching Resources

Quick Quizzes 62
Reteaching and Practice
Worksheets 62

■■■GETTING STARTED

Prerequisite Quiz

Multiply.

1. $\frac{1}{3} \cdot \frac{1}{5}$ $\frac{1}{15}$
2. $\frac{3}{8} \cdot \frac{5}{7}$ $\frac{15}{56}$
3. $4a^4 \cdot 5a^6$ $20a^{10}$
4. $(3a + 5)(2a - 1)$ $6a^2 + 7a - 5$
5. $(x - y)(x + y)$ $x^2 - y^2$

Motivator

Ask students to think about the methods they learned for adding and multiplying fractions in previous math courses. Then have them suggest procedures for multiplying two rational expressions, such as $\frac{x}{y} \cdot \frac{mx}{ny}$.   Answers will vary.

Additional Example 1

Multiply $\frac{8}{12} \cdot \frac{9c}{20}$. $\frac{3c}{10}$

Highlighting the Standards

Standards 7a, 7b: In the Examples and Exercises students apply what they have learned about rational expressions to two- and three-dimensional applications from geometry.

Lesson Note

Keeping the two fractions separated by the multiplication dot in Example 2 on page 298 makes it unnecessary to insert extra parentheses around $a + 3$. If students prefer to write one fraction, show them that parentheses would then be necessary. In Example 4 on page 299, point out the similarity of factoring out -2 to factoring out -1. Be sure the students understand why $-2(y - 5)$ is equivalent to $10 - 2y$.

Math Connections

Measurement: When multiplying 50 mph by 6 h the unit labels can be written as fractions so that $\frac{mi}{h} \cdot h$ cancels to give mi.

Critical Thinking Questions

Application: The Rule for Multiplication states that $\frac{a}{b} \cdot \frac{c}{d} = \frac{a \cdot c}{b \cdot d}$ so that $\frac{75}{52} \cdot \frac{91}{125} =$ $\frac{75 \cdot 91}{52 \cdot 125} = \frac{6825}{6500}$ which then needs to be simplified. Ask students to find a more efficient way for completing this problem.

$$\frac{\overset{3}{\cancel{75}}}{\underset{4}{\cancel{52}}} \cdot \frac{\overset{7}{\cancel{91}}}{\underset{5}{\cancel{125}}} = \frac{3 \cdot 7}{4 \cdot 5} = \frac{21}{20}$$

The same procedure can be used to multiply all rational expressions.

EXAMPLE 2 Multiply.

a. $\dfrac{6x^4}{5y^7} \cdot \dfrac{3y^5}{8x}$

b. $\dfrac{a + 5}{6a + 24} \cdot \dfrac{4a + 16}{a + 3}$

Plan Factor numerators and denominators. Divide out the common factors. Then multiply.

Solutions

a.
$$\dfrac{6x^4}{5y^7} \cdot \dfrac{3y^5}{8x}$$
$$= \dfrac{\overset{1}{\cancel{2}} \cdot 3 \cdot \overset{x^3}{\cancel{x^4}}}{5 \cdot \underset{y^2}{\cancel{y^7}}} \cdot \dfrac{3 \overset{1}{\cancel{y^5}}}{\underset{1}{\cancel{2}} \cdot 2 \cdot 2 \cdot \underset{1}{\cancel{x}}}$$
$$= \dfrac{3x^3 \cdot 3}{5y^2 \cdot 4}$$
$$= \dfrac{9x^3}{20y^2}$$

b.
$$\dfrac{a + 5}{6a + 24} \cdot \dfrac{4a + 16}{a + 3}$$
$$= \dfrac{a + 5}{6(a + 4)} \cdot \dfrac{4(a + 4)}{a + 3}$$
$$= \dfrac{(a + 5)}{\underset{3}{\cancel{6}}\cancel{(a+4)}} \cdot \dfrac{\overset{2}{4}\overset{1}{\cancel{(a + 4)}}}{a + 3}$$
$$= \dfrac{(a + 5) \cdot 2}{3 \cdot (a + 3)}$$
$$= \dfrac{2(a + 5)}{3(a + 3)}, \text{ or } \dfrac{2a + 10}{3a + 9}$$

EXAMPLE 3 Multiply $\dfrac{x^2 + 13x + 42}{x^2 - 3x - 40} \cdot \dfrac{x - 8}{x + 6}.$

Plan Factor numerators and denominators. Divide out the common factors. Then multiply.

Solution
$$\dfrac{(x + 7)(x + 6)}{(x - 8)(x + 5)} \cdot \dfrac{x - 8}{x + 6}$$
$$= \dfrac{(x + 7)\overset{1}{\cancel{(x + 6)}}}{\underset{1}{\cancel{(x - 8)}}(x + 5)} \cdot \dfrac{\overset{1}{\cancel{x - 8}}}{\underset{1}{\cancel{x + 6}}}$$
$$= \dfrac{x + 7}{x + 5}$$

TRY THIS 2. Multiply $\dfrac{a^2 - 81}{a^2 - 7a + 12} \cdot \dfrac{a - 4}{a - 9}.$ $\dfrac{a + 9}{a - 3}$

Sometimes the first step in multiplying rational expressions is to express one or more factors in convenient form.

298 Chapter 8 Rational Expressions

Additional Example 2

Multiply.

a. $\dfrac{4a^7}{3b^6} \cdot \dfrac{5b^3}{7a}$ $\dfrac{20a^6}{21b^3}$

b. $\dfrac{x - 5}{x + 3} \cdot \dfrac{2x + 6}{3x - 1}$ $\dfrac{2x - 10}{3x - 1}$

Additional Example 3

Multiply.

$\dfrac{x - 7}{x + 4} \cdot \dfrac{x^2 - 5x - 36}{x^2 - 49}$ $\dfrac{x - 9}{x + 7}$

EXAMPLE 4 Multiply $\dfrac{6a^2b^3}{10 - 2y} \cdot \dfrac{y^2 - y - 20}{4ab^8}$.

Plan First rewrite $10 - 2y$ in convenient form. Factor out -2 rather than -1: $10 - 2y = -2y + 10 = -2(y - 5)$.

Solution

$$\dfrac{6a^2b^3}{10 - 2y} \cdot \dfrac{y^2 - y - 20}{4ab^8} = \dfrac{6a^2b^3}{-2(y - 5)} \cdot \dfrac{(y + 4)(y - 5)}{4ab^8}$$

$$= \dfrac{\overset{3a}{\cancel{6a^2b^3}}}{-2\cancel{(y - 5)}} \cdot \dfrac{(y + 4)\overset{1}{\cancel{(y - 5)}}}{\underset{2b^5}{\cancel{4ab^8}}}$$
$${\underset{1}{}}$$

$$= \dfrac{3a(y + 4)}{-4b^5}, \text{ or } -\dfrac{3a(y + 4)}{4b^5}$$

TRY THIS 3. Multiply $\dfrac{9xy^4}{y^2 - 2y - 24} \cdot \dfrac{6 - y}{12x^4y^2}.$ $-\dfrac{3y^2}{4x^3(y + 4)}$

Rational expressions can be used to solve problems.

EXAMPLE 5 Find the area of the rectangle shown at the right.

Use the formula for the area of a rectangle, $A = lw$, where l is the length and w is the width.

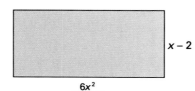

$x - 2$

$\dfrac{6x^2}{2x - 4}$

Plan Substitute the given dimension into the area formula. For convenience, write $x - 2$ as $\dfrac{x - 2}{1}$.

Solution

$A = lw$

$A = \dfrac{6x^2}{2x - 4} \cdot (x - 2)$

$A = \dfrac{6x^2}{2(x - 2)} \cdot \dfrac{x - 2}{1}$

$A = \dfrac{\overset{3}{\cancel{6x^2}}}{\underset{1}{\cancel{2}}\underset{1}{\cancel{(x - 2)}}} \cdot \dfrac{\overset{1}{\cancel{x - 2}}}{1} = 3x^2$ Thus, the area can be represented as $3x^2$.

TRY THIS 4. Find the area of the rectangle with length $\dfrac{9x - 6}{7x^3}$ and width $\dfrac{1}{3x - 2}.$ $\dfrac{3}{7x^3}$

Common Error Analysis

Error: When multiplying a fraction by $x - 2$, as in Example 5, some students may calculate or even write this as if it were $\dfrac{x - 2}{x - 2}$.

Show them that using this expression is equivalent to multiplying by 1, not by $x - 2$.

Checkpoint

1. $\dfrac{25}{24} \cdot \dfrac{12d}{10}$ $\dfrac{5d}{4}$

2. $\dfrac{5x^4}{3y^7} \cdot \dfrac{7y^3}{-4x^2}$ $\dfrac{-35x^2}{12y^4}$

3. $\dfrac{(3m + 1)}{m - 7} \cdot \dfrac{4}{6m + 2}$ $\dfrac{2}{m - 7}$

4. $\dfrac{-45ab^{12}}{a^2 - 100} \cdot \dfrac{40 - 4a}{3a^4b^6}$ $\dfrac{60b^3}{a^4 + 10a^3}$

5. $\dfrac{9 - 6x + x^2}{x^2 - 13x + 30} \cdot \dfrac{x^2 - 20x + 100}{x^2 - 100}$ $\dfrac{x - 3}{x + 10}$

Closure

Ask students to summarize the procedure for multiplying rational expressions.   Answers will vary.   Sample answer: Factor the numerators and denominators; cancel common factors; multiply.   Ask them what is meant by expressing factors in convenient form.   See page 294.

Additional Example 4

Multiply.

$\dfrac{36x^6y^4}{12 - 4y} \cdot \dfrac{2y^2 - 5y - 3}{9xy^3}$ $-2x^5y^2 - x^5y$

Additional Example 5

Find the area of the rectangle with length $2a - 1$ and width $\dfrac{3a^2}{6a - 3}.$ a^2

Guided Practice

Classroom Exercises 1–12
Try This   all

Independent Practice

A Ex. 1–30, **B** Ex. 31–41, **C** Ex. 42–44

Basic: WE 1–30 odd

Average: WE 11–40 odd

Above Average: WE 17–40 odd, 41–43 all

Additional Answers

Classroom Exercises

5. $\frac{y}{2}$

9. $\frac{2x-6}{x-1}$

Written Exercises

12. $\frac{x^2-2x+1}{10}$

14. $\frac{x^2-14x+49}{x+4}$

15. $\frac{y-5}{11}$

16. $\frac{-6p-9q}{-8p+12q}$

18. $\frac{4}{3}$

19. $5x+15$

21. $\frac{5}{y+1}$

22. $\frac{x+1}{2x-4}$

23. $\frac{2a-12}{a+5}$

24. $\frac{2}{3}$

26. $\frac{3p^3x-12p^3}{4q}$

27. $\frac{-5b^4}{2x+2}$

29. $\frac{y+5}{-9b}$

30. $\frac{3a^2}{-2b^5}$

33. $\frac{-x^2-6xy-5y^2}{x^2-3xy+2y^2}$

Classroom Exercises

Multiply.

1. $\frac{2}{3}\cdot\frac{5}{7}$ $\frac{10}{21}$

2. $\frac{8}{15}\cdot\frac{5}{12}$ $\frac{2}{9}$

3. $\frac{y}{9}\cdot\frac{7}{x}$ $\frac{7y}{9x}$

4. $\frac{x^3}{2}\cdot\frac{2}{x^5}$ $\frac{1}{x^2}$

5. $\frac{3}{y^4}\cdot\frac{y^5}{6}$

6. $\frac{a^3}{10}\cdot\frac{5}{a^2}$ $\frac{a}{2}$

7. $\frac{3x^7}{4y^5}\cdot\frac{2y^6}{3x^8}$ $\frac{y}{2x}$

8. $\frac{5a^3}{b^7}\cdot\frac{b^9}{15a}$ $\frac{a^2b^2}{3}$

9. $\frac{x-3}{4(x+2)}\cdot\frac{8(x+2)}{x-1}$

10. $\frac{12(a-2)}{a-3}\cdot\frac{a-3}{6(a+5)}$ $\frac{2a-4}{a+5}$

11. $\frac{a^3(x+1)}{x-4}\cdot\frac{x-4}{a^5}$ $\frac{x+1}{a^2}$

12. $\frac{3a+6}{a^2-4a+3}\cdot\frac{3-a}{a+2}$ $\frac{-3}{a-1}$

Written Exercises

Multiply.

1. $\frac{9}{11}\cdot\frac{5}{8}$ $\frac{45}{88}$

2. $\frac{3}{14}\cdot\frac{35}{33}$ $\frac{5}{22}$

3. $\frac{6a}{5}\cdot\frac{a^2}{7}$ $\frac{6a^3}{35}$

4. $\frac{3x^3}{14y^2}\cdot\frac{7y}{15x}$ $\frac{x^2}{10y}$

5. $\frac{-4a^5}{9b^2}\cdot\frac{12b^5}{22a^6}$ $\frac{-8b^3}{33a}$

6. $\frac{6m}{15n^4}\cdot\frac{30n^8}{9m^7}$ $\frac{4n^4}{3m^6}$

7. $\frac{a-3}{2}\cdot\frac{2a+10}{6}$ $\frac{a^2+2a-15}{6}$

8. $\frac{p+1}{5}\cdot\frac{3}{-2p-2}$ $\frac{-3}{10}$

9. $\frac{3x-12}{5y^2z}\cdot\frac{25yz^2}{x-4}$ $\frac{15z}{y}$

10. $\frac{y+1}{21}\cdot\frac{14}{y+1}$ $\frac{2}{3}$

11. $\frac{-5x}{7x-14}\cdot\frac{3x-6}{15}$ $\frac{-x}{7}$

12. $\frac{x-1}{5}\cdot\frac{x-1}{2}$

13. $\frac{-3a^2b}{c}\cdot\frac{1-c}{c-1}$ $\frac{3a^2b}{c}$

14. $\frac{x-7}{x+4}\cdot(x-7)$

15. $(y+5)\cdot\frac{y-5}{55+11y}$

16. $\frac{-(2p+3q)}{8}\cdot\frac{6}{-2p+3q}$

17. $\frac{35}{6y+8z}\cdot\frac{9y+12z}{7}$ $\frac{15}{2}$

18. $\frac{-a-b}{3x-1}\cdot\frac{12-36x}{9a+9b}$

19. $\frac{x^2+7x+12}{x+5}\cdot\frac{5x+25}{x+4}$

20. $\frac{a-7}{a-6}\cdot\frac{a^2-5a-6}{3a-21}$ $\frac{a+1}{3}$

21. $\frac{5y+30}{y^2-2y-3}\cdot\frac{y-3}{y+6}$

22. $(x+4)\cdot\frac{2x+2}{4x^2+8x-32}$

23. $\frac{4a-24}{4a^2+18a-10}\cdot(2a-1)$

24. $(3m-6)\cdot\frac{2m+4}{9m^2-36}$

25. $\frac{4a^3b^5}{x^2-9}\cdot\frac{6-2x}{6a^4b^2}$ $\frac{-4b^3}{3ax+9a}$

26. $\frac{x^2-5x+4}{10p^3q^2}\cdot\frac{15p^6q}{2x-2}$

27. $\frac{5-5x}{10b^2c}\cdot\frac{5b^6c}{x^2-1}$

28. $\frac{24x^4y^3}{4m-12}\cdot\frac{9-m^2}{18x^7y^4}$ $\frac{-m-3}{3x^3y}$

29. $\frac{y^2-y-30}{9ab^8}\cdot\frac{3ab^7}{18-3y}$

30. $\frac{15a^5b}{7-x}\cdot\frac{3x-21}{30a^2b^6}$

31. $\frac{2x^2-13x+20}{3x^2-10x-8}\cdot\frac{-3-x}{2x^2+x-15}$ $\frac{-1}{3x+2}$

32. $\frac{3x^2+20xy-7y^2}{x^2+5xy-14y^2}\cdot\frac{6y^2-xy-x^2}{3x^2+3xy-y^2}$ $\frac{-3x^2-8xy+3y^2}{3x^2+3xy-y^2}$

33. $\dfrac{x^2 - y^2}{3x^2 - 21xy + 30y^2} \cdot \dfrac{75y^2 - 3x^2}{x^2 - 2xy + y^2}$

34. $\dfrac{a^2 - ad - 6d^2}{2a^2 + 11ad + 5d^2} \cdot \dfrac{a^2 + 9ad + 20d^2}{a^2 + 6ad + 8d^2}$ $\dfrac{a - 3d}{2a + d}$

35. $\dfrac{10 - 3x - x^2}{x - 5} \cdot \dfrac{4x - 8}{2x^2 - 8x + 8}$ $\dfrac{-2x - 10}{x - 5}$

36. $\dfrac{3t^3 - 14t^2 + 8t}{20 + 3t - 2t^2} \cdot \dfrac{16t^2 + 34t - 15}{24t^2 - 25t + 6}$ $-t$

37. $\dfrac{3 - x}{x^2 - 2x - 15} \cdot \dfrac{x^2 - 4x - 5}{x^2 - x - 6}$ $\dfrac{-x - 1}{x^2 + 5x + 6}$

38. $\dfrac{2y^2 + 9y - 5}{2y^2 + 5y - 3} \cdot \dfrac{y^2 + y - 6}{y^2 + 7y + 10}$ $\dfrac{y - 2}{y + 2}$

39. $\dfrac{16 - x^2}{x^2 + 7x + 12} \cdot \dfrac{2x^2 + 3x - 9}{x^2 - 5x + 4}$ $\dfrac{-2x + 3}{x - 1}$

40. $\dfrac{2a^2 + 7a - 4}{a^3 - 16a} \cdot \dfrac{8 - 2a}{10a - 5}$ $\dfrac{-2}{5a}$

41. Use the formula $d = rt$ to find the distance traveled in t hours at a rate of r mi/h, given that $r = \dfrac{x - 5}{5}$ and $t = \dfrac{15}{13x - 65}$. $\dfrac{3}{13}$ mi

42. The formula for the volume of a rectangular solid is given by $V = lwh$, where l is the length, w is the width, and h is the height. Find the volume of the rectangular solid at the right. Give your answer in simplified form. $\dfrac{14a + 7}{6a^2 + 12a}$

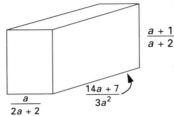

$\dfrac{a + 1}{a + 2}$

$\dfrac{a}{2a + 2}$ $\dfrac{14a + 7}{3a^2}$

The formula for the volume of a cube is $V = e^3$, where e is the length of one edge of the cube. Find the volume for the given length of an edge. Give your answer in simplified form.

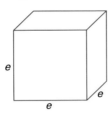

e e e

43. $e = \dfrac{2x - 1}{3}$

$\dfrac{8x^3 - 12x^2 + 6x - 1}{27}$

44. $e = \dfrac{a}{a + 1}$

$\dfrac{a^3}{a^3 + 3a^2 + 3a + 1}$

Mixed Review

Write an algebraic expression for each word description. *1.1*

1. 15 more than Myra's age m $m + 15$

2. 35 less than the cost c $c - 35$

3. the cost, in dollars, of m pounds of apples at \$.74 per pound $\$0.74m$

4. $2\frac{1}{2}$ times the regular team score s $\frac{5}{2}s$

Translate each problem into an equation and solve. *4.2*

5. Raul has 18 tapes. He has 6 fewer tapes than twice the number Marcia has. Find the number of tapes Marcia has. Let x = number of Marcia's tapes; $18 = 2x - 6$; $x = 12$

6. The selling price of a pair of skis is \$250. The selling price is \$20 less than twice the cost. Find the cost. Let c = cost; $250 = 2c - 20$; $c = \$135$.

8.3 Multiplying Rational Expressions **301**

Enrichment

Introduce *factorial notation*.

$5! = 5 \cdot 4 \cdot 3 \cdot 2 \cdot 1$

$7! = 7 \cdot 6 \cdot 5 \cdot 4 \cdot 3 \cdot 2 \cdot 1$

$n! = n(n - 1)(n - 2)\ldots(3)(2)(1)$ Then have students evaluate these expressions.

1. $4!$ 24

2. $(3!)(3!)$ 36

3. $\dfrac{5!}{4!}$ 5

4. $\dfrac{7!}{(5!)(2!)}$ 21

5. $\dfrac{10!}{(7!)(3!)}$ 120

6. $\dfrac{n!}{(n - 1)!}$ n

Teaching Resources

Manipulative Worksheet 8
Quick Quizzes 63
Reteaching and Practice
   Worksheets 63
Transparencies 20, 21

8.4 Dividing Rational Expressions

Objective

To divide rational expressions

You know that $\frac{2}{3} \div \frac{3}{5} = \frac{2}{3} \cdot \frac{5}{3}$. Since $\frac{5}{3}$ is the reciprocal

of $\frac{3}{5}$, $\frac{2}{3} \div \frac{3}{5}$ equals $\frac{2}{3}$ times the reciprocal of $\frac{3}{5}$.

Dividing rational expressions is similar to dividing fractions. That is,

$\frac{a}{b} \div \frac{c}{d}$ equals $\frac{a}{b}$ times the *reciprocal* of $\frac{c}{d}$.

Rule for Dividing Rational Expressions

If $\frac{a}{b}$ and $\frac{c}{d}$ are rational expressions, $b \neq 0$, $c \neq 0$, and $d \neq 0$, then

$$\frac{a}{b} \div \frac{c}{d} = \frac{a}{b} \cdot \frac{d}{c}.$$

GETTING STARTED

Prerequisite Quiz

Divide.

1. $\frac{18a^8}{27a^2}$ $\frac{2a^6}{3}$

2. $\frac{4m^2 - 11m - 3}{9 - 3m}$ $\frac{4m + 1}{-3}$

3. $\frac{3}{4} \div \frac{5}{7}$ $\frac{21}{20}$

4. $\frac{4}{9} \div \frac{8}{3}$ $\frac{1}{6}$

5. $12 \div \frac{2}{3}$ 18

6. $\frac{5}{8} \div 10$ $\frac{1}{16}$

Motivator

Ask students to give the reciprocal of
each expression.

1. $\frac{3}{8}$ $\frac{8}{3}$ 2. 1 1

3. 15 $\frac{1}{15}$ 4. $\frac{a}{2}$ $\frac{2}{a}$

5. $\frac{b + a}{c}$ $\frac{c}{b + a}$ 6. $a - 1$ $\frac{1}{a - 1}$

EXAMPLE 1   Divide: $\frac{12x^5}{15y^4} \div \frac{6x^2}{5y}$

Plan   Multiply $\frac{12x^5}{15y^4}$ by the reciprocal of $\frac{6x^2}{5y}$.

Solution $\frac{12x^5}{15y^4} \div \frac{6x^2}{5y} = \frac{12x^5}{15y^4} \cdot \frac{5y}{6x^2}$

$= \frac{\overset{2}{\cancel{12}}\overset{x^3}{\cancel{x^5}}}{\underset{3}{\cancel{15}}\underset{y^3}{\cancel{y^4}}} \cdot \frac{\overset{1}{\cancel{5}}\overset{1}{\cancel{y}}}{\underset{1}{\cancel{6}}\underset{1}{\cancel{x^2}}} = \frac{2x^3}{3y^3}$

EXAMPLE 2   Divide: $\frac{6x + 24}{7x - 49} \div \frac{3x - 6}{x - 7}$

Solution $\frac{6x + 24}{7x - 49} \div \frac{3x - 6}{x - 7} = \frac{6x + 24}{7x - 49} \cdot \frac{x - 7}{3x - 6}$

$= \frac{\overset{2}{\cancel{6}}(x + 4)}{7\cancel{(x - 7)}} \cdot \frac{\cancel{x - 7}^{1}}{3(x - 2)} = \frac{2(x + 4)}{7(x - 2)}$, or $\frac{2x + 8}{7x - 14}$

TRY THIS   1. Divide: $\frac{16a - 32}{21a + 21} \div \frac{4a + 12}{7a + 7}$ $\frac{4a - 8}{3a + 9}$

302   Chapter 8 Rational Expressions

Highlighting the Standards

Standards 5a, 5c, 7a: Here again students
operate on rational expressions and apply
these skills to three-dimensional geometric
figures.

Additional Example 1

Divide $\frac{36a^4}{22b^6} \div \frac{9a}{11b^5}$. $\frac{2a^3}{b}$

Additional Example 2

Divide $\frac{x + 2}{6x + 18} \div \frac{2x + 4}{3x + 15}$. $\frac{x + 5}{4(x + 3)}$

Sometimes expressions contain both multiplication and division. If there are no parentheses, multiply or divide in order from left to right. Write the reciprocal of a rational expression only if it immediately follows a division symbol.

EXAMPLE 3 Simplify $\dfrac{x + 2}{x - 3} \cdot \dfrac{(x - 3)(x - 4)}{x} \div \dfrac{(x + 2)(x - 4)}{x^2}$.

Plan To divide by the third expression, multiply by its reciprocal.

Solution $\dfrac{x + 2}{x - 3} \cdot \dfrac{(x - 3)(x - 4)}{x} \cdot \dfrac{x^2}{(x + 2)(x - 4)}$

$= \dfrac{\overset{1}{\cancel{x + 2}}}{\underset{1}{\cancel{x - 3}}} \cdot \dfrac{\overset{1}{\cancel{(x - 3)}}\overset{1}{\cancel{(x - 4)}}}{x} \cdot \dfrac{x^2}{\underset{1}{\cancel{(x + 2)}}\underset{1}{\cancel{(x - 4)}}} = x$

EXAMPLE 4 Simplify $\dfrac{x^2 + 3x - 10}{8x^4} \div \dfrac{x^2 - 8x + 12}{12x^3} \cdot \dfrac{x - 6}{3x^2 + 15x}$.

Plan To divide by the second expression, multiply by its reciprocal.

Solution $\dfrac{x^2 + 3x - 10}{8x^4} \cdot \dfrac{12x^3}{x^2 - 8x + 12} \cdot \dfrac{x - 6}{3x^2 + 15x}$

$= \dfrac{\cancel{(x + 5)}\cancel{(x - 2)}}{\underset{2x^2}{\cancel{8x^4}}} \cdot \dfrac{\overset{}{\cancel{12x^3}}}{\underset{1}{\cancel{(x - 6)}}\underset{1}{\cancel{(x - 2)}}} \cdot \dfrac{\overset{}{\cancel{x - 6}}}{\underset{1}{\cancel{3x}}\underset{1}{\cancel{(x + 5)}}} = \dfrac{1}{2x^2}$

TRY THIS **2.** Simplify $\dfrac{x^2 + 12x + 27}{4x^3 - 28x^2} \div \dfrac{x + 3}{x^2 - 13x + 42} \cdot \dfrac{24x^6}{x^2 + 3x - 54}$. $6x^4$

Classroom Exercises

Give the reciprocal.

1. 9 $\frac{1}{9}$

2. $\dfrac{1}{a - 3}$ $a - 3$

3. $x + 11$ $\frac{1}{x + 11}$

4. $\dfrac{b^2 - 7b + 6}{4b^3}$ $\dfrac{4b^3}{b^2 - 7b + 6}$

Divide.

5. $\dfrac{3}{8} \div \dfrac{4}{5}$ $\frac{15}{32}$

6. $\dfrac{5}{6} \div \dfrac{7}{8}$ $\frac{20}{21}$

7. $\dfrac{1}{3} \div \dfrac{5}{6}$ $\frac{2}{5}$

8. $\dfrac{-8}{9} \div \dfrac{2}{3}$ $\frac{-4}{3}$

9. $\dfrac{3}{8} \div 3$ $\frac{1}{8}$

10. $\dfrac{a}{b} \div \dfrac{3a^2}{4}$ $\frac{4}{3ab}$

11. $\dfrac{x^2}{3y} \div \dfrac{6x}{4y^3}$ $\frac{2xy^2}{9}$

12. $\dfrac{a + b}{2} \div \dfrac{a - b}{4}$ $\frac{2a + 2b}{a - b}$

13. $\dfrac{x^2 - 1}{y + 2} \div \dfrac{(x - 1)^2}{y - 2}$ $\frac{xy - 2x + y - 2}{xy + 2x - y - 2}$

14. $\dfrac{a^2 - 49}{5a^7b^6} \div \dfrac{a^2 + 4a - 21}{15ab^2}$ $\frac{3a - 21}{a^7b^4 - 3a^6b^4}$

Checkpoint

Divide.

1. $\dfrac{24x^4y^8}{3x^2 - 6x} \div \dfrac{16x^3y^2}{4x - 8}$ $2y^6$

2. $\dfrac{3y^2 - 5y + 2}{y^2 - 16} \div \dfrac{y - 1}{3y - 12}$ $\dfrac{3(3y - 2)}{y + 4}$

Simplify.

3. $\dfrac{2t^2 + 9t - 5}{t^5s^4} \cdot \dfrac{ts^6}{4t^2 - 1} \div \dfrac{t^2 + 4t - 5}{t^4s^2}$.

   $\dfrac{s^4}{(2t + 1)(t - 1)}$

Closure

Ask students to list the steps needed to divide rational expressions. Write the reciprocal of the divisor; factor, if possible; divide by common factors; multiply. Ask students whether it makes a difference if the order of the steps is changed. Yes, writing the reciprocal of the divisor must be the first step.

◢◣◤ FOLLOW UP

Guided Practice

Classroom Exercises 1–14
Try This   all

Independent Practice

A Ex. 1–33, **B** Ex. 34–44, **C** Ex. 45–51

Basic: WE 1–33 odd

Average: WE 9–41 odd

Above Average: WE 19–51 odd

Written Exercises

Divide.

1. $\dfrac{7}{9} \div \dfrac{2}{5}$ $\dfrac{35}{18}$

2. $\dfrac{6}{11} \div \dfrac{-8}{9}$ $-\dfrac{27}{44}$

3. $\dfrac{9x}{14} \div \dfrac{7x^2}{2}$ $\dfrac{9}{49x}$

4. $\dfrac{m^2}{p^2} \div \dfrac{p^4}{m^4}$ $\dfrac{m^6}{p^6}$

5. $\dfrac{3ab^2}{-4st} \div \dfrac{6b^3t}{5a}$ $\dfrac{-5a^2}{8bst^2}$

6. $\dfrac{(-x)^4}{y^3} \div \dfrac{-x^4}{y}$ $\dfrac{-1}{y^2}$

7. $\dfrac{y}{3x} \div \dfrac{2y^2}{9}$ $\dfrac{3}{2xy}$

8. $\dfrac{2m}{m + 2} \div (m + 2)$

9. $\dfrac{n^2}{n + 1} \div (n - 1)$ $\dfrac{n^2}{n^2 - 1}$

10. $\dfrac{x - 7}{3} \div \dfrac{x + 7}{6}$ $\dfrac{2x - 14}{x + 7}$

11. $\dfrac{a + 1}{a} \div \dfrac{-3a - 3}{ab}$ $-\dfrac{b}{3}$

12. $\dfrac{5x + 10}{x} \div \dfrac{x + 2}{y}$ $\dfrac{5y}{x}$

13. $\dfrac{a^2 - 2a - 8}{a^2 - 16} \div \dfrac{4a + 12}{a + 4}$

14. $\dfrac{3a + 21}{16} \div \dfrac{a^2 - 49}{4a - 8}$

15. $\dfrac{x^2 - 16}{y - 2} \div \dfrac{2x + 8}{7y - 14}$ $\dfrac{7(x - 4)}{2}$

16. $\dfrac{x^2 + 3x - 18}{m^5} \div \dfrac{2x + 12}{m^8}$ $\dfrac{m^3x - 3m^3}{2}$

17. $\dfrac{4x^5}{x - 7} \div \dfrac{16x^7}{3x^2 - 21x}$ $\dfrac{3}{4x}$

18. $\dfrac{3x + 21}{7x^5} \div \dfrac{3x^2 + 21x}{14x^2}$ $\dfrac{2}{x^4}$

19. $\dfrac{18a^5b^4}{5a - 10a^2} \div \dfrac{24a^4b^6}{6a - 3}$ $\dfrac{-9}{20b^2}$

20. $\dfrac{36 - x^2}{3x^4y^3} \div \dfrac{4x - 24}{15x^5y}$ $\dfrac{-5x^2 - 30x}{4y^2}$

21. $\dfrac{15 - 5m}{6m^3n^3} \div \dfrac{m^2 - 9}{14m^4n^2}$ $\dfrac{-35m}{3mn + 9n}$

22. $\dfrac{x^2 - 10x + 25}{x + 1} \div \dfrac{2x^2 - 10x}{x^2 + x}$ $\dfrac{x - 5}{2}$

23. $\dfrac{x^2 - 16}{6x} \div \dfrac{5x - 20}{3x^2 - 15x}$ $\dfrac{x^2 - x - 20}{10}$

24. $\dfrac{x^2 + 2x - 15}{x + 3} \div \dfrac{x^2 + 7x + 10}{x - 2}$ $\dfrac{x^2 - 5x + 6}{x^2 + 5x + 6}$

25. $\dfrac{x - 5}{x^2 + 3x - 10} \div \dfrac{9x^2}{3x^2 - 6x}$

26. $\dfrac{x^2 + 10x + 24}{3x^2 + 12x} \div (x^2 + 3x - 18)$ $\dfrac{1}{3x(x - 3)}$

27. $\dfrac{x^2 - 2x - 15}{5x^2 + 15x} \div (x^2 - 6x + 5)$

28. $\dfrac{x^2 - y^2}{6x} \div \dfrac{x^2y + xy^2}{3x^2y^2}$ $\dfrac{xy - y^2}{2}$

29. $\dfrac{4x^2 - 25y^2}{2x^2y + 5xy^2} \div \dfrac{6x^2 - 15xy}{9x^2y^2}$ $3y$

30. $\dfrac{3(x + y)^2}{x - y} \div 6(x + y)$ $\dfrac{x + y}{2x - 2y}$

31. $\dfrac{x - y}{x + y} \div \dfrac{5x^2 - 5y^2}{3x - 3y}$ $\dfrac{3(x - y)}{5(x + y)^2}$

32. $\dfrac{x^2 + 2x + 1}{3x} \div (x + 1)$ $\dfrac{x + 1}{3x}$

33. $\dfrac{a^3 - 6a^2 + 8a}{5} \div \dfrac{2a - 4}{10a - 40}$ $a(a - 4)^2$

34. $\dfrac{5x^2y^3}{2x + 6} \cdot \dfrac{x^2 - 16}{20x^7y} \div \dfrac{x + 4}{x^2 - 2x - 15}$

35. $\dfrac{k^2 + 8k + 15}{12k^2} \cdot \dfrac{9k^3}{k + 2} \div \dfrac{4k + 12}{k^2 + 2k}$

36. $\dfrac{a^2 - 2a - 15}{a^7b^2} \div \dfrac{a^2 - 25}{16a^7b^3} \cdot \dfrac{4a - 24}{3a + 9}$ $\dfrac{64b(a - 6)}{3(a + 5)}$

37. $\dfrac{b^2 - 3b}{b^6} \div \dfrac{b^2 + b - 12}{b^7} \cdot \dfrac{6b^2 + 24b}{4b^2}$ $\dfrac{3b}{2}$

38. $\dfrac{a^2 + 4a - 21}{a^2 + a - 20} \div \dfrac{a^2 + 8a + 7}{a^2 + 6a + 5}$ $\frac{a-3}{a-4}$

39. $\dfrac{2x^3 + 6x^2 - 20x}{15 - 2x - x^2} \div \dfrac{4x^2 + 24x - 64}{6x^2 - 54}$

40. $\dfrac{2x^2 + 5x + 3}{x^2 + 9x + 14} \div \dfrac{2x^2 - 3x - 9}{x^2 + 6x - 7}$ $\frac{x^2-1}{x^2-x-6}$

41. $\dfrac{3a^2 - a - 2}{a^2 - 3a} \div \dfrac{3a^2 + 11a + 6}{a^3}$

42. $\dfrac{9y^2 - b^2}{15y^2 + 6by} \div \dfrac{3y^2 + 2by - b^2}{25y^2 - 4b^2}$ $\frac{15y^2 - by - 2b^2}{3y^2 + 3by}$

43. $\dfrac{6a^2 + 11ab - 10b^2}{a^8} \div \dfrac{2a^2 + 11ab + 15b^2}{a^5 + 3a^4b}$ $\frac{3a-2b}{a^4}$

44. $\dfrac{xy - xz - x^2}{xyz} \cdot \dfrac{x^2y^2z^3}{yx - zx} \div x^3$ $\frac{yz^2(y-z-x)}{x^2(y-z)}$

45. $\dfrac{2a^3}{2a + b} \div \dfrac{10a^2}{4a^2 + 4ab + b^2} \cdot \dfrac{12a + 3b}{2a^2 + ab}$ $\frac{3(4a+b)}{5}$

46. $\dfrac{a^2 - 5a + 6}{a^2 + 3a} \cdot \dfrac{a^2 + 2a - 3}{a^2 + 4a + 3} \div \dfrac{2a^2 - 3a - 2}{2a^2 + a - 1}$ $\frac{2a^3 - 9a^2 + 10a - 3}{2a^3 + 7a^2 + 3a}$

47. $\dfrac{x^2 - 2x - 8}{x^3 - 9x} \div \dfrac{x^2 - 16}{x^2 + 3x} \cdot \dfrac{2x^2 - 7x - 4}{2x^2 + 5x + 2}$ $\frac{x-4}{x^2+x-12}$

A formula for the height of a rectangular solid is $h = \dfrac{V}{B}$, where V is the volume of the solid and B is the area of its rectangular base. Find an expression for the height of the rectangular solid.

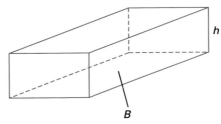

48. volume: $8x^3 + 8x^2$
area of base: $8x^2$ $x+1$

49. volume: $9x^3 + 45x^2 + 54x$
area of base: $9x^2 + 18x$ $x+3$

50. volume: $21x^2y + 105xy$
width of base: $7x$
length of base: $3y$ $x+5$

51. volume: $5x^2 + 5xy + 20x^2 + 20x$
width of base: $5x$
length of base: $x + 1$ $\frac{5x+y+4}{x+1}$

Mixed Review

Solve each equation or inequality. *5.2, 5.7, 7.8*

1. $x^2 + x - 12 = 0$ $\{-4, 3\}$

2. $x^2 + 2x - 15 = 0$ $\{-5, 3\}$

3. $a^2 + 6a - 7 = 0$ $\{-7, 1\}$

4. $x^2 - 25 = 0$ $\{-5, 5\}$

5. $a^2 - 7a - 8 = 0$ $\{-1, 8\}$

6. $x^2 - 5x = 0$ $\{0, 5\}$

7. $-6x + 8 \geq 20$ $x \leq -2$

8. $15 - 3y < 45$ $y > -10$

9. $|x - 1| \leq 3$ $-2 \leq x \leq 4$

Enrichment

Have students find the missing numerator or denominator.

1. $\dfrac{6}{x} \div \dfrac{72}{?} = \dfrac{x^2}{12}$ x^3

2. $\dfrac{1}{?} \div \dfrac{-5}{a^2} = \dfrac{a}{5}$ $-a$

3. $\dfrac{?}{8} \div \dfrac{15a^2}{12} = 1$ $10a^2$

4. $\dfrac{-6(a - 2)}{-3(a + 5)} \div \dfrac{?}{a(a + 5)} = \dfrac{2a}{3}$ $3(a - 2)$

305

Prerequisite Quiz

Add or subtract.

1. $\frac{4}{3} + \frac{7}{3}$ $\frac{11}{3}$
2. $\frac{7}{12} - \frac{3}{12}$ $\frac{1}{3}$

Simplify.

3. $a + 20 + a$ $2a + 20$
4. $2x^2 + 3 + 5x^2 - 4$ $7x^2 - 1$
5. $2m + n - (3m + n)$ $-m$

Motivator

Ask students to give examples from
everyday situations where fractions are
added or subtracted.   Answers will vary.
Sample answer: stock market, cooking,
measurements, and so on.

8.5 Adding and Subtracting Rational Expressions: Like Denominators

Objective   To add and subtract rational expressions with like denominators

One pizza weighs m ounces
and another weighs n ounces.
Each of the 2 pizzas will be
divided evenly among 8 people. How much pizza will
each person get? Each person
will get $\frac{1}{8}$ of each pizza.

In all, the fraction of both pizzas received by each person is $\frac{m}{8} + \frac{n}{8}$.
How can this be expressed as a single rational expression?

To answer this question, recall that two fractions are added or subtracted in the manner shown below.

$$\frac{4}{7} + \frac{2}{7} = \frac{4 + 2}{7} = \frac{6}{7} \qquad\qquad \frac{5}{7} - \frac{2}{7} = \frac{5 - 2}{7} = \frac{3}{7}$$

This suggests that an expression for each person's share of the pizzas is
$\frac{m + n}{8}$ ounces.

Rule for Adding and Subtracting Rational Expressions

If $\frac{a}{b}$ and $\frac{c}{b}$ are rational expressions, $b \neq 0$, then

$$\frac{a}{b} + \frac{c}{b} = \frac{a + c}{b} \text{ and } \frac{a}{b} - \frac{c}{b} = \frac{a - c}{b}.$$

EXAMPLE 1   Add or subtract.

   a. $\frac{7x}{4} + \frac{6x}{4}$ b. $\frac{5}{y} + \frac{6}{y}$ c. $\frac{4x}{x - 3} - \frac{3x}{x - 3}$

Plan   Use $\frac{a}{b} + \frac{c}{b} = \frac{a + c}{b}$ or $\frac{a}{b} - \frac{c}{b} = \frac{a - c}{b}$.

Solutions   a. $\frac{7x}{4} + \frac{6x}{4}$ b. $\frac{5}{y} + \frac{6}{y}$ c. $\frac{4x}{x - 3} - \frac{3x}{x - 3}$

     $= \frac{7x + 6x}{4} = \frac{13x}{4}$ $= \frac{5 + 6}{y} = \frac{11}{y}$ $= \frac{4x - 3x}{x - 3} = \frac{x}{x - 3}$

TRY THIS   1. Subtract $\frac{7x}{2x + 3} - \frac{5x}{2x + 3}$. $\frac{2x}{2x + 3}$

Highlighting the Standards

Standards 5a, 5c: The discussion on
rational expressions is expanded to
include addition. Geometric figures
show applications of these skills.

Additional Example 1

Add or subtract.

a. $\frac{4x}{3} + \frac{7x}{3}$ $\frac{11x}{3}$
b. $\frac{7}{t} + \frac{5}{t}$ $\frac{12}{t}$
c. $\frac{5a}{a - 4} - \frac{2a}{a - 4}$ $\frac{3a}{a - 4}$

The Rule for Adding Fractions can be extended to include three or more addends as illustrated in the next example.

EXAMPLE 2 Add or subtract.

a. $\dfrac{7a}{3} + \dfrac{4a}{3} + \dfrac{10a}{3}$

b. $\dfrac{3x}{2x-10} - \dfrac{9}{2x-10} + \dfrac{4-2x}{2x-10}$

Solutions

a. $\dfrac{7a}{3} + \dfrac{4a}{3} + \dfrac{10a}{3}$

$= \dfrac{7a + 4a + 10a}{3}$

$= \dfrac{\overset{7a}{\cancel{21a}}}{\underset{1}{\cancel{3}}}$

$= 7a$

b. $\dfrac{3x}{2x-10} - \dfrac{9}{2x-10} + \dfrac{4-2x}{2x-10}$

$= \dfrac{3x - 9 + (4 - 2x)}{2x - 10}$

$= \dfrac{3x - 9 + 4 - 2x}{2x - 10}$

$= \dfrac{\overset{1}{x-5}}{\underset{1}{2(x-5)}} = \dfrac{1}{2}$

EXAMPLE 3 Add $\dfrac{x^2}{x^2 - 7x + 12} + \dfrac{18 - 3x}{-x^2 + 7x - 12}$.

Plan Write $-x^2 + 7x - 12$ in convenient form: $-1(x^2 - 7x + 12)$.

Solution

$\dfrac{x^2}{x^2 - 7x + 12} + \dfrac{18 - 3x}{-1(x^2 - 7x + 12)}$

$= \dfrac{x^2}{x^2 - 7x + 12} - \dfrac{18 - 3x}{x^2 - 7x + 12}$ $\longleftarrow \dfrac{a}{-b} = -\dfrac{a}{b}$

$= \dfrac{x^2 - (18 - 3x)}{x^2 - 7x + 12}$

$= \dfrac{x^2 - 18 + 3x}{x^2 - 7x + 12}$

$= \dfrac{x^2 + 3x - 18}{x^2 - 7x + 12}$

$= \dfrac{(x + 6)\overset{1}{\cancel{(x - 3)}}}{\underset{1}{\cancel{(x - 3)}}(x - 4)} = \dfrac{x + 6}{x - 4}$

TRY THIS

2. Subtract $\dfrac{-x^2}{-x^2 + x + 56} - \dfrac{17x - 72}{x^2 - x - 56}$. $\frac{x-9}{x+7}$

3. Add $\dfrac{x}{x^2 + 2x} + \dfrac{2}{x^2 + 2x}$. $\frac{1}{x}$

8.5 Adding and Subtracting Rational Expressions: Like Denominators **307**

Additional Example 2

Add or subtract.

a. $\dfrac{13m}{9} + \dfrac{20m}{9} - \dfrac{6m}{9}$ $3m$

b. $\dfrac{8y - 9}{4y - 8} - \dfrac{y + 12}{4y - 8} + \dfrac{5y - 3}{4y - 8}$ 3

Additional Example 3

Add $\dfrac{x^2}{x^2 - 6x + 8} + \dfrac{20 - x}{-x^2 + 6x - 8}$. $\frac{x+5}{x-2}$

■■■**TEACHING SUGGESTIONS**

Lesson Note

In Example 3, point out the parentheses that enclose $18 - 3x$ in the third line of the solution. These are necessary for distributing the -1 factor over both terms of the numerator. In Example 4, you may wish to contrast finding the perimeter by addition with the area by multiplication.

Math Connections

History: The plus sign is thought to be a contraction of the Latin word "et," meaning "and." The word "et" was also often used to mean addition.

Critical Thinking Questions

Synthesis: Ask students to give a reason for each step in "developing" the Rule for Addition of Rational Expressions. Remind them of the Definition of Division and the Distributive Property.

$\dfrac{a}{b} + \dfrac{c}{b} = a \cdot \dfrac{1}{b} + c \cdot \dfrac{1}{b}$

$= (a + c) \cdot \dfrac{1}{b}$

$= \dfrac{a + c}{b}$

Def of div, Distrib Prop, Rule for mult.

Common Error Analysis

Error: When subtracting a rational expression whose numerator is a binomial, students sometimes subtract only the first term of the binomial rather than the entire binomial.

Have students use parentheses to show multiplication by -1 and the use of the Distributive Property.

Checkpoint

Add or subtract.

1. $\dfrac{5t}{t-3} + \dfrac{2t}{t-3}$ $\dfrac{7t}{t-3}$

2. $\dfrac{3m}{m+6} - \dfrac{m+3}{m+6}$ $\dfrac{2m-3}{m+6}$

3. $\dfrac{y^2}{2y+4} - \dfrac{4}{2y+4}$ $\dfrac{y-2}{2}$

4. $\dfrac{3x}{x-4} + \dfrac{12}{4-x}$ 3

Closure

Have students give the answer to each problem without using paper and pencil.

1. $\dfrac{7}{t} + \dfrac{9}{t}$ $\dfrac{16}{t}$

2. $\dfrac{10}{r} - \dfrac{3}{r}$ $\dfrac{7}{r}$

3. $\dfrac{m}{x+y} - \dfrac{n}{x+y}$ $\dfrac{m-n}{x+y}$

4. $\dfrac{3a}{2} + \dfrac{4a-7}{2}$ $\dfrac{7a-7}{2}$

5. $\dfrac{a+b}{9} - \dfrac{a-b}{9}$ $\dfrac{2b}{9}$

EXAMPLE 4   Find the perimeter of the rectangle at the right.

Plan   Add the lengths of the four sides.

Solution
$$\frac{3x+6}{3} + \frac{3x+6}{3} + \frac{2x}{3} + \frac{2x}{3}$$
$$= \frac{(3x+6)+(3x+6)+2x+2x}{3}$$
$$= \frac{3x+6+3x+6+2x+2x}{3}$$
$$= \frac{10x+12}{3} \quad \longleftarrow \text{simplified form}$$

Thus, a rational expression for the perimeter is $\dfrac{10x+12}{3}$.

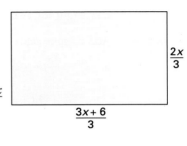

$\dfrac{2x}{3}$

$\dfrac{3x+6}{3}$

TRY THIS   4. Find the perimeter of a rectangle with length $\dfrac{4y+2}{9}$ and width $\dfrac{5y}{9}$. $\dfrac{18y+4}{9}$

Classroom Exercises

Add or subtract.

1. $\dfrac{2}{7} + \dfrac{3}{7}$ $\dfrac{5}{7}$

2. $\dfrac{7}{10} - \dfrac{3}{10}$ $\dfrac{2}{5}$

3. $\dfrac{5y}{8} + \dfrac{y}{8}$ $\dfrac{3y}{4}$

4. $\dfrac{14}{15a} - \dfrac{4}{15a}$ $\dfrac{2}{3a}$

5. $\dfrac{-3}{5y} + \dfrac{2}{5y}$ $-\dfrac{1}{5y}$

6. $\dfrac{6}{11m} + \dfrac{1}{-11m}$ $\dfrac{5}{11m}$

7. $\dfrac{1}{b-1} - \dfrac{b}{b-1}$ -1

8. $\dfrac{a}{a+1} - \dfrac{2}{1+a}$ $\dfrac{a-2}{a+1}$

9. $\dfrac{x}{9} + \dfrac{2x}{9} + \dfrac{3x}{9}$ $\dfrac{2x}{3}$

10. $\dfrac{m^2}{m-1} - \dfrac{2m}{m-1} - \dfrac{1}{1-m}$ $m-1$

Written Exercises

Add or subtract.

1. $\dfrac{8a}{5} + \dfrac{3a}{5}$ $\dfrac{11a}{5}$

2. $\dfrac{2x^2}{7} + \dfrac{4x^2}{7}$ $\dfrac{6x^2}{7}$

3. $\dfrac{7x^3}{3} - \dfrac{x^3}{3}$ $2x^3$

4. $\dfrac{b}{9} - \dfrac{10b}{9}$ $-b$

5. $\dfrac{4}{x^2} - \dfrac{2}{x^2}$ $\dfrac{2}{x^2}$

6. $\dfrac{12}{a} + \dfrac{5}{a}$ $\dfrac{17}{a}$

7. $\dfrac{9}{2x} + \dfrac{3}{2x}$ $\dfrac{6}{x}$

8. $\dfrac{10}{3b} - \dfrac{8}{3b}$ $\dfrac{2}{3b}$

9. $\dfrac{6}{x-1} - \dfrac{-5}{x-1}$

10. $\dfrac{7}{2-y} + \dfrac{5}{2-y}$

11. $\dfrac{a}{a-4} + \dfrac{4}{4-a}$ 1

12. $\dfrac{t}{t+1} - \dfrac{1}{1+t}$

13. $\dfrac{x+3}{4} + \dfrac{x-1}{4}$ $\dfrac{x+1}{2}$

14. $\dfrac{2y}{3} - \dfrac{y-1}{3}$ $\dfrac{y+1}{3}$

15. $\dfrac{a+9}{a} + \dfrac{8}{a}$ $\dfrac{a+17}{a}$

16. $\dfrac{3n}{2p} - \dfrac{n+2}{2p}$ $\dfrac{n-1}{p}$

Additional Example 4

Find the perimeter of a triangle whose sides have lengths $\dfrac{6x-4}{5}$, $\dfrac{x+10}{5}$, and $\dfrac{2x+5}{5}$.
$\dfrac{9x+11}{5}$

17. $\dfrac{5b}{21} + \dfrac{2b}{21} + \dfrac{8b}{21}$ $\dfrac{5b}{7}$ **18.** $\dfrac{2}{a} + \dfrac{3}{a} - \dfrac{4}{a}$ $\dfrac{1}{a}$ **19.** $\dfrac{5y}{2y-3} - \dfrac{3y}{2y-3}$ $\dfrac{2y}{2y-3}$

20. $\dfrac{6x}{x-4} + \dfrac{2x}{4-x}$ $\dfrac{4x}{x-4}$ **21.** $\dfrac{3x}{4x-20} - \dfrac{15}{4x-20}$ $\dfrac{3}{4}$ **22.** $\dfrac{2a^2}{a-1} + \dfrac{2}{1-a}$ $2a+2$

23. $\dfrac{x^2}{4x-12} + \dfrac{9}{12-4x}$ **24.** $\dfrac{x-y}{x+y} + \dfrac{2x+y}{x+y}$ $\dfrac{3x}{x+y}$ **25.** $\dfrac{3a}{a^2+6a} + \dfrac{18}{a^2+6a}$ $\dfrac{3}{a}$

26. $\dfrac{a}{a^2-9a+18} - \dfrac{6}{a^2-9a+18}$ $\dfrac{1}{a-3}$ **27.** $\dfrac{2a}{a^2-8a+12} - \dfrac{4}{a^2-8a+12}$ $\dfrac{2}{a-6}$

28. $\dfrac{3x}{x^2-25} - \dfrac{x}{x^2-25} + \dfrac{x+15}{x^2-25}$ $\dfrac{3}{x-5}$ **29.** $\dfrac{x^2}{2x-6} + \dfrac{5x}{2x-6} - \dfrac{24}{2x-6}$ $\dfrac{x+8}{2}$

30. $\dfrac{2x}{2x^2-3x-20} + \dfrac{5}{2x^2-3x-20}$ $\dfrac{1}{x-4}$ **31.** $\dfrac{3x}{3x^2+10x-8} - \dfrac{2}{3x^2+10x-8}$ $\dfrac{1}{x+4}$

32. $\dfrac{x^2}{x^2-7x+10} - \dfrac{4x+5}{x^2-7x+10}$ $\dfrac{x+1}{x-2}$ **33.** $\dfrac{2x^2}{x^2-6x-16} - \dfrac{6-x}{x^2-6x-16}$ $\dfrac{2x-3}{x-8}$

34. $\dfrac{16y^2-7}{5y^2+26y-24} - \dfrac{9y^2-9}{-5y^2-26y+24}$ **35.** $\dfrac{2b^2}{3b-6c} + \dfrac{7bc}{-3b+6c} + \dfrac{6c^2}{3b-6c}$

$\dfrac{5y+4}{y+6}$ $\dfrac{2b-3c}{3}$

Find the perimeters of the figures shown.

36.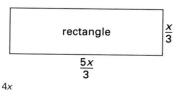

37.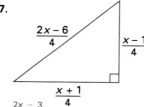

38.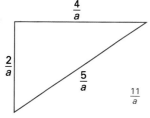

39. 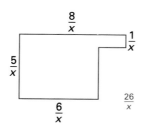 $\dfrac{26}{x}$

Add or subtract.

40. $\dfrac{7ax+4a-2b}{ay^2-by^2+3ay-3by+2a-2b} + \dfrac{-6ax-bx-2b}{ay^2-by^2+3ay-3by+2a-2b}$ $\dfrac{x+4}{y^2+3y+2}$

Guided Practice

Classroom Exercises 1–10
Try This   all

Independent Practice

A Ex. 1–25, **B** Ex. 26–39, **C** Ex. 40–42

Basic: WE 1–29 odd, Midchapter Review
Average: WE 9–37 odd, Midchapter Review
Above Average: WE 13–41 odd, Midchapter Review

Additional Answers

Written Exercises

9. $\dfrac{11}{x-1}$

10. $\dfrac{12}{2-y}$

12. $\dfrac{t-1}{t+1}$

23. $\dfrac{x+3}{4}$

Add or subtract.

41. $\dfrac{x^2}{x - 6 - 2y} - \dfrac{12x - 36}{x - 6 - 2y} - \dfrac{4y^2}{x - 6 - 2y}$ $x - 6 + 2y$

42. $\dfrac{a^2x}{ax^2 - bx^2 + cx^2} - \dfrac{2abx - b^2x}{ax^2 - bx^2 + cx^2} - \dfrac{c^2x}{ax^2 - bx^2 + cx^2}$ $\dfrac{a - b - c}{x}$

Midchapter Review

For what values of the variable is the rational expression undefined?   *8.1*

1. $\dfrac{a - 5}{a - 3}$ 3

2. $\dfrac{32}{3 - x}$ 3

3. $\dfrac{m + 3}{m - 2}$ 2

4. $\dfrac{t - 1}{t^2 - 2t - 15}$ $-3, 5$

5. $\dfrac{a + 1}{a^2 - 9}$ $-3, 3$

6. $\dfrac{4x - 8}{x^2 - 2x + 1}$ 1

Simplify each rational expression, if possible.   *8.1, 8.2*

7. $\dfrac{36a^6b^4}{-14ab^6}$ $\dfrac{18a^5}{-7b^2}$

8. $\dfrac{3p - 18}{2p - 12}$ $\dfrac{3}{2}$

9. $\dfrac{a^2 + 3a}{2a^2 + 7a + 3}$ $\dfrac{a}{2a + 1}$

10. $\dfrac{y - 3}{9 - y^2}$ $\dfrac{-1}{y + 3}$

11. $\dfrac{9 - a}{a^2 - 10a + 9}$ $\dfrac{-1}{a - 1}$

12. $\dfrac{x^2 + 7x + 10}{20 - 5x^2}$ $\dfrac{x + 5}{-5x + 10}$

Multiply.   *8.3*

13. $\dfrac{-15a^6}{9b^3} \cdot \dfrac{24b^6}{20a^7}$ $-\dfrac{2b^3}{a}$

14. $\dfrac{9y^2}{4 - 2y} \cdot \dfrac{5y - 10}{21y}$ $-\dfrac{15y}{14}$

Divide.   *8.4*

15. $\dfrac{-(x - 2)}{y} \div \dfrac{3x - 6}{y^2}$ $-\dfrac{y}{3}$

16. $\dfrac{12 - 4m}{18x^5y^3} \div \dfrac{m^2 - 9}{20x^3y}$ $\dfrac{-40}{9x^2y^2m + 27x^2y^2}$

Add or subtract.   *8.5*

17. $\dfrac{7a}{4} - \dfrac{3a}{4} + \dfrac{5a + 6}{4}$ $\dfrac{9a + 6}{4}$

18. $\dfrac{a^2}{3a + 12} - \dfrac{16}{3a + 12}$ $\dfrac{a - 4}{3}$

Enrichment

In preparation for the next lesson in this chapter, show the following pattern to the students.

$\dfrac{1}{3} + \dfrac{2}{5} = \dfrac{1}{3} \cdot \dfrac{5}{5} + \dfrac{2}{5} \cdot \dfrac{3}{3} = \dfrac{1 \cdot 5}{3 \cdot 5} + \dfrac{2 \cdot 3}{5 \cdot 3} =$
$\dfrac{1 \cdot 5 + 2 \cdot 3}{3 \cdot 5} = \dfrac{11}{15}$

Ask them to use this as a model for proving that $\dfrac{a}{b} + \dfrac{c}{d} = \dfrac{ad + bc}{bd}$.

$\dfrac{a}{b} + \dfrac{c}{d} = \dfrac{a}{b} \cdot \dfrac{d}{d} + \dfrac{c}{d} \cdot \dfrac{b}{b}$
$= \dfrac{ad}{bd} + \dfrac{bc}{bd}$
$= \dfrac{ad + bc}{bd}$

8.6 Adding and Subtracting: Unlike Denominators

Teaching Resources

Problem Solving Worksheet 8
Quick Quizzes 65
Reteaching and Practice
   Worksheets 65

Objective To add and subtract rational expressions with unlike monomial denominators

The fractions $\frac{10}{12}$ and $\frac{5}{6}$ are *equivalent*, that is, they have the same value.

$$\frac{10}{12} = \frac{5 \cdot 2}{6 \cdot 2} = \frac{5}{6}$$

The above steps can be reversed.

$$\frac{5}{6} = \frac{5 \cdot 2}{6 \cdot 2} = \frac{10}{12}$$

Notice that multiplying the numerator and denominator of $\frac{5}{6}$ by 2 does not change the value of the fraction $\frac{5}{6}$. This suggests the following alternate version of the *Cancellation Property*.

Multiplicative Identity Property for Rational Expressions
For all real numbers a, b, and c, if $b \neq 0$ and $c \neq 0$, then $\frac{a}{b} = \frac{a \cdot c}{b \cdot c}$.

The expression $\frac{5}{6} + \frac{1}{4}$ has unlike denominators. To write $\frac{5}{6} + \frac{1}{4}$ as a single fraction, begin by expressing it as a sum of two equivalent fractions with like denominators. Use the Multiplicative Identity Property.

$$\frac{5}{6} + \frac{1}{4} = \frac{5 \cdot 2}{6 \cdot 2} + \frac{1 \cdot 3}{4 \cdot 3} = \frac{10}{12} + \frac{3}{12} = \frac{10 + 3}{12} = \frac{13}{12}$$

In this example, the denominator 12 is the *least common multiple* of 6 and 4, which is also called the *least common denominator* (LCD) of the fractions.

The procedure for adding rational expressions with unlike denominators is the same as that just shown for fractions. Always begin by finding the LCD.

EXAMPLE 1 Add $\frac{3a}{20} + \frac{a}{6}$.

Plan Factor the denominators into prime factors. Find the number of times each prime factor must occur.

Additional Example 1
Add $\frac{4a}{75} + \frac{3a}{50}$. $\frac{17a}{150}$

GETTING STARTED

Prerequisite Quiz

Find the prime factorization of each number. Write the answer using exponents.

1. 24 $2^3 \cdot 3$
2. 45 $3^2 \cdot 5$
3. 64 2^6

Find the least common multiple of each pair of numbers.

4. 4; 8 8
5. 6; 10 30
6. 20; 24 120

Motivator

Model: Have students draw a large circle on their paper to represent a pizza. Then ask them how they would cut the pizza so that Bob would get $\frac{1}{3}$ of it, Cindy $\frac{1}{4}$ of it, and Sarah what was left over. Who got the most pizza?   Divide the pizza into 12 equal pieces so that Bob gets 4 pieces, Cindy 3 pieces, and Sarah 5 pieces (the most). Point out that this method demonstrates the usefulness of a common demoninator.

Highlighting the Standards

Standards 3a, 2c: The Focus on Reading asks students to make and test generalizations about the LCD. In the Exercises, they write their own explanations of mathematical procedures.

Lesson Note

In presenting Example 1, remind students that the LCM for 6 and 20 is the smallest number that is a multiple of both 6 and 20 and therefore, the smallest number that is divisible by both 6 and 20 without a remainder. The following alternate method for presenting or reteaching Example 2 may work well with some students.

$$\frac{1}{2x^3} - \frac{5}{6x} + \frac{2}{3x^2}$$

Identify the LCD: LCD $= 6x^3$

Write: $\dfrac{?}{6x^3} - \dfrac{?}{6x^3} + \dfrac{?}{6x^3}$

Then ask what $2x^3$ is multiplied by to get $6x^3$? 3 Multiply the numerator, 1, by the same factor, 3, and fill in the new numerator on the first fraction. Continue in like manner with the other two fractions. Then proceed as in the last line of Example 2.

Math Connections

Geometry: Although rays have no measure, two rays which have a *common* endpoint form an angle which can be measured.

Solution $20 = 2 \cdot 2 \cdot 5 \qquad 6 = 2 \cdot 3$

The LCD will have each of 2, 3, and 5 as its prime factors at least once. Since one denominator, 20, has 2 as a factor *twice*, the LCD must also have 2 as a factor twice. The LCD is $2 \cdot 2 \cdot 3 \cdot 5$, or 60.

$$\frac{3a}{20} + \frac{a}{6} = \frac{3a}{2 \cdot 2 \cdot 5} + \frac{a}{2 \cdot 3}$$

$$\underset{\text{needs }3}{} \qquad \underset{\text{needs 2 and 5}}{}$$

$$= \frac{3a \cdot 3}{2 \cdot 2 \cdot 5 \cdot 3} + \frac{a \cdot 2 \cdot 5}{2 \cdot 3 \cdot 2 \cdot 5}$$

$$= \frac{9a}{60} + \frac{10a}{60}$$

$$= \frac{9a + 10a}{60} = \frac{19a}{60}$$

EXAMPLE 2 Subtract and add $\dfrac{1}{2x^3} - \dfrac{5}{6x} + \dfrac{2}{3x^2}$.

Solution First, find the prime factors of each denominator.

$$2x^3 = 2 \cdot x \cdot x \cdot x \qquad 6x = 2 \cdot 3 \cdot x \qquad 3x^2 = 3 \cdot x \cdot x$$

There are at most three x's, one 3, and one 2 in any denominator. The LCD is $2 \cdot 3 \cdot x \cdot x \cdot x$, or $6x^3$.

$$\frac{1}{2x^3} - \frac{5}{6x} + \frac{2}{3x^2} = \frac{1}{2 \cdot x \cdot x \cdot x} - \frac{5}{2 \cdot 3 \cdot x} + \frac{2}{3 \cdot x \cdot x}$$

$$\underset{\text{needs 3}}{} \quad \underset{\text{needs } x \cdot x}{} \quad \underset{\text{needs } 2 \cdot x}{}$$

$$= \frac{1 \cdot 3}{2 \cdot x \cdot x \cdot x \cdot 3} - \frac{5 \cdot x \cdot x}{2 \cdot 3 \cdot x \cdot x \cdot x} + \frac{2 \cdot 2 \cdot x}{3 \cdot x \cdot x \cdot 2 \cdot x}$$

$$= \frac{3}{6x^3} - \frac{5x^2}{6x^3} + \frac{4x}{6x^3} = \frac{3 - 5x^2 + 4x}{6x^3}, \text{ or } -\frac{5x^2 - 4x - 3}{6x^3}$$

This result cannot be simplified further.

TRY THIS 1. Add and subtract $\dfrac{3}{4y^4} + \dfrac{5}{2y^2} - \dfrac{1}{8y}$. $-\dfrac{y^3 - 20y^2 - 6}{8y^4}$

The next example illustrates the process for adding or subtracting rational expressions that contain binomials in the numerators.

312 Chapter 8 Rational Expressions

Additional Example 2

Simplify.

$$\frac{-4}{16x^4} + \frac{9}{24x^3} - \frac{5}{36x^5} \qquad \frac{27x^2 - 18x - 10}{72x^5}$$

EXAMPLE 3 Subtract $\dfrac{5a + 4}{9a} - \dfrac{3a - 1}{12a}$.

Solution $9a = 3 \cdot 3 \cdot a \qquad 12a = 2 \cdot 2 \cdot 3 \cdot a$
The LCD is $2 \cdot 2 \cdot 3 \cdot 3 \cdot a$, or $36a$.

$$\dfrac{5a + 4}{9a} - \dfrac{3a - 1}{12a}$$

$$= \dfrac{(5a + 4) \cdot 2 \cdot 2}{3 \cdot 3 \cdot a \cdot 2 \cdot 2} - \dfrac{(3a - 1) \cdot 3}{2 \cdot 2 \cdot 3 \cdot a \cdot 3}$$

$$= \dfrac{20a + 16}{36a} - \dfrac{9a - 3}{36a}$$

$$= \dfrac{20a + 16 - (9a - 3)}{36a}$$

$$= \dfrac{20a + 16 - 9a + 3}{36a} = \dfrac{11a + 19}{36a}$$

The result cannot be simplified further.

TRY THIS   2. Subtract $\dfrac{2y - 1}{14y} - \dfrac{-3y - 2}{8y^2}$. $\dfrac{8y^2 + 17y + 14}{56y^2}$

 Focus on Reading

Suppose that two or more rational expressions are being added. Determine whether each statement is *always true*, *sometimes true*, or *never true*. Justify your answer.

1. The LCD is the product of the denominators.
2. The LCD is the sum of the denominators.
3. The factors of each denominator are also factors of the LCD.
4. The factors of the LCD are also factors of each denominator.

Classroom Exercises

Find the LCD of the fractions or rational expressions.

1. $\dfrac{5}{7} + \dfrac{3}{2 \cdot 5}$ 70

2. $\dfrac{3}{2 \cdot 3 \cdot 5} - \dfrac{5}{2 \cdot 3 \cdot 3}$ 90

3. $\dfrac{1}{3 \cdot 5 \cdot 5} + \dfrac{7}{2 \cdot 3 \cdot 5}$ 150

4. $\dfrac{8m}{3 \cdot 5} + \dfrac{13m}{2 \cdot 3 \cdot 3 \cdot 5}$ 90

5. $\dfrac{3x - 4}{2x^2} - \dfrac{4x - 1}{3x^3}$ $6x^3$

6. $\dfrac{3q - 1}{2 \cdot 3 \cdot 3 \cdot 7} - \dfrac{5q + 2}{2 \cdot 2 \cdot 5}$ 1,260

7. $\dfrac{3x + 2}{5} + \dfrac{2x - 1}{15}$ 15

8. $\dfrac{2x - 1}{3a} + \dfrac{x}{6a} + \dfrac{x - 4}{2a}$ $6a$

9. $\dfrac{3}{2a^2} - \dfrac{4}{a^3} + \dfrac{5}{7a}$ $14a^3$

Analysis: The GCF of 12 and 30 has been struck out in the factorization of 12.
$$12 = \cancel{2} \cdot 2 \cdot \cancel{3}$$
$$30 = 2 \cdot 3 \cdot 5$$
Then ask them to multiply all the factors that remain: $2 \cdot 2 \cdot 3 \cdot 5 = 60$. Ask the students how this number is related to 12 and 30.   LCM Have them try this with other pairs of numbers and ask whether this method always gives the LCM.   Yes

Common Error Analysis

Error: After identifying the common denominator, some students may fail to multiply one or more of the numerators by the appropriate factors.

Tell students that they should check mentally to be sure each fraction can be simplified to its original fraction.

Checkpoint

Simplify.

1. $\dfrac{9a}{16} - \dfrac{5a}{20}$ $\dfrac{5a}{16}$

2. $\dfrac{7t}{45} + \dfrac{6t}{60}$ $\dfrac{23t}{90}$

3. $\dfrac{-9m}{36m^4n^3} + \dfrac{4m}{24mn^7}$ $\dfrac{-3n^4 + 2m^3}{12m^3n^7}$

4. $\dfrac{3}{14xy} - \dfrac{xy - 1}{20x^2y^2}$ $\dfrac{23xy + 7}{140x^2y^2}$

Additional Example 3
Subtract $\dfrac{6x + 5}{24x^2} - \dfrac{x + 4}{18x^2}$. $\dfrac{14x - 1}{72x^2}$

Closure

Have students complete each expression that follows the " = " symbol.

1. $\frac{3}{3x} + \frac{2}{4x^2} = \frac{3 \cdot 4x + 2 \cdot 3}{?}$ $12x^2$

2. $\frac{3}{b} - \frac{11}{3b} = \frac{3 \cdot 3 - ?}{3b}$ 11

3. $\frac{a}{b} - \frac{c}{d} = \frac{? - bc}{bd}$ ad

4. $\frac{x + y}{y} - \frac{x + y}{x} = \frac{?}{xy}$ $x^2 - y^2$

◼FOLLOW UP

Guided Practice

Classroom Exercises 1–18
Try This   all, FR   all

Independent Practice

A Ex. 1–22, **B** Ex. 23–38, **C** Ex. 39–44

Basic: WE 1–22 all
Average: WE 1–10 all, 11–37 odd
Above Average: WE 1–43 odd

Add or subtract.

10. $\frac{3a}{15} + \frac{a}{3}$ $\frac{8a}{15}$

11. $\frac{3k}{10} - \frac{2k}{5} + \frac{k}{2}$ $\frac{2k}{5}$

12. $\frac{x}{12} - \frac{3x}{2} + \frac{x}{18}$ $-\frac{49x}{36}$

13. $\frac{2}{21x} + \frac{5}{7x} - \frac{1}{3x}$ $\frac{10}{21x}$

14. $\frac{8}{9a} + \frac{7}{15a^2} + \frac{1}{a^3}$

15. $\frac{2x + 1}{3} - \frac{x - 1}{2}$ $\frac{x + 5}{6}$

16. $\frac{1}{6a} - \frac{1}{4a} + \frac{1}{3a}$ $\frac{1}{4a}$

17. $\frac{3}{b} + \frac{5}{2b} - \frac{11}{6b}$ $\frac{11}{3b}$

18. $\frac{3x - 5}{4} + \frac{5x - 3}{3}$ $\frac{29x - 27}{12}$

Written Exercises

Add and subtract as indicated.

1. $\frac{5x}{14} + \frac{x}{7}$ $\frac{x}{2}$

2. $\frac{a}{2} - \frac{3a}{10}$ $\frac{a}{5}$

3. $\frac{2b}{9} + \frac{5b}{21}$ $\frac{29b}{63}$

4. $\frac{3x}{5} - \frac{4x}{15}$ $\frac{x}{3}$

5. $\frac{5a}{8} + \frac{a}{6}$ $\frac{19a}{24}$

6. $\frac{5b}{4} - \frac{3b}{10}$ $\frac{19b}{20}$

7. $\frac{2k}{3} + \frac{5k}{6}$ $\frac{3k}{2}$

8. $\frac{m}{5} + \frac{3m}{10} + \frac{m}{2}$ m

9. $\frac{x}{12} + \frac{x}{3} + \frac{5}{6}$ $\frac{5x + 10}{12}$

10. $\frac{5x}{12} + \frac{x}{4} - \frac{x}{6}$ $\frac{x}{2}$

11. $\frac{11}{14a} - \frac{3}{7a} + \frac{1}{2a}$ $\frac{6}{7a}$

12. $\frac{1}{3x^2} - \frac{2}{15x^3} + \frac{2}{5x}$

13. $\frac{7}{4m^2} + \frac{2}{8m} + \frac{1}{2m^3}$

14. $\frac{6y}{5} + \frac{3y}{4} + \frac{y}{10}$

15. $\frac{7}{4x^3} - \frac{5}{6x} + \frac{1}{3x^2}$

16. $\frac{3}{5x^3} + \frac{7}{10x} - \frac{5}{2x^2}$

17. $\frac{2y + 7}{4y} + \frac{2y - 1}{3y}$ $\frac{14y + 17}{12y}$

18. $\frac{2x - 5}{5x} - \frac{x + 2}{4x}$ $\frac{3x - 30}{20x}$

19. $\frac{a + 1}{3a} + \frac{2a + 3}{6a}$ $\frac{4a + 5}{6a}$

20. $\frac{3x - 1}{4} - \frac{5x + 1}{6}$ $-\frac{x + 5}{12}$

21. $\frac{b - 2}{9} + \frac{2b - 1}{4}$ $\frac{22b - 17}{36}$

22. $\frac{x + 2}{3x} - \frac{x + 4}{4x}$ $\frac{x - 4}{12x}$

23. $\frac{2x - 3}{6} + \frac{2}{3} - \frac{4x + 1}{2}$ $-\frac{5x + 1}{3}$

24. $\frac{4y - 3}{7} - \frac{2y + 1}{14} + \frac{3y - 4}{2}$ $\frac{27y - 35}{14}$

25. $\frac{2a - 1}{3a} + \frac{5a + 4}{2a} + \frac{a - 7}{2a}$ $\frac{22a - 11}{6a}$

26. $\frac{b + 3}{8b} + \frac{4b - 3}{4b} - \frac{b - 3}{2b}$ $\frac{5b + 9}{8b}$

27. $\frac{2k - 1}{15} - \frac{3k}{5} + \frac{3k - 1}{2}$ $\frac{31k - 17}{30}$

28. $\frac{7}{24b} + \frac{2b - 3}{6b} + \frac{b + 2}{4b}$ $\frac{14b + 7}{24b}$

29. $\frac{7y - 2}{6} - \frac{y}{5} + \frac{3y - 2}{10}$ $\frac{19y - 8}{15}$

30. $\frac{2z - 1}{9z} - \frac{3z + 4}{3z} + \frac{3z - 1}{4z}$

31. $\frac{5x - 1}{6} + \frac{2x - 1}{4} + \frac{4 - x}{3}$

32. $\frac{2a - 1}{5} - \frac{a + 3}{3} + \frac{a + 4}{10}$ $\frac{5a - 24}{30}$

33. Write in your own words how to find the LCD of two rational expressions.

34. Write in your own words how to add or subtract rational expressions with unlike denominators.

314   Chapter 8 Rational Expressions

Find the perimeters of the figures shown.

35.

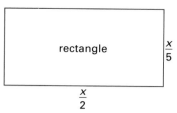

rectangle

$\frac{x}{5}$

$\frac{x}{2}$

36. $\frac{7x}{5}$

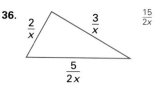

$\frac{2}{x}$ $\frac{3}{x}$ $\frac{15}{2x}$

$\frac{5}{2x}$

37. 2

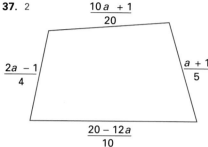

$\frac{10a+1}{20}$

$\frac{2a-1}{4}$

$\frac{a+1}{5}$

$\frac{20-12a}{10}$

38.

$\frac{x}{6}$ $\frac{x}{8}$ $\frac{11x+18}{4}$

$\frac{x+3}{4}$

$\frac{x+3}{2}$ $\frac{x}{3}$ $\frac{x+3}{2}$

Simplify.

39. $\dfrac{x-y}{3xy} + \dfrac{2x+y}{4x} + \dfrac{3x+2y}{6y}$

40. $\dfrac{a+3}{4a} - \dfrac{a+2}{3a^2} + \dfrac{a-4}{12a^2}$ $\dfrac{a^2+2a-4}{4a^2}$

41. $\dfrac{3a^2-4a+6}{4a} + \dfrac{3a-1}{5a} + \dfrac{a^2}{10a}$

42. $\dfrac{y-x}{7y} + \dfrac{x^2+3xy+y^2}{14xy} + \dfrac{3x-y}{4x}$

43. $\left(\dfrac{5a}{6} - \dfrac{1}{3}\right) \div \dfrac{4-25a^2}{12}$ $\dfrac{-2}{5a+2}$

44. $\left(\dfrac{2}{b} - \dfrac{1}{a}\right) \cdot \dfrac{bax+bay}{2ax+2ay-bx-by}$ 1

Mixed Review

Multiply. *6.8, 6.9, 8.3*

1. $(3x-1)(2x+5)$ $6x^2+13x-5$

2. $(4y+3)(2y-1)$ $8y^2+2y-3$

3. $(a-5)(a+5)$ a^2-25

4. $(3m+4)(3m-4)$

5. $(2x-t)(3x+2t)$

6. $(3m+4p)(2m-p)$

7. $\dfrac{4a^7}{9b^3} \cdot \dfrac{-2a}{5b^2}$ $\dfrac{-8a^8}{45b^5}$

8. $\dfrac{(-3a^2)^4}{5b^2} \cdot \dfrac{a^3}{-2b^4}$ $\dfrac{-81a^{11}}{10b^6}$

9. $\dfrac{9x^3y}{7xy^3} \cdot \dfrac{3x^4y}{5x^2y^7}$ $\dfrac{27x^4}{35y^8}$

Simplify. *2.10*

10. $5a - [3 - 2a(5-2a)]$ $-4a^2+15a-3$

11. $15x - 3[x - (10-2x)]$ $6x+30$

Prerequisite Quiz

Find the LCM for each pair of numbers.

1. $24a^3b^4$ and $36a^4b$ $72a^4b^4$
2. $2x^2 - 5x - 3$ and $x - 3$ $2x^2 - 5x - 3$
3. $x^2 - 3x$ and $x^2 - 9$ $x^3 - 9x$

Write each of the following in convenient form.

4. $2 - x$ $-1(x - 2)$
5. $25 - x^2$ $-1(x^2 - 25)$

Motivator

Ask students to give examples where they need to name the sum of a whole number and a fraction as a fraction.   Answers will vary. Sample answer: $3 + \frac{1}{4} = \frac{13}{4}$ to show the number of quarters in $3.25.

8.7 Adding and Subtracting: Polynomial Denominators

Objectives | To add and subtract rational expressions with unlike polynomial denominators
To add and subtract a polynomial and a rational expression

Sums and differences of rational expressions may contain denominators that are polynomials.

EXAMPLE 1 | Add $\dfrac{6}{a - 2} + \dfrac{5}{a - 4}$.

Plan | Find the LCD. Then replace the two rational expressions with equivalent expressions that have the LCD as the denominator.

Solution | The denominators $a - 2$ and $a - 4$ are irreducible. That is, their factored forms are $1 \cdot (a - 2)$ and $1 \cdot (a - 4)$. The LCD is $(a - 2)(a - 4)$.

$$\underset{\text{needs } a - 4}{\dfrac{6}{a - 2}} + \underset{\text{needs } a - 2}{\dfrac{5}{a - 4}}$$

$$= \dfrac{6 \cdot (a - 4)}{(a - 2) \cdot (a - 4)} + \dfrac{5 \cdot (a - 2)}{(a - 4) \cdot (a - 2)}$$

$$= \dfrac{6a - 24}{(a - 2)(a - 4)} + \dfrac{5a - 10}{(a - 4)(a - 2)}$$

$$= \dfrac{11a - 34}{(a - 2)(a - 4)} \quad \longleftarrow \text{THINK: Can the numerator be factored?}$$

Since the numerator and denominator have no common factors, the answer, in simplified form, is written as either $\dfrac{11a - 34}{(a - 2)(a - 4)}$ or $\dfrac{11a - 34}{a^2 - 6a + 8}$.

EXAMPLE 2 | Subtract $(a + 2) - \dfrac{a - 1}{2a - 3}$.

Plan | Write $a + 2$ as $\dfrac{a + 2}{1}$. Then find the LCD.

Solution | The binomial $2a - 3$ is irreducible. The LCD is $1 \cdot (2a - 3)$, or $2a - 3$.

Highlighting the Standards

Standards 5a, 5b: In the Exercises, students represent situations with expressions and operate on more complex rational expressions.

Additional Example 1

Add $\dfrac{3}{y + 4} + \dfrac{5}{y + 3}$. $\dfrac{8y + 29}{y^2 + 7y + 12}$

Additional Example 2

Subtract $(x + 3) - \dfrac{x + 4}{3x - 1}$. $\dfrac{3x^2 + 7x - 7}{3x - 1}$

$$(a + 2) - \frac{a - 1}{2a - 3}$$

$$= \frac{(a + 2)}{1} - \frac{a - 1}{2a - 3}$$

needs $2a - 3$

$$= \frac{(a + 2) \cdot (2a - 3)}{1 \cdot (2a - 3)} - \frac{a - 1}{2a - 3}$$

$$= \frac{2a^2 + a - 6}{2a - 3} - \frac{a - 1}{2a - 3}$$

$$= \frac{(2a^2 + a - 6) - (a - 1)}{2a - 3}$$

$$= \frac{2a^2 + a - 6 - a + 1}{2a - 3}$$

$$= \frac{2a^2 - 5}{2a - 3} \quad \longleftarrow \quad 2a^2 - 5 \text{ is irreducible and } 2a - 3$$
$$\text{is irreducible.}$$

The numerator and denominator are both irreducible, and they have no common factor. The answer, in simplified form, is $\frac{2a^2 - 5}{2a - 3}$.

TRY THIS

1. Subtract $\frac{x - 2}{x - 9} - \frac{x + 4}{x + 5}$. $\frac{8x + 26}{x^2 - 4x - 45}$

In the next two examples, it is convenient to change the form of one of the denominators in order to find the LCD.

EXAMPLE 3   Add $\frac{4x}{5x - 25} + \frac{2}{5 - x}$.

Plan   Use $\frac{a}{-b} = \frac{-a}{b}$ in order to obtain $\frac{2}{5 - x}$ in a more convenient form.

Solution
$$\frac{4x}{5x - 25} + \frac{2}{5 - x}$$

$$= \frac{4x}{5(x - 5)} + \frac{2}{-1(x - 5)} \quad \longleftarrow \quad 5 - x = -1(-5 + x) = -1(x - 5)$$

$$= \frac{4x}{5(x - 5)} + \frac{-2}{1(x - 5)} \quad \longleftarrow \quad \frac{a}{-b} = \frac{-a}{b}$$
needs 5

$$= \frac{4x}{5(x - 5)} + \frac{-2 \cdot 5}{(x - 5) \cdot 5} \quad \longleftarrow \quad 4x + (-2 \cdot 5) = 4x + (-10)$$

$$= \frac{4x - 10}{5(x - 5)}, \text{ or } \frac{2(2x - 5)}{5(x - 5)}, \text{ or } \frac{4x - 10}{5x - 25}$$

8.7 Adding and Subtracting: Polynomial Denominators   **317**

Lesson Note

Suggest to students that taking the time and the space on their paper to write the LCD on a line by itself will help them keep it in mind as they rewrite each rational expression. In Example 2, use a paralled procedure with an expression such as $8 - \frac{23}{20}$ to show how a fraction subtracted from a whole number can be expressed as a fraction.

$$8 - \frac{23}{20} = \frac{8 \cdot 20}{20} - \frac{23}{20} = \frac{160 - 23}{20} = \frac{137}{20}$$

Math Connections

Arithmetic: Rational expressions are similar to fractions in arithmetic. The procedures for adding, subtracting, multiplying, and dividing rational expressions are similar to the arithmetic procedures for carrying out these operations with fractions.

Critical Thinking Questions

Have students study this pattern.
$$1 \times \frac{1}{2} = 1 - \frac{1}{2}$$
$$2 \times \frac{2}{3} = 2 - \frac{2}{3}$$
$$3 \times \frac{3}{4} = 3 - \frac{3}{4}$$
Ask them to write the pattern in general form and to prove that their answer is correct.

Pattern: $n \times \frac{n}{n + 1} = n - \frac{n}{n + 1}$

$$n \times \frac{n}{n + 1} = \frac{n^2}{n + 1}; n - \frac{n}{n + 1} =$$
$$\frac{n(n + 1) - n}{n + 1} = \frac{n^2 + n - n}{n + 1} = \frac{n^2}{n + 1}$$

Ask students how many rational expressions that are equivalent to 5 can be found.   An infinite number

Additional Example 3
Add $\frac{3y + 5}{2y^2 - 5y - 3} + \frac{2}{3 - y}$. $\frac{-1}{2y + 1}$

Common Error Analysis

Error: Some students may rewrite $\frac{x-3}{3-y}$ as $\frac{x-3}{y-3}$ forgetting about the factor of -1 that has been introduced.

Remind them that $3 - y \neq y - 3$. Rather, $3 - y = -(y - 3)$.

Checkpoint

Add or subtract.

1. $\dfrac{7}{60a^3b} + \dfrac{11}{90a^5b^{10}}$ $\dfrac{21a^2b^9 + 22}{180a^5b^{10}}$

2. $\dfrac{-3}{5-a} - \dfrac{13}{a^2 - 25}$ $\dfrac{3a + 2}{a^2 - 25}$

3. $\dfrac{-9}{2a^2 + 5a - 3} + \dfrac{5}{4a - 2}$ $\dfrac{5a - 3}{4a^2 + 10a - 6}$

4. $\dfrac{x - 3}{x^2 - 6x + 9} - \dfrac{5}{x^2 - 3x}$ $\dfrac{x - 5}{x^2 - 3x}$

Closure

Ask students to identify the two steps in adding and subtracting rational expressions that depend on factoring.   Finding the LCD and simplifying the answer

◼◼◼FOLLOW UP

Guided Practice

Classroom Exercises 1–12
Try This   all

Independent Practice

A Ex. 1–25, **B** Ex. 26–37, **C** Ex. 38–41

Basic: WE 1–27 odd

Average: WE 11–37 odd

Above Average: WE 15–41 odd

EXAMPLE 4   Add and subtract $\dfrac{-2y - 10}{y^2 - 11y + 28} + \dfrac{1}{y - 4} - \dfrac{y + 1}{7 - y}$.

Solution

$\dfrac{-2y - 10}{y^2 - 11y + 28} + \dfrac{1}{y - 4} - \dfrac{y + 1}{7 - y}$

$= \dfrac{-2y - 10}{(y - 4)(y - 7)} + \dfrac{1}{y - 4} - \dfrac{y + 1}{-1(y - 7)}$

$= \dfrac{-2y - 10}{(y - 4)(y - 7)} + \dfrac{1}{y - 4} - \dfrac{-(y + 1)}{1(y - 7)}$ ⟵ $\dfrac{a}{-b} = \dfrac{-a}{b}$

$= \dfrac{-2y - 10}{(y - 4)(y - 7)} + \dfrac{1 \cdot (y - 7)}{(y - 4) \cdot (y - 7)} - \dfrac{-(y + 1) \cdot (y - 4)}{(y - 7) \cdot (y - 4)}$

$= \dfrac{(-2y - 10) + (y - 7) - [-(y^2 - 3y - 4)]}{(y - 4)(y - 7)}$

$= \dfrac{-2y - 10 + y - 7 + (y^2 - 3y - 4)}{(y - 4)(y - 7)}$ ⟵ $-(-y^2) = y^2$

$= \dfrac{y^2 - 4y - 21}{(y - 4)(y - 7)}$ ⟵ The numerator is factorable.

$= \dfrac{(y + 3)\overset{1}{\cancel{(y - 7)}}}{(y - 4)\underset{1}{\cancel{(y - 7)}}}$

$= \dfrac{y + 3}{y - 4}$

TRY THIS   2. Subtract and add $\dfrac{9 - 12y}{y^2 + 3y - 18} - \dfrac{y + 2}{y + 6} + \dfrac{-4y}{3 - y}$. $\dfrac{3y^2 + 13y + 15}{y^2 + 3y - 18}$

Classroom Exercises

Give the LCD for each sum or difference.

1. $\dfrac{3}{x - 2} + \dfrac{-2}{x + 1}$ $x^2 - x - 2$

2. $\dfrac{5}{2a + 1} - \dfrac{6}{a + 3}$ $2a^2 + 7a + 3$

3. $\dfrac{3}{a - 1} + \dfrac{2}{a + 5}$ $a^2 + 4a - 5$

4. $2 - \dfrac{4}{x}$ x

5. $15 - \dfrac{2}{m}$ m

6. $\dfrac{4}{x} + 1$ x

7. $(x + 3) + \dfrac{7}{x - 2}$ $x - 2$

8. $(a + 2) - \dfrac{2a}{a + 1}$ $a + 1$

9. $\dfrac{5}{a - 9} + \dfrac{3}{a(a - 9)}$ $a^2 - 9a$

Add or subtract.

10. $\dfrac{3}{x - 2} + \dfrac{2}{x}$ $\dfrac{5x - 4}{x^2 - 2x}$

11. $5 - \dfrac{1}{a + 5}$ $\dfrac{5a + 24}{a + 5}$

12. $\dfrac{2x}{x^2 - 25} - \dfrac{10}{5 - x}$ $\dfrac{12x + 50}{x^2 - 25}$

Additional Example 4

Subtract.

$\dfrac{a - 4}{a + 4} - \dfrac{3}{3 - a} - \dfrac{7a}{a^2 + a - 12}$ $\dfrac{a - 8}{a + 4}$

Written Exercises

1–9. Find the sum or difference for Classroom Exercises 1–9.

Add and subtract as indicated.

10. $\dfrac{x + 5}{x + 2} + \dfrac{4}{x}$

11. $\dfrac{5}{x} - \dfrac{2x - 1}{x + 1}$

12. $3 - \dfrac{4b - 1}{2b}$ $\dfrac{2b + 1}{2b}$

13. $\dfrac{x}{x - 1} + 2$ $\dfrac{3x - 2}{x - 1}$

14. $(a + 3) + \dfrac{a - 4}{3a - 2}$

15. $\dfrac{x + 4}{x - 3} + (x - 2)$

16. $\dfrac{2}{b - 4} - \dfrac{8}{b^2 - 4b}$ $\dfrac{2}{b}$

17. $\dfrac{y}{y - 3} - \dfrac{9}{y^2 + 3y}$

18. $\dfrac{3x - 4}{x^2 - 16} + \dfrac{2}{4 - x}$

19. $\dfrac{9x + 14}{x^2 + 7x} + \dfrac{x}{x + 7}$

20. $\dfrac{a}{a - 6} - \dfrac{a + 30}{a^2 - 6a}$

21. $\dfrac{4}{x - 6} - \dfrac{3}{x^2 - 2x - 24}$

22. $\dfrac{3}{a - 2} - \dfrac{12}{a^2 - 4} + \dfrac{2}{a + 2}$ $\dfrac{5}{a + 2}$

23. $\dfrac{44}{a^2 - 7a - 18} + \dfrac{3}{a - 9} + \dfrac{4}{a + 2}$

24. $\dfrac{2}{z - 5} + \dfrac{4}{z^2 + 2z - 35} + \dfrac{1}{z + 7}$

25. $\dfrac{3}{a - 4} + \dfrac{2}{2 - a} + \dfrac{2}{a^2 - 6a + 8}$

26. $\dfrac{x - 1}{x - 4} - \dfrac{2x - 3}{x^2 + x - 20}$ $\dfrac{x^2 + 2x - 2}{x^2 + x - 20}$

27. $\dfrac{14 - a^2}{a^2 - 9a + 20} + \dfrac{a + 6}{a - 5}$ $\dfrac{2}{a - 4}$

28. $\dfrac{-24}{a^2 - 7a + 10} - \dfrac{a + 3}{5 - a}$ $\dfrac{a + 6}{a - 2}$

29. $\dfrac{3b + 1}{b^2 - 25} - \dfrac{2b + 1}{5 - b}$ $\dfrac{2b^2 + 14b + 6}{b^2 - 25}$

30. $\dfrac{3x - 1}{x - 4} + \dfrac{2x + 1}{x^2 - 16}$ $\dfrac{3x^2 + 13x - 3}{x^2 - 16}$

31. $\dfrac{x + 1}{x^2 - 5x + 6} + \dfrac{x + 4}{x - 3}$ $\dfrac{x^2 + 3x - 7}{x^2 - 5x + 6}$

32. $\dfrac{x^2 + 2}{x^2 - 5x + 4} - \dfrac{x - 2}{x - 1}$ $\dfrac{6}{x - 4}$

33. $\dfrac{a^2 - 22}{a^2 - 9a + 20} - \dfrac{a - 2}{a - 5}$ $\dfrac{6}{a - 4}$

34. $\dfrac{m^2 - 8}{m^2 - 8m + 12} - \dfrac{m + 1}{m - 6}$ $\dfrac{1}{m - 2}$

35. $\dfrac{3}{2y^2 - 5y - 12} - \dfrac{y + 1}{2y + 3} + \dfrac{y - 5}{y - 4}$

36. $(4x - 3) - \dfrac{x^2 + 4}{2x + 1}$ $\dfrac{7x^2 - 2x - 7}{2x + 1}$

37. $(3x + 4) - \dfrac{2x^2 - x - 1}{5x - 2}$ $\dfrac{13x^2 + 15x - 7}{5x - 2}$

38. $(x^2 - x + 1) - \dfrac{x^3 + 1}{x + 1}$ 0

39. $\dfrac{-3a - 9}{a^2 - 7a + 10} + \dfrac{a + 3}{a - 5} + \dfrac{3}{a - 2}$

40. $\dfrac{2}{12 + a - a^2} - \dfrac{5}{a^4 - 25a^2 + 144}$ $\dfrac{-2a^2 - 2a + 19}{a^4 - 25a^2 + 144}$

41. $x^2 + 16 + \dfrac{5}{x - 4} + \dfrac{4}{x + 4}$ $\dfrac{x^4 + 9x - 252}{x^2 - 16}$

Mixed Review

Solve. **5.6**

1. $|x| = 4$ $4, -4$

2. $|x - 2| = 8$ $-6, 10$

3. $|x + 3| = 5$ $-8, 2$

4. $|2x - 1| = 13$ $-6, 7$

5. Find two consecutive integers whose sum is 89. **4.3** 44, 45

6. One number is 8 more than twice another. Represent the numbers. **4.1** $x, 2x + 8$

Enrichment

Have the students find the sum $\dfrac{1}{a} + \dfrac{1}{b}$ and use the answer as a formula to do the following problems mentally. $\dfrac{a + b}{ab}$

1. $\dfrac{1}{3} + \dfrac{1}{5}$ $\dfrac{8}{15}$

2. $\dfrac{1}{8} + \dfrac{1}{11}$ $\dfrac{19}{88}$

3. $\dfrac{1}{12} + \dfrac{1}{13}$ $\dfrac{25}{156}$

4. $\dfrac{1}{x + 1} + \dfrac{1}{x - 1}$ $\dfrac{2x}{x^2 - 1}$

■■GETTING STARTED

Prerequisite Quiz

Divide.

1. $\frac{x^3}{x}$ x^2

2. $\frac{8a^3}{2a}$ $4a^2$

Multiply.

3. $x(x + 3)$ $x^2 + 3x$

4. $2x^2(3x - 5)$ $6x^3 - 10x^2$

Subtract.

5. $(x^2 + 8x) - (x^2 - 3x)$ $11x$

6. $(x^3 + 0x^2) - (7x^3 + 9x^2)$ $-6x^3 - 9x^2$

Motivator

Ask students what is meant by the statement that one number is divisible by another.   There is no remainder.   Ask them to find the quotient and remainder for 39 ÷ 7.   Quotient: 5; remainder: 4   Ask them to express this relationship in two ways. 39 = (5 × 7) + 4; 39 = 7 × 5$\frac{4}{7}$ Ask them to express these two relationships in general form. $\frac{dividend}{divisor}$ = quotient + $\frac{remainder}{divisor}$, dividend = (divisor)(quotient) + remainder

8.8 Dividing Polynomials

Objectives   To divide polynomials by monomials and binomials

You know that when one whole number is divided by another, there may be a nonzero remainder, as illustrated below for 53 ÷ 13.

$$53 \div 13 = \frac{53}{13} = 4 + \frac{1}{13} \quad \longleftarrow \quad \frac{dividend}{divisor} = quotient + \frac{remainder}{divisor}$$

Another way to express the division example above is shown below.

$$53 = 4 \cdot 13 + 1 \quad \longleftarrow \quad dividend = quotient \cdot divisor + remainder$$

If the remainder happens to be zero, then the quotient and divisor are *factors* of the dividend.

$$52 \div 13 = 4, \text{ or } 52 = 4 \cdot 13 \quad \longleftarrow \quad dividend = quotient \cdot divisor$$

You can also divide polynomials.

EXAMPLE 1   Divide $(16x^4 - 12x^3 + 8x^2) \div (4x^2)$.

Plan   Think of the quotient as $\dfrac{16x^4 - 12x^3 + 8x^2}{4x^2}$ and use the Rule for Adding and Subtracting Rational Expressions.

Solution

$$\frac{16x^4 - 12x^3 + 8x^2}{4x^2}$$

Divide each term of the numerator by $4x^2$. $= \dfrac{16x^4}{4x^2} - \dfrac{12x^3}{4x^2} + \dfrac{8x^2}{4x^2}$

Simplify each term. $= 4x^2 - 3x + 2$

Thus, the quotient is $4x^2 - 3x + 2$, with a zero remainder.

TRY THIS   1. Divide $(6y^5 + 9y^8 - 21y^6) \div (3y^3)$. $3y^5 - 7y^3 + 2y^2$

In Example 1, the divisor $4x^2$ was a *monomial*. To understand the division of a polynomial by a *binomial* such as $2x - 5$, it is helpful to think of the steps of a numerical example such as 807 ÷ 32 as shown at the top of page 321 where Step 2 is described.

320   Chapter 8 Rational Expressions

Highlighting the Standards

Standards 4a, 7b, 1b: Here students connect what they know about dividing integers with dividing polynomials to determine whether the division processes are equivalent. In the Mixed Review they apply properties of geometric figures and, in Problem Solving, they learn another strategy.

Additional Example 1

Divide.

$(21a^5 + 15a^3 - 24a^2) \div 3a^2$
$7a^3 + 5a - 8$

Step 2

Bring down 7, the next digit of the dividend.

$$\begin{array}{r} 25 \\ 32\overline{)807} \\ 64 \\ \hline 167 \\ 160 \\ \hline 7 \end{array}$$

Divide. $32\overline{)167}^{\,5}$

Multiply. $5 \cdot 32 = 160$

Subtract. $167 - 160 = 7$

nonzero remainder

Check: $25 \cdot 32 + 7 = 807$ True

quotient · divisor + remainder = dividend

So, the quotient is 25 and the remainder is 7.

EXAMPLE 2 Divide $(6x^2 - 9x - 12)$ by $(2x - 5)$.

Plan First rewrite as $2x - 5\overline{)6x^2 - 9x - 12}$. Then repeat the *divide-multiply-subtract* steps shown for the numerical example until a remainder is obtained.

Solution $2x - 5\overline{)6x^2 - 9x - 12}$

Step 1

Divide. $2x\overline{)6x^2}^{\,3x}$

Multiply. $3x(2x - 5) = 6x^2 - 15x$

$$2x - 5\overline{)6x^2 - 9x - 12}^{\,3x}$$
$$6x^2 - 15x$$
$$\overline{ 6x}$$

Subtract. $(6x^2 - 9x) - (6x^2 - 15x)$
$= 6x^2 - 9x - 6x^2 + 15x$
$= 6x$

Step 2

Bring down -12.

Divide. $2x\overline{)6x}^{\,3}$

Multiply. $3(2x - 5) = 6x - 15$

Subtract. $(6x - 12) - (6x - 15) = 3$

$$2x - 5\overline{)6x^2 - 9x - 12}^{\,3x + 3}$$
$$6x^2 - 15x \quad\downarrow$$
$$\overline{6x - 12}$$
$$6x - 15$$
$$\overline{3}$$

nonzero remainder

Check

$(3x + 3)(2x - 5) + 3 = (6x^2 - 9x - 15) + 3 = 6x^2 - 9x - 12$

quotient · divisor + remainder = dividend

Thus, the quotient is $3x + 3$ and the remainder is 3.

The answer may also be represented as $3x + 3 + \dfrac{3}{2x - 5}$.

8.8 Dividing Polynomials **321**

Lesson Note

Work through Example 2 very slowly and carefully with students. In dividing $6x^2$ by $2x$ have students think, "What multiplied by $2x$ is equal to $6x^2$?" After multiplying, some students may find it easier to actually change each sign of $6x^2 - 15x$ so that they can *add* $-6x^2 + 15x$ to the terms above. Remind students that a $+$ sign is required when writing the quotient obtained in Example 2 as $3x + 3 + \dfrac{3}{2x - 5}$. When presenting Example 3, emphasize the necessity of inserting the $0x^2$ term to allow proper alignment of the terms in the rest of the problem.

Math Connections

Computer: Some versions of BASIC use the operators " \ " and "MOD" to give integral quotients and remainders. $22 \setminus 6 = 3$ and 22 MOD $6 = 4$

Critical Thinking Questions

Application: Given that $x - 1$ is a factor of $x^3 + x^2 - 17x + 15$, ask students how they can find the other two factors. Divide and factor the quotient found. Then have them find these factors. $x - 3$ and $x + 5$

Additional Example 2

Divide.

$(x^2 + 7x + 6) \div (x + 3)$ $x + 4 - \dfrac{6}{x + 3}$

Error: Students make errors in the subtraction step of the division process.

Remind them that to subtract a binomial, they must change the sign of both terms of the binomial before adding.

Checkpoint

Divide.

1. $(12x^4 - 15x^3 + 3x) \div (3x)$
   $4x^3 - 5x^2 + 1$
2. $(12x^2 + 20x - 17) \div (3x - 1)$
   $4x + 8 - \frac{9}{3x - 1}$
3. $(18x^3 - 9x^2 + 8x - 7) \div (2x - 5)$
   $9x^2 + 18x + 49 + \frac{238}{2x - 5}$
4. $(6a^3 + 16a^2 + 10) \div (3a - 1)$
   $2a^2 + 6a + 2 + \frac{12}{3a - 1}$

Closure

Write this example on the chalkboard. Then ask the questions that follow.

$$
\begin{array}{r}
b + 7 \\
2b - 3 \overline{)2b^2 + 11b - 15} \\
\underline{2b^2 - 3b} \\
14b - 15 \\
\underline{14b - 21} \\
6
\end{array}
$$

1. How was the b in the quotient obtained?
   $2b^2 \div 2b = b$
2. How was the $14b - 15$ obtained?
   $11b - (-3b) = 14b$; -15 is brought down
3. How was the 7 in the quotient obtained?
   $14b \div 2b = 7$
4. How was the remainder 6 obtained?
   $14b - 15 - 14b + 21 = 6$

The division of Example 2 can be performed using a more compact form, as shown below.

$$
\begin{array}{r}
3x + 3 \\
2x - 5 \overline{)6x^2 - 9x - 12} \\
\underline{6x^2 - 15x} \\
6x - 12 \\
\underline{6x - 15} \\
3
\end{array}
$$

EXAMPLE 3 Divide $(x^3 - 5x + 2)$ by $(x - 2)$.

Plan Rewrite as $x - 2 \overline{)x^3 + 0x^2 - 5x + 2}$. Then proceed with the *divide-multiply-subtract* cycle.

Solution
$$
\begin{array}{r}
x^2 + 2x - 1 \\
x - 2 \overline{)x^3 + 0x^2 - 5x + 2} \\
\underline{x^3 - 2x^2} \\
2x^2 - 5x \\
\underline{2x^2 - 4x} \\
-x + 2 \\
\underline{-x + 2} \\
0 \quad \longleftarrow \text{ zero remainder}
\end{array}
$$

Check $(x^2 + 2x - 1)(x - 2) = (x^3 + 2x^2 - x) + (-2x^2 - 4x + 2)$
$= x^3 - 5x + 2$

quotient · divisor = dividend

TRY THIS 2. Divide $(8x^3 - 12x^2 - 4)$ by $(2x - 1)$. $4x^2 - 4x - 2 - \frac{6}{2x - 1}$

Classroom Exercises

Divide.

1. $2x^2 \overline{)-8x^3}$ $-4x$
2. $-3x^2 \overline{)-27x^4}$ $9x^2$
3. $5x^3 \overline{)15x^3}$ 3
4. $-6x \overline{)12x^3}$ $-2x^2$

Divide.

5. $(9x^3 - 15x^2 + 27x) \div (3x)$
   $3x^2 - 5x + 9$
6. $(a^3 - 6a^2 + a - 6) \div (a - 6)$
   $a^2 + 1$

Additional Example 3

Divide.

$(8x^3 - 12x - 9) \div (2x - 3)$ $4x^2 + 6x + 3$

Written Exercises

Divide and check.

1. $(30a^5 + 12a^3 - 16a^2) \div (2a)$ $15a^4 + 6a^2 - 8a$
2. $(24t^8 - 12t^6 + 18t^4) \div (6t^2)$ $4t^6 - 2t^4 + 3t^2$
3. $(24n^3 - 18n^2 + 36n) \div (6n)$ $4n^2 - 3n + 6$ **4.** $(18b^4 - 24b^3 - 6b^2) \div (-3b^2)$
5. $(3x^2 - 5x - 2) \div (x - 2)$ $3x + 1$
6. $(10x^2 - 39x - 27) \div (5x + 3)$ $2x - 9$
7. $(3y^2 + 5y - 5) \div (y + 2)$ $3y - 1 - \dfrac{3}{y + 2}$
8. $(8a^2 - 22a + 3) \div (4a - 1)$ $2a - 5 - \dfrac{2}{4a - 1}$

Divide.

9. $(6a^2 - 5a - 2) \div (2a - 3)$ $3a + 2 + \dfrac{4}{2a - 3}$
10. $(15y^2 + 13y - 6) \div (3y - 1)$ $5y + 6$
11. $(12x^2 - 11x - 8) \div (3x + 1)$
12. $(12a^2 - 9a - 30) \div (4a + 5)$ $3a - 6$
13. $(12t^2 - 15t - 14) \div (4t + 3)$
14. $(4t^2 - 12t - 7) \div (2t + 1)$ $2t - 7$
15. $(10y^2 + 2y - 12) \div (2y - 2)$ $5y + 6$
16. $(12x^2 - x - 23) \div (3x - 4)$
17. $(6x^3 - x^2 - 8x + 4) \div (3x - 2)$
18. $(12a^3 - a^2 + 3a - 1) \div (4a + 1)$
19. $(6t^3 - 5t^2 + 16t - 5) \div (3t - 1)$
20. $(6y^3 + 14y^2 - 10y + 9) \div (2y + 6)$
21. $(4a^2 + 4a + 10) \div (2a - 1)$
22. $(4x^3 - 44x + 24) \div (x - 6)$
23. $(3t^3 + 16t^2 + 11) \div (3t + 4)$
24. $(4b^3 - 59b - 48) \div (2b + 6)$
25. Factor $a^3 - 13a - 12$ given that $a + 3$ is one of its factors. $(a + 3)(a - 4)(a + 1)$
26. Factor $a^3 - 8a^2 + 19a - 12$ completely, given that $a - 1$ is one of its factors. $(a - 1)(a - 3)(a - 4)$
27. Factor $8a^3 - 27$ given that $2a - 3$ is one of its factors. $(4a^2 + 6a + 9)(2a - 3)$
28. Divide $(x^{3c} + 3x^{2c} - x^{2c+1} - 3x^{c+1} + 2) \div (x^c + 3)$. $x^{2c} - x^{c+1} + \dfrac{2}{x^c + 3}$
29. Find the value of k for which $2x + 1$ is a factor of $2x^3 - 7x^2 + kx + 3$. $k = 2$

Mixed Review

Multiply.   **6.9**

1. $(x + 3)(x + 5)$
2. $(2x - 1)(x + 7)$ $2x^2 + 13x - 7$
3. $(x - 5)(x + 5)$ $x^2 - 25$

Simplify.   **8.1, 8.2**

4. $\dfrac{x^2 - x - 6}{x^2 + x - 2}$ $\dfrac{x - 3}{x - 1}$
5. $\dfrac{2x + 8}{6x - 6}$ $\dfrac{x + 4}{3x - 3}$
6. $\dfrac{x^2 - 25}{x^2 + 4x - 5}$ $\dfrac{x - 5}{x - 1}$
7. $\dfrac{5 - 5x}{x^2 + 6x - 7}$ $\dfrac{-5}{x + 7}$
8. $\dfrac{36 - x^2}{x^2 + 4x - 12}$ $\dfrac{-x + 6}{x - 2}$
9. $\dfrac{3x^2 + 3x - 6}{6x^2 - 6}$ $\dfrac{x + 2}{2x + 2}$
10. A triangle has angles measuring 48 and 52. What is the measure of the third angle of the triangle?   **4.2**   80
11. The length of a rectangle is 5 yd more than 3 times the width. The perimeter is 90 yd. Find the length and the width of the rectangle.   **4.2** $l = 35$ yd; $w = 10$ yd

■ FOLLOW UP

Guided Practice

Classroom Exercises 1–6
Try This   all

Independent Practice

A Ex. 1–16, **B** Ex. 17–22, **C** Ex. 23–29

Basic: WE 1–17 odd, Problem Solving Strategies

Average: WE 1, 3, 9–21 odd, Problem Solving Strategies

Above Average: WE 1,3 17–29 odd, Problem Solving Strategies

Additional Answers

Written Exercises

4. $-6b^2 + 8b + 2$
11. $4x - 5 - \dfrac{3}{3x + 1}$
13. $3t - 6 + \dfrac{4}{4t + 3}$
16. $4x + 5 - \dfrac{3}{3x - 4}$
17. $2x^2 + x - 2$
18. $3a^2 - a + 1 - \dfrac{2}{4a + 1}$
19. $2t^2 - t + 5$
20. $3y^2 - 2y + 1 + \dfrac{3}{2y + 6}$
21. $2a + 3 + \dfrac{13}{2a - 1}$
22. $4x^2 + 24x + 100 + \dfrac{624}{x - 6}$
23. $t^2 + 4t - \dfrac{16}{3} + \dfrac{97}{9t + 12}$
24. $2b^2 - 6b - \dfrac{23}{2} + \dfrac{21}{2b + 6}$

Mixed Review

1. $x^2 + 8x + 15$

Enrichment

Find the missing digits.

```
        *7***
124 )********
     ****
      ***
      ***
     ****
      ***
     ****
     ****
       0
```

```
          97809
124)12128316
    1116
     968
     868
    1003
     992
    1116
    1116
       0
```

HINTS: The first digit in the dividend is 1. (Why?) The divisor, when multiplied by 7, produces only a three-digit number. The **second to the last digit in the quotient is 0.** (Why?) The last two figures of the dividend were brought down together.

Problem Solving Strategies

Testing Conditions

Problems have conditions that affect which outcomes are reasonable. Testing different numbers can help you discover the conditions.

Example   A box of My-Grain cereal lists the ingredients in decreasing order by weight: oats, wheat, brown sugar, sugar, coconut oil, malt flavoring, salt, baking soda. What is the smallest possible percent of grain (oats and wheat) that could be in the cereal?

Think: The total of the percents by weight of the ingredients must equal 100%. Find the minimum percent for the first 2 ingredients.

Try some possible percents for the 8 ingredients, and each time try to make the percents for the first 2 items smaller.

(1) 30%, 25%, 15%, 10%, 5%, 5%, 5%, 5%
(2) 20%, 20%, 15%, 15%, 10%, 10%, 5%, 5%
(3) 15%, 15%, 15%, 15%, 15%, 15%, 5%, 5%

These examples suggest that the more there is of other ingredients, the less there will be of grain. But there can't be more of any other ingredient than there is of either grain. Therefore, the minimum for the first 2 items—the grains—is reached when each of the 8 items is the same percent. Under these conditions, each item would be $\frac{100}{8}$%, or 12.5% of the total. So, the cereal must contain *at least* 12.5% oats and 12.5% wheat for a total of 25% grain.

Exercises

Solve each problem, if possible.

1. Jolly O cereal has the same ingredients list as MyGrain but the sugars account for 30% of the cereal. What is the minimum percent of grain in Jolly O?   30%

2. Kim had an 89% average for 3 tests. If there was no extra credit for any test, what is the lowest score she could have received for any of the 3 tests?   67%

3. Mark claims that he walked straight for 5 mi, turned, and walked straight again for 3 mi, and then turned and walked 1 mi straight back to his starting point. Could Mark have done this?   Impossible

8.9 Complex Rational Expressions

Objective

To simplify complex fractions and complex rational expressions

The indicated division, $5 \div 6$, can be written as a fraction.

$$5 \div 6 = \frac{5}{6}$$

The division of two fractions can be written as a complex fraction.

$$\frac{3}{5} \div \frac{2}{3} = \frac{\frac{3}{5}}{\frac{2}{3}}$$

Definition

A **complex fraction** is a fraction that has at least one fraction in its numerator, denominator, or in both its numerator and denominator.

A complex fraction can be simplified by multiplying its numerator and denominator by the common denominator of the two fractions in its numerator and denominator.

EXAMPLE 1   Simplify $\dfrac{\frac{3}{5}}{\frac{2}{3}}$.

Plan   Find the LCD of $\frac{3}{5}$ and $\frac{2}{3}$. Then use the rule $\frac{a}{b} = \frac{a \cdot c}{b \cdot c}$.

Solution   The LCD of $\frac{3}{5}$ and $\frac{2}{3}$ is $3 \cdot 5$. Multiply by $3 \cdot 5$.

$$\frac{\frac{3}{5}}{\frac{2}{3}} = \frac{\frac{3}{5} \cdot 3 \cdot 5}{\frac{2}{3} \cdot 3 \cdot 5} \quad \longleftarrow \frac{a}{b} = \frac{a \cdot c}{b \cdot c} \text{ (Divide out common factors.)}$$

$$= \frac{3 \cdot 3 \cdot 1}{2 \cdot 1 \cdot 5}, \text{or } \frac{9}{10}$$

TRY THIS   Simplify $\dfrac{\frac{4}{7}}{\frac{4}{5}}$. $\frac{5}{7}$

The rational expression $\dfrac{\frac{4}{5} - \frac{3}{y}}{\frac{5y}{6}}$ is a complex rational expression.

Teaching Resources

Quick Quizzes 68
Reteaching and Practice Worksheets 68

▬▬ GETTING STARTED

Prerequisite Quiz

Simplify.

1. $\dfrac{4}{a + 5} - \dfrac{3}{a - 1}$ $\dfrac{a - 19}{a^2 + 4a - 5}$

2. $1 - \dfrac{9}{x^2}$ $\dfrac{x^2 - 9}{x^2}$

3. $\dfrac{-9}{x^2 + 7x - 8} + \dfrac{1}{x - 1}$ $\dfrac{1}{x + 8}$

Motivator

Have students complete this problem.

$$\frac{\frac{1}{2} \cdot 4}{\frac{3}{4} \cdot 4} = \frac{?}{} \quad \frac{2}{3}$$

Ask them if these three expressions are equivalent and why.

$$\frac{\frac{1}{2}}{\frac{3}{4}}, \quad \frac{\frac{1}{2} \cdot 4}{\frac{3}{4} \cdot 4}, \quad \frac{\frac{1}{2}}{\frac{3}{4}} \cdot \frac{4}{4}$$

Yes; $\frac{4}{4} = 1$, and 1 is the multiplicative identity.   Ask them to state the relationship between 4 and the denominators of $\frac{1}{2}$ and $\frac{3}{4}$.   LCD

Additional Example 1

Simplify $\dfrac{\frac{4}{9}}{\frac{7}{8}}$. $\frac{32}{63}$

Highlighting the Standards

Standards 4a, 7a: Again students are led to extend what they know about numbers to include polynomials. In the Exercises, they apply their knowledge of rational numbers to three-dimensional figures.

Lesson Note

Emphasize that if students use the LCD correctly, the result should be a rational expression that is *not* complex.

Math Connections

Previous Algebra: In Chapter 4, an equation such as $\frac{3}{4}x = \frac{3}{5}$ was solved by multiplying both sides of the equation by the LCD of the fractions. Since, in this example, $x = \frac{3}{5} \div \frac{3}{4}$, the same approach is used in the present lesson.

Critical Thinking Questions

Synthesis: Ask the students to explain how a common denominator could be used for dividing two fractions. As a hint, show them that $\frac{5}{6} \div \frac{1}{6} = 5$. If the two fractions have a common denominator, they have the form $\frac{a}{c} \div \frac{b}{c}$. Since this is equal to $\frac{a}{c} \cdot \frac{c}{b} = \frac{a}{b}$, the division can be completed by dividing the numerator of the first fraction by the numerator of the second.

Definition | A **complex rational expression** is a rational expression that has at least one rational expression in its numerator, denominator, or both numerator and denominator.

A complex rational expression can be simplified in the same way as a complex fraction.

EXAMPLE 2 Simplify $\dfrac{\dfrac{p}{14}}{\dfrac{p}{2} - \dfrac{3}{7}}$.

Plan Find the LCD of $\dfrac{p}{14}, \dfrac{p}{2}$, and $\dfrac{3}{7}$. Then, use the rule $\dfrac{a}{b} = \dfrac{a \cdot c}{b \cdot c}$.

Solution Multiply the numerator and denominator of the complex rational expression by 14, the *LCD* of $\dfrac{p}{14}, \dfrac{p}{2}$, and $\dfrac{3}{7}$.

$$\frac{\dfrac{p}{14}}{\dfrac{p}{2} - \dfrac{3}{7}} = \frac{\dfrac{p}{14} \cdot 14}{\left(\dfrac{p}{2} - \dfrac{3}{7}\right) \cdot 14} \qquad \longleftarrow \frac{a}{b} = \frac{a \cdot c}{b \cdot c}$$

$$= \frac{\dfrac{p}{\underset{1}{14}} \cdot \overset{1}{14}}{\dfrac{p}{\underset{1}{2}} \cdot \overset{7}{14} - \dfrac{3}{\underset{1}{7}} \cdot \overset{2}{14}} \qquad \longleftarrow (a - b)c = ac - bc$$

$$= \frac{p}{7p - 6} \qquad \longleftarrow \text{simplified form}$$

TRY THIS 2. Simplify $\dfrac{\dfrac{x}{12}}{\dfrac{5}{6} + \dfrac{x}{4}}$. $\quad \frac{x}{3x + 10}$

EXAMPLE 3 Simplify $\dfrac{1 - \dfrac{1}{n} - \dfrac{30}{n^2}}{1 - \dfrac{36}{n^2}}$.

Solution The *LCD* is n^2.

326 Chapter 8 Rational Expressions

Additional Example 2

Simplify $\dfrac{\dfrac{5x}{4}}{\dfrac{5x}{8} - \dfrac{5}{12}}$. $\quad \frac{6x}{3x - 2}$

Additional Example 3

Simplify $\dfrac{1 - \dfrac{4}{x} - \dfrac{12}{x^2}}{1 - \dfrac{4}{x^2}}$. $\quad \frac{x - 6}{x - 2}$

$$\frac{1 - \dfrac{1}{n} - \dfrac{30}{n^2}}{1 - \dfrac{36}{n^2}} = \frac{n^2\left(1 - \dfrac{1}{n} - \dfrac{30}{n^2}\right)}{n^2\left(1 - \dfrac{36}{n^2}\right)} \quad \longleftarrow \; \frac{a}{b} = \frac{c \cdot a}{c \cdot b}$$

$$= \frac{n^2 \cdot 1 - \overset{n}{\cancel{n^2}} \cdot \dfrac{1}{\cancel{n}\,1} - \overset{1}{\cancel{n^2}} \cdot \dfrac{30}{\cancel{n^2}\,1}}{n^2 \cdot 1 - \overset{1}{\cancel{n^2}} \cdot \dfrac{36}{\cancel{n^2}\,1}} \quad \longleftarrow \; \text{Distributive Property}$$

$$= \frac{n^2 - n - 30}{n^2 - 36} \quad \longleftarrow \; \text{THINK: Can the numerator or the denominator be factored?}$$

$$= \frac{(\overset{1}{\cancel{n-6}})(n + 5)}{(n + 6)(\underset{1}{\cancel{n-6}})} = \frac{n + 5}{n + 6}$$

EXAMPLE 4 Simplify $\dfrac{x - 1 + \dfrac{4}{x + 3}}{\dfrac{4}{x + 3} - 2}$.

Solution

$$\frac{x - 1 + \dfrac{4}{x + 3}}{\dfrac{4}{x + 3} - 2} = \frac{(x + 3)\left[(x - 1) + \dfrac{4}{x + 3}\right]}{(x + 3)\left[\dfrac{4}{x + 3} - 2\right]}$$

$$= \frac{(x + 3 \cdot (x - 1) + \overset{1}{\cancel{(x+3)}} \cdot \dfrac{4}{\cancel{x+3}\,1}}{\underset{1}{\cancel{(x+3)}} \cdot \dfrac{4}{\cancel{x+3}} - (x + 3) \cdot 2}$$

$$= \frac{(x + 3)(x - 1) + 4}{4 - 2(x + 3)}$$

$$= \frac{x^2 + 2x - 3 + 4}{4 - 2x - 6}$$

$$= \frac{x^2 + 2x + 1}{-2x - 2}$$

$$= \frac{(\overset{1}{\cancel{x+1}})(x + 1)}{-2(\underset{1}{\cancel{x+1}})} = \frac{x + 1}{-2}, \text{ or } -\frac{x + 1}{2}$$

TRY THIS

3. Simplify $\dfrac{\dfrac{x + 5}{x - 2} - 6}{7 - \dfrac{1}{x - 2}}$. $\frac{17 - 5x}{7x - 15}$

8.9 Complex Rational Expressions **327**

Common Error Analysis

Error: In a problem such as Example 4, some students will fail to multiply the non-fractional term , $x - 1$ and 2, by the LCD.

Remind them that they must multiply every term in the numerator and denominator by the LCD.

Checkpoint

Simplify.

1. $\dfrac{\dfrac{3}{5}}{\dfrac{6}{5}}$ $\frac{1}{2}$

2. $\dfrac{\dfrac{1}{n} - \dfrac{1}{5}}{\dfrac{-5}{n^2} + \dfrac{1}{5}}$ $-\frac{n}{n + 5}$

3. $\dfrac{\dfrac{12}{x - 3} + 10 + x}{-6 - \dfrac{6}{x - 3}}$ $-\frac{x + 9}{6}$

4. $\dfrac{3 - \dfrac{10}{x} + \dfrac{8}{x^2}}{\dfrac{4}{x^2} - 1}$ $\frac{-3x + 4}{x + 2}$

Closure

Ask students to define in their own words a complex fraction and a complex rational expression.   See pages 325 and 326. Then ask to give the first step for simplifying such expressions.   Multiplying all terms by the LCD of all their denominators

---

Additional Example 4

Simplify $\dfrac{n + 7 + \dfrac{-3}{n + 5}}{\dfrac{9}{n + 5} + 3}$. $\frac{n + 4}{3}$

327

Guided Practice

Classroom Exercises 1–16
Try This   all

Independent Practice

A Ex. 1–15, **B** Ex. 16–21, **C** Ex. 22–28
Basic: WE 1–12 all
Average: WE 1–23 odd
Above Average: WE 7–27 odd, 28

Classroom Exercises

Find the LCD that you would use to simplify each complex expression.

1. $\dfrac{\frac{2}{7}}{\frac{3}{5}}$ 35

2. $\dfrac{\frac{x}{6}}{\frac{3x}{6}}$ 6

3. $\dfrac{x + \frac{1}{3}}{5 - \frac{1}{9}}$ 9

4. $\dfrac{\frac{a}{9}}{\frac{a}{2} + \frac{a}{18}}$ 18

5. $\dfrac{8 + \frac{1}{5}}{4\frac{1}{4}}$ 20

6. $\dfrac{1\frac{2}{7}}{2\frac{5}{21}}$ 21

7. $\dfrac{2a + \frac{1}{3}}{\frac{7}{15} + \frac{a}{5}}$ 15

8. $\dfrac{1 + \frac{1}{n} - \frac{6}{n^2}}{1 - \frac{4}{n^2}}$ n^2

9–16. Simplify Classroom Exercises 1–8 above.

9. $\frac{10}{21}$ **10.** $\frac{1}{3}$ **11.** $\frac{9x + 3}{44}$ **12.** $\frac{1}{5}$ **13.** $\frac{164}{85}$ **14.** $\frac{27}{47}$ **15.** $\frac{30a + 5}{3a + 7}$ $\frac{n + 3}{n + 2}$

Written Exercises

Simplify.

1. $\dfrac{\frac{3}{16}}{\frac{4}{7}}$ $\frac{21}{64}$

2. $\dfrac{\frac{1}{5} + \frac{2}{9}}{\frac{5}{9} + \frac{3}{5}}$ $\frac{19}{52}$

3. $\dfrac{\frac{x}{5}}{\frac{x}{2} - \frac{1}{10}}$ $\frac{2x}{5x - 1}$

4. $\dfrac{\frac{3t}{7} + \frac{1}{14}}{\frac{t}{2} + \frac{5}{7}}$ $\frac{6t + 1}{7t + 10}$

5. $\dfrac{6 - \frac{3}{4}}{8 + \frac{5}{8}}$ $\frac{14}{23}$

6. $\dfrac{10\frac{2}{3}}{11\frac{1}{9}}$ $\frac{24}{25}$

7. $\dfrac{2a + \frac{1}{6}}{\frac{a}{6} + \frac{2}{3}}$ $\frac{12a + 1}{a + 4}$

8. $\dfrac{5a + \frac{1}{2}}{7a + \frac{1}{2}}$ $\frac{10a + 1}{14a + 1}$

9. $\dfrac{\frac{6}{m} - \frac{2}{9}}{\frac{1}{3} + \frac{4}{m^2}}$ $\frac{54m - 2m^2}{3m^2 + 36}$

10. $\dfrac{\frac{7}{2} - \frac{3}{b}}{\frac{5}{b^2} + \frac{11}{14}}$ $\frac{49b^2 - 42b}{11b^2 + 70}$

11. $\dfrac{\frac{8}{x^2} - \frac{3}{x}}{\frac{11}{x} + \frac{1}{x^2}}$ $\frac{-3x - 8}{11x + 1}$

12. $\dfrac{1 - \frac{5}{m} + \frac{\frac{m - 1}{m + 4}}{m^2}}{1 - \frac{16}{m^2}}$

13. $\dfrac{3 + \frac{6}{a} - \frac{9}{a^2}}{6 - \frac{6}{a^2}}$ $\frac{a + 3}{2a + 2}$

14. $\dfrac{4 - \frac{8}{x} - \frac{60}{x^2}}{2 - \frac{18}{x^2}}$ $\frac{2x - 10}{x - 3}$

15. $\dfrac{-2 - \frac{4}{y} + \frac{30}{y^2}}{4 - \frac{100}{y^2}}$ $-\frac{y - 3}{2y - 10}$

16. $\dfrac{\frac{3}{m - 8} + 4}{2 + \frac{1}{m - 8}}$ $\frac{4m - 29}{2m - 15}$

17. $\dfrac{3 + \frac{2}{a - 1}}{4 - \frac{3}{a + 1}}$ $\frac{3a^2 + 2a - 1}{4a^2 - 3a - 1}$

18. $\dfrac{5 - \frac{2}{x - 3}}{6 + \left(\frac{-1}{x + 5}\right)}$ $\frac{5x^2 + 8x - 85}{6x^2 + 11x - 87}$

19. $\dfrac{\frac{2}{a - 2} + \frac{7}{a^2 - 4}}{\frac{5}{a + 2} + \frac{6}{a - 2}}$ $\frac{2a + 11}{11a + 2}$

20. $\dfrac{\frac{4}{a^2 + 4a} - \frac{3}{a}}{\frac{2}{a + 4} + \frac{4}{a}}$ $-\frac{1}{2}$

21. $\dfrac{\frac{3}{x} - \frac{12}{x^2 + 2x}}{\frac{2}{x + 2} - \frac{1}{x}}$ 3

22. $\dfrac{\dfrac{3}{(x-5)}+\dfrac{-2}{x+3}}{\dfrac{4}{x^2-2x-15}+\dfrac{1}{x+3}}$ $\dfrac{x+19}{x-1}$

23. $\dfrac{\dfrac{-3}{a+5}-\dfrac{2}{a-2}}{\dfrac{4}{a^2+3a-10}+\dfrac{5}{a+5}}$ $\dfrac{-5a-4}{5a-6}$

24. $\dfrac{\dfrac{2x+4}{x+8}-\dfrac{x-1}{x-2}}{\dfrac{x^2-49}{x^2+6x-16}}$

25. $\dfrac{1-\dfrac{13}{x^2}+\dfrac{36}{x^4}}{\dfrac{1}{x^2}-\dfrac{1}{x^3}-\dfrac{6}{x^4}}$

26. $\dfrac{1-\dfrac{34}{x^2}-\dfrac{225}{x^4}}{1-\dfrac{8}{x}+\dfrac{15}{x^2}}$

In Exercises 27 and 28, S represents the area of the figure's surface and V represents its volume. Find the quotient $\dfrac{S}{V}$ for each figure and simplify.

27. Sphere
$S = 4\pi r^2$
$V = \frac{4}{3}\pi r^3$ $\dfrac{3}{r}$

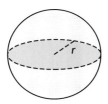

28. Right circular cone
$S = \pi r^2 + \pi r l$
$V = \frac{1}{3}\pi r^2 h$ $\dfrac{3r+3l}{rh}$

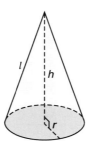

Mixed Review

Simplify. Then evaluate for the given value of the variable.   *1.7, 2.10*

1. $3(3x-18)+6x-1$ for $x = 3$ -10

2. $6s - (10 - s) + 18$ for $s = -4$ -20

3. $-4(6-4a)-(2a-10)$ for $a = 2$ 14

4. $x - 3(2-x)-(1-x)$ for $x = -2$
-17

Solve.   *3.2, 3.3, 3.4, 3.5, 4.5, 4.6*

5. $16x - 3 = -2x - 3 + 17x$ $x = 0$

6. $8x - 3(2-3x) = -2(2x-1)$ $x = \frac{8}{21}$

7. $\frac{3}{5}a = 9$ $a = 15$

8. $\frac{5}{3}x - 3 = 4$ $x = \frac{21}{5}$

9. $\frac{1}{2}+\frac{1}{3}x = \frac{4}{6}$ $x = \frac{1}{2}$

10. $1.3x + 0.05 = 0.31$ $x = 0.2$

8.9 Complex Rational Expressions   **329**

Enrichment

Have students complete the exercises for the three defined operations.

Operations

$a @ b = ab^2$ $a \$ b = \frac{1}{a}+\frac{1}{b}$
$a * b = 2a + b$

1. $(\frac{1}{6} @ 3) \$ 2$ $\frac{7}{6}$

2. $(4 @ \frac{1}{2}) * (6 \$ \frac{1}{2})$ $\frac{25}{6}$

3. $(x + y) \$ (y - x)$ $\dfrac{-2y}{x^2-y^2}$

4. $-\frac{1}{4x^2} @ 2x$ -1

Key Terms

complex fraction (p. 325)
complex rational expression (p. 326)
least common denominator (LCD)
   (p. 311)
least common multiple (LCM)
   (p. 311)
Multiplicative Identity Property for
   Rational Expressions (p. 311)

rational expression (p. 289)
Rule for Adding and Subtracting Rational
   Expressions (p. 306)
Rule for Dividing Rational Expressions
   (p. 302)
Rule for Multiplying Rational Expressions
   (p. 297)

Key Ideas and Review Exercises

8.1, 8.2 To find the value(s), if any, for which a rational expression is undefined, set the denominator equal to 0 and solve the equation.

To simplify a rational expression, factor the numerator and denominator. If necessary, rewrite the numerator and denominator in convenient form. Then divide out the common factor(s), noting all restrictions on the variables.

1. For what value or values of x is the expression undefined?

$$\frac{3x - 2}{2x^2 + 5x - 3}$$ $-3, \frac{1}{2}$

Simplify.

2. $\dfrac{36t^5}{14t}$ $\dfrac{18t^4}{7}$

3. $\dfrac{x^2 + 3x + 2}{x^2 - x - 6}$ $\dfrac{x + 1}{x - 3}$

4. $\dfrac{49 - a^2}{a + 7}$ $7 - a$

5. $-\dfrac{1}{x + 2}$

5. $\dfrac{x - 5}{10 + 3x - x^2}$

8.3 To multiply two rational expressions, factor the numerators and denominators, divide out common factors, and multiply.

Multiply.

6. $\dfrac{12a^2b^4}{5 - a} \cdot \dfrac{a^2 - 25}{3ab^3}$ $-4a^2b - 20ab$

7. $\dfrac{x^2 - 3x + 2}{x + 3} \cdot \dfrac{5x + 15}{x^2 + 4x - 5}$ $\dfrac{5x - 10}{x + 5}$

8.4 To divide two rational expressions, multiply the first expression by the reciprocal of the second expression.

Divide.

8. $\dfrac{36a^7b^3}{27 - 3a^2} \div \dfrac{12a^5b^4}{5a + 15}$ $\dfrac{5a^2}{3b - ab}$

9. $\dfrac{x^2 + 7x + 12}{x^2 - 16} \div \dfrac{x^2 + 4x - 21}{x - 4}$ $\dfrac{x + 3}{x^2 + 4x - 21}$

8.5   To add or subtract rational expressions with like denominators,

use $\dfrac{a}{b} + \dfrac{c}{b} = \dfrac{a + c}{b}$ or $\dfrac{a}{b} - \dfrac{c}{b} = \dfrac{a - c}{b}$.

Add or subtract.

10. $\dfrac{2a}{a - 4} - \dfrac{8}{a - 4}$ 2

11. $\dfrac{2x}{x^2 + 7x + 10} + \dfrac{10}{x^2 + 7x + 10}$ $\dfrac{2}{x + 2}$

8.6,   To add or subtract rational expressions with unlike denominators, find the
8.7   LCD. Then use the Multiplicative Identity Property to write equivalent
expressions with like denominators.

Add or subtract.

12. $\dfrac{3}{10x^2} + \dfrac{7}{15x^3}$ $\dfrac{9x + 14}{30x^3}$

13. $\dfrac{b + 2}{18} + \dfrac{3b - 1}{9} - \dfrac{b - 1}{6}$ $\dfrac{4b + 3}{18}$

14. $\dfrac{25}{5 - m} - \dfrac{5m}{m^2 - 25}$ $\dfrac{-30m - 125}{m^2 - 25}$

15. $\dfrac{35}{x^2 + x - 12} - \dfrac{5}{x - 3} + \dfrac{7}{x + 4}$ $\dfrac{2}{x + 4}$

16. Explain how to add or subtract a polynomial and a rational expression

Use $(x - 2) + \dfrac{6}{x + 3}$ as an illustration. Answers will vary.

8.8   To divide a polynomial by a monomial, divide each term of the polynomial
by the monomial.

To divide a polynomial by a binomial, arrange the terms in descending order
of exponents. If there is a missing term in the dividend, insert the term with
a coefficient of 0. Then divide as in Examples 2 and 3 in Lesson 8.8.

Divide.

17. $(36t^4 - 18t^3 + 12t) \div (6t)$ $6t^3 - 3t^2 + 2$ **18.** $(4x^3 - 19x + 15) \div (2x - 3)$
$2x^2 + 3x - 5$

8.9   To simplify a complex fraction or complex rational expression, multiply the
numerator and the denominator of the expression by the LCD of all the
terms.

Simplify.

19. $\dfrac{\dfrac{3t}{5} - \dfrac{2}{10}}{\dfrac{2t}{20} + \dfrac{5}{4}}$ $\dfrac{12t - 4}{2t + 25}$

20. $\dfrac{1 + \dfrac{2}{n} - \dfrac{15}{n^2}}{1 - \dfrac{25}{n^2}}$ $\dfrac{n - 3}{n - 5}$

For what values of the variable is the rational expression undefined?

1. $\dfrac{y - 3}{3y + 6}$ -2

2. $\dfrac{x + 4}{6 - x}$ 6

3. $\dfrac{m + 3}{m^2 + 7m + 6}$ $-1, -6$

Simplify.

4. $\dfrac{-54t^7}{6t^2}$ $-9t^5$

5. $\dfrac{6a^2 + a - 2}{4a^2 + 4a - 3}$ $\frac{3a + 2}{2a + 3}$

6. $\dfrac{8a^3 - 4a^4}{5a^3 - 10a^2}$ $-\frac{4a}{5}$

Multiply.

7. $\dfrac{5a^5}{3b^6} \cdot \dfrac{4a^2}{7b^7}$ $\frac{20a^7}{21b^{13}}$

8. $\dfrac{x^2 + 6x - 7}{x + 3} \cdot \dfrac{4x + 12}{x - 1}$ $4x + 28$

9. $\dfrac{3 - a}{a^2 + 3a - 4} \cdot \dfrac{a^2 - 1}{a - 3}$ $-\frac{a + 1}{a + 4}$

10. Divide. $\dfrac{a^2 + 2a - 15}{a^2 - 9} \div \dfrac{a^2 + 4a - 5}{3a - 9}$ $\frac{3a - 9}{a^2 + 2a - 3}$

11. Simplify. $\dfrac{x^2 + 4x + 3}{4x^4y^3} \div \dfrac{x^2 - 1}{12x^5y} \cdot \dfrac{x^2 - 6x + 5}{6x^2 + 18x}$ $\frac{x - 5}{2y^2}$

Add or subtract.

12. $\dfrac{3x}{x + 5} + \dfrac{15}{x + 5}$ 3

13. $\dfrac{-5}{18a^5} + \dfrac{7}{12a^3}$ $\frac{21a^2 - 10}{36a^5}$

14. $\dfrac{3t - 2}{2} + \dfrac{5t - 7}{7} - \dfrac{3t}{14}$ $2t - 2$

15. $(a + 5) - \dfrac{2}{a - 2}$ $\frac{a^2 + 3a - 12}{a - 2}$

16. $\dfrac{x + 3}{x + 5} + \dfrac{4x - 2}{x^2 - x - 30}$ $\frac{x - 4}{x - 6}$

17. $\dfrac{-6}{x - 3} + \dfrac{18}{x^2 - 9} + \dfrac{3}{x + 3}$ $\frac{-3}{x - 3}$

Divide.

18. $(36m^6 - 45m^3 + 81m^2) \div (9m^2)$ $4m^4 - 5m + 9$

19. $(6x^3 + 7x^2 - 5x + 15) \div (2x - 1)$ $3x^2 + 5x + \frac{15}{2x - 1}$

Simplify.

20. $\dfrac{\dfrac{2}{3} - \dfrac{1}{2}}{\dfrac{1}{5} + \dfrac{1}{10}}$ $\frac{5}{9}$

21. $\dfrac{1 - \dfrac{4}{x} - \dfrac{5}{x^2}}{1 - \dfrac{8}{x} + \dfrac{15}{x^2}}$ $\frac{x + 1}{x - 3}$

22. Factor $27x^3 - 8$ if one of its factors is $3x - 2$. $(3x - 2)(9x^2 + 6x + 4)$

23. Simplify $\dfrac{\dfrac{2}{x + 1} + \dfrac{-3}{x - 3}}{\dfrac{5}{x^2 - 2x - 3} + \dfrac{1}{x + 1}}$. $-\frac{x + 9}{x + 2}$

College Prep Test

In each Exercise, you are to compare a quantity in Column 1 with a quantity in Column 2. Write the letter of the correct answer from these choices.

A—The quantity in Column 1 is greater than the quantity in Column 2.
B—The quantity in Column 2 is greater than the quantity in Column 1.
C—The quantity in Column 1 is equal to the quantity in Column 2.
D—The relationship cannot be determined from the given information.

NOTE: Information centered over both columns refers to one or both of the quantities to be compared.

Sample Question and Answer	Answer: C, because
Column 1 $\qquad$ **Column 2** $x \neq 0$ and $y \neq 0$ $\dfrac{1}{x} - \dfrac{1}{y} \qquad\qquad \dfrac{y-x}{xy}$	$\dfrac{1}{x} - \dfrac{1}{y} = \dfrac{1}{x} + \dfrac{-1}{y}$ $= \dfrac{1 \cdot y}{x \cdot y} + \dfrac{-1 \cdot x}{y \cdot x}$ $= \dfrac{1y - 1x}{xy}, \text{ or } \dfrac{y-x}{xy}$

	Column 1	Column 2
	$a \neq 0, b \neq 0$	
1. C	$\dfrac{1}{a} + \dfrac{1}{b}$	$\dfrac{a+b}{ab}$
	$c > b > a > 0$	
2. B	$\dfrac{a}{b}$	$\dfrac{c}{a}$
	$-4 < x < 0$	
3. A	$\dfrac{1}{x^2}$	$\dfrac{1}{x^3}$
	$a = -100$	
4. B	$\dfrac{a^8}{a^3}$	$\dfrac{a^7}{a}$
	$z = \dfrac{1}{x+y}, x > 0, y > 0$	
5. C	$\dfrac{m}{z}$	$mx + my$

The 15 small rectangles below have the same length and width. The large rectangle has length x and width y.

	Column 1	Column 2
6. C	$\dfrac{7xy}{15}$	area of the shaded region
	$\dfrac{5}{a} = \dfrac{7}{b}$ and $a > 0, b > 0$	
7. B	a	b
	$\dfrac{1}{4} + \dfrac{2}{x} + \dfrac{2}{3} = \dfrac{13}{12}$	
8. A	x	3

For additional standardized test practice, see the SAT/ACT test booklet for cumulative tests Chapters 1–8.

Cumulative Review

24. $x = -\frac{1}{2}$ or $x = 4$

36. $-8a^3 + 6a^2 - 11$

49. $w : 6$ m; $l : 31$ m

50. $w : 20$ m; $l : 70$ m

55. 47, 47, and 86

56. 49.8 cm^2

Cumulative Review (Chapters 1–8)

Choose the one best answer.

1. Compute $(-3)^2 \cdot 2^3$. C   *6.2*
(A) 36   (B) -72 (C) 72
(D) 48   (E) None of these

2. Solve $-6 \le 9 + 5x$. D   *5.2*
(A) $x \le -3$ (B) $x \le 3$
(C) $x \ge 3$ (D) $x \ge -3$
(E) None of these

3. Simplify $(2a - 3)^2$. B   *6.10*
(A) $4a^2 + 12a + 9$
(B) $4a^2 - 12a + 9$
(C) $4a^2 + 6a - 9$
(D) $4a^2 - 6a + 9$
(E) None of these

4. Simplify.   D   *2.10*
$-3(2x - 4) + 2(6 - x)$
(A) $-4x$ (B) $-8x$
(C) $-4x + 24$
(D) $-8x + 24$
(E) None of these

5. Multiply $-4x^3y^5(-6xy^2)$. A   *6.1*
(A) $24x^4y^7$ (B) $24x^3y^7$
(C) $-24xy^{11}$ (D) $24xy^{10}$
(E) None of these

6. Solve $|3x - 5| = 13$. B   *5.6*
(A) -6 (B) $-\frac{8}{3}, 6$
(C) $\frac{8}{3}, 6$ (D) -3
(E) None of these

In Exercises 7 and 8, what property is illustrated?

7. $-8 + 10 = 10 + (-8)$ B   *1.5*
(A) Assoc Prop for Add
(B) Comm Prop for Add
(C) Add Inverse Prop
(D) Rule of Subt
(E) None of these

8. $-9a \cdot 1 = -9a$ C
(A) Add Identity Prop
(B) Closure Prop for Mult
(C) Mult Identity Prop
(D) Distr Prop
(E) None of these

Evaluate.

9. a^2 for $a = -\frac{1}{3}$ $\frac{1}{9}$ *2.6*
10. $4x^3$ for $x = -2$ -32
11. $2m - 3n + 6$ for $m = 3$ and   *2.8*
$n = -2$ 18
12. $-3a^2bc^3$ for $a = -1$, $b = 2$,   *2.6*
and $c = -2$ 48
13. $\dfrac{x + 12}{2x}$ for $x = 4$ 2   *8.1*

14. $\dfrac{x^2 - 2}{x}$ for $x = -5$ $\frac{-23}{5}$

Solve each equation or inequality.

15. $|x - 3| \le 4$ $-1 \le x \le 7$ *5.7*
16. $|x - 10| > 7$ $x < 3$ or $x > 17$
17. $|3x + 4| \le 17$ $-7 \le x \le \frac{13}{3}$
18. $a - 15 = -8$ $a = 7$ *3.1*
19. $3y + 8 - y = -2 + y$ $y = -10$ *3.4*
20. $9x = -45$ $x = -5$ *3.2*
21. $y - \frac{1}{5} = \frac{7}{6}$ $y = \frac{41}{30}$ *3.1*
22. $5(2x - 3) = 5x + 50$ $x = 13$ *3.5*
23. $|x - 3| = 8$ $x = -5, x = 11$ *5.6*
24. $2x^2 - 7x - 4 = 0$ *7.7*
25. $m^2 - 14m = 0$ $m = 0$ or $m = 14$
26. $5 + 14t = 3t^2$
$t = -\frac{1}{3}$ or $t = 5$

Factor.

27. $x^2 - 6x - 7$ $(x - 7)(x + 1)$ *7.3*
28. $6x^2 - x - 2$ $(3x - 2)(2x + 1)$ *7.4*

29. $m^2 - 36$ $(m + 6)(m - 6)$ **7.5**

30. $5y^2 + 30y + 40$ $5(y + 2)(y + 4)$ **7.6**

31. $6x^3 + 28x^2 - 10x$ $2x(3x - 1)(x + 5)$

32. $25x^2 - 81y^2$ $(5x + 9y)(5x - 9y)$ **7.5**

Simplify.   **33.** $4x^3 - x^2 + 3x$

33. $8x^3 - 2x^2 + 3x - 4x^3 + x^2$ **6.6**

34. $-8 + 2y^4 - 3y^3 - y^4 +$
$6y^3 - 5y^4$ $-4y^4 + 3y^3 - 8$

35. $3x - 2(8 - 9x)$ $21x - 16$ **2.10**

36. $-(8a^3 + 6 - 7a^2) - (a^2 + 5)$ **6.7**

37. $(-3x^2y^3)^2$ $9x^4y^6$ **6.2**

38. $\dfrac{18a^4b^6}{-6ab^7}$ $\frac{-3a^3}{b}$ **6.3**

39. $(8x^4y^5)(-4xy^7)$ $-32x^5y^{12}$ **6.1**

Write an algebraic expression.

40. $12x$ increased by 4 $12x + 4$ **1.1**

41. $4x$ decreased by 9 $4x - 9$

42. 9 times $6x$ $9 \cdot 6x$, or $54x$

43. 23 divided by x $23 \div x$

Solve each problem.

44. Use $p = 2l + 2w$ to find p for   **1.4**
$l = 8$ cm and $w = 12$ cm.   40 cm

45. Use $A = 4s^2$ to find A for
$s = 8$ m.   256 m²

46. Find two consecutive integers   **4.3**
whose sum is 87.   43 and 44

47. Find three consecutive odd integers whose sum is 57.   17, 19, 21

48. The second of three numbers is   **4.2**
2 less than 3 times the first.
The third is 8 more than twice
the first. If 4 times the second
is decreased by the third, the
result is 8 times the first. Find
the numbers.   8, 22, 24

49. The length of a rectangle is   **4.4**
5 m less than 6 times the
width. The perimeter is 74 m.
Find the length and the width.

50. The length of a rectangle is
10 m more than 3 times the
width. The perimeter is 180 m.
Find the length and the width.

51. The larger of two numbers is 9   **4.2**
less than twice the smaller. The
sum of the numbers is 36. Find
the two numbers.   15 and 21

52. A team won 16 games and lost   **4.7**
4 games. The number of games
won is what percent of the
games played?   80%

53. The selling price of a VCR is   **4.8**
$420. The profit is 40% of the
cost. Find the cost. $300

54. A college bookstore makes a
profit of $4 on each book sold.
How many books must be sold
to make a total profit of at
least $1,000?   250

55. The degree measure of the ver-   **4.4**
tex angle of an isosceles trian-
gle is 39 more than that of a
base angle. Find the measure
of each angle of the triangle.

56. Find the area of a trapezoid if   **1.4**
the height is 8.3 cm and the
bases are 5.4 cm and 6.6 cm.

57. Eighty students are separated   **4.2**
into two groups. The second
group is 7 times as large as the
first. How many students are
in each group?   10 and 70

58. In a collection of 30 nickels
and quarters, there are twice as
many nickels as quarters. Find
the number of coins of each
type.   20 nickels; 10 quarters

OVERVIEW

In this chapter, students solve equations which involve the rational expressions they worked with in Chapter 8. In Section 9.2, students will see that proportions are a special case of this type of equation. The lessons on work and motion problems (Sections 9.4 and 9.5) both extend and reinforce problem-solving skills. A lesson on dimensional analysis (see pages 364–366) extends these skills to measurement conversions, thus reducing the number of measurement facts to be memorized.

OBJECTIVES

- To solve rational equations
- To solve proportions
- To solve literal equations
- To solve motion and work problems
- To convert among various units of measure

PROBLEM SOLVING

Drawing a Diagram and Making a Table are demonstrated as useful problem solving techniques for the motion and work problems in this chapter. The focus of the lesson on pages 343–347 is on identifying problems that can be solved by Writing a Proportion. The Application on page 348 applies the strategy of Using a Formula to solve problems involving direct variation.

READING AND WRITING MATH

Students are asked to solve several types of word problems throughout this chapter. Exercise 35 on page 353 gives students a chance to write an explanation of an equation-solving technique. The Focus on Reading on page 346 requires that students identify equivalent equations.

TECHNOLOGY

Calculator: A calculator can be used for several of the problems involving calculations with decimals. Specifically, have students use a calculator for Exercise 3 on page 350, and Exercises 25 and 26 on page 352.

SPECIAL FEATURES

Mixed Review pp. 342, 348, 358, 363, 366
Focus on Reading p. 346
Application: Direct Variation p. 348
Brainteaser p. 353
Midchapter Review p. 353
Mixed Problem Solving p. 367
Key Terms p. 368
Key Ideas and Review Exercises pp. 368–369
Chapter 9 Test p. 370
College Prep Test p. 371

PLANNING GUIDE

Lesson	Basic	Average	Above Average	Resources
9.1 pp. 341–342	CE 1–8 WE 1–19 odd, 27	CE 1–8 WE 9–29 odd	CE 1–8 WE 11–31 odd	Reteaching 69 Practice 69
9.2 pp. 346–348	FR all CE 1–8 WE 1–33 odd, 34 Application	FR all CE 1–8 WE 13–45 odd Application	FR all CE 1–8 WE 17–49 odd Application	Reteaching 70 Practice 70
9.3 pp. 352–353	CE 1–12 WE 1–25 odd, 35 Midchapter Review Brainteaser	CE 1–12 WE 11–35 odd Midchapter Review Brainteaser	CE 1–12 WE 19–43 odd Midchapter Review Brainteaser	Reteaching 71 Practice 71
9.4 pp. 357–358	CE 1–8 WE 1–8 all	CE 1–8 WE 1–7 odd, 9–12 all	CE 1–8 WE 1–15 odd	Reteaching 72 Practice 72
9.5 pp. 362–363	CE 1–10 WE 1–10 all	CE 1–10 WE 5–14 all	CE 1–10 WE 11–20 all	Reteaching 73 Practice 73
9.6 pp. 366–367	CE 1–9 WE 1–13 all Mixed Problem Solving	CE 1–9 WE 4–16 all Mixed Problem Solving	CE 1–9 WE 6–18 all Mixed Problem Solving	Reteaching 74 Practice 74
Chapter 9 Review pp. 368–369	all	all	all	
Chapter 9 Test p. 370	all	all	all	
College Prep Test p. 371	all	all	all	

CE = Classroom Exercises   WE = Written Exercises   FR = Focus on Reading
Note: For each level, all students should be assigned all Try This and all Mixed Review Exercises.

Manipulative: In this *Investigation,* students use a manipulative to help them solve distance, percent, and interest problems.

Materials: Each student will need a copy of Manipulative Worksheet 9.

Students often have trouble with the three different forms of the distance formula: $d = rt$, $r = \frac{d}{t}$, and $t = \frac{d}{r}$.

To solve a problem using the triangle, have students place their thumbs over the unknown quantity. The position of the two remaining variables will show them which operation to use.

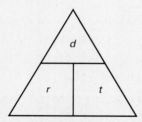

For example, have students find the time for a trip of 600 miles at a rate of 50 mph by placing their thumbs over the *t.* Ask them what the positions of the other two variables indicates.
To find the time, divide distance by rate.

The worksheet has the students use a similar triangle for percent problems, where percent = rate × base ($p = rb$), and for interest problems where $i = prt$.

When students have completed the worksheet, have them suggest other formulas from math or science that could be placed on a similar triangle.
Possible answers:
area: $A = bh$;
electricity: watts = volts · amps;
physics: mass = density · volume

9 APPLYING RATIONAL EXPRESSIONS

Susan Butcher has won the sled dog race over the Iditarod Trail in Alaska four times. Only one other person has done this! In 1990, she and her champion dogs completed the 1,158 miles from Anchorage to Nome in a little over 11 days, a record time.

More About the Photo

Susan Butcher raises Alaskan huskies and usually has about 150 at her property at an abandoned gold-mining camp. Each dog has its own plywood house where it likes to sit on the roof. For the sled-dog race, Susan uses a team of 16 or 18 of the strongest and smartest dogs. When the dogs are in training for a race, they eat five meals a day. Temperatures during the race may be as low as 60 degrees below zero. For the 11 or 12 days of the race, the dogs run for four hours and then sleep for four hours, with just one required 24-hour stopover.

9.1 Rational Equations

Teaching Resources

Manipulative Worksheet 9
Quick Quizzes 69
Reteaching and Practice
  Worksheets 69

Objective

To solve equations containing rational expressions

An equation such as $\frac{4}{x} + \frac{1}{2x} = \frac{1}{3}$ is called a **rational equation** because it contains one or more rational expressions. To solve a rational equation, first multiply each side of the equation by the least common denominator of the fractions.

EXAMPLE 1 Solve $\frac{4}{x} + \frac{1}{2x} = \frac{1}{3}$.

Plan Multiply each side of the equation by the LCD, $2 \cdot x \cdot 3$, or $6x$.

Solution

$$\frac{4}{x} + \frac{1}{2x} = \frac{1}{3}$$

$$6x\left(\frac{4}{x} + \frac{1}{2x}\right) = 6x \cdot \frac{1}{3}$$

$$6x \cdot \frac{4}{x} + 6x \cdot \frac{1}{2x} = 6x \cdot \frac{1}{3}$$

$$24 + 3 = 2x$$

$$27 = 2x$$

$$x = \frac{27}{2}, \text{ or } 13\frac{1}{2}$$

Check:

$$\frac{4}{x} + \frac{1}{2x} = \frac{1}{3}$$

$$\frac{4}{\frac{27}{2}} + \frac{1}{2\left(\frac{27}{2}\right)} \stackrel{?}{=} \frac{1}{3}$$

$$\frac{8}{27} + \frac{1}{27} \stackrel{?}{=} \frac{1}{3}$$

$$\frac{9}{27} \stackrel{?}{=} \frac{1}{3}$$

$$\frac{1}{3} = \frac{1}{3} \text{ True}$$

Thus, the solution is $\frac{27}{2}$.

TRY THIS 1. Solve $\frac{3}{5x} + \frac{2}{x} = \frac{1}{2}$. $\frac{26}{5}$

Sometimes, when you multiply by the LCD, you obtain a quadratic equation. Such an equation may have more than one root. Then, you need to check each root in the *original equation* to see whether it is a *solution* of the original equation.

EXAMPLE 2 Solve $\frac{a + 7}{a - 3} = \frac{9}{a} + \frac{30}{a^2 - 3a}$.

Plan Factor $a^2 - 3a$.
Then multiply each side of the equation by the LCD.

Solution Factor.

$$\frac{a + 7}{a - 3} = \frac{9}{a} + \frac{30}{a(a - 3)}$$

9.1 Rational Equations **337**

GETTING STARTED

Prerequisite Quiz

Solve each equation.

1. $\frac{x}{3} - 4 = \frac{1}{2}$ $\frac{27}{2}$
2. $\frac{a}{4} - \frac{3a}{2} = 5$ -4
3. $\frac{5}{2} - \frac{y}{8} = \frac{y}{4}$ $\frac{20}{3}$
4. $\frac{2a}{5} - \frac{a}{2} = \frac{1}{10}$ -1
5. $\frac{x}{24} - \frac{4}{3} = \frac{4x}{8}$ $-\frac{32}{11}$

Additional Example 1

Solve $\frac{1}{5} + \frac{3}{x} = \frac{2}{x}$. -5

Additional Example 2

Solve $\frac{x - 3}{x - 2} = \frac{5}{x} + \frac{10}{x^2 - 2x}$. 8

Highlighting the Standards

Standards 14a, 4d: Students see that a derived equation may not be an equivalent equation. In the Exercises, they apply mathematics to lens and image problems.

Motivator

Ask students to state the number or expression by which to multiply each side of the equation in order to have no denominator other than 1.

1. $\frac{x}{2} = 12$ 2

2. $\frac{a}{3} - \frac{a}{4} = 6$ 12

3. $\frac{3t}{5} + \frac{3}{2} = \frac{7t}{10}$ 10

4. $\frac{2}{3x} = \frac{1}{9}$ 9x

5. $\frac{y}{y-3} = 2$ $y - 3$

Ask them what value of x will make the denominator in Exercise 4 equal to 0. $x = 0$ Ask them what value of y in Equation 5 will make the denominator equal to 0. $y = 3$

Multiply by $a(a - 3)$.

$$a(a - 3)\frac{a + 7}{a - 3} = a(a - 3)\left[\frac{9}{a} + \frac{30}{a(a - 3)}\right]$$

$$a(a - 3)\frac{a + 7}{a - 3} = a(a - 3)\frac{9}{a} + a(a - 3)\frac{30}{a(a - 3)}$$

$$a(a + 7) = (a - 3)9 + 30$$

$$a^2 + 7a = 9a - 27 + 30$$

$$a^2 - 2a - 3 = 0$$

Factor.

$$(a - 3)(a + 1) = 0$$

$$a - 3 = 0 \ or \ a + 1 = 0$$

$$a = 3 \ or \qquad a = -1$$

Checks

Substitute 3 for a in the *original* equation.

$$\frac{a + 7}{a - 3} = \frac{9}{a} + \frac{30}{a^2 - 3a}$$

$$\frac{3 + 7}{3 - 3} \overset{?}{=} \frac{9}{3} + \frac{30}{3^2 - 3 \cdot 3}$$

$$\frac{10}{0} \overset{?}{=} \frac{9}{3} + \frac{30}{0}$$

The symbols $\frac{10}{0}$ and $\frac{30}{0}$ are undefined. Therefore, 3 is not a solution of the original equation.

Substitute -1 for a in the *original* equation.

$$\frac{a + 7}{a - 3} = \frac{9}{a} + \frac{30}{a^2 - 3a}$$

$$\frac{-1 + 7}{-1 - 3} \overset{?}{=} \frac{9}{-1} + \frac{30}{(-1)^2 - 3(-1)}$$

$$\frac{6}{-4} \overset{?}{=} -9 + \frac{30}{1 + 3}$$

$$-\frac{6}{4} \overset{?}{=} -9 + \frac{30}{4}$$

$$-\frac{6}{4} \overset{?}{=} \frac{-36}{4} + \frac{30}{4}$$

$$-\frac{6}{4} = -\frac{6}{4} \ \text{True}$$

Thus, -1 is the only solution of the original equation.

TRY THIS

2. Solve $\frac{6}{x} + \frac{x + 4}{x - 2} = \frac{12}{x^2 - 2x}$. -12

In Example 2, 3 is a solution of the *derived* equation, $a^2 - 2a - 3 = 0$. However, 3 is not a solution of the original equation. In this case, 3 is called an *extraneous solution,* or *extraneous root.*

An **extraneous solution** is a solution of a derived equation that is not a solution of the original equation.

EXAMPLE 3   Solve $\dfrac{-17}{x^2 + 5x - 6} = \dfrac{x + 2}{x + 6} + \dfrac{3}{1 - x}$.

Solution   Factor $x^2 + 5x - 6$ in the denominator.

Additional Example 3

Solve $\dfrac{-5}{x^2 - x - 6} = \dfrac{x + 3}{x + 2} + \dfrac{4}{3 - x}$.   6

$$\frac{-17}{(x + 6)(x - 1)} = \frac{x + 2}{x + 6} + \frac{3(-1)}{x - 1} \longleftarrow \text{Put } 1 - x \text{ in convenient form.}$$

$$\frac{-17}{(x + 6)(x - 1)} = \frac{(x + 2)}{(x + 6)} + \frac{-3}{x - 1} \longleftarrow \text{The LCD is } (x + 6)(x - 1).$$

$$(x + 6)(x - 1)\frac{-17}{(x + 6)(x - 1)} = (x + 6)(x - 1)\left[\frac{x + 2}{x + 6} + \frac{-3}{x - 1}\right]$$

$$-17 = (x + 6)(x - 1)\frac{x + 2}{x + 6} + (x + 6)(x - 1) \cdot \frac{-3}{x - 1}$$

$$-17 = (x - 1)(x + 2) + (x + 6)(-3)$$
$$-17 = x^2 + 2x - 1x - 2 - 3x - 18$$
$$-17 = x^2 - 2x - 20$$
$$0 = x^2 - 2x - 3$$

Factor. $\qquad 0 = (x - 3)(x + 1)$

$$x - 3 = 0 \text{ or } x + 1 = 0$$
$$x = 3 \qquad\qquad x = -1 \qquad \text{The check is left to you.}$$

EXAMPLE 4   The sum of a number and its reciprocal is $\frac{13}{6}$. Find the number and its reciprocal.

Solution   Let $x =$ the number. Then $\frac{1}{x} =$ the reciprocal of the number.

$$x + \frac{1}{x} = \frac{13}{6} \longleftarrow \text{Number + reciprocal} = \frac{13}{6}$$

$$6x\left[x + \frac{1}{x}\right] = 6x\left(\frac{13}{6}\right)$$

$$6x \cdot x + 6x \cdot \frac{1}{x} = 13x$$

$$6x^2 + 6 = 13x$$
$$6x^2 - 13x + 6 = 0$$
$$(3x - 2)(2x - 3) = 0$$
$$3x - 2 = 0 \text{ or } 2x - 3 = 0$$
$$3x = 2 \qquad\qquad 2x = 3$$

$$x = \frac{2}{3} \qquad\qquad x = \frac{3}{2}$$

If $x = \frac{2}{3}$, then $\frac{1}{x} = \frac{3}{2}$. If $x = \frac{3}{2}$, then $\frac{1}{x} = \frac{2}{3}$.

Check $\frac{2}{3}$ and $\frac{3}{2}$ are reciprocals since $\frac{2}{3} \times \frac{3}{2} = 1$. Also, $\frac{2}{3} + \frac{3}{2} = \frac{4}{6} + \frac{9}{6} = \frac{13}{6}$.

Thus, the number and its reciprocal are $\frac{2}{3}$ and $\frac{3}{2}$, or $\frac{3}{2}$ and $\frac{2}{3}$.

▰ TEACHING SUGGESTIONS

Lesson Note

When presenting Example 1 on page 337, tell students that after cancelling, the equation should contain no fractions. When presenting Example 2 on pages 337–338, tell them that if an equation contains a squared term after it has been simplified, it is a quadratic equation and must be solved by setting one member equal to 0 and factoring. Most students will need a little practice with checking these equations since they contain so many fractions. Emphasize also the importance of checking every apparent solution in the original equation.

Math Connections

Science: The total resistance R of two resistors R_1 and R_2 in a parallel circuit is determined by the rational equation

$$R = \frac{1}{\frac{1}{R_1} + \frac{1}{R_2}}$$ where resistance is measured in ohms.

Additional Example 4

The sum of a number and its reciprocal is $\frac{26}{5}$. Find the number and its reciprocal.

$x + \frac{1}{x} = \frac{26}{5}$; 5; $\frac{1}{5}$

Critical Thinking Questions

Application: Ask students how many solutions the equation $x = 5$ has.   1
Then have them multiply each side of this equation by x to get a new equation, $x^2 = 5x$. Have them solve this equation.   0, 5   Now ask if both of these roots are roots of the original equation.   No   Explain that the extraneous root was introduced when they multiplied each side of the original equation by the variable, x. Ask students to explain why $x = 5$ and $x^2 = 5x$ are not equivalent equations.   They do not have the same solution set.

Common Error Analysis

Error: Students do not use the Distributive Property correctly when solving equations such as

$$\frac{3}{x-1} + \frac{2}{x-4} = \frac{3x}{x^2 - 5x + 4}.$$

To avoid this, encourage students to write the product of each term by the LCD as follows.

$(x-4)(x-1) \cdot \dfrac{3}{x-1} + (x-4)(x-1) \cdot \dfrac{2}{x-4} = (x-4)(x-1) \cdot \dfrac{3x}{x^2 - 5x + 4}$

Checkpoint

Solve.

1. $\dfrac{5}{x} - \dfrac{1}{3} = \dfrac{2}{x}$ 9

2. $\dfrac{6}{x^2 - 5x} + \dfrac{3}{x} = \dfrac{x-3}{x-5}$ 3

3. $\dfrac{4}{x^2 + x - 2} = \dfrac{4}{1-x} + \dfrac{x-3}{x+2}$ 9, −1

4. The sum of a number and its reciprocal is $\dfrac{65}{8}$. Write an equation for this situation. Then find the number and its reciprocal.

$x + \dfrac{1}{x} = \dfrac{65}{8}$; 8; $\dfrac{1}{8}$

EXAMPLE 5 The denominator of a fraction is 1 less than twice the numerator. If the numerator and denominator are each increased by 3, the resulting fraction simplifies to $\frac{3}{4}$. Find the original fraction.

Solution Let a = the numerator of the original fraction.
Then $2a - 1$ = the denominator of the original fraction.
The original fraction is $\dfrac{a}{2a - 1}$.
If 3 is added to the original numerator and denominator, the resulting fraction is equal to $\frac{3}{4}$.

$$\frac{a + 3}{(2a - 1) + 3} = \frac{3}{4}$$

$$\frac{a + 3}{2a + 2} = \frac{3}{4}$$

$$\frac{a + 3}{2(a + 1)} = \frac{3}{2 \cdot 2} \quad \longleftarrow \text{LCD} = 2 \cdot 2(a + 1)$$

$$2 \cdot \overset{1}{\cancel{2}}(a \overset{1}{\cancel{+ 1}}) \left[\frac{a + 3}{\underset{1}{\cancel{2}}(a \underset{1}{\cancel{+ 1}})} \right] = \overset{1}{\cancel{2}} \cdot \overset{1}{\cancel{2}}(a + 1) \left[\frac{3}{\underset{1}{\cancel{2} \cdot \cancel{2}}} \right]$$

$$2(a + 3) = (a + 1)3$$
$$2a + 6 = 3a + 3$$
$$6 = a + 3$$
$$3 = a$$

Original fraction $= \dfrac{a}{2a - 1} = \dfrac{3}{2 \cdot 3 - 1} = \dfrac{3}{6 - 1} = \dfrac{3}{5}$

Check The denominator of $\frac{3}{5}$ is 1 less than twice the numerator. If the numerator and the denominator are each increased by 3, the new fraction simplifies to $\frac{3}{4}$.

$$\frac{3 + 3}{5 + 3} = \frac{6}{8} = \frac{3}{4}$$

Thus, the original fraction is $\frac{3}{5}$.

TRY THIS 3. The denominator of a fraction is 3 less than 3 times the numerator. If the numerator and denominator are each increased by 4, the resulting fraction simplifies to $\frac{1}{2}$. Find the original fraction. $\frac{7}{18}$

Additional Example 5

The denominator of a fraction is 1 more than twice the numerator. If the numerator and denominator are each increased by 4, the resulting fraction simplifies to $\frac{2}{3}$. Find the original fraction. $\frac{2}{5}$

Classroom Exercises

State the LCD for the denominators of each rational equation.

1. $\frac{1}{2} - \frac{1}{x} = \frac{1}{8}$ $8x$

2. $\frac{3}{y} - \frac{2}{5y} = 13$ $5y$

3. $\frac{2}{t^2 - 9} + \frac{5}{t - 3} = \frac{4}{t + 3}$ $t^2 - 9$

4. $\frac{6}{x^2 - 3x - 28} = \frac{3}{x - 7} - \frac{2}{x + 4}$
$x^2 - 3x - 28$

Solve. Check the extraneous solutions.

5. $\frac{1}{x} - \frac{1}{4} = \frac{1}{12}$ $x = 3$

6. $\frac{2}{a + 1} + \frac{1}{a - 1} = 1$ $a = 0 \text{ or } a = 3$

Write algebraic representations for the fractions described below.

7. The numerator of a fraction is twice the denominator. $d = \text{denominator}; \frac{2d}{d}$

8. The denominator of a fraction is 4 less than 3 times the numerator. $n = \text{numerator}; \frac{n}{3n - 4}$

Written Exercises

Solve. Check for extraneous solutions.

1. $\frac{1}{2} - \frac{1}{x} = \frac{1}{3}$ $x = 6$

2. $\frac{5}{2y} + \frac{1}{12} = \frac{3}{y}$ $y = 6$

3. $\frac{1}{2x} - \frac{1}{3x} = 1$ $x = \frac{1}{6}$

4. $\frac{5}{6} = \frac{1}{n} + \frac{2}{3n}$ $n = 2$

5. $\frac{5}{y + 3} = \frac{1}{y + 3} + 3$ $y = -\frac{5}{3}$

6. $\frac{1}{3x + 6} + 2 = \frac{3}{x + 2}$ $x = -\frac{2}{3}$

7. $\frac{2}{n^2 - n} + 1 = \frac{2}{n - 1}$ $n = 2$

8. $\frac{1}{x} + \frac{2}{x + 1} = \frac{7}{x^2 + x}$ $x = 2$

9. $\frac{1}{x - 4} = \frac{3}{x + 4} - \frac{2}{x^2 - 16}$ $x = 9$

10. $\frac{2}{y^2 - 9} = \frac{4}{y + 3} - \frac{5}{y - 3}$ $y = -29$

11. $\frac{6}{n^2 - 3n - 28} + \frac{2}{n + 4} = \frac{3}{n - 7}$

12. $\frac{4}{x^2 - 2x - 15} = \frac{2}{5 - x} + \frac{1}{x + 3}$

13. $\frac{5}{a^2 - 2a - 24} = \frac{2}{6 - a} + \frac{3}{a + 4}$

14. $\frac{4}{x - 5} + \frac{3}{x - 2} = \frac{4}{x^2 - 7x + 10}$

15. $\frac{2y - 1}{y^2 - 9y + 20} = \frac{4}{4 - y} + \frac{7}{y - 5}$

16. $\frac{-20}{a^2 - 4a - 45} = \frac{2}{a - 9} + \frac{a + 3}{a + 5}$

17. $\frac{x + 5}{x - 4} = \frac{3}{x} + \frac{36}{x^2 - 4x}$ $x = -6$

18. $\frac{16}{x^2 - 4x} = \frac{2}{x} + \frac{x}{x - 4}$ $x = -6$

19. $\frac{13}{x^2 - 4} + \frac{x}{2 - x} = \frac{-2}{x + 2}$ $x = \pm 3$

20. $\frac{x + 1}{x - 3} = \frac{3}{x} + \frac{12}{x^2 - 3x}$ $x = -1$

9.1 Rational Equations **341**

Closure

Have students explain *how* to solve each equation, but do not have them solve the equation.

1. $\frac{-3}{2x} + \frac{1}{x} = \frac{1}{4}$

2. $\frac{x - 3}{x - 2} + \frac{3}{x} = \frac{-15}{x^2 - 2x}$

a. Find the LCD of the denominators.
b. Multiply each side of the equation by the LCD. **c.** Simplify the resulting equation.
d. Solve the equation for x. **e.** Check in the original equation.   Ask what an extraneous root is and how to check for such roots.
See p. 338.

◼◼◼ FOLLOW UP

Guided Practice

Classroom Exercises 1–8
Try This   all

Independent Practice

A Ex. 1–20, **B** Ex. 21–29, **C** Ex. 30–31

Basic: WE 1–19 odd, 27

Average: WE 9–29 odd

Above Average: WE 11–31 odd

Additional Answers

Written Exercises

11. $n = -20$
12. $x = -15$
13. $a = 31$
14. $x = \frac{27}{7}$
15. $y = 7$
16. 1, 3
24. No solution

Sometimes, none of the solutions of a derived equation is a solution of the original equation. In that case, the original equation has no solution. Solve and check.

21. $\dfrac{6}{3y - 2} + \dfrac{6y}{9y^2 - 4} = \dfrac{1}{3y + 2}$ No solution

22. $\dfrac{x}{x - 2} = \dfrac{1}{x} + \dfrac{4}{x^2 - 2x}$ $x = -1$

23. $\dfrac{n}{n - 2} = \dfrac{4}{n + 3} + 1$ $n = 7$

24. $\dfrac{3}{9 - 4x^2} - \dfrac{2}{3 + 2x} = \dfrac{2x}{9 - 4x^2}$

The formula at the right refers to a convex lens and the way it forms an inverted image of a given object.

$$\dfrac{1}{D} + \dfrac{1}{d} = \dfrac{1}{F}$$

Here, D is the distance from the lens to the object; d is the distance from the lens to the image; and F is the distance from the lens to a point called the *focus*. F is called the *focal length* of the lens.

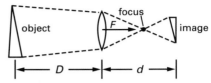

25. Find F for $D = 18$ cm and $d = 12$ cm. 7.2 cm

26. Find d for $D = 10$ cm and $F = 3$ cm. $4\frac{2}{7}$ cm

27. The sum of a number and its reciprocal is $\frac{25}{12}$. Find the number. $\frac{4}{3}, \frac{3}{4}$

28. Find two consecutive integers if the sum of their reciprocals is $\frac{11}{30}$. 5, 6

29. The denominator of a fraction is 2 less than twice the numerator. If the numerator is decreased by 2 and the denominator is increased by 3, the resulting fraction simplifies to $\frac{1}{3}$. Find the original fraction. $\frac{7}{12}$

Solve and check.

30. $\dfrac{5}{x^2 + 2x - 24} - \dfrac{2}{4x - x^2} = \dfrac{4}{x^2 + 6x}$

$x = -\frac{28}{3}$

31. $\dfrac{1}{x^2 + 5x + 6} = \dfrac{2}{x^2 - x - 6} - \dfrac{3}{9 - x^2}$

$x = -\frac{15}{4}$

Mixed Review

State the correct property for each equation. **1.5, 1.6, 1.7**

Assoc Prop for Add

1. $a + 6 = 6 + a$ Comm Prop for Add

2. $3 + (b + 9) = (3 + b) + 9$

3. $7(a + 6) = 7a + 7 \cdot 6$ Distr Prop for Mult Over Add

4. $1 \cdot x = x$ Mult Identity Prop

Simplify. **6.2, 6.10, 8.2, 8.3**

5. $(-3x^2y^4)^3$ $-27x^6y^{12}$

6. $(x + 3)^2$ $x^2 + 6x + 9$

7. $\dfrac{a - 3}{9 - a^2}$ $-\dfrac{1}{a + 3}$

8. $\dfrac{36a^4b^2}{9x - 27} \cdot \dfrac{9 - x^2}{12ab^5}$ $-\dfrac{a^3(x + 3)}{3b^3}$

Enrichment

Explain to students that it is helpful to know the resistance of household electrical appliances. The lower the resistance, the greater the amount of current used, and the higher the electricity bill. Two appliances with resistances R and r together have a total resistance T that can be found by using the equation, $\frac{1}{T} = \frac{1}{R} + \frac{1}{r}$. (Resistance is measured in ohms Ω.) Have the students find the missing resistance in each of the following.

1. $\dfrac{1}{T} = \dfrac{1}{49} + \dfrac{1}{14}$ $T = 10\frac{8}{9}\Omega$

2. $\dfrac{1}{T} = \dfrac{1}{2T} + \dfrac{1}{54}$ $T = 27\Omega, 2T = 54\Omega$

3. $\dfrac{1}{16} = \dfrac{1}{R} + \dfrac{1}{R + 24}$ $R = 24\Omega$

9.2 Ratios and Proportions

Objective

To solve proportions

A **ratio** is a comparison of two numbers by division. The equation $\frac{7}{2} = \frac{x}{3}$ states that two ratios are equal. Such an equation is called a *proportion*.

Definition

A **proportion** is an equation that states that two ratios are equal.

$$\frac{a}{b} = \frac{c}{d}, \; b \neq 0 \text{ and } d \neq 0 \quad \text{or} \quad a{:}b = c{:}d$$

In this proportion, a and d are the **extremes**; b and c are the **means**.

EXAMPLE 1

Identify the extremes and the means of the proportion $\frac{7}{2} = \frac{x}{3}$. Then solve for x.

Solution

The extremes are 7 and 3. The means are 2 and x.

A proportion such as $\frac{7}{2} = \frac{x}{3}$ can be solved by multiplying each side of the equation by the LCM of the denominators, $2 \cdot 3$.

$$\overset{1}{\cancel{2}} \cdot 3 \cdot \frac{7}{\underset{1}{\cancel{2}}} = 2 \cdot \overset{1}{\cancel{3}} \cdot \frac{x}{\underset{1}{\cancel{3}}}$$

$$3 \cdot 7 = 2 \cdot x$$

$$21 = 2x$$

$$10\frac{1}{2} = x, \text{ or } x = 10\frac{1}{2}$$

TRY THIS

1. Identify the extremes and the means of the proportion $\frac{x}{9} = \frac{4}{5}$. Then solve for x. x, 5; 9, 4; $7\frac{1}{5}$

Proportion Property

For all real numbers a, b, c, and d where $b \neq 0$, $d \neq 0$, if $\frac{a}{b} = \frac{c}{d}$, then $ad = bc$. That is, the product of the extremes equals the product of the means.

9.2 Ratios and Proportions **343**

Teaching Resources

Problem Solving Worksheet 9
Quick Quizzes 70
Reteaching and Practice
  Worksheets 70
Transparency 26

GETTING STARTED

Prerequisite Quiz

A football team won 8 games and lost 2 games. Find each of the following ratios.

1. wins to losses   8:2
2. losses to wins   2:8
3. wins to total games   8:10
4. losses to total games   2:10

Which of the following ratios are equal?

5. $\frac{8}{15}$ and $\frac{1}{2}$ $\neq$
6. $\frac{16}{20}$ and $\frac{4}{5}$ $=$

Additional Example 1

Identify the extremes and the means of the proportion $\frac{x}{10} = \frac{1}{9}$.   Extremes: x and 9;
Means: 10 and 1

Highlighting the Standards

Standards 4d, 1b, 10d: In the Examples and Exercises, students apply proportion to problems in several different areas. In the Application, they use direct variation and see how sampling is used in ecology.

Ask students to answer the questions based on this information.
A recent survey found that 9 out of 10 people preferred Whitewash Toothpaste.

1. Does this mean that exactly 10 people were surveyed?   No
2. Does the survey tell you exactly how many people were surveyed?   No
3. If the results of the survey are true, what is the least number of people who might have responded to it?   10
4. If the results of the survey are true, does this mean that 90% of the people who responded preferred Whitewash Toothpaste?   Yes
5. If 500 people actually responded to the survey, how many did *not* prefer Whitewash Toothpaste?   50

■■■TEACHING SUGGESTIONS

Lesson Note

Show students that writing a proportion as $a:b = c:d$ shows that a and d are the outermost terms (extremes) and b and c occupy the middle positions (means). When presenting Example 4 on page 345, emphasize the importance of checking whether an answer is reasonable. When explaining Example 5 on page 345, be sure students understand that $4x$ and $5x$ are in the ratio 4:5. That is, $\frac{4x}{5x} = \frac{4}{5}$.

The Proportion Property provides another method for solving proportions.

EXAMPLE 2   Solve $\frac{2}{x} = \frac{3}{x + 6}$.

Plan   Use the Proportion Property: product of extremes = product of means.

Solution

$$\frac{2}{x} = \frac{3}{x + 6} \qquad \text{Check:} \quad \frac{2}{x} = \frac{3}{x + 6}$$

$$2(x + 6) = x \cdot 3 \qquad\qquad\qquad \frac{2}{12} \stackrel{?}{=} \frac{3}{12 + 6}$$

$$2x + 12 = 3x$$

$$12 = x \qquad\qquad\qquad\qquad \frac{1}{6} \stackrel{?}{=} \frac{3}{18}$$

$$x = 12 \qquad\qquad\qquad\qquad \frac{1}{6} = \frac{1}{6} \ \text{True}$$

Thus, 12 is the solution of the proportion.

TRY THIS   2. Solve $\frac{4}{a} = \frac{6}{a + 3}$.   6

In solving a proportion, you may need to solve a quadratic equation.

EXAMPLE 3   Solve $\frac{y - 6}{7} = \frac{1}{y}$.

Plan   Use the fact that product of extremes = product of means.

Solution

$$(y - 6)y = 7 \cdot 1$$

$$y^2 - 6y = 7$$

$$y^2 - 6y - 7 = 0$$

Factor. $(y - 7)(y + 1) = 0$

$$y - 7 = 0 \quad or \quad y + 1 = 0$$

$$y = 7 \qquad\qquad y = -1$$

The checks are left for you. Both 7 and -1 make the proportion true.
Thus, the solutions are 7 and -1.

TRY THIS   3. Solve $\frac{1}{x} = \frac{(x + 5)}{24}$. -8, 3

Many practical problems involve ratio and proportion. Some of these are illustrated in the next three examples.

344   Chapter 9 Applying Rational Expressions

Additional Example 2

Solve $\frac{3}{a - 4} = \frac{-5}{a}$. $\frac{5}{2}$

Additional Example 3

Solve $\frac{y - 7}{10} = \frac{-1}{y}$.   5, 2

EXAMPLE 4 A recipe for $3\frac{1}{2}$ dozen muffins requires 700 g of flour. How many dozens of muffins can be made using 800 g of flour? (THINK: 800 g is a little more than 700 g, so the answer must be a little more than $3\frac{1}{2}$.)

Solution Let x = the number of dozens of muffins using 800 g of flour.

$$\frac{3\frac{1}{2}}{700} = \frac{x}{800}$$

Calculator Steps:
3.5 $\boxed{\times}$ 800 $\boxed{\div}$ 700 $\boxed{=}$ 4

$$\frac{7}{2} \cdot 800 = 700 \cdot x$$
$$2{,}800 = 700x$$
$$4 = x$$

(THINK: 4 is a little more than $3\frac{1}{2}$, so the answer is reasonable.)

Check The ratios $\frac{3\frac{1}{2}}{700}$ and $\frac{4}{800}$ are equal, since $3\frac{1}{2} \times 800 = 2{,}800$ and $4 \times 700 = 2{,}800$.
Thus, 4 dozen muffins can be made.

EXAMPLE 5 A recent poll found that 7 out of 8 people use No-Cavit toothpaste. How many people can be expected to use this brand in a city of 40,000?

Solution 7 *out of* 8 means $\frac{7}{8}$. Let x = the number of people who can be expected to use No-Cavit in a city of 40,000.

$$\frac{7}{8} = \frac{x}{40{,}000} \quad \begin{array}{l} \longleftarrow \text{No-Cavit users} \\ \longleftarrow \text{total number of people} \end{array}$$

$$7 \cdot 40{,}000 = 8 \cdot x$$
$$280{,}000 = 8x$$
$$35{,}000 = x$$

Calculator Steps:
7 $\boxed{\times}$ 40,000 $\boxed{\div}$ 8 $\boxed{=}$ 35,000

The check is left for you.
In a city of 40,000 people, 35,000 can be expected to use No-Cavit.

TRY THIS 4. A recipe for $2\frac{1}{2}$ dozen cookies requires 400 g of flour. How many dozens can be made using 600 g of flour? $3\frac{3}{4}$ dozen

Some problems involving *ratios* can be solved without writing proportions. If two numbers are in a given ratio, they can be represented in terms of a single variable.

Additional Example 4

Three cans of soda cost $1.20. Find the cost of 5 cans. $2.00

Additional Example 5

A councilwoman deduced from a poll that 4 out of 7 people will vote for her. How many votes will she get if 55,000 votes are cast? $\approx$31,429, or 31,400 to the nearest hundred

1. Identify the extremes and the means of the proportion $\frac{3}{y} = \frac{-4}{9}$. The extremes are 3 and 9; the means, y and -4.

2. Solve $\frac{x}{3} = \frac{x-5}{-2}$.   3

3. Solve $\frac{m}{3} = \frac{2}{m-1}$. $-2, 3$

4. In Martinsville, 2 out of 5 people belong to a union. How many union members are there if the population is 70,000?   28,000

5. It costs Tim $3.25 to bake 5 loaves of bread. How much will it cost him to bake 7 loaves? $4.55

6. Find the measures of two supplementary angles if they are in a ratio of 3:2.
108, 72

Closure

Ask students these questions.

1. What is a rational equation?
See p. 337.

2. What is the first step in solving a rational equation?   See p. 337.

3. Why must you always check the roots of a rational equation?   Sometimes there is an extraneous root

4. What is a proportion?   See p. 343.

5. How do you solve an equation involving a proportion?   Use the Proportion Property.

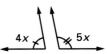

This figure shows two angles whose measures are in a ratio of 4:5.
Two angles are **supplementary** if the sum of their degree measures is 180.

EXAMPLE 6   Find the measures of two supplementary angles if they are in a ratio of 4:5.

Solution   Let $4x$ = measure of one angle. Let $5x$ = measure of second angle.

$$4x + 5x = 180 \quad \longleftarrow \text{ Sum of measures of supplementary angles is 180.}$$
$$9x = 180$$
$$x = 20$$

Then, $4x = 80$ and $5x = 100$.

Check   The ratio $\frac{80}{100}$ equals $\frac{4}{5}$. The angles are supplementary, because $80 + 100 = 180$.

Thus, the degree measures of the angles are 80 and 100.

TRY THIS   5. Find the measures of two supplementary angles if they are in a ratio of 2:1.   120, 60

Focus on Reading

Which equation is *not* equivalent to the others in the group? Explain.

1. a. $\frac{8}{3} = \frac{y}{9}$ b. $\frac{72}{3} = y$ c. $9y = 24$ d. $8 \cdot 9 = 3 \cdot y$

2. a. $\frac{3}{x} = \frac{4}{x-2}$ b. $4(x-2) = 3x$ c. $3(x-2) = 4x$ d. $-6 = x$

1. c; for a, b, and d, $y = 24$.
2. b; for a, c, and d, $x = -6$.

Classroom Exercises

Identify the extremes and means of each proportion. Then use the Proportion Property to write an equivalent equation without fractions.

3, $2x - 3$; x, 5; $6x - 9 = 5x$

1. $\frac{n}{5} = \frac{3}{4}$ 2. $\frac{4}{x} = \frac{3}{5}$ 3. $\frac{x}{7} = \frac{x-2}{10}$ 4. $\frac{3}{x} = \frac{5}{2x-3}$

n, 4; 5, 3; $4n = 15$ 4, 5; x, 3; $20 = 3x$ x, 10; 7, $x - 2$; $10x = 7x - 14$

Solve.

5. $\frac{n}{3} = \frac{4}{7}$ $n = \frac{12}{7}$ 6. $\frac{3}{a-2} = \frac{4}{a}$ $a = 8$ 7. $\frac{x-1}{3} = \frac{x}{4}$ $x = 4$ 8. $\frac{8}{n} = \frac{n}{18}$

$n = 12 \text{ or } n = -12$

Additional Example 6

Find the measures of two complementary angles if they are in the ratio of 8:1.   80, 10

Written Exercises

Identify the extremes and the means of each proportion.

10, 13; *m*, 52

1. $\dfrac{5}{2} = \dfrac{10}{x}$ **2.** $\dfrac{9}{12} = \dfrac{36}{n}$ **3.** $\dfrac{x}{3} = \dfrac{7}{4}$ *x*, 4; 3, 7 **4.** $\dfrac{10}{m} = \dfrac{52}{13}$

5. $\dfrac{3}{n+6} = \dfrac{2}{n}$ **6.** $\dfrac{x+3}{5} = \dfrac{x-2}{4}$ **7.** $\dfrac{n}{3} = \dfrac{2}{n-5}$ **8.** $\dfrac{x}{2} = \dfrac{2}{x+3}$

x, *x* + 3; 2, 2

9–12. Solve the proportions in Written Exercises 1–4.

9. $x = 4$ **10.** $n = 48$ **11.** $x = 5\frac{1}{4}$ **12.** $m = 2\frac{1}{2}$

Solve.

13. $\dfrac{x}{5} = \dfrac{3}{7}$ $x = 2\frac{1}{7}$ **14.** $\dfrac{11}{x} = \dfrac{3}{1}$ $x = 3\frac{2}{3}$ **15.** $\dfrac{4}{n} = \dfrac{5}{9}$ $n = 7\frac{1}{5}$ **16.** $\dfrac{8}{3} = \dfrac{m}{7}$ $m = 18\frac{2}{3}$

17. $\dfrac{2}{x-3} = \dfrac{5}{x}$ **18.** $\dfrac{x-1}{3} = \dfrac{x+1}{5}$ **19.** $\dfrac{n}{3} = \dfrac{n+4}{7}$ **20.** $\dfrac{1}{n-3} = \dfrac{3}{n-5}$

21. $\dfrac{x}{2} = \dfrac{30}{x-4}$ **22.** $\dfrac{12}{x} = \dfrac{x+4}{1}$ **23.** $\dfrac{n}{4} = \dfrac{10}{n-3}$ **24.** $\dfrac{y}{3} = \dfrac{56}{y+2}$

25. $\dfrac{8}{x} = \dfrac{x}{2} \pm 4$ **26.** $\dfrac{n}{3} = \dfrac{1}{12n} \pm \dfrac{1}{2}$ **27.** $\dfrac{y-5}{4} = \dfrac{y+3}{3}$ **28.** $\dfrac{m}{2} = \dfrac{36}{m+6}$

29. $\dfrac{x}{3} = \dfrac{6}{2x} \pm 3$ **30.** $\dfrac{12}{y} = \dfrac{y}{12} \pm 12$ **31.** $\dfrac{2x}{3} = \dfrac{16}{x+2}$ **32.** $\dfrac{1}{y} = \dfrac{6y-1}{1}$

33. If 6 out of 8 people use Clean-White toothpaste, how many people out of 60,000 can be expected to use Clean-White toothpaste? 45,000 people

34. Sal uses 4 skeins of yarn to make 3 scarfs. How many skeins of yarn will he need to make 15 scarfs?

20 skeins

35. $\dfrac{3}{x+2} = \dfrac{x-8}{13}$ **36.** $\dfrac{2}{n-3} = \dfrac{n-4}{1}$ **37.** $\dfrac{9}{m-5} = \dfrac{m+7}{-3}$ **38.** $\dfrac{b+2}{3} = \dfrac{13}{b-8}$

39. $\dfrac{n}{n+3} = \dfrac{n}{2}$ **40.** $\dfrac{y}{y+4} = \dfrac{y}{10}$ **41.** $\dfrac{n+3}{n-1} = \dfrac{5n}{2}$ **42.** $\dfrac{2x+2}{x+1} = \dfrac{x-2}{1}$

43. A recipe for making bran muffins requires 1,600 g of flour to make 40 muffins. How many muffins can be made using 1,000 g of flour? 25 muffins

44. Mr. Carmelito used 15 gallons of gasoline to drive 450 mi. How far can he drive on a full tank of 20 gallons of gasoline? 600 mi

45. Find the measures of two supplementary angles if their measures are in a ratio of 7:2. 140, 40

46. Find the measures of two supplementary angles if their measures are in a ratio of 1:3. 45, 135

47. Find the measures of two complementary angles if their measures are in a ratio of 5:13. (Recall that two angles are complementary if the sum of their measures is 90.) 25, 65

48. The area of a rectangle is 300 in². The sides are in the ratio 3:4. Find the perimeter of the rectangle. 70 in

49. If $3x = 4y$, find the ratio *y*:*x*. 3:4

50. If $pq = rs$, find the ratio *r*:*q*. *p*:*s*

FOLLOW UP

Guided Practice

Classroom Exercises 1–8
Try This all, Focus on Reading all

Independent Practice

A Ex. 1–34, **B** Ex. 35–46, **C** Ex. 47–50
Basic: WE 1–33 odd, 34, Application
Average: WE 13–45 odd, Application
Above Average: WE 17–49 odd, Application

Additional Answers

Written Exercises

1. E: 5, *x*; M: 10, 2
2. E: 9, *n*; M: 12, 36
5. E: 3, *n*; M: *n* + 6, 2
6. E: *x* + 3, 4; M: 5, *x* − 2
7. E: *n*, *n* − 5; M: 3, 2
17. $x = 5$ 18. $x = 4$
19. $n = 3$ 20. $n = 2$
21. −6, 10 22. −6, 2
23. −5, 8 24. −14, 12
27. −27 28. 6, −12
31. 4, −6 32. $-\frac{1}{3}, \frac{1}{2}$
35. 11, −5 36. 5, 2
37. 2, −4 38. 11, −5
39. 0, −1 40. 0, 6
41. $-\frac{3}{5}, 2$ 42. $x = 4$

Mixed Review

Solve.   *3.3, 5.6, 7.7*

1. $3x - 2 = -23$
$x = -7$

2. $|3x + 1| = 16$
$x = 5 \text{ or } x = -\frac{17}{3}$

3. $a^2 - 6a = 0$
$a = 0 \text{ or } a = 6$

4. $x^2 - 16 = 0$
$x = 4 \text{ or } x = -4$

Multiply.   *8.3*

5. $(a - 1) \cdot \dfrac{5a + 15}{a^2 + 2a - 3}$ 5

6. $\dfrac{x^2 - 9}{4x + 20} \cdot \dfrac{x^2 - 25}{x^2 - 8x + 15}$ $\frac{x + 3}{4}$

Divide.   *8.4*

7. $\dfrac{a^2 - 9a + 20}{4a - a^2} \div \dfrac{a^2 + 2a - 35}{a}$ $-\frac{1}{a + 7}$

8. $\dfrac{18x^4y^2}{a^2 + 4a - 5} \div \dfrac{6xy^4}{a^2 - 4a + 3}$ $\frac{3ax^3 - 9x^3}{ay^2 + 5y^2}$

9. The larger of two numbers is 8 more than 4 times the other. Their sum is 113. Find the two numbers.   *4.2* 21, 92

▰ Application: Direct Variation

A visitor to a sheep ranch asked a sheepherder how he could count the sheep so quickly. The herder answered, "Oh, it's easy. I just count the legs and divide by four!" This old joke may not describe an efficient way to count sheep, but it is an example of *direct variation*. The ratio of the number of legs to the number of heads is always 4:1.

In a **direct variation**, the ratio of y to x is always the same. If y *varies directly* as x, then

$$\frac{y}{x} = k, \text{ or } y = kx \text{ where } k \text{ is constant.}$$

Solve.

1. A conservationist catches 650 deer, tags them, and releases them. Later she catches 216 deer and finds that 54 of them are tagged. Estimate how many deer are in the forest. Use this proportion.

$$\frac{\text{Total tagged}}{\text{Total in forest}} = \frac{\text{Sample with tags}}{\text{Sample caught}}$$

(Assume that the number of tagged deer caught *varies directly* as the number of deer later caught.)   2,600 deer

2. A game warden catches 125 fish, tags them, and puts them back in the lake. Later, he catches 65 fish and finds that 13 of them are tagged. How many fish are in the lake? Assume that the number of tagged fish caught *varies directly* as the number of fish later caught.   625 fish

3. Emil's hourly earnings *vary directly* as the number of hours worked. For working 45 h, the earnings are $168.75. Find his earnings for 35 h of work. $131.25

348   Chapter 9 Applying Rational Expressions

Enrichment

The pattern 1, 2, 4, 8, 16, $\cdots$ is called a geometric sequence because there is a *common ratio* between each term and its preceeding term. In the sequence above, $\frac{2}{1} = \frac{4}{2} = \frac{8}{4} = 2$. Thus, the common ratio is 2. Ask students to find the common ratio in each of the following sequences and to use it to write the next 3 terms in the sequence.

1. 2, 6, 18, 54, $\cdots$ 3; 162, 486, 1458

2. 5, 50, 500, 5000, $\cdots$ 10; 50,000, 500,000, 5,000,000

3. 16, 4, 1, $\frac{1}{4}$, $\cdots$ $\frac{1}{4}$; $\frac{1}{16}, \frac{1}{64}, \frac{1}{256}$

4. 1, $\frac{1}{2}, \frac{1}{4}, \frac{1}{8}$, $\cdots$ $\frac{1}{2}$; $\frac{1}{16}, \frac{1}{32}, \frac{1}{64}$

Ask students to find the sum of the first 7 terms in the last sequence. $1\frac{63}{64}$

If the 11th term of this sequence is $\frac{1}{1024}$, ask what they think the sum of these 11 terms would be. $1\frac{1023}{1024}$

9.3 Literal Equations

Objectives

To solve a literal equation or formula for one of its variables

To evaluate a formula for one of its variables, given the value(s) of its other variables(s)

Teaching Resources

Application Worksheet 9
Quick Quizzes 71
Reteaching and Practice
  Worksheets 71

Simple interest on money invested can be expressed by the formula

$$i = prt,$$

where i is the amount of interest, p is the principal, r is the rate of interest, and t is the time in years. The formula $i = prt$ can also be solved for one of the other variables. For example, you can solve for r by dividing each side of the equation by pt.

$$i = prt$$

$$\frac{i}{pt} = \frac{\overset{1}{\cancel{p}}\,\overset{1}{\cancel{r}}t}{\underset{1}{\cancel{p}}\,\underset{1}{\cancel{t}}}$$

$$\frac{i}{pt} = r, \text{ or } r = \frac{i}{pt}$$

A formula contains more than one variable. Similarly, a *literal equation* such as $ax + b = 10c$ contains more than one letter or variable. In a **literal equation** any variable can be expressed in terms of the others.

EXAMPLE 1 Solve $ax + b = 10c$ for x.

Plan Solve for x in the same way you would solve the equation $2x + 18 = 16$.

Solution

$$\begin{aligned} ax + b &= 10c \\ ax + b - b &= 10c - b \\ ax &= 10c - b \\ x &= \frac{10c - b}{a} \end{aligned}$$

THINK: $2x + 18 = 16$

$$\begin{aligned} 2x + 18 - 18 &= 16 - 18 \\ 2x &= -2 \\ x &= -1 \end{aligned}$$

TRY THIS 1. Solve $ax - 3b = 8c$ for a. $\frac{8c + 3b}{x}$

Additional Example 1

Solve $c - bx = 8d$ for x. $\frac{c - 8d}{b}$

GETTING STARTED

Prerequisite Quiz

For $i = prt$, find each of the following.

1. Find i if $p = 100$, $r = 6\%$, and $t = 2$.
   12
2. Find p if $i = 30$, $r = 8\%$, and $t = 3$.
   125
3. Find r if $p = 400$, $i = 16$, and $t = 4$.
   1%
4. Find t if $r = 7\%$, $p = 1{,}000$, and $i = 140$.
   2

Factor out the GCF.

5. $mx + bx$ $x(m + b)$
6. $px - 4x$ $x(p - 4)$

Motivator

Draw a rectangle with given length, l, and width, w, on the chalkboard. Write the equation $P = 2l + 2w$ near the rectangle. Ask students to solve the formula for $2l$. $P - 2w$ Ask them to solve for l.
$\frac{P - 2w}{2}$ Ask them to solve for $2w$. $P - 2l$.
Ask them to solve for w. $\frac{P - 2l}{2}$
Now write the formula $A = lw$ on the chalkboard. Ask students to solve for l.
$l = \frac{A}{w}$ Ask them to solve for w. $w = \frac{A}{l}$

Highlighting the Standards

Standards 4b, 4d: Solving literal equations is shown to be the same procedure as solving equations with numerical coefficients. Applications of this are taken from banking and from geometry.

Lesson Note

Be sure that the students understand what a formula does. Take time to show an example using the interest formula. For example, if $800 is invested at a rate of 8% for 2 years, then the interest earned is $i = 800(.08)(2)$, or $128. When presenting the Examples of this lesson, emphasize that the goal is to "isolate" the required variable. Ask questions such as: What must be "undone" to get bx on the left side of the equation? (See Example 2). What must be "undone" to get x alone? Divide both sides by $(a + b)$. Remind students that the easiest way to solve an equation involving fractions is to clear fractions first by multiplying by the LCD of the denominator.

Math Connections

Measurement: The formula for changing a Celsius temperature to a Fahrenheit temperature is $F = \frac{9}{5}C + 32$. If this equation is solved for C it becomes $C = \frac{5}{9}(F - 32)$.

To solve an equation for a variable that appears in more than one term, get all of those terms by themselves on one side of the equation.

EXAMPLE 2   Solve $ax = 4c - bx$ for x.

Plan   Add bx to each side.

Solution

$$ax = 4c - bx$$
$$ax + bx = 4c - bx + bx$$

Factor out x. $ax + bx = 4c$

Divide each side by $(a + b)$. $x(a + b) = 4c$

$$x = \frac{4c}{a + b}$$

TRY THIS   **2.** Solve $ay = xy + 6c$ for y. $\frac{6c}{a - x}$

An important formula in banking is $A = p + prt$. The formula is used to find: (1) the amount of money in an account after interest has been credited, or (2) the total amount due on a loan including the interest.

EXAMPLE 3   Solve $A = p + prt$ for t. Then find the time t in years, for principal $p = \$500$, amount $A = \$560$, and rate $r = 6\%$ per year.

Plan   First solve for t.

Solution   Subtract p from each side.

$$A = p + prt$$
$$A - p = p - p + prt$$

Divide each side by pr. $A - p = prt$

$$\frac{A - p}{pr} = t$$

Now find t by substituting 560 for A, 500 for p, and 0.06 for r.

$$t = \frac{A - p}{pr}$$

$$= \frac{560 - 500}{500(0.06)}$$

Calculator Steps:

560 [−] 500 [=] [÷] 500 [÷] .06 [=]

$$= \frac{60}{30} = 2$$

Thus, $t = \frac{A - p}{pr}$, and $t = 2$ years for the given values of A, p, and r.

TRY THIS   **3.** Solve $A = p + prt$ for p. Then find the principal, p, when the amount $A = \$399$, the rate $r = 7\%$ per year, and the time $t = 2$ years. $p = \frac{A}{(1 + rt)}, p = \$350$

Additional Example 2

Solve $ax = -5b - cx$ for x. $\frac{-5b}{a + c}$

Additional Example 3

Solve $A = p + prt$ for r. $\frac{A - p}{pt}$

EXAMPLE 4 Solve $\frac{1}{a} = \frac{5}{b} + \frac{3}{c}$ for b.

Solution Multiply by the LCD, abc.

$$abc \cdot \frac{1}{a} = abc \cdot \frac{5}{b} + abc \cdot \frac{3}{c}$$

$$\cancel{a} \cdot b \cdot c \cdot \frac{1}{\cancel{a}} = a \cdot \cancel{b} \cdot c \cdot \frac{5}{\cancel{b}} + a \cdot b \cdot \cancel{c} \cdot \frac{3}{\cancel{c}}$$

$$bc = 5ac + 3ab$$

Subtract $3ab$. $$bc - 3ab = 5ac$$

Factor out b. $$b(c - 3a) = 5ac$$

$$b = \frac{5ac}{c - 3a}$$

EXAMPLE 5 A formula for the area of a trapezoid is

$$A = \tfrac{1}{2}h(b + c).$$

Solve this formula for c. Then find c for $A = 70$, $b = 6$, and $h = 10$.

Solution Multiply each side by 2.

Use the Distributive Property.
Subtract bh from each side to get ch alone.
Divide each side by h.

$$A = \tfrac{1}{2}h(b + c)$$
$$2A = h(b + c)$$
$$2A = bh + ch$$
$$2A - bh = ch$$
$$\frac{2A - bh}{h} = c$$

Substitute 70 for A, 6 for b and 10 for h.

$$c = \frac{2A - bh}{h}$$
$$= \frac{2 \cdot 70 - 6 \cdot 10}{10}$$
$$= \frac{140 - 60}{10}$$
$$= \frac{80}{10}, \text{ or } 8$$

Thus, $c = \frac{2A - bh}{h}$, and $c = 8$ for the given values of A, h, and b.

TRY THIS 4. Solve $\frac{2}{x} = \frac{5}{2y} - \frac{3}{z}$ for z. $z = \frac{6xy}{5x - 4y}$

Analysis: Have students solve the equation $A = p + prt$ for p. $p = \frac{A}{1 + rt}$ Ask them to explain how this formula can be used. To find the principal needed to invest at a given rate, r, in order to have a given amount A over a given time, t. Now have students solve the same formula for r and explain how it can be used. $r = \frac{A - p}{pt}$; to find the rate, r, at which a given principal, p, must be invested over a given amount of time, t, to yield a desired amount, A.

Common Error Analysis

Error: When the variable being solved for appears on both sides of the equation, some students will fail to get all these terms on one side and end up with a "solution" that contains the variable on both sides.

Show them that their formula would not be very useful, and remind them to collect all terms involving the variable on *one* side of the equation.

Checkpoint

1. Solve $ax + m = n$ for x. $\frac{n - m}{a}$
2. Solve $mx - 3p = nx$ for x. $\frac{3p}{m - n}$
3. Solve $A = ax - a$ for a. $\frac{A}{x - 1}$
4. Solve $\frac{1}{x} - \frac{1}{a} = \frac{1}{b}$ for x. $\frac{ab}{a + b}$
5. Solve $V = \frac{a(x + b)}{3}$ for x. $\frac{3V}{a} - b$

Additional Example 4

Solve $\frac{3}{a} = \frac{4}{b} - \frac{2}{c}$ for c. $\frac{2ab}{4a - 3b}$

Additional Example 5

Solve $A = \frac{h(b + c)}{2}$ for b.

Then find b for $A = 81$, $h = 9$, and $c = 11$.
$b = \frac{2A}{h} - c$; 7

351

Closure

Have students show every step in solving $6 = 2(4 + x)$. -1 Then have them use the same steps to solve $P = 2(l + w)$ for w.
$\frac{P - 2l}{2}$

■ FOLLOW UP

Guided Practice

Classroom Exercises 1–12
Try This   all

Independent Practice

A Ex. 1–24, **B** Ex. 25–34, **C** Ex. 35–43

Basic: WE 1–25 odd, 35, Midchapter Review, Brainteaser

Average: WE 11–35 odd, Midchapter Review, Brainteaser

Above Average: WE 19–43 odd, Midchapter Review, Brainteaser

Additional Answers

Written Exercises

9. $x = \frac{c - b}{6}$

12. $x = \frac{q - r}{p}$

15. $x = \frac{c}{a - b}$

18. $x = \frac{d}{m - r}$

Classroom Exercises

Solve for x.

1. $ax = 3$ $x = \frac{3}{a}$
2. $x + b = c$ $x = c - b$
3. $x - a = b$ $x = b + a$
4. $ax = b$ $x = \frac{b}{a}$
5. $\frac{x}{a} = b$ $x = ab$
6. $-5 = ax$ $x = -\frac{5}{a}$
7. $x + r = s$ $x = s - r$
8. $-4 = \frac{x}{t}$ $x = -4t$
9. $ax - b = 6$ $x = \frac{6 + b}{a}$
10. $-n = 12x + 5$ $x = -\frac{n + 5}{12}$
11. $ax - 3c = 2x$ $x = \frac{3c}{a - 2}$
12. $\frac{1}{x} = \frac{a}{b}$ $x = \frac{b}{a}$

Written Exercises

Solve for x.

1. $bx = 5$ $x = \frac{5}{b}$
2. $x - e = f$ $x = f + e$
3. $x + 2c = b$ $x = b - 2c$
4. $-ax = c$ $x = -\frac{c}{a}$
5. $\frac{x}{y} = z$ $x = yz$
6. $\frac{2}{x} = \frac{a}{c}$ $x = \frac{2c}{a}$
7. $\frac{r}{s} = \frac{3}{x}$ $x = \frac{3s}{r}$
8. $-5 = \frac{1}{cx}$ $x = -\frac{1}{5c}$
9. $6x + b = c$
10. $ax - b = c$ $x = \frac{c + b}{a}$
11. $n = ax + m$ $x = \frac{n - m}{a}$
12. $px - q = -r$
13. $ax + bx = c$ $x = \frac{c}{a + b}$
14. $c = ax - bx$ $x = \frac{c}{a - b}$
15. $ax = c + bx$
16. $ax = 4c - 3x$ $x = \frac{4c}{a + 3}$
17. $bx = 36 - dx$ $x = \frac{36}{b + d}$
18. $mx = d + rx$
19. $cx = dx + 18$ $x = \frac{18}{c - d}$
20. $7ax = 16 - cx$ $x = \frac{16}{7a + c}$
21. $\frac{x}{3} = \frac{a}{2} + \frac{t}{6}$ $x = \frac{3a + t}{2}$
22. $\frac{h}{5} + \frac{h}{3} = \frac{x}{15}$ $x = 8h$
23. $\frac{x}{5b} = \frac{k}{j}$ $x = \frac{5bk}{j}$
24. $\frac{x}{y} - g = v$ $x = y(v + g)$

25. The formula for the perimeter of a rectangle is $p = 2l + 2w$. Solve for l. Then find l for $p = 49.4$ cm and $w = 6.3$ cm.

$l = \frac{p - 2w}{2}$;
$l = 18.4$ cm

26. The formula for the perimeter of a model of a certain racetrack is $p = 4r + 2\pi r$. Solve for r. Then find r for $p \approx 30.84$.

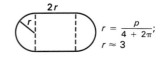

$r = \frac{p}{4 + 2\pi}$;
$r \approx 3$

Solve for x.

27. $-6x + c = -3x - 14c$ $x = 5c$
28. $16a - 5bx = bx - 4$ $x = \frac{2 + 8a}{3b}$
29. $kx - cd = 4e + f$ $x = \frac{4e + f + cd}{k}$
30. $5ax + b = 80ax - c$ $x = \frac{b + c}{75a}$
31. $px + q = rx - 6p$ $x = \frac{6p + q}{r - p}$
32. $-4a + 2bx = 12a + 10bx$ $x = \frac{-2a}{b}$
33. $\frac{1}{a} + \frac{1}{b} = \frac{1}{x}$ $x = \frac{ab}{a + b}$
34. $v = \frac{x - t}{s}$ $x = vs + t$

Enrichment

Have students research the "rule of 72" for compound interest, and illustrate it with some examples.   The rule of 72 states that if you divide 72 by the yearly interest rate, you will get the approximate time in years that it would take for an invested amount of money to double in value using compound interest. For example, at 8% this would take about 9 years, at 12% about 6 years, and so on.

35. Give an example of a literal equation and write in your own words a step-by-step description of how to solve it. **Answers will vary**

Solve each equation for the variable indicated.

36. $a^2 - ax + 12 = 4x - 7a$ for x $a + 3$ **37.** $rx + 2sx = r^2 - 7rs - 18s^2$ for x $r - 9s$

38. $\dfrac{rs}{a} + \dfrac{st}{b} = 1$ for s $\dfrac{ab}{br + at}$ **39.** $s = \dfrac{a - ar^2}{1 - r}$ for a $\dfrac{s}{1 + r}$

40. $W = -G\left(c - \dfrac{Q}{m}\right)$ for Q $\dfrac{mW}{G} + mc$ **41.** $ry^2 - b^2r - ty^2 + b^2t = y^2 - b^2$ for r $\dfrac{1}{1 + t}$

42. $a_1 = b\left(\dfrac{a_2}{b} + 4\right)$ for a_2 $a_1 - 4b$ **43.** $V = \dfrac{A}{x_1} - \dfrac{B}{x_2}$ for x_2 $\dfrac{Bx_1}{A - Vx_1}$

Midchapter Review

Solve. Check for extraneous solutions.   **9.1, 9.2**

1. $\dfrac{2}{3} - \dfrac{5}{x} = \dfrac{1}{4}$ $x = 12$ **2.** $\dfrac{1}{y} - \dfrac{2}{3y} = 2$ $y = \dfrac{1}{6}$

3. $\dfrac{3a - 1}{a^2 - a - 12} = \dfrac{5}{a + 3} - \dfrac{6}{4 - a}$ $a = \dfrac{1}{8}$ **4.** $\dfrac{3y - 1}{y - 4} = \dfrac{3}{y^2 - 16} + \dfrac{3}{y + 4}$

5. $\dfrac{x}{12} = \dfrac{3}{x}$ $x = 6$ or $x = -6$ **6.** $\dfrac{y - 5}{6} = \dfrac{y - 1}{3}$ $y = -3$

7. $\dfrac{a + 4}{-5} = \dfrac{2}{a - 3}$ $a = 1$ or $a = -2$ **8.** $\dfrac{n}{3} = \dfrac{150}{2n}$ $n = 15$ or $n = -15$

4. $y = -1\dfrac{2}{3}$ or $y = -1$

Solve for x.   **9.3**

9. $ax - 3 = 5b$ $x = \dfrac{5b + 3}{a}$ **10.** $n = \dfrac{x}{y}$ $x = ny$

11. $bx - c = ax$ $x = \dfrac{c}{b - a}$ **12.** $\dfrac{x}{4} = \dfrac{a}{3} + \dfrac{b}{6}$ $x = \dfrac{4a + 2b}{3}$

13. The sum of a number and its reciprocal is $\dfrac{61}{30}$. Find the number.   **9.1** $\dfrac{5}{6}$ or $\dfrac{6}{5}$

14. The formula for the perimeter of a rectangle is $p = 2l + 2w$. Solve for w. Then find w for $p = 62$ cm and $l = 7$ cm.   **9.3** $w = \dfrac{p - 2l}{2}$; $w = 24$ cm

◢◢◢/ Brainteaser

A young boy could not yet tell time but was counting the number of times the grandfather clock chimed. At 3:40 in the afternoon he told his mother that the clock had chimed 34 times. If the clock chimes the number of times of the hour on the hour and once on the half hour, what time was it when the boy began counting the chimes?   11 AM

Teaching Resources

Project Worksheet 9
Quick Quizzes 72
Reteaching and Practice
    Worksheets 72
Transparency 27

■■■GETTING STARTED

Prerequisite Quiz

1. Solve $d = rt$ for r. $\frac{d}{t}$
2. Solve $d = rt$ for t. $\frac{d}{r}$

Use $d = rt$ for the following.

3. Find d when $r = 50$ km/h and $t = 2$ h.
   100 km
4. Find r when $d = 800$ mi and $t = 4$ h.
   200 mi/h
5. Find t when $d = 1,000$ km and $r = 300$ km/h. $3\frac{1}{3}$ h

Motivator

Ask students how far they will travel if they drive 45 mi/h for 3 h.   135 mi   Ask them what was the rate if they drove 200 mi in 4 h?   50 mi/h   How long will it take them to drive 300 mi if they drive at an average rate of 60 mi/h?   5 h

9.4 Problem Solving: Motion Problems

Objective   To solve motion problems

Motion problems involve three variables: distance, rate, and time where distance = rate × time, or $d = rt$. The formula $d = rt$ can be solved for either r or t.

$$r = \frac{d}{t} \qquad \text{or} \qquad t = \frac{d}{r}$$

Each form of the distance formula is useful in solving problems.

EXAMPLE 1   *Traveling in Opposite Directions*
Two cyclists leave at the same time from the same point and travel in opposite directions. The rate of one cyclist is 7 km/h less than the rate of the other cyclist. After 10 h, the cyclists are 530 km apart. Find the rate of each cyclist.

Solution   Let x = the rate of the faster cyclist.
Then $x - 7$ = the rate of the slower cyclist
Make a sketch. Also, make a table to represent the distance traveled by each cyclist. Then write an equation and solve it.

	Rate	Time	Distance ($d = r \cdot t$)
Faster cyclist	x	10	$10x$
Slower cyclist	$x - 7$	10	$10(x - 7)$, or $10x - 70$

$$\frac{\text{distance of}}{\text{faster cyclist}} + \frac{\text{distance of}}{\text{slower cyclist}} = \frac{\text{total distance}}{\text{apart}}$$

$$
\begin{aligned}
10x \quad + \quad 10x - 70 \quad &= \quad 530 \\
20x &= 600 \\
x &= 30 \text{ (faster rate: 30 km/h)} \\
\text{So, } x - 7 &= 23 \text{ (slower rate: 23 km/h)}
\end{aligned}
$$

Check   The rates differ by 7 km/h. After 10 h, are the cyclists 530 km apart? Yes, since $(10 \cdot 30) + (10 \cdot 23) = 300 + 230 = 530$.

Thus, the two rates are 30 km/h and 23 km/h.

TRY THIS   1. Two cars leave at the same time from the same point and travel in opposite directions. The rate of one car is 4 km/h more than the rate of the other car. After 6 h, the cars are 624 km apart. Find the rate of each car. 50 km/h; 54 km/h

354   Chapter 9 Applying Rational Expressions

Highlighting the Standards

Standards 1b, 5b: Students apply problem-solving approaches, including making tables, to writing equations for, and solving, motion problems.

Additional Example 1

Two runners leave at the same time from the same point and travel in opposite directions. The rate of the first runner is 3 km/h more than the rate of the second runner. After 3 h the runners are 63 km apart. Find the rate of each runner.   9 km/h; 12 km/h

EXAMPLE 2   *Traveling in Same Direction*
Rae Marie and Fred were practicing for a long-distance boat race. Rae Marie started first and rowed at the rate 12 km/h. One hour later Fred left from the same point and rowed in the same direction at 16 km/h. How many hours did he row before he caught up with Rae Marie?

Solution   Let x = number of hours that Rae Marie rowed.

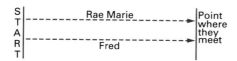

Then $x - 1$ = number of hours that Fred rowed.

	Rate	Time	Distance ($d = rt$)
Rae Marie	12	x	$12x$
Fred	16	$x - 1$	$16(x - 1)$, or $16x - 16$

Rae Marie's distance = Fred's distance
$$12x = 16x - 16$$
$$16 = 4x$$
$$x = 4 \quad \longleftarrow \text{Rae Marie's time}$$
$$x - 1 = 3 \quad \longleftarrow \text{Fred's time}$$

The check is left for you.

Thus, Fred rowed 3 h before he caught up with Rae Marie.

EXAMPLE 3   *Going and Returning at Different Rates*
Larry, Judy, and their daughters drove to the ski slope at 70 km/h. After a snowstorm, they managed to drive home at 40 km/h. It took 2 h longer to return home than it took going to ski. How long did it take for them to get home?

Solution   Let x = number of hours that it took to go.

Distances are equal.

Then $x + 2$ = number of hours that it took to return.

	Rate	Time	Distance ($d = rt$)
Going	70	x	$70x$
Returning	40	$x + 2$	$40(x + 2)$, or $40x + 80$

9.4 Problem Solving: Motion Problems   **355**

TEACHING SUGGESTIONS

Lesson Note

Encourage students to draw a diagram and complete a table for each problem. Then they should make the following decisions.

1. Are the distances traveled by each person or object the same? If not, how are they related?
2. Are the rates at which each travels the same? If not, how are they related?
3. Are the times each spends traveling the same? If not, how are they related?

Remind students to assign a variable to the quantity they must find and use this variable in the appropriate column of the chart. Then fill in another column of the chart with given information. Finally, the third column can be completed by expressing the relationship indicated in the other two columns.

Math Connections

Vocabulary The prefix "per" usually means "for every" or "divided by." Common usages include mph (miles per hour), mpg (miles per gallon), rpm (revolutions per minute), percent (per hundred).

Additional Example 2

James started canoeing in a river at the rate of 8 mi/h. A half hour later, Anne left the same point and paddled in the same direction at 10 mi/h. How many hours did it take Anne to overtake James?   2 h

Additional Example 3

Kevin, Candace, and their two children drove to the museum at 80 km/h. Because of traffic they drove home at 50 km/h. It took 3 h longer to return home than it took to get to the museum. How long did it take them to get home?   8 h

Critical Thinking Questions

Synthesis: Ask students to demonstrate why this problem has no answer. Town A and Town B are 40 miles apart. If a car is to make a round trip in 2 hours, and it travels at 20 mi/h going from A to B, at what speed must it make the return trip so that the average speed for both directions is 40 mi/h? Traveling at 20 mi/h it takes 2 hours to get from A to B. If the complete 80 mi trip is to be completed in 2 hours ($\frac{80}{40}$ mi/h = 2 h), all of the time has been used up just to travel one way.

Common Error Analysis

Error: In problems like Example 4, some students will try to assign a variable to the time as well as the rate.

Point out that in order to solve using one variable, they must have only one variable in the equation they set up.

$$\text{distance going} = \text{distance returning}$$
$$70x = 40x + 80$$
$$30x = 80$$
$$x = \frac{80}{30}, \text{ or } 2\frac{2}{3} \quad \longleftarrow \text{ time going}$$
$$\text{So, } x + 2 = 4\frac{2}{3}. \quad \longleftarrow \text{ time returning}$$

The check is left for you.

Thus, it took $4\frac{2}{3}$ h for them to return home.

EXAMPLE 4   *Traveling Different Distances in the Same Time*
It took Peter the same amount of time to drive 325 km as it took Mary to drive 275 km. Peter's speed was 10 km/h more than Mary's speed. How fast did each drive?

Solution   Let x = Mary's rate in kilometers per hour.
Then $x + 10$ = Peter's rate in kilometers per hour.
Since the distances are given, represent the time in terms of distance and rate.

	Rate	Time $\left(t = \dfrac{d}{r}\right)$	Distance
Mary	x	$\dfrac{275}{x}$	275
Peter	$x + 10$	$\dfrac{325}{x + 10}$	325

$$\text{Peter's time} = \text{Mary's time}$$
$$\frac{325}{x + 10} = \frac{275}{x}$$
$$325x = 275(x + 10)$$
$$325x = 275x + 2{,}750$$
$$50x = 2{,}750$$
$$x = 55 \quad \longleftarrow \text{ Mary's rate}$$
$$\text{So, } x + 10 = 65 \quad \longleftarrow \text{ Peter's rate}$$

The check is left for you.

Thus, Peter's rate was 65 km/h and Mary's rate was 55 km/h.

TRY THIS   2. Sue and Tim drove for the same amount of time. Sue's rate was 52 km/h and Tim's rate was 42 km/h. Sue drove 40 km farther than Tim. How far did each drive?   Tim: 168 km; Sue: 208 km

Additional Example 4

It took Andy the same time to drive 500 km as it took Sandy to drive 700 km. Andy's speed was 20 km/h less than Sandy's speed. How fast did each drive?   Andy, 50 km/h; Sandy, 70 km/h

Classroom Exercises

Use one of the formulas, $d = rt$, $r = \frac{d}{t}$, or $t = \frac{d}{r}$ to find the indicated value. (Exercises 1–4)

1. rate: 50 km/h
   time: 4 h
   distance: ____?____ 200 km

2. distance: 300 km
   rate: 60 km/h
   time: ____?____ 5 h

3. distance: 40 km
   time: 5 h
   rate: ____?____ 8 km/h

4. rate: 24 km/h
   time: 1.5 h
   distance: ____?____ 36 km

5. Bob drives 60 km/h for x hours. What is an expression for the distance traveled? $d = 60x$

6. Carlotta drives 55 km/h for $(x + 2)$ hours. What is an expression for the distance traveled? $d = 55x + 110$

7. Pierre drove 680 km at a rate of $(y + 10)$ km/h. Write an expression for the time that he spent traveling.
   $t = \frac{680}{y + 10}$

8. Two jets left from the same airport at the same time. One flew east at 600 km/h. The other flew west at 900 km/h. In how many hours were they 6,000 km apart? 4 h

Written Exercises

Solve each problem.

1. Two cyclists left from the same place at the same time and rode in opposite directions. One cyclist rode 5 km/h faster than the other. After 6 h, the cyclists were 78 km apart. How fast did each travel? Slow: 4 km/h; fast: 9 km/h

2. Two cars left at the same time and traveled in opposite directions from the same starting point. One car traveled 20 km/h faster than the other car. After 7 h the cars were 910 km apart. Find each car's rate. Slow: 55 km/h; fast: 75 km/h

3. Bea ran along a bicycle path at 6 km/h. One hour later, Roberto left from the same point and ran along the same path at 9 km/h. How many hours did he run before he caught up with Bea? 2 h

4. Nathan started out in his car at the rate of 42 km/h. Two hours later, Joelle left from the same point driving along the same road at 80 km/h. How long did it take her to catch up to Nathan? $2\frac{4}{19}$ h

5. The Smith family drove to the beach at 75 km/h. They returned later, in heavy traffic, at 50 km/h. The return trip took 2 h longer than the trip to the beach. How long did it take to get home? 6 h

6. Bart drove to a family reunion at the rate of 70 km/h. A few days later, the return trip took 1 h less because he drove at 80 km/h. How long did it take him to get home? 7 h

7. Miguel biked 14 km in the same amount of time that Linda hiked 10 km. Miguel's rate was 4 km/h more than Linda's. How fast did each travel? L: 10 km/h; M: 14 km/h

8. Chuck drove 240 km in the same amount of time that Millie drove 190 km. Chuck drove 20 km/h faster than Millie. How fast did each drive? C: 96 km/h; M: 76 km/h

9.4 Problem Solving: Motion Problems **357**

Checkpoint

Solve each problem.

1. Two trains left the same station at the same time and traveled in opposite directions. The E-train averaged 130 km/h. The A-train's speed was 110 km/h. In how many hours were they 480 km apart? 2 h

2. Leroy started out in his car at the rate of 50 km/h. One hour later, Tina left from the same point driving along the same road at 75 km/h. How long did it take her to catch up with Leroy? 2 h

Closure

Have students match each type of motion problem with the type of equation that might be used to solve the problem.

Motion Problems

1. traveling in the same direction
2. going and returning at different rates
3. traveling different distances in the same time
4. traveling in opposite directions

Equations

a. distance going = distance returning
b. time for one distance = time for second distance
c. first distance + second distance = total distance
d. first distance = second distance

1 d; 2 a; 3 b; 4 c

◼◼◼ FOLLOW UP

Guided Practice

Classroom Exercises 1–8
Try This   all

Independent Practice

A Ex. 1–8, **B** Ex. 9–12, **C** Ex. 13–16

Basic: WE 1–8 all
Average: WE 1–7 odd, 9–12 all
Above Average: WE 1–15 odd

Additional Answers

Mixed Review

4. $-5a^5b^2 + 15a^3b^3 + 10a^4b^2$
7. $a^3 - 3a^2 + 5a - 3$
8. $3ax + 3bx - 2ay - 2by$

---

9. Two cars drove away from the same place at the same time in opposite directions. One car drove 12.5 km/h faster than the other. After 4 h, the cars were 570 km apart. Find each car's rate.   Slow: 65 km/h; fast 77.5 km/h

10. Carol started running at 7 km/h. One half-hour later, Roberto left from the same point and ran along the same path at 9 km/h. How many hours did he run before he caught up with Carol? $1\frac{3}{4}$ h

11. Two trains left at the same time from the same station and traveled in the same direction. The first train averaged 65 km/h and the second train averaged 85 km/h. How long did it take the faster train to get 200 km ahead of the slower train?   10 h

12. Two cars traveled in opposite directions from the same starting point. The rate of one car was 18 km/h faster than the rate of the other car. After 2 h 20 min, the cars were 428 km apart. Find each car's rate.   Slow: $82\frac{5}{7}$ km/h; fast: $100\frac{5}{7}$ km/h

13. An aircraft carrier and a destroyer left the same port at 7:00 A.M. and sailed in the same direction. The destroyer traveled at a rate of 36 km/h and the carrier traveled at a rate of 24 km/h. At what time were they 300 km apart?   8:00 A.M., next day

14. Eight minutes after a bank was robbed, the police started a chase to pursue the thief. The thief was driving at 80 km/h while the police were driving at 95 km/h. How long did it take the police to catch up with the thief? $\frac{32}{45}$ h, or almost 43 min

15. Two trucks started toward each other at the same time from towns x km apart. One truck averaged y km/h and the other z km/h. Express algebraically the number of hours before they reached each other. $t = \frac{x}{y + z}$

16. Tony started driving at the rate of q km/h. Then r hours later, Mary left from the same point driving along the same road at s km/h. Assume that $s > q$. Express algebraically the number of hours before Mary caught up with Tony.
Let t = number of hours Mary drove. $t = \frac{qr}{s - q}$

Mixed Review

Multiply.   **6.4, 6.8, 6.9**

1. $3a^7b^2(-2a^{-2}b)$ $-6a^5b^3$
2. $-4a^3b^{-4}(-3a^2b^3)$ $\frac{12a^5}{b}$
3. $-3(x^4 + 2x^2 - 3x)$ $-3x^4 - 6x^2 + 9x$
4. $-5a^2b(a^3b - 3ab^2 - 2a^2b)$
5. $(2x - 1)(x + 8)$ $2x^2 + 15x - 8$
6. $(3x - 1)(3x + 1)$ $9x^2 - 1$
7. $(a - 1)(a^2 - 2a + 3)$
8. $(a + b)(3x - 2y)$
9. The length of a rectangle is 8 cm more than 4 times the width. The perimeter is 123 cm. Find the length and the width.   **4.4**
$l = 50.8$ cm; $w = 10.7$ cm
10. Maria has 14 coins, all in dimes and quarters. The number of dimes is 2 fewer than 3 times the number of quarters. How many coins of each type are there?   **4.2** 4 quarters, 10 dimes

---

Enrichment

Read this paragraph to students. Have them answer the four questions. Pat, Mary, and Tim went canoeing upstream together. Their floatable lunch box fell out when they hit a submerged rock. However, they did not realize it until 10 min later. At that time, they turned around and paddled downstream. They caught up with the lunch box 1 mi from the place where it had originally fallen overboard.

1. After they turned around, did the moving water affect the distance between the boat and the lunch?   No, the current moves boat and lunch in same direction at same speed.

2. Once they realized their lunch was missing, how long did it take for them to reach the lunch box if they paddled at the same speed as before?   10 min
3. How long did it take for the lunch box to travel 1 mi?   20 min
4. What is the rate of the stream?   3 mi/h

9.5 Problem Solving: Work Problems

Teaching Resources

Quick Quizzes 73
Reteaching and Practice
   Worksheets 73
Transparency 28

Objective   To solve work problems

If Linda Sue can paint her room in 6 h, it takes 1 h to paint $\frac{1}{6}$ of the room and 3 h to paint $\frac{3}{6}$, or $\frac{1}{2}$, of the room. The part of work done in 5 h is $\frac{5}{6}$.

$$5 \cdot \frac{1}{6} = \frac{5}{6}$$

$$\underset{\substack{\text{time} \\ \text{worked}}}{\text{time}} \cdot \underset{\substack{\text{rate of} \\ \text{work}}}{\text{rate of}} = \underset{\substack{\text{part of work} \\ \text{completed}}}{\text{part of work}}$$

This suggests a formula for solving work problems.

time · rate = part of work completed: $t \cdot r = w$, or $w = rt$

EXAMPLE 1   Frank can type a report in 5 h. What part can he type in 2 h? in 3 h? in n h?

Solution   To type the report in 5 h means $\frac{1}{5}$ of the report can be typed each hour. That is, $r = \frac{1}{5}$.

$w = rt$: In 2 h, $w = \frac{1}{5} \cdot 2 = \frac{2}{5}$ ←— $\frac{2}{5}$ of the report can be typed.

In 3 h, $w = \frac{1}{5} \cdot 3 = \frac{3}{5}$ ←— $\frac{3}{5}$ of the report can be typed.

In n h, $w = \frac{1}{5} \cdot n = \frac{n}{5}$ ←— $\frac{n}{5}$ of the report can be typed.

Thus, Frank can type $\frac{2}{5}$ of the report, $\frac{3}{5}$ of the report, and $\frac{n}{5}$ of the report in the given numbers of hours.

TRY THIS   1. Mary can wallpaper a room in 8 h. What part of the room can she wallpaper in 2h? in 5 h? in x h? $\frac{1}{4}, \frac{5}{8}, \frac{x}{8}$

Sometimes two or more people work together to complete a job. Suppose, in Example 1, Frank did $\frac{2}{5}$ of the job and Gail did $\frac{3}{5}$ of the job. Then, together, they completed the job. So, $\frac{2}{5} + \frac{3}{5} = 1$.

Prerequisite Quiz

Solve.

1. $\frac{x}{5} + \frac{x}{3} = 1$ $\frac{15}{8}$

2. $\frac{8}{x} + \frac{16}{80} = 1$ 10

3. $\frac{5}{x} + \frac{5}{3x} = 1$ $\frac{20}{3}$

4. $\frac{5}{15} + \frac{3}{x} = 1$ $4\frac{1}{2}$

Motivator

Have students answer the following questions given this information.
A painter takes 4 hours to paint a room.

1. What part of the job can the painter do in 1 hour? $\frac{1}{4}$

2. What part of the job can he do in 2 hours? $\frac{1}{4} \cdot 2 = \frac{2}{4}$, or $\frac{1}{2}$

3. What part of the job can he do in n hours? $\frac{1}{4} \cdot n$, or $\frac{n}{4}$

Then have students give examples of jobs that they do, the time it takes them to complete the job, and the part of the job they can complete in 1 minute.   Answers will vary.

Additional Example 1

Pierre can paint a house in 7 days. What part of the house can he paint in 3 days? $\frac{3}{7}$
in 4 days? $\frac{4}{7}$ in x days? $\frac{x}{7}$

Highlighting the Standards

Standards 1b, 5b: Students apply problem-solving approaches, including making tables, to writing equations for, and solving, work problems.

Lesson Note

Emphasize that if a job is completed, the *sum* of the fractional parts of the job done by each person must equal 1. In Example 4, point out to students that the reciprocal of $\frac{3}{2}x$ is $1 \div \frac{3x}{2} = 1 \cdot \frac{2}{3x} = \frac{2}{3x}$.

Math Connections

Calculator: Problems such as those in Example 2 can be solved by using a calculator. First, express

$\frac{x}{6} + \frac{x}{8} = 1$ as $\frac{1}{6} + \frac{1}{8} = \frac{1}{x}$.

Then use a calculator to evaluate $\frac{1}{6} + \frac{1}{8}$. Finally use the inverse key, $\frac{1}{x}$, to find the reciprocal of the sum of $\frac{1}{6}$ and $\frac{1}{8}$.

Critical Thinking Questions

Application: Ask students to solve this problem. If it takes 6 painters 6 days to paint 6 houses, how long will it take 2 painters to paint 2 houses?   6 days

Common Error Analysis

Error: Some students will forget to multiply the side of the equation equal to 1 by the common denominator.

Have students write the common multiplier next to each term of the equation before canceling.

EXAMPLE 2 Kim takes 6 days to prepare a boat for use. Dick takes 8 days to prepare the same boat for use. How long will it take to prepare the boat if they work together? (THINK: Will the number of days be less than 6 or more than 6?)

Plan Let x = number of days Kim and Dick work together. Make a table to represent the part of the job completed by each person.

Solution

	rate	×	time	=	work
	Part of job done in 1 day		Number of days working together		Part of job completed
Kim	$\frac{1}{6}$		x		$\frac{1}{6}x = \frac{x}{6}$
Dick	$\frac{1}{8}$		x		$\frac{1}{8}x = \frac{x}{8}$

Kim's part + Dick's part = the whole job

$$\frac{x}{6} + \frac{x}{8} = 1$$
$$24\left(\frac{x}{6} + \frac{x}{8}\right) = 24 \cdot 1$$
$$24 \cdot \frac{x}{6} + 24 \cdot \frac{x}{8} = 24$$
$$4x + 3x = 24$$
$$7x = 24$$
$$x = \frac{24}{7}, \text{ or } 3\frac{3}{7} \quad \longleftarrow \text{ number of days to prepare the boat}$$

Check In $3\frac{3}{7}$ days, will Kim and Dick complete the job?

Yes, since in $3\frac{3}{7}$ days, Kim will do $\frac{1}{6} \cdot \frac{24}{7}$, or $\frac{4}{7}$ of the job; Dick will do $\frac{1}{8} \cdot \frac{24}{7}$, or $\frac{3}{7}$ of the job, and $\frac{4}{7} + \frac{3}{7} = 1$.

Thus, the job will take $\frac{24}{7}$ days, or $3\frac{3}{7}$ days, if Kim and Dick work together.

TRY THIS 2. Dan paints a house in 6 days. It takes Joan 9 days to paint the same house. How long will it take them to paint the house if they work together? $3\frac{3}{5}$ days

Suppose that Kim can complete a job in x days, then $\frac{1}{x}$ is the rate (the part she can complete in 1 day). In 2 days, she can complete $\frac{2}{x}$ of the job; in n days she can complete $\frac{n}{x}$ of the job.

Additional Example 2

It takes Regina 3 h to prepare surgical equipment. Another nurse, Mort, can do it in 2 h. How long will it take to do the job if they work together?   1 h 12 min

EXAMPLE 3   Sue and Ed can clean their house in 7 h. It takes Sue 15 h to do the job alone. How long would it take Ed to do the job alone?

Solution   Let x = number of hours it would take Ed to do the job alone.

	Part of job done in 1 h	Number of hours working together	Part of job completed
Sue	$\frac{1}{15}$	7	$\frac{1}{15} \cdot 7 = \frac{7}{15}$
Ed	$\frac{1}{x}$	7	$\frac{1}{x} \cdot 7 = \frac{7}{x}$

Sue's part + Ed's part $= 1$

$$\frac{7}{15} + \frac{7}{x} = 1$$
$$15x\left(\frac{7}{15} + \frac{7}{x}\right) = 15x \cdot 1$$
$$7x + 105 = 15x$$
$$105 = 8x$$
$$x = \frac{105}{8}, \text{ or } 13\frac{1}{8}$$

The check is left for you. Ed would take $13\frac{1}{8}$ h to do the job alone.

EXAMPLE 4   Working together, Jodi and her father can deliver the Sunday papers in 3 h. It takes Jodi $1\frac{1}{2}$ times as long to deliver the papers alone. How long would it take each of them to deliver the papers alone?

Solution   Let x = number of hours it would take the father alone.
Then $\frac{3}{2}x$ = number of hours it would take Jodi alone.

	Part of job done in 1 h	Number of hours working together	Part of job completed
Father	$\frac{1}{x}$	3	$\frac{1}{x} \cdot 3 = \frac{3}{x}$
Jodi	$\frac{1}{\frac{3}{2}x} = \frac{2}{3x}$	3	$\frac{2}{3x} \cdot 3 = \frac{2}{x}$

Father's part + Jodi's part $= 1$ $\longrightarrow$ $\frac{3}{x} + \frac{2}{x} = 1$

The solution of this equation and the check are left for you.
It would take the father 5 h and Jodi $7\frac{1}{2}$ h to do the job alone.

Checkpoint

1. Kris can type her report in 12 h. What part of the report can she type in 3 h? $\frac{1}{4}$ in 6 h? $\frac{1}{2}$ in n h? $\frac{n}{12}$

2. It takes Jack 5 h and Joan 10 h to paint a house. How long will it take them to do the job if they work together?
   3 h 20 min

3. Working together, Salvatore and Marie can wallpaper an apartment in 14 h. It takes Marie 30 h to do it alone. How long will it take Salvatore to do it alone?
   26 h 15 min

Closure

Tell students that pipe A can empty a pool in 5 hours and pipe B can empty the same pool in 3 hours. Then ask:

1. What part of the pool can pipe B empty in 1 hour? $\frac{1}{3}$

2. What part of the pool can pipe A empty in 4 hours? $\frac{4}{5}$

3. What part of the pool can both pipes drain in 1 hour? $\frac{1}{5} + \frac{1}{3}$, or $\frac{8}{15}$

4. What part of the pool can both pipes drain in x hours? $\frac{x}{5} + \frac{x}{3}$, or $\frac{8x}{15}$

Additional Example 3

Working together, Hank and Pat can paint a house in 16 h. It will take Hank 40 h to do it alone. How long will it take Pat to do the job alone?   26 h 40 min

Additional Example 4

It takes Lois 3 times as long as Richie to mow a lawn. How long will it take each of them to mow the lawn alone, if together they can do it in 5 h?   Lois: 20 h; Richie: 6 h 40 min

Guided Practice

Classroom Exercises 1–10
Try This   all

Independent Practice

A Ex. 1–10, **B** Ex. 11–16, **C** Ex. 17–20

Basic: WE 1–10 all
Average: WE 5–14 all
Above Average: WE 11–20 all

Classroom Exercises

Maria can write a computer program in 12 h. What part can she write in:

1. 1 h? $\frac{1}{12}$ **2.** 3 h? $\frac{1}{4}$ **3.** 5 h? $\frac{5}{12}$ **4.** x h? $\frac{x}{12}$

Kris takes 5 days to construct the set for the school play. Susan takes 7 days to construct the same set. Solve Exercises 5–9 based on this information.

5. What part of the job can Kris do in 2 days? $\frac{2}{5}$

6. What part of the job can Susan do in 6 days? $\frac{6}{7}$

7. What part of the job can Kris do in $3\frac{1}{2}$ days? $\frac{7}{10}$

8. What part of the job can Susan do in $4\frac{1}{3}$ days? $\frac{13}{21}$

9. How long will it take to construct the set if they work together? $2\frac{11}{12}$ days

10. It takes Pam 5 h to complete a job. It takes Uri 10 h to complete the same job. How long will it take them to complete the job if they work together? $3\frac{1}{3}$ h

Written Exercises

Juan can paint his room in 8 h. What part of the room can he paint in:

1. 3 h? $\frac{3}{8}$ **2.** 4 h? $\frac{1}{2}$ **3.** 6 h? $\frac{3}{4}$ **4.** n h? $\frac{n}{8}$

Solve.

5. Mike takes 12 h to clean the basement. Jackie takes 14 h to clean the same basement. How long will it take them to clean the basement together? $6\frac{6}{13}$ h

6. A large pump can empty a pool in 8 h. A smaller pump can empty the pool in 10 h. How long will it take both pumps working together? $4\frac{4}{9}$ h

7. One machine can complete an order in 16 h while another machine takes 22 h to complete the same order. How long will it take both machines working together to complete the same order? $9\frac{5}{19}$ h

8. Working together, Audrey and Bill can address invitations in 5 h. If it takes Audrey 8 h to do it alone, how long would it take Bill to do the job alone? $13\frac{1}{3}$ h

9. It takes José 28 h to wallpaper two rooms of a house. Together José and Carol can finish the job in 16 h. How long would it take Carol to do the job alone? $37\frac{1}{3}$ h

10. Together Kathy and Mark can do their holiday baking in 5 h. If it takes Kathy 9 h to do the job alone, how long would it take Mark to do the job alone? $11\frac{1}{4}$ h

11. Working together, Anita and Dawn can install all the windows in a house in 7 days. It takes Anita twice as long as Dawn to install the windows alone. How long would it take each of them to install the windows alone? **D: $10\frac{1}{2}$ days, A: 21 days**

12. Jack and his son can build a house in 8 months. It takes the son twice as long as it takes his father to build the house alone. How long would it take each to do it alone?
J: 12 mos, son: 24 mos

13. One pipe can fill a pool in 8 h. A second pipe can do it in 7 h, and a third pipe can do it in 9 h. How long would it take to fill the pool if all three pipes are used? **$2\frac{122}{191}$ h**

14. Working together, Mel, Roz, and Abner can clean the Teen Center in 6 h. It takes Mel 18 h to do all the cleaning alone, and Roz 20 h to do it alone. How long would it take Abner to do it alone? **$16\frac{4}{11}$ h**

15. To clean the house, it would take Candy 6 h, Joe 7 h, and Jennifer 4 h. How long would it take to do the job if they all worked together? **$1\frac{37}{47}$ h**

16. To do a job alone, it takes Hannah 2 h. It takes Andy 1 h longer than Hannah, and it takes Bob 3 times as long as Hannah. How long would it take to do the job if they worked together? **1 h   S: $2\frac{1}{4}$ h; B: $4\frac{1}{2}$ h**

17. A swimming pool can be filled by a pipe in 18 h and emptied by an outlet pipe in 24 h. How long will it take to fill the empty pool if the outlet pipe is mistakenly left open at the same time as the inlet pipe is opened? **72 h**

18. Sandy and her little brother can clean the snow off the driveway in $1\frac{1}{2}$ h if they work together. It takes her little brother twice as long as it takes Sandy to do it alone. How long does it take each of them to do it alone?

19. It takes Dennis 16 h to plow a field. After he worked for 6 h, Dawn began to help him. Together they finished the job in 2 more hours. How long would it take Dawn to do the job alone? **4 h**

20. It takes Charles 12 h to clean his apartment. After he has worked 4 h, he is joined by Marie, and they finish cleaning in $2\frac{1}{2}$ more hours. How long would it take Marie to clean the apartment alone? **$5\frac{5}{11}$ h**

Mixed Review

Factor completely.   **7.2, 7.4, 7.5, 7.6**

1. $5x - 20$ **$5(x - 4)$**

2. $3a^2 + 15a$ **$3a(a + 5)$**

3. $2x^2 - 5x - 3$ **$(2x + 1)(x - 3)$**

4. $x^2 - 16$ **$(x - 4)(x + 4)$**

5. $4y^2 - 49$ **$(2y - 7)(2y + 7)$**

6. $a^2x^2 - x^2 - 4a^2 + 4$
   $(a - 1)(x + 2)(x - 2)(a + 1)$

Simplify each rational expression.   **8.1, 8.2**

7. $\dfrac{48x^6y^4}{-20xy^6}$ **$-\dfrac{12x^5}{5y^2}$**

8. $\dfrac{5n - 20}{8n - 32}$ **$\dfrac{5}{8}$**

9. $\dfrac{x - 4}{16 - x^2}$ **$-\dfrac{1}{x + 4}$**

9.5 Problem Solving: Work Problems   **363**

Enrichment

Explain to students that the ancient Egyptians represented most fractions as the sum of "unit fractions", or fractions with a numerator of 1.

Thus, $\frac{1}{2}$ was represented as $\frac{1}{3} + \frac{1}{6}$. Have students find a pair of unlike unit fractions whose sum is equal to each of these fractions.

1. $\frac{1}{3}$ $\frac{1}{4} + \frac{1}{12}$

2. $\frac{1}{4}$ $\frac{1}{5} + \frac{1}{20}$, $\frac{1}{12} + \frac{1}{6}$

3. $\frac{1}{6}$ $\frac{1}{7} + \frac{1}{42}$, $\frac{1}{8} + \frac{1}{24}$, $\frac{1}{9} + \frac{1}{18}$, $\frac{1}{10} + \frac{1}{15}$

9.6 Dimensional Analysis

Objective   To convert from one unit of measurement to another using dimensional analysis

The equation 3 ft = 1 yd states the relationship of feet to yards. By treating "ft" and "yd" as factors, this equation can be expressed in two ways as a fraction that is equal to 1.

Divide each side by 1 yd.

$$3 \text{ ft} = 1 \text{ yd}$$

$$\frac{3 \text{ ft}}{1 \text{ yd}} = 1$$

Divide each side by 3 ft.

$$3 \text{ ft} = 1 \text{ yd}$$

$$1 = \frac{1 \text{ yd}}{3 \text{ ft}}$$

Such fractions are used in converting one unit of measurement to another. The fractions are called **conversion fractions** or *conversion factors*. Their use in computations is called **dimensional analysis.**

To convert 29 ft to yards, multiply 29 ft by $\frac{1 \text{ yd}}{3 \text{ ft}}$.

The common factor, ft, divides out.

$$\frac{29 \text{ ft}}{1} \cdot \frac{1 \text{ yd}}{3 \text{ ft}} = \frac{29 \text{ yd}}{3}$$

$$= 9\frac{2}{3} \text{ yd}$$

To convert 42 yd to feet, multiply 42 yd by $\frac{3 \text{ ft}}{1 \text{ yd}}$.

The common factor, yd, divides out.

$$\frac{42 \text{ yd}}{1} \cdot \frac{3 \text{ ft}}{1 \text{ yd}} = 42 \cdot 3 \text{ ft}$$

$$= 126 \text{ ft}$$

Some measures, such as speed, involve two types of measurements. Speed involves distance and time. In science, measurements of speed often need to be converted into different units.

EXAMPLE 1   Convert 1,100 feet per second (ft/s) into feet per minute (ft/min).

Plan   Write 1,100 ft/s as $\frac{1,100 \text{ ft}}{1 \text{ s}}$. Since 60 s = 1 min, multiply by $\frac{60 \text{ s}}{1 \text{ min}}$.

Solution $\frac{1,100 \text{ ft}}{1 \text{ s}} \cdot \frac{60 \text{ s}}{1 \text{ min}} = \frac{1,100 \text{ ft} \cdot 60}{1 \text{ min}} = \frac{66,000 \text{ ft}}{1 \text{ min}}$, or 66,000 ft/min

Thus, 1,100 ft/s is equivalent to 66,000 ft/min

Prerequisite Quiz

Complete.

1. ___?___ cm = 1 m   100
2. 1 ft = ___?___ in   12
3. 1 lb = ___?___ oz   16
4. 1 mi = ___?___ ft   5,280
5. ___?___ m = 1 km   1,000
6. 1 kg = ___?___ g   1,000

Motivator

Ask students these questions.

1. What units are generally used to give the area of a room?   ft²
2. What units are generally used to determine how much carpeting to buy for a room?   yd²
3. Which unit is larger, 1 ft² or 1 yd²?   1 yd²
4. What do you need to know to convert from ft² to yd²?   3 ft = 1 yd, or 9 ft² = 1 yd²

Highlighting the Standards

Standards 4b, 1b, 7a: Examples from geometry in two and three dimensions, as well as from the world outside the classroom, are used to illustrate the analysis of units for problem solving.

Additional Example 1

Convert 800 meters per second (m/s) into meters per minute (m/min).   48,000 m/min

EXAMPLE 2   Convert 60 miles per hour (mi/h) to feet per second (ft/s).

Plan   Write 60 mi/h as $\dfrac{60 \text{ mi}}{1 \text{ h}}$. Then use conversion fractions based on these equivalent measures: 5,280 ft = 1 mi, 1 h = 60 min, and 1 min = 60 s

Solution $\dfrac{60 \text{ mi}}{1 \text{ h}} \cdot \dfrac{5{,}280 \text{ ft}}{1 \text{ mi}} \cdot \dfrac{1 \text{ h}}{60 \text{ min}} \cdot \dfrac{1 \text{ min}}{60 \text{ s}} = \dfrac{60 \cdot 5{,}280 \text{ ft}}{60 \cdot 60 \text{ s}}$

$= \dfrac{88 \text{ ft}}{1 \text{ s}}$, or 88 ft/s

Thus, 60 mi/h is equivalent to 88 ft/s.

TRY THIS   1. Convert 10 miles per hour (mi/h) to inches per second (in./s).   176 in./s

Conversion fractions can be used with square units. Refer to the figure of 1 yard square. Since 3 ft = 1 yd, you can square each side of the equation to show that $9 \text{ ft}^2 = 1 \text{ yd}^2$.

Then, $\dfrac{9 \text{ ft}^2}{1 \text{ yd}^2} = 1$ and $\dfrac{1 \text{ yd}^2}{9 \text{ ft}^2} = 1$

Similarly, 12 in. = 1 ft, so $144 \text{ in}^2 = 1 \text{ ft}^2$.

Then, $\dfrac{144 \text{ in}^2}{1 \text{ ft}^2} = 1$ and $\dfrac{1 \text{ ft}^2}{144 \text{ in}^2} = 1$.

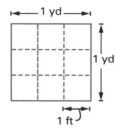

EXAMPLE 3   A rectangular rug measures 14.5 ft by 12 ft. What is its area in square yards? Give your answer to the nearest integer.

Plan   Use $A = lw$ to find the area in square feet. Then multiply by $\dfrac{1 \text{ yd}^2}{9 \text{ ft}^2}$.

Solution   The area of the rug is 14.5 ft × 12 ft = 174 ft².

$\dfrac{174 \text{ ft}^2}{1} \cdot \dfrac{1 \text{ yd}^2}{9 \text{ ft}^2} = \dfrac{174}{9} \text{ yd}^2$, or $\approx 19 \text{ yd}^2$

Thus, the area is 19 yd² to the nearest square yard.

TRY THIS   2. A rectangular patio measures 84 in. by 65 in. What is its area to the nearest square foot?   38 ft²

Lesson Note

Suggest to students that they must multiply by a conversion factor that enables them to cancel unwanted units and retain the desired units. For example, to change ft/sec to ft/min, the conversion factor must have units in the form sec/min.

$$\dfrac{\text{ft}}{\text{sec}} \cdot \dfrac{\text{sec}}{\text{min}} = \dfrac{\text{ft}}{\text{min}}$$

Math Connections

Astronomy: Since light travels at approximately 186,000 mi/sec, the distance light travels in one year (light year) is approximately 5,865,696,000,000 miles, or close to 6 trillion miles.

Critical Thinking Questions

Application: Remind students that in the United States, 1 million is a thousand thousands, 1 trillion is a thousand billions, and so on, with each new name representing a number 1000 times the former. In England, a billion is written with 12 zeros and a trillion with 18 zeros. Ask students what the common multiple is in the English system of naming large numbers?   1,000,000

Common Error Analysis

Error: Some students may use the reciprocal of the correct conversion unit, i.e., 3 ft/1 yd instead of 1yd/3 ft.

Remind them that the units must cancel to give the desired result.

Additional Example 2

Convert 500 mi/h to feet per second.

$733\frac{1}{3}$ ft/s

Additional Example 3

The dimensions of a rectangular floor are 18.6 ft by 14 ft. What is its area in square yards? Give your answer to the nearest integer.   29 yd²

Checkpoint

1. Convert 80 mi/h to feet per minute. **7,040 ft/min**
2. Convert 800 ft/min to inches per second. **160 in/s**
3. The dimensions of a rectangular rug are 9 ft by 12 ft. What is its area in square yards? **12 yd²**

Closure

Have students do the following conversions and explain the method used. **Explanations will vary.**

1. 29 ft to yards $9\frac{2}{3}$ yd
2. 60 mi/h to feet per second   **88 ft/s**
3. 400 ft² to square yards $44\frac{4}{9}$ yd²
4. 300 ft/s to feet per minute **18,000 ft/min**

◼◼◼FOLLOW UP

Guided Practice

Classroom Exercises 1–9
Try This   all

Independent Practice

A Ex. 1–13, **B** Ex. 14–17, **C** Ex. 18
Basic: WE 1–13 all, Mixed Problem Solving
Average: WE 4–16 all, Mixed Problem Solving
Above Average: WE 6–18 all, Mixed Problem Solving

Additional Answers

Classroom Exercises

4. $\frac{1,000 \text{ mg}}{1 \text{ g}}$ 6. $\frac{1 \text{ kg}}{1,000 \text{ g}}$

Determine the conversion fraction or fractions that can be used for the conversion indicated. Use the Table of Equivalent Measures if necessary.

1. 20 yd to feet $\frac{3 \text{ ft}}{1 \text{ yd}}$
2. 81 in. to feet $\frac{1 \text{ ft}}{12 \text{ in}}$
3. 36 oz to pounds $\frac{1 \text{ lb}}{16 \text{ oz}}$
4. 300 g to milligrams
5. 250 cm to meters $\frac{1 \text{ m}}{100 \text{ cm}}$
6. 750 g to kilograms
7. 3 mi/h to miles per minute $\frac{1 \text{ hr}}{60 \text{ min}}$
8. 40 mi/h to feet per minute $\frac{5,280 \text{ ft}}{\text{mi}} \cdot \frac{1 \text{ h}}{60 \text{ min}}$
9. 30 yd² to square feet $\frac{9 \text{ ft}^2}{1 \text{ yd}^2}$

Written Exercises

1–9. Convert each measure in Classroom Exercises 1–9 as indicated.

Use dimensional analysis to convert each measure.

10. 75 mi/h to feet per second 110 ft/s
11. 24 tons/h to pounds per minute 800 lb/min
12. 352 ft/s to miles per hour 240 mi/h
13. 12 gallons/min to quarts per second 0.8 qt/s
14. 6.5 ft/min to inches per second 1.3 in/s
15. 4.5 lb/min to tons per hour 0.135 T/h
16. A satellite is traveling around the earth at an average speed of about 4.1 mi/s. How many miles per hour is this? 14,760 mi/h
17. A rectangular rug measures 16 ft by 11.5 ft. What is its area in square yards? $20\frac{4}{9}$ yd²
18. The figure at the right shows a cube with a volume of 1 m³. What is its volume in cubic decimeters? In cubic centimeters? 1,000 dm³; 1,000,000 cm³

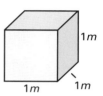

Mixed Review

Evaluate.   *1.1, 1.3*

1. $10 - 5a$ for $a = \frac{1}{5}$ 9
2. $64x^4$ for $x = \frac{2}{3}$ $12\frac{52}{81}$

Solve.   *3.3, 3.5*

3. $\frac{x}{4} - 3 = 8$ x = 44
4. $8x - \frac{2}{3}(15 - 9x) = 12$ $x = 1\frac{4}{7}$

Add or subtract.   *8.5*

5. $\frac{2n^2}{n-1} + \frac{3}{1-n}$ $\frac{2n^2 - 3}{n-1}$
6. $\frac{y}{y+4} - \frac{3}{4+y}$ $\frac{y-3}{y+4}$

Solve.   *4.3, 4.7*

7. Find two consecutive integers whose sum is 137. 68, 69
8. If 40% of a number is 78, what is the number? 195

Enrichment

Ask students to find, to the nearest year, how long it would take to count a billion dollars, at the rate of one dollar per second.   32 yr

If you find that you often make careless errors as you write the equation that models a problem, pay more attention to the step where you *define the variables*. Remember that the equation is often easier to write if you let x represent the smaller quantity.

For example, in Exercise 13, let x represent the amount of this month's electric bill. Then $x + 18.70$ represents the amount of last month's electric bill.

1. Find two consecutive odd integers whose sum is -84. $-43, -41$

2. One number is 12 times another number. Their sum is 117. Find the numbers. 9, 108

3. One number is 2 more than 4 times a second. The sum of the numbers is 22. Find the numbers. 4, 18

4. Find three consecutive even integers such that the sum, decreased by 52, is equal to the second integer. 24, 26, 28

5. Kristin's age is 6 less than $\frac{1}{2}$ of Slim's age. The sum of their ages is 108 years. Find their ages. Kristin: 32; Slim: 76

6. A triangle has an area of 36 in² and a height of 12 in. Find the length of the base. 6 in.

7. Each base angle of an isosceles triangle has degree measure 8 less than half that of the vertex angle. Find all three angle measures. 98, 41, 41

8. Membership fees at the health club are $80 less than twice the original cost. The fees this year are $460. What was the original cost? $270

9. One side of a triangle is 7 in. longer than a second side. The remaining side is 4 in. shorter than the second side. The perimeter is 78 in. Find the length of each side. 21 in., 25 in., 32 in.

10. Sam Ferris lost 25% of the value of his stock portfolio when the stock market collapsed. His portfolio is now worth $86,125. How much money did he lose? $28,708.33

11. The regular price of a ski suit is $210. It is on sale for $140. Find the rate of discount. $\frac{1}{3}$

12. The selling price of a compact disc player is $243. The profit is 35% of the cost. Find the cost. $180

13. Last month's electric bill was $18.70 more than this month's bill. The bill for the two months totaled $73.98. What was this month's bill? $27.64

14. Mary lives 2 mi from the store. She walks at a rate of 6 mi/h. How long must she walk to get from home to the store? 20 min

15. Separate 87 people into two groups so that the first group has 9 fewer than 3 times the number of people in the second group. first: 63; second: 24

16. The product of the first and third of three consecutive integers is 37 less than 20 times the second. Find the three integers. 1, 2, 3 or 17, 18, 19

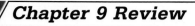

Key Terms

conversion fractions   (p. 364)
dimensional analysis   (p. 364)
direct variation   (p. 348)
extraneous solution   (p. 338)
extremes   (p. 343)
literal equation   (p. 349)

means   (p. 343)
proportion   (p. 343)
Proportion Property   (p. 343)
ratio   (p. 343)
rational equation   (p. 337)
supplementary angles   (p. 346)

Key Ideas and Review Exercises

9.1   To solve an equation containing rational expressions, multiply each side by the LCM of all the denominators.

Solve. Check for extraneous solutions.

1. $\dfrac{3}{a - 6} - \dfrac{4}{a + 2} = \dfrac{5}{a^2 - 4a - 12}$ $a = 25$

2. $\dfrac{-2}{x^2 - 9} + \dfrac{3}{x + 3} = \dfrac{5}{x - 3}$ $x = -13$

9.2   To solve a proportion, set the product of the extremes equal to the product of the means.

Solve.

3. $\dfrac{4}{a + 1} = \dfrac{5}{2a - 4}$ $a = 7$

4. $\dfrac{n - 3}{5} = \dfrac{2n + 3}{6}$ $n = -8\frac{1}{4}$

5. $\dfrac{2x - 1}{-3} = \dfrac{1}{x + 2}$ $x = -\frac{1}{2}\ or\ x = -1$

6. $\dfrac{y}{5} = \dfrac{y}{y - 1}$ $y = 0\ or\ y = 6$

9.3   To solve a literal equation or formula for one of its variables, follow the same steps you use in solving a nonliteral equation.

Solve for x.

7. $7x + b = c$ $x = \frac{c - b}{7}$

8. $3x - b = cd$ $x = \frac{cd + b}{3}$

9. $ax + bx = 6c$ $x = \frac{6c}{a + b}$

10. $-7x + b = 3x - 10b$ $x = \frac{11b}{10}$

9.4   To solve problems about motion, begin with a sketch and a table. Represent the distance, the rate, and the time traveled by each person or moving object. Write an equation and solve it. Determine whether the answer is reasonable. Check it in the original problem.

Solve.

11. Two cars traveled in opposite directions from the same starting point. The rate of one car was 30 km/h faster than the rate of the second car. After 3 h, the cars were 300 km apart. Find the rate of each car.
First: 65 km/h; second: 35 km/h

12. Mike jogged 15 km in the same time that it took Jan to jog 20 km. Jan's rate was 3 km/h faster than Mike's. How fast did each travel?
Mike: 9 km/h; Jan: 12 km/h

9.5 To solve problems about work, complete a table such as the one below. For each person, rate of work × time worked = part of job completed.

	Part of job done in 1 h	Number of hours working together	Part of job completed
First name	?	?	?
Second name	?	?	?

If the job is completed, then the parts of the job completed by each person must have a sum of 1.

Solve.

13. Working alone, it takes Milt 8 h and Ellen 10 h to paint a garage. How long will it take them to do the job if they work together? $4\frac{4}{9}$ h

14. Together, Howard and Willie can assemble a TV kit in 6 h. If it takes Howard 9 h to do the job alone, how long would it take Willie to do the job alone? 18 h

9.6 To convert from one unit of measurement to another using dimensional analysis, multiply by conversion fractions such as the following.

$$\frac{60 \text{ s}}{1 \text{ min}} = 1, \frac{1 \text{ mi}}{5,280 \text{ ft}} = 1, \text{ and } \frac{1 \text{ yd}^2}{9 \text{ ft}^2} = 1$$

For example, $660 \text{ ft} = \frac{660 \text{ ft}}{1} \cdot \frac{1 \text{ mi}}{5,280 \text{ ft}} = \frac{660}{5,280} \text{ mi} = \frac{1}{8} \text{ mi}$

Convert each measure as indicated.

15. 180 in. to feet 15 ft

16. 27 ft² to square yards 3 yd²

17. 75 cm to meters 0.75 m

18. 15 lb/in² to pounds per square foot

19. 66 ft/s to miles per hour 45 mi/h

20. 3 km/h to meters per minute 50 m/min

21. Explain how to convert 4.5 gallons per day to quarts per hour.

18. 2,160 lb/ft²

21. $\frac{4.5 \text{ gal}}{\text{day}} \cdot \frac{4 \text{ qt}}{\text{gal}} \cdot \frac{1 \text{ day}}{24 \text{ h}} = \frac{0.75 \text{ qt}}{\text{h}}$, or $\frac{3}{4}$ qt/h

Solve. Check for extraneous solutions.

1. $\dfrac{1}{x} + \dfrac{1}{4x} = 5$ $x = \frac{1}{4}$

2. $\dfrac{1}{x+2} = \dfrac{1}{2x+4} + \dfrac{1}{2}$ $x = -1$

3. $\dfrac{x+1}{x-4} = \dfrac{4x+4}{x^2-4x}$ $x = -1$

4. $\dfrac{n+2}{n-4} - n = \dfrac{12}{4-n}$ $n = -2 \ or \ n = 7$

Solve.

5. $\dfrac{m}{9} = \dfrac{5}{3}$ $m = 15$

6. $\dfrac{4}{7} = \dfrac{x}{5}$ $x = 2\frac{6}{7}$

7. $\dfrac{3x-5}{6} = \dfrac{2x+1}{4}$ No Solution

8. $\dfrac{b-3}{4} = \dfrac{9}{b+6}$ $b = 6 \ or \ b = -9$

Solve for x.

9. $ax + 9 = -6$ $x = -\dfrac{15}{a}$

10. $ax = b - c$ $x = \dfrac{b-c}{a}$

11. $px + qx = 3$ $x = \dfrac{3}{p+q}$

12. $kx - bc = 5d + fx$ $x = \dfrac{5d+bc}{k-f}$

Convert each measure as indicated.

13. 180 in² to square feet 1.25 ft²

14. 1,320 ft/s to miles per minute 15 mi/min

Solve.

15. If 8 out of 9 people buy an AM/FM radio, how many buy this type of radio in a city with a population of 81,000? 72,000 people

16. Sarah started out on her bike traveling at 9 mi/h. One hour later, Carlos left from the same point on his bike and pedaled along the same route at 12 mi/h. How many hours had he pedaled before he caught up with Sarah? 3 h

17. The Saltzmans drove 65 km/h to visit their daughter. They returned later at the rate of 80 km/h. It took them 8 h less time to return home than it did to go. How long did it take to return? $34\frac{2}{3}$ h

18. It takes Edith 4 h to paint her room. Her brother, Fred, can do it in 5 h. How long will it take to do the job if they work together? $2\frac{2}{9}$ h

19. It takes Rico twice as long as Woody to build a bridge over a stream. How long would it take each of them to do it alone if they can build the bridge together in 9 h? Rico: 27 h; Woody: $13\frac{1}{2}$ h

20. The perimeter of a triangle is 45 cm. Find the length of the shortest side if the lengths of the sides are in the ratio 2:3:4. 10 cm

21. Solve $\dfrac{xy}{a} - \dfrac{tx}{b} = \dfrac{4}{ab}$ for x. $x = \dfrac{4}{by-at}$

Choose the *one* best answer to each question or problem.

1. In 1986 a moving company transported 32,000 tons of cargo. In 1990 the same company handled 48,800 tons of cargo. The average annual increase in tons handled was ___?___.
(A) 4,200   (B) 5,600
(C) 8,000   (D) 12,200
(E) 16,800

2. The product of 798 and 694 is ___?___.
(A) 553,813   (B) 553,814
(C) 553,812   (D) 553,815
(E) 553,817

3. Find the value closest to the sum.
$$70 + 60 + 30 + \frac{3}{4} + \frac{4}{3}$$
(A) 161.55   (B) 167.80
(C) 240.75   (D) 235.80
(E) 315

4. The sum of three consecutive odd integers is 21. The smallest of these integers is ___?___.
(A) 3   (B) 5
(C) 7   (D) 9
(E) 11

5. The perimeter of a rectangle is 54 in. Its length is 15 in.

15 in.

The area of the rectangle is ___?___.
(A) 27 in^2 (B) 39 in^2
(C) 144 in^2 (D) 180 in^2
(E) 225 in^2

6. Which of the following is not a multiple of 11?
(A) 5,555   (B) 968
(C) 1,111   (D) 11,111
(E) 2,805

7. How many numbers are there between 1 and 101 that are divisible by either 2 or 7, but not both?
(A) 50   (B) 58
(C) 63   (D) 69
(E) 7

8. Let $A \cdot B$ be defined by the equation $A \cdot B = (7 - B)(8 - A)$. Find the numerical value of $3 \cdot 5$.
(A) 18   (B) 20
(C) 15   (D) 12
(E) 10

9. The average of 20 students' test scores is 84. If the two highest and the two lowest scores are eliminated, the average of the remaining scores is 88. What was the average of the test scores eliminated?
(A) 86   (B) 68   (C) 77
(D) 93   (E) 34

10. How many centimeters are there in p meters?
(A) 10p (B) 100p
(C) 1,000p (D) 0.10p
(E) 0.100p

11. If $n > \frac{n^2}{10}$, then n may take on which of the following values?
(A) -40 (B) 40
(C) 0   (D) 10
(E) 0.5

12. Find the sum of all the distinct whole number factors of 80.
(A) 66   (B) 87
(C) 105   (D) 186
(E) 204

10 RELATIONS, FUNCTIONS, AND VARIATIONS

OVERVIEW

This chapter introduces the student to coordinate graphing and uses ordered pairs to define the concepts of relation and function. The student sees that equations in two variables have many solutions and that those of the form $Ax + By = C$ can be graphed as lines. Then direct and inverse variation are explored as practical examples of functions in two variables.

OBJECTIVES

- To graph a point given its coordinates
- To find the domain and range of a relation and determine whether it is a function
- To find values of a function using the notation $f(x)$
- To solve a linear equation with two variables and then to graph it
- To determine whether a relation is direct or inverse variation and to solve for missing values

PROBLEM SOLVING

This chapter introduces graphing in the coordinate plane as a way to identify a function and demonstrate its values. Using a Formula and Making a Table of Values are techniques used to graph functions and solve for unknown quantities. On page 390, the strategy of Making a Table is used with that of Making a Graph to solve problems related to distance, rate, and time. In the Application on page 391, students use a formula to estimate fixed and variable costs. The strategy of Restating the Problem is presented on page 405 as an aid in writing arguments to prove general statements related to integers and rational numbers.

READING AND WRITING MATH

Students are asked to give written explanations of newly introduced concepts in Exercise 40 on page 377, Exercise 21 on page 380, and Exercise 32 on page 407. The Focus on Reading Exercises on pages 375, 395, 399 review vocabulary and concepts covered in the preceding lessons.

TECHNOLOGY

Calculator: Using function keys on a scientific calculator is explained on page 381. Exercises 29 and 30 on page 385 and Exercises 11 and 12 on page 403 provide students with opportunities to use a calculator for the solution of a problem.

SPECIAL FEATURES

Mixed Review pp. 377, 381, 385, 396, 400, 404
Focus on Reading pp. 375, 395, 399
Using the Calculator p. 381
Midchapter Review p. 389
Application: Fixed and Variable Costs p. 391
Problem Solving Strategies: Making a Graph p. 390
Problem Solving Strategies: Restating the Problem p. 405
Key Terms p. 406
Key Ideas and Review Exercises pp. 406–407
Chapter 10 Test p. 408
College Prep Test p. 409
Cumulative Review (Chapters 1–10) pp. 410–411

PLANNING GUIDE

Lesson	Basic	Average	Above Average	Resources
10.1 pp. 376–377	FR all CE 1–15 WE 1–28 all, 40	FR all CE 1–15 WE 1–16 all, 17–39 odd, 40	FR all CE 1–15 WE 1–49 odd	Reteaching p. 75 Practice p. 75
10.2 pp. 379–381	CE 1–10 WE 1–17 all Using the Calculator	CE 1–10 WE 5–21 all Using the Calculator	CE 1–10 WE 9, 10, 15–29 all Using the Calculator	Reteaching p. 76 Practice p. 76
10.3 pp. 384–385	CE 1–8 WE 1–25 odd	CE 1–8 WE 9–33 odd	CE 1–8 WE 15–39 odd	Reteaching p. 77 Practice p. 77
10.4 pp. 388–391	CE 1–9 WE 1–21 odd Midchapter Review Problem Solving Strategies Application	CE 1–9 WE 7–27 odd Midchapter Review Problem Solving Strategies Application	CE 1–9 WE 11–31 odd Midchapter Review Problem Solving Strategies Application	Reteaching p. 78 Practice p. 78
10.5 pp. 395–396	FR all CE 1–8 WE 1–27 odd	FR all CE 1–8 WE 9–35 odd	FR all CE 1–8 WE 17–43 odd	Reteaching p. 79 Practice p. 79
10.6 pp. 399–400	FR all CE 1–12 WE 1–15 odd	FR all CE 1–12 WE 5–17 odd, 18	FR all CE 1–12 WE 7–21 odd	Reteaching p. 80 Practice p. 80
10.7 pp. 403–405	CE 1–10 WE 1–15 odd Problem Solving Strategies	CE 1–10 WE 5–19 odd Problem Solving Strategies	CE 1–10 WE 9–23 odd Problem Solving Strategies	Reteaching p. 81 Practice p. 81
Chapter 10 Review pp. 406–407	all	all	all	
Chapter 10 Test p. 408	all	all	all	
College Prep Test p. 409	all	all	all	
Cumulative Review pp. 410–411	all	all	all	

CE = Classroom Exercises   WE = Written Exercises   FR = Focus on Reading
Note: For each level, all students should be assigned all Try This and Mixed Review Exercises.

▰▰▰ INVESTIGATION

Project: In this *Investigation,* students explore the use of polar coordinates to locate cities in Colorado.

Materials: Each student will need a copy of Project Worksheet 10, a protractor, and a ruler.

Remind students that the usual way to locate points on a map is with latitude and longitude lines. In this project, they will use an alternate method to locate cities in Colorado.

Have students think of an imaginary east-west line running through Denver and have them draw a ray pointing east with Denver as the initial point. Explain that the **direction angle** is the angle, measured in a counterclockwise direction, from Denver to Colorado Springs. Since the angle is greater than 180°, have them use a protractor to measure the smaller of the two angles formed, and subtract this measure from 360°. (Remind students that a circle contains 360°). Then have students use a scale to find the distance from Denver to Colorado Springs, and tell them to write the coordinates as approximately (40, 276°), where 40 is the distance in miles and 276° is the direction angle.

Be sure that students understand that they are assuming air distance, not actual highway distance and that direction angles less than 180° can be measured directly without subtracting from 360°.

Have students complete the worksheet, working in pairs or in small groups.

Divers that explore the inner world of the oceans can avoid the painful and dangerous "bends" by carefully calculating the amount of oxygen in the diving tank, the duration of the dive, and the speed of rise to the surface.

More About the Photo

Underwater, the pressure increases by almost a half pound per square inch for each foot of depth. The pressure on a diver who is 33 feet deep is twice as great as the air pressure at the surface. A diver breathing compressed air (which is more than three-fourths nitrogen) absorbs nitrogen into the blood. If a diver rises too quickly, bubbles of nitrogen may form in the blood and cause the "bends" as the pressure decreases rapidly. A compression table gives a safe rate of ascent so that the pressure in the lungs stays equal to the decreasing water pressure.

10.1 Coordinates of Points in a Plane

Teaching Resources

Project Worksheet 10
Quick Quizzes 75
Reteaching and Practice
  Worksheets 75
Transparency 29

Objectives   To give the coordinates of a point in a plane
To identify the quadrant in which a given point is located
To graph a point, given its coordinates

You already know that only one number is needed to locate a point on a number line. To locate a point in a plane, two numbers, called **coordinates**, are needed.

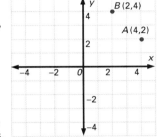

The figure at the right shows a **coordinate plane**. It has two number lines, called **axes**, that intersect in a point called the **origin**.

The horizontal number line is the **x-axis** with the positive direction to the right. The vertical number line is the **y-axis** with the positive direction upward.

For each point in a coordinate plane, there is exactly one **ordered pair** of real numbers. For each ordered pair of real numbers, there is exactly one point that is its graph. The figure shows point A, the graph of $(4, 2)$, and point B, the graph of $(2, 4)$.

To graph or **plot** a point given its coordinates, follow these steps: Start at the origin and move to the right if the first coordinate is positive or to the left if it is negative. Then move up if the second coordinate is positive or down if it is negative.

Examples 1 and 2 show how to graph ordered pairs of numbers.

EXAMPLE 1   Point P has coordinates $(5, -3)$.
Graph the point.

Solution   Start at the origin. Move 5 units to the right. Then move 3 units down.

Label the point as $P(5, -3)$.

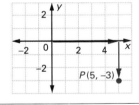

TRY THIS   1. Plot each point on the same graph. Write the ordered pair next to each point.
$A(-3, 1)$ $B(-2, -3)$ $C(4, 1)$ $D(1, -2)$

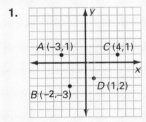
Additional Example 1

Point Q has coordinates $(-2, 2)$. Graph the point.

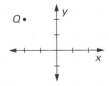

Lesson Note

Demonstrate why the coordinates of a point are called an *ordered* pair by showing the difference between the points with coordinates (2, −3) and (−3, 2). Show students that points such as (4, 0) and (0, −3) are not in any of the quadrants, but rather on one of the axes.

Math Connections

History: The concept of coordinates of points in a plane is sometimes referred to as the Cartesian Coordinate System in honor of its developer, Rene Descartes (1596–1650), a French mathematician. Legend has it that the idea first came to him as he watched a fly moving about the ceiling in his room. He noted that at any moment the position of the fly could be described in terms of its distances from two adjacent walls.

Critical Thinking Questions

Analysis: Ask students what is true of all points that have an abscissa of 4. They all lie on a vertical line parallel to, and 4 units to the right of, the y-axis. Ask what is true of all points that have an ordinate of −2. They all lie on a horizontal line parallel to, and 2 units below, the y-axis. Ask what all points which have an abscissa equal to the ordinate have in common. They all lie on line which goes through the origin and which makes a 45°-angle with the positive x-axis.

Common Error Analysis

Error: Students sometimes write the coordinates of a point in the wrong order.

Emphasize that the first coordinate describes movement to the left or right of the origin.

Sometimes, a point such as P is shown but its coordinates are not given. To determine its coordinates, find the number on the x-axis that is directly above P and the number on the y-axis that is directly to the left of P.

In the figure, the number 3 is on the x-axis directly above P. It is called the **x-coordinate** of the point. The number −2 is on the y-axis directly to the left of P. It is called the **y-coordinate** of the point. Thus, $(3, -2)$ are the coordinates of point P.

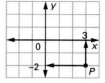

The x-coordinate of a point is also called the *abscissa*. The y-coordinate is also called the *ordinate*.

EXAMPLE 2 Find the coordinates of point A.

Solution Point A is 2 units to the left of the y-axis and 1 unit above the x-axis.

Therefore, the coordinates of point A are $(-2, 1)$.

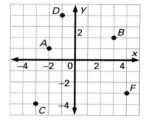

TRY THIS 2. Find the coordinates of points B, C, D, and F on the graph above. $B(3, 2)$; $C(-3, -4)$; $D(-1, 4)$; $F(4, -3)$

The axes divide a coordinate plane into four **quadrants,** as shown. Notice that the quadrants are numbered in a counterclockwise direction, beginning at the upper right. Each point is located in one of the quadrants or on one of the axes. The origin has the coordinates (0, 0) and is located on both axes.

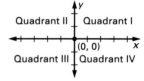

EXAMPLE 3 Give the quadrant for each point.

    a. $P(3, 2)$ **b.** $Q(-2, 4)$ **c.** $R(-1, -4)$ **d.** $S(4, -4)$

374   Chapter 10 Relations, Functions, and Variations

Additional Example 2

Give the coordinates for each of the points A, B, C, and D. $A(1, 2)$; $B(-2, 3)$; $C(-1, -2)$; $D(2, -1)$

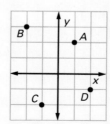

Solution   Plot each point.

    a. $P(3, 2)$ is in Quadrant I.

    b. $Q(-2, 4)$ is in Quadrant II.

    c. $R(-1, -4)$ is in Quadrant III.

    d. $S(4, -4)$ is in Quadrant IV.

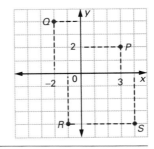

TRY THIS   Give the quadrants for each point.
3. $W(-3, 4)$ II   4. $X(4, -1)$ IV   5. $Y(2, 5)$ I   6. $Z(-1, -1)$ III

The next example illustrates how to determine the fourth vertex of a rectangle if the other three are known. Recall that the opposite sides of a rectangle are equal in length, and its adjacent sides are perpendicular.

EXAMPLE 4   Three vertices of rectangle $ABCD$ are $A(1, 1)$, $B(6, 1)$, and $C(6, 4)$. Find the coordinates of the fourth vertex D.

Solution   Graph the points A, B, and C. Point D must be directly above A and directly to the left of C. D has the same x-coordinate as A and the same y-coordinate as C.

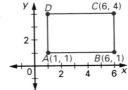

    Thus, the coordinates of the fourth vertex are $D(1, 4)$.

TRY THIS   7. $A(1, -3)$, $B(1, 2)$, and $C(4, 2)$ are the vertices of a rectangle. Find the coordinates of the fourth vertex. $(4, -3)$

Focus on Reading

Determine whether each statement is true or false. If false, give a reason for your answer.

1. There is exactly one ordered pair of numbers that locates a given point in the coordinate plane. T
2. Every point is in one of four quadrants. F; a point could lie on an axis.
3. The abscissa of an ordered pair is always positive. F; abscissa is negative if the point is in Quadrant II or III.
4. The horizontal axis is the x-axis. T
5. The ordinate of every point on the x-axis is zero. T

1. Give the coordinates and the quadrant for each of the points P, Q, R, S, and T.

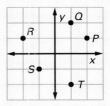

    $P(2, 1)$, I; $Q(1, 2)$, I; $R(-2, 1)$, II; $S(-1, -1)$, III; $T(1, -2)$, IV

2. Three vertices of rectangle $ABCD$ are at $A(-2, 1)$, $B(6, 1)$, and $C(6, 5)$. Find the coordinates of the fourth vertex D. $(-2, 5)$

Closure

Ask students to define the words axes, coordinates, origin, abscissa, ordinate and quadrant. See pp. 373–374. Then have them describe how to locate a point, given its coordinates. Answers will vary.

◼◼◼ FOLLOW UP

Guided Practice

Classroom Exercises 1–15
Try This all FR all

Independent Practice

A Ex. 1–30, **B** Ex. 31–40, **C** Ex. 41–49

Basic: WE 1–28 all, 40

Average: WE 1–16 all, 17–39 odd, 40

Above Average: WE 1–49 odd, 40

Additional Example 3

Name the quadrant for each point, A, B, C, and D in Additional Example 2. A, I; B, II; C, III, D, IV

Additional Example 4

Three vertices of rectangle $ABCD$ are $A(2, 3)$, $B(6, 3)$, and $C(6, 7)$. Find the coordinates of the fourth vertex, D. $(2, 7)$

Additional Answers

Classroom Exercises

10–15.

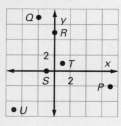

Written Exercises

17–24.

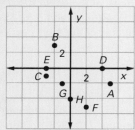

25.

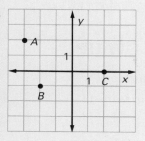

26.

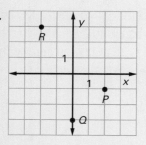

Classroom Exercises

Tell what point has the given coordinates.

1. (3, 1) B
2. (−5, 1) F
3. (0, 3) D
4. (−4, 0) E
5. (3, −4) I
6. (−4, −2) G

For each point, tell which number is the abscissa and which is the ordinate.

7. A a:2, o:0
8. C a:1, o:4
9. H a: −1, o: −3

Graph each point in the same coordinate plane.

10. P(7, −2)
11. Q(−2, 7)
12. R(0, 5)
13. S(−1, 0)
14. T(1, 1)
15. U(−5, −5)

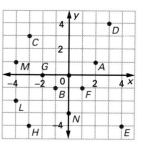

Written Exercises

Give the coordinates of each point.

1. A (2, 1)
2. B (−1, −1)
3. C (−3, 3)
4. D (3, 4)
5. E (4, −4)
6. F (1, −1)
7. G (−2, 0)
8. H (−3, −4)
9. L (−4, −2)
10. M (−4, 1)
11. N (0, −3)
12. O (0, 0)

Which points in the diagram above are in the given quadrant?

13. Quadrant I A, D
14. Quadrant II C, M
15. Quadrant III B, H, L
16. Quadrant IV E, F

Graph each point in the same coordinate plane.

17. A(5, −2)
18. B(−2, 3)
19. C(−3, −1)
20. D(4, 0)
21. E(−3, 0)
22. F(2, −5)
23. G(−1, −2)
24. H(0, −4)

Graph the three points in the same coordinate plane.

25. A(−3, 2), B(−2, −1), C(2, 0)
26. P(2, −1), Q(0, −3), R(−2, 3)
27. M(5, −1), N(−4, −2), P(−4, 3)
28. R(0, 2), S(−4, −5), T(3, −4)
29. D(−3, −2), E(1, 4), F(5, 0)
30. A(−2, −3), B(5, −2), C(0, 0)

376   Chapter 10 Relations, Functions, and Variations

A, B, and **C** are three vertices of a rectangle. Find the coordinates of the fourth vertex of the rectangle.

31. $A(-1, 2)$, $B(4, 2)$, $C(4, 4)$ $(-1, 4)$ **32.** $A(3, -1)$, $B(7, -1)$, $C(7, 3)$ $(3, 3)$
33. $A(3, 0)$, $B(5, 0)$, $C(5, 5)$ $(3,5)$ **34.** $A(-4, -3)$, $B(3, -3)$, $C(3, 3)$ $(-4, 3)$

Using the graph at the right, find the point(s) whose coordinates satisfy the given condition. **37.** A, B, C, G

35. The abscissa is -2. D; F   **36.** The ordinate is 1. C
37. The abscissa is positive.   **38.** The ordinate is negative.
39. The sum of the abscissa and the ordinate is 5. A, B   E, F, G
40. Write a short paragraph to explain the construction of a coordinate system. Mention how the axes are located, how they are labeled, how direction is determined, and the relation between ordered pairs of real numbers and points. Give examples.
Answers will vary.

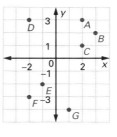

A, B, and **C** are three vertices of parallelogram **ABCD.**
Find the coordinates of the fourth vertex of the parallelogram.

41. $A(1, 1)$, $B(5, 1)$, $C(7, 3)$ $(3, 3)$ **42.** $A(2, -1)$, $B(7, -1)$, $C(8, 1)$ $(3, 1)$
43. $A(-1, 1)$, $B(4, 1)$, $C(6, 4)$ $(1, 4)$ **44.** $A(-1, -3)$, $B(4, -3)$, $C(5, -1)$ $(0, -1)$

45–48. In Exercises 41–44, suppose that the order of the vertices (clockwise or counterclockwise) is not in the alphabetical order "ABCD." Find all possible positions for vertex D.

49. Find two ordered pairs (x,y) such that $2|x - 6| - 4 = 8$
and $\dfrac{10}{y^2 + 2y} = \dfrac{5}{y + 2}$. $(12, 2)$, $(0, 2)$
45. (11, 3), (3, 3), (−1, −1)
46. (13, 1), (3, 1), (1, −3)
47. (11, 4), (1, 4), (−3, −2)
48. (10, −1), (0, −1), (−2, −5)

Mixed Review

Solve and check. **3.3, 3.5, 4.6**

1. $4x - 9 = 39$ $x = 12$ **2.** $\frac{2}{3}x + 15 = -5$ $x = -30$
3. $10a + 2(1 - a) = 6a - 4$ $a = -3$ **4.** $1.8 - 3.1n = 48.6 + 7.3n$ $n = -4.5$

Graph the solution set on a number line. **5.4**

5. $5x - 7 \geq 8$ **6.** $2x - 18 \leq 34$

7. The square of a number is 6 less than 5 times the number. Find the number. **7.9** 2 or 3

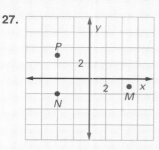

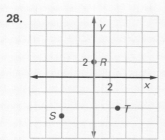

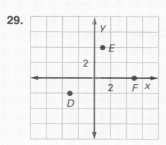

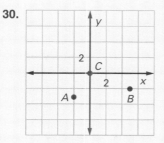

See page 385 for the answers to Mixed Review Ex. 5–6.

Enrichment

Have students locate a point in each of the four quadrants and label the coordinates of each point. Then have them draw a quadrilateral with these points as vertices. Next have them identify (estimating if necessary) and label the midpoints of the four sides of the quadrilateral. Finally, have them draw a new quadrilateral having as its vertices the midpoints of the sides of the original quadrilateral. Ask: What type of quadrilateral does the new figure appear to be? Have students repeat the experiment a few times with other sets of points in the four quadrants. Parallelogram.

■GETTING STARTED

Prerequisite Quiz

Given set $A = \{(-1, 2), (3, -2), (0, -5),$
$(-7, -3), (4, 2)\}$.

1. List the abscissas for the ordered
   pairs. $-1, 3, 0, -7, 4$
2. List the ordinates for the ordered
   pairs. $2, -2, -5, -3, 2$
   Given points $A(-1, 6), B(2, 6), C(0, 2)$,
   and $D(-1, 5)$:
3. Which points have the same abscissa?
   A and D
4. Which points have the same ordinate?
   A and B

Motivator

Remind students that two numbers are often
related in some way. Have them write
ordered pairs to represent the data in the
following.

1. It takes Mike 4 h to type a 40-page
   report, 9 hr to type a 100-page report,
   and $1\frac{1}{2}$ h to type a 6-page letter.
   $\{(4, 40), (9, 100), (1\frac{1}{2}, 6)\}$
2. The cost of 3 different sizes of a boxed
   detergent is $2.25 for 24 oz, $3.60 for
   36 oz, and $5.20 for 60 oz. $\{(24, \$2.25),$
   $(36, \$3.60), (60, \$5.20)\}$ Ask students
   to name other ways to represent this
   paired data.   tables, graphs, or
   equations.

10.2 Relations and Functions

Objectives

To graph relations and functions
To determine the domain and the range of a relation
To determine whether a relation is a function

Anne worked as a lifeguard for $7 per hour. In 1 hour she earned $7,
in 2 hours she earned $14, in 3 hours she earned $21, and in 4 hours
she earned $28. The number of hours she worked and the amount she
earned can be described by a set of ordered pairs.

$$S = \{(1, 7), (2, 14), (3, 21), (4, 28)\}$$

A set of ordered pairs is called a **relation**. The set S is a relation. The
set of all first coordinates is called the **domain** of a relation. For S, the
domain is $\{1, 2, 3, 4\}$. The set of all second coordinates is called the
range of a relation. For S, the range is $\{7, 14, 21, 28\}$. For the relation
$N = \{(\text{James}, 19), (\text{Kris}, 23), (\text{Scott}, 19)\}$, the domain is $\{\text{James}, \text{Kris},$
Scott$\}$, and the range is $\{19, 23\}$. Notice that 19 is listed only once.

EXAMPLE 1

Graph the relation $B = \{(3, 0), (-1, 2), (4, 1), (-2, -1), (1, -2)\}$.
Then give the domain and range of B.

Solution

The domain is the set of all first coordinates.

The range is the set of all second coordinates.

The domain of B is $\{-2, -1, 1, 3, 4\}$.

The range of B is $\{-2, -1, 0, 1, 2\}$.

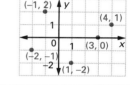

TRY THIS

1. Give the domain and range of the relation   Domain: $\{2, 0, -1, 3\}$
   $M = \{(2, 1), (0, 0), (-1, -2), (3, 6)\}$.   Range: $\{1, 0, -2, 6\}$

Notice that for the relation $B = \{(3, 0), (-1, 2), (4, 1), (-2, -1),$
$(1, -2)\}$, no first coordinates are the same. This kind of relation is
called a *function*. In a function some of the second coordinates may be
the same, but all the first coordinates must be different.

Definition

A **function** is a relation in which no two ordered pairs have the same
first coordinate.

Highlighting the Standards

Standards 6a, 6c: In the Exercises,
students see how to recognize relations and
functions and their graphs and to use
functions in problems. In Using the
Calculator, they explore functions with a
scientific calculator.

Additional Example 1

Graph the relation
$T = \{(-1, -3), (3, 2), (-4, 1)\}$.

Give the domain and the range of
T. $D = \{-4, -1, 3\}, R = \{-3, 1, 2\}$

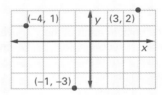

EXAMPLE 2 Given the relation $A = \{(1, -3), (-1, 3), (2, 5), (-2, -3), (3, -1)\}$, determine the domain and range. Is A a function?

Plan Determine whether any two first coordinates are the same.

Solution The domain of A is $\{-2, -1, 1, 2, 3\}$.
The range of A is $\{-3, -1, 3, 5\}$.
Since no ordered pairs have the same first coordinate, A is a function.

EXAMPLE 3 List the set of ordered pairs in relation G. Is G a function?

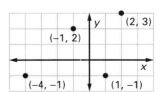

Plan List the set of ordered pairs. Then determine whether any two first coordinates are the same.

Solution $G = \{(-2, 1), (3, 2), (-2, 3), (-1, -3)\}$.
Since the ordered pairs $(-2, 1)$ and $(-2, 3)$ have the same x-coordinate, G is not a function.

TRY THIS 2. Is $D = \{(-2, 4), (-1, 1), (0, 0), (1, 1), (1, 2)\}$ a function? No

Classroom Exercises

Determine which relations are functions and which are not functions.

1. $\{(1, 2), (-3, 2), (4, -1), (2, 1), (2, 3)\}$ No 2. $\{(-3, 1), (4, 2), (3, -1), (2, -4)\}$ Yes
3. $\{(4, 2), (4, -1), (3, -2), (1, 4), (-2, 4)\}$ No 4. $\{(3, -2), (-3, 2), (5, -2), (-1, -1)\}$ Yes

State the domain and the range of each relation.

5. $\{(2, 1), (-3, 2), (-1, 4)\}$ 6. $\{(-4, 0), (3, -1), (2, 3)\}$
7. $\{(0, -3), (-2, 5), (1, -6)\}$ 8. $\{(5, 6), (7, 0), (-8, 1)\}$
9. $\{(35, \text{John}), (43, \text{Hilda}), (10, \text{Dawn})\}$ 10. $\{(\text{red}, 1), (\text{blue}, 10), (\text{red}, 3)\}$
D: {35, 43, 10}; R: {John, Hilda, Dawn}   D: {red, blue}; R: {1, 3, 10}

Written Exercises

Graph. Give the domain and range of the relation.

1. $\{(3, 1), (0, -2), (5, -3), (-3, -1)\}$ 2. $\{(4, 0), (-3, -2), (-2, 1), (2, -1)\}$
3. $\{(-1, 2), (4, 3), (-1, 4), (2, -3)\}$ 4. $\{(3, -2), (3, -4), (0, 2), (2, -2)\}$

10.2 Relations and Functions   **379**

Checkpoint

1. Graph the relation $S = \{(-3, 0),$ $(5, -3), (2, 2)\}$. Give the domain and the range of S. $D = \{-3, 2, 5\}$, $R = \{-3, 0, 2\}$

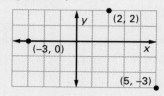

2. Is S a function?   Yes
3. List the set of ordered pairs in relation T graphed below. Is T a function? $\{(1, 2),$ $(1, -1), (-3, 1), (0, -2)\}$; No

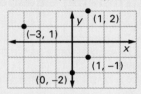

Closure

Ask students to state the difference between a relation and a function.   (See p. 378).
Ask them to classify each of the following as a function, or not a function.

1. $\{(0, 0), (2, 22), (3, 33)\}$ Function
2. $\{(10, 20), (10, 50), (11, 30)\}$ Not a function
3. $\{(1, 7), (1, 14), (1, 21)\}$ Not a function

State the domain and the range of each relation. Is it a function?

5. $\{(-6, -3), (7, -3), (-4, 7), (5, 10)\}$

6. $\{(0, 1), (3, 0), (0, 0), (-5, 0)\}$

7. $\{(-3, -4), (3, -2), (2, -3), (-3, 5)\}$
   D: $\{-3, 2, 3\}$; R: $\{-4, -3, -2, 5\}$; no

8. $\{(9, 6), (6, 9), (5, 4), (4, 5)\}$
   D: $\{4, 5, 6, 9\}$; R: $\{4, 5, 6, 9\}$; yes

List the set of ordered pairs of each relation. Is the relation a function?

9.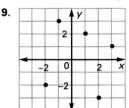

10.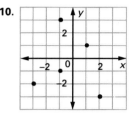

$\{(1, 2), (3, 1), (-1, 3), (-2, -2), (2, -3)\}$; yes

$\{(1, 1), (-1, 3), (-1, -1), (-3, -2), (2, -3)\}$; no

Determine whether or not the relation is a function.

11. $\{(-3, 1), (4, 0), (2, -3), (5, -1), (3, -3)\}$ A function

12. $\{(4, 1), (5, 0), (-2, 3), (4, -2), (3, 2)\}$ Not a function

13. $\{(0, -3), (-1, 2), (-1, -3), (5, 0), (2, 1)\}$ Not a function

14. $\{(3, -4), (-4, 2), (5, 1), (-2, 3), (0, 3)\}$ A function

15. $\{(-5, -1), (4, 3), (3, -1), (-5, 5), (0, 2)\}$ Not a function

16. $\{(-3, -2), (-1, 4), (3, -1), (-5, 2), (4, -1)\}$ A function

17. The life expectancy of a monkey is 15 years, an African elephant 35 years, a beaver 5 years, and a grizzly bear 25 years. Write a relation that describes this information. Determine the domain and the range.

18. An elephant can run 25 mi/h, a reindeer 32 mi/h, a greyhound 39 mi/h, and an elk 45 mi/h. Write a relation that describes this information. Determine the domain and the range.

19. For what value of k will the relation $R = \{(2k+1, 3), (3k-2, -6)\}$ *not* be a function?
    (HINT: R will not be a function if $2k + 1 = 3k - 2$.)   3

20. For what value of k will the relation $S = \{(2k+3, 1), (3k-1, -2)\}$ *not* be a function?   4

21. Write a paragraph explaining what is meant by a relation and a function. Be sure to include these features: examples of functions, examples of relations that are not functions, and examples of functions that show a specific relationship between domain and range.   Answers will vary.

For what value(s) of k will the relation *not* be a function?

22. $A = \{(k^2, 16), (4k, 32)\}$ 0, 4
23. $B = \{(k^2 - 5k, 10), (k + 7, 4)\}$ $-1, 7$
24. $R = \{(k^3 - 5k^2 + 3k, -5), (-k, 4)\}$ 0, 1, 4
25. $S = \{(|k + 1| + 2, 4), (8, 7)\}$ $-7, 5$

Graph the function described below. The domain of each function is $\{-2, -1, 0, 1, 2\}$.

26. In each ordered pair, the second number is equal to twice the absolute value of the first number.

27. In each ordered pair, the second number is the opposite of the absolute value of the first number.

28. In each ordered pair, the second number is the square of the first number.

29. In each ordered pair, the second number is either 2 or -2. If the first number is nonnegative, then the second number is 2. If the first number is negative, then the second number is -2.

Mixed Review

Evaluate. *1.1, 2.6*

1. $x + 9$ for $x = 12$ 21
2. $4y - 3$ for $y = 5$ 17
3. $5m^2$ for $m = -4$ 80
4. x^2y for $x = -3$ and for $y = -4$ -36

Solve. *7.7, 9.1*

5. $\dfrac{3x}{2} + 1 = \dfrac{6x}{5} - \dfrac{14}{10}$ $x = -8$
6. $3x^2 + 8x - 3 = 0$ $-3, \frac{1}{3}$

◢◣/ Using the Calculator

Scientific calculators are often compared by the number of "functions" they can perform. Select several function keys and test one at a time. Form an ordered pair by first listing the entry, and then the result displayed. You may try making the same entry several times. For example,

$\boxed{x^2}$ key:	$\boxed{\sqrt{x}}$ key:
(19, 361)	(361, 19)
(27, 729)	(729, 27)
(-27, 729)	(729, 27)

1. Examine the set of relations obtained by using the $\boxed{x^2}$ key. Is this relation a function? Yes
2. Is the set of relations displayed by the $\boxed{\sqrt{x}}$ key a function? Yes
3. Do the function keys on a calculator give sets of values that are functions? Yes

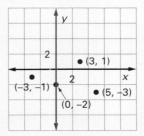
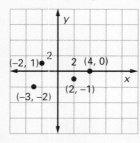

■■■GETTING STARTED

Prerequisite Quiz

Evaluate for the given value of the variable.

1. $a - 3$ for $a = 5$ 2
2. $3x + 8$ for $x = -2$ 2
3. $m^2 + 9$ for $m = -3$ 18
4. $5y^2 - 6$ for $y = 4$ 74
5. $a^2 - 3a + 5$ for $a = -2$ 15
6. $\frac{t^2 - 4}{t + 1}$ for $t = 4$ $2\frac{2}{5}$

Motivator

There are many practical examples of a function. For example, if a car is driven at a constant rate of 50 mi/h, then the distance traveled is a function of the time t. Write the ordered pairs for this function: (t,d) and (2 h, 100 mi) on the chalkboard. Ask students these questions.

1. What is the value of d when t is 3 h?
   150 mi
2. What is the value of d when t is 7 h?
   350 mi
3. Does it make sense for the value of t in this function to be a negative?   No, because it represents time in hours.

10.3 Values of a Function

Objectives

To find values of a function using the notation $f(x)$
To determine the range of a function for a given domain

A **function** is a relation in which no two ordered pairs have the same first coordinate. Consider the set of ordered pairs below.

$$\{(2, 5), (6, 9), (-1, 2), (-3, 0)\}$$

Since no two of the ordered pairs have the same first coordinate, the relation is a function. You know that you can graph this function in a coordinate plane. Another way to picture this function is in a **mapping** as illustrated below.

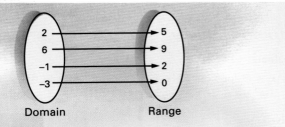

You can see that the domain is $\{-3, -1, 2, 6\}$. The range is $\{0, 2, 5, 9\}$. From the mapping, note that for each number in the domain, there is just one number in the range.

EXAMPLE 1   Which relations are functions?

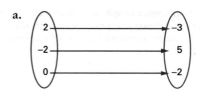
a.

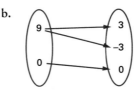
b.

Solutions

a. For each number in the domain, there is exactly one number in the range. So the relation *is* a function.

b. For the number 9 in the domain, there are two numbers in the range. So the relation *is not* a function.

Highlighting the Standards

Standards 4a, 6c: This lesson represents a function as an equivalent mapping and students translate between different representations of a function.

Additional Example 1

Which relations are functions?

a.

b.

a. Not a function
b. Function

Which relations are functions?

1. $S = \{(-1, 1), (2, 1), (3, 2)\}$ 2. $T = \{(1, 3), (-1, -3), (0, 0)\}$
   Both S and T

This mapping shows the function
$f = \{(-2, -8), (1, 4), (3, 12)\}$.

The -2 in the domain corresponds to -8 in the range. Similarly, the 1 in the domain corresponds to 4 in the range.

Each number in the *range* of a function f is called a **value of the function.** Thus, when x is -2, the value of the function shown is -8. This can be represented as

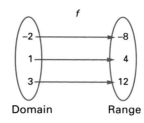

$$f(-2) = -8$$

which is read "the value of f at -2 is -8." (Here, f is *not* a variable and $f(-2)$ does *not* mean "f times negative 2.")

Similarly, the values of f at 1 and 3 are represented as

$$f(1) = 4 \text{ and } f(3) = 12.$$

EXAMPLE 2   Find the indicated value of the function.
   a. $f(-1)$ b. $f(-3)$

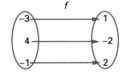

Solutions   a. The value of f b. The value of f
   at -1 is 2.   at -3 is 1.

   $f(-1) = 2$ $f(-3) = 1$

TRY THIS   Find the indicated value of the function for each of the following.

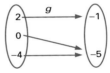

3. $g(2)$ -1 4. $g(-4)$ -5

Sometimes a function can be defined by an algebraic expression or rule. Consider the function with domain $D = \{-12, -5, -2, 0, 2, 5, 12\}$, where $f(x) = x + 2$.

This indicates that for each number x in the domain, the value of the function is $x + 2$. Thus, when x is 5 or 12 or any other number in the domain, the value of the function can be computed as shown.

$f(5) = 5 + 2 = 7$ $f(12) = 12 + 2 = 14$

Thus, $f(5) = 7$.   Thus, $f(12) = 14$.

Lesson Note

The notation $f(x)$ or $g(x)$ is new for many students and will need careful discussion. Emphasize that $f(x)$ means the value of the function for a given value x and not f times x. Be sure the students understand that the values of the function make up the *range* of the function.

Math Connections

Calculator: Many of the keys on a scientific calculator, such as the x^2 or $\sqrt{\ }$ keys, are referred to as function keys. Each function key requires one or more numbers as input to give a value of the function.

Critical Thinking Questions

Analysis: Begin by explaining that the domain and range of a function are sometimes restricted by either real-life or mathematical considerations. Ask these questions.

1. If the distance d is a function of the time t, and $d = 50t$, are there any restrictions on the domain (the values of t) or the range (the values of d)? Each must consist of non-negative numbers.

2. If $g(x)$ is a function of x and $g(x) = \frac{1}{x}$, are there any restrictions on the domain (the values of x) or the range (the values of $g(x)$)? Neither can include zero.

Additional Example 2

Find the indicated value of the function.
a. $f(-2)$ -8
b. $f(8)$ 27

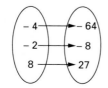

383

Common Error Analysis

Error: Errors sometimes occur when substituting negative values. For example, if $g(x) = -x^2 + 4$, then $g(-2) = -(-2)^2 + 4 = -(4) + 4$, or 0. A student may erroneously write this as $g(-2) = (-2)^2 + 4 = 4 + 4$, or 8.

Suggest that students use parentheses when they substitute elements of the domain.

Checkpoint

Given $f(x) = -3x + 4$, find the indicated value.

1. $f(-4)$ 16   **2.** $f(0)$ 4
3. $f(9)$ -23 **4.** $f(12)$ -32

Given $g(x) = 3x^2 - 8$, find the indicated value.

5. $g(-2)$ 4   **6.** $g(1)$ -5

Closure

Have each student specify a domain of 3 elements and a function rule in the form $f(x) =$ algebraic expression on a piece of paper. Then have students exchange papers and give the range for each other's functions. Have students exchange papers again and check the results.

EXAMPLE 3 Given $f(x) = 4x^2 - 1$ and the domain $D = \{-2, 1, 3\}$, find the indicated values of the function and the range.

    a. $f(3)$ **b.** $f(1)$ **c.** $f(-2)$

Solutions

a. $f(x) = 4x^2 - 1$ **b.** $f(x) = 4x^2 - 1$ **c.** $f(x) = 4x^2 - 1$
    $f(3) = 4(3^2) - 1$ $f(1) = 4(1^2) - 1$ $f(-2) = 4(-2)^2 - 1$
        $= 4(9) - 1$ $= 4(1) - 1$ $= 4(4) - 1$
        $= 36 - 1$ $= 4 - 1$ $= 16 - 1$
        $= 35$ $= 3$ $= 15$

Thus, the range of f is $\{3, 15, 35\}$.

TRY THIS **5.** Given $g(x) = -3x^2 + 6$ and the domain $D = \{-3, 0, 3\}$, find the range of g. $\{-21, 6\}$

Classroom Exercises

Find the indicated value of the function shown.

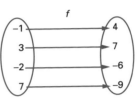

1. $f(-1)$ 4 **2.** $f(7)$ -9 **3.** $f(3)$ 7 **4.** $f(-2)$ -6

Given a function $g(x) = 3x + 6$, find the indicated value of the function.

5. $g(8)$ 30 **6.** $g(-2)$ 0 **7.** $g(0)$ 6 **8.** $g(-5)$ -9

Written Exercises

Which relations are functions?

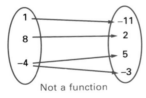

Not a function   **2.**   A function

Find the indicated value of the function shown at the right.

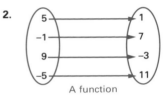

3. $f(0)$ 4 **4.** $f(-1)$ -2 **5.** $f(1)$ 3 **6.** $f(-3)$ -5

Given $f(x) = 2x - 6$, find the indicated value.

7. $f(5)$ 4 **8.** $f(3)$ 0 **9.** $f(1)$ -4 **10.** $f(-5)$ -16

Given $g(x) = -4x + 5$, find the indicated value.

11. $g(0)$ 5 **12.** $g(3)$ -7 **13.** $g(-3)$ 17 **14.** $g(7)$ -23

Additional Example 3

Given $f(x) = -2x^2 + 4$ and the domain $D = \{-1, 2, 4\}$. Find the indicated value and the range of the function.

a. $f(-1)$ 2
b. $f(2)$ -4
c. $f(4)$ -28 range: $\{-28, -4, 2\}$

Given $f(x) = x^2 - 9$, find the indicated values.

15. $f(2)$ −5   **16.** $f(5)$ 16   **17.** $f(-1)$ −8   **18.** $f(-4)$ 7

Given $g(x) = -2x^2 + 1$, find the indicated values.

19. $g(0)$ 1   **20.** $g(-3)$ −17   **21.** $g(2)$ −7   **22.** $g(-4)$ −31

Use the given domain D to find the range of each function. **23.** {−8, −5, 4}
{7, −2, −11}

23. $f(x) = 3x - 5, D = \{-1, 0, 3\}$ **24.** $g(x) = -3x + 1, D = \{-2, 1, 4\}$
25. $f(x) = 2x + 6, D = \{-5, -1, 4\}$ **26.** $g(x) = 5x - 3, D = \{-3, -2, -1\}$
27. $f(x) = x - 4,$ {−4, 4, 14}   **28.** $g(x) = -3x^2 - 4,$ {−18, −13, −8}
  $D = \{-2, -1, 3\}$ {−6, −5, −1} $D = \{-1, 2, 3\}$ {−31, −16, −7}
29. $f(x) = x^2 - 4,$ **30.** $g(x) = -4x + 9,$
  $D = \{-14.1, 0.04, 21.06\}$ $D = \{-2.01, 3.65, 5.98\}$
  {−3.9984, 194.81, 439.5236}   {−14.92, −5.6, 17.04}

If a machine produces 280 appliances per hour, the total number of
appliances produced in x hours can be represented as a function
$f(x) = 280x$.

31. How many appliances can be pro-   **32.** How many more appliances can be
duced in 3.5 h? 980   produced in 7 h than in $4\frac{1}{4}$ h? 770

Use the given domain to find the range of the function.

33. $f(x) = -x^3 + x^2 + 5,$ **34.** $g(x) = 2x^3 - 3x + 1,$
  $D = \{-2, -1, 2\}$ {1, 7, 17} $D = \{-3, -1, 4\}$ {−44, 2, 117}
35. $f(x) = |x|, D = \{-4, 0, -3\}$ {0, 3, 4}   **36.** $g(x) = (x - 3)^2, D = \{-3, 2, 3\}$
37. $f(x) = (2x - 1)^2, D = \{-5, 3, 4\}$ **38.** $g(x) = |x - 3|^2, D = \{-3, 2, 3\}$
  {25, 49, 121}   {0, 1, 36}

Write an equation with $f(x)$ to show how the numbers in the domain
and the range are related.

39. $\{(1, 2), (-2, -4), (3, 6), (0, 0)\}$ $f(x) = 2x$ **40.** $\{(1, 5), (3, 17), (-2, -13), (-4, -25)\}$
    $f(x) = 6x - 1$

Mixed Review

Simplify. *2.2, 2.8*

1. $|7| + |-7|$ 14   **2.** $|-1.7| + |1.8|$ 3.5   **3.** $|8| \cdot |0|$ 0   **4.** $|-2| \cdot |-4|$ 8
5. $-3 \cdot 5 + 4(-2) - 9(-5)$ 22   **6.** $8(-1) + (-2)^2 - 4 \cdot 3$ −16

Factor. *7.3, 7.4* **7.** $(x - 1)(x - 3)$ $(x - 2)(x + 4)$ $(2x - 1)(x + 2)$ $(3x + 2)(x - 1)$
7. $x^2 - 4x + 3$ **8.** $x^2 + 2x - 8$ **9.** $2x^2 + 3x - 2$ **10.** $3x^2 - x - 2$
11. Find two consecutive integers whose product is 342. *7.9* 18, 19 *or* −19, −18

10.3 Values of a Function   **385**

Guided Practice

Classroom Exercises 1–8
Try This   all

Independent Practice

A Ex. 1–22, **B** Ex. 23–34, **C** Ex. 35–40

Basic: WE 1–25 odd
Average: WE 9–33 odd
Above Average: WE 15–39 odd

Additional Answers

Written Exercises

36. {0, 1, 36}

Additional Answers, page 377

Mixed Review

5.

6.

Enrichment

Give students the following functions.
$f(x) = 2x + 3$
$g(x) = x^2 - 3x$
$h(x) = \frac{x - 3}{2}$
Then show an example of a composite
function, such as $f(g(2))$, where you first
find $g(2) = -2$ and then $f(-2) = -1$; so
$f(g(2)) = -1$.

Have students find each of the following
values.

1. $f(5) - g(-4)$ −15
2. $f(g(8))$ 83
3. $f(h(11))$ 11
4. $h(f(-4))$ −4
5. $f(h(x))$ x

10.4 Equations with Two Variables

Objective   To solve an equation with two variables for a given replacement set

The perimeter of each rectangle below can be represented by an equation. In each case, begin with the perimeter formula, $p = 2l + 2w$.

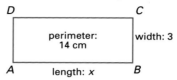

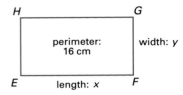

Equation: $14 = 2x + 2 \cdot 3$
Solution: *one* solution
$x = 4$
(Verify that 4 *is* a solution by
substituting it into the equation.)

Equation: $16 = 2x + 2y$
Solutions: *many* solutions
$x = 2$ when $y = 6$,
$x = 5$ when $y = 3$,
$x = 7$ when $y = 1$,
and so on
(Use substitution to verify that
these numbers *are* solutions.

EXAMPLE 1   Find two solutions of the equation $5x - y = 9$.

Plan   Choose any value for x and solve for y.

Solution

Let $x = -1$.
$$5(-1) - y = 9$$
$$-5 - y = 9$$
$$-y = 14$$
$$y = -14$$

Let $x = \frac{1}{5}$.
$$5(\tfrac{1}{5}) - y = 9$$
$$1 - y = 9$$
$$-y = 8$$
$$y = -8$$

Thus, two solutions are $(-1, -14)$ and $(\frac{1}{5}, -8)$.

TRY THIS

1. Find two solutions of the equation $2x - 3y = 5$.
   Answers may vary. Possible answers: $(1, -1), (0, -\frac{5}{3})$

The solution set for an equation with two variables is a set of ordered pairs. To find the solution set for an equation such as $2x + 3y = 15$:

1. Solve the equation for y.
2. Then substitute numbers for x.

GETTING STARTED

Prerequisite Quiz

For $x = 6$, what value of y makes each equation true?

1. $x + y = 5$ -1
2. $2x + 3y = 6$ -2
3. $5x - 2y = -4$ 17

Solve for y in terms of x.

4. $x + 2y = 8$ $y = -\frac{1}{2}x + 4$
5. $-3x + 4y = 12$ $y = \frac{3}{4}x + 3$
6. $x - 3y = 2$ $y = \frac{1}{3}x - \frac{2}{3}$

Motivator

Tell students that they have 26 feet of fencing to surround a rectangular flower garden. Ask them to give possible dimensions of the garden.   Answers will vary.   Point out that their answers are solutions of the equation $2l + 2w = 26$ and that such an equation in two variables has many solutions.

Highlighting the Standards

Standards 5d, 6c, 1b: Here students practice functional notation and relate a table to a function. In Problem Solving Strategies, they see how to use the strategy of Making a Graph.

Additional Example 1

Find two solutions of the equation
$3x - y = 0$. $(0, 0), (\frac{1}{3}, 1)$; Answers will vary.

386

EXAMPLE 2   Find the solution set for $2x + 3y = 15$ if the replacement set for each variable is $\{0, 3, 5\}$.

Plan   Solve the equation for y in terms of x. Substitute values from $\{0, 3, 5\}$ for x. Accept only numbers from $\{0, 3, 5\}$ for y.

Solution
$$2x + 3y = 15$$
$$3y = 15 - 2x$$
$$y = 5 - \frac{2}{3}x$$

Let $x = 0$.
$y = 5 - \frac{2}{3}(0)$
$y = 5$
So, $(0, 5)$ *is a* solution.

Let $x = 3$.
$y = 5 - \frac{2}{3}(3)$
$= 5 - 2$
$= 3$
So, $(3, 3)$ *is a* solution.

Let $x = 5$.
$y = 5 - \frac{2}{3}(5)$
$= 5 - \frac{10}{3}$
$= 1\frac{2}{3}$ ← Not in the replacement set
$(2, 1\frac{2}{3})$ is *not* a solution.

The solution set is $\{(0, 5), (3, 3)\}$.

EXAMPLE 3   The area of a rectangle is 20 cm². Find all possible whole-number values of l and w.

20 cm²   w

l

Solution
$$A = lw$$
$$20 = lw$$
$$l = \frac{20}{w}$$

Both w and l must be whole numbers. So, use values for w that are factors of 20.

Check each ordered pair (w, l) in $20 = lw$.

The solution set is $\{(1, 20), (2, 10), (4, 5), (5, 4), (10, 2), (20, 1)\}$.

w	$\frac{20}{w}$	l	ordered pair (w,l)
1	$\frac{20}{1}$	20	$(1, 20)$
2	$\frac{20}{2}$	10	$(2, 10)$
4	$\frac{20}{4}$	5	$(4, 5)$
5	$\frac{20}{5}$	4	$(5, 4)$
10	$\frac{20}{10}$	2	$(10, 2)$
20	$\frac{20}{20}$	1	$(20, 1)$

TRY THIS   2. The perimeter of a rectangle is 24 in. Find all the possible whole number values of its sides.
$\{(1, 11), (2, 10), (3, 9), (4, 8), (5, 7), (6, 6), (7, 5), (8, 4), (9, 3), (10, 2), (11, 1)\}$

10.4 Equations with Two Variables   **387**

Additional Example 2

Find the solution set for $x + 2y = 9$ if the replacement set for each variable is $\{0, 1, 2, 3, 4, 5\}$ $\{(1, 4), (3, 3), (5, 2)\}$

Additional Example 3

The area of a rectangle is 18 cm². Find all the possible values of l and w if they must be whole numbers.

18 cm²   w

l

$\{(1, 18), (2, 9), (3, 6), (6, 3), (9, 2), (18, 1)\}$

387

Checkpoint

Find two solutions for the given equation. Use {1, 2, 3, 4, 5, 6} as the replacement set for each variable.

1. $3x - y = 1$ (1, 2), (2, 5)
2. $2x + y = 5$ (1, 3), (2, 1)
3. $x - 2y = 2$ (4, 1), (6, 2)

Find the solution set if the replacement set is the set of whole numbers.

4. $xy = 7$ {(1, 7), (7, 1)}
5. $x^2 + y^2 = 25$ {(0, 5), (3, 4), (4, 3), (5, 0)}

Closure

Have students give the steps for finding the solution set for an equation with two variables. (1) Solve the equation for y in terms of x; (2) Substitute values from the replacement set for x.

■■■FOLLOW UP

Guided Practice

Classroom Exercises 1–9
Try This   all

Independent Practice

A Ex. 1–15, **B** Ex. 16–28, **C** Ex. 29–32

Basic: WE 1–21 odd, Midchapter Review, Application, Problem Solving Strategies

Average: WE 7–27 odd, Midchapter Review, Application, Problem Solving Strategies

Above Average: WE 11–31 odd, Midchapter Review, Application, Problem Solving Strategies

Note that if the replacement set in Example 3 were the set of rational numbers or the set of real numbers, the solution set would consist of an unlimited number of ordered pairs. For example, if $w = \frac{1}{2}$, then $l = 40$; if $w = \frac{1}{3}$, $l = 60$, and so on.

Classroom Exercises

Solve for y in the ordered pair so that it shows a solution for $3x + y = 5$.

1. $(0, y)$ $y = 5$ **2.** $(4, y)$ $y = -7$ **3.** $(-2, y)$ $y = 11$

Find the solution set if the replacement set for each variable is {1, 2, 3, 4, 5}.

{(3, 2) (4, 4)}
4. $-x + 3y = 6$ {(3, 3)}   **5.** $3x - 2y = 8$ {(4, 2)}   **6.** $2x - y = 4$
7. $2x + y = 3$ {(1, 1)}   **8.** $5x - 3y = 1$ {(2, 3)}   **9.** $-7x + y = 2$ $\varnothing$

Written Exercises

Find two solutions for each given equation. Use {0, 1, 2, 3, 4, 5, 6} as the replacement set for each variable.

(2, 0), (4, 5)
1. $4x - y = 2$ (1, 2), (2, 6)   **2.** $2x + y = 16$ (5, 6), (6, 4)   **3.** $5x - 2y = 10$
4. $x - 3y = 3$ (3, 0), (6, 1)   **5.** $2x + 4y = 12$ **6.** $3x - 2y = -6$
                           (0, 3), (2, 2), (4, 1), (6, 0)   (0, 3), (2, 6)

Find the solution set if the replacement set is the set of whole numbers. If there are many solutions, find two.

{(3, 2), (4, 4)}
7. $-3x + 2y = 6$ {(0, 3), (2, 6)}   **8.** $5x - 3y = -8$ **9.** $6x - 3y = 12$
10. $5x - y = -3$ {(0, 3), (1, 8)}   **11.** $-2x + 3y = 9$ **12.** $-x - y = -4$
13. $x + 3y = 5$ {(5, 0), (2, 1)}   **14.** $-x + 4y = -3$ **15.** $-2x - 3y = -9$
                              {(3, 0), (7, 1)}   {(0, 3), (3, 1)}

Find the solution set if the replacement set is the set of whole numbers.

16. $xy = 10$ **17.** $ab = 12$ **18.** $mn = 18$
19. $3xy = 48$ **20.** $2pq = 72$ **21.** $5ab = 125$
22. $x^2 + y^2 = 13$ **23.** $x^2 + y^2 = 16$ **24.** $3x^2 + y^2 = 7$
      {(2, 3), (3, 2)}   {(0, 4), (4, 0)}   {(1, 2)}

For Exercises 25–28, write an equation in two variables. Then find the solution set if the replacement set is the set of whole numbers.

25. The area of a rectangle is 32 cm².
26. The area of a triangle is 24 in².
(HINT: Area $= \frac{1}{2}bh$)
27. The sum of two even numbers is 10.
28. The sum of two odd numbers is 12.

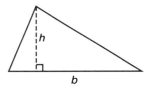

388   Chapter 10 Relations, Functions, and Variations

29. If m and n are positive single-digit numbers, find all the possible solutions of $3m + 6n = 33$.

30. If p and q are positive two-digit numbers, find all possible solutions of $2p - 3q = 11$.

31. If p and q are two-digit numbers and $q > 0$, find all possible solutions of $3p + 2q = -36$.

32. If a and b are integers, how many solutions of $-3ab = 30$ are there? 8

Midchapter Review

Give the coordinates of each point. *10.1*

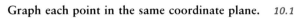

1. A (1, 2)
2. B $(-2, -3)$
3. C $(-1, 1)$
4. D $(4, -2)$
5. E $(3, 0)$
6. F $(0, -1)$
7. G $(-3, 3)$

Graph each point in the same coordinate plane. *10.1*

8. $P(-1, 2)$
9. $Q(5, 0)$
10. $R(-3, -1)$
11. $S(4, -4)$

Give the domain and the range of each relation. Is it a function? *10.2*

12. $\{(-2, 3), (6, 3), (-1, -2), (5, -1)\}$
D: $\{-2, -1, 5, 6\}$; R: $\{-2, -1, 3\}$; yes

13. $\{(1, 3), (-1, 6), (1, -2), (3, -2)\}$
D: $\{-1, 1, 3\}$; R: $\{-2, 3, 6\}$; no

Given $f(x) = -5x + 1$, find each indicated value. *10.3*

14. $f(-3)$ 16
15. $f(0)$ 1
16. $f(2)$ -9
17. $f(-1)$ 6

Use the given domain D to find the values of each function. *10.3*

18. $g(x) = 3x - 4$, $D = \{-3, 0, 2\}$
$g(-3) = -13$; $g(0) = -4$; $g(2) = 2$

19. $f(x) = 3x^2 - x + 2$, $D = \{-2, 0, 3\}$
$f(-2) = 16$; $f(0) = 2$; $f(3) = 26$

Find the solution set if the replacement set for each variable is $\{0, 1, 2, 3, 4, 5\}$. *10.4*

20. $x + 3y = 10$ (4, 2), (1, 3)
21. $2x - 3y = 1$ (2, 1), (5, 3)

22. The area of a rectangle is 36 cm². Find all possible values of l and w if they must be whole numbers. *10.4*

23. For the possible pairs of whole numbers for the length and the width of a rectangle when the area is 36 cm², predict which pair results in a rectangle with the least perimeter. Check your prediction. (6, 6)

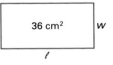

10.4 Equations with Two Variables **389**

Enrichment

Using the set of whole numbers as a replacement set, have students find an ordered pair that is a solution for both $2x + 3y = 24$ and $3x + 2y = 31$. (9, 2)

Midchapter Review

8–11.

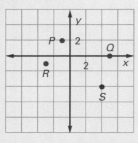

22. {(1, 36), (2, 18), (3, 12), (4, 9), (6, 6), (9, 4), (12, 3), (18, 2), (36, 1)}

Additional Answers, pp. 379–380

3. D: {−1, 2, 4}; R: {−3, 2, 3, 4}

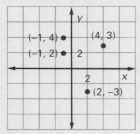

4. D: {0, 2, 3}; R: {−4, −2, 2}

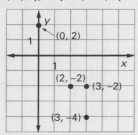

5. D: {−6, −4, 5, 7}; R: {−3, 7, 10}; yes
6. D: {−5, 0, 3}; R: {0, 1}; no
17. {(monkey, 15), (elephant, 35), (beaver, 5), (bear, 25)}; D: {monkey, elephant, beaver, bear}; R: {15, 35, 5, 25}
18. {(elephant, 25), (deer, 32), (greyhound, 39), (elk, 45)}; D: {elephant, deer, greyhound, elk}; R: {25, 32, 39, 45}

Problem Solving Strategies

Making a Graph

Many problems can be solved by using graphs. One type is the motion problem, which is related to the distance formula, $d = rt$. If the rate traveled is constant, then the distance traveled increases as the time increases. The graph is a straight line.

Mrs. Zemora left her office at 9:00 A.M. and drove 50 mi/h to attend a business meeting in a town 400 mi away. One hour later her assistant left the same office with a package that Mrs. Zemora needed and drove 60 mi/h. At what time did he catch up to her? Write the equations and make a graph. Find where the lines intersect.

	Mrs. Zemora		Assistant	
	t	$d = (50t)$	t	$d = 60t$
9 A.M.	0	0	t	$(60t)$
10 A.M.	1	50	0	0
11 A.M.	2	100	1	60
12 NOON	3	150	2	120
1 P.M.	4	200	3	180
2 P.M.	5	250	4	240
3 P.M.	6	300	5	300
4 P.M.	7	350	6	360

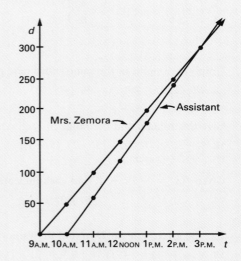

The graphs of $d = 50t$ and $d = 60t$ intersect at 3:00 P.M.

Therefore, Mrs. Zemora's assistant caught up to her at 3:00 P.M.

Exercises

1. How many hours did the assistant drive to catch Mrs. Zemora? 5 h

2. How far did each drive? 300 mi

3. If Mrs. Zemora had driven 60 mi/h, how fast would the assistant have to drive to catch up to her by 3:00 P.M.? Make a graph. 72 mi/h

390 Problem Solving Strategies

390

Application: *Fixed and Variable Costs*

In order to make a profit, Pete's Pizza Parlor prices each pizza higher than the costs of making and delivering the pizzas. Certain costs for making a pizza remain fixed for pizzas of all sizes. These fixed costs include amounts paid for rent, taxes, electricity and other utilities, equipment, and personnel. On the other hand, the cost of the ingredients changes with the size of each pizza.

Pete uses this formula to estimate the cost of the ingredients for his SuperDuper Special pizza.

$$I = 0.045d^2$$

In the formula, I represents the cost in dollars of the ingredients, and d represents the diameter of a pizza in inches. Note that the size of a pizza is expressed in terms of its diameter.

Exercises

Solve each problem. Assume that all problems refer to Pete's SuperDuper Special pizza.

1. Find the cost of the ingredients for a 10-inch pizza and for a 12-inch pizza. $4.50; $6.48

2. The fixed costs for each pizza amount to $5.00. Find the total cost of producing a 10-inch pizza and for producing a 12-inch pizza. $9.50; $11.48

3. If Pete wants to make a 10% profit on each pizza, what should be the selling price of a 10-inch pizza? Of a 12-inch pizza? $10.45; $12.63

4. Pete wants to make a new SuperSize pizza that will sell for $18.20. What size should the pizza be? Round your answer to the nearest inch. 16 inches

Application: Fixed and Variable Costs **391**

26.
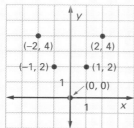

27.
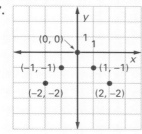

28.
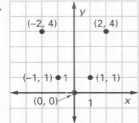

29.
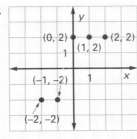

Teaching Resources
Manipulative Worksheet 10
Problem Solving Worksheet 10
Quick Quizzes 79
Reteaching and Practice
    Worksheets 79
Transparency 31
Teaching Aid 4

10.5 Graphing Linear Equations

Objectives   To graph linear equations

To determine whether a relation is a function, by the vertical-line test

The solution set of $x + y = 3$ is a set of ordered pairs. Some of the solutions are $(0, 3)$, $(2, 1)$, and $(4, -1)$. None of the ordered pairs in the solution set has the same first coordinate. Thus, the equation $x + y = 3$ defines a function which is called a **linear function.** Such a function can be described by a *linear equation in two variables*. The equations $x + y = 3$ and $4x - 5y = 6$ are examples of such equations.

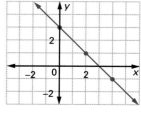

Definition   An equation of the form $Ax + By = C$, with A and B not both 0, is called the **standard form** of a linear equation in two variables.

EXAMPLE 1   Graph $-2x + y = 1$.

Plan   Solve for y in terms of x. Then find three ordered pairs that are solutions for the equation. You can draw the graph through two points. The third point is used as a check.

Solution
$$-2x + y = 1$$
$$y = 2x + 1$$

Make a table of values. Choose -1, 0, and 1 as values for x, find the corresponding values for y, and draw the graph through the points.

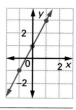

x	$2x + 1$	y	(x, y)
-1	$2 \cdot (-1) + 1$	-1	$(-1, -1)$
0	$2 \cdot 0 + 1$	1	$(0, 1)$
1	$2 \cdot 1 + 1$	3	$(1, 3)$

TRY THIS   1. Graph $x - y = 3$.   2. Graph $x = 2y + 1$.

Sometimes you are asked to graph a linear function given in $f(x)$ notation. $f(x) = 3x$ indicates that for each number x, the value of the function is $3x$. The same relationship can be represented as $y = 3x$.

392   Chapter 10 Relations, Functions, and Variations

GETTING STARTED

Prerequisite Quiz

List the ordered pairs and give the domain and range of the relation in the graph below. Is the relation a function?

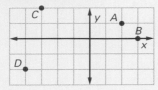

$A(2, 1)$, $B(3, 0)$, $C(-3, 2)$, $D(-4, -2)$;
$D = \{2, 3, -3, -4\}$, $R = \{1, 0, 2, -2\}$; yes

Motivator

Ask these questions to help students understand that a line can represent *all* the solutions to an equation.

1. Ask students if the ordered pair $(0, 3)$ is a solution of $y - 2x = 3$.   Yes
2. Ask them if $(2, 7)$ is a solution of $y - 2x = 3$.   Yes
3. Ask students to give three other solutions of $y - 2x = 3$.   Answers will vary.
4. Ask students to plot all the solutions they found in Exercises 1–3. What shape does this graph suggest?   A line
5. Ask them if they can find *all* the solutions of $y - 2x = 3$. Why or why not?   No, too many

---

Highlighting the Standards

Standards 4a, 6b: This lesson introduces the relationship between an equation and its equivalent expression as a graph.

Additional Example 1

Graph $3x - y = 4$.

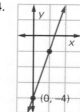

Additional Answers

Try This

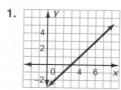

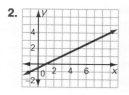

EXAMPLE 2   Graph the linear function $f(x) = 3x$.

Solution   Make a table of values that shows three ordered pairs, $(x, f(x))$.

x	3x	f(x)	(x, f(x))
−1	3 · (−1)	−3	(−1, −3)
0	3 · 0	0	(0, 0)
2	3 · 2	6	(2, 6)

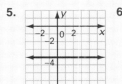

Plot the points and draw the graph.

TRY THIS   **3.** Graph $f(x) = x - 1$.   **4.** Graph $f(x) = -2x + 1$.

These equations are special cases of the standard form, $Ax + By = C$.

$y = 3$ This equation is equivalent to $0 \cdot x + 1 \cdot y = 3$.

$x = -2$ This equation is equivalent to $1 \cdot x + 0 \cdot y = -2$.

EXAMPLE 3   Graph the equation $y = 3$.

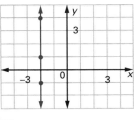

Solution   Any ordered pair that has 3 as its ordinate is a solution of $y = 3$. Three such ordered pairs are listed below.

$(-2, 3)$ $(1, 3)$ $(4, 3)$

The graph is a horizontal line three units above the x-axis.

TRY THIS   **5.** Graph $y = -4$.   **6.** Graph $y = 0$.

The graph of a linear equation of the form $y = c$ is a *horizontal line*. A function whose graph is a horizontal line is a **constant linear function**.

EXAMPLE 4   Graph the equation $x = -2$. Is the line the graph of a function?

Solution   Any ordered pair that has -2 as its abscissa is a solution of $x = -2$. Three such ordered pairs are listed below:

$(-2, -1)$ $(-2, 1)$ $(-2, 4)$

The graph is a vertical line two units to the left of the y-axis.

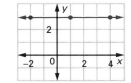

Since each ordered pair has the same first coordinate, the line is not the graph of a function

10.5 Graphing Linear Equations   **393**

■ TEACHING SUGGESTIONS

Lesson Note

Show students that the line for $x + y = 3$ includes such points as $(\frac{1}{2}, 2\frac{1}{2})$ and (1.7, 1.3). Point out that the graph in Example 1 extends indefinitely in two directions. Emphasize that every point on the line has coordinates that are solutions of $-2x + y = 1$. Emphasize also that, every solution for $-2x + y = 1$ corresponds to a point on the line.

Tell the students that the word "linear" refers to something that is a straight line. When presenting Examples 3 and 4, show students that in the equation $0 \cdot x + 1 \cdot y = 3$, x can have any value, while in the equation $0 \cdot y + 1 \cdot x = -2$, y can have any value.

Math Connections

Science: A linear accelerator is one in which the paths of the particles (electrons, protons, or heavy ions) accelerated are essentially straight lines.

Try This

3. 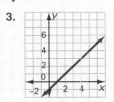   4.

5.   6.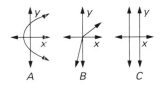

Additional Example 2

Graph the equation $y = -2$.

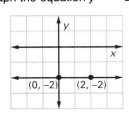

Additional Example 3

Graph the equation $x = 3$.

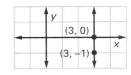

Additional Example 4

Determine whether each of the following is the graph of a function.   *B is; A and C are not.*

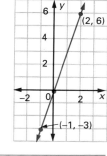

393

Critical Thinking Questions

Synthesis: Have students think about the ways two lines in the same plane can relate to each other. Ask them to describe the possibilities.   parallel or intersecting Then ask them to graph $x + y = 6$ and $x + y = 10$ on the same set of axes. Ask if they can explain why these two lines seem to be parallel.   There is no ordered pair of x and y that is a solution for both of these equations.

Common Error Analysis

Error: Some students will graph equations such as $x = 4$ and $y = -2$ as the points $(4, 0)$ and $(0, -2)$. Have students rewrite the equation $x = 4$ as $0 \cdot y + 1 \cdot x = 4$ and rewrite the equation $y = -2$ as $0 \cdot x + 1 \cdot y = -2$. Then have them substitute values for the variables and graph several points.

Checkpoint

Graph each equation. Which relations are functions?   1, 3, 4   Which relations are linear functions?   1, 3, 4   Which functions are constant linear functions?   3

1. $-3x + y = 3$
2. $x = -1$
3. $y = 3$
4. $4x + 2y = 6$

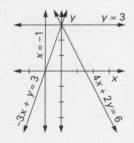

The graph of a linear equation of the form $x = c$ is a *vertical line*. On a vertical line, all the points have the same first coordinate. This suggests that a vertical line can be used to determine whether a graph is the graph of a function.

The Vertical Line Test
If no vertical line can be found that intersects a graph more than once, then the relation is a function. If there is a vertical line that intersects the graph more than once, then the relation is not a function.

EXAMPLE 5   Determine whether each of the following is the graph of a function.

a. 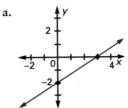   b. 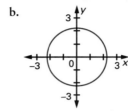   c.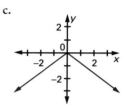

Plan   Draw vertical lines. Check to see whether any vertical line intersects the graph in more than one point.

Solutions   a. 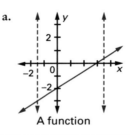   b. 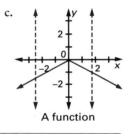   c.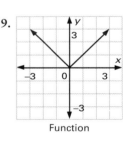

A function   Not a function   A function

TRY THIS   Which of the following is the graph of a function?

7. 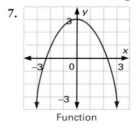   8. 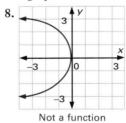   9.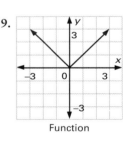

Function   Not a function   Function

394   Chapter 10 Relations, Functions, and Variations

Additional Example 5

Graph the linear function; $f(x) = -2x + 1$.

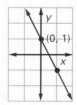

394

Focus on Reading

Match each phrase at the left with one or more graphs at the right. Justify your answer.

1. linear function a, c
2. horizontal line a
3. vertical line b
4. constant linear function a
5. not a function b

a.   **b.** 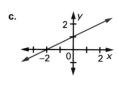 **c.**

Classroom Exercises

Graph each equation.

1. $x + y = 7$ **2.** $y = 5$ **3.** $x = -4$ **4.** $8x + 3y = 12$

Use your graphs in Exercises 1–4 to answer these questions.

5. Which graphs represent relations? 1, 2, 3, 4
6. Which graphs represent functions? 1, 2, 4
7. Which graphs represent constant linear functions? 2
8. Which graphs do not represent functions? 3

Written Exercises

Graph each equation.

1. $x + y = 9$ **2.** $x - y = 3$ **3.** $x + y = -5$ **4.** $x - y = -1$
5. $x = 6$ **6.** $y = 2$ **7.** $x + 4 = 0$ **8.** $y = -5$

9. $y = 2x$

Solve for y in terms of x. Then graph the equation.   **10.** $y = -3x + 9$

                                                  $y = -2x + 4$

9. $-2x + y = 0$ **10.** $3x + y = 9$ **11.** $4x = 4y$ $y = x$ **12.** $6x + 3y = 12$
13. $4x - 2y = 6$ **14.** $6x - 3y = 6$ **15.** $-4x + y = -1$ **16.** $-6x + 2y = 10$
   $y = 2x - 3$ $y = 2x - 2$ $y = 4x - 1$ $y = 3x + 5$

Graph each equation. Which relations are linear functions? Which functions are constant linear functions?

17. $2x - y = 3$ **18.** $y = -3$ **19.** $x = 2$ **20.** $-x = 4$
21. $4x + 2y = 10$ **22.** $x - 6 = 0$ **23.** $y = -4$ **24.** $0 \cdot x + 1 \cdot y = 7$

Linear: Ex. 17, 18, 21, 23, 24; constant linear: Ex. 18, 23, 24

10.5 Graphing Linear Equations **395**

 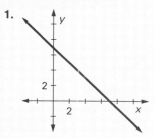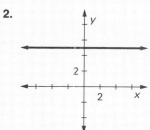

Enrichment

Have students graph the equations $y = 2x - 4$ and $y = -3x + 1$ on the same set of axes and find their point of intersection. $(1, -2)$ Then have them write a pair of equations which would intersect at the point $(-3, 2)$. Answers will vary.

3.

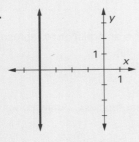

4.

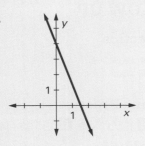

Written Exercises

1.

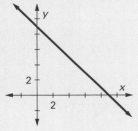

See pages 410–411 answers to Ex. 2–8 and pages 652–654 for the answers to Ex. 9–24, 29–44, and Mixed Review Ex. 1, 3, and 7.

396

Use the vertical line test to determine which relations are functions.

25.

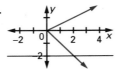

No

26.

Yes

27.

Yes

28.

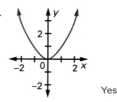

Yes

Graph each function.

29. $f(x) = 4x$ **30.** $f(x) = 3x - 5$ **31.** $f(x) = -x$ **32.** $f(x) = -3x - 2$

Write an equation. Then draw the graph. **33.** $y = -2x + 10$

33. The sum of $2x$ and y is 10.   **34.** y is 5 less than x. $y = x - 5$

                                                         $y = 2|x|$

35. The opposite of $3x$ equals $\frac{1}{2}y$. $y = -6x$ **36.** y equals twice the absolute value of x.

37. $\frac{2}{3}x$ is 1 more than $\frac{1}{6}y$. $y = 4x - 6$ **38.** No matter what x is, twice y equals 9.

                                                           $y = \frac{9}{2}$

Graph each function.

39. $f(x) = \frac{x}{4}$ **40.** $f(x) = 2 - \frac{1}{3}x$ **41.** $\frac{f(x) - 4}{2} = x - 2$

42. $3x - 2f(x) = 4$ **43.** $3x - [2x - f(x)] = 4$ **44.** $4 - [6 - (f(x)] = 8$

Mixed Review

Add, subtract, multiply, or divide.   *6.7, 6.8, 6.9, 8.4, 8.7*

                                                                           $-y^2 + y$

1. $(2x^2 + 3x + 4) + (x^2 - 7x - 6)$ **2.** $(5y^2 - 3y + 7) - (6y^2 - 4y + 7)$

3. $-8n^2(2n^2 - 4n + 5)$ **4.** $(x - 7)(x + 8)$ $x^2 + x - 56$

5. $(2y + 4)(y - 6)$ $2y^2 - 8y - 24$ **6.** $(y + 3)(y - 3)$ $y^2 - 9$

7. $\dfrac{x^2 - x - 6}{x^2 - 36} \div \dfrac{2x^2 - 7x + 3}{x^2 + 8x + 12}$ **8.** $\dfrac{2}{x + 3} - \dfrac{15}{x^2 - 9} + \dfrac{2}{x - 3}$ $\frac{4x - 15}{x^2 - 9}$

9. The selling price of a turntable is \$156. The profit is 30% of the cost. Find the cost.   *4.8* \$120

10.6 Direct Variation

Objectives
To determine whether a relation is a direct variation
To determine the constant of variation
To find missing values in direct variations

Peter earns $20 per hour playing in a band. If he works 1 h, he earns $20; if he works 2 h, he earns $40, and so on. The amount of money he earns is directly related to the number of hours he works. The relationship can be expressed by the equation $e = 20n$, where e represents the total earnings, 20 is the number of dollars earned per hour, and n represents the number of hours worked. The equation $e = 20n$ is an example of *direct variation*.

Definitions

A **direct variation** is a linear function defined by an equation of the form $y = kx$, where k is a nonzero real number.

The constant k is called the **constant of variation**.

The equation $y = kx$ is read "y varies directly as x."

EXAMPLE 1
y varies directly as x. y is 45 when x is 9. Find the constant of variation. Then find y when x is 7.

Solution

Write an equation for direct variation.	$y = kx$
Substitute 45 for y and 9 for x.	$45 = k \cdot 9$
	$9k = 45$
	$k = 5$

Thus, the constant of variation is 5.

Replace k with 5 in $y = kx$.	$y = 5x$
Substitute 7 for x.	$y = 5 \cdot 7$
	$y = 35$

Thus, y is 35 when $x = 7$.

TRY THIS
1. y varies directly as x and y is 18 when x is 27. Find x when y is 8. 12

Sometimes a table of ordered pairs (x, y) is given, and you need to determine whether the relationship is a direct variation.

Teaching Resources

Quick Quizzes 80
Reteaching and Practice
Worksheets 80

◼◼◼GETTING STARTED

Prerequisite Quiz

Solve each equation for the variable indicated.

1. $C = 2\pi r$ for r $r = \frac{C}{2\pi}$
2. $d = rt$ for t $t = \frac{d}{r}$
3. $A = lw$ for l $l = \frac{A}{w}$
4. $v = \pi dl$ for l $l = \frac{v}{\pi d}$

Motivator

Explain to students that if record albums are on sale at $5 each, the price for buying several albums varies directly as the number of albums you buy. Ask them to write an equation to represent this relationship where y = the price paid and x = the number of albums bought. $y = 5x$. Ask students what happens to y as x increases in value. y increases Ask them the value of the ratio $\frac{y}{x}$. 5 Ask them if the value of the ratio will ever be any other number than 5. No, because $y \div x = 5$.

Additional Example 1

If y varies directly as x and y is 60 when x is 10, find the constant of variation. Then find y when x is 9. $k = 6$; 54

Highlighting the Standards

Standards 2d, 1b: In the Focus on Reading, students demonstrate their understanding of direct variation by answering questions on the reading. In the Exercises, they translate between a table and an equation, and apply their skills to solving various problems.

Lesson Note

After presenting Example 1, tell students that if the replacement set is all real numbers, the graph of a direct variation is a straight line that includes the origin. Demonstrate this with the graph of $y = 5x$. In presenting Example 2, emphasize that solving $y = kx$ for k gives the *constant of variation,* which is the constant ratio between y and x. After showing direct variation written as a proportion, tell students that $y = kx$ can also be read as "y is directly proportional to x."

Math Connections

Previous Algebra: Direct variation is a special case of the linear function $Ax + By = C$, where C is 0 and the constant of variation is $-\frac{A}{B}$.

Critical Thinking Questions

Analysis: If the value of the constant of variation, k, is positive, then direct variation could be described with the phrase "as the value of x increases, the value of y increases." Ask students how this explanation of direct variation could be altered to include the case where k is negative.   As the absolute value of x increases, the absolute value of y increases.

Common Error Analysis

Error: When solving a word problem, some students will chose the wrong values to use in $y = kx$.

Suggest that they use the proportion form (see page 398) so that all values can be used in the same equation.

EXAMPLE 2   From the table, determine whether y varies directly as x. If so, find the constant of variation.

x	y
2	-12
5	-30
-2	12

Plan   Use $\frac{y}{x} = k$. See whether $\frac{y}{x} = k$ for all pairs (x, y) such that $x \neq 0$ and $y \neq 0$.

Solution $\dfrac{-12}{2} = -6 \qquad \dfrac{-30}{5} = -6 \qquad \dfrac{12}{-2} = -6$

Thus, y varies directly as x, and the constant of variation is -6

TRY THIS   2. From the set of ordered pairs, determine the constant of variation.
$$\{(-3, 5), (6, -10), (-15, 25)\} \quad -\tfrac{5}{3}$$

In Example 2, the ratios $\frac{-12}{2}, \frac{-30}{5}$, and $\frac{12}{-2}$ have the same value, -6. Therefore, the following *proportions* are true.

$$\frac{-12}{2} = \frac{-30}{5} \qquad \frac{-30}{5} = \frac{12}{-2} \qquad \frac{-12}{2} = \frac{12}{-2}$$

In general, if two ratios, $\dfrac{y_1}{x_1}$ and $\dfrac{y_2}{x_2}$, have the same value, k, then the *direct variation* $y = kx$ can be written as the proportion $\dfrac{y_1}{x_1} = \dfrac{y_2}{x_2}$.

EXAMPLE 3   The distance d that a plane travels varies directly as the time t that it travels if the rate of travel is constant. If a plane travels 2,600 mi in 4 h, how far does it travel in 6 h?

Plan   Write the direct variation as a proportion.

Solution   Since $\dfrac{d}{t}$ is constant, write: $\dfrac{d_1}{t_1} = \dfrac{d_2}{t_2}$

Substitute 2,600 for d_1, 4 for t_1, and 6 for t_2. $\dfrac{2,600}{4} = \dfrac{d_2}{6}$

Solve for d_2. $2,600(6) = 4d_2$

$d_2 = \dfrac{2,600(6)}{4}$, or $3,900$

Thus, the distance traveled in 6 h is $3,900$ mi

TRY THIS   3. The number of widgets a machine can make varies directly as the time it operates. The machine can make 1,275 widgets in 2 hours. How many can it make in 7 hours? 4462.5

Additional Example 2

From the table, determine whether y varies directly as x. If so, find the constant of variation.

a.
x	y
-4	40
5	-50
10	-100

Yes, -10

b.
x	y
6	36
8	64
-2	4

No

Additional Example 3

The bending of a beam varies directly as its mass. A beam is bent 30 mm by a mass of 90 kg. How much will the beam bend with a mass of 120 kg?   40 mm

Focus on Reading

Which statements do not apply to $\dfrac{d}{t} = \dfrac{7}{5}$? Explain. Statements 3 and 5 do not apply.

1. d varies directly as t.
2. d is directly proportional to t.
3. The constant of variation is 7.
4. t varies directly as d.
5. d is directly proportional to $\dfrac{7}{5}$.
6. $d = \dfrac{7}{5}t$.

Statement 3: Constant of variation is $\frac{7}{5}$, not 7.
Statement 5: Variable cannot be directly proportional to a constant.

Classroom Exercises

Find the constant of variation, if y varies directly as x.

1. $y = -3x$ -3
2. $\dfrac{y}{x} = \dfrac{4}{3}$ $\frac{4}{3}$
3. $7s = t$ 7
4. $c = 2\pi r$ 2π

For Exercises 5–8, y varies directly as x.

5. y is 27 when x is 6.
Find y when x is 12. $y = 54$

6. y is 100 when x is 40.
Find y when x is 16. $y = 40$

7. y is 6 when x is 9.
When y is 32, what is x? $x = 48$

8. y is 25 when x is 20.
When y is 35, what is x? $x = 28$

Determine whether y varies directly as x. If so, find the constant of variation.

9.

x	y
8	4
18	9
20	10

Yes; $k=\frac{1}{2}$

10.

x	y
30	2
90	6
40	3

No

11.

x	y
-1	-7
-2	-14
-3	-21

Yes; $k=7$

12.

x	y
-10	5
10	-5
20	-10

Yes; $k=-\frac{1}{2}$

Written Exercises

In each of the following, y varies directly as x.

1. y is 54 when x is 9.
Find y when x is 3. $y = 18$

2. y is -36 when x is -4.
Find y when x is 7. $y = 63$

3. y is 27 when x is -3.
Find y when x is 18. $y = -162$

4. y is -10 when x is -4.
Find y when x is -6. $y = -15$

5. y is 100 when x is 60.
When y is 80, what is x? $x = 48$

6. y is 9 when x is 12.
When y is 48, what is x? $x = 64$

Determine whether y varies directly as x. If so, find the constant of variation.

7.

x	y
5	20
6	24
-3	-12

Yes; $k = 4$

8.

x	y
-75	-15
-60	-12
50	10

Yes; $k = 0.2$

9.

x	y
7	15
10	18
15	23

No

10.

x	y
-30	3
-20	2
50	-5

Yes; $k = -0.1$

10.6 Direct Variation **399**

Checkpoint

1. y varies directly as x and y is 20 when x is -5. Find y when x is 7. -28
2. y varies directly as x and y is 80 when x is 4. When y is -60, what is x? -3
3. The cost of a certain chemical varies directly as its weight. If 12 g cost \$65, find the cost of 18 g. \$97.50

Closure

Have students give an example from everyday life that illustrates direct variation and then write an equation to state this relationship. Answers will vary. Then ask them to think of a situation involving two variable quantities that is *not* an example of direct variation. Possible answer: The time required to make a 200 mile trip does not vary directly as the speed at which you travel.

◼◼◼ FOLLOW UP

Guided Practice

Classroom Exercises 1–12
Try This all

Independent Practice

Ⓐ Ex. 1–10, Ⓑ Ex. 11–18, Ⓒ Ex. 19–22
Basic: WE 1–15 odd
Average: WE 5–17 odd, 18
Above Average: WE 7–21 odd

Solve each problem.

11. The cost of chocolates varies directly as the number of pounds. If 2 lb of chocolates cost $4.60, find the cost of 5 lb of chocolates. $11.50

12. The weight of a metal rod varies directly as its length. If a 12-ft rod weighs 18 lb, how much does a 20-ft rod weigh? 30 lb

13. The distance a car travels varies directly as the time traveled, but only if the rate of travel is constant. If a car travels 330 mi in 6 h, how far does it travel in 8 h? (Assume rate is constant.) 440 mi

14. The property tax on a house varies directly as the assessed value of the house. The tax on a house assessed at $20,000 is $4,000. Find the taxes on a house assessed at $35,000. $7,000

15. At a given time and place, the height of a vertical pole varies directly as the length of its shadow. A 6-m pole casts a 9-m shadow. Find the height of a building that casts a 60-m shadow. 40 m

16. The number of kilograms of water in a person's body varies directly as a person's mass. A person with a mass of 100 kg contains 75 kg of water. How many kilograms of water are in a person with a mass of 96 kg? 72 kg

17. Gas consumption of a car is approximately proportional to the distance traveled. A car uses 40 liters of gas to travel 240 km. About how much gas will the car use to travel 300 km? 50 liters

18. The weight of an object on the moon varies directly as its weight on earth. On earth, an object weighs 125 lb. But on the moon it weighs 20 lb. What would an 80-pound crate weigh on the moon? 12.8 lb

19. y varies directly as x^2. If y is 64 when x is 7, find y when x is 3. $y = 11\frac{37}{49}$

20. y varies directly as x^2. If y is 49 when x is 12, find y when x is 9. $y = 27\frac{9}{16}$

21. The distance that a falling object travels varies directly as the square of the time it falls. A ball falls 320 m in 8 s. How far will it fall in 18 s? 1,620 m

22. The distance needed to stop a car varies directly as the square of its speed. It requires 173 m to stop a car traveling at 82 km/h. What distance is required to stop a car traveling 88 km/h? $199\frac{409}{1,681}$ m

Mixed Review

Solve. *3.2, 3.3, 4.6*

1. $5y = -80$ $y = -16$

2. $-6x = 72$ $x = -12$

3. $-\frac{2}{3}x = 16$ $x = -24$

4. $-\frac{3}{5}y = -45$ $y = 75$

5. $4n + 15 = 7$ $n = -2$

6. $3.2x = 0.75 - 0.7x$ $x = \frac{5}{26}$

7. Two cars traveled in opposite directions from the same starting point. The rate of one car was 15 km/h faster than the other car. After 2 h, the cars were 240 km apart. Find the rate of each car. *9.4* 52.5 km/h; 67.5 km/h

Enrichment

Have students make lists of applications of direct variation in everyday life. Answers will vary. Possible answer: The amount of interest on a savings account is directly proportional to the amount of money in the account. Have students work in groups to write equations and draw graphs for these relationships.

10.7 Inverse Variation

Objectives

To determine whether a relation is an inverse variation
To determine the constant of variation in an inverse variation
To find missing values in inverse variations

A train is traveling to a city 240 mi away. The faster it goes, the less time it will take. In the table, notice that rate × time is a constant, 240.

Rate in miles per hour (r)	20	30	40	60	80	120
Time in hours (t)	12	8	6	4	3	2

The equation, $r \cdot t = 240$, or $t = \dfrac{240}{r}$, is an example of *inverse variation*. We say that the time, t, varies inversely as the rate, r.

Definition

An **inverse variation** is a function defined by an equation of the form $xy = k$, or $y = \dfrac{k}{x}$ where k is a nonzero real number. The constant k is called the **constant of variation.**

EXAMPLE 1

Given that y varies inversely as x and $y = 12$ when $x = 5$, find the constant of variation. Then find y when x is 20.

Solution

Write an equation for inverse variation.	$xy = k$
Substitute 5 for x and 12 for y.	$5 \cdot 12 = k$
	$k = 60$

So, k is 60.

Replace k by 60 in $xy = k$.	$xy = 60$
Substitute 20 for x.	$20y = 60$
	$y = \dfrac{60}{20}$
	$y = 3$

Thus, y is 3 when x is 20.

TRY THIS

1. Given that y varies inversely as x, and $y = 6$ when $x = 3$, find y when $x = -36$. $-\dfrac{1}{2}$

10.7 Inverse Variation　**401**

Additional Example 1

y varies inversely as x and y is 15 when x is 4. Find the constant of variation. Then find y when x is 30. $k = 60; 2$

▰▰▰ GETTING STARTED

Prerequisite Quiz

Solve each equation for the variable indicated.

1. $lw = 36$ for w $w = \dfrac{36}{l}$
2. $T = \dfrac{c}{s}$ for s $s = \dfrac{c}{T}$
3. $v = \dfrac{k}{p}$ for p $p = \dfrac{k}{v}$
4. $V = Bh$ for h $h = \dfrac{V}{B}$

Motivator

Ask students to give possible lengths and widths for a rectangular carpet that has an area of 12 square yards. Then have them graph the (l, w) pairs on a coordinate graph and determine whether the relationship is a function, and if it is a function, whether it is a linear function.   Possible (l, w) pairs: (1, 12), (2, 6), (3, 4), (4, 3), (6, 2), (12, 1). The graph is a function; it is not linear.

▰▰▰ TEACHING SUGGESTIONS

Lesson Note

When presenting Example 1, tell students that "varies inversely as" is sometimes stated as "is inversely proportional to." On page 402, note that $x_1y_1 = x_2y_2$ could also be written as $\dfrac{x_1}{y_2} = \dfrac{x_2}{y_1}$

Highlighting the Standards

Standards 4d, 1b: In the Exercises, students apply inverse variation to a number of problems from physics. In Problem Solving Strategies they see the strategy of Restating the Problem.

Math Connections

Physics: When two turning gear wheels mesh, the one with fewer teeth rotates more rapidly. If t stands for the number of teeth and n for the number of revolutions per minute (rpm) then $t_1 n_1 = t_2 n_2$, or $\frac{t_1}{n_2} = \frac{t_2}{n_1}$

Critical Thinking Questions

Application: Tell the students that you are thinking of two numbers whose product is 6. Ask them to write an equation for this, using x and y as the two numbers. $xy = 6$ Ask if this is a direct or inverse variation. Inverse Have students sketch the graph of the equation, using: (1) the positive x-values: 1, 2, 3, 6; (2) the negative x-values: $-1, -2, -3, -6$.

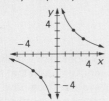

Explain that this graph is called a *hyperbola*, a figure that students will encounter in future study.

Common Error Analysis

Error: Some students may confuse $xy = k$ with $y = kx$.

Suggest that they use the form $x_1 y_1 = x_2 y_2$ for inverse variation.

For a table of ordered pairs (x, y) such as the one at the right, you can tell whether y varies inversely as x by finding the product $x \cdot y$ for each pair.

x	y
3	16
12	4
-6	-8

$$3 \cdot 16 = 48 \qquad 12 \cdot 4 = 48 \qquad (-6) \cdot (-8) = 48$$

The product $x \cdot y$ is 48 for each pair. Therefore, y varies inversely as x, and the constant of variation is 48.

The graph of an inverse variation is *not* a straight line. One part of the graph of $xy = 48$, or $y = \frac{48}{x}$, is shown below. This figure shows the graph when both x and y are positive. Another part of the graph is located in Quadrant III (where both x and y are negative).

x	$\frac{48}{x}$	y	(x, y)
3	$\frac{48}{3}$	16	$(3, 16)$
4	$\frac{48}{4}$	12	$(4, 12)$
6	$\frac{48}{6}$	8	$(6, 8)$
8	$\frac{48}{8}$	6	$(8, 6)$
12	$\frac{48}{12}$	4	$(12, 4)$
16	$\frac{48}{16}$	3	$(16, 3)$

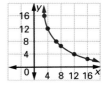

The inverse variation $xy = k$ can be expressed using subscripts, as shown below.

$$x_1 y_1 = k \text{ and } x_2 y_2 = k, \text{ so } x_1 y_1 = x_2 y_2$$

EXAMPLE 2 In a closed container, the volume V of a gas varies inversely as the pressure p that is applied to the gas. If the volume is 100 m³ under 5 atmospheres of pressure, find the volume under 8 atmospheres of pressure.

Plan Since V varies inversely as p, use $p_1 V_1 = p_2 V_2$.

Solution

Substitute 100 for V_1, 5 for p_1, and 8 for p_2.

Solve for V_2.

$$p_1 V_1 = p_2 V_2$$
$$5 \cdot 100 = 8 \cdot V_2$$
$$V_2 = \frac{5 \cdot 100}{8}$$
$$= 62.5$$

Thus, the volume is 62.5 m³ when the pressure is 8 atmospheres.

Additional Example 2

Determine from the table whether y varies inversely as x. If so, find the constant of variation.

a.

x	y
20	5
10	10
-25	-4

Yes, $k = 100$

b.

x	y
-3	12
4	-16
5	-20

No

Classroom Exercises

Find the constant of variation for each inverse variation.

1. $xy = 30$ $k = 30$ **2.** $lw = 16$ $k = 16$ **3.** $rt = 400$ $k = 400$ **4.** $y = \dfrac{-48}{x}$
                                                                             $k = -48$

For Exercises 5–8, y varies inversely as x.

5. y is 3 when x is 8. Find y when x is 2. $y = 12$

6. y is 12 when x is 9. Find y when x is 36. $y = 3$

7. y is 8 when x is 6. When y is 24, what is x? $x = 2$

8. y is 32 when x is 3. When y is 16, what is x? $x = 6$

For Exercises 9 and 10, let $y = \dfrac{36}{x}$.

9. If x increases, what happens to y? y decreases.

10. If x decreases, what happens to y? y increases.

Written Exercises

In each of the following, y varies inversely as x.

1. y is 13 when x is 4. Find y when x is 26. $y = 2$

2. y is 18 when x is 6. Find y when x is 9. $y = 12$

3. y is 3 when x is 8. Find y when x is 6. $y = 4$

4. y is 4 when x is -16. Find y when x is -8. $y = 8$

5. y is 20 when x is 4. When y is 80, what is x? $x = 1$

6. y is -60 when x is $\frac{3}{5}$. When y is 2, what is x? $x = -18$

7. y is 3 when x is -8. When y is -6, what is x? $x = 4$

8. y is 12 when x is $\frac{3}{4}$. When y is 27, what is x? $x = \frac{1}{3}$

Determine whether y varies inversely as x. If so, find the constant of variation.

9.

x	y
3	4
2	6
-4	-3
12	2 No

10.

x	y
6	4
-8	-3
-12	2
3	-8 No

11.

x	y
3.2	20
6.4	10
12.8	5
21.9	3 No

12.

x	y
3.2	15.0
6.4	7.5
10.0	4.8
19.2	2.5

Yes; $k = 48$

Solve each problem.

13. The current in an electric circuit varies inversely as the resistance. When the current is 40 amps, the resistance is 25 ohms. Find the current when the resistance is 15 ohms. $66\frac{2}{3}$ amps

14. The time to travel a fixed distance varies inversely as the rate of travel. When the time traveled is 6 h, the rate of travel is 90 km/h. Find the time when the rate of travel is 80 km/h. 6.75h

1. y varies inversely as x and y is 50 when x is -3. Find y when x is 15. -10

2. y varies inversely as x and y is 35 when x is 5. When y is 25, what is x? 7

3. In a closed container, the volume v of gas varies inversely as the pressure p that is applied to the gas. If the volume is 100 m^2 under 6 atmospheres of pressure, find the volume under 30 atmospheres of pressure. 20 m^2

Closure

The formula $d = rt$ can be used to illustrate both direct and inverse variation. Have the students use this formula to develop two problems, one for each type of variation. Possible answers: direct: If you can travel 200 miles in 4 hours, how far could you travel at the same speed in 7 hours?; inverse: If you can travel to Smithville in 3 hours at a speed of 50 mph, how long would it take at 60 mph?

◼◼◼ FOLLOW UP

Guided Practice

Classroom Exercises 1–10
Try This   all

Independent Practice

A Ex. 1–12, **B** Ex. 13–20, **C** Ex. 21–24

Basic: WE 1–15 odd

Average: WE 5–19 odd, Problem Solving Strategies

Above Average: WE 9–23 odd, Problem Solving Strategies

Additional Answers, page 405

Problem Solving Strategies

Answers will vary. Sample answers are given.

1. Let n be any integer. Add 1 to n, to get $n + 1$, which is always larger than n. (NOTE: Students may add any positive integer to n, for example 3, 100, or k where k is any positive integer.)

2. Restatement: For any rational number, you can always find a larger rational number. Let $\frac{p}{q}$ be any rational number, where p and q are integers. For any $\frac{p}{q}$, you can find a larger rational number $\frac{p}{q} + 1$.

3. Restatements: For any rational number, you can always find a smaller rational number. For any $\frac{p}{q}$, you can find a smaller rational number $\frac{p}{q} - 1$.

4. Restatement: Any pair of numbers that satisfied the given conditions does not consist of integers. Let x and y be numbers that satisfy the conditions and let x be the larger of the pair:
$x + y = 10 \quad x - y = 5$, or $x = 5 + y$
Substitute the second equation in the first:

$5 + y + y = 10 \qquad x = 2\frac{1}{2} + 5$
$5 + 2y = 10 \qquad\qquad = 7\frac{1}{2}$
$2y = 5$
$y = 2\frac{1}{2}$

The numbers that satisfy the conditions are not integers.

15. The number of vibrations a string makes under constant tension is inversely proportional to its length. If a 32-cm string vibrates 420 times per second, what length string vibrates 640 times per second? **21 cm**

16. When two meshed gears revolve, their speeds are inversely proportional to the number of teeth they have. If a gear with 60 teeth revolves at a speed of 2,500 rev/min, at what speed should a gear with 90 teeth revolve?

17. The length of a rectangle with a constant area varies inversely as the width. When the length is 24 in., the width is 8 in. Find the length when the width is 12 in. **16 in**

18. The volume of gas varies inversely as the pressure. If the volume is 60 m³ under 6 atmospheres of pressure, find the volume under 4 atmospheres of pressure. **90 m³**

19. The base of a triangle with constant area varies inversely as the height. When the base is 22 cm, the height is 6 cm. Find the length of the base when the height is 12 cm. **11 cm**

20. The frequency of a radio wave is inversely proportional to its wave-length. If a 300-m wave has a frequency of 2,000 kilocycles, what length wave has a frequency of 1,000 kilocycles? **600 m**

21. The height of a cylinder of constant volume varies inversely as the square of the radius of the base. The height of a cylinder is 12 m and the radius of the base is 5 m. Find the height of the cylinder of the same volume with a base radius of 6 m. **$8\frac{1}{3}$ m**

22. The brightness of the illumination of an object varies inversely as the square of the distance of the object from the source of illumination. If a light meter reads 45 luxes at a distance of 3 m from a light source, find the reading at 5 m from the source. **16.2 luxes**

23. The weight of a body at, or above, the earth's surface varies inversely as the square of the body's distance from the earth's center. An object has a weight of 350 lb when it is at the earth's surface. What is its weight when it is 250 mi above the earth's surface? (Use 4,000 mi as the earth's radius.) **Approximately 310 lb**

24. How far above the earth's surface would the object in Exercise 23 have to be for its weight to be 290 lb? **Approximately 394.35 mi**

Mixed Review

Simplify. *6.1, 6.3*

1. $-3y(-2y^3)$ **6y^4**

2. $-a^2 \cdot (-a)^2$ **$-a^4$**

3. $-ab(-2ab)$ **$2a^2b^2$**

4. $\dfrac{25ab}{5a^3b}$ **$\frac{5}{a^2}$**

5. $\dfrac{36a^3bc^4}{-6ab^4c^5}$ **$-\frac{6a^2}{b^3c}$**

6. $\dfrac{(3a^3b^2)^3}{9ab^7}$ **$\frac{3a^8}{b}$**

Solve. *7.7*

7. $x^2 - 8x + 15 = 0$ **3, 5**

8. $x^2 - 3x - 4 = 0$ **$-1, 4$**

9. $y^2 + 5y = 0$ **$-5, 0$**

10. $y^2 - 16 = 0$ **± 4**

11. $2y^2 = 2 - 3y$ **$-2, \frac{1}{2}$**

12. $-y = 6y^2 - 1$ **$-\frac{1}{2}, \frac{1}{3}$**

13. Five is what percent of 8? *4.7* **62.5%**

14. Eighty is 125% of what number? *4.7* **64**

Enrichment

Explain that a formula of the form $y = kxz$ is an example of *combined* variation and one of the form $y = \frac{kz}{z}$ is a *joint* variation. Have students look through their science books to find formulas that involve different types of variation. Have students list the formulas or equations, identify the variables, the constants of variation, and the type of variation.

Problem Solving Strategies

Restating the Problem

When you are asked to show that a general statement is true, it sometimes helps to restate the problem first.

Example Show that there is no smallest positive rational number.

Restating the problem may give you an idea of how to solve it. In this case, restate the problem as shown below.

> Show that for any positive rational number, you can always find another positive rational number that is smaller.

Represent any rational number as $\dfrac{p}{q}$, where p and q are positive integers.

Then you can always find a positive rational number smaller than $\dfrac{p}{q}$ by adding 1 to the denominator:

Original Rational Number   **Smaller Rational Number**

$$\dfrac{p}{q} \qquad\qquad\qquad\qquad \dfrac{p}{q+1}$$

To help see that $\dfrac{p}{q+1} < \dfrac{p}{q}$, think of some examples.

$$\dfrac{1}{4} < \dfrac{1}{3} \qquad \dfrac{1}{11} < \dfrac{1}{10} \qquad \dfrac{1}{1{,}001} < \dfrac{1}{1{,}000}$$

You can see that for any positive rational number, no matter how small, you can always find a smaller one. In other words, there is no *smallest* positive rational number.

Show that each statement is true.

1. There is no largest integer. (Restate. Show that for any integer, you can always find a larger integer.)

2. There is no largest rational number.

3. There is no smallest rational number.

4. There is no pair of integers whose sum is 10 and whose difference is 5.

5. There is no smallest element of the set of rational numbers greater than 3 and less than 4.

6. There is an infinite number of rational numbers between any two rational numbers.

5. Restatement: For any number greater than 3 and less than 4, you can always find a smaller number in that range. Let x be a number greater than 3 but less than 4. You can always find a number smaller than x but greater than 3 by calculating the average of 3 and x: $\dfrac{3+x}{2}$. Proof that the average of any two numbers (a and b, where a is the larger) is less than a and greater than b:

$$\begin{aligned} a &> b & a &> b \\ 2a &> a+b & a+b &> 2b \\ a &> \tfrac{a+b}{2} & \tfrac{a+b}{2} &> b \end{aligned}$$

6. Restatement: For any 2 rational numbers, no matter how close one is to the other, you can keep finding another rational number that is even closer. Let x and y be any 2 rational numbers, where y is the larger of the pair. The average of x and y, which is $\dfrac{x+y}{2}$ is between x and y. That is, $\dfrac{x+y}{2}$ is closer to x than y is. Next you can find the average of x and $\dfrac{x+y}{2}$, which will be closer to x than $\dfrac{x+y}{2}$ is. You can keep repeating this process for as long as you like.

1–8.

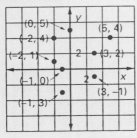

18.

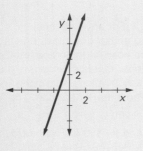

19.

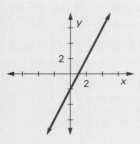

20.

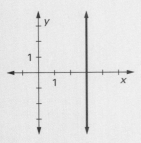

Chapter 10 Review

Key Terms

constant linear function   (p. 393)
constant of variation   (p. 397, 401)
coordinate plane   (p. 373)
direct variation   (p. 397)
domain   (p. 378)
function   (p. 378)
inverse variation   (p. 401)
linear function   (p. 392)

mapping   (p. 382)
ordered pair   (p. 373)
origin   (p. 373)
range   (p. 378)
relation   (p. 378)
value of a function   (p. 383)
x- and y-axis   (p. 373)
x- and y-coordinates   (p. 374)

Key Ideas and Review Exercises

10.1   To graph a point given its coordinates, start at the origin and move right if the abscissa is positive, or left if it is negative. Then move up if the ordinate is positive, or down if it is negative.

Graph each point in the same coordinate plane.

1. $(3, -1)$ **2.** $(-2, 4)$ **3.** $(-1, -3)$ **4.** $(3, 2)$

5. $(0, 5)$ **6.** $(5, 4)$ **7.** $(-1, 0)$ **8.** $(-2, 1)$

10.2   To determine the domain of a relation, find the set of first coordinates. To determine the range of a function, find the set of second coordinates.

To show that a relation is a function, show that no two ordered pairs have the same first coordinate.

Given the relation $G = \{(-1, 3), (1, 5), (-6, -3), (7, 3), (-3, 1)\}$.

9. Determine the domain of G.   **10.** Determine the range of G. $\{-3, 1, 3, 5\}$

11. Is G a function? Explain your answer.   **9.** $\{-6, -3, -1, 1, 7\}$

   Yes; no two ordered pairs have the same first coordinate.

10.3   To find the value of a function $f(x)$, substitute a given value for x in the expression that defines the function.

To determine the range of a function, find the set of function values for all elements in the domain.

Given $f(x) = x^2 - 8$, find the indicated value.

12. $f(1)$ -7 **13.** $f(-2)$ -4 **14.** $f(5)$ 17 **15.** $f(-6)$ 28

16. Use the domain $D = \{-1, 0, 5\}$ to find the range of $g(x) = 3x - 5$. $\{-8, -5, 10\}$

10.4   To solve an equation with two variables x and y, substitute values of x from the replacement set and solve for y. If the value of y is in the replacement set, then (x, y) is a solution of the equation.

17. Find two solutions of the equation $2x - 3y = 0$. Use $\{0, 1, 2, 3\}$ as the replacement set. (0, 0), (3, 2)

10.5   To graph a function described by a linear equation, find three ordered pairs that are solutions of the equation. If the function is given in $f(x)$ form, find three ordered pairs $(x, f(x))$. Then plot the ordered pairs and draw the graph.

The vertical line test can be used to determine whether a relation is a function. If no vertical line can be found that intersects the graph of the relation in more than one point, then the relation is a function.

Graph each equation or function.

18. $y = 3x + 4$ **19.** $4x - 2y = 4$ **20.** $x = 3$ **21.** $y = -2$

22. $-2x - y = 3$ **23.** $f(x) = 2x + 3$ **24.** $f(x) = -3x - 1$ **25.** $f(x) = -x + 2$

Use the vertical line test to determine which relations are functions.

26.
Yes

27.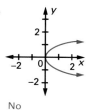
No

28.
Yes

29.
Yes

10.6,
10.7   To solve problems with constant variation:

1. Use $y = kx$ or $\dfrac{y}{x} = k$ for direct variation;

use $xy = k$ or $y = \dfrac{k}{x}$ for inverse variation.

2. Substitute for x and y to find the constant of variation, k.

3. Using the value of k, substitute for one variable to find the other.

30. y varies directly as x, and y is 48 when x is 6. Find y when x is 8. $y = 64$

31. y varies inversely as x, and y is 15 when x is 3. Find y when x is 9. $y = 5$

32. Given the set of ordered pairs $\{(8, 6), (-12, -4), (-16, -3)\}$, write in your own words how to determine whether y varies directly as x, or whether y varies inversely as x, or neither.

Chapter 10 Review   **407**

21.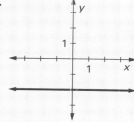

22.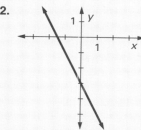

23.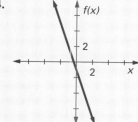

24.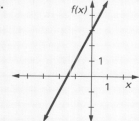

See page 409 for the answers to Ex. 25 and 32.

1.

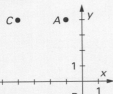

11.

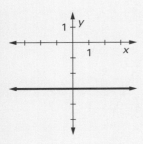

12.

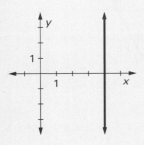

13.

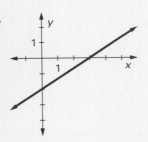

See page 409 for the answer to Ex. 14.

Chapter 10 Test

A Exercises: 1–3, 5–8, 10–14, 17–19
B Exercises: 4, 9, 15–16, 20   C Exercises: 21–22

1. Graph the points $A(-1, 4)$, $B(0, -1)$, and $C(-4, 4)$ in the same coordinate plane.

   2. D: $\{-1, 3, 4\}$; R: $\{-3, 0, 2, 5\}$; no

Give the domain and range of the relation. Is it a function?

2. $\{(4, -3), (3, 0), (-1, 5), (4, 2)\}$ 3. $\{(3, -3), (1, -3), (0, 4), (-1, 2)\}$

   D: $\{-1, 0, 1, 3\}$;
4. For what value of k will the relation $R = \{(-2k+3, 4),$ R: $\{-3, 2, 4,\}$; yes
   $(5k-4, -3)\}$ *not* be a function? $k = 1$

Given $g(x) = 5x^2 - 4$, find the indicated value.

5. $g(-2)$ 16   6. $g(3)$ 41   7. $g(0)$ -4 8. $g(-3)$ 41

9. Use the domain $D = \{-3, 0, 3\}$ to find the range of $f(x) = (x - 3)^2$. $\{0, 9, 36\}$

10. Find two solutions for $4x - y = 3$. Use $\{0, 1, 2, 3, 4, 5\}$ as the replacement set for each variable. $(1, 1)$; $(2, 5)$

Graph each equation.

11. $y = -3$ 12. $x = 4$ 13. $2x - 3y = 6$ 14. $x = 2y$

Use the vertical line test to determine which relations are functions.

15. 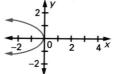 Not a function   16. 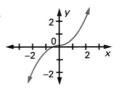 A function

17. Determine whether y varies directly as x, or whether y varies inversely as x, or neither. If there is a variation, find the constant of variation. Directly; $k = -4$

x	y
-3	12
2	-8
5	-20

                                    $y = -40$
18. y varies directly as x, and y is 30   19. y varies inversely as x, and y is -4
    when x is -6. Find y when x is 8.   when x is 4. Find x when y is 8. $x = -2$

20. The width of a rectangle of constant area varies inversely as the length. If the width is 4 ft when the length is 16 ft, find the width when the length is 32 ft. 2 ft

21. If y varies directly as the square of x, and y is 50 when x is 5, find y when x is 6. $y = 72$

22. If $f(x) = 3x - 1$ and $g(x) = x^2$, find $f(g(-2))$. 11

Choose the one best answer to each question or problem.

1. A train traveling at 40 mi/h is stopped $3\frac{1}{3}$ mi from its destination at 7:00 A.M. At what time would the train have arrived if it had not been delayed? D
   (A) 7:10 A.M.   (B) 6:55 A.M.
   (C) 7:03 A.M.   (D) 7:05 A.M.
   (E) 7:53 A.M.

2. Connect points $P(2, 0)$, $Q(6, 0)$, and $R(4, 5)$ with line segments. Which of the following is true? C
   (A) $RQ < PQ$ (B) $RQ < RP$
   (C) $RP = RQ$ (D) $RQ = PQ$
   (E) $RP = PQ$

3. Inside the square, the area of each circle is 16π. What is the area of the shaded region? A

   (A) $256 - 64\pi$ (B) $256 - 16\pi$
   (C) $64\pi - 256$ (D) $64 - 64\pi$
   (E) $64\pi - 64$

4. For what value of k will $x^2 + 8x + k = 0$ have one solution?
   (A) 4   (B) -4 (C) 8   E
   (D) -16 (E) 16

5. If the operation $*$ for positive numbers is defined as $x * y = x^2 + xy$, then $3 * (3 * 3)$ is ___?___. A
   (A) 63   (B) 18   (C) 27
   (D) 36   (E) 21

6. The vertices of triangle ABC are $A(3, 1)$, $B(3, 5)$, $C(5, 1)$. The area of triangle ABC is ___?___. B
   (A) 2   (B) 4   (C) 8
   (D) 12   (E) 16

7. Six consecutive odd integers are given. The sum of the first three is 33. What is the sum of the last 3? E
   (A) 24   (B) 39   (C) 42
   (D) 45   (E) 51

8. Mary Alice drove 300 km at 50 km/h.
   E If she had driven 10 km/h faster, how many hours would be saved?
   (A) 11   (B) 10   (C) 6
   (D) 5   (E) 1

9. If the operation $*$ for positive numbers is defined as $m * n = \dfrac{mn}{m + n}$, then $(3 * 3) * 3$ is ___?___. B
   (A) 3   (B) 1   (C) $\dfrac{3}{2}$
   (D) $\dfrac{9}{4}$ (E) $\dfrac{9}{2}$

10. If $\dfrac{2}{3} \cdot \dfrac{3}{4} \cdot \dfrac{4}{5} \cdot \dfrac{5}{6} \cdot \dfrac{6}{7} \cdot n = 5$, then what is the value of n? D
    (A) $\dfrac{2}{35}$ (B) 1   (C) $\dfrac{35}{7}$
    (D) $\dfrac{35}{2}$ (E) 35

11. If $\dfrac{3}{4}$ of a number is 54, then $\dfrac{1}{3}$ of the number is ___?___. C
    (A) $13\frac{1}{2}$ (B) 18   (C) 24
    (D) 32   (E) $40\frac{1}{2}$

12. If $4x - 4y = 1$, what is the value of $\sqrt{x - y}$? C
    (A) $\dfrac{1}{16}$ (B) $\dfrac{1}{4}$ (C) $\dfrac{1}{2}$
    (D) 2   (E) 4

14.

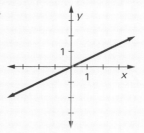

25.

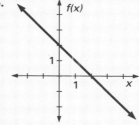

32. y varies directly as x if for every ordered pair in the set $\frac{y}{x} = k$, where k is a nonzero real number. In the given set, y doesn't vary directly as x. y varies inversely as x if for every ordered pair in the set $xy = k$, where k is a nonzero real number. In the given set, y varies inversely as x since for each ordered pair in the set $xy = 48$.

2.

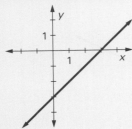

3.

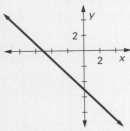

4.

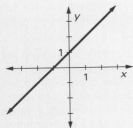

5.

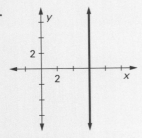

410

Cumulative Review (Chapters 1–10)

Add or subtract.

1. $\dfrac{2x}{14} + \dfrac{x}{7} + \dfrac{5x}{6}$ $\dfrac{47x}{42}$ *8.6*

2. $\dfrac{p-4}{10} - \dfrac{2p-3}{15}$ $-\dfrac{p+6}{30}$

3. $\dfrac{-a-3}{a^2+a-6} + \dfrac{6}{2-a}$ $\dfrac{7}{2-a}$ *8.7*

Simplify. Then give the degree of the polynomial. $4a^3 - 3a^2 - a\,;\,3$

4. $8a^3 - 4a^2 - a - 4a^3 + a^2$ *6.6*

5. $-6 + 3y^4 - 2y^2 + 8y - y^4 - y^2$ $2y^4 - 3y^2 + 8y - 6;\,4$

6. $-5x + 3 - 4x^2 - x + x^2 + 3x$ $-3x^2 - 3x + 3;\,2$

Divide and simplify.   **8.** $-6a^3 + 8a^2 + 3$

7. $24y^5 \div (-6y^2)$ $-4y^3$ *6.3*

8. $(-18a^4 + 24a^3 + 9a) \div 3a$ *8.8*

9. $\dfrac{64m^9 - 32m^6 + 48m^5}{-4m^2}$ $-16m^7 + 8m^4 - 12m^3$

10. $\dfrac{3a^2 + 13a - 10}{3a - 2}$ $a + 5$ *8.1*

11. $(x^3 - 12x^2 - 6) \div (x - 2)$ *8.8*
    $x^2 - 10x - 20 - \dfrac{46}{x-2}$

Multiply and simplify.

12. $-3x^5 \cdot (6x^4)$ $-18x^9$ *6.1*

13. $-2a^4b \cdot (-3ab^4)$ $6a^5b^5$

14. $\dfrac{8x^3y^2}{5xy^4} \cdot \dfrac{-3xy^5}{4x^2y}$ $-\dfrac{6xy^2}{5}$ *8.3*

15. $2x^{-3}y^2 \cdot 5x^5y^0$ $10x^2y^2$ *6.4*

Solve.

16. $|3m - 2| = 13$ $-\dfrac{11}{3},\,5$ *5.6*

17. $\dfrac{x-3}{2} + \dfrac{x}{4} = 6$ $x = 10$ *4.5*

18. $9a - (6 - a) = 18 - 2a$ $a = 2$ *3.5*

19. $5m - 3 = -2m^2$ $m = -3$ or $m = \frac{1}{2}$ *7.7*

20. $y^2 - 8y = 0$ $y = 0$ or $y = 8$

21. $x^3 - x^2 - 6x = 0$

22. $6x^2 + x - 1 = 0$ $x = -\frac{1}{2}$ or $x = \frac{1}{3}$

23. $25 - m^2 = 0$ $m = 5$ or $m = -5$

21. $x = 0$ or $x = -2$ or $x = 3$

For each of Exercises 24–27, choose the one best answer.

24. The value of x for which   *8.1*
   $\dfrac{2x-6}{5x+10}$ is undefined is __?__.
   (A) 2   (B) -2 (C) 3   B
   (D) -3 (E) none of these

25. Solve $-5x - 3 \le 12$.   A   *5.2*
   (A) $x \ge -3$ (B) $x > -3$
   (C) $x \le -3$ (D) $x < -3$
   (E) None of these

26. Evaluate $-3x - 4y^2$ for   *2.8*
   $x = -1$ and $y = -2$.   C
   (A) 19   (B) -19
   (C) -13 (D) 13
   (E) None of these

27. Subtract $-3a^2 + 4a$ from   *6.7*
   $a^3 + 5a^2 - 3a - 10$.   A
   (A) $a^3 + 8a^2 - 7a - 10$
   (B) $a^3 + 2a^2 + a - 10$
   (C) $-a^3 - 8a^2 + 7a + 10$
   (D) $a^3 + 8a^2 + 4a - 10$
   (E) None of these

32. Comm Prop Add

Which property is illustrated?

28. $-5(2x - 3) = -5 \cdot 2x +$ *1.6*
   $(-5)(-3)$ Distr Prop

29. $a + b = b + a$ Comm Prop Add   *1.5*

30. $x \cdot 1 = x$ Mult Identity Prop   *1.7*

31. $a + 0 = a$ Add Identity Prop   *2.3*

32. $a + (b + c) = (b + c) + a$ *1.5*

33. $x \cdot y = y \cdot x$ Comm Prop for Mult

Factor completely.

34. $6a^2 + a - 1$ $(3a - 1)(2a + 1)$ *7.4*

35. $25x^2 - 1$ $(5x + 1)(5x - 1)$ *7.5*

36. $3m^3 + 12m^2 - 15m$ *7.6*

37. $-9a^2 + 9$ $-9(a + 1)(a - 1)$

38. $6x^2 - 54$ $6(x + 3)(x - 3)$

36. $3m(m + 5)(m - 1)$

Simplify. $x^4 + 4x^3 - 7x - 10$

39. $(x^4 - x) - (6x + 10 - 4x^3)$ *6.7*

40. $-(8a^3 + 6 - 7a^2) - (a^2 + 5)$ $-8a^3 + 6a^2 - 11$

41. $(-3x^2y^3)^2$ $9x^4y^6$ *6.2*

42. $\dfrac{18a^4b^6}{-6ab^7}$ $-\dfrac{3a^3}{b}$ *6.3*

43. $(8x^4y^5)(-4xy^7)$ $-32x^5y^{12}$ *6.1*

Write an algebraic expression for each word expression.

44. y more than the product of 12 and x $12x + y$ *4.1*

45. Six more than $\frac{1}{5}$ of n $\frac{1}{5}n + 6$

46. t more than triple the cost of d dollars $3d + t$

47. Seven less than twice a number, increased by 3 more than 4 times the number $(2n - 7) + (4n + 3)$

Solve each word problem.

48. If 3 times Dan's age is decreased by 10 more than twice his age, the result is 22. How old is he? 32 yrs old *3.3*

49. Marina leases a car for $45 plus $.28/mi. How far can she travel, to the nearest mile, on a budget of $310? 946 mi

50. The second of three numbers is 6 times the first. The third is 5 more than the second. If the second is decreased by twice the third, the result is -40. Find the three numbers. 5, 30, 35 *4.2*

51. Forty-two percent of 80 is what number? 33.6 *4.7*

52. Twelve percent of what number is 10.8? 90

53. Twenty-seven is what percent of 90? 30%

54. The first side of a triangle is 4 cm longer than the second side. The third side is 3 times the second side. The perimeter is 44 cm. How long is each side? 8 cm, 12 cm, and 24 cm *4.4*

55. Eight more than 4 times a number is at most 32. Find the numbers. $x \le 6$ *5.2*

56. The selling price of a video player is $420. The profit is 25% of the cost. Find the cost. $336 *4.8*

57. If the square of a number is increased by 12, the result is the same as 8 times the number. Find the number. 2 or 6 *7.8*

58. Find two consecutive even integers such that the sum of the squares is 244. 10, 12; $-10, -12$ *7.9*

59. The size of next year's graduating class will be 98% of this year's. There are 320 in this year's class. How many students will be in the graduating class? 314 students *4.8*

60. Alex averages 625 mi per week on his car. His car averages 26 mi/gal. If gasoline sells for $1.40/gallon, how much does he spend on gasoline per week? $33.65 *9.6*

61. The sum of three consecutive odd integers is -129. Find the integers. $-45, -43, -41$ *4.3*

Additional Answers, page 395

6.
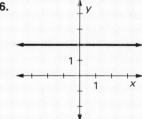

7.
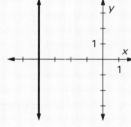

8.
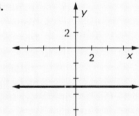

OVERVIEW

In this chapter, students use what they have learned about coordinate graphing to find the slope of a line. By observing patterns involving slopes and intercepts, the students will be able to graph lines. They will also be able to write the equation of a line, given any two points on the line or given one point and the slope of the line. Students then study the relationships between the slopes of parallel lines and perpendicular lines. Finally, students apply these skills in graphing linear inequalities in two variables.

OBJECTIVES

- To find the slope of a line
- To write an equation of a line, given one of its points and its slope or two of its points
- To find the slope and y-intercept of a line, given its equation, and to graph it
- To determine whether two lines are parallel or perpendicular
- To graph linear inequalities in two variables

PROBLEM SOLVING

The emphasis in Chapter 11 is on using the strategy of Making a Graph to show the solution sets of linear equations. Then this process is reversed so that the student is asked to write an equation or inequality to describe a table of values or a given graph. The Application on page 423 applies logical reasoning and the strategy of Writing a Formula to solving problems that relate altitude and temperature. Using Logical Reasoning is the strategy applied in the problem-solving exercises on page 438.

READING AND WRITING MATH

In Exercise 35 on page 422, students are asked to write equations that meet certain conditions. Then they compare the graphs of these equations. Exercises 19–24 on page 432 explore the relationships between lines and the effect that changes in slope and y-intercept will have on the graph of an equation.

TECHNOLOGY

Calculator: You may wish to have students use a calculator to calculate coordinates that will satisfy an equation or inequality. A graphing calculator would be an interesting and useful tool to use throughout this chapter.

SPECIAL FEATURES

PLANNING GUIDE

Lesson	Basic	Average	Above Average	Resources
11.1　pp. 416–417	CE 1–12 WE 1–18 all	CE 1–12 WE 10–28 all	CE 1–12 WE 16–34 all	Reteaching 82 Practice 82
11.2　pp. 420–423	CE 1–9 WE 1–21 odd, 27 Application	CE 1–9 WE 5–23 odd, 27, 28 Application	CE 1–9 WE 11–35 odd Application	Reteaching 83 Practice 83
11.3　pp. 426–428	CE 1–9 WE 1–27 odd, 40–43 all Midchapter Review	CE 1–9 WE 11–45 odd Midchapter Review	CE 1–9 WE 19–51 odd Midchapter Review	Reteaching 84 Practice 84
11.4　pp. 431–432	CE 1–8 WE 1–9 odd, 10–15 all, 19	CE 1–8 WE 1–23 odd	CE 1–8 WE 1, 5, 9, 19–24 all, 25–29 odd	Reteaching 85 Practice 85
11.5　pp. 436–439	CE 1–12 WE 1–17 odd, 25–28 all Problem Solving Strategies Mixed Problem Solving	CE 1–12 WE 5–29 odd Problem Solving Strategies Mixed Problem Solving	CE 1–12 WE 9–33 odd Problem Solving Strategies Mixed Problem Solving	Reteaching 86 Practice 86
Chapter 11 Review pp. 440–441	all	all	all	
Chapter 11 Test p. 442	all	all	all	
College Prep Test p. 443	all	all	all	

CE = Classroom Exercises　　WE = Written Exercises　　FR = Focus on Reading
Note: For each level, all students should be assigned all Try This and all Mixed Review Exercises.

◼️ INVESTIGATION

Application: In this *Investigation,* students look for patterns in graphing equations of the form $y = mx + b$.

Materials: Each student will need a copy of Application Worksheet 11, some graph paper and a straightedge.

Before presenting technical terms, such as slope, give the students a chance to use what they know about graphing points to find patterns involving graphs of $y = mx + b$.

Have the students complete the first set of charts for the y values. Then have them decide how the change in y compares to the change in x in the first chart. *y changes twice as fast as x.* Have students graph the points for this chart and draw a line through the points. Point out that y changes twice as fast as x everywhere on the line, for any two points they choose. When they have finished this first set of lines, they should see how the coefficient of x affects the graph of the line.

In the second set of exercises, students determine what is the effect of a change in the value of b. *It raises or lowers the line vertically.* Students should also be able to determine what connection b has with where the line crosses the y-axis. *The line crosses the y-axis at point $(0, b)$.*

When students have completed the worksheet, ask them to use what they have learned to graph the following equations.

1. $y = 4x$
2. $y = x - 2$
3. $y = \frac{1}{2}x + 3$

The work of Dr. Selma Burke may be in your pocket right now! She is the sculptor who carved the profile of President Franklin D. Roosevelt that appears on every dime.

More About the Photo

As a child, Selma Burke began sculpting with the creek clay used to make whitewash for her North Carolina home. She wanted to study art, but her mother, whose other children included four lawyers, two preachers, and a doctor, insisted that she go to nursing school. A private nursing job took her to New York City where she became part of the art scene. She then traveled and studied art in Europe. She was 40 when she completed her master of fine arts degree at Columbia University and 70 when she earned her doctorate. Her personal art collection is housed in the Selma Burke Gallery at Winston-Salem State University.

11.1 Slope of a Line

Objective

To find the slope of a line

When walking or pedaling a bike up a hill, you can notice the steepness of the hill. The steepness, or *slope*, is defined as the ratio of the vertical *rise* of the hill to the horizontal *run* of the hill. The steeper the hill, the greater the slope.

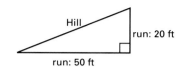

The slope of the hill at the right above is $\frac{20}{50}$, or $\frac{2}{5}$. This means that the hill rises 2 ft vertically for every 5 ft of horizontal distance.

In mathematics, you find slopes of lines in a coordinate plane. The slope of a line is found by forming the ratio $\frac{\text{rise}}{\text{run}}$. The *rise* represents *vertical* change from one point to another. The *run* represents *horizontal* change between the same two points. If a line *slants upward* as you move from left to right, the line has a *positive* slope. If a line *slants downward* as you move from left to right, its slope is *negative*.

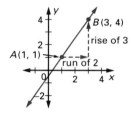

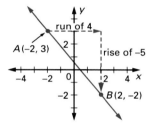

Line *AB* slants upward.

$$\text{slope} = \frac{\text{rise}}{\text{run}} = \frac{3}{2}$$

Line *AB* slants downward.

$$\text{slope} = \frac{\text{rise}}{\text{run}} = \frac{-5}{4} = -\frac{5}{4}$$

For two points such as $A(2, 1)$ and $B(3, 5)$, the run is the difference in the *x*-coordinates and the rise is the *corresponding* difference in the *y*-coordinates.

change in *x*-coordinates: $3 - 2 = 1$

change in *y*-coordinates: $5 - 1 = 4$

Therefore, the slope is $\frac{4}{1}$, or 4.

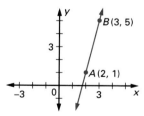

11.1 Slope of a Line **413**

Teaching Resources

Application Worksheet 11
Manipulative Worksheet 11
Quick Quizzes 82
Reteaching and Practice Worksheets 82
Transparency 32

◼◼◼GETTING STARTED

Prerequisite Quiz

Compute.

1. $-8 - (-2)$ -6
2. $-9 - 5$ -14
3. $12 - (-5)$ 17
4. $1 - 6$ -5
5. $\frac{4 - 3}{0 - (-2)}$ $\frac{1}{2}$
6. $\frac{-3 - (-7)}{2 - 5}$ $-\frac{4}{3}$

Motivator

Ask students to give examples of the use of the word *slope*.
Possible answers: Ski slope, slope of a roof.   Ask them what special characteristic the word describes.   steepness

Highlighting the Standards

Standards 8a, 8b: Students use a coordinate definition of the slope of a line and translate between graphs and coordinates.

Lesson Note

Be sure that the students understand the meaning of a "negative" rise. A rise of -5 is really a drop of 5 units. When presenting Example 2, show students that, if the order of the points is reversed, then the slope of line PQ is

$$\frac{y_1 - y_2}{x_1 - x_2} = \frac{1 - 4}{3 - 0} = \frac{-3}{3} = -1.$$

Emphasize that, when calculating slope, either point may be considered as the first point. The slope remains the same.

Math Connections

Trigonometry: The slope of a line is the same as the tangent of the angle that a line makes with the positive x-axis. In calculus, the slope of a curve at a given point is defined as the slope of the tangent line at that point.

Additional Answers

Try This

1. 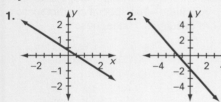 2.

EXAMPLE 1 Graph the line passing through each pair of points. Then find the slope of each line.

a. $A(2, 3)$ and $B(7, 6)$ b. $R(2, 5)$ and $S(6, 3)$

Solutions a. 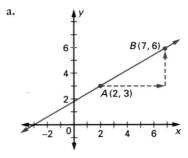   b.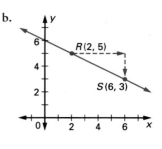

slope $= \dfrac{\textit{change in y-coordinates}}{\textit{change in x-coordinates}}$ slope $= \dfrac{\textit{change in y-coordinates}}{\textit{change in x-coordinates}}$

$= \dfrac{6 - 3}{7 - 2} = \dfrac{3}{5}$ $= \dfrac{3 - 5}{6 - 2} = \dfrac{-2}{4} = -\dfrac{1}{2}$

TRY THIS Graph the line passing through each pair of points. Then find the slope of each line.

1. $C(2, -1)$ and $D(5, -3)$ $-\dfrac{2}{3}$
2. $E(-4, 3)$ and $F(1, -3)$ $-\dfrac{6}{5}$

In the figure at the right, $A(x_1, y_1)$ and $B(x_2, y_2)$ represent any two points. The subscripts are used to distinguish between the coordinates of the two points. Note that the coordinates are subtracted in the same order,
$y_2 - y_1$ and $x_2 - x_1$.

Slope is usually represented by the letter m, as shown in the definition below.

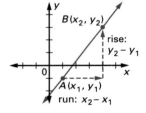

Definition

For any two points $A(x_1, y_1)$ and $B(x_2, y_2)$, on a nonvertical line, the **slope** m of the line is defined as follows:

$$m = \frac{\textit{change in y-coordinates}}{\textit{change in x-coordinates}}, \text{ or } m = \frac{y_2 - y_1}{x_2 - x_1}, \text{ where } x_1 \neq x_2$$

414 Chapter 11 Analytic Geometry

Additional Example 1

Find the slope of each line.

a. 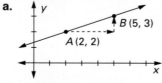 $\dfrac{1}{3}$ b. 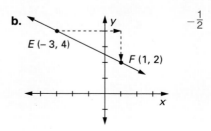 $-\dfrac{1}{2}$

EXAMPLE 2   Find the slope of the line determined by $P(0, 4)$ and $Q(3, 1)$.

Solution   Slope of $\overleftrightarrow{PQ} = \dfrac{y_2 - y_1}{x_2 - x_1}$

$= \dfrac{4 - 1}{0 - 3}$

$= \dfrac{3}{-3}$, or -1

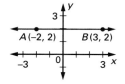

Thus, the slope of $\overleftrightarrow{PQ}$ is -1.

EXAMPLE 3   Find the slope of each line.

a.

b.

Solutions   a. slope of $\overleftrightarrow{AB} = \dfrac{y_2 - y_1}{x_2 - x_1}$ b. slope of $\overleftrightarrow{CD} = \dfrac{y_2 - y_1}{x_2 - x_1}$

$= \dfrac{2 - 2}{3 - (-2)}$ $= \dfrac{3 - (-1)}{-2 - (-2)}$

$= \dfrac{0}{5}$ $= \dfrac{4}{0}$, undefined.

$= 0$ The slope is undefined.

The slope is zero.

TRY THIS   Find the slope of the line determined by each pair of points.

3. $A(-3, 5)$ and $B(2, 8)$ $\frac{3}{5}$ 4. $H(-1, 4)$ and $K(-1, 5)$
5. $S(4, -2)$ and $T(-4, 2)$ $-\frac{1}{2}$ undefined

As Example 3 illustrates, the slope of a horizontal line is 0, since $\dfrac{y_2 - y_1}{x_2 - x_1} = \dfrac{0}{x_2 - x_1} = 0$. The slope of a vertical line is undefined,

since $\dfrac{y_2 - y_1}{x_2 - x_1} = \dfrac{y_2 - y_1}{0}$ (with a zero denominator) is undefined.

11.1 Slope of a Line   **415**

Additional Example 2

Find the slope of the line determined by $P(-5, -3)$ and $Q(-3, 7)$.   5

Additional Example 3

Find the slope of each line.

a.

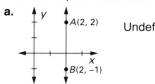

Undef

b.

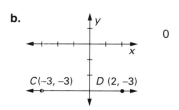

0

415

Guided Practice

Classroom Exercises 1–12
Try This   all

Independent Practice

A Ex. 1–18, **B** Ex. 19–28, **C** Ex. 29–34

Basic: WE 1–18 all
Average: WE 10–28 all
Above Average: WE 16–34 all

Written Exercises

13. $-\frac{1}{4}$; down to right
14. 1; up to right
15. 0; horiz
16. $-\frac{6}{7}$; down to right
17. Undef; vert

26.

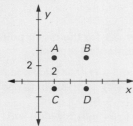

Information about the slope of a line can be determined by the direction in which it slants.

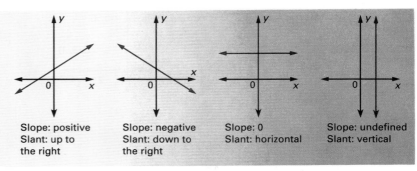

Slope: positive
Slant: up to
the right

Slope: negative
Slant: down to
the right

Slope: 0
Slant: horizontal

Slope: undefined
Slant: vertical

Classroom Exercises

Use this diagram for Exercises 1–3.

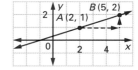

1. What is the rise? 1   2. What is the run? 3
3. What is the slope of the line? $\frac{1}{3}$

For points A and B, is the slope of $\overleftrightarrow{AB}$ positive, negative, zero, or undefined?

4. $A(2, 3)$, $B(6, 8)$ Pos   5. $A(1, 4)$, $B(10, 6)$ Pos   6. $A(0, 1)$, $B(4, 5)$ Pos
7. $A(6, 4)$, $B(-8, 4)$ Zero   8. $A(5, -3)$, $B(5, 7)$ Undef   9. $A(-3, 2)$, $B(6, -4)$
                                                                                                Neg

Find the slope of $\overleftrightarrow{PQ}$ for the given points, and describe the slant of the line.

10. $P(-2, 1)$, $Q(3, 4)$
$\frac{3}{5}$; up to right

11. $P(-1, 6)$, $Q(1, -4)$
-5; down to right

12. $P(4, 1)$, $Q(4, -6)$
Undef; vert

Written Exercises

Find the slope of each line.

1. 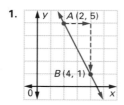 -2

2. 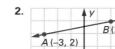 $\frac{1}{5}$

3. 0

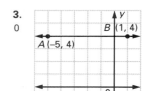

Find the slope of the line determined by the given points M and N.

4. $M(2, 5)$, $N(4, 6)$ $\frac{1}{2}$ 5. $M(6, 5)$, $N(4, 3)$ 1   6. $M(1, 0)$, $N(2, 8)$ 8
7. $M(4, 8)$, $N(0, 8)$ 0   8. $M(-1, 3)$, $N(-3, 9)$ -3 9. $M(-2, 3)$, $N(-2, -4)$
                                                                                              Undef

416   Chapter 11 Analytic Geometry

Enrichment

Challenge students to solve this problem.
Given the four points $A(0, 5)$, $B(3, q)$,
$C(r, 4)$, and $D(2, s)$, where the slope of
$\overline{AB}$ = slope of $\overline{CD}$ = $\frac{2}{3}$, and the slope of
$\overline{BC}$ = slope of $\overline{AD}$ = $-\frac{3}{2}$, find the values of
q, r, and s. $q = 7$; $r = 5$, $s = 2$ Then ask
students to describe the figure formed by
joining the points in order.   square

Find the slope of $\overleftrightarrow{MN}$ for the given points, and describe the slant of
the line. $\frac{1}{2}$; up to the right

0; horiz   −1; down to the right

10. $M(3, -4)$, $N(7, -2)$ **11.** $M(4, 7)$, $N(-3, 7)$ **12.** $M(7, -6)$, $N(4, -3)$

13. $M(-3, -2)$, $N(-7, -1)$ **14.** $M(4, -2)$, $N(3, -3)$ **15.** $M(7, -3)$, $N(-2, -3)$

16. $M(0, -6)$, $N(-7, 0)$ **17.** $M(-9, -3)$, $N(-9, 6)$ **18.** $M(0, -4)$, $N(5, -4)$

0; horiz

19. The vertices of a triangle are at $A(1, 3)$, $B(3, 5)$ and $C(4, 1)$. Find
the slope of each side of the triangle. AB:1; BC:−4; AC:−$\frac{2}{3}$

20. The vertices of a rectangle are at $R(-2, 4)$, $V(-2, -4)$, $S(3, -4)$,
and $T(3, 4)$. Find the slope of each side of the rectangle. RT, VS, 0; TS, RV: undef

21. Find the slope of a line passing through $A(a, 0)$ and $B(0, a)$.   −1

Use the given expressions to find the slope of $\overleftrightarrow{AB}$.

22. $A(4n, 3r)$, $B(6n, 2r)$ $-\frac{r}{2n}$ **23.** $A(-8n, -2r)$, $B(-n, -r)$ $\frac{r}{7n}$

24. $A(2n - 1, 3r - 2)$ **25.** $A(-3n + 4, 2r - 1)$
$B(4n + 3, 5r - 1)$ $\frac{2r + 1}{2n + 4}$ $B(8n, 5r + 2)$ $\frac{3r + 3}{11n - 4}$

26. Plot the four points $A(2, 3)$, $B(6, 3)$, $C(6, -1)$, and $D(2, -1)$.
Which lines determined by pairs of these points have a slope of 0? $\overleftrightarrow{AB}$, $\overleftrightarrow{DC}$

27. In Exercise 26, which of the lines determined by pairs of the points
have an undefined slope? $\overrightarrow{AD}$, $\overrightarrow{BC}$

28. In Exercise 26, which of the lines determined by pairs of the points
have a negative slope? $\overleftrightarrow{AC}$

Determine the value(s) of the variable so that $\overleftrightarrow{XY}$ has the given slope.

29. $X(4, -3)$, $Y(a, 9)$; slope = 1   16   **30.** $X(-3, 2)$, $Y(7, a)$; slope = $\frac{2}{3}$ $8\frac{2}{3}$

31. $X(-4, 5a)$, $Y(-1, a^2)$; slope = 0   **32.** $X(6, a^2 - 5a)$, $Y(-3, 2a^2 - a)$;
0 or 5   slope = 0   0 or −4

33. $X(5, -8a)$, $Y(10, a^2)$; slope = −3   **34.** $X(-2, a)$, $Y(3, 6a^2)$; slope = $\frac{2}{5}$
−3 or −5 $-\frac{1}{2}$ or $\frac{2}{3}$

Mixed Review

Graph the solution set on a number line.   **5.7**

1. $|x| \le 3$ **2.** $|x| > 2$ **3.** $|-x| \ge 4$ **4.** $|x - 3| < 5$

Multiply.   **6.9, 6.10**   **5.** $6a^2 + 7a - 3$

$14x^2 + 3x - 5$ $4m^2 - n^2$

5. $(3a - 1)(2a + 3)$ **6.** $(7x + 5)(2x - 1)$ **7.** $(2m - n)(2m + n)$ **8.** $(2x - 3)^2$

9. Keith started out in his car at the rate of 58 km/h. One hour later $4x^2 - 12x + 9$
Shirley left from the same point driving along the same road at 70
km/h. How long did it take her to catch up to Keith?   **9.4** $4\frac{5}{6}$ h

10. The square of a number is 11 more than 10 times the number. Find
the number(s).   **7.8** −1, 11

1.

2.

3.

4.

11.2 Equation of a Line: Point-Slope Form

Objectives

To write an equation of a line, given a point on the line and its slope
To write an equation of a line, given two points on the line
To write an equation of a line, given a table or graph

◤◥ GETTING STARTED

Prerequisite Quiz

Find the slope of $\overleftrightarrow{PQ}$ for the given points.

1. $P(3,-1)$, $Q(4,-2)$ -1
2. $P(-8,-2)$, $Q(4,-2)$ 0
3. $P(5,-1)$, $Q(6,5)$ 6
4. $P(0,1)$, $Q(7,-4)$ $-\dfrac{5}{7}$

Motivator

Show students the graph of $y = 2x$ and mark several points including their coordinates. Then have students calculate the slope using several different pairs of these points. Ask students to draw a conclusion based on their calculations. The slope of a straight line is the same everywhere.

This figure shows a line, $\overleftrightarrow{HG}$, through $H(2,-4)$. The slope of the line is -2. Let $G(x,y)$ be any other point on the line. The slope m of a line is

$$\frac{y_2 - y_1}{x_2 - x_1}.$$

So, for the line through H and G,

$$\frac{y - (-4)}{x - 2} = -2$$
$$y + 4 = -2(x - 2)$$

Thus, $y + 4 = -2(x - 2)$ is an equation of $\overleftrightarrow{HG}$.

This derivation suggests the following *point-slope* form of an equation.

Equation of a Line: Point-Slope Form
The **point-slope form** of the equation of a line with slope m and containing point $P(x_1, y_1)$ is $y - y_1 = m(x - x_1)$.

EXAMPLE 1   Find an equation of the line passing through point $P(2,-3)$ with slope $\frac{4}{3}$. Write your answer in the form $Ax + By = C$.

Solution   Use $y - y_1 = m(x - x_1)$. $y - (-3) = \frac{4}{3}(x - 2)$

Multiply each side by 3. $3(y + 3) = 4(x - 2)$
$$3y + 9 = 4x - 8$$

Rewrite the equation in the form $Ax + By = C$. $-4x + 3y = -17$, or $4x - 3y = 17$

Thus, $4x - 3y = 17$ is an equation of the line passing through $P(2,-3)$ with slope $\frac{4}{3}$.

TRY THIS   1. Find the equation of the line passing through $R(-1, 4)$ and with slope $-\frac{2}{3}$. Write your answer in the form $Ax + By = C$.

$2x + 3y = 10$

418   Chapter 11 Analytic Geometry

Highlighting the Standards

Standards 8a, 4c, 4d: Here students continue to move between points, graphs, and equations. In the Application, they apply these skills to problems about the formation of ice on airplane wings.

Additional Example 1

Find an equation of the line passing through point $P(-1, 2)$ with slope $-\frac{2}{3}$. Write your answer in the form $Ax + By = C$. $2x + 3y = 4$

You can also use slope to write an equation of a line that passes through two given points.

EXAMPLE 2   Write an equation of $\overleftrightarrow{PQ}$ passing through points $P(4, -1)$ and $Q(2, 1)$.

Solution $m = \dfrac{y_2 - y_1}{x_2 - x_1} = \dfrac{1 - (-1)}{2 - 4} = \dfrac{2}{-2} = -1$

Substitute in $y - y_1 = m(x - x_1)$, using -1 for m and either of the given points for (x_1, y_1).

$$y - (-1) = -1(x - 4)$$
$$y + 1 = -x + 4$$
$$x + y = 3$$

Thus, $x + y = 3$ is an equation, in standard form, of $\overleftrightarrow{PQ}$ passing through points $P(4, -1)$ and $Q(2, 1)$.

EXAMPLE 3   Write an equation in the form $Ax + By = C$.

a.

x	y
1	2
3	1
5	0

b.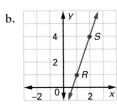

Solutions

a. Choose any two ordered pairs in the table. For example, use $(1, 2)$ and $(5, 0)$.

$$m = \frac{0 - 2}{5 - 1} = \frac{-2}{4} = -\frac{1}{2}$$

Next, use $y - y_1 = m(x - x_1)$.

$$y - 2 = -\frac{1}{2}(x - 1)$$
$$2(y - 2) = -1(x - 1)$$
$$2y - 4 = -x + 1$$
$$x + 2y = 5$$

b. Choose any two convenient points on the graph: R and S. Determine the ordered pairs and use them.

$$R(1, 1) \text{ and } S(2, 4)$$
$$m = \frac{4 - 1}{2 - 1} = \frac{3}{1} = 3$$

Next, use $y - y_1 = m(x - x_1)$.

$$y - 1 = 3(x - 1)$$
$$-3x + y = -2$$
$$3x - y = 2$$

TRY THIS   2. Write an equation of the line passing through the points $A(-3, 3)$ and $B(-1, -1)$. Write the equation in the form $Ax + By = C$.

$2x + y = -3$

Lesson Note

In using the point-slope form, point out that x and y are "stand-in" variables that will be part of the equation of the line, whereas, x_1 and y_1 are replaced with the coordinates of the given point. Emphasize that $Ax + By = C$ is called the *standard form* of a linear equation in two variables. In Example 2, show students that using point $Q(2, 1)$ in the point-slope formula will give the same equation.

Math Connections

Life Skills: When driving through the mountains, often a road sign will warn of a steep grade. The steepness, or *gradient*, of a road is expressed as a percent. A 5% grade would indicate a rise of 5 feet for every 100 feet of distance traveled.

Critical Thinking Questions

Synthesis: On a coordinate graph, draw a random line that crosses both axes. Ask students to state, without counting, what they know about the point where this line crosses the x-axis.   Its y-coordinate is 0.   Ask them what they know about the point where it crosses the y-axis. Its x-coordinate is 0.

Additional Example 2

Write an equation of $\overleftrightarrow{AB}$ passing through points $A(3, -3)$ and $B(-4, 1)$.
$4x + 7y = -9$

Additional Example 3

Write an equation in the form $Ax + By = C$.

a.

x	y
2	3
4	7
6	11

$2x - y = 1$

Common Error Analysis

Error: When calculating slope, some students may subtract the x-coordinates in a different order than the y-coordinates; for example,

$$\frac{y_2 - y_1}{x_1 - x_2}.$$

Suggest that they designate one of the points as the first point and then mark it clearly as such to avoid this error.

Checkpoint

Write an equation, in the form $Ax + By = C$, of the line passing through the given point and having the given slope.

1. $P(4, -6)$; slope $= -3$ $3x + y = 6$
2. $P(3, 5)$; slope $= -\frac{1}{3}$ $x + 3y = 18$
3. $P(-9, 8)$; slope $= 0$ $y = 8$

Write an equation, in the form $Ax + By = C$, of the line passing through the given points.

4. $A(-3, -7)$, $B(-4, 7)$ $14x + y = -49$
5. $A(0, -4)$, $B(7, -5)$ $x + 7y = -28$
6. $A(7, -1)$, $B(-3, -1)$ $y = -1$

Closure

Have students state how to write an equation of a line given:

1. one of its points and its slope
   See p. 418.
2. two of its points   See p. 419.
3. its graph   See p. 419.

EXAMPLE 4

When a weight of 2 oz is attached to a spring, the length of the spring is 3 in. When a weight of 4 oz is attached, the length of the spring is 4 in. Assume that the relation between the attached weight and the spring length is linear for $(0 \le$ weight $\le 8)$. Write an equation for the relation in the form $Ax + By = C$, and draw its graph.

Solution

Let x represent the weight attached to the spring: $0 \le x \le 8$.

Let y represent the resulting length of the spring.

Use the ordered pairs $(2, 3)$ and $(4, 4)$ to find m.

$$m = \frac{4 - 3}{4 - 2} = \frac{1}{2}$$

Now substitute in $y - y_1 = m(x - x_1)$, using either ordered pair.

$$y - 3 = \frac{1}{2}(x - 2)$$

$$2(y - 3) = 1(x - 2)$$

$$2y - 6 = x - 2$$

Thus, $-x + 2y = 4$, or $x - 2y = -4$

The graph is shown at the right.

Since the values of x are restricted to $0 \le x \le 8$, the graph is a line segment, not a line.

TRY THIS

3. Use the graph in Example 4 to determine the length of the spring when a weight of 6 oz is attached.   5 in.

Classroom Exercises

Find an equation for the line passing through the given point with the given slope. Give the equation in the form $y - y_1 = m(x - x_1)$. $y - 1 = -2(x - 3)$

1. $P(2, 3)$; slope $= 3$ $y - 3 = 3(x - 2)$ 2. $P(3, 1)$; slope $= -2$

3. $P(5, 4)$; slope $= \frac{2}{3}$ $y - 4 = \frac{2}{3}(x - 5)$ 4. $P(-1, 2)$; slope $= \frac{1}{2}$ $y - 2 = \frac{1}{2}(x + 1)$

Find an equation, in the form $Ax + By = C$, of the line passing through the given point and having the given slope.

5. $P(-1, -2)$; slope $= -3$ $3x + y = -5$ 6. $P(5, -4)$; slope $= \frac{2}{3}$ $2x - 3y = 22$

Additional Example 4

When a weight of 1 oz is attached to a spring, the length of the spring is 2 in. long. When a weight of 3 oz is attached, the length of the spring is 4 in. Assume that the relation between weight x and length y is a linear one for $0 \le x \le 6$. Write an equation for the relation in the form $Ax + By = C$, and draw its graph. $-x + y = 1$, or $x - y = -1$

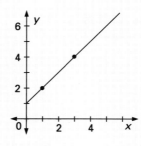

Find an equation, in the form $Ax + By = C$, of the line passing through the given points.

7. $P(4, 0)$, $Q(3, -4)$
$4x - y = 16$

8. $P(-3, -1)$, $Q(1, 5)$
$3x - 2y = -7$

9.

x	y
1	2
2	0
4	-4

$2x + y = 4$

Written Exercises

Write an equation, in the form $Ax + By = C$, of the line passing through the given point and having the given slope.

1. $P(-3, 4)$; slope $= -\frac{1}{3}$ $x + 3y = 9$ **2.** $P(4, 1)$; slope $= 4$ $4x - y = 15$

3. $P(-2, 8)$; slope $= -\frac{1}{4}$ $x + 4y = 30$ **4.** $P(3, -4)$; slope $= 3$ $3x - y = 13$

5. $P(0, 0)$; slope $= -5$ $5x + y = 0$ **6.** $P(-1, -5)$; slope $= 0$ $y = -5$

7. $P(-9, -3)$; slope $= \frac{3}{4}$ $3x - 4y = -15$ **8.** $P(8, 0)$; slope $= -\frac{5}{6}$ $5x + 6y = 40$

Write an equation, in the form $Ax + By = C$, of the line passing through the given points.

9. $P(3, 1)$, $Q(2, 4)$ $3x + y = 10$ **10.** $P(-1, -5)$, $Q(3, -6)$ $x + 4y = -21$
11. $P(2, 0)$, $Q(-1, 4)$ $4x + 3y = 8$ **12.** $P(-1, -6)$, $Q(3, -7)$ $x + 4y = -25$
13. $P(-1, 6)$, $Q(-3, -4)$ $5x - y = -11$ **14.** $P(5, 4)$, $Q(-5, -3)$ $7x - 10y = -5$
15. $P(0, -3)$, $Q(5, 0)$ $3x - 5y = 15$ **16.** $P(-3, -4)$, $Q(-1, -4)$ $y = -4$

Write an equation, in the form $Ax + By = C$, for each graph or table.

17. 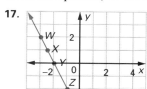 $2x + y = -4$ **18.** 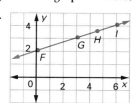 $x - 3y = -6$

19.

x	y
2	4
3	6
4	8

$2x - y = 0$

20.

x	y
2	0
6	2
12	5

$x - 2y = 2$

21.

x	y
0	2
2	-4
4	-10

$3x + y = 2$

22.

x	y
0	3
6	-1
9	-3

$2x + 3y = 9$

Guided Practice

Classroom Exercises 1–9
Try This   all

Independent Practice

A Ex. 1–22, **B** Ex. 23–28, **C** Ex. 29–35
Basic: WE 1–21 odd, 27, Application
Average: WE 5–23 odd, 27, 28, Application
Above Average: WE 11–35 odd, Application

27.

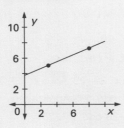

33. $(2 + 2r)x + (3 - r)y = 2r^2 + 6$

35.

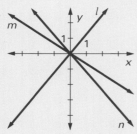

Write an equation, in the form $Ax + By = C$, of the line passing through the given point and having the given slope.

23. $P(\frac{1}{2}, \frac{1}{4})$; slope $= 3$ $\quad 12x - 4y = 5$

24. $P(\frac{3}{4}, \frac{2}{3})$; slope $= -2$ $\quad 12x + 6y = 13$

25. $P(\frac{5}{4}, -\frac{5}{6})$; slope $= \frac{1}{2}$ $\quad 12x - 24y = 35$

26. $P(-\frac{1}{2}, -\frac{3}{5})$; slope $= \frac{3}{4}$ $\quad 30x - 40y = 9$

27. When a weight of 3 oz is attached to a spring, the length of the spring is 5 in. When a weight of 8 oz is attached, the length of the spring is 7.5 in. Assume that the relationship between weight and length is linear for $0 \leq$ weight ≤ 10. Write an equation for the relation in the form $Ax + By = C$, and draw its graph. $x - 2y = -7$

28. From the graph drawn in Exercise 27, determine the length of the spring when a weight of 9 oz is attached. 8 in

Write an equation, in the form $Ax + By = C$, of the line passing through the given point and having the given slope.

29. $P(0, d)$; slope $= 4$ $\quad 4x - y = -d$

30. $P(\frac{2}{3}, 0)$; slope $= 6k$ $\quad 6kx - y = 4k$

31. $P(\frac{1}{3}, 0)$; slope $= -4g$ $\quad 12gx + 3y = 4g$

32. $P(-s, t)$; slope $= 5s$
$\qquad\qquad\qquad\qquad 5sx - y = -5s^2 - t$

Write an equation of the line passing through the given points.

33. $P(r, 2), \; Q(3, -2r)$

34. $P(s, t), \; Q(5s, -6t)$ $7tx + 4sy = 11st$

35. Write three equations, each in the form $Ax + By = C$, of lines that meet these conditions:

Line l passes through $(3, 4)$ and $(4.5, 6)$. $4x - 3y = 0$
Line m passes through $(-6, 4)$ and $(-7.5, 5)$. $2x + 3y = 0$
Line n passes through $(8, -10)$ and $(4, -5)$. $5x + 4y = 0$

What do you notice about the value of C in each equation? Graph all the lines in the same plane. Do the three lines intersect in the same point? If so, give the coordinates of the point.

$C = 0$; the three lines intersect at the origin.

Mixed Review

Evaluate for $x = -2$ and $y = 3$. *1.1, 2.6*

1. $y + 15$ 18

2. $3x - 2y$ -12

3. $xy - 3y$ -15

4. $-4xy + 5y - x$ 41

5. $x^3 - 3y^2 + 2xy$ -47

6. $x^2 - y - 3xy + y^2$ 28

7. The length of a rectangle is 6 more than 3 times the width. The perimeter is 44 cm. Find the length and the width. *4.4* $l = 18$ cm; $w = 4$ cm

8. The retail price of a camera is \$399. The profit is 40% of the cost. Find the cost. *4.8* \$285

Enrichment

Have students plot these points in a coordinate plane: $A(-3, 1)$, $B(3, 1)$, $C(0, -2)$. Draw $\overleftrightarrow{AB}$, $\overleftrightarrow{CB}$, and $\overleftrightarrow{AC}$. Ask: Do any of these lines have the same slope? No

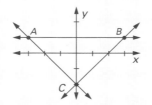

Ask: What is the relationship of the slopes of $\overleftrightarrow{CB}$ and $\overleftrightarrow{AC}$? They are opposites.
Ask: What is the area of triangle ABC? 9 sq units

Application: *Temperature and Altitude*

When flying at certain altitudes, pilots of small planes must be concerned with the possible icing of moisture on their planes' wings.

As the altitude of the plane increases, the outside air temperature cools about one degree Fahrenheit for each 273 feet.

Exercises

Refer to these additional facts to solve each problem.

- Water freezes at 32°F.
- One mile equals 5,280 feet.

1. If the ground temperature is 70°F, what will the air temperature be at an altitude of one mile? Round your answer to the nearest degree. 51°F

2. If the outside temperature at the altitude at which the plane is flying is 24°F below the ground temperature, what is the altitude of the plane? Round your answer to the nearest tenth of a mile.

   1.2 miles

3. A pilot notices ice on the wings of his plane, and decides to descend to an altitude with a temperature of 37°F. How many feet does the pilot descend? 1,365 feet

4. If the ground temperature is 68°F, how high in feet can a small plane fly before the outside temperature is at the freezing point?

   9,828 feet

5. Write a formula that a pilot could use to compute the temperature, t, in degrees Fahrenheit at a height of h feet above the ground when the ground temperature is G° Fahrenheit. Answers will vary.

6. Test your formula to see whether it produces the same data as shown in the table at the right. If it does not, rewrite your formula.

t	G	h (in feet)
58	80	6,000
30	56	7,000

7. A pilot knows that she can fly without ice forming on the plane's wings if the outside temperature is 37°F or higher. If she wishes to fly at 10,000 feet, what is the minimum ground temperature needed? About 74° F

◼◼GETTING STARTED

Prerequisite Quiz

Solve for y.

1. $3x + 4y = 12$ $y = -\frac{3}{4}x + 3$
2. $2x - y = 8$ $y = 2x - 8$
3. $-x + 3y = 10$ $y = \frac{1}{3}x + \frac{10}{3}$
4. $4x - 3y = -15$ $y = \frac{4}{3}x + 5$
5. $x = 2y + 3$ $y = \frac{1}{2}x - \frac{3}{2}$

Additional Answers

Try This

3.
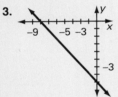

Objectives
To write an equation of a line, given its slope and y-intercept
To find the slope and y-intercept of a line, given its equation
To graph a line using its slope and y-intercept or both of its intercepts
To write the slope-intercept form of an equation of a line, given its graph

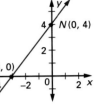

The line $\overleftrightarrow{MN}$ at the right intersects the x-axis at $(-3, 0)$ and the y-axis at $(0, 4)$. The number -3 is the *x-intercept* of the line. The number 4 is the *y-intercept* of the line.

In general, the **x-intercept** of a line is the abscissa of the point where the line intersects the x-axis. The **y-intercept** is the ordinate of the point where the line intersects the y-axis.

The slope of the line above is $\frac{4}{3}$. To write an equation of the line, you can use $m = \frac{4}{3}$ and the point $(0, 4)$ in $y - y_1 = m(x - x_1)$.

$$y - 4 = \tfrac{4}{3}(x - 0), \text{ or } y - 4 = \tfrac{4}{3}x, \text{ or } y = \tfrac{4}{3}x + 4$$

        slope y-intercept

Now consider any line that has slope m and passes through $(0, b)$. The y-intercept of the line is b.

To write an equation of the line, substitute 0 for x_1 and b for y_1 in $y - y_1 = m(x - x_1)$.

$$y - b = m(x - 0)$$
$$y - b = mx$$
$$y = mx + b$$

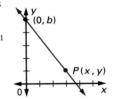

Since the slope of the line is m and the y-intercept is b, $y = mx + b$ is called the *slope-intercept form* of the equation.

Equation of a Line: Slope-Intercept Form
If a line has slope m and y-intercept b, the **slope-intercept form** of the equation of the line is $y = mx + b$.

---

424   Chapter 11 Analytic Geometry

Highlighting the Standards

Standards 8a, 5b, 5c: In the Exercises, students use the slope-intercept form to write equations and to graph them.

Additional Example 1

Write an equation of the line with slope $-\frac{1}{4}$ and y-intercept 6. $y = -\frac{1}{4}x + 6$

EXAMPLE 1   Write an equation of a line with slope $\frac{1}{2}$ and y-intercept -5.

Plan   Substitute $\frac{1}{2}$ for m and -5 for b in $y = mx + b$.

Solution $y = mx + b$

$y = \frac{1}{2}x + (-5)$

$y = \frac{1}{2}x - 5$, or $x - 2y = 10$ ⟵ Standard form: $Ax + By = C$

EXAMPLE 2   Write each equation in slope-intercept form. Identify the slope and y-intercept.

   a. $2x + 3y = 9$ **b.** $y + 3x = 0$

Plan   Solve each equation for y. The coefficient of x is the slope, and the y-intercept is the constant term.

Solutions   **a.** $3y = -2x + 9$ **b.** $y = -3x + 0$

$y = -\frac{2}{3}x + 3$ The slope is -3
                                        The y-intercept is 0.

The slope is $-\frac{2}{3}$

The y-intercept is 3

TRY THIS   1. Write an equation of a line with slope $-\frac{2}{3}$ and y-intercept 3. $y = -\frac{2}{3}x + 3$
                2. Write the equation $3x - 2y = 12$ in slope-intercept form. $y = \frac{3}{2}x - 6$

To graph a line given its y-intercept and its slope, plot the point $(0, b)$. Then use the slope to plot a second point. (A third point can be plotted as a check.)

EXAMPLE 3   Write the slope m and the y-intercept b of the line with equation $y = \frac{3}{4}x + 2$. Then graph the equation.

Solution   The slope is $\frac{3}{4}$. The y-intercept is 2.

Plot $(0, 2)$, the point of intersection of the line and the y-axis. Using the slope $\frac{3}{4}$, go right 4 and up 3. Plot $(4, 5)$. Draw the line through P and Q.

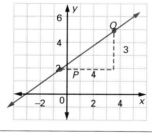

TRY THIS   3. Graph the equation $y = -\frac{1}{2}x - 4$.

11.3 Equation of a Line: Slope-Intercept Form   **425**

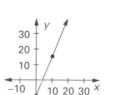

Critical Thinking Questions

Analysis: Remind students that the following are equivalent. $-\frac{2}{3} = \frac{-2}{3} = \frac{2}{-3}$
Then ask how they would interpret a slope of $\frac{2}{-3}$.   A rise of 2 and a run of -3 (3 units to the left)   Have a student demonstrate graphing the line $y = -\frac{2}{3}x + 2$ by starting at the y-intercept and finding two other points, first by using $\frac{-2}{3}$ as the slope and then by using $\frac{2}{-3}$ as the slope. Ask them what all three points have in common.   They all lie on the same line.

Common Error Analysis

Error: When solving for y in Example 2, students sometimes get $y = -\frac{2}{3}x + 9$ because they forget to divide the 9 by 3.

Emphasize that both $-2x$ and $+9$ must be divided by 3.

Try This

4.

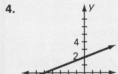

EXAMPLE 4   Find the slope-intercept form of $2x + 3y = -3$. Graph the equation.

Solution
$$2x + 3y = -3$$
$$3y = -2x - 3$$
$$y = -\frac{2}{3}x - 1 \quad \longleftarrow y = mx + b$$

Plot $P(0, -1)$, the point of intersection of the line and the y-axis. Using the slope $-\frac{2}{3}$, or $\frac{-2}{3}$, go right 3 and down 2. Plot $Q(3, -3)$. Draw the line through P and Q.

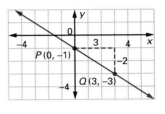

EXAMPLE 5   Write an equation, in slope-intercept form, of the line shown.

Plan   Determine the y-intercept and the slope.

Solution   The y-intercept is -4. Using $(0, -4)$ and another convenient point such as $(1, -2)$, find the slope. Go right 1, up 2. The slope is $\frac{2}{1}$, or 2. So, in slope-intercept form, an equation of the line is $y = 2x - 4$.

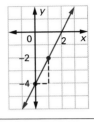

TRY THIS   4. Write $2x - 5y = -10$ in slope-intercept form. Then graph the equation. $y = \frac{2}{5}x + 2$

Equations such as $3x + 6y = 12$ can also be graphed by using intercepts.

To find the x-intercept, let $y = 0$. Then $x = 4$. To find the y-intercept, let $x = 0$. Then $y = 2$. The graph is the line through $(4, 0)$ and $(0, 2)$.

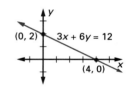

Classroom Exercises

State the slope and the y-intercept of the line with the given equation.

1. $y = 3x + 4$ $m = 3, b = 4$

2. $y = -2x + \frac{1}{3}$ $m = -2, b = \frac{1}{3}$

3. $y = -6x$ $m = -6, b = 0$

4. $y = 5$ $m = 0, b = 5$

5. $3x - y = 4$ $m = 3, b = -4$

6. $2x - 5y = -10$ $m = \frac{2}{5}, b = 2$

Write an equation of the line with the given slope m and y-intercept b.

7. $m = -\frac{2}{3}, b = 1$
$y = -\frac{2}{3}x + 1$

8. $m = 3, b = \frac{1}{2}$
$y = 3x + \frac{1}{2}$

9. $m = \frac{1}{2}, b = -2$
$y = \frac{1}{2}x - 2$

426   Chapter 11 Analytic Geometry

Additional Example 4

Find the slope-intercept form of $6x + 12y = 12$. Then graph the equation.

$y = -\frac{1}{2}x + 1$

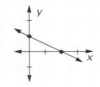

Additional Example 5

Write an equation, in slope-intercept form, for the line shown.

$y = -\frac{1}{3}x + 1$

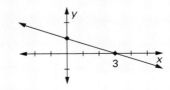

Written Exercises

Write an equation of the line with the given slope m and y-intercept b.

1. $m = 5, b = 2$ $y = 5x + 2$ **2.** $m = -3, b = 4$ **3.** $m = \frac{1}{2}, b = -4$ $y = \frac{1}{2}x - 4$

$y = -3x + 4$

4. $m = -\frac{1}{3}, b = -\frac{1}{5}$ **5.** $m = -\frac{1}{9}, b = -7$ **6.** $m = \frac{1}{5}, b = -\frac{1}{4}$

$y = -\frac{1}{3}x - \frac{1}{5}$ $y = -\frac{1}{9}x - 7$

7. $m = -4, b = \frac{4}{3}$ **8.** $m = -\frac{6}{5}, b = -\frac{1}{3}$ **9.** $m = 0, b = \frac{1}{3}$

$y = -4x + \frac{4}{3}$

$y = -\frac{6}{5}x - \frac{1}{3}$

Find the slope m and the y-intercept b.

6. $y = \frac{1}{5}x - \frac{1}{4}$

9. $y = \frac{1}{3}$ $-\frac{1}{4}; 3$

10. $3x - 2y = 8$ $\frac{3}{2}; -4$ **11.** $-5x - 3y = 12$ $-\frac{5}{3}; -4$ **12.** $x + 4y = 12$

13. $-3y + 4x = 15$ $\frac{4}{3}; -5$ **14.** $2x + 5y = -15$ $-\frac{2}{5}; -3$ **15.** $-x - y = -9$

16. $y - 3x = -10$ $3; -10$ **17.** $3x - 5y = 20$ $\frac{3}{5}; -4$ **18.** $7x - 8y = 16$

15. $-1; 9$

18. $\frac{7}{8}; -2$

Write the slope and y-intercept. Then graph the equation.

19. $y = 2x - 3$ $2; -3$ **20.** $y = -4x + 5$ $-4; 5$ **21.** $y = -3x + 1$ $-3; 1$

22. $y = \frac{2}{3}x + 2$ $\frac{2}{3}; 2$ **23.** $y = -\frac{4}{3}x - 2$ $-\frac{4}{3}; -2$ **24.** $y = \frac{3}{5}x - 4$ $\frac{3}{5}; -4$

25. $y = 3x$ $3; 0$ **26.** $y = -\frac{2}{3}x - \frac{2}{3}$ 0 **27.** $y = -4x$ $-4; 0$

Graph by using the x-intercept and the y-intercept.

28. $2x - 3y = 6$ **29.** $5x + 2y = 10$ **30.** $3y + 4x = -12$

31. $y + x = -2$ **32.** $3y - x = -6$ **33.** $3x - 5y = 15$

Write the equation in slope-intercept form. Then graph the line.

34. $3x - y = 0$ **35.** $2y + 5x = 0$ **36.** $-x + y = 0$

37. $4x - 3y = 2$ **38.** $3x + 2y = -3$ **39.** $y + 3x = 0$

For the line shown, write its equation in slope-intercept form.

40.
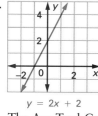
$y = 2x + 2$

41.
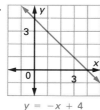
$y = -x + 4$

42.
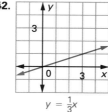
$y = \frac{1}{3}x$

43. The Ace Tool Company rents out a floor waxer for $2 per hour plus $4 for the wax. Suppose you rent a floor waxer for x hours. Write an equation that shows what the total cost y will be. $y = 2x + 4$

Checkpoint

Write an equation for the line with given slope m and y-intercept b.

1. $m = -\frac{4}{5}, b = 4$ $y = -\frac{4}{5}x + 4$

2. $m = 2, b = -\frac{2}{3}$ $y = 2x - \frac{2}{3}$

Find the slope m and the y-intercept b.

3. $2x - 5y = 10$
Slope $= \frac{2}{5}$, y-intercept: -2

4. $-7x + 3y = 14$
Slope $= \frac{7}{3}$, y-intercept: $\frac{14}{3}$

Write the equation in slope-intercept form, and graph.

5. $4x - 2y = 8$ $y = 2x - 4$

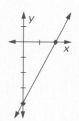

Closure

Ask students to write two equations in the slope-intercept form. Have them graph these equations on a separate piece of paper. Have each student exchange graphs with another student. After students write the equations of the exchanged graphs, have them return the graphs with equations to the original writer for checking.

Enrichment

Have students take the standard form of a linear equation, $Ax + By = C$, and solve for y. $y = \frac{-Ax}{B} + \frac{C}{B}$ Ask students to explain how this can help them identify the slope and y-intercept of an equation without changing it to another form.

The slope is found by taking the opposite of the coefficient of x and dividing by the coefficient of y. The y-intercept is the constant term divided by the coefficient of y.   Ask them to name at sight the slope and y-intercept of each of the following equations.

1. $3x + 2y = 8$ slope: $\frac{-3}{2}$; y-intercept: 4

2. $x - 4y = 11$ slope: $\frac{1}{4}$; y-intercept: $\frac{-11}{4}$

3. $x + y = 0$ slope: -1; y-intercept: 0

Guided Practice

Classroom Exercises 1–9
Try This   all

Independent Practice

A Ex. 1–27, **B** Ex. 28–46, **C** Ex. 47–51

Basic: WE 1–27 odd, 40–43 all,
Midchapter Review

Average: WE 11–45 odd, Midchapter
Review

Above Average: WE 19–51 odd,
Midchapter Review

Written Exercises

19.

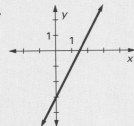

20.

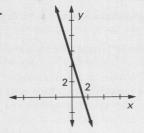

See pages 654–656 for the answers to
Written Ex. 21–39, 44, 47–51 and
Midchapter Review 13–14.

44. Graph the equation for Exercise 43. Assume that $1 \le x \le 6$. Use the graph to determine what the cost for 5 h will be. $14

Write an equation of the line described.

45. The line has a y-intercept of -5 and the same slope as the line with equation $y = -3x + 2$. $y = -3x - 5$

46. The line has a y-intercept of 3 and the same slope as the line with equation $4x + 6y = -12$. $y = -\frac{2}{3}x + 3$

Write the equation in slope-intercept form. Then graph the line.   **11.3** $y = \frac{5}{2}x - \frac{1}{2}$

47. $2x - (3 - 2y) = 8$ $y = -x + \frac{11}{2}$

48. $6 - (4 - 2y) = 5x + 1$

49. $4x - 2(3 - 3x) = 3y - 6$ $y = \frac{10}{3}x$

50. $8(-y + 2) + 6(y + 3) = 2(x - \frac{1}{2})$

51. The equation $F = \frac{9}{5}C + 32$ shows the relationship between Fahrenheit temperature F and Celsius temperature C. Draw a graph of the relation. Let Celsius readings be indicated along the horizontal axis, and Fahrenheit readings along the vertical axis. $y = -x + \frac{35}{2}$

Midchapter Review

Find the slope of $\overleftrightarrow{PQ}$ for the given points, and describe the slant of the line.   **11.1**

1. $P(-2, 3)$, $Q(8, -1)$ $-\frac{2}{5}$; down to right

2. $P(2, -1)$, $Q(-3, 9)$ -2; down to right

3. $P(6, 1)$, $Q(3, -4)$ $\frac{5}{3}$; up to right

4. $P(-4, -2)$, $Q(-5, 3)$ -5; down to right

Find the slope of the line $\overleftrightarrow{MN}$ for the given points.   **11.1**

5. $M\left(\frac{1}{4}, \frac{1}{5}\right)$, $N\left(\frac{5}{4}, \frac{4}{5}\right)$ $\frac{3}{5}$

6. $M\left(\frac{2}{3}, \frac{1}{4}\right)$, $N\left(\frac{5}{6}, -\frac{3}{4}\right)$ -6

Write an equation, in the form $Ax + By = C$, of the line passing through the given point and having the given slope.   **11.2**

7. $P(-1, 6)$; slope $= -\frac{2}{3}$ $2x + 3y = 16$

8. $P(2, 0)$; slope $= -3$ $3x + y = 6$

Write an equation, in the form $Ax + By = C$, of the line passing through the given points.   **11.2**

9. $P(0, -5)$, $Q(6, 7)$ $2x - y = 5$

10. $P(1, 3)$, $Q(7, 1)$ $x + 3y = 10$

Write an equation, in slope-intercept form, of the line with the given slope m and y-intercept b.   **11.3**

11. $m = -5$, $b = \frac{4}{3}$ $y = -5x + \frac{4}{3}$

12. $m = 0$, $b = \frac{4}{5}$ $y = 0x + \frac{4}{5}$, or $y = \frac{4}{5}$

Write the equation in slope-intercept form. Then graph the line.   **11.3**

13. $4x - 5y = 8$

14. $9x + 3y = 12$

11.4 Line Relationships

Objective To determine whether two lines are parallel or perpendicular

This photograph shows a painting by the famous artist Piet Mondrian. Notice how Mondrian used geometric shapes formed by parallel and perpendicular lines to create his art. In this lesson, you will explore relationships of lines in a coordinate plane.

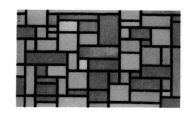

EXAMPLE 1 Graph $y - 5 = 2x$ and $y - 2x = -1$ in the same coordinate plane. Will the two graphs intersect?

Solution Write the equations in slope-intercept form. Then graph the equations using the slope and y-intercept.

$$y - 5 = 2x \qquad y - 2x = -1$$
$$y = 2x + 5 \qquad y = 2x - 1$$

The graphs are shown at the right.

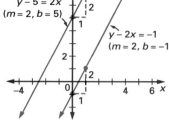

Since the lines have the same slope ($m = 2$) but different y-intercepts, the lines do not intersect. The two lines are parallel.

TRY THIS

1. Graph $4x + 2y = 6$ and $2x + y = 1$ in the same coordinate plane. Describe the graphs. The two lines are parallel.

Parallel Lines and Equal Slopes
If two lines have equal slopes, then the two lines are parallel.

Vertical lines are parallel also, as the figure at the right illustrates. Recall that the slope of a vertical line is undefined (see page 415).

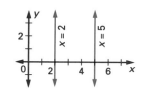

Teaching Resources

Quick Quizzes 85
Reteaching and Practice
  Worksheets 85

■■■ **GETTING STARTED**

Prerequisite Quiz

Name the slope of each line.

1. $y = 3x - 7$ 3
2. $y = -2x + 4$ -2
3. $5y = 4x$ $\frac{4}{5}$

Find the value of n.

4. $5 \times -\frac{1}{5} = n$ -1
5. $-\frac{7}{8} \times \frac{8}{7} = n$ -1
6. $-9 \times n = -1$ $\frac{1}{9}$
7. $\frac{14}{15} \times n = -1$ $-\frac{15}{14}$

Motivator

Have students graph the four points $A(3, 1)$, $B(8, 3)$, $C(6, 8)$, and $D(1, 6)$. Have them connect the points with segments to form what appears to be a square. Then have them find the slopes of segments AB, BC, CD, and DA. $\frac{2}{5}, \frac{-5}{2}, \frac{2}{5}, \frac{-5}{2}$ Show them that the opposite sides are parallel and have the same slope. Then show them that the segments that appear to be perpendicular to each other have slopes that have a product of -1.

Additional Example 1

Graph $y = \frac{1}{2}x - 2$ and $y = \frac{1}{2}x + 1$ in the same coordinate plane. Will the two graphs intersect? What is the relationship of the two lines?

Since the two graphs have the same slope ($m = \frac{1}{2}$), they will not intersect. The two lines are parallel.

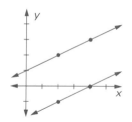

Highlighting the Standards

Standards 5b, 5c, 8b: Again in this lesson, students translate between various representations for lines and test whether or not lines are parallel or perpendicular.

429

Lesson Note

When explaining Example 1, show students that any other line with a slope of 2 will be parallel to the lines shown in Example 1. After presenting Example 2, emphasize that the slopes of non-vertical perpendicular lines are *negative reciprocals* of each other.

Math Connections

Architecture: Parallel and perpendicular lines are used in the design of buildings and other structures both to enhance the beauty of the design and for utilitarian purposes, such as outlining space for windows, storage space, and so on.

Critical Thinking Questions

Application: Have students graph the lines with equations $y = -2x - 1$ and $y = \frac{1}{2}x - 6$ in the same coordinate plane. Then have them find the equations of the two lines that pass through the point $(2, 5)$ and form a rectangle when they intersect the two given lines. $y = -2x + 9$ and $y = \frac{1}{2}x + 4$

Common Error Analysis

Error: Some students will forget that the slopes of perpendicular lines are *negative reciprocals* of each other.

Show students that the positive reciprocal of 1 is 1 and that two lines with these slopes would be parallel rather than perpendicular.

EXAMPLE 2 Graph $y = -\frac{3}{2}x + 4$ and $y = \frac{2}{3}x - 6$ in the same coordinate plane. Do the two graphs intersect? What is the relationship of the two lines?

Solution To graph $y = -\frac{3}{2}x + 4$, use the slope $-\frac{3}{2}$ and the y-intercept 4. To graph $y = \frac{2}{3}x - 6$, use the slope $\frac{2}{3}$ and the y-intercept -6.

The graphs are shown at the right. The two lines intersect. They seem to be perpendicular.

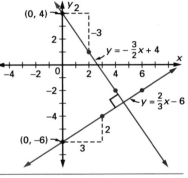

TRY THIS 2. Graph $x + 2y = 1$ and $2x - y = 4$ in the same coordinate plane. Describe the graphs. The two lines intersect. They seem to be perpendicular.

In Example 2, the lines appear to be perpendicular, and the product of their slopes is -1: $-\frac{3}{2} \cdot \frac{2}{3} = -1$. This suggests the following generalization.

> **Perpendicular lines**
> If two lines have slopes whose product is -1, then the two lines are perpendicular.

One other case of perpendicular lines occurs when one line is horizontal ($m = 0$) and the other line is vertical (m undefined).

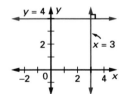

Some lines are neither parallel nor perpendicular. Lines with equations $y = -x + 1$ and $y = -\frac{1}{5}x + 1$ intersect but are not perpendicular. The product of their slopes is not -1.

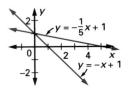

Additional Example 2

Graph $y = 2x$ and $y = -\frac{1}{2}x$ in the same coordinate plane. Do the two graphs intersect? What is the relationship of the two lines?

The two lines intersect. They seem to be perpendicular.

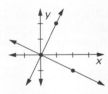

Classroom Exercises

Graph the pair of equations in the same coordinate plane. Tell whether the lines are parallel, perpendicular, or neither.

1. $y = -3x + 5$
$y = \frac{1}{3}x$
Perpendicular

2. $y = 2x - 1$
$y = -2x + 2$
Neither

3. $y = \frac{2}{5}x + 4$
$y = \frac{2}{5}x + 1$
Parallel

4. $y = 4$
$x = -2$
Perpendicular

Without drawing the graphs, tell whether the lines are parallel, perpendicular, or neither.

5. $y = -\frac{5}{3}x + 3$
$y = \frac{3}{5}x$
Perpendicular

6. $y = 9$
$y = 5$
Parallel

7. $x = 4$
$y = 1$
Perpendicular

8. $y = \frac{1}{2}x - 1$
$y = \frac{1}{2}x + 3$
Parallel

Written Exercises

Graph the pair of lines in the same coordinate plane. Tell whether the lines are parallel, perpendicular, or neither parallel nor perpendicular.

1. $y = \frac{1}{2}x$
$y = -2x$
Perpendicular

2. $y = 3x - 2$
$y = -3x + 4$
Neither

3. $y = \frac{2}{3}x - 1$
$y = \frac{2}{3}x + 5$
Parallel

4. $y = \frac{5}{4}x - 3$
$y = \frac{4}{5}x + 2$
Neither

5. $y = -\frac{5}{6}x + 1$
$y = -\frac{5}{6}x + 4$
Parallel

6. $y = -\frac{3}{4}x + 2$
$y = \frac{4}{3}x - 2$
Perpendicular

7. $y - x = 2$
$y + x = 2$
Perpendicular

8. $x - y = 5$
$x - y = 7$
Parallel

9. $x - 3y = 4$
$3x + y = 1$
Perpendicular

Without drawing the graphs, tell whether the lines are parallel, perpendicular, or neither.

10. $y = 4x$
$y = \frac{1}{4}x$ Neither

11. $y = -\frac{1}{3}x + 2$
$y = 3x - 4$ Perpendicular

12. $y = \frac{1}{5}x + 1$
$y = \frac{1}{5}x - 2$ Parallel

13. $x + 2y = 5$
$x + 3y = 5$ Neither

14. $8x - 4y = 12$
$8x + 4y = 8$ Neither

15. $y = 7$
$y + 2 = 0$ Parallel

16. $\frac{y}{10} - \frac{x}{4} = 2$
$\frac{x}{2} - \frac{y}{5} = 1$ Parallel

17. $0.02x - 0.03y = 4$
$1.5x + y = 2$
Perpendicular

18. $x - (3y + 2) = 4$
$6x + 2(y - 1) = 1$
Perpendicular

Checkpoint

Tell whether the lines are parallel, perpendicular, or neither. Do not draw the graphs.

1. $y = 5x + 4$
$y = 4x + 5$ Neither

2. $y = -\frac{2}{3}x + 3$
$y = -\frac{2}{3}x - 1$ $\parallel$

3. $y = 4x + 5$
$y = -\frac{1}{4}x + 5$ $\perp$

Closure

Ask students to write an equation of a line whose graph is parallel to the graph of $y = 3x + 8$. Any equation of the form $y = 3x + b$ Then ask them to give the equation of a line whose graph is perpendicular to the graph of $y = 3x + 8$. Any equation of the form $y = \frac{-1}{3}x + b$

▬▬FOLLOW UP

Guided Practice

Classroom Exercises 1–8
Try This   all

Independent Practice

A Ex. 1–15, **B** Ex. 16–23, **C** Ex. 24–30

Basic: WE 1–9 odd, 10–15 all, 19

Average: WE 1–23 odd

Above Average: WE 1, 5, 9, 19–24 all, 25–29 odd

Classroom Exercises

1.
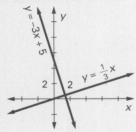

2.
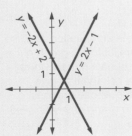

3.
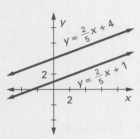

4.
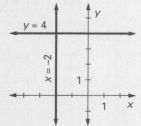

See pages 438 and 439 for the answers to Written Ex. 1–8. See page 656 for the answers to Written Ex. 9, 19 and 22–24.

432

19. Write an explanation in your own words of how to tell whether two lines in a coordinate plane are parallel, perpendicular, or neither. Answers may vary.
20. Write three equations of lines that are parallel to $\overleftrightarrow{AB}$.
21. Write three equations of lines that have the same y-intercept as $\overleftrightarrow{AB}$. Answers may vary.
22. Consider the equation $-2x + ry = 10$, where r is a positive number. Choose a value such as 5 for r and write the equation. Then let the value of r decrease and write several other equations. What happens to the slopes of the graphs as r decreases? Draw several graphs to check your answer.
23. Consider the equation $4x + 3y = k$, where k is any real number. Choose a value such as -3 for k and write the equation. Then let the value of k increase and write several other equations. What happens to the slopes of the graphs as k increases? What happens to the y-intercepts? Draw several graphs to check your answers.
24. If two or more points are points of the same line, then the points are called **collinear points**. Explain why the points A, B, and C of a coordinate plane must be collinear if $\overleftrightarrow{AB}$ and $\overleftrightarrow{BC}$ have the same slope. (HINT: If $\overleftrightarrow{AB}$ and $\overleftrightarrow{BC}$ have the same slope, then either they are parallel or they are the same line. Show that one of these cases cannot be true. Then the other case must be true.)

Use the results of Exercise 24 to determine whether or not the given points are collinear.

25. $P(0, -1)$, $Q(5, -3)$, $R(-5, 1)$ Collinear
26. $P(-1, -1)$, $Q(1, 0)$, $R(3, 1)$ Collinear
27. $P(7, 1)$, $Q(10, -1)$, $R(-7, -1)$ Not collinear
28. $P(0, 0)$, $Q(1, 2)$, $R(2, 4)$ Collinear
29. $P(-6, 1)$, $Q(-3, 1)$, $R(0, 1)$ Collinear
30. $P(8, 2)$, $Q(8, 7)$, $R(8, -1)$ Collinear

Mixed Review

Solve. 3.1, 3.3, 9.2

1. $3x - 9 = 3$ $x = 4$
2. $a + 4 = 1$ $a = -3$
3. $5y + 2 = -8$ $y = -2$
4. $-\frac{1}{2}x - 4 = 6$ $x = -20$
5. $\frac{x}{8} = \frac{5}{2}$ $x = 20$
6. $\frac{x-3}{2} = \frac{x+4}{-3}$ $x = \frac{1}{5}$
7. A track team won 8 out of 12 meets. What percent of its meets did the team win? 4.7 $66\frac{2}{3}\%$
8. The regular price of a T-shirt is $12. It is on sale for 30% off. What is the sale price? 4.7 $8.40

Enrichment

Have students write the equation of each line as described.

1. The line passes through the origin and is parallel to the graph of $2x + 3y = 5$.

   $y = -\frac{2}{3}x$

2. The line contains the point $T(3, 2)$ and is parallel to the x-axis. $y = 2$
3. The line is parallel to the y-axis and contains the point $H(3, 2)$. $x = 3$
4. The line contains the point $S(-4, -7)$ and is perpendicular to the graph of $2x - 3y = 6$. $y = -\frac{3}{2}x - 13$

11.5 Graphing Linear Inequalities

Objectives

To graph linear inequalities in two variables
To write inequalities for given graphs

You have been graphing linear equations such as $y = \frac{2}{3}x - 2$. The graph of every linear equation separates a coordinate plane into two regions called **half-planes**. Every point in the plane is either on the line or in one of the half-planes. This figure shows the graph of $y = \frac{2}{3}x - 2$ and the half-planes related to it.

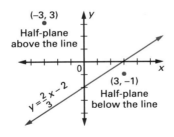

(−3, 3)
Half-plane above the line

$y = \frac{2}{3}x - 2$

(3, −1)
Half-plane below the line

For any point with coordinates (x, y), one of these relationships must be true.

$$y = \frac{2}{3}x - 2 \qquad y > \frac{2}{3}x - 2 \qquad y < \frac{2}{3}x - 2$$

For example, for $(-3, 3)$, $y > \frac{2}{3}x - 2$ is true.

$3 > \frac{2}{3}(-3) - 2$
$3 > -2 - 2$
$3 > -4$ True

For $(3, -1)$, $y < \frac{2}{3}x - 2$ is true.

$-1 < \frac{2}{3}(3) - 2$
$-1 < 2 - 2$
$-1 < 0$ True

The set of all points for which $y > \frac{2}{3}x - 2$ is true is called the **open half-plane** above the line. The set of all points for which $y < \frac{2}{3}x - 2$ is true is called the open half-plane below the line. The set of all points for which $y = \frac{2}{3}x - 2$ is true is the **boundary line**.

A half-plane that includes the boundary line is called a **closed half-plane**. In the graph of an inequality, a solid boundary line indicates that the boundary is part of the graph. A dashed line is used to indicate that the boundary is *not* part of the graph.

11.5 Graphing Linear Inequalities **433**

■■■GETTING STARTED

Prerequisite Quiz

Tell whether the inequality is true for $x = -2$ and $y = 5$.

1. $y < 3x + 7$ F
2. $y = 3x + 7$ F
3. $y > 3x + 7$ T

Solve for y.

4. $x + 3y > 15$ $y > -\frac{1}{3}x + 5$
5. $2x - y \le 3$ $y \ge 2x - 3$

Motivator

Ask students to write an inequality for this statement, using x to represent amount of income and y for the amount spent on entertainment.

Many teenagers spend at least $\frac{1}{3}$ of their income on entertainment. $y \ge \frac{1}{3}x$

Have students graph the line $y = \frac{1}{3}x$, and then locate the points (10, 4), (15, 5) and (20, 5) on the line. Suggest that the coordinates of these points could stand for amounts in dollars and have students check which of these points make the original inequality true. Have them tell where these points are in relationship to the line they have drawn.   (10, 4): true; above line; (15, 5): true; on line; (20, 5): false; below line

Highlighting the Standards

Standards 5b, 5c, 6c: This lesson adds inequalities and half-planes to the topic of graphing.

Lesson Note

As you discuss the graph of $y = \frac{2}{3}x - 2$ and the half-planes associated with it on page 433, use the Trichotomy Property, $a < b$, $a = b$, or $a > b$. When discussing the choice of test points in Example 1 in order to determine whether a shaded region is the correct graph of an inequality, emphasize that at least two test points must be used— one inside the shaded region and the other outside the shaded region. Be sure that students understand which check is false and which is true and why.

Additional Answers

Try This

1.

2.

3.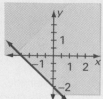

4.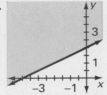

5.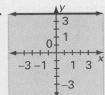

6.

Summary

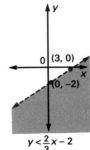

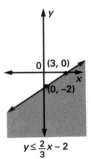

$y > \frac{2}{3}x - 2$ $y \geq \frac{2}{3}x - 2$ $y < \frac{2}{3}x - 2$ $y \leq \frac{2}{3}x - 2$

EXAMPLE 1 Graph $y < \frac{1}{2}x + 3$.

Plan Use the slope-intercept method to graph the boundary line whose equation is $y = \frac{1}{2}x + 3$. Then shade the appropriate half-plane, and check.

Solution The boundary line has the equation $y = \frac{1}{2}x + 3$. Using the point $(0, 3)$ and the slope $\frac{1}{2}$, draw the boundary as a dashed line since it is not part of the graph. Shade the half-plane below the boundary line.

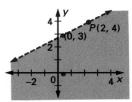

Check To check that the correct half-plane is shaded, test two points, one above the line and one below it. Choose each point so that the coordinates are easily checked.

Check $P(-2, 4)$: $y < \frac{1}{2}x + 3$ Check $O(0, 0)$: $y < \frac{1}{2}x + 3$

$4 \overset{?}{<} \frac{1}{2}(-2) + 3$ $0 \overset{?}{<} \frac{1}{2}(0) + 3$

                                                                $0 \overset{?}{<} 0 + 3$

$4 \overset{?}{<} -1 + 3$ $0 \overset{?}{<} 3$

$4 \overset{?}{<} 2$ $0 < 3$ True

$4 < 2$ False

So, the graph of $y < \frac{1}{2}x + 3$ is the half-plane below the boundary line.

TRY THIS   1. Graph $y < -2x + 1$.   2. Graph $y > \frac{2}{3}x - 1$.

Note that when you graph an inequality such as $y < \frac{1}{2}x + 3$, you are graphing the *solution set* of the inequality. The ordered pair $(0, 0)$ is

Additional Example 1

Graph $y > 2x - 1$.

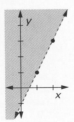

one solution of $y = \frac{1}{2}x + 3$. For every point in the half-plane below the dashed line, there is an ordered pair of numbers that is a solution of $y < \frac{1}{2}x + 3$.

EXAMPLE 2   Graph $2x - y \le 3$.

Plan   Rewrite the inequality so that y is alone on one side. Then the equation for the boundary line will be in slope-intercept form.

Solution   Multiply each side by -1 and simplify.

$$2x - y \le 3$$
$$-y \le -2x + 3$$
$$-1(-y) \ge -1(-2x + 3) \quad \longleftarrow \text{Reverse the order of the inequality.}$$
$$y \ge 2x - 3$$

Graph $y = 2x - 3$. Since $y \ge 2x - 3$ means $y > 2x - 3$ *or* $y = 2x - 3$, the boundary line is part of the graph. So, use a solid line for the boundary. Shade the half-plane above the line.

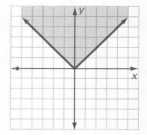

Check   Check the coordinates of two test points in the *original inequality*.

Check $O(0, 0)$.

$$2x - y \le 3$$
$$2(0) - 0 \overset{?}{\le} 3$$
$$0 - 0 \overset{?}{\le} 3$$
$$0 \overset{?}{\le} 3$$
$$0 < 3 \quad \text{True}$$

Check $Q(3, -2)$.

$$2x - y \le 3$$
$$2(3) - (-2) \overset{?}{\le} 3$$
$$6 + 2 \overset{?}{\le} 3$$
$$8 \overset{?}{\le} 3$$
$$8 \le 3 \quad \text{False}$$

So, the graph of $2x - y \le 3$ is the boundary line and the half-plane above the line.

TRY THIS   3. Graph $x + y \ge -2$.   4. Graph $2y - x \ge 4$.

In the next example, you will consider a horizontal line and a vertical line, and the half-planes associated with each line. Observe that in the case of a vertical line, the half-planes are to the right and left of the line, not above and below it.

11.5 Graphing Linear Inequalities   **435**

Math Connections

Life Skills: When qualifying for a loan to buy a house, many banking institutions require that the resulting house payment be no more than a certain percent of one's gross income. For example, it may be required that the payment be no more than 8% of gross monthly income.

Critical Thinking Questions

Application: Have students draw the graph of $y = |x|$ by substituting several negative and positive values for x. Then have them shade the graph appropriately to show $y \ge |x|$ and check several points to verify their solution.

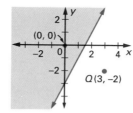

Common Error Analysis

Error: Students shade the incorrect side of the line when graphing an inequality.

Have students always check two points as a minimum.

Additional Example 2

Graph $3x - y < 3$.

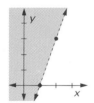

435

Graph.

1. $y \geq \frac{1}{2}x - 1$
2. $2x + 3y > 6$
3. $x \leq 2$
4. $y \geq -1$

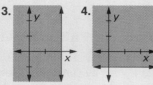

Closure

Have students tell whether the solutions of each inequality will lie to the left or to the right of the graph of $x = 3$, and whether the points on the graph of $x = 3$ are also included.

1. $x \leq 3$ To the left of, and including
2. $3 < x$ To the right of, not including
3. $3 \leq x$ To the right of, and including
4. $x < 3$ To the left of, and not including

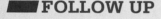**FOLLOW UP**

Guided Practice

Classroom Exercises 1–12
Try This   all

Independent Practice

A Ex. 1–16, **B** Ex. 17–29, **C** Ex. 30–34

Basic: WE 1–17 odd, 25–28 all, Problem Solving Strategies, Mixed Problem Solving

Average: WE 5–29 odd, Problem Solving Strategies, Mixed Problem Solving

Above Average: WE 9–34 odd, Problem Solving Strategies, Mixed Problem Solving

---

EXAMPLE 3   Graph each inequality.

a. $y \geq -2$ b. $x < 3$

Solutions   a. Graph $y = -2$. Since the inequality symbol is $\geq$, the boundary line is part of the graph. So, use a solid line for the boundary. Shade the half-plane above the line.

b. Graph $x = 3$ with a dashed line. Since the x-coordinate of every point to the left of the line satisfies $x < 3$, shade the half-plane to the left of the line.

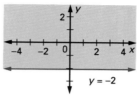
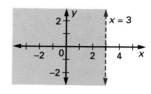

So, the graph of $y \geq -2$ is the line and the half-plane above the line.

So, the graph of $x < 3$ is the half-plane to the left of the boundary line.

EXAMPLE 4   Write an inequality of the graph shown at the right.

Solution   The boundary line passes through $(0, 4)$ and has slope 1. So, an equation for the line is $y = x + 4$. Since the boundary line is solid, it is part of the graph. The region above the line is shaded. Thus, $y \geq x + 4$ is an inequality for the graph.

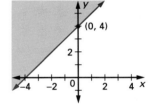

TRY THIS   5. Graph $y \leq 4$.   6. Graph $x \geq -1$.

Classroom Exercises

State whether the shaded half-plane will be above, below, to the left, or to the right of the boundary line. State whether the boundary line will be solid or dashed.

Above; solid	Below; dashed	Left; dashed	Below; dashed
1. $y \geq 2x - 1$	2. $y < 3x - 4$	3. $x < -3$	4. $y < 2$
5. $y > 4$	6. $x \geq -2$	7. $y \leq 8x - 7$	8. $y < -x + 3$
Above; dashed	Right; solid	Below; solid	Below; dashed

Graph each inequality.

9. $y \geq -2x + 1$ 10. $y > -3x - 2$ 11. $x \leq 1$ 12. $y \geq -1$

Additional Example 3

Graph each inequality.

a. $x \geq 2$
b. $y < -1$

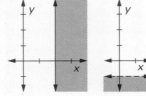

Additional Example 4

Write an inequality for the graph shown below.

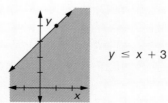

$y \leq x + 3$

Written Exercises

Graph each inequality.

1. $y > x + 3$ **2.** $y < x + 2$ **3.** $y \geq x + 1$ **4.** $y \leq x - 3$

5. $y \geq 2x - 1$ **6.** $y \leq 2x + 3$ **7.** $y < 4$ **8.** $y > 5$

9. $y \leq -3$ **10.** $x > 1$ **11.** $x \geq -2$ **12.** $x \leq 3$

13. $y < -\frac{3}{2}x$ **14.** $y > \frac{1}{2}x - 3$ **15.** $y \geq -\frac{1}{5}x - 1$ **16.** $y < -\frac{2}{5}x + 2$

17. $2x + 3y < 9$ **18.** $3x - 2y \leq 8$ **19.** $-3x - 5y \leq 5$ **20.** $-2x + 4y > 8$

21. $-6x - y \leq -1$ **22.** $8x - 3y < -4$ **23.** $-x - y \geq -3$ **24.** $2x - 4y > 10$

Write an inequality for the graph.

25.

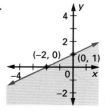

26.

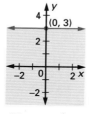

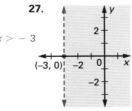

$y \leq \frac{1}{2}x + 1$

$y \leq 3$

27.

$x > -3$

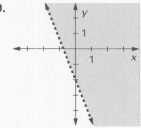

For each sentence, write an inequality. Use x and y as the variables. Tell what each variable represents.

28. The majority of American families save less than $\frac{1}{10}$ of their income. Let y = savings and x = income. $y < \frac{1}{10}x$.

29. The perimeter of a quadrilateral is greater than or equal to 4 times the length of its shortest side. Let y = perimeter and x = shorter side. $y \geq 4x$

Graph the inequality in a coordinate plane.

30. $|x| > 2$ **31.** $|y| > 3$ **32.** $|x| < 1$ **33.** $|y| \leq 1$ **34.** $|y - 2| > 0$

Mixed Review

Simplify.   **6.2, 6.3**

1. $(-3x^2y)^4$ $81x^8y^4$ **2.** $-5a^3(-2a^4b^2)^3$ $40a^{15}b^6$ **3.** $\dfrac{-18a^3b^2c^6}{9a^5bc^2}$ $-\frac{2bc^4}{a^2}$

Factor completely.   **7.5, 7.6**

4. $a^2 - 36$ **5.** $81x^2 - 16$ **6.** $x^3 + x^2 - 2x$

$(a - 6)(a + 6)$ $(9x + 4)(9x - 4)$ $x(x + 2)(x - 1)$

Solve for x.   **9.3**

7. $-5x + b = x - 4b$ $x = \frac{5b}{6}$ **8.** $\dfrac{1}{b} - \dfrac{1}{a} = \dfrac{1}{x}$ $\frac{ab}{a - b}$

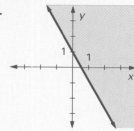
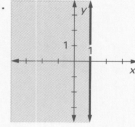
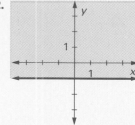
Enrichment

Have students experiment with various values of x and y to draw a graph of $|x| + |y| \leq 4$.

Then have them write equations for the four lines which intersect to form the boundary (square) of this figure. $y = x + 4$, $y = x - 4$, $y = -x + 4$, $y = -x - 4$

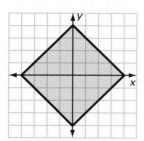

1.
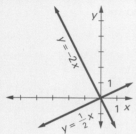

2.
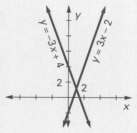

3.
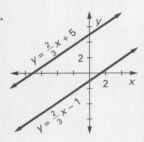

4.
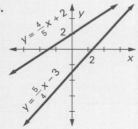

Problem Solving Strategies

Using Logical Reasoning

Ronald mows lawns for three families. The three families live in houses at points A, B, and C as shown on the street grid at the right. Ronald plans to move into a house that is as close as possible to all three houses.

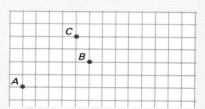

Notice that Ronald doesn't necessarily want to be at an equal distance from each of the three houses. Rather, he wants the shortest possible total distance for the three trips he makes from his house to the lawns he mows. The route to the houses must be along horizontal or vertical streets. No diagonal shortcuts are allowed! That is, if Ronald's house is at R, he wants T, the total distance, to be as small as possible.

$T = $ (dist. from R to A) + (dist. from R to B) + (dist. from R to C)

To help Ronald decide where to live, you might use the strategy of *Guessing and Checking* or you might try this strategy.

Place R at some point. Then find T. Next consider moving one block north. Will this shorten the distance from R to A? From R to B? From R to C? Whenever you get more *yes* answers than *no* answers, make the move. From the new position of R, consider the same questions. If there are two *yes* answers to the one *no* answer, make the move. Eventually, when there are only *yes* answers, you will find the best possible location for R.

Problems such as this have many practical applications, such as finding locations for schools, airports, and telephone lines.

Exercises

1. Copy the grid and find the best location for Ronald's house. Call the lower left corner the origin, $(0, 0)$ and give the ordered pair that names the position of R. (5, 3)

2. After Ronald moves to his new house, his customer at point B moves 7 blocks away. Ronald thought that he might have to move again, but he checked the map and decided that it wasn't necessary. Give the ordered pair that names B's new location. (13, 3)

3. This new strategy will not work if you are trying to find the shortest possible distance to an even number of points. Explain why. You could fail to get more *yes* than *no* answers.

Mixed Problem Solving

Making a table is an effective strategy to use in solving some problems. Making a table helps to organize the given information and also helps you to identify the relationships that can be expressed as an equation.

Solve each problem.

1. How fast must a bus travel to cover a distance of 150 mi in $2\frac{1}{2}$ h? 60 mi/h

2. Flying at 420 mi/h, how long does it take a plane to travel 1,420 mi? 3.38 h

3. The width of a rectangle is 10 cm less than the length. The perimeter is 52 cm. Find the length and width. 18 cm, 8 cm

4. Juan worked 12 h longer than Jeanine. Together they worked 42 h. How long did each work? Jeanine: 15 h; Juan: 27 h

5. In a basketball game, Kyle's score was 6 less than 3 times Craig's score. Together they scored 42 points. Find their scores. Craig: 12 pts; Kyle: 30 pts

6. A television repair person charges $30 to make a house call plus $42 an hour for each hour she works. Her bill for a job was $198. How many hours did she work? 4 h

7. The sum of two consecutive odd integers is 144. Find the integers. 71, 73

8. Find three consecutive even integers whose sum is 90. 28, 30, 32

9. If $10,000 is invested at a simple interest rate of 7.4% per year for 4 years, how much interest is earned? $2,960

10. Lincoln Memorial High School has 2,750 students. If 6% are absent, how many students are absent? 165

11. Mary's wage was increased by 0.12 of last year's wage, which was $12.50 per hour. What does she earn now? $14/h

12. The sum of a number and 0.6 of the number is 38.88. Find the number. 24.3

13. Six less than 7 times a number is the same as the square of the number. Find the number. 1 or 6

14. The length of a rectangle is 2 cm more than 3 times its width. The area is 33 cm². Find its dimensions. 11 cm, 3 cm

15. Find two consecutive even integers whose product is 168. 12, 14 or −12, −14

16. Find two consecutive odd integers whose product is 255. 15, 17 or −15, −17

17. Jim walked along a path at 4 mi/h. Rachel left from the same point as Jim $\frac{1}{2}$ h later, and ran along the same path at 9 mi/h. How long did she run before catching up with Jim? 24 min

18. It would take Steve 14 h to build some stairs. Helen would take 18 h to build the same stairs. How long would it take them to build the stairs if they worked together? $7\frac{7}{8}$ h

19. The cost of gasoline varies directly as the number of gallons. If 3 gallons cost $3.66, find the cost of 14 gallons of gasoline. $17.08

20. The volume of a gas varies inversely as the pressure. If the volume is 55 in³ under 25 lb/in² of pressure, find the volume under 6 lb/in² of pressure. 229.17 in³

5.
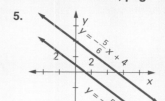

6.
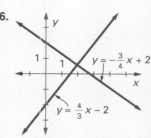

7.
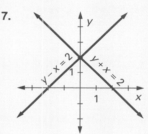

8.
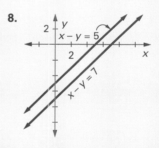

4. $-\frac{5}{9}$; down to right

5. -1; down to right

6. 0; horiz

14.

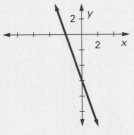

15.

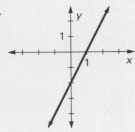

Key Terms

boundary line (p. 433)
closed half-plane (p. 433)
half-plane (p. 433)
open half-plane (p. 433)
parallel lines (p. 429)
perpendicular lines (p. 430)
point-slope form (p. 418)

rise (p. 413)
run (p. 413)
slope (p. 414)
slope-intercept form (p. 424)
x-intercept (p. 424)
y-intercept (p. 424)

Key Ideas and Review Exercises

11.1 To find the slope of a line, use

$$\text{slope} = \frac{\text{change in } y\text{-coordinates}}{\text{change in } x\text{-coordinates}}, \text{ or } m = \frac{y_2 - y_1}{x_2 - x_1}, \text{ where } x_1 \neq x_2.$$

If the slope is positive, the line slants up to the right.
If the slope is negative, the line slants down to the right.
If the slope is zero, the line is horizontal.
If the slope is undefined, the line is vertical.

Find the slope of the line determined by the given points A and B.

1. $A(3, 4)$, $B(-7, 6)$ $-\frac{1}{5}$ **2.** $A(-2, -3)$, $B(8, -5)$ $-\frac{1}{5}$ **3.** $A(5, -1)$, $B(5, 4)$ Undef

Find the slope of $\overleftrightarrow{MN}$ for the given points, and describe the slant of the line.

4. $M(4, -2)$, $N(-5, 3)$ **5.** $M(7, 4)$, $N(6, 5)$ **6.** $M(-6, -5)$, $N(7, -5)$

11.2 To write an equation of a line, given one of its points and its slope, use $y - y_1 = m(x - x_1)$, where the given point is (x_1, y_1) and the given slope is m.

To write an equation of a line, given two of its points, find the slope m. Then use $y - y_1 = m(x - x_1)$, where (x_1, y_1) is either of the two given points.

Write an equation, in the form $Ax + By = C$, of the line passing through the given point and having the given slope.

7. $P(4, 3)$; slope $= \frac{2}{5}$
$2x - 5y = -7$

8. $P(5, -3)$; slope $= -3$
$3x + y = 12$

9. $P(1, 6)$; slope $= 0$
$0x + y = 6$, or $y = 6$

Write an equation, in the form $Ax + By = C$, of the line passing through the given points.

10. $P(7, 1)$, $Q(8, -6)$
$7x + y = 50$

11. $P(-5, 6)$, $Q(-2, 7)$
$x - 3y = -23$

12. $P(4, 1)$, $Q(-5, -2)$
$x - 3y = 1$

11.3   To write an equation of a line, given its slope and y-intercept, use the form $y = mx + b$, where m is the slope and b is the y-intercept.

To find the slope m and the y-intercept b, given the equation of a line, write it in the form $y = mx + b$. Then identify m and b.

To graph an equation using its intercepts, let $y = 0$ and solve for x, and let $x = 0$ and solve for y. Then draw a line through the two points.

To graph a line using its slope and y-intercept, write the equation in slope-intercept form, $y = mx + b$. Plot the point $(0, b)$. Use the slope to plot a second point. Then draw a line through the two points.

13. Write an equation of the line with slope $-\frac{3}{4}$ and y-intercept 6.

Find the slope m and the y-intercept b. $y = -\frac{3}{4}x + 6$

14. Graph the equation $3x + y = -6$ using the x-intercept and the y-intercept.

15. Graph the equation $-2x + y = -2$ using the slope and y-intercept.

11.4   Two lines are parallel if their slopes are the same and their y-intercepts differ, or if both lines are vertical (slope undefined).

Two lines are perpendicular if their slopes have a product of -1, or if one line is vertical (slope undefined) and the other is horizontal (slope $= 0$).

Tell whether the lines are parallel, perpendicular, or neither.

16. $y = \frac{1}{4}x + 6$ Perpendicular
$y = -4x + 1$

17. $y = \frac{1}{2}x + 3$ Parallel
$y = \frac{1}{2}x$

18. Perpendicular
$y = x - 3$
$y = -x + 5$

19. $y + 3x = -3$ Neither
$y - 3x = 0$

20. $2x + 3y = 4$ Perpendicular
$3x - 2y = 6$

21. $y - 3 = 0$
$y = 8$ Parallel

11.5   To graph a linear inequality in two variables, first graph the related equation as a boundary line. Draw this as a dashed line if it is not part of the graph. Then shade the half-plane to the left or right, or above or below the boundary line.

Graph the inequality.

22. $y > -x + 1$

23. $3x - 4y \geq 12$

24. $y < -1$

25. Write an explanation in your own words of how to write an inequality for the graph at the right. Refer to the equation of the boundary line and to the shaded half-plane.

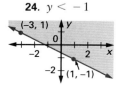

Chapter 11 Review   **441**

22.

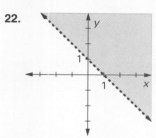

23.

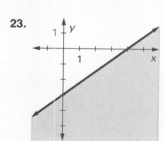

24.

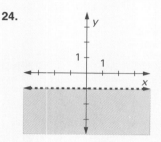

25. Sample answer: The slope of the line is $m = \frac{-1-1}{1+3} = -\frac{1}{2}$. By using the equation $y - y_1 = m(x - x_1)$ the equation of the boundary line can be found: $y + 1 = -\frac{1}{2}(x - 1)$, or $y = -\frac{1}{2}x - \frac{1}{2}$. The shaded half-plane is above the boundary line. Thus, the inequality for the graph is $y \geq -\frac{1}{2}x - \frac{1}{2}$.

10.

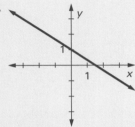

11.

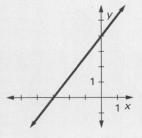

12.

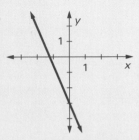

17.

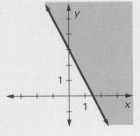

See page 443 for the answer to Ex. 18.

Chapter 11 Test

A Exercises: 1–10, 13–15, 17   B Exercises: 11, 12, 15, 16, 18, 19

C Exercises: 20–21

Find the slope of the line determined by the given points M and N.

1. $M(-1, 8)$, $N(7, -9)$ $-\frac{17}{8}$
2. $M(10, -2)$, $N(1, -4)$ $\frac{2}{9}$
3. $M(4, 1)$, $N(-6, 1)$ 0
4. $M(3, 5)$, $N(3, -2)$ Undef
5. Find the slope of $\overleftrightarrow{AB}$ for $A(-1, 4)$ and $B(5, 6)$. $\frac{1}{3}$

Write an equation, in the form $Ax + By = C$, for the line passing through the given point and having the given slope.

6. $P(-4, -3)$; slope $= -\frac{5}{4}$ $5x + 4y = -32$
7. $P(1, -1)$; slope $= \frac{4}{3}$ $4x - 3y = 7$

Write an equation, in the form $Ax + By = C$, of the line passing through the given points.

8. $P(-3, 4)$, $Q(-1, 8)$ $2x - y = -10$
9. $P(-2, -3)$, $Q(9, -4)$ $x + 11y = -35$
10. Write the slope and the y-intercept of the line with equation $y = -\frac{2}{3}x + 1$. Then graph the equation. $m = -\frac{2}{3}$; y-intercept: 1
11. Graph $4x - 3y = -12$ using the x-intercept and the y-intercept.
12. Write the equation $2y + 5x = -6$ in slope-intercept form, and graph. $y = -\frac{5}{2}x - 3$

Tell whether the lines are parallel, perpendicular, or neither.

13. $y = 6x - 12$ Parallel
    $y = 6x - 3$
14. $y - 2x = 8$ Neither
    $y = -2x - 2$
15. $y = \frac{1}{5}x$ Perpendicular
    $y = -5x + 2$
16. $5y = 3x - 5$ Perpendicular
    $3y + 5x = 2$

Graph the inequality.

17. $y \geq -2x + 3$
18. $3x - 2y > 8$
19. Write an inequality for the graph at the right. $y < \frac{2}{3}x - \frac{1}{3}$
20. Write an equation, in the form $Ax + By = C$, of the line with y-intercept 3 and slope the same as that of the line with equation $2x - 3y = 6$. $2x - 3y = -9$
21. Use slopes to determine whether points $P(-1, 6)$, $Q(8, -9)$, and $R(5, -4)$ are collinear.
    P, Q, and R are collinear.

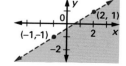

18.

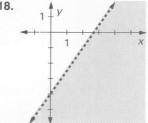

In each Exercise, you are to compare a quantity in Column 1 with a quantity in Column 2. Write the letter of the correct answer from these choices.

A—The quantity in Column 1 is greater than the quantity in Column 2.
B—The quantity in Column 2 is greater than the quantity in Column 1.
C—The quantity in Column is equal to the quantity in Column 2.
D—The relationship cannot be determined from the given information.

Note: Information centered over both columns refers to one or both of the quantities to be compared.

Column 1	Column 2	Column 1	Column 2

1. $a \neq 0$, $b \neq 0$

$$\frac{a}{b} - 1 \qquad \frac{b}{a} - 1 \quad \text{D}$$

4. Triangle ABC is a right triangle.

x $\qquad$ 30 $\quad$ C

2. $x < 0$

$$x^3 + x - 3 \qquad 0 \quad \text{B}$$

5.

$$x(y + x) \qquad xy + z \quad \text{D}$$

3. $\dfrac{10^n}{10^5} > 1$

6. $a > 0$, $b < 0$

$n \qquad 5 \quad \text{A}$

$b - a \qquad |b| + a \quad \text{B}$

For Exercises 7–11, choose the letter of the correct answer.

7. If $y = x + 3$, what is the value of $(x - y)^3$? A
(A) -27 (B) 27 (C) 9
(D) -9 (E) 6

8. If $x = 3$, which of the following is not an odd integer? E
(A) $3x + 4$ (B) $x - 4$ (C) $-x$
(D) $x^2 + 2$ (E) $6x - 10$

9. If $2 - 3y = -13$ and $x + y = 3$, what is the value of x^3? D
(A) 125 (B) -125 (C) 8
(D) -8 (E) 3

10. Solve $\dfrac{1 - \dfrac{y}{x}}{8} = 3$ for x. C
(A) $-y - 24$ (B) $y + 24$
(C) $-\dfrac{y}{23}$ (D) $23y$ (E) None of these

11. If a wheel rotates 180 times each minute, how many degrees does it rotate in 10 s? (HINT: There are 360 degrees in one rotation.) E
(A) 30 (B) 360 (C) 1,080
(D) 5,400 (E) 10,800

OVERVIEW

This chapter gives students three different methods for solving a system of linear equations. Although the graphing method is presented first, the emphasis in the rest of the chapter is on the substitution and addition methods of solution. The student then has the opportunity to apply these skills to the solution of digit, age, coin, mixture, and motion problems. In the final lesson, students graph systems of inequalities which are then used in an application on linear programming.

OBJECTIVES

- To solve linear systems of equations in two variables by graphing, the substitution method, and the addition method
- To solve word problems using two equations and two variables
- To solve linear systems of inequalities in two variables by graphing

PROBLEM SOLVING

The primary goal of this chapter is to have students apply systems of equations in problem solving. In the first lesson, students apply the strategy of Making a Graph to solve a system. Then two other methods of solution are presented in order to encourage students to Find More Than One Way to solve problems. Making a Table is particularly useful for solving age and motion problems. Drawing a Diagram and Using a Formula are two other strategies that are useful in solving motion problems. All of these types of problems require the student to Write a Pair of Equations for their solution. In this chapter, students also apply the strategy of Working Backwards to solving problems involving distance. In the Application on pages 486–487, students use the strategies of Making a Table, Writing Equations, and Testing Conditions to solve linear programming problems.

READING AND WRITING MATH

In Exercises 33 and 41 on page 450, students are asked to explain how graphing can be used to determine the number of solutions of a system and when graphing is not a good method of solution. In Exercise 28 on page 463, students are asked to explain why the addition method will not work for some types of systems. The Focus on Reading on page 465 asks students to list the steps for solving a system in the correct sequence.

TECHNOLOGY

Calculator: The system of equations in Exercise 37 on page 467 provides a good opportunity for using a calculator in solving a system. Throughout the chapter, the calculator can be used to check student solutions to many types of problems.

SPECIAL FEATURES

Mixed Review pp. 450, 455, 459, 463, 471, 474, 478, 482, 485
Problem Solving: Working Backwards p. 451
Focus on Reading p. 465
Midchapter Review p. 467
Application: Linear Programming pp. 486–487
Key Terms p. 488
Key Ideas and Review Exercises pp. 488–489
Chapter 12 Test p. 490
College Prep Test p. 491
Cumulative Review (Chapters 1–12) pp. 492–493

PLANNING GUIDE

Lesson	Basic	Average	Above Average	Resources
12.1 pp. 448–451	CE 1–12 WE 1–8 all, 9–23 odd Problem Solving Strategies	CE 1–12 WE 1–33 odd Problem Solving Strategies	CE 1–12 WE 9–41 odd Problem Solving Strategies	Reteaching 87 Practice 87
12.2 pp. 454–455	CE 1–8 WE 1–19 odd	CE 1–8 WE 9–27 odd	CE 1–8 WE 17–39 odd	Reteaching 88 Practice 88
12.3 pp. 457–459	CE 1–4 WE 1–15 odd	CE 1–4 WE 7–21 odd	CE 1–4 WE 11–25 odd	Reteaching 89 Practice 89
12.4 pp. 462–463	CE 1–9 WE 1–19 odd	CE 1–9 WE 11–29 odd	CE 1–9 WE 17–37 odd	Reteaching 90 Practice 90
12.5 pp. 466–467	FR all, CE 1–8 WE 1–19 odd, 25, 27 Midchapter Review	FR all, CE 1–8 WE 9–31 odd Midchapter Review	FR all, CE 1–8 WE 13–35 odd, 36, 37 Midchapter Review	Reteaching 91 Practice 91
12.6 pp. 470–471	CE 1–8 WE 1–10 all	CE 1–8 WE 4–13 all	CE 1–8 WE 1–13 odd, 15–18 all	Reteaching 92 Practice 92
12.7 pp. 473–474	CE 1–8 WE 1–8 all	CE 1–8 WE 5–12	CE 1–8 WE 5–15 odd	Reteaching 93 Practice 93
12.8 pp. 477–478	CE 1–8 WE 1–6 all	CE 1–8 WE 1–11 odd	CE 1–8 WE 5–15 odd	Reteaching 94 Practice 94
12.9 pp. 481–482	CE 1–8 WE 1–6 all	CE 1–8 WE 1–11 odd	CE 1–8 WE 5–15 odd	Reteaching 95 Practice 95
12.10 pp. 485–487	CE 1–9 WE 1–13 odd	CE 1–9 WE 5–17 odd Application	CE 1–9 WE 11–23 odd Application	Reteaching 96 Practice 96
Chapter 12 Review pp. 488–489	all	all	all	
Chapter 12 Test p. 490	all	all	all	
College Prep Test p. 491	all	all	all	
Cumulative Review (Chapters 1–12) pp. 492–493	all	all	all	

CE = Classroom Exercises   WE = Written Exercises   FR = Focus on Reading
Note: For each level, all students should be assigned all Try This and all Mixed Review Exercises.

■■ **INVESTIGATION**

Manipulative: In this *Investigation,* students locate points that satisfy given inequalities.

Materials: Each student will need a copy of Manipulative Worksheet 12, two pens or pencils of different colors, and a straightedge.

This manipulative allows students to combine what they have learned about inequalities, compound conditional statements, and graphing lines to determine the location of all points in the coordinate plane that satisfy a system of inequalities. The emphasis is on finding specific points that satisfy the given conditions and then on locating the region that contains these points and the lines that act as boundaries.

Encourage students to work slowly and carefully so that their results will give them a clear picture of the desired results.

As a follow-up activity, ask students to consider the following two inequalities.

$$y < x + 5 \text{ and } y > x + 1$$

1. Name 3 ordered pairs that will satisfy one inequality *or* the other. Characterize these points.   Sample points: $(-2, 2)$, $(0, 4)$, $(2, 3\frac{1}{2})$; Points between the parallel lines $y = x + 5$ and $y = x + 1$.

2. Name 3 ordered pairs that will satisfy both the first and the second inequalities. Characterize these points.   Points between the parallel lines $y = x + 5$ and $y = x + 1$.

Navaho Indians, famous for handwoven blankets and beautifully crafted silver and turquoise jewelry, originally learned to weave from the neighboring Pueblo Indians. Today, however, they create their own unique geometric designs.

More About the Photo

In the Navaho designs for weaving, you can find various lines of symmetry, tessellations, proportional figures, and reflected patterns. There are about 140,000 Navahos, making them the largest group of Native Americans. The Navaho reservation in Arizona, New Mexico, and Utah, consists of 14 million acres, about the size of the state of West Virginia. This homeland, selected by their ancestors, is of great importance to the Navaho, as is their large extended family or clan, defined as all the people related through the female relatives.

12.1 Systems of Equations— Graphing

Objectives

To solve systems of two linear equations by graphing
To determine the number of solutions of a system of two linear equations

A storekeeper has \$120 to give as bonus money to his two workers, Jean and Karl. If Jean gets j dollars and Karl gets k dollars, then

$$j + k = 120.$$

That year, Jean worked 3 times as many hours as Karl, so Jean will get 3 times as much bonus money as Karl. Thus,

$$j = 3k.$$

These two linear equations in two variables form a **system of linear equations.**

$$j + k = 120$$
$$j = 3k$$

Each equation has a solution set of ordered pairs. For example, $(20, 100)$ is a solution of $j + k = 120$, since $20 + 100 = 120$, and $(30, 10)$ is a solution of $j = 3k$, since $30 = 3 \cdot 10$.

The ordered pair that makes *both* equations true is the **solution of the system.**

The solution can be found by graphing both equations in the same coordinate plane. The graphs of $j + k = 120$ and $j = 3k$ meet at a point with coordinates that appear to be $(90, 30)$. If this ordered pair satisfies both equations, then it is a solution of the system.

$$j + k = 120$$
$$j = 3k$$

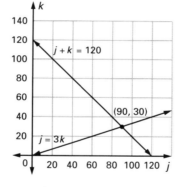

Check:

$j + k = 120$	$j = 3k$
$90 + 30 \overset{?}{=} 120$	$90 \overset{?}{=} 3(30)$
$120 = 120$ True	$90 = 90$ True

Thus, Jean gets \$90 and Karl gets \$30.

Teaching Resources

Application Worksheet 12
Manipulative Worksheet 12
Quick Quizzes 87
Reteaching and Practice
  Worksheets 87
Transparency 35
Computer Investigation,
  p. 635

GETTING STARTED

Prerequisite Quiz

Give the slope, *m*, and the *y*-intercept, *b*, of the graph of each equation.

1. $y = 6x + 7$ $m = 6; b = 7$
2. $y = -\frac{1}{2}x - 3$ $m = -\frac{1}{2}; b = -3$
3. $y = -\frac{2}{3}x$ $m = -\frac{2}{3}; b = 0$
4. $y = -4$ $m = 0; b = -4$
5. $2x + y = 1$ $m = -2; b = 1$
6. $-3x + 2y = 4$ $m = \frac{3}{2}; b = 2$

Motivator

Ask students to give ordered pairs of numbers that are solutions for the equation $y = 2x - 3$. Then ask what is true of all of these solutions when they are graphed.   They all lie on the same line. Ask students to give ordered pairs that are solutions for $y = -\frac{1}{2}x + 2$. Then ask them to find an ordered pair that is a solution for both of these equations.   (2, 1)

Highlighting the Standards

Standards 5c, 6b, 6c, 1a: Students solve a pair of equations, represent them as graphs, and translate between the two forms. In Problem Solving Strategies, they practice the strategy of working backwards.

Lesson Note

Point out that the axes in the graph on page 445 are marked in intervals of 20 rather than 1. This is to accommodate the larger numbers. When presenting Example 1, emphasize that the two lines intersect because they have different slopes. In Example 3, be sure students understand that there is an infinite number of solutions for the pair of equations, and not just one pair of numbers. Remind students that one solution to a system of linear equations indicates intersecting lines; no solution indicates parallel lines; and infinitely many solutions indicate that the graphs are the same line.

Math Connections

Analytic Geometry: Determining the number of solutions of a system of equations by deciding in how many points they can intersect, especially when the graphs are circles, ellipses, parabolas, and hyperbolas, is often helpful in finding the solutions.

Critical Thinking Questions

Synthesis: Ask students to consider the equations $y = 2x - 3$ and $y = -3x + 2$. Since y is equal to two different expressions, ask what would happen if they set these two expressions equal to each other? They will get an equation in one variable. Ask how this could be used to find where the two lines intersect. Solve the new equation for x, and substitute this value into one of the original equations to obtain the y-coordinate of the point of intersection.

EXAMPLE 1 Solve the system $\begin{array}{l} y = x - 1 \\ y = -x + 3 \end{array}$ by graphing.

Plan Use the slope–intercept method to graph each equation.

Solution

$y = x - 1$
slope: 1
y-intercept: -1

$y = -x + 3$
slope: -1
y-intercept: 3

The graphs intersect at $P(2, 1)$.
Check to see whether $(2, 1)$
satisfies both equations.

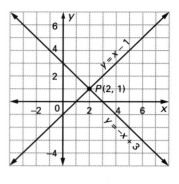

Check

$y = x - 1$
$1 \stackrel{?}{=} 2 - 1$
$1 = 1$ 　　True

$y = -x + 3$
$1 \stackrel{?}{=} -2 + 3$
$1 = 1$ 　　True

Thus, the solution is $(2, 1)$

TRY THIS　**1.** Solve the system $\begin{array}{l} y = 2x - 1 \\ y = -3x + 4 \end{array}$ by graphing.　(1, 1)

When two lines have the *same* slope and *different* y-intercepts, the lines will not intersect.

EXAMPLE 2 Solve the system $\begin{array}{l} y = 2x + 4 \\ y = 2x - 3 \end{array}$ by graphing.

Solution

Graph each equation.
$y = 2x + 4$
slope: 2
y-intercept: 4

$y = 2x - 3$
slope: 2
y-intercept: -3

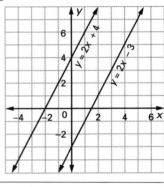

The graphs of the equations have the same slope but different y-intercepts. They are *parallel* lines.

Because there is no point of intersection, there is no ordered pair that satisfies both equations.
Thus, the system has *no solution*.

When two lines have the same slope and the same y-intercepts, their graphs coincide.

Additional Example 1

Solve by graphing.
$y = x + 1$
$y = -x - 1$　$(-1, 0)$

Additional Example 2

Solve by graphing.
$y = x - 2$
$y = x + 1$　No solution

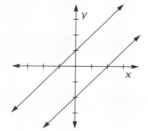

EXAMPLE 3   Solve the system $\begin{array}{l} y = 3x + 2 \\ -6x + 2y = 4 \end{array}$ by graphing.

Solution   Write $-6x + 2y = 4$ in slope-intercept form and compare the two equations.

$$-6x + 2y = 4$$
$$2y = 6x + 4$$
$$y = 3x + 2$$

Note that $-6x + 2y = 4$ is *equivalent* to $y = 3x + 2$. Their graphs are the same line, with slope 3 and y-intercept 2. Since the coordinates of every point on the line satisfy both equations, the system has *infinitely many* solutions.

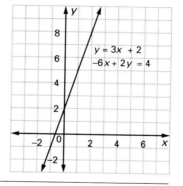

TRY THIS   Solve each system by graphing.

2. $\begin{array}{l} 2x + y = -1 \\ 2x + y = 4 \end{array}$ no solution

3. $\begin{array}{l} x - y = 3 \\ 2y - 2x = -6 \end{array}$ infinitely many solutions

The number of solutions of a system of two linear equations can be determined by comparing the slopes and y-intercepts of the equations.

EXAMPLE 4   Determine the number of solutions of the system: $\begin{array}{l} y = -\frac{1}{2}x + 3 \\ x + 2y = 5 \end{array}$

Plan   Write the second equation in slope-intercept form. Then compare the slopes and the y-intercepts.

Solution $x + 2y = 5$
$$2y = -x + 5$$
$$y = -\tfrac{1}{2}x + \tfrac{5}{2}$$

Equation	Slope	y-intercept
(1) $y = -\frac{1}{2}x + 3$	$-\frac{1}{2}$	3
(2) $y = -\frac{1}{2}x + \frac{5}{2}$	$-\frac{1}{2}$	$\frac{5}{2}$

Since the slopes are the same and the y-intercepts are different, the system has *no solution*.

TRY THIS   Determine the number of solutions for each system.

4. $\begin{array}{l} 2x - y = 4 \\ 2y = 4x - 3 \end{array}$ no solution

5. $\begin{array}{l} y = 7 \\ y = 3x + 1 \end{array}$ one solution

Common Error Analysis

Error: Since lines with different slopes intersect, students may think that lines with the same slope must be parallel.

Remind them that the two equations may represent the same line, so they must check the y-intercepts.

Additional Answers

Try This

1.

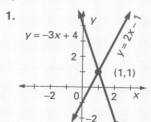

2.

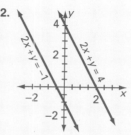

3.

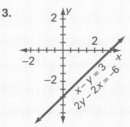

Additional Example 3

Solve by graphing.
$y = -x + 3$
$2x + 2y = 6$ Infinitely many solutions

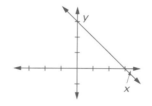

Additional Example 4

Determine the number of solutions for the system.

$y = \frac{1}{3}x - 3$

$2x - 6y = 4$ No solution

Checkpoint

Solve each system by graphing.

1. $y = x - 2$
   $x + y = 0$ $(1, -1)$

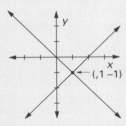

2. $2y = x + 2$
   $x - 2y = 2$ No solution

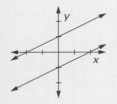

3. $x - y = 1$
   $y = x - 1$ Infinitely many solutions

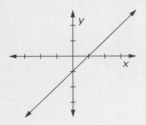

Closure

Have students graph the line that goes through the points (0, 1) and (3, 3) and the line through the points (0, 4) and (5, 7) on the same set of axes. Ask them how they can determine whether these two lines are parallel?   Find the slope of each line.

The chart below summarizes how to determine the number of solutions of a system of two linear equations.

Slopes	y-intercepts	Graph	Number of solutions
different⟶	same or different⟶	two intersecting lines→	1
same ⟨	different⟶	two parallel lines⟶	0
	same⟶	same line⟶	infinitely many

Classroom Exercises

Tell the number of solutions for each system graphed below.

1.
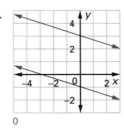
0

2.
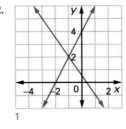
1

3.
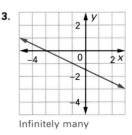
Infinitely many

Give the solution of each system graphed below.

4.
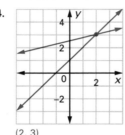
(2, 3)

5.
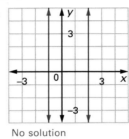
No solution

6.
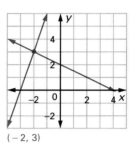
(-2, 3)

Is the given ordered pair a solution of the given system?

7. $(2, 1)$; $y = -3x + 7$
   $y = 2x - 3$ Yes

8. $(-1, -7)$; $y = x - 6$
   $y = 3x + 5$ No

9. $(5, 1)$; $x - y = 4$ Yes
   $x + 2y = 7$

Solve each system by graphing.

10. $y = 2x - 3$
    $y = -1$ $(1, -1)$

11. $y = -\frac{1}{3}x - 2$
    $x - 3y = -12$ $(-9, 1)$

12. $x - 2y = -6$
    $y = -\frac{3}{2}x - 1$ $(-2, 2)$

448   Chapter 12 Systems of Linear Equations

Written Exercises

Use the graph to find the solution of each system of equations.

1. $y = x - 4$
$y = -x - 2$

2. $y = 2x + 4$
$y = -\frac{1}{3}x + 4$

3. $y = -x - 2$
$y = x + 8$

4. $y = x - 4$
$y = -\frac{1}{3}x + 4$

5. $y = x + 8$
$y = x - 4$

6. $y = 2x + 4$
$y = -x - 2$

7. $y = x + 8$
$y = -\frac{1}{3}x + 4$

8. $y = x - 4$
$y = 0$

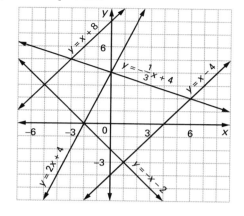

1. $(1, -3)$ **2.** $(0, 4)$ **3.** $(-5, 3)$ **4.** $(6, 2)$
5. No solution **6.** $(-2, 0)$ **7.** $(-3, 5)$ **8.** $(4, 0)$

Solve each system by graphing.

9. $y = x + 6$
$y = -2x$ $(-2, 4)$

10. $y = -x + 4$
$y = 2x - 2$ $(2, 2)$

11. $y = \frac{1}{3}x + 1$
$y = -x + 3$ $\left(\frac{3}{2}, \frac{3}{2}\right)$

12. $x = -3$
$y = 2x + 1$
$(-3, -5)$

13. $x - y = 6$
$y = -2$ $(4, -2)$

14. $y = 2x + 3$
$y = -5x - 4$ $(-1, 1)$

15. $y = -2x + 9$
$y = 4x - 3$ $(2, 5)$

16. $y = 3x - 9$
$y = -4x + 5$
$(2, -3)$

For each system, determine the number of solutions.

Infinitely many

17. $y = \frac{1}{3}x + 2$ 0
$y = \frac{1}{3}x - 7$

18. $y = 6x + 1$ 1
$y = -x + 1$

19. $y = 3x - 4$ 1
$y = -\frac{1}{3}x + 6$

20. $y = \frac{1}{2}x - 4$
$2y = x - 8$

21. $x - 2y = 6$
$2x - 4y = 12$
Infinitely many

22. $x = 3$ 1
$y = -2$

23. $x + y = 4$ 1
$x = 4$

24. $x - 3y = 1$ 0
$x - 3y = -9$

Solve by graphing. If necessary, estimate answers to the nearest half-unit.

$(-6, 8)$

25. $x - y = 2$ $(2, 0)$
$3x - 2y = 6$

26. $x + 3y = 6$ $(6, 0)$
$x - 3y = 6$

27. $y = -2x + 5$ $(2, 1)$
$3x - 2y = 4$

28. $2x + y = -4$
$5x + 3y = -6$

29. $2x - 2y = 1$
$x + 2y = 8$ $\left(3, \frac{5}{2}\right)$

30. $3x + 2y = 10$
$-x + 6y = 0$ $\left(3, \frac{1}{2}\right)$

31. $2x + 3y = 9$
$2x - 3y = -3$ $\left(\frac{3}{2}, 2\right)$

32. $3x + 5y = -5$
$2y + 5 = 0$
$\left(\frac{5}{2}, -\frac{5}{2}\right)$

12.1 Systems of Equations—Graphing **449**

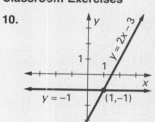
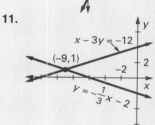
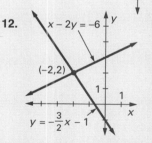

9.

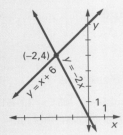

10.

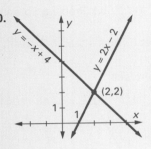

11.

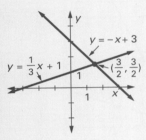

12.

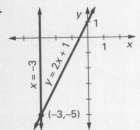

See page 451 for the answers to Ex. 13–16, page 455 for the answers to 25–28, and page 463 for the answers to Ex. 29–32.

33. Explain why a system whose solution is $\left(-\frac{7}{8}, \frac{5}{3}\right)$, cannot be solved accurately by graphing. Fractions are difficult to read on a graph.

34. If two equations in a system have the same slope, how many solutions could the system have? Infinitely many or 0

35. If two equations in a system have the same y-intercept, how many solutions could the system have? Infinitely many or 1

36. If two equations in a system have different slopes, how many solutions could the system have? 1

37. If two equations in a system have different y-intercepts, how many solutions could the system have? 1 or 0

38. A triangular region is enclosed by the x-axis and the lines whose equations are $-3x + y = 0$ and $y = -3x + 6$. Give the coordinates of the three vertices of the triangle. (0, 0), (2, 0), (1, 3)

39. A quadrangular region is enclosed by the x-axis, the y-axis, and the lines whose equations are $y = x + 3$ and $y = -2x + 6$. Give the coordinates of the four vertices of the region. (0, 0), (0, 3), (3, 0), (1, 4)

40. If $(3, -1)$ is a solution of the system $\begin{array}{c} Ax + 2y = 10 \\ x - By = 8 \end{array}$, find A and B. $A = 4$, $B = 5$

41. Write in your own words how to determine the number of solutions of a system of two linear equations in two variables.

Answers will vary.

Mixed Review

Solve for x. 3.5, 9.2, 9.3

1. $-15 = 3(x + 7)$ -12

2. $\frac{x + 3}{7} = \frac{4}{6}$ $\frac{5}{3}$

3. $2x + a = c$ $\frac{c - a}{2}$

Simplify. 2.10, 6.1, 6.4

4. $5(3x - y) - 2(x + y)$ $13x - 7y$

5. $5ab^2(-2a^3b)$ $-10a^4b^3$

6. $2x(4x^2y)^{-1}$ $\frac{1}{2xy}$

Factor completely. 7.3, 7.5, 7.6

7. $x^2 - 7x + 12$ $(x - 3)(x - 4)$

8. $a^2 + 24a + 144$ $(a + 12)^2$

9. $2y^3 - 32y$ $2y(y - 4)(y + 4)$

Solve. 7.8, 7.9

10. Ten times a number is 24 less than the square of the number. Find the number. -2 or 12

11. The area of a rectangular parking lot is 1056 m². The length of the lot is 20 meters greater than its width. Find the dimensions (length and width) of the lot. length: 44 m; width: 24 m

Enrichment

Challenge students to find two solutions for each of the following systems of equations. The students should find that the equations with absolute value bars are not linear equations.

1. $y = 2|x|$
   $y = x + 3$ (3,6), $(-1,2)$
2. $y = |x - 1|$
   $3y + x = 9$ (3,2), $(-3,4)$

Problem Solving Strategies

Working Backwards

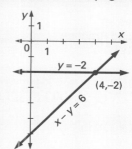

One Saturday morning, Peter and Jerry leave their homes at the same time and bicycle to meet each other. Peter bicycles at 15 miles per hour and Jerry travels at 17 miles per hour. They meet at 11:00 A.M. If they live 50 miles apart, how far apart are they at 10:30 A.M.?

You could begin to solve this problem by making a table based on the distance formula, $d = rt$. However, you can also arrive at the answer by applying the strategy of *Working Backwards*. Start from the moment the two boys meet. Think about what happens as the hands of a clock move backwards from the moment of meeting.

Time	Miles Left to Travel
10 A.M. (1 h before meeting)	Peter: 15 mi Jerry: 17 mi
10:30 A.M. ($\frac{1}{2}$ h before meeting)	Peter: 7.5 mi Jerry: 8.5 mi

Since $7.5 + 8.5 = 16$, Peter and Jerry are 16 miles apart at 10:30 A.M.

Exercises

Use the Working Backwards strategy to solve these problems.

1. Two trains start toward each other at the same time on the same track. At 8 P.M., they are 400 miles apart. One is traveling at 50 miles per hour, and the other is traveling at 36 miles per hour. Fifteen minutes before they might crash, the engineer on one train sees the light of the second train. How far apart are the trains? 21.5 mi

2. What information in Exercise 1 was not needed to solve the problem? At 8:00 P.M., they are 400 mi apart.

3. The dough from a batch of raisin bread doubles in bulk every 90 minutes. If the bowl is competely filled at 4 o'clock, at what time was the bowl half full? 2:30

13.

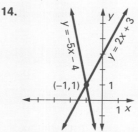

14.

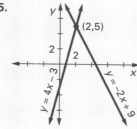

15.

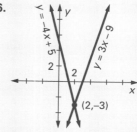

16.

12.2 The Substitution Method

Objective | To solve systems of equations using the substitution method

GETTING STARTED

Prerequisite Quiz

Simplify.

1. $3y + 5y + (2y - 7)$ $10y - 7$
2. $4(3x - 1) - 6x + 2$ $6x - 2$
3. $-5(4 - 3y) - 1 - 3y$ $12y - 21$
4. $3(2x + 4) - x - (2x + 4)$ $3x + 8$
5. $y - (-3y - 1) + 2(-3y - 1)$ $-2y - 1$

Motivator

Ask students to name some disadvantages to finding the solution of a system of equations by graphing. Some fractional or decimal values are difficult to read from a graph. Sometimes the lines meet at a point not on the graph. The *substitution method* presented in this lesson will facilitate the solution of systems of linear equations and always give an exact answer.

■■■**TEACHING SUGGESTIONS**

Lesson Note

For checking the solution to Example 1, you might suggest that since the first equation was used to solve for the second variable, the second equation could then be used as a "quick check". However, the solution should be checked in both of the original equations. After presenting Example 2, demonstrate that the same solution can be obtained by first solving the second equation for *y*.

In the previous lesson, the following system of equations was solved by graphing.

$$j + k = 120$$
$$j = 3k$$

For some systems, the graphing method will give only approximate solutions, while the *substitution method* will always give an exact solution. The **substitution method** is based on the property: If $a = b$, then a can be replaced by b and b can be replaced by a.

Since $j = 3k$ in the system above, you can replace j by $3k$ in the first equation, $j + k = 120$, as shown at the right.

$$3k + k = 120$$
$$4k = 120$$
$$k = 30$$

Then, since $j = 3k$,
$$j = 3 \cdot 30 = 90.$$

EXAMPLE 1 | Solve the system $\begin{matrix} y = -3x + 5 \\ -6x + y = -1 \end{matrix}$ using the substitution method.

Solution | The first equation gives the value of y in terms of x.

$$y = -3x + 5 \quad (1)$$
$$-6x + y = -1 \quad (2)$$

Substitute $-3x + 5$ for y in Equation (2) and solve for x.

$$-6x - 3x + 5 = -1$$
$$-9x = -6$$
$$x = \frac{2}{3}$$

To find the value of y, substitute $\frac{2}{3}$ for x in either equation. Equation (1) is used here.

$$y = -3x + 5$$
$$y = -3 \cdot \frac{2}{3} + 5$$
$$y = -2 + 5$$
$$y = 3$$

Thus, the solution is $\left(\frac{2}{3}, 3\right)$. The check is left for you.

TRY THIS | Solve each system by substitution.

1. $y = 2x + 1$ $(-1, -1)$
　　$3x + y = -4$

2. $2x - y = 5$ $(3, 1)$
　　$y = -x + 4$

Highlighting the Standards

Standards 5c, 4d: Here students solve systems of equations algebraically and apply this method to practical problems.

Additional Example 1

Solve by the substitution method.
$$y = 6x + 2$$
$$-4x + y = 3 \quad \left(\tfrac{1}{2}, 5\right)$$

If neither original equation is solved for one variable, look for a variable whose coefficient is 1 or -1. Then solve the equation for that variable. If no variable has a coefficient of 1 or -1, you can still use the substitution method to solve the system.

EXAMPLE 2   Solve the system $\begin{array}{l} x - 4y = 17 \\ 2x + y = -2 \end{array}$ by the substitution method.

Plan   The coefficient of x is 1 in the first equation. So, solve the first equation for x in terms of y. Then substitute in the second equation.

Solution

$$x - 4y = 17 \quad (1)$$
$$x = 4y + 17$$

Substitute $4y + 17$ for x in Equation (2) and solve for y.

$$2x + y = -2 \quad (2)$$
$$2(4y + 17) + y = -2$$
$$8y + 34 + y = -2$$
$$9y + 34 = -2$$
$$9y = -36$$
$$y = -4$$

Substitute -4 for y in $x = 4y + 17$. (This equation is used since it is already solved for x.)

$$x = 4y + 17$$
$$x = 4(-4) + 17$$
$$x = -16 + 17$$
$$x = 1$$

Thus, the solution is $(1, -4)$. The check is left for you.

EXAMPLE 3   Solve using the substitution method: $\begin{array}{ll} 3x + 2y = 5 & (1) \\ 5x + 3y = 9 & (2) \end{array}$

Plan   Solve Equation (1) for y. Then substitute for y in Equation (2).

Solution

$$3x + 2y = 5 \quad (1)$$
$$2y = -3x + 5$$
$$y = -\tfrac{3}{2}x + \tfrac{5}{2}$$

Now substitute $-\tfrac{3}{2}x + \tfrac{5}{2}$ for y in Equation (2).

$$5x + 3y = 9 \quad (2)$$
$$5x + 3\left(-\tfrac{3}{2}x + \tfrac{5}{2}\right) = 9$$
$$5x - \tfrac{9}{2}x + \tfrac{15}{2} = 9$$
$$10x - 9x + 15 = 18$$
$$x = 3$$

Now substitute 3 for x in either equation and solve for y: $y = -2$

Thus, the solution is $(3, -2)$. The check is left for you.

TRY THIS   Solve each equation by substitution.

3. $\begin{array}{l} x + 2y = 3 \\ 3x - y = -1 \end{array}$ $\left(\tfrac{1}{7}, \tfrac{10}{7}\right)$
     4. $\begin{array}{l} 5x + 2y = 1 \\ 3x - 2y = 7 \end{array}$ $(1, -2)$

12.2 The Substitution Method   **453**

453

Closure

Have students summarize the substitution method for solving a system of equations in their own words. Be sure that they name all the steps in the correct sequence. See the Summary on p. 454.

See the Summary on p. 454.

■■■FOLLOW UP

Guided Practice

Classroom Exercises 1–8
Try This all

Independent Practice

A Ex. 1–20, **B** Ex. 21–28, **C** Ex. 29–40

Basic: WE 1–19 odd

Average: WE 9–27 odd

Above Average: WE 17–39 odd

Additional Answers

Written Exercises

8. $(-1, -1)$
12. $(0, 1)$
16. $\left(\frac{9}{2}, -\frac{3}{2}\right)$
20. $(-10, -14)$
23. $\left(4, \frac{10}{3}\right)$
24. $\left(\frac{2}{3}, \frac{22}{3}\right)$
30. $\left(-\frac{24}{7}, -\frac{12}{7}\right)$
31. $\left(\frac{13}{12}, \frac{7}{12}\right)$

▬▬▬ Summary

To solve a system of equations using substitution:

1. Solve one of the equations for one of its variables (if neither equation is already solved for one of the variables).
2. Substitute for that variable in the other equation.
3. Solve the resulting equation.
4. Find the value of the second variable by substituting the result from Step (3) in either of the equations.
5. Check the solution in *both* of the original equations.

Classroom Exercises

For each system, solve one of the equations for one of the variables.

1. $x + 8 = 5y$
$3x = 7y - 8$
$x = 5y - 8$

2. $2x + y = 14$
$3x + 2y = -1$
$y = -2x + 14$

3. $2x + 5y = 7$
$x + 3y = 0$
$x = -3y$

4. $2x - 3y = 4$
$13 + y = \frac{1}{4}x$
$y = \frac{1}{4}x - 13$

Solve using the substitution method.

5. $x - 3y = 9$
$y = 2$
$(15, 2)$

6. $3a + b = 7$
$b = a - 1$
$(2, 1)$

7. $5c + 3d = -1$
$c + d = 3$
$(-5, 8)$

8. $x + y = 3$
$x - y = 9$
$(6, -3)$

Written Exercises

Solve using the substitution method.

1. $y = x$ $(-2, -2)$
$3x - y = -4$

2. $x + y = 8$
$y = x - 2$ $(5, 3)$

3. $x + y = 12$
$y = 2x$ $(4, 8)$

4. $y = 2x - 6$ $(2, -2)$
$x - y = 4$

5. $x = y - 8$ $(-4, 4)$
$-3x = -y + 16$

6. $x - y = -1$ $(2, 3)$
$-2x + 3y = 5$

7. $y = 3x - 14$ $(2, -8)$
$y = -5x + 2$

8. $y = 5x + 4$
$y = -2x - 3$

9. $3x - 5y = -9$
$4x + y = -12$ $(-3, 0)$

10. $5x + 4y = 0$
$x - y = 9$ $(4, -5)$

11. $y + x = -1$
$3x - 4y = 4$ $(0, -1)$

12. $4x + 3y = 3$
$x + 2y = 2$

13. $3x + y = 2$
$x - y = 0$ $\left(\frac{1}{2}, \frac{1}{2}\right)$

14. $5x - y = 1$
$y = -3x + 1$ $\left(\frac{1}{4}, \frac{1}{4}\right)$

15. $2x - y = -1$
$x - 2y = -11$ $(3, 7)$

16. $x - y = 6$
$x + y = 3$

17. $3x = y + 12$
$-5y = -4x + 16$ $(4, 0)$

18. $2x + y = 1$
$10x - 4y = 2$ $\left(\frac{1}{3}, \frac{1}{3}\right)$

19. $3y - 2 + x = 0$
$-2x = 4y + 2$ $(-7, 3)$

20. $y = \frac{1}{2}x - 9$
$y = \frac{3}{2}x + 1$

21. $3x - 2y = 8$
$2x + 3y = 14$ $(4, 2)$

22. $3y = 5 - 2x$
$1 = 4x + 3y$ $(-2, 3)$

23. $-3x + 6y = 8$
$-2x + 6y = 12$

24. $2c - d = -6$
$c + d = 8$

25. $3(x + y) = -9$
$5x - 9y = -1$ $(-2, -1)$

26. $\dfrac{x + y}{4} = 6$
$x - \frac{3}{2}y = -6$ $(12, 12)$

27. $\dfrac{y}{2} = \dfrac{12 - 3x}{3}$
$3x - y = 7$ $(3, 2)$

28. $3x = 4 - y$
$\frac{x}{3} + \frac{y}{5} = \frac{8}{15}$ $(1, 1)$

Enrichment

Have students use the substitution method to solve the following systems for *x*, *y*, and *z*.

1. $x + 2y + z = 4$
$4y + 9z = -1$
$x = y - 1$ $(1, 2, -1)$

2. $4x - 3y + z = -7$
$-4x + 2y = 4$
$z + y = 13$ $(3, 8, 5)$

3. $x + y = -3$
$y + z = -5$
$x + z = 6$ $(4, -7, 2)$

29. $3g = \frac{1}{2}h + 3$
$h - 2 = -g$ $\left(\frac{8}{7}, \frac{6}{7}\right)$

30. $x + 12 = -5y$
$\frac{y}{2} = \frac{x}{4}$

31. $p - q = \frac{1}{2}$
$\frac{1}{7}p = q - \frac{3}{7}$

32. $2a + b = 1$
$\frac{a}{5} - 1 = \frac{b}{2}$
$\left(\frac{5}{4}, -\frac{3}{2}\right)$

33. $\frac{x + y}{2} = 10$
$3y = 2x$ $(12, 8)$

34. $4a + b - 8 = 0$
$5a + 3b - 3 = 0$ $(3, -4)$

35. $2u - r + 2 = 0$
$6u + 12r = 1$ $(u, r) = \left(-\frac{23}{30}, \frac{7}{15}\right)$

36. $4s + 3t - 1 = 0$
$2 = 8s + 6t$ Infinite number
of solutions

Try to solve each system using the substitution method. Explain the results.

37. $y = -\frac{1}{3}x + 5$
$3y = -x - 9$ No solution

38. $y = \frac{1}{2}x + 3$
$-x + 2y = 6$ Infinite number
of solutions

Solve each system of three equations for x, y, and z. Use the substitution method.

39. $3x + 2y + z = 7$ $(2, 1, -1)$
$4y + 5z = -1$
$x = y + 1$

40. $-5x - 2y + z = 7$ $(5, -12, 8)$
$3x + 2y = -9$
$z - x = 3$

Mixed Review

1. 35 is 5% of what number? *4.7* 700

Factor. *7.3, 7.5*

2. $y^2 - 2y - 35$ $(y - 7)(y + 5)$

3. $x^2 + 10x + 25$ $(x + 5)^2$

4. $x^2 - 16$
$(x - 4)(x + 4)$

Solve. *3.5, 5.2, 7.7*

5. $9 - (7 - x) = 5x - 6$ 2

6. $4x - 9 \le 1 + 3x$ $x \le 10$

7. $3a^2 + 7a - 6 = 0$
$\frac{2}{3}, -3$

8. Subtract $\frac{3y + 2}{y - 8}$ from 2. Simplify, if possible. *8.7* $\frac{-y - 18}{y - 8}$

Solve. *10.6, 10.7*

9. The cost, c, of cleaning a carpet varies directly as its area, A. The cost of cleaning a 9-ft by 12-ft carpet is $21.60. At the same rate per square foot, how much will it cost to clean a carpet 18 ft long by 10 ft wide? $36

10. The amount of time it takes Jenny to travel to the seashore varies inversely as the rate of travel. At 30 mph, it takes Jenny 4 hours to make the trip from her home to the shore. How long will it take if she travels at a rate of 60 mph? 2 h

25.

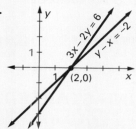

26.

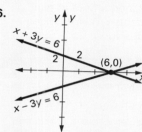

27.

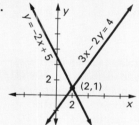

28.

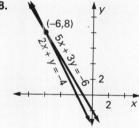

▬ GETTING STARTED

Prerequisite Quiz

Using the underlined letters in each sentence as variables, rewrite these sentences as equations.

1. Janet is three years younger than Cora. $J = C - 3$, or $J + 3 = C$
2. The larger of two numbers is 10 more than half the smaller. $r = \frac{1}{2}s + 10$
3. The area of a rectangle is 4 less than twice the area of a square. $r = 2s - 4$
4. In 5 years Henry will be twice as old as Burt will be in 5 years. $H + 5 = 2(B + 5)$

Motivator

Present this puzzle: I am thinking of two numbers. If you add them, the sum is 20. If you subtract them, the difference is 8. Then ask students to write the puzzle as two algebraic statements. $x + y = 20$, $x - y = 8$. Ask them to find the numbers. 14, 6

12.3 Problem Solving: Using Two Variables

Objective | To solve word problems using systems of equations

The problems in this chapter will be solved by writing two equations in two variables and solving the resulting system. Although some of the problems can be solved using one equation and one variable, it is often easier to use two variables.

EXAMPLE 1 | A pencil and 3 erasers cost 50¢. Two pencils and 4 erasers cost 86¢. Find the cost of 1 pencil and the cost of 1 eraser.

Solution | Let p = cost of 1 pencil, in cents.
Let e = cost of 1 eraser, in cents.

One pencil and 3 erasers cost 50¢.	$p + 3e = 50 \qquad (1)$
Two pencils and 4 erasers cost 86¢.	$2p + 4e = 86 \qquad (2)$
Solve Equation (1) for p.	$p = 50 - 3e$
Substitute $50 - 3e$ for p in (2) and solve for e.	$2(50 - 3e) + 4e = 86$
	$100 - 6e + 4e = 86$
	$-2e = -14$
	$e = 7$
Substitute 7 for e in $p = 50 - 3e$.	$p = 50 - 3 \cdot 7$
Solve for p.	$= 50 - 21$, or 29

Check | One pencil and 3 erasers cost 50¢: $1 \cdot 29 + 3 \cdot 7 = 29 + 21 = 50$

Two pencils and 4 erasers cost 86¢: $2 \cdot 29 + 4 \cdot 7 = 58 + 28 = 86$

Thus, one pencil costs 29¢ and one eraser costs 7¢.

EXAMPLE 2 | The perimeter of a rectangle is 54 cm. The length is 1 cm shorter than 3 times the width. Find the length and the width.

Solution | What are you to find? → length and width of a rectangle

Draw and label a figure.

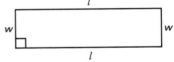

What is given? → perimeter = 54 cm
length = 1 cm less than 3 times the width
(THINK: $\frac{1}{2}$ the perimeter is 27 cm.)
$l + w = 27$

Highlighting the Standards

Standards 4c, 4d: In this lesson, students use systems of equations to solve problems from situations within and outside mathematics.

Additional Example 1

The cost of a ruler and 2 pads is $1.60. Three rulers and 5 pads cost $4.20. Find the cost of 1 ruler and the cost of 1 pad. One ruler costs 40¢; one pad costs 60¢.

Additional Example 2

The perimeter of a rectangle is 76 cm. The length is 6 cm longer than 3 times the width. Find the length and the width. $l = 30$ cm; $w = 8$ cm

Choose two variables. What do they represent?	Let l = the length. Let w = the width.
Write two equations.	$2l + 2w = 54$ (1) $l = 3w - 1$ (2)
Solve the system. Substitute $3w - 1$ for l in Equation (1).	$2(3w - 1) + 2w = 54$ $6w - 2 + 2w = 54$ $8w - 2 = 54$ $8w = 56$ $w = 7$
Now substitute in Equation (2).	$l = 3w - 1$ (2) $= 3(7) - 1$ $= 21 - 1$ $l = 20$ The answers are reasonable. $\frac{1}{2}$ the perimeter is 27, and $20 + 7 = 27$.
Check Check in the original problem.	Is the perimeter 54 cm? Yes, because $2 \cdot 20 + 2 \cdot 7$ $= 40 + 14 = 54$. Is the length 1 cm shorter than 3 times the width? Yes, because $3 \cdot 7 - 1 = 21 - 1 = 20$
State the answer.	The length is 20 cm and the width is 7 cm.

TRY THIS

1. The sum of two numbers is 53. One number is 3 less than the other. Find the numbers.   28, 25
2. The perimeter of a rectangle is 80 ft. One side is $\frac{2}{3}$ the length of the other. Find the dimensions of the rectangle.   24, 16

Classroom Exercises

For each problem, use the given information to write two equations in two variables. State what each variable represents.

1. Mary is 3 times as old as Jean. The difference in their ages is 18 years. Find each girl's age. Let m = Mary's age; j = Jane's age; $m = 3j$; $m - j = 18$

2. A 73-ft length of wire is cut into two pieces. How long is each piece if one is 9 ft longer than the other. Let x = shorter and y = longer; $x + y = 73$; $y = x + 9$

12.3 Problem Solving: Using Two Variables   **457**

Common Error Analysis

Error: Some students have difficulty because they fail to write *two* equations for each problem.

Remind them that in order to solve for two variables, they must have two equations.

Checkpoint

The sum of two numbers is 20. What are the numbers if:

1. the larger number is 3 times the smaller?   15 and 5
2. one number is 4 smaller than the other?   8 and 12
3. the smaller number is $\frac{1}{4}$ of the larger number?   4 and 16
4. the larger number is 1 less than twice the smaller?   13 and 7
5. the smaller number is 1 less than half the larger?   14 and 6

Closure

Suppose a word problem has two unknown quantities that are to be found. Ask students how many variables and how many equations are used to solve such a problem.   2; 2   Then ask what kind of problem lends itself to being solved by the substitution method.   A problem where one equation has one variable by itself on one side of the equation.

---

3. The sum of two numbers is 121. The first number is 11 less than the second number. Find the two numbers.
Numbers: x and y; $x + y = 121$; $x = y - 11$

4. One notebook and 3 packs of paper cost $6.20. Two notebooks and 1 pack of paper cost $8.65. Find the cost of a notebook and of a pack of paper.
Let $n =$ cost of notebook and p cost of paper; $n + 3p = 620$; $2n + p = 865$

Written Exercises

Solve each problem by using two equations and two variables.
Marcia: 18 yr; Kate: 6 yr

1. Bob is 11 years older than Hank. The sum of their ages is 27 years. Find the age of each. Bob: 19 yr; Hank: 8 yr

2. Marcia is 3 times as old as Kate. The difference between their ages is 12 years. Find the age of each.

3. The perimeter of a rectangle is 46 m. The length is 8 m more than 4 times the width. Find the length and the width. l: 20 m; w: 3 m

4. The difference of two numbers is 5. Three times the larger, decreased by 5 times the smaller, is 7. Find the numbers. 9, 4
Paul: 40 mi; Tim: 8 mi

5. Separate $74 into two parts so that one amount is $7 less than twice the other amount. $27, $47

6. Paul drove 5 times as far as Tim. The sum of their distances was 48 mi. How far did each drive?

7. Mr. Hernandez weighs 3 times as much as his son, Carlos. Together, they weigh 280 lb. How much does each weigh? Mr. H: 210 lb; C: 70 lb

8. Two hamburgers and 1 salad cost $2.95. Five hamburgers and 3 salads cost $8.00. Find the cost of a hamburger and the cost of a salad.

9. Jane is 4 times as old as Karen. Jane's age decreased by Karen's age is 21 years. Find the age of each.
Jane: 28 yr; Karen: 7 yr

10. A piece of ribbon 52 cm long is cut into two pieces. The first piece is 7 cm longer than twice the length of the second. How long is each piece?

11. Jack scored 6 more points than twice the number of points Igor scored. Their combined score was 27 points. How many points did each score? J: 20;
I: 7

12. Sun Lee has a total of 33 tapes and records. The number of tapes is 3 less than twice the number of records. How many of each does he have?

13. The length of a rectangle is 5 in. more than twice the width. The perimeter is 52 in. Find the length and the width of the rectangle. l: 19 in; w: 7 in

14. The perimeter of a rectangle is 68 cm. The length is 2 cm less than 3 times the width. Find the length and width. l: 25 cm; w: 9 cm
24,11

15. The sum of two numbers is 20. The first number decreased by the second number is 2. Find the numbers. 11, 9

16. The product of 4 and the sum of two numbers is 140. The difference of the two numbers is 13. Find the numbers.

17. There are twice as many students on Mr. Sneed's team as on Mr. Johnson's team. Together, they want to raise $360 to buy uniforms. According to the size of the teams, how much money should each team raise?
Sneed: $240; Johnson: $120

18. A hotel charges less for weekend nights than for weekday nights. Three weekday nights and 1 weekend night cost $286. Four weekend nights and 2 weekday nights cost $414. Find the cost of a weekend night and the cost of a weekday night.
Weekend: $67; weekday: $73

19. The formula $F = \frac{9}{5}C + 32$ is used to change from degrees Celsius to degrees Fahrenheit. At what temperature do both scales show the same number? –40

20. One number is 4 less than 3 times a second number. If 3 more than twice the first number is decreased by twice the second number, the result is 11. Find the numbers. 8, 4

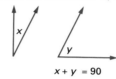

21. Two angles are *complementary*; that is, the sum of their degree measures is 90. The measure of one angle is 6 less than twice the measure of the other. Find the measure of each angle. 32, 58

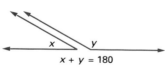

$x + y = 90$

22. Two angles are *supplementary*; that is, the sum of their degree measures is 180. The measure of one angle is 12 more than 3 times the measure of the other. Find the measure of each angle. 42, 138

$x + y = 180$

23. A number and twice another number are added. Twice this sum is 346. The sum of the original two numbers is 118. Find the numbers. 63, 55

24. Half the sum of two numbers is equal to 91. Twice the larger number subtracted from 3 times the smaller number is 51. Find the numbers. 83, 99

25. Half the sum of the distances traveled by Nicole and Melissa is 640 km. If $\frac{1}{4}$ of the distance Nicole traveled is 5 km more than $\frac{1}{6}$ of the distance Melissa traveled, how far did each travel? Nicole: 524 km; Melissa: 756 km

26. During the holiday weekend, the number of campers admitted to a park was 5 less than $\frac{1}{2}$ the number of trailers admitted. If 4 more trailers had been admitted, there would have been a total of 89 trailers and campers. How many trailers and how many campers were admitted? 25 campers, 60 trailers

Mixed Review

Solve. If there is no solution, so indicate. (Exercises 1–6) *4.5, 4.6, 5.6, 7.7, 9.2*

No solution

1. $\frac{2}{5}x + 4 = 3 + \frac{1}{2}x$ $x = 10$

2. $0.04x = 0.1$ $x = 2.5$

3. $-8 - 2|4x - 2| = 16$

4. $x^2 - x - 30 = 0$ (–5, 6)

5. $6y^2 + 28y = -32$ $-\frac{8}{3}, -2$

6. $\frac{a - 6}{4} = \frac{2}{a - 4}$ 2, 8

7. Multiply $(2c + 7)(c - 9)$. *6.9*
$2c^2 - 11c - 63$

8. Add $\dfrac{2}{x - 2} + \dfrac{3}{x - 5} + \dfrac{7}{x^2 - 7x + 10}$. *8.7*

8. $\dfrac{5x - 9}{x^2 - 7x + 10}$

12.3 Problem Solving: Using Two Variables **459**

Enrichment

Have students make a poster comparing the Fahrenheit and Celsius scales. The poster should show the comparable temperatures for the boiling point of water, the freezing point of water and normal body temperature. It should also include an explanation of where the fractions $\frac{5}{9}$ and $\frac{9}{5}$ and the number 32 come from in the conversion formulas.

■GETTING STARTED

Prerequisite Quiz

Simplify each expression.

1. $3x + 2y - 3x + 1$ $2y + 1$
2. $4 - 6x + 2y + 6x - y$ $y + 4$
3. $7x - 3y - 5x + 3y + 1$ $2x + 1$
4. $6y + 2x - 4y - 2x - 2y$ 0
5. $-8x + 3y - 5x - 7y + 4y$ $-13x$

Motivator

Show students the system of equations $2x + 3y = -8$ and $5x - 3y = 1$ and ask them to solve it by the substitution method. $(-1, -2)$ Then show them that there is a way to avoid the fractions this method will involve by solving the system with the addition method shown in this lesson.

12.4 The Addition Method

Objectives
To solve systems of equations using the addition method
To explain why a given system cannot be solved using a given method

You have solved systems of equations using graphs and using the substitution method. In this lesson, you will use the **addition method.** It is based on the following form of the Addition Property of Equality.

$$
\begin{array}{ll}
\text{If} & a = b \\
\text{and} & \underline{c = d} \\
\text{then} & a + c = b + d
\end{array}
$$

Adding the left sides and the right sides of two equations is often referred to as "adding the equations."

EXAMPLE 1 Use the addition method to solve $\begin{array}{l} x + y = 4 \\ -x + y = -2 \end{array}$.

Plan Since x and $-x$ are opposites, eliminate the x terms by adding the equations. Then solve for y.

Solution
$$
\begin{array}{ll}
x + y = 4 & (1) \\
\underline{-x + y = -2} & (2) \\
2y = 2 \\
y = 1
\end{array}
$$

Substitute 1 for y in either equation; then solve for x.
$$
\begin{array}{l}
x + y = 4 \\
x + 1 = 4 \\
x = 3
\end{array}
$$

Check
$$
\begin{array}{ll}
x + y = 4 & -x + y = -2 \\
3 + 1 \stackrel{?}{=} 4 & -3 + 1 \stackrel{?}{=} -2 \\
4 = 4 \ \text{True} & -2 = -2 \ \text{True}
\end{array}
$$

Thus, the solution is $(3, 1)$.

TRY THIS Use the addition method to solve each system.

1. $\begin{array}{l} x - 2y = 2 \\ 2x + 2y = 1 \end{array}$ $\left(1, -\frac{1}{2}\right)$ 2. $\begin{array}{l} \frac{1}{2}r + \frac{2}{5}s = 15 \\ -\frac{1}{2}r + \frac{4}{5}s = 45 \end{array}$ $(-10, 50)$

Highlighting the Standards

Standards 1b, 4d: Students practice and apply another strategy for solving systems of equations.

Additional Example 1

Solve by the addition method.
$x + 3y = 1$
$-x + y = 7$ $(-5, 2)$

Sometimes a system is easier to solve if you write the equations in standard form first.

EXAMPLE 2   Use the addition method to solve $\begin{array}{l} 2x = 3y - 5 \\ 3y = 4x + 1 \end{array}$.

Solution

1. Write in standard form and solve for one variable.

$$\begin{array}{rcl} 2x - 3y &=& -5 \\ -4x + 3y &=& 1 \\ \hline -2x &=& -4 \\ x &=& 2 \end{array}$$

2. Substitute for x and solve.

$$\begin{array}{rcl} 3y &=& 4(2) + 1 \\ 3y &=& 8 + 1 \\ 3y &=& 9 \\ y &=& 3 \end{array}$$

Check $(2, 3)$ in both original equations.
Thus, the solution is $(2, 3)$.

EXAMPLE 3   Two packages have a total weight of 39 lb. One package weighs 7 lb more than the other package. Find the weight of each package.

What are you to find?   The weights of two packages

What is given?   The total weight of two packages is 39 lb. One package weighs 7 lb more than the other package.

Choose two variables.
What do they represent?   Let x = weight of heavier package.
Let y = weight of lighter package.

Write two equations. $x + y = 39$ (1)
$x = y + 7$ (2)

Solve the system using the addition method.

$$\begin{array}{rcll} x - y &=& 7 & \quad(2) \\ x + y &=& 39 & \quad(1) \\ \hline 2x &=& 46 \\ x &=& 23 \\ 23 + y &=& 39 \\ y &=& 16 \end{array}$$

Check in the original problem.   Is the total weight 39 lb?
Yes, since $23 + 16 = 39$.

Does one package weigh 7 lb more than the other?
Yes, since $23 = 16 + 7$.

State the answer.   The packages weigh 23 lb and 16 lb.

TRY THIS   **3.** Lupe has $9 less than Juanita. Together they have $53. How much does each person have?   Juanita: $31; Lupe: $22

■TEACHING SUGGESTIONS

Lesson Note

Remind students that the Addition Property of Equality states that if $a = b$, then $a + c = b + c$. In the present case, since $c = d$, this becomes $a + c = b + d$. For each system in this lesson, the coefficients of one pair of like terms are the additive inverses of each other. Thus, that pair of terms will be eliminated when the two equations are added. More general systems will be treated in the next lesson. You may wish to point out that Example 3 could also be solved by the substitution method.

Math Connections

Geometry: To find the sum of the measures of the angles of any quadrilateral, draw one of the diagonals to divide the quadrilateral into two triangles. Since the sum of the measures of the angles of each triangle is 180°, addition shows that the sum of the angles of a quadrilateral is 360°.

Critical Thinking Questions

Application: Show students the following arrangement for adding all the integers from 1 through 100.

$$\begin{array}{rrrrrr} 1 + & 2 + & 3 + & & + 49 + & 50 \\ 100 + & 99 + & 98 + & \ldots + & 52 + & 51 \end{array}$$

Then ask how this sum can be found quickly.   There will be 50 sums of 101 each. So $(50)(101) = 5050$

Common Error Analysis

Error: Some students may still have trouble with "more than" and "less than." Compare "a number is 3 more than a second number" ($x = y + 3$) and "a number is more than 3 times a second number" ($x > 3y$).

Additional Example 2

Solve by the addition method.
$\frac{1}{3}r + \frac{3}{5}s = 19$
$\frac{4}{3}r - \frac{3}{5}s = 31$ (30, 15)

Additional Example 3

Two cuts of roast beef have a total weight of 14 lb. One piece weighs 2 lb more than the other. Find the weight of each piece.
6 lb and 8 lb

Checkpoint

Solve each system by the addition method.

1. $2x - y = -6$
$-2x + 2y = 4$ $\quad(-4, -2)$
2. $2x - y = 15$
$x + y = 0$ $\quad(5, -5)$
3. $x + 6y = 19$
$-x + 4y = 6$ $\quad\left(4, \frac{5}{2}\right)$

Closure

Ask students to consider the following system: $3x + 2y = 12$, $4y - 5x = 2$. Ask whether the system can be solved easily by using the addition method. If not, why not? No; neither variable will be eliminated if the two equations are added.

■■■FOLLOW UP

Guided Practice

Classroom Exercises 1–9
Try This all

Independent Practice

A Ex. 1–19, **B** Ex. 20–33, **C** Ex. 34–37

Basic: WE 1–19 odd
Average: WE 11–29 odd
Above Average: WE 17–37 odd

Classroom Exercises

Add each pair of equations. State the resulting equation. Then solve the system.

$\qquad\qquad\qquad\qquad\qquad x = -2; (-2, -1) \qquad -2n = -6; (-3, 3)$

1. $x + y = 15$
$-x + 3y = -7$ $\;4y = 8; (13, 2)$
2. $2x - 5y = 1$
$-x + 5y = -3$
3. $4m + n = -9$
$-4m - 3n = 3$

4. $2x + y = 3$
$-x - y = 3$ $\;x = 6; (6, -9)$
5. $\frac{1}{2}p - q = 9$
$-\frac{1}{2}p + 2q = -3$ $\;q = 6; (30, 6)$
6. $\frac{1}{4}x + 2y = 10$
$\frac{1}{4}x - 2y = -2$

7. $x - 0.5y = 10$
$x + 0.5y = 8$ $\;2x = 18; (9, -2)$
8. $0.4x + y = 5$
$0.6x - y = 0$ $\;x = 5; (5, 3)$
9. $3x + 5y = 4$
$-3x + 4y = 14$
$9y = 18; (-2, 2)$

6. $\frac{1}{2}x = 8; (16, 3)$

Written Exercises

Solve using the addition method. **9.** $\left(-\frac{4}{3}, 2\right)$ **12.** $\left(-9, \frac{19}{2}\right)$

1. $x - 3y = -3$
$x + 3y = 9$ $(3, 2)$
2. $-2x + 3y = 2$
$2x + 7y = 18$ $(2, 2)$
3. $3x + 7y = 3$
$x - 7y = 1$ $(1, 0)$

4. $-7c - d = 19$
$3c + d = -7$ $(-3, 2)$
5. $-2x + 8y = 14$
$2x + 7y = 16$ $(1, 2)$
$(1, 1)$ **6.** $5p - 4q = 1$
$7p + 4q = 11$

7. $2x + 9y = 4$
$-5x - 9y = 17$ $(-7, 2)$
8. $5x - y = 13$
$y + x = -1$ $(2, -3)$
9. $-3x + 5y = 14$
$3x + 8y = 12$

10. $5x - 3y = -5$
$-5x + 2y = 1$ $\left(\frac{7}{5}, 4\right)$
11. $3r - 7s = 24$
$3r + 7s = 6$ $\left(5, -\frac{9}{7}\right)$
12. $s + 2t = 10$
$-2t = 3s + 8$

13. $4x - 3y = 3$
$7x = 19 - 3y$ $\left(2, \frac{5}{3}\right)$
14. $y = -2x + 3$
$-2x + y = -4$ $\left(\frac{7}{4}, -\frac{1}{2}\right)$
15. $0.2x + y = 6$
$0.7x - y = 3$
$(10, 4)$

Solve each problem using a system of two equations and the addition method.

16. The sum of two numbers is 40. Their difference is 16. Find the two numbers. 12, 28

17. Jane weighs 8 lb less than Pat. Their total weight is 202 lb. How much does each weigh? Jane: 97 lb; Pat: 105 lb

18. If Jean and Joe combined their weekly allowances, they would have $12. Jean's allowance is $2 more than Joe's. Find the weekly allowances for Jean and Joe. Jean: $7; Joe: $5

19. Together, a house and a lot cost $78,500. The difference between the cost of the house and the cost of the lot is $35,900. Find the cost of the house and the cost of the lot.

20. Mary's salary is $50 more than Sue's salary. If 3 times Mary's salary is added to Sue's salary, the result is $1,550. Find Mary's salary and Sue's salary. Mary: $400; Sue: $350

21. The total cost of two books is $32.90. One book costs $3 more than the other book. Find the cost of each book. $14.95, $17.95

19. $h = \$57,200; l = \$21,300$

462 Chapter 12 Systems of Linear Equations

Enrichment

This enrichment section and those in Lessons 12.5 and 12.7 form a sequence which introduces students to the determinant method of solving two simultaneous linear equations. Explain to students that a 2 by 2 determinant is defined as follows.

$$\begin{vmatrix} a & b \\ c & d \end{vmatrix} = ad - bc$$

For example,

$$\begin{vmatrix} 2 & -4 \\ 1 & 3 \end{vmatrix} = (2)(3) - (-4)(1) = 10$$

Then have them evaluate the following determinants.

1. $\begin{vmatrix} 5 & 3 \\ 4 & 7 \end{vmatrix}$ $\;23$
2. $\begin{vmatrix} 3 & -2 \\ -6 & -1 \end{vmatrix}$ $\;-15$

3. $\begin{vmatrix} 24 & 3 \\ -8 & 0 \end{vmatrix}$ $\;24$
4. $\begin{vmatrix} 13 & 4 \\ 13 & 4 \end{vmatrix}$ $\;0$

Solve each system using the addition method.

22. $-\frac{1}{3}p + \frac{3}{5}q = 1$
$\frac{4}{3}p - \frac{3}{5}q = 5$ (6, 5)

23. $\frac{2}{3}r + \frac{4}{5}s = 4$
$-\frac{2}{3}r + \frac{3}{5}s = 10$
(−6, 10)

24. $2(x + y) = 3x$
$x + 6y = 0$ (0, 0)

25. $-3(x + 5y) = 0$
$15y - 4x = 35$
(−5, 1)

26. $4x - y = 7$ (2, 1)
$3x - (6 - y) = 1$

27. $0.18x - 0.13y = -3.6$ $\left(\frac{5}{3}, 30\right)$
$-0.18x + 0.25y = 7.2$

28. Write an explanation of why the system shown below cannot be solved using the addition method. What can you do to the second equation so that you can use the addition method?

$3x - 4y = 13$ Neither the *x*-terms nor the *y*-terms are opposites.
$-2x + y = -2$ Multiply Equation 2 by 4 to obtain 4*y*.

What values of *r* and *s* will eliminate the *y*-term? Sample answers are given.

29. $r(-2x + y - 5) + s(x + 3y - 1) = 0$
$r = -3, s = 1$

30. $r(2x + 3y - 3) + s(3x - 5y + 2) = 0$
$r = 5, s = 3$

Write each equation of the system in the form $Ax + By = C$. Then solve using the addition method.

31. $\frac{x + y}{3} = 2$
$\frac{x - y}{2} = -1$ (2, 4)

32. $\frac{2c + d}{5} = c + d$
$\frac{4d + 3c}{7} = d + 3$ (4, −3)

33. $\frac{x + 2y}{3} = 2$
$1 = \frac{5x - 2y}{6}$ (2, 2)

34. $\frac{x + 3}{2} = x + 4y$
$x = 5y$ $\left(\frac{15}{13}, \frac{3}{13}\right)$

35. $16x - 6y = -1 + 6y$
$\frac{2}{3}x = 4y + 3$ $\left(-\frac{5}{7}, -\frac{73}{84}\right)$

36. $2(5x - 3y) = 2x + 1$
$6y + \frac{9}{2} = x$ $\left(-\frac{1}{2}, -\frac{5}{6}\right)$

37. Solve the system $\begin{array}{l} Ax + By = C \\ -Ax + Dy = E \end{array}$ for *y*, where $B + D \neq 0$. $y = \frac{C + E}{B + D}$

Mixed Review

Solve.   **3.5, 5.2, 5.7, 7.7**

1. $6 = 7 - (4 - x)$
$x = 3$

2. $5x - 3 < 9 + 3x$
$x < 6$

3. $|x| < 2$
$-2 < x < 2$

4. $25 = y^2$
$y = 5$ or $y = -5$

Divide.   **8.4, 8.8**

5. $7x\overline{)49x^3 - 21x^2 + 14x}$ $7x^2 - 3x + 2$

6. $\frac{15a^4b}{9a^2 - 18b} \div \frac{5a^3b^3}{6a - 12}$ $\frac{2a^2 - 4a}{a^2b^2 - 2b^3}$

7. Find three consecutive integers whose sum is -54.   **4.3** $-19, -18, -17$

8. It takes James 10 h and Bob 15 h to plow a field. How long will it take them if they work together?   **9.5** 6 h

12.4 The Addition Method   **463**

Additional Answers, page 449

29.
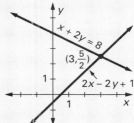

30.
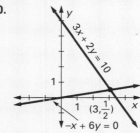

31.
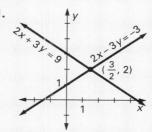

32.
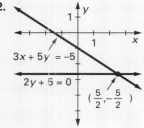

12.5 The Multiplication with Addition Method

Objective To solve systems of equations using multiplication with addition

In the system at the right, adding the two equations does not eliminate either variable. In such a case, you can use the Multiplication Property of Equality, as shown in Example 1.

$$\begin{array}{r} -x + y = -3 \\ 4x + y = 2 \\ \hline 3x + 2y = -1 \end{array}$$

EXAMPLE 1 Solve the system $\begin{array}{l} -x + y = -3 \\ 4x + y = 2 \end{array}$ using multiplication with addition.

Plan Multiply each side of the first equation by -1 to make the y terms opposites. Then solve the system using the addition method.

Solution

$$
\begin{array}{ll}
-1(-x + y) = -1(-3) & \\
x - y = 3 & (1) \\
4x + y = 2 & (2) \\
\hline
\text{Add (1) and (2).} \quad 5x \quad\quad = 5 & \\
x = 1 &
\end{array}
$$

Substitute 1 for x in (1).
$-x + y = -3$
$-1 + y = -3$
$y = -2$

Thus, the solution is $(1, -2)$. The check is left for you.

EXAMPLE 2 Solve the system. $\begin{array}{ll} -y = 7 - 4x & (1) \\ 5x + 3y = 13 & (2) \end{array}$

Plan First add $4x$ to each side of (1) to get the x and y terms on one side. Then multiply each side by 3 to make the y terms opposites.

Solution

$$
\begin{array}{ll}
(1) & -y = 7 - 4x \\
(3) & 4x - y = 7 \\
 & 3(4x - y) = 3 \cdot 7 \\
(4) & 12x - 3y = 21 \\
(2) & 5x + 3y = 13 \\
\hline
\text{Add (4) and (2).} & 17x \quad\quad = 34 \\
 & x = 2
\end{array}
$$

Substitute 2 for x in (2).
$5x + 3y = 13$
$5 \cdot 2 + 3y = 13$
$10 + 3y = 13$
$3y = 3$
$y = 1$

Thus, the solution is $(2, 1)$. The check is left for you.

TRY THIS Solve each system.

1. $x - 3y = 18$ $(12, -2)$
   $x + 5y = 2$

2. $-4y = x - 3$ $(-9, 3)$
   $2x + 3y = -9$

Prerequisite Quiz

Find the value of *a* and *b* in each statement that will make the expression equal to zero.

1. $4x - ax + 6y + by$ $a = 4, b = -6$
2. $4x - 3y + 5x + ax - by$
   $a = -9, b = -3$
3. $by + 2x - 5y + ax - 2x$ $a = 0, b = 5$
4. $2x - by + 3x - 5y - y + ax$
   $a = -5, b = -6$

Motivator

Show students the following system.
   $x + 3y = 9$
   $2x - y = 4$
Ask if the addition method can be used to solve this system. No Ask students how many solutions there are for the equation $2x - y = 4$. An infinite number Then ask how the solutions for $6x - 3y = 12$ compare with those for $2x - y = 4$. Same Ask whether the original system could be written equivalently as
   $x + 3y = 9$
   $6x - 3y = 12$ Yes
Ask them whether this equivalent system can be solved by the addition method.
Yes What is the solution? (3, 2)

Highlighting the Standards

Standards 1b, 4d, 2d: Here another strategy is added for solving systems of equations. The Focus on Reading verifies that students understand the procedure.

Additional Example 1

Solve by using multiplication with addition.
$x - y = 2$
$3x - y = 10$ (4, 2)

Additional Example 2

Solve the system.
$-y = 10 - 2x$
$3x + 2y = 22$ (6, 2)

This method of using addition with multiplication, if necessary, is sometimes called the *linear-combination* method of solving a system. You may need to use the Multiplication Property with both equations.

EXAMPLE 3 Solve the system $\begin{array}{ll} 5x + 6y = 14 & (1) \\ 3x - 4y = 16 & (2) \end{array}$

Solution Multiply Equation (1) by 3 and Equation (2) by -5.

$$3(5x + 6y) = 3 \cdot 14 \qquad \text{Substitute } -1 \text{ for } y \text{ in (1).}$$
$$-5(3x - 4y) = -5 \cdot 16 \qquad 5x + 6y = 14$$
$$15x + 18y = 42 \quad (3) \qquad 5x + 6(-1) = 14$$
$$-15x + 20y = -80 \quad (4) \qquad 5x - 6 = 14$$

Add (3) and (4).
$$38y = -38 \qquad 5x = 20$$
$$y = -1 \qquad x = 4$$

Thus, the solution is $(4, -1)$. The check is left for you.

EXAMPLE 4 At a fruit-packing plant, fruit of uniform size is shipped in gift baskets. For a large basket, 8 grapefruit and 14 oranges weigh 15 lb. For a small basket, 4 grapefruit and 10 oranges weight 9 lb. Find the approximate weight of a grapefruit and of an orange.

Solution Let g = the weight of a grapefruit, in pounds.
Let r = the weight of an orange, in pounds.

Solve the system: $\begin{array}{l} 8g + 14r = 15 \\ 4g + 10r = 9 \end{array}$. You will find that $r = \frac{1}{2}$ and $g = 1$.

Thus, a grapefruit weighs about 1 lb and an orange weighs about $\frac{1}{2}$ lb.

The solution and check are left for you.

TRY THIS 3. Two baseballs and a bat cost $41. Five baseballs and two bats cost $94. Find the cost of one baseball. $12

Focus on Reading

List five steps, in order, to solve the system: $\begin{array}{ll} (1) & 7x + 2y = 11 \\ (2) & -2x + 5y = 8 \end{array}$

b, e, c, d, a

a. Solve for x.
b. Multiply each side of Equation (1) by 2.
c. Add Equations (1) and (2) and solve for y.
d. Substitute the value for y in (1) or (2).
e. Multiply each side of Equation (2) by 7.

12.5 The Multiplication with Addition Method **465**

■■■■**TEACHING SUGGESTIONS**

Lesson Note

Both equations should be written in standard form, $Ax + By = C$, before using the Multiplication with Addition Method. The following arrangement for Example 3 may be easier for the students to understand.

$$5x + 6y = 14 \longrightarrow 15x + 18y = 42$$
$$3x - 4y = 16 \longrightarrow -15x + 20y = -80$$

In Example 3, you may want to show students that the system can be rewritten as

$$20x + 24y = 56$$
$$18x - 24y = 96$$

to eliminate the y terms first.

Math Connections

Linear Algebra: The solution of systems of linear equations in any number of variables is an important part of a college-level course called linear algebra. Systems having many equations are usually best solved by numerical methods which can be programmed into a computer.

Additional Example 3

Solve the system.
$4x + 2y = -2$
$3x - 5y = -34$ $(-3, 5)$

Additional Example 4

A melon and 8 apples weigh 4 lb. Two melons and 4 apples weigh 5 lb. Find the approximate weight of a melon and an apple.   A melon weighs about 2 lb and an apple weighs about $\frac{1}{4}$ lb.

Critical Thinking Questions

Application: Have students try to use the multiplication-addition method on the following systems and tell what their results indicate about these systems.

1. $2x + 4y = 20$
   $x + 2y = 10$ $0 = 0$ indicates an infinite number of solutions. These are equations of the same line.
2. $6x - 3y = 15$
   $2x - y = 6$ Empty set. These are equations of parallel lines.

Common Error Analysis

Error: Students sometimes forget to multiply both sides of an equation by the same number.

Remind them to apply the Distributive Property and multiply each term on each side of the equation by the same number.

Checkpoint

Solve each system using the multiplication–with–addition method.

1. $2x + y = -10$
   $-x + 2y = 0$ $(-4, -2)$
2. $2x + 6y = 9$
   $x - 2y = 2$ $(3, \frac{1}{2})$
3. $4x + y = -1$
   $x - 3y = 16$ $(1, -5)$

Classroom Exercises

Find a number by which you would multiply one equation before eliminating variables using addition.

1. $2x - 3y = 8$
   $-5x + y = 3$
   Mult (2) by 3.

2. $x + 2y = -1$
   $x - 3y = 2$
   Mult either by -1.

3. $7x - 6y = 8$
   $-2x - 2y = 1$
   Mult (2) by -3.

4. $3x - 2y = 5$
   $2x + 4y = 14$
   Mult (1) by 2.

Solve each system using multiplication with addition.

5. $x + 2y = 7$
   $x + 3y = 10$
   $(1, 3)$

6. $2x - y = -8$
   $3x + 6y = -12$
   $(-4, 0)$

7. $6p - 2q = -8$
   $3p + 7q = 4$
   $(-1, 1)$

8. $2x - 3y = 2$
   $3x - 4y = -1$
   $(-11, -8)$

Written Exercises

8. $(3, -1)$ 16. $(5, -2)$ 20. $\left(\frac{2}{3}, \frac{3}{4}\right)$

Solve each system using multiplication with addition.

1. $x + y = 4$
   $2x + 3y = 9$ $(3, 1)$

2. $5x + 2y = 7$
   $x + 4y = 5$ $(1, 1)$

3. $x + 4y = 14$
   $6x - 2y = 6$ $(2, 3)$

4. $3x + 5y = 11$
   $2x - 3y = 1$
   $(2, 1)$

5. $3x + 2y = 12$
   $2x + 5y = 8$ $(4, 0)$

6. $2s - 5t = 22$
   $2s - 3t = 6$ $(-9, -8)$

7. $2x - 7y = 3$
   $5x - 4y = -6$ $(-2, -1)$

8. $3x + 9y = 0$
   $11x - 2y = 35$

9. $3x - 7y = 8$
   $2x - 5y = 7$ $(-9, -5)$

10. $7p + 3q = 13$
    $3p - 2q = -1$ $(1, 2)$

11. $3x - 2y = 6$
    $5x + 7y = 41$ $(4, 3)$

12. $5x + 2y = 7$ $(1, 1)$
    $3x + 7y = 10$

13. $5x - 2y = 8$
    $3x = 5y + 1$ $(2, 1)$

14. $3c = 2d + 5$
    $2c + 5d = 16$ $(3, 2)$

15. $4x - 3y = 1$
    $y = -2x + 3$ $(1, 1)$

16. $2x + 7y = -4$
    $6x = 24 - 3y$

17. $10x = 3y + 2$
    $5x = 2y + 3$ $(-1, -4)$

18. $2x = 4y + 1$
    $x = y$ $(-\frac{1}{2}, -\frac{1}{2})$

19. $9x + 2y = 2$
    $21x + 6y = 4$ $(\frac{1}{3}, -\frac{1}{2})$

20. $3x - 4y = -1$
    $6x + 8y = 10$

21. $7y - \frac{1}{2}x = 2$
    $-2y = x + 4$
    $(-4, 0)$

22. $r + \frac{3}{2}s = 0$
    $s - 2 = -r$
    $(6, -4)$

23. $2.5c - d = 5$
    $4d = 3c - 6$
    $(2, 0)$

24. $x - 1.5y = 4$
    $0.8x = -0.4y$
    $(1, -2)$

Solve each problem.

25. Three pairs of sneakers and 2 pairs of socks cost $78. Two pairs of sneakers and 4 pairs of socks cost $60. Find the cost of a pair of sneakers and a pair of socks. Socks: $3; sneakers: $24

26. Harry bought 2 mints and 5 apples for $2.36. Craig bought 4 mints and 3 apples for $1.78. Find the cost of a mint and the cost of an apple.
   Mint: 13¢; apple: 42¢

27. When buying string for a kite, Jay found that 3 small spools of string and 1 large spool would give him a total of 115 yd of string. Two small spools and 2 large spools would give him a total of 130 yd. How many yards of string are in the small spool and in the large spool? 25 yd, 40 yd

28. When Jane babysat for 8 h and did odd jobs for 3 h, she made a total of $39. When she babysat for 2 h and did odd jobs for 5 h, she made a total of $31. How much does she charge per hour for babysitting and for doing odd jobs?
   Jobs: $5/h; babysitting: $3/h

29. A jacket with 1 pair of trousers costs \$114. A jacket with 2 pairs of trousers costs \$153. Find the cost of a jacket and the cost of a pair of trousers. Jacket: \$75; trousers: \$39

Solve each system using multiplication with addition.

30. $\dfrac{2(x + 6y)}{3} = 2$
$x + 10y = -1$ (9, −1)

31. $\dfrac{x - 2y}{8} = \dfrac{1}{2}$
$3x + 2y = 4$ (2, −1)

32. $x - \dfrac{3}{4}y = \dfrac{1}{4}$
$\dfrac{1}{2}x + \dfrac{1}{4}y = \dfrac{3}{4}$ (1, 1)

33. $\dfrac{6}{7}x - y = \dfrac{3}{7}$
$\dfrac{9}{7}x + 2y = \dfrac{1}{7}$ $\left(\dfrac{1}{3}, -\dfrac{1}{7}\right)$

34. $\dfrac{c}{6} + \dfrac{d}{4} = \dfrac{3}{2}$ (3, 4)
$\dfrac{2}{3}c - \dfrac{d}{2} = 0$

35. $6x - 3(y - 2x) = 3x + 15$ (2, 1)
$y - (4y - x) = 7x - 15$

36. Solve the system $\begin{array}{l} Ax + By = C \\ Dx + Ey = F \end{array}$ for x and y, where $AE \neq BD$. $x = \dfrac{CE - BF}{AE - BD}$; $y = \dfrac{AF - CD}{AE - BD}$

37. Use your answer to Exercise 36 and a calculator to solve the system $\begin{array}{l} 4x + 5y = 24.05 \\ 2x + 3y = 13.15 \end{array}$ (3.2, 2.25)

Midchapter Review

Solve each system of equations. Use the most convenient method.

1. $y = 6x + 8$ (−1, 2)
$y = -4x - 2$

2. $3x + y = 7$
$x - y = 5$ (3, −2)

3. $x = y$ (1, 1)
$3x + 4y = 7$

4. $5x - 4y = 13$
$-3x + 2y = 1$ (−15, −22)

5. $x = 4y - 3$
$3x + 5y = -1$ $\left(-\dfrac{19}{17}, \dfrac{8}{17}\right)$

6. $\dfrac{1}{3}x - \dfrac{3}{4}y = 5$
$\dfrac{2}{3}x - \dfrac{3}{4}y = 4$ (−3, −8)

7. $-5c + d = -13$
$c + d = -1$ (2, −3)

8. $-2x + 8y = -1$
$4x - 6y = 7$ $\left(\dfrac{5}{2}, \dfrac{1}{2}\right)$

Solve each problem using a system of two equations with two variables.

9. The sum of two numbers is 15. Five times the first number minus 2 times the second number is 12. Find the two numbers. 6, 9

10. At a resort, 1 round of golf and 2 tennis lessons cost \$55. Three rounds of golf and 5 tennis lessons cost \$150. At this rate, what is the cost of 1 round of golf and 1 tennis lesson? Golf: \$25; tennis: \$15

11. At an arcade game, you get 5 points each time you hit the target, but you lose 2 points each time you miss. After 26 tries, Jason's score was −3. How many hits and misses did he have? 7 hits, 19 misses

12. One egg and 2 slices of bacon cost \$1.40. Two eggs and 3 slices of bacon cost \$2.55. At this rate, what is the cost of 1 egg and the cost of 1 slice of bacon? Egg: 90¢; bacon: 25¢

13. Mary worked 2 h more than 3 times the number of hours that Victor worked. If they worked a total of 86 h, how long did each work? Mary: 65 h; Victor: 21 h

12.5 The Multiplication with Addition Method **467**

Closure

Have students suggest four different ways to use the multiplication-addition method to solve the following system of equations.
(1) $3x + 2y = 12$
(2) $2x + 5y = 8$
   Multiply (1) by 2 and (2) by −3.
   Multiply (1) by −2 and (2) by 3.
   Multiply (1) by 5 and (2) by −2.
   Multiply (1) by −5 and (2) by 2.

■■■FOLLOW UP

Guided Practice

Classroom Exercises 1–8
Try This   all; FR   all

Independent Practice

A Ex. 1–16, **B** Ex. 17–29, **C** Ex. 30–37

Basic: WE 1–19 odd, 25, 27, Midchapter Review all

Average: WE 9–31 odd, Midchapter Review

Above Average: WE 13–35 odd, 36, 37 Midchapter Review

Enrichment

Show students how to solve for x in the system, $ax + by = c$
$ dx + ey = f$
by going through the following steps.

$aex + bey = ce$
$-bdx - bey = -bf$
_____
$aex - bdx = ce - bf$
$x(ae - bd) = ce - bf$
$ x = \dfrac{ce - bf}{ae - bd}$

Have them use this result as a "formula" to solve for x in selected systems from the Written Exercises on page 466.

Prerequisite Quiz

Solve each system without using pencil and paper.

1. $x - 3y = 4$
   $x + 3y = 6$ $(5, \frac{1}{3})$
2. $2x - y = 4$
   $-2x + 2y = 1$ $(\frac{9}{2}, 5)$
3. $-x + y = -1$
   $x - 2y = 3$ $(-1, -2)$
4. $-8x + 2y = 4$
   $8x - y = 2$ $(1, 6)$

Motivator

Remind students that in the number 58, the 5 actually has a value of 50 since it is in the tens place. Then ask how they could write an expression using t and u to represent 58 if $t = 5$ and $u = 8$. $10t + u$ Ask how the number 85 can be represented by an expression in t and u. $10u + t$

12.6 Problem Solving: Digit Problems

Objective To solve two-digit number problems using systems of equations

Any two-digit number can be written in expanded form as

tens digit   units digit
↓
$36 = 10 \cdot 3 + 6$

The chart below introduces vocabulary that will be used in the examples and exercises.

Tens digits	Units digit	The two-digit number	Sum of the digits	Number with digits reversed
4	7	$10 \cdot 4 + 7$, or 47	$4 + 7$, or 11	$10 \cdot 7 + 4$, or 74
t	u	$10t + u$	$t + u$	$10u + t$

EXAMPLE 1 The sum of the digits of a two-digit number is 11. The number is 13 times the units digit. Find the number.

Solution

What are you to find?	a two-digit number
What is given?	The sum of the digits is 11. The number is 13 times the units digit.
Choose two variables. What do they represent?	Let t = the tens digit. Let u = the units digit. Then $10t + u$ = the two-digit number.
Write two equations.	(1) $t + u = 11$ (2) $10t + u = 13u$, or $10t - 12u = 0$

Solve the system.   Multiply (1) by -10.   Replace u by 5 in (1).

$$-10(t + u) = -10 \cdot 11 \qquad t + u = 11$$
$$\underline{\begin{array}{l} -10t - 10u = -110 \quad (3) \\ 10t - 12u = 0 \qquad\quad (2) \end{array}} \qquad t + 5 = 11$$
$$\begin{array}{r} -22u = -110 \\ u = 5 \end{array} \qquad\qquad t = 6$$

So, $10t + u = 65$.

Check   Check in the original problem.

Is the sum of the digits 11?
Yes, $6 + 5 = 11$

Is the number 13 times the units digit?
Yes, $65 = 13 \cdot 5$

State the answer.   The number is 65.

Highlighting the Standards

Standards 1b, 1c: Students apply their strategies for solving systems of equations as they formulate problems from stated situations.

Additional Example 1

The sum of the digits of a two-digit number is 16. The number is 11 times the units digit. Find the number.   88

EXAMPLE 2

Three times the tens digit of a two-digit number, increased by the units digit, is 16. If the digits are reversed, the new number is 1 less than twice the original number. Find the original number.

Solution

| What are you to find? | a two-digit number |

What is given?

3 · tens digit + units digit = 16
number with digits reversed =
2 · original number − 1

Choose two variables.
What do they represent?

Let t = the tens digit.
Let u = the units digit.
$10t + u$ = original two-digit number
$10u + t$ = new number (digits reversed)

Write two equations.

(1) $\quad 3t + u = 16$
(2) $\quad 10u + t = 2(10t + u) - 1$

Solve the system.

Solve Equation (1)
for u.

$$3t + u = 16$$
$$u = 16 - 3t$$

Write Equation (2)
with variables on
one side.

$$10u + t = 2(10t + u) - 1$$
$$10u + t = 20t + 2u - 1$$
(3) $\quad 8u - 19t = -1$

Replace u by $16 - 3t$
in Equation (3) and
solve for t.

$$8(16 - 3t) - 19t = -1$$
$$128 - 24t - 19t = -1$$
$$-43t = -129$$
$$t = 3$$

To find u, replace t by
3 in $u = 16 - 3t$.

$$u = 16 - 3 \cdot 3$$
$$u = 7$$

So, $10t + u = 10(3) + 7$, or 37.

Thus, the number is 37.

The check is left for you.

TRY THIS

1. The units digit of a two-digit number is 4 less than 6 times the tens digit. If the digits are reversed, the new number is 2 less than 3 times the original number. Find the original number. 28

Lesson Note

Give several examples like those in the chart until students become comfortable with the difference between t, which stands for the tens digit, and $10t$, which stands for the value of the tens digit. Also give several examples showing that $10t + u$ is the value of the original number and $10u + t$ is the value of the number with its digits reversed. Example 2 uses the substitution method to solve the system of equations. Remind students that they have two methods for solving systems and that they may use either method for any given problem, but that often one method is more convenient than the other.

Math Connections

History: A major difference between our present-day numerals (Hindu-Arabic) and Roman numerals is the importance of place value. Compare a number in the Roman system, such as XXX, with a number in our system, such as 222. In the Roman system, each X has the same value of 10; in our system, 222 has the value of $2(100) + 2(10) + 2(1)$.

Additional Example 2

Twice the tens digit of a two-digit number increased by the units digit is 13. If the digits are reversed, the new number is 5 more than 3 times the original number. Find the original number. 29

Critical Thinking Questions

Application: Have each student compile a list of ten two-digit numbers in a chart, such as the following.

Two Digit Number	Digits Reversed	Difference
85	58	27
46	64	−18

Then ask if they see any pattern in the last column of this chart.   Each entry is a positive or negative multiple of 9. Also, the difference between the tens and units digit of the original number tells what multiple of 9 the number in the last column is.

Common Error Analysis

Error: Some students may try to write the expression *tu*, rather than $10t + u$, for a two-digit number.

Remind them that *tu* would indicate the product of *t* and *u* rather than the appropriate value of the two-digit number.

Checkpoint

Find these two-digit numbers.

1. The units digit is 1 less than the tens digit, and the sum of the digits is 11.   65
2. The tens digit is 8 less than the units digit, and the sum of the digits is 10.   19
3. The tens digit is 6 less than 4 times the unit digit, and the units digit is half the tens digit.   63

Classroom Exercises

1. What is the sum of the digits of 48? 12
2. In 80, what is the units digit? 0
3. In the number 63, what is the tens digit? 6
4. What new number is formed when the digits of 25 are reversed? 52
5. Does reversing the digits of a two-digit number affect the sum of the digits? No

Solve each problem.

6. The sum of the digits of a two-digit number is 11. The tens digit is 7 more than the units digit. Find the number. 92
7. The sum of the digits of a two-digit number is 9. The number is 6 times the units digit. Find the number. 36
8. The units digit of a two-digit number is 5 times the tens digit. If the digits are reversed, the new number is 36 more than the original number. Find the original number. 15

Written Exercises

Solve each problem.

1. The tens digit of a two-digit number is 3 times the units digit. The difference between the digits is 6. Find the number. 93
2. The tens digit of a two-digit number is 4 more than the units digit. The number is 7 less than 8 times the sum of the digits. Find the number. 73
3. The sum of the digits of a two-digit number is 7. If the digits are reversed, the new number is 27 more than the original number. Find the original number. 25
4. The sum of the digits of a two-digit number is 10. If the digits are reversed, the new number is 18 less than the original number. Find the original number. 64
5. The units digit of a two-digit number is 4 times the tens digit. If the digits are reversed, the new number is 54 more than the original number. Find the original number. 28
6. The tens digit of a two-digit number is 3 times the units digit. If the digits are reversed, the new number is 36 less than the original number. Find the original number. 62
7. The tens digit of a two-digit number is 3 more than twice the units digit. If the digits are reversed, the new number is 54 less than the original number. Find the original number. 93
8. Four times the units digit of a two-digit number is 1 less than the tens digit. If the digits are reversed, the new number is 63 less than the original number. Find the original number. 92

9. The units digit of a two-digit number is 12 less than twice the tens digit. If the digits are reversed, the new number is 3 less than 8 times the tens digit of the original number. Find the original number. 96

10. The units digit of a two-digit number is 1 more than 4 times the tens digit. If the digits are reversed, the new number is 5 more than 3 times the original number. Find the original number. 29

11. Use h, t, and u to represent a three-digit number. Then represent the number with its digits reversed.

12. When the digits of a three-digit number are reversed, what is true of the tens digit? It remains the same.

13. In a three-digit number, the tens digit is twice the units digit, and the hundreds digit is 1 more than the units digit. The sum of the digits is 17. Find the number. 584

14. In a three-digit number, the hundreds digit is 6 less than the units digit. The tens digit is 1 more than the hundreds digit. The sum of the digits is 13. Find the number. 238

15. Show that if a two-digit number is added to the number with its digits reversed, the result is divisible by 11.

16. Show that if a two-digit number is subtracted from the number with its digits reversed, the result is divisible by 9.

17. Find all possible two-digit numbers such that twice the tens digit increased by the units digit is 19.
59, 67, 75, 83, 91
11. $100h + 10t + u$, $100u + 10t + h$

18. In a three-digit number, the hundreds digit and the tens digit are the same. The sum of the digits is 18. If the digits are reversed, the number is decreased by 297. Find the number. 774

Mixed Review

Factor completely. 7.4, 7.5, 7.6

1. $5a^2 - 19a - 4$
$(5a + 1)(a - 4)$

2. $36x - 4x^3$
$-4x(x + 3)(x - 3)$

3. $-9x^2 + 15x + 3x^3$
$3x(x^2 - 3x + 5)$

Simplify if possible. 8.5, 8.6, 8.7

4. $\dfrac{6}{5x} + \dfrac{9}{5x}$ $\dfrac{3}{x}$

5. $\dfrac{6}{y + 5} - \dfrac{4}{y - 5} + \dfrac{8}{y^2 - 25}$ $\dfrac{2y - 42}{y^2 - 25}$

6. $\dfrac{2}{3a^2} + \dfrac{4}{5a} - \dfrac{1}{9a^3}$ $\dfrac{36a^2 + 30a - 5}{45a^3}$

7. Find the slope of $\overleftrightarrow{AB}$ determined by points $A(-4, -2)$ and $B(2, 3)$. 11.1 $\dfrac{5}{6}$

8. Write an equation for $\overleftrightarrow{CD}$ determined by points $C(-1, 5)$ and $D(0, 2)$. 11.2 $y = -3x + 2$

9. Write an equation, in the form $Ax + By = C$, of a line with a slope of $-\dfrac{2}{5}$ and a y-intercept of -3. 11.2, 11.3 $2x + 5y = -15$

12.6 Problem Solving: Digit Problems **471**

471

◼◼◼GETTING STARTED

Prerequisite Quiz

Based on the given information, how old will each person be in 10 years?

1. Cora was 14 last year.   25
2. Brad will be 14 next year.   23
3. Gail was 5 five years ago.   20
4. Carlos will be 15 in six years.   19
5. Deb was 3 nine years ago.   22

Motivator

Give students the following puzzle. Janice Brown is three times as old as her son Mike. In 10 years, she will be twice as old as Mike. How old is Janice now? Allow students to use trial and error to find the answer.   30 Then suggest that writing a pair of equations would be a more efficient way to solve the puzzle.

◼◼◼TEACHING SUGGESTIONS

Lesson Note

Some students may find it helpful to make a chart such as the following for Example 1.

	Age now	8 yrs from now
Cathy:	c	$c + 8$
Paul:	p	$p + 8$
Equations:	$p = 3c$	$p + 8 = 2(c + 8)$

12.7 Problem Solving: Age Problems

Objective   To solve problems about ages using systems of equations

Problems about ages may involve ages in the past or in the future as well as in the present time. It is customary to let a variable represent the present age of a person. For example,

$$\text{Let } t = \text{Tim's age now.}$$
$$\text{Then } t + 6 = \text{Tim's age 6 years from now,}$$
$$t - 1 = \text{Tim's age last year,}$$

and so on.

EXAMPLE 1   Paul is 3 times as old as Cathy. In 8 years, he will be twice as old as she will be. How old is each now?

Solution

What are you to find?   Paul's and Cathy's ages now

What is given?   Paul's age now = $3 \cdot$ Cathy's age now
Paul's age in 8 years = $2 \cdot$ Cathy's age in 8 years

Choose two variables. What do they represent?   Let $c =$ Cathy's age now.
Then $c + 8 =$ her age in 8 years.
Let $p =$ Paul's age now.
Then $p + 8 =$ his age in 8 years.

Write two equations. $p = 3c$ (1)
$p + 8 = 2(c + 8)$ (2)

Solve the system.

Substitute $3c$ for p in (2). Solve for c.	Substitute 8 for c in (1). Solve for p.
$3c + 8 = 2(c + 8)$	$p = 3 \cdot 8$
$3c + 8 = 2c + 16$	$p = 24$
$c + 8 = 16$	
$c = 8$	

Check   Check in the original problem.   Is Paul 3 times as old as Cathy? Yes, $24 = 3 \cdot 8$. In 8 years, will Paul be twice as old as Cathy will be? Yes, since Paul will be 32 years old, Cathy will be 16 years old, and $32 = 2 \cdot 16$.

State the answer.   Cathy is 8 years old.
Paul is 24 years old.

Highlighting the Standards

Standards 5a, 5c, 1c: This chapter continues the application of systems of equations to problem solving.

Additional Example 1

Sue is 5 times as old as Gary. In 5 years she will be 3 times as old as he will be. How old is each now?   Sue: 25 yr old, Gary: 5 yr old

EXAMPLE 2 Elaine is 15 years younger than Al. Ten years ago, Al was 4 times as old as Elaine was then. How old is each now?

Solution Let e = Elaine's age now.
Let a = Al's age now.

Al's age now:	$a = e + 15$	(1)
Al's age 10 years ago:	$a - 10 = 4(e - 10)$	(2)

Substitute $e + 15$
for a in (2) and
solve for e.

$$e + 15 - 10 = 4e - 40$$
$$e + 5 = 4e - 40$$
$$45 = 3e$$
$$e = 15$$

Substitute 15 for e in (1). $a = 15 + 15$, or 30

Thus, Elaine is 15 years old now, and Al is 30 years old.
The check is left for you.

TRY THIS

1. David is 4 years younger than Joan. Six years ago, Joan was 3 times as old as David was then. How old is each now? David: 8 yr; Joan: 12 yr

Classroom Exercises

Give an expression for each age if j = Janet's present age.

1. Janet's age 5 years from now $j + 5$
2. twice Janet's age 4 years ago $2(j - 4)$

Refer to the problem below for Exercises 3–6. 3. Darlene's and Kara's ages now

Darlene is 4 times as old as Kara. In 10 years, she will be twice as old as Kara will be then. How old is each now?

3. What are you to find?
4. What is given?
5. Write two equations. Let k = Kara's age now. Let d = Darlene's age now. $d = 4k$, $d + 10 = 2(k + 10)$
6. Solve the system of equations and answer the question. Darlene: 20 yr; Kara: 5 yr

Solve each problem.

7. Leah is 8 years younger than Walt. Twelve years ago Walt was twice as old as Leah was then. How old is each now? Leah: 20 yr; Walt: 28 yr
8. Roberto is 30 years younger than Kate. In 12 years, Kate will be 3 times as old as Roberto will be. Find the present ages of Roberto and Kate. Kate: 33 yr; Roberto: 3 yr

Written Exercises
4. Darlene's age now = 4 · Kara's age now
Darlene's age in 10 years = 2 · Kara's age in 10 years

For each problem, find the age of each person now.

1. Mrs. O'Malley is 28 years older than her son Sean. She was 5 times as old as her son 14 years ago. Sean: 21; Mrs. O'Malley: 49
2. In 4 years, Mike will be 3 times as old as Chris will be. The sum of their ages now is 56. Chris: 12; Mike: 44

Math Connections

Life Skills: It has been suggested that dogs and cats age approximately seven years for each human year. This type of comparison gives you a convenient way to judge the age of pets and to make decisions about their health and care.

Critical Thinking Questions

Synthesis: Tell students that Cindy was born when Scott was 3 years old. Then have them use the x-axis to represent years in Cindy's age and the y-axis for Scott's age. After they have graphed several ordered pairs (Cindy's age, Scott's age) and connected them with a line, ask them how they could use the slope of some other line(s) to show when Scott is 2 times as old as Cindy, and to show when he is $1\frac{1}{2}$ times as old.   The slope of the line through the origin and a particular point of the line they have drawn will give the relative age of Scott to Cindy at that point.   slope $\overline{OA} = 2$

slope $\overline{OB} = \frac{3}{2}$

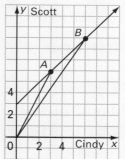

Common Error Analysis

Error: Some students may forget to add or subtract the appropriate number of years for *both* people involved in the problem. Remind them that in a years both ages will be increased by a.

Additional Example 2

Phil is 6 years older than Sandra. Eight years ago he was twice as old as she was. How old is each now? Phil: 20 yr old, Sandra: 14 yr old

Checkpoint

Lisa's age and Marissa's age total 30 years. Give the age of each for the additional given data in each exercise.

1. Marissa is 5 times as old as Lisa.
   Li: 5; M: 25
2. Lisa is twice as old as Marissa.
   Li 20; M: 10
3. Lisa is 2 years younger than Marissa.
   Li: 14; M: 16
4. Marissa is 6 years younger than twice Lisa's age.   Li: 12; M: 18
5. In 12 years, Lisa will be $\frac{4}{5}$ Marissa's age.   Li: 12; M: 18

Closure

Suggest to students that Mary is presently twice as old as Tony. Ask them to compare their ages 5 years ago.   Mary was more than twice as old as Tony.   Ask students to compare Mary's and Tony's ages 10 years from now.   Mary will be less than twice as old as Tony.

FOLLOW UP

Guided Practice

Classroom Exercises 1–8
Try This   all

Independent Practice

A Ex. 1–8, **B** Ex. 9–12, **C** Ex. 13–16
Basic: WE 1–8 all
Average: WE 5–12 all
Above Average: WE 1–15 odd

---

3. Four years ago, Shirley's age was 2 years more than 4 times Kim's age. Four years from now, she will be 3 times as old as Kim will be.   Kim: 18; Shirley: 62
4. A grandfather is 5 times as old as his granddaughter. In 6 years, he will be 4 times as old as his granddaughter will be.   90 yr, 18 yr
5. The sum of Mary's and Al's ages is 48. Twelve years ago, Mary was twice as old as Al.   Al: 20; Mary: 28
6. Kathy's father is 6 times as old as Kathy. Two years ago, he was 8 times as old as Kathy was.   K: 7; father: 42
7. Denise is 14 years older than Jack. In 10 years, the sum of their ages will be 60.   Denise: 27; Jack: 13
8. Wilma is 8 years younger than Josh. In 6 years, the sum of their ages will be 46.   Wilma: 13; Josh: 21
9. Regina is $\frac{3}{5}$ as old as Mort. Ten years from now, she will be $\frac{4}{5}$ as old as he will be.   Regina: 6; Mort: 10
10. Four years ago, Rich was $\frac{2}{3}$ as old as Marie. Six years from now, he will be $\frac{4}{5}$ as old as she will be.   R: 14; M: 19
11. In 16 years, Ian's age will be 4 years less than $\frac{1}{2}$ Judy's age then. The sum of their ages now is 36.   I: 4; J: 32
12. Elias is $2\frac{1}{2}$ times as old as his daughter. In $8\frac{1}{2}$ years, he will be twice as old as his daughter will be.
13. Penelope is $\frac{1}{2}$ as old as William. In $6\frac{1}{2}$ years, she will be $\frac{2}{3}$ as old as William will be.   Penelope: $6\frac{1}{2}$; William: 13
14. Carl is $1\frac{2}{3}$ times as old as his friend. In $\frac{1}{2}$ year, Carl will be $1\frac{3}{5}$ as old as his friend.   Carl: $7\frac{1}{2}$; friend: $4\frac{1}{2}$
15. Clint is 8 years older than Fern. Fern is $1\frac{1}{2}$ times as old as Geri. Clint's age will be twice Geri's age 4 years from now.   Clint: 20; Fern: 12; Geri: 8
16. Enid is 9 years older than Alice. Cathy was $\frac{3}{4}$ as old as Alice was 2 years ago. Next year, Enid will be twice as old as Cathy will be.   Enid: 23; Alice: 14; Cathy: 11

12. Elias: $42\frac{1}{2}$; daughter: 17

Mixed Review

Use the given formulas and values to find the perimeter.   *1.4*

1. $p = a + b + c$ 21 m
   $a = 6$ m, $b = 9$ m, $c = 6$ m

2. $p = 4s$ 48 cm
   $s = 12$ cm

3. Find the perimeter of a rectangle if the length is 6.2 cm and the width is 4.7 cm. Use the formula $p = 2l + 2w$.   *1.4*   21.8 cm
4. What is 15% of 42?   *4.7*   6.3
5. Six is what percent of 30?   *4.7*   20%
6. Write an equation for a line with slope $\frac{2}{3}$ and y-intercept -5.   *11.3* $y = \frac{2}{3}x - 5$

Enrichment

First, have students review the Enrichment lesson on page 467. Then ask them to rewrite the solution for x in determinant form by using coefficients of the first equation of the system for the top row of each determinant (numerator and denominator) and coefficients of the second equation for the second row in each determinant.

$$\frac{\begin{vmatrix} c & b \\ f & e \end{vmatrix}}{\begin{vmatrix} a & b \\ d & e \end{vmatrix}}$$

Now ask them to find a solution for y in determinant form.
(Hint: See the Enrichment on page 467.)

$$\frac{\begin{vmatrix} a & c \\ d & f \end{vmatrix}}{\begin{vmatrix} a & b \\ d & e \end{vmatrix}}$$

Then have students use these determinant forms to solve selected systems of equations.

12.8 Problem Solving: Coin and Mixture Problems

Objective   To solve coin and mixture problems using systems of equations

When solving problems about coins, you often need to express the value of the coins in cents, as shown in the chart below.

Coins	Number of coins	Value of coins in cents
7 nickels	7	$5 \cdot 7 = 35$
4 dimes	4	$10 \cdot 4 = 40$
3 dimes and 5 quarters	8	$10 \cdot 3 + 25 \cdot 5 = 155$
n nickels and q quarters	$n + q$	$5n + 25q$

Cents are used rather than dollars in order to avoid decimals.

EXAMPLE 1   Jane has a collection of nickels and quarters worth $3.05. She has 7 more nickels than quarters. How many coins of each type does she have?

Solution

What are you to find?	the number of nickels and the number of quarters
What is given?	value of nickels + value of quarters = 305¢ number of nickels = number of quarters + 7
Choose two variables. What do they represent?	Let n = the number of nickels. Let q = the number of quarters.
Write two equations.	(1) $5n + 25q = 305$ (2) $n = q + 7$

Solve the system.

Substitute $q + 7$ for n in (1). Solve for q.

$5(q + 7) + 25q = 305$
$5q + 35 + 25q = 305$
$30q = 270$
$q = 9$

Substitute 9 for q in (2). Solve for n.

$n = 9 + 7$
$n = 16$

Check

Check in the original problem.	Are 9 quarters and 16 nickels worth $3.05? Yes, because $9 \cdot 25 + 16 \cdot 5 = 225 + 80 = 305$ cents. Are there 7 more nickels than quarters? Yes, because $16 = 9 + 7$.
State the answer.	Jane has 16 nickels and 9 quarters.

▬ GETTING STARTED

Prerequisite Quiz

Attempt to solve these problems mentally without using pencil and paper.

1. What is 20% of 60?   12
2. Four is what percent of 8?   50%
3. Three is what percent of 12?   25%
4. Six is 10% of what number?   60
5. Four is 1% of what number?   400
6. What is 3% of 600?   18

Motivator

Model: Show students a handful of change (quarters, nickels, and dimes), tell them how many of each you have, and ask them to quickly calculate the total value in cents. Take time to discuss the difference between the *number* of coins you have and their total *value*. Then ask them to give the value of a collection of q quarters, n nickels, and d dimes. $25q + 5n + 10d$ Ask how many coins this is. $q + n + d$

Additional Example 1

Ruby has a collection of nickels and quarters worth $3.70. She has 14 more nickels than quarters. How many coins of each type does she have?   24 nickels and 10 quarters

Highlighting the Standards

Standards 1a, 1b, 4c: These problems relate to such fields as chemistry, nursing, and pharmacy.

Lesson Note

Emphasize that the first equation in Example 1 is a "value of" equation, while the second equation is a "number of" equation. This distinction should help the students to write the necessary two equations to model each problem. When a problem has been analyzed and a system of equations has been found, students should determine on their own which method of solution would be easier, addition with multiplication, substitution, or graphing. In Example 3, point out that the first equation represents the *number* of gallons of each solution in the mixture. The second equation represents the amount of alcohol (value) in each container. Remind students that a 20%-alcohol solution contains 20% alcohol and 80% water.

Math Connections

Chemistry: A chemist who needs to dilute a 10%-alcohol solution to a 5%-alcohol solution must know exactly how much water to add to get the correct ratio of alcohol to water.

TRY THIS

1. John has a collection of dimes and quarters worth $4.50. He has 4 more quarters than dimes. How many coins of each type does he have? dimes: 10; quarters: 14

The same method that is used for solving coin problems can be used to solve other problems about mixtures.

EXAMPLE 2   A coffee wholesaler mixes Colombian beans selling at $1.20/lb with Venezuelan beans selling at $1.60/lb. He wants a mixture of 90 pounds to sell at $1.24/lb. How many pounds of each should he use?

Solution

| What are you to find? | The number of pounds of each kind of coffee bean needed for a 90-lb mixture worth $1.24/lb. |

What is given?   Colombian beans sell at $1.20/lb.
Venezuelan beans sell at $1.60/lb.
90 lb are wanted in all.

Choose two variables.   Let c = the number of pounds of
What do they represent?   Colombian beans.
Let v = the number of pounds of
Venezuelan beans.

Write two equations.
$$c + v = 90 \qquad (1)$$
$$120c + 160v = 124 \cdot 90 \quad (2)$$

Solve the system.
Solve (1) for c. $c = 90 - v$
Substitute in (2). $120(90 - v) + 160v = 124 \cdot 90$
Solve for v. $10{,}800 - 120v + 160v = 11{,}160$
$$v = 9$$

Substitute in (1). $c + 9 = 90$
$$c = 81$$

Check in the original problem.   Does the mixture weigh 90 lb?
Yes, $9 + 81 = 90$.

Does the value of 81 lb of Colombian beans plus the value of 9 lb of Venezuelan beans equal the value of the mixture? Yes, since $81 \times \$1.20 + 9 \times \$1.60 = \$97.20 + \$14.40$ and $90 \times \$1.24 = \$111.60$.

State the answer.   The wholesaler should use 81 lb of Colombian beans and 9 lb of Venezuelan beans.

EXAMPLE 3   A 20% alcohol solution is mixed with a 30% alcohol solution to obtain 25 gal of a 24% solution. How many gal of each are needed?

Additional Example 2

A packager of frozen foods mixes shelled peas that sell for $1.80/lb with diced carrots that sell for 90 ¢/lb. He wants a mixture of 200 pounds to sell at a $1.20/lb. How many pounds of each should he use?   The packager should use $133\frac{1}{3}$ lb of carrots and $66\frac{2}{3}$ lb of peas.

Additional Example 3

A 40% alcohol solution is mixed with a 70% alcohol solution to obtain 60 gallons of a 50% solution. How many gallons of each are needed?   40 gal of 40% solution and 20 gal of 70% solution

Solution

Let l = the number of gal of the 20% solution.
Let h = the number of gal of the 30% solution.

$$\underset{\text{of 20\% solution}}{\text{no. of gal}} + \underset{\text{of 30\% solution}}{\text{no. of gal}} = 25$$
$$l \quad + \quad h \quad = 25$$

$$\underset{\text{20\% solution}}{\text{alcohol in}} + \underset{\text{30\% solution}}{\text{alcohol in}} = \underset{\text{24\% solution}}{\text{alcohol in}}$$
$$0.20\,l \quad + \quad 0.30h \quad = \quad 0.24(25)$$

Solve the system.

$$l + h = 25 \qquad (1)$$
$$0.20l + 0.30h = 0.24(25) \qquad (2)$$

Multiply (2) by 100. $\qquad 20l + 30h = 600 \qquad (3)$
Multiply (1) by -20. $\qquad -20l - 20h = -500 \qquad (4)$

Add equations (3) and (4). $\qquad 10h = 100$
$$h = 10$$

Substitute 10 for h in (1). $\qquad l + 10 = 25$
$$l = 15$$

Thus, 15 gal of the 20% solution and 10 gal of the 30% solution are needed. The check is left for you.

TRY THIS

2. A 36% salt solution is mixed with a 42% salt solution to obtain 21 gallons of a 40% solution. How many gallons of each are needed?
14 gallons of 42% solution; 7 gallons of 36% solution

Classroom Exercises

Express the total value in cents. (Exercises 1–6)

1. q quarters and d dimes $25q + 10d$
2. n nickels and q quarters $5n + 25q$
3. b dollars and d dimes $100b + 10d$
4. n nickels and k dollars $5n + 100k$
5. c tickets at \$1.75 each and r tickets at \$3.00 each $175c + 300r$
6. r lb of oranges at 45¢/lb and q lb of grapefruit at 49¢/lb $45r + 49q$
7. Sarah has \$2.30 in quarters and dimes. She has 5 fewer dimes than quarters. How many coins of each type does she have? 8 quarters, 3 dimes
8. A cash register contains 15 coins in dimes and nickels. The total value is \$1.25. How many coins of each type are there? 5 nickels, 10 dimes

Written Exercises

Solve.

1. José has 6 more dimes than quarters. He has \$1.65 total. How many coins of each type does he have?
9 dimes, 3 quarters

2. If 24 coins in half-dollars and dimes are worth \$3.60, how many coins of each type are there?
21 dimes, 3 half-dollars

Critical Thinking Questions

Application: Pose this problem to students: You have 2 gallons of a 25% salt solution, and you need a 30% solution in order to complete a scientific experiment before class tomorrow. It is late, so all the stores are closed, and you have no more salt. How can you make your solution stronger without adding salt? Boil off some of the water. How could you use linear equations to determine when your solution is the correct strength? Let x = amount of water removed; $.75(2) - x = .7(2 - x)$; $x = \frac{1}{3}$

Common Error Analysis

Error: To represent a quantity such as "7 more nickels than quarters," students will often write $q = n + 7$ rather than $n = q + 7$.

Ask them to decide whether there are fewer quarters or nickels. Then have them add 7 to the number of quarters to get the number of nickels.

Checkpoint

A gourmet food shop sells a 5 lb box of fruit and nuts for $15/lb. As the costs of fruits and nuts vary, the shop has to vary the percents of each ingredient in the 5-lb box. Based on these facts, complete the following table.

	COST PER LB		AMOUNT (LB)	
	Fruits	Nuts	Fruits	Nuts
1.	$12	$20	$3\frac{1}{8}$	$1\frac{7}{8}$
2.	$18	$12	$2\frac{1}{2}$	$2\frac{1}{2}$
3.	$5	$30	3	2
4.	$8	$20	$2\frac{1}{12}$	$2\frac{11}{12}$

Closure

Ask students to state how the two equations in a coin problem are related to the two equations in a mixture problem.   Both have a "value" equation and a "number" equation.

▉▉▉ FOLLOW UP

Guided Practice

Classroom Exercises 1–8
Try This   all

Independent Practice

A Ex. 1–6, **B** Ex. 7–11, **C** Ex. 12–15

Basic: WE 1–6 all

Average: WE 1–11 odd

Above Average: WE 5–15 odd

---

3. A box of walnuts mixed with pecans costs $15.75. Walnuts cost $3.75/lb and pecans cost $4.50/lb. The number of pounds of walnuts is 3 times the number of pounds of pecans. How many pounds of each type are there? p: 1 lb; w: 3 lb

4. Heidi sold 56 tickets to a play and collected $215.00. Adult tickets cost $4.50 each and student tickets cost $3.50 each. How many tickets of each kind did she sell? 19 adults, 37 students

5. A 10% salt solution is mixed with an 18% salt solution to obtain 32 oz of a 15% solution. How many ounces of each are needed? 12 oz of 10%; 20 oz of 18%

6. A 15% acid solution is mixed with a 25% acid solution to obtain 20 gallons of a 21% acid solution. How many gallons of each are needed? 8 g of 15%; 12 g of 25%

7. Ed has 95¢ in dimes and nickels. The total number of coins is 1 more than twice the number of dimes. How many coins of each type are there? n: 7; d: 6

8. Angelo has $1.90 in dimes and nickels. If he has 4 fewer nickels than 5 times the number of dimes, how many dimes does he have? 6

9. Milk that is 2% butterfat is mixed with milk that is 4% butterfat to make 10 gallons that is 3.25% butterfat. How many gallons of each type are needed? $6\frac{1}{4}$ g of 4%; $3\frac{3}{4}$ g of 2%

10. Thirty quarts of a 24% iodine solution were mixed with a 52% solution to make a 40% iodine solution. How many quarts of the 52% solution were needed? 40

11. Ms. Garcia wants to sell a box of fruit that contains cherries and plums for $19.20. The cherries sell for $2.40/kg. The plums sell for $3.60/kg. The total number of kilograms of fruit is 3 kg more than twice the number of kilograms of plums. How many kilograms of each type of fruit are in the box? 2 kg of plums, 5 kg of cherries

12. A pharmacist wants to add water to a solution that contains 80% medicine. She wants to obtain 12 oz of a solution that is 20% medicine. How much water and how much of the 80% solution should she use? 9 oz water; 3 oz of 80% solution

13. Brine is a solution of salt and water. If a tub contains 50 lb of a 5% salt solution of brine, how much water must evaporate to change it to an 8% solution? $18\frac{3}{4}$, or 18.75 lb

14. How many liters of water must be evaporated from 60 liters of a 12% acid solution to make it a 36% acid solution? 40 liters

15. How many liters of a 72% alcohol solution must be added to 15 liters of an 18% solution to obtain a 25% alcohol solution? $\frac{105}{47}$, or $2\frac{11}{47}$ liters

Mixed Review

Solve for x.   3.4, 5.6, 5.7, 9.1, 9.3

1. $-14 + 3x = 6 + 2x - 5$ x = 15

2. $4x + 2a = b$ $x = \frac{b-2a}{4}$

3. $3|x + 2| = 12$ x = 2 or x = −6

4. $|x - 5| < -4$ No solution

5. $\frac{14}{x^2 + 7x - 18} = \frac{x + 5}{x + 9} + \frac{4}{x - 2}$ x = −4 or x = −3

Give the slope and the y-intercept of the line with the given equation.   11.3

6. $y = 4x - 2$ m = 4; b = −2

7. $y = x$ m = 1; b = 0

8. $2x - 4y = 7$ $m = \frac{1}{2}$; $b = -\frac{7}{4}$

Enrichment

Challenge students to devise a strategy to solve this problem. In a class of 36 students, 18 belong to the Art Club, 18 belong to the Drama Club, and 19 belong to the Outdoor Club. Three students belong to all 3 clubs. Seven belong to both the Art and Outdoor Clubs. Eight belong to both the Art and Drama Clubs, and 8 belong to both the Drama and Outdoor Clubs.

1. How many belong to the Art Club only? 6

2. How many belong to the Drama Club only? 5

3. How many belong to the Outdoor Club only? 7

4. How many belong to none of the three clubs? 1

A Venn diagram explains the answers.

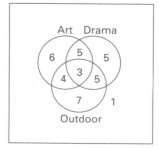

12.9 Problem Solving: Motion Problems

Teaching Resources

Project Worksheet 12
Quick Quizzes 95
Reteaching and Practice
  Worksheets 95

Objective   To solve motion problems using systems of equations

Problems related to distance, rate, and time are based on the formula $d = rt$. It is often helpful to draw a diagram and organize the data in a chart before writing the equations.

EXAMPLE 1   A freight train left Pennsylvania Station traveling at 35 mi/h. Two hours later, a high-speed train left Pennsylvania Station on parallel tracks traveling 55 mi/h. In how many hours after the slow train starts will the two trains meet?

Solution   Let t = time (hours) for the slow train and
$t - 2$ = time for the fast train.
Let d = distance (miles) each train travels before they meet.

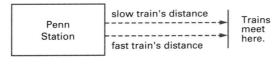

distance	= rate	× time
Slow train	d	35	t
Fast train	d	55	$t - 2$

$$d = 35t \quad (1)$$
$$d = 55(t - 2) \quad (2)$$

Substitute $35t$ for d in (2).

$$35t = 55(t - 2)$$
$$35t = 55t - 110$$
$$-20t = -110$$
$$t = 5\frac{1}{2}$$
$$\text{and } t - 2 = 3\frac{1}{2}$$

Check   Is the distance traveled at 35 mi/h for $5\frac{1}{2}$ h the same as the distance traveled at 55 mi/h for $3\frac{1}{2}$ h? Yes, since
$$35\left(\frac{11}{2}\right) = 55\left(\frac{7}{2}\right) = \frac{385}{2}, \text{ or } 192\frac{1}{2} \text{ mi.}$$

Thus, the trains will meet $5\frac{1}{2}$ h after the slow train starts.

Prerequisite Quiz

Answer each question without using pencil and paper.

1. How fast do you have to drive in order to travel 120 mi in 3 h?   40 mi/h
2. At an average rate of 45 mi/h, how far will you travel in 4 h?   180 mi
3. What distance will you walk in 4 h if you walk at an average rate of 3.5 mi/h?   14 mi
4. What is your average rate of speed if you walk 15 mi in $2\frac{1}{2}$ h?   6 mi/h

Motivator

Review with students the meaning of the formula $d = rt$. Use simple examples such as those given in the Prerequisite Quiz to help them see the relationship between the three variables. Given r and t, they have a multiplication problem. Any other combination gives a division problem.

Additional Example 1

A propeller-driven airplane leaves Kennedy International Airport traveling 200 mi/h. Three hours later, a jet leaves the airport on a parallel route traveling at 600 mi/h. In how many hours will the jet overtake the propeller–driven plane? $1\frac{1}{2}$ h

Highlighting the Standards

Standards 1a, 1b: These applications focus on using systems of equations to solve motion problems.

Lesson Note

Encourage students to set up tables similar to the one used in Example 1 as a means of finding a system of equations. When presenting Example 2, remind students that the wind affects the speed of an airplane, just as a current affects the speed of a boat. For example, if the speed (rate) of a motorboat is 20 mi/h and the rate of the current is 2 mi/h, then the rate of the boat going downstream is 20 + 2, or 22 mi/h. The rate of the boat going upstream is 20 − 2, or 18 mi/h.

Math Connections

Advanced Math: The expression $\frac{2ab}{a+b}$ is called the harmonic mean of the numbers a and b. This formula can be used to find an average speed for a complete trip when the speed going is different than the returning speed.

Critical Thinking Questions

Evaluation: Bob drove to a city across the state at a rate of 40 mi/h. On the return trip, he drove at a rate of 60 mi/h. Ron decided that Bob's average rate for the whole trip was 50 mi/h. Ask students to explain or demonstrate whether Ron's conclusion is valid or not.   Not valid. Suppose the city is 240 miles away. At a rate of 40 mi/h the trip there would take 6 hours; the return trip at a rate of 60 mi/h would take 4 hours. The complete trip would be a distance of 480 miles and would take 10 hours to complete. This is at an average rate of 48 mi/h.

EXAMPLE 2   The distance between Chicago and New York is 735 mi. A plane left Chicago flying with the wind and landed in New York in 1 h 45 min. Then the plane left New York flying against the same wind, and landed in Chicago after 2 h. Find the rate of the plane in calm air and the rate of the wind.

Solution   Let w = rate of wind.   Let r = rate of plane in calm air.
Then $r + w$ = rate of the plane with the wind and
$r - w$ = rate of the plane against the wind.

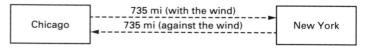

distance	=	rate	×	time
with wind	735	$r + w$	$1\frac{3}{4}$	← 1 h 45 min
against wind	735	$r - w$	2	

Solve the system.

$$735 = (r + w) \cdot \tfrac{7}{4} \quad (1)$$
$$735 = (r - w) \cdot 2 \quad (2)$$

Multiply each side of (1) by $\frac{4}{7}$.

$$\tfrac{4}{7} \cdot 735 = r + w$$
$$420 = r + w \quad (3)$$

Multiply each side of (3) by 2. Then add Equation (2).

$$840 = 2r + 2w$$
$$\underline{735 = 2r - 2w \quad (2)}$$
$$1{,}575 = 4r$$
$$393\tfrac{3}{4} = r$$

Substitute $393\frac{3}{4}$ for r in (3).

$$420 = 393\tfrac{3}{4} + w$$
$$26\tfrac{1}{4} = w$$

Thus, the rate of the plane in calm air is $393\frac{3}{4}$ mi/h and the rate of the wind is $26\frac{1}{4}$ mi/h. The check is left for you.

TRY THIS   A car left Dodge City traveling at 50 mi/h. One hour 30 min later, another car left Dodge City using the same highway traveling at 60 mi/h. In how many hours after the slow car starts will the two cars meet? 9 h

Additional Example 2

The distance between New York City and Houston is about 1,600 mi. A plane left N.Y.C. flying with the wind, and landed in Houston in 4 h. Then the plane left Houston flying against the same wind, and landed in N.Y.C. in 4.5 h. Find the rate of the plane in still air and the rate of the wind. $377\frac{7}{9}$ mi/h and $22\frac{2}{9}$ mi/h

Classroom Exercises

Complete the table. All rates are in miles per hour.

	Rate of boat in still water	Rate of current	Rate of boat upstream	Rate of boat downstream
1.	10	2	8	12
2.	25	3	22	28
3.	14	1	13	15
4.	19	5	14	24
5.	18	6	12	24
6.	30	5	25	35
7.	r	c	$r - c$	$r + c$

8. On a bike hike, Jim rode his bike at a rate of 5 mi/h. Barry followed Jim, leaving from the same place 1 hour later. He rode at a rate of 8 mi/h. In how many hours did Barry catch up with Jim? $1\frac{2}{3}$

Written Exercises

Solve each problem.

1. A freight train left Columbus traveling 40 mi/h. Two hours later a passenger train left the same station traveling 50 mi/h on a parallel track. How many hours after the freight train left will the trains meet? 10 h

2. A car left Detroit traveling 45 mi/h. A second car left from the same place $1\frac{1}{2}$ h later. It traveled on the same road and met the first car after driving $4\frac{1}{2}$ h. Find the rate of the second car. 60 mi/h

3. A motorboat went 12 mi upstream in 40 min. It made the return trip downstream with the same current in 15 min. Find the rate of the boat in still water. 33 mi/h

4. A rowing team rowed their boat 12 mi upstream in $1\frac{3}{4}$ h. They made the return trip downstream with the same current in 1 h. Find the rate of the current. $2\frac{4}{7}$ mi/h

5. A pilot flew a distance of 960 mi in 8 h against the wind. On the return trip he made the flight in 6 h with the same wind behind him. What was the rate of the plane in still air? 140 mi/h

6. A carrier pigeon flew against the wind for 1 h to deliver a message. It made the return trip with the same wind in 45 min. The trip was 6 mi each way. Find the rate of the pigeon in still air and the rate of the wind.
Pigeon: 7 mi/h; wind: 1 mi/h

12.9 Problem Solving: Motion Problems   **481**

Common Error Analysis

Error: When working on word problems, students often fail to answer the question after they have solved the system of equations. Remind students that a simple statement such as, "The rate of the plane is 400 mi/h and of the wind is 20 mi/h," is much more meaningful as an answer.

Checkpoint

1. A boy leaves home at 12 noon on his bicycle and travels at 10 mi/h. An hour later his father leaves home and travels the same direction in his car at 30 mi/h. At what time will the father catch up to the boy?   1:30 P.M.
2. Refer to Exercise 1. At what time would the father have to leave the house to catch up to the boy at 1:00 P.M.?   12:40 P.M.

Closure

Ask students to write three simple problems using the distance formula that can be solved without using paper and pencil. Each of the three problems should require solving for a different variable. Have several students read out their problems for others to solve.

◼️ FOLLOW UP

Guided Practice

Classroom Exercises 1–8
Try This   all

Independent Practice

🅰 Ex. 1–7, 🅱 Ex. 8–11, 🅲 Ex. 12–15
Basic: WE 1–6 all
Average: WE 1–11 odd
Above Average: WE 5–15 odd

Additional Answers

Mixed Review

1.
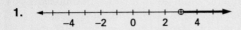

2.
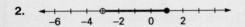

3.
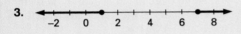

---

7. Jean and José started bicycling from the same spot in the same direction. Jean rode at a rate of 12 mi/h. José rode at a rate of 10 mi/h. After how many hours will they be 10 mi apart? 5 h

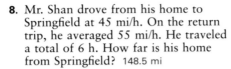

8. Mr. Shan drove from his home to Springfield at 45 mi/h. On the return trip, he averaged 55 mi/h. He traveled a total of 6 h. How far is his home from Springfield? 148.5 mi

9. At what rate did a car travel to catch up with a motorcycle in 6 h if the motorcycle traveled at 30 mi/h and left 3 h before the car? 45 mi/h

10. The current of the Wahakee River runs 2 mi/h. It took Marcia and Jessie 45 min to paddle upstream and 30 min to come back with the current. What was their rate of paddling in still water? 10 mi/h

11. A fish swam upstream for 1 h and then returned downstream with the current. It took only 40 min to return to its starting point. The rate of the current is 2 mi/h. How far did the fish travel each way? 8 mi

12. The rate of a boat in still water is 12 mi/h. The boat can travel 18 mi upstream in the same time that it can travel 24 mi downstream. Find the rate of the current. $\frac{12}{7}$ mi/h

13. A plane flew 900 mi. If it had flown 75 mi/h faster, the plane could have traveled 1,350 mi in the same time. Find the speed of the plane during the 900-mile trip. 150 mi/h

14. A bus traveled 180 mi. If weather conditions had been better, it could have driven 5 mi/h faster and completed the trip in 30 min less time. How fast did the bus travel? 40 mi/h

15. Let r be the rate of a boat in still water and c be the rate of the current. Show that the rate of the boat going downstream is equal to the rate of the boat going upstream plus twice the rate of the current. Let $r + c$ = rate downstream. Then $r + c = r + c + c - c = (r - c) + (c + c) = (r - c) + 2c$ = rate upstream + 2 (rate of current).

Mixed Review

Graph the solution set. *5.2, 5.3, 5.7*

1. $2y - 15 > -9$

2. $x > -3$ *and* $x \le 1$

3. $|y - 4| \ge 3$

Multiply. *6.8, 6.9, 6.10*

4. $3a(2a^2 + 6a + 5)$ $6a^3 + 18a^2 + 15a$

5. $(x + 5)(x - 9)$ $x^2 - 4x - 45$

6. $(y + 11)^2$ $y^2 + 22y + 121$

7. Use the formula $A = 4s^2$ to find A for $s = 7$. *1.4* $A = 196$

8. Jack's hourly wage jumped from \$5 to \$7. What was the percent increase in his wage? *4.7* 40%

9. Separate 80 people into two groups such that one group is 4 times as large as the other. *4.2* 64, 16

482 Chapter 12 Systems of Linear Equations

Enrichment

Tell students that if you travel in one direction at a rate of r_1 mph and return at a rate of r_2 mph, the average rate of the trip will be the harmonic mean of r_1 and r_2. To find the harmonic mean of two numbers "take the reciprocal of the mean (average) of the reciprocals of the two numbers." Have students use this definition to find the harmonic mean of r_1 and r_2.

$$\frac{2r_1r_2}{r_1 + r_2}$$

Then have them find the average rate for the following round trips, where r_1 is the rate going and r_2 is the rate returning. Assume all rates are in mi/h.

1. $r_1 = 30$, $r_2 = 50$ 37.5
2. $r_1 = 30$, $r_2 = 60$ 40
3. $r_1 = 40$, $r_2 = 60$ 48

12.10 Systems of Inequalities

Objective   To solve systems of inequalities by graphing

In Chapter 11 you graphed linear inequalities such as $y > x - 1$. In this lesson you will graph *systems* of linear inequalities.

The solution set of $y > x - 1$ consists of the coordinates of every point in the shaded region above the line $y = x - 1$.

To solve a system of inequalities such as

$$y > x - 1$$
$$3x + 2y < 6$$

by graphing, locate all points whose coordinates satisfy *both* inequalities. Graph both inequalities in the same coordinate plane. The intersection of the two regions will contain all points whose coordinates are solutions of the system.

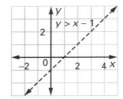

EXAMPLE 1   Solve the system $\begin{array}{l} y > x - 1 \\ 3x + 2y \leq 6 \end{array}$ by graphing.

Solution   Graph the boundary for each half-plane by the slope-intercept method.

Graph $y = x - 1$ with a dashed line since the boundary is not included. Shade the open half-plane above the line.

$3x + 2y \leq 6$, or $y \leq -\frac{3}{2}x + 3$

Graph $y = -\frac{3}{2}x + 3$ with a solid line since the boundary is included. Shade the closed half-plane *below* the line.

To check, choose a point inside the double-shaded region. Try $(0, 0)$.

$y > x - 1$	$3x + 2y \leq 6$
$0 \overset{?}{>} 0 - 1$	$3 \cdot 0 + 2 \cdot 0 \overset{?}{\leq} 6$
$0 \overset{?}{>} -1$	$0 + 0 \overset{?}{\leq} 6$
$0 > -1$ True	$0 \leq 6$ True

Thus, the double-shaded region contains all points whose coordinates are the solutions of the system. There are infinitely many solutions.

12.10 Systems of Inequalities   **483**

Additional Example 1

Solve by graphing.
$y < x + 2$
$2x + y \geq -2$

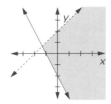

Teaching Resources

Quick Quizzes 96
**Reteaching and Practice
Worksheets** 96

GETTING STARTED

Prerequisite Quiz

Solve for x.

1. $5x - 1 > 14$ $x > 3$
2. $6 - x \geq -4$ $x \leq 10$
3. $2 - 3x < x - 10$ $x > 3$
4. $x - 5 \leq 2x + 1$ $x \geq -6$
5. $3x - 4 < -4x + 10$ $x < 2$

Motivator

Ask students to describe the location of all points in the coordinate plane that are solutions of the equation $y = 2x + 3$. On a straight line which is the graph of $y = 2x + 3$.   Ask them to determine whether the point $(0, 0)$ is above or below this line.   Below   Ask them to determine whether the point $(-4, 2)$ is above or below the line.   Above

TEACHING SUGGESTIONS

Lesson Note

Show students the graph of the system of Example 2 with the inequality signs reversed: $2x < 4 - y$ and $y > -2x - 1$. The solution will be the points which lie between the two parallel lines. Before presenting Example 3, have students describe the graphs of $x = -3$ (vertical line) and $y = 2$ (horizontal line).

Highlighting the Standards

Standards 4b, 4c, 4d: The skills of the previous lessons are extended to include solving systems of inequalities. The Application then relates this to the use of linear programming to maximize profit.

Math Connections

Computers: Computers are used to solve complex problems using linear programming with many variables and constraints (see the Application on page 486). There are numerous applications for these in business and industry.

Critical Thinking Questions

Analysis: Show the following shaded region to the students and ask them to write a system of linear inequalities whose graphs would intersect to give this region.

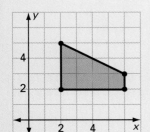

$y \geq 2$
$x \geq 2$
$x \leq 6$
$y \leq -\frac{1}{2}x + 6$

Common Error Analysis

Error: Some students will shade the incorrect side of the line when graphing an inequality. Have them check at least two points to verify each graph before continuing with the other graphs for the problem.

EXAMPLE 2   Solve the system $\begin{array}{c} 2x > 4 - y \\ y < -2x - 1 \end{array}$ by graphing.

Solution   Find the equation of the boundary line of $2x > 4 - y$ in slope-intercept form. Solve for y.

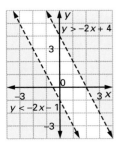

$2x > 4 - y$
$2x + y > 4$
$\quad\quad y > -2x + 4$ ⟵ boundary: $y = -2x + 4$

The equations of the boundaries of the half-planes are $y = -2x - 1$ and $y = -2x + 4$. Use dashed lines for both boundaries. The regions do not intersect.
Thus, the system has no solution.

TRY THIS   Solve each system by graphing.

1. $y \leq x + 4$
$\quad 3y - 2x > 6$

2. $3y + 9x \leq 6$
$\quad y + 3 \geq \frac{1}{3}x$

In Example 2, note that if the inequality signs were reversed, the resulting system would have infinitely many solutions. They would be the coordinates of points lying between the two parallel lines.

EXAMPLE 3   Solve the system $\begin{array}{c} x > -3 \\ y \leq 2 \\ y \geq 2x - 1 \end{array}$ by graphing.

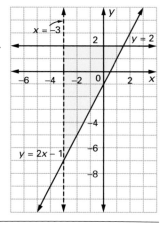

Plan   Graph each inequality. Locate the region that is the intersection of all three graphs.

Solution   When each inequality is graphed, the intersection of the three graphs is the triangular region shown at the right.

To check, show that the coordinates of a point in the triangular region satisfy all three inequalities. The check is left for you. Try $(0, 0)$.

TRY THIS   **3.**   Solve the system consisting of these three inequalities.

a. $x + y \geq 2$ **b.** $2x - y \leq 3$ **c.** $y > 0$

Additional Answers

Try This

1.
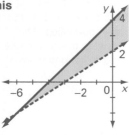

2.
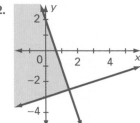

3.
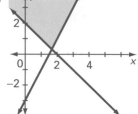

Classroom Exercises

To graph each inequality, would you use a dashed line or a solid line?
Would you shade the region above or below the line?

1. $y > x - 6$ Dashed, above   **2.** $y \le -2x + 7$ Solid, below   **3.** $y \ge -x + 5$ Solid, above

4. $2x + y < -8$ Dashed, below   **5.** $3x - y > -2$ Dashed, below   **6.** $4x + 3y \le 1$ Solid, below

Tell whether or not the given coordinates are a solution of the system
of inequalities.

7. $(4,3)$ Yes
$y < x + 1$
$y > -x + 5$

8. $(-1,-6)$ No
$y \le 2x - 4$
$y > -x + 2$

9. $(2,-3)$ Yes
$3x - 2y \ge 12$
$4x + 2y \le 2$

Written Exercises

Solve by graphing.

1. $y \ge -5$
$x > 2$

2. $y < 4$
$x \ge -3$

3. $y > 3$
$y \le -2$

4. $x \le 4$
$x \ge -1$

5. $y < x + 2$
$x \ge -1$

6. $y > 3x + 2$
$y \le 3x - 1$

7. $y > -2x - 3$
$y < -2x + 4$

8. $2x + y < 3$
$3y > -6x - 3$

9. $3x - 2y < 2$
$3x + 2y > -2$

10. $y - 2x > -3$
$x - 3y < 9$

11. $y - x < 0$
$y \ge 2x - 3$

12. $x > y + 1$
$2x + y \ge -4$

Solve by graphing.

13. $y \le 2$
$x < 2$
$y > -1$

14. $x \ge -2$
$y \ge 3$
$x < 1$

15. $x \ge 1$
$y < -2$
$x - 2y \le 8$

16. $y < 4$
$x \ge 0$
$x + y > 2$

17. $y \ge 0$
$x + 4y < 8$
$3x - 2y > -4$

18. $y - 2x > 1$
$x + y - 6 < 0$
$y > -2$

19. $x \ge -3$
$x < 4$
$y < -1$
$y \ge -5$

20. $x \le 3$
$x < -2$
$y \le 7$
$y > -1$

21. $x > -1$
$x \le 2$
$y > 4$
$y \le -3$

22. $|x| > 3$
$|x| \le 5$

23. $|y| \ge 2$
$|y| < 6$

24. $|x| \le 4$
$|y| \le 2$

Mixed Review

1. Evaluate $3x^4$ for $x = -2$. *2.6* 48

2. Add $-7x^2yz + 10x^2yz$. *6.6* $3x^2yz$

3. Simplify $(2a^2b^3)^4$. *6.2* $16a^8b^{12}$

4. Multiply $(3y - 8)(y + 1)$. *6.9*
   $3y^2 - 5y - 8$

5. Factor $2x^2 - 3x - 5$. *7.4*
   $(2x - 5)(x + 1)$

6. Solve $x^2 - 11x + 28 = 0$. *7.7*
   $x = 4 \text{ or } x = 7$

Checkpoint

Solve each system by graphing.

1. $y \ge x$
$y \ge -x$

2. $y \le 1$
$x \le 1$
$y > 1 - x$

Closure

Ask students whether a system
of inequalities is the union or the intersection
of the graphs of the separate inequalities.
intersection   Then ask them to describe
how to graph a such a system.   Graph each
inequality and double-shade the intersection.

◼◼◼FOLLOW UP

Guided Practice

Classroom Exercises 1–9
Try This   all

Independent Practice

A Ex. 1–12, **B** Ex. 13–18, **C** Ex. 19–24

Basic: WE 1–13 odd

Average: WE 5–17 odd, Application

Above Average: WE 11–23 odd,
Application

See pages 486–488 for the answers to
Written Ex. 1–13. See pages 491–493
for the answers to Written Ex. 14–24.

Additional Example 2

Solve by graphing.
$x < 2 + y$
$x - y > -1$

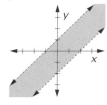

Additional Example 3

Solve by graphing.
$x < 1$
$y \le -1$
$y \ge -x - 1$

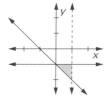

Enrichment

Have students use the library to research
the uses of linear programming in the areas
of industry, business, transportation, and so
on. Have them present their findings to the
class.

485

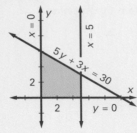
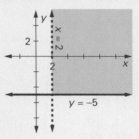
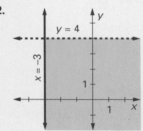
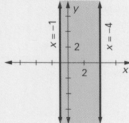

Application: *Linear Programming*

A technique called *linear programming* is used to solve a variety of problems in business, industry, and government. The problems involve decisions that will maximize or minimize certain quantities, such as profit. Linear inequalities are used to represent given conditions, or *constraints*.

Example

A company produces two kinds of special handmade bolts: zero-bolts and one-bolts. Boltmaker *A* takes 2 min to make a zero-bolt and 4 min to make a one-bolt. Boltmaker *B* takes 3 min to make a zero-bolt and 1 min to make a one-bolt. Working a maximum of 3 h per day, each boltmaker makes the same number of zero-bolts and the same number of one-bolts. The profit is $3 on each zero-bolt and $4 on each one-bolt. How many of each type should be made for a maximum daily profit?

Solution

1. Use two variables to represent the data.

   Let x = the total number of zero-bolts produced each day.
   Let y = the total number of one-bolts produced each day.

   Then the daily profit in dollars is $3x + 4y$.

2. Write a system of inequalities to represent the constraints.

   THINK: Each boltmaker makes $\frac{1}{2}$ of the zero-bolts and $\frac{1}{2}$ of the one-bolts produced each day. $\longrightarrow \frac{1}{2}x + \frac{1}{2}y$

   Each boltmaker works a maximum of 3 hours, or 180 minutes per day.

   Time Boltmaker *A* works $\longrightarrow$ $2 \cdot \frac{1}{2}x + 4 \cdot \frac{1}{2}y \le 180$, or $x + 2y \le 180$

   Time Boltmaker *B* works $\longrightarrow$ $3 \cdot \frac{1}{2}x + 1 \cdot \frac{1}{2}y \le 180$, or $3x + y \le 360$

   The number of bolts made cannot be negative. $\longrightarrow$ $x \ge 0, y \ge 0$

3. Graph the system of inequalities. The shaded region is the graph of the solution set. Each vertex is found by solving a system of equations.

System	Solution (vertex)
$x = 0$ $y = 0$	$(0,0)$
$x + 2y = 180$ $x = 0$	$(0,90)$
$3x + y = 360$ $y = 0$	$(120,0)$
$x + 2y = 180$ $3x + y = 360$	$(108,36)$

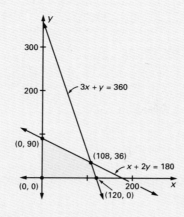

4. Since $3x + 4y$ represents the profit, find the ordered pair for which $3x + 4y$ has a maximum value. It can be proved that a maximum or minimum value of such an expression occurs at a vertex.

Vertex	$(0,0)$	$(0,90)$	$(120,0)$	$(108,36)$
$3x + 4y$	$3 \cdot 0 + 4 \cdot 0$	$3 \cdot 0 + 4 \cdot 90$	$3 \cdot 120 + 4 \cdot 0$	$3 \cdot 108 + 4 \cdot 36$
Value	0	360	360	468 (maximum)

Thus, each day the bolt company should make 108 zero-bolts and 36 one-bolts. The maximum profit is $468 a day.

EXERCISES

Use the following system of inequalities for Exercises 1–3.
$$x \geq 0, \ y \geq 0, \ x \leq 5, \ 5y + 3x \leq 30$$

1. Graph the system of inequalities and shade the solution set.

2. Find the maximum and the minimum value of $x + 5y$. 30, 0

3. Find the maximum and the minimum value of $x - y$. 5, −6

4. The Econo-Company manufactures two sizes of TV screens, size A and size B. The screens are made by two machines. The old machine makes a size-A in 4 min and size-B in 1 min. The new machine makes a size-A in 2 min and a size-B in 8 min. Each machine makes the same number of size-A screens and the same number of size-B screens and operates, at most, 2 h per day. If the profit is $2 on a size-A and $3 on a size-B, how many of each size should be made per day to make the maximum profit? 56 size A, 16 size B

5.

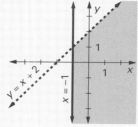

6. No solution

7.

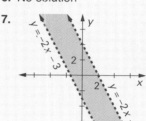

8.

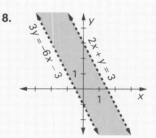

9.

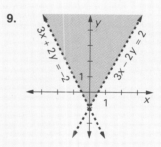

10.

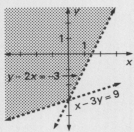

11.

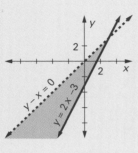

12.

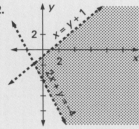

13.

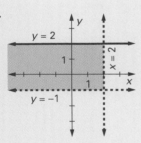

Chapter 12 Review

Key Terms

addition method   (p. 460)
linear combination method   (p. 465)
multiplication with addition method   (p. 464)
substitution method   (p. 452)
system of linear equations   (p. 445)
tens digit   (p. 468)
units digit   (p. 468)

Key Ideas and Review Exercises

12.1   To solve a system of two linear equations by graphing, graph both equations in the same coordinate plane. If the lines intersect in a point, the coordinates of the point are the solution.

Solve by graphing.

1. $2x - y = -5$
$y = -x + 2$ $(-1, 3)$

2. $y = 2x - 1$
$x - 3y = -7$ $(2, 3)$

3. $2x + 3y = 5$
$x - 2y = 6$
$(4, -1)$

12.2   To solve a system of linear equations using the substitution method, solve one equation for one of the variables. Substitute the value for that variable in the other equation. Solve the equation. Find the value of the second variable by substituting the value of the first variable in either equation of the system and then solve the equation.

Solve by using the substitution method. $\left(\frac{1}{2}, \frac{1}{2}\right)$

4. $y = 2x + 1$
$3x - y = 7$ $(8, 17)$

5. $3x - 4y = 19$
$x + y = 4$ $(5, -1)$

6. $5x - 3y = 1$
$x + y = 1$

12.3   To solve word problems using systems of equations:

1. Choose two variables to represent two unknowns.

2. Use two facts from the problem to write two equations with the two variables.

3. Solve the resulting system. Check in the original problem.

Solve each problem using two equations and two variables.

7. The length of a rectangle is 6 m more than 3 times the width. The perimeter is 44 m. Find the length and the width. $l = 4$ m; $w = 18$ m

8. Barbara's age is 5 years less than her brother's age. If her age is increased by 3 times her brother's age, the result is 51 years. Find each of their ages.
Barbara: 9; brother: 14

488   Chapter 12 Review

12.4, To solve a system of linear equations using addition, the equations must
12.5 contain opposite terms in one variable. If necessary, multiply one or both equations by a number to obtain this result. Add the equations to eliminate the variable. Then proceed as in the substitution method.

Solve using the addition method. Use multiplication first, if necessary.

$(1, -1)$

9. $3x + 4y = 11$
$-3x + y = -16$ $(5, -1)$

10. $x - 3y = 1$
$2x + 6y = 14$ $(4, 1)$

11. $5x - 3y = 8$
$2x = 7y + 9$

12. Write a description of two algebraic methods of solving the
system: $\begin{array}{l} 2x + 3y = 16 \\ x - y = 3 \end{array}$. Answers will vary.

12.6, To solve special types of word problems, choose variables as follows:
12.7,
12.8, • For two-digit numbers, let t = the tens digit, u = the units digit, and
12.9 $10t + u$ = the number. ($10u + t$ = the number with the digits reversed)

 • For age problems, choose variables to represent the present ages.

 • For coin problems, represent both the number and value of each coin. For example, if q = the number of quarters and d = the number of dimes, then $25q$ and $10d$ are the values of the quarters and dimes.

 • For mixture problems, represent the total amount of each solution and the amount of a particular ingredient in each solution.

 • For motion problems, draw a diagram to represent distance, rate, and time in a chart. Use the equation $d = rt$.

11 quarters; 5 dimes

13. Jim is 6 years older than Pete. Two years ago, Jim was twice as old as Pete. How old is each now? J: 14; P: 8

14. Tad has 6 more quarters than dimes. He has $3.25 total. How many coins of each type does he have?

15. The sum of the digits of a two-digit number is 9. If the digits are reversed, the new number is 27 less than the original number. Find the original number. 63

16. Mr. and Mrs. Lee drove home from Canton, he at 40 mi/h and she at 50 mi/h. If she left 1 h after he did, and they reached home at the same time, how far did they drive? 200 mi

17. A butterfly flew 4 mi with the wind in 1 h. It returned against the same wind in 1 h 15 min. Find the rate of the wind. 0.4 mi/h

18. Milk that is 5% butterfat is mixed with milk that is 2% butterfat. How much of each is needed to obtain 60 gallons that is 3% butterfat?
5%; 20 gal; 2%; 40 gal

12.10 To solve a system of inequalities by graphing, graph the inequalities in the same coordinate plane. The intersection of the graphs is the set of points whose coordinates are the solution.

19. Solve the system $\begin{array}{l} y > x + 1 \\ y < -x - 7 \end{array}$ by graphing.

1.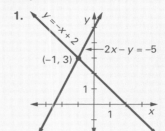

2.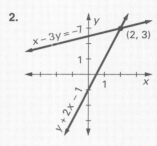

3.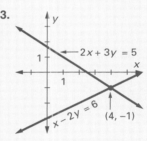

19.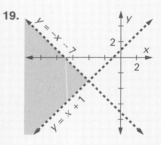

1.

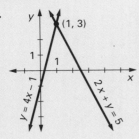

6.

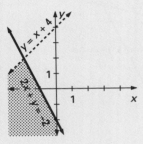

7.

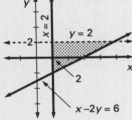

490

Chapter 12 Test A Exercises: 1, 2, 4, 6, 8–12
B Exercises: 3, 5, 7, 13   C Exercises: 14–15

1. Solve the system $\begin{array}{l} 2x + y = 5 \\ y = 4x - 1 \end{array}$ by graphing. (1, 3)

Solve using the substitution method.

2. $y = 3x + 1$ (1, 4)
$-2x + 3y = 10$

3. $2(y - 8) + 3x = 0$ (2, 5)
$7x = 19 - y$

Solve using the addition method. Use multiplication first if necessary.

4. $3x - 2y = 17$ (5, −1)
$5x + 2y = 23$

5. $-7c + 2d = -5$ $(\frac{3}{5}, -\frac{2}{5})$
$-9c + 4d = -7$

Solve by graphing.

6. $y < x + 4$
$2x + y \geq -2$

7. $x \geq 2$
$y < 2$
$x - 2y \leq 6$

Solve each problem using two equations and two variables.

8. The sum of the digits of a two-digit number is 15. If the digits are reversed, the new number is 27 less than the original number. Find the original number. 96

9. A parking meter contains $3.20 in nickels and quarters. There are 3 times as many nickels as quarters. How many coins of each type are in the parking meter? 24 nickels, 8 quarters

10. Paul is three times as old as Kate. Fourteen years from now, Paul will be twice as old as Kate. How old is each now? Kate: 14; Paul: 42

11. A robin flew 3 mi against the wind in 40 min. It returned the same distance, flying with the same wind in 30 min. Find the rate of the wind. $\frac{3}{4}$ mi/h

12. A freight train left Paddington Station traveling 35 mi/h. A passenger train left 2 h later on parallel tracks at 40 mi/h. How far from the station will the passenger train meet the freight train? 560 mi

13. A science teacher wants to obtain 6.25 qt of a 35% acid solution by mixing a 25% acid solution with a 75% acid solution. How much of each type of solution should be in the mixture? 5 qt of 25%, $1\frac{1}{4}$ qt of 75%

Solve using the most convenient method.

14. $\frac{x}{3} - \frac{y}{2} = \frac{4}{3}$ (7, 2)
$3x - 7y = 7$

15. $\frac{4}{3}x - y = \frac{5}{3}$ $(\frac{7}{5}, \frac{1}{5})$
$\dfrac{-2x + 9y + 1}{6} = 0$

In each item, you are to compare the quantity in Column 1 with the quantity in Column 2. Write the letter of the correct answer from the following choices.

A—The quantity in Column 1 is greater than the quantity in Column 2.
B—The quantity in Column 2 is greater than the quantity in Column 1.
C—The quantity in Column 1 is equal to the quantity in Column 2.
D—The relationship cannot be determined from the given information.

NOTE: Information centered over both columns refers to one or both of the quantities to be compared.

Column 1	Column 2
1. $x = 5$ and $y = 7$	
B	
$2x - y$	$2y - x$
2. $x < 0$	
B	
$x^3 - 3$	1
3. $a^2 - b^2$	$(a + b)(a - b)$
C	
4. $\lvert x \rvert$	$\lvert x - 2 \rvert$
D	
5. $y = -x$	
D	
x	y
6. area of	half the area
B circle A	of circle B

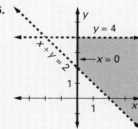

Column 1	Column 2
7. $x > 0$ and $y > 0$	
A	
$2x + 3y$	$2x - 3y$

Column 1	Column 2
8. x	y
A	
	$2x - y = 4$
	$3x + y = 6$
9. slope of line	slope of line
A	
	$y = -x + 1$ $y = -2x - 2$
10. $3(4x - y)$	$-3(y - 4x)$
C	
11. x	y
D	
	$0 < x < 7$
	$0 < y < 5$
12. $3y - x$	$3x - y$
B	
	$2x + 9y = 24$
	$5y = 2x + 4$
13. $y \neq 0$	
C	
$\dfrac{x + y}{y}$	$\dfrac{x}{y} + 1$

Additional Answers, page 485

14.

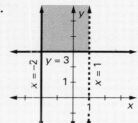

15.

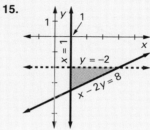

16.

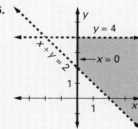

17.

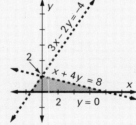

For additional standardized test practice, see the SAT/ACT test booklet for cumulative tests Chapters 1–12.

4. $9x + 2$

16.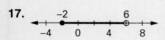
0 2 4

17.
-2 6
-4 0 4 8

36. $D: \{2, 1, 3\}$; $R: \{-2, -1, 1, 3\}$; no

38.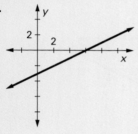

Additional Answers, page 485

18.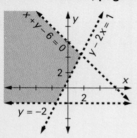

19.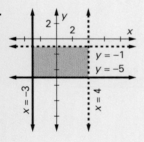

---

1. Evaluate $2y^2 - 4y - 1$ for $y = 3$. **5** **1.3**

2. Simplify $\dfrac{3 + 5}{7 - 2}$. $\dfrac{8}{5}$, or $1\dfrac{3}{5}$ **1.2**

3. Which property is illustrated? **1.6**
$7(4x + 1) = 7 \cdot 4x + 7 \cdot 1$ **Dist**

4. Simplify $8 + 2(4x - 3) + x$. **1.7**

5. Evaluate $4c^3 - 7c$ for $c = -2$. **-18** **2.8**

6. Simplify: $-10y + x + 4$ **2.9**
$-3y + 2x - 7y + 4 - x$

7. Simplify: **2.10**
$7(8 - x) - 2(3x + 4)$ **$-13x + 48$**

Solve.

8. $8 = r + 17$ $r = -9$ **3.1**

9. $-4 = \dfrac{2}{3}c$ $c = -6$ **3.2**

10. $7x + 4 - 2x = 19$ $x = 3$ **3.3**

11. $-2(7y + 9) = 4(3 - y)$ $y = -3$ **3.5**

12. $\dfrac{1}{2}y - 2 = \dfrac{3}{5}y$ $y = -20$ **4.5**

13. $0.5x + 7.2 = -0.7x$ $x = -6$ **4.6**

14. $-3x + 2 \le -13$ $x \ge 5$ **5.2**

15. $|x - 3| = 9$ $x = -6 \text{ or } x = 12$ **5.6**

Graph the solution set.

16. $|x - 2| < 2$ **5.7**

17. $x < 6$ and $x \ge -2$ **5.4**

Simplify. (Exercises 18–21)

18. $-\dfrac{2}{3}ab^2(3ab^3)$ $-2a^2b^5$ **6.1**

19. $(-2x^2yz^3)^3$ $-8x^6y^3z^9$ **6.2**

20. $\dfrac{-6a^2b^3c}{3abc^4}$ $-\dfrac{2ab^2}{c^3}$ **6.3**

21. $x^2 - 8x^3 + 3 - x - x^2 + 2x^3 - 4$ $-6x^3 - x - 1$ **6.6**

22. Subtract $a^2 - a + 2$ from $3a^2 - a - 5$. $2a^2 - 7$ **6.7**

23. Multiply $(x + 5)(x^2 + 2x + 1)$. **6.9**
$x^3 + 7x^2 + 11x + 5$

24. Multiply $(3b + 4c)(3b - 4c)$. **6.10**
$9b^2 - 16c^2$

Factor completely. (Exercises 25–27)

25. $y^2 - 4y - 21$ $(y - 7)(y + 3)$ **7.3**

26. $4x^2 - 20x + 25$ $(2x - 5)(2x - 5)$ **7.5**

27. $4x^3 - 12x^2 + 8x$ $4x(x - 1)(x - 2)$ **7.6**

28. Solve $y^2 + 6y - 16 = 0$. **7.7**
$y = -8 \text{ or } y = 2$

Simplify. (Exercises 29–31)

29. $\dfrac{9 - y^2}{y^2 - 5y - 24}$ $\dfrac{3 - y}{y - 8}$ **8.2**

30. $\dfrac{x - 7}{x} \div \dfrac{x^2 - 3x - 28}{x^2 + 4x}$ **1** **8.4**

31. $3x - 3 + \dfrac{2}{x + 1}$ $\dfrac{3x^2 - 1}{x + 1}$ **8.7**

32. Divide $(8m^3 - 4m + 12)$ by $(2m - 1)$. $4m^2 + 2m - 1 + \dfrac{11}{2m - 1}$ **8.8**

Solve for x. (Exercises 33–35)

33. $\dfrac{1}{3} - \dfrac{2}{x} = \dfrac{5}{2x}$ $x = 13\dfrac{1}{2}$ **9.1**

34. $\dfrac{x - 1}{5} = \dfrac{x + 2}{3}$ $x = -6\dfrac{1}{2}$ **9.2**

35. $px - 7 = 5a$ $x = \dfrac{5a + 7}{p}$ **9.3**

36. Give the domain and the range of the relation: $\{(2, -1), (2, 1), (3, -2), (1, 3)\}$. Is the relation a function? **10.2**

37. If $f(x) = 4x^2 - x$, find $f(-1)$. **5** **10.3**

38. Graph $x - 2y = 6$. **10.5**

39. Find the slope of the line determined by $C(4, -3)$ and $D(2, 1)$. **-2** **11.1**

40. Write an equation for the line passing through $R(1, -3)$ and having a slope of -2. $y = -2x - 1$ **11.2**

41. Solve: $\begin{array}{l} 2x - 4y = 8 \\ y = 2x + 1 \end{array}$ $(-2, -3)$ **12.2**

42. Solve: $\begin{array}{l} x - y = 7 \\ 3x + 2y = 11 \end{array}$. $(5, -2)$ **12.5**

43. The perimeter of a rectangle is 56 m. The width is 2 m less than the length. Find the length and width. $l = 15$ m, $w = 13$ m   **4.4**

44. One number is 8 more than 3 times the other. If the larger is decreased by twice the smaller, the result is 10. Find the numbers. 14 and 2   **4.2**

45. The formula $l_1w_1 = l_2w_2$ can be used to balance two weights, w_1 and w_2, on a see-saw. Their distances from the fulcrum are l_1 and l_2, respectively. If a 120-lb boy sits on a seesaw 5 ft from the fulcrum, how far from the fulcrum must a 100-lb boy sit to balance the seesaw? 6 ft   **3.6**

46. The first side of a triangle is 4 ft shorter than twice the second side. The third side is 3 times as long as the first side. Represent the three lengths in terms of one variable. s, $2s - 4$, $3(2s - 4)$ **4.1**

47. 18 is what percent of 24? 75%   **4.7**

48. What is 42% of 76? 31.92

49. The price of a tape increased from \$8 to \$9. What was the percent increase in the price? $12\frac{1}{2}\%$

50. How much simple interest is earned when \$1,500 is invested at 6.5% for 3 years? \$292.50

51. The selling price of a microwave oven is \$162. The profit is 80% of the cost. Find the cost. \$90   **4.8**

52. A freight elevator can carry 2,500 lb safely. A shipping crate weighs 80 lb. At most, how many crates can be safely carried on the elevator? 31   **5.5**

53. The area of a rectangle is 65 ft². The length is 2 ft less than 3 times the width. Find the length and the width. $l = 13$ ft; $w = 5$ ft   **7.9**

54. A computer can execute an instruction in an average of 2 microseconds (1 microsecond = 1 millionth of a second). A certain program requires 6,372 instructions. How many seconds will it take the program to run? Round your answer to two digits. Then give your answer in scientific notation. 1.3×10^{-2} **6.5**

55. The degree measures of the two smaller angles of a right triangle are in a ratio of 2:3. Find the measure of each angle of the triangle. 36, 54, 90   **9.2**

56. It took Jane the same time to drive 270 mi as it took Jake to drive 300 mi. Jane's speed was 5 mi/h slower than Jake's speed. How fast did each drive? Jane: 45 mi/h; Jake: 50 mi/h   **9.4**

57. Working together, Mrs. Kalb and her daughter can paint a room in 4 h. It takes Mrs. Kalb twice as long as it takes her daughter to do it alone. How long would it take each to do it alone? Mrs. K: 12 h; d: 6 h   **9.5**

58. The current in an electrical circuit varies inversely as the resistance. When the current is 50 amps, the resistance is 32 ohms. Find the current when the resistance is 20 ohms. 80 amps   **10.7**

Additional Answers, page 485

20.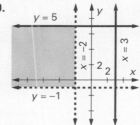

21. No solution

22.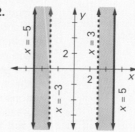

23.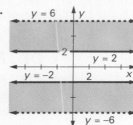

24.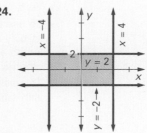

13 RADICALS

OVERVIEW

This chapter connects with and extends the concept of real numbers presented in earlier chapters. After irrational numbers are defined, students are introduced to the Product Property of Square Roots and to methods for approximating their values. The usefulness of radicals is demonstrated in the lesson on the Pythagorean Theorem and its converse. Then students learn how to add, subtract, multiply and divide radicals in lessons which emphasize the importance of expressing answers in simplest form. Finally, Lesson 13.9 allows students to apply previously-learned skills to solving equations that involve radicals.

OBJECTIVES

- To identify irrational numbers and give approximate values of square roots
- To add, subtract, multiply, and divide radical expressions
- To simplify radical expressions
- To use the Pythagorean Theorem to identify right triangles and to find the length of one side given the lengths of the other sides
- To solve problems leading to equations with radicals

PROBLEM SOLVING

When working with square roots, students are encouraged to Estimate Before Solving to pinpoint the integers between which solutions will lie. Using a Formula is the key to solving problems involving right triangles (Pythagorean Theorem) as well as in the application involving the resistance of electrical currents on page 507. Looking Back by checking solutions is useful for identifying extraneous roots to radical equations.

READING AND WRITING MATH

Students are encouraged to explore the real numbers by answering questions concerning rational and irrational numbers in Exercises 43–48 on page 499. In Exercise 34 on page 506, students are asked to explain the position of an irrational number on the number line.

TECHNOLOGY

Calculator: The calculator allows students to investigate patterns of repeating digits such as those in Exercises 7–12 on page 498. The calculator can also be used to compare radical expressions before and after they have been simplified, as illustrated in Example 3 on page 501. Approximate decimal values for irrational solutions to problems can be found by using a calculator.

SPECIAL FEATURES

Mixed Review pp. 499, 503, 507, 512, 518, 521, 525, 528
Focus on Reading p. 502
Application: Electricity p. 507
Midchapter Review p. 515
Mixed Problem Solving p. 529
Key Terms p. 530
Key Ideas and Review Exercises pp. 530–531
Chapter 13 Test p. 532
College Prep Test p. 533

PLANNING GUIDE

Lesson	Basic	Average	Above Average	Resources
13.1 pp. 498–499	CE 1–6 WE 1–19 odd, 33–42 all	CE 1–6 WE 1–41 odd	CE 1–6 WE 7–41 odd, 43, 46, 47	Reteaching 97 Practice 97
13.2 pp. 502–503	FR, CE 1–10 WE 1–39 odd	FR, CE 1–10 WE 9–47 odd	FR, CE 1–10 WE 13–51 odd	Reteaching 98 Practice 98
13.3 pp. 506–507	CE 1–12 WE 1–23 odd Application	CE 1–12 WE 9–31 odd Application	CE 1–2 WE 13–35 odd Application	Reteaching 99 Practice 99
13.4 pp. 510–512	CE 1–9 WE 1–33 odd, 50–53 all	CE 1–9 WE 15–57 odd	CE 1–9 WE 23–57 odd, 58–61 all	Reteaching 100 Practice 100
13.5 pp. 514–515	CE 1–16 WE 1–39 odd Midchapter Review	CE 1–16 WE 11–49 odd Midchapter Review	CE 1–16 WE 21–59 odd Midchapter Review	Reteaching 101 Practice 101
13.6 pp. 517–518	CE 1–10 WE 1–20 all	CE 1–10 WE 1–29 odd, 31–34 all	CE 1–10 WE 1–33 odd 35–37 all	Reteaching 102 Practice 102
13.7 pp. 520–521	CE 1–14 WE 1–31 odd	CE 1–14 WE 7–39 odd	CE 1–14 WE 13–43 odd	Reteaching 103 Practice 103
13.8 pp. 524–525	CE 1–8 WE 1–20 all	CE 1–8 WE 1–35, 45–51 odd	CE 1–8 WE 13–55 odd	Reteaching 104 Practice 104
13.9 pp. 527–529	CE 1–10 WE 1–23 odd Mixed Problem Solving	CE 1–10 WE 11–33 odd Mixed Problem Solving	CE 1–10 WE 23–45 odd Mixed Problem Solving	Reteaching 105 Practice 105
Chapter 13 Review pp. 530–531	all	all	all	
Chapter 13 Test p. 532	all	all	all	
College Prep Test p. 533	all	all	all	

CE = Classroom Exercises WE = Written Exercises FR = Focus on Reading

Note: For each level, all students should be assigned all Try This and all Mixed Review Exercises.

INVESTIGATION

Project: This *Investigation* allows students to explore the difference between fractions that are equivalent to terminating decimals and those that are equivalent to repeating decimals.

Materials: Each student will need a copy of Project Worksheet 13. Each group of students will need at least one calculator.

Since students are asked to find patterns in this project, they are likely to be more successful if they work in groups of 3 or 4 students.

This project should lead students to two major conclusions.

1. Fractions in lowest terms whose denominators can be factored into powers of only 2 and/or 5 can be expressed as terminating decimals. All other fractions can be expressed as repeating decimals.
2. The maximum number of digits possible in the repeating decimal is one less than the denominator of the fraction in lowest terms.

After students have completed the worksheet, have them answer the following questions.

1. If a fraction is in lowest terms, how can you determine whether the equivalent decimal will terminate or repeat?   See conclusion 1 above.
2. How can you quickly determine the maximum possible number of digits in a repeating decimal?   See conclusion 2 above.
3. Must a repeating decimal always be written with a maximum possible number of digits that repeat? Give an example.   No; $\frac{1}{6} = 0.1\overline{6}$

Studio photographers may work in areas such as the fashion, publishing, and advertising industries. They make precise adjustments to camera lenses and measure the light carefully with a meter to produce imaginative and dramatic effects.

More About the Photo

One of the most common kinds of camera, the 35 mm, gets its name from the fact that the film it uses has an image area that is 35 by about 23 millimeters. When a roll of 36 exposures is printed directly on photographic paper to make contact prints, the whole roll can fit on a sheet of paper that measures 8 by 10 inches. Enlarging an image creates a picture that is proportional to the image on the film. Color photography uses three basic dyes, yellow, magenta (a deep red), and cyan (an intense blue). Filters in these primary colors for light are used in combination to create all the colors.

13.1 Rational Numbers and Irrational Numbers

Objectives

To express rational numbers as decimals
To write repeating or terminating decimals in fractional form
To determine whether a decimal is a rational or an irrational number
To explain how to write a repeating decimal in fractional form

On a number line, each point corresponds to some *real* number. Some of these numbers are *rational* numbers.

Numbers such as -2, $-1\frac{3}{4}$, -1, 0, $\frac{1}{2}$, $\frac{7}{8}$, and 1 are rational numbers.

Definition

A **rational number** is a real number that can be expressed in the form $\frac{a}{b}$, where a and b are integers and $b \neq 0$.

Note that every integer a is a rational number because $a = \frac{a}{1}$.

EXAMPLE 1

Show that each number is a rational number.

   a. 3   **b.** 0   **c.** $-6\frac{5}{8}$ **d.** -5 **e.** 1.67

Solutions

   a. $3 = \frac{3}{1}$ **b.** $0 = \frac{0}{1}$ **c.** $-6\frac{5}{8} = \frac{-53}{8}$ **d.** $-5 = \frac{-5}{1}$ **e.** $1.67 = \frac{167}{100}$

TRY THIS

Show that each number is a rational number.

   1. 2.4 $\frac{24}{10}$ **2.** -1 $\frac{-1}{1}$ **3.** $2\frac{1}{5}$ $\frac{11}{5}$ **4.** $-1\frac{1}{4}$ $\frac{-5}{4}$ **5.** 4 $\frac{4}{1}$

Rational numbers can be expressed as decimals. Some rational numbers can be expressed as *terminating decimals*. For example, $\frac{1}{2} = 0.5$ and $\frac{5}{8} = 0.625$. Others can be expressed as *repeating decimals*. For example, $\frac{2}{3} = 0.666 \cdots = 0.\overline{6}$ and $\frac{5}{11} = 0.4545 \cdots = 0.\overline{45}$. The bars mean that 6 and 45 repeat indefinitely.

Teaching Resources

Project Worksheet 13
Quick Quizzes 97
Reteaching and Practice
  Worksheets 97

▰▰ GETTING STARTED

Prerequisite Quiz

Multiply.

  1. 9.666 × 10 96.66
  2. 0.4545 × 100 45.45
  3. 47.258258 × 1,000 47,258.258

Solve. Leave your answer in simplified–fraction form.

  4. $44n = 4,634$ $\frac{2,317}{22}$
  5. $9n = 12$ $\frac{4}{3}$
  6. $99n = 1,425$ $\frac{475}{33}$

Additional Example 1

Show that each number is a rational number.

  a. 4 $\frac{4}{1}$
  b. $7\frac{2}{5}$ $\frac{37}{5}$
  c. -2 $\frac{-2}{1}$
  d. -3.48 $\frac{-348}{100}$

Highlighting the Standards

Standards 14a, 4a, 4b: In this lesson, students examine rational numbers expressed as decimals and look again at the real number system.

Motivator

Have students draw two squares, each 1 cm on a side. Ask them to give the area of each square. 1 cm² Then have them draw the diagonal of each square, cut out each square, and cut the squares along the diagonal. Challenge them to reassemble the pieces to form a larger square with each side having the length of a diagonal of the two smaller squares. Ask them to give the area of the larger square. 2 cm² Ask them to give the length of a side of the larger square. Answers will vary; ≈ 1.4 cm. Tell students that this length is actually √2 cm, that √2 is an irrational number, and that they will learn about irrational numbers in this chapter.

■ TEACHING SUGGESTIONS

Lesson Note

Tell students that the word *rational* comes from the word *ratio*. Emphasize that every rational number can be written as the ratio, or quotient, of two integers. After presenting Example 4, point out that the number 2.525525552 . . . does follow a pattern, but this is not the same as having a repeating block of digits of a specific length.

Math Connections

History: The followers of Pythagoras (ca. 540 B.C.) were called Pythagoreans and one of their greatest achievements was the discovery of irrational numbers. They were dismayed when they realized that the diagonal of a unit square could not be expressed as a rational number.

EXAMPLE 2 Express each rational number as a decimal. Then classify the decimal as either terminating or repeating.

a. $\frac{3}{4}$　　　b. $-\frac{5}{8}$　　　c. $\frac{7}{9}$　　　d. $-1\frac{5}{11}$

Plan Divide the numerator of each fraction by the denominator.

Solutions a. $\frac{3}{4}$　　　b. $-\frac{5}{8}$　　　c. $\frac{7}{9}$　　　d. $-1\frac{5}{11}$

$\begin{array}{r} 0.75 \\ 4\overline{)3.00} \end{array}$　　$\begin{array}{r} 0.625 \\ 8\overline{)5.000} \end{array}$　　$\begin{array}{r} 0.777\ldots \\ 9\overline{)7.000} \end{array}$　　$\begin{array}{r} 0.4545\ldots \\ 11\overline{)5.0000} \end{array}$

$\frac{3}{4} = 0.75$　　$-\frac{5}{8} = -0.625$　　$\frac{7}{9} = 0.\overline{7}$　　$-1\frac{5}{11} = -1.\overline{45}$

　terminating　　　terminating　　　repeating　　　repeating

EXAMPLE 3 For each repeating decimal, write a fraction $\frac{a}{b}$, where a and b are integers.

a. $0.\overline{5}$　　　　　　　　　　b. $3.7\overline{2}$

Plan Let n = the decimal. Multiply each side of the equation by 10^1, or 10, since exactly one digit repeats. Subtract the first equation from the second.

Solutions

a. Let $n = 0.555\ldots$

$\begin{array}{r} 10n = 5.555\ldots \\ -\quad n = 0.555\ldots \\ \hline 9n = 5.000\ldots \\ n = \frac{5}{9} \end{array}$

Multiply by 10 ⟶ to eliminate the decimal.

Thus, $0.\overline{5} = \frac{5}{9}$.

b. Let $n = 3.7222\ldots$

$\begin{array}{r} 10n = 37.222\ldots \\ -\quad n = 3.722\ldots \\ \hline 9n = 33.5 \\ 90n = 335 \\ n = \frac{335}{90}, \text{ or } \frac{67}{18} \end{array}$

Thus, $3.7\overline{2} = \frac{67}{18}$.

TRY THIS 6. Write $\frac{3}{7}$ and $-2\frac{1}{5}$ as decimals. $0.\overline{428571}$; -2.2

7. Write $0.0\overline{9}$ and $-2.\overline{3}$ as fractions. $\frac{1}{10}$; $-2\frac{1}{3}$

When a repeating decimal contains a block of two or more repeating digits, the process for finding the corresponding fraction is similar. However, if a is the number of repeating digits, multiply by 10^a in order to obtain a terminating decimal after the subtraction.

Additional Example 2

Express each rational number as a decimal. Then classify the number as either repeating or terminating.

a. $\frac{1}{6}$　$0.1\overline{6}$, repeating

b. $\frac{-3}{7}$　$-0.\overline{428571}$, repeating

c. $\frac{11}{8}$　1.375, terminating

d. $-\frac{8}{9}$　$-0.\overline{8}$, repeating

Additional Example 3

Write a fraction for each repeating decimal.

a. $0.\overline{4}$　$\frac{4}{9}$

b. $5.6\overline{7}$　$\frac{511}{90}$

EXAMPLE 4   Write a fraction $\frac{a}{b}$, where a and b are integers, for the repeating decimal $12.\overline{47}$.

Solution   Multiply each side of $n = 12.4747$ by 10^2, or 100, since two digits repeat.

$$100n = 1{,}247.4747$$
$$- \quad n = \quad\;\; 12.4747$$
$$99n = 1{,}235$$
$$n = \frac{1{,}235}{99}$$

Thus, $12.\overline{47} = \frac{1{,}235}{99}$.

TRY THIS   8. Write $-0.\overline{45}$ and $4.\overline{15}$ as fractions. $-\frac{5}{11}$; $4\frac{5}{33}$

There are some points on the number line that do not correspond to rational numbers. These points correspond to **irrational numbers**.

The decimals for irrational numbers are *nonterminating* and *nonrepeating*. In 2.525525552 . . . , each string of 5s has one more 5 than the preceding string. Therefore, no block of digits repeats indefinitely.

Other irrational numbers have no pattern. One such irrational number is π. To the nearest millionth, π is 3.141593. The table below shows examples of rational and irrational numbers.

Number	Rational or Irrational	Expressed as $\frac{a}{b}$ (a and b integers, $b \neq 0$)
0.373373337 . . .	Irrational	Not possible
0.88888 . . .	Rational	$\frac{8}{9}$
-0.423	Rational	$-\frac{423}{1{,}000}$
0.28293031 . . .	Irrational	Not possible

If you combine the set of rational numbers and the set of irrational numbers, the result is the set of *real numbers*. In other words, a number is a real number if it is a rational number or an irrational number.

You have previously worked with the following sets of numbers.

Set of natural or counting numbers, N $N = \{1, 2, 3, \cdots\}$

Set of whole numbers, W $W = \{0, 1, 2, 3, \cdots\}$

Set of integers, I $I = \{\cdots, -3, -2, -1, 0, 1, 2, 3, \cdots\}$

13.1 Rational Numbers and Irrational Numbers   **497**

Additional Example 4

Write a fraction for the repeating decimal $16.\overline{32}$. $\frac{1{,}616}{99}$

Closure

Have the students copy the following Venn diagram.

Then have them place each of the following numbers inside the *smallest* circle within which it belongs.

$\frac{2}{3}$, 5, -3, $\frac{-8}{3}$, $.16$, $\sqrt{3}$, $\sqrt{16}$
$.\overline{16}$, 3.2, π, $.\overline{7}$, 0,
$\sqrt{4+5}$, $.121231234\ldots$, $\frac{9}{25}$,
$\sqrt{1}$, $\frac{0}{5}$, $\sqrt{.36}$ Nat.: 5, $\sqrt{16}$, $\sqrt{4+5}$,
$\sqrt{1}$; Wh.: 0, $\frac{0}{5}$; Int.: -3; Rat.: $\frac{2}{3}$, $\frac{-8}{3}$, $.16$,
$.\overline{16}$, 3.2, $.\overline{7}$, $\frac{9}{25}$,
$\sqrt{.36}$; Real: $\sqrt{3}$, π, $.121231234\ldots$

◼◼◼ FOLLOW UP

Guided Practice

Classroom Exercises 1–6
Try This all

Independent Practice

A Ex. 1–20, **B** Ex. 21–44, **C** Ex. 45–48

Basic: WE 1–19 odd, 33–42 all
Average: WE 1–41 odd
Above Average: WE 7–41 odd, 43, 46, 47

If all members of a set X are also members of a set Y, then X is said to be a **subset** of Y. The diagram below shows how subsets of real numbers are related. Notice that N is a subset of W. Also, W is a subset of I.

Real Numbers

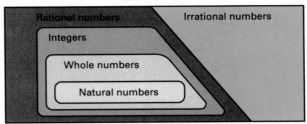

Classroom Exercises

Identify each set of numbers as natural numbers, whole numbers, integers, rational numbers, irrational numbers, or real numbers.

1. the set of numbers that can be expressed as terminating or repeating decimals Rational

2. the set of numbers that are expressed as nonterminating, nonrepeating decimals Irrational

3. the set of real numbers that are not rational Irrational

4. the set of whole numbers together with the set of negative integers Integers

5. $\{0, 1, 2, 3, \ldots\}$ Whole

6. $\{1, 2, 3, 4, \ldots\}$ Natural

Written Exercises

Show that each number is rational by expressing it in the form $\frac{a}{b}$, where a and b are integers.

1. $1\frac{1}{1}$ $\frac{1}{1}$
2. $13\frac{2}{5}$ $\frac{67}{5}$
3. $-2\frac{7}{8}$ $\frac{-23}{8}$
4. 0.7 $\frac{7}{10}$
5. -7 $\frac{-7}{1}$
6. -4.53 $\frac{-453}{100}$

Express each number as a decimal. Then classify the decimal as either terminating or repeating.

7. $\frac{2}{3}$ $0.\overline{6}$, rep
8. $\frac{4}{5}$ 0.8, term
9. $-\frac{1}{6}$ $-0.1\overline{6}$, rep
10. $\frac{3}{7}$ $0.\overline{428571}$, rep
11. $-2\frac{7}{9}$ $-2.\overline{7}$, rep
12. $-1\frac{1}{4}$ -1.25, term

Write a fraction $\frac{a}{b}$, where a and b are integers, for each repeating decimal.

13. $5.\overline{6}$ $\frac{17}{3}$
14. $15.5\overline{6}$ $\frac{467}{30}$
15. $4.\overline{3}$ $\frac{13}{3}$
16. $8.1414\ldots$ $\frac{806}{99}$
17. $0.\overline{62}$ $\frac{62}{99}$
18. $3.\overline{9}$ $\frac{4}{1}$
19. $0.7\overline{5}$ $\frac{68}{90}$
20. $5.8181\ldots$ $\frac{64}{11}$

21. $0.\overline{182}$ $\frac{182}{999}$ **22.** $10.\overline{351}$ $\frac{383}{37}$ **23.** $2.\overline{8642}$ $\frac{28,640}{9,999}$ **24.** $5.139139\ldots$

25. $52.2\overline{735}$ $\frac{522,683}{9,999}$ **26.** $476.\overline{438}$ $\frac{158,654}{333}$ **27.** $5.\overline{257}$ $\frac{5,252}{999}$ **28.** $15.\overline{3534}$ $\frac{51,173}{3,333}$

                                              24. $\frac{5,134}{999}$

Express each rational number as a decimal. Then classify the decimal as either terminating or repeating.

                                 0.1875; terminating

29. $\frac{435}{333}$ $1.\overline{306}$; repeating   **30.** $\frac{159}{99}$ $1.\overline{60}$; repeating   **31.** $\frac{231}{1,232}$ **32.** $\frac{492}{450}$

                                       $1.09\overline{3}$; repeating

True or false? Justify your answer. (Exercises 33–42)

33. The set of integers is a subset of the set of rational numbers.

34. Every nonterminating, nonrepeating decimal is a rational number.

35. Every irrational number is real.

36. A rational number cannot be negative.

37. Zero is a real number.

38. 0.3 is an irrational number.

39. Zero is a natural number.

40. Every real number is rational.

41. Every rational number is real.

42. Some integers are irrational.

43. Write an explanation of why every integer is a rational number.

44. Explain how to write a repeating decimal in fractional form.

45. When two repeating decimals are added, the sum is the decimal for the sum of the corresponding fractions. Give an illustration to show this. (HINT: Use $0.\overline{3}$ and $0.\overline{6}$.)

46. Describe the pattern found in the decimals for $\frac{1}{7}$, $\frac{2}{7}$, $\frac{3}{7}$, $\frac{4}{7}$, $\frac{5}{7}$, and $\frac{6}{7}$.

47. Form a decimal by using the following procedure. First, write the decimal point. Toss a coin. If "heads" comes up, write 2 at the right of the decimal point in the tenths place. Otherwise, write a different digit. Use the coin-tossing procedure to select either 2 or another digit for the hundredths place, the thousandths place, and so on, indefinitely. Is the constructed number rational or irrational? Explain your reasoning.

48. Explain how to construct an irrational number by using a pair of dice. Why would the number be irrational?

Mixed Review

Evaluate. *1.1, 1.3*

1. $x + 7.2$ for $x = 5.9$ 13.1

2. $5.3w$ for $w = 4$ 21.2

3. $x - y$ for $x = 8.1$ and $y = 4.7$ 3.4

4. $6x^2 - x$ for $x = 5$ 145

Simplify, if possible. If not possible, so indicate. *1.2, 6.2, 8.1, 8.2*

5. $39 - 3 \cdot 6 - 6 \cdot 2$ 9

6. $\frac{2}{3}\left(\frac{3}{5} - \frac{1}{5}\right)$ $\frac{4}{15}$

7. $4 + (7 - 3)6$ 28

8. $(3x^4y^2)^3$ $27x^{12}y^6$

9. $\frac{3x - 15}{x - 5}$ 3

10. $\frac{x - 9}{9 - x^3}$ Not possible

Enrichment

Explain to the students that perfect square integers such as 1, 4, 9, 16, . . . are numbers that are squares of positive integers. Have students express each perfect square integer through 144 as a product of prime factors. Then have them notice that, within each factorization, each prime number occurs an *even* number of times. Since any number that has a factor of 2 is *even,* what can be said about an integer whose square is an even number? The integer is even. This result will be used in the proof of the irrationality of $\sqrt{2}$ in the Enrichment for Lesson 13.2.

■■■■**GETTING STARTED**

Prerequisite Quiz

Factor into primes.

1. 36 $2 \cdot 2 \cdot 3 \cdot 3$
2. 63 $3 \cdot 3 \cdot 7$
3. 48 $2 \cdot 2 \cdot 2 \cdot 2 \cdot 3$
4. 225 $3 \cdot 3 \cdot 5 \cdot 5$
5. 224 $2^5 \cdot 7$
6. 1,764 $2 \cdot 2 \cdot 3 \cdot 3 \cdot 7 \cdot 7$

Motivator

Have students give the prime factorization of the square of each number given in the Prerequisite Quiz. In other words, $36^2 = 2 \cdot 2 \cdot 2 \cdot 2 \cdot 3 \cdot 3 \cdot 3 \cdot 3$. Then point out that each factor occurs an even number of times in their answers.

13.2 Square Roots

Objective   To find square roots of positive numbers

The inverse of squaring a number is finding its *square root*, that is, finding one of two equal factors of the number.

Since $49 = 7^2$ and $49 = (-7)^2$, the number 49 has two square roots, 7 and -7. The *principal square root* is the positive square root, which is written as $\sqrt{49}$. Therefore, $\sqrt{49} = 7$.

In general, for any real number a, a^2 is positive, and $\sqrt{a^2} = |a|$.

Definition
> The **principal square root** of a positive real number is the positive square root of the number.

The symbol $\sqrt{49}$ is called a **radical**, $\sqrt{}$ is called a **radical sign**, and the number under the radical sign is called the **radicand**. So, 49 is the radicand of $\sqrt{49}$.

Notice that $-\sqrt{49} = -7$, but $\sqrt{-49}$ is not a real number, since there is no real number whose square is -49.

EXAMPLE 1   Simplify.
  a. $\sqrt{36}$ b. $\sqrt{81}$ c. $-\sqrt{81}$

Solutions   a. $36 = 6^2$ b. $81 = 9^2$ c. $81 = 9^2$
            Thus, $\sqrt{36} = 6$.   Thus, $\sqrt{81} = 9$.   Thus, $-\sqrt{81} = -9$.

TRY THIS   Simplify: **1.** $\sqrt{121}$ 11   **2.** $-\sqrt{64}$ -8

An important property used in multiplying and simplifying square roots is illustrated below.

$$\sqrt{4} \cdot \sqrt{100} = 2 \cdot 10 = 20 \text{ and } \sqrt{4 \cdot 100} = \sqrt{400} = 20$$
$$\text{So, } \sqrt{4} \cdot \sqrt{100} = \sqrt{4 \cdot 100}.$$

Product Property for Square Roots
For all real numbers a and b, where $a \geq 0$ and $b \geq 0$,
$\sqrt{a} \cdot \sqrt{b} = \sqrt{a \cdot b}$ and $\sqrt{a \cdot b} = \sqrt{a} \cdot \sqrt{b}$.

500   Chapter 13 Radicals

Highlighting the Standards

Standards 4c, 4d, 5d: Square roots are found by algebra and with the calculator, and are applied to problems in geometry and physics.

Additional Example 1

Simplify.

  a. $\sqrt{49}$ 7
  b. $\sqrt{100}$ 10
  c. $-\sqrt{121}$ -11

EXAMPLE 2   Simplify.
  a. $\sqrt{5} \cdot \sqrt{7}$
  b. $\sqrt{8} \cdot \sqrt{8}$

Plan   Use the Product Property for Square Roots.

Solutions
  a. $\sqrt{5} \cdot \sqrt{7} = \sqrt{5 \cdot 7}$
       $= \sqrt{35}$
  b. $\sqrt{8} \cdot \sqrt{8} = \sqrt{8^2}$
       $= 8$

TRY THIS   Simplify: 3. $\sqrt{3} \cdot \sqrt{5}$ $\sqrt{15}$ 4. $-\sqrt{4} \cdot -\sqrt{4}$ 4

Part **b** of Example 2 shows that $\sqrt{8} \cdot \sqrt{8} = \sqrt{8^2} = 8$, or $(\sqrt{8})^2 = 8$.

In general, for $a \geq 0$, $(\sqrt{a})^2 = \sqrt{a} \cdot \sqrt{a} = \sqrt{a \cdot a} = \sqrt{a^2} = a$.

The numbers 4, 9, 16, and $\frac{1}{25}$ are *perfect squares* since $\sqrt{4} = \sqrt{2^2} = 2$, $\sqrt{9} = \sqrt{3^2} = 3$, $\sqrt{16} = \sqrt{4^2} = 4$, and $\sqrt{\frac{1}{25}} = \sqrt{\left(\frac{1}{5}\right)^2} = \frac{1}{5}$.

Definition   A **perfect square** is a positive rational number whose principal square root is a rational number.

If the radicand is not a perfect square, the square root is an *irrational number*. To simplify such a radical, find the greatest perfect square factor of the radicand. Then use $\sqrt{a \cdot b} = \sqrt{a} \cdot \sqrt{b}$.

EXAMPLE 3   Simplify.
  a. $\sqrt{18}$
  b. $\sqrt{72}$

Solutions
  a. $\sqrt{18} = \sqrt{9 \cdot 2}$
       $= \sqrt{9} \cdot \sqrt{2}$
       $= 3\sqrt{2}$
  b. $\sqrt{72} = \sqrt{36 \cdot 2}$
       $= \sqrt{36} \cdot \sqrt{2}$
       $= 6\sqrt{2}$

Calculator check: (Compare the approximations in the displays.)
  a. 18 $\boxed{\sqrt{}}$ 4.2426407 and 2 $\boxed{\sqrt{}}$ $\boxed{\times}$ 3 $\boxed{=}$ 4.2426407
  b. 72 $\boxed{\sqrt{}}$ 8.4852814 and 2 $\boxed{\sqrt{}}$ $\boxed{\times}$ 6 $\boxed{=}$ 8.4852814

TRY THIS   Simplify: 5. $\sqrt{27}$ $3\sqrt{3}$ 6. $-\sqrt{50}$ $-5\sqrt{2}$

Sometimes it is difficult to recognize the greatest perfect square factor. In that case, factor the radicand into its prime factors.

Additional Example 2

Simplify.

  a. $\sqrt{3} \cdot \sqrt{5}$ $\sqrt{15}$
  b. $\sqrt{7} \cdot \sqrt{7}$ 7

Additional Example 3

Simplify.

  a. $\sqrt{27}$ $3\sqrt{3}$
  b. $\sqrt{32}$ $4\sqrt{2}$

Common Error Analysis

Error: Sometimes students do not find the greatest perfect square factor and then do not simplify sufficiently. Remind students that $\sqrt{12}$ still contains a perfect square integer, and so is not completely simplified.

Checkpoint

Simplify.

1. $\sqrt{81}$ 9
2. $\sqrt{64}$ 8
3. $\sqrt{196}$ 14
4. $\sqrt{75}$ $5\sqrt{3}$
5. $-2\sqrt{243}$ $-18\sqrt{3}$
6. $9\sqrt{576}$ 216
7. $\sqrt{5} \cdot \sqrt{7}$ $\sqrt{35}$
8. $\sqrt{29} \cdot \sqrt{29}$ 29

Closure

Have students represent each of the following as a product of a perfect and a nonperfect square.

1. 50 $25 \cdot 2$
2. 18 $9 \cdot 2$
3. 32 $16 \cdot 2$

Then ask students what is meant by the principal square root of a number. (See p. 500). Ask them to simplify $\sqrt{a} \cdot \sqrt{a}$. $\sqrt{a^2} = |a|$.

EXAMPLE 4   Simplify.

     a. $6\sqrt{84}$ b. $-3\sqrt{990}$

Plan   First, factor into primes to find the greatest perfect square factor.

Solutions

a. $6\sqrt{84} = 6\sqrt{2 \cdot 2 \cdot 3 \cdot 7}$ b. $-3\sqrt{990} = -3\sqrt{3 \cdot 3 \cdot 2 \cdot 5 \cdot 11}$

$\phantom{a. 6\sqrt{84}} = 6\sqrt{4} \cdot \sqrt{21}$ $= -3\sqrt{9} \cdot \sqrt{110}$

$\phantom{a. 6\sqrt{84}} = 6 \cdot 2 \cdot \sqrt{21}$ $= -3 \cdot 3 \cdot \sqrt{110}$

$\phantom{a. 6\sqrt{84}} = 12\sqrt{21}$ $= -9\sqrt{110}$

EXAMPLE 5   The area of a circular garden is 100π ft². Find its diameter.

Solution

$A = \pi r^2$

$100\pi = \pi r^2$

$r^2 = 100$

$r = \sqrt{100}$, or 10 ft   So, the diameter is 20 ft.

TRY THIS   Simplify: 7. $5\sqrt{60}$ $10\sqrt{15}$ 8. $-2\sqrt{125}$ $-10\sqrt{5}$

                9. The area of a square is 169 cm². Find the length of each side.   13

Focus on Reading

Tell whether the statement is true or false. If false, explain why.

1. $\sqrt{36} = 6$ T   2. $-\sqrt{25} = -5$ T   3. $\sqrt{-1} = -1$ 4. $\sqrt{10} \cdot \sqrt{10} = 10^2$

           3. F; no real number has -1 for its square.   4. F; $\sqrt{10} \cdot \sqrt{10} = \sqrt{10 \cdot 10} = 10$

Classroom Exercises

State each radicand as its greatest perfect square factor times another factor.

1. $\sqrt{8}$ $\sqrt{4 \cdot 2}$ 2. $\sqrt{50}$ $\sqrt{25 \cdot 2}$ 3. $\sqrt{27}$ $\sqrt{9 \cdot 3}$ 4. $\sqrt{48}$ $\sqrt{16 \cdot 3}$ 5. $\sqrt{300}$

                                                                            $\sqrt{100 \cdot 3}$

Simplify.

6. $\sqrt{9}$ 3   7. $-\sqrt{81}$ -9 8. $\sqrt{3,600}$ 60   9. $\sqrt{500}$ $10\sqrt{5}$ 10. $\sqrt{28}$ $2\sqrt{7}$

Written Exercises

Simplify.

1. $\sqrt{4}$ 2   2. $\sqrt{100}$ 10   3. $\sqrt{16}$ 4   4. $\sqrt{25}$ 5   5. $-\sqrt{18}$ $-3\sqrt{2}$

6. $\sqrt{121}$ 11   7. $\sqrt{144}$ 12   8. $-\sqrt{49}$ -7 9. $\sqrt{64}$ 8   10. $\sqrt{169}$ 13

11. $\sqrt{27}$ $3\sqrt{3}$ 12. $\sqrt{20}$ $2\sqrt{5}$ 13. $\sqrt{32}$ $4\sqrt{2}$ 14. $\sqrt{75}$ $5\sqrt{3}$ 15. $\sqrt{108}$ $6\sqrt{3}$

16. $\sqrt{24}$ $2\sqrt{6}$ 17. $-\sqrt{8}$ 18. $\sqrt{50}$ $5\sqrt{2}$ 19. $-\sqrt{72}$ 20. $\sqrt{98}$ $7\sqrt{2}$

                  $-2\sqrt{2}$ $-6\sqrt{2}$

Additional Example 4

Simplify.

   a. $8\sqrt{98}$ $56\sqrt{2}$

   b. $-5\sqrt{360}$ $-30\sqrt{10}$

Additional Example 5

The area of a circular garden is 900π ft². What is the diameter of the garden?   60 ft

21. $\sqrt{3} \cdot \sqrt{7}$ $\sqrt{21}$ **22.** $\sqrt{7} \cdot \sqrt{6}$ $\sqrt{42}$ **23.** $\sqrt{11} \cdot \sqrt{7}$ $\sqrt{77}$ **24.** $\sqrt{15} \cdot \sqrt{2}$ $\sqrt{30}$

25. $\sqrt{15} \cdot \sqrt{15}$ 15 **26.** $\sqrt{10} \cdot \sqrt{10}$ 10 **27.** $\sqrt{7} \cdot \sqrt{7}$ 7 **28.** $\sqrt{35} \cdot \sqrt{35}$ 35

29. $-4\sqrt{180}$ $-24\sqrt{5}$ **30.** $9\sqrt{243}$ $81\sqrt{3}$ **31.** $-6\sqrt{600}$ $-60\sqrt{6}$ **32.** $15\sqrt{343}$ $105\sqrt{7}$

33. $10\sqrt{224}$ $40\sqrt{14}$ **34.** $-7\sqrt{275}$ $-35\sqrt{11}$ **35.** $12\sqrt{252}$ $72\sqrt{7}$ **36.** $-20\sqrt{363}$

37. $\sqrt{18} \cdot \sqrt{14}$ $6\sqrt{7}$ **38.** $-\sqrt{30} \cdot \sqrt{40}$ **39.** $\sqrt{72} \cdot \sqrt{\frac{4}{9}}$ $4\sqrt{2}$ **40.** $\sqrt{\frac{9}{25}} \cdot \sqrt{50}$ $3\sqrt{2}$
$\quad\quad\quad\quad\quad\quad -20\sqrt{3}$

36. $-220\sqrt{3}$

Solve each problem.

41. The area of a square garden plot is 256 ft². What is the length of each side? 16 ft

42. The area of a circle is 121π cm². What is the radius of the circle? 11 cm

43. The length of a rectangle is 3 times its width. The area of the rectangle is 192 cm². What is the length and the width of the rectangle? $l = 24$ cm, $w = 8$ cm

Simplify.

$\quad\quad\quad\quad\quad\quad\quad -12\sqrt{14}\quad\quad\quad\quad\quad -34\sqrt{11}\quad\quad\quad\quad 45\sqrt{3}$

44. $\sqrt{686}$ $7\sqrt{14}$ **45.** $-\sqrt{2,016}$ **46.** $-2\sqrt{3,179}$ **47.** $3\sqrt{675}$

The time T (in seconds) it takes for a pendulum to swing back and forth is given by the formula

$$T = 2\pi\sqrt{\frac{l}{g}}$$

where l is the length of the pendulum in meters and g is the acceleration due to gravity, which is 9.8 meters per second per second (m/s²).

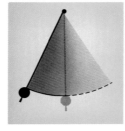

Find the time T for each value of l. Give answers in terms of π and in simplest radical form.

48. $l = 0.8$ $\frac{4}{7}\pi$ s **49.** $l = 1.8$ $\frac{6}{7}\pi$ s **50.** $l = 5.0$ $1\frac{3}{7}\pi$ s **51.** $l = 9.8$ 2π s

Mixed Review

Factor completely. *7.3, 7.6*

$\quad\quad\quad\quad\quad\quad\quad\quad\quad\quad\quad\quad\quad\quad\quad\quad\quad\quad\quad\quad (y-6)(y-3)$

1. $x^2 + 8x + 12$ $(x+2)(x+6)$ **2.** $x^2 - x - 6$ $(x-3)(x+2)$ **3.** $y^2 - 9y + 18$

4. $2y^3 - 2y$ $2y(y+1)(y-1)$ **5.** $3x^3 - 27x$ $3x(x+3)(x-3)$ **6.** $2x^3 + 16x^2 + 30x$
$\quad\quad\quad\quad\quad\quad\quad\quad\quad\quad\quad\quad\quad\quad\quad\quad\quad\quad\quad\quad\quad\quad 2x(x+3)(x+5)$

Find the slope of $\overline{PQ}$ determined by the given points. *11.1*

7. $P(9,1), Q(-6,3)$ $-\frac{2}{15}$ **8.** $P(0,-8), Q(7,-2)$ $\frac{6}{7}$ **9.** $P(-1,-3), Q(8,-4)$
$\quad\quad\quad\quad\quad\quad\quad\quad\quad\quad\quad\quad\quad\quad\quad\quad\quad\quad\quad\quad\quad\quad\quad\quad -\frac{1}{9}$

FOLLOW UP

Guided Practice

Classroom Exercises 1–10
Try This all, FR all

Independent Practice

A Ex. 1–40, **B** Ex. 41–47, **C** Ex. 48–51

Basic: WE 1–39 odd
Average: WE 9–47 odd
Above Average: WE 13–51 odd

Additional Answers, page 499

47. In general, the decimal will be irrational since the sequences of blocks of 2's will not all be the same length as the decimal is extended indefinitely.

48. Sample answer: First, write a decimal point. Toss a die. If the die shows a 1, write 1 at the right of the decimal point in the tenths place; otherwise write a different digit. Continue as in Ex. 47. The number will be irrational since, in general, 1 will not be part of a block of repeating digits.

Enrichment

Present the following argument for the irrationality of $\sqrt{2}$ to the students. Suppose $\sqrt{2}$ is rational. It can then be written as a fraction of integers $\frac{a}{b}$ in lowest terms. Since $\sqrt{2} = \frac{a}{b} \cdot 2 = \frac{a^2}{b^2}$ and $2b^2 = a^2$. This implies that a^2 is even which implies that a is even (see Enrichment for Lesson 13.1).

Let $a = 2n$, for some integer n. Then $2b^2 = (2n)^2$; $2b^2 = 4n^2$; and $b^2 = 2n^2$. This implies that b^2 is even which implies that b is even. But now we have both a and b even, which contradicts the fact that the ratio $\frac{a}{b}$ was in lowest terms. Therefore, $\sqrt{2}$ cannot be written as a ratio of two integers

and $\sqrt{2}$ is not rational. So $\sqrt{2}$ must be irrational. Have students develop an argument for the irrationality of $\sqrt{3}$.

Prerequisite Quiz

Divide. Round answer to the nearest hundredth.

1. $84 \div 9.5$ 8.84
2. $75 \div 8.7$ 8.62
3. $125 \div 11.12$ 11.24
4. $26 \div 5.13$ 5.07

Find the average between each pair.

5. 6.4 and 6.7   6.55
6. 8.3 and 8.11   8.205
7. 28.45 and 29.33   28.89

Motivator

Tell the students that an acre of land has an area of 4,840 yd^2. Then ask if the plot of land were square, how long would each side be. Lead students to a discussion of approximating a square root. Since $70^2 = 4,900$ and $69^2 = 4,761$, each side must be between 69 and 70 yards long.

13.3 Approximating Square Roots

Objectives
To approximate square roots by using a square root table
To approximate square roots after simplifying radicals
To approximate square roots by the divide-and-average method

The value of the square root of a whole number that is not a perfect square can be approximated in different ways. One way is to use a calculator. Another way is to use a Table of Roots and Powers.

EXAMPLE 1 Find an approximation of $\sqrt{15}$ by using a square root table. A portion of the table is shown below.

Table of Roots and Powers		
Number	Square	Square Root
1	1	1.000
2	2	1.414
⋮	⋮	⋮
14	196	3.742
15	225	3.873

Solution Read down to the number in the number column. Then read across to the square root column.

So, $\sqrt{15} \approx 3.873$. (Recall that the symbol $\approx$ means "is approximately equal to.")

TRY THIS 1. Find an approximation of $\sqrt{2}$ using a square root table. 1.414

When possible, simplify a radical before using the square root table.

EXAMPLE 2 Approximate $\sqrt{500}$ to the nearest tenth.

Solution $\sqrt{500} = \sqrt{100 \cdot 5} = \sqrt{100} \cdot \sqrt{5} = 10\sqrt{5} \approx 10(2.236) \approx 22.36$
Thus, $\sqrt{500} \approx 22.4$.

TRY THIS 2. Approximate $\sqrt{108}$ to the nearest tenth. 10.4

504   Chapter 13 Radicals

Additional Example 1

Find an approximation of $\sqrt{24}$ using a square root table.   4.899

Additional Example 2

Approximate $\sqrt{810}$ to the nearest tenth.   28.5

The *divide-and-average* method can help you find square roots.

$$\begin{array}{r} 6 \leftarrow \text{quotient} \\ \text{divisor} \longrightarrow 6\overline{)36} \end{array}$$

When 36 is divided by 6, the quotient is also 6, since $\sqrt{36} = 6$.

$$\begin{array}{r} 5 \leftarrow \text{quotient} \\ \text{divisor} \longrightarrow 3\overline{)15} \end{array}$$

When 15 is divided by 3, the quotient is 5, so $\sqrt{15}$ is *between* 3 and 5. (From the table you know that $\sqrt{15} \approx 3.873$.)

EXAMPLE 3 Approximate $\sqrt{32}$ to the nearest tenth by using the divide-and-average method.

Solution

1. Locate 32 between two perfect squares.

   $25 < 32 < 36$
   $5^2 < 32 < 6^2$
   So, $5 < \sqrt{32} < 6$.

2. Divide 32 by any number in the interval from 5 through 6. (Use 5.5.)

   $$\begin{array}{r} 5.8\ 1 \leftarrow \text{two decimal places} \\ 5.5\ \overline{)3\ 2.0\ 0\ 0} \end{array}$$

3. Average 5.5 and 5.81 to find a new divisor.

   one decimal place
   $$\frac{5.5 + 5.81}{2} = \frac{11.31}{2} = 5.65$$

4. Divide 32 by 5.65. The divisor and quotient agree to the nearest tenth.

   $$\begin{array}{r} 5.6\ 6\ 3 \\ 5.65\ \overline{)3\ 2.0\ 0\ 0\ 0\ 0} \end{array}$$

   divisor ≈ 5.7; quotient ≈ 5.7

   Thus, $\sqrt{32} \approx 5.7$. This is closer to 6 than to 5. This is reasonable since 32 is closer to 6^2 than to 5^2.

   (This procedure can be continued to find more accurate values for $\sqrt{32}$.)

EXAMPLE 4 The area of a square field is 53 m². Find the length of a side of the field correct to the nearest tenth of a meter.

$$A = s^2 \quad \boxed{\ s\ }$$

Solution

$A = s^2$
$53 = s^2$
$\quad s = \sqrt{53} \leftarrow$ (THINK: $7^2 = 49$ and $8^2 = 64$, so $\sqrt{53}$ is between 7 and 8.)
$\quad s \approx 7.280 \leftarrow$ From the table of square roots

Thus, the length of a side of the field is approximately 7.3 m.

TRY THIS 3. The area of a circle is 71π cm². Find the length of the radius to the nearest tenth. Use $A = \pi r^2$. 8.4

Lesson Note

Although a calculator can be used to find square roots, it is still valuable for students to be able to estimate the approximate square root of a number. For example, $\sqrt{200}$ is between 14 and 15 since 200 is between 196 and 225.

Math Connections

Life Skills: The least amount of fencing needed to enclose a rectangular garden of 300 square feet would be that needed to enclose a square area each side of which would be $\sqrt{300}$ ft. Since $\sqrt{300}$ is approximately 17.3, this would require about 4(17.3), or 70 feet of fencing.

Critical Thinking Questions

Application: Since $2^4 = 16$, the fourth root of 16 is 2. Ask students how the square root table could be used to find the fourth root of a number. Find the square root of the square root of the number.

Common Error Analysis

Error: Some students assume that the square root table gives an exact square root for every number.

Remind students that all square roots, except those of perfect squares, are irrational and as such cannot be expressed exactly as a decimal number.

Additional Example 3

Use the divide-and-average method to approximate $\sqrt{52}$ to the nearest tenth. 7.2

Additional Example 4

The area of a square tile is 28 cm². Find the length of a side of the tile to the nearest tenth of a centimeter. 5.3 cm

505

Approximate, to the nearest tenth, using the square root table.

1. $\sqrt{17}$ 4.1
2. $\sqrt{60}$ 7.7
3. $\sqrt{700}$ 26.5

Approximate, to the nearest tenth, using the divide-and-average method.

4. $\sqrt{70}$ 8.4
5. $\sqrt{45}$ 6.7
6. $\sqrt{130}$ 11.4

Closure

Call out numbers between 1 and 100 and have students give the two consecutive integers between which the square root of the number lies. Then have a student summarize the divide-and-average method (see page 505).

Classroom Exercises

Between what two consecutive whole numbers is the square root?

1. $\sqrt{74}$ 8–9 2. $\sqrt{38}$ 6–7 3. $\sqrt{43}$ 6–7 4. $\sqrt{111}$ 10–11 5. $\sqrt{52}$ 7–8 6. $\sqrt{12}$ 3–4

Approximate to the nearest tenth. Use a square root table.

7. $\sqrt{22}$ 4.7 8. $\sqrt{86}$ 9.3 9. $\sqrt{43}$ 6.6 10. $\sqrt{93}$ 9.6 11. $\sqrt{12}$ 3.5 12. $\sqrt{71}$ 8.4

Written Exercises

Approximate to the nearest tenth. Use a square root table.

1. $\sqrt{19}$ 4.4 2. $\sqrt{97}$ 9.8 3. $\sqrt{48}$ 6.9 4. $\sqrt{51}$ 7.1 5. $\sqrt{62}$ 7.9 6. $\sqrt{90}$ 9.5
7. $\sqrt{72}$ 8.5 8. $\sqrt{33}$ 5.7 9. $\sqrt{84}$ 9.2 10. $\sqrt{21}$ 4.6 11. $\sqrt{13}$ 3.6 12. $\sqrt{35}$ 5.9

Approximate to the nearest tenth. Use the divide-and-average method.

13. $\sqrt{38}$ 6.2 14. $\sqrt{27}$ 5.2 15. $\sqrt{65}$ 8.1 16. $\sqrt{44}$ 6.6 17. $\sqrt{98}$ 9.9 18. $\sqrt{71}$ 8.4

Approximate to the nearest tenth. Use either a square root table or the divide-and-average method.

19. $\sqrt{245}$ 15.7 20. $\sqrt{124}$ 11.1 21. $\sqrt{248}$ 15.7 22. $\sqrt{775}$ 27.8 23. $\sqrt{550}$ 23.5 24. $\sqrt{463}$ 21.5

25. Find all the integers between 1 and 100 whose square roots are integers.

26. Find all the integers between 120 and 250 whose square roots are integers.

Solve each problem. Round answers to the nearest tenth.

27. The area of a square is 18 cm². Find the length of a side. 4.2 cm

28. The area of a square is 47 cm². Find the length of a side. 6.9 cm

29. The area of a circle is 31π. Find the length of the radius. Use $A = \pi r^2$. 5.6

30. The area of a circle is 134π. Find the length of the radius. Use $A = \pi r^2$. 11.6

31. The length of a rectangle is 4 times the width. The area is 440 m². Find the length and the width of the rectangle. $l = 42.0$ m, $w = 10.5$ m

32. The area of a rectangle is 90 cm². The length is 3 times the width. Find the length and the width of the rectangle. $l = 16.4$ cm, $w = 5.5$ cm

33. Approximate $\sqrt{0.40}$ to the nearest tenth without using a calculator. 0.6

34. Write an explanation of why $\sqrt{57}$ is a number between 7 and 8.

A centripetal force keeps an object moving in a circular path at a constant velocity. The formula used to calculate centripetal force F is $F = \dfrac{mv^2}{r}$, where r is the radius, m is the mass, and v is the velocity.

35. A discus thrower swings a discus, then lets it fly. Before it flies off, what is its velocity in meters per second if its mass is 4 kg, the radius of its path is 0.8 m, and the force on the arm of the thrower is 80 newtons? 4 m/s

36. A centripetal force of 78 newtons keeps a ball swinging on a chain 3 m in length. If the mass of the ball is 2 kg, find the velocity of the ball in meters per second. Give the answer to one decimal place. 10.8 m/s

Mixed Review

Simplify. *2.2, 2.10, 6.1, 6.3*

1. $-|-3| \cdot |-4|$ -12 **2.** $-1(-24)$ 24 **3.** $-7(3^2)$ -63 **4.** $-(3a - 2b) - 2a$ $-5a + 2b$

5. $x^5 \cdot x^2$ x^7 **6.** $a \cdot a^4$ a^5 **7.** $(-3a^5)(4a^3)$ $-12a^8$ **8.** $\dfrac{-4x^3y^2}{16xy^5}$ $-\dfrac{x^2}{4y^3}$

Application: *Electricity*

Appliances that produce high amounts of heat (hairdryers, heaters, toasters) cost more to run than other kinds of appliances. Wires in these appliances oppose, or resist, the flow of electrons so the wires turn red and become hot. More power is required to push electrons through a wire that has high resistance. This formula gives the electrical current, I, in amperes, A, for an appliance that uses W watts of power and has a resistance of R ohms.

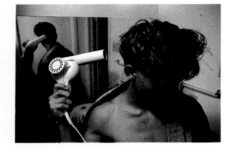

$$I = \sqrt{\dfrac{W}{R}}$$

Use the formula to find the amps for each of these appliances. Round your answers to the nearest tenth.

1. A toaster that uses 575 watts of power and has a resistance of 20 ohms 5.4 A

2. An electric heater that uses 790 watts of power and has a resistance of 16 ohms 7.0 A

3. An iron that uses 632.5 watts of power and has a resistance of 20.9 ohms 5.5 A

13.3 Approximating Square Roots **507**

Enrichment

The **Golden Ratio** is a number that occurs frequently in mathematics, as well as in nature, astronomy, and architecture. The value of this number is $\dfrac{1 + \sqrt{5}}{2}$.

1. Have students approximate this number to the nearest thousandth using the divide-and-average method.   1.618

The **Fibonacci Series** is a famous series whose first two numbers are 1 and 1. Each subsequent number is found by adding the two preceding numbers to get: 1, 1, 2, 3, 5, 8, . . .

2. Have students compute the ratio of the larger to smaller number in each successive adjacent pair of numbers for 10 ratios.

$\dfrac{1}{1} = 1$; $\dfrac{2}{1} = 2$; $\dfrac{3}{2} = 1.5$; $\dfrac{5}{3} \approx 1.667$; $\dfrac{8}{5} = 1.6$; $\dfrac{13}{8} = 1.625$; $\dfrac{21}{13} \approx 1.615$; $\dfrac{34}{21} \approx 1.619$; $\dfrac{55}{34} \approx 1.618$; $\dfrac{89}{55} \approx 1.618$

3. What conclusion can be drawn from the results of the two above exercises?
The ratios using the successive numbers in the Fibonacci series grow closer to the value of the Golden Ratio.

Teaching Resources

Manipulative Worksheet 13
Problem Solving Worksheet 13
Quick Quizzes 100
Reteaching and Practice
  Worksheets 100
Transparency 39

■■ GETTING STARTED

Prerequisite Quiz

Simplify.

1. $\sqrt{289}$ 17
2. $\sqrt{60}$ $2\sqrt{15}$
3. $\sqrt{189}$ $3\sqrt{21}$
4. $\sqrt{875}$ $5\sqrt{35}$
5. 15^2 225
6. $(\sqrt{3})^2$ 3
7. $(4\sqrt{5})^2$ 80
8. $(2\sqrt{2})^2$ 8

Motivator

Model: Use three straws of lengths 6 in., 8 in., and 10 in. to form a right triangle. Then draw this triangle on the board and build a square on each of its sides. Show that the area of the larger square is equal to the sum of the areas of the two smaller squares. Point out that if the lengths 6, 8, and 10 are replaced by the variables a, b, and c, we have $a^2 + b^2 = c^2$.

13.4 The Pythagorean Theorem

Objectives

To find the length of one side of a right triangle, given the lengths of the other two sides

To determine whether a triangle is a right triangle, given the lengths of its three sides

A surveyor walked 8 m east, then 6 m north. How far was she from the starting point?

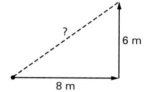

This problem can be solved by using a relationship between the sides of a **right triangle**, that is, a triangle with one right angle. A right angle has a degree measure of 90 and is indicated in diagrams by the symbol ⌐, as in the figure below.

In the sixth century B.C., Pythagoras, a Greek philosopher and mathematician, discovered a relationship between the hypotenuse and the two legs of a right triangle. It is known as the *Pythagorean Theorem*.

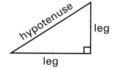

Pythagorean Theorem
If triangle ABC is a right triangle with c the length of the hypotenuse and a and b the lengths of the legs, then $c^2 = a^2 + b^2$.

In this lesson, c will represent the length of the hypotenuse of a right triangle and a and b the lengths of the legs.

For the problem about the surveyor, it is necessary to find the length of the hypotenuse, given legs of lengths 6 m and 8 m.

Use the Pythagorean Theorem.

$$\begin{aligned} c^2 &= a^2 + b^2 \\ &= 6^2 + 8^2 \\ &= 36 + 64 \\ &= 100 \\ c &= \sqrt{100}, \text{ or } 10 \end{aligned}$$

So, the surveyor was 10 m from her starting point.

Highlighting the Standards

Standards 4c, 7d: This lesson connects geometry and algebra as the students apply the Pythagorean Theorem.

EXAMPLE 1 For each right triangle, find the missing length. If the answer is not a rational number, give it in simplest radical form.

  a. $a = 2, b = 6$ **b.** $a = 8, c = 17$

Plan Use the Pythagorean Theorem, $c^2 = a^2 + b^2$.

Solutions

a.
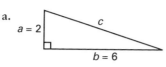

$$c^2 = a^2 + b^2$$
$$= 2^2 + 6^2$$
$$= 4 + 36$$
$$= 40$$
$$c = \sqrt{40}, \text{ or } c = 2\sqrt{10}$$

b.

$$c^2 = a^2 + b^2$$
$$17^2 = 8^2 + b^2$$
$$289 = 64 + b^2$$
$$225 = b^2$$
$$b = \sqrt{225} = 15$$

TRY THIS 1. The lengths of the legs of a right triangle are 5 and 12. Find the length of the hypotenuse. 13

The *converse of the Pythagorean Theorem* is also true.

Converse of the Pythagorean Theorem
If $c^2 = a^2 + b^2$, where a, b, and c are the lengths of the sides of a triangle, then the triangle is a right triangle.

EXAMPLE 2 Given the lengths of three sides of a triangle, determine whether the triangle is a right triangle.

  a. 3, 3, $3\sqrt{2}$ **b.** 4, 9, 7

Plan Let c = the length of the longest side. Determine if $c^2 = a^2 + b^2$.

Solutions

a.

$$c^2 \stackrel{?}{=} a^2 + b^2$$
$$(3\sqrt{2})^2 \stackrel{?}{=} 3^2 + 3^2$$
$$9(2) \stackrel{?}{=} 9 + 9$$
$$18 = 18$$
So, $c^2 = a^2 + b^2$.

The triangle *is* a right triangle.

b.

$$c^2 \stackrel{?}{=} a^2 + b^2$$
$$9^2 \stackrel{?}{=} 4^2 + 7^2$$
$$81 \stackrel{?}{=} 16 + 49$$
$$81 \neq 65$$
So, $c^2 \neq a^2 + b^2$.

The triangle *is not* a right triangle.

13.4 The Pythagorean Theorem   **509**

Additional Example 1

Find the missing length of the right triangle. If the answer is not a rational number, give it in simplest radical form.

  a. $a = 10, b = 10$ $c = 10\sqrt{2}$
  b. $b = 20, c = 29$ $a = 21$

Additional Example 2

Given the lengths of three sides of a triangle, determine whether the triangle is a right triangle.

  a. 7, 24, 25   Yes
  b. 2, $2\sqrt{3}$, $4\sqrt{3}$ No

Common Error Analysis

Error: Students sometimes substitute numbers incorrectly in the formula for the Pythagorean Theorem. Remind them that *c* always stands for the hypotenuse, the longest side of the triangle. Values for the legs of the triangle, *a* and *b*, may be assigned arbitrarily.

Checkpoint

1. Find the length of the hypotenuse of a right triangle if the lengths of the legs are 7 and 8. $\sqrt{113}$
2. For right triangle *ABC*, $a = 6$ and $b = 8$. Find *c*. 10
3. The lengths of three sides of a triangle are 4, 16 and $\sqrt{272}$. Determine whether the triangle is a right triangle. Yes
4. A car was driven 75 mi south, then 100 mi west. How far was the car from the starting point? 125 mi

EXAMPLE 3   A 10-ft ladder is leaning against a building. Its base is 3 ft from the base of the building. How high up the building will the ladder reach? Round the answer to the nearest tenth of a foot.

Solution   Draw a diagram and label the parts.

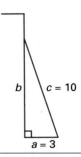

$$c^2 = a^2 + b^2$$
$$10^2 = 3^2 + b^2$$
$$100 = 9 + b^2$$
$$91 = b^2$$
$$b = \sqrt{91}$$
$$b \approx 9.539, \text{ or } 9.5 \text{ to the nearest tenth}$$

Thus, the ladder will reach approximately 9.5 ft up the wall.

TRY THIS   2. A rectangle is 3 m long and 8 m wide. Find the length of a diagonal to the nearest tenth. 8.5

Classroom Exercises

For each right triangle, find the missing length. If the answer is not a rational number, write it in simplest radical form. Use $c^2 = a^2 + b^2$.

1. $a = 3, b = 4$ $c = 5$
2. $a = 1, b = 3$ $c = \sqrt{10}$
3. $a = 6, c = 9$ $b = 3\sqrt{5}$
4. $b = 7, c = 7\sqrt{2}$ $a = 7$
5. $b = 40, c = 50$ $a = 30$
6. $a = 5, b = 5\sqrt{3}$ $c = 10$

Given the lengths of three sides of a triangle, determine whether the triangle is a right triangle.

7. 10, 15, 20 No
8. $\sqrt{6}, \sqrt{15}, 3$ Yes
9. 3, $3\sqrt{3}$, 6 Yes

Written Exercises

Solve each problem.

1. Find the length of the hypotenuse if the lengths of the legs of a right triangle are 3 m and 4 m. 5 m
2. Find the length of the hypotenuse if the lengths of the legs of a right triangle are 5 cm and 12 cm. 13 cm
3. The length of the hypotenuse of a right triangle is 4 m. The length of one leg is 2 m. Find the length of the other leg. Give the answer in simplest radical form. $2\sqrt{3}$ m
4. The length of the hypotenuse of a right triangle is 6 cm. The length of one leg is 4 cm. Find the length of the other leg. Give the answer in simplest radical form. $2\sqrt{5}$ cm

510   Chapter 13 Radicals

Additional Example 3

A 12–ft ladder is 4 ft from the base of a house. How high up the house will the ladder reach? $8\sqrt{2}$ ft ≈ 11.3 ft

For each right triangle, find the length of the hypotenuse. If the answer is not a rational number, write it in simplest radical form.

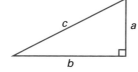

$2\sqrt{85}$

5. $a = 2, b = 4$ $2\sqrt{5}$ **6.** $a = 7, b = 9$ $\sqrt{130}$ **7.** $a = 12, b = 14$

8. $a = 24, b = 10$ 26   **9.** $a = 3, b = 5$ $\sqrt{34}$ **10.** $a = 9, b = 13$

11. $a = 5, b = 7$ $\sqrt{74}$ **12.** $a = \sqrt{3}, b = \sqrt{6}$ 3   **13.** $a = \sqrt{5}, b = 10$

10. $5\sqrt{10}$ **13.** $\sqrt{105}$

For each right triangle, find the missing length. If the answer is not a rational number, write it in simplest radical form.

$8\sqrt{7}$

14. $a = 3, b = 7$ $c = \sqrt{58}$ **15.** $a = 5, c = 8$ $b = \sqrt{39}$ **16.** $b = 9, c = 23$

17. $a = 9, c = 41$ $b = 40$ **18.** $b = 10, c = 12$ $a = 2\sqrt{11}$ **19.** $a = 5, c = 13$ $b = 12$

20. $a = 7, b = 11$ $c = \sqrt{170}$ **21.** $a = 3, c = 9$ $b = 6\sqrt{2}$ **22.** $b = 4, c = 7$ $a = \sqrt{33}$

23. $b = 9, c = 15$ $a = 12$ **24.** $b = 5, c = 11$ $a = 4\sqrt{6}$ **25.** $a = 7, b = 10$

$c = \sqrt{149}$

Given the lengths of three sides of a triangle, determine whether it is a right triangle.

26. 9, 12, 15 Yes   **27.** 4, 5, 3 Yes   **28.** 2, 3, 4 No

29. 20, 21, 29 Yes   **30.** 14, 48, 50 Yes   **31.** 15, 39, 36 Yes

32. 21, 72, 75 Yes   **33.** 20, 30, 40 No   **34.** $\sqrt{3}, \sqrt{4}, \sqrt{5}$ No

35. 7, 9, $\sqrt{130}$ Yes   **36.** $\sqrt{7}, 8, \sqrt{71}$ Yes   **37.** 5, 12, $\sqrt{119}$ Yes

38. $\sqrt{5}, 12, 13$ No   **39.** $\sqrt{3}, 4, \sqrt{19}$ Yes   **40.** 10, 12, $\sqrt{22}$ No

For each right triangle, find the missing length. If the answer is not a rational number, write it in simplest radical form.

$a = 2\sqrt{17}$ $c = 6\sqrt{5}$

41. $a = 3\sqrt{3}, c = 7$ **42.** $b = 4\sqrt{2}, c = 10$ **43.** $a = \sqrt{5}, b = 5\sqrt{7}$

44. $a = 4\sqrt{7}, b = 3\sqrt{5}$ **45.** $a = 5\sqrt{6}, c = 14$ **46.** $b = 8\sqrt{2}, c = 6\sqrt{8}$

47. $b = 3\sqrt{7}, c = 4\sqrt{7}$ **48.** $a = 3\sqrt{11}, b = 4\sqrt{10}$ **49.** $a = 11\sqrt{3}, b = 4\sqrt{5}$

$a = 7$ $c = \sqrt{259}$ $c = \sqrt{443}$

Solve each problem. If an answer is not rational, round it to the nearest tenth of the unit.

50. A 6-m ladder is leaning against a building. Its base is 1 m from the base of the building. How high up the building will the ladder reach? 5.9 m

51. Find the length of the diagonal across a television screen that is 18 cm wide and 18 cm long. 25.5 cm

52. A carpenter wants to build a brace for a gate that is 4 ft wide and 5 ft high. Find the length of the brace. 6.4 ft

53. A rectangular field is 50 yd wide and 120 yd long. How long is the diagonal path connecting two opposite corners of the field? 130 yd

The lengths of three sides of a triangle are given below. For each triangle, have students tell whether it is a right triangle. If it is a right triangle, have them give the length of the hypothenuse.

1. 10, 20, 30 No
2. 12, 13, 5 Yes; 13
3. 40, 30, 50 Yes; 50
4. 15, 9, 12 Yes; 15

◢◢◢◢ FOLLOW UP

Guided Practice

Classroom Exercises 1–9
Try This all

Independent Practice

A Ex. 1–34, **B** Ex. 35–57, **C** Ex. 58–61

Basic: WE 1–33 odd, 50–53 all
Average: WE 15–57 odd
Above Average: WE 23–57 odd, 58–61 all

Additional Answers

Written Exercises

41. $b = \sqrt{22}$
44. $c = \sqrt{157}$
45. $b = \sqrt{46}$
46. $a = 4\sqrt{10}$

58. The area of a square equals the square of the length of a side. Hence, the area of the large square is $(a + b)^2$. Another way to find the area of the large square is to first notice that the large square consists of 4 congruent triangles and one smaller square. By adding the total area of the 4 triangles and the area of the small square, one finds the area of the large square. The area of a triangle is $\frac{1}{2}$ (base)(height). In this case, the area of the 4 congruent triangles is $4(\frac{1}{2}ab)$. The area of the smaller square is c^2. Hence, $(a + b)^2 = 4(\frac{1}{2}ab) + c^2$.

59. In Exercise 58, $(a + b)^2 = 4(\frac{1}{2}ab) + c^2$ is given where c is the length of the hypotenuse and a and b are the lengths of the legs. Thus, $a^2 + 2ab + b^2 = 2ab + c^2$. Subtract $2ab$ from each side of this equation to obtain $a^2 + b^2 = c^2$, the Pythagorean Theorem.

61. An equilateral triangle can be separated into 2 congruent right triangles. In one of these triangles let s equal the length of the hypotenuse, and h equal the length of one leg. Then $\frac{1}{2}s$ equals the length of the other leg. Then apply the Pythagorean Theorem: $h^2 + (\frac{1}{2}s)^2 = s^2$. Solve this equation for h. The result is $h = \frac{1}{2}s\sqrt{3}$.

54. A 15-m loading ramp is connected to a dock 3 m high. How far is the foot of the ramp from the loading dock? 14.7 m

55. A surveyor walked 9 km east, then 6 km north. How far was the surveyor from his starting point? 10.8 km

56. A baseball diamond is a square with sides 90 ft in length. How far is it from second base to home plate? 127.3 ft

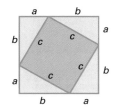

57. A mover must try to fit a thin circular mirror, 2 m in diameter, through a doorway measuring 1 m by 1.8 m. Will the mirror fit through the doorway? Yes

Use the diagram at the right for Exercises 58 and 59. The diagram shows a square with sides of length $a + b$. The four-sided figure inside the square is a square with sides of length c.

58. Use the concept of area to explain why the following equation is true: $(a + b)^2 = 4(\frac{1}{2}ab) + c^2$

59. Use the Equation in Exercise 58 to verify the Pythagorean Theorem.

60. Find the greatest possible length of a ski to be placed flat in a box that is 4 ft by 5 ft. Give the answer to the nearest inch. 77 in

61. Derive a formula for finding the altitude h of an equilateral triangle in terms of s, the length of a side.

Mixed Review

Simplify. *6.1*

1. $x^5 \cdot x^4$ x^9

2. $(5a^2)(-6a^3)$ $-30a^5$

3. $(-3x^4)(2x)$ $-6x^5$

4. $(-a^3b^2)(-3a^4b^5)$ $3a^7b^7$

5. $(-3b^3c^4)(-b^4c^4)$ $3b^7c^8$

6. $(4a^6b^7)(-2a^7b^{10})$ $-8a^{13}b^{17}$

7. The price of a bus ticket was $80. The price was increased by 12%. What is the new price? *4.7* $89.60

8. A baseball team won 16 games and lost 4 games. What percent of its games did the team win? *4.7* 80%

9. A rectangle is 8 cm long and 6 cm wide. If the length and the width are increased by the same amount, the area is increased by 32 cm². Find the amount by which each side must be increased. *7.9* 2 cm

Enrichment

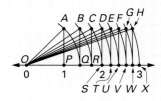

Find the numbers for points P through X on the number line at the left. Since each triangle OPA, OQB, etc., is a right triangle, students can use the Pythagorean Theorem. Each vertical segment, $\overline{PA}$, $\overline{QB}$, etc., is 1 unit long. $P(1)$, $Q(\sqrt{2})$, $R(\sqrt{3})$, $S(2)$, $T(\sqrt{5})$, $U(\sqrt{6})$, $V(\sqrt{7})$, $W(2\sqrt{2})$, $X(3)$

13.5 Simplifying Radicals

Objective To simplify square roots in which the radicands contain variables raised to even or odd powers

Square roots such as $\sqrt{2^6}$ and $\sqrt{10^4}$ can be simplified using the Product of Powers Property $b^{m+n} = b^m \cdot b^n$, as shown below.

$$\sqrt{2^6} = \sqrt{2^3 \cdot 2^3} = 2^3 \qquad\qquad \sqrt{10^4} = \sqrt{10^2 \cdot 10^2} = 10^2$$

These examples illustrate a property of square roots of the form $\sqrt{a^{2n}}$, where $2n$ is an even number and a is positive. For the examples and exercises of this lesson, assume that each base is a positive number.

Square Root Property of Even Powers
For every even number $2n$, $\sqrt{a^{2n}} = \sqrt{a^{n+n}} = \sqrt{a^n \cdot a^n} = a^n$, where $a > 0$.

EXAMPLE 1 Simplify $4\sqrt{25a^8b^{18}}$.

Solution
$$\begin{aligned}
4\sqrt{25a^8b^{18}} &= 4 \cdot \sqrt{25} \cdot \sqrt{a^8} \cdot \sqrt{b^{18}} \\
&= 4 \cdot 5 \cdot a^4 \cdot b^9 \\
&= 20a^4b^9
\end{aligned}$$

Square Root Property of Odd Powers
For every odd number $2n + 1$, $\sqrt{a^{2n+1}} = \sqrt{a^{2n} \cdot a^1} = a^n\sqrt{a}$, where $a > 0$.

EXAMPLE 2 Simplify $-3\sqrt{28x^3y^7}$.

Solution
$$\begin{aligned}
-3\sqrt{28x^3y^7} &= -3\sqrt{4 \cdot 7 \cdot x^2 \cdot x^1 \cdot y^6 \cdot y^1} \\
&= -3\sqrt{4 \cdot 7} \cdot \sqrt{x^2 \cdot x} \cdot \sqrt{y^6 \cdot y} \\
&= -6\sqrt{7} \cdot x\sqrt{x} \cdot y^3\sqrt{y} \\
&= -6xy^3 \cdot \sqrt{7} \cdot \sqrt{x} \cdot \sqrt{y} \\
&= -6xy^3\sqrt{7xy}
\end{aligned}$$

TRY THIS Simplify: **1.** $4\sqrt{27x^5y^3}$ $\quad 12x^2y\sqrt{3xy}$ **2.** $-2\sqrt{32x^7y^4}$ $\quad -8x^3y^2\sqrt{2x}$

13.5 Simplifying Radicals **513**

Prerequisite Quiz

Multiply.

1. $x^5 \cdot x^5$ $\quad x^{10}$
2. $x \cdot x^8$ $\quad x^9$

Complete to make true statements.

3. $a^8 = a^4 \cdot \underline{\ ?\ }$ $\quad a^4$
4. $a^9 = a \cdot \underline{\ ?\ }$ $\quad a^8$
5. $a^{14} = a^7 \cdot \underline{\ ?\ }$ $\quad a^7$

Motivator

Have students consider this question. If $\sqrt{2} \approx 1.4$, will $\sqrt{8}$ be closer to 2×1.4, 3×1.4, or 4×1.4? $\quad 2 \times 1.4$ Have them use a calculator to check their answers.

■■■ **TEACHING SUGGESTIONS**

Lesson Note

Since $\sqrt{x^2} = |x|$, simplifying an expression such as $\sqrt{a^{2n}}$ really depends on the value of a. To avoid such problems in this book, we will assume that $a > 0$ unless otherwise stated.

Additional Example 1

Simplify.

a. $\sqrt{a^4}$ $\quad a^2$
b. $\sqrt{36m^{10}n^{12}}$ $\quad 6m^5n^6$

Additional Example 2

Simplify.

a. $\sqrt{y^{11}}$ $\quad y^5\sqrt{y}$
b. $-7\sqrt{12x^5y^9}$ $\quad -14x^2y^4\sqrt{3xy}$

Highlighting the Standards

Standard 5c, 5d: Students become familiar with square root notation as they practice combining and simplifying radicals.

513

Math Connections

Algebra 2: When exponents are extended to the set of rational numbers, $x^{\frac{1}{2}} = \sqrt{x}$. For example,

$\sqrt{16x^6y^3} = (16x^6y^3)^{\frac{1}{2}}$

$= 16^{\frac{1}{2}}x^{\frac{6}{2}}y^{\frac{3}{2}} = 4x^3y \cdot y^{\frac{1}{2}}$

$= 4x^3y\sqrt{y}$

Critical Thinking Questions

Synthesis: Ask students to consider what would happen if the Square Root Properties are not restricted to $a > 0$. Ask them whether the expression $\sqrt{a^6} = a^3$ is true if a has a value of -2; in other words, is $\sqrt{(-2)^6} = (-2)^3$? No Then ask how the statement of this property might be changed to include values less than zero.

$\sqrt{a^{2n}} = |a^n|$

Common Error Analysis

Error: In a problem such as $\sqrt{a^{16}}$, students may try to take the square root of the exponent giving an incorrect result of a^4. Point out that taking the square root of a variable to an even power actually is accomplished by finding half of the exponent.

Checkpoint

Simplify.

1. $\sqrt{25a^6}$ $5a^3$
2. $-\sqrt{16m^{18}}$ $-4m^9$
3. $-15\sqrt{100x^9}$ $-150x^4\sqrt{x}$
4. $24\sqrt{24a^{10}b^{19}}$ $48a^5b^9\sqrt{6b}$
5. $-6\sqrt{63x^{19}y^{31}}$ $-18x^9y^{15}\sqrt{7xy}$

EXAMPLE 3   Simplify $-7x^5y\sqrt{12x^9y^6z^{11}}$.

Plan   Use $\sqrt{a^{2n}} = a$ and $\sqrt{a^{2n+1}} = a^n\sqrt{a}$.

Solution $-7x^5y\sqrt{12x^9y^6z^{11}} = -7x^5y\sqrt{4 \cdot 3 \cdot x^8 \cdot x \cdot y^6 \cdot z^{10} \cdot z}$

$= -7x^5y \cdot 2x^4y^3z^5\sqrt{3xz}$

$= -14x^9y^4z^5\sqrt{3xz}$

TRY THIS   Simplify: **3.** $-6x^3y^2z\sqrt{9x^5y^9z^3}$ $-18x^5y^6z^2\sqrt{xyz}$

4. $4x^3z^4\sqrt{8y^9z^8}$ $8x^3y^4z^8\sqrt{2y}$

Classroom Exercises

Simplify.

1. $\sqrt{a^8}$ a^4
2. $\sqrt{x^9}$ $x^4\sqrt{x}$
3. $\sqrt{a^5}$ $a^2\sqrt{a}$
4. $\sqrt{b^{12}}$ b^6
5. $\sqrt{y^{15}}$ $y^7\sqrt{y}$
6. $\sqrt{m^{36}}$ m^{18}
7. $\sqrt{y^{21}}$ $y^{10}\sqrt{y}$
8. $\sqrt{a^{19}}$ $a^9\sqrt{a}$
9. $\sqrt{y^{14}}$ y^7
10. $\sqrt{m^7}$ $m^3\sqrt{m}$
11. $\sqrt{a^2}$ a
12. $\sqrt{b^{11}}$ $b^5\sqrt{b}$
13. $\sqrt{16x^{14}}$ $4x^7$
14. $\sqrt{81b^{17}}$ $9b^8\sqrt{b}$
15. $-2\sqrt{18m^5n^6}$ $-6m^2n^3\sqrt{2m}$
16. $5\sqrt{50a^{24}b^{25}}$ $25a^{12}b^{12}\sqrt{2b}$

Written Exercises

Simplify.

1. $\sqrt{4x^{24}}$ $2x^{12}$
2. $-\sqrt{9a^{14}}$ $-3a^7$
3. $-\sqrt{16y^{16}}$ $-4y^8$
4. $-\sqrt{16a^{20}}$ $-4a^{10}$
5. $14\sqrt{25y^{28}}$ $70y^{14}$
6. $-9\sqrt{36x^{32}}$ $-54x^{16}$
7. $5\sqrt{81b^{30}}$ $45b^{15}$
8. $-10\sqrt{49y^{44}}$ $-70y^{22}$
9. $\sqrt{9a^7}$ $3a^3\sqrt{a}$
10. $-3\sqrt{x^5}$ $-3x^2\sqrt{x}$
11. $2\sqrt{4y^3}$ $4y\sqrt{y}$
12. $5\sqrt{15x^{15}}$ $5x^7\sqrt{15x}$
13. $-2\sqrt{36c^5}$
14. $10\sqrt{3x^5}$ $10x^2\sqrt{3x}$
15. $-2\sqrt{12a^9}$
16. $5\sqrt{24x^{13}}$
17. $\sqrt{x^9y^7}$ $x^4y^3\sqrt{xy}$
18. $-2\sqrt{20a^5b^{11}}$
19. $-\sqrt{45m^7n^5}$
20. $8\sqrt{18x^{15}y^9}$
21. $-\sqrt{40xy^5}$
22. $-4\sqrt{50x^3y^7}$
23. $4\sqrt{18a^{17}b^{15}}$
24. $-\sqrt{32a^{11}b}$
25. $\sqrt{4a^{10}b^{12}}$ $2a^5b^6$
26. $-\sqrt{25x^4y^6}$
27. $-\sqrt{49a^{12}b^{10}}$
28. $\sqrt{81m^{16}n^{20}}$
29. $8\sqrt{9x^{14}y^{40}}$
30. $-12\sqrt{4a^{18}b^{16}}$
31. $12\sqrt{36x^{24}y^{36}}$
32. $-\sqrt{100m^{10}n^{30}}$
33. $9\sqrt{27a^{15}b^9}$
34. $5\sqrt{12a^{11}b^{13}}$
35. $10\sqrt{24x^{21}y^{33}}$
36. $-4\sqrt{18x^{15}y^9}$
37. $-3a^3b\sqrt{75a^4b^8}$
38. $4xy^6\sqrt{12x^4y^{12}}$ $8x^3y^{12}\sqrt{3}$
39. $-5a^3b^4\sqrt{18a^6b^{16}}$
40. $-5x^2y\sqrt{18x^5y^7}$ $-15x^4y^4\sqrt{2xy}$
41. $9a^5b^{10}\sqrt{72ab^9}$ $54a^5b^{14}\sqrt{2ab}$
42. $10xy\sqrt{24x^5y^3}$ $20x^3y^2\sqrt{6xy}$

514   Chapter 13 Radicals

Additional Example 3

Simplify.

$-6a^4b^3\sqrt{18a^7b^{10}c^{15}}$ $-18a^7b^8c^7\sqrt{2ac}$

43. $7a^4b^{10}\sqrt{36a^{13}b^{11}}$ 44. $-10xy^5\sqrt{75x^{22}y^{13}}$ 45. $8e^4f^3\sqrt{27e^4f^3}$

46. $-\sqrt{100a^4b^2c^6}$ $-10a^2bc^3$ 47. $\sqrt{144x^{10}y^6z^{10}}$ $12x^5y^3z^5$ 48. $-\sqrt{81a^{24}b^{12}c^8}$ $-9a^{12}b^6c^4$

49. $5\sqrt{44x^3yz^4}$ $10xz^2\sqrt{11xy}$ 50. $6ab^2c\sqrt{90a^2bc^7}$ 51. $-4x^3y^3z\sqrt{60x^7y^9z^6}$

52. $4\sqrt{48x^8y^{14}z^{36}}$ 53. $10\sqrt{32a^{41}b^{38}c^{22}}$ 54. $4\sqrt{128x^4y^{16}z^{80}}$

55. $-3\sqrt{28a^{30}b^{25}c^{36}}$ 56. $9\sqrt{98x^{44}y^{36}z^{18}}$ 57. $-3\sqrt{243a^{40}b^{80}c^{100}}$

58. $-6\sqrt{a^{2n+1}}$ $-6a^n\sqrt{a}$ 59. $-3\sqrt{x^{4n+1}}$ $-3x^{2n}\sqrt{x}$ 60. $9x\sqrt{y^{16n}z^{5n}}$ $9xy^{8n}z^{2n}\sqrt{z^n}$

52. $16x^4y^7z^{18}$ 53. $40a^{20}b^{19}c^{11}\sqrt{2a}$ 54. $32x^2y^8z^{40}\sqrt{2}$
55. $-6a^{15}b^{12}c^{18}\sqrt{7b}$ 56. $63x^{22}y^{18}z^9\sqrt{2}$ 57. $-27a^{20}b^{40}c^{50}\sqrt{3}$

Midchapter Review

Express each rational number as a decimal. Then classify the decimal as either terminating or repeating.   13.1

1. $\frac{1}{5}$
   0.2, term
2. $-\frac{5}{6}$
   $-0.8\overline{3}$, rep
3. $3\frac{2}{3}$
   $3.\overline{6}$, rep
4. $-\frac{3}{11}$
   $-0.\overline{27}$, rep

For each repeating decimal, find a fraction $\frac{a}{b}$ where a and b are integers.   13.1

5. $4.\overline{5}$ $\frac{41}{9}$ 6. $13.4\overline{3}$ $\frac{403}{30}$ 7. $3.8282\cdots$ $\frac{379}{99}$ 8. $0.\overline{153}$ $\frac{17}{111}$

Simplify.   13.2

9. $\sqrt{81}$ 9   10. $\sqrt{27}$ $3\sqrt{3}$ 11. $\sqrt{50}$ $5\sqrt{2}$ 12. $-6\sqrt{500}$ $-60\sqrt{5}$

Approximate to the nearest tenth. Use a square root table.   13.3

13. $\sqrt{15}$ 3.9   14. $\sqrt{43}$ 6.6   15. $\sqrt{80}$ 8.9   16. $\sqrt{94}$ 9.7

17. Approximate $\sqrt{135}$ to the nearest tenth. Use the divide-and-average method.   13.3 11.6

18. The area of a rectangle is 216 cm². The length is twice the width. Find the length and the width of the rectangle.   13.3 l = 20.8 cm, w = 10.4 cm

For each right triangle, find the missing length. If the answer is not a rational number, give it in simplest radical form.   13.4

19. $a = 5, b = 5$ 20. $a = 3, c = 7$ 21. $b = 10, c = 20$ 22. $b = 21, c = 29$
   $c = 5\sqrt{2}$ $b = 2\sqrt{10}$ $a = 10\sqrt{3}$ $a = 20$

Given the lengths of the sides of a triangle, determine if it is a right triangle.   13.4

23. 18, 24, 30 Yes   24. 2, 3, $\sqrt{5}$ Yes   25. 10, 11, 15 No   26. 20, 48, 52 Yes

Simplify.   13.5

27. $-4\sqrt{36a^{18}}$ 28. $\sqrt{24x^{11}}$ 29. $6\sqrt{16x^{12}y^{19}}$ 30. $a^2b\sqrt{75a^{21}b^{24}}$
   $-24a^9$ $2x^5\sqrt{6x}$ $24x^6y^9\sqrt{y}$ $5a^{12}b^{13}\sqrt{3a}$

Have students do the following exercises orally to demonstrate understanding.

Simplify.

1. $\sqrt{x^{10}}$ x^5
2. $\sqrt{x^7}$ $x^3\sqrt{6}$
3. $\sqrt{16x^{12}y}$ $4x^6\sqrt{y}$
4. $\sqrt{36x^6y^7}$ $6x^3y^3\sqrt{y}$

◼◼ FOLLOW UP

Guided Practice

Classroom Exercises 1–16
Try This   all

Independent Practice

A Ex. 1–39, **B** Ex. 40–51, **C** Ex. 52–60

Basic: WE 1–39 odd, Midchapter Review

Average: WE 11–49 odd, Midchapter Review

Above Average: WE 21–59 odd, Midchapter Review

Additional Answers

Written Exercises

13. $-12c^2\sqrt{c}$ 15. $-4a^4\sqrt{3a}$
16. $10x^6\sqrt{6x}$ 18. $-4a^2b^5\sqrt{5ab}$
19. $-3m^3n^2\sqrt{5mn}$ 20. $24x^7y^4\sqrt{2xy}$
21. $-2y^2\sqrt{10xy}$ 22. $-20xy^3\sqrt{2xy}$
23. $12a^8b^7\sqrt{2ab}$ 24. $-4a^5\sqrt{2ab}$
26. $-5x^2y^3$ 27. $-7a^6b^5$
28. $9m^8n^{10}$ 29. $24x^7y^{20}$

See page 518 for the answers to Ex. 30–37, 39, 43–45, and 50–51.

Enrichment

Ask students to find the distance between points $P(1, 1)$ and $Q(7, 1)$ as shown in the graph.   6

Then have them find the distance between $Q(7, 1)$ and $R(7, 9)$.   8   Ask them how they might find the distance between points P and R.   Use the Pythagorean Theorem.

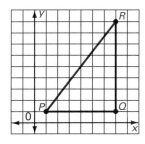

Have them find this distance.   10
Students should now be able to find the distance between the following pairs of points by locating them on a graph, drawing a right triangle, and using the Pythagorean Theorem.

1. $A(-5, -2)$, $B(7, 3)$ 13
2. $C(2, -4)$, $D(7, 1)$ $5\sqrt{2}$

◼ GETTING STARTED

Prerequisite Quiz

Simplify, if possible.

1. $4x - y$ Not possible
2. $18a - 7b + 6a - b$ $24a - 8b$
3. $\sqrt{36a^4b^2}$ $6a^2b$
4. $\sqrt{81a^5b^3}$ $9a^2b\sqrt{ab}$
5. $-3\sqrt{49x^9y^{10}}$ $-21x^4y^5\sqrt{x}$

Motivator

Give students the approximate value of
$\sqrt{3}$ as 1.732 and ask them to find an
approximate decimal value for the following.
$7\sqrt{3} - 2\sqrt{3} + 5\sqrt{3}$ 17.32   Then show
them how these terms can be combined to
give $10\sqrt{3}$ which is easier to evaluate.

◼ TEACHING SUGGESTIONS

Lesson Note

Although $\sqrt{2}$ is not a variable, suggest to
the students that it can be treated somewhat
like a variable when combining radicals.
Since $5a + 6a = 11a$, in like manner
$5\sqrt{2} + 6\sqrt{2} = 11\sqrt{2}$. Just as the
expression $5a + 5b$ cannot be simplified,
there is no way to combine $5\sqrt{2} + 6\sqrt{3}$.

13.6 Adding and Subtracting Radicals

Objective   To add and subtract expressions containing radicals

Recall that like terms such as $5a$ and $6a$ can be
combined by using the Distributive Property.

5	6	
$5\sqrt{7}$	$6\sqrt{7}$	$\sqrt{7}$

$$5a + 6a = (5 + 6)a = 11a$$

Similarly, $5\sqrt{7} + 6\sqrt{7} = (5 + 6)\sqrt{7} = 11\sqrt{7}$

11	
$11\sqrt{7}$	$\sqrt{7}$

Therefore, you can combine expressions with radicals by adding or subtracting when the radicands are the same.

EXAMPLE 1   Simplify $8\sqrt{2} - 7\sqrt{2} + 4\sqrt{2}$.

Solutions

Method 1
Group the positive terms.
$8\sqrt{2} - 7\sqrt{2} + 4\sqrt{2}$
$= (8\sqrt{2} + 4\sqrt{2}) - 7\sqrt{2}$
$= 12\sqrt{2} - 7\sqrt{2}$
$= 5\sqrt{2}$

Method 2
Combine using the given order.
$8\sqrt{2} - 7\sqrt{2} + 4\sqrt{2}$
$= (8\sqrt{2} - 7\sqrt{2}) + 4\sqrt{2}$
$= 1\sqrt{2} + 4\sqrt{2}$
$= 5\sqrt{2}$

EXAMPLE 2   Simplify $9\sqrt{5} + 3\sqrt{3} - 2\sqrt{5}$.

Solution $9\sqrt{5} + 3\sqrt{3} - 2\sqrt{5} = (9\sqrt{5} - 2\sqrt{5}) + 3\sqrt{3}$
$= 7\sqrt{5} + 3\sqrt{3}$ ⟵ unlike radicands

TRY THIS   Simplify: 1. $6\sqrt{3} + 4\sqrt{3} - 2\sqrt{3}$ 2. $5\sqrt{6} - 5\sqrt{2} + 3\sqrt{6}$
$8\sqrt{3}$ $8\sqrt{6} - 5\sqrt{2}$

You may need to simplify radicals before combining like terms.

EXAMPLE 3   Simplify $5\sqrt{8} - 6\sqrt{2} + 4\sqrt{72}$.

Plan   Simplify $\sqrt{8}$ and $\sqrt{72}$ first. Then combine like terms.

Solution $5\sqrt{8} - 6\sqrt{2} + 4\sqrt{72} = 5\sqrt{4 \cdot 2} - 6\sqrt{2} + 4\sqrt{36 \cdot 2}$
$= 5 \cdot 2\sqrt{2} - 6\sqrt{2} + 4 \cdot 6\sqrt{2}$
$= 10\sqrt{2} - 6\sqrt{2} + 24\sqrt{2}$
$= 28\sqrt{2}$

516   Chapter 13 Radicals

Highlighting the Standards

Standards 5c, 4c: Students use radicals to
solve problems relating to geometric figures.

Additional Example 1

Simplify.
$12\sqrt{3} - 8\sqrt{3} + 5\sqrt{3}$ $9\sqrt{3}$

Additional Example 2

Simplify.
$10\sqrt{7} + 2\sqrt{5} - 3\sqrt{7}$ $7\sqrt{7} + 2\sqrt{5}$

EXAMPLE 4 Simplify $3b\sqrt{36a^3b} - 14a\sqrt{ab^3} - 4\sqrt{81a^3b^3}$.

Solution
$3b\sqrt{36a^3b} - 14a\sqrt{ab^3} - 4\sqrt{81a^3b^3}$
$= 3b\sqrt{36} \cdot \sqrt{a^2 \cdot ab} - 14a\sqrt{ab \cdot b^2} - 4\sqrt{81} \cdot \sqrt{a^2 \cdot b^2 \cdot ab}$
$= 3b \cdot 6 \cdot a\sqrt{ab} - 14a \cdot b\sqrt{ab} - 4 \cdot 9 \cdot a \cdot b\sqrt{ab}$
$= 18ab\sqrt{ab} - 14ab\sqrt{ab} - 36ab\sqrt{ab}$
$= -32ab\sqrt{ab}$

EXAMPLE 5 Find the perimeter of a rectangle with a length of $9 - \sqrt{3}$ and a width of $3\sqrt{5}$.

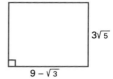
$3\sqrt{5}$
$9 - \sqrt{3}$

Plan Use $P = l + w + l + w$ or $P = 2l + 2w$.

Solution
$P = l + w + l + w$ $\qquad\qquad$ $P = 2l + 2w$
$= 9 - \sqrt{3} + 3\sqrt{5} + 9 - \sqrt{3} + 3\sqrt{5}$ $\qquad$ $= 2(9 - \sqrt{3}) + 2(3\sqrt{5})$
$= 18 - 2\sqrt{3} + 6\sqrt{5}$ $\qquad\qquad\qquad$ $= 18 - 2\sqrt{3} + 6\sqrt{5}$

Thus, the perimeter of the rectangle is $18 - 2\sqrt{3} + 6\sqrt{5}$.

TRY THIS
3. Simplify $2x\sqrt{25xy^5} + 8y\sqrt{9x^3y^3} - 7xy^2\sqrt{49xy}$. $-15xy^2\sqrt{xy}$
4. Find the perimeter of a rectangle with a length of $2 + \sqrt{3}$ and a width of $1 + \sqrt{2}$. $6 + 2\sqrt{3} + 2\sqrt{2}$

Classroom Exercises

Which terms are like terms?
1. $-8\sqrt{2}, \sqrt{2}, \sqrt{7}, \sqrt{14}$ $-8\sqrt{2}, \sqrt{2}$
2. $9, 4\sqrt{5}, -\sqrt{5}, \sqrt{3}$ $4\sqrt{5}, -\sqrt{5}$
3. $5\sqrt{3}, 4\sqrt{3}, 4\sqrt{2}, -3\sqrt{3}$
   All except $4\sqrt{2}$
4. $-5\sqrt{7}, \sqrt{7}, -3\sqrt{7}, -5$
   All except -5

Simplify.
5. $5\sqrt{3} + 7\sqrt{3}$ $12\sqrt{3}$ $\quad$ 6. $8\sqrt{5} - 2\sqrt{5}$ $6\sqrt{5}$ $\quad$ 7. $14\sqrt{7} + 6\sqrt{7}$ $20\sqrt{7}$ $\quad$ 8. $9\sqrt{2} - 3\sqrt{2}$ $6\sqrt{2}$
9. $35\sqrt{44} - 15\sqrt{99}$ $25\sqrt{11}$ $\quad$ 10. $-15\sqrt{54} - 3\sqrt{150} + 10\sqrt{6}$ $-50\sqrt{6}$

Written Exercises

Add or subtract. Simplify.
1. $18\sqrt{2} - 7\sqrt{2} + 12\sqrt{2}$ $23\sqrt{2}$ $\qquad\qquad$ 2. $12\sqrt{3} - 7\sqrt{3} + 2\sqrt{3}$ $7\sqrt{3}$
3. $10\sqrt{5} - 2\sqrt{5} - 9\sqrt{5}$ $-\sqrt{5}$ $\qquad\qquad$ 4. $-4\sqrt{7} + 5\sqrt{7} + 3\sqrt{7}$ $4\sqrt{7}$

Previous Math: The fact that $\sqrt{a} + \sqrt{b} \neq \sqrt{a + b}$ is comparable to the fact that $a^2 + b^2 \neq (a + b)^2$. In fact, if $\sqrt{a}$ is written as $a^{\frac{1}{2}}$, the comparison is even more evident. $a^{\frac{1}{2}} + b^{\frac{1}{2}} \neq (a + b)^{\frac{1}{2}}$.

Critical Thinking Questions

Application: Ask students if there are any values of x and y that make the following statement true. $\sqrt{a} + \sqrt{b} = \sqrt{a + b}$. $a = 0$ or $b = 0$

Common Error Analysis

Error: Students may try to simplify $4\sqrt{3} + 7\sqrt{5}$ to $11\sqrt{15}$.

Tell them to think of this as $4x + 7y$, which cannot be simplified.

Checkpoint

Simplify.
1. $-15\sqrt{7} - 3\sqrt{7} + 20\sqrt{7}$ $2\sqrt{7}$
2. $2\sqrt{32} + 3\sqrt{18} - \sqrt{50}$ $12\sqrt{2}$
3. $-4\sqrt{48} + 5\sqrt{192} - 6\sqrt{108}$
   $-12\sqrt{3}$
4. $12\sqrt{xy} - 2\sqrt{xy} + \sqrt{xy}$ $11\sqrt{xy}$
5. $10\sqrt{18ab} - \sqrt{50ab} - 8\sqrt{72ab}$
   $-23\sqrt{2ab}$

Additional Example 3

Simplify.
$6\sqrt{12} - 3\sqrt{48} + 2\sqrt{27}$ $6\sqrt{3}$

Additional Example 4

Simplify.
$12b\sqrt{16a^5b} - 6a\sqrt{9a^3b^3} + 14ab\sqrt{a^3b}$
$44a^2b\sqrt{ab}$

Additional Example 5

Find the perimeter of the square whose side has length $5 + 2\sqrt{5}$. $20 + 8\sqrt{5}$

Ask students what must be true of two radicals before they can be combined. They must have like radicands. Ask them how to express $\sqrt{8}$ and $\sqrt{32}$ with like radicands. $2\sqrt{2}$ and $4\sqrt{2}$

■■■FOLLOW UP

Guided Practice

Classroom Exercises 1–10
Try This all

Independent Practice

A Ex. 1–20, **B** Ex. 21–34, **C** Ex. 35–37
Basic: WE 1–20 all
Average: WE 1–29 odd, 31–34 all
Above Average: WE 1–33 odd, 35–37 all

Additional Answers, page 515

30. $-24a^9b^8$
31. $72x^{12}y^{18}$
32. $-10m^5n^{15}$
33. $27a^7b^4\sqrt{3ab}$
34. $10a^5b^6\sqrt{3ab}$
35. $20x^{10}y^{16}\sqrt{6xy}$
36. $-12x^7y^4\sqrt{2xy}$
37. $-15a^5b^5\sqrt{3}$
39. $-15a^6b^{12}\sqrt{2}$
43. $42a^{10}b^{15}\sqrt{ab}$
44. $-50x^{12}y^{11}\sqrt{3y}$
45. $24e^6f^4\sqrt{3f}$
50. $18a^2b^2c^4\sqrt{10bc}$
51. $-8x^6y^7z^4\sqrt{15xy}$

---

5. $7\sqrt{8} - \sqrt{18}$ $11\sqrt{2}$
6. $4\sqrt{75} + 6\sqrt{27}$ $38\sqrt{3}$
7. $-3\sqrt{20} + 2\sqrt{45} - \sqrt{7}$ $-\sqrt{7}$
8. $7\sqrt{18} - 2\sqrt{50} - \sqrt{12}$ $11\sqrt{2} - 2\sqrt{3}$
9. $6\sqrt{11} + \sqrt{99} + 2\sqrt{44}$ $13\sqrt{11}$
10. $-7\sqrt{98} + 6\sqrt{18} - \sqrt{32}$ $-35\sqrt{2}$
11. $8\sqrt{a} - 3\sqrt{a}$ $5\sqrt{a}$
12. $7\sqrt{x} - 3\sqrt{x} + \sqrt{x}$ $5\sqrt{x}$ $21\sqrt{xy}$
13. $-16\sqrt{ab} + 5\sqrt{ab} - \sqrt{ab}$ $-12\sqrt{ab}$
14. $3\sqrt{25xy} + 4\sqrt{36xy} - 2\sqrt{81xy}$
15. $4\sqrt{cd} - 2\sqrt{cd} - \sqrt{cd} + 10\sqrt{cd}$
16. $3x\sqrt{100x^2} - 2x\sqrt{25x^2} + \sqrt{36x^2}$
17. $-2\sqrt{12xy} + 8\sqrt{27xy} - \sqrt{3xy}$
18. $9\sqrt{49ab} + 3\sqrt{81ab} - \sqrt{4ab}$ $88\sqrt{ab}$
19. $9\sqrt{32mn} + 2\sqrt{18mn} - 3\sqrt{50mn}$
20. $-10\sqrt{45xy} + 2\sqrt{20xy} + 9\sqrt{80xy}$
21. $\sqrt{4x^3y^2} + xy\sqrt{36x}$ $8xy\sqrt{x}$
22. $a\sqrt{ab^3} + b\sqrt{a^3b}$ $2ab\sqrt{ab}$
23. $-2\sqrt{a^3b^3} + 3b\sqrt{a^3b} - \sqrt{25ab^3}$
24. $\sqrt{16cd^3} + 3\sqrt{cd^3} - 5d\sqrt{25cd}$
25. $\sqrt{18x^4y} + 3x^2\sqrt{2y}$ $6x^2\sqrt{2y}$
26. $4mn\sqrt{49n} + 6\sqrt{m^2n^3}$ $34mn\sqrt{n}$
27. $5y\sqrt{125x^2} + 8\sqrt{80x^3y^3}$
28. $y\sqrt{12x^3y} - x\sqrt{6xy^3} + \sqrt{54x^3y^3}$
29. $x^2\sqrt{2y} + \sqrt{18x^4y} + 3\sqrt{8x^4y}$
30. $2\sqrt{75x^3y} + x\sqrt{48xy} + 2\sqrt{3x \cdot 3y}$

31. Find the perimeter of a rectangle with a length of $5\sqrt{2}$ and a width of $3 + \sqrt{5}$. $6 + 10\sqrt{2} + 2\sqrt{5}$

32. The length of a rectangle is $2 + 3\sqrt{5}$ and the width is $1 + 5\sqrt{2}$. Find the perimeter. $6 + 10\sqrt{2} + 6\sqrt{5}$

Find the perimeter. Write your answer in simplest radical form.

33. 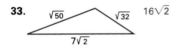 $16\sqrt{2}$

34. 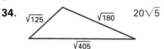 $20\sqrt{5}$

Solve each problem. Give answers in simplest radical form.

35. The area of a square quilt is 104 cm^2. What is the length? What is its perimeter? $l = 2\sqrt{26}$ cm, $p = 8\sqrt{26}$ cm

104 cm²

36. The length of a rectangle is $6 + \sqrt{3}$ and the perimeter is $8 + 10\sqrt{3}$. Find the width. $-2 + 4\sqrt{3}$

37. Find the perimeter of a right triangle with legs that measure $\sqrt{5}$ and $\sqrt{2}$. $\sqrt{2} + \sqrt{5} + \sqrt{7}$

15. $11\sqrt{cd}$ 16. $20x^2 + 6x$ 17. $19\sqrt{3xy}$ 19. $27\sqrt{2mn}$ 20. $10\sqrt{5xy}$ 23. $(ab - 5b)\sqrt{ab}$
24. $-18d\sqrt{cd}$ 27. $25xy\sqrt{5} + 32xy\sqrt{5xy}$ 28. $2xy\sqrt{3xy} + 2xy\sqrt{6xy}$ 29. $10x^2\sqrt{2y}$
30. $14x\sqrt{3xy} + 6\sqrt{xy}$

Mixed Review

Solve each equation or system.   *3.4, 3.5, 12.2, 12.5*

1. $3(2x - 5) - 3 = 2x - 8$ $x = \frac{5}{2}$
2. $0.4x - 3 = 0.1x + 0.09$ $x = 10.3$
3. $y = 2x$
   $x + y = 9$ $(3, 6)$
4. $2x - 3y = 8$
   $x = 4 - y$ $(4, 0)$
5. $5x + 2y = 12$
   $3x - 2y = 4$ $(2, 1)$
6. $2x - 7y = -3$
   $9y = 4x + 11$ $(-5, -1)$

Enrichment

If only the lengths of the sides of a triangle are known, the area of the triangle can be found by using Heron's formula.

Area of a triangle $=$
$\sqrt{s(s - a)(s - b)(s - c)}$ where a, b, and c are the lengths of the three sides and

$s = \frac{1}{2}(a + b + c)$. Have students use Heron's formula to compute the area of each triangle with the given lengths of sides.

1. 5 cm, 10 cm, 13 cm $6\sqrt{14}$ cm²
2. 9 m, 40 m, 41 m   180 m²

13.7 Multiplying Radicals

Objective   To multiply and simplify expressions containing radicals

The rectangle at the right is
$6\sqrt{3}$cm long and $2\sqrt{2}$ cm wide.
To find the area, use the formula
$A = lw$.

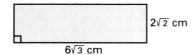

2$\sqrt{2}$ cm

6$\sqrt{3}$ cm

$$A = lw$$
Substitute. $A = 6\sqrt{3} \cdot 2\sqrt{2}$
$$= 6 \cdot 2 \cdot \sqrt{3} \cdot \sqrt{2}$$
Use $\sqrt{a} \cdot \sqrt{b} = \sqrt{ab}$. $= 12 \cdot \sqrt{3 \cdot 2} = 12\sqrt{6}$

The area is $12\sqrt{6}$ cm².

Thus, to multiply two expressions that contain square roots, use the
property $\sqrt{a} \cdot \sqrt{b} = \sqrt{ab}$ to multiply the radicals. Then simplify.

EXAMPLE 1   Simplify.

     a. $3\sqrt{2x^5} \cdot 4\sqrt{8x}$ **b.** $-12\sqrt{12a^3} \cdot 4\sqrt{5a^2}$

Solutions   **a.** $3\sqrt{2x^5} \cdot 4\sqrt{8x}$ **b.** $-12\sqrt{12a^3} \cdot 4\sqrt{5a^2}$
           $= 12\sqrt{16x^6}$ $= -48\sqrt{60a^5}$
           $= 12 \cdot 4 \cdot x^3$ $= -48\sqrt{4 \cdot 15 \cdot a^4 \cdot a}$
           $= 48x^3$ $= -48 \cdot 2 \cdot \sqrt{15} \cdot a^2 \cdot \sqrt{a}$
                               $= -96a^2\sqrt{15a}$

EXAMPLE 2   Simplify $-3\sqrt{7}(5\sqrt{3} - 2\sqrt{5})$.

Plan   Use the Distributive Property, $a(b + c) = ab + ac$.

Solution $-3\sqrt{7}(5\sqrt{3} - 2\sqrt{5}) = -3\sqrt{7}(5\sqrt{3}) + (-3\sqrt{7})(-2\sqrt{5})$
                                     $= -15\sqrt{21} + 6\sqrt{35}$
Thus, $-3\sqrt{7}(5\sqrt{3} - 2\sqrt{5}) = -15\sqrt{21} + 6\sqrt{35}$.

TRY THIS   Simplify.

     1. $2\sqrt{24a^7}(-5\sqrt{9a^3})$ $-60a^5\sqrt{6}$ **2.** $\sqrt{2}(2\sqrt{3} - 3\sqrt{4})$ $2\sqrt{6} - 6\sqrt{2}$

     3. $2\sqrt{3}(-4\sqrt{5} - 6\sqrt{2})$ **4.** $\sqrt{8x^3} \cdot 2\sqrt{18x^5}$ $24x^4$
        $-8\sqrt{15} - 12\sqrt{6}$

GETTING STARTED

Prerequisite Quiz

Multiply.

   1. $15x^4 \cdot 4x^7$ $60x^{11}$
   2. $(x + 3)(x - 2)$ $x^2 + x - 6$
   3. $(4y + 3)(4y - 3)$ $16y^2 - 9$
   4. $(x - 3)^2$ $x^2 - 6x + 9$

Motivator

Have students recall the FOIL method for
multiplying two binomials and ask one
of them to give the meaning of each
letter.   Sum of products of First terms,
Outer terms, Inner terms, and Last
terms   Then ask what happens when the
two binomials differ only by the sign between
the two terms as in $(2x + 7)(2x - 7)$.
The Outer and Inner terms are opposites
and have a sum of 0.   Tell them that these
patterns will be applied to multiplying
radicals.

Additional Example 1

Simplify.

   a. $14\sqrt{32x^4} \cdot 3\sqrt{2x^3}$ $336x^3\sqrt{x}$
   b. $9\sqrt{9a^6} \cdot -5\sqrt{16a^3}$ $-540a^4\sqrt{a}$

Additional Example 2

Simplify $-5\sqrt{3}(-3\sqrt{5} + \sqrt{3})$.

$15\sqrt{15} - 15$

Highlighting the Standards

Standards 5c, 4c: Expressions involving
the multiplication of radicals are practiced
and applied to geometric figures.

Lesson Note

In addition to $\sqrt{a} \cdot \sqrt{b} = \sqrt{ab}$, remind students that $\sqrt{a} \cdot \sqrt{a} = a$. Tell students that the two factors in Example 4, ($2\sqrt{7} + 3\sqrt{5}$ and $2\sqrt{7} - 3\sqrt{5}$) are called *conjugates*. These will be very important for dividing radicals in the next lesson.

Math Connections

Algebra 2: Conjugates are used to simplify expressions containing complex numbers.
$(4 + i)(4 - i) = 16 - i^2$
Since $i = \sqrt{-1}$, $i^2 = -1$. So
$(4 + i)(4 - i) = 16 + 1$, or 17.

Critical Thinking Questions

Application: Ask students if they can find an efficient way to find the following product without using a method that involves multiplying the two trinomials.
$(a + b)^2(a - b)^2$ $(a + b)^2(a - b)^2$
$\quad = (a + b)(a + b)(a - b)(a - b)$
$\quad = (a + b)(a - b)(a + b)(a - b)$
$\quad = (a^2 - b^2)(a^2 - b^2)$
$\quad = (a^2 - b^2)^2$
$\quad = a^4 - 2a^2b^2 + b^4$

In Example 3, the FOIL method is used to multiply the two binomials.

$$\begin{array}{ccccccc} & \text{First terms} & \text{Outer terms} & \text{Inner terms} & \text{Last terms} \\ (a + b)(c + d) = & ac & + & ad & + & bc & + & bd \end{array}$$

EXAMPLE 3   Simplify $(4\sqrt{5} + 3\sqrt{2})(2\sqrt{5} - 9\sqrt{2})$.

Solution $(4\sqrt{5} + 3\sqrt{2})(2\sqrt{5} - 9\sqrt{2})$

$$\begin{array}{cccc} \text{First terms} & \text{Outer terms} & \text{Inner terms} & \text{Last terms} \end{array}$$
$= 4\sqrt{5}(2\sqrt{5}) + 4\sqrt{5}(-9\sqrt{2}) + 3\sqrt{2}(2\sqrt{5}) + 3\sqrt{2}(-9\sqrt{2})$
$= \quad 8 \cdot 5 \quad - \quad 36\sqrt{10} \quad + \quad 6\sqrt{10} \quad - \quad 27 \cdot 2$
$= 40 - 30\sqrt{10} - 54$
$= -14 - 30\sqrt{10}$

EXAMPLE 4   Simplify $(2\sqrt{7} + 3\sqrt{5})(2\sqrt{7} - 3\sqrt{5})$.

Plan   Use the formula for the special product, $(a + b)(a - b) = a^2 - b^2$.

Solution $(2\sqrt{7} + 3\sqrt{5})(2\sqrt{7} - 3\sqrt{5}) = (2\sqrt{7})^2 - (3\sqrt{5})^2$
$= 2^2 \cdot (\sqrt{7})^2 - 3^2 \cdot (\sqrt{5})^2$
$= 4 \cdot 7 - 9 \cdot 5$
$= 28 - 45$
$= -17$

TRY THIS   Simplify.

5. $(6\sqrt{2} - 2\sqrt{3})(3\sqrt{2} - 5\sqrt{3})$ $66 - 36\sqrt{6}$

6. $(2\sqrt{3} + 5)^2$ $37 + 20\sqrt{3}$

7. $(4\sqrt{11} - 3\sqrt{5})(4\sqrt{11} + 3\sqrt{5})$ 131

Classroom Exercises

Simplify.

1. $\sqrt{5} \cdot \sqrt{5}$ 5   2. $\sqrt{5} \cdot \sqrt{3}$ $\sqrt{15}$ 3. $\sqrt{8} \cdot \sqrt{8}$ 8   4. $2 \cdot 4\sqrt{7}$ $8\sqrt{7}$

5. $\sqrt{3} \cdot 2\sqrt{7}$ $2\sqrt{21}$ 6. $\sqrt{2} \cdot 3\sqrt{2}$ 6   7. $5\sqrt{7} \cdot 2\sqrt{7}$ 70   8. $9\sqrt{5a^4} \cdot 3\sqrt{6a^2}$ $27a^3\sqrt{30}$

9. $-3\sqrt{3}(\sqrt{2} - \sqrt{3})$ $-3\sqrt{6} + 9$ 10. $(\sqrt{7} + \sqrt{6})(\sqrt{7} - \sqrt{6})$ 1

11. $(2\sqrt{3} + 4\sqrt{2})(2\sqrt{3} - 4\sqrt{2})$ -20 12. $(3\sqrt{5} - \sqrt{2})(3\sqrt{5} + \sqrt{2})$ 43

13. $(3\sqrt{2} - 1)(3\sqrt{2} - 1)$ $19 - 6\sqrt{2}$ 14. $(4\sqrt{7} + 3)(4\sqrt{7} + 3)$ $121 + 24\sqrt{7}$

Additional Example 3

Simplify.
$(3\sqrt{7} - 2\sqrt{3})(2\sqrt{7} + 3\sqrt{3})$
$24 - 5\sqrt{21}$

Additional Example 4

Simplify.
$(5\sqrt{3} + 2\sqrt{2})(5\sqrt{3} - 2\sqrt{2})$ 67

Written Exercises

Simplify.

14. $-16a^8\sqrt{6}$ **17.** $9\sqrt{35} - 15\sqrt{14}$ **18.** $-35\sqrt{6} + 28\sqrt{3}$
19. $-9\sqrt{30} + 135\sqrt{2}$ **20.** $48\sqrt{6} - 48$ **21.** $40\sqrt{3} - 24$
24. $-25 + 14\sqrt{6}$ **25.** $-36 - 21\sqrt{2}$ **30.** $-28\sqrt{3} - 18$

1. $3\sqrt{4} \cdot 5\sqrt{5}$ $30\sqrt{5}$
2. $9\sqrt{3} \cdot 3\sqrt{5}$ $27\sqrt{15}$
3. $4\sqrt{7} \cdot 4\sqrt{5}$ $16\sqrt{35}$

4. $3\sqrt{2} \cdot 7\sqrt{2}$ 42
5. $10\sqrt{5} \cdot 3\sqrt{2}$ $30\sqrt{10}$
6. $5\sqrt{7} \cdot 2\sqrt{7}$ 70

7. $4\sqrt{x} \cdot 3\sqrt{x}$ $12x$
8. $7\sqrt{a} \cdot \sqrt{a}$ $7a$
9. $10\sqrt{2x} \cdot 3\sqrt{2x}$ $60x$

10. $3\sqrt{7a^3} \cdot 3\sqrt{14a^4}$ $63a^3\sqrt{2a}$ **11.** $4\sqrt{10x^3} \cdot 3\sqrt{5x^4}$
12. $5\sqrt{3y} \cdot 7\sqrt{6y}$ $105y\sqrt{2}$

13. $-5\sqrt{6x^5} \cdot 7\sqrt{2x^2}$
14. $-2\sqrt{2a^{10}} \cdot 4\sqrt{12a^6}$
15. $7\sqrt{8y^9} \cdot \sqrt{2y^5}$ $28y^7$

16. $4\sqrt{3}(\sqrt{2} + \sqrt{3})$ $4\sqrt{6} + 12$ **17.** $3\sqrt{7}(3\sqrt{5} - 5\sqrt{2})$
18. $-7\sqrt{2}(5\sqrt{3} - 2\sqrt{6})$

19. $-9\sqrt{6}(\sqrt{5} - 5\sqrt{3})$
20. $6\sqrt{8}(4\sqrt{3} - 2\sqrt{2})$
21. $4\sqrt{2}(5\sqrt{6} - 3\sqrt{2})$

22. $(3\sqrt{6} + \sqrt{5})(2\sqrt{6} - \sqrt{5})$ $31 - \sqrt{30}$
23. $(3\sqrt{5} + \sqrt{2})(\sqrt{5} - 2\sqrt{2})$ $11 - 5\sqrt{10}$

24. $(-2\sqrt{2} + 3\sqrt{3})(4\sqrt{2} - \sqrt{3})$
25. $(-5\sqrt{3} - \sqrt{6})(2\sqrt{3} + \sqrt{6})$

26. $(3\sqrt{7} + \sqrt{5})(3\sqrt{7} - \sqrt{5})$ 58
27. $(4\sqrt{6} + \sqrt{2})(4\sqrt{6} - \sqrt{2})$ 94

28. $(2\sqrt{11} - 3\sqrt{3})(2\sqrt{11} + 3\sqrt{3})$ 17
29. $(6\sqrt{3} - 7\sqrt{2})(6\sqrt{3} + 7\sqrt{2})$ 10

30. $(3\sqrt{2} - 2\sqrt{24})(5\sqrt{2} + \sqrt{24})$
31. $(3\sqrt{12} - 2\sqrt{2})(\sqrt{12} + 5\sqrt{2})$

32. $(7\sqrt{6} - \sqrt{12})(2\sqrt{6} + 2\sqrt{12})$
$60 + 72\sqrt{2}$
33. $(\sqrt{3} - 4\sqrt{20})(-3\sqrt{3} + 2\sqrt{20})$
$-169 + 28\sqrt{15}$

For Exercises 34–39, use $(a + b)^2 = a^2 + 2ab + b^2$.

34. $(2\sqrt{3} + \sqrt{2})^2$ $14 + 4\sqrt{6}$
35. $(4\sqrt{2} + 3\sqrt{5})^2$
36. $5(\sqrt{10} - 2\sqrt{2})^2$ $90 - 40\sqrt{5}$

37. $(-2\sqrt{3} - \sqrt{5})^2$
38. $(5\sqrt{5} - 2\sqrt{3})^2$
39. $(-7\sqrt{3} - 2\sqrt{5})^2$
$167 + 28\sqrt{15}$

Find the area of each rectangle. Give answers in simplest radical form.

40.
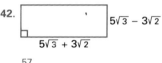
$\sqrt{3}$
$3\sqrt{2}$
$3\sqrt{6}$

41.
$\sqrt{10}$
$5\sqrt{5} - 3\sqrt{2}$
$25\sqrt{2} - 6\sqrt{5}$

42.
$5\sqrt{3} - 3\sqrt{2}$
$5\sqrt{3} + 3\sqrt{2}$
57

43. Find the area of a triangle with a base of $8\sqrt{5}$ cm and altitude of $2\sqrt{3}$ cm.
$8\sqrt{15}$ cm²

44. Find the area of a right triangle if the length of each leg is $\sqrt{5}$ in. $\frac{5}{2}$ in²

31. $16 + 26\sqrt{6}$ **35.** $77 + 24\sqrt{10}$
37. $17 + 4\sqrt{15}$ **38.** $137 - 20\sqrt{15}$

Mixed Review

Factor completely.   7.4, 7.6

1. $3y^2 - 14y - 24$
$(3y + 4)(y - 6)$
2. $8a^2 + 12a - 80$
$4(2a - 5)(a + 4)$
3. $3a^3 - 5a^2 - 8a$
$a(3a - 8)(a + 1)$

Solve.   7.7

4. $3b^2 + 11b + 6 = 0$ $-\frac{2}{3}, -3$ **5.** $2x^2 + 5x + 2 = 0$ $-\frac{1}{2}, -2$ **6.** $6y^2 - y - 1 = 0$ $-\frac{1}{3}, \frac{1}{2}$

7. Write an equation for the line that passes through points $P(-3, -1)$ and $Q(6, 2)$.   11.2 $y = \frac{1}{3}x$

13.7 Multiplying Radicals   **521**

■■■GETTING STARTED

Prerequisite Quiz

Divide and simplify.

1. $\frac{a^8}{a}$ a^7
2. $\frac{16x^7}{4x^3}$ $4x^4$
3. $\frac{64b^5}{6b^5}$ $\frac{32}{2}$
4. $\frac{-18a^6b^3}{9a^2b}$ $-2a^4b^2$

Multiply.

5. $(5 - \sqrt{2})(5 + \sqrt{2})$ 23
6. $(3\sqrt{5} + 1)(3\sqrt{5} - 1)$ 44
7. $(\sqrt{3} - \sqrt{5})(\sqrt{3} + \sqrt{5})$ -2
8. $(-3\sqrt{5} - 2\sqrt{3})(-3\sqrt{5} + 2\sqrt{3})$ 33

Motivator

Tell the students that $\sqrt{2} \approx 1.414$. Ask them to use this value to find an approximate value of $\frac{1}{\sqrt{2}}$. .707   Then ask them to find an approximate value of $\frac{\sqrt{2}}{2}$ .707   After they notice that the answers are the same, ask them which calculation would be easier to do mentally. Then show them that $\frac{1}{\sqrt{2}} = \frac{1}{\sqrt{2}} \cdot \frac{\sqrt{2}}{\sqrt{2}} = \frac{\sqrt{2}}{2}$

13.8 Dividing Radicals

Objectives
To divide and simplify expressions containing radicals
To simplify radical expressions by rationalizing the denominator

The *Quotient Property* for square roots is similar to the product property, $\sqrt{a} \cdot \sqrt{b} = \sqrt{ab}$. Consider the following illustration.

$$\frac{\sqrt{49}}{\sqrt{100}} = \frac{7}{10} \text{ and } \sqrt{\frac{49}{100}} = \frac{7}{10}, \text{ so } \frac{\sqrt{49}}{\sqrt{100}} = \sqrt{\frac{49}{100}}$$

Quotient Property for Square Roots
For all positive numbers a and b, $\frac{\sqrt{a}}{\sqrt{b}} = \sqrt{\frac{a}{b}}$.

EXAMPLE 1   Simplify.

a. $\frac{\sqrt{45}}{\sqrt{5}}$ b. $\frac{\sqrt{100a^7}}{\sqrt{2a^3}}$

Solutions   a. $\frac{\sqrt{45}}{\sqrt{5}} = \sqrt{\frac{45}{5}} = \sqrt{9} = 3$

b. $\frac{\sqrt{100a^7}}{\sqrt{2a^3}} = \sqrt{\frac{100a^7}{2a^3}} = \sqrt{50a^4} = \sqrt{25 \cdot 2 \cdot a^4} = 5a^2\sqrt{2}$

TRY THIS   Simplify: 1. $\frac{\sqrt{108}}{\sqrt{3}}$ 6   2. $\frac{\sqrt{144x^5}}{\sqrt{3x^2}}$ $4x\sqrt{3x}$

A radical in simplest form contains no radical in the denominator. In Example 1, the radicand in the numerator was exactly divisible by the radicand in the denominator. If this is not the case, it is necessary to *rationalize* the denominator, as shown in Example 2.

EXAMPLE 2   Simplify $\frac{-5\sqrt{7}}{\sqrt{12}}$.

Plan   Multiply the numerator and the denominator by the least square root needed to make the radicand in the denominator a perfect square.

522   Chapter 13 Radicals

Highlighting the Standards

Standard 5c, 4c: Students use conjugates to simplify radicals, and again apply these skills to problems relating to geometry.

Additional Example 1

Simplify.

a. $\frac{\sqrt{60}}{\sqrt{3}}$ $2\sqrt{5}$

b. $\frac{\sqrt{75x^6}}{\sqrt{3x^2}}$ $5x^2$

Additional Example 2

Simplify.

$\frac{11\sqrt{3}}{\sqrt{20}}$ $\frac{11\sqrt{15}}{10}$

Solution

$$\frac{-5\sqrt{7}}{\sqrt{12}} = \frac{-5\sqrt{7} \cdot \sqrt{3}}{\sqrt{12} \cdot \sqrt{3}} \quad \longleftarrow \frac{a}{b} = \frac{a \cdot c}{b \cdot c}$$

$$= -\frac{5\sqrt{21}}{\sqrt{36}} = -\frac{5\sqrt{21}}{6}$$

TRY THIS   Simplify: **3.** $\dfrac{-6\sqrt{5}}{\sqrt{8}}$ $-\frac{3\sqrt{10}}{2}$ **4.** $\dfrac{4\sqrt{5}}{\sqrt{27}}$ $\frac{4\sqrt{15}}{9}$

A radical expression is not simplified if the radicand contains a fraction. To simplify a radical with a fraction in the radicand, use the property $\sqrt{\dfrac{a}{b}} = \dfrac{\sqrt{a}}{\sqrt{b}}$. Then rationalize the denominator.

EXAMPLE 3   Simplify $\sqrt{\dfrac{12}{8a^3}}$.

Solution

$$\sqrt{\frac{12}{8a^3}} = \sqrt{\frac{3}{2a^3}} = \frac{\sqrt{3} \cdot \sqrt{2a}}{\sqrt{2a^3} \cdot \sqrt{2a}} = \frac{\sqrt{6a}}{\sqrt{4a^4}} = \frac{\sqrt{6a}}{2a^2}$$

TRY THIS   Simplify: **5.** $\sqrt{\dfrac{18}{21x^5}}$ $\frac{\sqrt{42x}}{7x^3}$ **6.** $\sqrt{\dfrac{10}{6x}}$ $\frac{\sqrt{15x}}{3x}$

Radical expressions such as $-2\sqrt{7} + 3\sqrt{5}$ and $-2\sqrt{7} - 3\sqrt{5}$ that differ only in the sign of one term are called **conjugates**. To simplify an expression such as $\dfrac{3}{2 - \sqrt{3}}$, multiply the numerator and the denominator by $2 + \sqrt{3}$, which is the conjugate of $2 - \sqrt{3}$. Then use $(a + b)(a - b) = a^2 - b^2$. The resulting denominator will be a rational number as shown in Example 4.

EXAMPLE 4   Simplify $\dfrac{3}{2 - \sqrt{3}}$.

Plan   Multiply the numerator and the denominator by $2 + \sqrt{3}$, the conjugate of $2 - \sqrt{3}$.

13.8 Dividing Radicals   **523**

TEACHING SUGGESTIONS

Lesson Note

Tell students that to *rationalize* the denominator means to change the denominator of a fraction to a rational number. In Example 2, emphasize that multiplying by $\frac{\sqrt{3}}{\sqrt{3}}$ is really multiplication by 1. You could also multiply by $\frac{\sqrt{12}}{\sqrt{12}}$, but this would require more work in simplifying the final answer. Show this method of solution so that students can compare it with the solution given in the textbook. After presenting Example 3, remind students that simplifying the expressions under the radicand first is a good procedure to follow.

Math Connections

Calculation: Before the invention of the calculator, the most efficient way to calculate with radical expressions was to use logarithms. However, it is still an advantage to be able to calculate some expressions at sight, hence the "simplification" of $\frac{1}{\sqrt{2}}$ to $\frac{\sqrt{2}}{2}$.

Critical Thinking Questions

Synthesis: Since neither a radical in the denominator of a fraction nor a fraction under the radical sign is in simplest form, ask students to give a simplified equivalent form for $\frac{\sqrt{a}}{\sqrt{b}}$. $\frac{\sqrt{ab}}{b}$

Additional Example 3

Simplify $\sqrt{\dfrac{18}{4x^3}}$. $\frac{3\sqrt{2x}}{2x^2}$

Additional Example 4

Simplify $\dfrac{-5}{3 + \sqrt{2}}$. $\frac{-15 + 5\sqrt{2}}{7}$

Common Error Analysis

Error: Some students may try to reduce $\frac{10}{\sqrt{15}}$ to $\frac{2}{\sqrt{3}}$.

Show them that $\sqrt{15}$ is an irrational number between 3 and 4 and certainly not a factor of 10. In general, fractions can be reduced only if both the numerator and denominator are inside, or both are outside, the radical symbol.

Checkpoint

Simplify.

1. $\frac{\sqrt{40}}{\sqrt{5}}$ $2\sqrt{2}$

2. $\frac{\sqrt{64x^7}}{\sqrt{8x^3}}$ $2x^2\sqrt{2}$

3. $\frac{7}{\sqrt{3}}$ $\frac{7\sqrt{3}}{3}$

4. $\frac{5}{3 \div \sqrt{2}}$ $\frac{15 - 5\sqrt{2}}{7}$

Closure

Give the students the following rules for expressing a radical answer in simplest form and have them give an example to illustrate each. A radical is in simplest form if:

1. there are no perfect square factors in the radicand;
2. there are no radicals in the denominator of a fraction;
3. there are no fractions in the radicand.

Solution

$$\frac{3}{2 - \sqrt{3}} = \frac{3 \cdot (2 + \sqrt{3})}{(2 - \sqrt{3}) \cdot (2 + \sqrt{3})} = \frac{3(2 + \sqrt{3})}{2^2 - (\sqrt{3})^2} = \frac{6 + 3\sqrt{3}}{4 - 3}$$
$$= 6 + 3\sqrt{3}$$

So, $\frac{3}{2 - \sqrt{3}} = 6 + 3\sqrt{3}$. To check, multiply $(6 + 3\sqrt{3})(2 - \sqrt{3})$.

TRY THIS   Simplify: 7. $\frac{4}{2 - \sqrt{5}}$ $-(8 + 4\sqrt{5})$ 8. $\frac{8}{2 + \sqrt{6}}$ $4\sqrt{6} - 8$

Classroom Exercises

By what factor should the denominator be multiplied to give a perfect square radicand? Give the least possible square root.

1. $\frac{\sqrt{3}}{\sqrt{5}}$ $\sqrt{5}$

2. $\frac{\sqrt{7y}}{\sqrt{5y^3}}$ $\sqrt{5y}$

3. $\frac{-3a^4}{\sqrt{6b^5}}$ $\sqrt{6b}$

4. $\frac{3\sqrt{7a}}{2\sqrt{8c^7}}$ $\sqrt{2c}$

Simplify.

5. $\frac{\sqrt{27x^6}}{\sqrt{3x^2}}$ $3x^2$

6. $\frac{\sqrt{12}}{\sqrt{18a}}$ $\frac{\sqrt{6a}}{3a}$

7. $\frac{9}{\sqrt{27a^5}}$ $\frac{\sqrt{3a}}{a^3}$

8. $\frac{24xy}{\sqrt{x^6y^3}}$ $\frac{24\sqrt{y}}{x^2y}$

Written Exercises

Divide and simplify.

1. $\frac{\sqrt{35}}{\sqrt{5}}$ $\sqrt{7}$

2. $\frac{\sqrt{48a^5}}{\sqrt{6a^3}}$ $2a\sqrt{2}$

3. $\frac{\sqrt{50x^3}}{\sqrt{2x}}$ $5x$

4. $\frac{\sqrt{28y^7}}{\sqrt{7y^3}}$ $2y^2$

5. $\frac{\sqrt{54a^5}}{\sqrt{6a}}$ $3a^2$

6. $\frac{3}{\sqrt{5}}$ $\frac{3\sqrt{5}}{5}$

7. $\frac{-2}{\sqrt{3}}$ $\frac{-2\sqrt{3}}{3}$

8. $\frac{1}{\sqrt{7}}$ $\frac{\sqrt{7}}{7}$

9. $\frac{14}{\sqrt{2}}$ $7\sqrt{2}$

10. $\frac{-10}{\sqrt{5}}$ $-2\sqrt{5}$

11. $\frac{7\sqrt{2}}{\sqrt{6}}$ $\frac{7\sqrt{3}}{3}$

12. $\frac{-18\sqrt{5}}{\sqrt{9}}$ $-6\sqrt{5}$

13. $\frac{12\sqrt{4}}{-\sqrt{3}}$ $-8\sqrt{3}$

14. $\frac{16\sqrt{7}}{\sqrt{8}}$ $4\sqrt{14}$

15. $\frac{-36\sqrt{12}}{\sqrt{2}}$ $-36\sqrt{6}$

16. $\frac{\sqrt{20}}{\sqrt{30x^3}}$ $\frac{\sqrt{6x}}{3x^2}$

17. $\frac{\sqrt{14}}{\sqrt{21a}}$ $\frac{\sqrt{6a}}{3a}$

18. $\frac{\sqrt{35}}{\sqrt{15a^5}}$ $\frac{\sqrt{21a}}{3a^3}$

19. $\frac{\sqrt{48}}{\sqrt{16x^3}}$ $\frac{\sqrt{3x}}{x^2}$

20. $\frac{\sqrt{75}}{\sqrt{50a^7}}$ $\frac{\sqrt{6a}}{2a^4}$

21. $\frac{4}{\sqrt{a^3}}$ $\frac{4\sqrt{a}}{a^2}$

22. $\frac{-10}{\sqrt{y^5}}$ $-\frac{10\sqrt{y}}{y^3}$

23. $\frac{16}{\sqrt{20a^7}}$ $\frac{8\sqrt{5a}}{5a^4}$

24. $\frac{-81}{\sqrt{3a^3}}$ $-\frac{27\sqrt{3a}}{a^2}$

524   Chapter 13 Radicals

25. $\dfrac{6}{3 - \sqrt{7}}$ $9 + 3\sqrt{7}$ **26.** $\dfrac{4}{1 + \sqrt{2}}$ $-4 + 4\sqrt{2}$ **27.** $\dfrac{-12}{\sqrt{7} - 4}$ $\dfrac{4\sqrt{7} + 16}{3}$ **28.** $\dfrac{-7}{5 + \sqrt{2}}$

29. $\dfrac{24ab}{\sqrt{a^4b^3}}$ $\dfrac{24\sqrt{b}}{ab}$ **30.** $\dfrac{-16y}{\sqrt{24y}}$ $-\dfrac{4\sqrt{6y}}{3}$ **31.** $\dfrac{48c^3d^2}{\sqrt{c^2d}}$ $48c^2d\sqrt{d}$ **32.** $\dfrac{30x}{\sqrt{xy^4}}$ $\dfrac{30\sqrt{x}}{y^2}$

33. $\dfrac{18x^4y^3}{\sqrt{xy}}$ $18x^3y^2\sqrt{xy}$ **34.** $\dfrac{-12x^4y^3}{\sqrt{20x^5y^2}}$ **35.** $\dfrac{3a^4b^5}{\sqrt{6a^2b^3}}$ $\dfrac{a^3b^3\sqrt{6b}}{2}$ **36.** $\dfrac{36c^9d^3}{\sqrt{12cd^3}}$

37. $\dfrac{-15a^4b}{\sqrt{5a^6b^7}}$ $\dfrac{-3a\sqrt{5b}}{b^3}$ **38.** $\dfrac{100x^7y^4}{\sqrt{xy^2}}$ $100x^6y^3\sqrt{x}$ **39.** $\dfrac{\sqrt{18b^4c^8}}{\sqrt{3a^3b^6}}$ $\dfrac{c^4\sqrt{6a}}{a^2b}$ **40.** $\dfrac{\sqrt{72x^5y^7}}{\sqrt{8xy^{10}}}$

41. $\dfrac{\sqrt{6\frac{1}{2}}}{\sqrt{4\frac{1}{3}}}$ $\dfrac{\sqrt{6}}{2}$ **42.** $\dfrac{\sqrt{6\frac{2}{5}}}{\sqrt{5\frac{1}{3}}}$ $\dfrac{\sqrt{30}}{5}$ **43.** $\dfrac{\sqrt{9\frac{2}{3}}}{\sqrt{5\frac{4}{5}}}$ $\dfrac{\sqrt{15}}{3}$ **44.** $\dfrac{\sqrt{15\frac{1}{3}}}{\sqrt{9\frac{2}{3}}}$ $\dfrac{\sqrt{1,334}}{29}$

45. $\dfrac{5}{3\sqrt{5} - 1}$ **46.** $\dfrac{14}{3 - 2\sqrt{3}}$ **47.** $\dfrac{12}{5 + 3\sqrt{2}}$ **48.** $\dfrac{135}{7 + 2\sqrt{7}}$

49. $\dfrac{-15}{3\sqrt{5} - 1}$ $\dfrac{-45\sqrt{5} - 15}{44}$ **50.** $\dfrac{100}{5 - \sqrt{2}}$ $\dfrac{500 + 100\sqrt{2}}{23}$ **51.** $\dfrac{3}{\sqrt{2} + 2\sqrt{3}}$ $\dfrac{3\sqrt{2} - 6\sqrt{3}}{-10}$ **52.** $\dfrac{-7\sqrt{2}}{-3\sqrt{5} - 2\sqrt{3}}$ $\dfrac{21\sqrt{10} - 14\sqrt{6}}{33}$

Solve.

53. A rectangle has an area of $6\sqrt{10}$ in². If its length is $3\sqrt{5}$ in., what is its width? $2\sqrt{2}$ in

54. A rectangle is $2\sqrt{5}$ cm long. If its area is $(6\sqrt{15} - 6\sqrt{10})$ cm², what is its width? $3\sqrt{3} - 3\sqrt{2}$ cm

55. The area of a triangle is 72 cm² and its altitude is $6\sqrt{2}$ cm. Find the length of the base. $12\sqrt{2}$ cm

56. The area of a rectangle is $(\sqrt{10} + 3\sqrt{5})$ cm² and its length is $(\sqrt{2} + 3)$ cm. Find its width. $\sqrt{5}$ cm

Mixed Review

Evaluate for the given values of the variables. *1.3, 2.6*

1. x^2 for $x = -\dfrac{2}{3}$ $\dfrac{4}{9}$ **2.** $2a^4$ for $a = -2$ 32

3. $\dfrac{x^2 - 2x}{3x}$ for $x = 3$ $\dfrac{1}{3}$ **4.** $5ab^2c^3$ for $a = -1$, $b = 2$, and $c = -2$ 160

Factor. *7.2, 7.4–7.6*

5. $2m^5n + 16m^4$ $2m^4(mn + 8)$ **6.** $6x^2 - x - 2$ $(3x - 2)(2x + 1)$

7. $m^2 - n^2$ $(m - n)(m + n)$ **8.** $ac + bc + ad + bd$ $(c + d)(a + b)$

9. Wanda's age is 12 years more than twice Carlos's age. The sum of their ages is 36. How old is each? *12.3* Carlos: 8, Wanda: 28

Guided Practice

Classroom Exercises 1–8
Try This all

Independent Practice

A Ex. 1–20, **B** Ex. 21–52, **C** Ex. 53–56

Basic: WE 1–20 all

Average: WE 1–35 odd, 45–51 odd

Above Average: WE 13–55 odd

Additional Answers

Written Exercises

28. $\dfrac{-35 + 7\sqrt{2}}{23}$

34. $-\dfrac{6xy^2\sqrt{5x}}{5}$

36. $6c^8d\sqrt{3cd}$

40. $\dfrac{3x^2\sqrt{y}}{y^2}$

45. $\dfrac{15\sqrt{5} + 5}{44}$

46. $\dfrac{42 + 28\sqrt{3}}{-3}$

47. $\dfrac{60 - 36\sqrt{2}}{7}$

48. $\dfrac{315 - 90\sqrt{7}}{7}$

Enrichment

Einstein's theory of Special Relativity predicts that the speed of an object affects its length. When an object is stationary, it has a certain length, known as its proper length; however, when the object moves, it has a different length. Einstein related these two lengths in the equation: $L = L'\sqrt{1 - \dfrac{v^2}{c^2}}$, where L is the length of the object in motion

L' is the length of the object in a stationary position; v is the speed of the object; and c is the speed of light $(1.86 \times 10^5$ mi/s$)$.

1. If a stationary bar is 1 m in length, what is its length when it moves at half the speed of light? $\dfrac{\sqrt{3}}{2}$ m ≈ 0.866 m

2. Before takeoff, a spaceship is 20 m long. What is its length when it moves at two-thirds the speed of light? $\dfrac{20\sqrt{5}}{3}$ m ≈ 14.907

3. A student standing by her desk is 5 ft tall. What is her height if she runs at a speed of 6 mi/h, or 1.67×10^{-3} mi/s? ≈ 5.000 ft

13.9 Radical Equations

Objective   To solve equations containing radicals

An equation with a variable in the radicand, such as $\sqrt{x} = 6$, is called a **radical equation**. To solve $\sqrt{x} = 6$, square each side.

$$\sqrt{x} = 6 \qquad \text{Check:} \quad \sqrt{x} = 6$$
$$(\sqrt{x})^2 = 6^2 \qquad\qquad \sqrt{36} \stackrel{?}{=} 6$$
$$x = 36 \qquad\qquad\qquad 6 = 6 \ \text{(True)}$$

In this case the original equation ($\sqrt{x} = 6$) is equivalent to the squared equation ($x = 36$), since both have the same solution.

To solve an equation such as $\sqrt{3x - 5} - 4 = 3$, get the radical alone on one side of the equation. Then square each side and solve for x.

EXAMPLE 1   Solve $\sqrt{2x - 5} - 4 = 3$. Check.

Plan   Add 4 to each side of the equation to get the radical alone.

Solution

$$\sqrt{2x - 5} - 4 = 3 \qquad\qquad \text{Check:} \quad \sqrt{2x - 5} - 4 = 3$$
$$\sqrt{2x - 5} = 7 \qquad\qquad\qquad\qquad \sqrt{2(27) - 5} - 4 \stackrel{?}{=} 3$$
$$2x - 5 = 49 \qquad\qquad\qquad\qquad\quad \sqrt{54 - 5} - 4 \stackrel{?}{=} 3$$
$$2x = 54 \qquad\qquad\qquad\qquad\qquad\qquad \sqrt{49} - 4 \stackrel{?}{=} 3$$
$$x = 27 \qquad\qquad\qquad\qquad\qquad\qquad\quad 7 - 4 \stackrel{?}{=} 3$$
$$3 = 3 \ \text{(True)}$$

Thus, the solution is 27.

TRY THIS   1. Solve $\sqrt{4x - 3} - 6 = 3$. 21

Squaring each side of an equation does not always result in an equivalent equation. Let $x = 9$. Then $x^2 = 81$.

The squared equation has two solutions, 9 and -9, but only 9 checks in the original equation. The root, -9, that was introduced by squaring is called an **extraneous** solution.

When you square both sides of an equation, you must always check *all* solutions in the *original* equation. Eliminate extraneous solutions.

GETTING STARTED

Prerequisite Quiz

Square and simplify.

1. $3\sqrt{7}$ 63
2. $4\sqrt{6b}$ 96b
3. $\sqrt{x + 1}$ $x + 1$
4. $3\sqrt{x + 5}$ $9x + 45$
5. $-4\sqrt{y - 1}$ $16y - 16$
6. $-\sqrt{3b^2 + 4}$ $3b^2 + 4$

Motivator

Most students will agree that if $a = b$, then $a^2 = b^2$. So have them work along with you on the following. Start with $x + 3 = 5$ and square each side of the equation to get $x^2 + 6x + 9 = 25$. Solve this equation for x by setting equal to 0 and factoring. $x = 2, -8$ Now ask students to check these solutions in the original equation, $x + 3 = 5$. Will they both check?   No   Tell them that squaring both sides of an equation will not necessarily give an equivalent equation.

Highlighting the Standards

Standards 5c, 4a, 1a, 1b: Students see how the familiar process of solving equations can be applied to equations involving radicals, but may also introduce extraneous solutions. In the Mixed Problem Solving, they apply various strategies to many different kinds of problems.

Additional Example 1

Solve and check.
$\sqrt{2x - 3} - 2 = 1$ 6

EXAMPLE 2   Solve $\sqrt{x + 2} + 3\sqrt{x - 6} = 0$. Check.

Plan   First, get one radical expression on each side of the equation. Then square each side.

Solution

$$\sqrt{x + 2} + 3\sqrt{x - 6} = 0$$
$$\sqrt{x + 2} = -3\sqrt{x - 6}$$
$$x + 2 = 9(x - 6)$$
$$x + 2 = 9x - 54$$
$$56 = 8x$$
$$7 = x$$

Check $x = 7$.
$$\sqrt{x + 2} + 3\sqrt{x - 6} = 0$$
$$\sqrt{7 + 2} + 3\sqrt{7 - 6} \overset{?}{=} 0$$
$$\sqrt{9} + 3\sqrt{1} \overset{?}{=} 0$$
$$3 + 3 \overset{?}{=} 0$$
$$6 \neq 0$$

There is no solution since 7 is extraneous.

EXAMPLE 3   Solve $\sqrt{y + 1} = y - 5$. Check.

Solution

$$\sqrt{y + 1} = y - 5$$
$$y + 1 = (y - 5)^2$$
$$y + 1 = y^2 - 10y + 25$$
$$0 = y^2 - 11y + 24$$
$$0 = (y - 8)(y - 3)$$
$$y - 8 = 0 \quad or \quad y - 3 = 0$$
$$y = 8 \qquad\qquad y = 3$$

Check $y = 8$.
$$\sqrt{y + 1} = y - 5$$
$$\sqrt{8 + 1} \overset{?}{=} 8 - 5$$
$$\sqrt{9} \overset{?}{=} 3$$
$$3 = 3 \quad \text{True}$$

Check $y = 3$.
$$\sqrt{y + 1} = y - 5$$
$$\sqrt{3 + 1} \overset{?}{=} 3 - 5$$
$$\sqrt{4} \overset{?}{=} -2$$
$$2 \neq -2 \quad \text{Thus, the solution is } 8.$$

TRY THIS   **2.** Solve $\sqrt{6y - 8} = y - 4$. $y = 12$

Classroom Exercises

What equation results from squaring each side of the given equation?

1. $\sqrt{y} = 2$ $y = 4$ **2.** $\sqrt{2x} = 8$ $2x = 64$ **3.** $\sqrt{3y - 2} = 7$ **4.** $\sqrt{3x - 1} = \sqrt{2x + 4}$

$3y - 2 = 49$ $3x - 1 = 2x + 4$

Solve each equation. Check.

5. $\sqrt{2x - 1} = 7$ $x = 25$ **6.** $\sqrt{3y} + 2 = 11$ $y = 27$ **7.** $7 + \sqrt{5a} = 9$ $a = \frac{4}{5}$

8. $2\sqrt{5y} = 10$ $y = 5$ **9.** $\sqrt{4x} = 2\sqrt{5}$ $x = 5$ **10.** $\sqrt{3x - 1} = \sqrt{5x + 2}$

No solution

13.9 Radical Equations   **527**

Analysis: Have students analyze the equation $\sqrt{2x + 3} + 8 = 5$ and tell how they can decide if its solution is the empty set without having to work through the steps presented in the lesson. Since $-3 + 8 = 5$, $\sqrt{2x + 3}$ must be equal to -3. But this is not possible, since the radical sign indicates a positive result.

Common Error Analysis

Error: When squaring both sides of $\sqrt{y + 1} = y - 5$, some students will obtain the incorrect equivalent equation $y + 1 = y^2 + 25$.

Remind them that each *side* must be squared and that the square of a binomial is a trinomial.

Checkpoint

Solve and check.

1. $\sqrt{5x} = 10$ 20
2. $\sqrt{3x + 1} = 7$ 16
3. $\sqrt{3a - 8} - \sqrt{a - 2} = 0$ 3
4. $y + 3 = \sqrt{-y - 1}$ -2

Closure

Ask students for the first step in solving the equation $\sqrt{7x - 2} - 5 = 3$. Isolate the radical on one side of the equal sign. Ask for the second step. Square both sides. Then ask why it is essential to check all solutions in the original equation? To find all possible extraneous solutions.

Written Exercises

17. $x = 9$ 18. $a = -\dfrac{4}{5}$ 19. $y = 1$
20. No sol   21. No sol   27. No sol

Solve each equation. Check.

1. $\sqrt{x} = 5$ $x = 25$
2. $8 = \sqrt{y}$ $y = 64$
3. $\sqrt{z} = -3$ No solution
4. $\sqrt{a} + 2 = 3$ $a = 1$
5. $7 = \sqrt{y} - 2$ $y = 81$
6. $\sqrt{c} - 4 = 2$ $c = 36$
7. $\sqrt{2x + 3} = 7$ $x = 23$
8. $13 = \sqrt{3y - 2} + 9$ $y = 6$
9. $\sqrt{7x - 2} + 5 = 3$ No sol
10. $0 = \sqrt{3x + 4} - 7$ $x = 15$
11. $4 - \sqrt{x - 3} = 9$ No sol
12. $\sqrt{x} = 3\sqrt{3}$ $x = 27$
13. $\sqrt{2y} = 2\sqrt{5}$ $y = 10$
14. $\sqrt{44} = 2\sqrt{y}$ $y = 11$
15. $2\sqrt{5x - 4} = 12$ $x = 8$
16. $4\sqrt{x - 5} = 15$ $x = \dfrac{305}{16}$
17. $\sqrt{4x - 2} = \sqrt{3x + 7}$
18. $\sqrt{1 - 7a} = \sqrt{5 - 2a}$
19. $\sqrt{10y + 2} = 2\sqrt{4y - 1}$
20. $\sqrt{3x + 1} + 2\sqrt{x} = 0$
21. $3\sqrt{1 + a} = \sqrt{5a + 1}$
22. $\sqrt{x^2 - 6x} = 4$ $-2, 8$
23. $\sqrt{x^2 - 12} = -1$ No sol
24. $\sqrt{3y + 1} = y - 3$ $y = 8$
25. $1 + y = \sqrt{y^2 + 5}$ $y = 2$
26. $c - 5 = \sqrt{c + 7}$ $c = 9$
27. $\sqrt{x^2 + 2} = x - 2$
28. $\sqrt{2x^2 - 12} = x$ $x = 2\sqrt{3}$
29. $\sqrt{17 - 4y} = y + 1$ $y = 2$
30. $\sqrt{4x^2 + 5} = 3x$ $x = 1$
31. $a - 2 = \sqrt{2a - 1}$ $a = 5$
32. $\sqrt{23 - x} + 3 = x$ $x = 7$
33. $6 = \sqrt{2y + 3} + y$ $y = 3$

Solve each problem.

34. If the square root of a number is increased by 9, the result is 16. Find the number.   49
35. The square root of 1 less than 5 times a number is equal to 3. Find the number.   2
36. Find a number if half the square root of the number is equal to 5.   100

Solve each equation. Check. (HINT: Square each side two different times.) $a = 9$

37. $\sqrt{x} + 1 = \sqrt{x + 11}$
38. $\sqrt{x} + 4 = \sqrt{x + 40}$
39. $\sqrt{a + 27} = 3 + \sqrt{a}$
40. $\sqrt{y} + 1 = \sqrt{3y - 3}$
41. $\sqrt{x - 7} = 2 + \sqrt{x}$
42. $\sqrt{2x + 4} = 2\sqrt{x + 3}$
37. $x = 25$ 40. $y = 4$
38. $x = 9$ 41. No solution
42. No solution

Solve each formula for the given variable.

43. $r = \sqrt{\dfrac{A}{\pi}}$ for A $A = \pi r^2$
44. $T = 2\pi\sqrt{\dfrac{l}{g}}$ for l $l = \dfrac{T^2 g}{4\pi^2}$
45. $r = \sqrt{\dfrac{3v}{\pi h}}$ for h $h = \dfrac{3V}{\pi r^2}$

Mixed Review

Simplify.   *2.2, 6.2, 6.7, 8.3, 8.7*

1. $-|-14|$ -14
2. $-(7a^2 + 6a^3 - 5) - (a^2 + 3)$ $-6a^3 - 8a^2 + 2$
3. $(-3x^5y^3)^2$ $9x^{10}y^6$
4. $\dfrac{8x^4y^2}{5xy^6} \cdot \dfrac{15x^3y^5}{16x^2y^8}$ $\dfrac{3x^4}{2y^7}$
5. $\dfrac{-4}{n - 3} + \dfrac{2}{n + 4}$ $\dfrac{-2n - 22}{n^2 + n - 12}$
6. $\dfrac{a - 1}{2a + 1} + 3$ $\dfrac{7a + 2}{2a + 1}$
7. Four is what percent of 12?   *4.7* $33\frac{1}{3}\%$
8. Thirty-six percent of what number is 52.2?   *4.7* 145
9. A sweater sells for \$95. If the tax rate is $7\frac{1}{2}\%$, what is the total cost?   *4.8* \$102.13

528   Chapter 13 Radicals

Mixed Problem Solving

When solving problems that involve geometric figures, drawing a diagram will help you to visualize the situation and to understand the problem. Remember to label the diagram carefully, and to express given dimensions in terms of one variable, when possible.

Solve each problem.

1. An isosceles triangle has a base angle of 38°. What is the measure of the vertex angle? 104

2. How fast must a bicycle travel to cover a distance of 105 mi in $3\frac{1}{2}$ h? 30 mi/h

3. The length of a rectangle is 8 in. more than 5 times the width. The perimeter is 40 in. Find the length and the width of the rectangle. l: 18 in., w: 2 in.

4. A plumber charges $50 for a house call plus $48 for each hour she works. Her bill for a job was $194. How many hours did she work? 3 h

5. Find the height of a triangle with an area of 240 cm² and a base of 13 cm. Use the formula $A = \frac{1}{2}bh$. $36\frac{12}{13}$ cm

6. Use the formula $A = \frac{1}{2}(b + c)h$ to find the area of a trapezoid if the height is 8.7 cm and the bases are 6.4 cm and 12.6 cm long. 82.65 cm²

7. When a number is decreased by 0.6 of the number, the result is 128. Find the number. 320

8. A company earned $12,500 more in the second year than in the first. What did it earn each year if the total for the two years was $197,500?

9. A collection of dimes and and quarters is worth $5.00. There are $2\frac{1}{2}$ times as many dimes as quarters. Find the number of coins of each type. 10 quarters; 25 dimes

10. Kim has 16 coins, all in nickels and quarters. The total value of the coins is $2.80. How many coins of each type are there? 10 quarters, 6 nickels

11. Nine is what percent of 45? 20%

12. Sixteen percent of what number is 36? 225

13. Find the interest earned on $1,400 invested at a simple interest rate of 8% per year for 9 years. $1,008

14. A new car was priced at $16,000. Then the price was increased by 3%. What was the new price? $16,480

15. Find two consecutive even integers whose product is 168. 12, 14 or $-14, -12$

16. Find two consecutive odd integers whose product is 63. 7, 9 or $-7, -9$

17. Find three consecutive odd integers such that the square of the third minus the square of the second is 9 less than the square of the first.

18. The cost of chocolates varies directly as the weight. If 4 lb of chocolates cost $26, find the cost of 6 lb of chocolates. $39

19. The volume of a gas varies inversely as the pressure. If the volume of a gas is 48 m³ at 4.5 atmospheres of pressure, find the volume at 7 atmospheres of pressure. $30\frac{6}{7}$ m³

20. For a triangle of constant area, the height h varies inversely as the length of the base b. If $b = 16$ cm when $h = 36$ cm, find b when $h = 48$ cm. 12 cm

17. 7, 9, 11 or $-3, -1, 1$

8. First: $92,500; second: $105,000

FOLLOW UP

Guided Practice

Classroom Exercises 1–10
Try This all

Independent Practice

A Ex. 1–21, **B** Ex. 22–36, **C** Ex. 37–45

Basic: WE 1–23 odd, Mixed Problem Solving

Average: WE 11–33 odd, Mixed Problem Solving

Above Average: WE 23–45 odd, Mixed Problem Solving

Enrichment

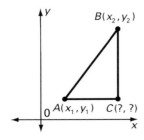

Refer to the Enrichment Exercise for Lesson 13.5 where the distance between two points in the coordinate plane is calculated by using the Pythagorean Theorem. Show students the graph at the left. Have them complete the following.

1. Give the coordinates of C. (x_2, y_1)
2. Find the distance from A to C.
   $\sqrt{x_2 - x_1}$

3. Find the distance from C to B. $|y_2 - y_1|$
4. Use the Pythagorean Theorem to find the distance from A to B.
   $\sqrt{(x_2 - x_1)^2 + (y_2 - y_1)^2}$
5. Use the answer to question 4 to find the distance between $(-4, -3)$ and $(2, 3)$. $6\sqrt{2}$

Chapter 13 Review

Key Terms

conjugates (p. 523)
irrational number (p. 497)
perfect square (p. 501)
principal square root (p. 500)
Product Property for Square Roots
  (p. 500)
Pythagorean Theorem (p. 508)
Quotient Property for Square Roots
  (p. 522)
radical (p. 500)

radical sign (p. 500)
radicand (p. 500)
rational number (p. 495)
rationalize the denominator (p. 522)
right triangle (p. 508)
Square Root Property of Even Powers
  (p. 513)
Square Root Property of Odd Powers
  (p. 513)
subset (p. 498)

Key Ideas and Review Exercises

13.1   To write repeating decimals in fractional form, let $n =$ the decimal. Multiply the equation by 10^a, where a is the number of digits in the block of repeating digits. Subtract the first equation from the second, and solve for n.

For each repeating decimal, write a fraction $\frac{a}{b}$ where a and b are integers.

1. $13.\overline{42}$ $\frac{443}{33}$ **2.** $9.\overline{7}$ $\frac{88}{9}$ **3.** $41.\overline{17}$ $\frac{4,076}{99}$ **4.** $12.\overline{241}$ $\frac{12,229}{999}$ **5.** $7.9191\ldots$ $\frac{784}{99}$

13.2   To simplify a square root when the radicand is not a perfect square, find the greatest perfect square factor of the radicand. Then use $\sqrt{a \cdot b} = \sqrt{a} \cdot \sqrt{b}$.

Simplify, if possible. If not possible, so indicate.

6. $\sqrt{18}$ $3\sqrt{2}$ **7.** $\sqrt{79}$ **8.** $\sqrt{99}$ $3\sqrt{11}$ **9.** $4\sqrt{500}$ **10.** $-10\sqrt{180}$
       Irreducible $40\sqrt{5}$ $-60\sqrt{5}$

13.3   To approximate the square root of a number, use a square root table or the divide-and-average method.

Approximate to the nearest tenth. Use a square root table.

11. $\sqrt{41}$ 6.4   **12.** $\sqrt{79}$ 8.9   **13.** $\sqrt{99}$ 9.9   **14.** $\sqrt{125}$ 11.2   **15.** $\sqrt{234}$ 15.3

16. Approximate $\sqrt{54}$ to the nearest tenth. Use the divide-and-average method. 7.3

13.4   To find the length of one side of a right triangle given the lengths of the other two sides, use the Pythagorean Theorem $c^2 = a^2 + b^2$.

To determine whether or not a triangle is a right triangle, use $c^2 = a^2 + b^2$ where c is the length of the longest side.

For each right triangle, find the missing length. Give answers in simplest radical form.

17. $a = 5, b = 7$ $\sqrt{74}$ **18.** $a = 3, c = 10$ $\sqrt{91}$ **19.** $b = 8, c = 10\sqrt{3}$ $2\sqrt{59}$

20. Write an explanation of why a triangle with sides 3 cm, 5 cm, and $\sqrt{34}$ cm is a right triangle, but one with sides 8 cm, 10 cm, and 12 cm is not a right triangle.

13.5   To simplify square roots with variables in the radicand, use $\sqrt{a^{2x}} = a^x$ for even powers and $\sqrt{a^{2x+1}} = a^x\sqrt{a}$ for odd powers.

Simplify.

21. $\sqrt{25y^{14}}$ $5y^7$ **22.** $-3\sqrt{49a^{22}}$ $-21a^{11}$ **23.** $\sqrt{16x^7}$ $4x^3\sqrt{x}$

24. $-\sqrt{36y^{15}}$ $-6y^7\sqrt{y}$ **25.** $-\sqrt{18x^5y^{10}}$ $-3x^2y^5\sqrt{2x}$ **26.** $\sqrt{121xy^8}$ $11y^4\sqrt{x}$

13.6   To add or subtract expressions containing radicals, simplify the radicals, if possible. Then combine like terms.

Add or subtract. Simplify.

27. $9\sqrt{5} - 3\sqrt{5}$ $6\sqrt{5}$ **28.** $-5a\sqrt{4a} + \sqrt{9a^3}$ **29.** $\sqrt{80} - 3\sqrt{45}$ $-5\sqrt{5}$
                                                $-7a\sqrt{a}$

13.7   To multiply and simplify expressions containing radicals, use $\sqrt{a} \cdot \sqrt{b} = \sqrt{ab}$ to multiply the radicals. Then simplify.

Simplify.

30. $5\sqrt{9} \cdot 3\sqrt{3}$ **31.** $9\sqrt{3}(\sqrt{3} + 4\sqrt{6})$ **32.** $(2\sqrt{3} + 3)(3\sqrt{3} - 2)$
    $45\sqrt{3}$ $27 + 108\sqrt{2}$ $12 + 5\sqrt{3}$

13.8   To divide expressions containing radicals, use $\dfrac{\sqrt{a}}{\sqrt{b}} = \sqrt{\dfrac{a}{b}}$ if the radicand in the numerator is divisible by the radicand in the denominator. Otherwise, rationalize the denominator.

Divide and simplify.

33. $\dfrac{\sqrt{95}}{\sqrt{5}}$ $\sqrt{19}$ **34.** $\sqrt{\dfrac{3}{75x}}$ $\dfrac{\sqrt{x}}{5x}$ **35.** $\dfrac{8}{3 + \sqrt{5}}$ $6 - 2\sqrt{5}$

13.9   To solve radical equations, get a radical alone on one side of the equation. Then square each side. Check all solutions in the original equation. Eliminate any extraneous solutions.

Solve each equation. Check.

36. $\sqrt{x - 2} + 4 = 9$ **37.** $3 + \sqrt{x + 6} = 0$ **38.** $\sqrt{x^2 + 8x} = 3$
    27   No solution   1, -9

20. In the first triangle, let $a = 3$, $b = 5$, and $c = \sqrt{34}$. Since $9 + 25 = 34$, $a^2 + b^2 = c^2$. Then, by the converse of the Pythagorean Theorem, the triangle is a right triangle. In the second triangle, let $a = 8$, $b = 10$, and $c = 12$. Since $64 + 100 \neq 144$, $a^2 + b^2 \neq c^2$. Then, by the Pythagorean Theorem, the triangle is not a right triangle.

![Chapter 13 Test banner]

Chapter 13 Test

A Exercises: 1–10, 13–15, 17–20, 22–24, 26–32
B Exercises: 11, 12, 16, 21, 25, 33, 34
C Exercises: 35, 36

For each repeating decimal, write a fraction $\frac{a}{b}$ where a and b are integers.

1. $0.\overline{7}$ $\frac{7}{9}$

2. $0.\overline{145}$ $\frac{145}{999}$

3. Approximate $\sqrt{13}$ to the nearest tenth. Use a square root table. 3.6

4. Approximate $\sqrt{38}$ to the nearest tenth. Use the divide-and-average method. 6.2

Simplify.

        12. $15m^8n^{14}\sqrt{3m}$ 15. $\sqrt{3ab} + 5\sqrt{ab}$

5. $\sqrt{40}$ $2\sqrt{10}$

6. $\sqrt{99}$ $3\sqrt{11}$

7. $\sqrt{48} \cdot \sqrt{48}$ 48

8. $3\sqrt{64y^{16}}$ $24y^8$

9. $\sqrt{4x^5y^7}$ $2x^2y^3\sqrt{xy}$

10. $-\sqrt{28a^{10}b^{24}}$ $-2a^5b^{12}\sqrt{7}$

11. $3c\sqrt{16c^{12}d^9}$

12. $5m\sqrt{27m^{15}n^{28}}$

13. $9\sqrt{5} - 4\sqrt{5}$ $5\sqrt{5}$

14. $8\sqrt{2} - \sqrt{18} + \sqrt{8}$ $7\sqrt{2}$

15. $3\sqrt{27ab} - 4\sqrt{12ab} + 5\sqrt{ab}$

16. $3xy\sqrt{36y} + 7\sqrt{x^2y^3}$ $25xy\sqrt{xy}$

17. $5\sqrt{2} \cdot 3\sqrt{8}$ 60

18. $6\sqrt{8x} \cdot \sqrt{5x^5}$ $12x^3\sqrt{10}$

19. $-3\sqrt{2}(\sqrt{7} - 2\sqrt{5})$ $-3\sqrt{14} + 6\sqrt{10}$

20. $(3\sqrt{7} - \sqrt{2})(3\sqrt{7} + \sqrt{2})$ 61

21. $(\sqrt{2} - 3\sqrt{5})^2$ $47 - 6\sqrt{10}$

22. $\dfrac{\sqrt{40}}{\sqrt{8}}$ $\sqrt{5}$

23. $\dfrac{-5}{\sqrt{3}}$ $\dfrac{-5\sqrt{3}}{3}$

24. $\dfrac{\sqrt{6}}{\sqrt{8y^5}}$ $\dfrac{\sqrt{3y}}{2y^3}$

25. $\dfrac{3x^2y^3}{\sqrt{6x^4y^5}}$ $\dfrac{\sqrt{6y}}{2}$

26. $\dfrac{-3}{5 + \sqrt{3}}$ $\dfrac{-15 + 3\sqrt{3}}{22}$

For each right triangle, find the missing length. Give answers in simplest radical form.

27. $a = 6, b = 5$ $\sqrt{61}$

28. $b = 4, c = 9$ $\sqrt{65}$

Given the lengths of three sides of a triangle, determine whether it is a right triangle.

29. 48, 14, 50 Yes

30. 5, 10, 15 No

Solve each equation. Check.

31. $\sqrt{7x} + 1 = 8$ $x = 7$

32. $x - 3 = \sqrt{2x - 6}$ $x = 5$ or $x = 3$

Solve each problem. Round answers to the nearest tenth.

33. The area of a rectangle is 96 cm². Find the length and the width if the length is 4 times the width.
$l = 19.6$ cm, $w = 4.9$ cm

34. A surveyor walked 10 km north, then 8 km east. How far was the surveyor from his starting point? 12.8 km

Simplify.

35. $-14\sqrt{a^{6m}b^{4m+3}}$ $-14a^{3m}b^{2m+1}\sqrt{b}$

36. $(\sqrt{2} - 2\sqrt{3} + \sqrt{5})^2$
$19 - 4\sqrt{6} + 2\sqrt{10} - 4\sqrt{15}$

College Prep Test

For each Exercise, you are to compare a quantity in Column 1 with a quantity in Column 2. Select the correct answer from the following choices.

A—The quantity in Column 1 is greater than the quantity in Column 2.
B—The quantity in Column 2 is greater than the quantity in Column 1.
C—The quantity in Column 1 is equal to the quantity in Column 2.
D—The relationship cannot be determined from the given information.

NOTE: Information centered over both columns refers to one or both of the quantities to be compared.

Column 1	Column 2	Column 1	Column 2
1. $a > 0$ A		**9.** C	
$\sqrt{a}$	$-\dfrac{a}{\sqrt{a}}$		
2. $x \geq 8$ D			
$\sqrt{x + 8}$	4	*ABCD* is a square.	
3. $4 - \sqrt{6}$ A	$\sqrt{5} - 4$	$9x^2$	$4y^2$
4. $\sqrt{\dfrac{1}{4} - \dfrac{1}{36}}$ A	$\sqrt{\dfrac{1}{4}} - \sqrt{\dfrac{1}{36}}$	**10.** $a > 0, b > 0$ B	
		$(\sqrt{a + b})^2$	$(\sqrt{a} + \sqrt{b})^2$
5. $x > y > 0$ D			
$\sqrt{x}$	$\sqrt{xy}$	**11.** $\sqrt{7} + \sqrt{13}$ A	$\sqrt{20}$
6. $\sqrt{0.49}$	$(0.7)^2$ A		
7. $\sqrt{\dfrac{1}{0.50}}$	$\sqrt{4}$ B	**12.** $3 - \sqrt{x + 12} = 0$ A	
8. $a > 0, b < 0, x > 0, y > 0$ A			
$\dfrac{a}{\sqrt{x}}$	$\dfrac{b}{\sqrt{y}}$	x	$2x$

College Prep Test **533**

14 QUADRATIC EQUATIONS AND FUNCTIONS

OVERVIEW

The focus in this chapter is on the solution of quadratic equations and their applications in problem solving. In addition to the factoring method covered in previous lessons, students use the Square Root Property, completing the square, and the quadratic formula to solve these equations. Students are shown how these equations can be used to solve problems in physics, business, and geometry. Finally, students graph quadratic functions to find maximum and minimum values and use the discriminant to determine the number of real solutions.

OBJECTIVES

- To solve quadratic equations by using the Square Root Property, by completing the square, or by using the quadratic formula
- To solve word problems involving quadratic equations
- To graph a quadratic function, find the coordinates of its vertex, the equation of the axis of symmetry, and to determine the maximum or minimum value of the parabola
- To determine the number of solutions of a quadratic equation using the discriminant

PROBLEM SOLVING

In this chapter, students are shown that there is More Than One Way to solve quadratic equations, and decide which is the most effective way for a given problem. Drawing a Diagram and Writing an Equation help in the solution of geometric problems, while Using a Formula is the strategy used in the Application on page 552 and in determining the number of solutions of a quadratic equation. The usefulness of Checking Assumptions as a strategy in problem solving is introduced on page 557.

READING AND WRITING MATH

Students are asked to explain the difference between a solution to an equation and a solution to a problem in Exercise 7 on page 555. They are asked how to use the discriminant to decide what kind of solutions an equation will have in Exercise 29 on page 567. Exercises 19 and 20 on page 562 guide students in explaining what happens to the graph of a parabola when different parameters are changed.

TECHNOLOGY

Calculator: The calculator can be used instead of the tables at the back of the book for finding approximate decimal solutions to quadratic equations. This is demonstrated at the bottom of page 545. The calculator is also valuable for checking decimal solutions to word problems as illustrated in Example 2 on page 554.

SPECIAL FEATURES

PLANNING GUIDE

Lesson	Basic	Average	Above Average	Resources
14.1 pp. 537–538	CE 1–8 WE 1–33 odd	CE 1–8 WE 17–49 odd	CE 1–8 WE 23–55 odd	Reteaching 106 Practice 106
14.2 pp. 541–542	CE 1–18 WE 1–31 odd	CE 1–18 WE 13–41 odd, 49	CE 1–18 WE 21–51 odd	Reteaching 107 Practice 107
14.3 pp. 546–547	CE 1–18 WE 1–12 all	CE 1–18 WE 9–27 odd, 36, 37	CE 1–18 WE 17–35 odd, 36, 37	Reteaching 108 Practice 108
14.4 pp. 550–552	FR all, CE 1–18 WE 1–21 odd, 28–31 all Midchapter Review Application	FR all, CE 1–18 WE 11–27 odd, 32–37 all Midchapter Review Application	FR all, CE 1–18 WE 17–27 odd, 35–43 all Midchapter Review Application	Reteaching 109 Practice 109
14.5 pp. 555–557	CE 1–8 WE 1–7 all Problem Solving Strategies	CE 1–8 WE 1–9 odd, 10,11 Problem Solving Strategies	CE 1–8 WE 3–13 odd, 14 Problem Solving Strategies	Reteaching 110 Practice 110
14.6 pp. 561–563	CE 1–9 WE 1–17 odd	CE 1–9 WE 7–17 odd 19–21 all	CE 1–9 WE 13–19 odd 20–24 all, 28	Reteaching 111 Practice 111
14.7 pp. 566–567	CE 1–9 WE 1–18 all Brainteaser	CE 1–9 WE 1–9 all, 11–27 odd Brainteaser	CE 1–9 WE 1–29 odd, 30–33 all Brainteaser	Reteaching 112 Practice 112
Chapter 14 Review pp. 568–569	all	all	all	
Chapter 14 Test p. 570	all	all	all	
College Prep Test p. 571	all	all	all	
Cumulative Review Chapters 1–14 pp. 572–573	all	all	all	

CE = Classroom Exercises WE = Written Exercises FR = Focus on Reading

Note: For each level, all students should be assigned all Try This and all Mixed Review Exercises.

Project: For this *Investigation,* groups of students use a function to encode a message which another group must decode by using an inverse function.

Materials: Each student will need a copy of Project Worksheet 14.

Before handing out the worksheet, have students make a list of the letters of the alphabet and number them in order from 1 to 26. Then give them the following coded message:

20 8 27 15 16 26 13 28 21

Tell them that this message was formed by using the function $y = x + 7$. In other words, seven was added to the number of each letter in the original message. Have them decode the message. They should quickly see that they need only subtract seven from each number in the code to find the original numbers and the message: MATH IS FUN. Show them that if the y and x are exchanged in $y = x + 7$, you get $x = y + 7$, and if you solve this equation for y, you get $y = x - 7$. This is an inverse function.

Follow the same procedure with this message.

36 441 196 9 400 81 225 196

Tell them that the encoding function was $y = x^2$. Have them decode the message. FUNCTION   Show the students that since we are dealing with positive numbers only, we are only concerned with positive square roots. Therefore, an inverse function could be written as $y = \sqrt{x}$. (Solving $x = y^2$ for y.)

Hand out the worksheets and have students work in groups to encode messages, exchange papers with another group, and then decode messages.

Ben Livingston, an artist in neon, is shown here with his Neon Mural #1, a computer-animated cartoon. Two friends of his, a computer programmer and a neon glass bender, helped him turn his ideas into reality.

More About the Photo

Neon Mural #1 is a computer-animated neon cartoon that measures 14 by 40 feet. The mural presents an action-filled sequence that begins with a flower, shows a rocket traveling through the sky, and returns to the flower. The animation, or movement, in the gas-filled glass tubes is sequenced by a computer that activates 10 relays and transformers. This method of operation is something like an electronic player piano. Livingston's mural has been the subject of various articles and has won the Illuminating Engineering Society Paul Waterbury International Design Award of Excellence.

14.1 The Square Root Property

Objective

To solve quadratic equations by using the Square Root Property

Teaching Resources

Project Worksheet 14
Quick Quizzes 106
Reteaching and Practice
   Worksheets 106

Suppose that a ball falls 80 ft from the roof of a building. How long will it take the ball to reach the ground?

The time t in seconds can be found by using the formula $d = 16t^2$, where d is the distance in feet. Since the ball travels 80 ft, $80 = 16t^2$, or $16t^2 = 80$.

So, $\qquad t^2 = 5$ (Only the positive square root
$\qquad\qquad t = \sqrt{5} \approx 2.236$ can be the answer.)

Thus, it will take the ball about 2.2 s to reach the ground.

Consider the quadratic equation $x^2 = c$ for different values of c.

$c > 0$ (c is positive)	$c = 0$	$c < 0$ (c is negative)
$x^2 = 36$	$x^2 = 0$	$x^2 = -9$
There are *two* solutions, 6 and -6, since $6^2 = 36$ and $(-6)^2 = 36$.	There is *one* solution, 0, since $0^2 = 0$.	There is *no* real number solution, since there is no real number that has -9 as its square.

The Square Root Property is stated below for $c > 0$.

The Square Root Property

If $x^2 = c$, then $x = \sqrt{c}$ *or* $x = -\sqrt{c}$, for $c > 0$.

EXAMPLE 1 Find the solution set of $x^2 = 52$.

Plan Use the Square Root Property.

Solution
$$x^2 = 52$$
$$x = \sqrt{52} \ or \ x = -\sqrt{52} \quad\longleftarrow \sqrt{52} = \sqrt{4} \cdot \sqrt{13} = 2\sqrt{13}$$
$$x = 2\sqrt{13} \ or \ x = -2\sqrt{13} \quad\longleftarrow \text{This can be written as } x = \pm 2\sqrt{13}.$$

Check Since $(2\sqrt{13})^2 = 52$ and $(-2\sqrt{13})^2 = 52$, both solutions check.

Thus, the solution set is $\{2\sqrt{13}, -2\sqrt{13}\}$.

TRY THIS Find the solution set: **1.** $x^2 = 50$ **2.** $x^2 = 27$
$\qquad\qquad\qquad\qquad\qquad \{-5\sqrt{2}, 5\sqrt{2}\} \qquad\qquad \{-3\sqrt{3}, 3\sqrt{3}\}$

Additional Example 1

Find the solution set of $x^2 = 63$.
$\{3\sqrt{7}, -3\sqrt{7}\}$

535

Lesson Note

When using the shorthand method of expressing two roots, such as $\pm 2\sqrt{13}$, be sure the students understand that this indicates two distinct solutions, $-2\sqrt{13}$ and $2\sqrt{13}$. When presenting Example 4, emphasize the strategy of isolating $(2x - 5)^2$ by first adding 8 to each side and then dividing each side by 3.

Math Connections

Science: The distance an object falls in t seconds is given by the equation $s = \frac{1}{2}gt^2$, where g is acceleration due to gravity (32 ft/sec²).

Critical Thinking Questions

Analysis: Have students explain why $a(x - b)^2 = c$ has no real solutions if $a > 0$ and $c < 0$. If a and c have different signs, their quotient will be negative and the square root of a negative number is not a real number.

Common Error Analysis

Error: Students sometimes take the square root of an equation such as $(x - 5)^2 = 81$, and forget that there are *two* roots.

Remind them that every positive number has two square roots.

Sometimes you may need to use the Addition Property of Equality first. Then apply the Square Root Property.

EXAMPLE 2   Solve $5x^2 - 6 = 3$.

Plan   Write the equation in the form $x^2 = c$. Then solve.

Solution

$$5x^2 - 6 = 3$$
$$5x^2 = 9$$
$$x^2 = \frac{9}{5}$$
$$x = \sqrt{\frac{9}{5}} \quad or \quad x = -\sqrt{\frac{9}{5}}$$
$$x = \frac{3}{\sqrt{5}} \qquad x = -\frac{3}{\sqrt{5}}$$
$$\frac{3 \cdot \sqrt{5}}{\sqrt{5} \cdot \sqrt{5}} = \frac{3\sqrt{5}}{5} \longrightarrow x = \frac{3\sqrt{5}}{5} \qquad x = -\frac{3\sqrt{5}}{5}$$

Thus, the solutions are $\frac{3\sqrt{5}}{5}$ and $-\frac{3\sqrt{5}}{5}$. The check is left for you.

TRY THIS   Solve.

3. $3x^2 - 2 = 10 \quad \pm 2$

4. $2x^2 + 1 = 51 \quad \pm 5$

5. $x^2 + 4 = 10 \quad \pm\sqrt{6}$

6. $3x^2 - 4 = 7 \quad \pm\frac{\sqrt{33}}{3}$

More complex equations such as $(x + 3)^2 = 49$ can also be solved by using the Square Root Property.

EXAMPLE 3   Find the solution set of $(x + 3)^2 = 49$.

Solution

$(x + 3)^2 = 49$
$x + 3 = \sqrt{49} \quad or \quad x + 3 = -\sqrt{49}$
$x + 3 = 7 \qquad\qquad x + 3 = -7$
$x = 4 \qquad\qquad\quad x = -10$

Checks

$x = 4: (x + 3)^2 = 49$ $x = -10: \quad (x + 3)^2 = 49$
$(4 + 3)^2 \stackrel{?}{=} 49$ $(-10 + 3)^2 \stackrel{?}{=} 49$
$7^2 \stackrel{?}{=} 49$ $(-7)^2 \stackrel{?}{=} 49$
$49 = 49$ True $49 = 49$ True

Thus, the solution set is $\{4, -10\}$

Additional Example 2

Solve $3x^2 - 2 = 5$. $\frac{\sqrt{21}}{3}$ and $-\frac{\sqrt{21}}{3}$

Additional Example 3

Find the solution set of
$(x + 5)^2 = 4$. $\{-3, -7\}$

EXAMPLE 4   Solve $3(2x - 5)^2 - 8 = 10$.

Plan   First add 8 to each side. Then divide each side by 3.

Solution
$$3(2x - 5)^2 - 8 = 10$$
$$3(2x - 5)^2 = 18$$
$$(2x - 5)^2 = 6$$

$$2x - 5 = \sqrt{6} \qquad \text{or} \qquad 2x - 5 = -\sqrt{6}$$
$$x = \frac{5 + \sqrt{6}}{2} \qquad\qquad x = \frac{5 - \sqrt{6}}{2}$$

Thus, the solutions are $\dfrac{5 + \sqrt{6}}{2}$ and $\dfrac{5 - \sqrt{6}}{2}$, or $\dfrac{5 \pm \sqrt{6}}{2}$.

The check is left for you.

TRY THIS   Solve.

7. $(x - 2)^2 = 25$ $-3, 7$

8. $2(3x + 4)^2 - 5 = 9$ $\dfrac{-4 \pm \sqrt{7}}{3}$

Classroom Exercises

Solve each equation. Give irrational solutions in simplest radical form.

1. $x^2 = 4$ $x = \pm 2$
2. $y^2 = 0$ $y = 0$
3. $c^2 = 15$ $c = \pm\sqrt{15}$
4. $2x^2 = 32$ $x = \pm 4$
5. $y^2 - 21 = 0$ $y = \pm\sqrt{21}$
6. $(y - 8)^2 = 81$ $y = 17 \text{ or } y = -1$
7. $2(a - 5)^2 = 98$ $a = 12 \text{ or } a = -2$
8. $(2x + 1)^2 - 4 = 7$ $x = \dfrac{-1 \pm \sqrt{11}}{2}$

Written Exercises
22. $\pm\dfrac{3\sqrt{2}}{2}$ 23. $\pm\dfrac{3\sqrt{2}}{2}$ 24. $\pm\dfrac{3\sqrt{5}}{5}$

Solve each equation. Give irrational solutions in simplest radical form.

1. $x^2 = 49$ ± 7
2. $x^2 = 100$ ± 10
3. $a^2 = 0$ 0
4. $x^2 = 11$ $\pm\sqrt{11}$
5. $y^2 = 26$ $\pm\sqrt{26}$
6. $z^2 = 30$ $\pm\sqrt{30}$
7. $c^2 = 24$ $\pm 2\sqrt{6}$
8. $x^2 = 72$ $\pm 6\sqrt{2}$
9. $y^2 = 75$ $\pm 5\sqrt{3}$
10. $x^2 = 128$ $\pm 8\sqrt{2}$
11. $y^2 = 300$ $\pm 10\sqrt{3}$
12. $x^2 = 250$ $\pm 5\sqrt{10}$
13. $x^2 - 64 = 0$ ± 8
14. $y^2 - 15 = 0$ $\pm\sqrt{15}$
15. $z^2 - 32 = 0$ $\pm 4\sqrt{2}$
16. $-2x^2 = -50$ ± 5
17. $3x^2 - 75 = 0$ ± 5
18. $-5y^2 = -10$ $\pm\sqrt{2}$
19. $5a^2 = 3$ $\pm\dfrac{\sqrt{15}}{5}$
20. $7y^2 = 1$ $\pm\dfrac{\sqrt{7}}{7}$
21. $3x^2 = 16$ $\pm\dfrac{4\sqrt{3}}{3}$
22. $-4z^2 = -18$
23. $2x^2 = 9$
24. $5y^2 - 8 = 1$
25. $7x^2 + 4 = 19$
26. $3y^2 - 2 = 1$ $y = \pm 1$
27. $4a^2 - 2 = 0$
28. $(x + 5)^2 = 36$ $-11, 1$
29. $(c - 4)^2 = 100$ $-6, 14$
30. $(y + 1)^2 = 6$
31. $2(x - 8)^2 = 8$ $6, 10$
32. $-4(y - 9)^2 = -36$ $6, 12$
33. $5(c + 1)^2 = 80$ $-5, 3$

25. $\pm\dfrac{\sqrt{105}}{7}$ 27. $\pm\dfrac{\sqrt{2}}{2}$ 30. $-1 \pm \sqrt{6}$

14.1 The Square Root Property   **537**

Additional Example 4

Solve $2(3x - 7)^2 - 10 = 4$. $\dfrac{7 \pm \sqrt{7}}{3}$

Guided Practice

Classroom Exercises 1–8
Try This   all

Independent Practice

A Ex. 1–27, **B** Ex. 28–50, **C** Ex. 51–56

Basic: WE 1–33 odd

Average: WE 17–49 odd

Above Average: WE 23–55 odd

---

34. $(r - 2)^2 = 19$ $2 \pm \sqrt{19}$ **35.** $(x + 8)^2 = 12$ $-8 \pm 2\sqrt{3}$ **36.** $(z - 4)^2 = 30$

37. $(3x - 4)^2 = 18$ **38.** $(2y + 1)^2 = 5$ **39.** $(4c - 3)^2 = 2$

40. $(2y + 3)^2 - 1 = 5$ **41.** $(6a - 5)^2 + 2 = 8$ **42.** $(2d + 3)^2 - 4 = -2$

43. $\left(x + \frac{1}{2}\right)^2 = 0$ $-\frac{1}{2}$ **44.** $\left(y + \frac{1}{4}\right)^2 = \frac{1}{16}$ $-\frac{1}{2}, 0$ **45.** $\left(c - \frac{1}{3}\right)^2 = \frac{4}{9}$ $-\frac{1}{3}, 1$

In Exercises 46 and 47, use the formula $d = 16t^2$, where d is the distance in feet traveled by a freely falling object dropped from rest, and t is the time of the fall in seconds.

46. Find the time to the nearest tenth of a second for an object to fall 1,500 ft. 9.7 s

47. An object dropped from the top of a tower hit the ground in 4 s. If the tower were twice as tall, how long would it take the object to hit the ground? 5.7 s

48. The area of the floor of a square room is 42 ft². Find the length of each side, to the nearest tenth of a foot. 6.5 ft

49. If the area of a square is doubled, the result is 72 m². Find the length of each side of the square. 6 m

50. When the length of each side of a square is doubled, the area is 100 cm². Find the length of each side of the square. 5 cm

36. $4 \pm \sqrt{30}$

37. $\frac{4 \pm 3\sqrt{2}}{3}$

38. $\frac{-1 \pm \sqrt{5}}{2}$

39. $\frac{3 \pm \sqrt{2}}{4}$

40. $\frac{-3 \pm \sqrt{6}}{2}$

41. $\frac{5 \pm \sqrt{6}}{6}$

42. $\frac{-3 \pm \sqrt{2}}{2}$

Solve. If the equation has no real number solution, write *no solution*.

No solution

51. $y^2 + 14y + 49 = 6$ **52.** $c^2 - 10c + 25 = 25$ $\pm \sqrt{2}$ **53.** $m^2 + 16m + 64 = -1$

54. $4x^2 + 12x + 9 = 3$ **55.** $49y^2 - 42y = -9$ $\frac{3}{7}$ **56.** $2x^2 + 20x + 50 = 2$

51. $-7 \pm \sqrt{6}$ $\frac{-3 \pm \sqrt{3}}{2}$ $-6, -4$

Mixed Review

Multiply. 6.10

1. $(x + 7)^2$ $x^2 + 14x + 49$ **2.** $(y - 4)^2$ $y^2 - 8y + 16$ **3.** $(2r + 1)^2$ $4r^2 + 4r + 1$

Factor. 7.5

4. $x^2 + 16x + 64$ $(x + 8)^2$ **5.** $z^2 - 20z + 100$ $(z - 10)^2$ **6.** $9y^2 - 30y + 25$ $(3y - 5)^2$

Divide. 8.4

7. $\frac{y}{4x} \div \frac{3y^3}{8}$ $\frac{2}{3xy^2}$ **8.** $\frac{n + 2}{n} \div \frac{-4n - 8}{nm}$ $-\frac{m}{4}$ **9.** $\frac{x^2}{x - 1} \div (x + 1)$ $\frac{x^2}{x^2 - 1}$

Solve. 9.1

10. The sum of a number and its reciprocal is $-\frac{13}{6}$. Find the numbers. $-\frac{3}{2}, -\frac{2}{3}$

11. Find two consecutive integers such that the sum of their reciprocals is $\frac{17}{72}$. 8, 9

Enrichment

Have students look up Newton's Universal Law of Gravitation and write a short explanation of how it makes use of direct and inverse variation.

14.2 Completing the Square

Objective

To solve quadratic equations by completing the square

Teaching Resources

Manipulative Worksheet 14
Quick Quizzes 107
Reteaching and Practice
Worksheets 107

A square vegetable garden measures $(x + 3)$ meters on each side. It has several sections, as shown in this figure. The areas of most of the sections are given. The sum of the given areas is $x^2 + 3x + 3x$, or $x^2 + 6x$. What is the area, in meters, of the section shown in blue?

3	area: 3x	?
x	area: x²	area: 3x
	x	3

You know that the area of the garden can be represented as $(x + 3)^2$, or $x^2 + 6x + 9$. Thus, the area of the section shown in blue is 9 m^2.

In each case below, a binomial of the form $(x + a)$ is squared. The result is a *perfect square trinomial*. In each trinomial, observe the relationship between the coefficient of the x term and the third term.

$$(x + 3)^2 = x^2 + 6x + 9 \qquad (x - 4)^2 = x^2 - 8x + 16$$
$$\left(\tfrac{1}{2} \cdot 6\right)^2 \nearrow \qquad\qquad \left[\tfrac{1}{2} \cdot (-8)\right]^2 \nearrow$$

In both cases, the third term $= \left(\tfrac{1}{2} \cdot \text{coefficient of } x\right)^2$.

To complete the square of a binomial such as $x^2 + 12x$ means to add a third term that will form a perfect square trinomial. To complete the square of a binomial $x^2 + bx$, first take half of b and square it. Then add the square to $x^2 + bx$. The result is a perfect square trinomial.

EXAMPLE 1

Complete the square. Show the result as the square of a binomial.

a. $x^2 + 12x$

b. $y^2 - 5y$

Solutions

a. $\left(\tfrac{1}{2} \cdot 12\right)^2 = 6^2 = 36$
Add 36.
$x^2 + 12x + 36 = (x + 6)^2$

b. $\left[\tfrac{1}{2}(-5)\right]^2 = \left(-\tfrac{5}{2}\right)^2 = \tfrac{25}{4}$
Add $\tfrac{25}{4}$.
$y^2 - 5y + \tfrac{25}{4} = \left(y - \tfrac{5}{2}\right)^2$

TRY THIS

Complete the square. Write the result as the square of a binomial.

1. $r^2 - 20r$
$r^2 - 20r + 100 = (r - 10)^2$

2. $t^2 + t$ $\quad t^2 + t + \tfrac{1}{4} = (t + \tfrac{1}{2})^2$

14.2 Completing the Square **539**

GETTING STARTED

Prerequisite Quiz

Multiply.

1. $(x + 3)(x + 3)$ $\quad x^2 + 6x + 9$
2. $(y - 6)(y - 6)$ $\quad y^2 - 12y + 36$
3. $(r + 11)^2$ $\quad r^2 + 22r + 121$
4. $(c - 8)^2$ $\quad c^2 - 16c + 64$

Factor.

5. $x^2 - 8x + 16$ $\quad (x - 4)^2$
6. $x^2 + 2x + 1$ $\quad (x + 1)^2$

Motivator

Ask students to name numbers which are perfect squares. 1, 4, 9, 16, 25, . . . Then have them square some binomials such as, $(x + 5)^2$ and $(x - 2)^2$, at sight. Tell them that their results are perfect square trinomials. Write down each answer as it is completed and ask students how the coefficient of the middle term of the trinomial relates to the binomial. It is double the constant term of the binomial. Ask how the last term of the trinomial relates to the binomial. It is the square of the constant term.

Additional Example 1

Complete the square. Show the result as the square of a binomial.

a. $x^2 + 18x$ $\quad x^2 + 18x + 81; (x + 9)^2$
b. $y^2 - 7y$ $\quad y^2 - 7y + \tfrac{49}{4}; (y - \tfrac{7}{2})^2$

Highlighting the Standards

Standards 4a, 4b: A gardening problem is related to a geometric figure which then leads to a method for solving equations.

Lesson Note

When presenting Example 1b, tell students to leave fractions in the form shown. In this case, fractions are much easier to work with than mixed numbers or decimals. Before presenting Example 2, ask the students why $x^2 - 2x - 15$ is not a perfect square trinomial.   Last term is not the square of half the middle term coefficient.

Math Connections

Vocabulary: The equations in this chapter are called *quadratic* equations, from the Latin *quadratus* or square. The word quadratic refers to the fact that the highest degree of the variable in these equations is 2 (square).

Critical Thinking Questions

Evaluation: Ask students if they can use the solution to Example 2 to write the factors of $x^2 - 2x - 15$. $(x - 5)(x + 3)$ Ask them to use the solution of Example 3 to write the factors of $y^2 + 20y + 17$.
$[y - (-10 + \sqrt{83})][y - (-10 - \sqrt{83})]$, or $(y + 10 - \sqrt{83})(y + 10 + \sqrt{83})$.

You can use the process of completing the square to solve quadratic equations. Be sure that the quadratic equation is written in the form $ax^2 + bx + c = 0$ before starting this process.

EXAMPLE 2   Find the solution set of $x^2 - 2x - 15 = 0$ by completing the square.

Plan   Add 15 to each side. Complete the square for $x^2 - 2x$ and add the result to each side. Then solve by using the Square Root Property.

Solution
$$x^2 - 2x - 15 = 0$$
$$x^2 - 2x = 15$$
$$x^2 - 2x + 1 = 15 + 1 \longleftarrow \text{Add } \left[\tfrac{1}{2}(-2)\right]^2 = (-1)^2 = 1.$$
$$(x - 1)^2 = 16$$
$$x - 1 = \sqrt{16} \quad \text{or} \quad x - 1 = -\sqrt{16}$$
$$x = 1 + 4 \qquad\qquad x = 1 - 4$$
$$x = 5 \qquad\qquad\quad x = -3$$

Check

$x = 5$:

$$x^2 - 2x - 15 = 0$$
$$5^2 - 2 \cdot 5 - 15 \overset{?}{=} 0$$
$$25 - 10 - 15 \overset{?}{=} 0$$
$$0 = 0 \ \text{(True)}$$

$x = -3$:

$$x^2 - 2x - 15 = 0$$
$$(-3)^2 - 2(-3) - 15 \overset{?}{=} 0$$
$$9 + 6 - 15 \overset{?}{=} 0$$
$$0 = 0 \ \text{(True)}$$

Thus, the solution set is $\{5, -3\}$.

EXAMPLE 3   Solve $y^2 + 20y + 17 = 0$ by completing the square.

Solution
$$y^2 + 20y + 17 = 0$$
$$y^2 + 20y = -17$$
$$y^2 + 20y + 100 = -17 + 100 \longleftarrow \text{Add } \left(\tfrac{1}{2} \cdot 20\right)^2 = 10^2 = 100.$$
$$(y + 10)^2 = 83$$
$$y + 10 = \sqrt{83} \qquad \text{or} \quad y + 10 = -\sqrt{83}$$
$$y = -10 + \sqrt{83} \qquad\qquad y = -10 - \sqrt{83}$$

Thus, the solutions are $-10 + \sqrt{83}$ and $-10 - \sqrt{83}$, which can be written as $-10 \pm \sqrt{83}$

TRY THIS   3. Solve $a^2 - 16a + 13 = 0$ by completing the square. $8 \pm \sqrt{51}$

Additional Example 2

Find the solution set of $x^2 - 5x - 6 = 0$ by completing the square. $\{6, -1\}$

Additional Example 3

Solve $y^2 + 12y - 6 = 0$ by completing the square. $-6 \pm \sqrt{42}$

EXAMPLE 4   Solve $3c^2 = 4c + 1$ by completing the square.

Plan   Subtract $4c$ from each side. Then divide each side by 3 so that the coefficient of c^2 is 1.

Solution
$$3c^2 = 4c + 1$$
$$3c^2 - 4c = 1$$
$$c^2 - \frac{4}{3}c = \frac{1}{3}$$
$$c^2 - \frac{4}{3}c + \frac{4}{9} = \frac{1}{3} + \frac{4}{9} \quad \longleftarrow \text{Add } \left[\frac{1}{2}\left(-\frac{4}{3}\right)\right]^2 = \left(-\frac{2}{3}\right)^2 = \frac{4}{9}.$$
$$\left(c - \frac{2}{3}\right)^2 = \frac{7}{9}$$

$$c - \frac{2}{3} = \sqrt{\frac{7}{9}} \qquad or \qquad c - \frac{2}{3} = -\sqrt{\frac{7}{9}}$$

$$c = \frac{2}{3} + \frac{\sqrt{7}}{3} \qquad\qquad c = \frac{2}{3} - \frac{\sqrt{7}}{3}$$

$$c = \frac{2 + \sqrt{7}}{3} \qquad\qquad c = \frac{2 - \sqrt{7}}{3}$$

Thus, the solutions are $\dfrac{2 \pm \sqrt{7}}{3}$.

TRY THIS
4. Solve $5x^2 = 2 - 6x$ by completing the square. $\dfrac{-3 \pm \sqrt{19}}{5}$
5. Solve $2 = 3a^2 - a$ by completing the square. $\left\{-\frac{2}{3}, 1\right\}$

Classroom Exercises

What number should be added to make a perfect square trinomial?

1. $x^2 + 2x$ 1   2. $y^2 - 10y$ 25   3. $z^2 - 4z$ 4   4. $a^2 + 12a$ 36

5. $r^2 + 9r$ $\frac{81}{4}$ 6. $y^2 - y$ $\frac{1}{4}$ 7. $y^2 - \frac{1}{4}y$ $\frac{1}{64}$ 8. $n^2 + \frac{3}{4}n$ $\frac{9}{64}$

Simplify, if possible.

9. $1 + \sqrt{25}$ 6   10. $-3 - \sqrt{36}$ -9 11. $-\frac{2}{5} + \frac{\sqrt{11}}{5}$ $\frac{-2 + \sqrt{11}}{5}$ 12. $\frac{3 - \sqrt{49}}{2}$ -2

Solve by completing the square.

13. $x^2 + 12x - 28 = 0$ 2, -14 14. $c^2 - 8c - 84 = 0$ 14, -6

15. $x^2 - 7x + 10 = 0$ 2, 5   16. $b^2 + 10b + 16 = 0$ $-8, -2$

17. $a^2 + 6 = -5a$ $-3, -2$ 18. $c^2 - c = 30$ $-5, 6$

14.2 Completing the Square   **541**

Common Error Analysis

Error: When completing the square by adding the appropriate number to the left side of the equation, some students forget to add the same quantity to the right side of the equation.

Remind students that the same number must be added to both sides of the equation if the resulting equation is to be equivalent.

Checkpoint

Solve by completing the square.

1. $x^2 - 4x - 21 = 0$ $7, -3$
2. $y^2 + 6y - 27 = 0$ $3, -9$
3. $r^2 - 10r + 16 = 0$ $8, 2$
4. $c^2 - 8c + 3 = 0$ $\frac{8 \pm 2\sqrt{13}}{2}$, or $4 \pm \sqrt{13}$
5. $x^2 + 12x - 2 = 0$ $-6 \pm \sqrt{38}$
6. $y^2 - 3y - 7 = 0$ $\frac{3 \pm \sqrt{37}}{2}$

Closure

Ask students why some method other than factoring is needed to solve quadratic equations.   Some equations cannot be factored.   Then write the equation $x^2 - 8x + 5 = 0$ on the chalkboard and have different students each give one step in the correct sequence for solving this equation by the method of completing the square.

Additional Example 4

Solve $5c^2 = 3c + 5$ by completing the square. $\dfrac{3 \pm \sqrt{109}}{10}$

Guided Practice

Classroom Exercises 1–18
Try This   all

Independent Practice

A Ex. 1–32, **B** Ex. 33–54, **C** Ex. 55–60

Basic: WE 1–31 odd

Average: WE 13–41 odd, 49

Above Average: WE 21–51 odd

Additional Answers

Written Exercises

29. $3 \pm \sqrt{13}$

31. $-3 \pm \sqrt{19}$

32. $2 \pm \sqrt{11}$

40. $\dfrac{3 \pm \sqrt{21}}{6}$

42. No real-number solution

Mixed Review

2. $n < -9$

Written Exercises

Complete the square. Show the result as the square of a binomial. $(x + \frac{11}{2})^2$

1. $x^2 + 14x$ $(x + 7)^2$ **2.** $y^2 - 2y$ $(y - 1)^2$ **3.** $x^2 + 18x$ $(x + 9)^2$ **4.** $x^2 + 11x$

5. $y^2 - 15y$ $_{(y - \frac{15}{2})^2}$ **6.** $n^2 - \frac{1}{2}n$ $(n - \frac{1}{4})^2$ **7.** $y^2 + \frac{3}{5}y$ $(y + \frac{3}{10})^2$ **8.** $x^2 + \frac{2}{3}x$
$(x + \frac{1}{3})^2$

Solve by completing the square. Give irrational solutions in simplest
radical form.   **18.** $-5, -11$ **20.** $-4, -8$ **26.** $-1 \pm \sqrt{6}$ **27.** $-7 \pm \sqrt{43}$

9. $x^2 - 2x - 24 = 0$ $6, -4$ **10.** $c^2 - 4c - 12 = 0$ $6, -2$ **11.** $z^2 - 6z + 5 = 0$ $5, 1$

12. $y^2 - 10y + 21 = 0$ $7, 3$ **13.** $a^2 + 2a - 3 = 0$ $1, -3$ **14.** $x^2 - 6x - 27 = 0$ $9, -3$

15. $x^2 - 2x - 15 = 0$ $5, -3$ **16.** $y^2 + 4y + 3 = 0$ $-1, -3$ **17.** $a^2 - 4a - 21 = 0$ $7, -3$

18. $x^2 + 16x + 55 = 0$ **19.** $r^2 + 6r - 7 = 0$ $1, -7$ **20.** $c^2 + 12c + 32 = 0$

21. $c^2 - 18c + 77 = 0$ $11, 7$ **22.** $x^2 - 24x + 80 = 0$ $4, 20$ **23.** $y^2 - 18y + 72 = 0$ $6, 12$

24. $x^2 + 4x + 2 = 0$ $-2 \pm \sqrt{2}$ **25.** $y^2 - 8y + 5 = 0$ $4 \pm \sqrt{11}$ **26.** $c^2 + 2c - 5 = 0$

27. $a^2 + 14a + 6 = 0$ **28.** $x^2 + 2x - 6 = 0$ $-1 \pm \sqrt{7}$ **29.** $r^2 - 6r - 4 = 0$

30. $c^2 - 10c + 22 = 0$ $5 \pm \sqrt{3}$ **31.** $y^2 + 6y - 10 = 0$ **32.** $y^2 - 4y - 7 = 0$

33. $x^2 - 3x + 2 = 0$ $1, 2$ **34.** $y^2 - 7y - 8 = 0$ $-1, 8$ **35.** $a^2 + a = 30$ $-6, 5$

36. $c^2 + 3c = 40$ $5, -8$ **37.** $d^2 - 15d + 56 = 0$ $7, 8$ **38.** $x^2 + 5x = 84$ $-12, 7$

39. $2y^2 + 7y - 4 = 0$ $-4, \frac{1}{2}$ **40.** $3x^2 = 3x + 1$ **41.** $6r^2 = r + 1$ $\frac{1}{2}, -\frac{1}{3}$

42. $2a^2 + 6a + 5 = 0$ **43.** $4c^2 + 4 = 17c$ $4, \frac{1}{4}$ **44.** $2x + 8 = 3x^2$ $2, -\frac{4}{3}$

45. $2y^2 + 3y = 17$ $_{\frac{-3 \pm \sqrt{145}}{4}}$ **46.** $2x^2 - 2x - 1 = 0$ **47.** $3a^2 + 2a = 3$ $\frac{-1 \pm \sqrt{10}}{3}$
$\frac{1 \pm \sqrt{3}}{2}$

Solve each problem.

48. If 3 times the square of a number is decreased by twice the number, the result is 1. Find the number. $1, -\frac{1}{3}$

49. Twice the square of a number is equal to the number increased by 3. Find the number. $\frac{3}{2}, -1$

50. The square of a number is 2 less than 6 times the number. Find the number.
$3 \pm \sqrt{7}$

51. The square of a number equals twice the number plus 4. Find the number.
$1 \pm \sqrt{5}$

Mixed Review

Solve each open inequality.   5.1, 5.2, 5.7

1. $x + 5 < 2$ $x < -3$ **2.** $-2n + 10 > 28$ **3.** $16 - y \le 4$ $y \ge 12$ **4.** $|x - 2| < 7$
$-5 < x < 9$

5. Given that $f(x) = -2x^2 + 7$, find $f(-3)$.   10.3 -11

Solve.   9.2, 9.3

6. If 3 out of 5 people use Drain-Fix, how many use Drain-Fix in a city of 42,000 people? 25,200

7. Solve the formula, $I = prt$, for t. Then find the time t in years for $I = \$200$, $p = \$8,000$, and $r = 5\%$.
$t = \frac{I}{pr}; \frac{1}{2}$ yr

Enrichment

Challenge students to solve this equation:
$x^4 - 7x^2 = -10$. Tell them to start by letting
$y = x^2$ and remind them that a fourth-degree
equation will have 4 solutions.   By letting
$y = x^2$, the equation becomes
$y^2 - 7y = -10$, having solutions $y = 5$
and $y = 2$. Thus, $x^2 = 5$ or $x^2 = 2$,
$x = \pm \sqrt{5}$ or $x = \pm \sqrt{2}$

14.3 The Quadratic Formula

Teaching Resources

Application Worksheet 14
Quick Quizzes 108
Reteaching and Practice
   Worksheets 108

Objective   To solve quadratic equations by using the quadratic formula

The Parthenon, a building in ancient Greece, has the shape of a rectangle. The width of this rectangle is about six-tenths of the length. Such a rectangle is called a *golden rectangle* because the ratio of the width to the length is considered "most pleasing to the eye." The exact ratio is found by using the quadratic formula (see Written Exercises 36 and 37).

The **quadratic formula** can be used to find the solutions of *any* quadratic equation. Every quadratic equation can be written in the standard form, $ax^2 + bx + c = 0$, where $a > 0$. This form can be solved for x by completing the square. Begin by adding $-c$ to each side.

■■■ **GETTING STARTED**

Prerequisite Quiz

Rewrite each quadratic equation in the form $ax^2 + bx + c = 0$, where $a > 0$.

1. $7 + 9x^2 = 5x$ $9x^2 - 5x + 7 = 0$
2. $3y^2 = -2 - 4y$ $3y^2 + 4y + 2 = 0$
3. $8r - r^2 = 3$ $r^2 - 8r + 3 = 0$

Simplify.

4. $\sqrt{25 - (4)(1)(4)}$ 3
5. $\sqrt{81 - 4(-2)(3)}$ $\sqrt{105}$
6. $\sqrt{6^2 - 4(3)(1)}$ $2\sqrt{6}$

$$ax^2 + bx + c = 0$$

Add $-c$ to each side.
$$ax^2 + bx = -c$$

Divide each side by a.
$$x^2 + \frac{b}{a}x = -\frac{c}{a}$$

Complete the square.
$$x^2 + \frac{b}{a}x + \frac{b^2}{4a^2} = -\frac{c}{a} + \frac{b^2}{4a^2}$$

Factor the left side.
$$\left(x + \frac{b}{2a}\right)^2 = -\frac{c}{a} + \frac{b^2}{4a^2}$$

$$\left(x + \frac{b}{2a}\right)^2 = -\frac{4ac}{4a^2} + \frac{b^2}{4a^2}$$

Simplify the right side.
$$\left(x + \frac{b}{2a}\right)^2 = \frac{b^2 - 4ac}{4a^2}$$

$$x + \frac{b}{2a} = \pm\sqrt{\frac{b^2 - 4ac}{4a^2}}$$

$$x = \frac{\pm\sqrt{b^2 - 4ac}}{2a} - \frac{b}{2a}$$

$$x = \frac{-b \pm \sqrt{b^2 - 4ac}}{2a}$$

Motivator

Ask students which would be easier and more accurate, to use a formula for finding the area of a triangle or to count squares. Use a formula. Tell them that the method of completing the square that they learned in the previous lesson can be used to develop a formula for finding the solution of all quadratic equations of the form $ax^2 + bx + c = 0$, $a \neq 0$.

Highlighting the Standards

Standards 2d, 5d: As students read the derivation of the quadratic formula, they can see the power and economy of such mathematical generalizations.

Lesson Note

It is helpful to work through a parallel concrete example as you go through the steps of the derivation of the quadratic formula. For example, the steps for solving $3x^2 + 4x - 5$ by completing the square can be compared with the general steps in the derivation of the formula. For practical applications, students need to be able to express the answers in approximate decimal form as shown at the bottom of page 545.

Math Connections

History: Solutions for third and fourth degree equations were discovered in the sixteenth century. In the early nineteenth century, it was proved that there is no general solution for polynomial equations of degree higher than four.

Critical Thinking Questions

Analysis: Ask students to describe the solutions of a quadratic equation when $b^2 - 4ac = 0$. There will be one solution, $-\frac{b}{2a}$. Ask them to describe the solutions when $b^2 - 4ac$ is a perfect square. There will be two rational-number solutions. Ask them to describe the solution when $b^2 - 4ac$ is an irrational number. There will be two solutions, both irrational.

The Quadratic Formula

The solutions of a quadratic equation in standard form, $ax^2 + bx + c = 0$, where $a > 0$, are given by this formula.

$$x = \frac{-b \pm \sqrt{b^2 - 4ac}}{2a}$$

EXAMPLE 1   Find the solution set of $2x^2 + x - 10 = 0$ by the quadratic formula.

Plan   The equation is in standard form. Determine the values of a, b, and c, and substitute them into the quadratic formula.

Solution $2x^2 + x - 10 = 0 \quad\longleftarrow a = 2, b = 1, c = -10$

$$x = \frac{-b \pm \sqrt{b^2 - 4ac}}{2a}$$

$$x = \frac{-1 \pm \sqrt{1^2 - 4 \cdot 2 \cdot (-10)}}{2 \cdot 2}$$

$$x = \frac{-1 \pm \sqrt{1 + 80}}{4}$$

$$x = \frac{-1 \pm \sqrt{81}}{4}$$

$$x = \frac{-1 \pm 9}{4}$$

$$x = \frac{-1 + 9}{4} \quad or \quad x = \frac{-1 - 9}{4}$$

$$x = \frac{8}{4} \qquad\qquad\qquad x = \frac{-10}{4}$$

$$x = 2 \qquad\qquad\qquad x = -\frac{5}{2}$$

Each solution checks in the original equation. Thus the solution set is $\left\{2, -\frac{5}{2}\right\}$. The check is left for you.

TRY THIS   1. Use the quadratic formula to find the solution set of $3x^2 - 2x - 5 = 0$. $\left\{-1, \frac{5}{3}\right\}$

You could solve the equation in Example 1 by factoring, since $2x^2 + x - 10 = (2x + 5)(x - 2)$. However, some quadratic

---

Additional Example 1

Find the solution set of $2x^2 - 5x - 3 = 0$ by the quadratic formula. $\left\{3, -\frac{1}{2}\right\}$

equations cannot be solved by factoring, or they may be difficult to factor. In such cases, the formula is the best method for solution.

EXAMPLE 2 Solve $2y + 1 = 2y^2$ by the quadratic formula.

Plan First put the equation in standard form. Then use the quadratic formula.

Solution
$$2y + 1 = 2y^2$$
$$2y^2 - 2y - 1 = 0 \longleftarrow a = 2, b = -2, c = -1$$
$$y = \frac{-b \pm \sqrt{b^2 - 4ac}}{2a}$$
$$y = \frac{-(-2) \pm \sqrt{(-2)^2 - 4 \cdot 2 \cdot (-1)}}{2 \cdot 2}$$
$$y = \frac{2 \pm \sqrt{4 + 8}}{4}$$
$$y = \frac{2 \pm \sqrt{12}}{4}$$
$$y = \frac{2 \pm 2\sqrt{3}}{4} \longleftarrow \sqrt{12} = 2\sqrt{3}$$
$$y = \frac{\overset{1}{\cancel{2}}(1 \pm \sqrt{3})}{\underset{1}{\cancel{2}} \cdot 2} = \frac{1 \pm \sqrt{3}}{2}$$

Thus, the solutions are $\dfrac{1 \pm \sqrt{3}}{2}$

TRY THIS 2. Solve $6x = 2 - 5x^2$ by the quadratic formula. $\dfrac{-3 \pm \sqrt{19}}{5}$

The solutions in Example 2 are given in *simplest radical form*. The solutions can be approximated to the nearest tenth by using a table of square roots or the square root key on a calculator.

$$\frac{1 + \sqrt{3}}{2} \approx \frac{1 + 1.732}{2} = \frac{2.732}{2} = 1.366 \longleftarrow 1.4 \text{ to the nearest tenth}$$

$$\frac{1 - \sqrt{3}}{2} \approx \frac{1 - 1.732}{2} = \frac{-0.732}{2} = -0.366 \longleftarrow -0.4 \text{ to the nearest tenth}$$

Thus, the solutions are 1.4 and -0.4, to the nearest tenth.
The portion of the quadratic formula that is under the radical sign, $b^2 - 4ac$, is called the **discriminant**.

Common Error Analysis

Error: Since in many of the problems a has a value of one, some students will forget that the denominator of the fraction in the formula is $2a$, and will use 2 regardless of the value of a.

Have students write out the formula before working each problem.

Checkpoint

Solve by using the quadratic formula. Give irrational solutions in simplest radical form.

1. $x^2 + 4x - 21 = 0$ $-7, 3$
2. $y^2 - 4y = -3$ $1, 3$
3. $2c^2 + c = 3$ $1, -\frac{3}{2}$
4. $4x^2 - 3x + 6 = 0$ $\varnothing$
5. $x^2 + 5x + 2 = 0$ $\dfrac{-5 \pm \sqrt{17}}{2}$
6. $y^2 = 3 - 8y$ $-4 \pm \sqrt{19}$

Closure

Have students identify a, b, and c in each of the equations. Then have them substitute these values in the "quadratic" formula.

1. $3x^2 + 2x - 3 = 0$
   $a = 3, b = 2, c = -3;$
   $x = \dfrac{-2 \pm \sqrt{(-2)^2 - 4(3)(-3)}}{2(3)}$
2. $2x^2 - 7x + 2 = 0$
   $a = 2, b = -7, c = 2;$
   $x = \dfrac{7 \pm \sqrt{(-7)^2 - 4(2)(2)}}{2(2)}$
3. $2y^2 - 3y = 4$
   $a = 2, b = -3, c = -4;$
   $x = \dfrac{3 \pm \sqrt{(-3)^2 - 4(2)(-4)}}{2(2)}$

Additional Example 2

Solve $3y + 4 = 2y^2$ by the quadratic formula. $\dfrac{3 \pm \sqrt{41}}{4}$

Guided Practice

Classroom Exercises 1–18
Try This   all

Independent Practice

A Ex. 1–12, **B** Ex. 13–29, **C** Ex. 30–38

Basic: WE 1–12 all
Average: WE 9–27 odd, 36, 37
Above Average: WE 17–35 odd, 36, 37

Additional Answers

Written Exercises

8. $\frac{7 \pm \sqrt{37}}{2}$

9. $\frac{7 \pm \sqrt{53}}{2}$

10. $\frac{-1 \pm \sqrt{5}}{2}$

11. No real–number solution

12. $\frac{-5 \pm \sqrt{37}}{2}$

13. $\frac{1 \pm \sqrt{73}}{12}$

14. $\frac{11 \pm \sqrt{181}}{10}$

15. $\frac{5 \pm \sqrt{73}}{6}$

19. $-1 \pm \sqrt{3}$

20. $\frac{2 \pm \sqrt{10}}{2}$

21. $4, \frac{2}{3}$

33. $\frac{2 \pm 2\sqrt{13}}{3}$

37. From Exercise 36, $\frac{w}{l} \approx 0.618$, where w and l are the width and length of a golden rectangle. Thus, $w \approx 0.618\,l$. Since $0.618 > 0.6$, it follows that $0.618\,l > 0.6\,l$, or $w > 0.6\,l$.

38. If $4ac > b^2$, the discriminant of the quadratic equation is negative, or $b^2 - 4ac < 0$. However, the square root of a negative number does not exist as a real number. Thus, a quadratic equation has no real-number solution if $4ac > b^2$.

Before using the quadratic formula, it is wise to see whether $b^2 - 4ac$ is negative. If the discriminant is negative, the equation will have no real number solutions. In other words, if $b^2 - 4ac < 0$, the solution set of the quadratic equation is the empty set, or $\varnothing$.

EXAMPLE 3   Solve $x^2 = 3x - 8$.

Solution $x^2 - 3x + 8 = 0$ ⟵ standard form; $a = 1, b = -3, c = 8$
$b^2 - 4ac = (-3)^2 - 4 \cdot 1 \cdot 8 = 9 - 32 = -23$

The discriminant is negative, so there are no real number solutions.

TRY THIS   3. Solve $6 = 4y - y^2$. $\varnothing$

Classroom Exercises

For each equation, give the value of a, b, and c.

1. $3x^2 - 2x + 1 = 0$ $3, -2, 1$ 2. $y^2 - y - 1 = 0$ $1, -1, -1$ 3. $2y^2 + 3 = 4y$ $2, -4, 3$

4. $d^2 = 5d - 2$ $1, -5, 2$ 5. $p^2 - 2p = 0$ $1, -2, 0$ 6. $-3x^2 = 5x$ $3, 5, 0$

Find the value of $b^2 - 4ac$, for the given values of a, b, and c.

7. $a = 1, b = 5, c = 5$ 5 8. $a = 1, b = -5, c = 5$ 5 9. $a = 1, b = 2, c = 4$ -12

Each expression represents two numbers. Simplify, if possible.

10. $4 \pm \sqrt{12}$ $4 \pm 2\sqrt{3}$ 11. $-2 \pm \sqrt{18}$ $-2 \pm 3\sqrt{2}$ 12. $0 \pm \sqrt{9}$ ± 3

13. $\frac{8 \pm 4\sqrt{3}}{2}$ $4 \pm 2\sqrt{3}$ 14. $\frac{-4 \pm 8\sqrt{2}}{4}$ $-1 \pm 2\sqrt{2}$ 15. $\frac{-7 \pm \sqrt{11}}{3}$ Does not simplify.

Solve by using the quadratic formula.

16. $x^2 + 7x + 6 = 0$ $-1, -6$ 17. $y^2 - y - 12 = 0$ $4, -3$ 18. $3c - 8 = -2c^2$
$\frac{-3 \pm \sqrt{73}}{4}$

Written Exercises

Solve by using the quadratic formula. Give irrational solutions in simplest radical form.

1. $x^2 - 7x + 10 = 0$ $5, 2$ 2. $c^2 + 6c + 8 = 0$ $-2, -4$ 3. $x^2 - 9x + 14 = 0$ $7, 2$

4. $d^2 + 2d - 3 = 0$ $1, -3$ 5. $y^2 + 4y - 3 = 0$ $-2 \pm \sqrt{7}$ 6. $x^2 - 9x + 8 = 0$ $8, 1$

7. $2x^2 + 1 = 3x$ $1, \frac{1}{2}$ 8. $x^2 - 7x = -3$ 9. $x^2 - 1 = 7x$

10. $x^2 = 1 - x$ 11. $2x + x^2 + 7 = 0$ 12. $3 = y^2 + 5y$

13. $y = 6y^2 - 3$ 14. $11x = 5x^2 - 3$ 15. $4 = 3y^2 - 5y$

16. $-x^2 - 4 = 2x$ 17. $x^2 + 4x + 2 = 8$ 18. $-y^2 + 2 = -6y$
    No real sol $-2 \pm \sqrt{10}$ $3 \pm \sqrt{11}$

Additional Example 3

Solve $2x^2 = 5x - 4$. $\varnothing$

19. $2x - 1 + x^2 = 1$ **20.** $2x(x - 2) = 3$ **21.** $(x + 1)^2 = (2x - 3)^2$

22. $x^2 + (x - 1)^2 = 2$ $\frac{1 \pm \sqrt{3}}{2}$ **23.** $3x^2 = (3x)^2 - 2x + 1$ **24.** $(x + 2)^2 = 7 - x^2$
No real number solution $\frac{-2 \pm \sqrt{10}}{2}$

Solve, to the nearest tenth, by using the quadratic formula.

25. $3x^2 + 2x + 1 = -7$ **26.** $4d^2 = 5d + 2$ 1.6, -0.3 **27.** $4x = -x^2 - 2$
No real solution $-0.6, -3.4$

Solve each problem.

28. Twice the square of a number is 4 more than the opposite of the $\frac{-1 \pm \sqrt{33}}{4}$
number. Find the number.

29. If twice a number is decreased by 1, the result is equal to 3 times
the square of the number. Find the number.   No real number solution

Solve by using the quadratic formula.

30. $\frac{3}{2}y + y^2 = \frac{5}{2}$ $1, -\frac{5}{2}$ **31.** $\frac{1}{2}x^2 - 2 = -\frac{1}{2}x$ $\frac{-1 \pm \sqrt{17}}{2}$ **32.** $1 + \frac{x - 4}{x - 3} = \frac{3}{x - 1}$ 4, 2

33. $\frac{4}{3y + 4} = -1 + \frac{4}{y}$ **34.** $2r\sqrt{2} = r^2 - 6$ $3\sqrt{2}, -\sqrt{2}$ **35.** $4x + 2\sqrt{6} = x^2\sqrt{6}$
                                                           $\sqrt{6}, -\frac{\sqrt{6}}{3}$

36. In a golden rectangle whose length is 1 unit and
whose width is x units, the following proportion
is always true. Find x correct to three decimal
places.   0.618

$$\frac{x}{1} = \frac{1 - x}{x}$$

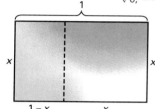

37. Write an explanation of why Exercise 36 enables
you to conclude that in any golden rectangle, the
width is a little more than six-tenths of the length.

38. Explain why a quadratic equation has no real solutions if $4ac > b^2$.

Mixed Review

Add or subtract.   8.5, 8.6, 8.7

1. $\frac{7x}{3} + \frac{2x}{3}$ $3x$ **2.** $\frac{6y^2}{7} - \frac{y^2}{7}$ $\frac{5y^2}{7}$ **9.** $\frac{5y + 19}{y^2 - 2y - 24}$

3. $\frac{x^2}{3x - 9} + \frac{6x}{3x - 9}$ $\frac{x^2 + 6x}{3x - 9}$

4. $\frac{x + 3}{2x} + \frac{2x + 1}{4x}$ $\frac{4x + 7}{4x}$ **5.** $\frac{y + 2}{3y} - \frac{y + 3}{2y}$ $\frac{-y - 5}{6y}$ **6.** $\frac{7}{10n} - \frac{5}{2n^2}$ $\frac{7n - 25}{10n^2}$

7. $\frac{3}{n - 4} - \frac{5}{n^2 - 4n}$ $\frac{3n - 5}{n^2 - 4n}$ **8.** $\frac{x}{x - 4} + \frac{3}{x + 4}$ $\frac{x^2 + 7x - 12}{x^2 - 16}$ **9.** $\frac{5}{y - 6} - \frac{1}{y^2 - 2y - 24}$

10. y varies directly as x. y is 12 when x is 4. Find x when y is 14.   **10.6** $4\frac{2}{3}$

11. One machine can complete an order in 10 h. Another takes 15 h.
How long will it take both machines working together?   **9.5** 6 h

Enrichment

Write the sum of the roots of the general
quadratic equation on the chalkboard in this
form.

$$\frac{-b + \sqrt{b^2 - 4ac}}{2a} + \frac{-b - \sqrt{b^2 - 4ac}}{2a}$$

1. Now have the students find, in simplest
form, the sum of the roots of a quadratic
equation. $-\frac{b}{a}$

2. Next, have the students multiply the two
roots of the general quadratic equation,
and write the result in simplest form. $\frac{c}{a}$

Explain that these two results provide a
good way to check solutions of quadratic
equations. Have students verify these sum
and product formulas with some of the
solutions of examples and exercises in this
lesson.

■■■ **GETTING STARTED**

Prerequisite Quiz

1. Solve $x^2 = 48$ by using the Square Root Property. $\pm 4\sqrt{3}$
2. Solve $x^2 - 15x + 54 = 0$ by using the factoring method. 6, 9
3. Solve $x^2 - 6x + 4 = 0$ by completing the square. $3 \pm \sqrt{5}$
4. Solve $5x^2 - 19x - 4 = 0$ by using the quadratic formula. $4, -\frac{1}{5}$

Motivator

Ask students why it would be very inefficient to solve the equation $x^2 = 25$ by using the quadratic formula. It can be solved more quickly by applying the Square Root Property. Then ask what would probably be the fastest way to solve the equation $x^2 + 7x + 10 = 0$. By factoring Ask students what is the advantage of using the quadratic formula. It works with any quadratic equation.

14.4 Choosing a Method of Solution

Objective   To solve quadratic equations using the most appropriate method

You have learned four methods for solving quadratic equations.

(1) Square Root Property   (3) Completing the Square
(2) Factoring   (4) Quadratic Formula

You can use Methods 3 and 4 for any quadratic equation. However, Methods 1 and 2 are used when an equation has no middle term.

EXAMPLE 1   Solve $x^2 = 98$.

Solutions

Method 1: Square Root Property

$$x^2 = 98$$
$$x = \sqrt{98} \quad or \quad x = -\sqrt{98}$$
$$x = 7\sqrt{2} \qquad\qquad x = -7\sqrt{2}$$

Method 2: Factoring

$$x^2 = 98$$
$$x^2 - 98 = 0$$
$$(x - \sqrt{98})(x + \sqrt{98}) = 0$$
$$x = \sqrt{98} \quad or \quad x = -\sqrt{98}$$
$$x = 7\sqrt{2} \qquad\qquad x = -7\sqrt{2}$$

Thus, the solutions are $7\sqrt{2}, -7\sqrt{2}$.

If a quadratic equation can be solved by factoring, then that is usually a simpler method than completing the square or using the formula.

EXAMPLE 2   Solve $2x^2 = 9x + 5$.

Solutions   Method 2: Factoring

$$2x^2 - 9x - 5 = 0$$
$$(2x + 1)(x - 5) = 0$$
$$2x + 1 = 0 \quad or \quad x - 5 = 0$$
$$2x = -1 \qquad\qquad x = 5$$
$$x = -\frac{1}{2}$$

Method 4: Quadratic Formula

For $2x^2 - 9x - 5 = 0$, $a = 2$, $b = -9$, and $c = -5$.

$$x = \frac{-(-9) \pm \sqrt{(-9)^2 - 4 \cdot 2 \cdot (-5)}}{2 \cdot 2}$$
$$x = \frac{9 \pm \sqrt{121}}{4} = \frac{9 \pm 11}{4}$$
$$x = 5 \quad or \quad x = -\frac{1}{2}$$

Thus, the solutions are 5 and $-\frac{1}{2}$.

TRY THIS   1. Solve $3x^2 - 2x = 1$. $-\frac{1}{3}, 1$

Highlighting the Standards

Standards 1a, 3c: This lesson shows that there is more than one way to solve a quadratic equation. In the Focus on Reading, students construct a logical argument.

Additional Example 1

Solve $x^2 = 128$ $\{8\sqrt{2}, -8\sqrt{2}\}$

Additional Example 2

Solve $3x^2 = 7x + 6$. $\{3, -\frac{2}{3}\}$

When neither the Square Root Property nor factoring can be applied, then completing the square or the quadratic formula should be used.

EXAMPLE 3 A pellet is to be shot upward from the earth's surface at an initial velocity of 25 m/s. Use the formula $h = 25t - 5t^2$ to find the height h in meters of the pellet t seconds after it is shot upward. After how many seconds will the pellet be 10 m above the ground? Give answers correct to the nearest tenth of a second.

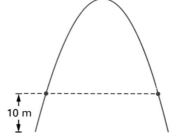

10 m

Plan Substitute 10 for h in $h = 25t - 5t^2$. Solve for t.

There will be two solutions, one for the number of seconds for the flight upward and one for the number of seconds downward.

Solution

	$h = 25t - 5t^2$
Substitute 10 for h.	$10 = 25t - 5t^2$
Write in standard form.	$5t^2 - 25t + 10 = 0$
Factoring won't work.	$t = \dfrac{-b \pm \sqrt{b^2 - 4ac}}{2a}$
Use the quadratic formula.	

$$a = 5, b = -25, c = 10 \qquad t = \frac{-(-25) \pm \sqrt{(-25)^2 - 4 \cdot 5 \cdot 10}}{2 \cdot 5}$$

$$t = \frac{25 \pm \sqrt{625 - 200}}{10} = \frac{25 \pm \sqrt{425}}{10}$$

$$t \approx \frac{25 - 20.616}{10} \quad or \quad t \approx \frac{25 + 20.616}{10}$$

$$t \approx \frac{4.384}{10} \qquad\qquad t \approx \frac{45.616}{10}$$

$$t \approx 0.4 \qquad\qquad t \approx 4.6$$

THINK: 0.4 s seems reasonable for the pellet to rise to 10 m, and 4.6 s seems reasonable for it to peak and fall to 10 m.

Thus, the pellet will rise to 10 m after 0.4 s and will fall to 10 m after 4.6 s

TRY THIS 2. After how many seconds (nearest tenth of a second) will the pellet in Example 3 be 5 m above the ground? Use the formula $h = 25t - 5t^2$. Rise to 5 m after 0.2 s; fall after 4.8 s

549

Lesson Note

Point out to students that the factoring method for Example 1 is extended to include factors with irrational numbers. Before writing an equation for Example 3, have students look at the diagram and ask them how many times the pellet will be at a height of 10 m. Twice Then tell them that, since a quadratic equation has two solutions, they will be able to find both values of t.

Math Connections

Computers: The most efficient way for a computer to solve quadratic equations would be to use the formula rather than to factor. In fact, the name for the computer language FORTRAN comes from *for*mula *tran*slation.

Critical Thinking Questions

Application: Write this equation on the chalkboard.

$$s = \frac{n}{2}(n + 1)$$

Explain that this formula is used to find the sum of the first n counting numbers. For example, when $n = 5$, $1 + 2 + 3 + 4 + 5 = \frac{5}{2}(5 + 1) = 15$. Have students solve for the formula, n, in terms of s, keeping in mind that $n > 0$. $n = \dfrac{-1 + \sqrt{1 + 8s}}{2}$

Evaluation: Ask students which of the four methods for solving quadratic equations would probably be used the least often and why? Completing the square, because the quadratic formula actually replaces it and is usually more efficient.

Additional Example 3

A pellet is to be shot upward from the earth's surface at an initial velocity of 100 m/s. The formula $h = 100t - 5t^2$ can be used to find the height h (in meters) of the pellet, t seconds after it is shot upward. After how many seconds will the pellet be 10 m above the ground? The pellet will rise to 10 m after 0.1 s and fall to 10 m after 19.9 s.

Common Error Analysis

Error: When evaluating $-4ac$ in the formula, some students will make an error in the sign of this product.

Advise them to write out each factor rather than trying to multiply mentally.

Checkpoint

Solve each equation. Give irrational roots in simplest radical form.

1. $c^2 - 60 = 0$ $\pm 2\sqrt{15}$
2. $2r^2 - 7r + 3 = 0$ $3, \frac{1}{2}$
3. $8 + 5x = x^2$ $\frac{5 \pm \sqrt{57}}{2}$
4. $3x^2 + 4x = 1$ $\frac{-2 \pm \sqrt{7}}{3}$
5. The formula $h = 45t - 5t^2$ can be used to find the height h (in meters) of a pellet shot upward at an initial velocity of 45 m/s after t seconds. After how many seconds will the pellet be 20 m above the ground? Give answers to the nearest tenth of a second.   0.5 s, 8.5 s

---

Below are four steps for solving a quadratic equation. Put the steps in the most logical order.

1. Use the quadratic formula.   4
2. Write the equation in standard form.   1
3. Try factoring the equation.   3
4. Check to see whether the discriminant is negative.   2

Classroom Exercises

Which method would you use to solve each equation?   Answers may vary.

1. $x^2 = 49$ 7, -7
2. $x^2 - 5x + 6 = 0$ 3, 2
3. $(y - 6)^2 = 81$ 15, -3
4. $x^2 + 4x - 9 = 0$ $-2 \pm \sqrt{13}$
5. $(x + 2)^2 = 13$ $-2 \pm \sqrt{13}$
6. $3c^2 + 11c = 4$ $\frac{1}{3}, -4$
7. $-2a - 10 = -a^2$ $1 \pm \sqrt{11}$
8. $y^2 - 12y + 36 = 0$ 6
9. $2r^2 + 12r = -9$ $\frac{-6 \pm 3\sqrt{2}}{2}$

10–18. Solve each equation in Classroom Exercises 1–9.

Solutions near exercises above.

Written Exercises

9. $\frac{2}{3}, -\frac{5}{2}$ 17. $\frac{3 \pm \sqrt{7}}{2}$ 18. $\frac{1 \pm \sqrt{17}}{4}$

Solve each equation. Give irrational solutions in simplest radical form.

1. $x^2 = 64$ ± 8
2. $a^2 + 8a + 7 = 0$ $-1, -7$
3. $3x^2 + 7x + 3 = 0$ $\frac{-7 \pm \sqrt{13}}{6}$
4. $2x^2 - 3x - 4 = 0$ $\frac{3 \pm \sqrt{41}}{4}$
5. $2d^2 - 5d - 12 = 0$ $4, -\frac{3}{2}$
6. $(x - 2)^2 = 64$ 10, -6
7. $y^2 - 3y = 28$ 7, -4
8. $y^2 - 50 = 0$ $\pm 5\sqrt{2}$
9. $6x^2 + 11x - 10 = 0$
10. $4x^2 + x = 2$ $\frac{-1 \pm \sqrt{33}}{8}$
11. $r^2 - 8r + 16 = 0$ 4
12. $y^2 - 2y = 5$ $1 \pm \sqrt{6}$
13. $(y - 8)^2 = 17$ $8 \pm \sqrt{17}$
14. $-3 = y - 4y^2$ $1, -\frac{3}{4}$
15. $x^2 - 3x = 0$ 0, 3
16. $r^2 - 81 = 0$ 9, -9
17. $6x - 1 = 2x^2$
18. $2 = 2y^2 - y$
19. $4x^2 + 12x + 9 = 0$ $-\frac{3}{2}$
20. $5x + x^2 = 0$ 0, -5
21. $5c^2 + 2c - 1 = 0$ $\frac{-1 \pm \sqrt{6}}{5}$

Solve each equation. Approximate irrational solutions to the nearest tenth.

22. $-2x - 2 = -x^2$ 2.7, -0.7
23. $y^2 + 22 = 10y$ 6.7, 3.3
24. $25n^2 = 4,225$ 13, -13
25. $7 - x^2 = 4x + 4$ 0.6, -4.6
26. $12 - 4y^2 = 0$ 1.7, -1.7
27. $3y^2 + 2 = 5y + 6$ 2.3, -0.6

In the polygon with n sides, the number of diagonals, d, is given by the formula $d = \dfrac{n^2 - 3n}{2}$. Use this formula for Exercises 28–31.

28. Find the number of diagonals in a polygon with 8 sides.   20
29. Find the number of diagonals in a polygon with 27 sides.   324

30. If a polygon has 14 diagonals, how many sides does it have? 7

31. If a polygon has 35 diagonals, how many sides does it have? 10

The formula $h = 40t - 5t^2$ can be used to find the height in meters of an object shot upward at an initial velocity of 40 m/s after t seconds. Use this formula for Exercises 32–34. Approximate irrational answers to the nearest tenth of a second.

A baseball pitcher can throw a ball with an initial velocity of 40 m/s.

32. If the pitcher throws the ball straight up, when will its height be 20 m from where he threw it? 7.5 s, 0.5 s

33. When will the height of the ball be 80 m from where the pitcher threw it? 4 s

34. After how many seconds will the ball be at the height from which the pitcher threw it? 0 s, 8 s

The sum S of the first n positive integers is given by the formula $S = \dfrac{n^2 + n}{2}$.

For example, for the first 10 positive integers $(1, 2, 3, \cdots, 8, 9, 10)$,

$$S = \frac{10^2 + 10}{2} = \frac{100 + 10}{2} = \frac{110}{2} = 55.$$

Use this formula for Exercises 35–37.

35. Find the sum of the first 25 positive integers. 325

36. Starting with 1, how many consecutive integers must be used to obtain a sum of 136? 16

37. Starting with 1, how many consecutive integers must be used to obtain a sum of 210? 20

The formula $h = 60t - 5t^2$ can be used to find the height in meters of an object shot upward at an initial velocity of 60 m/s after t seconds. Use this formula for Exercises 38–41.

From the top of a building 10 m tall, a pellet is launched at an initial velocity of 60 m/s.

38. When will the pellet be 110 m above the ground? 2 s, 10 s

39. When will the pellet fall to the level of the top of the building? 12 s

40. When will the pellet be 190 m above the ground? 6 s

41. Why do you think there is only one answer to Exercise 40?

42. Keith wrote a quadratic equation that had $\sqrt{7}$ and $2\sqrt{7}$ as its roots. What was the equation? $x^2 - 3x\sqrt{7} + 14 = 0$

43. Eva wrote a quadratic equation that had $2\sqrt{11}$ as its only root. What was the equation? $x^2 - 4x\sqrt{11} + 44 = 0$

Guided Practice

Classroom Exercises 1–18
Try This all, FR all

Independent Practice

A Ex. 1–21, **B** Ex. 22–37, **C** Ex. 38–43

Basic: WE 1–21 odd, 28–31 all,
Midchapter Review, Application

Average: WE 11–27 odd, 32–37 all,
Midchapter Review, Application

Above Average: WE 17–27 odd, 35–43
all, Midchapter Review, Application

Midchapter Review

Give all irrational solutions in simplest radical form.

Solve each equation. *14.1*

1. $x^2 = 34$ $\pm\sqrt{34}$ 　　**2.** $(r-3)^2 = 64$ $11, -5$ 　　**3.** $6(y+2)^2 = 30$
$-2 \pm \sqrt{5}$

Complete the square. Show the result as the square of a binomial. *14.2*

4. $y^2 - 14y$ $(y-7)^2$ 　**5.** $x^2 + 24x$ $(x+12)^2$ 　**6.** $n^2 + \frac{1}{2}n$ $\left(n+\frac{1}{4}\right)^2$

Solve by completing the square. *14.2*

7. $x^2 + 6x + 7 = 0$ $-3 \pm \sqrt{2}$ 　**8.** $y^2 - 8y - 48 = 0$ $12, -4$ 　**9.** $5x^2 = 2x + 1$
$\frac{1 \pm \sqrt{6}}{5}$

Solve by using the quadratic formula. *14.3*

10. $x^2 - 9x + 14 = 0$ $7, 2$ 　**11.** $2y^2 - 2y - 4 = 0$ $2, -1$ 　**12.** $x^2 - 5x - 2 = 0$
$\frac{5 \pm \sqrt{33}}{2}$

13. $5 = c^2 + 3c$ $\frac{-3 \pm \sqrt{29}}{2}$ 　**14.** $y^2 - 4y = 6$ $2 \pm \sqrt{10}$ 　**15.** $-x^2 - 4 = 3x$
No real number solution

Solve each equation. *14.4*

16. $x^2 - 14 = 0$ $\pm\sqrt{14}$ 　**17.** $a^2 - 4a - 45 = 0$ $9, -5$ 　**18.** $3y^2 - 2y = 1$ $1, -\frac{1}{3}$

19. $r^2 + 7r = 0$ $0, -7$ 　**20.** $x^2 - 8x = 5$ $4 \pm \sqrt{21}$ 　**21.** $2c^2 = -6c - 1$
$\frac{-3 \pm \sqrt{7}}{2}$

Solve. *14.1, 14.2*

22. If the length of each side of a square is doubled, then the area is 144 cm². Find the length of each side of the original square. 6 cm

23. The square of a number is 9 more than 5 times the number. Find the number. $\frac{5 \pm \sqrt{61}}{2}$

Application: *Stopping Distance for a Car*

You can estimate the stopping distance, s, needed for a car going x mph by adding the reaction distance (x feet) and the braking distance $\left(\frac{x^2}{20}\right)$.

$$s = x + \frac{x^2}{20}$$

1. You are driving at 30 mph and you see an animal in the road. How many feet will the car travel before you can stop? 75 ft

2. A police officer estimates the stopping distance for a car as 206.25 ft. Is the car traveling at a speed greater than, or less than, 60 mph?
less than

Enrichment

Remind students (from the Enrichment for Lesson 14.3) that the sum of the roots of a quadratic equation is $\frac{-b}{a}$ and the product is $\frac{c}{a}$. Have the students write a quadratic equation whose roots have a sum of $\frac{2}{3}$ and a product of $\frac{-3}{2}$, and then solve the equation to check for this sum and product. $6x^2 - 4x - 9$; $\frac{2 \pm \sqrt{58}}{6}$

14.5 Problem Solving: Quadratic Equations and Geometry

Objective To solve geometric problems that lead to quadratic equations

You are now able to choose from four methods when you solve problems that lead to quadratic equations.

EXAMPLE 1 The rectangular floor of a tree house is constructed so that its perimeter is 42 ft. and its area is 104 ft². Find the width and the length of the floor.

Plan Draw a diagram and represent the data. If you need to solve a quadratic equation, try factoring first. If that is not successful, use the quadratic formula.

Solution Let w = the width and l = the length.
The perimeter formula is $p = 2l + 2w$.
$$42 = 2l + 2w$$
$$21 = l + w$$
So, $l = 21 - w$.

$$A = lw$$
$$104 = (21 - w)w \quad \longleftarrow \text{The area is 104 ft}^2.$$
$$104 = 21w - w^2$$
$$w^2 - 21w + 104 = 0$$
$$(w - 13)(w - 8) = 0$$
$$w = 13 \text{ or } w = 8 \quad \longleftarrow \text{Width: 13 ft or 8 ft}$$

If $w = 13$, $l = 21 - 13$, or 8.
If $w = 8$, $l = 21 - 8$, or 13.

Check Is the perimeter 42 ft? Yes, since $2 \cdot 13 + 2 \cdot 8 = 26 + 16 = 42$.
Is the area 104 ft²? Yes, since $13 \cdot 8 = 104$.
Thus, the width of the floor is 8 ft and the length is 13 ft

TRY THIS 1. The perimeter of a rectangular garden is 80 m and its area is 76 m². Find the width and length. width: 2 m; length: 38 m

Teaching Resources

Quick Quizzes 110
Reteaching and Practice Worksheets 110

■■■ GETTING STARTED

Prerequisite Quiz

Match the numbers in the first list with the letters in the second list.

1. area of a rectangle   e
2. perimeter of a rectangle   c
3. area of a triangle   g
4. perimeter of a triangle   f
5. area of a square   d
6. perimeter of a square   b
7. Pythagorean Theorem   a

a. $a^2 + b^2 = c^2$ **b.** $4s$
c. $2l + 2w$ **d.** s^2
e. lw **f.** $a + b + c$
g. $\frac{1}{2}bh$

Motivator

Have students sketch a square with one of its diagonals on a piece of paper. Tell them that the length of the diagonal is 10 units. Then have them suggest a method for finding the perimeter of the square.   Use the Pythagorean Theorem to find the length of a side. Then multiply the length by 4 to find the perimeter.

Additional Example 1

The area of a rectangular bulletin board is 21 ft² and its perimeter is 20 ft. Find the length and the width of the bulletin board.
$l = 7$ ft, $w = 3$ ft

Highlighting the Standards

Standards 4c, 1b: This lesson focuses on problem solving, and connects algebra with geometry. The Problem Solving Strategies section teaches the strategy of checking assumptions.

553

Lesson Note

Point out to students that these problems are solved by using one variable rather than two. Remind them to draw a diagram and label it carefully for each problem. In Example 2, students could use a calculator to find $\sqrt{388}$. Remind them that a decimal solution is usually needed for a practical application.

Math Connections

Geometry: In a golden rectangle the length is the geometric mean between the width and the sum of the length and the width.
$$\frac{w}{l} = \frac{l}{l+w}$$

Critical Thinking Questions

Application: Ask students to find the dimensions and area of the largest possible rectangular flower garden that could be enclosed by 24 feet of fencing.   A 6 by 6 square; area = 36 ft^2 Then ask if they can show that this is the maximum area possible.   Possible answer: Let the width = x and the length = $12 - x$. Then, $x(12 - x) = A$. Solving for x in terms of A, $x = 6 \pm \sqrt{36 - A}$. From this it is clear that A could not be greater than 36. The maximum area is 36 cm^2.

Common Error Analysis

Error: Some students may set the length plus the width of a rectangle equal to the perimeter.

Have them make a sketch of each figure to help avoid this type of error.

Some applications of the Pythagorean Theorem lead to quadratic equations. When these applications involve geometric figures, it is important to *draw a diagram* to represent the given information as you begin to solve each problem.

EXAMPLE 2   The distance between two opposite corners of a rectangular garden is 14 m. The length of the garden is 2 m longer than the width. Find the length and the width of the garden.

Solution

Let w = the width.
Then $w + 2$ = the length.
Use the Pythagorean Theorem.

$$a^2 + b^2 = c^2$$
$$w^2 + (w + 2)^2 = 14^2$$
$$w^2 + w^2 + 4w + 4 = 196$$
$$2w^2 + 4w - 192 = 0 \quad \longleftarrow \text{Divide each side by 2.}$$
$$w^2 + 2w - 96 = 0$$
$$w = \frac{-2 \pm \sqrt{2^2 - 4 \cdot 1 \cdot (-96)}}{2 \cdot 1} \quad \longleftarrow \begin{array}{l}\text{Use the quadratic formula.} \\ a = 1, b = 2, c = -96\end{array}$$
$$w = \frac{-2 \pm \sqrt{4 + 384}}{2}$$
$$w = \frac{-2 \pm \sqrt{388}}{2}$$
$$w = \frac{-2 \pm 2\sqrt{97}}{2} \quad \longleftarrow \sqrt{388} = \sqrt{4} \cdot \sqrt{97} = 2\sqrt{97}$$
$$w = -1 \pm \sqrt{97} \quad \longleftarrow \text{Reject the negative solution.}$$
$$w = -1 + 9.8 \quad \longleftarrow \sqrt{97} \approx 9.849 \approx 9.8$$
$$w = 8.8 \text{ and } w + 2 = 10.8$$

Calculator check: (Check that $\sqrt{8.8^2 + 10.8^2} \approx 14$.)

8.8 ⌈x^2⌉ ⌈+⌉ 10.8 ⌈x^2⌉ ⌈=⌉ ⌈$\sqrt{x}$⌉   display ⟶ 13.931259 ≈ 14

Thus, the length is about 10.8 m, and the width is about 8.8 m.

TRY THIS

2. The diagonal of a rectangle is 8 mm. The length is 5 mm longer than the width. Find the width and the length. 2.6 mm, 7.6 mm

3. The distance between the opposite corners of a rectangular pool is 12 yd. The length of the pool is 4 yd longer than the width. Find the width and the length. 6.2 yd, 10.2 yd

Additional Example 2

The distance between two opposite corners of a rectangular game court is 20 m. The length of the court is 1 m longer than the width. Find the length and the width of the game court. $l \approx 14.6$ m, $w \approx 13.6$ m

Classroom Exercises

Use the data from each figure to write an equation in one variable.

1.
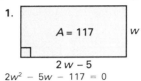
$A = 117$, w

2w − 5

$2w^2 − 5w − 117 = 0$

2.
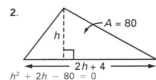
h, $A = 80$

2h + 4

$h^2 + 2h − 80 = 0$

3.
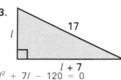
17, l

l + 7

$l^2 + 7l − 120 = 0$

4–6. Solve the equations in Classroom Exercises 1–3. **4.** $w = 9$ **5.** $h = 8$ **6.** $l = 8$

Solve each problem.

7. Opposite corners of a rectangle are 30 ft apart. The length is 6 ft longer than the width. Find the length and the width. $l = 24$ ft, $w = 18$ ft

8. The area of a rectangle is 72 ft² and its perimeter is 34 ft. Find the length and the width of the rectangle. $l = 9$ ft, $w = 8$ ft

Written Exercises

Solve each problem.

1. Jaime's bedroom floor is the shape of a rectangle. The distance between two opposite corners is 15 ft. The length is 3 ft longer than the width. Find the length and the width of the floor. $l = 12$ ft; $w = 9$ ft

2. Opposite sides of a square are increased by 3 cm. The other sides are increased by 2 cm. The area of the resulting rectangle is 72 cm². Find the length of a side of the original square. 6 cm

3. The perimeter of a rectangular floor is 38 ft and area of the floor is 90 ft². Find the length and the width of the floor. $l = 10$ ft, $w = 9$ ft

4. The number of square centimeters in the area of a square is 42 more than the number of centimeters in the length of each side. Find the length of each side. 7 cm

5. The distance between opposite corners of a rectangular deck is 25 ft. The length is 3 ft shorter than 3 times the width. Find the length and the width of the deck. $l = 23.4$ ft, $w = 8.8$ ft

6. A patio is in the shape of a triangle. The area of the patio is 28.5 m². The base of the triangle is 2.5 m less than twice the height. Find the base and the height. $h = 6$ m, $b = 9.5$ m

7. Refer to Exercises 1–6 to give an example of a number that is a solution to an equation used to solve a problem, even though it is not a solution to that problem. Explain in writing why the number is not a solution to the problem.

Checkpoint

1. Jill's room has the shape of a square. If one side of the square is increased by 5 ft and an adjacent side is decreased by 3 ft, the area of the resulting rectangular room would be 105 ft². Find the length of each side of the square–shaped room. 10 ft

2. The distance between two opposite corners of a rectangular lot is 13 ft. The length is 2 ft longer than twice the width. Find the length and the width of the lot. 12 ft, 5 ft

Closure

Ask students to give three examples of word problems related to geometry that can be solved by using a quadratic equation. Answers will vary.

▬▬ FOLLOW UP

Guided Practice

Classroom Exercises 1–8
Try This all

Independent Practice

A Ex. 1–7, **B** Ex. 8–12 **C** Ex. 13–14

Basic: WE 1–7 all, Problem Solving Strategies

Average: WE 1–9 odd, 10, 11, Problem Solving Strategies

Above Average: WE 3–13 odd, 14, Problem Solving Strategies

Additional Answers

Written Exercises

7. In Exercise 1, the quadratic equation $w^2 + 3w - 108 = 0$ has two solutions, 9 and -12. Since the width of a rectangle cannot be negative, -12 is rejected as a solution.

Mixed Review

1–4.

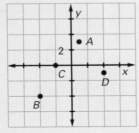

5.

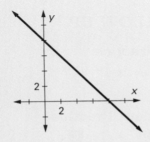

6.

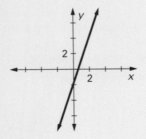

Solve. Give irrational answers to the nearest tenth.

8. The length of a rectangular piece of cloth is 8 cm more than the width. The area is 50 cm². Find the length and the width. l = 12.1 cm, w = 4.1 cm

9. A windowpane is in the shape of a right triangle. The area of the glass is 4 ft². One leg of the triangle measures 1 ft more than 3 times the other leg. Find the length of each leg. 1.5 ft, 5.5 ft

10. A wire 28 cm long can be bent to form a rectangle with an area of 42 cm². Find the length and the width of the rectangle. l = 9.6 cm w = 4.4 cm

11. A rectangular plot of ground measures 14 m by 10 m. A path is to be constructed within the plot so that a 72 m² garden can be planted with the path surrounding it. Find the width of the path. 1.6 m

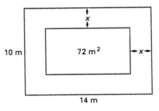

12. The dimensions of a rectangle are 16 ft by 12 ft. If each dimension of the rectangle is decreased by the same amount, the area of the resulting rectangle is 32 ft². Find the length and the width of the resulting rectangle. l = 8 ft, w = 4 ft

13. The diagonal of a square piece of glass is 5 in. longer than the length of a side. Find the length of a side. 12.1 in

14. A ladder is leaning against a tree so that it touches the tree 8 ft above the ground and touches the ground 6 ft from the base of the tree. How much should the ladder be lowered so that the distance from the top of the ladder to the ground will equal the distance from the bottom of the ladder to the base of the tree? 0.9 ft

Mixed Review

Graph each point in the same coordinate plane. **10.1**

1. $A(1, 3)$
2. $B(-4, -4)$
3. $C(-2, 0)$
4. $D(4, -1)$

Graph each equation. **10.5, 11.3**

5. $x + y = 8$
6. $y = 3x - 2$

Solve. **4.7**

7. 48 is what percent of 150? 32%
8. 96 is 120% of what number? 80
9. y varies inversely as x. y is 8 when x is 3. Find y when x is -2. **10.7** -12

Simplify. **13.2**

10. $\sqrt{98}$ $7\sqrt{2}$
11. $\sqrt{196}$ 14
12. $8\sqrt{12}$ $16\sqrt{3}$
13. $-2\sqrt{32}$ $-8\sqrt{2}$
14. $\sqrt{5} \cdot \sqrt{2}$ $\sqrt{10}$
15. $\sqrt{10} \cdot \sqrt{20}$ $10\sqrt{2}$
16. $\sqrt{50} \cdot \sqrt{2}$ 10
17. $\sqrt{7} \cdot \sqrt{7}$ 7

Enrichment

Have students solve the following puzzle.

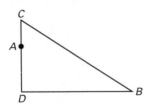

Two motorists start at point A in the diagram at the left. One travels 6 miles to D and another 15 miles to B. The other motorist travels in the opposite direction by going to point C and then to point B. The distance either way is the same. Find the distance from A to C. $\frac{10}{3}$ mi.

Checking Assumptions

Carla and Joann have two large identical buckets. Carla fills hers with glass marbles that are $\frac{1}{2}$ inch in diameter. Joann fills hers with larger marbles that are 1 inch in diameter. Whose bucket has the most marbles in it?

Then Carla and Joann each fill their buckets (with the marbles in them) with water to the brim. Whose bucket has more water in it? Whose bucket has more glass in it?

Many people assume that one bucket will have more water (or glass) than the other. This is not so. In theory, the proportion of water to glass is the same in both buckets.

Mathematics provides a way to check out some assumptions such as this one to determine the truth of what appears to be so.

Exercises

Use the strategy of Checking Assumptions to solve these problems.

1. Betsy uses a square piece of paper, 11 inches on a side, to paint the design on the left. Rafael uses paper of the same size to paint the design on the right. Whose paper had more of its area covered with paint? Both have the same area.

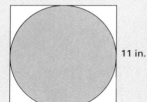 11 in.

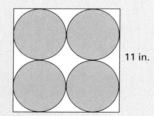 11 in.

2. Bob sold half his hamsters, plus an additional one-half a hamster to George. Then Bob sold half of the hamsters that remained plus one-half a hamster to Joe. Bob had exactly one hamster left. Did Bob have to divide a hamster in half? No

3. How many hamsters did Bob have to start with? 7

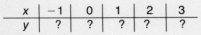
▰▰▰ GETTING STARTED

Prerequisite Quiz

Complete the charts for the given functions.

1. $y = x^2$

x	−2	−1	0	1	2
y	?	?	?	?	?

4, 1, 0, 1, 4

2. $y = x^2 − 2x + 1$

x	−1	0	1	2	3
y	?	?	?	?	?

4, 1, 0, 1, 4

Additional Answers

Try This

1.

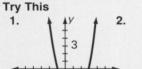

2.

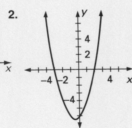

3.

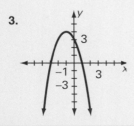

14.6 Quadratic Functions

Objectives

To graph quadratic functions
To find the coordinates of the vertex and the equation of the axis of symmetry of a parabola
To find the minimum or maximum value of a quadratic function

Below are the graphs of two functions with domain $\{-2, -1, 0, 1, 2\}$.

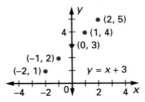
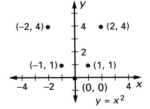

The equation $y = x + 3$ is a linear equation. The equation $y = x^2$ is a quadratic equation.

An equation of the form $y = ax^2 + bx + c$ $(a \neq 0)$ defines a **quadratic function**. To graph a quadratic function, select several values of x, find the corresponding values of y, graph the ordered pairs, and draw a smooth curve through the points. In all examples in this lesson, use the set of real numbers as the domain of x.

EXAMPLE 1   Graph $y = x^2 + 2x - 3$.

Plan   Make a table of values. Then graph the ordered pairs and draw the curve.

Solution

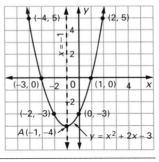

x	$x^2 + 2x - 3$	y	Ordered pairs
0	$0^2 + 2 \cdot 0 - 3$	−3	(0, −3)
1	$1^2 + 2 \cdot 1 - 3$	0	(1, 0)
2	$2^2 + 2 \cdot 2 - 3$	5	(2, 5)
−1	$(-1)^2 + 2(-1) - 3$	−4	(−1, −4)
−2	$(-2)^2 + 2(-2) - 3$	−3	(−2, −3)
−3	$(-3)^2 + 2(-3) - 3$	0	(−3, 0)
−4	$(-4)^2 + 2(-4) - 3$	5	(−4, 5)

TRY THIS   1. Graph $y = x^2 - 2x - 3$.   2. Graph $y = x^2 + x - 6$.

Highlighting the Standards

Standards 5b, 6a: Students investigate the graph of a parabola and relate it to applications outside the classroom.

Additional Example 1

Graph $y = x^2 - x - 2$.

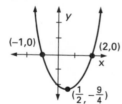

Any equation of the form $y = ax^2 + bx + c$ $(a \neq 0)$ has a graph that is a **parabola**. In Example 1, the parabola opens upward. The point $(-1, -4)$ is the **vertex**, or the *turning point*, of the parabola. The y-coordinate of the vertex, -4, is the **minimum value** (least value) of the function. Thus, the vertex A is called a **minimum point**.

Notice that the parabola described by $y = x^2 + 2x - 3$ is symmetric with respect to the dashed vertical line that contains the vertex, $(-1, -4)$. This line is called the **axis of symmetry** of the parabola and its equation is $x = -1$.

EXAMPLE 2   Graph $y = -x^2 + 6x - 5$.

Solution

x	y	Ordered pair
0	-5	$(0, -5)$
1	0	$(1, 0)$
2	3	$(2, 3)$
3	4	$(3, 4)$
4	3	$(4, 3)$
5	0	$(5, 0)$
6	-5	$(6, -5)$

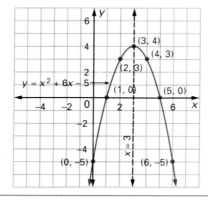

TRY THIS   3. Graph $y = -x^2 - 2x + 3$.

For the function in Example 2, $y = -x^2 + 6x - 5$, the coefficient of x^2 is a negative number, -1. This causes the parabola to open downward. The vertex $(3, 4)$ is a **maximum point**, and the y-coordinate of the vertex, 4, is the **maximum value** (greatest value) of the function. The axis of symmetry with equation $x = 3$ contains the vertex, $(3, 4)$.

The graph of a quadratic function $y = ax^2 + bx + c$ is made up of pairs of points with the same y-coordinate (except for the vertex). In Example 2, notice that pairs of points such as $(2, 3)$, $(4, 3)$, and $(1, 0)$, $(5, 0)$ have the same y-coordinate.

The axis of symmetry contains the vertex and lies halfway between any two points with the same y-coordinate. Thus, the x-coordinate of the vertex is the *arithmetic mean*, or average, of the x-coordinates of any two points with the same y-coordinate.

14.6 Quadratic Functions   **559**

Additional Example 2

Graph $y = -x^2 + 2x + 3$.

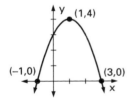

Critical Thinking Questions

Synthesis: To develop the formula for the x-coordinate of the vertex of a parabola, have students consider a pair of points with a y-value of zero. Substitute 0 for y in the quadratic function $y = ax^2 + bx + c$. The result is the standard form of a quadratic equation, $0 = ax^2 + bx + c$. Then have them find the average (arithmetic mean) of the two solutions for this equation.

Common Error Analysis

Error: Some students may give the x-coordinate of the vertex as the maximum or minimum value of the function.

Remind students that the value of the function is represented by the variable y.

Additional Answers

Try This

4.

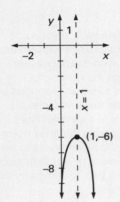

(1, −6)

Pairs of points	Mean of x-coordinates
(1, 0), (5, 0)	$\frac{1}{2}(1 + 5) = \frac{1}{2} \cdot 6 = 3$
(2, 3), (4, 3)	$\frac{1}{2}(2 + 4) = \frac{1}{2} \cdot 6 = 3$

x-coordinate of vertex

The values of a and b in $y = ax^2 + bx + c$ can be used to find the x-coordinate of the vertex of its graph as well as the equation of the axis of symmetry of its graph.

x-coordinate of the vertex: $\quad -\dfrac{b}{2a}$

equation of the axis of symmetry: $\quad x = -\dfrac{b}{2a}$

EXAMPLE 3 For the function $y = 2x^2 - 8x + 5$:
a. find the coordinates of the vertex of the parabola,
b. find the equation of the axis of symmetry,
c. find the minimum or maximum value, and
d. graph the function.

Solution

$a = 2, b = -8$

$x\text{-coordinate of vertex} = -\dfrac{b}{2a} = -\left(\dfrac{-8}{2 \cdot 2}\right) = -\left(\dfrac{-8}{4}\right) = 2$

Substitute 2 for x and solve for y.
$\quad y = 2x^2 - 8x + 5$
$\quad y = 2 \cdot 2^2 - 8 \cdot 2 + 5$
$\quad y = -3$

Thus, the coordinates of the vertex are $(2, -3)$. The equation of the axis of symmetry is $x = 2$. Since $a = 2$ and $2 > 0$, the parabola opens upward, and the minimum value is -3.

x	y	Ordered pairs
2	−3	(2, −3)
3	−1	(3, −1)
4	5	(4, 5)
1	−1	(1, −1)
0	5	(0, 5)

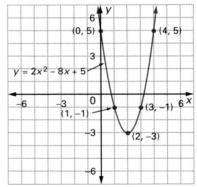

TRY THIS 4. For the function $y = -3x^2 + 6x - 9$, find the coordinates of the vertex, the equation of the axis of symmetry, the maximum or minimum value, and graph the function.

vertex: $(1, -6)$; axis of symmetry: $x = 1$; max. value: -6

Additional Example 3

For the function $y = 2x^2 + 12x + 16$:

a. Find the coordinates of the vertex. $(-3, -2)$
b. Find the equation of the axis of symmetry. $x = -3$
c. Find the maximum or the minimum value. minimum $= -2$
d. Graph the function.

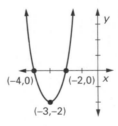

(−4,0) (−2,0) (−3,−2)

EXAMPLE 4   Mrs. Truro sells packages of towels. She uses the following formula to approximate her profits per day.

$$p = -x^2 + 50x - 350$$

In the formula, p is the profit derived from selling x packages of towels. How many packages of towels must she sell to make a maximum profit? What is the maximum profit?

Plan   Use the characteristics of the graphs of quadratic functions. The graph of $p = -x^2 + 50x - 350$ will have a maximum since the coefficient of x^2 is negative. Use the formula $x = -\frac{b}{2a}$ to find the value of x that makes p a maximum.

Solution $a = -1, b = 50, c = -350$

$$x = -\frac{b}{2a} = \frac{-50}{2(-1)} = \frac{-50}{-2} = 25 \longleftarrow \begin{array}{l}\text{number of packages per day}\\ \text{to give maximum profit}\end{array}$$

To find the maximum profit, let $x = 25$ in $p = -x^2 + 50x - 350$.

$p = -1x^2 + 50x - 350$
$p = -1(25)^2 + 50 \cdot 25 - 350$
$p = -625 + 1,250 - 350$
$p = 275$

Thus, she must sell 25 packages per day to make a maximum profit of $275.

TRY THIS   5. Use the formula $p = x^2 - 14x + 89$ to find how many packages of towels must be sold to make a minimum profit and to find the minimum profit.   7 packages; $40

Classroom Exercises

Tell whether the function has a maximum or a minimum value.

1. $y = -x^2 + 5x + 2$ Max
2. $y = 3x^2 + 7x - 1$ Min
3. $y = 12 - x^2$ Max
4. $2 + 3x^2 = y$ Min
5. $y = x^2 + x$ Min
6. $2x - 4x^2 + 1 = y$ Max

For each quadratic function: (a) find the coordinates of the vertex of the parabola, (b) find the equation of the axis of symmetry, (c) find the minimum or maximum value, and (d) graph the function.

7. $y = 3x^2$
(0, 0); $x = 0$; min: 0

8. $y = -x^2 + 2$
(0, 2); $x = 0$; max: 2

9. $y = x^2 - 5x + 4$
$\left(\frac{5}{2}, -\frac{9}{4}\right)$; $x = \frac{5}{2}$; min: $-\frac{9}{4}$

14.6 Quadratic Functions   **561**

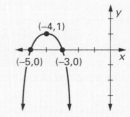
Additional Example 4

Mrs. Jackson makes bead necklaces to sell at craft fairs. She uses the following formula to estimate her profits per day:
$p = -x^2 + 10x + 40$, where p is the profit for selling x necklaces. How many necklaces must she sell to make a maximum profit? What is the maximum profit?
5 necklaces; $65 maximum profit

Closure

Ask students the following questions.

1. What is the shape of the graph of any equation of the form $y = ax^2 + bx + c$? parabola
2. What is the turning point called? vertex
3. What expression gives the x-coordinate of the vertex? $\frac{-b}{2a}$
4. What does the y-coordinate of the vertex indicate? A minimum or maximum value of the function
5. What determines whether the function has a minimum or a maximum value? $a > 0$ indicates that the function has a minimum value; $a < 0$ indicates it has a maximum value.
6. What is the equation of the axis of symmetry? $x = \frac{-b}{2a}$

■■■ FOLLOW UP

Guided Practice

Classroom Exercises 1–9
Try This   all

Independent Practice

A Ex. 1–18, **B** Ex. 19–21, **C** Ex. 22–28

Basic: WE 1–17 odd

Average: WE 7–17 odd, 19–21 all

Above Average: WE 13–19 odd, 20–24 all, 28

Written Exercises

Graph each function.

1. $y = x^2$
2. $y = -2x^2$
3. $y = \frac{1}{2}x^2$
4. $y = -x^2 + 1$
5. $y = 3x^2 + 2x$
6. $y = x^2 + 2x + 4$

For each quadratic function, (a) find the coordinates of the vertex of the parabola, (b) find the equation of the axis of symmetry, (c) find the minimum or maximum value, and (d) graph the function.

7. $y = 3x^2 + 2$
8. $y = -x^2 - 6x - 8$
9. $y = x^2 - 4x + 3$
10. $y = x^2 - 8x + 7$
11. $y = \frac{1}{2}x^2 - x - 2$
12. $y = -4x^2 + 16x - 6$

13. The formula for the cost of running a taco stand is $c = x^2 - 12x + 60$. How many units x of tacos must be sold to keep costs at a minimum? Find the minimum cost. 6; 24

14. The cost c of producing x units of radios is $c = x^2 - 12x + 72$. How many radios should be made to produce the minimum cost? What is the minimum cost? 6; 36   400; $160, 200

15. The formula for the height s reached by a rocket fired straight up from the ground with an initial velocity of 96 ft/s is $s = -16t^2 + 96t$. Find the time t for the rocket to reach a maximum height. Find this height. 3 s; 144 ft

16. The owner of a new company finds that the profit p is related to the number x of items sold by $p = 200 + 800x - x^2$. How many items must be sold for the maximum profit? What is the maximum profit?

17. A hair dryer manufacturer determines that the total profit p (in dollars) of manufacturing x hair dryers is $p = -0.18x^2 + 36x + 4,000$. How many must be sold for the maximum profit? 100

18. A biologist's formula to predict the number of impulses fired after stimulation of a nerve is $i = -x^2 + 30x - 50$, where i is the number of impulses per millisecond and x is the number of milliseconds after stimulation. Find the time for the maximum number of impulses. 15 milliseconds

19. Graph each function on the same set of axes.
   a. $y = x^2$
   b. $y = x^2 + 2$
   c. $y = x^2 - 3$
   d. For the function $y = x^2 + k$, where k is constant, explain what happens to the resulting parabolas when k increases. The parabolas move up.
   e. For the function $y = x^2 + k$, where k is a constant, give the coordinates of the vertex of the parabola. Give the minimum value. (0,k); k

20. Graph each function on the same set of axes.
   a. $y = x^2$
   b. $y = (x - 1)^2$
   c. $y = (x + 3)^2$
   d. For the function $y = (x - h)^2$, where h is a constant, explain what happens to resulting parabolas as h increases. The parabolas move right.
   e. For the function $y = (x - h)^2$, where h is a constant, give the coordinates of the vertex of the parabola and the minimum value. (h,0); 0

21. Graph each function on the same set of axes.
    a. $y = (x + 2)^2 - 3$ **b.** $y = (x - 1)^2 - 1$ **c.** $y = (x - 3)^2 + 2$
    d. For the function $y = (x - h)^2 + k$, where h and k are constants, give the coordinates of the vertex. Give the minimum value. $(h, k); k$

Consider the quadratic function $y = ax^2 + bx + c$ $(a \neq 0)$ for Exercises 22–23.

22. Give the coordinates of the vertex of the parabola. $-\frac{b}{2a}, -\frac{b^2}{4a} + c$
23. Give the maximum or minimum value. $-\frac{b^2}{4a} + c$
24. Use your answer to Exercise 23 to give the maximum value of
    $y = 4 - \frac{1}{2}x^2 + x$. $4\frac{1}{2}$

Refer to the relation $x = y^2$ for Exercises 25–27.

25. Make a table of values and graph the relation.
26. Do any two ordered pairs have the same first coordinate? If your answer is yes, give an example to support it. Yes; (4, 2) and (4, −2)
27. Is the equation a function? Explain your answer. No; more than one y-value for a given x-value.
28. Write an inequality to describe each shaded region below.

a.
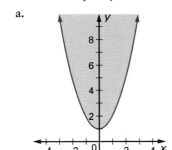
$y \geq x^2 + 1$

b.
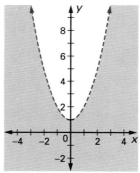
$y < x^2 + 1$

Mixed Review

Solve. *3.3, 3.4*

1. $-2x + 5 = -7$ $x = 6$
2. $8n + 19 = -4$ $n = -2\frac{7}{8}$
3. $y + 4 = 7y + 5 - 10y$ $y = \frac{1}{4}$

Graph each equation. Is the relation a function? *10.5*

4. $y = 3$ Yes
5. $x = 4$ No
6. $x - y = 3$ Yes
7. Find the slope of the line determined by $A(-2, 3)$ and $B(5, -1)$. *11.1* $-\frac{4}{7}$
8. For the relation $\{(-3, 1), (6, 1), (3, -1), (4, 2)\}$, give the domain and range. Is the relation a function? *10.2* $D = \{-3, 3, 4, 6,\}; R = \{-1, 1, 2\};$ yes

14.6 Quadratic Functions **563**

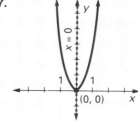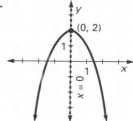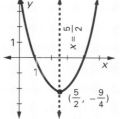
Enrichment

If a farmer harvests his melons today, he will average 5 melons per plant, and he will be paid $1/melon at the market. For every week he waits to harvest the melons, he will average an additional 2 melons per plant, but will receive 20¢ less for each melon at the market.

1. Write an equation for profit p in terms of the number x of weeks that the farmer waits to harvest the melons.
   $p = (5 + 2x)(1 - 0.2x)$
2. About how long should the farmer wait to harvest the melons in order to make a maximum profit per plant?
   About $1\frac{1}{4}$ weeks, or about 9 days
3. What will be his average profit per plant? About $5.63

■■■ GETTING STARTED

Prerequisite Quiz

Solve by using the quadratic formula.

1. $x^2 + 6x + 9 = 0$ -3
2. $x^2 - 11x + 18 = 0$ $9, 2$
3. $5x^2 + x + 7 = 0$ $\varnothing$
4. $3x^2 - 2x - 4 = 0$ $\dfrac{1 \pm \sqrt{13}}{3}$

Motivator

Remind students that every non-horizontal line has an x-intercept. Ask them how they can determine where the graph of $3x + 2y = 15$ crosses the x-axis without having to graph the line. Set $y = 0$ and solve for x.   Ask them how they might find the x-intercepts, if they exist, of the graph of $y = x^2 - 2x - 3$.   Set $y = 0$ and solve for x.   Ask them to find the x intercepts. $-1, 3$ Ask them how they might use the x-intercepts to find the vertex of the parabola. x-coordinate: Find the average of the x-intercepts; y-coordinate: Substitute this value into the equation and solve for y. Ask them to find the coordinates of the vertex. $(1, -4)$

14.7 Quadratic Functions and the Discriminant

Objectives   To find the x-intercepts of a quadratic function

To determine the number of real-number solutions of a quadratic equation by using the discriminant

You have learned that the quadratic equation $x^2 - 4x + 3 = 0$ is related to the quadratic function $y = x^2 - 4x + 3$. The solutions of the equation can be found by factoring.

$$x^2 - 4x + 3 = 0$$
$$(x - 3)(x - 1) = 0$$
$$x = 3 \ or \ x = 1$$

Notice that the graph of the function $y = x^2 - 4x + 3$ crosses the x-axis at points $(3, 0)$ and $(1, 0)$. The x-coordinates of these two ordered pairs are the two solutions of the equation $x^2 - 4x + 3 = 0$.

Recall that the x-coordinate of a point where a graph crosses the x-axis is called an x-*intercept* of the graph. Thus, the x-intercepts of the parabola, 3 and 1, are the solutions of the equation.

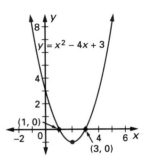

EXAMPLE 1   Find the x-intercepts of the graph of $y = 2x^2 - 3x - 5$.

Plan   Find the solutions of $2x^2 - 3x - 5 = 0$.
Try factoring.

Solution
$$2x^2 - 3x - 5 = 0$$
$$(2x - 5)(x + 1) = 0$$
$$x = \tfrac{5}{2} \ or \ x = -1$$

Thus, the x-intercepts are $\tfrac{5}{2}$ and -1.

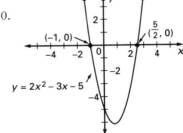

TRY THIS   1. Find the x-intercepts of the graph of $y = -3x^2 + 14x - 8$. $\tfrac{2}{3}, 4$

564   Chapter 14 Quadratic Equations and Functions

Additional Example 1

Find the x-intercepts of the graph of
$y = 3x^2 + 11x - 4$. $\tfrac{1}{3}, -4$

Examine the relationships in the table below.

Function	$y = x^2 - 6x + 5$	$y = x^2 - 6x + 9$	$y = x^2 - 6x + 11$
Graph			
Number of x-intercepts	2	1	0
Related equation	$x^2 - 6x + 5 = 0$	$x^2 - 6x + 9 = 0$	$x^2 - 6x + 11 = 0$
Discriminant $b^2 - 4ac$	$(-6)^2 - 4 \cdot 1 \cdot 5$ $= 36 - 20$ $= 16$ (positive)	$(-6)^2 - 4 \cdot 1 \cdot 9$ $= 36 - 36$ $= 0$ (zero)	$(-6)^2 - 4 \cdot 1 \cdot 11$ $= 36 - 44$ $= -8$ (negative)
Number of real number solutions	2	1	0

The relationship between the discriminant and the number of real number solutions of a quadratic equation is summarized below.

If $b^2 - 4ac > 0$, there are two solutions. They are $\dfrac{-b \pm \sqrt{b^2 - 4ac}}{2a}$.

If $b^2 - 4ac = 0$, there is one solution. The solution is $-\dfrac{b}{2a}$.

If $b^2 - 4ac < 0$, there is no real number solution.

EXAMPLE 2 Use the discriminant to determine the number of real number solutions.
a. $x^2 - 7x - 2 = 0$ **b.** $5x^2 + 4x + 1 = 0$ **c.** $-x^2 + 10x - 25 = 0$

Solutions
a. $b^2 - 4ac$
$= (-7)^2 - 4 \cdot 1 \cdot (-2)$
$= 49 + 8$
$= 57 \leftarrow$ positive
Two solutions

b. $b^2 - 4ac$
$= 4^2 - 4 \cdot 5 \cdot 1$
$= 16 - 20$
$= -4 \leftarrow$ negative
No solution

c. $b^2 - 4ac$
$= 10^2 - 4(-1)(-25)$
$= 100 - 100$
$= 0 \leftarrow$ zero
One solution

TRY THIS 2. Use the discriminant to determine the number of real number solutions of $-4x^2 - 3x + 6 = 0$. two solutions

■■■ **TEACHING SUGGESTIONS**

Lesson Note

Emphasize the difference between $y = 2x^2 - 3x - 5$, which is true for an infinite number of ordered pairs (x, y), which lie on a parabola, and $2x^2 - 3x - 5 = 0$, which is true for only two values of x. You may want to point out that the discriminant also reveals whether real solutions are rational or irrational, depending on whether it is, or is not, a perfect square.

Math Connections

Vocabulary: To discriminate means to distinguish or recognize as distinct. The discriminant in the quadratic formula can be used to distinguish or differentiate between the number and type of roots of a quadratic equation.

Critical Thinking Questions

Application: Ask students to determine, without graphing, whether the graphs of $y = x^2 - 3x - 10$ and $y = -x^2 + 4x + 5$ intersect 0, 1, or 2 times. Set $x^2 - 3x - 10 = -x^2 + 4x + 5$ and solve for x. There are two solutions; therefore, the parabolas intersect at two points.

Additional Example 2

Use the discriminant to determine the number of solutions of each equation.

a. $x^2 + x + 1 = 0$ No real-number solution
b. $x^2 - 6x + 9 = 0$ 1
c. $x^2 - x - 1 = 0$ 2

565

Common Error Analysis

Error: When asked to state the value of the discriminant some students may give the value of $\sqrt{b^2 - 4ac}$. Remind them that the discriminant is the quantity $b^2 - 4ac$ before the square root is taken.

Checkpoint

Use the related quadratic equation to find the *x*-intercepts of each quadratic function.

1. $y = x^2 - 25$ ± 5
2. $y = x^2 - 7x + 6$ 6, 1
3. $y = 2x^2 + x - 1$ $\frac{1}{2}, -1$

Use the discriminant to determine the number of solutions of each equation. Do not solve.

4. $4x^2 + 4x + 1 = 0$ 1
5. $8 = 5x + x^2$ 2
6. $3x^2 - 4x + 2 = 0$ 0

Closure

Ask students how they can determine the *x*-intercepts of a parabola before graphing it.   Set $y = 0$ and solve for *x*.   Ask students to state the expression for the discriminant. $b^2 - 4ac$ Then ask how the discriminant can be used to determine of number of solutions for a quadratic equation.   (See page 565).

Classroom Exercises

For each quadratic function graphed below, give the related quadratic equation and its real number solution(s), if they exist. Tell whether the discriminant is positive, negative, or zero.

1. $y = x^2 - 4x + 7$

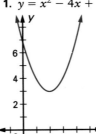

$x^2 - 4x + 7 = 0$; no real solutions; neg

2. $y = -x^2 - 2x + 3$

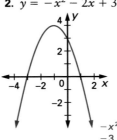

$-x^2 - 2x + 3 = 0$;
-3, 1; pos

3. $y = 4x^2 - 4x + 1$

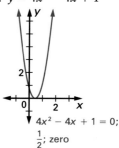

$4x^2 - 4x + 1 = 0$;
$\frac{1}{2}$; zero

Use the related quadratic equation to find the *x*-intercepts of each quadratic function.

4. $y = x^2 - 8x + 16$ 4
5. $y = x^2 + 9x - 2$ $\frac{-9 \pm \sqrt{89}}{2}$
6. $y = 3 - 2x^2 + x$ $-1, \frac{3}{2}$

Use the discriminant to determine the number of real number solutions of each equation. Do not solve.

7. $y = x^2 - 8x + 4$ 2
8. $y = 2x^2 + 5x + 3$ 2
9. $y = 1 + 4x^2 - 4x$ 1

Written Exercises

Use the related quadratic equation to find the *x*-intercepts of each quadratic function.

1. $y = x^2 - 7x - 18$ $-2, 9$
2. $y = x^2 + 12x + 36$ -6
3. $y = x^2 + 5x - 24$ $-8, 3$
4. $y = 64 - x^2$ $-8, 8$
5. $y = 3x^2 + 11x - 4$ $-4, \frac{1}{3}$
6. $y = -4x^2 - 4x + 3$ $-\frac{3}{2}, \frac{1}{2}$
7. $y = x^2 + 7x - 4$ $\frac{-7 \pm \sqrt{65}}{2}$
8. $y = x - 2x^2 + 5$ $\frac{1 \pm \sqrt{41}}{4}$
9. $-6 + 3x^2 - 2x = y$ $\frac{1 \pm \sqrt{19}}{3}$

Use the discriminant to determine the number of real number solutions of each equation. Do not solve.

10. $x^2 - 5x + 7 = 0$ 0
11. $x^2 - 8x + 16 = 0$ 1
12. $x^2 + 4x + 2 = 0$ 2
13. $x^2 - 2x + 9 = 0$ 0
14. $-3x^2 + 2x - 4 = 0$ 0
15. $4x^2 - 6x - 1 = 0$ 2
16. $-9 = 6x + x^2$ 1
17. $-x^2 + 40 = 0$ 2
18. $3x^2 - 7x = 0$ 2
19. $\frac{1}{2}x^2 + x + 1 = 0$ 0
20. $0 = 2x^2 - \frac{1}{2}x + 3$ 0
21. $x^2 + \frac{1}{4} = x$ 1
22. $x^2 + 2.1x = -1.1025$ 1
23. $0.5x = x^2 - 0.14$ 2
24. $x^2 + 2.25 = 0$ 0

25. For what value(s) of c will the graph of $y = x^2 - 18x + c$ have exactly one x-intercept? $c = 81$

26. For what value(s) of c will the vertex of $y = 16x^2 + 24x + c$ lie on the x-axis? $c = 9$

27. For what value(s) of c will the graph of $y = x^2 - 6x + c$ have no x-intercepts? $c > 9$

28. For what value(s) of c will the graph of $y = x^2 - 8x + c$ have two x-intercepts? $c < 16$

29. Explain in writing how the discriminant can be used to determine whether the solutions of $ax^2 + bx + c = 0$ are rational or irrational. Give examples to illustrate your statements. Answers will vary.

30. What is the relationship between b^2 and $4ac$ if $ax^2 + bx + c = 0$ has just one solution? $b^2 = 4ac$

31. Explain why the sign of b has no effect on the number of solutions of $ax^2 + bx + c = 0$.

32. Show that the quadratic function $y = ax^2 + bx + c$ $(a \neq 0)$ will have two x-intercepts if $a > 0$ and $c < 0$.

33. In Exercise 32, what conclusion can you draw about the number of x-intercepts if $a < 0$ and $c > 0$? There are two.

Mixed Review

Simplify. *6.1, 6.2*

1. $(4x^2)^2$ $16x^4$

2. $(-2y^2)^3$ $-8y^6$

3. $(-3xy^2)(5x^4y^2)$ $-15x^5y^4$

Solve. Check for extraneous solutions. *9.1, 9.2*

4. $\dfrac{3}{4} - \dfrac{1}{x} = \dfrac{5}{12}$ $x = 3$

5. $\dfrac{x}{9} = \dfrac{4}{x}$ $x = 6 \text{ or } x = -6$

6. $\dfrac{x - 3}{2} = \dfrac{-5}{x + 4}$ $x = 1 \text{ or } x = -2$

Solve. *4.4, 12.8*

7. The width of a rectangle is 6 cm less than the length. The perimeter of the rectangle is 54 cm. Find the length and the width. $l = 16.5$ cm; $w = 10.5$ cm

8. Abu has 18 coins, all dimes and quarters. The total value of the coins is $3.30. How many of each type are there? 8 dimes, 10 quarters

9. Given that $g(x) = 4x^2 - 3$, find $g(-2)$. *10.3* 13

Brainteaser

When $\frac{1}{4}$ of the adults left a beach party, the ratio of adults to children was 1:2. When 30 of the children left, the ratio of the children to adults was then 4:3. How many people remained at the beach? 105 (45 adults, 60 children)

FOLLOW UP

Guided Practice

Classroom Exercises 1–9
Try This   all

Independent Practice

A Ex. 1–18, **B** Ex. 19–29, **C** Ex. 30–33
Basic: WE 1–18 all, Brainteaser
Average: WE 1–9 all, 11–27 odd, Brainteaser
Above Average: WE 1–29 odd, 30–33 all, Brainteaser

Additional Answers

Written Exercises

31. To determine the number of real–number solutions of a quadratic equation, only the value of the discriminant, $b^2 - 4ac$, is needed. Since b is squared, the sign of b has no effect on the value of the discriminant. Hence the sign of b doesn't affect the number of solutions.

32. If $a > 0$ and $c < 0$, then $ac < 0$. Then $4ac < 0$ and thus $-4ac > 0$. So, $b^2 - 4ac > b^2 + 0 \geq 0$, and $b^2 - 4ac > 0$. The function has two x–intercepts.

Enrichment

Challenge the students to find a quadratic equation whose roots are $\dfrac{3 \pm \sqrt{17}}{4}$. $2x^2 - 3x - 1 = 0$
Then have the students challenge each other by making up similar problems of their own.

28.

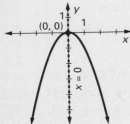

(0, 0) $x = 0$

29.

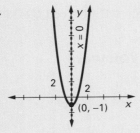

$x = 0$ (0, −1)

30.

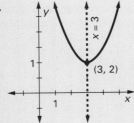

$x = 3$ (3, 2)

37. The graph will have exactly one
x–intercept when the discriminant $= 0$
(when $b^2 - 4ac = 0$). When $c = \frac{b^2}{4a}$, the
graph will have exactly one x–intercept.
Since $y = x^2 - 20x + c$, $a = 1$ and
$b = -20$. If $c = \frac{(-20)^2}{4(1)} = \frac{400}{4} = 100$, the
graph will have exactly one x–intercept.

Chapter 14 Review

Key Terms

axis of symmetry   (p. 559)
completing the square   (p. 539)
discriminant   (p. 545)
maximum point   (p. 559)
maximum value   (p. 559)
minimum point   (p. 559)

minimum value   (p. 559)
parabola   (p. 559)
quadratic formula   (p. 544)
quadratic function   (p. 558)
Square Root Property   (p. 535)
vertex of a parabola   (p. 559)

Key Ideas and Review Exercises

14.1   To solve an equation of the form $x^2 = c$, use the Square Root Property. If
$x^2 = c$, where $c > 0$, then $x = \sqrt{c}$ or $x = -\sqrt{c}$.

Solve each equation. Give irrational solutions in simplest radical form.

1. $y^2 = 14$ $\pm\sqrt{14}$
2. $2x^2 = 36$ $\pm 3\sqrt{2}$
3. $-z^2 = -121$ ± 11
4. $-4 = 19 - x^2$ $\pm\sqrt{23}$
5. $(r - 2)^2 = 81$ $11, -7$
6. $(2a + 1)^2 - 3 = 4$ $\frac{-1 \pm \sqrt{7}}{2}$

14.2   To solve an equation by completing the square, write it in the form
$x^2 + bx = -c$, take $\frac{1}{2}$ of b, square it, and add the result to each side
of the equation. The left side is now a perfect square, $\left(x + \frac{b}{2}\right)^2$. Then
solve by using the Square Root Property.

Solve by completing the square. Give irrational solutions in simplest
radical form.

7. $x^2 + 10x - 24 = 0$ $2, -12$
8. $y^2 - 8y = -16$ 4
9. $x^2 + 2x - 35 = 0$ $5, -7$
10. $x^2 - 6x - 2 = 0$ $3 \pm \sqrt{11}$
11. $10 - 6y = y^2$ $-3 \pm \sqrt{19}$
12. $x^2 + 7x + 10 = 0$ $-2, -5$

14.3   To solve a quadratic equation by using the quadratic formula, write the
equation in standard form, $ax^2 + bx + c = 0$, where $a > 0$. Then use the
formula, $x = \dfrac{-b \pm \sqrt{b^2 - 4ac}}{2a}$. If the discriminant $b^2 - 4ac$ is negative,
the equation has no real number solutions.

Solve by using the quadratic formula. Give irrational solutions in
simplest radical form.

13. $c^2 - 9c + 20 = 0$ $5, 4$
14. $x^2 + 8x = 12$ $-4 \pm 2\sqrt{7}$
15. $2x^2 = -4x - 1$ $\frac{-2 \pm \sqrt{2}}{2}$
16. $a^2 - 5a + 1 = 0$ $\frac{5 \pm \sqrt{21}}{2}$
17. $-y^2 + 6y = 4$ $3 \pm \sqrt{5}$
18. $2x - 4 = -x^2$ $-1 \pm \sqrt{5}$

14.4 Solve a quadratic equation by (1) using the Square Root Property, (2) factoring, (3) completing the square, or (4) by using the quadratic formula.

Solve. Give irrational solutions in simplest radical form.

19. $5x^2 = 100$ $\pm 2\sqrt{5}$ 　　**20.** $y^2 + 10y = -24$ $-4, -6$ 　　**21.** $r^2 - 12r + 36 = 0$ 6

22. $4x^2 + x - 3 = 0$ $\frac{3}{4}, -1$ 　　**23.** $x^2 + 5x = 3$ $\frac{-5 \pm \sqrt{37}}{2}$ 　　**24.** $x^2 - 2 = 6x$

$3 \pm \sqrt{11}$

25. If a ball is thrown straight up at an initial velocity of 40 m/s, when will its height be 50 m from where it was thrown? Use the formula $h = 40t - 5t^2$. Give answers correct to the nearest tenth of a second. 6.4 s, 1.6 s

14.5 To solve a geometric problem that leads to a quadratic equation, draw and label a figure. Then solve an equation by any appropriate method.

26. The distance between two opposite corners of a rectangular patio is 13 m. The length of the patio is 3 m shorter than 3 times the width. Find the length and the width of the patio. $w = 5$ m, $l = 12$ m

27. The perimeter of a picture frame is 38 in. and its area is 84 in². Find the length and the width of the frame. $w = 7$ in, $l = 12$ in

14.6 For the parabola $y = ax^2 + bx + c$ $(a \neq 0)$:

(1) The x-coordinate of the vertex is $-\frac{b}{2a}$. To find the y-coordinate of the vertex, substitute the x-value in the equation and solve for y.

(2) The equation of the axis of symmetry is $x = -\frac{b}{2a}$.

(3) The maximum or minimum value is the y-coordinate of the vertex.

For each quadratic function: (a) find the coordinates of the vertex of the parabola, (b) find the equation of the axis of symmetry, (c) find the minimum or maximum value, and (d) graph the function.

28. $y = -x^2$ 　　　　　**29.** $y = 2x^2 - 1$ 　　　　　**30.** $y = x^2 - 6x + 11$
(0, 0); $x = 0$; max: 0 　　(0, -1); $x = 0$; min: -1 　　(3, 2); $x = 3$; min: 2

14.7 The x-intercepts of the graph of $y = ax^2 + bx + c$ are the solutions of $ax^2 + bx + c = 0$. For the equation $ax^2 + bx + c = 0$, there are two solutions if $b^2 - 4ac > 0$; there is one solution if $b^2 - 4ac = 0$; and there are no real number solutions if $b^2 - 4ac < 0$.

Use the related quadratic equation to find the x-intercepts.

31. $y = x^2 - 4x - 21$ $-3, 7$ 　　**32.** $y = 3x^2 - 14x + 8$ $\frac{2}{3}, 4$ 　　**33.** $-3x + 2x^2 - 1 = y$

Use the discriminant to find the number of solutions. Do not solve. $\frac{3 \pm \sqrt{17}}{4}$

34. $x^2 - 6x + 3 = 0$ 2 　　　**35.** $-2x^2 + 3x + 5 = 0$ 2 　　**36.** $25 + 10x = -x^2$ 1

37. Given the function $y = x^2 - 20x + c$, write an explanation of how to find the value(s) of c such that the graph will have exactly one x-intercept.

1.

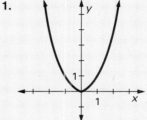

2.

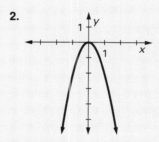

3.

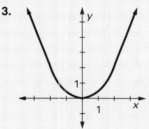

4.

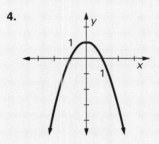

Chapter Test

16.

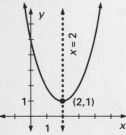

Additional Answers, page 562

5.

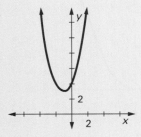

6.

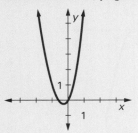

7. $(0, 2); x = 0;$ min: 2

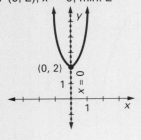

8. $(-3, 1); x = -3;$ max: 1

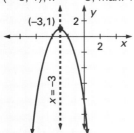

570

Solve each equation. Give irrational solutions in simplest radical form.

1. $x^2 = 35$ $\pm\sqrt{35}$

2. $3x^2 - 48 = 0$ ± 4

3. $(c - 2)^2 = 17$ $2 \pm \sqrt{17}$

4. $-(2x - 3)^2 + 9 = 2$ $\dfrac{3 \pm \sqrt{7}}{2}$

Solve by completing the square. Give irrational solutions in simplest radical form.

5. $x^2 + 2x - 3 = 0$ $1, -3$

6. $x^2 + 8x - 20 = 0$ $2, -10$

7. $y^2 - 8y + 48 = 0$
No real number solution

8. $3z^2 = 3 - 2z$ $\dfrac{-1 \pm \sqrt{10}}{3}$

Solve by using the quadratic formula. Give irrational solutions in simplest radical form.

9. $x^2 - 7x + 12 = 0$ $4, 3$

10. $2c^2 = 5c + 1$ $\dfrac{5 \pm \sqrt{33}}{4}$

11. $3 - x^2 = 5x$ $\dfrac{-5 \pm \sqrt{37}}{2}$

12. $2y + y^2 = -4$
No real number solution

Use the quadratic function $y = x^2 - 4x + 5$ for Items 13–16.

13. Find the coordinates of the vertex of the parabola. $(2, 1)$

14. Write the equation of the axis of symmetry of the parabola. $x = 2$

15. Find the maximum or minimum value.   Min: 1

16. Graph the function.

Use the discriminant to determine the number of solutions for each equation. Do not solve.

17. $x^2 + 3x + 3 = 0$ 0

18. $25x^2 - 10x - 1 = 0$ 2

19. $4x^2 + 9 = 12x$ 1

20. $-\frac{1}{2}x^2 + 3x - 1 = 0$ 2

21. The formula $h = 50t - 5t^2$ can be used to find the height in meters of an object shot upward at an initial velocity of 50 m/s after t seconds. If a pellet is shot upward at an initial velocity of 50 m/s, when will its height be 70 m from where it was shot? Give answers to the nearest tenth of a second.   8.3 s, 1.7 s

22. The distance between two opposite corners of a rectangular deck is 13 ft. The length is 7 ft longer than the width. Find the length and the width. $w = 5$ ft, $l = 12$ ft

23. Solve $\dfrac{x^2 - 4}{2} - 1 = -x - 2$ $x = -1 \pm \sqrt{3}$

24. What is the solution of the quadratic equation $ax^2 + bx + c = 0$ $(a \neq 0)$ if the discriminant is equal to zero? $\dfrac{-b}{2a}$

Choose the *one* best answer to each question or problem.

1. If x and y are positive integers with a difference of 9, what is the least possible value of $x + y$? D
   (A) 8 (B) 9 (C) 10 (D) 11
   (E) 18

2. If $4y + 3y = 35$, then $(2y - 1)^2 = \underline{\ ?\ }$. A
   (A) 81 (B) 25 (C) 9 (D) 6
   (E) 5

3. If three-thirds of $5\frac{1}{3}$ is subtracted from $5\frac{1}{3}$, the result is $\underline{\ ?\ }$. D
   (A) $10\frac{2}{3}$ (B) $5\frac{1}{3}$ (C) 1
   (D) 0 (E) -1

4. For what value(s) of y is the statement $\frac{5^y}{5^6} > 1$ true? B
   (A) $y > 1$ (B) $y > 6$
   (C) $y > 0$ (D) $y = 6$
   (E) $y < 6$

5. Solve $10 = \dfrac{5 - \frac{x}{y}}{4}$ for x. D
   (A) $35y$ (B) $\dfrac{5 - \frac{x}{y}}{40}$
   (C) $\dfrac{5y - x}{40}$ (D) $-35y$
   (E) $35 - \dfrac{x}{y}$

6. If $\frac{2}{3}$ of a number is 12, what is $\frac{5}{6}$ of the number? A
   (A) 15 (B) 10 (C) 8 (D) $\frac{20}{3}$
   (E) None of these

7. Carrie cycled 10 mi at 20 mi/h. How much time would she have saved if she had cycled at 25 mi/h? E
   (A) $\frac{1}{2}$ h (B) $\frac{2}{5}$ h (C) 10 min
   (D) 5 min (E) 6 min

8. What is the maximum possible area of the rectangle (in square units)? C

   (A) 7 (B) 14 (C) 49
   (D) 196 (E) It cannot be determined from the information given.

9. If $0 < c < 1$, then $\dfrac{1}{\frac{1}{c}}$ is $\underline{\ ?\ }$. C
   (A) greater than 1 (B) less than 0
   (C) between 0 and 1 (D) equal to 0
   (E) It cannot be determined from the information given.

10. If $x + y = 15$ and $x - y = 7$, then what is the value of y^2? B
    (A) 22 (B) 16 (C) 11 (D) 4
    (E) 121

11. The equation $x^2 - 12x + c = 0$ has just one solution if $c = \underline{\ ?\ }$. E
    (A) 0 (B) -24 (C) 24
    (D) -36 (E) 36

12. For what value(s) of x is it true that $4 + 5x \neq 6x$? B
    (A) 4 only (B) Every value except 4
    (C) $\frac{4}{11}$ only (D) $\frac{11}{4}$ only
    (E) -4 only

13. If $\frac{2}{3}x = \frac{3}{2}y$ and $y \neq 0$, then $\frac{x}{y} = \underline{\ ?\ }$. B
    (A) 1 (B) $\frac{9}{4}$ (C) $\frac{4}{9}$
    (D) $\frac{2}{3}$ (E) $\frac{3}{2}$

14. The area of a square is quadrupled if each side is $\underline{\ ?\ }$. C
    (A) increased by 2
    (B) increased by 4 (C) doubled
    (D) quadrupled (E) None of these

9. $(2, -1)$; $x = 2$; min: -1

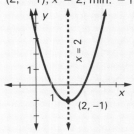

10. $(4, -9)$; $x = 4$; min: -9

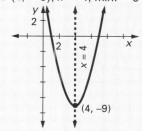

11. $(1, -\frac{5}{2})$; $x = 1$; min: $-\frac{5}{2}$

12. $(2, 10)$; $x = 2$, max: 10

571

19. a–c.

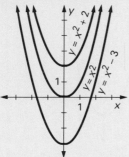

20. a–c.

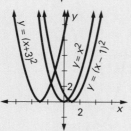

21. a–c.

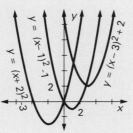

25.

x	0	1	1	4	4	9	9
y	0	1	−1	2	−2	3	−3

Cumulative Review *(Chapters 1–14)*

1. Evaluate $16 - 5x$ for $x = 3$. 1   **1.2**
2. Which property is illustrated?   **1.5**
   $(6 + 3x) + x = 6 + (3x + x)$
3. Simplify $8y + 2(6 + 3y) + 8$.   **1.7**
4. Simplify $-|10 - 8|$. −2   **2.2**
5. Evaluate $-2x^3 + 7x$ **2.8**
   for $x = -1$. −5
6. Simplify.   **2.10**
   $7(3 - 2x) - 4(x + 6)$ −3 − 18x

Solve.

7. $y - 9 = -4$ 5   **3.1**
8. $-\frac{3}{5}c = 21$ −35   **3.2**
9. $(r - 6)3 = 4(2r - 1) + 1$ −3   **3.5**
10. $\frac{2}{3}x - \frac{1}{2} = \frac{3}{4}$ $\frac{15}{8}$ or $1\frac{7}{8}$ **4.5**
11. $2.4 - 1.6y = 3.2y$ 0.5   **4.6**
12. $-5a + 15 > -20$ $a < 7$ **5.2**
13. $|14 - x| = -2$ No solution   **5.6**

Graph the solution set.

14. $y \le 4$ *and* $y > -1$ **5.4**
15. $7 - |2x - 5| \le -2$ **5.7**

Simplify. (Exercises 16–18)

16. $2x^2y(-3xy^3)^2$ $18x^4y^7$ **6.2**
17. $\dfrac{5a^2bc^3}{-15ab^4c}$ $\frac{ac^2}{-3b^3}$ **6.3**
18. $3a^3 - 2a + 7 - 4a^3 - a^2 + 9a$ −a³ − a² + 7a + 7   **6.6**
19. Subtract $-3x^2 + 5x - 2$ **6.7**
    from $6x^2 - 7x - 1$. 9x² − 12x + 1
20. Multiply $(2x - 8)(3x + 1)$.   **6.9**
    6x² − 22x − 8

Factor completely. (Exercises 21–23)

21. $x^2 - 3x - 28$ (x − 7)(x + 4)   **7.3**
22. $c^2 - 81$ (c − 9)(c + 9)   **7.5**
23. $2y^2 + 10y - 48$ 2(y + 8)(y − 3)   **7.6**
24. Solve $r^2 + 7r = -10$. −5, −2   **7.7**

Simplify.

25. $\dfrac{y + 3}{y - 1} \div \dfrac{y^2 - 9}{y^2 - 6y + 5}$ $\frac{y - 5}{y - 3}$ **8.4**
26. $\dfrac{-3x - 9}{x^2 - 3x} + \dfrac{2x}{x - 3}$ $\frac{2x + 3}{x}$ **8.7**
27. $\dfrac{1 + \frac{1}{x}}{\frac{x}{2} - \frac{1}{2x}}$ $\frac{2}{x - 1}$ **8.9**

Solve for x.

28. $\dfrac{x + 1}{x - 3} = \dfrac{3}{x} + \dfrac{12}{x^2 - 3x}$ −1   **9.1**
29. $\dfrac{y + 7}{5} = \dfrac{y - 3}{4}$ 43   **9.2**
30. $3x - 3b = 5x - a$ $\frac{a - 3b}{2}$ **9.3**

Give the domain and the range of each relation. Is the relation a function?

31.   **10.2**

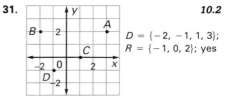

$D = \{-2, -1, 1, 3\};$
$R = \{-1, 0, 2\};$ yes

32. $\{(-1, 0), (0, 1), (-1, 2), (2, 3)\}$
33. If $f(x) = -2x^2 + 9$, find   **10.3**
    $f(-3)$. −9
34. Graph $2x - 3y = 9$.   **10.5**
35. Find the slope of the line determined by $A(2, -6)$ and   **11.1**
    $B(3, 2)$. 8
36. Write an equation determined   **11.2**
    by points $R(4, 3)$ and
    $S(0, -1)$. y = x − 1
37. Find the slope of the line.   **11.3**
    $y = -\frac{4}{5}x - 3$ $-\frac{4}{5}$

Solve each system. (Exercises 38 and 39).

38. $3x - 2y = 8$
$\quad y = -2$ $\left(\frac{4}{3}, -2\right)$ *12.2*

39. $x - 2y = 9$
$\quad -3x + 2y = 5$ $(-7, -8)$ *12.4*

Solve by graphing.

40. $y > x - 2$
$\quad y \le -x + 4$ *12.10*

41. Write a fraction $\frac{a}{b}$, where a *13.1*
and b are integers, for $0.6\overline{2}$. $\frac{28}{45}$

Simplify. (Exercises 42–44) **47.** No real sol

42. $\sqrt{28}$ $2\sqrt{7}$ *13.2*
43. $-\sqrt{27x^3 y^8}$ $-3xy^4\sqrt{3x}$ *13.5*
44. $8\sqrt{2} - \sqrt{50}$ $3\sqrt{2}$ *13.6*
45. Approximate $\sqrt{87}$ to the   *13.3*
nearest tenth. 9.3
46. Solve $3x^2 - 2 = 5$. $\frac{\pm\sqrt{21}}{3}$ *14.1*
47. Solve $x^2 - 2x + 7 = 0$.   *14.3*
48. Solve $3x^2 + x - 5 = 0$. $\frac{-1 \pm \sqrt{61}}{6}$
49. Find the coordinates of the   *14.6*
vertex of the parabola
$y = x^2 + 4x - 10$. Graph
the function. $(-2, -14)$

50. If 142 is 7 more than 3 times
a number, find the number. 45   *3.3*

51. A train went 147 mi in $3\frac{1}{2}$h.   *3.6*
What was the average rate in
miles per hour? 42 mi/h

52. The sum of two consecutive   *4.3*
odd integers is -76. Find the
integers. $-39, -37$

53. The base of an isosceles tri-   *4.4*
angle is 8 cm longer than a
leg. The perimeter is 104 cm.
Find the length of each side. 32, 32, 40

54. In a class of 180 students,   *4.7*
45% are boys. Find the num-
ber of boys and the number
of girls in the class. 81 boys; 99 girls

55. The price of a stock dropped
from $12\frac{1}{2}$ to $11\frac{1}{2}$. What was
the percent decrease? 8%

56. Forty-eight is 12% of what
number? 400

57. Jane earns $250 a week, plus   *4.8*
a 5% commission on her to-
tal sales. Last week she
earned $671. What were her
total sales for the week? $8,420

58. The length of a rectangular   *7.9*
garden is 3 ft less than twice
the width. The area is 54 ft².
Find the length and the width
of the garden. 6 ft, 9 ft

59. Separate a 72-cm board into   *9.2*
two parts with lengths in the
ratio 3:5. Find the lengths of
the two parts. 27 cm, 45 cm

60. One machine can complete an   *9.5*
order in 6 h while another
machine takes 9 h. How long
will it take both machines to
complete this order working
together? $3\frac{3}{5}$ h, or 3 h 36 min

61. If y varies inversely as x and   *10.7*
y is 24 when x is 9, find x
when y is 6. 36

62. One pencil and 3 erasers cost   *12.3*
79¢. Two pencils and 5 eras-
ers cost $1.48. Find the cost
of a pencil and the cost of an
eraser. pencil: 49¢; eraser: 10¢

63. The tens digit of a two-digit   *12.6*
number is 3 more than the
units digit. If the digits are reversed, the new number is 27
less than the original number.
Find the original number.

64. A rectangular lot measures 12   *13.4*
yd by 16 yd. How many yards
farther is it to walk the length
and the width of the lot than
to walk from one corner to
the opposite corner? 8 yd

Cumulative Review

2. Assoc Prop Add
3. $14y + 20$
14.
15.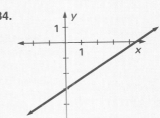

32. $D: \{-1, 0, 2\}$; $R: \{0, 1, 2, 3\}$; no

34.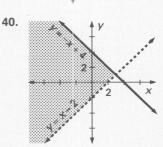

40.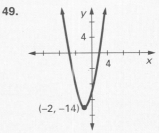

49.

$(-2, -14)$

63. 96, 85, 74, 63, 52, 41

15 TRIGONOMETRY

OVERVIEW

Similar triangles are used in the first lesson of this chapter to introduce students to indirect measurement. Other ratios involving the length of sides of triangles are presented in the following lesson as the definitions of the sine, cosine, and tangent of an acute angle. After learning to read trigonometric tables, students use these to solve for unknown parts of right triangles. In the final lesson, students apply these concepts and skills to solving word problems.

OBJECTIVES

- To find the lengths of sides of similar triangles
- To find the value of the sine, cosine, and tangent of given angles and to find angle measures given one of these ratios
- To find the lengths of the sides and the measures of the acute angles of a right triangle
- To solve problems involving trigonometric ratios

PROBLEM SOLVING

The student is asked to Use Formulas and Definitions to Write Equations for solving problems involving similar triangles and right triangles. In most of the problems, students Look for Patterns in order to use the correct ratio to solve each problem. For all of the problems, students are encouraged to Draw a Diagram.

READING AND WRITING MATH

The Focus on Reading exercises on page 576 help students to clarify the meaning of corresponding sides and corresponding angles of similar triangles. In Exercises 38–40 on page 585, students use patterns they have observed to justify basic trigonometric identities.

TECHNOLOGY

Calculator: This chapter offers an excellent opportunity for student use of the scientific calculator. Using a calculator to find the values of trigonometric ratios and to solve triangle problems allows the student to concentrate on the processes involved rather than on routine calculations.

SPECIAL FEATURES

PLANNING GUIDE

Lesson	Basic	Average	Above Average	Resources
15.1 pp. 577–578	FR all, CE 1–8 WE 1–12 all	FR all, CE 1–8 WE 1–11 odd, 13–18 all	FR all, CE 1–8 WE 1–21 odd	Reteaching 113 Practice 113
15.2 pp. 581–582	CE 1–11 WE 1–24 all	CE 1–11 WE 7–30 all	CE 1–11 WE 19–41 all	Reteaching 114 Practice 114
15.3 pp. 584–585	CE 1–9 WE 1–14 all Midchapter Review	CE 1–9 WE 1–27 odd, 28–36 all Midchapter Review	CE 1–9 WE 1–31 odd, 32–40 all Midchapter Review	Reteaching 115 Practice 115
15.4 pp. 587–589	CE 1–8 WE 1–19 odd Brainteaser	CE 1–8 WE 7–25 odd Brainteaser	CE 1–8 WE 15–33 odd Brainteaser	Reteaching 116 Practice 116
15.5 pp. 591–593	CE 1–8 WE 1–6 all Mixed Problem Solving	CE 1–8 WE 1–11 odd Mixed Problem Solving	CE 1–8 WE 5–11 odd,13,14 Mixed Problem Solving	Reteaching 117 Practice 117
Chapter 15 Review pp. 594–595	all	all	all	
Chapter 15 Test p. 596	all	all	all	
College Prep Test p. 597	all	all	all	

CE = Classroom Exercises   WE = Written Exercises   FR = Focus on Reading
Note: For each level, all students should be assigned all Try This and all Mixed Review Exercises.

INVESTIGATION

Project: This project can involve the whole class in building a table of values for the sines, cosines, and tangents of acute angles in increments of 10 degrees. The students may or may not use the words sin, cos, and tan, according to the discretion of the teacher.

Materials: Each group of 3 or 4 students will need a large piece of paper to construct triangles, a meter stick, a protractor, and a calculator. Each student will need a copy of Project Worksheet 15.

Each group of students should use a compass and ruler to construct a right triangle with an acute angle of 10°, of 20°, of 30° . . . of 80°. Assign one or two of these measures to each group. Have students draw the triangle as large as they can on the paper provided. After each side of the triangle is measured, the necessary calculations can be done with a calculator. If there are enough students and enough time, you may wish to have more than one group do the same angles so that results can be compared.

Have a large sheet of paper available to record the results for all three ratios of each of the angles so that the whole class can see the compiled results.

When this chart has been completed, have students look for patterns by asking questions such as these.

1. What is the range of values for the sine?   0 to 1

2. As the measure of the angle increases, what happens to the value of the cosine? It becomes smaller.

If more than one group has drawn a triangle with the same specified acute angle, have students compare the two triangles and ask them if the size of the triangle has any effect on the answers.   No

Bill Grove, of Whitenoise, plays the saxophone, which is a wind instrument. The music made by wind instruments, such as the clarinet, flute, and trombone, is produced by sound waves vibrating inside a tube.

More About the Photo

Sound travels in waves of compressed air. The amplitude (or maximum height) of the wave determines the loudness; the frequency (or cycles per second) of the wave determines the pitch (high or low). These waves may be represented mathematically by sine functions. With stringed instruments, such as the violin, guitar, or string bass, the sound produced by plucking the string is inversely proportional to the length of the string (L), inversely proportional to the square root of the thickness or weight (W), and directly proportional to the square root of its tension or how tight it is (T).

15.1 Similar Triangles

Objective

To find the lengths of sides of similar triangles

Some distances are impossible to measure directly. In such cases, *similar triangles* can often be used to measure distances indirectly. An ancient method of measuring heights is called *shadow reckoning*. This makes use of similar triangles that involve the sun's rays, two vertical objects, and the shadows of these objects.

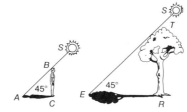

Similar triangles have the same shape. If all pairs of **corresponding angles** have the same measure, then the two triangles are **similar triangles**. In the figure at the right below, $\triangle ABC$ is similar to $\triangle DEF$, or $\triangle ABC \sim \triangle DEF$.

The symbol "$\angle$" means *angle*. The symbol "m $\angle$" means *measure of an angle*.

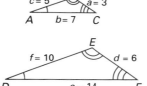

Notice the same markings for pairs of corresponding angles, $\angle A$ and $\angle D$, $\angle B$ and $\angle E$, $\angle C$ and $\angle F$. This indicates that m $\angle A$ = m $\angle D$, m $\angle B$ = m $\angle E$, and m $\angle C$ = m $\angle F$.

The sides opposite pairs of corresponding angles are called **corresponding sides**. Thus, $\overline{AC}$ (opposite $\angle B$) and $\overline{DF}$ (opposite $\angle E$) are corresponding sides.

In $\triangle ABC$, a, b, and c denote the lengths of the sides. Compare the ratios of the lengths of corresponding sides in $\triangle ABC$ and $\triangle DEF$.

$$\frac{a}{d} = \frac{3}{6}, \text{ or } \frac{1}{2}$$

$$\frac{b}{e} = \frac{7}{14}, \text{ or } \frac{1}{2}$$

$$\frac{c}{f} = \frac{5}{10}, \text{ or } \frac{1}{2}$$

Each ratio is equal to $\frac{1}{2}$. This illustrates an important property of similar triangles.

In *similar triangles, the lengths of all three pairs of corresponding sides have the same ratio.*

15.1 Similar Triangles **575**

Teaching Resources

Manipulative Worksheet 15
Project Worksheet 15
Quick Quizzes 113
Reteaching and Practice Worksheets 113
Transparencies 43A, 43B, 43C

GETTING STARTED

Prerequisite Quiz

Solve each proportion.

1. $\frac{n}{8} = \frac{21}{28}$ 6

2. $\frac{9}{6} = \frac{15}{n}$ 10

3. $\frac{6}{22} = \frac{n}{55}$ 15

4. $\frac{1.2}{n} = \frac{2.0}{1.5}$ 0.9

Motivator

Manipulative: Have students draw a triangle on a piece of paper. Then have them use a ruler to measure and extend two sides of the triangle to double their original lengths. Have them draw the third side of the longer triangle. Ask students to measure this third side and compare its length to the corresponding side of the smaller triangle. It is twice as long.   Now have students use a protractor to measure the angles of both triangles and compare them.   Corresponding angles will have the same measure.

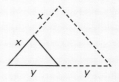

Highlighting the Standards

Standards 7b, 7c: In this lesson students use similar triangles to represent problem situations. They then solve the problems by applying the properties of similar triangles.

Lesson Note

Use the triangles on page 575 to show students that corresponding angles are those angles that occur in the same relative position when the two triangles have the same orientation. Tell students that the shape of a triangle is determined by the size of its angles. Therefore, two triangles whose three pairs of corresponding angles have equal measures, have the same shape. When assigning exercises for homework, suggest to the students that it may help to identify corresponding parts if they redraw the figures with the same orientation.

Math Connections

Drawing: A tool called a pantograph can be used to draw a figure similar to a given figure by enlarging or reducing its size.

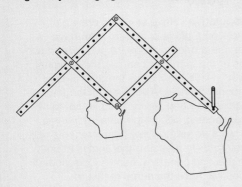

EXAMPLE 1 $\triangle ABC \sim \triangle RST$, $a = 8$, $r = 12$, and $c = 10$. Find t.

Plan   Write a proportion using a, r, c, and t, since a and r are lengths of corresponding sides, and c and t are lengths of corresponding sides.

Solution $\dfrac{a}{r} = \dfrac{c}{t}$

$\dfrac{8}{12} = \dfrac{10}{t}$

$8t = 12 \cdot 10$ ⟵ If $\dfrac{a}{b} = \dfrac{c}{d}$, then $ad = bc$.

$8t = 120$

$t = 15$

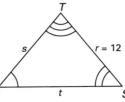

EXAMPLE 2   A pole that is 2 m high casts a shadow that is 3 m long. At the same time, a nearby building casts a shadow that is 36 m long. Find the height of the building.

Plan   Draw a diagram to show two similar right triangles. Let x represent the height of the building. Write a proportion to find the height.

Solution $\dfrac{x}{2} = \dfrac{36}{3}$

$3x = 72$

$x = 24$

(The triangles are not shown in true relative size.)

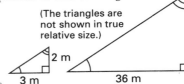

The height of the building is 24 m.

TRY THIS   1. Jane is 5 ft tall. She casts a 7-ft shadow at the same time that a nearby pole casts a 15-ft shadow. How high is the pole? 10.7 ft

---

■ Focus on Reading

$\triangle EFG \sim \triangle KLM$. **Tell whether each statement is true or false.**

1. $\angle E$ corresponds to $\angle K$. True
2. $\angle G$ corresponds to $\angle L$. False
3. $\overline{EG}$ corresponds to $\overline{KL}$. False
4. $\overline{FG}$ corresponds to $\overline{LM}$. True
5. $\dfrac{e}{k}$ and $\dfrac{f}{l}$ must be equal. True
6. $\dfrac{m}{g}$ and $\dfrac{l}{f}$ must be equal. True

Additional Example 1

$\triangle ABC \sim \triangle UVW$ and $a = 8$, $u = 10$, and $c = 20$. Find w.   25

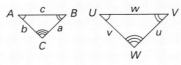

Additional Example 2

A 4-ft stick casts a shadow 6 ft long. At the same time, a nearby building casts a shadow 27 ft long. Find the height of the building.   18 ft

Classroom Exercises

$\triangle ABC \sim \triangle DEF$. Complete the proportion.

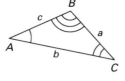

1. $\frac{c}{f} = \frac{a}{?} d$

2. $\frac{c}{f} = \frac{?}{e} b$

3. $\frac{e}{b} = \frac{d}{?} a$

4. $\frac{?}{c} = \frac{d}{a} f$

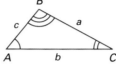

Find the missing length.

5. $c = 12$, $f = 8$, and $b = 9$. Find e. 6
6. $a = 16$, $d = 12$, and $c = 8$. Find f. 6
7. $d = 8$, $a = 16$, and $f = 9$. Find c. 18
8. $e = 10$, $b = 15$, and $d = 12$. Find a. 18

Written Exercises

In Exercises 1–8, $\triangle ABC \sim \triangle XZY$.
Find the missing length.

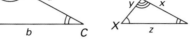

1. $b = 8$, $z = 4$, $a = 12$. Find x. 6
2. $a = 12$, $x = 6$, and $b = 8$. Find z. 4
3. $z = 16$, $b = 4$, and $x = 12$. Find a. 3
4. $y = 5$, $b = 6$, and $z = 15$. Find c. 2
5. $b = 250$, $y = 200$, and $c = 150$. Find z. 333.3
6. $a = 40$, $x = 30$, and $b = 36$. Find z. 27
7. $b = 12.5$, $y = 10.0$, and $c = 8.5$. Find z. 14.7
8. $y = 4.5$, $b = 6.0$, and $z = 6.6$. Find c. 4.1

In Exercises 9–12, $\triangle DEF \sim \triangle RST$.

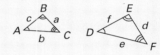

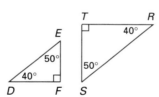

9. $\angle E$ corresponds to which angle in $\triangle RST$? $\angle S$
10. $\angle D$ corresponds to which angle in $\triangle RST$? $\angle R$
11. $\angle F$ corresponds to which angle in $\triangle RST$? $\angle T$
12. What is the degree measure of $\angle F$? 90

Solve each problem.

13. A 2-m stick casts a 10-m shadow. At the same time, a nearby building casts a 40-m shadow. Find the height of the building. 8 m

14. Mr. Forbes is 6 ft tall. He casts a 4-ft shadow at the same time that a nearby flagpole casts an 18-ft shadow. How high is the flagpole? 27 ft

15. A 30-m ladder touches the side of a building at a height of 25 m. At what height would a 12-m ladder touch the building if it makes the same angle with the ground? 10 m

15.1 Similar Triangles **577**

Critical Thinking Questions

Analysis: Tell students that similar polygons have the measures of their corresponding angles equal and the ratios of corresponding sides equal. Then ask if all rectangles are similar. Why or why not? No; the ratios of corresponding sides are not necessarily equal. Are all squares similar? Why or why not? Yes; both parts of the definition are true.

Common Error Analysis

Error: If triangles are drawn in different positions, some students will match corresponding sides incorrectly.

Suggest that they redraw the figures in the same orientation.

Checkpoint

In Exercises 1–3, $\triangle ABC \sim \triangle DEF$. Find the missing length.

1. $c = 32$, $f = 20$, and $b = 10$. Find e. $6\frac{1}{4}$
2. $f = 6$, $c = 15$, and $d = 8$. Find a. 20
3. $b = 3.0$, $e = 4.5$, and $f = 6.0$. Find c. 4.0
4. A 2-m pole casts a shadow 3 m long. At the same time, a nearby tree casts a shadow 18 m long. Find the height of the tree. 12 m

Ask students what these terms mean.
1. similar triangles
2. corresponding angles
3. corresponding sides   See page 575.
   Have students draw two similar triangles
   and show corresponding angles and
   corresponding sides.

■■■FOLLOW UP

Guided Practice

Classroom Exercises 1–8
Try This   all, FR   all

Independent Practice

A Ex. 1–12, **B** Ex. 13–18, **C** Ex. 19–22
Basic: WE 1–12 all
Average: WE 1–11 odd, 13–18 all
Above Average: WE 1–21 odd

Additional Answers

Written Exercises

21. Yes. In $\triangle ABC$, m$\angle C$ = 180 − (62 +
28) = 90. In $\triangle XYZ$, m$\angle Y$ = 180 −
(90 + 62) = 28. Since the three pairs
of corresponding angles have the same
measures, the triangles are similar.

22. No. In $\triangle DEF$, the measures of the
angles are 90, 36, and 180 − (90 + 36),
or 54. In $\triangle JKL$, the measures of
the angles are 90, 53, and 180 −
(90 + 53), or 37. Since all three pairs
of corresponding angles do not have
the same measures, the triangles are
not similar.

---

16. Tony stands so that his shadow just
reaches the tip of the shadow of a
tree. His height is 5 ft and the length
of his shadow is 8 ft. The length of
the shadow of the tree is 40 ft. What
is the height of the tree? 25 ft

17. Ms. Carter wants to find the height of
one of her trees. When she holds a
yardstick in a vertical position, touch-
ing the ground, it casts a shadow 2 ft
long. At the same time, the tree's
shadow is 32 ft long. What is the
height of the tree? 48 ft

18. In the figure at the right, BC = 180 m,
CD = 50 m, and ED = 40 m.
Angles B and D are right angles, and
$\triangle ABC \sim \triangle EDC$. Use this information to find x, the distance across the
river. 144 m

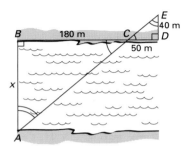

Round lengths to the nearest tenth.

22.5 cm; 26.3 cm

19. The lengths of three sides of a triangle are 5 cm, 8 cm, and 9 cm. The
shortest side of a similar triangle is
9 cm. Find the other lengths. 14.4 cm;
16.2 cm

20. The lengths of three sides of a triangle are 12 cm, 14 cm, and 16 cm.
The longest side of a similar triangle
is 30 cm. Find the other lengths.

21. In $\triangle ABC$, m $\angle A$ = 62 and
m $\angle B$ = 28. In $\triangle XYZ$, m $\angle X$ = 62
and m $\angle Z$ = 90. Are the two trian-
gles similar? Explain.

22. In right $\triangle DEF$, m $\angle D$ = 36.
In right $\triangle JKL$, m $\angle K$ = 53.
Are the two right triangles similar?
Explain.

Mixed Review

Divide. **8.4**

1. $\dfrac{x^2 - 1}{y + 2} \div \dfrac{(x - 1)^2}{2y - 8}$ $\dfrac{2xy + 2y - 8x - 8}{xy + 2x - y - 2}$

2. $\dfrac{n^2 + 4n - 21}{n^2 + n - 20} \div \dfrac{n^2 + 8n + 7}{n^2 + 6n + 5}$ $\dfrac{n - 3}{n - 4}$

Simplify. **13.2, 13.7**

3. $\sqrt{3} \cdot \sqrt{11}$ $\sqrt{33}$

4. $3\sqrt{8y^5} \cdot 5\sqrt{2y}$ $60y^3$ $106 - 17\sqrt{10}$

5. $\sqrt{2}(\sqrt{8} + 3\sqrt{2})$ 10

6. $(4\sqrt{5} - \sqrt{2})(5\sqrt{5} - 3\sqrt{2})$

7. Find two consecutive odd integers
whose product is 399. **7.9**
19 and 21 or −19 and −21

8. The sum of a number and its reciprocal is $\frac{25}{12}$. Find the number. **9.1**
$\frac{4}{3}$ or $\frac{3}{4}$

Enrichment

Have each student choose three points, A,
B, and C, with coordinates such that $\overline{AB}$,
$\overline{BC}$, and $\overline{CA}$ form a triangle. Then have them
follow the following directions and answer
the questions.
  1. Multiply the coordinates of A, B, and C
     by 2 to obtain the coordinates of three
     new points, D, E, and F, respectively.
     Plot D, E, and F, and draw $\overline{DE}$, $\overline{EF}$, and

$\overline{FD}$. Is $\triangle DEF$ similar to $\triangle ABC$? Is it
smaller, larger, or the same size as
$\triangle ABC$?   Yes; larger
  2. Multiply the coordinates of A, B, and C
     by $\frac{1}{2}$ to obtain the coordinates of three
     new points, G, H, and I, respectively. Is
     $\triangle GHI$ similar to $\triangle ABC$? Is it smaller,
     larger, or the same size as $\triangle ABC$?
     Yes; smaller

15.2 Trigonometric Ratios

Objectives   To compute or find the value of the sine, cosine, and tangent of acute angles of right triangles

To explain why corresponding angles of similar right triangles have the same trigonometric ratios

Recall that in a right triangle, the side opposite the right angle is called the *hypotenuse*. The other sides are called the *legs*.

In right triangle ABC, leg a is opposite $\angle A$; leg b is opposite $\angle B$. $\angle A$ and $\angle B$ are acute angles, since each has a degree measure between 0 and 90. Each leg is also referred to as *adjacent to an acute angle*. Thus, leg a is adjacent to $\angle B$ and leg b is adjacent to $\angle A$.

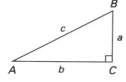

The figure below shows two similar right triangles.

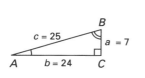 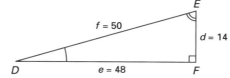

Corresponding sides of the two similar triangles are proportional.

$$\frac{a}{d} = \frac{c}{f} \qquad \frac{b}{e} = \frac{c}{f} \qquad \frac{a}{d} = \frac{b}{e}$$

In $\frac{a}{d} = \frac{c}{f}$, the means ($c$ and d) can be interchanged, as shown below.

$$\frac{a}{d} = \frac{c}{f} \qquad \left(\frac{7}{14} = \frac{25}{50}\right)$$

Use the Proportion Property. $a \cdot f = d \cdot c$

Divide each side by $c \cdot f$. $\dfrac{a \cdot f}{c \cdot f} = \dfrac{d \cdot c}{c \cdot f}$

A new proportion is obtained. $\dfrac{a}{c} = \dfrac{d}{f} \qquad \left(\dfrac{7}{25} = \dfrac{14}{50}\right)$

Notice that the left side of the above proportion involves sides of $\triangle ABC$ and the right side involves sides of $\triangle DEF$. In a similar way, you can rewrite $\frac{b}{e} = \frac{c}{f}$ as $\frac{b}{c} = \frac{e}{f}$, and $\frac{a}{d} = \frac{b}{e}$ as $\frac{a}{b} = \frac{d}{e}$.

15.2 Trigonometric Ratios   **579**

Teaching Resources
Problem Solving Worksheet 15
Quick Quizzes 114
Reteaching and Practice Worksheets 114
Transparencies 44A, 44B, 44C

◼◼ GETTING STARTED

Prerequisite Quiz

Find each of the following values to three decimal places.

1. $\frac{3}{5}$ 0.600
2. $\frac{12}{13}$ 0.923
3. $\frac{7}{9}$ 0.778

Find the missing lengths.

4. $a = 9$, $b = 12$, $c = \underline{\ ?\ }$ 15
5. $a = 3$, $b = \underline{\ ?\ }$, $c = 4$ $\sqrt{7}$

Simplify. Rationalize the denominator.

6. $\frac{1}{\sqrt{2}}$ $\frac{\sqrt{2}}{2}$
7. $\frac{3}{\sqrt{3}}$ $\frac{\sqrt{3}}{3}$

Motivator

On the chalkboard, draw and label right triangle ABC as shown at the top of page 579. Have students identify each of the following parts.

hypotenuse c
side opposite $\angle A$ a
side opposite $\angle B$ b
side adjacent to $\angle A$ b
side adjacent to $\angle B$ a

Highlighting the Standards

Standard 9a: Here students apply trigonometry to problem situations involving triangles.

Lesson Note

After giving the definitions of the trigonometric ratios, refer the students back to triangles *ABC* and *DEF* on page 579 so that students can notice that m∠*A* = m∠*D* and that sin *A* = sin *D*. In other words, the value of the sine does not depend on the lengths of the sides of the triangle, but rather on the measure of the angle. If a larger or smaller right triangle with an acute angle having the same measure were drawn, the value of the sine of that angle would remain the same. You may wish to have students use a calculator for all exercises that call for a three-place decimal answer so that they can concentrate on the definitions rather than on routine calculations.

Math Connections

History: Trigonometry has its basis in early studies of the heavenly bodies and of the angles that these bodies make from an observation point on the earth. It also has its basis in the study of astronomy and geography. Ptolemy (ca. 150 AD) and Hipparchus (ca. 160 BC) created much of the work leading to the mathematics known as trigonometry.

Critical Thinking Questions

Analysis: Have students look at the solutions to Example 1 (p. 580) and notice that sin *A* = cos *B* and cos *A* = sin *B*. Ask them why this happens.   The ratios are the same.   Have them notice that ∠*A* and ∠*B* are complementary (acute angles of a right triangle are complementary) and ask if the relationships just observed will be true for any pair of complementary angles.   Yes

---

The ratios $\frac{a}{c}$, $\frac{b}{c}$, and $\frac{a}{b}$ in right triangle *ABC* are called, respectively, the *sine ratio* of angle *A*, the *cosine ratio* of angle *A*, and the *tangent ratio* of angle *A*. These ratios, abbreviated as *sin A*, *cos A*, and *tan A*, are defined as follows.

Definition

For all right triangles *ABC* with right angle *C* and sides *a*, *b*, and *c*:

$$\sin A = \frac{\text{length of opposite side}}{\text{length of hypotenuse}} = \frac{a}{c}$$

$$\cos A = \frac{\text{length of adjacent side}}{\text{length of hypotenuse}} = \frac{b}{c}$$

$$\tan A = \frac{\text{length of opposite side}}{\text{length of adjacent side}} = \frac{a}{b}$$

EXAMPLE 1   For right triangle *ABC*, find the trigonometric ratios for ∠*A* and ∠*B* to three decimal places.

Plan   Use the definitions of sine, cosine, and tangent.

Solutions

$\sin A = \frac{\text{opp}}{\text{hyp}} = \frac{6}{10}$, or 0.600 $\qquad$ $\sin B = \frac{\text{opp}}{\text{hyp}} = \frac{8}{10}$, or 0.800

$\cos A = \frac{\text{adj}}{\text{hyp}} = \frac{8}{10}$, or 0.800 $\qquad$ $\cos B = \frac{\text{adj}}{\text{hyp}} = \frac{6}{10}$, or 0.600

$\tan A = \frac{\text{opp}}{\text{adj}} = \frac{6}{8}$, or 0.750 $\qquad$ $\tan B = \frac{\text{opp}}{\text{adj}} = \frac{8}{6}$, or 1.333

EXAMPLE 2   For right triangle *PQR*, find sin *P*, cos *P*, and tan *P*. Find sin *Q*, cos *Q*, and tan *Q*. Leave the answers in fraction form.

Plan   Use the definitions of sine, cosine, and tangent.

Solutions

$\sin P = \frac{5}{13}$ $\qquad$ $\sin Q = \frac{12}{13}$

$\cos P = \frac{12}{13}$ $\qquad$ $\cos Q = \frac{5}{13}$

$\tan P = \frac{5}{12}$ $\qquad$ $\tan Q = \frac{12}{5}$

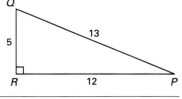

580   Chapter 15 Trigonometry

Additional Example 1

For right triangle *ABC*, find the basic trigonometric ratios for ∠*A* to three decimal places.

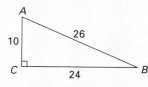

$\sin A = \frac{24}{26}$, or 0.923;

$\cos A = \frac{10}{26}$ or 0.385;

$\tan A = \frac{24}{10}$, or 2.4;

EXAMPLE 3 For right triangle *RST*, find sin *R*, cos *R*, and tan *R*.

Plan First use the Pythagorean Theorem to find the length, *t*.

Solution
$$t^2 = r^2 + s^2$$
$$t^2 = 2^2 + 1^2 = 5$$
$$t = \sqrt{5}$$

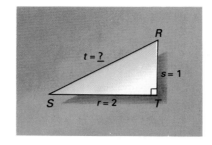

Thus, $\sin R = \dfrac{2}{\sqrt{5}}$, or $\dfrac{2\sqrt{5}}{5}$

$\cos R = \dfrac{1}{\sqrt{5}}$, or $\dfrac{\sqrt{5}}{5}$

$\tan R = \dfrac{2}{1}$, or 2

TRY THIS For right triangle *RST* in Example 3, find sin *S*, cos *S*, and tan *S* if *s* = 2 and *r* = 6. $\dfrac{\sqrt{10}}{10}; \dfrac{3\sqrt{10}}{10}; \dfrac{1}{3}$

Classroom Exercises

Refer to the figure at the right.
1. Name the hypotenuse. *p*
2. Name the leg opposite ∠*M*. *m*
3. Name the leg adjacent to ∠*M*. *n*
4. Name the leg adjacent to ∠*N*. *m*
5. Name the leg opposite ∠*N*. *n*

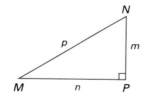

Find each of the following to three decimal places.
6. sin *D* 0.280
7. cos *D* 0.960
8. tan *D* 0.292
9. sin *E* 0.960
10. cos *E* 0.280
11. tan *E* 3.429

Written Exercises

Find each of the following to three decimal places.
1. sin *A* 0.385
2. cos *A* 0.923
3. tan *A* 0.417
4. sin *B* 0.923
5. cos *B* 0.385
6. tan *B* 2.400

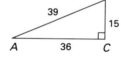

15.2 Trigonometric Ratios **581**

Checkpoint

In Exercises 1–4, use right triangle *ABC*. Find each of the following values to three decimal places.

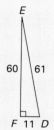

1. sin *A* 0.976
2. cos *A* 0.220
3. sin *B* 0.220
4. tan *B* 0.225

In Exercises 5–7, use right triangle *DEF*. Find each of the following. Leave the answer in fraction form.

5. sin *E* $\dfrac{11}{61}$
6. cos *E* $\dfrac{60}{61}$
7. tan *E* $\dfrac{11}{60}$

Additional Example 2

For right triangle *XYZ*, find sin *Y*, cos *Y*, and tan *Y*. Leave the answers in fraction form.

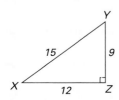

$\sin Y = \dfrac{12}{15}$, or $\dfrac{4}{5}$;

$\cos Y = \dfrac{9}{15}$, or $\dfrac{3}{5}$;

$\tan Y = \dfrac{12}{9}$, or $\dfrac{4}{3}$

Additional Example 3

For right triangle *JKL*, find sin *J*, cos *J*, and tan *J*.

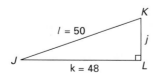

$\sin J = \dfrac{7}{25}$;

$\cos J = \dfrac{24}{25}$;

$\tan J = \dfrac{7}{24}$

Ask students to give the definitions of the sine, cosine, and tangent of an acute angle of a right triangle. See page 580. Draw a right triangle on the chalkboard with sides $a = 3$, $b = 4$, and $c = 5$. Have students give these values orally: sin A, cos A, tan A, sin B, cos B, and tan B. $\frac{3}{5}, \frac{4}{5}, \frac{3}{4}, \frac{4}{5}, \frac{3}{5}, \frac{4}{3}$

◤◤◤FOLLOW UP

Guided Practice

Classroom Exercises 1–11
Try This   all

Independent Practice

A Ex. 1–24, **B** Ex. 25–30, **C** Ex. 31–41

Basic: WE 1–24 all
Average: WE 7–30 all
Above Average: WE 19–41 all

Find each of the following to three decimal places.

7. sin J 0.280

8. cos J 0.960

9. tan J 21   0.292

10. sin K 0.960

11. cos K 0.280

12. tan K 3.429

Find each of the following values. Leave answers in fraction form.

13. sin Y $\frac{4}{5}$

14. cos Y $\frac{3}{5}$

15. tan Y $\frac{4}{3}$

16. sin X $\frac{3}{5}$

17. cos X $\frac{4}{5}$

18. tan X $\frac{3}{4}$

Find the missing length. Then find the trigonometric ratios.

19. b 48

20. sin A 0.385

21. cos A 0.923

22. tan A 0.417

23. sin B 0.923

24. cos B 0.385

Find the missing length. Then find the trigonometric ratios. Rationalize denominators if necessary.

25. x $\sqrt{13}$

26. sin X $\frac{\sqrt{13}}{7}$

27. cos X $\frac{6}{7}$

28. tan X $\frac{\sqrt{13}}{6}$

29. cos Y $\frac{\sqrt{13}}{7}$

30. tan Y $\frac{6\sqrt{13}}{13}$

Refer to right triangle *DEF* at the right. Find the trigonometric ratios. Rationalize denominators if necessary.

31. sin D $\frac{\sqrt{14}}{7}$

32. cos D $\frac{\sqrt{35}}{7}$

33. tan D $\frac{\sqrt{10}}{5}$

34. sin E $\frac{\sqrt{35}}{7}$

35. cos E $\frac{\sqrt{14}}{7}$

36. tan E $\frac{\sqrt{10}}{2}$

In right triangle QRS, angles Q and R are acute angles. The leg adjacent to $\angle Q$ is 12 cm long. The leg opposite $\angle Q$ is 9 cm long.

37. Find the length of $\overline{QR}$.   15 cm

38. Find tan Q. 0.750

39. Find cos R. 0.600

Triangle *ABC* is an *isosceles* right triangle. The two legs have the same length, and the measure of each acute angle is 45.

40. Find the length of the hypotenuse. $\sqrt{2}$

41. Find sin 45, cos 45, and tan 45. Rationalize denominators if necessary. $\frac{\sqrt{2}}{2}, \frac{\sqrt{2}}{2}, 1$

Mixed Review

Factor.   *7.4, 7.5*

1. $2x^2 + x - 6$
$(2x - 3)(x + 2)$

2. $y^2 - 24y + 144$ $(y - 12)^2$

3. $n^2 - 25$
$(n - 5)(n + 5)$

Solve each equation.   *7.7*

4. $x^2 - 9 = 0$ $x = \pm 3$

5. $x^2 - 7x + 10 = 0$
$x = 2$ *or* $x = 5$

6. $2x^2 - 8x = 0$
$x = 0$ *or* $x = 4$

Enrichment

Have students draw a 30°-60°-90° triangle and show them that the hypotenuse is twice as long as the leg opposite the 30°-angle. Have them label the leg opposite the 30°-angle with a length of 1. Therefore, the hypotenuse has a length of 2. Tell them to use this triangle to find sin 30, cos 30, tan 30, sin 60, cos 60, and tan 60 to three decimal places. Then have them check their answers with those in the Trigonometric Table at the back of the book.

15.3 Trigonometric Tables

Objectives

To find the value of the sine, cosine, and tangent of an acute angle using a trigonometric table or a calculator

To find the measure of an angle using a trigonometric table or a calculator

To find the values of trigonometric ratios for angles with degree measures between 0 and 90, use the trigonometric table on page 643. Although the table gives *approximations* to four decimal places, the symbol = is used instead of ≈ for convenience.

EXAMPLE 1

Find each value.

a. sin 12   **b.** tan 47

Plan

Find the appropriate number in the Angle Measure column. Then read across to the correct column.

Solutions

a.
Angle Measure	sin	cos	tan
10	.1736	.9848	.1763
11	.1908	.9816	.1944
12 →	.2079	.9781	.2126
13	.2250	.9744	.2309

Thus, sin 12 = 0.2079.

b.
Angle Measure	sin	cos	tan
46	.7193	.6947	1.036
47 —	.7314	.6820 →	1.072
48	.7431	.6691	1.111
49	.7547	.6561	1.150

Thus, tan 47 = 1.072.

TRY THIS

Find each value: **1.** tan 10 .1763   **2.** cos 49 .6561

You can also use a trigonometric table to find the measure of an angle, given a trigonometric ratio of the angle.

EXAMPLE 2

Find m∠B if cos B = 0.2419.

Solution

Read down the cos column until you find 0.2419. Then read across to the Angle Measure column to find the correct angle measure.

Thus, m ∠B = 76.

Angle Measure	sin	cos	tan
75	.9659	.2588	3.732
76 ←	.9703	.2419	4.011
77	.9744	.2250	4.332
78	.9781	.2079	4.704

TRY THIS

3. Find m∠A if sin A = .9744. 77

◼◼◼ GETTING STARTED

Prerequisite Quiz

Find each square root to the nearest tenth. Use the table on page 642.

1. $\sqrt{2}$ 1.4   **2.** $\sqrt{12}$ 3.5
3. $\sqrt{23}$ 4.8   **4.** $\sqrt{5}$ 2.2
5. $\sqrt{17}$ 4.1   **6.** $\sqrt{33}$ 5.7

Motivator

Have students draw 5 right triangles with one angle having these measures: 15°, 30°, 45°, 60°, and 75°. Have them measure the sides of each triangle and use a calculator to evaluate sin 15, sin 30, sin 45, sin 60, and sin 75. Then ask them to complete this generalization: As the measure of an angle increases, the value of the sine ratio __?__ . Increases

◼◼◼ TEACHING SUGGESTIONS

Lesson Note

If students are using calculators, care must be taken to see that the calculator is set to the proper mode for working in degrees. A calculator set for radians or gradients will display an incorrect function for an angle measured in degrees. To find the measure of an angle whose sine is given, they must enter the value given followed by INV SIN or 2nd sin or $\sin^{-1}$.

Additional Example 1

Find each value. Use the table in Example 1.

a. cos 10   0.9848
b. tan 48   1.1111

Additional Example 2

Find the measure of ∠B if sin B = 0.9744. Use the table in Example 2.   77

Highlighting the Standards

Standards 9a, 6a, 6b: The application of trigonometry continues as students use tables and calculators to find values for various functions.

Math Connections

History: In the early 16th century, Copernicus suggested that the sun was at the center of a solar system in which earth was just one of the orbiting planets. To prove his theory, very accurate trigonometric tables were needed. By 1700 the enormous task of finding these values to at least ten decimal places for every ten-second interval had been accomplished entirely by hand calculations.

Critical Thinking Questions

Analysis: Ask students within what range of numbers must the sine of an acute angle lie. Between 0 and 1 Why? In a right triangle, the length of the side opposite the acute angle must be greater than 0 and less than the length of the hypotenuse.

Common Error Analysis

Error: Some students make careless errors in reading the tables.

Emphasize the need for care in reading the tables. Rulers, or the edges of index cards, can be used to help students read down along the correct column or across the correct row.

If the trigonometric ratio of an angle does not appear in the table, use the closest value to find the measure of the angle.

EXAMPLE 3   Find m $\angle A$ to the nearest degree if cos $A = 0.9710$.

Plan   Locate the value closest to 0.9710 in the cos column. Then read across to find the angle measure.

Solution

Angle Measure	sin	cos	tan	
11	.1908	.9816	.1944	
12	.2079	.9781	.2126	
13	.2250	.9744	.2309	
14 ←	.2419	.9703	.2493	← .9710 is closer to .9703 than to .9744.

Thus, m $\angle A = 14$, to the nearest degree.

TRY THIS   4. Find m $\angle B$ to the nearest degree if sin $B = .1937$. 11

Classroom Exercises

Use the tables in Example 1 to find the value of the trigonometric ratio.

1. sin 10 0.1736
2. cos 13 0.9744
3. tan 11 0.1944
4. cos 48 0.6691
5. tan 49 1.150
6. sin 46 0.7193

Use the table in Example 2 to find the measure of angle A.

7. tan $A = 4.704$ 78
8. cos $A = 0.2588$ 75
9. sin $A = 0.9781$ 78

Written Exercises

Find the value of the trigonometric ratio to four decimal places. Use the table on page 643.

1. sin 36 0.5878
2. cos 8 0.9903
3. tan 73 3.271
4. cos 15 0.9659
5. tan 31 0.6009
6. cos 54 0.5878
7. tan 12 0.2126
8. sin 81 0.9877
9. tan 43 0.9325
10. sin 48 0.7431
11. tan 4 0.0699
12. sin 79 0.9816
13. cos 21 0.9336
14. sin 2 0.0349
15. cos 86 0.0698

Find m $\angle A$ to the nearest degree. Use the table on page 643.

16. cos $A = 0.9877$ 9
17. tan $A = 0.5543$ 29
18. sin $A = 0.9925$ 83
19. tan $A = 0.0875$ 5
20. sin $A = 0.7193$ 46
21. cos $A = 0.5000$ 60

Additional Example 3

Find the measure of $\angle A$ to the nearest degree if sin $A = 0.2380$. Use the table in Example 3.   14°

22. sin A = 0.9205 67
23. cos A = 0.4067 66
24. tan A = 3.732 75
25. cos A = 0.8740 29
26. tan A = 2.500 68
27. sin A = 0.6460 40
28. tan A = 9.001 84
29. sin A = 0.0299 2
30. cos A = 0.1351 82
31. cos 60 + sin 30 1
32. sin 70 − cos 20 0
33. sin 25 + cos 65 0.8452
34. (sin 30)² + (cos 30)² 1
35. (sin 40)² + (cos 40)² 1
36. (sin 70)² + (cos 70)² 1

37. Determine the degree measure of ∠A if sin A = cos A. 45
38. Describe the relationship between m ∠A and m ∠B if sin A = cos B.
39. Show by choosing three measures of ∠A from the table that
sin A = cos (90 − A). Sample answer: sin 14 = 0.2419 = cos (90 − 14)
40. Show by choosing three measures of ∠A from the table that
tan A = $\frac{\sin A}{\cos A}$. Sample answer: $\frac{\sin 40}{\cos 40} = \frac{0.6428}{0.7660} = 0.8392 = \tan 40$

Midchapter Review

In Exercises 1–4, △ABC ~ △DEF. Find the missing length. *15.1*

1. a = 10, d = 6, c = 15. Find f. 9
2. b = 9, e = 5, and a = 18. Find d. 10
3. a = 15, b = 9, and d = 10. Find e. 6
4. e = 14, d = 16, and b = 7. Find a. 8

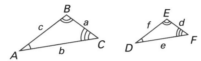

Find each of the following to three decimal places. *15.2*

5. sin A 0.280
6. cos A 0.960
7. tan A 0.292
8. sin B 0.960
9. cos B 0.280
10. tan B 3.429

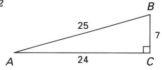

Find each of the following. Leave the answer in the fraction form. *15.2*

11. sin A $\frac{1}{3}$
12. cos A $\frac{2\sqrt{2}}{3}$
13. tan A $\frac{\sqrt{2}}{4}$
14. sin B $\frac{2\sqrt{2}}{3}$

Find the value of the trigonometric ratio. *15.3*

15. sin 38 0.6157
16. cos 9 0.9877
17. tan 41 0.8693

Find the measure of ∠A to the nearest degree. *15.3*

18. cos A = 0.8192 35
19. tan A = 4.103 76
20. sin A = 0.8301 56

21. A 3-m stick casts a 5-m shadow. At the same time, a nearby tree
casts a 15-m shadow. Find the height of the tree. *15.1* 9 m

15.3 Trigonometric Tables **585**

■■■**GETTING STARTED**

Prerequisite Quiz

Solve. Find x to the nearest tenth.

1. $\frac{x}{12} = 0.4532$ 5.4

2. $1.3561 = \frac{x}{9}$ 12.2

3. $0.3176 = \frac{10}{x}$ 31.5

4. $0.3821 = \frac{x}{5.1}$ 1.9

Motivator

Have students sketch a 3-4-5 right triangle on a piece of paper and ask them how they could use what they have learned about trigonometric ratios to find the measures of the acute angles of the triangle without using a protractor. Guide them through using the table to find one of the angles and then use the fact that the acute angles of a right triangle are complementary to find the measure of the other acute angle.   About 37° and 53°

15.4 Right-Triangle Solutions

Objectives

To find the lengths of the sides of a right triangle to the nearest tenth

To find the measures of the acute angles of a right triangle to the nearest degree

In this lesson, you will see how trigonometric ratios can be used to find the lengths of sides or the measures of angles in a right triangle.

EXAMPLE 1   In $\triangle ABC$, if m $\angle A = 24$ and $b = 32.0$, find a to the nearest tenth.

Plan   You know m $\angle A$ and the length of b, the leg adjacent to $\angle A$. To find the length a of the leg opposite $\angle A$, use the definition $\tan A = \dfrac{\text{opp}}{\text{adj}}$.

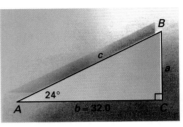

Solution

$$\tan 24 = \frac{a}{32}$$

$$32 \tan 24 = a$$

$$32(0.4452) = a$$

$$14.2464 = a$$

So, $a = 14.2$ to the nearest tenth.

EXAMPLE 2   In $\triangle ABC$, if m $\angle B = 50$ and $b = 37.0$, find c to the nearest tenth.

Plan   You know the measure of $\angle B$ and the length of the leg opposite $\angle B$. To find the length of the hypotenuse, use $\sin B = \dfrac{\text{opp}}{\text{hyp}}$.

Solution

$$\sin 50 = \frac{37}{c}$$

$$c(\sin 50) = 37$$

$$c = \frac{37}{\sin 50}$$

$$c = \frac{37}{0.7660} = 48.30$$

So, $c = 48.3$ to the nearest tenth.

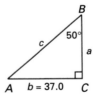

TRY THIS   In right triangle ABC in Example 2, if m $\angle A = 17$ and $c = 23.0$, find b to the nearest tenth. 22.0

Highlighting the Standards

Standards 7b, 9a: Students apply trigonometric functions to solving problems with geometric figures.

Additional Example 1

If the measure of $\angle A$ of $\triangle ABC$ is 38 and $b = 18.0$, find a to the nearest tenth.   14.1

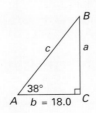

Additional Example 2

If the measure of $\angle B$ of $\triangle ABC$ is 64 and $a = 52.0$, find c to the nearest tenth.   118.6

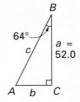

In Examples 1 and 2, a trigonometric ratio was used to find *the length of a side* of a right triangle when the following facts were known:

the measure of an acute angle in the triangle

the length of another side in the triangle

In the next example, you will see how a trigonometric ratio can be used to find *the measure of an acute angle* in a right triangle when the following facts are known:

lengths of *any two sides* of the triangle

EXAMPLE 3   If $a = 2.0$ and $b = 3.0$, find m $\angle A$ to the nearest degree.

Plan   You know the length of the leg opposite $\angle A$ and the length of the leg adjacent to $\angle A$. Use $\tan A = \dfrac{\text{opp}}{\text{adj}}$ to find the measure of $\angle A$.

Solution $\tan A = \dfrac{2}{3} = 0.6667$

The value closest to 0.6667 in the tangent column is $\tan 34 = 0.6745$.

So, m $\angle A = 34$, to the nearest degree.

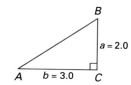

Classroom Exercises

For Exercises 1–5, refer to $\triangle ABC$.

1. Identify the leg opposite $\angle A$. *a*
2. Identify the leg adjacent to $\angle A$. *b*
3. Identify the hypotenuse of the triangle. *c*
4. What trigonometric ratio would you use to find *b*?
5. Find *b* to the nearest tenth. 9.2 $\cos 40 = \dfrac{b}{12}$

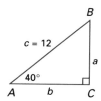

In Exercises 6–8, find the length of the indicated side to the nearest tenth, or the measure of the indicated angle to the nearest degree.

6.

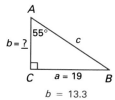

$b = 13.3$

7.

$b = 13.2$

8.

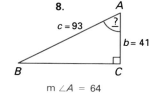

m $\angle A = 64$

15.4 Right-Triangle Solutions   **587**

Additional Example 3

If $a = 6.0$ and $b = 8.0$, find the measure of $\angle A$ to the nearest degree.   37

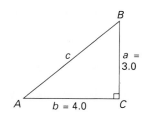

Lesson Note

You may wish to have students use calculators to solve the exercises for this lesson. The important element in these problems is the choice of the correct trigonometric ratio, *not* the routine computations. Show students that they can solve each problem by using the ratio made up of the side whose length is given and the side whose length must be found.

Math Connections

Language: The word *trigonometry* comes from two Greek words, *trigonon* and *metria*, meaning "triangle" and "measure," respectively. Trigonometry is a method of finding unknown measures of parts of a triangle by using the relationships that exist between the sides and angles of any triangle.

Critical Thinking Questions

Analysis: A triangle has 6 parts: 3 sides and 3 angles. In a right triangle, one part is always a right angle. Ask the students if it is always possible to find the missing measurements if you are given the measures of any two parts in addition to the right angle of a right triangle.   No; the measures of the two acute angles of a right triangle will not give any information about the lengths of the sides.

Common Error Analysis

Error: Some students may use the wrong ratio; for example, they might use sin instead of cos for $\dfrac{\text{adjacent side}}{\text{hypotenuse}}$.

Emphasize the importance of learning the definitions of the three trigonometric ratios presented in this chapter.

Checkpoint

Use right triangle *ABC* for these exercises.

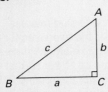

1. If $a = 12.0$, and m $\angle A = 15$, find b to the nearest tenth. **44.8**
2. If $c = 72.0$, and m $\angle A = 37$, find a to the nearest tenth. 43.3
3. If $a = 15.0$ and $b = 25.0$, find m $\angle B$ to the nearest degree. 59
4. If $b = 16$ and $c = 38$, find m $\angle A$ to the nearest degree. 65

Closure

Draw the following right triangle on the chalkboard and ask students how to use the given information to find the required measurement.

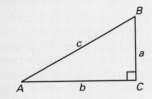

Given:	Find:	
1. a, A	c	$\sin A = \frac{a}{c}$
2. a, b	B	$\tan B = \frac{b}{a}$
3. b, c	A	$\cos A = \frac{b}{c}$
4. A, c	B	$B = 90 - A$
5. a, c	b	$a^2 + b^2 = c^2$
6. B, c	b	$\sin B = \frac{b}{c}$

Written Exercises

Find the length of the indicated side to the nearest tenth.

1.

2.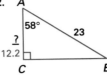

3.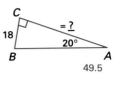

Find the measure of the indicated angle to the nearest degree.

4.

5.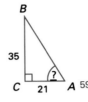

6.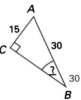

Find the missing lengths of sides to the nearest tenth and the missing measures of angles to the nearest degree.

7. If $a = 14$ and m $\angle A = 60$, find b. 8.1
8. If $b = 32$ and m $\angle B = 15$, find c. 123.6
9. If $a = 93$ and m $\angle B = 52$, find c. 151.1
10. If $c = 18$ and m $\angle B = 33$, find a. 15.1
11. If $b = 72$ and $c = 96$, find m $\angle A$. 41
12. If $a = 31$ and $b = 46$, find m $\angle B$. 56
13. If $a = 44$ and $c = 77$, find m $\angle A$. 35
14. If $a = 15$ and $b = 16$, find m $\angle B$. 47
15. If $a = 33.6$ and m $\angle B = 12$, find c. 34.4
16. If $a = 76.3$ and $c = 98.7$, find m $\angle A$. 51
17. If $b = 86.8$ and m $\angle A = 46$, find a. 89.9
18. If $b = 49.6$ and $c = 76.4$, find m $\angle B$. 40

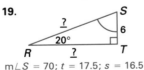

Find all the missing measures (lengths of sides to nearest tenth and measures of angles to nearest degree).

19.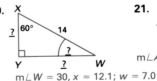

m$\angle S = 70$; $t = 17.5$; $s = 16.5$

20.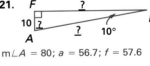

m$\angle W = 30$, $x = 12.1$; $w = 7.0$

21.

m$\angle A = 80$; $a = 56.7$; $f = 57.6$

Solve.

22. Anna drew a right triangle ABC with legs 4.0 cm and 5.0 cm long. What is the measure of $\angle A$? 39

23. Georgio drew a right triangle DEF with $\angle F$ the right angle. The degree measure of $\angle D$ was 38 and DF was 12.0 cm. What was the length of the hypotenuse? 15.2 cm

24. Shoshana drew a right triangle GHI with $\angle I$ the right angle. GI was 3.6 cm and IH was 4.8 cm. What was the measure of $\angle G$? 53

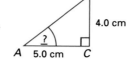

Figure $ABCD$ is a rhombus. A rhombus is a parallelogram whose diagonals bisect each other and are perpendicular to each other. The diagonals of $ABCD$ are $\overline{BD}$ and $\overline{AC}$. m $\angle DAC = 25$ and $DE = 4.0$.

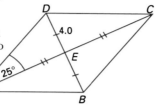

Find the indicated lengths to the nearest tenth and angle measures to the nearest degree.

25. AD 9.5	**26.** BC 9.5	**27.** EC 8.6
28. AE 8.6	**29.** BE 4.0	**30.** m $\angle DCE$ 25
31. m $\angle EBA$ 65	**32.** AB 9.5	**33.** DC 9.5

Mixed Review

Add or subtract. 8.5, 8.6, 8.7

1. $\dfrac{8y}{5} + \dfrac{3y}{5}$ $\frac{11y}{5}$

2. $\dfrac{9}{10n} - \dfrac{5}{2n^2}$ $\frac{9n-25}{10n^2}$

3. $\dfrac{4}{x-6} - \dfrac{1}{x^2 - 2x - 24}$ $\frac{4x+15}{x^2-2x-24}$

Solve by using the quadratic formula. Write irrational solutions in simplest radical form. 14.3

4. $x^2 - 7x + 3 = 0$ $\frac{7 \pm \sqrt{37}}{2}$

5. $2x + 7 = -x^2$ No real number

6. $3 = 5x^2 - 11x$ $\frac{11 \pm \sqrt{181}}{10}$

Solve. 4.7

7. Ninety-two percent of 150 is what number? 138

8. Thirty-eight percent of what number is 19? 50

9. Thirty-two is what percent of 80? 40%

▰▰/Brainteasers

1. If x is a real number, what is the least value of this product?
$(x + \sqrt{2} + \sqrt{3})(x + \sqrt{2} - \sqrt{3})(x - \sqrt{2} + \sqrt{3})(x - \sqrt{2} - \sqrt{3})$ −24

2. Solve for x if $x^2 + |x| = 30$. ±5

3. Find all values of x such that $|x| + 2 = |x - 2|$. $x \leq 0$

15.4 Right-Triangle Solutions **589**

◼◼◼FOLLOW UP

Guided Practice

Classroom Exercises 1–8
Try This   all

Independent Practice

A Ex. 1–14, **B** Ex. 15–24, **C** Ex. 25–33

Basic: WE 1–19 odd, Brainteaser,

Average: WE 7–25 odd, Brainteaser

Above Average: WE 15–33 odd, Brainteaser

Enrichment

Introduce these three additional trigonometric ratios: cosecant (csc), secant (sec), and cotangent (cot).

$\csc A = \dfrac{\text{hyp}}{\text{opp}}$

$\sec A = \dfrac{\text{hyp}}{\text{adj}}$

$\cot A = \dfrac{\text{adj}}{\text{opp}}$

Have students find the relationship between these three ratios and the sine, cosine and tangent of $\angle A$.

$\csc A = \dfrac{1}{\sin A}$, $\sec A = \dfrac{1}{\cos A}$, $\cot A = \dfrac{1}{\tan A}$

◼◼◼GETTING STARTED

Prerequisite Quiz

Solve. Use the table on page 643.
Give each answer to the nearest
whole number.

1. $x \cos 35 = 18$ 22
2. $\tan 18 = \frac{49}{x}$ 151
3. $\sin 73 = \frac{x}{16}$ 15
4. $\cos 59 = \frac{103}{x}$ 200

Motivator

Ask students whether they elevate or
depress their line of vision when they stand
on the ground and look up to see an airplane
flying overhead.   Elevate   Ask them if the
pilot of a plane elevates or depresses his
line of vision when he looks downward
toward the airport runway when landing.
Depresses

15.5 Problem Solving: Applying Trigonometry

Objective   To solve problems using trigonometric ratios

The following examples illustrate applications of trigonometric ratios.
All answers are rounded and given as approximations.

EXAMPLE 1   Find the distance across the pond to the nearest meter.

Plan   Use $\tan D = \dfrac{\text{opp}}{\text{adj}}$.

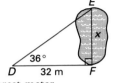

Solution $\tan 36 = \dfrac{x}{32}$

$x = 32(\tan 36) = 32(0.7265)$, or 23.25

So, the distance across the pond is 23 m , to the nearest meter.

An **angle of elevation** is an angle formed by
a horizontal line and the line of sight to a
point at a higher elevation. An **angle of
depression** is an angle formed by a horizontal line and the line of sight to a point at a
lower elevation. For any given line of sight,
the measure of the angle of elevation equals
the measure of the angle of depression.

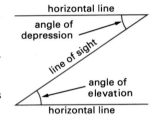

EXAMPLE 2   A ship is 335 m from the base of a cliff. The degree measure of the
angle of elevation from the ship to the top of the cliff is 34. Find the
distance from the ship to the top of the cliff. Give the answer to the
nearest 10 meters.

Plan   Use $\cos S = \dfrac{\text{adj}}{\text{hyp}}$.

Solution $\cos 34 = \dfrac{335}{x}$

$x(\cos 34) = 335$

$x = \dfrac{335}{\cos 34} = \dfrac{335}{0.8290}$, or 404.1013

So the distance from the ship to the top of the cliff is about 400 m .

Highlighting the Standards

Standards 1d, 9a: Students use
trigonometric relationships to solve problems
from the world outside the classroom.

Additional Example 1

How long a ladder is needed to reach the
top of a building 39 ft high if the ladder
makes an angle that measures 67 with the
ground? Give the answer to the nearest
foot.   42 ft

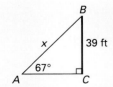

Classroom Exercises

The figure at the right shows a pond and right triangle *XYZ*. Identify each given side as the leg opposite ∠*X*, the leg adjacent to ∠*X*, or the hypotenuse.

1. $\overline{XY}$ Hyp

2. $\overline{XZ}$ Adj to ∠*X*

3. $\overline{YZ}$ Opp ∠*X*

4. In right triangle *XYZ*, which trigonometric ratio would you use to find *x*, the distance across the lake? $\tan X = \frac{x}{56}$

5. In right triangle *XYZ*, find *x*. 47 m

For Classroom Exercises 6–7, use right triangle *ABC*.

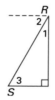

6. What is the degree measure of the angle of elevation? 71

7. Find the height of the tree to the nearest foot. 17 ft

8. In the figure at the right, identify which angle is the angle of depression from point *R* to point *S*. ∠2

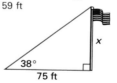

Written Exercises

1. Find the length *x* of the ladder to the nearest foot. 11 ft

2. Find the height *x* of the flagpole to the nearest foot. 59 ft

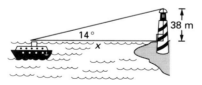

3. Find the length *x* of the ladder to the nearest meter. 20 m

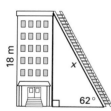

4. Find the distance *x* from the ship to the lighthouse to the nearest meter. 152 m

15.5 Problem Solving: Applying Trigonometry **591**

Additional Example 2

Find *y*, the distance across the lake. Give the answer to the nearest meter. 60 m

591

Checkpoint

1. Find the length of the ladder to the nearest foot. **22 ft**

2. Find the height of the building to the nearest meter. **9 m**

Closure

Have each student draw a sketch similar to one of those in the lesson with enough information to solve for an unknown measurement. Then have students exchange papers and work each others problems.

▰▰▰FOLLOW UP

Guided Practice

Classroom Exercises 1–8
Try This   all

Independent Practice

A Ex. 1–4, **B** Ex. 5–12, **C** Ex. 13–14
Basic: WE 1–6 all, Mixed Problem Solving
Average: WE 1–11 odd, Mixed Problem Solving
Above Average: WE 5–11 odd, 13, 14, Mixed Problem Solving

Solve each problem. Find lengths to the nearest integer and angle measures to the nearest degree.

5. A kite is flying at the end of a 220-m string. How high above the ground is the kite if the string has an angle of elevation of measure 55? **180 m**

6. A tree casts a 30-m shadow when the degree measure of the angle of elevation of the sun is 24. How tall is the tree? **13 m**

7. The degree measure of the angle of depression from the top of a lighthouse 120 ft high to an object in the water is 63. How far from the base of the lighthouse is the object? The foot of the lighthouse is at sea level. **61 ft**

8. A straight road up a hill is 430 m long and has an angle of elevation of degree measure 12. Find the height of the hill. **89 m**

9. A plane flying at an altitude of 9,400 m makes an angle of depression with its carrier that has degree measure 28. How far is the plane from its carrier? **20,021 m**

10. A ramp is 156 m long. It rises a vertical distance of 31 m. Find the measure of the angle of elevation. **11**

11. An airplane flies in a northeasterly direction making an angle with the north line that has degree measure 35. If the plane flies a distance of 275 mi, how many miles east will it fly? **158 mi**

12. In Exercise 11, suppose the plane flies in a northeasterly direction at an angle of degree measure 42 from the north line. How many miles would it have to fly to be 215 mi east of the starting position? **321 mi**

Suppose that a rectangular park is labeled _ABCD_. A straight diagonal path from _B_ to _D_ is 85 ft long and makes an angle of degree measure 22 with each of the longer sides of the rectangle.

13. Find the length of the park to the nearest foot. **79 ft**

14. Find the area of the park to the nearest hundred square feet. **2,500 ft²**

Mixed Review

Simplify if possible. _6.1, 6.2, 6.3_

1. $-4xy^2z(5x^3y^5z)$ $-20x^4y^7z^2$

2. $(-3a^4)^2$ $9a^8$

3. $-(4m^9)^4$ $-256m^{36}$

4. $\dfrac{-18a^4b^2}{3ab^2}$ $-6a^3$

Solve the system of equations. _12.4, 12.5_

5. $x + y = 12$
   $x - y = 6$ (9, 3)

6. $-3x + 2y = 1$
   $-x - y = 2$ (−1, −1)

7. $5x - 4y = 1$
   $3x + 2y = 5$
   (1, 1)

You have now used a number of problem-solving strategies. Here are some of them.

Checking Assumptions   Making a Graph
Defining the Variables   Making a Table
Drawing a Diagram   Solving a Simpler Problem
Estimating Before Solving   Using Formulas
Guessing and Checking   Working Backwards

Use one or more of these strategies to solve these problems.

6. 4.32×10^4

1. Twenty-four is what percent of 60? 40% **2.** Eighty is 32% of what number? 250

3. If $8,000 is invested at a simple interest rate of 8.4% per year for 3 years, how much interest is earned? $2,016

4. An isosceles triangle has a vertex angle with degree measure 42. What is the degree measure of a base angle? 69

5. Find three consecutive even integers whose sum is 72. 22, 24, 26

6. Express the number of seconds in a 12-hour day in scientific notation.

7. The area of a garden that is circular is 64π ft². What is the diameter of the garden? 16 ft

8. The area of a square is 27 cm². Find the length of a side to the nearest tenth. 5.2 cm

9. A hiker walked 6 mi south and then 8 mi east. How far was the hiker from the starting point? 10 mi

10. Find the area of a rectangle if its length is $8\sqrt{2}$ cm and its width is $5\sqrt{3}$ cm. $40\sqrt{6}$ cm²

11. The bending of a beam varies directly as its mass. A beam is bent 20 mm by a mass of 40 kg. How much will the beam bend with a mass of 100 kg? 50 mm

12. It takes Karen 3 h to cut the grass. Scott takes $3\frac{1}{2}$ h to cut the same area. How long will it take them to cut the grass together? $1\frac{8}{13}$ h

13. Write an equation of the line with slope $\frac{1}{2}$ and y-intercept the same as that of the line described by the equation $y = -\frac{2}{3}x + 5$. $y = \frac{1}{2}x + 5$

14. Walt has $3.30 in dimes and quarters. The number of dimes is 2 less than the number of quarters. How many coins of each type does he have?
8 dimes, 10 quarters

15. Cashews cost $11.25/kg. Pecans cost $13/kg. A box contains a mixture that will sell for $136.00. The number of kilograms of cashews in the box is 3 less than the number of kilograms of pecans. How many kilograms of each are in the box?
4 kg cashews, 7 kg pecans

16. Two cities are 1,630 mi apart. Flying with the wind, a jet took 2 h 30 min to travel from one city to the other. On the return trip, the jet flew against the same wind and took 3 h. Find the rate of the jet in still air and the rate of the wind. $597\frac{2}{3}$ mi/h, $54\frac{1}{3}$ mi/h

Enrichment

Have students research and write a short paper on how trigonometry is used in navigation or surveying.

Key Terms

angle of depression   (p. 590)
angle of elevation   (p. 590)
corresponding angles   (p. 575)
corresponding sides   (p. 575)

cosine ratio   (p. 580)
similar triangles   (p. 575)
sine ratio   (p. 580)
tangent ratio   (p. 580)

Key Ideas and Review Exercises

15.1   In similar triangles, all pairs of corresponding sides have the same ratio. To find the length of a side in one of two similar triangles, use a proportion involving the known lengths and the unknown length.

In the figures below, $\triangle ABC \sim \triangle DEF$.

1. If $a = 12$, $c = 10$, and $d = 6$, find f. 5
2. If $d = 12$, $e = 18$, and $b = 24$, find a. 16

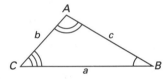 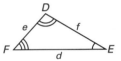

15.2   Three trigonometric ratios are defined.

$$\text{sine} = \frac{\text{opposite}}{\text{hypotenuse}} \qquad \text{cosine} = \frac{\text{adjacent}}{\text{hypotenuse}} \qquad \text{tangent} = \frac{\text{opposite}}{\text{adjacent}}$$

Find each of the following to three decimal places.

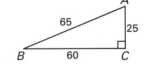

3. $\sin A$ 0.923
4. $\cos A$ 0.385
5. $\tan A$ 2.400

Find the missing length. Then find the trigonometric ratios. Rationalize the denominator if necessary.

6. x 4
7. $\cos Y$ $\frac{1}{2}$
8. $\tan X$ $\frac{\sqrt{3}}{3}$

15.3 Use trigonometric tables or a calculator to find the sine, cosine, or tangent of an acute angle, or to find the measure of an acute angle when its trigonometric ratio is given.

Find the value of the trigonometric ratio.

9. sin 38 0.6157

10. tan 42 0.9004

11. cos 62 0.4695

12. sin 28 0.4695

Find m $\angle C$ to the nearest degree.

13. cos C = 0.7193 44

14. sin C = 0.8387 57

15. tan C = 0.5310 28

16. cos C = 0.4700 62

15.4 To find the length of a side or the measure of an acute angle in a right triangle, use trigonometric ratios.

Find the length of the indicated side to the nearest tenth, or the measure of the indicated angle to the nearest degree.

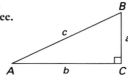

17. If a = 16 and m $\angle A$ = 32, find c. 30.2

18. If b = 45 and a = 17, find m $\angle B$. 69

19. Write in your own words how to find the measure of an acute angle of a right triangle when
a. the length of a leg and the length of the hypotenuse are known.
b. the lengths of the two legs are known. Answers will vary.

15.5 Use trigonometric ratios to solve problems involving angle measures and lengths of sides in a right triangle.

20. Find the height of the tent pole to the nearest foot. Refer to the figure at the right. 11 ft

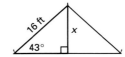

21. The height of a building is 40 m. How far from the building would you be if the degree measure of your angle of elevation to the top of the building were 38? Give your answer to the nearest meter. 51 m

21. $\sin A = \frac{a}{c}$ and $\cos A = \frac{b}{c}$

$\dfrac{\sin A}{\cos A} = \dfrac{\frac{a}{c}}{\frac{b}{c}} = \dfrac{a}{b} = \tan A$

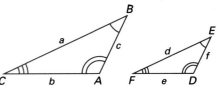

Chapter 15 Test

A Exercises: 1, 2, 4–9, 11–15, 17, 18   B Exercises: 3, 10, 16, 19, 20   C Exercise: 21

In Exercises 1 and 2, $\triangle ABC \sim \triangle DEF$.

1. If $a = 22$, $c = 8$, and $f = 4$, find d. 11

2. If $b = 24$, $c = 12$, and $e = 6$, find f. 3

3. A 3-m stick casts a 12-m shadow while a tree casts a 16-m shadow. How high is the tree? 4 m

Find each of the following to three decimal places.

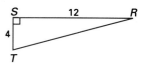

4. $\sin A$ 0.923

5. $\cos A$ 0.385

6. $\tan B$ 0.417

Find each of the following. Leave the answers in fraction form.

7. $\tan M$ $\sqrt{3}$

8. $\cos M$ $\frac{1}{2}$

9. $\sin N$ $\frac{1}{2}$

10. Refer to $\triangle RST$ at the right. Find the missing length. Then find $\cos R$. Rationalize the denominator if necessary.

$$s = 4\sqrt{10}, \ \cos R = \frac{3\sqrt{10}}{10}$$

Find each value.

11. $\sin 34$ 0.5592

12. $\cos 81$ 0.1564

13. $\tan 14$ 0.2493

Find m $\angle B$ to the nearest degree.

14. $\sin B = 0.6018$ 37

15. $\cos B = 0.3907$ 67

16. $\tan B = 9.500$ 84

Find the length of the indicated side to the nearest tenth, or the measure of the indicated angle to the nearest degree.

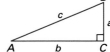

17. If $a = 8$ and m $\angle A = 21$, find c. 22.3

18. If $b = 14$ and $a = 7$, find m $\angle A$. 27

19. If $b = 4.8$ and $c = 11.2$, find m $\angle B$. 25

20. A plane flying at an altitude of 7,000 ft makes an angle of depression of degree measure 32 with its carrier. To the nearest hundred feet, how far is the plane from the carrier? 13,200 ft

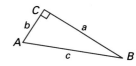

21. Use the figure at the right to show that $\tan A = \dfrac{\sin A}{\cos A}$.

College Prep Test

Choose the *one* best answer to each question or problem. (The figures in this test may not be drawn to scale.)

1. x, y, and z are in the ratio of 3:1:2.
B Find z.

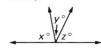

(A) 30   (B) 60   (C) 90
(D) 15   (E) 45

Find the true statement.

2.
C

(A) $m \angle 1 = m \angle 3$
(B) $m \angle 3 = m \angle 4$
(C) $m \angle 1 + m \angle 2 = m \angle 3 + m \angle 4$
(D) $m \angle 2 = m \angle 4$
(E) $m \angle 1 = m \angle 4$

3.
C

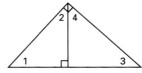

(A) $x > y$ (B) $x < y$
(C) $x + y = 150$ (D) $x > 30$
(E) $x + y = 90$

4. $\tan W =$ _____
E

(A) $\frac{\sqrt{6}}{2}$ (B) $\frac{\sqrt{6}}{6}$ (C) $\frac{\sqrt{3}}{3}$
(D) $\frac{\sqrt{2}}{2}$ (E) 1

5. $\sin T =$ _____
B

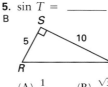

(A) $\frac{1}{2}$ (B) $\frac{\sqrt{5}}{5}$ (C) $\frac{\sqrt{3}}{3}$
(D) $\frac{\sqrt{2}}{2}$ (E) 1

6.
A

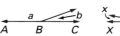

(A) $\sin A = \cos B$
(B) $\sin A = \cos A$
(C) $\tan A = \tan B$
(D) $\sin B = \cos B$
(E) $\cos A = \cos B$

7. $\overline{AC}$ and $\overline{XZ}$ are straight lines;
C $b = 10°$, $y = 170°$

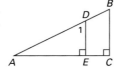

(A) $x + y = x + b$
(B) $y + b = y + a$
(C) $a + b = x + a$
(D) $y + a = 180$ (E) $x + b = 90$

8.
A

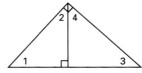

(A) $m \angle 1 = m \angle B$
(B) $m \angle A = m \angle B$
(C) $\frac{AB}{AD} = \frac{AE}{AC}$ (D) $\frac{BC}{AD} = \frac{DB}{AE}$
(E) $m \angle A = m \angle 1$

Additional Answers, page 585

38. Sample answer:
   sin 45 = 0.7071 = cos 45
   = cos(90 − 45);
   sin 30 = 0.5000 = cos 60
   = cos(90 − 30);
   sin 14 = 0.2419 = cos 76
   = cos(90 − 14).
   m ∠A = 90 − m ∠B

PROBABILITY AND STATISTICS

OVERVIEW

This chapter introduces students to probability and statistics, two topics that have become increasingly important in the study of mathematics. Students calculate probabilities with common objects such as coins and dice. The mean, median, and mode are presented as important statistical averages. Then students have a chance to interpret two important types of graphs, the scatter plot and the histogram. Finally, students are asked to find the standard deviation of a set of data.

OBJECTIVES

- To determine the probability of a simple and a compound event
- To find the mean, median, and mode for a set of data
- To read and interpret a scatter plot and a histogram
- To find the range and standard deviation for a set of data

PROBLEM SOLVING

Drawing a Diagram is the technique used to introduce the counting principle that is used to find probabilities. The students then Interpret Key Words in order to find the mean, median, and mode of a set of data and Construct a Table to find the same measures. To find the standard deviation, students must be able to Use the Formula for this measure of data.

READING AND WRITING MATH

The Focus on Reading exercises on page 611 review the important terms related to finding different types of averages. On page 618, Exercise 15 asks the students to write a short description of a scatter plot.

TECHNOLOGY

Calculator: The calculator is a valuable tool in statistical applications. Finding the mean of a large set of data, as in the exercises on page 612, and the calculations needed to find standard deviations on page 620 become less tedious when a calculator is used.

SPECIAL FEATURES

Mixed Review pp. 602, 607, 618, 621
Brainteaser p. 607
Midchapter Review p. 613
Focus on Reading p. 611
Key Terms p. 622
Key Ideas and Review Exercises
   pp. 622–623
Chapter 16 Test p. 624
College Prep Test p. 625
Cumulative Review (Chapters 1–16)
   pp. 626–627

PLANNING GUIDE

Lesson	Basic	Average	Above Average	Resources
16.1 pp. 601–602	CE 1–8 WE 1–25 odd	CE 1–8 WE 5–29 odd	CE 1–8 WE 11–35 odd	Reteaching 118 Practice 118
16.2 pp. 606–607	CE 1–8 WE 1–12 all Brainteaser	CE 1–8 WE 1–23 odd Brainteaser	CE 1–8 WE 5–27 odd, 28–30 all Brainteaser	Reteaching 119 Practice 119
16.3 pp. 611–613	FR all, CE 1–9 WE 1–7 odd, 9–20 all Midchapter Review	FR all, CE 1–9 WE 1–19 odd, 21–26 all Midchapter Review	FR all, CE 1–9 WE 9–20 odd, 21–24 all, 25–33 odd Midchapter Review	Reteaching 120 Practice 120
16.4 pp. 616–618	CE 1–8 WE 1–10 all	CE 1–8 WE 1–4, 9–14 all	CE 1–8 WE 9–18 all	Reteaching 121 Practice 121
16.5 pp. 620–621	CE 1–9 WE 1–9 all	CE 1–9 WE 1–17 odd	CE 1–9 WE 5–17 odd 19–22 all	Reteaching 122 Practice 122
Chapter 16 Review pp. 622–623	all	all	all	
Chapter 16 Test p. 624	all	all	all	
College Prep Test p. 625	all	all	all	
Cumulative Review (Chapters 1–16) pp. 626–627	all	all	all	

CE = Classroom Exercises   WE = Written Exercises   FR = Focus on Reading

Note: For each level, all students should be assigned all Try This and all Mixed Review Exercises.

◼◼◼ INVESTIGATION

Project: This *Investigation* shows students some surprising results when taking averages of a group of data.

Materials: Each student will need a copy of Project Worksheet 16. Each group of students will need at least one calculator.

Have students draw a picture as you give the following information. A grey bag has 7 yellow and 9 blue marbles. The chances of drawing a yellow marble is $\frac{7}{16}$.

Have them use a calculator to give this result as a two-place decimal. 0.44 A red bag has 5 yellow and 7 blue marbles. Ask for the chances of drawing a yellow marble. $\frac{5}{12} = 0.42$

If drawing a yellow marble will win a prize, ask students from which bag should they draw in order to have a better chance of winning. grey

Now have the students put all the marbles in the two grey bags into one grey bag, and in like manner, all the other marbles in one red bag. What are the chances of drawing a yellow marble from the grey bag? $\frac{10}{20} = 0.50$; the red bag? $\frac{16}{27} = 0.59$ Which bag gives you the better chance of winning? red

Now hand out the worksheet for students to complete in small groups.

Mae Jamison, an astronaut, is pictured here during ejection seat training at Vance Air Force Base. Before joining the space shuttle program, Dr. Jamison worked in engineering, with computers, and as a medical officer in Africa.

More About the Photo

Mae Jemison, M.D., has traveled far from her birthplace in Decatur, Alabama. Jemison not only received her degree in Chemical Engineering from Stanford, she also fullfilled the requirements for a second degree in African and Afro-American Studies. Her interest in Africa led her, after she became a doctor in 1981, to serve in Sierra Leone and Liberia with the Peace Corps. She was working as a general practitioner in Los Angeles when NASA (National Aeronautics and Space Administration) selected her as an astronaut candidate.

16.1 Probability of an Event

Teaching Resources

Project Worksheet 16
Quick Quizzes 118
Reteaching and Practice Worksheets 118
Transparencies 46A, 46B

Objectives   To determine the probability of an event
To determine the probability of an event that is certain to happen and one that cannot happen

Probability is the study of the chances that particular events will occur.

When a coin is tossed, there are 2 *possible outcomes*: heads or tails. These are *equally likely* outcomes if the coin is "fair." Assume that you want tails as the outcome. Then tails is called a *favorable* or *successful* outcome. Out of 2 possible outcomes, there is one way to get the outcome *tails*. So the probability of getting tails is $\frac{1}{2}$. A way to determine the probability of an event is given in the following formula.

$$\textbf{Probability of an event} = \frac{\text{number of successful outcomes}}{\text{total number of possible outcomes}}$$

In symbols, $P(E) = \frac{s}{t}.$

EXAMPLE 1   In one throw of a *die* (singular for *dice*), what is the probability of each event?
a. The upper face will show 4 dots.
b. The upper face will show an even number of dots.

Since there are 6 possible outcomes, $t = 6$.
Find the successful outcomes s for the event.

Then use $P(E) = \frac{s}{t}$.

Solutions   **a.** The upper face will show 4 dots.
1 successful outcome (4); so, $s = 1$.
$P(4) = \frac{s}{t} = \frac{1}{6}$

b. The upper face will show an even number of dots.
3 successful outcomes (2, 4, or 6); so, $s = 3$.
$P(\text{even number}) = \frac{s}{t} = \frac{3}{6} = \frac{1}{2}$

TRY THIS   In one throw of a die, what is the probability of each event?
1. The upper face will show one dot. $\frac{1}{6}$
2. The upper face will show an odd number of dots. $\frac{1}{2}$
3. The upper face will show three or more dots. $\frac{2}{3}$

Additional Example 1

In one throw of a die, what is the probability of each event?
a. The upper face will show 1 dot. $\frac{1}{6}$
b. The upper face will show more than 3 dots. $\frac{1}{2}$

Highlighting the Standards

Standard 11b: Students use probability to represent and solve problems involving uncertainty.

Motivator

Activity: Use the following activity to illustrate that the probability for a certain outcome in an experiment is a reasonably good prediction rather than an exact prediction. Have each student toss a coin 10 times, list the results, and compute the ratio $\frac{\text{number of heads}}{\text{number of tosses}}$. Compare the results. The ratio of the total number of heads over the total number of tosses should be close to $\frac{1}{2}$.

Have them repeat the experiment with 20 tosses and then 30 tosses and compare the results. Generally, as the number of tosses increases, the closer the ratio of the number of heads to the total number of tosses gets to $\frac{1}{2}$.

◼◼◼ TEACHING SUGGESTIONS

Lesson Note

Be sure that students understand that $P(E)$ does not indicate multiplication but rather the probability of an event. Emphasize that most probabilities are expressed as fractions. All probabilities are in the range $0 \leq P \leq 1$.

Math Connections

Meteorology: If a television weather report indicates a 20% chance of rain, this is the probability $\frac{1}{5}$ expressed as a percent. It means that about 20% of the viewing area can expect to have some rain and that about 80% can expect no rain.

EXAMPLE 2   A bag contains 12 marbles, 10 are white marbles and 2 are blue marbles. One marble is drawn from the bag. Find the probability of each event.

a. A white marble is drawn.   b. A blue marble is drawn.

Plan   There are 12 possible outcomes; so, $t = 12$. Find the successful outcomes s for the event. Then use $P(E) = \frac{s}{t}$.

Solutions   a. A white marble is drawn.   b. A blue marble is drawn.
10 successful outcomes;   2 successful outcomes;
so, $s = 10$.   so, $s = 2$.

$P(\text{white}) = \frac{s}{t} = \frac{10}{12}$, or $\frac{5}{6}$ $P(\text{blue}) = \frac{s}{t} = \frac{2}{12}$, or $\frac{1}{6}$

EXAMPLE 3   A bag contains 8 dimes. One coin is drawn from the bag. Find each probability.

a. $P(\text{dime})$ b. $P(\text{quarter})$

Plan   There are 8 possible outcomes; so, $t = 8$. Use $P(E) = \frac{s}{t}$.

Solutions   a. $P(\text{dime})$ b. $P(\text{quarter})$
8 successful outcomes;   no successful outcomes;
so, $s = 8$.   so, $s = 0$.

$P(\text{dime}) = \frac{s}{t} = \frac{8}{8}$, or 1 $P(\text{quarter}) = \frac{s}{t} = \frac{0}{8}$, or 0

TRY THIS   One marble is drawn from a bag containing two yellow and six red marbles. Find the probability of each event.

4. A red marble is drawn. $\frac{3}{4}$
5. A yellow marble is drawn. $\frac{1}{4}$
6. A white marble is drawn. 0

Example 3 illustrates two important ideas.
If an event is *certain* to happen, its probability is 1: $P(E) = 1$.
If an event *cannot happen*, its probability is 0: $P(E) = 0$.
All probabilities fall in the interval from 0 to 1. Notice that in Example 2, the probability of drawing a white marble is $\frac{5}{6}$. The probability of *not drawing* a white marble is $\frac{1}{6}$. Thus,

$$P(\text{white}) + P(\text{not white}) = \frac{5}{6} + \frac{1}{6} = 1$$

600   Chapter 16 Probability and Statistics

Additional Example 2

A bag contains 20 marbles; 15 are white marbles and 5 are red marbles. One marble is drawn from the bag. Find the probability of each event.

a. A red marble is drawn. $\frac{1}{4}$
b. A white marble is drawn. $\frac{3}{4}$

Additional Example 3

A drawer contains only black socks. One sock is pulled from the drawer. Find each probability.

a. $P(\text{black socks})$ 1
b. $P(\text{white socks})$ 0

In general, if $P(E)$ is the probability of an event E occurring and $P(\text{not } E)$ is the probability of E not occurring, then

$$P(E) + P(\text{not } E) = 1 \quad \text{and} \quad P(\text{not } E) = 1 - P(E).$$

Classroom Exercises

A bag contains 4 white marbles and 6 blue marbles. One marble is drawn. Find the probability of each event.

1. A white marble is drawn. $\frac{2}{5}$
2. A blue marble is drawn. $\frac{3}{5}$
3. A red marble is drawn. 0
4. A red marble is not drawn. 1

A person spins the pointer of this spinner. Find each probability.

5. $P(4)$ $\frac{1}{6}$
6. $P(1)$ 0
7. $P(\text{a number less than 6})$ $\frac{1}{3}$
8. $P(\text{a number less than 14})$ 1

Written Exercises

In one throw of a die, what is the probability of each event?

1. The upper face will show 3 dots. $\frac{1}{6}$
2. The upper face will show 5 dots. $\frac{1}{6}$
3. The upper face will show an odd number of dots. $\frac{1}{2}$
4. The upper face will show less than 5 dots. $\frac{2}{3}$
5. The upper face will show 2 dots. $\frac{1}{6}$
6. The upper face will not show 2 dots. $\frac{5}{6}$

Refer to the spinner at the right. Find the probability of each event.

7. $P(\text{white})$ $\frac{3}{8}$
8. $P(\text{red})$ $\frac{1}{2}$
9. $P(\text{blue})$ $\frac{1}{8}$
10. $P(\text{not white})$ $\frac{5}{8}$
11. $P(\text{not red})$ $\frac{1}{2}$
12. $P(\text{not blue})$ $\frac{7}{8}$

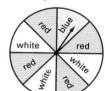

13. $P(3)$ $\frac{1}{2}$
14. $P(4)$ $\frac{1}{3}$
15. $P(2)$ 0
16. $P(1)$ $\frac{1}{6}$
17. $P(\text{even number})$ $\frac{1}{3}$
18. $P(\text{not 5})$ 1

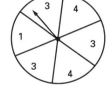

A bag contains 20 marbles; 4 are white, 10 are blue, and 6 are red. One marble is drawn. Find the probability of each event.

19. $P(\text{white})$ $\frac{1}{5}$
20. $P(\text{blue})$ $\frac{1}{2}$
21. $P(\text{red})$ $\frac{3}{10}$
22. $P(\text{green})$ 0
23. $P(\text{not white})$ $\frac{4}{5}$
24. $P(\text{not blue})$ $\frac{1}{2}$
25. $P(\text{not red})$ $\frac{7}{10}$
26. $P(\text{not green})$ 1

16.1 Probability of an Event **601**

Application: Pose the following problem to the students. Suppose you throw a "fair" die 10 times and 6 dots land facing up each time. If you throw the die once more, is the probability that 6 dots will again land facing up greater than $\frac{1}{6}$, equal to $\frac{1}{6}$, or less than $\frac{1}{6}$? Equal to $\frac{1}{6}$

Common Error Analysis

Error: In problems like Exercise 14, some students may think that $P(4) = \frac{4}{6}$.

Remind them that the definition of probability on page 599 states that the numerator is the *number* of successful outcomes.

Checkpoint

1. What is the probability that, in one throw of a die, a 4 will appear? $\frac{1}{6}$ a 5 will not appear? $\frac{5}{6}$

2. A bag contains 16 red balls and 8 blue balls. What is the probability of drawing a red ball? $\frac{2}{3}$ a blue ball? $\frac{1}{3}$ a white ball? 0

Refer to the spinner at the right below. Find the probability of each event.

3. $P(\text{red})$ $\frac{1}{4}$
4. $P(\text{blue})$ $\frac{1}{4}$
5. $P(\text{white})$ $\frac{1}{2}$
6. $P(\text{not red})$ $\frac{3}{4}$

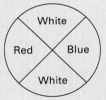

Closure

Have students answer the following questions without paper and pencil.

In one throw of a die, what is the probability of each event?

1. The upper face will show 1 dot. $\frac{1}{6}$
2. The upper face will show an even number of dots. $\frac{1}{2}$
3. The number of dots on the upper face will be greater than 4. $\frac{1}{3}$
4. The number of dots on the upper face will be greater than 6. 0
5. The upper face will *not* show 3 dots. $\frac{5}{6}$

◼◼◼FOLLOW UP

Guided Practice

Classroom Exercises 1–8
Try This   all

Independent Practice

A Ex. 1–26, **B** Ex. 27–30, **C** Ex. 31–35

Basic: WE 1–25 odd

Average: WE 5–29 odd

Above Average: WE 11–35 odd

Additional Answers

Written Exercises

35. Let s = successful outcomes and t = total outcomes. Then $t - s$ = the number of unsuccessful outcomes.

$$\frac{P(E)}{P\,(not\ E)} = \frac{\frac{s}{t}}{\frac{t-s}{t}} = \frac{s}{t-s} =$$

$$\frac{\text{successful outcomes}}{\text{unsuccessful outcomes}}$$

---

27. If the probability that it will snow is $\frac{3}{10}$, what is the probability that it will not snow? $\frac{7}{10}$

In a scientific experiment, a gerbil is placed in a large space with 5 exits labeled A, B, C, D, and E.

28. What is the probability that the gerbil will use Exit C? $\frac{1}{5}$
29. What is the probability that the gerbil will *not* use Exit C? $\frac{4}{5}$
30. If the gerbil uses an exit, what is the probability that it will use one of the five labeled exits? 1

The relative chances that an event E will occur are often expressed in terms of the *odds* in favor of E. The odds that an event E will occur are defined by the ratio $\dfrac{\text{number of successful outcomes}}{\text{number of unsuccessful outcomes}}$.

Example: What are the odds of spinning a prime number on this spinner? Five numbers are prime (2, 3, 7, 11, 13); three numbers are not prime (4, 9, 21).

So, the odds of spinning a prime number are $\frac{5}{3}$.

Refer to the spinner. Find each of the following.

31. the odds of spinning an even number $\frac{1}{3}$
32. the odds of spinning a number that is not even $\frac{3}{1}$
33. the odds of spinning a number that is less than 15 $\frac{7}{1}$
34. the odds of spinning a number that is a factor of 42 $\frac{1}{1}$
35. Show that the odds that an event E will occur can be found by using the ratio $\dfrac{P(E)}{P(\text{not } E)}$.

Mixed Review

Solve. Check the solutions.   *4.5, 5.6, 5.7, 7.7, 14.1, 14.3, 14.4*

1. $|x + 3| = 2$ $x = -1\ or\ x = -5$
2. $|x - 5| = -1$ No solution
3. $|x - 3| \le 9$ $-6 \le x \le 12$
4. $x^2 + x - 12 = 0$ $-4, 3$
5. $\frac{x}{6} + \frac{x}{2} = 9$ 13.5
6. $(x - 2)^2 = 41$ $2 \pm \sqrt{41}$
7. $4x^2 = 30$ $\frac{\sqrt{30}}{2}$
8. $y^2 - 1 = 7y$ $\frac{7 \pm \sqrt{53}}{2}$
9. $4 = 2x^2 - x + 2$ $\frac{1 \pm \sqrt{17}}{4}$
10. If the length of each side of a square is tripled, then the area is 225 cm². Find the length of each side of the original square.   *14.1* 5 cm
11. The square of a number is 22 more than 9 times the number. Find the number.   *7.8* -2 or 11

Enrichment

Probability is applied to many problems in economics, medicine, business, social sciences, biology, and in determining insurance premiums. Have students research applications in one of the disciplines above, or have them find other practical examples of probability. Some possible topics are the Mendelian theory of heredity, theory of epidemics, theory of earthquakes, risks in using certain drugs, and predicting election results.

16.2 Probability: Compound Events

Teaching Resources

Manipulative Worksheet 16
Problem Solving Worksheet 16
Quick Quizzes 119
Reteaching and Practice
  Worksheets 119
Teaching Aid 5

Objectives

To determine the probability of independent events
To determine the probability of compound events

In many situations, you need to make a systematic listing of all the outcomes. The set of all possible outcomes is called a **sample space**. Below are the 6 outcomes in the sample space for rolling a die.

$$\{1, 2, 3, 4, 5, 6\}$$

An event is a subset of the sample space. The event "roll a 3, 4, or 6" is $\{3, 4, 6\}$. Note that *outcome* and *event* have different meanings.

There are 2 different roads from East-vale to Centerton. There are 4 different roads from Centerton to Westvale. Suppose you want to list all the possible routes from Eastvale to Westvale by way of Centerton.

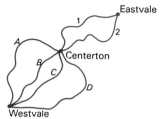

A diagram called a **tree diagram** can be used to determine all possible routes (outcomes).

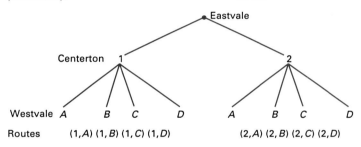

If you use Road 1 to Centerton, you can use any of the 4 roads to Westvale: *A*, *B*, *C*, or *D*.

If you use Road 2 to Centerton, you can use any of the 4 roads to Westvale: *A*, *B*, *C*, or *D*.

The sample space of possible routes from Eastvale to Westvale is $\{(1,A), (1,B), (1,C), (1,D), (2,A), (2,B), (2,C), (2,D)\}$.

Note that each route is shown as an *ordered pair*.

16.2 Probability: Compound Events **603**

GETTING STARTED

Prerequisite Quiz

Simplify.

1. $\frac{1}{3} \cdot \frac{1}{4}$ $\frac{1}{12}$
2. $\frac{2}{6} \cdot \frac{3}{6}$ $\frac{1}{6}$
3. $\frac{1}{12} \cdot \frac{3}{4}$ $\frac{1}{16}$
4. $\frac{3}{15} + \frac{2}{15}$ $\frac{1}{3}$
5. $\frac{6}{20} + \frac{4}{20}$ $\frac{1}{2}$
6. $\frac{3}{4} + \frac{5}{36}$ $\frac{8}{9}$

Motivator

Remind students that they have worked with "and" and "or" statements earlier in this course when studying compound inequalities. Ask students when the statement, "Today is Tuesday *and* it is raining." is true.   Only when both parts are true.   Then ask when the following statement is true: "Next week is the last week of school *or* next week is the first week of June."   When either or both parts are true.

Highlighting the Standards

Standards 11a, 11d: Students construct the sample space and find the probability for a compound event.

Lesson Note

When presenting Example 1, ask students how many ways there are to "travel" from the top to the bottom of the tree diagram. 12   This shows that there are 12 possible outcomes. Show the sample space for Example 3 on the chalkboard or overhead and have students follow as you circle all the ordered pairs whose first number, *s,* is less than or equal to 3. Then circle those where the second number, *b,* is less than or equal to 2. Count each set and point out the 6 pairs that belong to both sets that should not be counted twice.

Math Connections

Games: Games of chance, such as those with dice or cards, are based on the laws of probability. The less the chance of getting a particular result, the bigger the payoff. Likewise, the chances or odds in favor of winning a large amount of money in a contest are expressed as a very small probability.

Critical Thinking Questions

Analysis: Ask students what operation is associated with *P(A and B)* and with *P(A or B).*   Multiplication; addition

---

EXAMPLE 1   Suppose you toss one coin and roll one die. What is the probability of getting a tail on the coin and a 6 on the die?

Plan   Make a tree diagram and list the sample space.

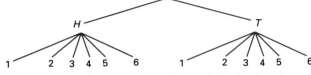

Solution   Sample space: $\{(H,1), (H,2), (H,3), (H,4), (H,5), (H,6), (T,1), (T,2), (T,3), (T,4), (T,5), (T,6)\}$

There are 12 outcomes. In the sample space, there is only one successful outcome: $(T,6)$. Since all the outcomes are equally likely,

$$P[(T,6)] = \tfrac{1}{12}, \text{ or } P(T \text{ and } 6) = \tfrac{1}{12}.$$

So, the probability of getting a tail on the coin and a 6 on the die is $\tfrac{1}{12}$.

The outcome of rolling a die does not affect the outcome of tossing a coin. Similarly, the outcome of tossing a coin does not affect the outcome of rolling a die. Such events are called **independent events**.

In Example 1, you can see that $P(T) = \tfrac{1}{2}$ and that $P(6) = \tfrac{1}{6}$. Note that $\tfrac{1}{2} \cdot \tfrac{1}{6} = \tfrac{1}{12},$ which is equal to $P(T \text{ and } 6)$. This suggests the following rule.

Probability of *A and B*
If *A* and *B* are independent events, then $P(A \text{ and } B) = P(A) \cdot P(B)$.

EXAMPLE 2   A red die and a white die are rolled. What is the probability of getting a 3 on the red die and a 5 on the white die?

Plan   The two events are independent. Use $P(A \text{ and } B) = P(A) \cdot P(B)$.

Solution $P(r = 3) = \tfrac{1}{6}$ $P(w = 5) = \tfrac{1}{6}$ $P(r = 3 \text{ and } w = 5) = \tfrac{1}{6} \cdot \tfrac{1}{6} = \tfrac{1}{36}$

So, the probability of getting a 3 on the red die and a 5 on the white die is $\tfrac{1}{36}$.

TRY THIS   A blue die and a red die are rolled. What is the probability of getting an even number on the blue die and a 3 on the red die? $\tfrac{1}{12}$

604   Chapter 16 Probability and Statistics

Additional Example 1

Roll one die and toss 1 coin. What is the probability of getting a 3 on the die and a head on the coin? $\tfrac{1}{12}$

Additional Example 2

A coin is tossed and a die is rolled. What is the probability of getting a tail on the coin and a 6 on the die? $\tfrac{1}{12}$

EXAMPLE 3

In a roll of two dice, a small one and a big one, find $P(s \leq 3 \text{ or } b \leq 2)$.

Plan

Make a sample space of ordered pairs. Count the number of outcomes in which $s \leq 3$ and the number of outcomes in which $b \leq 2$, and then subtract the number of ordered pairs common to both sets. They cannot be counted twice.

Solution

The sample space is shown at the right. There are 36 possible outcomes.

There are 18 outcomes in which $s \leq 3$ and 12 outcomes in which $b \leq 2$.

Big						
6	(1,6)	(2,6)	(3,6)	(4,6)	(5,6)	(6,6)
5	(1,5)	(2,5)	(3,5)	(4,5)	(5,5)	(6,5)
4	(1,4)	(2,4)	(3,4)	(4,4)	(5,4)	(6,4)
3	(1,3)	(2,3)	(3,3)	(4,3)	(5,3)	(6,3)
2	(1,2)	(2,2)	(3,2)	(4,2)	(5,2)	(6,2)
1	(1,1)	(2,1)	(3,1)	(4,1)	(5,1)	(6,1)
	1	2	3	4	5	6

Small

However, 6 ordered pairs are common to both sets and must not be counted twice. So, the number of these common ordered pairs must be subtracted. For $s \leq 3$ or $b \leq 2$, there are $18 + 12 - 6$, or 24 successful outcomes.

$P(s \leq 3 \text{ or } b \leq 2) = \frac{24}{36}$, or $\frac{2}{3}$

In Example 3, $P(s \leq 3 \text{ or } b \leq 2) = \frac{18}{36} + \frac{12}{36} - \frac{6}{36} = \frac{24}{36}$, or $\frac{2}{3}$.

This suggests the following.

Probability of A or B

For two events A and B, $P(A \text{ or } B) = P(A) + P(B) - P(A \text{ and } B)$.

Note that if the two sets are *mutually exclusive*, or do not overlap, then $P(A \text{ or } B) = P(A) + P(B)$.

The events A and B and A or B are called *compound events* since each contains more than one element of the sample space.

EXAMPLE 4

A bag contains 4 white, 8 green, and 6 red beads. A bead is drawn. Find the probability that the bead is white or green.

Plan

Use $P(A \text{ or } B) = P(A) + P(B)$ since the two sets are mutually exclusive.

Solution

$P(\text{white or green}) = P(w) + P(g) = \frac{4}{18} + \frac{8}{18} = \frac{12}{18}$, or $\frac{2}{3}$

So, the probability that the bead is white or green is $\frac{2}{3}$.

Error: Some students forget to subtract the overlap in situations such as Example 3.

Suggest that students count the pairs that are left over (not part of either separate outcome) and subtract this number from the total number of outcomes to find the number of favorable outcomes.

Checkpoint

1. Suppose that you toss a coin and roll a 12-sided die with numbers 1 through 12. What is the probability of getting a tail on the coin and a 6 on the die? $\frac{1}{24}$

2. A blue die and a pink die are rolled. Find the probability of a 6 on the blue die and a 3 on the pink die. $\frac{1}{36}$

Refer to the two spinners above. Find the probability of each event.

3. 100 *and* black $\frac{1}{4}$

4. 200 *and* black $\frac{1}{6}$

5. 100 *and* yellow $\frac{1}{8}$

6. 300 *and* green $\frac{1}{24}$

Use the sample space in Example 3. Find each of the following.

7. $P(b \leq 4 \text{ or } s > 2)$ $\frac{8}{9}$

8. $P(b \leq 3 \text{ or } s \leq 3)$ $\frac{3}{4}$

Additional Example 3

In the roll of two dice, a red one and a white one, find $P(r \leq 4 \text{ or } w < 3)$. $\frac{7}{9}$

Additional Example 4

A bag contains 2 white, 3 red, and 5 green beads. A bead is drawn. Find $P(\text{red or green})$. $\frac{4}{5}$

Closure

Have students answer the following orally. If A and B are independent events, find $P(A$ and $B)$.

1. $P(A) = \frac{1}{2}$ and $P(B) = \frac{1}{6}$? $\frac{1}{12}$
2. $P(A) = \frac{1}{3}$ and $P(B) = \frac{1}{5}$? $\frac{1}{15}$

Have students find the probability of each event, given that one coin and one die are tossed. Have students refer to the sample space for Example 1.

1. $P(H$ and $1)$ $\frac{1}{12}$
2. $P(T$ and $6)$ $\frac{1}{12}$
3. $P(H$ or $4)$ $\frac{7}{12}$
4. $P(T$ or $7)$ $\frac{1}{2}$

■■FOLLOW UP

Guided Practice

Classroom Exercises 1–8
Try This   all

Independent Practice

A Ex. 1–10, **B** Ex. 11–26, **C** Ex. 27–30
Basic: WE 1–12 all, Brainteaser
Average: WE 1–23 odd, Brainteaser
Above Average: WE 5–27 odd, 28–30 all, Brainteaser

Additional Answers

Mixed Review

10. $13\frac{1}{7}$ in; $11\frac{1}{7}$ in; $3\frac{5}{7}$ in

---

Classroom Exercises

Use the tree diagram in Example 1. Find the probability of each event.

1. $P(H$ and $2)$ $\frac{1}{12}$
2. $P(T$ and $5)$ $\frac{1}{12}$
3. $P(T$ and $1)$ $\frac{1}{12}$
4. $P(H$ and $8)$ 0

A red die and a blue die are rolled.

5. Find the probability of a 4 on the red die and a 2 on the blue die. $\frac{1}{36}$
6. Find the probability of a 2 on the red die and a 2 on the blue die. $\frac{1}{36}$
7. Find $P(r > 4$ and $b > 4)$. $\frac{1}{9}$
8. Find $P(r > 4$ or $b > 4)$. $\frac{5}{9}$

Written Exercises

1. {(H, 1), (H, 2), (H, 3), (H, 4), (T, 1), (T, 2), (T, 3), (T, 4)}

Suppose you toss a coin and spin this spinner.

1. Make a tree diagram and list the sample space.
2. What is the probability of getting a head on the coin and a 3 on the spinner? $\frac{1}{8}$
3. What is the probability of getting a tail on the coin and a 5 on the spinner? 0
4. What is the probability of getting a tail on the coin and a 2 on the spinner? $\frac{1}{8}$

Refer to the two spinners at the right. Find the probability of each event.

5. $P(\text{blue and } 4)$ $\frac{1}{64}$
6. $P(\text{red and } 1)$ $\frac{3}{32}$
7. $P(\text{white and } 4)$ $\frac{1}{16}$
8. $P(\text{white and } 1)$ $\frac{1}{8}$
9. $P(\text{red and } 2)$ $\frac{3}{32}$
10. $P(\text{blue and } 3)$ $\frac{3}{64}$

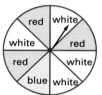 

A small die and a big die are rolled. Find each of the following. (Use the sample space in Example 3.)

11. $P(s \geq 4$ or $b \leq 2)$ $\frac{2}{3}$
12. $P(s \leq 2$ or $b \leq 2)$ $\frac{5}{9}$
13. $P(s \geq 5$ or $b \geq 5)$ $\frac{5}{9}$
14. $P(s \leq 3$ or $b \geq 5)$ $\frac{2}{3}$
15. $P(s < 3$ or $b < 3)$ $\frac{5}{9}$
16. $P(s > 5$ or $b > 5)$ $\frac{11}{36}$
17. $P(s < 1$ or $b > 6)$ 0
18. $P(s < 7$ or $b > 0)$ 1

A bag contains 3 white, 6 green, and 9 red beads. One bead is drawn.

19. Find $P(\text{white or red})$. $\frac{2}{3}$
20. Find $P(\text{green or red})$. $\frac{5}{6}$

Enrichment

Use the following exercises to have students explore what happens when asked to find $P(A$ and $B)$ if events A and B are dependent. A bag contains 4 blue, 3 green, and 5 red marbles.

1. What is the probability that one marble drawn is red? $\frac{5}{12}$

2. Suppose the first marble drawn is red. If this marble is not replaced, what is the probability that the second marble drawn will be red? $\frac{4}{11}$

3. If two marbles are drawn from the bag, one at a time without replacing the first one, what is the probability that both marbles are red? $\frac{5}{12} \cdot \frac{4}{11} = \frac{20}{132} = \frac{5}{33}$

4. If two marbles are drawn, one at a time without replacement, what is the probability that both marbles are green? $\frac{1}{22}$

This set of cards is mixed and placed in a bag. Find the probability that the number drawn is

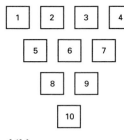

21. even *and* greater than 7. $\frac{1}{5}$

22. even *or* greater than 7. $\frac{3}{5}$

23. odd *and* prime. $\frac{3}{10}$

24. odd *or* prime. $\frac{3}{5}$

25. a factor of 3 *or* a factor of 7. $\frac{3}{10}$

26. a factor of 6 *and* a factor of 9. $\frac{1}{5}$

The Gallon Organization is taking a survey of families with 3 children. The order of birth of boys (*B*) and girls (*G*) in a given family can be shown as *GBG*, which indicates this outcome: a girl was born first, a boy was born second, and a girl was born third.

27. Make a tree diagram and list all the possible outcomes in a family with three children. How many possible outcomes are there?

28. Assume that the outcomes are equally likely. Find the probability of three girls *or* three boys. $\frac{1}{4}$

29. Find the probability of two girls *and* a boy. (HINT: This does *not* mean that the girls were born before the boy. The order is *not* significant in this problem.) $\frac{3}{8}$

30. Find the probability that the oldest child is a boy *or* the youngest child is a girl. $\frac{3}{4}$

27. {(GGG), (GGB), (GBG), (GBB), (BGG), (BGB), (BBG), (BBB)}; 8 possible outcomes

Mixed Review

Simplify. *13.2, 13.5, 13.7*

1. $\sqrt{12} \cdot \sqrt{12}$ 12

2. $-8\sqrt{800}$ $-160\sqrt{2}$

3. $\sqrt{30} \cdot \sqrt{50}$ $10\sqrt{15}$

4. $\sqrt{x^6}$ x^3

5. $2\sqrt{x^5y^7}$ $2x^2y^3\sqrt{xy}$

6. $8\sqrt{36x^6y^{12}z^7}$

7. $5\sqrt{3} \cdot 7\sqrt{3}$ 105

8. $3\sqrt{5} \cdot 6\sqrt{2}$ $18\sqrt{10}$

9. $-5\sqrt{12a} \cdot 2\sqrt{3a}$ $-60a$

10. The first side of a triangle is 2 in. longer than the second. The second side is 3 times as long as the third side. The perimeter is 28 in. How long is each side? *4.4*

6. $48x^3y^6z^3\sqrt{z}$

11. The selling price of a compact disc player is $889. The profit is 40% of the cost. Find the cost. *4.8* $635

Brainteaser

The mean (average) of the numbers in a set of 3*n* numbers is 50. The set is subdivided into two disjoint sets, one containing 2*n* of the numbers, the other containing the remaining *n* numbers. The mean of the numbers in the larger subset is twice the mean of the numbers in the smaller. Find the mean of the numbers in the smaller subset. 30

16.3 Mean, Median, and Mode

Objectives
To find the mean, the median, and the mode for a set of data
To construct a frequency table and find the mean, the median, and the mode for the data in the table

---

■■■GETTING STARTED

Prerequisite Quiz

Add.

1. $98 + 95 + 93 + 92 + 91$ 469
2. $17 + 24 + 16 + 31 + 43 + 36$ 167
3. $100 + 105 + 95 + 90 + 110$ 500

Divide.

4. $\frac{496}{5}$ 99.2
5. $\frac{87}{4}$ 21.75
6. $\frac{103}{6}$ $17.1\overline{6}$

Motivator

Present the following scenario to the students. The Acme Company has five employees, four office workers and the president. The salaries of the five people are $12,000, $12,000, $17,000, $24,000 and $135,000. Ask students to find the average of these salaries. $40,000 Ask students to arrange the salaries in order from least to greatest and find the middle salary. $17,000 Ask them which amount of salary occurs most frequently. $12,000 Tell them that each of these amounts, $40,000, $17,000, and $12,000 represents a different kind of "average."

In statistics, the average of a set of numbers is called the *arithmetic mean*, or simply *mean* of the numbers, or *data*. Often the data in statistics are called *scores*.

The **mean** of a set of data is found by adding the scores and dividing the sum by the number of scores.

$$\text{mean of a set of data} = \frac{\text{sum of the scores}}{\text{number of scores}}$$

The symbol $\overline{x}$ (read "x-bar") is often used to represent the mean.

One advantage of using the mean is that it provides a way of using a single number to represent a set of data. However, the mean may be misleading if there are some extreme scores. Extreme scores are scores that are much greater or much less than most of the other scores.

EXAMPLE 1
Carol scored 98, 96, 88, 92, and 86 on her mathematics tests. Find the mean of her scores.

Solution
$$\overline{x} = \frac{\text{sum of the scores}}{\text{number of scores}}$$
$$= \frac{98 + 96 + 88 + 92 + 86}{5}$$
$$= \frac{460}{5}$$
$$= 92$$

TRY THIS
Susan, José, and Lucia received these scores on five tests. Compute the mean score for each student.

1. Susan: 75, 83, 91, 94, 87 86
2. José: 84, 86, 70, 85, 90 83
3. Lucia: 84, 96, 96, 70, 79 85

There are other numbers, apart from the mean, that can be used to represent a set of data. One such number is called the *median*.

608   Chapter 16 Probability and Statistics

Highlighting the Standards

Standard 10c: Here students use and compare several measures of central tendency.

Additional Example 1

Will scored 90, 92, 98, 96, and 100 on his exams. Find the mean of his scores.   95.2

The **median** of a set of data is found by arranging the data in order and choosing the middle score. If the data contain an even number of items, the mean of the two middle scores is the median.

One advantage of using the median is that it is not influenced by extreme scores.

EXAMPLE 2   Find the median 3:00 P.M. temperature for a week when the Fahrenheit temperatures at 3:00 P.M. were 48, 43, 51, 63, 49, 50, and 68.

Plan   Arrange the temperatures in order and choose the middle temperature.

Solution   43   48   49   50   51   63   68

So, the median is 50.

TRY THIS   **4.** This list shows the number of students attending the seven high schools in Orange County. Find the median.

   539   625   517   525   415   560   478 525

To find the median of an *even number of scores*, you find the mean of the two middle scores.

EXAMPLE 3   Find the median of these six temperatures: 97, 94, 86, 89, 92, 90.

Plan   Arrange the temperatures in order. Since there is an even number of scores, find the mean of the two middle numbers.

Solution   86   89   90   92   94   97

Median $= \dfrac{90 + 92}{2}$

$= \dfrac{182}{2}$, or 91

So, the median is 91.

TRY THIS   **5.** The price in cents of a pint-size carton of 10 brands of orange juice is given below. Find the median price.

   69   71   68   70   65   71   69   75   72   67
                                                          69.5

Sometimes it is important to know the most frequent score, or *mode*, of a set of data.

The **mode** of a set of data is the score that occurs most frequently.

Lesson Note

When presenting Example 2, remind students that the median of a road is the section in the middle. When presenting Example 4 on page 610, tell students that if each entry in a set of data occurs only once, then the data has no mode.

Math Connections

Life Skills: Many banks have a service charge on a checking account based on the average daily balance in the account. The more days a deposit is left without withdrawals, the higher the average daily balance will be and the lower the service charge.

Critical Thinking Questions

Synthesis: Have students find a method for solving the following type of problem. If Lupita makes test grades of 89, 95, and 82, what grade must she make on the fourth test in order to have a test average of 90?
$(4 \cdot 90) - (89 + 95 + 82) = 360 - 266 = 94$

Additional Example 2

Find the median 2 A.M. temperature for a week when the Celsius temperatures at 2 A.M. were 8°, 14°, 4°, 0°, −1°, −2°, −2°.
0

Additional Example 3

Find the median of the set of data:
93, 92, 89, 94, 96, 90.   92.5

Common Error Analysis

Error: Some students may choose the middle score for the median without having first put the data in order.

Remind students that the median remains the same no matter the order of listing the data. However, there must be as many scores with a *value* below the median value as scores with a value above it.

Checkpoint

Find the mean of each set of data.

1. 14, 12, 16, 18, 15   15
2. 98, 97, 97, 96, 92   96

Find the median of each set of data.

3. 98, 99, 96, 94, 90   96
4. 14, 18, 16, 12, 17, 19   16.5

Find the mode of each set of data, if it exists.

5. 10, 12, 16, 14, 12, 10, 9, 12, 16   12
6. 98, 99, 100, 97, 96   Does not exist

Closure

Have students list a set of data (no more than 10 entries) and exchange papers with a classmate who must find the mean, median, and mode of this data. Then have students check each others work.

To find the mode, list each number as many times as it occurs and then choose the number or numbers that occur most frequently. If each item appears an equal number of times, there is no mode.

EXAMPLE 4 The test scores for a student were 90, 95, 90, 85, 100, 95, 80, 90, 100, 80. Find the mode.

Plan Arrange the scores in order. Group the same scores together.

Solution

80, 80,	85,	90, 90, 90,	95, 95,	100, 100
twice	once	three times	twice	twice

So, the mode is 90.

Notice, in Example 4, that if the test scores had included three 80s as well as three 90s, there would have been two modes, 80 and 90.

The three types of statistical measures you have just studied—mean, median, and mode—are called **measures of central tendency**. They help tell about the data and identify its "middle" points or representative numbers.

A **frequency table** is often used to summarize and represent a set of data. The table below shows the frequency of runs scored per game by a baseball team in its 50-game schedule.

In this table, the runs are the scores s; the Tally column is used to record the times each score occurred; and the Frequency column gives the total number of times each score occurred.

The column headed $f \cdot s$ (frequency × score) is not an essential part of every frequency table, but it is useful in calculating the mean, as shown in Example 5. To calculate the mean, find the sum of the products $f \cdot s$, then divide by the sum of the frequencies.

Runs (s)	Tally	Frequency (f)	$f \cdot s$
0	II	2	0
1	JHT	5	5
2	III	3	6
3	JHT I	6	18
4	JHT JHT II	12	48
5	JHT IIII	9	45
6	I	1	6
7	IIII	4	28
8	II	2	16
9	IIII	4	36
10	II	2	20
		Sum: 50	Sum: 228

Additional Example 4

The test scores for a student were 75, 70, 85, 80, 85, 75, 90, 80, 85, and 70. Find the mode.   85

EXAMPLE 5 Find the mean, the median, and the mode for the runs scored per game by the baseball team. Use the table preceding this example.

Solution
$$\text{mean} = \frac{\text{sum of the products } f \cdot s}{\text{sum of the frequencies}} = \frac{228}{50}, \text{ or } 4.56$$

So, the mean of the scores is 4.56 runs per game.

The median is the middle score. Count to find the 25th and 26th scores. They occur in the row that has frequency 12—the row for 4 runs. So, the median is 4 runs per game.

The mode is the most frequently occurring score. So, the mode is 4 runs per game.

Summary Note that a frequency table is helpful in determining the median and the mode as well as the mean. An $f \cdot s$ column is also useful in many situations.

To construct a frequency table:

1. label the columns;
2. tally the data; and
3. find the frequency of each score.

◼️ Focus on Reading

True or false?

1. The mean is always a number that appears in the set of data. F
2. The median is always the middle score in a set of scores. F
3. There is no mode in this set of data: 1, 2, 3, 4, 6, 8, 10. T
4. The mode and the median can never be the same number. F

Classroom Exercises

Find the mean of each set of data.

1. 3, 4, 5, 5, 8 5
2. 4, 3, 4, 2, 7 4
3. 6, 9, 8, 10, 7, 3 $7\frac{1}{6}$
4-6. Find the median of each set of data in Exercises 1–3. 5; 4; 7.5
7-9. Find the mode of each set of data in Exercises 1–3. 5; 4; none

Guided Practice

Classroom Exercises 1–9
Try This all, FR all

Independent Practice

🅰 Ex. 1–20, 🅱 Ex. 21–30, 🅲 Ex. 31–34
Basic: WE 1–7 odd, 9–20 all, Midchapter Review
Average: WE 1–19 odd, 21–26 all, Midchapter Review
Above Average: WE 9–20 odd, 21–24 all, 25–33 odd, Midchapter Review

Additional Example 5

The runs scored by a baseball team in its last 6 games were 2, 4, 3, 4, 13, 1 respectively. Find the mean, median, and mode, for the runs scored over the last 6 games.   mean: 4.5, median: 3.5, mode: 4

21.

Test Scores(s)	Frequency(f)	f · s
70	3	210
75	4	300
80	2	160
85	4	340
90	6	540
95	5	475
100	6	600
	30	2,625

Written Exercises

Find the mean of each set of data.

1. 18, 15, 12, 14, 16 15

2. 5, 10, 12, 13, 10 10

3. 16, 24, 19, 36, 35 26

4. 96, 94, 97, 93, 95 95

5. 46, 48, 87, 95, 85 72.2

6. 77, 76, 70, 72, 68, 69 72

7. 91, 72, 73, 89, 73, 83 $80\frac{1}{6}$

8. 87, 99, 100, 96, 99, 95 96

Find the median.

9. 98, 93, 97, 100, 92 97

10. 49, 43, 40, 41, 48 43

11. 105, 100, 98, 104, 110 104

12. 12, 13, 9, 8, 14, 15 12.5

13. 91, 93, 98, 87, 90 91

14. 100, 98, 103, 95, 96, 100 99

Find the mode.

15. 91, 90, 91, 98, 97, 100 91

16. 38, 36, 37, 36, 39 36

17. 50, 48, 50, 48, 55 48 *and* 50

18. 100, 101, 98, 103, 105 No mode

19. 98, 97, 98, 100, 96, 98 98

20. 85, 80, 85, 100, 80 80 *and* 85

A class had the following test scores:
90, 95, 85, 100, 100, 95, 80, 95, 100, 75, 70, 75, 90, 95, 100, 85, 80, 70, 85, 90, 100, 90, 85, 75, 70, 75, 95, 100, 90, **and 90. (Exercises 21–24)**

21. Make a frequency table.

22. Find the mean. 87.5

23. Find the median. 90

24. Find the mode. 90 *and* 100

Solve.

25. Frankie scored 48.5, 43.2, 38.7, 53.1, and 49.3 on a battery of tests. What is the mean of these scores? 46.56

26. In her music auditions, Linda Sue received scores of 98.3, 89.1, 96.5, 93.4, 91.9, and 95.8. Find the mean of these scores. 94.1$\overline{6}$

27. Kris has a bowling average of 210. One day she bowled 190 three times and 200 twice. What must she bowl in her next game to maintain her average? 290

28. Wally's test grades so far are 85, 80, 75, and 90. He expects to get the same grade on each of the remaining tests and to have an average of 87 on a total of ten tests. What is that grade? 90

29. Find the mean of this set of data.
$x + 3, x + 2, x + 5, 2 + x$ $x + 3$

30. Find the median and the mode of the set of data in Exercise 29.

31. Find the value(s) of x for the set of data if the mean of $8x^2$, $4x$, $2x$, 7, -8, $-7x^2$, $-3x$, 2, $-x$ is 1. 2 *or* -4

32. Find the value(s) of t for the set of data if the mean of t^2, $-3t$, t, $5t^2$ is 0. 0 *or* $\frac{1}{3}$

30. median: $x + 2.5$; mode: $x + 2$

612 Chapter 16 Probability and Statistics

33. The mean of 20 students' scores is 88. If the 2 lowest and the 2 highest scores are removed, the mean of the remaining scores is 84. What is the mean of the removed scores? 104

34. The mean of 48 students' test scores is 94. If the 4 lowest and the 4 highest scores are removed, the mean of the remaining scores is 96. What is the mean of the removed scores? 84

Midchapter Review

A bag contains 8 red marbles and 10 white marbles. One marble is drawn. Find the probability of each event.

1. P(red) $\frac{4}{9}$ **2.** P(green) 0   **3.** P(not yellow) 1

Refer to the spinner at the right. Find the probability of each event.

4. P(1) $\frac{1}{4}$
5. P(4) $\frac{1}{8}$
6. P(even number) $\frac{3}{8}$
7. P(5) 0

Suppose you spin both of these spinners. Find the indicated probability.

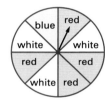 

8. P(white *and* 4) $\frac{1}{8}$ **9.** P(red *and* 3) $\frac{1}{6}$
10. P(blue *and* 1) $\frac{1}{48}$ **11.** P(white *and* 2) $\frac{1}{16}$

A blue die and a green die are rolled. Make a sample space of ordered pairs. Find the probability of each event.

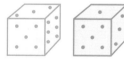

12. $P(b < 2 \text{ or } g < 2)$ $\frac{11}{36}$ **13.** $P(b \geq 3 \text{ or } g \leq 3)$ $\frac{5}{6}$

A football team scored the following points in its games: 13, 6, 14, 20, 14, 7, and 10. Use this data for Exercises 14–16.

14. Find the mean. 12   **15.** Find the median. 13   **16.** Find the mode. 14

Enrichment

Have students find examples of the use of statistical averages in newspaper and magazine articles, graphs, and advertisements. Have them bring these items to class and hold group discussions of how these items may, or may not, be misleading.

Teaching Resources

Application Worksheet 16
Quick Quizzes 121
Reteaching and Practice
   Worksheets 121
Transparency 48
Computer Investigation, p. 641

▰▰▰▰GETTING STARTED

Prerequisite Quiz

If $f(x) = \frac{1}{2}x + 2$, find each of the following values.

1. $f(0)$ 2
2. $f(1)$ $2\frac{1}{2}$
3. $f(-1)$ $1\frac{1}{2}$
4. $f(2)$ 3
5. $f(-2)$ 1
6. $f(3)$ $3\frac{1}{2}$
7. $f(-3)$ $\frac{1}{2}$
8. Graph $f(x) = \frac{1}{2}x + 2$ for $-3 \le x \le 3$.

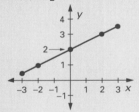

Motivator

This material will have more interest for most students if you can obtain the height and weight of each member of one of the teams in your school. (You could assign this task to one of your students for extra credit.) Use this information to make a scatter plot similar to the one on page 614 in the text. Show the students how the points are plotted and let them help you decide where to draw the line.

Highlighting the Standards

Standards 10a, 10b: Students use various kinds of graphs to summarize data and make predictions.

16.4 Statistical Graphs

Objectives   To read and interpret scatter plots and histograms

Although the mean, median, and mode of a set of data help to characterize the data, it is often convenient to use a graph to show how two sets of data vary relative to each other.

The list below gives ordered pairs (heights, weights) for twenty young male adults (*A* to *T*) with medium frames. The heights are in inches and the weights are in pounds.

A: (67,146)   E: (70,157)   I: (67,145)   M: (69,151)   Q: (68,147)

B: (68,151)   F: (70,156)   J: (69,153)   N: (71,160)   R: (70,154)

C: (66,145)   G: (67,148)   K: (68,149)   O: (66,141)   S: (67,145)

D: (71,158   H: (69,155)   L: (66,142)   P: (66,144)   T: (71,159)

You can probably see that the greater the height, the greater the weight seems to be. However, a graph called a **scatter plot** can be used to learn more about the relationship and to make predictions about data.

To make a scatter plot, plot each ordered pair as a point in the coordinate plane. Then you see whether or not the points suggest a clear relationship.

In this figure, the points seem to suggest a linear relation, although they obviously do not lie exactly on a line. Some points are above the line that has been drawn; other points are below it. In general, they cluster around the line. So, the line can be used to make judgments or predictions about data for weights and heights of male adults with medium frames.

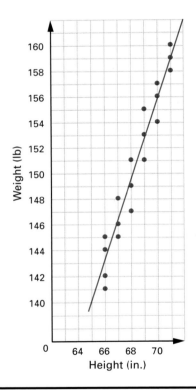

EXAMPLE 1   Use the graph on the preceding page to answer the questions.

a. Approximately what weight would you predict for a young male adult (medium frame) whose height is 65 in?

b. Approximately what weight would you predict for a young male adult (medium frame) whose height is 72 in?

Solutions   **a.** approximately 140 pounds   **b.** approximately 162 pounds

TRY THIS   Approximately what weight would you predict for a young male adult (medium frame) whose height is 70 inches?   Approximately 156 lb

Note in Example 1 that the answers are approximate. Data obtained from a graph are usually approximate.

Another type of useful graph is a *histogram*. A histogram is often used to represent the data in a frequency table. The table below represents the distribution of semester grades in English for 460 juniors in a high school. The scores are grouped in intervals (50–60, 60–70, 70–80, 80–90, 90–100) with each boundary score assigned to the lesser interval. For example, a score of 80 would be assigned to the interval 70–80.

Interval	Frequency
90–100	98
80–90	150
70–80	100
60–70	83
50–60	29

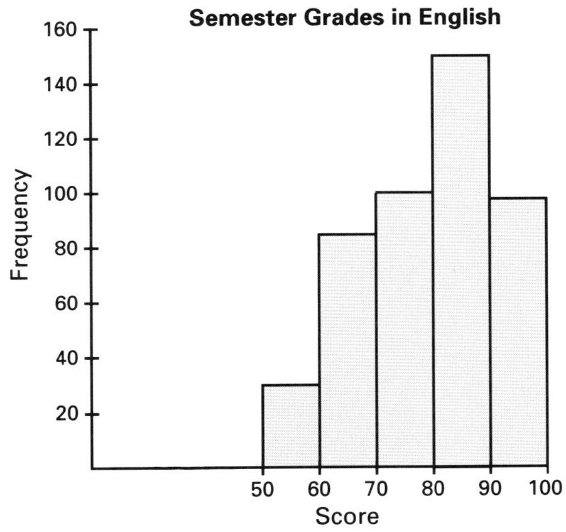

Semester Grades in English

The **histogram** for the set of data is shown above at the right. Notice that a histogram is a kind of bar graph. The horizontal axis is used to indicate the intervals for the scores. The vertical axis is used to indicate the frequencies of the scores in the given interval.

16.4 Statistical Graphs   **615**

Common Error Analysis

Error: Some students fail to read the instructions concerning the placement of boundary scores in a histogram.

Remind them that the peripheral information on a graph is very important to an understanding of the graph.

Checkpoint

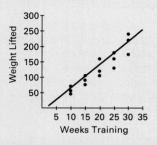

Weight Lifted vs. Weeks Training

Refer to the graph above to answer to questions.

1. Approximately what weight would you predict a student who trained 5 weeks could lift? Approximately 25 lb
2. Approximately what weight would you predict a student who trained 35 weeks could lift? Approximately 250 lb

Refer to the histogram in Additional Example 2 to answer the questions.

3. Approximately how many students scored below 500? Approximately 470
4. Approximately how many students scored between 400 and 600?
   Approximately 540

EXAMPLE 2 From the histogram for the 460 students, answer each question.

a. Approximately how many scores were above 90?

b. Approximately how many scores were above 80?

Plan Find the correct interval or intervals on the horizontal axis. Then determine the appropriate frequency or frequencies.

Solutions

a. For all scores above 90, use the interval 90–100. Approximately 100 scores were in this interval.

b. There are two appropriate intervals: 80–90 and 90–100. The number of scores in these two intervals was approximately 100 + 150, or 250.

Classroom Exercises

Use the scatter plot that precedes Example 1 for Exercises 1–2.

1. What point on the line has 67 as its *x*-coordinate? (67, 146)
2. Complete this ordered pair for the point on the line: (69, _?_). 153
   Tell what the ordered pair means in the problem situation.
   A young male adult 5 ft 9 in tall weighs 153 pounds.

The histogram below represents a number of students and their scores on a social studies exam. Each boundary score is assigned to the lesser interval.

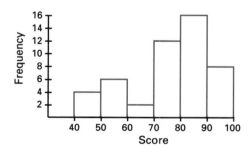

3. How many students had scores above 90? 8
4. How many students had scores above 80? 24
5. How many students had scores of 50 or less? 4
6. How many students had scores of 60 or less? 10
7. In which interval did the greatest number of scores occur? 80–90
8. In which interval did the least number of scores occur? 60–70

Additional Example 2

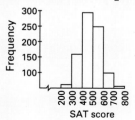

SAT score

The histogram at the left represents the scores on an aptitude test and the number of students taking the test.

a. Approximately how many students scored above 600? Approximately 110
b. Approximately how many students scored above 400? Approximately 650

Written Exercises

A group of scientists kept a record of the number of chirps that crickets made at various temperatures. The scatter plot at the right shows points for the ordered pairs (temperature Fahrenheit, number of chirps per minute); it also shows the line around which the points cluster. Use the line to answer the questions in Exercises 1–4.

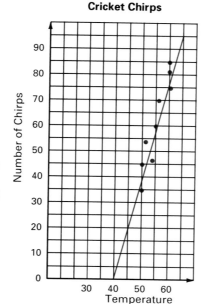

Cricket Chirps

1. When the temperature is 50°F, what is the number of chirps per minute? About 40
2. When the temperature is 60°F, what is the number of chirps per minute? About 75
3. Approximately what number of chirps per minute would you predict for a temperature of 65°F? 95
4. Approximately what number of chirps per minute would you predict for a temperature of 40°F? 0

The histogram below represents the numbers of hours that 250 bulbs burned. Each boundary score is assigned to the lesser interval.

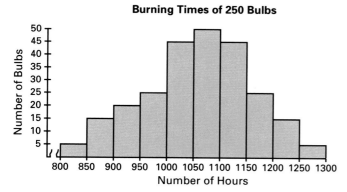

Burning Times of 250 Bulbs

5. How many bulbs burned 800–850 h? 5
6. How many bulbs burned 1,250–1,300 h? 5

7. Approximately how many bulbs burned longer than 1,200 h? 20
8. In which interval did the greatest number of hours occur? 1,050–1,100

16.4 Statistical Graphs **617**

Closure

Ask students how the position of the line in the scatter plot on page 614 might be determined. It is a line that seems to approximate closely the position of most of the points in the graph. Ask what is true of all the intervals shown in the histogram on page 615. Each interval has the same range.

◼️◼️FOLLOW UP

Guided Practice

Classroom Exercises 1–8
Try This   all

Independent Practice

🅰 Ex. 1–8, 🅱 Ex. 9–15 🅲 Ex. 16–18
Basic: WE 1–10 all
Average: WE 1–4, 9–14 all
Above Average: WE 9–18 all

9. Approximately how many bulbs burned longer than 1,050 h? 140
10. Approximately how many bulbs burned 1,050 h or less? 110
11. What percent of the bulbs burned 1,050–1,100 h? 20
12. What percent of the bulbs burned longer than 1,150 h? 18
13. What percent of the bulbs burned 900 h or less? 8
14. Suppose 625 bulbs had been used and the same percents of the total number occurred in the same intervals. How many bulbs would have burned 1,050–1,100 h? 125
15. Write a short description of a scatter plot. Answers will vary.

Suppose that many families with 4 children were surveyed to find the sexes of the 4 children. The results can be summarized by using a "representative sample" of 16 families and displaying the data in this histogram. Note that order of birth is not significant in these data. For example, a family that has 3 girls and 1 boy includes the following possible outcomes (in terms of orders of birth): GGGB, GGBG, GBGG, BGGG.

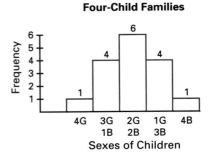

Four-Child Families

Use the histogram to answer the questions in Exercises 16–18.

16. What is the probability that a 4-child family will have 4 girls? $\frac{1}{16}$
17. What is the probability that a 4-child family will have 2 girls and 2 boys? $\frac{3}{8}$
18. If 8,000 families with 4 children were included in the survey, approximately how many would you expect to have more than 2 girls? 2,500

Mixed Review

Factor completely. 7.2–7.6
$4n(n + 1)(n - 1)$

1. $x^2 - 9x + 8$ $(x - 1)(x - 8)$
2. $y^2 + 6y + 9$ $(y + 3)^2$
3. $4n^3 - 4n$
4. $6x^3 - 15x$ $3x(2x^2 - 5)$
5. $x^2 - x - 12$ $(x - 4)(x + 3)$
6. $6x^2 + 19x + 10$
   $(3x + 2)(2x + 5)$

Solve by factoring. 7.7

7. $8y^2 - 32 = 0$ ± 2
8. $9t^3 - 16t = 0$ $0, \pm 1\frac{1}{3}$
9. $3a^2 - 2a = 5$
10. Doug's age is 6 years less than his brother's age. If his age is increased by twice his brother's age, the result is 39 years. Find each of their ages. 12.3 Doug: 9 yr old; brother: 15 yr old
    $(-1, 1\frac{2}{3})$

Enrichment

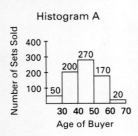

Histogram A

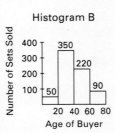

Histogram B

The two histograms illustrate the number of encyclopedias a salesperson has sold to people of various age groups. Have students refer to the histograms to answer the following questions.

1. The salesperson makes a sales pitch to a couple in their forties who have children in high school. Which histogram is the salesman more likely to show

them? A Why? It suggests that many more people in their forties like the encyclopedias.
2. Which histogram is the salesperson more likely to show a couple in their twenties? B Why? It suggests that many more people in their twenties like the encyclopedias.

16.5 Range and Standard Deviation

Teaching Resources

Quick Quizzes 122
Reteaching and Practice
Worksheets 122

Objective   To calculate and interpret the range and the standard deviation

Scores in a set of data may vary widely or they may cluster close to the mean. Therefore, a study of the spread, or *dispersion* of the data, is important. The simplest measure of dispersion is the **range**, which is the difference between the highest and lowest scores.

For example, the range of this set of data is 95–60, or 30.

$$95, 85, 75, 70, 60$$

The *standard deviation* measures the extent to which scores deviate from the mean. A small standard deviation indicates that most scores are close to the mean. A large standard deviation indicates that most scores are not close to the mean.

The **standard deviation** of a set of data, $x_1, x_2, x_3, \cdots, x_n$, with a mean $\bar{x}$, is

$$s = \sqrt{\frac{(x_1 - \bar{x})^2 + (x_2 - \bar{x})^2 + (x_3 - \bar{x})^2 + \cdots + (x_n - \bar{x})^2}{n}}.$$

EXAMPLE 1   Find the standard deviation for the set of data: 12, 10, 8, 13, 9, 8.

Plan   First find the mean $\bar{x}$. Then find the standard deviation s.

Solution $\bar{x} = \dfrac{12 + 10 + 8 + 13 + 9 + 8}{6} = \dfrac{60}{6}$, or 10

$$s = \sqrt{\frac{(x_1 - \bar{x})^2 + (x_2 - \bar{x})^2 + (x_3 - \bar{x})^2 + \cdots + (x_n - \bar{x})^2}{n}}$$

$$s = \sqrt{\frac{(12 - 10)^2 + (10 - 10)^2 + (8 - 10)^2 + (13 - 10)^2 + (9 - 10)^2 + (8 - 10)^2}{6}}$$

$$s = \sqrt{\frac{2^2 + 0^2 + (-2)^2 + 3^2 + (-1)^2 + (-2)^2}{6}}$$

$$s = \sqrt{\frac{4 + 0 + 4 + 9 + 1 + 4}{6}} = \sqrt{\frac{22}{6}} \approx 1.91$$

TRY THIS   Find the standard deviation for this set of data:
14   18   16   15   12 2

Math Connections

Calculators: Many scientific and business calculators have functions to calculate mean, variance, standard deviation, and other statistical values.

Critical Thinking Questions

Evaluation: Ask students to play the role of a teacher and decide what might be called for if the standard deviation on a set of test scores is very large.   Possible answer: a teacher may consider assigning enrichment exercises or extra help to students obtaining either very high or very low scores relative to the average score for the class.

Common Error Analysis

Error: Some students will have difficulty setting up the calculations for a problem in standard deviation because they are not familiar with the symbols used.

Remind them that the first step is to find the *mean* of the data which is represented by $\bar{x}$.

Checkpoint

Find the range and standard deviation for each set of data.

1. 5, 9, 6, 8, 7   Range: 4; s: 1.41
2. 40, 45, 52, 37   Range: 15; s: 5.68

Classroom Exercises

Find the range for each set of data.

1. 18, 16, 21, 10, 14 11
2. 99, 100, 98, 97, 80 20
3. 50, 65, 60, 70, 55 20

Find the standard deviation for each set of data.

4. 6, 8, 10, 12 2.24
5. 10, 5, 14, 3 4.30
6. 3, 2, 4, 5, 2 1.17
7. 2, 4, 6, 4 1.41
8. 1, 8, 7, 4 2.74
9. 4, 5, 6, 3, 2 1.41

Written Exercises

Find the range for each set of data.

1. 21, 16, 19, 27, 25 11
2. 60, 58, 59, 56, 57 4
3. 75, 70, 80, 90, 100 30

Find the standard deviation for each set of data.

4. 15, 13, 19, 17 2.24
5. 3, 2, 4, 6, 5 1.41
6. 72, 79, 93, 70, 82, 61 7.73
7. 24, 42, 36, 30 6.71
8. 22, 25, 28, 32, 18 4.82
9. 98, 97, 86, 75, 91, 92 10.09
10. 5.68, 2.84, 4.26 1.16
11. 43.1, 52.6, 48.4, 36.3, 49.7 5.75
12. 110.7, 98.6, 101.3, 99.3, 102.8 4.34

The weights of 9 objects used in a scientific experiment were:

15 g, 6 g, 9 g, 8 g, 15 g, 11 g, 6 g, 9 g, 11 g

13. Find the range of the set of data. 9 g
14. Find the standard deviation. 3.16 g

The heights of 10 players on a basketball team were:

5.8 ft, 6.2 ft, 5.7 ft, 5.9 ft, 6.3 ft, 6.1 ft, 6.4 ft, 5.7 ft, 6.2 ft, 6.0 ft

15. Find the range of the set of data. 0.7 ft
16. Find the standard deviation. 0.24 ft
17. The lengths (in centimeters) of a set of line segments were:
$$x + 2.2, x + 2.5, x - 0.5, x - 1.5, x + 1.8, x + 0.9$$
Find the range, the mean, and the standard deviation for this set of data.
18. Elena kept a record of the number of errors she made in each softball game. After 7 games, the range for the set of data, the mean, the median, the mode, and the standard deviation were all the same number. What was that number? 0

There is a special bell-shaped graph that happens to fit the plots of some kinds of statistical measures—for example, the heights of pea plants, test scores on an examination, and lifespans of organisms of a particular species—where large numbers of data are used. This curve is called a **normal curve**. The highest point of the curve corresponds to the mean of the measures. For data that are normally distributed:

- 50% of the data lie on each side of the mean;
- about 68.2% of the measures fall within 1 standard deviation from the mean;
- about 95.4% of the measures fall within 2 standard deviations from the mean; and
- about 99.6% of the measures fall within 3 standard deviations from the mean.

The life expectancy of a certain automobile approximately fits the normal curve. The mean life is 80,000 mi with one standard deviation of 12,000 mi. Thus, the probability that a car of this kind will last between 80,000 mi and 92,000 mi is 34.1%, or 0.341.

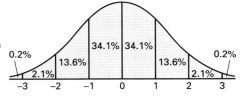

19. What is the probability that a car will last between 68,000 and 92,000 mi? 68.2%

20. What is the probability that a car will not last 80,000 mi? 50%

21. What is the probability that a car will last more than 92,000 mi? 15.9%

22. What is the probability that a car will last less than 68,000 mi? 15.9%

Mixed Review

Solve each system of equations. *12.2, 12.4, 12.5*

1. $3y = x$ (3, 1)
   $5x - 2y = 13$

2. $4x - 3y = 14$ (2, -2)
   $5x + 3y = 4$

3. $-x - 5y = 3$ (2, -1)
   $2x - y = 5$

Simplify. *13.8*

4. $\dfrac{\sqrt{18}}{\sqrt{2}}$ 3

5. $\dfrac{\sqrt{32a^5}}{\sqrt{8a}}$ $2a^2$

6. $\dfrac{16x^2y}{\sqrt{8x^5y^3}}$ $\dfrac{4\sqrt{2xy}}{xy}$

7. Four times the tens digit of a two-digit number, increased by the units digit, is 16. If the digits are reversed, the new number is 2 less than 3 times the original number. Find the original number. *12.6* 28

Closure

Have students summarize the steps for finding the standard deviation of a set of data. (1) Find the mean of the data; (2) Find the difference between each score and the mean; (3) Square each difference; (4) Add the squares of the differences; (5) Divide by the number of the terms in the set; (6) Take the square root.

◼◼◼ FOLLOW UP

Guided Practice

Classroom Exercises 1–9
Try This   all

Independent Practice

🅰 Ex. 1–9, 🅱 Ex. 10–17, 🅲 Ex. 18–22
Basic: WE 1–9 all
Average: WE 1–17 odd
Above Average: WE 5–17 odd, 19–22 all

Additional Answers

Written Exercises

17. R: 4.0 cm; M: $x + 0.9$ cm; SD: 1.46 cm

Enrichment

The standard deviation of a set of data is a measure of the consistency of the data. Have students use the standard deviation to answer the following questions.

	Bill	Rich		Bill	Rich
J	165	150	A	175	155
F	168	148	M	173	157
M	170	149	J	168	160

	Savings 1	Savings 2
J	$1,000	$2,000
F	1,500	1,500
M	1,750	1,250
A	1,600	1,200
M	1,400	1,600
J	1,500	1,500

1. Which person, Bill or Rich, has the more consistent weight in pounds over a six-month period?   Bill

2. Which bank account, Savings 1 or Savings 2, has the more consistent balance over a six-month period?   Savings 1

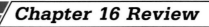

Chapter 16 Review

Key Terms

event   (p. 603)
frequency table   (p. 610)
histogram   (p. 615)
independent events   (p. 604)
mean   (p. 608)
measures of central tendency   (p. 610)
median   (p. 609)

mode   (p. 609)
probability   (p. 599)
range   (p. 619)
sample space   (p. 603)
scatter plot   (p. 614)
standard deviation   (p. 619)
tree diagram   (p. 603)

Key Ideas and Review Exercises

16.1   To find the probability of an event E, use

$$P(E) = \frac{s}{t} = \frac{\text{number of favorable outcomes}}{\text{total number of possible outcomes}}.$$

Ten cards are numbered consecutively from 1 to 10 and placed in a box. One card is drawn from the box. Find the probability of each event.

1. P(even number) $\frac{1}{2}$ **2.** P(odd number) $\frac{1}{2}$ **3.** $P(5)$ $\frac{1}{10}$ **4.** P(not 8) $\frac{9}{10}$

16.2   To find the probability of $P(A \text{ and } B)$, use
$P(A \text{ and } B) = P(A) \cdot P(B)$, if A and B are independent events.
To find the probability of $P(A \text{ or } B)$, use
$P(A \text{ or } B) = P(A) + P(B) - P(A \text{ and } B)$. If A and B are mutually exclusive events, use $P(A \text{ or } B) = P(A) + P(B)$.

A red die and a white die are rolled.

5. Give a sample space of ordered pairs. See Example 3, p. 605.

6. Find $P(r = 5 \text{ and } w = 1)$. $\frac{1}{36}$

7. Find $P(r < 4 \text{ and } w \geq 2)$. $\frac{5}{12}$

8. Find $P(r \leq 3 \text{ or } w = 6)$. $\frac{7}{12}$

A pouch contains 6 red marbles, 4 white marbles, and 5 pink marbles. A marble is drawn.

$\frac{3}{5}$

9. Find P(red marble *or* white marble). $\frac{2}{3}$ **10.** Find P(pink marble *or* white marble).

16.3   To find the mean of a set of data, add the scores and divide the sum by the number of scores.
To find the median of a set of data, arrange the scores in order and choose the middle score. If the data contains an even number of scores, use the mean of the two middle scores as the median.
To find the mode of a set of data, find the score that occurs most frequently. A set of data may have more than one mode.

Find the mean, the median, and the mode of each set of data.

11. 8, 12, 10, 8, 16, 9
mean: 10.5; median: 9.5; mode: 8

12. 98, 99, 97, 98, 96, 100
mean: 98; median: 98; mode: 98

The ages of 20 students on a camping trip were: 17, 15, 16, 15, 16, 15, 15, 16, 17, 15, 18, 15, 17, 16, 17, 16, 15, 18, 15, 16.

13. Make a frequency table for the data.

14. Find the mean. 16

15. Find the median. 16

16. Find the mode. 15

17. In a short paragraph, explain the advantages and the disadvantages of using the mean, the median, and the mode. Use examples to illustrate your explanation. Answers will vary.

16.4   To interpret a scatter plot, see p. 614.
To construct or to use a histogram, see p. 615.

The histogram at the right represents the weights of a freshman football team. Each boundary score is assigned to the lesser interval.

18. How many players weigh between 195 lb and 205 lb? 4

19. How many players weigh more than 205 lb? 6

The scatter plot at the right represents the number of hours students studied for their final exam and their scores.

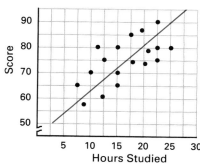

20. A student studies about 5 h. What score did this student probably receive? About 55

21. A student's score was close to 95. About how many hours did this student study for the exam? About 27 h

16.5   To find the range of a set of data, find the difference between the highest and lowest scores.

To find the standard deviation of a set of data, use

$$s = \sqrt{\frac{(x_1 - \bar{x})^2 + (x_2 - \bar{x})^2 + \cdots + (x_n - \bar{x})^2}{n}}$$

For Exercises 22–23, use the set of data: 18, 24, 26, 13, 19.

22. Find the range. 13

23. Find the standard deviation. 4.60

13.

Ages(s)	Frequency(f)	f · s
18	2	36
17	4	68
16	6	96
15	8	120
	20	320

A bag contains 15 red marbles, 10 white marbles, and 5 yellow marbles. One marble is drawn. Find the probability of each event.

1. $P(\text{red})$ $\frac{1}{2}$
2. $P(\text{yellow})$ $\frac{1}{6}$
3. $P(\text{not white})$ $\frac{2}{3}$
4. $P(\text{not yellow})$ $\frac{5}{6}$

5. $P(\text{blue})$ 0
6. $P(\text{not red})$ $\frac{1}{2}$
7. $P(\text{white})$ $\frac{1}{3}$
8. $P(\text{not blue})$ 1

A white die and a black die are rolled.

9. Give a sample space of ordered pairs. See Example 3, p. 605.

10. Find $P(w = 6 \text{ and } b = 3)$. $\frac{1}{36}$

11. Find $P(b \geq 3 \text{ and } w = 1)$. $\frac{1}{9}$

12. Find $P(w = 4 \text{ or } b > 2)$. $\frac{13}{18}$

Find the mean, the median, and the mode of each set of data.

13. 12, 10, 10, 20, 8

mean: 12; median: 10; mode: 10

14. 96, 95, 96, 97, 94, 90

mean: $94\frac{2}{3}$; median: $95\frac{1}{2}$; mode: 96

Two columns of a frequency table are shown at the right. Use the scores in this table for Exercises 15–17.

Scores	Frequency
8	6
9	4
10	5
11	3
12	2

15. Find the mean. 9.55

16. Find the median. 9.5

17. Find the mode. 8

The histogram at the right represents the SAT scores of a college class. Each boundary score is assigned to the lesser interval.

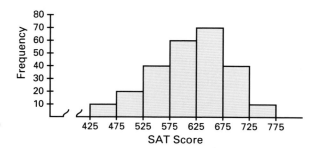

18. How many scores are between 675 and 725? 40

19. How many scores are between 425 and 525? 30

20. How many scores are 575 or less? 70

For Exercises 21–22, use the following set of data: 16, 14, 23, 27, 30.

21. Find the range. 16

22. Find the standard deviation. 6.16

23. A green die and a red die are rolled. Find $P(g + r = 7 \text{ or } g + r = 11)$. $\frac{2}{9}$

24. When a die is rolled, what are the odds that the upper face will show 5 dots? $\frac{1}{5}$

Choose the one best answer to each question or problem.

1. What is the decimal for
   $60 + 4 + \frac{3}{5} + \frac{2}{50}$? B
   (A) 70.04   (B) 64.64
   (C) 64.46   (D) 64.44
   (E) 64.05

2. What is the decimal for
   $90 + 5 + \frac{3}{10} + \frac{6}{25}$? E
   (A) 95.90   (B) 95.84
   (C) 95.98   (D) 95.27
   (E) 95.54

3. What is the sum of all the whole
   number factors of 48? A
   (A) 124   (B) 75   (C) 48
   (D) 26   (E) 14

4. How many numbers between 1 and
   100 inclusive are divisible by either
   2 or 3 or both? C
   (A) 99   (B) 98   (C) 67
   (D) 66   (E) 16

5. How many numbers between 1 and
   100 inclusive are divisible by either
   5 or 12 but not both? D
   (A) 30   (B) 28   (C) 27
   (D) 26   (E) 1

6. How many numbers between 1 and
   100 inclusive are divisible by both 3
   and 8? D
   (A) 45   (B) 33   (C) 12   (D) 4
   (E) 0

7. If two dice are rolled, the probability that neither lands with 4 dots up
   is __?__. A
   (A) $\frac{25}{36}$ (B) $\frac{11}{12}$ (C) $\frac{5}{6}$ (D) $\frac{1}{36}$
   (E) $\frac{1}{12}$

8. In the formula $A = \frac{bh}{2}$, for $A = 48$
   and $b = 8$, find the mean of b and h. D
   (A) 52   (B) 28   (C) 12
   (D) 10   (E) 7

9.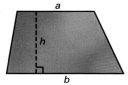

   The formula for the area of the trapezoid is $A = \frac{1}{2}h(a + b)$, where a
   and b are the lengths of the bases
   and h is the height. Find the mean
   of the bases if the area is 198 in^2
   and the height is 22 in. D
   (A) 4,356   (B) 36   (C) 18
   (D) 9   (E) 4.5

10. In 1984, a company transported
    15,000 boxes of fruit per month. In
    1990, it transported 75,000 boxes of
    fruit per month. What was the aver-
    age increase in monthly shipments
    per year? A
    (A) 10,000   (B) 12,000
    (C) 15,000   (D) 5,000   (E) 857

11. The average of 22 students' scores is
    86. If the two highest and the two
    lowest scores are removed, the mean
    of the remaining scores is 82. What
    is the mean of the removed
    scores? D
    (A) 472   (B) 416   (C) 105
    (D) 104   (E) 4

12. If $-1 < x < 0$ and $x < y$, then the
    mean of x and y is __?__. B
    (A) less than x
    (B) greater than x
    (C) greater than y
    (D) less than -1
    (E) It cannot be determined from
        the information given.

For additional standardized test practice, see
the SAT/ACT test booklet for the cumulative
tests Chapters 1–16.

625

1. Evaluate $8 - 2x$ for $x = 2$. 4 **1.2**
2. Which property is illustrated? **1.5**
   $6 + (-a) = -a + 6$ Comm P. Add
3. Simplify $6m + 3(5 + 2m) + 4$. **1.7** $12m + 19$
4. Simplify $-6 \cdot 9 - 4^2$. -70
5. Evaluate $-4y + 6y^3$ for **2.8**
   $y = -2$. -40
6. Simplify $6(p - 3) - 2(1 + 4p)$. **2.10** $-2p - 20$
7. Pam plans to ride her bike $4\frac{1}{3}$ **3.6**
   mi for 40 min. At what rate
   must she travel? $6\frac{1}{2}$ mi/h

Solve each equation or inequality.

8. $r + 15 = -15$ -30 **3.1**
9. $2x - 8 = 17 + 3x$ -25 **3.4**
10. $\frac{3}{4}x + \frac{1}{3} = \frac{2}{3}x - 1\frac{2}{3}$ -24 **4.5**
11. $|x - 4| = 7$ -3 or 11 **5.6**
12. $-2x - 5 < 13$ $x > -9$ **5.7**
13. $6x^2 + 13x - 5 = 0$ $\frac{1}{3}$ or $-\frac{5}{2}$ **7.7**

Simplify.

14. $(2a^3b^2)(-5ab^5)$ $-10a^4b^7$ **6.1**
15. $(-3x^2yz^4)^4$ $81x^8y^4z^{16}$ **6.2**
16. $\dfrac{16x^{-2}y^{-1}}{-4xy^{-2}}$ $-\frac{4y}{x^3}$ **6.4**
17. $\dfrac{a - 3}{a^2 - 3a}$ $\frac{1}{a}$ **8.1**

Find each answer.

18. $(5x^2 - 5x - 3)$
    $+ (-3x^2 - x + 5)$ $2x^2 - 6x + 2$ **6.7**
19. $(4x^2 - 9x + 3)$
    $- (-2x^2 + 8x - 1)$ $6x^2 - 17x + 4$
20. $-3x(2x^3 - 5x + 1)$ **6.8**
21. $(2x - 3y)(3x + 4y)$ **6.9**
22. $\dfrac{x - 2}{x} \div \dfrac{3x - 6}{x^4}$ $\frac{x^3}{3}$ **8.4**

20. $-6x^4 + 15x^2 - 3x$ 21. $6x^2 - xy - 12y^2$ 26. $4(x^2 - 3x + 7)$

23. $\dfrac{a + 3}{9a} - \dfrac{4a + 2}{3a}$ $\frac{-11a - 3}{9a}$ **8.6**
24. $\dfrac{x}{x + 5} + \dfrac{8x + 15}{x^2 + 5x}$ $\frac{x + 3}{x}$ **8.7**
    $\frac{x+3}{2a^2 + a - 3}$
25. $(4a^3 - 7a + 3) \div (2a - 1)$ **8.8**
26. Factor $4x^2 - 12x + 28$. **7.2**
27. Factor $3a^2 - 9a - 30$. **7.6**
    $3(a - 5)(a + 2)$

Solve for x. Check for extraneous solutions, if necessary.

28. $\dfrac{x + 5}{x - 4} = \dfrac{3}{x} + \dfrac{36}{x^2 - 4x} - 6$ **9.1**
29. $\dfrac{x + 7}{6} = \dfrac{3}{x}$ -9 or 2 **9.2**
30. $7 - 2x = mx$ $\dfrac{7}{m + 2}, m \neq -2$ **9.3**
31. Is the relation $\{(-2, 3),$ **10.2**
    $(-1, 5), (1, 3), (7, 1)\}$ a func-
    tion? Explain your reasoning. yes
32. Given $f(x) = 2x^2 - 9$, find **10.3**
    $f(-1)$. -7
33. Find the slope of the line de- **11.1**
    termined by the points $P(2,3)$
    and $Q(7, -6)$. $-\frac{9}{5}$
34. Write an equation in the form **11.2**
    $Ax + By = C$ for the line
    that includes the point
    $A(-4, 3)$ and with a slope of $-\frac{2}{5}$.
    $2x + 5y = 7$
35. Solve. $\begin{array}{l} x + y = 4 \\ 3x - 2y = 2 \end{array}$ $x = 2, y = 2$ **12.5**

Simplify.

36. $-\sqrt{18x^7y^{12}}$ $(x > 0, y > 0)$ $-3x^3y^6\sqrt{2x}$ **13.5**
37. $4\sqrt{3} \cdot 2\sqrt{75}$ 120 **13.7**
38. $\sqrt{2}(\sqrt{2} - 3\sqrt{6})$ $2 - 6\sqrt{3}$
39. $\dfrac{-5}{3 - \sqrt{3}}$ $\frac{-15 - 5\sqrt{3}}{6}$ **13.8**

40. In right triangle ABC, $a = 4$ **13.4**
and $b = 6$. Find the length of
the hypotenuse c. $2\sqrt{13}$

41. Use the quadratic formula to   **14.3**
solve $x^2 + 4x + 2 = 0$. $-2 \pm \sqrt{2}$

42. If $\triangle ABC \sim \triangle XYZ$, find z for   **15.1**
$a = 14, c = 10, x = 7$.   5

**Refer to the figure to name each
trigonometric ratio in fraction form.**

43. $\sin A$ $\frac{3}{5}$ **15.2**

44. $\cos B$ $\frac{3}{5}$

45. $\tan A$ $\frac{3}{4}$

**Find the median and the mode,
if any, for the given data.**

46. 16, 14, 36, 32, 38   median: 32;   **16.3**
no mode

47. 98, 97, 98, 96, 97, 98
Median: 97.5, mode 98

**For the set of data 12, 8, 6, 10,
16, 8, find each of the following.**

48. the range $16 - 6$, or 10   **16.5**

49. the standard deviation   3.27

Solve each problem.

50. The length of a rectangle is   **4.4**
twice the width. The perime-
ter is 45 cm. Find the length
and the width. l: 15 cm; w: 7.5 cm

51. Twenty percent of 45 is what   **4.7**
number?   9

52. Six is what percent of 30?   20%

53. Ten percent of what number
is 20?   200

54. The selling price of a tele-   **4.8**
vision set is $260. The profit
is 30% of the cost. Find the
cost. $200

55. The square of a number is 12   **7.8**
less than 8 times the number.
Find the number.   2 or 6

56. Two cars started toward each   **9.4**
other at the same time from
towns 170 km apart. One
car's rate was 45 km/h, and
the other car's rate was 40
km/h. After how many hours
did they meet?   2 h

57. To do a job alone takes Rose   **9.5**
4 h, Bob 3 h, and Shelly 5 h.
How long would the job take
if they all worked together? $1\frac{13}{47}$ h

58. The cost of a metal varies di-   **10.6**
rectly as its mass. If 5 kg cost
$15, find the cost of 8 kg. $24

59. y varies inversely as x. y is 20   **10.7**
when x is -4. Find y when x
is -16.   5

60. The cost of an adult ticket to   **12.3**
a school play was $2.00. A
student ticket cost $1.50. The
number of student tickets
sold was 50 more than twice
the number of adult tickets.
The total income from tickets
was $800. How many tickets
of each type were sold?

61. The sum of the digits of a   **12.6**
two-digit number is 8. If the
digits are reversed, the new
number is 36 less than the
original number. Find the
original number.   62

62. Mark has 6 more quarters   **12.8**
than dimes. He has $3.25 in
all. How many coins of each
type does he have?   5 d; 11 q

63. What is the probability that if   **16.1**
a letter is picked from the al-
phabet, it will be a vowel? $\frac{5}{26}$

64. A bag contains 20 colored   **16.2**
disks. There are 12 red disks
and 3 blue disks. What is the
probability of picking a disk
that is either red or blue? $\frac{3}{4}$

60. 145 adult; 340 student

Computer Investigations: Algebra 1

The following is a listing of the computer investigations that are included in *Holt Algebra 1*. The related textbook page for each investigation is also indicated. For each investigation, you will need the *Investigating Algebra with the Computer* software.

Title	Appendix Page	Related Text Page
SLOPE–INTERCEPT FORM OF A LINE	629	424
GRAPHING $y = |Ax + B|$	631	424
GRAPHING $y = |Ax + B| + C$	633	424
GRAPHING SYSTEMS OF EQUATIONS	635	445
GRAPHING QUADRATIC FUNCTIONS	637	558
MAXIMUM AND MINIMUM	638	558
THE DISCRIMINANT	639	564
USING STATISTICS: PREDICTING	641	614

Investigation: Slope-Intercept Form of a Line

Program: SLOPEINT

Objectives: To use the computer to graph equations of the form $y = Mx + B$

To predict the slope and y intercept of a line before graphing

Part 1

1. **a.** With BASIC loaded into the computer and the disk in drive 1 (or drive A), run the program SLOPEINT by typing the appropriate command.
   Apple: RUN SLOPEINT   IBM: RUN "SLOPEINT"
   Be sure to press the RETURN key after typing the command.
   b. Read the opening messages of the program.

2. Use the program to draw the graph of $y = 2x + 3$.
   a. Type 2 for M and press RETURN.
   b. Enter 3 for B (and press RETURN).

3. With the graph on the screen, answer the following questions.
   a. Is the graph a straight line?
   b. What is the y coordinate of the point where the graph crosses the y axis? This value is called the y intercept of the graph.
   c. Use the point $(-1,1)$ and the point where the graph crosses the y axis to compute the slope. What is the slope of the graph?

4. **a.** Answer Y (yes) to the questions ANOTHER GRAPH? and CLEAR THE SCREEN?.
   b. Graph the equation $y = 4x - 1$. Enter 4 for M and -1 for B
   c. What is the y intercept?
   d. Use the y intercept and the point $(1,3)$ to compute the slope. What is the slope?

5. Answer Y to ANOTHER GRAPH? and Y to CLEAR THE SCREEN?. Then, before having the computer graph $y = -3x + 2$, predict the y intercept and the slope of the graph.
   a. y intercept?   **b.** slope?

6. Now have the computer graph $y = -3x + 2$. (M $= -3$ and B $= 2$.)
   a. Was your prediction for the y intercept correct?
   b. Was your prediction for the slope correct?

7. Repeat steps 5 and 6 for the equation $y = .5x - 3$. Do not leave the previous graph on the screen. Were your predictions correct?

8. *Complete*: For the graph of the equation $y = Mx + B$, the slope equals __?__ and the y intercept equals __?__.

Try These

Have the computer graph the following equations. In each case, predict the slope and y intercept before seeing the graph. Clear the screen before each graph is drawn.

9. $y = -.5x - 4$

10. $y = x + 5$

11. $y = 1.5x - 2$

12. $y = -x + 2.5$

Part 2

1. Clear the screen and graph $y = 3x$. That is, enter 3 for M and 0 for B.
   a. Does the graph go through the origin?
   b. What is the y intercept of the graph?
   c. Does the y intercept equal the value of B in the equation?
   d. What is the slope of the line?
   e. Does the slope equal the value of M in the equation?

2. a. Before graphing $y = -2x$, predict the y intercept and the slope.
   b. Now clear the screen and graph $y = -2x$. Were your predictions correct?

Try These

Have the computer graph the following equations. In each case, predict the slope and y intercept before seeing the graph. Clear the screen before each graph is drawn.

3. $y = 2x$ 4. $y = -3x$ 5. $y = 2.5x$ 6. $y = -4x$

Part 3

1. Clear the screen and graph $y = 4$. In this case, M = 0 and B = 4.
   a. The graph is what kind of straight line?
   b. What is the y intercept of the graph?
   c. Does the y intercept equal the value of B?
   d. What is the slope of the line?
   e. Does the slope equal the value of M for this equation?

2. a. Before graphing $y = -2$, predict the y intercept and the slope.
   b. Now graph $y = -2$. (Enter 0 for M.) Were your predictions correct?

Try These

Have the computer graph these equations. In each case, predict the slope and y intercept before seeing the graph. Clear the screen before each graph is drawn.

3. $y = 1$ 4. $y = -5$ 5. $y = 2.5$ 6. $y = -1$

Investigation: Graphing $y = |Ax + B|$

Program: ABSVAL1

Objectives: To use the computer to graph equations of the form $y = |Ax + B|$

To predict the x intercept of the graph of an equation of the form $y = |Ax + B]$

To predict the slopes of the arms of the graph of an equation of the form $y = |Ax + B|$

Part 1

1. **a.** With BASIC loaded into the computer and the disk in drive 1 (or drive A), run the program ABSVAL1 by typing the appropriate command.
   Apple: RUN ABSVAL1   IBM: RUN "ABSVAL1"
   Be sure to press the RETURN key after typing the command.

2. Have the computer graph the equation $y = |x|$. Enter 1 for A and 0 for B. With the graph on the screen, answer the following questions.
   a. The graph is shaped like what letter of the alphabet?
   b. The graph lies in which quadrants?
   c. What is the x intercept of the graph?
   d. What are the slopes of the two "arms" or "branches" of the graph?

3. **a.** Answer Y (yes) to ANOTHER GRAPH? and N (no) to CLEAR THE SCREEN?.
   b. Have the computer graph the equation $y = |2x|$. That is, enter 2 for A and 0 for B.
   c. The graph is shaped like what letter?
   d. The graph lies in which quadrants?
   e. What is the x intercept?   **f.** What are the slopes of the arms?

4. Before having the computer graph $y = |3x|$, predict the x intercept and the slopes of the arms of the graph.
   a. x intercept?   **b.** slopes?

5. Without clearing the screen, graph $y = |3x|$. (Use A = 3 and B = 0.)
   a. Was your prediction for the x intercept correct?
   b. Was your prediction for the slopes correct?

6. Answer Y to ANOTHER GRAPH? and Y to CLEAR THE SCREEN?. Then graph $y = |-2x|$. (A = -2, B = 0)
   a. What is the x intercept?   **b.** What are the slopes of the arms?
   c. The graph is the same as that of what equation graphed earlier?

7. *Complete:* For the graph of $y = |ax|$, the x intercept is __?__ and the slopes of the arms are __?__ and __?__.

Try These Clear the screen each time and have the computer graph the following equations. In each case, predict the slopes of the arms before seeing the graph.

8. $y = |.5x|$ **9.** $y = |-4x|$ **10.** $y = |-x|$ **11.** $y = |5x|$

Part 2
1. Clear the screen and graph $y = |x + 1|$. Enter 1 for A and 1 for B.
   a. The graph is shaped like what letter?
   b. The graph is in which quadrants?
   c. What is the x intercept?   d. What are the slopes of the arms?
2. Graph $y = |x - 2|$. Do not clear the screen. (Enter 1 for A and -2 for B.)
   a. What is the x intercept?   b. What are the slopes of the arms?
3. Before graphing $y = |x + 3|$, predict the x intercept and the slopes of the arms.   a. x intercept?   b. slopes?
4. Without clearing the screen, graph $y = |x + 3|$. Were your predictions for the x intercept and the slopes correct?

Try These Clear the screen and have the computer graph the following equations on the same axes. In each case, predict the x intercept before seeing the graph.

5. $y = |x - 1|$ **6.** $y = |x - 3|$
7. $y = |x + 2|$ **8.** $y = |x + 4|$

Part 3
1. Clear the screen. Then graph $y = |2x - 4|$. (Enter 2 for A and -4 for B.)
   a. The graph is shaped like what letter?
   b. The graph is which quadrants?
   c. What is the x intercept?   d. What are the slopes of the arms?
2. Repeat step 1 for $y = |3x + 6|$.   a. x intercept?   b. slopes?
3. Before seeing the graph of $y = |2x + 8|$, predict the x intercept and the slopes of the arms.   a. x intercept?   b. slopes?
4. Clear the screen and graph $y = |2x + 8|$. Were your predictions correct?
5. Repeat steps 3–4 for $y = |-2x + 6|$.   a. x intercept?   b. slopes?
6. Repeat steps 3–4 for $y = |-x + 4|$.   a. x intercept?   b. slopes?
7. *Complete:* The graph of $y = |Ax + B|$ (with $a \neq 0$) is shaped like the letter __?__ and lies in quadrants __?__ and __?__. The x intercept is __?__ and the slopes of the arms are __?__ and __?__.

Try These Predict the x intercept and the slopes of the arms of the graph of each equation. Then clear the screen and have the computer graph the equation. Correct your predictions if they were wrong.

8. $y = 2x + 1$ **9.** $y = 3x = 2$ **10.** $y = -5x - 4$

Investigation: Graphing $y = |Ax + B| + C$

Program: ABSVAL2

Objectives: To use the computer to graph equations of the form $y = |Ax + B| + C$
To predict the coordinates of the vertex and the slopes of the arms of the graph.

Part 1

1. **a.** With BASIC loaded into the computer and the disk in drive 1 (or drive A), run the program ABSVAL2 by typing the appropriate command.
   Apple: RUN ABSVAL2   IBM: RUN "ABSVAL2"
   Be sure to press the RETURN key after typing the command.
   b. Read the opening messages of the program.

2. **a.** Have the computer graph the equation $y = |x|$. Enter 1 for A, 0 for B, and 0 for C.
   b. The vertex of the graph is the endpoint where the two arms meet. What are the coordinates of the vertex of the graph? (_?_, _?_)
   c. What are the slopes of the arms of the graph?

3. **a.** Answer Y (yes) to ANOTHER GRAPH and N (no) to CLEAR THE SCREEN?.
   b. Have the computer graph $y = |x| + 3$. That is, enter 1 for A, 0 for B, and 3 for C.
   c. What are the coordinates of the vertex?
   d. What are the slopes of the arms?

4. **a.** Before having the computer graph $y = |x| - 2$, predict the coordinates of the vertex and the slopes of the arms.
   b. Without clearing the screen, graph $y = |x| - 2$. (A = 1, B = 0, C = −2)
   c. Were your predictions for the vertex and the slopes correct?

5. Answer Y to ANOTHER GRAPH? and Y to CLEAR THE SCREEN?. Then graph $y = |2x| + 4$. (A = 2, B = 0, C = 4) Give the following.
   a. coordinates of the vertex?   **b.** slopes of the arms?

6. Repeat step 5 for $y = |-3x| - 1$. (A = −3, B = 0, C = −1)

7. **a.** Before graphing $y = |4x| - 6$, predict the coordinates of the vertex and the slopes of the arms of the graph.
   b. Without clearing the screen, graph $y = |4x| - 6$. Were your predictions correct?

8. Repeat step 7 for $y = |-2x| + 5$.

Try These

Clear the screen and graph the following equations on the same screen. In each case, predict the coordinates of the vertex and the slopes of the arms before seeing the graph. Then check your predictions from the graph.

9. $y = |x| + 2$ **10.** $y = |x| + 5$ **11.** $y = |x| - 3$ **12.** $y = |x| - 7$

13. $y = |5x| + 6$ **14.** $y = |-4x| + 1$ **15.** $y = |-5x| - 7$ **16.** $y = |.5x| - 8$

Part 2

1. Clear the screen and graph $y = |x + 1| + 2$. ($A = 1$, $B = 1$, and $C = 2$.)

  a. What are the coordinates of the vertex?

  b. What are the slopes of the arms?

2. Repeat step 1 for $y = |x - 2| + 3$. Do not clear the screen. ($A = 1$, $B = -2$, $C = 3$)

3. a. Before having the computer graph $y = |x + 3| - 4$, predict the coordinates of the vertex and the slopes of the arms.

  b. Without clearing the screen, graph $y = |x + 3| - 4$. Were your predictions for the intercepts and the slopes correct?

4. Clear the screen and graph $y = |2x - 4| - 5$. Give the following.

  a. coordinates of the vertex? **b.** slopes of the arms?

5. Without clearing the screen, repeat step 4 for $y = |3x - 9| + 1$.

6. a. Before graphing $y = |4x + 12| - 2$, predict the coordinates of the vertex and the slopes of the arms.

  b. Without clearing the screen, graph $y = |4x + 12| - 2$. Were your predictions correct?

7. Repeat step 6 for $y = |-3x + 6| - 7$.

8. *Complete*: For the graph of $y = |Ax + B| + C$ ($A \neq 0$), the vertex is (_?_, _?_) and the slopes of the arms are _?_ and _?_.

Try These

Clear the screen and graph the following equations on the same screen. In each case, predict the coordinates of the vertex and the slopes of the arms before seeing the graph. Then check your predictions from the graph.

9. $y = |x - 1| + 4$ **10.** $y = |x + 6| + 2$

11. $y = |2x + 8| - 1$ **12.** $y = |3x - 15| - 6$

13. $y = |-4x - 12| + 5$ **14.** $y = |-.5x + 4| - 8$

Investigation: Graphing Systems of Equations

Program: SYSTEMS

Objective: To use the computer to solve systems of two linear equations by graphing

1. **a.** With BASIC loaded into the computer and the disk in drive 1 (or drive A), run the program SYSTEMS by typing the appropriate command.
   Apple: RUN SYSTEMS   IBM: RUN "SYSTEMS"
   Be sure to press the RETURN key after typing the command.
   b. Read the opening messages of the program.

2. Use the program to draw the graph of these two equations on the same axes.

$$\begin{cases} x + 2y = 5 \\ 3x - y = 1 \end{cases}$$

   a. For the first equation enter 1 for A, 2 for B, and 5 for C.
   b. Is the graph of the first equation a straight line?
   c. For the second equation enter 3 for A, −1 for B, and 1 for C.
   d. Is the graph of the second equation a straight line?
   e. At what point do the two graphs intersect? (_?_ , _?_)
   f. Substitute the x and y coordinates of the point of intersection into the first equation. Does a true sentence result?
   g. Substitute the coordinates of the intersection point into the second equation. Does a true sentence result?

3. **a.** Enter Y (yes) to the question ANOTHER SYSTEM?.
   b. Have the computer graph these equations on the same axes.

   $$\begin{cases} 3x - 2y = 14 \\ 2x + 3y = -8 \end{cases} \qquad \begin{array}{l} A = 3, B = -2, C = 14 \\ A = 2, B = 3, \quad C = -8 \end{array}$$

   c. What is the point of intersection of the two lines? (_?_ , _?_)
   d. Substitute the x and y coordinates of this point into the first equation. Do you get a true sentence?
   e. Substitute the x and y coordinates of this point into the second equation. Do you get a true sentence?

4. Repeat step 3 for this system. $$\begin{cases} -3x + y = 4 \\ 10x - 7y = 5 \end{cases}$$

   a. What is the point of intersection of the two lines?
   b. Substitute the x and y coordinates of the intersection point into both equations. Are both sentences true?

5. Repeat step 3 for this system. $\begin{cases} 6x - 2y = 12 \\ 3x - y = -4 \end{cases}$

a. Do the two lines intersect?

b. Do the two equations have any common solution?

6. Repeat step 3 for this system. $\begin{cases} x + y = 6 \\ 2x + 2y = 12 \end{cases}$

a. What do you notice about the graphs of the two equations?

b. How many common solutions do the two equations have?

Try These

Have the computer graph each pair of equations on the same axes. Write the coordinates of the point of intersection of each pair or, if the lines do not intersect, write <u>no common point</u>. Check by substitution that the coordinates of the point of intersection make both equations true.

7. $\begin{cases} 3x + y = 11 \\ 2x - 3y = 11 \end{cases}$ **8.** $\begin{cases} x + y = 7 \\ 2x - 3y = 14 \end{cases}$

9. $\begin{cases} 4x - 3y = 18 \\ -8x + 6y = 11 \end{cases}$ **10.** $\begin{cases} -9x + 7y = -59 \\ 11x + 12y = 31 \end{cases}$

11. $\begin{cases} 2x = 14 \\ x - 2y = 3 \end{cases}$ (NOTE: B = 0)   **12.** $\begin{cases} 10y = -30 \\ 4x + 2y = -2 \end{cases}$ (NOTE: A = 0)

Put each equation in the form $Ax + By = C$. Then use the program to determine the common solution, if any, of each system. Check each solution by substitution.

13. $\begin{cases} y = 3x + 2 \\ y = 6x - 1 \end{cases}$ **14.** $\begin{cases} 4x = 9y - 7 \\ 3 = 3x - 4y \end{cases}$

15. $\begin{cases} 2y = 11x + 15 \\ y + 3x = 0 \end{cases}$ **16.** $\begin{cases} 16 = -2x \\ 20 + x = 2y \end{cases}$

17. $\begin{cases} 3x - y = 7 \\ 4y = 12x - 15 \end{cases}$ **18.** $\begin{cases} y = 5x - 2 \\ 15x - 3y = 6 \end{cases}$

Investigation: Graphing Quadratic Functions

Program: PARA1

Objectives: To use the computer to graph equations of the form $y = Ax^2 + Bx + C$ and discover that the graph is a parabola

To read the zeros of a quadratic function from its graph

1. **a.** With BASIC loaded into the computer and the disk in drive 1 (or drive A), run the program PARA1 by typing the appropriate command.
   Apple: RUN PARA1   IBM: RUN "PARA1"
   Be sure to press the RETURN key after typing the command.

   b. Read the opening messages of the program.

2. Use the program to draw the graph of $y = x^2$, as follows.

   a. Enter 1 for A and press RETURN.   **b.** Enter 0 for B (and press RETURN).

   c. Enter 0 for C.   **d.** Is the graph a straight line?

3. The graph of $y = x^2$ is called a **parabola.** Where does the parabola touch the x axis; that is, what is its x intercept?

4. **a.** Enter Y to ANOTHER GRAPH? and N to CLEAR THE SCREEN?.

   b. Graph $y = -3x^2$. Enter -3 for A, 0 for B, and 0 for C.

5. Answer these questions about the graph of $y = -3x^2$.

   a. Is the graph a parabola?   **b.** What is the x intercept?

   c. The **zeros** of a function are the values for x which make the function equal zero. List any zeros of the function $y = -3x^2$.

6. **a.** Answer Y to ANOTHER GRAPH? and Y to CLEAR THE SCREEN?.

   b. Graph $y = x^2 + 2x$. Enter 1 for A, 2 for B, and 0 for C.

7. Answer these questions about the graph of $y = x^2 + 2x$.

   a. Is the graph a parabola?   **b.** How many zeros (x intercepts) does this function have?

   c. List the zeros.

8. Clear the screen and graph $y = x^2 - 3x$. Enter $A = 1$, $B = -3$, and $C = 0$. What are the zeros of this function?

9. Clear the screen and graph $y = x^2 - x - 2$. ($A = 1$, $B = -1$, $C = -2$)

   a. How many zeros does this function have?   **b.** List the zeros.

Try These   Have the computer graph the following functions, each on a separate screen. List the zeros of each function. Estimate any zeros that are not integers to the nearest tenth.

10. $y = x^2 + 2x - 8$ 11. $y = -x^2 + 4x$

12. $y = 2x^2 - 11x - 6$ 13. $y = x^2 - 12x + 36$

Investigation: Maximum and Minimum

Program: PARA3

Objectives: To use the computer to graph functions of the form $y = Ax^2 + Bx + C$ and draw the axis of symmetry

To discover that the parabola opens upward when $A > 0$ and downward when $A < 0$

To discover that the x coordinate of the vertex is $\frac{-B}{2A}$ and the equation of the axis of symmetry is $x = \frac{-B}{2A}$

1. **a.** With BASIC loaded into the computer and the disk in drive 1 (or drive A), run the program PARA3 by typing the appropriate command.
   Apple: RUN PARA3   IBM: RUN "PARA3"
   Be sure to press the RETURN key after typing the command.
   b. Read the opening messages of the program.

2. **a.** Have the computer graph $y = x^2$. Enter 1 for A, 0 for B, and 0 for C.
   b. The computer graphs the parabola and draws the **axis of symmetry.** It also prints the equation of the axis of symmetry and the coordinates of the turning point or **vertex** of the parabola.
   c. Does the parabola open upward or downward?
   d. What are the coordinates of the vertex?   (_?_, _?_)
   e. What is the equation of the axis of symmetry? $x = $ _?_

3. **a.** Enter Y (yes) for ANOTHER GRAPH?.
   b. Have the computer graph $y = -2x^2$. Enter -2 for A, 0 for B, and 0 for C.
   c. Does the parabola open upward or downward?
   d. What are the coordinates of the vertex?
   e. What is the equation of the axis of symmetry?

4. Repeat step 3 for $y = x^2 - 2x + 3$. (A = 1, B = -2, C = 3)
   a. open upward or downward?   **b.** coordinates of the vertex?
   c. equation of the axis of symmetry?

5. **a.** For the function $y = x^2 - 2x + 3$ above, compute $\frac{-B}{2A}$.
   b. What connection does this value have to the vertex of the parabola?
   c. What connection does this value have to the axis of symmetry equation?

6. **a.** For $y = 2x^2 - 16x + 35$, compute $\frac{-b}{2a}$.

   b. Before graphing the function, predict whether it will open up or down.

   c. Predict the equation of the axis of symmetry.

7. Have the computer graph $y = 2x^2 - 16x + 35$.

   a. Does the parabola open the way you predicted?

   b. Was your prediction of the equation of the axis of symmetry correct?

   c. What are the coordinates of the vertex?

8. Before graphing $y = -3x^2 + 6x - 2$, make predictions about the graph.

   a. open upward or downward?   **b.** equation of the axis of symmetry?

   c. x coordinate of the vertex?

9. Have the computer graph $y = -3x^2 + 6x - 2$. Were your predictions correct?

10. Complete for the parabola $y = Ax^2 + Bx + C$. $(A \neq 0)$

   a. The parabola opens upward if $A > \underline{\ ?\ }$.

   b. The parabola opens downward if $\underline{\ ?\ }$.

   c. The x coordinate of the vertex of the parabola equals $\underline{\ ?\ }$.

   d. The equation of the axis of symmetry is $\underline{\ ?\ }$.

Try These   Before having the computer graph each parabola, predict whether it will open upward or downward. Also predict the x coordinate of the vertex and the equation of the axis of symmetry. Check your predictions from the computer's graph. Also list the y coordinate of the vertex.

11. $y = x^2 + 6x + 7$ 12. $y = -2x^2 - 4x - 3$

13. $y = -5x^2 - 30x - 45$ 14. $y = x^2 + 2$

15. $y = 4x^2 - 8x$ 16. $y = -3x^2 + 4$

Investigation: The Discriminant

Program: PARA2

Objectives: To use the computer to graph equations of the form $y = Ax^2 + Bx + C$

To use the discriminant to predict the number of zeros of the function before seeing its graph

1. **a.** With BASIC loaded into the computer and the disk in drive 1 (or drive A), run the program PARA2 by typing the appropriate command.

        Apple: RUN PARA2   IBM: RUN "PARA2"

      Be sure to press the RETURN key after typing the command.

   b. Read the opening messages of the program.

2. **a.** Use the program to graph the function $y = 3x^2 - 8x - 3$. Enter 3 for A, -8 for B, and -3 for C.

   b. After the graph is displayed, count the number of **zeros** of the function. That is, count the number of points where the graph crosses the x axis. How many zeros does it have?

   c. For a quadratic function $y = Ax^2 + Bx + C$, the value of $B^2 - 4AC$ is called the **discriminant** of the function. Compute the value of $B^2 - 4AC$ for $y = 3x^2 - 8x - 3$. Enter this value into the computer. If your value is incorrect, the computer will tell you to try again. What is the correct value?

   d. Based on the discriminant, the computer prints the number of zeros of this function. Does this number agree with your answer to part **b** above?

3. Answer Y (yes) to ANOTHER GRAPH? . Then repeat step 2 for the function $y = x^2 + x + 2$. (A = 1, B = 1, C = 2). Be sure to count the zeros from the graph before computing the discriminant.

   a. number of zeros?   **b.** discriminant?

4. Repeat step 2 for the function $y = x^2 + 4x + 4$. (A = 1, B = 4, C = 4)

   a. number of zeros?   **b.** discriminant?

5. **a.** What is the discriminant of $y = 2x^2 - x - 6$? (A = 2, B = -1, C = -6)

   b. Predict the number of zeros of the graph of $y = 2x^2 - x - 6$.

   c. Have the computer graph the function. Was your prediction correct?

   d. Enter your value for the discriminant. Was it correct?

6. Repeat step 5 for the function $y = 4x^2 - 4x + 1$.

   a. discriminant?   **b.** number of zeros?

7. Repeat step 5 for the function $y = -x^2 + 3x - 4$. Notice that A = -1.

   a. discriminant?   **b.** number of zeros?

8. *Complete*: For the quadratic function $y = Ax^2 + Bx + C$,

   a. if the discriminant is positive, the function has __?__ zero(s);

   b. if the discriminant is negative, the function has __?__ zero(s);

   c. if the discriminant is zero, the function has __?__ zero(s).

Try These   Compute the discriminant of each function and predict the number of zeros. Then have the computer graph the function. Check your predictions.

9. $y = x^2 + 2x + 7$ 10. $y = 2x^2 - 5x$ (NOTE: C = 0.)

11. $y = 3x^2 + 5x - 2$ 12. $y = 6x^2 + 5$ (NOTE: B = 0.)

13. $y = -3x^2 + 4x + 6$ 14. $y = -x^2 - 6x - 9$

Investigation: Using Statistics: Predicting

Program: SCATPLOT

Objectives: To use the computer to draw scatterplots and draw the median line of fit
To have the computer predict a Y value on the median line of fit for any given X value

1. **a.** With BASIC loaded into the computer and the disk in drive 1 (or drive A), run the program SCATPLOT by typing the appropriate command.
    Apple: RUN SCATPLOT   IBM: RUN "SCATPLOT"
    Be sure to press the RETURN key after typing the command.
   b. Read the opening messages of the program.

2. Enter the pairs of values at the right into the computer. Be sure to type the comma between the pairs. Also press RETURN after entering each pair. After typing the last pair, answer PAIR #11? by typing 999, 999 (and pressing RETURN).

0.2, 70	1.1, 75
1.2, 80	1.5, 79
1.6, 83	1.9, 84
2.2, 86	2.4, 86
2.7, 89	3.0, 91

3. **a.** The computer draws the **scatterplot.** After studying the plot, press any key to see the **median line of fit.**
   b. What is the equation of the line of fit?
   c. To the question PREDICT Y VALUE FOR WHICH X VALUE? enter 3.5 (and press RETURN).
   d. What is the predicted Y value on the line of fit for X = 3.5?
   e. To ANOTHER GRAPH? enter Y (yes).

4. **a.** Enter the pairs of values below. Enter 999,999 to end the list.
   1,4   2,7   3,6   5,9   6,10   7,12   9,12   10,13   11,15   12,16
   b. Press any key to see the median line of fit.
   c. What is the equation of the line of fit?
   d. To PREDICT Y VALUE FOR WHICH X VALUE? enter 15.
   e. What is the predicted Y value for X = 15?

Try These

Use the computer to draw the scatterplot of each set of data. In each case record the equation of the line of fit. Then enter the X value and record the predicted Y value for that X value.

5.

X	10	20	20	25	30	35	40	45	50	60
Y	10	12	11	14	18	23	30	29	31	35

Predict Y for X = 53.

6.

X	17.5	18.0	19.0	19.5	20.5	21.0	22.5	23.5	25.0	27.0
Y	8.7	8.5	8.4	7.9	7.8	8.0	7.4	7.1	6.8	6.1

Predict Y for X = 18.0.

Table of Roots and Powers

No.	Sq.	Sq. Root	Cube	Cu. Root	No.	Sq.	Sq. Root	Cube	Cu. Root
1	1	1.000	1	1.000	51	2,601	7.141	132,651	3.708
2	4	1.414	8	1.260	52	2,704	7.211	140,608	3.733
3	9	1.732	27	1.442	53	2,809	7.280	148,877	3.756
4	16	2.000	64	1.587	54	2,916	7.348	157,564	3.780
5	25	2.236	125	1.710	55	3,025	7.416	166,375	3.803
6	36	2.449	216	1.817	56	3,136	7.483	175,616	3.826
7	49	2.646	343	1.913	57	3,249	7.550	185,193	3.849
8	64	2.828	512	2.000	58	3,364	7.616	195,112	3.871
9	81	3.000	729	2.080	59	3,481	7.681	205,379	3.893
10	100	3.162	1,000	2.154	60	3,600	7.746	216,000	3.915
11	121	3.317	1,331	2.224	61	3,721	7.810	226,981	3.936
12	144	3.464	1,728	2.289	62	3,844	7.874	238,328	3.958
13	169	3.606	2,197	2.351	63	3,969	7.937	250,047	3.979
14	196	3.742	2,744	2.410	64	4,096	8.000	262,144	4.000
15	225	3.875	3,373	2.466	65	4,225	8.062	274,625	4.021
16	256	4.000	4,096	2.520	66	4,356	8.124	287,496	4.041
17	289	4.123	4,913	2.571	67	4,489	8.185	300,763	4.062
18	324	4.243	5,832	2.621	68	4,624	8.246	314,432	4.082
19	361	4.359	6,859	2.668	69	4,761	8.307	328,509	4.102
20	400	4.472	8,000	2.714	70	4,900	8.357	343,000	4.121
21	441	4.583	9,261	2.759	71	5,041	8.426	357,911	4.141
22	484	4.690	10,648	2.802	72	5,184	8.485	373,248	4.160
23	529	4.796	12,167	2.844	73	5,329	8.544	389,017	4.179
24	576	4.899	13,824	2.884	74	5,476	8.602	405,224	4.198
25	625	5.000	15,625	2.924	75	5,625	8.660	421,875	4.217
26	676	5.099	17,576	2.962	76	5,776	8.718	438,976	4.236
27	729	5.196	19,683	3.000	77	5,929	8.775	456,533	4.254
28	784	5.292	21,952	3.037	78	6,084	8.832	474,552	4.273
29	841	5.385	24,389	3.072	79	6,241	8.888	493,039	4.291
30	900	5.477	27,000	3.107	80	6,400	8.944	512,000	4.309
31	961	5.568	29,791	3.141	81	6,561	9.000	531,441	4.327
32	1,024	5.657	32,768	3.175	82	6,724	9.055	551,368	4.344
33	1,089	5.745	35,937	3.208	83	6,889	9.110	571,787	4.362
34	1,156	5.831	39,304	3.240	84	7,056	9.165	592,704	4.380
35	1,225	5.916	42,875	3.271	85	7,225	9.220	614,125	4.397
36	1,296	6.000	46,656	3.302	86	7,396	9.274	636,056	4.414
37	1,369	6.083	50,653	3.332	87	7,569	9.327	658,503	4.431
38	1,444	6.164	54,872	3.362	88	7,744	9.381	681,472	4.448
39	1,521	6.245	59,319	3.391	89	7,921	9.434	704,969	4.465
40	1,600	6.325	64,000	3.420	90	8,100	9.487	729,000	4.481
41	1,681	6.403	68,921	3.448	91	8,281	9.539	753,571	4.498
42	1,764	6.481	74,088	3.476	92	8,464	9.592	778,688	4.514
43	1,849	6.557	79,507	3.503	93	8,649	9.644	804,357	4.531
44	1,936	6.633	85,184	3.530	94	8,836	9.695	830,584	4.547
45	2,025	6.708	91,125	3.557	95	9,025	9.747	857,375	4.563
46	2,116	6.782	97,336	3.583	96	9,216	9.798	884,736	4.579
47	2,209	6.856	103,823	3.609	97	9,409	9.849	912,673	4.595
48	2,304	6.928	110,592	3.634	98	9,604	9.899	941,192	4.610
49	2,401	7.000	117,649	3.659	99	9,801	9.950	970,299	4.626
50	2,500	7.071	125,000	3.684	100	10,000	10.000	1,000,000	4.642

Trigonometric Ratios

Angle Measure	Sin	Cos	Tan	Angle Measure	Sin	Cos	Tan
0°	0.000	1.000	0.000	46°	.7193	.6947	1.036
1°	.0175	.9998	.0175	47°	.7314	.6820	1.072
2°	.0349	.9994	.0349	48°	.7431	.6691	1.111
3°	.0523	.9986	.0524	49°	.7547	.6561	1.150
4°	.0698	.9976	.0699	50°	.7660	.6428	1.192
5°	.0872	.9962	.0875	51°	.7771	.6293	1.235
6°	.1045	.9945	.1051	52°	.7880	.6157	1.280
7°	.1219	.9925	.1228	53°	.7986	.6018	1.327
8°	.1392	.9903	.1405	54°	.8090	.5878	1.376
9°	.1564	.9877	.1584	55°	.8192	.5736	1.428
10°	.1736	.9848	.1763	56°	.8290	.5592	1.483
11°	.1908	.9816	.1944	57°	.8387	.5446	1.540
12°	.2079	.9781	.2126	58°	.8480	.5299	1.600
13°	.2250	.9744	.2309	59°	.8572	.5150	1.664
14°	.2419	.9703	.2493	60°	.8660	.5000	1.732
15°	.2588	.9659	.2679	61°	.8746	.4848	1.804
16°	.2756	.9613	.2867	62°	.8829	.4695	1.881
17°	.2924	.9563	.3057	63°	.8910	.4540	1.963
18°	.3090	.9511	.3249	64°	.8988	.4384	2.050
19°	.3256	.9455	.3443	65°	.9063	.4226	2.145
20°	.3420	.9397	.3640	66°	.9135	.4067	2.246
21°	.3584	.9336	.3839	67°	.9205	.3907	2.356
22°	.3746	.9272	.4040	68°	.9272	.3746	2.475
23°	.3907	.9205	.4245	69°	.9336	.3584	2.605
24°	.4067	.9135	.4452	70°	.9397	.3420	2.747
25°	.4226	.9063	.4663	71°	.9455	.3256	2.904
26°	.4384	.8988	.4877	72°	.9511	.3090	3.077
27°	.4540	.8910	.5095	73°	.9563	.2924	3.271
28°	.4695	.8829	.5317	74°	.9613	.2756	3.487
29°	.4848	.8746	.5543	75°	.9659	.2588	3.732
30°	.5000	.8660	.5774	76°	.9703	.2419	4.010
31°	.5150	.8572	.6009	77°	.9744	.2250	4.331
32°	.5299	.8480	.6249	78°	.9781	.2079	4.704
33°	.5446	.8387	.6494	79°	.9816	.1908	5.145
34°	.5592	.8290	.6745	80°	.9848	.1736	5.671
35°	.5736	.8192	.7002	81°	.9877	.1564	6.314
36°	.5878	.8090	.7265	82°	.9903	.1392	7.115
37°	.6018	.7986	.7536	83°	.9925	.1219	8.144
38°	.6157	.7880	.7813	84°	.9945	.1045	9.514
39°	.6293	.7771	.8098	85°	.9962	.0872	11.43
40°	.6428	.7660	.8391	86°	.9976	.0698	14.30
41°	.6561	.7547	.8693	87°	.9986	.0523	19.08
42°	.6691	.7431	.9004	88°	.9994	.0349	28.64
43°	.6820	.7314	.9325	89°	.9998	.0175	57.29
44°	.6947	.7193	.9657	90°	1.000	0.000	
45°	.7071	.7071	1.000				

Glossary

absolute value: The distance between x and 0 on a number line. (p. 46)

acute angle: An angle with a degree measure less than 90. (p. 579)

addition method: A method for solving a system of linear equations in two variables, based on the Addition Property. (p. 460)

algebraic expression: An expression that contains one or more operations with one or more variables. (p. 1)

angle of elevation: An angle formed by a horizontal line and the line of sight to a point at a higher elevation. (p. 590)

angle of depression: An angle formed by a horizontal line and the line of sight to a point at a lower elevation. (p. 590)

area: A measurement of the number of square units that a geometric figure contains. (p. 13)

arithmetic mean: The average of a set of numbers. (p. 69)

average: The sum of the members of the set divided by the number of members. (p. 69)

axis of symmetry: An imaginary line about which a curve is symmetric. (p. 559)

binomial: A polynomial with two terms. (p. 218)

boundary line: A line that separates a coordinate plane into two regions. (p. 433)

central tendency: The "middle points" of statistical data, measured by the *mean, mode,* and *median.* (p. 610)

coefficient: See *numerical coefficients.* (p. 26)

completing the square: A method for solving quadratic equations. (p. 539)

complex fraction: A fraction with at least one fraction in its numerator, denominator, or in both. (p. 325)

complex rational expression: A rational expression with at least one rational expression in its numerator, denominator, or in both. (p. 326)

composite number: A positive integer with two or more positive factors other than 1 (not prime). (p. 243)

conclusion: The *then* clause in a conditional statement. (p. 116)

conditional: A statement written in *if-then* form. (p. 116)

conjugates: Binomial expressions that differ only in the sign of the second term. (p. 523)

conjunction: The logical union of two statements, which are connected by the word *and.* (p. 175)

constant linear function: A function of the form $f(x) = c$, where c is constant and whose graph is a horizontal line. (p. 393)

constant of variation: The constant k in the equation $y = kx$ or $xy = k$, where k is a nonzero real number. (pp. 397, 401)

conversion fractions: Fractions equal to 1 that are used to convert units of measurement. (p. 364)

coordinate: The number, or number pair, that corresponds to a point on a number line. (pp. 41, 373)

coordinate plane: A plane that has a vertical axis and a horizontal axis that intersect in a point called the origin. (p. 373)

corresponding angles: Angles of two similar figures, which have the same measure. (p. 575)

corresponding sides: Sides, proportional in length, that lie opposite the pairs of corresponding angles in similar figures. (p. 575)

cosine ratio: The cosine of an acute angle of a right triangle is the ratio of the length of the side that is adjacent to the acute angle to the length of the hypotenuse. (p. 580)

degree of a monomial: The sum of the exponents of all its variables. (p. 219)

degree of a polynomial: The degree of a polynomial is the same as that of the greatest degree term. (p. 219)

difference of two squares: A perfect square minus another perfect square: $a^2 - b^2$. (p. 259)

dimensional analysis: The use of conversion fractions to convert units of measurement. (p. 364)

direct variation: A linear function defined by an equation of the form $y = kx$, where k is a nonzero real number. (p. 397)

discriminant: A portion of the quadratic formula that is under a radical sign, $b^2 - 4ac$. (p. 545)

disjunction: A logical intersection of two statements connected by the word *or*. (p. 176)

domain: The set of all first coordinates of a relation. (p. 378)

double root: A root which appears twice as a factor. (p. 269)

empty set: The only set containing no members, symbolized by $\varnothing$. (p. 32)

equation: Two expressions with an equals symbol between them. (p. 30)

equivalent equations: Equations that have the same solution set. (p. 89)

equivalent expressions: Expressions that are equal for all values of their variables for which the expressions have meaning. (p. 19)

even integer: An integer divisible by 2. (p. 135)

event: A specific outcome which is a subset of the sample space. (p. 603)

exponent: The number which tells how many times the base of a power, such as 3^2, is used as a factor. (p. 9)

extraneous solution: A solution of a derived equation that is not a solution of the original equation. (pp. 338, 526)

extremes: The numbers a and d in the equation $\frac{a}{b} = \frac{c}{d}$, where $b \neq 0$ and $d \neq 0$. (p. 343)

factoring by grouping: A method of grouping terms to find a common binomial factor. (p. 264)

factors: Numbers or groups of numbers that are multiplied. (p. 9)

formula: In a formula, variables and symbols are used to show how quantities are related. (p. 9)

frequency table: A representation used to summarize statistical data. (p. 610)

function: A relation in which no two ordered pairs have the same first coordinate. (p. 378)

graph: The set of points on a number line or in a coordinate plane that correspond to a number, an ordered pair, or a relation. (p. 41)

greatest common factor (GCF): The largest integer that is a common factor of two or more integers. (p. 244)

grouping symbols: Parentheses ( ) and brackets [ ] that are used to clarify or to change the order of operations. (p. 5)

half-planes: Two regions of a coordinate plane that are separated by a boundary line. (p. 433)

histogram: A type of bar graph that is used to illustrate a set of data. (p. 615)

hypothesis: The *if* clause in a conditional statement. (p. 116)

hypotenuse: The side that is opposite the right angle in a right triangle. (p. 579)

identity: An equation that is true for any value from the replacement set of the variable. (p. 103)

independent events: Separate events whose outcomes do not affect each other. (p. 604)

inequality: Two expressions with an inequality symbol between them. (p. 30)

integers: The set of whole numbers and their opposites: $\{..., -3, -2, -1, 0, 1, 2, 3, ...\}$. (p. 41)

inverse variation: A function defined by an equation of the form $xy = k$, or $y = \dfrac{k}{x}$, where k is a nonzero real number. (p. 401)

irrational number: A number that cannot be expressed as the quotient of two integers. (p. 497)

isosceles triangle: A triangle with two congruent sides. (p. 140)

least common denominator (LCD): The LCM of all denominators in a group of fractions by which the fractions can be multiplied to eliminate the denominators. (p. 144)

least common multiple (LCM): The smallest number that is divisible by a group of numbers. (p. 144)

like terms: Terms that include the same variables and their corresponding exponents. (p. 26)

linear combination method: A method for solving a system of linear equations in two variables using addition with multiplication. (p. 464)

linear function: A function of one variable, which has a line as its graph. (p. 392)

literal equation: An equation with more than one variable in which one of the variables can be expressed in terms of the other variables. (p. 349)

mathematical sentence: A statement that indicates a relationship between numerical or variable expressions. (p. 30)

maximum point: A point on a curve whose y-coordinate is greater than or equal to the y-coordinate of every other point on the curve. (p. 559)

maximum value: The y-coordinate of the maximum point. (p. 559)

mean: See *Arithmetic mean.* (p. 608)

means: In the equation $\dfrac{a}{b} = \dfrac{c}{d}$, where $b \neq 0$ and $d \neq 0$, the means are b and c. (p. 343)

median: The middle score or the mean of two middle scores of an ordered set of data. (p. 609)

minimum point: A point on a curve whose y-coordinate is less than or equal to the y-coordinate of every other point on the curve. (p. 559)

minimum value: The y-coordinate of the minimum point. (p. 559)

mode: The value that occurs most frequently in a set of data. (p. 609)

monomial: A variable, or a product of a numeral and one or more variables. (p. 201)

null set: See *empty set*. (p. 31)

numerical coefficient: A number by which a variable or a product of variables is multiplied; also simply called a *coefficient*. (p. 26)

numerical expression: An expression that contains one or more numbers involving one or more operations. (p. 1)

odd integer: An integer that is not divisible by 2. (p. 136)

open half-plane: A half-plane that does not include the boundary line. (p. 433)

open sentence: A mathematical sentence that contains at least one variable. (p. 31)

opposites: Like numbers with opposite signs whose sum is zero. (p. 45)

ordered pair: A pair of numbers that corresponds to a point in a coordinate plane. (p. 373)

origin: The point at which the axes intersect in a coordinate plane. (p. 373)

parabola: A graph of any equation of the form $y = ax^2 + bx + c$, where $a \neq 0$. (p. 559)

parallel lines: Coplanar lines with equal slopes. The lines do not intersect. (p. 429)

percent: *per hundred*, or *hundredths*. For example, 6% means six per hundred, or six hundredths. (p. 150)

perfect square: A positive rational number whose principal square root is rational. (p. 501)

Perfect Square Trinomial: A trinomial that can be factored as a perfect square of binomials.
$a^2 + 2ab + b^2 = (a + b)^2$
$a^2 - 2ab + b^2 = (a - b)^2$. (p. 260)

perimeter: A measure of the distance around the boundary of a figure. (p. 13)

perpendicular lines: Lines that have slopes whose product is -1, and intersect at an angle of measure 90. (p. 430)

point-slope form: The form of an equation of a line with slope m and containing point $P(x_1, y_1)$: $y - y_1 = m(x - x_1)$. (p. 418)

polynomial: A monomial or the sum of two or more monomials. (p. 218)

power: The third power of 5 is written as 5^3. The raised three is called an exponent. An exponent indicates how many times a number is used as a factor. (p. 9)

prime factorization: A factorization in which all the factors are prime numbers. (p. 244)

prime number: An integer greater than 1 whose only positive factors are 1 and itself. (p. 243)

principal square root: The positive *square root* of a number. (p. 500)

probability of an event: A ratio between 0 and 1 that describes how likely it is that a particular event will occur. (p. 599)

proportion: An equation which states that two ratios are equal. (p. 343)

Proportion Property: For all real numbers a, b, c, and d, if $\frac{a}{b} = \frac{c}{d}$, then $ad = bc$. The product of the *extremes* equals the product of the *means*. (p. 343)

Pythagorean Theorem: If triangle ABC is a right triangle with c the length of the hypotenuse and a and b the lengths of the legs, then $c^2 = a^2 + b^2$. (p. 508)

quadrant: One-fourth of a coordinate plane, as divided by the x- and y-axes. (p. 374)

quadratic equation: A second degree equation of the form $ax^2 + bx + c = 0$, where $a \neq 0$. (p. 268)

quadratic formula: The solutions of a quadratic equation of the form $ax^2 + bx + c = 0$, where $a \neq 0$, are given by this formula: $x = \frac{-b \pm \sqrt{b^2 - 4ac}}{2a}$. (p. 544)

quadratic function: An equation of the form $y = ax^2 + bx + c$, where $a \neq 0$. (p. 558)

quadratic polynomial: A polynomial of degree 2. (p. 268)

radical: The expression $\sqrt{49}$, is called a *radical*. The $\sqrt{}$ is a *radical sign,* and the number under the radical sign is called the *radicand.* (p. 500)

range of a relation: The set of all second coordinates of the relation. (p. 378)

range of a set of data: The difference between the greatest and least values in a set of data. (p. 619)

ratio: the comparison of two numbers by division. (p. 343)

rational equation: An equation that contains one or more rational expressions. (p. 337)

rational number: A real number that can be expressed in the form $\frac{a}{b}$, where a and b are integers and $b \neq 0$. (p. 495)

real number: The set of positive and negative numbers including all rational numbers, all irrational numbers, and 0. (pp. 42, 497)

reciprocal: Two numbers are called reciprocals, or *multiplicative inverses,* of each other if their product is 1; 4 and $\frac{1}{4}$ are reciprocals, since $4 \cdot \frac{1}{4} = 1$. (p. 68)

relation: A set of ordered pairs. (p. 378)

replacement set: The set of numbers that can replace a variable. (p. 30)

right triangle: A triangle with one angle of measure 90. (p. 508)

rise: The vertical change from one point to another in the slope of a line. (p. 413)

roots: The solutions of an open sentence. (p. 268)

run: The horizontal change from one point to another in the slope of a line. (p. 413)

sample space: The set of all possible outcomes. (p. 603)

scatter plot: A graph of ordered pairs that indicates the kind of relation that may exist between variables. When the relation is linear, a line approximating the points is drawn and then used to make predictions. (p. 614)

scientific notation: A number in the form $a \times 10^n$, where $1 < a < 10$ and n is an integer. (p. 214)

similar triangles: Triangles that have the same shape based on corresponding congruent angles. (p. 575)

sine ratio: The sine of an acute angle of a right triangle is the ratio of the length of the side that is opposite the acute angle to the length of the hypotenuse. (p. 580)

slope: A description of the steepness of a line, defined by the ratio of the vertical distance to the horizontal distance between any two points on the line. (p. 414)

slope-intercept form: If a line has slope m and y-intercept b, then the slope-intercept form of an equation of the line is $y = mx + b$. (p. 424)

solution set: The set of all numbers from a given replacement set that make an open sentence true. (p. 31)

square root: One of two equal factors of a number. (p. 500)

standard deviation: A measure s of the dispersion of a set of data, $x_1, x_2, x_3, \ldots x_n$ about the mean x for that set, defined by

$$s = \sqrt{\frac{(x_1 - x)^2 + (x_2 - x)^2 + (x_3 - x)^2 + \ldots + (x_n - x)^2}{n}}$$

where n is the number of values in the set. (p. 619)

subset: If all members of a set X are also members of a set Y, then X is said to be a subset of Y. (p. 498)

substitution method: A method for solving a system of linear equations in two variables, based on the Substitution Property. (p. 452)

supplementary: When the sum of the measures of two angles is 180, the angles are supplementary. (p. 346)

system of linear equations: A set of two or more linear equations in the same variables. (p. 445)

tangent ratio: The tangent of an acute angle of a right triangle is the ratio of the length of the side that is opposite the acute angle to the length of the side adjacent to the angle. (p. 580)

term: In the expression $7x^2 + 8xy + 9$, $7x^2$, $8xy$, and 9 are the terms of the expression. (p. 26)

then-clause: The conclusion in a conditional statement. (p. 116)

tree diagram: A diagram that can be used to determine or illustrate a *sample space*. (p. 603)

trinomial: A polynomial with three terms. (p. 218)

truth table: A summary of the truth values of a conjunction or a disjunction. (p. 176)

value of a function: A member of the range of a function. (p. 383)

variable: A symbol that is used to represent one or more numbers. (p. 1)

vertex of a parabola: The maximum or minimum point of the graph of a parabola. (p. 559)

volume: The number of cubic units contained in a solid figure. (p. 14)

x- and y-axes: The horizontal and vertical number lines in a coordinate plane which intersect at the *origin*. (p. 373)

x- and y-coordinates: In an ordered pair of numbers, the x-coordinate is the first number in the pair and the y-coordinate is the second number. (p. 374)

x-intercept: The x-coordinate of the point where a line intersects the x-axis. (p. 424)

y-intercept: The y-coordinate of the point where a line intersects the y-axis. (p. 424)

zero exponent: For each nonzero real number b, $b^0 = 1$ (0^0 is undefined). (p. 210)

Additional Answers

11.

Statements	Reasons
1. $(ax + b) + ay = (b + ax) + ay$	1. Comm Prop Add
2. $= b + (ax + ay)$	2. Assoc Prop Add
3. $= b + a(x + y)$	3. Dist Prop
4. $= a(x + y) + b$	4. Comm Prop Add
5. $(ax + b) + ay = a(x + y) + b$	5. Trans Prop of Eq

12.

Statements	Reasons
1. $x + (3 + y) = (3 + y) + x$	1. Comm Prop Add
2. $3 + (y + x)$	2. Assoc Prop Add
3. $x + (3 + y) = 3x + (y + x)$	3. Trans Prop of Eq

13.

Statements	Reasons
1. $mx + (a + x) = mx + (x + a)$	1. Comm Prop Add
2. $= (mx + x) + a$	2. Assoc Prop Add
3. $= (m + 1)x + a$	3. Dist Prop
4. $= a + (m + 1)x$	4. Comm Prop Add
5. $mx + (a + x) = a + (m + 1)x$	5. Trans Prop of Eq

14.

Statements	Reasons
1. $a(b - c) = a[b + (-c)]$	1. Def of subtr
2. $a \cdot b + a \cdot (-c)$	2. Dist Prop
3. $ab + (-ac)$	3. Mult
4. $ab - ac$	4. Def of subtr
5. $a(b - c) = ab - ac$	5. Trans Prop of Eq

26.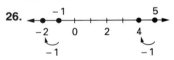

27. By Add Prop Ineq, if $a < b$, then $a + c < b + c$ and if $a > b$, then $a + c > a + b$. Let $c = (-d)$. If $a < b$, then $a + (-d) < b + (-d)$. Thus, $a - d < b - d$. If $a > b$, then $a + (-d) > b + (-d)$. Thus, $a - d > b - d$.

13. $x < -27$;

14. $y < 5$;

15. $x > 3$;

16. All real numbers;

17.

18.

19.

20.

21.

22.

23.

24.

25.

26.

42. If $a > 0$, then $x < \dfrac{b}{a}$; if $a < 0$, then $x > \dfrac{b}{a}$.

43. If $(a + b) > 0$ and $c > 0$ then $x < \dfrac{cd}{a + b}$;

if $(a + b) > 0$ and $c < 0$ then $x > \dfrac{cd}{a + b}$;

if $(a + b) < 0$ and $c > 0$ then $x > \dfrac{cd}{a + b}$;

if $(a + b) < 0$ and $c < 0$ then $x < \dfrac{cd}{a + b}$.

23.

24.

25.

26.

27. −4   0   4

28. 5; 0   10   20

29. 3   5; 0   2   4   4

30. −10   0   10

31. −3   6⅔; −4   0   4   8

32. −4   0   4

33. −15   −13   −11

34. 3; −6   0   6

38. 5; 0   2   4   6

39. 5; 0   10   20

40. −6   0   6

41. −8   0   8

42. 2½; 0   2   4   8

43. −9; −12   −6   0   6

44. 0   4   8

45. 3; 0   2   4

46. −a   0   a

47. a   0   −a

48. b   a

49. b   a

50. a   b

51. a   0   b

Midchapter Review

1. −6   0   6

2. −7; −14   0   14

3. −2   0   2

4. 0   12   24

5. −13; −26   0   26

6. −3; −12   −6   0

11. 7; 0   14   28

12. −3   5; −6   0   6

13. −11   −5; −12   −8   −4   0

Classroom Exercises

1. $x > 6$ or $x < -6$
2. $x \leq 3$ and $x \geq -3$
3. $y \geq 1$ or $y \leq -1$
4. $x < \frac{2}{7}$ or $x > \frac{6}{7}$
5. $c = -\frac{1}{3}$
6. −6   0   6

7. −3   3; −6   0   6

8. −1   1; −2   0   2

9. $\frac{2}{7}$ $\frac{6}{7}$; 0 $\frac{4}{7}$ $\frac{8}{7}$

10. $-\frac{1}{3}$; −2   0   2

Written Exercises

1. −4   0   4

2. −4   0   4

3. −4   0   4

4. 2; 0   4   8

5. −11   3; −12   −6   0   6

6. −4   0   4

7. 1   7; 0   4   8

8. 2; −4   0   4

9. −3   7; −4   0   4   8

10. 1   3; 0   2   4

11. −7   3; −8   −4   0   4

12. 1   5; 0   2   4   6

13. 1; −6   0   6

651

14.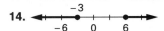
−3
−6 0 6

15.
−5   9
−6 0 6 12

16.
−4 0 4

17.
−1/2   5 1/2
−4 0 4 8

18.
−2/3
−2 0 2

19.
1/2
0 2 4

20.
−2 0 2 4

21.
−8 0 8

22.
72 78 84

23.
10,000
8,000   12,000

24.
8 7/8   11 1/8
8 10 12

25.
0 2 4

26.
−7   3
−8 −4 0 4

27.
−5   −1
−6 −4 −2 0

28.
−9 −3
−12 −6 0

29.
−3   9
−6 0 6 12

30.
−4 0 4

31.
2   6
0 4 8

32.
1 1/2   2 1/2
0 1 2 3

33.
−5 1/2 −1 1/2
−8 −4 0 4

9.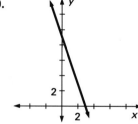

10.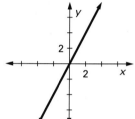

11.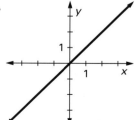

12.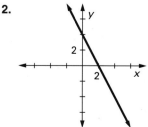

13.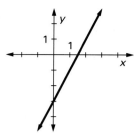

14.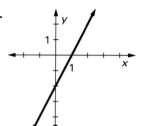

15.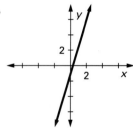

16.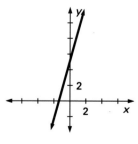

17.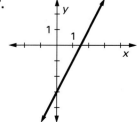

652

18.
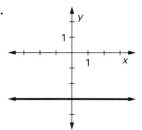

23.
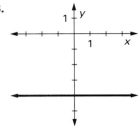

32.
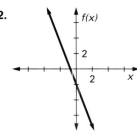

19.
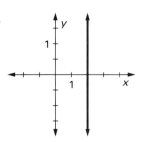

24.
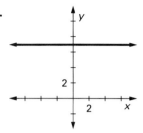

33.
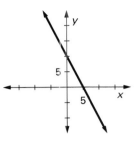

20.
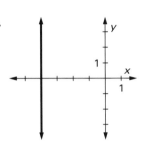

29.
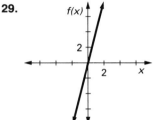

34.
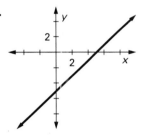

21.
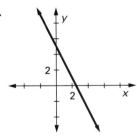

30.
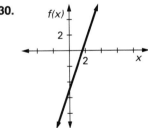

35.
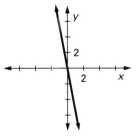

22.
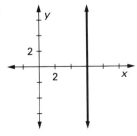

31.
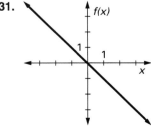

36.
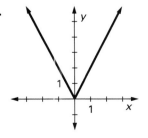

653

37.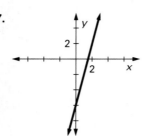

42.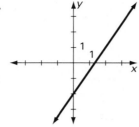

22.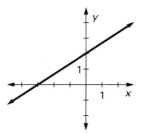

38.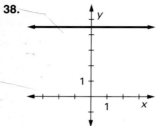

43.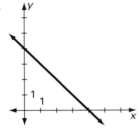

23.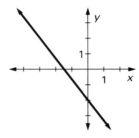

39.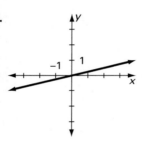

44.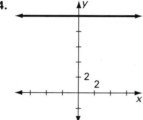

24.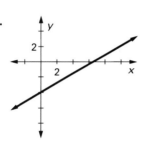

40.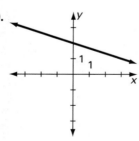

Mixed Review

1. $3x^2 - 4x - 2$

3. $-16n^4 + 32n^3 - 40n^2$

7. $\frac{x^2 + 4x + 4}{2x^2 - 13x + 6}$

PAGE 427

25.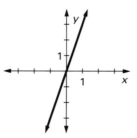

41.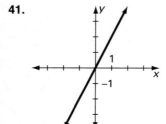

21.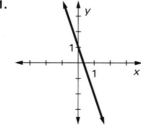

26.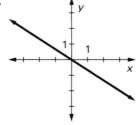

654

27.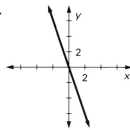

28.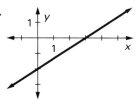

29.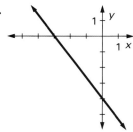

30.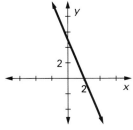

31.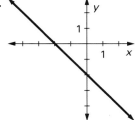

32.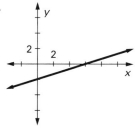

33.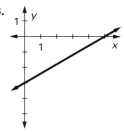

34. $y = 3x$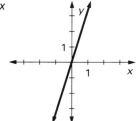

35. $y = -\frac{5}{2}x$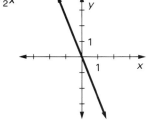

36. $y = x$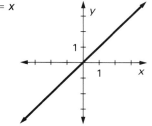

37. $y = \frac{4}{3}x - \frac{2}{3}$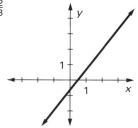

38. $y = -\frac{3}{2}x - \frac{3}{2}$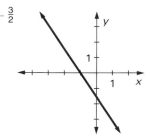

39. $y = -3x$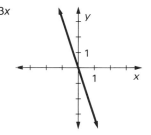

44.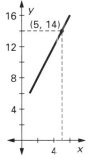

47.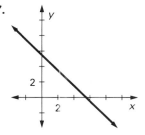

48.

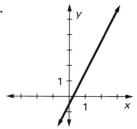

49.

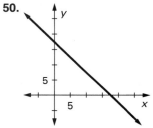

50.

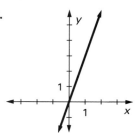

51.

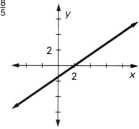

Midchapter Review

13. $y = \frac{4}{5}x - \frac{8}{5}$

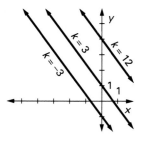

14. $y = -3x + 4$

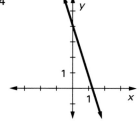

PAGES 431–432

9.

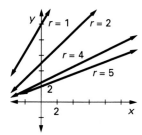

$3x + y = 1$

$x - 3y = 4$

19. Sample answer: Two lines in a coordinate plane are parallel if the two lines have equal slopes or are vertical. Two lines are perpendicular if the product of the slopes of the two lines is -1, or if one line is horizontal ($m = 0$) and the other line is vertical (m is undefined). If two lines do not possess one of the characteristics above, then the lines are neither parallel nor perpendicular.

22. As r decreases, the slope increases;

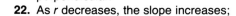

$r = 1$ $r = 2$

$r = 4$

$r = 5$

23. As k increases, the slopes of the graphs remain equal and the y-intercepts increase.

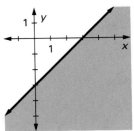

$k = 3$ $k = 12$

$k = -3$

24. Given that $\overleftrightarrow{AB}$ and $\overleftrightarrow{BC}$ have the same slope, then either $\overleftrightarrow{AB}$ and $\overleftrightarrow{BC}$ are parallel or they are the same line. Since $\overleftrightarrow{AB}$ and $\overleftrightarrow{BC}$ have point B in common they cannot be parallel. Hence, $\overleftrightarrow{AB}$ and $\overleftrightarrow{BC}$ are the same line. Since A, B, and C are points on the same line, A, B, and C are collinear.

PAGE 437

1.

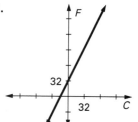

2.

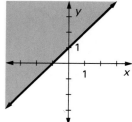

3.

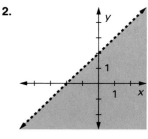

4.

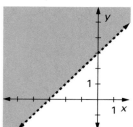

5.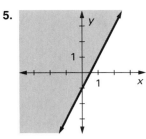

6.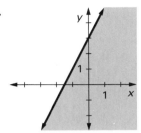

7.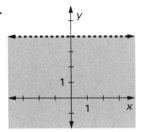

8.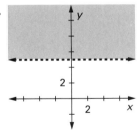

9.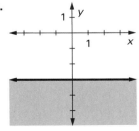

10.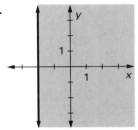

11.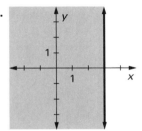

12.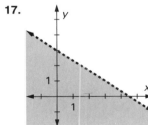

13.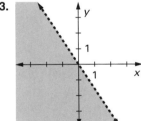

14.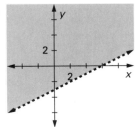

15.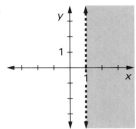

16.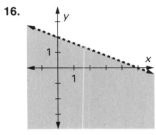

17.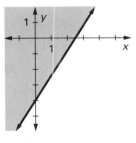

18.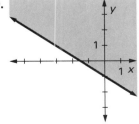

19.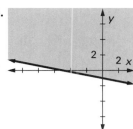

20.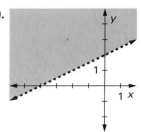

21.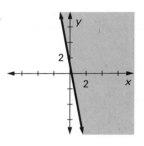

22.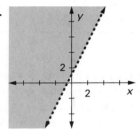

23.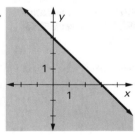

24.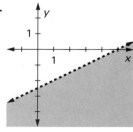

30.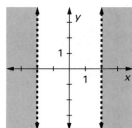

31.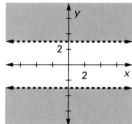

32.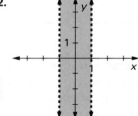

33.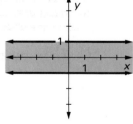

34.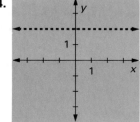

III NCTM Evaluation Standards //// Grades 9–12

STANDARD 1:
Mathematics as Problem Solving

In grades 9–12, the mathematics curriculum should include the refinement and extension of the methods of mathematical problem solving so that all students can—

1a. use, with increasing confidence, problem solving approaches to investigate and understand mathematical content;

1b. apply integrated mathematical problem-solving strategies to solve problems from within and outside mathematics;

1c. recognize and formulate problems from situations within and outside mathematics;

1d. apply the process of mathematical modeling to real-world problem situations.

STANDARD 2:
Mathematics as Communication

In grades 9–12, the mathematics curriculum should include the continued development of language and symbolism to communicate mathematical ideas so that all students can—

2a. reflect upon and clarify their thinking about mathematical ideas and relationships;

2b. formulate mathematical definitions and express generalizations discovered through investigations;

2c. express mathematical ideas orally and in writing;

2d. read written presentations of mathematics with understanding;

2e. ask clarifying and extending questions related to mathematics they have read or heard about;

2f. appreciate the economy, power, and elegance of mathematical notation and its role in the development of mathematical ideas.

STANDARD 3:
Mathematics as Reasoning

In grades 9–12, the mathematics curriculum should include numerous and varied experiences that reinforce and extend logical reasoning skills so that all students can—

3a. make and test conjectures;

3b. formulate counterexamples;

3c. follow logical arguments;

3d. judge the validity of arguments;

3e. construct simple valid arguments;

and so that, in addition, college-intending students can—

3f. construct proofs for mathematical assertions, including indirect proofs and proofs by mathematical induction.

STANDARD 4:
Mathematical Connections

In grades 9–12, the mathematical curriculum should include investigation of the connections and interplay among various mathematical topics and their applications so that all students can—

4a. recognize equivalent representations of the same concept;

4b. relate procedures in one representation to procedures in an equivalent representation;

4c. use and value connections among mathematical topics;

4d. use and value the connections between mathematics and other disciplines.

659

In grades 9–12, the mathematics curriculum should include the continued study of algebraic concepts and methods so that all students can—

5a. represent situations that involve variable quantities with expressions, equations, inequalities, and matrices;

5b. use tables and graphs as tools to interpret expressions, equations, and inequalities;

5c. operate on expressions and matrices, and solve equations and inequalities;

5d. appreciate the power of mathematical abstraction and symbolism;

and so that, in addition, college-intending students can—

5e. use matrices to solve linear systems;

5f. demonstrate technical facility with algebraic transformations, including techniques based on the theory of equations.

STANDARD 6:
Functions

In grades 9–12, the mathematics curriculum should include the continued study of functions so that all students can—

6a. model real-world phenomena with a variety of functions;

6b. represent and analyze relationships using tables, verbal rules, equations, and graphs;

6c. translate among tabular, symbolic, and graphical representations of functions;

6d. recognize that a variety of problem situations can be modeled by the same type of function;

6e. analyze the effects of parameter changes on the graphs of functions;

and so that, in addition, college-intending students can—

6f. understand operations on, and the general properties and behavior of, classes of functions.

STANDARD 7:
Geometry from a Synthetic Perspective

In grades 9–12, the mathematics curriculum should include the continued study of the geometry of two and three dimensions so that all students can—

7a. interpret and draw three-dimensional objects;

7b. represent problem situations with geometric models and apply properties of figures;

7c. classify figures in terms of congruence and similarity and apply these relationships;

7d. deduce properties of, and relationships between, figures from given assumptions;

and so that, in addition, college-intending students can—

7e. develop an understanding of an axiomatic system through investigating and comparing various geometries.

STANDARD 8:
Geometry from an Algebraic Perspective

In grades 9–12, the mathematics curriculum should include the study of the geometry of two and three dimensions from an algebraic point of view so that all students can—

8a. translate between synthetic and coordinate representations;

8b. deduce properties of figures using transformations and using coordinates;

8c. identify congruent and similar figures using transformations;

8d. analyze properties of Euclidean transformations and relate translations to vectors;

and so that, in addition, college-intending students can—

8e. deduce properties of figures using vectors;

8f. apply transformations, coordinates, and vectors in problem solving.

STANDARD 9:
Trigonometry

In grades 9–12, the mathematics curriculum should include the study of trigonometry so that all students can—

9a. apply trigonometry to problem situations involving triangles;

9b. explore periodic real-world phenomena using sine and cosine functions;

and so that, in addition, college-intending students can—

9c. understand the connection between trigonometric and circular functions;

9d. use circular functions to model periodic real-world phenomena;

9e. apply general graphing techniques to trigonometric functions;

9f. solve trigonometric equations and verify trigonometric identities

9g. understand the connections between trigonometric functions and polar coordinates, complex numbers, and series.

STANDARD 10:
Statistics

In grades 9–12, the mathematics curriculum should include the continued study of data analysis and statistics so that all students can—

10a. construct and draw inferences from charts, tables, and graphs that summarize data from real-world situations;

10b. use curve fitting to predict from data;

10c. understand and apply measures of central tendency, variability, and correlation;

10d. understand sampling and recognize its role in statistical claims;

10e. design a statistical experiment to study a problem, conduct the experiment, and interpret and communicate the outcomes;

10f. analyze the effects of data transformations on measures of central tendency and variability;

and so that, in addition, college-intending students can—

10g. transform data to aid in data interpretation and prediction;

10h. test hypotheses using appropriate statistics.

STANDARD 11:
Probability

In grades 9–12, the mathematics curriculum should include the continued study of probability so that all students can—

11a. use experimental or theoretical probability, as appropriate, to represent and solve problems involving uncertainty;

11b. use simulations to estimate probabilities;

11c. understand the concept of a random variable;

11d. create and interpret discrete probability distributions;

11e. describe, in general terms, the normal curve and use its properties to answer questions about sets of data that are assumed to be normally distributed;

and so that, in addition, college-intending students can—

11f. apply the concept of a random variable to generate and interpret probability distributions including binomial, uniform, normal, and chi square.

STANDARD 12:
Discrete Mathematics

In grades 9–12, the mathematics curriculum should include topics from discrete mathematics so that all students can—

12a. represent problem situations using discrete structures such as finite graphs, matrices, sequences, and recurrence relations;

12b. represent and analyze finite graphs using matrices;

12c. develop and analyze algorithms;

12d. solve enumeration and finite probability problems;

and so that, in addition, college-intending students can—

12e. represent and solve problems using linear programming and difference equations;

12f. investigate problem situations that arise in connection with computer validation and the application of algorithms.

STANDARD 13:
Conceptual Underpinnings of Calculus

In grades 9–12, the mathematics curriculum should include the informal exploration of calculus concepts from both a graphical and a numerical perspective so that all students can—

13a. determine maximum and minimum points of a graph and interpret the results in problem situations;

13b. investigate limiting processes by examining infinite sequences and series and areas under curves;

and so that, in addition, college-intending students can—

13c. understand the conceptual foundations of limit, the area under a curve, the rate of change, and the slope of a tangent line, and their applications in other disciplines;

13d. analyze the graphs of polynomial, rational, radical, and transcendental functions.

STANDARD 14:
Mathematical Structure

In grades 9–12, the mathematics curriculum should include the study of mathematical structure so that all students can—

14a. compare and contrast the real number system and its various subsystems with regard to their structural characteristics;

14b. understand the logic of algebraic procedures;

14c. appreciate that seemingly different mathematical systems may be essentially the same;

and so that, in addition, college-intending students can—

14d. develop the complex number system and demonstrate facility with its operations;

14e. prove elementary theorems within various mathematical structures, such as groups and fields;

14f. develop an understanding of the nature and purpose of axiomatic systems.

Correlation of Holt Algebra 1 to the NCTM Evaluation Standards

NCTM Standard	Holt Algebra 1 Lesson	
1a	1.8, 4.2, 4.8, 5.7, 7.4, 7.8, 12.1, 12.8, 12.9, 13.9, 14.4	
1b	1.8, 3.2, 4.2, 4.8, 5.5, 5.7, 7.4, 7.7, 7.8, 7.9, 8.8, 9.2, 9.4, 9.5, 9.6, 10.4, 10.6, 10.7, 12.4, 12.5, 12.6, 12.8, 12.9, 13.9, 14.1, 14.5	
1c	1.1, 3.3, 3.5, 4.2, 12.6, 12.7	
1d	1.3, 1.7, 2.3, 2.4, 3.5, 3.6, 4.2, 7.6, 15.5	
2a	1.3, 2.6, 2.8, 7.1, 7.7	
2b	2.3, 2.6, 3.6, 6.4, 7.5, 8.5	
2c	1.2, 2.2, 3.2, 6.8, 8.6	
2d	1.7, 6.1, 6.7, 7.6, 7.7, 7.8, 8.2, 10.6, 12.5, 14.3	
2f	2.2, 5.1, 5.3, 5.6, 6.1, 6.2	
3a	1.6, 5.1, 8.6	
3b	1.6, 5.1, 6.1	
3c	2.10, 3.7, 14.4	
3d	2.1, 3.7	
3e	3.7, 4.3, 5.1	
4a	2.7, 5.1, 5.4, 5.6, 6.4, 6.10, 7.3, 7.5, 8.2, 8.8, 8.9, 10.3, 10.5, 13.1, 13.9, 14.2	
4b	1.5, 3.4, 4.7, 6.9, 7.3, 9.3, 9.6, 12.10, 13.1, 14.2	
4c	4.1, 4.4, 6.5, 6.7, 6.8, 6.9, 7.2, 7.5, 7.9, 11.2, 12.3, 12.8, 12.10, 13.2, 13.4, 13.6, 13.7, 13.8, 14.5	
4d	4.7, 6.5, 6.7, 9.1, 9.2, 9.3, 10.7, 11.2, 12.2, 12.3, 12.4, 12.5, 12.10, 13.2, 13.3	
5a	1.4, 3.1, 3.3, 4.7, 5.2, 5.7, 8.1, 8.4, 8.5, 8.7, 12.7	
5b	2.5, 6.10, 8.7, 9.4, 9.5, 11.3, 11.4, 11.5, 13.3, 14.6	
5c	5.2, 5.3, 5.5, 5.6, 5.7, 6.2, 6.3, 8.1, 8.4, 8.5, 11.3, 11.4, 11.5, 12.1, 12.2, 12.7, 13.5, 13.6, 13.7, 13.8, 13.9, 14.1	
5d	10.4, 13.2, 13.5, 14.3	
6a	6.6, 10.2, 14.6, 15.3	
6b	1.4, 2.9, 7.5, 8.5, 10.5, 12.1, 14.7, 15.3	
6c	10.2, 10.3, 10.4, 11.5, 12.1	
6e	14.7	
7a	1.4, 2.9, 6.6, 6.8, 8.1, 8.3, 8.4, 8.9, 9.6	
7b	4.4, 8.3, 8.8, 15.1, 15.4	
7c	15.1	
7d	4.4, 5.5, 7.9, 13.4	
8a	10.1, 11.1, 11.2, 11.3	
8b	10.1, 11.1, 11.4	
9a	15.2, 15.3, 15.4, 15.5	
10a	16.4	
10b	16.4	
10c	4.5, 16.3, 16.5	
10d	9.2	
11a	16.2	
11b	16.1	
11d	16.2	
14a	2.1, 2.5, 4.3, 4.6, 7.1, 9.1, 13.1	
14b	1.5, 5.4, 7.2	
14c	4.3	
14d	2.4	

Index

Boldfaced numerals indicate the pages that contain definitions.